Handbook of Research on Global Environmental Changes and Human Health

Kholoud Kahime
Cadi Ayyad University, Morocco

Moulay Abdelmonaim El Hidan
Ibn Zohr University, Morocco

Omar El Hiba
Chouaib Doukkali University, Morocco & Cadi Ayyad University, Morocco

Denis Sereno
University of Montpellier, France

Lahouari Bounoua
NASA, USA

A volume in the Advances in Environmental Engineering and Green Technologies (AEEGT) Book Series

IGI Global
DISSEMINATOR OF KNOWLEDGE

Published in the United States of America by
IGI Global
Engineering Science Reference (an imprint of IGI Global)
701 E. Chocolate Avenue
Hershey PA, USA 17033
Tel: 717-533-8845
Fax: 717-533-8661
E-mail: cust@igi-global.com
Web site: http://www.igi-global.com

Library of Congress Cataloging-in-Publication Data

Names: Kahime, Kholoud, 1986- editor.
Title: Handbook of research on global environmental changes and human health / Kholoud Kahime, Moulay Abdelmonaim El Hidan, Omar El Hiba, Denis Sereno, and Lahouari Bounoua, editors.
Description: Hershey, PA : Engineering Science Reference, [2019] | Includes bibliographical references.
Identifiers: LCCN 2018035057| ISBN 9781522577751 (h/c) | ISBN 9781522577768 (eISBN)
Subjects: LCSH: Climatic changes--Health aspects. | Global environmental change--Health aspects. | Environmental health.
Classification: LCC RA793 .H36 2019 | DDC 616.9/88--dc23 LC record available at https://lccn.loc.gov/2018035057

This book is published in the IGI Global book series Advances in Environmental Engineering and Green Technologies (AEEGT) (ISSN: 2326-9162; eISSN: 2326-9170)

British Cataloguing in Publication Data
A Cataloguing in Publication record for this book is available from the British Library.

Advances in Environmental Engineering and Green Technologies (AEEGT) Book Series

Sang-Bing Tsai
University of Electronic Science and Technology of China Zhongshan Institute, China
Ming-Lang Tseng
Lunghwa University of Science and Technology, Taiwan
Yuchi Wang
University of Electronic Science and Technology of China Zhongshan Institute, China

ISSN:2326-9162
EISSN:2326-9170

Mission

Growing awareness and an increased focus on environmental issues such as climate change, energy use, and loss of non-renewable resources have brought about a greater need for research that provides potential solutions to these problems. Research in environmental science and engineering continues to play a vital role in uncovering new opportunities for a "green" future.

The **Advances in Environmental Engineering and Green Technologies (AEEGT)** book series is a mouthpiece for research in all aspects of environmental science, earth science, and green initiatives. This series supports the ongoing research in this field through publishing books that discuss topics within environmental engineering or that deal with the interdisciplinary field of green technologies.

Coverage

- Industrial Waste Management and Minimization
- Renewable Energy
- Biofilters and Biofiltration
- Green Technology
- Air Quality
- Sustainable Communities
- Radioactive Waste Treatment
- Cleantech
- Green Transportation
- Electric Vehicles

IGI Global is currently accepting manuscripts for publication within this series. To submit a proposal for a volume in this series, please contact our Acquisition Editors at Acquisitions@igi-global.com or visit: http://www.igi-global.com/publish/.

The Advances in Environmental Engineering and Green Technologies (AEEGT) Book Series (ISSN 2326-9162) is published by IGI Global, 701 E. Chocolate Avenue, Hershey, PA 17033-1240, USA, www.igi-global.com. This series is composed of titles available for purchase individually; each title is edited to be contextually exclusive from any other title within the series. For pricing and ordering information please visit http://www.igi-global.com/book-series/advances-environmental-engineering-green-technologies/73679. Postmaster: Send all address changes to above address.

Titles in this Series

For a list of additional titles in this series, please visit: www.igi-global.com/book-series

Building Sustainability Through Environmental Education
Lynn A. Wilson (SeaTrust Institute, USA) and Carolyn N. Stevenson (Purdue University Global, USA)
Engineering Science Reference • copyright 2019 • 335pp • H/C (ISBN: 9781522577270) • US $195.00 (our price)

Unique Sequence Signatures in Plant Lipolytic Enzymes Emerging Research and Opportunities
Nihed Ben Halima (University of Sfax, Tunisia)
Engineering Science Reference • copyright 2019 • 197pp • H/C (ISBN: 9781522574828) • US $145.00 (our price)

Handbook of Research on the Adverse Effects of Pesticide Pollution in Aquatic Ecosystems
Khursheed Ahmad Wani (ITM University Gwalior, India) and Mamta (Jiwaji University, India)
Engineering Science Reference • copyright 2019 • 538pp • H/C (ISBN: 9781522561118) • US $295.00 (our price)

Harnessing Marine Macroalgae for Industrial Purposes in an Australian Context Emerging Research and Opportunities
Göran Roos (Economic Development Board of South Australia, Australia) Anthony Cheshire (Balance Carbon Pty Ltd, Australia) Sasi Nayar (South Australian Aquatic Sciences Centre, Australia) Steven M. Clarke (South Australian Research and Development Institute, Australia) and Wei Zhang (Flinders University, Australia)
Engineering Science Reference • copyright 2019 • 291pp • H/C (ISBN: 9781522555773) • US $195.00 (our price)

Advanced Methodologies and Technologies in Engineering and Environmental Science
Mehdi Khosrow-Pour, D.B.A. (Information Resources Management Association, USA)
Engineering Science Reference • copyright 2019 • 506pp • H/C (ISBN: 9781522573593) • US $295.00 (our price)

Transitioning Island Nations Into Sustainable Energy Hubs Emerging Research and Opportunities
Catalina Spataru (UCL Energy Institute, UK)
Engineering Science Reference • copyright 2019 • 172pp • H/C (ISBN: 9781522560029) • US $135.00 (our price)

Recent Challenges and Advances in Geotechnical Earthquake Engineering
T.G. Sitharam (Indian Institute of Science, India) and J. S. Vinod (University of Wollongong, Australia)
Engineering Science Reference • copyright 2019 • 315pp • H/C (ISBN: 9781522569480) • US $195.00 (our price)

Agrifood Economics and Sustainable Development in Contemporary Society
Gabriel Popescu (Bucharest Academy of Economic Studies, Romania)
Engineering Science Reference • copyright 2019 • 401pp • H/C (ISBN: 9781522557395) • US $195.00 (our price)

IGI Global
DISSEMINATOR OF KNOWLEDGE

701 East Chocolate Avenue, Hershey, PA 17033, USA
Tel: 717-533-8845 x100 • Fax: 717-533-8661
E-Mail: cust@igi-global.com • www.igi-global.com

To my parents
To my friends
Kahime Kholoud
To my parents
To my wonderful wife Naima
To my sisters
To all my friends without whom this book wouldn't have been completed
Moulay Abdelmonaim El Hidan
To my parents
To my wonderful wife Maryam
To my brother
To the soul of my friend Youssef
To all my friends without whom this book wouldn't have been completed
Omar El Hiba

Editorial Advisory Board

List of Contributors

Table of Contents

Detailed Table of Contents

(SeVI) that measures socioeconomic vulnerability by regions. Result suggest that three southern and one north region are relatively more vulnerable, which are the most exposed to natural hazard. Furthermore, significant adaptive capacity scores are recorded in in the remaining regions, and average sensitivity for all. Therefore, societies and economies at all levels and on every region have to prepare for and adapt to impacts of climate change.

Abdelmohcine Aimrane, Cadi Ayyad University, Morocco
Hasna Lahouaoui, Cadi Ayyad University, Morocco
Youssef Khamsi, University of Arizona, USA
Ahmed Draoui, Cadi Ayyad University, Morocco
Hassan Alahyane, Cadi Ayyad University, Morocco
Nadia Zouhairi, Cadi Ayyad University, Morocco
Hind Benammi, Cadi Ayyad University, Morocco

Climate change is an actual fact setting off an imbalance in many living systems. Among these affected systems, water is a major essential element in the globe and in every existing living being. Therefore, several complications have been stated to occur, following water scarcity and water flood in many regions of the world, which make of them a global major threat of water security. The global disease burden is an additional factor that appeals to serious interventions worldwide in order to alleviate the water scarcity and water flood-related effects.

Mohamed Ait El Mokhtar, Cadi Ayyad University, Morocco
Mohamed Anli, Cadi Ayyad University, Morocco
Raja Ben Laouane, Cadi Ayyad University, Morocco
Abderrahim Boutasknit, Cadi Ayyad University, Morocco
Hanane Boutaj, Cadi Ayyad University, Morocco
Ahmed Draoui, Cadi Ayyad University, Morocco
Lamia Zarik, Cadi Ayyad University, Morocco
Abdessamad Fakhech, Cadi Ayyad University, Morocco

Climate change may potentially disrupt progress toward a world without hunger. Today, a clear and consistent global pattern is perceptible of the different impacts of climate change on crop productivity that could have repercussions on food security. Consequently, the stability of the whole food systems may be at risk under climate change because of its unpredictable variations. Indeed, agricultural production is highly vulnerable even to 2°C predictions augmentation for global mean temperatures in 2100, with major implications for poverty and for food security. The climate change impacts seem to be clear in areas currently affected by hunger and undernutrition, which will heighten food insecurity in these parts of the world. Therefore, adapting food systems both to increase food security and to prevent future negative impacts from climate change will require attention to more than just agricultural production. The evidence sustains the need for thoughtful investment in adaptation and mitigation actions toward an efficient management of climate change influences on food security.

breast cancer. Furthermore, it has been reported that the excessive use of pesticides in agriculture in order to improve the productivity contaminates food materials and can be responsible to induce breast cancer in women. It is also noted that some fungi produce several type of mycotoxins such us zearalenone, aflatoxin, and ochratoxin that are dangerous for human health and can especially cause breast cancer. Thus, the objective of this chapter is to discuss the experimental data regarding the involvement of heavy metals, pesticides, and mycotoxins as well as the recent insights on the molecular mechanisms involved in the progress of breast cancer.

Leukemia is a cancer that starts in blood forming cells which occurs in several forms of chronic or acute diseases. It is the most common cause of pediatric malignancy, accounting for approximately 25% of all cancers occurring before age 20. It represents 32% of all cancer cases occurring among children younger than 15 years of age, with an annual incidence rate of 43 cases per million. In the last decade of the 20th century, the occurrence of childhood leukemia has shown a rise. This disease, like most cancers, has a multifactorial etiological causal mechanism and a heterogeneous biological composition involving the interaction between different aspects originating from the environment as well as human genetics. This chapter discusses, through the current published literature, the relationship between cancer, particularly childhood leukemia, and environmental exposures to heavy metals, pesticides, and mycotoxins.

The liver is the main detoxifier organ of the body. When normal liver function is compromised, other systems in the body can be affected, including the brain. Hepatocerebral disorder is the term used to describe some neuropsychiatric conditions that result from liver failure and characterized by the accumulation of these toxic metals in brain. Examples of such disorders are Wilson's disease (WD), an autosomal recessive disorder that is characterized by the deposition of copper in liver and brain tissues and acquired (non-Wilsonian) hepatocerebral degeneration (AHCD), a complication that occurs most frequently in patients with hepatic coma or that suffered multiple episodes of severe HE. AHCD is characterized by accumulation in brain of manganese. This chapter will focus on the crucial importance of relationship between liver and brain functioning and on the effects produced when this relationship is compromised. Specifically, the chapter will discuss on the physiopathology of WD and AHCD and on the role that toxic metals play on neurological symptoms in such disorders.

This chapter presents a literature review on the effect of environmental changes factors exposure in the etiology of Alzheimer's and Parkinson's diseases. The use of pesticides is more intense and somehow erratic as it aims to face climate change consequences like drought and water scarcity. The rural population is getting to be more vulnerable to have these neurodegenerative diseases. However, intense food production and economic models mean also the use of heavy metals in many stages as well during the production and the consumption processes and practices. Evidence from experimental studies shows that such heavy metals may also be a factor for the occurrence of Parkinson's and Alzheimer's diseases. At least, the environmental lifestyle and, likely, genetic factors, individually and collectively, play a significant role in the etiology of the diseases.

Substantial epidemiological studies have established a link between environmental factors and prevalence of neurodegenerative diseases. Increasingly, some studies report a direct cause-and-effect relation between environmental agents such as heavy metals and pesticides, and some neurodegenerative disorders including multiple sclerosis (MS) and Huntington's disease (HD). Interestingly, high blood level of heavy metals

Panel on Climate Change (IPCC) reports and notably susceptible to such drastic climate-related health consequences. Climatic parameter change would directly affect disease transmission by acting on the vector's geographic range, activity, or reproduction and by reduction the period of pathogen incubation. This chapter will discuss the increasing risk of some vector-borne diseases in hazard-prone localities. It further identifies the severe challenges both of health adaptation to climate change by highlighting Moroccan adaptive capacity to such crises.

Leishmaniasis is a complex disease comprised of multiple organisms in association. Each of these organisms responds differently to external factors. The environmental and socio-economic associations contribute to the dynamics and emergence of leishmaniasis across the globe. Ecological dynamics of the vector-parasite-host system of leishmaniasis influenced directly and indirectly both human and animal health. The transitions and rapid climate and socio-economic changes caused a transition of emergence and re-emergence of leishmaniasis outbreaks. The pattern of changes is influenced by the distribution and abundance as well as the spatial dynamics of vector and reservoirs species, which in turn disrupts ecosystem structure of vector and parasite. In Morocco, leishmaniasis are endemic and constitute a major public health threat. The observations showed significant variations in its spatial distribution and forms through Morocco with increase in the number of recorded cases during the last couple of years. Here, the authors discuss disease change related to climate and socio-economic influence.

As in most arid and pre-Saharan North Africa regions, the local population is especially rural, socioeconomically vulnerable, and exposed to vector-borne disease like cutaneous leishmaniasis. Information on geographical and seasonal distribution of this disease can be helpful in the control of this disease. In this context, four sites were used to identify the seasonality of zoonotic cutaneous leishmaniasis (ZCL) in Middle Draa Valley (MDV). The seasonal occurrence of this disease was correlated with some climatic and hydrologic variables. The findings show that the most part of patients (86%) were from rural areas with a bit high rate of affected people is female gender (54%) and children less than 9 years (51%). The distribution of cases shows a prevalence of affected in the upstream area than in downstream where the climate is more and more arid. The highest number of cases were detected in the winter season (from December to March).

Malaria is a mosquito-borne infectious disease caused by obligate intraerythrocytic protozoa of the genus Plasmodium. As one of the most devastating global health issues, malaria is a sensitive disease to weather and climate conditions, in such a way the ongoing trends of increasing temperature and more variable weather could lead to malaria transmission spreading. Spatial and temporal variations in precipitation, temperature, and humidity that are projected to take place under different climate change scenarios will impact the biology and ecology of malaria vectors and subsequently the risk of disease transmission. Here, the authors review how climate and climate change may impact malaria transmission. They contrast ecological and behavioral characteristics of malaria vectors and parasites and how weather, climate, climate change, and socioeconomic factors may have very different impacts on their spatiotemporal occurrence and abundance and the resulting malaria risk.

As a result of increased frequency and intensity of heat waves, increased floods and droughts, change in climate will affect biological, physical, and chemical components of water through different paths thus enhancing the risk of waterborne diseases. Identifying the role of weather in waterborne infection is a priority public health research issue as climate change is predicted to increase the frequency of extreme precipitation and temperature events. This chapter provides evidence that precipitation and temperature can affect directly or indirectly water quality and consequently affect the health human. This chapter also highlights the complex relationship between precipitation or temperature and transmission of waterborne disease such as diarrheal disease, gastroenteritis, cryptosporidiosis, giardiasis, and cholera.

Preface

Among global environmental threats to human health, and according to the World Health Organization (WHO), are climate change, ozone depletion, changes in ecosystems services, alteration of hydrological systems and freshwater scarcity, land degradation, urbanization, and stresses on food-producing systems. Global environmental changes are intimately linked to increase and spread of human diseases. The WHO considers that nearly one third of worldwide diseases could be directly associated to environmental changes risk factors. Actually, environmental changes play a main role in the occurrence and spread of diseases related to food, water quality and environmental exposures to chemicals and toxins. Furthermore, extreme temperatures, climate-related disturbances have a direct influence on the emergence or re-emergence of vector-borne diseases and scourge associated to the distribution of venomous animals particularly in developing countries with poor access to sanitation, safe and sufficient water supply. This may be exacerbated as the human population grows and its density increases.

This book focuses on the global environmental changes and their effects on human health. The authors tried to cover and emphasize pesticides and heavy metals exposure in association with human diseases such as cancers, water and vector-borne diseases, neurodegenerative and neuropsychiatric disorders; food and water security, and climate changes and their effects on envenomation. They broadened the research horizon based of the latest findings with a purpose to help provide guidance towards identifying gaps and addressing future knowledge challenges?

The architecture of the book includes 24 chapters, with each focused on a specific topic to cover diverse perspectives. Four of these chapters focus on food and water security and exploring different aspects: the first chapter, “Space Weather and Link to Climate Change,” provides a brief description of the Sun-Earth connection and the Sun’s variability and the link between the Space Weather and Climate change. Chapter 2, “Vulnerability of Human Populations to Climate Change: Focus on Socio-Economic Factors,” highlights the vulnerability and adaptation”. It evaluates the vulnerability of human population to climate change through the introduction and analysis of the socioeconomic vulnerability index (SeVI) that measures regional socioeconomic vulnerability. In Chapter 3, “Climate Change and Water Security: Water Scarcity and Water Flood,” the authors review the effect of climate change on water scarcity and flood together with human’s leading behavior to water scarcity exacerbation including possible interventions that could mitigate water imbalance. Chapter 4, “Food Security and Climate Change,” focuses on climate change status in North Africa and the possible impact of the projected global climate change on agriculture with its different components and consequently on food security. Chapter 5, “Food Safety and Climate Change: Case of Mycotoxins,” provides an overview of the major mycotoxins toxicity and their negative economic impact together with the effect of climate change on their production as well as a the different strategies to reduce or otherwise eliminate the toxic effects of mycotoxins in contaminated foods and feeds by using chemical, physical, and biological/biotechnological methods and other innovative approaches. In Chapter 6, "Climate Change Outcomes on the Environmental Ecotoxicology," the author shows the likely influence of climate change on chemical exposure. The emphasis is on expected increases in the utilization of pesticides, heavy metals and hydrocarbons in the context of global warming. In the same context, Chapter 7, "Study of the Effect of Climate Changes on Well Water Contamination by Heavy Metals at a Mining Site in the City of Marrakech, Morocco," studies the spatial and seasonal variations of heavy metal deposition in groundwater in a mining site near the city of Marrakech (Morocco) to assess the extent of pollution generated by the mining activity and to identify the key mechanism responsible for this contamination and its relation to this mining activity. Chapters 8, "Prostate Cancer and Environmental Exposure: An Overview on Heavy Metals and Pesticides Involvement," develops another aspect of the possible impact of climate change on human health, while Chapter 9, "Breast Cancer With Relevance for Heavy Metals, Mycotoxins, and Pesticides," and Chapter 10, "Childhood Leukemia and Environmental Risk Factors," describe the role of heavy metals, pesticides and mycotoxins exposures in the pathogenesis of three different kind of highly frequent types of human cancers such as prostate adenocarcinoma, breast cancer and leukemia. The authors of these chapters provide a rich literature review of the possible carcinogenetic potential of those chemicals based on epidemiological as well as experimental investigations.

In addition to the possible relevance of climate change to human cancers, the book also investigates the ultimate outcomes of human contamination arising from climate change, on the neuropathogenesis of different mental as well as neurological disorders. Hence, in Chapter 11, "Metal Toxicity and Brain-Liver Axis: The Good, the Bad, and the Neurodegenerated," the authors focus on the crucial importance of the relationship between liver and brain functioning and on side effects produced when this relationship is compromised. Specifically the physiopathology of Wislson disease and Acquired hepatocerebral degeneration (AHCD) and on the role that toxic metals play on neurological symptoms in such disorders. Chapter 12, "Environmental Aspects of Alzheimer's and Parkinson's Diseases Neuropathologies: A Focus on Heavy Metals and Pesticides," aims at describing the effects of these heavy metals pesticides exposures and their relationship to the pathogenesis of the most common neurodegenerative diseases such as Parkinson's (PD) and Alzheimer's (AD), the chapter presents recent studies including experimental as well as epidemiological approaches supporting the involvement of environmental contamination by those chemicals, providing in some cases the molecular underlying mechanisms. Similarly, Chapter 13, "Neuropathogenesis of Multiple Sclerosis and Huntington's Disease: An Overview of Environment Patterns," and Chapter 14, "Amyotrophic Lateral Sclerosis Disease and Environmental Risk Factors: Role of Heavy Metals and Pesticides," shed light on the relevance of heavy metals and pesticides to the pathogenesis of rare neurodegenerative diseases such as multiple sclerosis, Huntington disease and Amyotrophic lateral sclerosis. These two chapters provide substantial evidence to support the non-genetic side of those neuropathologies and outline the usually neglected environmental part of the disease. Likewise, in Chapter 15, " Depression and Anxiety Emerging From Heavy Metals: What Relationship?" and Chapter 16, "Epilepsy and Stroke Emerging From Climate Change-Related Neurotoxicity: Involvement of Food and Water Contaminations," the authors try to cover other aspects of environmental contamination with heavy metals and pesticides as key elements in the pathogenesis of neuropsychiatric disorders such as depression, anxiety and epilepsy as well as other brain pathologies with increased incidence including stroke. A particular interest was given to bacterial and parasitic diseases arising from water and food contamination by microorganisms as a consequence of climate change. Chapter 17, "Vector Borne Diseases and Climate Change," highlights some diseases sensitive to climate change with particular attention in Morocco, while Chapter 18, "Climatic Factors Impacting Leishmaniasis Risk in a Global View: The Case of Morocco," addresses the environmental and socioeconomic factors influencing the leishmaniasis trend in Morocco in relation to climate change and suggests a linkage between change in leishmaniasis incidence over time across different areas of Morocco and the increase in population. In Chapter 19, "Climate Impact on Seasonal Distribution of Zoonotic Cutaneous Leishmaniasis in Southern Morocco," the authors conducted a study in the Middle Draa Valley (MDV), known to be among the most endemic areas for Zoonotic Cutaneous Leishmaniasis (ZCL), to assess the distribution and physical and anthropogenic characteristics of ZCL as well as explore its seasonal transmission rate and examine its transmission following the upstream-downstream situation of the sub-basin. Another aspect of vector born diseases is highlighted in Chapter 20, "Biology, Epidemiology, and Public Health Significance of Malaria Disease Linked to Climate Changes," in which the authors review the characteristics of Malaria transmission and their response to changes in weather and climate. In Chapter 21, "Waterborne Diseases Arising From Climate Change: An Overview on The Possible Link," and Chapter 22, "Climate Change and Fecal Peril: Possible Impacts and Emerging Trends," the authors provide evidence that precipitation and temperature can directly or indirectly affect water quality and consequently human health. They also highlight the complex relationship between precipitation or temperature and transmission of water-borne diseases such as diarrheal disease, gastroenteritis, cryptosporidiosis, giardiasis and cholera. Furthermore, the authors examine the health effects of climate change in developed and developing countries as well as water shortage associated with rapid urbanization and high population density. This is believed to have contributed to increased pressure on regional water resources leading to the use of urban wastewater as an alternative for irrigation, which is likely responsible of exacerbating the transmission of infectious diseases, including those caused by intestinal protozoa and helminths parasitic worms.

The last two chapters of the book, Chapter 23, "Climate Change, Scorpion Ecology, and Envenomation: What Are the Links?" and Chapter 24, "Climate Change Effects on Venomous Snakes: Distribution and Snakebite Epidemiology," address the link between climate and climate change and scorpion activity as well as scorpion envenomation and highlights how weather, climate, climate change and socioeconomic factors may have very different impacts on the spatiotemporal occurrence and abundance of scorpions. They also describe the resulting scorpion envenomation, along with a review of the possible relationship between climate changes and the increase of snakebites incidences.

Acknowledgment

Throughout this experience, we discovered that writing a book is harder than we thought and more rewarding than we could have ever imagined. Indeed, none of this would have been possible without the contribution of different parts. We are eternally grateful to the authors of the 24 chapters of our book; we would like to express our deepest appreciation for their seriousness, persistence and patience during all the steps of the book process.

A special and particular acknowledgement to the Editorial Advisory Board of our book for their generous help, and pertinent advices and guidance given throughout our book processing.

The quality of a book depends to a large extent on proper reviewing. We gratefully acknowledge the time and expertise devoted to reviewing all the book chapters by the reviewing board, who gave freely of their precious acknowledge and energy to the different book chapters meticulously, with particular attention to the scientific as well as the linguistic aspects.

The editors acknowledge IGI Global book development team for their confidence in our potential and supervision, especially to Amanda Fanton the Assistant Development Editor and Courtney Tychinski; Special Projects Coordinator for their continuous availability and assistance during the all steps of the book preparation.

We should not forget all scientists; authors of the numerous papers and reviews cited in the our book.

We are grateful to our professors and our PhD supervisors for the quality of their guidance and continuous encouragement.

Thank you all for your contribution.

Section 1

Sun–Earth Connection and the Sun's Variability in Relationship With Climate Change

Chapter 1
Space Weather and Link to Climate Change

Hamid Nebdi
Chouaïb Doukkali University, Morocco

ABSTRACT

Our nearest shining star, the Sun, source of radiations and energy, sometimes generates severe events and phenomena in space which can affect our technology and biosphere. On the other hand, space weather, as defined by National Aeronautics and Space Administration (NASA), is conditions on the Sun and in the solar wind, magnetosphere, ionosphere, and thermosphere that can influence the performance and reliability of space-borne and ground-based technological systems and can endanger human life or health. A brief description of the Sun-Earth connection is firstly presented. Secondly, a particular attention is given to highlight the Sun's variability and the link between the space weather and climate change by means of some recent studies.

INTRODUCTION

The Sun, our source of heat and light, continue shining and providing free energy to sustain life in our planet Earth. From photosynthesis, which is a radiation-matter natural interaction process and the principal generator of the oxygen, to the celestial dynamical processes such as earth tides, changing seasons and climate, the Sun is always present.

The Sun had also a great and important interest and significance in mythology and for different old civilizations such as for Egyptians who founded Heliopolis (City of the Sun) which it is now located in Ayn Shams, the Aztec Sun Stone (the Calendar Stone) which indicated that the Aztecs believed they were living in the fifth and last creation of the world. They called each creation a sun, because movement of the sun maintained human life. It is also cited in Holy books (Torah, Bible and Qur'an) of the three monotheistic religions; for example in the Noble Qur'an is mentioned 33 times and a Sûrah is especially dedicated to the Sun (Ash-Shams, Sûrah 91) with 15 Verse.

DOI: 10.4018/978-1-5225-7775-1.ch001

So, studying the behavior of our generous star and its caused phenomena was, is and will be a big and honorific both multidisciplinary and interdisciplinary challenge.

The first section is devoted to the Sun-Earth connection by introducing some necessary notions and definitions of key processes considered as drivers of the Space Weather. The second one is related to the variability of the Sun and a clarification of the link between the Space Weather and Climate change. A conclusion is given in the third section.

THE SUN-EARTH CONNECTION

In fact, our star is a giant sphere of electrically-charged hot gas. This charged gas moves, generating a powerful magnetic field, which goes through a cycle, called the solar magnetic cycle or simply, solar cycle. Its periodicity is 11 years or so, and during this magnetic field of the Sun completely flips. To flip back again it takes about another 11 years. The solar cycle affects activity on the surface of the Sun, such as sunspots which are caused by the Sun's magnetic fields.

Sunspots are sites of very strong magnetic fields that are cooler than the rest of the photosphere (see Figure 1). This is why they appear dark against the photosphere. A medium size sunspot is bigger than the Earth's diameter. Small sunspots may only last days, larger sunspots and sunspot groups may last several months. Sunspots usually come in groups with two sets of spots. One set will have positive or north magnetic field while the other set will have negative or south magnetic field. The field is strongest in the darker parts of the sunspots (the umbra), and is weaker and more horizontal in the lighter part (the penumbra). Sunspots have been, historically, important manifestation of variable solar activity.

It's known that solar activity is a very complex process, involving the whole Sun and its atmosphere, but sunspot numbers retain their value as a simple measure of solar activity. As the magnetic fields change, so does the amount of activity on the Sun's surface.

In addition, the Sun impacts not only our planet but all the solar system with electromagnetic radiations, from gamma rays to radio waves which takes about eight minutes to reach our land (only Infra-Red (IR), Visible and near Ultra-Violets (UV) rays) after all the dangerous ones have been absorbed by the different layers of our atmosphere (see Figure 2), and with particles intercepted by the magnetosphere.

These particles and radiations can be generated by different processes (Eddy, 2009):

1. **Solar Flares:** Solar flares appear as sudden and intense brightening in highly-localized regions on the surface of the Sun. The initial and brightest part of the ensuing whoosh of light usually lasts but a few minutes, and the sudden event will run its course, on average, in less than half an hour. Some of the effects on the Earth are almost immediate. Others last for several days. In the largest solar flares, the amount of energy released on the Sun from a relatively small region and in so short a span of time is far beyond all earthly experience.
2. **Solar Coronal Mass Ejections (CMEs):** CMEs are the product of impulsive changes in the corona, in which entire segments of the outer corona are driven outward from the Sun and ejected into the solar wind stream. They originate in regions of closed magnetic field, most often from the coronal streamers that extend outward from the disk of the Sun like the petals of a flower.
3. **Solar Wind:** In addition to heat and light the Sun also releases a continuous flow, called the solar wind, of atomic particles—protons, neutrons, electrons and ions of all the solar elements—that expands outward, night and day, in all directions everywhere. This never-ending flow of coronal

Figure 1. Illustration depicts Sun-Earth interactions that influence space weather (Credit: NASA's Goddard Space Flight Center)

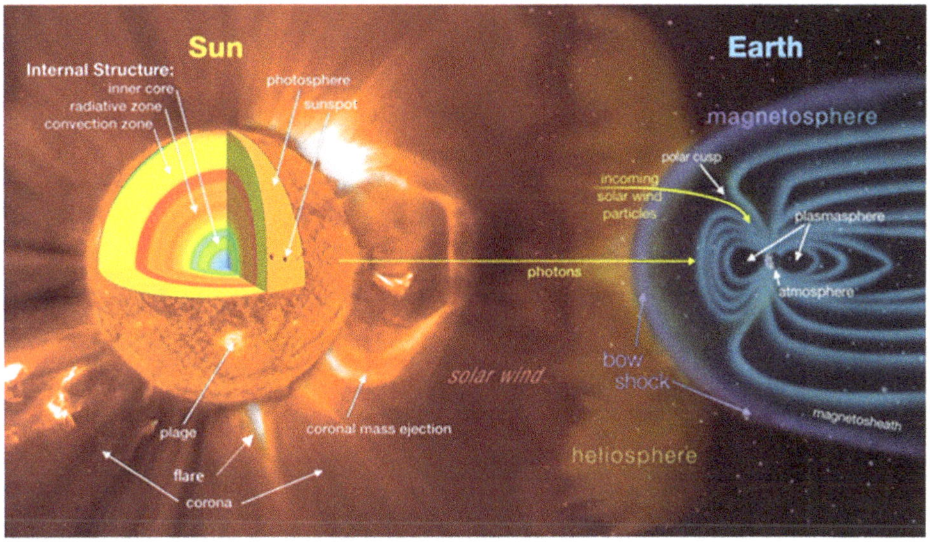

Figure 2. Layers of Earth's upper atmosphere (Credit: John Emmert/NRL, https://3c1703fe8d.site.internapcdn.net/newman/gfx/news/hires/thermosphere.jpg)

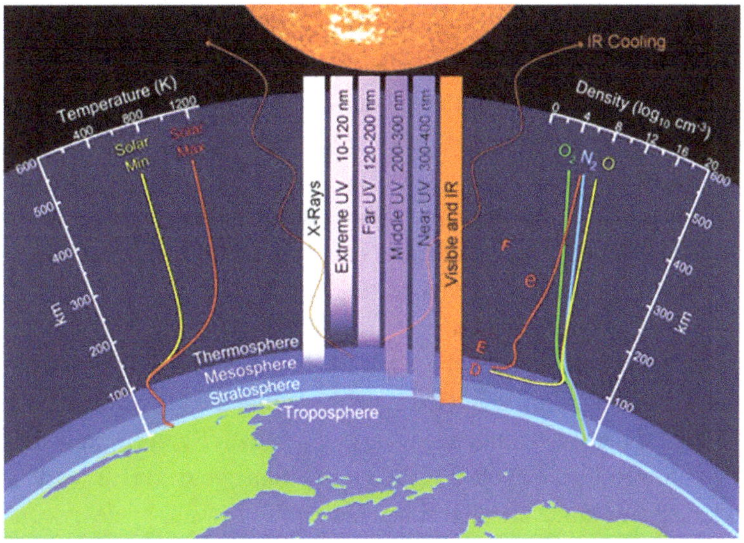

plasma is of considerable significance for the Earth. The solar wind expands out into interplanetary space and carves out a region surrounding the Sun called the heliosphere.

The different particles, generated by the above components of the solar activity, are magnetically driven in different regions of the space and particularly towards the Earth. Fortunately, our planet has a self-protection by means of the geomagnetic field lines which constitute the magnetosphere (see Figure

1); in simple words it's the earth's umbrella anti- particles. This natural shielding is deformed due to the huge impact of the solar wind which provokes geomagnetic storms at Earth and induces extra currents in the ionosphere and the ground.

THE VARIABILITY OF THE SUN AND THE LINK BETWEEN THE SPACE WEATHER AND CLIMATE CHANGE

The Sun's Variability and Space Weather

The processes discussed above constitute the drivers of the Space Weather and their generated elements are linked to the variability of the Sun. In other hand, solar variability can be divided into three components according to their influence on the structure and composition of our atmosphere (Hanslmeier 2004):

- Variation of the solar constant
- UV variation
- Energetic particle variation

Variation of the Solar Constant

The solar constant or the total solar irradiance (TSI) describes the radiant energy emitted by the sun over all wavelengths that falls vertically each second on one square meter outside the earth's atmosphere. In order to taking account of the global earth radiation budget, the solar constant is introduced in the different meteorological models, and is one of the drivers of the Climate change. The average value of the solar constant at the mean distance of the Earth from the Sun (referred to as the Astronomical Unit and equal to $1.496 x 10^8$ km) is about 1365 W/m^2.

Because of the influences of the Earth's atmosphere this constant is extremely difficult to measure on the surface and the most reliable measurements can only be done from space. TSI variations were undetectable until satellite observations began in late 1978 (Figure 3).

The 11-year solar cycle is explicit in irradiance measurements made by individual instruments and in the composite time series, according to which monthly averaged TSI values increase approximately 1.6 $W.m^{-2}$ (0.12%) between solar minima and maxima. Note that the TSI variations during cycles 22 and 23 are similar even though the peak annual mean sunspot number, Rz, a common index of solar activity shown in Figure 3c, was lower during cycle 23 (Rz = 119) than cycle 22 (Rz = 159). Rapid solar irradiance variations with larger amplitude are superimposed on the 11-year cycles; decreases on time scales of days to weeks can be as large as 4.6 $W.m^{-2}$ (0.34%). Dark sunspots and bright faculae are the two primary sources of solar irradiance variations. Both are magnetic features that occur in varying number and size on the solar disk, altering the Sun's radiative output affecting the Earth by respectively decreasing and enhancing the local emission. Because the solar activity cycle increases both the dark sunspot and bright facular contributions to TSI, the observed variability is the net effect of these two opposing influences, and thus is not a direct linear function of a single solar activity index such as sunspot number.

A stable record of solar irradiance variations combined with reliable global surface temperature observations can provide a tool for quantifying climate response processes to radiative forcing on decadal time scales.

Figure 3. (a) Space borne total solar irradiance (TSI) measurements are shown on "native" scales with offsets attributable to calibration errors. Instrument overlap allows corrections for offsets and the creation of a composite TSI record. (b) The average of three different reported composites [ACRIM, PMOD, and RMIB] adjusted to match the SORCE/TIM absolute scale. The grey shading indicates the standard deviation of the three composites. (c) The daily sunspot numbers indicate fluctuating levels of solar activity for the duration of the database. (Credit: Adapted from Kopp et al., 2011)

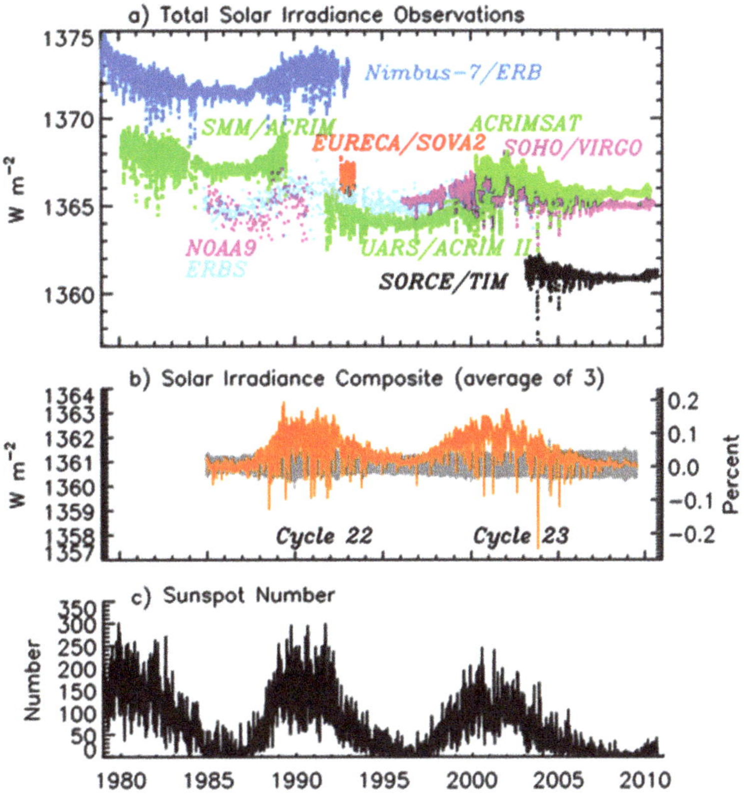

Variation of the UV Radiation

The primary energy source driving the earth's thermosphere and ionosphere is solar soft X-ray (XUV; 0.1-10 nm) and extreme ultra violet (EUV; 10-121 nm) radiation. It produces ionization, dissociation, excitation and heating, and its accurate knowledge is of prime importance for space weather applications. Solar irradiance varies with solar longitude on solar cycle, solar flare time scales, and solar rotation.

Solar radiation with a wavelength about 320 nm represents only 2% of the total solar irradiance; 0.01% of the incident flux is absorbed at about 80 km and 0.2% above 50 km. This radiation is extremely important since the thermal structure and photochemical processes above the troposphere are controlled by it. The stratosphere is controlled by absorption and dissociation of O_2 in the 175 to 240 nm range. The 205 to 295 nm range is predominantly absorbed by ozone O_3.

The short term variation of UV radiation is attributed to the evolution and rotation of plage regions on the solar disk. The XUV induced thermospheric temperature changes are depending to low and high solar activity.

Floyd et al. (2002) studied the solar UV irradiance variation during a period including the decline of solar cycle 22 followed by the rise of cycle 23, by using the daily solar measurements from the Solar Ultraviolet Spectral Irradiance Monitor (SUSIM) aboard the Upper Atmosphere Research Satellite (UARS). These measurements include scans over the wavelength range 115–410 nm at 1.1 nm resolution. As expected, the measured time series of UV irradiances reveal strong periodicities in solar cycle and solar rotation. For all wavelengths, the UV irradiance time series are similar to that of the Mg II core-to-wing ratio index: as implied by its name, the Mg II core-to-wing ratio index is the ratio of the UV spectral irradiance at the center of the feature to those of its adjoining wings, it has been shown (Heath and Schlesinger, 1986) to be an effective proxy for the variations in UV irradiance over a wide range of wavelengths). During solar cycle 22, the irradiance of the strong Ly-α line (Lyman-alpha line is a spectral line of hydrogen or H-like at 121,5 nm) varied by more than a factor of two. The peak-to-peak irradiance variation declined with increasing wavelength, reaching ~10% just below the Al edge at 208 nm. Between the Al edge and 250 nm the variation was ~6–7%. Above 250 nm, the variation declines further until none is observed above ~290 nm. Preliminary results for the first portion of cycle 23 indicate that the far UV below the Al edge is rising at about the same rate as the Mg II index while the irradiances in the Ly-α emission line and for wavelengths longer than the Al edge are rising more slowly — even after accounting for the lower level of activity of cycle 23.

On the other hand, Lilenstein et al. (2008) highlighted that the accurate knowledge of this flux is of prime importance for space weather applications that require nowcasting and forecasting of the solar XUV-EUV flux, such as satellite drag, telecommunication, positioning and classical weather. They reviewed present models and discussed how they account for the variability of the solar spectrum. They showed why the measurement of the full spectrum is difficult, and why it is illusory to retrieve it from its atmospheric effects. They also addressed the problem of determining a set of observations that are adapted for space weather purposes in the frame of ionospheric studies, and they reviewed the existing and future space experiments that are devoted to the observation of the solar XUV-EUV spectrum.

Variation of the Energetic Particles

There are three main contributions:

- **Electrons:** They reach the high latitude thermosphere after interaction with the geomagnetic field and acceleration;
- **High Energy Solar Protons:** Their flux is enhanced during periods of large flares;
- **Galactic Cosmic Rays:** They originate from outside the heliosphere but their input on Earth is partly controlled by solar activity.

During large flares, intense fluxes of energetic protons penetrate the Earth's polar cap regions. They produce ionization between 100 and 20 km. Such an event can last for a few hours to a few days by generating auroras borealis in the north regions and australis ones in the south regions. Large numbers of molecules are produced leading to a subsequent ozone depletion. Relativistic electron precipitation are possible sources for ionization and odd nitrogen production at altitudes above 80 km, thus well above the ozone layer.

To provide a long-term global picture of the radiation belt electrons and protons for understanding physical processes during different parts of the solar cycle, Li et al. (2001) have analyzed the long-term and continuous measurements of the solar wind and magnetosphere from the low-altitude and polar orbiting satellite SAMPEX (The Solar, Anomalous, and Magnetospheric Particle Explorer).

In Figure 4, they showed SAMPEX's measurements of radiation belt electrons and protons from launch on 3rd July 1992 to the end of 2000 together with the sunspot number and the Dst index (Disturbance storm time index is a major indicator of magnetic storm level, it has long been used as an indirect measure of the ring current around Earth caused by solar protons and electrons). They remarked that the inner proton belt ($L<2$, where L is the radial distance in R_E at the equator if the Earth's magnetic field is approximated as a dipole), is relatively stable and varies over the time scale of the solar cycle or longer and is anti-correlated with sunspot numbers as mentioned by Walt (1996) and Miyoshi et al. (2000). The slow variation of the inner proton belt is in contrast to the outer electron belt ($L>2$), which varies on a range of time scales. Several distinct features are notable in the long term electron measurements. The outer belt exhibits a strong seasonal and solar cycle variation. It was most intense, on average, during the descending phase of the sunspot cycle (1993-1995), weakest during sunspot minimum (1996-1997) and then became more intense again during the ascending phase of the solar cycle (1997-1999). Interestingly, the electrons are not most intense approaching or at sunspot maximum conditions. Seasonally, the outer belt is most intense as indicated by Baker et al. (1999) and also penetrates the deepest around the equinoxes. Equinox periods are marked by the vertical bars along the horizontal axis in Figure 4. Another remarkable feature of this Figure is the correlation of the inward extent of MeV electrons with the Dst index. For this interval of more than eight years, the two weakest electron periods corresponded to the only periods, summer of 1996 and 1999, when the averaged Dst was above zero. They concluded there work by the fact that long-term observations such as shown here demonstrate that the magnetosphere is strongly controlled by the solar wind. Large-scale magnetospheric features, such as currents, which

Figure 4. Selected SAMPEX measurements of protons of 19-27.4 MeV (#/cm²-s-sr-MeV) and electrons of 2-6 MeV (#/cm²-s-sr in logarithm, in bins of 0.1 L) since launch (July 3, 1992) and sunspot number and Dst index for the same period. The protons and sunspot numbers are window-averaged over a 9-month period and the electron and Dst index is window-averaged over a 30-day period in order to show the overall feature. The vertical bars on the horizontal axis are marks of equinoxes. (Credit: Li et al., 2001)

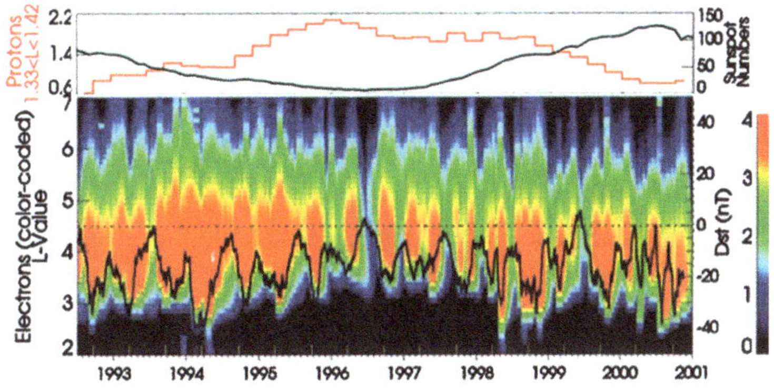

determine the Dst index, and radiation belt particles respond systematically to variations in the solar wind. The semiannual variation of the Dst index and MeV electrons deep in the inner magnetosphere can be attributed mostly to the equinoctial effect (orientation of the Earth's dipole axis relative to solar wind flow) with the axial (helio- graphic latitude) and Russell-McPherron (a larger z-component of the interplanetary magnetic field (IMF) in GSM coordinate) effects also contributing while the semi-annual variation of MeV electrons at geostationary orbit is attributed mostly to the semiannual variation of solar wind velocity as seen by Earth.

Space Weather and Its Effects

For the European Space Agency (ESA), Space Weather deals with phenomena involving ambient plasma, magnetic fields, radiation, particle flows in space and how these phenomena may influence man-made systems. In addition to the Sun, non-solar sources such as galactic cosmic rays can be considered as Space Weather since they alter space environment conditions near the Earth.

The Space Weather, as described above, generates sever events which have impact on technology, life and climate.

Effects to Technology

Different types of space weather events can affect different technologies in space and at Earth, and consequently, engender many big economical lose. For example:

- Solar flares can turn out strong x-rays that degrade or block high-frequency radio waves used for radio communication during events known as Radio Blackout Storms.

Nebdi et al. (2002) identified for the first time a lot of unknown particular phenomena (see Figure 5) observed on the SAFIR system (Système d'Alerte Foudre par Interférométrie Radioélectrique); it is the most accurate system used up to day in order to forecast thunderstorms and to alert for ground flash danger. By analyzing measurements from different locations in Belgium (four stations: Dourbes, La Gileppe, Mourcourt and Oelegem), in France (three stations: Arbois, Pic de Bure and Saint-Chaptes) and in Netherlands (three stations: Valkenburg, De Kooy and Deelen), and using the collected data from the GEOS satellites (Geosynchronous Operational Environmental Satellites) concerning the detected solar flares, they find that the origin of the observed phenomena on the collected data from the SAFIR system measurements is due to the effect of the solar flares on the very high frequencies (VHF), and the typical morphologies of the phenomena and their durations are related to the magnitude and duration of the solar flare.

- Solar Energetic Particles (energetic protons) can penetrate satellite electronics and cause electrical failure. These energetic particles also block radio communications at high latitudes in during Solar Radiation Storms.
- Coronal Mass Ejections (CMEs) can cause Geomagnetic Storms at Earth and induce extra currents in the ground that can degrade power grid operations, and accelerate erosion in pipelines and contribute to the environmental pollution by producing oil spills.

Figure 5. The phenomena observed in four Belgian stations in 21 August 1999 (Credit: Nebdi et al., 2002)

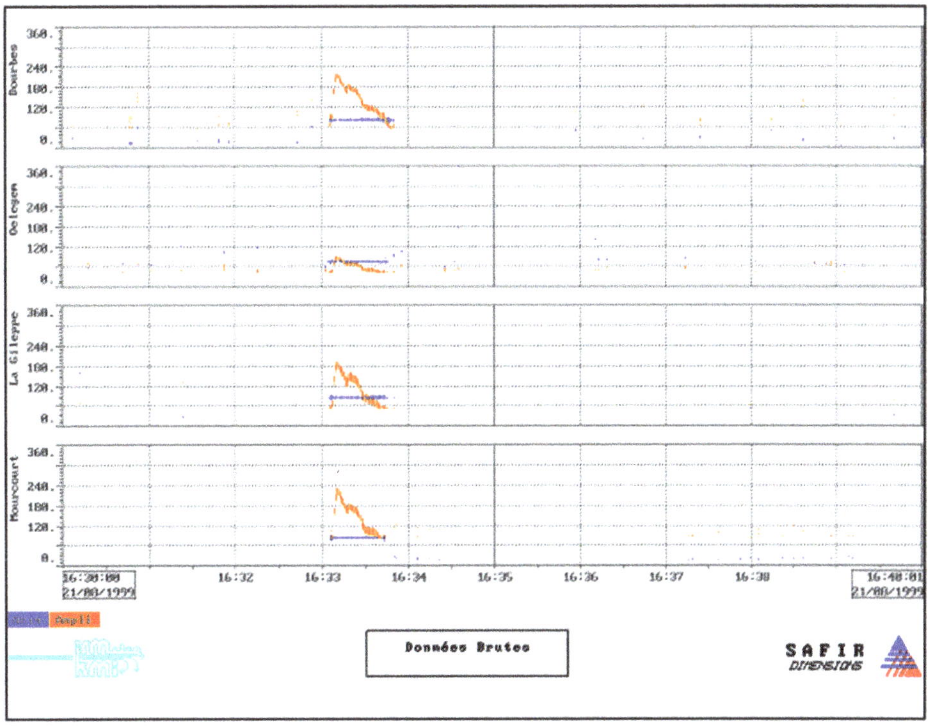

- Geomagnetic storms can also modify the signal from radio navigation systems (GPS, Galileo, …) causing degraded accuracy.

In the frame of the ESA Space Weather Pilot Project, Nebdi et al. (2004) established a correlation between highly disturbed geomagnetic activity and the occurrence of black or red conditions for Real-Time Kinematic (RTK) positioning. In fact, the RTK is a technique which allows a mobile user to measure its position in real-time with a centimetre-level precision by making use of differential corrections broadcast by a reference station. Black conditions mean extreme small-scale variability where RTK applications are severely degraded or even are impossible, and red ones mean severe to extreme small-scale (a few kilometres) ionospheric variability and severe degradations of RTK applications are expected. Based on the above correlation and on the forecasts of geomagnetic activity, the authors developed forecasts of RTK positioning conditions.

Effects to Life

Life on Earth for human or animals are also affected by the space weather events. Fortunately, the Earth's atmosphere and the geomagnetic field protect local life by shielding us against solar particle flows and dangerous radiations. Generally, magnetic fields can affect terrestrial life such as migrating animals and/or confuse and disorientate them like birds, bees, whales and dolphins by exploiting their internal 'compass,' which they use for orientation when traveling long distances. Thus, terrestrial life is connected to astronomical interrelations between different magnetic fields, particle flows and radiation.

Mass strandings or Multiple strandings of sperm whales (defined as two or more animals stranded at about the same time, in or near the same place) are infrequent but worldwide occurrences. Vanselow et al. 2016 investigated the possible reasons for this phenomenon based on a series of strandings of 29 male, mostly bachelor, sperm whales (Physeter macrocephalus) in the southern North Sea in early 2016. They conclude that whales'magnetic sense may play an important role in orientation and migration, and strandings may thus be triggered by geomagnetic storms. Ferrari (2017) focused his study on how cetacean life is influenced by disturbances in its environment that involves solar flares and cetacean beachings. Solar storms are caused by major coronal eruptions on the Sun. Upon reaching Earth, they cause disturbances in Earth's normally stable magnetosphere. Unable to follow an accurate magnetic bearing under such circumstances, cetaceans lose their compass reading while travelling and, depending on their juxtaposition and nearness to land, eventually beach themselves. His hypothesis was supported by six separate, independent surveys of beachings: in the Mediterranean Sea, the northern Gulf of Mexico, the east and west coasts of the USA and two surveys from around the world. When the six surveys were pooled (1614 strandings), a highly significant correlation (R^2= 0.981) of when strandings occurred with when major geomagnetic disturbances in Earth's magnetosphere occurred was consistent with his hypothesis. Whale and dolphin strandings in the northern Gulf of Mexico and the east coast of the USA were correlated (R^2= 0.919, R^2= 0.924) with the number of days before and after a geomagnetic storm. Yearly strandings were correlated with annual geomagnetic storm days. Annual beachings of cetaceans from 1998 to 2012 were linearly correlated (R^2= 0.751) with frequency of annual sunspot numbers. Thus, consistently strong statistical correlation evidence indicates that solar flares can cause cetaceans to change their behavior and become disoriented, which eventually causes them to swim onto a shore and beach themselves.

In other hand, the ionizing radiations are well known by their destructive influence on the human body. Of primary importance within the industry at the moment is the radiation hazard to aircrew and passengers. Increased radiation doses can generate many irreversible health problems.

In fact, the additional dose contribution to the radiation exposure at aviation altitudes during Solar Particle Events has been a matter of concern for many years. After the Halloween storms in 2003 several airlines began to implement mitigation measures such as rerouting and lowering flight altitudes in response to alerts on the National Oceanic and Atmospheric Administration (NOAA) S-scale (The S-scale for solar storm warnings was established in 1999 as a part of the NOAA Space Weather Scales, in order to provide information for the non-expert) regarding solar radiation storms. These alerts are based on the integral proton flux above 10 MeV measured aboard the corresponding GOES-satellite which is operated outside the Earth's atmosphere in a geosynchronous orbit. This integral proton flux has, however, been proved to be an insufficient parameter to apply to the radiation field at aviation altitudes without an accompanying analysis of the shape of the energy spectrum. Consequently, false alarms and corresponding disproportionate reactions ensued. Since mitigating measures can be quite cost-intensive, there has been a demand for appropriate space weather information among responsible airline managers for about a decade. Against this background, Meier et al. (2014) proposed the introduction of a new Space Weather index D, based on dose rates at aviation altitudes produced by solar protons during solar radiation storms, as the relevant parameter for the assessment of corresponding radiation exposure. The

Space Weather index D is a natural number given by a graduated table of ranges of dose rates in ascending order which is derived by an equation depending on the dose rate of solar protons.

In their ecological/populational retrospective study, Vieira et al. (2018) analyzed the correlation between the annual flux of local secondary galactic cosmic rays induced ionization (CRII) and mortality rates in the city of Sao Paulo, Brazil, between 1951–2012. The multivariate linear regression analyses adjusted by demographic and weather parameters showed that CRII are significantly correlated with total mortality, infectious disease mortality, maternal mortality, and perinatal mortality rates ($p<0.001$). But, the underlying mechanisms are still unclear. So, further cross-sectional and experimental cohort studies are necessary to understand the biophysical mechanisms of the association found.

Effects to Climate

There are other types of space weather that can impact the atmosphere. Energetic particles penetrate into the atmosphere and change the chemical constituents. These changes in minor species such as Nitrous Oxide (NO) can have long lasting consequences in the upper and middle atmosphere.

The duration of solar minimum may also have an impact on Earth's climate. During solar minimum there is a maximum in the amount of Cosmic rays, high energy particles, reaching earth.

Galactic Cosmic Rays Effects

Galactic cosmic rays (GCRs) are the main source of ionizing radiation in the lower troposphere, in which secondary products can penetrate the ground and underground layers. GCRs affect the physical–chemical properties of the terrestrial atmosphere, as well as the biosphere. GCRs are modulated by solar activity and latitudinal geomagnetic field distribution. However, these rays are not originated from the sun it's considered one of the drivers of the Space Weather.

Figure 6. Variations of low-altitude cloud cover (less than about 3 km), cosmic rays, and total solar irradiance between 1984 and 1994. (Credit: Carslaw et al., 2002)

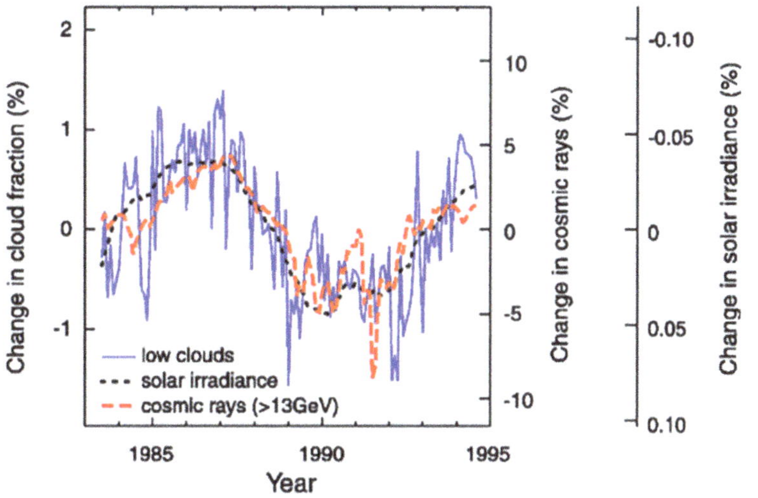

In Figure 6, the correlation between the variations of low-altitude cloud cover (less than about 3 km), cosmic rays and total solar irradiance, is presented. Carslaw et al. (2002) were able to show the very strong correlation between galactic cosmic rays, solar irradiance, and low cloud cover: When solar activity decreases, with a consequent small decrease in irradiance, the number of galactic cosmic rays entering the earth's atmosphere increases as does the amount of low cloud cover. This increase in cloud cover results in an increase in the earth's albedo, thereby lowering the average temperature. The sun's 11 year cycle is therefore not only associated with changes in irradiance, but also with changes in the solar wind, which in turn affect cloud cover by modulating the cosmic ray flux. This, it is argued, constitutes the strong positive feedback needed to explain the significant impact of small changes in solar activity on climate.

Solar Wind and Geomagnetic Effects

Prikryl et al. (2017) explored the influence of the tropospheric weather by solar wind through atmospheric vertical coupling downward control, by examining the occurrence of severe weather in the context of solar wind coupling to the magnetosphere-ionosphere-atmosphere (MIA) system. They observed that significant snowfall, wind and heavy rain, particularly if caused by low pressure systems in winter, tend to follow arrivals of high-speed solar wind. Previously published statistical evidence that explosive extratropical cyclones in the northern hemisphere tend to occur within a few days after arrivals of high-speed solar wind streams from coronal holes (Prikryl et al., 2009, 2016) is corroborated for the southern hemisphere. Cases of severe weather events are examined in the context of MIA coupling. Physical mechanism to explain these observations was proposed. The leading edge of high-speed solar wind streams is a locus of large-amplitude magneto-hydrodynamic waves that modulate Joule heating and/or Lorentz forcing of the high-latitude lower thermosphere generating medium-scale atmospheric gravity waves that propagate upward and downward through the atmosphere. Simulations of gravity wave propagation in a model atmosphere using the Transfer Function Model reveal that propagating waves originating in the lower thermosphere can excite a spectrum of gravity waves in the lower atmosphere. In spite of significantly reduced amplitudes but subject to amplification upon reflection in the upper troposphere, these gravity waves can provide a lift of unstable air to release instabilities in the troposphere and initiate convection to form cloud/precipitation bands. It is primarily the energy provided by release of latent heat that leads to intensification of storms. Their results indicated that vertical coupling in the atmosphere exerts downward control from solar wind to the lower atmospheric levels influencing tropospheric weather development.

Vieira and da Silva (2006) studied the geomagnetic modulation of clouds effects in the Southern Hemisphere Magnetic Anomaly through lower atmosphere cosmic ray effects. They showed that in the southern Pacific Ocean cloud effects on the net radiative flux in the atmosphere are related to the intensity of the Earth's magnetic field through lower atmosphere cosmic ray effects. In the inner region of the Southern Hemisphere Magnetic Anomaly (SHMA) it was observed a cooling effect of approximately 18 W/m^2 while in the outer region it was a heating effect of approximately 20 W/m^2. The variability in the inner region of the SHMA of the net radiative flux is correlated to galactic cosmic rays flux observed in Huancayo, Peru ($r = 0.73$). It was also observed in the correlation map that the correlation increases in the inner region of the SHMA. The geomagnetic modulation of cloud effects in the net radiative flux in the atmosphere in the SHMA is, therefore, unambiguously due to GCRs and/or highly energetic solar proton particles effects.

Elsner and Kavlakov (2001) explored the Hurricane intensity changes associated with geomagnetic variation. To analyze the possible relationship with North Atlantic hurricane intensification, they examined the geomagnetic data for ten days prior to 298 hurricanes (154 tropical-only hurricanes, 73 baroclinically-enhanced hurricanes and 71 baroclinically-initiated hurricanes) over 50 years (1950-1999). In fact, Hurricanes are mature tropical cyclones. Formed from low pressure areas over the warm oceanic waters, a developing tropical cyclone typically moves with the trade winds from east to west at low latitudes. A tropical cyclone reaches hurricane intensity when the rotational wind speed at 10 m above the ground exceeds 33 m/sec sustained over 1 min (maximum sustained wind speed). The authors identified a significant positive correlation between the averaged Kp index of global geomagnetic activity and hurricane intensity as measured by maximum sustained wind speed for baroclinically-initiated hurricanes (tropical cyclones that reach hurricane intensity having originated as baroclinic disturbances, such as a middle latitude trough; the term ``baroclinic'' refers to atmospheric processes that derive energy from thermal gradients on constant pressure surfaces. Their results were consistent with a mechanism whereby ionization processes trigger glaciation at cloud top which leads to hurricane intensification through upper tropospheric latent heat release. In 2008, Kavlakov et al. expanded their earlier work (Elsner & Kavlakov, 2001) by focusing on intensifcation rather than intensity and by examining hourly data. Their results appear to be more general in that there is no need to separate the tropical cyclones by type.

In other work, Hodges and Elsner (2010) investigated empirically the evidence linking solar variability with US hurricanes. First, a relationship between the probability of a US hurricane and the solar cycle were shown conditional on sea surface temperatures (SST). For years of above normal SST, the probability of three or more US hurricanes decreases from 40 to 20% as sunspot numbers (SSN) increase from lower to upper quartile amounts. Second, since SST is in phase with the 11-year total solar irradiance cycle but upper-air temperature is in phase with ultraviolet radiation changes on the monthly time scale, an anomaly index of SSN has been constructed. The index is significantly correlated with US hurricanes and major US hurricanes over the period 1866-2008. The chances of at least one hurricane affecting the United States in the lowest and highest SSN anomaly seasons are 68 and 91%, respectively. A similar relationship is noted using hurricane records spanning the period 1749-1850, providing independent corroborating evidence linking solar variability to the probability of a US hurricane.

According to El-Borie et al. (2012) study about the solar and geomagnetic activity effects on Global Surface Temperatures (GST), they mentioned that greenhouse gases (GHGs) and solar activity changes, the general main factors of the Earth's climate change, are not all-independent. They showed to what degree of connection, in the longer time (1880-2011), between changer of global surface temperature, and solar geomagnetic activist represented by sunspot number (Rz) and geomagnetic indices (aa, Kp).

Their results of spectral analysis (see Figure 8) revealed strong 21.3 year peak in GST than the 10.6 year peak, which is related to the changes in the polarity of main solar magnetic field; this obtained result demonstrate that the interplanetary magnetic field (IMF) effect is more powered on GST than the solar activity cycle. Significant peaks at 10.6 year are appear in aa, Kp and Rz series which is the most established cycle of solar activity. They also found that 10.6 year peak in Rz series is larger than the same peak in aa series this indicate the geomagnetic activity predominate over the solar activity in GST.

In order to investigate the solar-geomagnetic activity influence on Earth's climate, Mufti and Shah (2011) analyzed a long uninterrupted homogeneous data set on the annual mean Sea Surface Temperature (SST) anomaly records as a representative of the Earth's climatic parameter in conjunction with 158

Figure 7. Hodges and Elsner found that the probability of three or more hurricanes hitting the United States goes up drastically during low points of the 11-year sunspot cycle. Years with few sunspots and above-normal ocean temperatures spawn a less stable atmosphere and consequently more hurricanes. Years with more sunspots and above-normal ocean temperatures yield a more stable atmosphere and thus fewer hurricanes (Adopted from: D'Aleo, 2016)

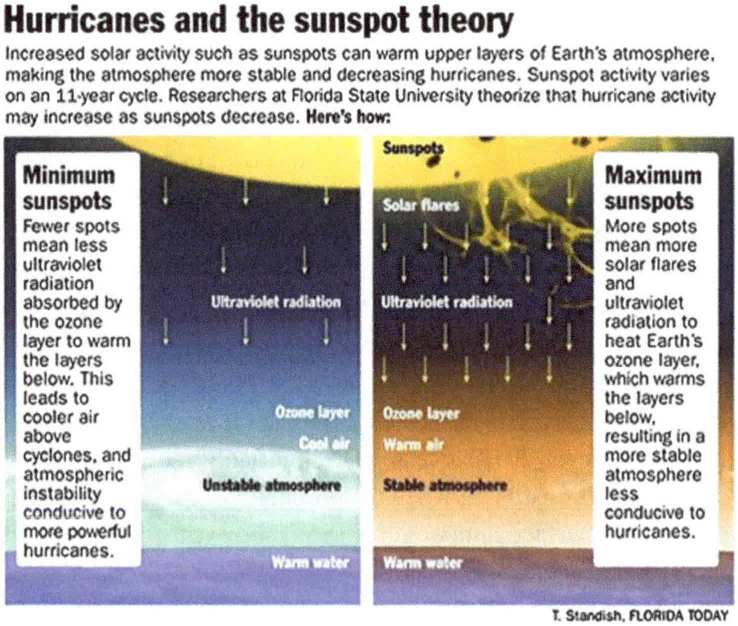

year (covering about 15 solar cycles) long time series on the annual sunspot indices, R_z and geomagnetic activity indices, aa for the period 1850–2007. The 11-year and 23-year overlapping means of global (δt_g) as well as northern (δt_n) and southern (δt_s) hemispheric SST anomalies reveal significant positive correlation with both Rz and aa indices. Rz, aa and δt_g depict a similar trend in their long-term variation and both seem to be on increase after attaining a minimum in the early 20th century (~1905). Whereas the results on the power spectrum analysis by the Multi-Taper Method (MTM) on δt_g, R_z and aa reveal periodicities of ~79-80 years (Gleissberg's cycle) and ~ 9-11 years (Schwabe solar cycle) consistent with earlier findings, MTM spectrum analysis also reveals fast cycles of 3–5 years. A period of 4.2 years in aa at 99% confidence level appears recorded in δt_g at ~ 4.3 years at 90% confidence level. A period of ~3.6-3.7 years at 99% confidence level found in δt_g is correlating with a similar periodic variation in sector structure of Interplanetary Magnetic Field (IMF). They found that this fast cycle parallelism is new and is supportive of a possible link between the solar-modulated geomagnetic activity and Earth's climatic parameter i.e. SST. The results point to the fact that geomagnetic forcing pre-dominates over the solar forcing on the climate.

Nebdi (2010) was interested by the study of the occurrence of the geomagnetic storms during the very geomagnetic activity days which characterized by a value more then 6 for the geomagnetic K index of Dourbes; in other words he considered the strong, severe and extreme geomagnetic days. A data base of 12045 days (three solar cycles: 20th, 21st and 22nd) was analyzed.

Figure 8. The normalized power spectra density for the GST, the geomagnetic indices aa and Kp and the sunspot number Rz (Credit: El-Borie et al., 2012)

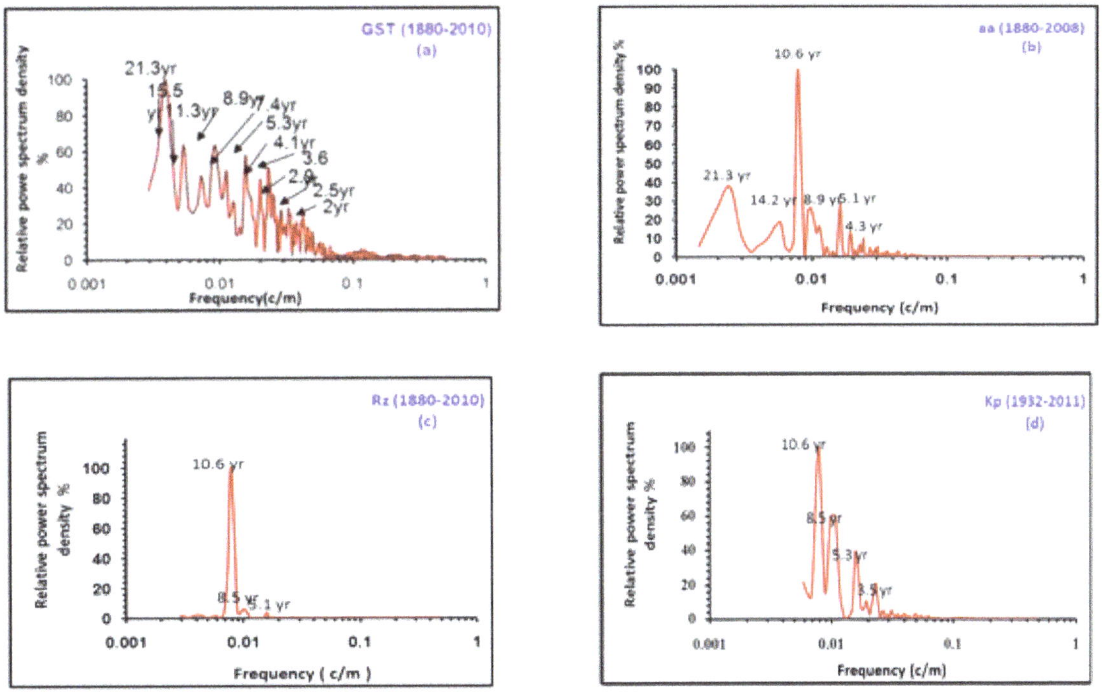

He showed that the occurrence of these very geomagnetic activity days was increasing cycle by cycle and there were 26, 29 and 37 very geomagnetic activity days for the three solar cycles, respectively. He concluded that 87% of the very geomagnetic activity days are located at the increasing and declining phases (25% in the increasing phase and 62% in the declining phase towards the minimum) of the solar cycles and 13% at the maximum (see Figure 9). In addition, he noted that the very geomagnetic activity months during the studied period are April, February and November. The "quiet" one is January.

The above effects indicate clearly that the space weather is not only a robust forcing driver for the climate change but, it is also for our technology and our biosphere.

CONCLUSION

From the arguments presented above, it's clear that the space weather is linked, without any doubt, to the climate change within the galactic cosmic rays, solar wind and geomagnetic activity as potential drivers and forcing factors.

On the other hand, multidisciplinary and interdisciplinary collaborations must be established, as they are the future for the well understanding of the above link, which will be more highlighted by using the new generation of super computers, artificial intelligence, big data collected from the past, present and

Figure 9. Distribution of the very geomagnetic activity days over the monthly sunspot numbers during the 20th, 21st and 22nd solar cycles; the triangles indicate the Very Active Days and the solid line the Sunspot Numbers (Credit: Nebdi, 2010)

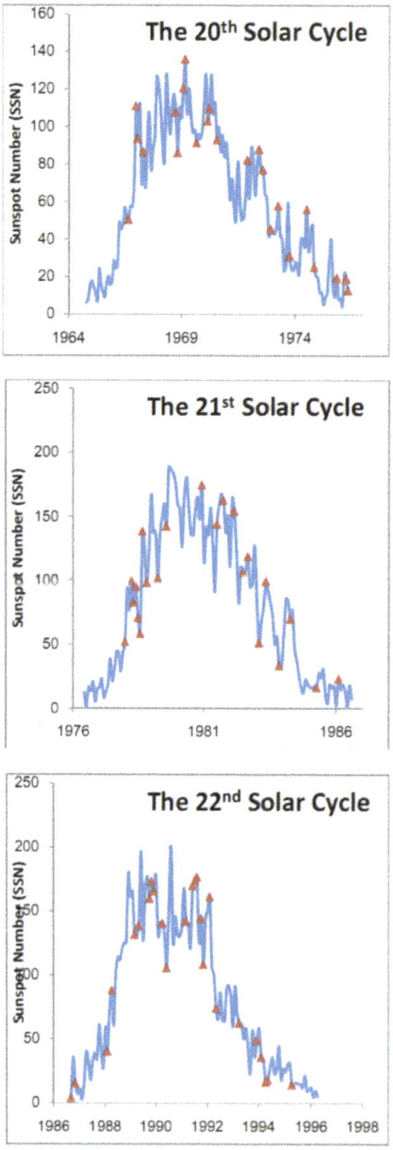

the future spatial missions and international programs such as Living with a Star, and Variability of the Sun and Its Terrestrial Impact (VarSITI).

Who knows, perhaps the connection between the Sun's kernel and the Earth's one will be possible to understand the physics behind!

REFERENCES

Baker, D. N., Kanekal, S. G., Pulkkinen, T. I., & Blake, J. B. (1999). Equinoctial and Solstitial Averages of Magnetospheric Relativistic Electrons: A Strong Semiannual Modulation. *Geophysical Research Letters*, *26*(20), 3193–3196. doi:10.1029/1999GL003638

Carslaw, K. S., Harrison, R. G., & Kirkby, J. (2002). Cosmic Rays, Clouds, and Climate. *Science*, *298*(5599), 1732–1737. doi:10.1126cience.1076964 PMID:12459578

D'Aleo, S. (2016). Solar Changes and the Climate. In D. J. Easterbrook (Ed.), *Evidence-Based Climate Science: Data Opposing CO_2 Emissions as the Primary Source of Global Warming* (pp. 263–282). Elsevier. doi:10.1016/B978-0-12-804588-6.00015-X

Eddy, J. A. (2010). *The Sun, the Earth, and Near-Earth Space: A Guide to the Sun-Earth System*. NASA.

El-Borie, M. A., Abd-Elzaher, M., & Al Shenawy, A. (2012). Solar and Geomagnetic Activity Effects on Global Surface Temperatures. *American Journal of Environmental Engineering*, *2*(4), 80–85. doi:10.5923/j.ajee.20120204.02

Elsner, J. B., & Kavlakov, S. P. (2001). Hurricane intensity changes associated geomagnetic variation. *Atmospheric Science Letters*, *2*(1-4), 86–93. doi:10.1006/asle.2001.0040

Ferrari, T. E. (2017). Cetacean beachings correlate with geomagnetic disturbances in Earth's magnetosphere: An example of how astronomical changes impact the future of life. *International Journal of Astrobiology*, *16*(02), 163–175. doi:10.1017/S1473550416000252

Floyd, L. E., Prinz, D. K., Crane, P. C., & Herring, L. C. (2002). Solar UV irradiance variation during cycles 22 and 23. *Advances in Space Research*, *29*(12), 1957–1962. doi:10.1016/S0273-1177(02)00242-9

Hanslmeier, A. (2004). *The Sun and Space Weather.* Dordrecht: Springer Netherlands.

Heath, D. F., & Schlesinger, B. M. (1986). The Mg 280-nm doublet as a monitor of changes in solar ultraviolet irradiance. *Journal of Geophysical Research*, *91*(D8), 8672–8682. doi:10.1029/JD091iD08p08672

Hodges, R. E., & Elsner, J. B. (2011). Evidence linking solar variability with US hurricanes. *International Journal of Climatology*, *31*(13), 1897–1907. doi:10.1002/joc.2196

Kavlakov, S., Elsner, J. B., & Pérez-Peraza, J. (2008). A statistical link between tropical cyclone intensification and major geomagnetic disturbances. *Geofísica Internacional*, *47*(3), 207–213.

Kopp, G., & Lean, J. L. (2011). A new, lower value of total solar irradiance: Evidence and climate significance. *Geophysical Research Letters*, *38*(1), L01706. doi:10.1029/2010GL045777

Li, X., Baker, D. N., Kanekal, S. G., Looper, M., & Temerin, M. (2001). Long Term Measurements of Radiation Belts by SAMPEX and Their Variations. *Geophysical Research Letters*, *28*(20), 3827–3830. doi:10.1029/2001GL013586

Lilensten, L., Dudok de Wit, T., Kretzschmar, L., Amblard, P.-O., Moussaoui, S., Aboudarham, J., & Auchère, F. (2008). Review on the solar spectral variability in the EUV for space weather purposes. *Annales Geophysicae*, *26*(2), 269–279. doi:10.5194/angeo-26-269-2008

Meier, M. M., & Matthia, D. (2014). A space weather index for the radiation field at aviation altitudes. *Journal of Space Weather Space Climate*, *4*, A13. doi:10.1051wsc/2014010

Miyoshi, Y., Morioka, A., & Misawa, H. (2000). Long Term Modulation of Low Altitude Proton Radiation Belt by the Earth's Atmosphere. *Geophysical Research Letters*, *27*(14), 2169–2172. doi:10.1029/1999GL003721

Mufti, S., & Shah, G. N. (2011). Solar-geomagnetic activity influence on Earth's climate. *Journal of Atmospheric and Solar-Terrestrial Physics*, *73*(13), 1607–1615. doi:10.1016/j.jastp.2010.12.012

Nebdi, H. (2010). Study of geomagnetic activity during three solar cycles using a K-index data base. *Umm Al-Qura University Journal of Applied Science*, *2*(1), 33–41.

Nebdi, H., Jodogne, J.-C., Crabbé, M., & Malcorps, H. (2002), *On the Identification of New Phenomena Observed on Safir System Measurements*. Paper presented at 2002' SAFIR Workshop, Budapest, Hungary.

Nebdi, H., Warnant, R., & Lejeune S. (2004). *Study of the correlation between small-scale TEC variability and geomagnetic activity.* Technical Report of Work Package 222. Solar Influences Data Centre Space Weather Pilot Project, ESA contract 16913/03/NL/LvH.

Prikryl, P., Bruntz, R., Tsukijihara, T., Iwao, K., Muldrew, D. B., Rusin, V., ... Stastný, P. (2017). Tropospheric weather influenced by solar wind through atmospheric vertical coupling downward control. *Journal of Atmospheric and Solar-Terrestrial Physics*, *171*, 94–110. doi:10.1016/j.jastp.2017.07.023

Prikryl, P., Iwao, K., Muldrew, D., Rusin, V., Rybansky, M., & Bruntz, R. (2016). A link between high-speed solar wind streams and explosive extratropical cyclones. *Journal of Atmospheric and Solar-Terrestrial Physics*, *149*, 219–231. doi:10.1016/j.jastp.2016.04.002

Prikryl, P., Muldrew, D. B., & Sofko, G. J. (2009b). The influence of solar wind on extratropical cyclones – Part 2: A link mediated by auroral atmospheric gravity waves? *Annales Geophysicae*, *27*(1), 31–57. doi:10.5194/angeo-27-31-2009

Prikryl, P., Rusin, V., & Rybanský, M. (2009a). The influence of solar wind on extratropical cyclones – Part 1: Wilcox effect revisited. *Annales Geophysicae*, *27*(1), 1–30. doi:10.5194/angeo-27-1-2009

Vanselow, K. H., Jacobsen, S., Hall, C., & Garthe, S. (2017). Solar storms may trigger sperm whale strandings: Explanation approaches for multiple strandings in the North Sea in 2016. *International Journal of Astrobiology*, 1–9. doi:10.1017/S147355041700026X

Vieira, C. L. Z., Janot-Pacheco, E., Lage, C., Pacini, A., Koutrakis, P., Cury, P. R., ... Saldiva, P. H. N. (2018). Long-term association between the intensity of cosmic rays and mortality rates in the city of Sao Paulo. *Environmental Research Letters*, *13*(2). doi:10.1088/1748-9326/aaa27a

Vieira, L. E. A., & da Silva, L. A. (2006). Geomagnetic modulation of clouds effects in the Southern Hemisphere Magnetic Anomaly through lower atmosphere cosmic ray effects. *Geophysical Research Letters*, *33*(14), L14802. doi:10.1029/2006GL026389

Walt, M. (1996). Source and loss processes for radiation belt particles. In Radiation Belts: Model & Standards. Academic Press.

KEY TERMS AND DEFINITIONS

Albedo: In Latin meaning "whiteness," it is the measure of the diffuse reflection of solar radiation out of the total solar radiation received by an astronomical body (e.g. a planet like Earth). It is dimensionless and measured on a scale from 0 (corresponding to a black body that absorbs all incident radiation) to 1 (corresponding to a body that reflects all incident radiation).

Cloud Cover: Known also as cloudiness or cloud amount, it refers to the fraction of the sky obscured by clouds when observed from a particular place.

Cosmic Rays: High energy particles, generally originating outside the Solar System. They are mainly 89% of protons (nuclei of hydrogen), they also include nuclei of helium (10%) and heavier nuclei (1%). When they arrive at Earth, they collide with the nuclei of atoms in the upper atmosphere, creating more particles.

Geomagnetic Storms: A momentary disturbance of the Earth's magnetosphere, which is associated with solar coronal mass ejections, coronal holes, or solar flares. It's caused by a solar wind shock wave when it strikes the geomagnetic field 24 to 36 hours after the event.

Hurricane: Tropical cyclone with winds of 119 kilometers per hour or greater that occurs particularly in the western Atlantic, that is usually accompanied by rain, thunder, and lightning, and that sometimes moves into temperate latitudes.

Ionosphere: The ionized part of Earth's (or any planet) upper atmosphere, from about 50 km 5 to 1,000 km altitude. The ionosphere is ionized by solar radiation. It plays an important role in atmospheric electricity and forms the inner edge of the magnetosphere. It has practical importance because, among other functions, it influences radio propagation to distant places on the Earth.

K-Index: Quantifies disturbances in the horizontal component of the geomagnetic field with an integer from 0 to 9, with 1 being calm and 5 or more indicating a geomagnetic storm. It is derived from the maximum fluctuations of horizontal components observed during a three-hour interval.

Kp-Index: Derived by calculating a weighted average of *K*-indices from a network of geomagnetic observatories.

Magnetosphere: The region around a planet dominated by the planet's magnetic field. Other planets in our solar system have magnetospheres, but Earth has the strongest one of all the rocky planets: Earth's magnetosphere is a vast, comet-shaped bubble. Life on Earth initially developed and continues to be sustained under the protection of this magnetic environment. The magnetosphere shields our planet from solar and cosmic particle radiation, as well as erosion of the atmosphere by the solar wind. It is generated by the convective motion of charged, molten iron, far below the surface in Earth's outer core. Constant bombardment by the solar wind compresses the sun-facing side of our magnetic field. The sun-facing side, or dayside, extends a distance of about six to 10 times the radius of the Earth. The side of the magnetosphere facing away from the sun - the night side - stretches out into an immense magnetotail, which fluctuates in length and can measure hundreds of Earth radii, far past the moon's orbit at 60 Earth radii.

Solar Coronal Mass Ejections: Large solar expulsions of plasma and magnetic field from the Sun's corona. CMEs travel outward from the Sun at speeds ranging from 250 km/s to 3000 km/s. The fastest Earth-directed CMEs can reach our planet in as little as 15-18 hours. Slower CMEs can take several days to arrive. They expand in size as they propagate away from the Sun and larger CMEs can reach a size comprising nearly a quarter of the space between Earth and the Sun by the time it reaches our planet.

Solar Cycle: Known also as solar magnetic activity cycle which have an average duration of about 11 years characterizing changes in the Sun's activity (including changes in the levels of solar radiation and ejection of solar material) and appearance (changes in the number and size of sunspots, flares, and other manifestations). Solar maximum and solar minimum refer respectively to periods of maximum and minimum sunspot counts. Cycles cover from one minimum to the next.

Solar Flare: An abrupt flash of increased brightness on the Sun, usually observed near its surface. The flare ejects clouds of electrons, ions, and atoms along through the Sun's corona into outer space, and also emits radio waves. According to the peak flux in watts per square metre (W/m^2) of X-rays with wavelengths 0.1 to 0.8 nanometer, as measured at the Earth by the Geostationary Operational Environmental Satellite (GOES) spacecraft, the classification system for solar flares uses the letters A ($< 10^{-7}$ W/m^2), B ($10^{-7} - 10^{-6}$ W/m^2), C ($10^{-6} - 10^{-5}$ W/m^2), M ($10^{-5} - 10^{-4}$ W/m^2) or X ($> 10^{-4}$ W/m^2).

Solar Wind: A flow of charged particles released from the upper atmosphere of the Sun, called the corona. This plasma consists of mostly electrons, protons and alpha particles with kinetic energy between 0.5 and 10 keV. Embedded within the solar-wind plasma is the interplanetary magnetic field. The solar wind varies in density, temperature and speed over time and over solar latitude and longitude. Its particles can escape the Sun's gravity because of their high energy resulting from the high temperature of the corona, which in turn is a result of the coronal magnetic field.

Sunspots: Areas where the magnetic field is about 2500 times stronger than Earth's one, much higher than anywhere else on the Sun. Because of the strong magnetic field, the magnetic pressure increases while the surrounding atmospheric pressure decreases. This in turn lowers the temperature relative to its surroundings because the concentrated magnetic field inhibits the flow of hot, new gas from the Sun's interior to the surface. One interesting aspect of the Sun is its sunspots.

Thermosphere: The layer of the Earth's atmosphere directly above the mesosphere and below the exosphere. Within this layer of the atmosphere, ultraviolet radiation causes photoionization/photodissociation of molecules, creating ions in the ionosphere.

Total Solar Irradiance: A measure of the solar power over all wavelengths per unit area incident on the Earth's upper atmosphere. It is measured perpendicular to the incoming sunlight. The solar constant is a conventional measure of mean TSI at a distance of one astronomical unit (AU).

Section 2

Climate Change Outcomes on Human Population Resources

Chapter 2
Vulnerability of Human Populations to Climate Change:
Focus on Socio-Economical Factors

Abdelkrim Ben Salem
Cadi Ayyad University, Morocco

Souad Ben Salem
Cadi Ayyad University, Morocco

Kholoud Kahime
Cadi Ayyad University, Morocco

Mohammed Messouli
Cadi Ayyad University, Morocco

Mohammed Yacoubi Khebiza
Cadi Ayyad University, Morocco

ABSTRACT

Moroccan ecosystems are considered endangered due to climate change that affects directly or indirectly different key features (biodiversity, snow cover, run-off processes, and water availability). The chapter describes the strategy for achieving collaboration between natural and social scientists, stakeholders, decision makers, and other societal groups in order to carry out an integrated assessment of climate change in the 12 Moroccan regions, with an emphasis on vulnerability and adaptation, and evaluate the vulnerability of human population to climate change applying the socioeconomic vulnerability index (SeVI) that measures socioeconomic vulnerability by regions. Result suggest that three southern and one north region are relatively more vulnerable, which are the most exposed to natural hazard. Furthermore, significant adaptive capacity scores are recorded in in the remaining regions, and average sensitivity for all. Therefore, societies and economies at all levels and on every region have to prepare for and adapt to impacts of climate change.

DOI: 10.4018/978-1-5225-7775-1.ch002

INTRODUCTION

Climate change is one of the key future challenges for human population, and the most serious announcements threatening the sustainability of development, with potential adverse effects on human health, food security, and socio-economic activity. The interface between exposure to the physical threats to human well-being and the capacity of people and communities to cope with those threats is vulnerability. Africa's vulnerability to climate change has been analyzed and projected in key sectors of the continent that protect human life, livelihoods and ecosystems. For example, the population at risk of increased water stress is projected to be 75 to 250 million by 2020 and 350 to 600 million by 2050. In some countries, yield reductions could reach 50% by 2020 (Arnell, 2004).

The vulnerability of water resources to climate change is dramatic (Immerzeel, et al., 2011). In terms of unmet demand, Morocco will see water shortages increase from 10 to 20 km3 in 2020 to 2030 and up to 40 km3 in 2040-2050. In addition, vulnerability in the water sector is characterized by climatic extremes, since drought sequences can alternate with flood sequences, as was the case, for example, in 1995 (Ourika), 1996 (El Jadida, Mohammedia, Beni Mellal, Casablanca, Kenitra), 2002 (Mohammedia), 2011 (Khenifra), 2010-2014 (Casablanca and Rabat) and. These floods are sometimes devastating, causing loss of life and significant economic damage (Messouli, 2014).

The annual cost of adaptation in the case of Morocco can range from 71 to 92 million USD for the specific needs of research, irrigation efficiency and rural roads (Rosegrant, 2009). Morocco gave a new impetus to the process of regionalization of the country. This project advanced in Morocco envisages the distribution of the country in 12 areas instead of 16, a reduction of 25%. This new regional configuration was made according to certain criteria based amongst other things on the principles of efficiency, homogeneity, proportionality and balance, but also of accessibility and proximity (CCR, 2011). If the questions are now more than answers, the position defended in this chapter is that the approach of the strengths and limitations of the territories by analyzing their vulnerability is particularly relevant, provided, however, to design the evaluation the level of vulnerability as a process rather than just a one-time operation. For this purpose, it is necessary to have elements of objective evaluation which will make it possible to better consider the vulnerability in the strategies and action plans for the adaptation. These elements of evaluation will be able to guide the decision makers and the government in its decisions. Vulnerability assessment methods have been developed in recent decades for natural disasters, food security, poverty analysis, livelihood sustainability and related areas. These approaches, each with its own nuances, provide a set of practices for use in climate change vulnerability and adaptation studies.

BACKGROUND

Vulnerability has received growing international recognition as an issue of central concern to the sustainable development of all countries. The vulnerability of a country, or its converse resilience, is the result of an interplay of factors, which can result in damage to social, economic or environmental systems (SOPAC, 2004). Therefore, development of vulnerability research and consequent adaptation policy has become top priority (Hinkel, 2011). The need for a socio-economic vulnerability index comes from the lack of practical recommendations on vulnerability to climate change that motivated the realization of this chapter. While there is significant literature on the impacts of climate change, this is not the case for policies and strategies for assessing vulnerability for information is limited. In order to demonstrate that planned activities will facilitate adaptation, the key vulnerabilities to climate changes need to be identified. It therefore seems essential to study the vulnerability of territories to future climate changes in order to determine a medium and long-term strategy. Adaptation policies are not intended to accept the inevitable, but to reduce the vulnerability of territories to the impacts of climate changes and put them in a position to take advantage of their beneficial effects (Messouli, Ben Salem, & Rochdane, 2013). Today, the need for a clear understanding of vulnerability and climate adaptation exists on many levels, from the local decision-making process to national development planning. Although efforts to develop vulnerability indices for countries are not new, popular focus has been on economic and social vulnerabilities, giving only a limited understanding of the overall problem. Various climate change assessment studies explore the vulnerability status for the poor whose livelihood is natural resource dependent (Ribot & Norton, 2009), which often leads to socioeconomic discrimination in the society (Ahmed, Diffenbaugh, & Hertel, 2009). For others vulnerability despite its conceptual applicability, questions remain about how to adequately integrate this forward-looking dimension into socio-economic vulnerability assessments of communities, sectors or regions. Notwithstanding this challenge, common among current assessment approaches is a reliance upon population statistics and the identification of variables that co-vary with increased vulnerability (e.g. the development of composite indexes) (Smith, Scott, Keys, & Smith, 2017). Despite the many challenges that exist in quantifying vulnerability, several parameters have been proposed and applied. These can be classified into two broad approaches: the indicator approach and the vulnerability variable assessment approach. In this paper, the indicator approach is used to identify vulnerable regions in Morocco for several reasons:

- Indices allow for an early approach and understanding of phenomena that can be very complex, such as poverty. This allows for adequate decision-making by the decision makers who are the first users of this type of tool (Eriksen & Kelly, 2007).
- At the global level, these indices allow comparisons between regions and countries and thus serve to determine area's most in need of assistance. At the local level, they reveal heterogeneities, which are masked on a small scale (Sullivan & Meigh, 2007), and target the priority areas once again.
- Finally, they allow situations to be compared at different time periods and thus to see an evolution and are therefore useful for evaluating the effectiveness of actions undertaken in the context of adaptation strategies (Eriksen & Kelly, 2007).

Need and Usefulness to Measure the Vulnerability of the Human Population to Climate Change in Morocco

National communications as well as all the reports consulted in relation to climate change in Morocco did not address the concept of vulnerability as defined by the IPCC (exposure, sensitivity, adaptive capacity). The term vulnerability is a priori often advanced subjectively to designate different meanings: impact, damage, resilience, sensitivity etc. (Messouli, Ben Salem, & Rochdane, 2013). Vulnerability assessment is needed to: i) identify the importance and location of short - and long - term threats, ii) respond to the impacts of natural disasters and climate change, iii) facilitate early warning to natural disasters and adaptation to climate change, iv) understanding adaptation vulnerability, v) helping identify and classify current and potential adaptation needs, and vi) guiding and supporting appropriate interventions.

As vulnerability assessment has not yet been discussed in detail, all of these findings imply new challenges for Morocco and suggests several actions to implement, including:

- Create a vulnerability scale to predict which regions are most vulnerable to climate change and that could, as a result, contribute significantly to making the country more and more disorganized and potentially unstable;
- Identify tools to assess vulnerability and adopt "no regrets or remorse" strategies such as increasing water management capacity; prepare adaptive responses; explore the local implications.

The Socioeconomic Vulnerability Index (SeVI)

The SeVI was developed by using a Composite Indicator Framework method, which consists of three main dimensions: adaptive capacity, sensitivity and exposure (Krishnamurthy, Choularton, Betts, & Lewis, 2011 and IPCC, 2007). These dimensions were further divided into five domains: demographic, social, economic, physical, and exposure to natural hazards. Five domains comprised of 24 indicators (for details see Table 1) formed the SeVI. Since each of the indicator was measured on different scale, it was necessary to standardise each as index value (Ahsan & Warner, 2014).

The "Vulnerability Sourcebook", which was commissioned by the Deutsche Gesellschaft für Internationale Zusammenarbeit (GIZ) and developed jointly by adelphi and the European Academy of Bozen (EURAC), offers a comprehensive tool for conducting regular vulnerability assessments. The Sourcebook provides a standardised approach to vulnerability assessments covering a broad range of sectors and topics (e.g. water sector, agriculture, fisheries, different ecosystems) as well as different spatial levels (community, sub-national, national) and time horizons (e.g. current vulnerability or vulnerability in the

Figure 1. Impact chain of vulnerability assessment on the potential impact, socio-economic factors

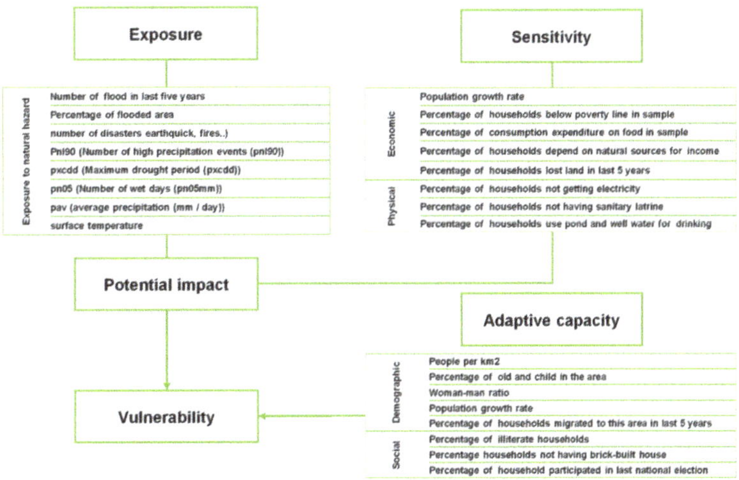

medium- to long-term (Fritzsche & al, 2017). The Vulnerability Sourcebook is based on the most widely used definition provided by the AR4 of the IPCC.

DOMAIN OF STUDY

The study is carried out over domain of Morocco region. This country is divided into three topographical regions: i) an open area in the northwest which is enclosed by Rif Mountains. ii) the mountains and uplands in the eastern and southern areas which include the Atlas Mountains. iii) The coastal plains of the west which include the fertile plains characteristic to agricultural zone. These distributed topographical units are largely a function of climatic and edaphic variables and disturbance regimes. It can exhibit some various local micro-climates (Driouech, Déqué, & Mokssit, 2009) and consequently a different vegetal covert (Benabid, 1982).

Climate Change: Global Appraisal

Climate change refers to a change in the state of the climate that can be identified (e.g., by using statistical tests) by changes in the mean and/or the variability of its properties, and that persists for an extended period, typically decades or longer (IPCC AR4, 2007). Climate change may be due to natural internal processes or external forcings such as modulations of the solar cycles, volcanic eruptions, and persistent anthropogenic changes in the composition of the atmosphere or in land use. Note that the Framework Convention on Climate Change (UNFCCC), in its Article 1, defines climate change as: "a change of climate which is attributed directly or indirectly to human activity that alters the composition of the global atmosphere and which is in addition to natural climate variability observed over comparable time periods." The UNFCCC thus makes a distinction between climate change attributable to human activities altering the atmospheric composition, and climate variability attributable to natural causes (UNFCCC, 1992).

Climate change events can reduce agrifood productivity, which in turn affects food security and raises food prices. Associated with other socio-economic factors, such as population growth, these consequences can lead to social unrest, riots and even civil war. Climate change can also affect local and transnational water availability. And water scarcity can lead to public unrest. In addition, transboundary watersheds may experience competition among states as water becomes an increasingly valuable resource in a warming world. Environment-related migration is expected to intensify as populations move away from drought-stricken areas, resulting food insecurity, and are forced to adapt to a changing climate.

Vulnerability of Human Populations to Climate Change

According to the 4th IPCC Report, vulnerability refers to the degree to which a system is likely to be negatively affected or adversely affected by the adverse effects of climate change, including climate variability and extreme events. Vulnerability depends on the character, magnitude, and rate of climate change a system is exposed to, its sensitivity, and its adaptive capacity (IPCC AR4, 2007).

Figure 2. Map of regions in Morocco

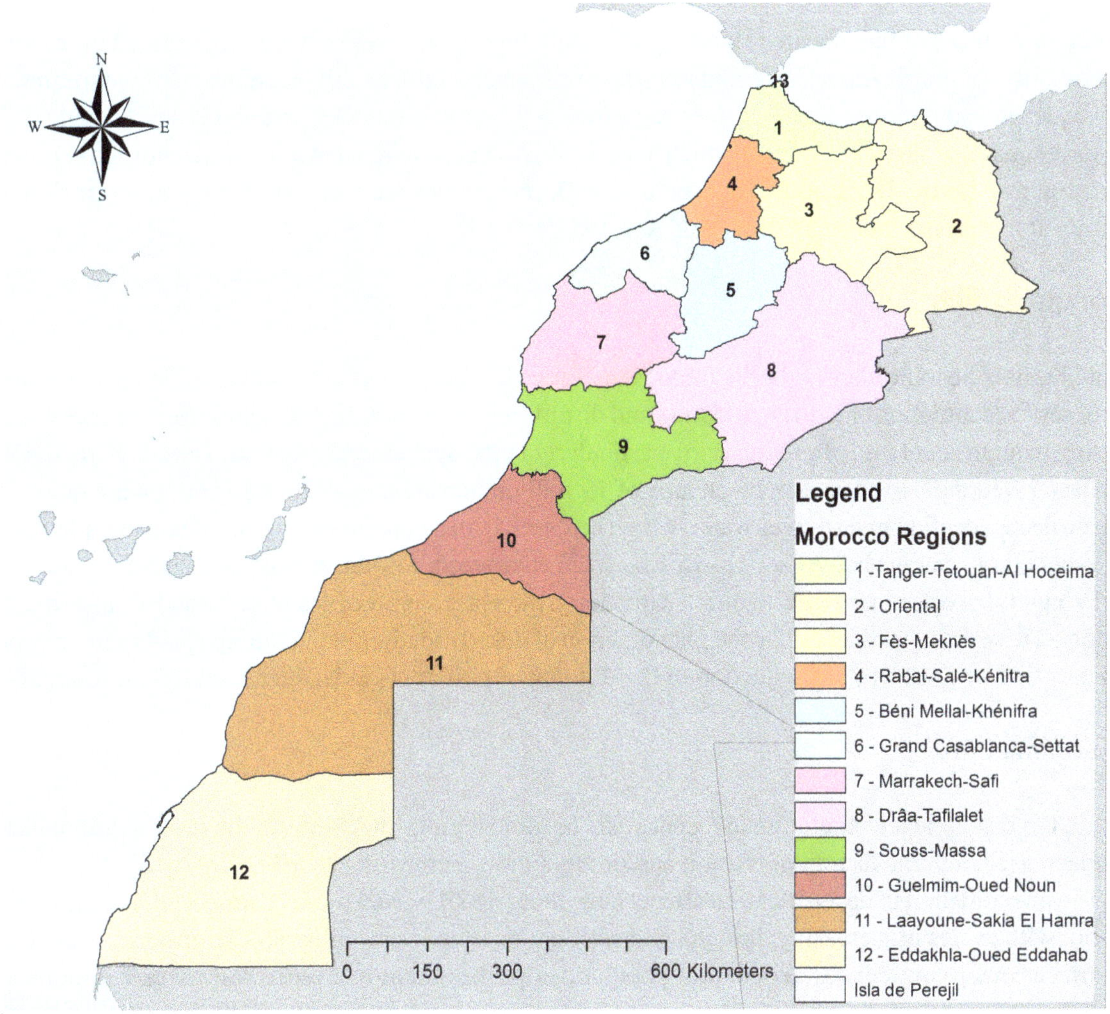

To present the observed effects of climate change, researchers have drawn on the four key components of vulnerability analysis that determine whether, and to what extent, a system is likely to be impacted by climate change: exposure, sensitivity, impact potential and adaptability.

Main Indicators of Exposition

Of all the components which contribute to vulnerability, exposure is the only one directly linked to climate parameters, that is, the character, magnitude, and rate of change and variation in the climate. Typical exposure factors include temperature, precipitation, evapotranspiration and climatic water balance, as well as extreme events such as heavy rain and meteorological drought (IPCC, 2001). Morocco is exposed to climatic, meteorological, geological or biological phenomena: major risks such as floods, torrential floods, earthquakes and landslides (Messouli, 2014).

Inundations

The consequences of the repeated floods observed in the region in recent years highlighted the extreme vulnerability of North American countries (the importance of the health, economic and environmental consequences) and the weakness of their response capacity. The expected rise in sea levels could affect the lower coastal cities of the region, which have the highest population densities and important economic infrastructure (agricultural, industrial and tourism), these areas are threatened by floods, erosion and submersion. (Ezzine, May 2015, and ZIC & TARGA, 2014)

Drought Period

Droughts have become more and more frequent in recent decades, accentuating desertification (advance of the desert, accentuation of land degradation and disturbing the already fragile equilibrium of ecosystems, leading to significant loss of productivity, particularly in the agricultural, oasis and forestry and the loss of coastal wetlands. In Morocco, on an area of 19 million hectares, more than 17 million are degraded. The projected decline in resources water, which are already in a situation of significant overexploitation, is evaluated on average between 10 to 15% by 2020 (Driouech, Déqué, & Mokssit, 2009).

Droughts have become with a significant impact on the water supply of populations and socio-economic sectors. These droughts accentuate the phenomenon of desertification, salinization of soils and increased water pollution caused by the reduction of the dilution of pollutants in watercourses. (Rosegrant, 2009)

Desertification

Due to the rise of heat temperature in recent years in some regions of Morocco, the level of groundwater has decreased, and the factors of erosion and drought have increased. (HCEFLCD, 2018)

The phenomenon of desertification affects large areas and it is even more pronounced that the climate is arid, with cycles of increasing drought, and soils are poor and very vulnerable to erosion. In addition, the precarious living conditions of rural populations pushes them to overexploit natural resources to meet their growing needs, which further amplifies environmental degradation. The factors that favor the triggering and accentuation of desertification are arid climate, which affects nearly 93% of the national territory. In addition to these natural constraints, there is the continued degradation of the vegetation cover, which is constantly being solicited to satisfy the needs of the populations in agricultural land, wood and fodder resources for the livestock. Both factors lead to poverty and soil fragility. This fragility is characterized by human uses incompatible with the principle of conservation (Mhirit, et al, 2011)

Main Indicators of Sensitivity

Sensitivity determines the degree to which a system is adversely or beneficially affected by a given climate change exposure. Sensitivity is typically shaped by natural and/or physical attributes of the system including topography, the capacity of different soil types to resist erosion, land cover type…

The different aspects of climate change sensitivity can be addressed by a thematic classification, according to the sector on which effects are observed. In this work, researchers have limited themselves to the study of economic and social indicators. (Fritzsche & al, 2017)

Economic Sector

Agriculture is highly sensitive to climate, both in terms of longer-term trends in the average conditions of rainfall and temperature, which determine the productivity and spatial distribution of crops, but also in terms of year-to-year variability and the occurrence of droughts, floods, heat waves, frosts and other extreme events. Direct effects are primarily expected from higher CO2 levels resulting in increased biomass production and water use efficiency. Indirect effects come through changes in climatic variables, such as temperature, precipitation, radiation, humidity and extreme weather events, which affect plant water uptake, occurrence of weeds, pests and diseases, soil moisture, and ultimately influence crop growth.

Water is already scarce, is also subject to the continuous increase of the needs, due to the fast evolution of the population, the improvement of the standard of living, the industrial development and the extension of the irrigated agriculture. Pressure on water resources is accompanied by increasing and increasingly serious deterioration of their quality.

Under the effects of climate change, the situation of the water resources of the country already critical may then become a problem impeding any further development, under the effect of expected climate change. These climate changes will have direct and indirect negative consequences on the water resource potential, both in terms of quantity and quality, of water needs and the efficiency of use of this resource for different uses (Messouli, Ben Salem, & Rochdane, 2013).

There are many weather-related effects on infrastructure performance which differ very much between geographical locations. Moreover the severity of the effects depends on infrastructure design, age and usage. Consequences of climate change will both be negative and positive for transportation infrastructure such as for rail, road, shipping and aviation, but will differ from region to region. In particular, the projected increase in frequency and intensity of extreme weather, such as heavy rain (e.g. causing floods), heavy snowfall, extreme heat and cold, drought and reduced visibility can enhance negative impacts on the transport infrastructure, causing injuries and damages as well as economic losses. (CSWD, 2013).

Social Sector

The IPCC concludes, with a high degree of confidence, that climate change will lead to increased heat-related morbidity and mortality, a decrease in cold-related mortality in temperate zones, a greater frequency of outbreaks of infectious diseases as a result of floods and storms, and will have a significant impact on the health of populations displaced by sea level rise and increased storm surges (WHO, 2004).

In addition, many important diseases are highly sensitive to changes in temperature and precipitation patterns. These are, for example, common vector-borne diseases such as malaria and leishmaniasis, as well as other major killers such as malnutrition and chronic diseases. Climate change is already contributing to the global burden of disease and this phenomenon is expected to increase in the future.

Among the consequences also on the populations and the social bond, the isolation of people living alone following a flood, or the difficulties met by the fragile old people during the episodes of heat wave (ADEM, 2012). Climate change will feed into existing conflicts with dwindling resources, especially when access to these resources is political power (HREC, 2008).

Main Indicators of Adaptability

Adaptive capacity refers to 'the ability of a system to adjust to climate change (including climate variability and extremes) to moderate potential damages, to take advantage of opportunities, or to cope with the consequences' (Parry et al. 2007). It is used to describe the various socio-economic, structural, institutional and technological abilities of a human system to produce adaptation measures.

In the face of the consequences of climate change, Morocco intends to preserve its territory and its civilization in the most appropriate way, anticipating an adaptation policy that prepares the whole of its population and its economic actors to face these vulnerabilities, by helping cities to become more resilient, the Bank will strive to replicate Morocco's successful disaster risk management experience through the introduction of early warning systems, flood protection infrastructure and national insurance against disasters (World Bank, 2016).

Socio-Economical Factors Increasing Climate Change Impacts

Pollution

Our planet is more and more polluted by human activities (industry, transport, agriculture ...) climate change is a consequence of this pollution. It is reflected by an increase in the temperature on the surface of the earth, but also by the intensification of the natural disasters and their magnitude (flood, drought, cyclones ...) the human activities, and in particular the gas emissions with effect of greenhouse are one of the main causes of the current climate change

Poverty and Precociousness

The poor - who suffer from poverty and live just above the poverty line - are already the most vulnerable to climate change. They have little to adapt to or quickly recover from climate shocks and often live in the most vulnerable areas because of low prices, such as houses on the sides of flooded or flooded water tables or on the edges of hills subject to Landslides or agricultural lands not supplied with clean drinking water (World Bank 2015).

The lack of rain and the difficult agricultural conditions have contributed significantly to the migration of the village to the city or to European countries due to the inability to adapt to the drought-causing climatic changes.

Demography Trend

Urbanization is one of the most important and sustainable forms of land transformation, and the extent of its growth is at least proportional to population growth and economic development.

As more people live in cities, more land is converted into buildings, sidewalks and roads, which in turn warm the air and increase the risk of sudden floods.

By giving the example of the city of Marrakech in Morocco, urbanization resulted in a warming of 1.60 ° C during the summer days and 0.87 ° C during the night. During the tourist peak of the summer months, the population of the Mediterranean region of Morocco doubles with seasonal tourism, which is also favored by the urbanization of the coasts. This urbanization has contributed to the erosion of Morocco's sandy beaches and increased the region's vulnerability to coastal flooding (Bounoua, 2017)

Moroccan demographic change is believed to be in urban areas, due in part to rural exodus and urbanization of rural areas. In 2050, Moroccan cities were home to 73.6% of the country's inhabitants, compared to 60.3% in 2014, with revenues of 32.1 million and 20.4 million. The rural population, on the other hand, would experience a slight decline in its workforce; 13.4 million people in 2014, it would be 11.5 million by 2050 (MHCP 2017).

RESULTS

Domain-Wise Vulnerability

Demographic Vulnerability

Grand Casablanca-Settat was found as demographically the most vulnerable region with a weighted average score of 0,81 (± 0,246) whereas Guelmim-Oued Noun was found as the least vulnerable region with a score of 0.36 (± 0,42) (see detail in Table 1 and Figure 3).

Study findings indicated Casablanca-Settat as the most densely populated region with 353 persons/ km2 (population density in casablanca city is 14200 inhabitants per square kilometer, fifth in the world and first in Africa) and Eddakhla-Oued Eddahab as the least densely populated region with 1persons/ km2. Regarding the reasons that drive people to migrate to cities, it is usually easier to find better-paid work in cities than in rural locations. Drâa-Tafilalet occupied the highest percentage (40,58) of elderly people and children (i.e. highest dependency ratio) while Eddakhla-Oued Eddahab occupied the lowest percentage (31,75) (average percentage: 37,13 ± 2,37). Eddakhla-Oued Eddahab possessed highest male-female ratio of 1.38 whereas Guelmim-Oued Noun possessed the lowest ratio of 0.95 (average ratio: 1,2

Figure 3. Domain-wise vulnerability scorss

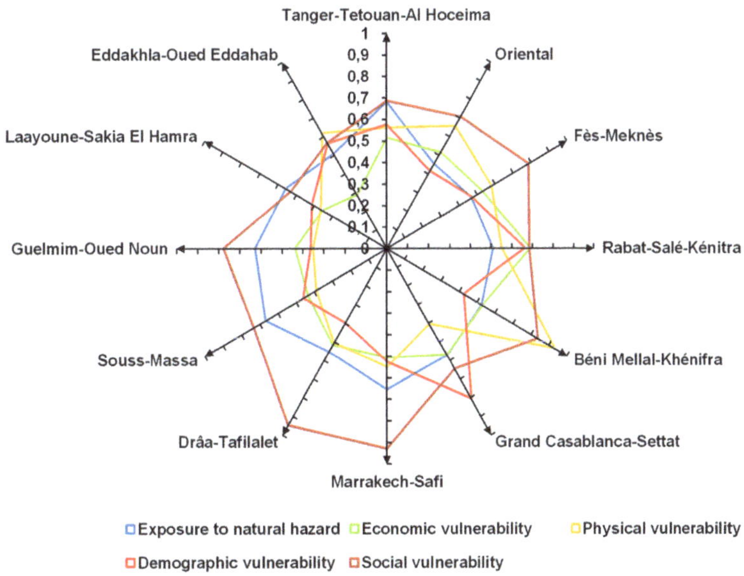

Table 1. Indicator index scores and overall SeVI scores of regions in Morocco

Domain		Indicators	Tanger-Tetouan-Al Hoceima	Oriental	Fès-Meknès	Rabat-Salé-Kénitra	Béni Mellal-Khénifra	Grand Casablanca-Settat	Marrakech-Safi	Drâa-Tafilalet	Souss-Massa	Guelmim-Oued Noun	Laayoune-Sakia El Hamra	Eddakhla-Oued Eddahab
Exposure	Exposure to natural	Number of flood in last five years	0,73	0,09	0,36	0,91	0,55	1,00	1,00	0,27	0,45	0,09	0,09	0,09
		Percentage of flooded area	1,00	0,00	0,01	0,14	0,14	0,02	0,00	0,00	0,00	0,00	0,00	0,00
		number of disasters earthquick, fires..)	1,00	0,37	0,52	0,33	0,17	0,23	0,11	0,02	0,07	0,00	0,00	0,00
		Pnl90 (Number of high precipitation events (pnl90))	0,41	0,58	0,51	0,41	0,59	0,51	0,71	0,81	0,91	1,00	0,81	0,73
		pxcdd (Maximum drought period (pxcdd))	0,35	0,50	0,49	0,45	0,58	0,62	0,80	0,67	0,98	0,98	0,84	0,69
		pn05 (Number of wet days (pn05mm))	0,53	0,64	0,55	0,48	0,65	0,61	0,81	0,89	1,00	1,00	0,90	0,79
		pav (average precipitation (mm / day))	0,53	0,51	0,46	0,52	0,60	0,67	0,85	0,78	1,00	1,00	0,87	0,78
		sensitivity index to desertification	0,89	0,93	0,85	0,85	0,93	0,93	0,98	1,00	0,96	0,96	0,96	0,96
		Total exposure	**0,681**	**0,453**	**0,469**	**0,511**	**0,525**	**0,574**	**0,657**	**0,555**	**0,671**	**0,629**	**0,558**	**0,506**
		Std Dev	**0,261**	**0,299**	**0,234**	**0,258**	**0,260**	**0,328**	**0,385**	**0,399**	**0,433**	**0,496**	**0,439**	**0,402**
Sensitivity	Economic	Population growth rate	0,53	0,77	0,59	0,60	0,56	0,58	0,47	0,55	0,51	0,97	1,00	0,36
		Percentage of households below poverty line in sample	0,73	0,87	0,89	0,71	1,00	0,49	0,72	0,89	0,66	0,59	0,44	0,61
		Percentage of consumption expenditure on food in sample	0,47	0,28	0,48	0,62	0,23	1,00	0,47	0,15	0,29	0,05	0,05	0,03
		Percentage of households depend on natural sources for income	0,81	0,58	0,66	0,50	0,85	0,78	0,88	1,00	0,73	0,59	0,28	0,43
		Percentage of households lost land in last 5 years	0,03	0,07	0,02	1,00	0,02	0,00	0,00	0,00	0,00	0,00	0,00	0,00
		Economic vulnerability	**0,51**	**0,51**	**0,53**	**0,69**	**0,53**	**0,57**	**0,51**	**0,52**	**0,44**	**0,44**	**0,35**	**0,29**
		Std Dev	**0,303**	**0,332**	**0,319**	**0,190**	**0,411**	**0,374**	**0,335**	**0,441**	**0,296**	**0,409**	**0,402**	**0,264**
	Physical	Percentage of households not getting electricity	0,42	0,73	0,51	0,63	0,97	0,41	0,58	0,50	0,46	0,41	0,50	1,00
		Percentage of households not having sanitary latrine	0,25	0,36	0,39	0,27	1,00	0,13	0,32	0,53	0,24	0,20	0,08	0,42
		Percentage of households use pond and well water for drinking	1,00	0,88	0,84	0,75	0,83	0,68	0,75	0,50	0,45	0,45	0,51	0,43
		Physical vulnerability	**0,56**	**0,65**	**0,58**	**0,55**	**0,93**	**0,41**	**0,65**	**0,51**	**0,39**	**0,35**	**0,36**	**0,62**
		Std Dev	**0,391**	**0,269**	**0,230**	**0,248**	**0,091**	**0,276**	**0,218**	**0,017**	**0,133**	**0,136**	**0,245**	**0,332**
		Total sensitivity	**0,503**	**0,500**	**0,509**	**0,583**	**0,585**	**0,483**	**0,477**	**0,478**	**0,384**	**0,381**	**0,319**	**0,311**
	Demographic	People per km2	0,58	0,07	0,30	0,71	0,17	1,00	0,33	0,04	0,14	0,03	0,01	0,00
		Percentage of old and child in the area	0,90	0,90	0,93	0,90	0,97	0,90	0,98	1,00	0,93	0,93	0,85	0,78
		Woman-man ratio	0,75	0,71	0,71	0,72	0,70	0,73	0,74	0,70	0,70	0,69	0,77	1,00
		Population growth rate	0,40	0,28	0,24	0,35	0,24	0,42	0,34	0,25	0,38	0,16	0,41	1,00
		Percentage of households migrated to this area in last 5 years	0,24	0,12	0,18	0,63	0,04	1,00	0,25	0,02	0,16	0,01	0,02	0,03
		Demographic vulnerability	**0,57**	**0,41**	**0,47**	**0,66**	**0,42**	**0,81**	**0,53**	**0,40**	**0,46**	**0,36**	**0,41**	**0,56**
		Std Dev	**0,265**	**0,373**	**0,331**	**0,199**	**0,394**	**0,246**	**0,316**	**0,432**	**0,347**	**0,420**	**0,399**	**0,508**
	Social	Percentage of illiterate households	0,79	0,68	0,89	0,75	1,00	0,67	0,97	0,88	0,65	0,84	0,56	0,57
		Percentage households not having brick-built house	0,46	0,44	0,57	0,50	0,61	0,56	0,92	1,00	0,48	0,50	0,14	0,31
		Percentage of household participated in last national election	0,80	0,79	0,90	0,80	0,91	0,71	0,91	0,96	0,88	1,00	0,90	0,81
		Social vulnerability	**0,68**	**0,70**	**0,79**	**0,68**	**0,84**	**0,65**	**0,93**	**0,95**	**0,74**	**0,78**	**0,53**	**0,57**
		Std Dev	**0,197**	**0,230**	**0,186**	**0,165**	**0,206**	**0,075**	**0,032**	**0,061**	**0,226**	**0,258**	**0,378**	**0,249**
		Total adaptive capacity	**0,616**	**0,523**	**0,590**	**0,671**	**0,579**	**0,747**	**0,678**	**0,606**	**0,564**	**0,520**	**0,457**	**0,564**
		POTENTIAL IMPACT	**0,592**	**0,477**	**0,489**	**0,547**	**0,555**	**0,529**	**0,567**	**0,517**	**0,527**	**0,505**	**0,438**	**0,408**
		IMPACT TO VULNERABILITY	**0,604**	**0,500**	**0,539**	**0,609**	**0,567**	**0,638**	**0,623**	**0,561**	**0,546**	**0,512**	**0,448**	**0,486**

± 0,12). Eddakhla-Oued Eddahab, despite the most natural hazard affected area, had the highest yearly population growth rate of 3,7 percent versus Guelmim-Oued Noun occupied the lowest percentage of 0,61 (average percentage:1,38 ± 0,79).

Social Vulnerability

Socially, the most vulnerable region in the study region was Drâa-Tafilalet with an average score of 0,95 (±0,06) and Laayoune-Sakia El Hamra was identified as socially the least vulnerable with a score of 0.53 (±0,38) (see detail in Table1). In second classe Marrakech-Safi with an average score of 0,93 (±0,03) and Béni Mellal-Khénifra with an average score of 0,84 (±0,2) (see detail in Table 1 and Figure 3).

Indicators of this domain showed that highest percentage (39,1) of illiterate households lived in Béni Mellal-Khénifra; whereas the lowest percentage (21,5) was found in Laayoune-Sakia El Hamra rgion (average percentage 31,34 ± 5,61).

Considering shelter security, very few households were found living with brick-built places. This indicates most households resided in places made of weak materials. The maximum percentage (48,4) households found in this category lived in Drâa-Tafilalet versus the minimum percentage (6,9) in Laayoune-Sakia El Hamra (average percentage: 26,13 ± 11,23). In rural areas of Drâa-Tafilalet, it is rural housing that predominates, sheltering 73.8% of households, followed by the Moroccan house, which concerns 24.4% of households. Summary or slum dwellings account for only 0.4% of households. These percentages are respectively 64.1%, 30.7% and 3.1% at the national level (HCP, 2014). in regards to household participated in last national election, an exceptionally positive response was found in case of households' voting-participation in national-level elections, where the lowest percentage of households (45,26) was found in Grand Casablanca-Settat and the maximum (64,18) was found in Guelmim-Oued-Noun (average percentage: 55,47).

Economic Vulnerability

Rabat-Salé-Kénitra was found economically the most vulnerable region (score: 0,69 ± 0,19), and Eddakhla-Oued Eddahab (score: 0,29 ± 0,26) was found the least vulnerable region in this category (see detail in Table 1 and Figure 3). Indicators within this domain showed in Drâa-Tafilalet the highest percentage (60) of households did depend on natural sources, especially on oasis for their income; whereas, the lowest percentage was in Laayoune-Sakia El Hamra (16,6) (average percentage (in study region): 40,45 ± 12,47). Again, highest percentage (28) of unemployed households (HCP, 2018) was found in Laayoune-Sakia-El Hamra whereas, the lowest (10,1) in Eddakhla-Oued Eddahab (average percentage:17,48 ± 5,39). Joblessness is a structural problem with several factors driving it. Firstly, population is rising beyond the economy's capacity to create jobs. Secondly, there is insufficient decent-job creation in the private sector. Finally, the inadequate supply of skills by the education system is also a major hindrance to youth accessing the labour market. (Seddiki, 2014). The highest percentage (31.47) of households living below the poverty line was found in Béni Mellal-Khénifra region and the lowest percentage was in Laayoune-Sakia El Hamra (13.7) (average percentage: 22.54 17.48 ± 5.41). Considering the land loss issue, we found the highest percentage (23.41) households in Rabat-Salé-Kénitra had lost their land due to natural hazards in last five years whereas all other region had low percentages less than (1.71) (average percentage 2.24 ± 6.68).

Physical Vulnerability

Results indicated Beni Mellal-Khenifra as physically the most vulnerable region with a score of (0.93 ±0.09); whereas, Guelmim-Oued Noun was the least vulnerable with a score of (0.35 ±0.14) (see detail in Table 1 and Figure 3). Indicators within this domain showed that the highest percentage (15.3) of households not getting electricity lived in Eddakhla-Oued Eddahab region and the lowest percentage (6.30) in both Guelmim-Oued Noun and Grand Casablanca-Settat regions (average percentage: 9.1± 3.11). The low rate of electrification is due to the geographical dispersion and elongation of buildings and the large area of the region. The highest percentage of households (19.60) not using sanitary latrine was found in Beni Mellal-Khenifra and the lowest (1.6) was found in Laayoune-Sakia El Hamra (average percentage: 6.9 ± 4.7). Percentage of households using pond, river and well water for drinking and cooking, varies between (31.1) in Tanger-Tetouan-Al Hoceima region and (15.5) in Eddakhla-Oued Eddahab (average percentage: 24.7 ± 6.8).

Exposure to Natural Hazards

Based on the weighted average scores, Tanger-Tetouan-Al Hoceima region was found as the most vulnerable region due to natural hazards with a score: 0.681 (±0.26)), followed by: Souss-Massa, Marrakech-Safi and Guelmim-Oued Noun with the following scores (succinctly) (0.671 (±0.43)), (0.657 (±0.39)) and (0.629 (±0.49)) ; and Oriental (score: 0.453(±0.29)) as the least vulnerable region(see detail in Table 1 and Figure 3). Indicators within this domain revealed that the highest number of floods in last five years (18) was in Grand Casablanca-Settat and Marrakech-Safi regions and 16.39 in Rabat-Sale-Kenitra. Floods have caused considerable material damage, loss of agricultural land, livestock and human lives. The floods are average in the regions of Tangier-Tetouan-Al Hoceima (13.11), Beni Mellal-Khenifra

(9.83) and Souss-Massa (8.19) but they are weak in the other regions of Morocco. During the period 1960 to 2014, 732 disasters were recorded in Morocco. including 84 flood that causes the death of 1360 people and distrcution 4039 house and damage to 9455 others (ZIC & TARGA, 2014).

Overall Vulnerability Score

Based on final weighted average score of SeVI, Grand Casablanca-Settat was identified socioeconomically the most vulnerable region (weighted average score: 0.64(±0.06)), followed by Marrakech-Safi (0.62(±0.06)), Rabat-Salé-Kénitra and Tanger-Tetouan-Al Hoceima (0.61(±0.06)), Béni Mellal-Khénifra (0.57(±0.06)), Drâa-Tafilalet and Souss-Massa (0.56(±0.06)). On the contrary, the least vulnerable region is Laayoune-Sakia El Hamra (0.44(±0.06)), followed by Eddakhla-Oued Eddahab and Oriental (0.48(±0.06) and 0.49(±0.06) succinctly) (Table 1). A supporting graph is also prepared (Figure 5) to show the magnitude of concerned indicators (separately) across regions and within the specific dimension having a scale of 0 to 12 (closer to 12 indicates more vulnerable).

IPCC-Dimension Wise Vulnerability and Magnitude of Contributing Indicators Within Dimensions

'IPCC-dimension wise' vulnerability scores, show that in study region both 'sensitivity' and 'exposure' were more dominant dimensions than 'adaptive capacity' (Figure 4).

Three categories of regions were identified, according to the dimensions " Adaptive Capacity " (AC), "Exposure" (E) and "Sensitivity" (S).

The first category, where exposure is more than adaptive capacity more than sensitivity (E> AC> S). It includes four regions of Laayoune-Sakia El Hamra, Guelmim-Oued Nun, Souss-Massa and Tangier-Tetouan-Al Hoceima.

The second category, in which adaptive capacity is more than exposure more than sensitivity (AC> E> S). It is represented by the Eddakhla-Oued Eddahab, Drâa-Tafilalet, Marrakech-Safi and Grand Casablanca-Settat regions.

The last category, whose adaptive capacity is more than sensitivity more than exposure (AC > S> E). It includes the remaining regions, Beni Mellal-Khénifra, Rabat-Salé-Kenitra, Fez-Meknes and the Oriental.

Results show that the highest average exposure score was recorded for Tanger-Tetouan-Al Hoceima region 0.681 (±0.261), followed by Souss-Massa 0.671 (±0.433), Marrakech-Safi 0.657 (±0.385), Guelmim-Oued Noun 0.629 (±0.496), Laayoune-Sakia El Hamra 0.558 (±0.439), Drâa-Tafilalet 0.555 (±0.399), Béni Mellal-Khénifra 0.525 (±0.260), Rabat-Salé-Kénitra 0.511 (±0.258), and Eddakhla-Oued Eddahab 0,506 (±0.385); while the lowest scores were found for Oriental 0.453 (±0.299) and Fès-Meknès 0.469 (±0.234) (Table 1, Figure 4).

Considering 'sensitivity' dimension Béni Mellal-Khénifra was found the region with the average highest score 0.558 (±0.377) followed by Rabat-Salé-Kénitra 0.553 (±0.208), moderate sensitivity was found in Tanger-Tetouan-Al Hoceima, Oriental and Fès-Meknès 0.500 (±0.298). On the contrary, the average lowest score was recorded for Eddakhla-Oued Eddahab 0.311 (±0.317) followed by Laayoune-Sakia El Hamra 0.319 (±0.331) and Guelmim-Oued Noun 0.381 (±0.321).

Finally, for adaptive capacity scores, Grand Casablanca-Settat have the average highest score 0.747 (±0.208) followed by Marrakech-Safi 0.678 (±0.318) and Rabat-Salé-Kénitra 0.678 (±0.318). In contrast, the lowest score for adaptive capacity was found with Laayoune-Sakia El Hamra 0.457 (±0.368), regions that remain a medium adaptability (0.5).

DISCUSSION

The importance of the vulnerability analysis to climate change is reflected in the fact that it highlights the imbalances and their origins, it allows on the one hand to qualitatively assess the vulnerability of the territory, and on the other hand to prioritize the "crucial points" of vulnerability. The issues related to the social and economic activities of the territory and its management should raise awareness of the importance of the consequences of climate change and adaptation policy in the development of Moroccan society.

The literacy of the population aged 15-24 years is in the process of being universalized, with a most significant improvement among girls than boys. However, the literacy rate of the age group 10 years and older is below the target set by the Millennium Development Goals. At all educational levels, gender parity is almost achieved through positive discrimination, with a higher growth rate in rural areas (HCP, 2015).

Infant and child mortality decreased by 60% during the last 20 years and will be between 26.2‰ and 27.6‰ as estimated by the Inter-agency Group of the United Nations system and the HCP experts for a target of 25‰ in 2015 (HCP, 2015).

- **A Context of Advanced Demographic**: Transition With a population where people under 40 years account for more than 62% and those less than 30 years for more than 54%, Morocco benefits from the bonus of a particularly advanced demographic transition. Under the effect of a

Figure 4. IPCC –dimension-wise vulnerability scores

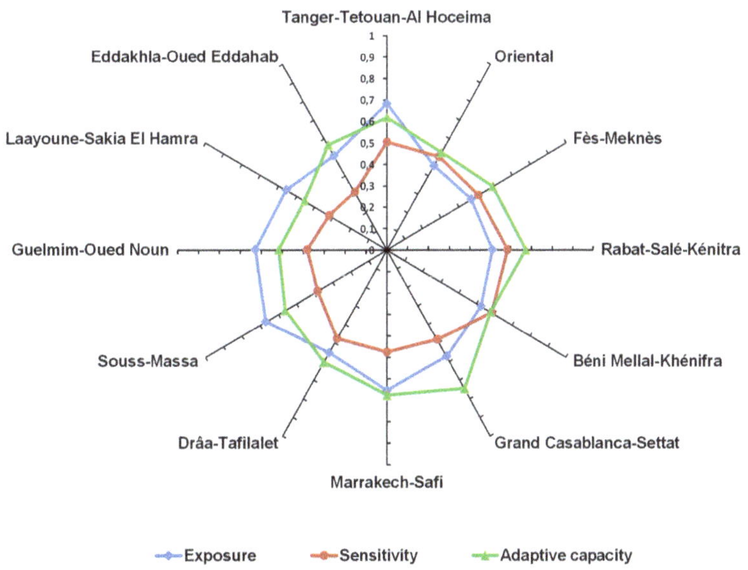

Figure 5. Magnitude of indicators within IPCC- dimensions across unions

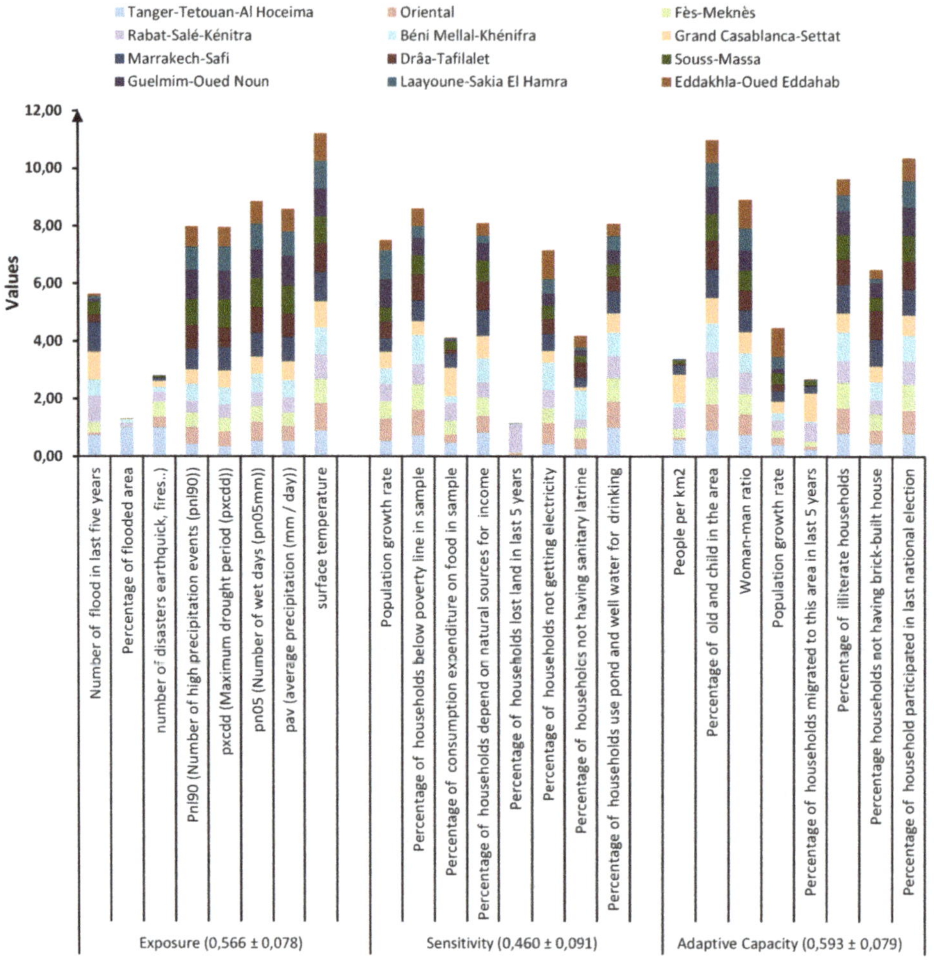

decline in mortality and fertility, the sharp decrease in the relative weight of the population under 15 years, from 31.2% in 2004 to 28% in 2014 and to 20.9% expected in 2030, is one source of this bonus (HCP, 2015).

Population growth, the strong movement of rural exodus following drought years, the weak integration capacity of Moroccan cities, the low purchasing power of a large segment of the population, the mismatch between supply and demand for housing, the depletion of land reserves, the complexity of land tenure and property taxation, the lack of a coherent housing intervention strategy, the legal weakness of housing town planning, etc. are all factors that explain the explosion of the phenomenon of substandard housing in Morocco (Adidi, 2009).

Socio-ecological systems in different regions of Morocco are vulnerable to climate variability and change. Vulnerable farm households were found in most rural livelihood systems. Major vulnerabilities identified by the local populations include: shortage of water for animals, insufficient food for people, conflicts/insecurity, crop failure, animal diseases, human diseases, limited land for cultivation, shortage of crops for cultivation, lack of employment, low prices for animals, insufficient pasture for animals, and shortage of water for domestic use. Most of these vulnerabilities can be traced to the impacts of climate variability (Messouli, et al., 2011).

Variables that influenced this vulnerability include household factors such as household size, availability of labour, health of household members, size of land cultivated, proportion of land under irrigation, size of cattle holdings, education of household members, combined income of household members from all sources, and gender of household head. Other factors include distance to market, proximity to motorable road, membership in community organizations, availability of storage facilities, and interaction with extension workers.

CONCLUSION

Climate in Morocco has been changing and will change further in the future. The major problems to climate change and variability are related to the pollution of air, the decline in average yearly rainfall and the increase of temperatures. Agriculture, energy and wastes are the main sources of greenhouse gas (GHG) emissions. The intense use of oils for transportation, the mismanagement of livestock and the decomposition of solid and liquid wastes are human activities generating GHG. The current level of contribution of Morocco to GHG emission is low. However, the population is projected to double in 22 years and the people living in urban areas will increase tremendously. This will have an impact on food production. Current land use practices such as putting more land under cultivation for crop production, extensive livestock systems, firewood and extension of irrigated crops will all contribute to the increase of GHG emissions. In contrast, forestry activities and someland use changes (land cover regeneration for example) have a potential of sequestering carbon.

Farm households in Morocco are vulnerable to drought or other climate change related negative impacts because the biophysical vulnerability (soil degradation, land cover degradation, water availability for development), the socio-economic vulnerability (rainfed subsistence and weather dependant agriculture, low literacy, lack of diversity, etc.) and technological vulnerability (low development of irrigation, lack of infrastructure, etc).

The case studies offered a valuable component to the profile by revealing insights about the determinants of vulnerability at the farm household or community levels. Numerous physical and socioeconomic factors come into play in enhancing or constraining the current capacity of farmers to cope with adverse changes. Rural communities have a rich experience of autonomous adaptation strategies, which can feed into future policy of adaptations. Prominent among the physical factors are the cropping patterns and the shifts to drought resistant varieties. The most important socio-economic factors include ownership of assets (like agricultural implements, cattle, sheep and goats), access to services (like credit, health and education), and infrastructural support (like irrigation, markets, and transport/communication networks).

Policies that are designed to fortify current coping strategies also have the power to strengthen long-term adaptive capacity. This best exemplified by measures such as rural infrastructure program, agricultural services support programs, alternative (off-farm) employment options, and enhanced access to inputs and markets). Another set of policy-relevant insights offered by the case study approach relates to the understanding of how certain factors change the vulnerability of a given community or region over time.

REFERENCES

Adidi, A. (2009). *La question de l'intégration sociale dans les programmes de lutte contre l'habitat insalubre au Maroc.* Retrieved from Ministère de la Réforme de l'Administration et de la Fonction Publique: http://www.mmsp.gov.ma

Ahmed, S., Diffenbaugh, N., & Hertel, T. (2009). Climate volatility deepens poverty vulnerability in developing countries. *Environmental Research Letters*, *4*(3), 034004. doi:10.1088/1748-9326/4/3/034004

Ahsan, M. N., & Warner, J. (2014). The socioeconomic vulnerability index: Apragmatic approach for assessing climate change led risks –A case study in thesouth-western coastal Bangladesh. *International Journal of Disaster Risk Reduction*, *8*, 32–49. doi:10.1016/j.ijdrr.2013.12.009

Arnell, W. N. (2004). Climate change and global water resources: SRES emissions and socio-economic scenarios. *Global Environmental Change*, *14*(1), 31–52. doi:10.1016/j.gloenvcha.2003.10.006

Benabid, A. (1982). *Bref aperçu sur la zonation altitudinale de la végétation climacique du Maroc.* Marseille: Centre National de Documentation Pedagogique.

CCR, L. C. (2011). *Rapport sur la régionalisation avancée.* Royaume du Maroc, La Commission Consultative de la Régionalisation (CCR).

Direction Regionale de l'Oriental, H. C. (2017). *La région de l'Oriental en.* Haut Commissariat Au Plan Direction Regionale de l'Oriental.

Driouech, F., Déqué, M., & Mokssit, A. (2009). Numerical simulation of the probability distribution function of precipitation over Morocco. *international journal. Climate Dynamics*, *32*(7-8), 1055–1063. doi:10.100700382-008-0430-6

Elhafi, A. (2012). *Changement Climatique et impact sur la sécurité alimentaire de la région d'Afrique du Nord.* Academic Press.

Eriksen, S., & Kelly, P. (2007). Developing Credible Vulnerability Indicators for Climate Adaptation Policy Assessment. *Mitigation and Adaptation Strategies for Global Change*, *12*(4), 495–524. doi:10.100711027-006-3460-6

Ezzine, H. (2015). *Regional Analysis of Desaster Loss Databases in Arab States.* Cairo, Egypt: UNISDR.

Fritzsche, K. (2017). *Vulnerability Sourcebook – Concept and Guidelines for standardised Vulnerability Assessments.* Bonn: Deutsche Gesellschaft für Internationale Zusammenarbeit (GIZ) GmbH.

HCEFLCD. (2018, June). *Le Haut Commissariat aux Eaux et Forêts et à la Lutte Contre la Désertification (HCEFLCD).* Retrieved from eaux et forets: http://www.eauxetforets.gov.ma

HCP. (2014). *Caractéristiques Démographiques et Socio-Economiques: REGION DRAA-TAFILALET. Errachidia: HCP.* Direction Régionale Drâa-Tafilalet.

HCP. (2015). *Morocco between Millennium Development Goals and Sustainable Development Goals Achievements and Challenges.* Kingdom of Morocco High Commission For Planning.

HCP. (2018, May). *High Commission for Planning (Morocco)*. Retrieved from Haut commissariat au plan: https://www.hcp.ma/

Hinkel, J. (2011). "Indicators of vulnerability and adaptive capacity": Towards a clarification of the science–policy interface. *Global Environmental Change*, *21*(1), 198–208. doi:10.1016/j.gloenvcha.2010.08.002

Immerzeel, W., Droogers, P., Terink, W., Hoogeveen, J., Hellegers, P., Bierkens, M., & Beek, R. v. (2011). *Middle-East and Northern Africa Water Outlook*. World Bank, Future Water.

IPCC. (2001). *IPCC Third Assessment Report. Synthesis Report*. Cambridge, UK: Cambridge University.

IPCC. (2007). *Climate change 2007-impacts, adaptation and vulnerability: Working group II contribution to the fourth assessment report of the IPCC* (Vol. 4). Cambridge University Press.

IPCC AR4. (2007). Climate Change 2007: Synthesis Report. Contribution of Working Groups I, II and III to the Fourth Assessment. Geneva, Switzerland: IPCC.

Krishnamurthy, P., Choularton, R., Betts, R., & Lewis, K. (2011). Assessing the impacts of climate risk on food security through a vulnerability index. *Vulnerability Workshop of GI Forum*. Salzburg.

Messouli, M. (2014). *Anticipation et gestion des risques d'evenements climatiques extremes et de catastrophes naturelles au maroc*. Rabat: Institut royal des études stratégiques (IRES).

Messouli, M., Ben Salem, A., & Rochdane, S. (2013). *Développement d'un concept d'Indice de Vulnérabilité au Changement Climatique et Environnementale au Niveau Régional*. Rabat: Département du développement durable (MNSSD), Région Moyen-Orient et Afrique du Nord, Banque Mondiale.

Mhirit, O., Khattabi, A., Rouchdi, M., Yessef, M., & Chouraichi, M. (2011). *Adaptation du Programme d'Action National de Lutte Contre la Désertification aux Spécificités Zonales*. Rabat: Le Haut Commissariat aux Eaux et Forêts et à la Lutte Contre la Désertification (HCEFLCD).

Ministère de l'Aménagement du Territoire, l. (2005). *Rapport national sur la prevention des desastres conference de kobe- hyogo. Royaume du Maroc, Ministère de l'Aménagement du Territoire, de l'Eau et de l'Environnement*. Département de l'Environnement.

Ministère de l'Equipement, d. T. (2014). *Impact de la Houle Exceptionnelle du 06-07 Janvier 2014 sur les infrastructures Portuaires*. Royaume du Maroc, Ministère de l'Equipement, du Transport, de la Logistique.

Mokssit, A. (2012). *Impact des Changements Climatiques au Maroc*. IRES-Rabat: Ministère des Energies, des Mines, de l'Eau et de l'Environnement.

Nissen, K. M., Leckebusch, G. C., Pinto, J. G., & Ulbrich, U. (2014). Mediterranean cyclones and windstorms in a changing climate. *Regional Environmental Change*, *14*(5), 1873–1890. doi:10.100710113-012-0400-8

Ribot, J., & Norton, A. (2009). Social dimensions of climate change: equity and vulnerability in a warming world. In Social Dimensions of Climate Change (pp. 47-75). Washington, DC: WB.

Rosegrant, M. (2009). Changement climatique et sécurité alimentaire au Maroc. Rencontre internationale « Le changement climatique: enjeux et perspectives d'adaptation pour le Maroc ». Rabat: Institut Royal des Etudes Stratégiques (IRES).

Seddiki, A. (2014, August 20). In Morocco youth unemployment is driving up inequality. The Guardian.

Smith, F. E., Scott, N. L., Keys, N., & Smith, F. T. (2017). The socio-economic vulnerability of the Australian east coast grazing sector to the impacts of climate change. *Regional Environmental Change*, 1185–1199.

SOPAC. (2004). *The Environmental Vulnerability Index (EVI), Technical Report 384.* South Pacific Applied Geoscience Commission.

Stour, L., & Agoumi, A. (2009). *Climatic drought in Morocco during the last decades [Sécheresse climatique au Maroc Sécheresse climatique au Maroc].* Paris: Hydroécologie Appliquée.

Sullivan, A. C., & Meigh, J. (2007). Integration of the biophysical and social sciences using an indicator approach: Addressing water problems at different scales. *Water Resources Management*, 111–128. doi:10.100711269-006-9044-0

UNFCCC. (1992). *United Nations framework convention on climate change.* New York: United Nations.

Zic, L., & Targa, A. (2014). Risk Nexus: Inondations au Maroc de 2014: quels enseignements tirer de Guelmim et Sidi Ifni. Zurich: Zurich Insurance Company Ltd.

Chapter 3
Climate Change and Water Security:
Water Scarcity and Water Flood

Abdelmohcine Aimrane
Cadi Ayyad University, Morocco

Hasna Lahouaoui
Cadi Ayyad University, Morocco

Youssef Khamsi
University of Arizona, USA

Ahmed Draoui
Cadi Ayyad University, Morocco

Hassan Alahyane
Cadi Ayyad University, Morocco

Nadia Zouhairi
Cadi Ayyad University, Morocco

Hind Benammi
Cadi Ayyad University, Morocco

ABSTRACT

Climate change is an actual fact setting off an imbalance in many living systems. Among these affected systems, water is a major essential element in the globe and in every existing living being. Therefore, several complications have been stated to occur, following water scarcity and water flood in many regions of the world, which make of them a global major threat of water security. The global disease burden is an additional factor that appeals to serious interventions worldwide in order to alleviate the water scarcity and water flood-related effects.

DOI: 10.4018/978-1-5225-7775-1.ch003

INTRODUCTION

Water is an essential commodity for the existence of life on earth. It is a vital element indispensable for humans, animals and plants. Water is obtained from two principal natural sources: Ground water like well water and borehole water and Surface water such as fresh water lakes, rivers and streams (McMurry & Fay, 2004; Mendie, 2005).It is the mediator of all the principal biochemical events in the body including cell survival, birth and death due to its unique chemical properties including its polarity and hydrogen bonds, for this water is able to dissolve, absorb, adsorb or suspend many different compounds(Lipinski, 2001).

Global climate change is expected to affect the frequency, the intensity and the duration of extreme water-related weather events such as excessive precipitation, floods, and drought (Evans, 2019). A wide range of water problems face nations and individuals around the world (Rijsberman, 2004). Water problems consist of water scarcity and contamination, ecological degradation, and the threat of climate change (Cann, Thomas, Salmon, Wyn-Jones, & Kay, 2013).

Water scarcity has been reported to promote drought and aridity in many parts of the world and could set off complications in the ecosystem (Santos Pereira, Cordery, & Iacovides, 2009). Furthermore, human welfare and health are the most threatened by water insufficiency (Patz et al., 1997 ; Rijsberman, 2004). The extent of water scarcity has been stated to affect roughly four billion people around the world, including rich and poor populations (Mekonnen & Hoekstra, 2016). The global climate changes as well as human overconsumption of fresh available water are among the essential causes leading to water scarcity, a fact which is forecasted to amplify in the future (Abahussain, Abdu, Al-Zubari, El-Deen, & Abdul-Raheem, 2002; Evans, 2019; World Health Organization, 2012).

Water flood is another case of water imbalance that takes part in many regions around the world (Evans, 2019 ; Paterson, Wright, & Harris, 2018). Adverse affects on the material and environment are in fact linked to this disaster (Patz et al., 1997). Also, a lot of floods related injuries, infections and diseases worldwide have been stated to alter human welfare (Vinet, 2017).

In this review, we present an overview on the effect of climate change on water scarcity and water flood. We also put forward an outline on human behavior leading to water scarcity exacerbation, in addition to the possible interventions that could mitigate water imbalance.

WATER SCARCITY

Definitions

Insufficiency of fresh available water, or water scarcity as it is called, results either from a basic lack of water or from a lack of suitable infrastructure. The basic lack of water is distinguished as physical water scarcity while the lack of suitable infrastructure refers to economic water scarcity. Both of these factors can have an important impact on water availability. Nevertheless, physical water scarcity has been reported to cause greater impact to ecological systems mainly because of the influence of the climate change as well as human influence (Bond, Burrows, Kennard, & Bunn, 2019). Two main scenarios of water reduction have been defined in literature; aridity and drought. Drought is a natural and temporal water shortage consisting of lower precipitation and leading to reduced carrying capacity of the ecosystems (Santos Pereira et al., 2009). Aridity is another manifestation of water shortage where the consequences on human and ecosystem are long lasting, irreversible, and may lead to desertification (Water, 2006). The International Convention on Desertification has explained it as 'the degradation of land in arid, semi-arid and sub- humid dry areas caused by climatic changes and human activities'(Kassas, 1995). Accordingly, the linked impacts of any sort of water unavailability has a great impact on human being and the environment.

Evidences Revealing the Extent of Water Scarcity

The extent of water withdrawal and the strength of its consequences are dependant specifically on climate change as well as the increase of human demand of affordable water rather than many other hydrologic and ecologic natural factors (Gosling & Arnell, 2016). The outcomes of these influences indeed present a threat not only for the biodiversity, but also for human beings' welfare, health and life. Consequently, its hazardous results may play a huge impact in most parts of the world (Mekonnen & Hoekstra, 2016 ; Rijsberman, 2004). Thus in water-scarce region, the hydrologic arising impacts on low flows and water storage infrastructures alters the function of rivers and to induce loss of biodiversity, which provoke the alteration of the ecological system (Rolls, Leigh, & Sheldon, 2012). On the other hand, the lack of access to fresh drinking water and sanitation cause massive health problems, specifically diarhoeal diseases, which have been estimated to annually take the lives of 2.18 million people among whom three quarters are children under the age of 5 years old, in addition with a measured annual increase of disease reaching 82 million Disability Adjusted Life Years (Pruss, Kay, Fewtrell, & Bartram, 2002).

Indeed, water scarcity has long been established as one of climate change and human influence that impacts the world to a great extent. Mekonnen and Hoekstra (2016), in a diachronic study on the annual measures of water scarcity for the period of 1996-2005, distinguished two main levels of geographic and temporal spreads of fresh water scarcity; a year-round low water scarcity and a high water scarcity. The high water scarcity is especially reported in regions under human influence, either in regions with high population density or in much irrigated areas, or both. The previous study stated that 4.3 billion people, roughly two-thirds (66%) of the world's population, suffer from severe water scarcity at least 1 month per year, 1.8 to 2.9 billion during four to six months per year and half a billion people for the whole year, every year (Mekonnen & Hoekstra, 2016). The negative impact on humans and the environment is preponderant in a major part of the world,several instances illustrate the extantof the dangers. In the Arab regions, for instance, fast urbanization and rapid population growth have lead to an increase in the number of countries suffering water scarcity from 5 countries in 1950 to 16 countries in 1993.Furthermore, 50% of the Arab countries are forecasted to be touched by water insufficiency by the year 2025 (Abahussain et al., 2002). Moreover, recent studies have stated that water scarcity in the Arab desert is worse than that in other deserts. For instance Saudi Arabia and Yemen are classified under countries facing severe water scarcity where all the population is affected at least 6 months per year (Mekonnen & Hoekstra, 2016).

The previous examples present the water scarcity situation in many parts of the world. Currently, the situation is becoming a global concern especially with the influence of climate change. Although drought and desertification could be produced by a natural phenomena as in the arid world areas, human influence has been reported to be strong and may provoke long lasting effects (Abahussain et al., 2002; World Health Organization, 2005).

Human's Behavior and Water Scarcity

The patterns in water over exploitation have recently become more and more important, especially during the21st century (Veldkamp et al., 2017). In fact, human behavior and demand for water are among the main factors leading to water insufficiency and exacerbating effects of climate changes. Natural aridity may indeed provoke a decrease on the per capita available water, a fact which may have an impact on the socioeconomic development plans. However, the over exploitation of water and irrational water demand are stated to be among the leading reasons to water scarcity and food decrease (Abahussain et al., 2002).

The most relevant facts related to such overconsumption are not only associated to populations suffering from fragility, conflict, poverty and low educational levels, but also to the lack of valuable planning in efficient land and sustainable water management. Increased wealth places extra pressure on water resources and domestic water usage (World Health Organization, 2012). Furthermore, the strong human consumption of water for economic and social use, in addition to its exploitation for leisure and recreational reasons have been stated to lead to aridity especially in semi-arid and sub-humid areas (Santos Pereira et al., 2009). At the same time, several African, South American and Arab populations have been reported to need supply of basic safe drinking water and sanitation (World Health Organization, 2012), a fact which reveals the need to make inquiries about the managerial tendencies for global sustainable development, especially on behalf of rich and developed countries.

The tendencies of human management of agriculture and food production is an additional factor and influence the demand of water (Chiarelli, Rosa, Rulli, & D'Odorico, 2018 ; Piao et al., 2010). The combination of climate change and the need to meet achievements on the economic scale, as well as the necessity of maintaining self sufficiency in food may be the main reason behind this fact. However, water scarcity actually occurs in many agricultural regions. For instance, the high agricultural activity in High Plains of the United States, in India, Eastern China and the Nile delta have been linked to severe water scarcity (Mekonnen & Hoekstra, 2016). In China, a study conducted by Piao and collaborators (2010) have stated that the increased heat waves have lead to drought in most parts of the country. Although the study did not meet a conclusion about the existence of a dangerous climate threshold, it is declared interesting and challenging to identify the interaction by which human management and climate change influence water availability and food production (Piao et al., 2010).

Knowing the importance of human influence on water availability, his integration in the right ways might be the key solution to mitigate water scarcity. Water management in fact, under variable conditions of drought and aridity, requires the adoption of politics and policy that implies rational and efficient planning for water use, appropriate strategies to alleviate the high demand and to increase water production and conservation. Furthermore, human conduct may be a crucial factor in sustainable development by valuing water resources as important good and recruiting educated and trained water managers, operators and users (Santos Pereira et al., 2009).

FLOOD WATER

Definitions

Flood is an overflow of water in a land beyond the usual limits. Several types of floods exist which are classified in function of the origin and according the generating process of water; river floods, flash floods, dam-break floods, ice-jam floods, glacial-lake floods, urban floods, coastal floods, and hurricane-related floods. The common factor driving all kinds of floods is climate through precipitation and temperature along with several other factors. Landscape, soil, geology, vegetation and channel characteristics are for instance factors determining the hydrological flood process like runoff generation and routing. The processes driving flood water vary widely around the world, also they vary between events in the same region. Additionally, climate characteristics are expected to change in the future, a fact which may have an influence on flood processes around the world (Few, Ahern, Matthies, & Kovats, 2004).

Flood Impact to Human and the Ecosystem

Flood is a natural disaster that is a threat to life and the health of human beings, wild animals and the ecosystem (Paterson et al., 2018 ; Vinet, 2017). Flood frequency is in fact stated to be affected by year-to-year changes in precipitation variability and the short-term rainfall properties (for example storm rainfall frequency). Thus, the seasonal distribution of precipitation being expected to change, the frequency of low flows is affected as well which may have a notable effect on prolonged droughts (Douglas, Vogel, & Kroll, 2000 ; Marx et al., 2018). Floods are in fact associated with good news as well as bad news in terms of water resources. However, the good news is often confusing. Although water scarcity may be alleviated in a region by getting more precipitations, the amount of rain can also increase the flood related risks. Furthermore, the high frequency of storms is stated to play a beneficial effect in lakes and estuaries by reducing the risk of macrophytes eutrophication and algal blooms proliferation (Dodds, 2006), the amount of storm runoff carries an extra charge of these nutrient to sea, which set off many adverse impacts in the marine ecosystem where the worst is the risk of eutrophication and hypoxia (Justić et al., 2018).

Climate change is expectd to alter the frequency of floods and exacerbate its related negative effects (Few et al., 2004). Water temperature increments, subsequent to the global warming, could influence water quality and intensify water pollution, such as thermal pollution, the increase of sediment, nutrients, dissolved organic carbon, pathogens, pesticides, amplification of salt (Vodela, Renden, Lenz, Mc Elhenney, & Kemppainen, 1997 ; G. Yang, Sau, Lai, Cichon, & Li, 2015 ; Yuan, Liu, Chen, & Yang, 2011). Such a fact does not only alter the ecosystem, but also human health and the reliability of the water system and operating costs. Moreover, the rise of sea level is anticipated to extend the affected area with salinization and therefore may result in a decrease of the available freshwater for humans and the ecosystem in coastal area (Salinas, Delaune, & Patrick, 1986).

Human health and life has been reported to be under direct effect of water overflow. Leading to death reasons are linked to flood disasters, where drowning is the first one, followed by heart attacks, hypothermia, blunt trauma caused by wind-borne objects and vehicle-related accidents (L. D. Wright, Elia, & Nichols, 2019). Moreover, the amount of fatalities caused by these issues has no notable improvement. For instance, a diachronic analysis was discussed in a historical comparison that flood caused mortality that occurred in 1850 or 1930 is equivalent to these caused in 2015. Nevertheless, a plethora of fatal injuries from electrocution, wound infections and bites from wild animals, such as snakebites, are also causes that could lead to death during evacuation (Vinet, 2017). In brief, the number of flood related dangers have been well studied, categorized and discussed as being a major factor, hence many studies have stated the necessity of preventive actions, such as proposing simulative models (Ogawa & Male, 1986), or using new technologies (Montz & Gruntfest, 2002) to alleviate these negative effects. Although significant improvements have been made in the accuracy of forecasting and in the technologies to get relevant data (Montz & Gruntfest, 2002), the amount of deaths caused and damage remain until this day unchanged (Vinet, 2017). Accordingly, more serious studies on the ways flood hazards can be alleviated must be undertaken.

Climate change is in actual fact forecasted to alter both, drought and flood with different intensities around the world (Evans, 2019). Therefore, the effect of floods and water scarcity on the natural system are reconciled, the question that may come to mind is about the main effects of these natural factors of climate change might have on human health.

WATER DISEASES

Water diseases remain a major concern in much of the developing world and are typically placed in four classes: waterborne, water-washed, water-based, and water-related insect vectors.

- **Waterborne Diseases**: Represent the major threat to global human health and are carried by pathogenic representatives of the three principle groups of organisms; bacteria, viruses and protozoa(Woodall, 2009).
- **Water-Washed Diseases**: Those whose transmission is due to poor personal and/or domestic hygiene as a result of lack of appropriate water (K. Yang et al., 2012).

Waterborne and water-washed diseases include infectious diarrhea. Infectious diarrhea consists of cholera, shigellosis, salmonellosis, amoebiasis, and other protozoal and viral intestinal infections. These micro-organisms are transmitted by water, contact (person to person and animal to human), foodborne and aerosol routes (Hutton, 2013).

- **Water-Based Diseases**: The widespread sources of water-based diseases are the exotoxins produced by several marine and freshwater algae species. Marine algal toxins come from dinoflagellates and cyanobacteria (blue-green algae) (Carmichael, Jones, Mahmood, & Theiss, 1985).
- **Water-Related Diseases**: The diseases spread by vectors and insects that live in or close to water. Stagnant ponds of water provides the breeding place for the disease spreading vectors such as mosquitoes, flies and insects (Ashbolt, 2004). Another kind of water-associated disease, schistosomiasis, is caused by a worm or blood fluke whose life cycle involves particular aquatic snails and human contact with infected water (Batterman et al., 2009).

In the developing world, 80% of illness and death is water-related and the number of deaths from water-related disease approaches 5 million annually, most of them children. These deaths occur among the estimated 1.2 billion people worldwide without access to safe and reliable drinking water and the 2.5 billion without access to sanitation services (Batterman et al., 2009).

WATER CONTAMINATION

In nature, water is not pure and different sources of contamination arise from humans and animals as well as other biological activities (Mendie, 2005). One of the most important environmental issues today is ground water contamination (Vodela et al., 1997). Untreated wastewater has microbial and chemical hazards and influences negatively on human health (Matthys et al., 2006). Bacterial and viral micro-organisms induce diarrhoea, respiratory tract infections, skin and eye diseases as well as epidemic disease outbreaks like cholera and typhoid fever (Becker et al., 2013 ; Blumenthal & Peasey, 2002 ; Stenström, Seidu, Nelson, & Christian, 2011). Furthermore, the bioaccumulation of heavy metals and pesticides may cause chronic diseases and cancer (Ackah et al., 2014 ; Järup, 2003 ; Marcussen, Dalsgaard, & Holm, 2008).

Water contamination represents an important environmental issue and the wide variety of contaminants influencing water resources, especially heavy metals and pesticides because of their toxicity, even at low concentrations (Marcovecchio et al., 2007).

Metal contamination in the aquatic environment has a attracted attention due to its environmental toxicity (Sin, Chua, Lo, & Ng, 2001 ; Yuan et al., 2011). Large quantities of heavy metals have been found in rivers worldwide due to intensive domestic activities (Islam, Han, Ahmed, & Masunaga, 2014 ; Srebotnjak, Carr, De Sherbinin, & Rickwood, 2012).

The main threats to human health from heavy metals are linked to the exposure to lead, cadmium, mercury and arsenic. Heavy metals are present in the environment through a wide range of processes and pathways, including surface waters (via runoff and releases from storage and transport). Atmospheric emissions tend to be of greatest concern in regards to human health, both because of the quantities involved and the widespread dispersion and potential exposure that often ensues (Järup, 2003).

Recently, in Kerala there was news about ground water contamination with heavy metals. The global environmental action group urged the state government to close down a bottling plant in Palakkad district for supplying inferior waste from the plant as fertilizer. The government reports say that ground water samples collected contained 65.7 μg of Pb and more than 10 μg of Cd per litre. Heavy mortality in cattle in Maharashtra, Punjab, Rajasthan and Delhi, was caused due to industrial lead poisoning and the death of animals was due to leakage of MIC gas during the Bhopal tragedy are some examples of industrial pollution catastrophe affecting domestic animals in India (Rajaganapa, Xavier, Sreekumar, & Mandal, 2011).

Waterborne Disease

Climate change has a great impact on waterborne diseases. These kinds of diseases are one of the major contributors to diseases burden and mortality (Pruss et al., 2002). Different viral, bacterial and parasitic diseases have been linked to waterborne transmission (Hunter, 1997). Diseases could spread by recreational water contact or via drinking water. The distribution of waterborne pathogens depends on countries. It was shown that Vibrio cholerae, Hepatitis E virus and schistosomiasis are restricted to certain tropical countries, others, like cryptosporidiosis and campylobacteriosis, are widespread (Hunter, 2003).

In the UK, there has been some 65 recorded outbreaks of infection associated to water affecting 4112 people during the years 1991–2000. Of these outbreaks, 25 were associated with public water supplies, 16 with private water supplies, 23 with swimming pools and one with recreational contact with surface waters (Hunter, 2003).

Waterborne diseases are the major reason for 4 billion cases of diarrhea annually that causes 1.6 million deaths worldwide (J. Wright & Gundry, 2009). Diarrhea represents the five most common disease causes of death (Kosek, Bern, & Guerrant, 2003). The pathogen waterborne parasitic infections inducing diarrhea are cryptosporidiosis and giardiasis (Lane & Lloyd, 2002). Other parasitic protozoa with waterborne transmission that cause human infections are *Toxoplasma gondii*, *Entamoebahistolytica*, *Acanthamoeba* spp., *Cyclosporacayetanensis*, *Microsporidia*, *Isospora*, *Blastocystishominis*, *Sarcocystis* spp., *Naegleria* spp. and *Balantidium coli*. Since most of these protozoa use the faecal-oral way of transmission they can infect humans through sewage and contamination of land and rivers by animal or human feces (Lanata, 2003).

CONCLUSION

In conclusion, we found several evidences supporting the effect of climate change on the health and the environment. However, few studies have been developed to mitigate water scarcity and flood water related hazards, which refers on the pertinence of developing more investigations on that issue.

REFERENCES

Abahussain, A. A., Abdu, A. S., Al-Zubari, W. K., El-Deen, N. A., & Abdul-Raheem, M. (2002). Desertification in the Arab Region: Analysis of current status and trends. *Journal of Arid Environments*, *51*(4), 521–545. doi:10.1016/S0140-1963(02)90975-4

Ackah, M., Anim, A. K., Gyamfi, E. T., Zakaria, N., Hanson, J., Tulasi, D., ... Osei, J. (2014). Uptake of heavy metals by some edible vegetables irrigated using wastewater: A preliminary study in Accra, Ghana. *Environmental Monitoring and Assessment*, *186*(1), 621–634. doi:10.100710661-013-3403-0 PMID:24114276

Ashbolt, N. J. (2004). Microbial contamination of drinking water and disease outcomes in developing regions. *Toxicology*, *198*(1-3), 229–238. doi:10.1016/j.tox.2004.01.030 PMID:15138046

Batterman, S., Eisenberg, J., Hardin, R., Kruk, M. E., Lemos, M. C., Michalak, A. M., ... Wilson, M. L. (2009). Sustainable control of water-related infectious diseases: A review and proposal for interdisciplinary health-based systems research. *Environmental Health Perspectives*, *117*(7), 1023–1032. doi:10.1289/ehp.0800423 PMID:19654908

Becker, S. L., Vogt, J., Knopp, S., Panning, M., Warhurst, D. C., Polman, K., ... Utzinger, J. (2013). Persistent digestive disorders in the tropics: Causative infectious pathogens and reference diagnostic tests. *BMC Infectious Diseases*, *13*(1), 37. doi:10.1186/1471-2334-13-37 PMID:23347408

Blumenthal, U. J., & Peasey, A. (2002). *Critical review of epidemiological evidence of the health effects of wastewater and excreta use in agriculture*. Repéré à http://www.who.int/water_sanitation_health/wastewater/whocriticalrev.pdf

Bond, N. R., Burrows, R. M., Kennard, M. J., & Bunn, S. E. (2019). *Water Scarcity as a Driver of Multiple Stressor Effects. Multiple Stressors in River Ecosystems. (S.l.)*. Elsevier Inc.; doi:10.1016/B978-0-12-811713-2.00006-6

Cann, K. F., Thomas, D. R., Salmon, R. L., Wyn-Jones, A. P., & Kay, D. (2013). Extreme water-related weather events and waterborne disease. *Epidemiology and Infection, 141*(4), 671–686. doi:10.1017/S0950268812001653 PMID:22877498

Carmichael, W. W., Jones, C. L. A., Mahmood, N. A., Theiss, W. C., & Krogh, P. (1985). Algal toxins and water-based diseases. *Critical Reviews in Environmental Control, 15*(3), 275–313. doi:10.1080/10643388509381734

Chiarelli, D. D., Rosa, L., Rulli, M. C., & D'Odorico, P. (2018). The water-land-food nexus of natural rubber production. *Journal of Cleaner Production, 172*, 1739–1747. doi:10.1016/j.jclepro.2017.12.021

Dodds, W. K. (2006). Eutrophication and trophic state in rivers and streams. *Limnology and Oceanography, 51*(1part2), 671–680. doi:10.4319/lo.2006.51.1_part_2.0671

Douglas, E. M., Vogel, R. M., & Kroll, C. N. (2000). Trends in floods and low fows in the United States : Impact of spatial correlation. *Journal of Hydrology (Amsterdam), 240*(1-2), 90–105. doi:10.1016/S0022-1694(00)00336-X

Evans, G. W. (2019). Projected Behavioral Impacts of Global Climate Change. *Annual Review of Psychology, 70*(1). doi:10.1146/annurev-psych-010418-103023

Few, R., Ahern, M., Matthies, F., & Kovats, S. (2004). Floods, health and climate change: a strategic review. *East*. Repéré à http://www.tyndall.ac.uk/publications/working_papers/wp63_summary.shtml%5Cnhttp://tyndall.ac.uk/sites/default/files/wp63.pdf

Gosling, S. N., & Arnell, N. W. (2016). A global assessment of the impact of climate change on water scarcity. *Climatic Change, 134*(3), 371–385. doi:10.100710584-013-0853-x

Hunter, P. (1997). *Waterborne disease: epidemiology and ecology*. John Wiley & Sons.

Hunter, P. R. (2003). Climate change and waterborne and vector-borne disease. *Journal of Applied Microbiology, 94*(s1), 37–46. doi:10.1046/j.1365-2672.94.s1.5.x PMID:12675935

Hutton, G. (2013). Global costs and benefits of reaching universal coverage of sanitation and drinking-water supply. *Journal of Water and Health, 11*(1), 1–12. doi:10.2166/wh.2012.105 PMID:23428544

Islam, M. S., Han, S., Ahmed, M. K., & Masunaga, S. (2014). Assessment of Trace Metal Contamination in Water and Sediment of Some Rivers in Bangladesh. *Journal of Water and Environment Technology, 12*(2), 109–121. doi:10.2965/jwet.2014.109

Järup, L. (2003). Hazards of heavy metal contamination. *British Medical Bulletin, 68*(1), 167–182. doi:10.1093/bmb/ldg032 PMID:14757716

Justić, D., Turner, R. E., Rabalais, N. N., Justic, D., Turner, R. E., & Rabalais, N. N. (2018). *Climatic Influences on Riverine Nitrate Flux : Implications for Coastal Marine Eutrophication and Hypoxia*. Retrieved from https://www.jstor.org/stable/1353186

Kassas, M. (1995). Desertification: A general review. *Journal of Arid Environments, 30*(2), 115–128. doi:10.1016/S0140-1963(05)80063-1

Kosek, M., Bern, C., & Guerrant, R. L. (2003). The global burden of diarrhoeal disease, as estimated from studies published between 1992 and 2000. *Bulletin of the World Health Organization, 81*(3), 197–204. PMID:12764516

Lanata, C. F. (2003). Studies of food hygiene and diarrhoeal disease. *International Journal of Environmental Health Research, 13*(sup1), 175–183. doi:10.1080/0960312031000102921 PMID:12775394

Lane, S., & Lloyd, D. (2002). Current trends in research into the waterborne parasite Giardia. *Critical Reviews in Microbiology, 28*(2), 123–147. doi:10.1080/1040-840291046713 PMID:12109771

Lipinski, C. (2001, April). Avoiding investment in doomed drugs. *Current Drug Discovery*, 17-19.

Marcovecchio, J. E., Botté, S. E., & Freije, R. H. (2007). Heavy metals, major metals, trace elements. Handbook of Water Analysis, 2, 275-311.

Marcussen, H., Dalsgaard, A., & Holm, P. E. (2008). Content, distribution and fate of 33 elements in sediments of rivers receiving wastewater in Hanoi, Vietnam. *Environmental Pollution, 155*(1), 41–51. doi:10.1016/j.envpol.2007.11.001 PMID:18164523

Marx, A., Kumar, R., Thober, S., Rakovec, O., Wanders, N., Zink, M., ... Samaniego, L. (2018). Climate change alters low flows in Europe under global warming of 1.5, 2, and 3 °C. *Hydrology and Earth System Sciences, 22*(2), 1017–1032. doi:10.5194/hess-22-1017-2018

Matthys, B., Vounatsou, P., Raso, G., Tschannen, A. B., Becket, E. G., Gosoniu, L., ... Utzinger, J. (2006). Urban farming and malaria risk factors in a medium-sized town in Côte d'Ivoire. *The American Journal of Tropical Medicine and Hygiene, 75*(6), 1223–1231. doi:10.4269/ajtmh.2006.75.1223 PMID:17172397

McMurry, J., & Fay, R. C. (2004). Hydrogen, Oxygen and Water. In K. P. Hamann (Ed.), *McMurry Fay Chemistry* (4th ed.; pp. 575–599). Pearson Education.

Mekonnen, M. M., & Hoekstra, Y. A. (2016). Four Billion People Experience Water Scarcity. *Science Advances, 2*(2), 1–7. doi:10.1126ciadv.1500323 PMID:26933676

Mendie, U. E. (2005). *The theory and practice of clean water production for domestic and industrial use: Purified and package water*. Lagos: Lacto-Medals Publishers.

Montz, B. E., & Gruntfest, E. (2002). Flash flood mitigation: Recommendations for research and applications. *Environmental Hazards, 4*(1), 15–22. doi:10.3763/ehaz.2002.0402

Ogawa, H., & Male, J. W. (1986). Simulating the Flood Mitigation Role of Wetlands. *Journal of Water Resources Planning and Management, 112*(1), 114–128. doi:10.1061/(ASCE)0733-9496(1986)112:1(114)

Paterson, D. L., Wright, H., & Harris, P. N. A. (2018, July). Health Risks of Flood Disasters. *Clinical Infectious Diseases*, 1–5. doi:10.1093/cid/ciy227

Patz, J. A., Githeko, A. K., McCarty, J. P., Hussein, S., Confalonieri, U., & De Wet, N. (1997). Climate Change and Human Health. *Population and Development Review, 23*(1), 205. doi:10.2307/2137486

Piao, S., Ciais, P., Huang, Y., Shen, Z., Peng, S., Li, J., ... Fang, J. (2010). The impacts of climate change on water resources and agriculture in China. *Nature*, *467*(7311), 43–51. doi:10.1038/nature09364 PMID:20811450

Pruss, A., Kay, D., Fewtrell, L., & Bartram, J. (2002). Estimating the Burden of Disease from Water, Sanitation, Hygene at a Global Level. *Environmental Health Perspectives*, *110*(5), 537–542. doi:10.1289/ehp.02110537 PMID:12003760

Rajaganapa, V., Xavier, F., Sreekumar, D., & Mandal, P. K. (2011). Heavy Metal Contamination in Soil, Water and Fodder and their Presence in Livestock and Products : A Review. *Journal of Environmental Science and Technology*, *4*(3), 234–249. doi:10.3923/jest.2011.234.249

Rijsberman, F. R. (2004). Water scarcity: Fact or fiction? *Proceedings of the 4th International Crop Science Congress*, *80*, 5-22. 10.1016/j.agwat.2005.07.001

Rolls, R. J., Leigh, C., & Sheldon, F. (2012). Mechanistic effects of low-flow hydrology on riverine ecosystems: Ecological principles and consequences of alteration. *Freshwater Science*, *31*(4), 1163–1186. doi:10.1899/12-002.1

Salinas, L. M., Delaune, R. D., & Patrick, W. H. (1986). Changes occurring along a rapidly submerging coastal area: Louisiana, USA. *Journal of Coastal Research*, *2*(3), 269–284.

Santos Pereira, L., Cordery, I., & Iacovides, I. (2009). *Coping with water scarcity: Addressing the challenges. Coping with Water Scarcity: Addressing the Challenges.* doi:10.1007/978-1-4020-9579-5

Sin, S. N., Chua, H., Lo, W., & Ng, L. M. (2001). Assessment of heavy metal cations in sediments of Shing Mun River, Hong Kong. *Environment International*, *26*(5-6), 297–301. doi:10.1016/S0160-4120(01)00003-4 PMID:11392742

Srebotnjak, T., Carr, G., De Sherbinin, A., & Rickwood, C. (2012). A global Water Quality Index and hot-deck imputation of missing data. *Ecological Indicators*, *17*, 108–119. doi:10.1016/j.ecolind.2011.04.023

Stenström, T. A., Seidu, R., Nelson, E., & Christian, Z. (2011). *Microbial Exposure and Health Assessments in Sanitation Technologies and Systems*. (Repéré à www.ecosanres.org

Veldkamp, T. I. E., Wada, Y., Aerts, J. C. J. H., Döll, P., Gosling, S. N., Liu, J., ... Ward, P. J. (2017). Water scarcity hotspots travel downstream due to human interventions in the 20th and 21st century. *Nature Communications*, *8*, 15697. doi:10.1038/ncomms15697 PMID:28643784

Vinet, F. (2017). Flood Impacts on Loss of Life and Human Health. *Floods*, 33-51. doi:10.1016/B978-1-78548-268-7.50003-1

Vodela, J. K., Renden, J. A., Lenz, S. D., Mc Elhenney, W. H., & Kemppainen, B. W. (1997). Drinking Water Contaminants (Arsenic, Cadmium, Lead, Benzene, and Trichloroethylene). Interaction of Contaminants with Nutritional Status on General Performance and Immune Function in Broiler Chickens. *Poultry Science*, *76*(11), 1474–1492. doi:10.1093/ps/76.11.1474 PMID:9355140

Water, U. N. (2006). *Coping with Water Scarcity–A Strategic Issue and Priority for System-Wide Action.* Geneva, Switzerland: UN-Water.

Woodall, C. J. (2009). Waterborne diseases - What are the primary killers? *Desalination*, *248*(1-3), 616–621. doi:10.1016/j.desal.2008.05.110

World Health Organization. (2005). *Ecosystems and human well-being : health synthesis : a report of the Millennium Ecosystem Assessment*. Author.

World Health Organization. (2012). UN-Water Global Analysis and Assessment of Sanitation and Drinking-Water: The Challenge of Extending and Sustaining Services. *UN Water Report*, 1-112.

Wright, J., & Gundry, S. W. (2009). Household characteristics associated with home water treatment: An analysis of the Egyptian demographic and health survey. *Journal of Water and Health*, *7*(1), 21–29. doi:10.2166/wh.2009.056 PMID:18957772

Wright, L. D., Elia, C. F. D., & Nichols, C. R. (2019). *Tomorrow's Coasts: Complex and Impermanent* (Vol. 27). Springer International Publishing. doi:10.1007/978-3-319-75453-6

Yang, G., Sau, C., Lai, W., Cichon, J., & Li, W. (2015). Article. *HHS Public Access*, *344*(6188), 1173–1178. doi:10.1126cience.1249098.Sleep

Yang, K., LeJeune, J., Alsdorf, D., Lu, B., Shum, C. K., & Liang, S. (2012). Global distribution of outbreaks of water-associated infectious diseases. *PLoS Neglected Tropical Diseases*, *6*(2), e1483. doi:10.1371/journal.pntd.0001483 PMID:22348158

Yuan, G. L., Liu, C., Chen, L., & Yang, Z. (2011). Inputting history of heavy metals into the inland lake recorded in sediment profiles: Poyang Lake in China. *Journal of Hazardous Materials*, *185*(1), 336–345. doi:10.1016/j.jhazmat.2010.09.039 PMID:20932644

Chapter 4
Food Security and Climate Change

Mohamed Ait El Mokhtar
https://orcid.org/0000-0001-8096-9927
Cadi Ayyad University, Morocco

Mohamed Anli
Cadi Ayyad University, Morocco

Raja Ben Laouane
Cadi Ayyad University, Morocco

Abderrahim Boutasknit
Cadi Ayyad University, Morocco

Hanane Boutaj
Cadi Ayyad University, Morocco

Ahmed Draoui
Cadi Ayyad University, Morocco

Lamia Zarik
Cadi Ayyad University, Morocco

Abdessamad Fakhech
Cadi Ayyad University, Morocco

ABSTRACT

Climate change may potentially disrupt progress toward a world without hunger. Today, a clear and consistent global pattern is perceptible of the different impacts of climate change on crop productivity that could have repercussions on food security. Consequently, the stability of the whole food systems may be at risk under climate change because of its unpredictable variations. Indeed, agricultural production is highly vulnerable even to 2°C predictions augmentation for global mean temperatures in 2100, with major implications for poverty and for food security. The climate change impacts seem to be clear in areas currently affected by hunger and undernutrition, which will heighten food insecurity in these parts of the world. Therefore, adapting food systems both to increase food security and to prevent future negative impacts from climate change will require attention to more than just agricultural production. The evidence sustains the need for thoughtful investment in adaptation and mitigation actions toward an efficient management of climate change influences on food security.

DOI: 10.4018/978-1-5225-7775-1.ch004

INTRODUCTION

Approaching hunger is one of the substantial challenges of our time (Ban, 2012). Hunger has many aspects and causes, including, increasing food demands, changes in diet and extreme climatic events, among other factors. Moreover, further pressures on the global food system are expected to increase in the future. For example, demand for agricultural products is estimated to heighten by about 50% by 2030 as the global population increases (Bruinsma, 2003), which will demand a shift toward sustainable intensification of food systems (Garnett et *al.*, 2013). Though present-day global food supplies achieve current global calorie requirements to feed the world population, food insecurity prevails in several parts of the world. Two billion people around the world are food insecure at least during a period of the year (Gleadow et *al.*, 2009) and around 815 million people in developing countries, 60% of which live in Sub-Saharan Africa and South Asia, have calorie-deficient diets (FAO, 2017).

Climate change, which is due in large part to economic activity, is directly affecting agricultural production and thereby food security (Easterling et *al.*, 2007; Schmidhuber and Tubiello, 2007; Thornton et *al.*, 2009). Even so, adaptation to climate change through interventions in agricultural systems or through changing social and institutional structures has potential to reduce food insecurity (Easterling et *al.*, 2007; Howden et *al.*, 2007). It is estimated that agriculture contributes to about 14% of annual greenhouse gas (GHG) emissions (Johnson et *al.*, 2007; Cohen et *al.*, 2008; Nelson, 2009; Sarris, 2009), while land use change, including forest loss, contributes by another 19% (Cohen et *al.*, 2008; Nelson, 2009). Accommodations in the agricultural sector thus could play a significant role in climate change mitigation and may as a result influence food security.

In this chapter, we will examine climate change status in North Africa because of the scarcity of studies about climate variations in this area and we will describe the possible impact of the projected global climate change on agriculture with its different components and consequently on food security.

CLIMATE CHANGE IN NORTH AFRICA

Climate change poses a significant challenge for North Africa, which emits low levels of greenhouse gases (between 1.5 and 3.5 emission tonnes of CO_2/inhabitant/year) (Radhouane, 2013). It affects environmental systems, major socio-economic and agricultural developments in the region. The severity of climate change affecting these countries is related to the ecological and geographic particularity of the region. The biophysical and socio-economic conditions, as well as the state of technology in the region are the major factors behind the extreme vulnerability of the region to climate change (Scott, 2008). Rising the sea level, increasing annual mean temperatures and decreasing precipitation, which has been observed for the second half of the 20th century in North Africa are likely to continue and to cause heater and drier conditions. The dynamical regional model REMO showed that temperatures are likely to rise between 2 and 3°C in North Africa, while precipitation is likely to decrease between 10% and 20% by 2050 (Paeth et *al.*, 2009). Furthermore, the sea level could rise by 23-47 cm by the end of the 21st century. Then, many Mediterranean regions would run a major risk of being submerged and eroded. The main economic and social activities in North Africa are concentrated along the coastal zones. Population within 100 km of coast is 68.8% in Algeria, 78.7% in Libya, 65.1% in Morocco, and 84% in Tunisia. Thus, sea level rise could result in major population movements and adversely affect many economic activities like tourism; a major source of employment and income in Morocco and Tunisia (Schilling et *al.*, 2012).

The rise in temperature and the decrease in annual precipitation decline the land areas suitable for agriculture, shorten the length of growing seasons, reduce crop yields and deteriorate food security. In these countries, it is estimated that a 1°C rise in temperature in a given year reduces economic growth in that year by about 1.1%. These effects will exacerbate in semiarid and arid regions that rely on irrigation for crop growth (Radhouane, 2013). As one of the world's most water-scarce regions with a high dependency on climate-sensitive agriculture, the economic and social conditions in North Africa are likely to deteriorate in the future. Therefore, the African countries face numerous environmental challenges and have to reconcile many conflicting priorities, from promoting economic diversification, ensuring water supply and food security, and furthering environmental.

CLIMATE CHANGE FORCING ON THE SPATIAL DISTRIBUTION OF USEFUL AGRICULTURAL AREA

Agriculture is the predominant activity of using lands on the planet today, because it provides a vital service for human societies. Moreover, agricultural activities, particularly, have been responsible for the big majority of these land use associated consequences (Tilman et *al.*, 2001; Green et *al.*, 2005). Almost 40% of free land surface of the planet is now being employed for agriculture, and a lot of this land has supersede grasslands, savannas and forests. Human activities of land use are a force of global importance. However, humans have widely modified the Earth's land surface, reducing the ability of ecosystems to continue rendering valuable resources such as freshwater, forest resources and food, and services such as control of water quality, air quality, climate and soil resources (Foley et *al.*, 2005). Freeing of tropical forests for culture or grazing is responsible for 12 to 26% of the total emissions of CO_2 to the atmosphere (Houghton, 2003), and land use changes can considerably modify regional and global climate (Pielke et *al.*, 2002). Furthermore, 20 to 30% of the total obtainable surface water on the planet is used for irrigation, and nitrogen fixation by way of crop cultivation and fertilizer production now equals or even transcends natural biotic fixation (Cassman & Wood, 2005).

Agriculture is a part or entirely responsible for environmental problems such as biodiversity loss, tropical deforestation, loss of habitats, fragmentation, losses of soil quality by erosion and salinization, emissions of important greenhouse gases, alteration of regional climates, reduction in air quality, decreases in quality of water resources, and increases in infectious diseases. On the other hand, agricultural intensification and expansion had furnished a crucial service to humanity by assembly the food needs of a speedily growing population, and thereby includes a trade-off between environmental deterioration and food production (Cassman & Wood, 2005). Despite of the differences in intended temperatures by different General Circulation Models (GCM), it exists some consensus on the probably changes in the climate, which can affected agriculture. Changes in productivity and crop yields resulting of climate change might vary significantly across regions and between localities, thusly ever-changing production patterns. Whereas, an enhance of less than 2°C in the average global temperature in the next 100 years might bring some advantages to the developed countries with temperate climates in the from milder winters, extended growing seasons and heighten yields of some crops, the ability to manage warming depends highly on access to technology and economic resources. In the subtropics and tropics regions, where non-irrigated agriculture is dominated by dry land, and where crops are close to their maximum temperature tolerance, yields are probably to reduce slightly with increment in atmospheric temperature. In general, agricultural productivity might decrease for the next century, leading vulnerable zones to malnutrition and hunger, particularly in drought regions of Africa. Until recently, most researches have demonstrated that agricultural technology and management will not improve and adapt. If adaptive strategies are taken into account, the management of climate variations might change. Otherwise, higher concentration of CO_2 can increase crop productivity. The study of agricultural effects of trace gas including climate change is complex by the fact that enhancement of atmospheric CO_2 has other impacts on crop plants besides its alteration of their climate regime (Choudhary, 2018).

Many studies have explored already the global distribution of potential land arable, under current and future climate change, in term of biophysical circumstances (Cramer & Solomon 1993; Xiao, 1997; Bot et *al*., 2000; Ramankutty et *al*., 2002). Nevertheless, in those studies, neither the ambiguity inherent in land classification nor the uncertainty engaged in climate change projections has been definitely addressed (Ahamed et *al*., 2000). Land availability evaluations must be concerned with the challenges engaged in land classification in relation to multiple criteria, and particularly, land evaluation at the global scale employing international datasets must considering the uncertainty engaged in those datasets. Recently, Cai et *al*. (2011) suggested a learning based on logic modeling approach and employed this framework to assess the global agricultural land availability, performed by probability based on land suitability values. In order to evaluate the consequences of agriculture on Earth system, the positive economic and social benefits and the frequently negative environmental effect, it is necessary to establish global data series of the spatial distribution of useful agricultural land and vegetation cover change (Bauer et *al*., 2003; Donner & Kucharik, 2003; Cassman & Wood, 2005). Current advances have led to the emersion of newly data series on land cover of agricultural, by combining satellite derived vegetation cover data series and soil based agricultural inventory data series (Donner, 2003; Leff et *al*., 2004; Ramankutty, 2004). Mendelsohn et *al*. (2000) presented in Figure 1 the results by country as a part of 2100 agricultural Gross domestic product (GDP). Accompanied by the experimental climate of response function and the 14 climate predictions, which 7 countries are predicted to be affected by important average losses in the agricultural area in down warding order (Algeria, Morocco, Nigeria, Cameroon, South Africa, Sudan, and Zaire). At the same time, these 7 countries constitute for 47% of the detriments in Africa. Moreover, those countries have great agricultural areas. The examination of impacts as a part of agricultural GDP therefore gives a different perspectives. The following countries, in descending order, were most severely affected (Zambia, Niger, Chad, Burkina Faso, Togo, Botswana, Guinea Bissau, and Gambia). These 8 countries beard losses greater than 70% of their agricultural areas, with the cross sectional climate of response functions perhaps, national effects significant although only Niger, Chad Togo and Burkina Faso had great impacts than 10% of agricultural GDP. Whereas, Zambia, Chad and Niger practically losing their complete farming sector.

Several approaches exist to derive large-scale climate change impacts on crop yields, such as the use of empirical relationships between crop growth and climate. Schlenker and Lobell (2010) combined crop production and weather data into a panel analysis, and showed that by the mid-21st century, the mean estimates of aggregate production changes in Sub-Saharan Africa under the preferred model specification are -22%, -17%, -17%, -18% and -8% for maize, sorghum, millet, groundnut and cassava, respectively. Figure 2 shows the R-square of different models which is presented as a colored bar. A model that has only fixed effects and quadratic time trend is presented as a black solid line. Considering the large difference in mean yields, these fixed effects take up on a large fraction of the global variation. Count on the crop, various models result in the upmost R-square, with nearly all presenting significant progress over a model with no weather. Just for cassava that the weather variables didn't added a lot, that is not astonishing as it is a root crop with much variable increasing season and it is therefrom empirically

Figure 1. Percentage variation in agricultural GDP (PAG) in African countries in 2100 because of climate change using the average values (AVG) of 14 global climate change models and cross sectional (CRS) coefficient (CRS_AVG_PAG) (Mendelsohn, 2000)

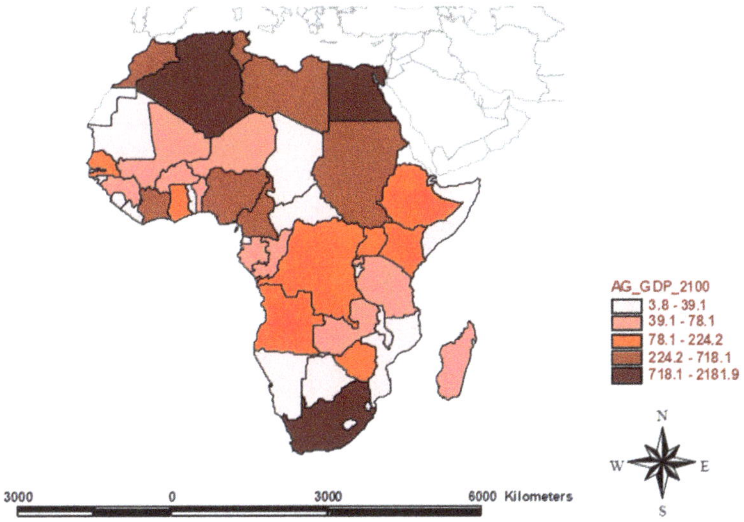

tough to set weather data for the increasing season to a particularly yield. Schlenker and Lobell (2010) choose the degree-days model (10 to 30 °C, beyond 30 °C) as their basic model. In general it gives lower predicted harms and they are hence conservative in their harm assess.

IMPACT OF CLIMATE CHANGE ON AGRICULTURAL INDUSTRY

Impact on Soil

Soil is an essential component of a terrestrial ecosystem (Singh and Gupta, 1977). It is characterized by significant physicochemical and biological properties. These properties are influenced by the functioning, composition, fertility and quality of the soil (Gros, 2002). The stability of the soil structure is an important physical property at the interface of many agronomic and environmental functions of the soil (Bulluck lii et *al*., 2002). The incorporation of organic matter is an agricultural practice to improve the stability of the soil structure, especially through the microbial activity generated (Six et *al*., 2004). Any disturbances or changes lead to a decline in the productivity of organic matter and can lead to imbalances in the functioning of the soil (Schoenholtz et *al*., 2000). However, climate change is among the major factors that lead to major problems of vulnerability to land degradation (Verchot et *al*., 2007; IPCC, 2014). On the other hand, soils in arid and semi-arid areas are particularly sensitive to climatic conditions that have become increasingly difficult as a result of prolonged climate change (Lioubimtseva et *al*., 2009). Precipitation, increased temperature, and decreased soil moisture influence the availability, decomposition, and mineralization of organic matter, and negatively affect the abundance and functioning of soil microorganisms (McDanie et *al*., 2013). In addition, increasingly frequent climatic events, such as drought, salinity, erosion and silting, are likely to aggravate the degradation and reduction of biologi-

Figure 2. R-square of different model specifications excepting all flagged yields: for each crop Wolfram Schlenker and David B Lobell, (2010) run four model specifications (model 1-4) using two different datasets (CRU 2.1 and NCC) and mean weather over all country of only crop growing zone. Maize, Groundnuts and Sorghum enclose results when an isolate regression is assessed for high fertilizer countries (shown in blue). The R-square for a model using frozen impacts as well as time trends is added as a black line (Schlenker & Lobell, 2010)

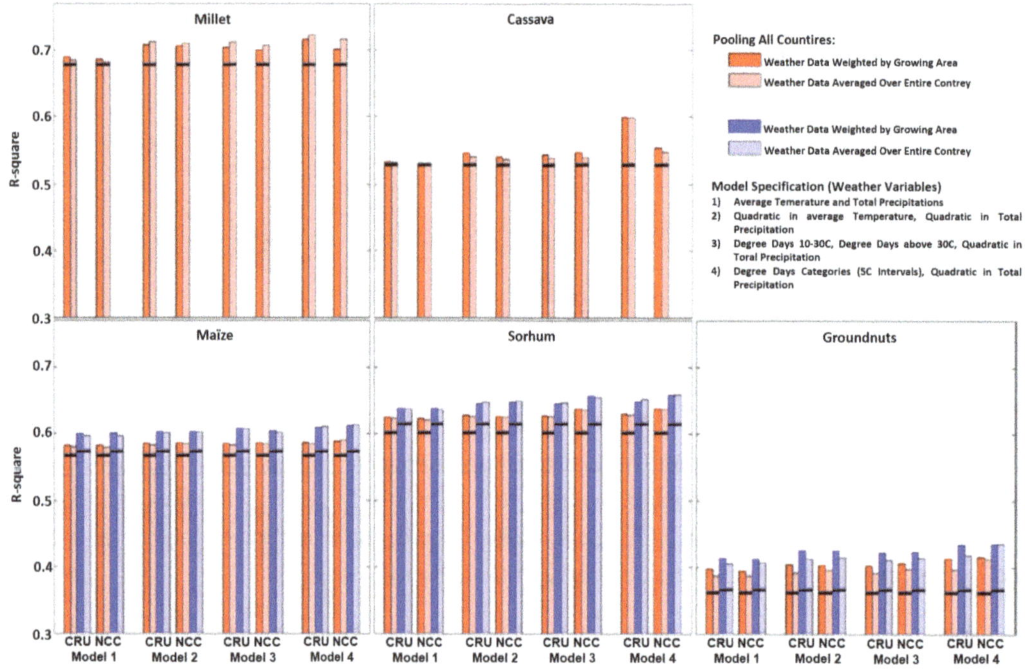

cal biomass, the physical and chemical degradation of soils (Zhang et *al.*, 2005; Waldrop & Firestone, 2006; Frey et *al.*, 2008; IPCC, 2014).

Climate change can be enhanced by the loss of organic matter reserves in soils and vegetation. In addition, degradations caused by reductions in vegetation cover influence the local microclimate by decreasing the humidity of the air and increasing the temperature of the soil (Stabler et *al.*, 2007). However, reduced vegetation cover usually increases the amount of solar radiation reflected by the earth's surface (albedo effect), which has a cooling effect on the overall temperature of the soil. It's very clear that climate change has the potential to disturb terrestrial ecological systems, which can lead to a failure of the supply of important elements in the rehabilitation and have significant negative consequences such as desertification, erosion, silting, increasing salinity and soil acidity (Sati, 2017).

Impact on Water

Issues of climate change and variability have been a concern for scientists and policy makers in the last decades. Climate variability can be manifested by long periods of drought with negative effects on the hydrological cycle, the environment and socio-economic activities. Water resources are considered one of the most important resources for all components of an ecosystem (Bouziani, 2006). In fact, areas characterized by arid, semi-arid and sub-Saharan climate are subject to adverse climatic variations, accentuated by periods of drought, and longer and drier episodes. Climate change in these areas is marked by large fluctuations, including temperature increase and a declining trend in rainfall patterns (Hulme et *al*, 2000). Therefore, decreasing precipitation, increasing CO_2 concentration and temperatures, and increasing their variability implies a reduction in water resources, and thus an acceleration of land degradation. Like a situation accentuates the risks of drought by the degradation of the plant cover. The decline in vegetation cover leads to a decrease in the infiltration of water into the groundwater and leads to soil erosion. The succession of years with a deficit rainfall also marks the flow negatively. For some authors (Mahé & Olivry, 1995, Bricquet et *al*., 1997, Mahé et *al*., 2000), the explanation for this difference between rainfall and hydrological deficit comes from a reduced contribution of inputs from underground sources. Since the 1970s, groundwater supplies were decreasing as the rainfall deficit continued.

Impact on Air

The air we breathe every day is a vital part of our survival on earth. Air pollution is a limiting factor for its purity. However, air quality is probably one of the first health and environmental concerns of this new century. Air consists essentially of gases including nitrogen, oxygen and carbon dioxide. These elements play a key role in physiology (Loreto et *al*., 1996), growth, yield and maturity and quality of fruits and seeds of crops. In addition, air quality disruption can directly or indirectly affects plant requirements (Zhu et *al*., 2018).

Currently, climate variability increases the concentration of greenhouse gases (N_2O) (Stocker et *al*., 2013). The intense use of nitrogenous chemical fertilizers contributes to the increase in atmospheric emissions of N_2O (Cai et *al*., 1997). Air pollution from human activities and natural disasters has been one of the major environmental challenges in recent decades. Human activities and population growth alter the atmospheric composition and damage the Earth's atmosphere (Aghamohammadi et *al*., 2018). Today, climate change and air pollution pose significant risks in the short and long term (Cairncross et *al*., 2018). Air pollution continues to increase, causing unpredictable circulatory and respiratory diseases (Langkulsen et *al*., 2018).

Impact on Seeds Quality

Agriculture is one of the sectors that contribute to the economy and development of the country. This sector is made up of several essential components that affect the quality and quantity of agricultural production. Indeed, these components in particular, soil, air, water, organic matter, nutrients ... must be ensured by favorable conditions on the one hand. On the other hand, the climatic conditions, which intervene in the germination of the seeds until the death of the plant, must be favorable also.

The main climatic conditions that influence seed germination are two sets including internal conditions, good physiological and genetic state of the seed and external conditions related to the physical factors of the surrounding environment that are moisture, substrate (Bonner, 2008, Costa et *al*., 2018), temperature (Bonner, 2008, Houang et *al*., 2018, Lindow-Lopez et *al*., 2018), light and aeration around the seed (Bonner, 2008). Lindow-Lopez et *al*. (2018) showed that temperature plays a big role in seed germination. Temperatures between 20 and 25 °C seem to be the suitable for germination unlike temperatures between 15 and 30 °C, which record low proportions of germination (Figure 3). These climatic conditions can be very variable, in particular temperature and humidity. In addition, climate change has become unpredictable from one year to another, particularly in arid, semi-arid, and sub-Saharan regions. These regions are known by the increase in temperature marked by significant periods of drought (Salem et *al.*, 2011). The agricultural soils of these regions are subjected to biotic and abiotic constraints as well as to low fertility which has led to an intense destruction of these environments, necessarily limiting agricultural production in these severe and fragile equilibrium environments (Zougari-Elwedi et *al.*, 2012, Meddich et *al.*, 2018). In many arid and semi-arid regions of the world, drought limits crop productivity (Al-karaki et *al.*, 2004). However, drought and salinity are major constraints that severely limit crop production (Leye et *al.*, 2012). Otherwise, climate has a great influence on plant regeneration (Lloret et *al.*, 2004; Fay and Schultz, 2009; Dalgleish et *al.*, 2010). For example, with reduction of winter period (Dong et *al.*, 2010), seeds could remain partially dormant in spring and need more time to germinate (Walck et *al.*, 1997). The alteration of temperature and water supply due to global climate change may prevent, delay or increase regeneration from seeds (Walck et *al.*, 2011).

It appears that unpredictable climate changes may lead to greater climatic variability (Kar and Das, 2015), greater irregularity of rainy periods and dry periods, greater rainfall violence and thus greater sensitivity to disturbance of factors of germination and growth of plants. However, seed germination would depend on the degree of dormancy between species in relation to global warming (Mondoni et *al.*, 2012).

EFFECT OF CLIMATE CHANGE ON FOOD SECURITY

Food security is one of the highest priority concerns on the development programs of many developing countries. As defined by the Food and Agriculture Organization (FAO), food security is a "situation that exists when all people, at all times, have physical, social and economic access to sufficient, safe and nutritious food that meets their dietary needs and food preferences for an active and healthy life"(FAO, 2012). It includes components of food availability, food stability, food access and food utilization. Several factors can affect food security including population increase; climate change; urbanization and globalization; land use shifts and water availability; income growth and nutritional trends; and tendencies in global energy supply and food trading (Von Braun, 2007; Premanandh, 2011). It is really challenging to evaluate precisely the present status of global food security from such a wide concept. However, the big picture is clear: About 2 billion of the global population of over 7 billion are food insecure because they fall short of one or many of FAO's components of food security (FAO, 2017). Great geographic differences in the percentage of undernourishment exist within this global estimate, with almost all countries in the most extreme "alarming" food security status located in sub-Saharan Africa, East Africa, middle Africa or South Asia (von Grebmer et *al.*, 2012; FAO, 2017) (Table 1). These areas have large rural populations, wide spread poverty and extensive lands of low agricultural productivity due to steadily degrading resource bases, fragile markets and high climatic risks (Misra et *al.*, 2014).

Among all factors, climate change will have a significant impact on food system activities and various aspects of food security (Wheeler and Von Braun, 2013). In order to understand the effects of climate change on food security it is necessary to understand the nexus between climate change, food security, and its drivers as shown in Figure 4. The drivers, that can all be affected by climate variations in some way, can include the environmental system, its cycles and its management as well as agricultural, socio-economic, demographic, cultural and political variables that can be directly or indirectly impacted by climate. Changes in climate variables might also affect science and technology demands and outcomes

Figure 3. Germination proportion (mean ± standard error) of six cactus species at constant and alternating temperatures. Different letters indicate significant differences in the mean value of the transformed variable for the combination of species x temperature (Lindow-López et al., 2018)

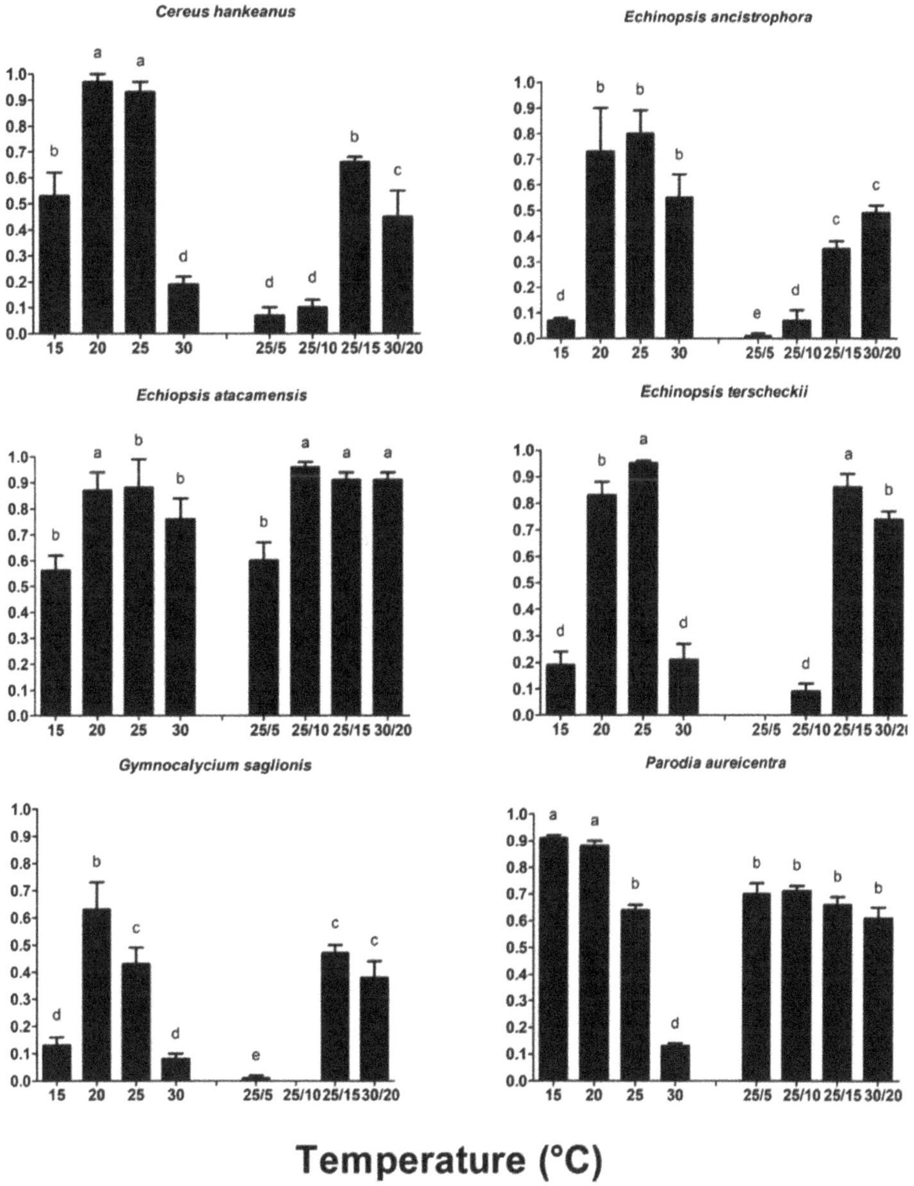

that in turn may affect food security outcomes. These drivers affect the four aspects of food security. However, there are also feedbacks from the outcomes of food security to the drivers. For example, if there is reduced rainfall and this is not controlled by agricultural management strategies, crop yields will decrease and reduce food availability. This might therefore affect the price of the crop. This economic driver might then affect accessibility to food.

Table 1. Prevalence of undernourishment in the world by region, 2000–2016 (FAO, 2017)

	2000	**2005**	**2010**	**2011**	**2012**	**2013**	**2014**	**2015**	**2016[1]**
	Pourcentage								
WORLD	**14.7**	**14.2**	**11.5**	**11.2**	**11.0**	**10.8**	**10.7**	**10.6**	**11.0**
AFRICA	**24.3**	**20.8**	**18.3**	**17.9**	**17.8**	**17.8**	**18.1**	**18.5**	**20.0**
Northern Africa	**6.8**	**6.3**	**5.1**	**4.8**	**8.5**	**8.4**	**8.3**	**8.3**	**8.3**
Sub-Saharan Africa	**28.1**	**23.7**	**20.6**	**20.2**	**20.0**	**20.0**	**20.4**	**20.8**	**22.7**
Eastern Africa	39.3	34.3	30.9	30.2	30.6	30.6	30.9	31.1	33.9
Middle Africa	37.4	29.4	23.8	23.1	22.5	22.3	23.0	24.4	25.8
Southern Africa	7.1	6.4	6.7	6.3	6.2	6.2	6.5	6.6	8.0
Western Africa	15.1	12.0	10.0	9.9	9.9	9.8	9.8	10.4	11.5
ASIA	**16.7**	**17.0**	**13.2**	**12.8**	**12.5**	**12.2**	**11.9**	**11.6**	**11.7**
Central Asia and Southern Asia	**17.6**	**20.1**	**15.7**	**15.7**	**15.6**	**15.4**	**15.1**	**14.7**	**14.2**
Central Asia	15.7	14.2	10.6	9.9	9.1	8.4	8.2	8.2	8.4
Southern Asia	17.7	20.4	15.9	15.9	15.9	15.7	15.3	14 .9	14.4
Eastern Asia and South-Eastern Asia	**16.6**	**15.2**	**11.6**	**10.9**	**10.4**	**9.9**	**9.6**	**9.2**	**9.7**
Eastern Asia	14.6	14.1	11.3	10.7	10.3	9.9	9.5	9.1	9.0
South-Eastern Asia	22.0	18.1	12.4	11.3	10.7	10.0	9.7	9.4	11.5
Western Asia	**11.3**	**10.5**	**9.4**	**9.1**	**8.9**	**8.7**	**8.9**	**9.3**	**10.6**
LATIN AMERICA AND THE CARIBBEAN	**12.0**	**9.1**	**6.8**	**6.6**	**6.4**	**6.3**	**6.3**	**6.3**	**6.6**
Latin America	**11.1**	**8.0**	**5.9**	**5.7**	**5.5**	**5.4**	**5.4**	**5.5**	**5.9**
Central America	8.1	8.3	7.1	7.2	7.1	7.1	6.9	6.7	6.5
South America	12.2	7.9	5.4	5.1	4.8	4.7	4.8	5.0	5.6
Caribbean	23.8	23.3	19.9	19.3	19.4	19.2	18.9	18.4	17.7
OCEANIA	5.3	5.3	5.0	5.2	5.3	5.7	6.0	6.4	6.8
NORTHERN AMERICA AND EUROPE	<2.5	<2.5	<2.5	<2.5	<2.5	<2.5	<2.5	<2.5	<2.5
Other country group: Western Asia and Northern Africa	9.3	8.7	7.6	7.3	8.7	8.5	8.6	8.8	9.5

[1]Projected values

Climate change shows profound threats to global food security due to variation in water supply and demand (Döll and Siebert, 2002; Barnett et *al.*, 2005; Alcamo et *al.*, 2007; Spash, 2008a), impacts on crop productivity (Droogers, 2004; Droogers & Aerts, 2005), impacts on food supply (Rosenzweig and Parry, 1994; Arnell et *al.*, 2004), and high costs of adaptation to climate change (Kandlikar & Risbey, 2000).

Climate change could influence agriculture and food security by disrupting the spatial and temporal distribution of precipitations, and the availability of water, land, capital, biodiversity and terrestrial resources. It may increase vulnerabilities all through the food chain, and yield to trading dynamics, and

Figure 4. Linkages between climate change and food security
(Ziervogel & Ericksen, 2010)

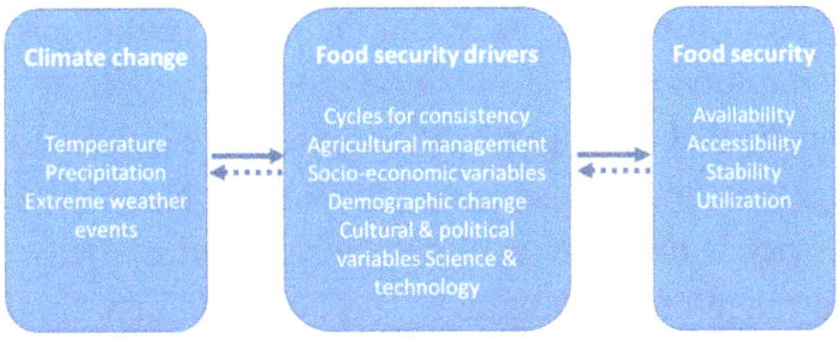

finely affect the global economy, food security and the ability to secure food to nine billion people by 2050. Modelling by the International Institute for Applied Systems Analysis (IIASA) indicates that future socioeconomic development and climate change may impact on regional and global irrigation requirements and therefore on agricultural water scarcity (Fischer et *al.*, 2007). The irrigation requirements may enhance by 45% by 2080. Even with improvements in irrigation efficiency, gross water scarcity may increase by 20%. Global irrigation requirements under climate change will increase by 20% above the reference case scenario (without climate change). The modelling simulation shows that the global effects of climate change on irrigation water demands could be as large as the projected increase in irrigation due to socioeconomic development.

The impacts of climate change on global food production are small but geographically very unequally distributed, where the most affected countries are in arid and sub-humid tropics in Africa and South Asia (Parry et *al.*, 2001) and particularly in developing countries with low capacity for adaptation (Kurukulasuriya et *al.*, 2006).

Climate change will impact on crop productivity, with implications for food security (Spash, 2008a,b). Global warming has been suspected to heighten yields due to the rising atmospheric carbon, but the impacts seemed to be negative for poor countries. For example, global warming will decrease food production in countries in the equator regions (Droogers & Aerts, 2005). African countries will experience issues of droughts and further food deficiency. Likewise, the Pacific Islands and Indonesia will be more dependent on imports and look toward more poverty and other social problems. Recently, a study by The International Water Management Institute (IWMI) expects a 50% decline in South Asian wheat production by 2050 – equal to about 7% of the global crop production (de Fraiture et *al.*, 2008). The Peterson Institute states that agricultural production in developing countries may fall between 10% and 25% and even by 40% if global warming is unabated (Cline, 2007).

Climate change may affect rainfall and the supply of water for irrigation in many areas and countries in the world. A decline in rainfall along with an increase in temperature will heighten crop water requirement due to high evapotranspiration while less rainfall will increase crop net irrigation water demands. Thus, the already existing water withdrawal problem will be intensified in many regions and countries, and impact food production. The most affected will be the areas with intense water scarcity and food security issues, such as the arid countries of sub-Saharan Africa and parts of South Asia, which are already subjected to malnutrition, poverty, and more episodes of hunger (Brown & Lall, 2006; Brown & Funk, 2008; Funk et *al.*, 2008).

CONCLUSION

Managing food security and its sustainability is one of the major challenges worldwide. Most countries are not able to supply sufficient quantities of nutritious food to people so that they can live healthily. There is an emerging consensus that climate variations can have imminent impacts on the food security of the most vulnerable people, in the absence of adaptation. Nevertheless, the aggregate impact of climate change on food security is not fully explored. In particular, many of the impacts are difficult to quantify and depend on a range of presumption. The available quantitative researches indicate that climate change will negatively affect food security at the global level in the long term. The studies suggest that, at the global scale, climate change will reduce crop yields and the land suitable for agricultural activities with the greatest impacts in tropical regions where the greatest food security issues persist. It is strongly believed that there is enough food in the world to feed everyone adequately except that the problem is distribution and management. Consequently, development and implementation of food security strategies are necessary that must include procedures for handling threats, product tampering, and product storage and distribution plan along with a monitoring procedure. This paper indicates that climate change is only one of multiple changes affecting food systems and that its relative importance varies between regions. Adaptations of food systems through interventions in availability, access, stability and utilization are possible to cope with climate change at different scales although their feed-backs to the planet system have yet to be fully assessed.

REFERENCES

Aghamohammadi, N., & Isahak, M. (2018). Climate Change and Air Pollution in Malaysia. In *Climate Change and Air Pollution* (pp. 241–254). Cham: Springer. doi:10.1007/978-3-319-61346-8_15

Ahamed, T. N., Rao, K. G., & Murthy, J. S. R. (2000). GIS-based fuzzy membership model for crop-land suitability analysis. *Agricultural Systems*, *63*(2), 75–95. doi:10.1016/S0308-521X(99)00036-0

Al-Karaki, G., McMichael, B. Z. A. K. J., & Zak, J. (2004). Field response of wheat to arbuscular mycorrhizal fungi and drought stress. *Mycorrhiza*, *14*(4), 263–269. doi:10.100700572-003-0265-2 PMID:12942358

Alcamo, J., Dronin, N., Endejan, M., Golubev, G., & Kirilenko, A. (2007). A new assessment of climate change impacts on food production shortfalls and water availability in Russia. *Global Environmental Change*, *17*(3-4), 429–444. doi:10.1016/j.gloenvcha.2006.12.006

Arnell, N. W., Livermore, M. J., Kovats, S., Levy, P. E., Nicholls, R., Parry, M. L., & Gaffin, S. R. (2004). Climate and socio-economic scenarios for global-scale climate change impacts assessments: Characterising the SRES storylines. *Global Environmental Change*, *14*(1), 3–20. doi:10.1016/j.gloenvcha.2003.10.004

Ban, K. (2012). *A message from the UN Secretary General for the opening session of the 39th Session of the Committee on World Food Security, Rome, 15 to 20 October 2012*. Retrieved from www.unfoodsecurity.org/node/1356)

Barnett, T. P., Adam, J. C., & Lettenmaier, D. P. (2005). Potential impacts of a warming climate on water availability in snow-dominated regions. *Nature*, *438*(7066), 303–309. doi:10.1038/nature04141 PMID:16292301

Bauer, E., Claussen, M., Brovkin, V., & Huenerbein, A. (2003). Assessing climate forcings of the Earth system for the past millennium. *Geophysical Research Letters*, *30*(6). doi:10.1029/2002GL016639

Bindi, M., & Moriondo, M. (2005). Impact of a 2°C global temperature rise on the Mediterranean region: Agriculture analysis assessment In C. Giannakopoulos, M, Bindi, M. Moriondo, P. Le Sager, & T. Tin (Eds.), Climate change impacts in the Mediterranean resulting from a 2°C global temperature rise (pp. 54-66). WWF Report.

Bonner, F. T. (2008). Storage of seeds. *Woody plant seed manual*, 85-96.

Bot, A., Nachtergaele, F., & Young, A. (2000). *Land resource potential and constraints at regional and country levels (No. 90)*. Food & Agriculture Org.

Bouziani, M. (2006). L'eau dans tous ses états. Ed. Dar El Gharb.

Bricquet, J., Bamba, F., Mahe, G., Toure, M., & Olivry, J. (1997). Evolution récente des ressources en eau de l'Afrique atlantique. *Revue des sciences de l'eau/Journal of Water Science*, *10*(3), 321-337.

Brown, C., & Lall, U. (2006). Water and economic development: The role of variability and a framework for resilience. *Natural Resources Forum*, *30*(4), 306–317. doi:10.1111/j.1477-8947.2006.00118.x

Brown, M. E., & Funk, C. C. (2008). Food security under climate change. *Science*, *319*(5863), 580–581. doi:10.1126cience.1154102 PMID:18239116

Bruinsma, J. (2003). *World Agriculture: Towards 2015/2030—an FAO perspective.* London: Earthscan.

Bulluck, L.R., III, Brosius, M., Evanylo, G.K., & Ristaino, J. B. (2002). Organic and synthetic fertility amendments influence soil microbial, physical and chemical properties on organic and conventional farms. *Applied Soil Ecology, 19*(2), 147-160.

Cai, X., Zhang, X., & Wang, D. (2010). Land availability for biofuel production. *Environmental Science & Technology*, *45*(1), 334–339. doi:10.1021/es103338e PMID:21142000

Cai, Z., Xing, G., Yan, X., Xu, H., Tsuruta, H., Yagi, K., & Minami, K. (1997). Methane and nitrous oxide emissions from rice paddy fields as affected by nitrogen fertilisers and water management. *Plant and Soil*, *196*(1), 7–14. doi:10.1023/A:1004263405020

Cairncross, E., Dalvie, A., Euripidou, R., Irlam, J., & Naidoo, R. N. (2018). Climate Change, Air Pollution and Health in South Africa. In *Climate Change and Air Pollution* (pp. 327–347). Cham: Springer. doi:10.1007/978-3-319-61346-8_20

Cassman, K. G., & Wood, S. (2005). Cultivated systems. In R. Hassan & ... (Eds.), *Ecosystems and Human Well-being: Current State and Trends* (Vol. 1, pp. 745–794). Washington, D. C.: Island Press.

Choudhary, D. (2018). *An Assignment on Climate Change and Agriculture Agmet-602, Fundamentals of Climatology (3+0).* Agrometeorologist, Dept. of Agricultural Meteorology B.A.C.A., AAU, ANAND.

Cline, W. R. (2007). *World agriculture faces serious decline from global warming*. Washington, DC: Centre for Global Development.

Cohen, M. J., Tirado, C., Aberman, N. L., & Thompson, B. (2008). Impact of climate change and bioenergy on nutrition. Rome: Food and Agricultural Organisations of the United Nations (FAO) and International Food Policy Research Institute (IFPRI).

Costa, C. R. X., Pivetta, K. F. L., de Souza, G. R. B., Mazzini-Guedes, R. B., Pereira, S. T. S., & Nogueira, M. R. (2018). Effects of Temperature, Light and Seed Moisture Content on Germination of Euterpe precatoria Palm. *American Journal of Plant Sciences*, *9*(01), 98–106. doi:10.4236/ajps.2018.91009

Cramer, W. P., & Solomon, A. M. (1993). Climatic classification and future global redistribution of agricultural land. *Climate Research*, *3*, 97–110. doi:10.3354/cr003097

Dalgleish, H. J., Koons, D. N., & Adler, P. B. (2010). Can life-history traits predict the response of forb populations to changes in climate variability? *Journal of Ecology*, *98*(1), 209–217. doi:10.1111/j.1365-2745.2009.01585.x

de Fraiture, C., Smakhtin, V., Bossio, D., McCornick, P., Hoanh, C., Noble, A., ... Turral, H. (2008). *Facing Climate Change by Securing Water for Food, Livelihoods and Ecosystems*. Colombo, Sri Lanka: International Water Management Institute.

Döll, P., & Siebert, S. (2002). Global modeling of irrigation water requirements. *Water Resources Research*, *38*(4), 8-1–8-10. doi:10.1029/2001WR000355

Dong, W., Jiang, Y., & Yang, S. (2010). Response of the starting dates and the lengths of seasons in Mainland China to global warming. *Climatic Change*, *99*(1-2), 81–91. doi:10.100710584-009-9669-0

Donner, S. (2003). The impact of cropland cover on river nutrient levels in the Mississippi River Basin. *Global Ecology and Biogeography*, *12*(4), 341–355. doi:10.1046/j.1466-822X.2003.00032.x

Donner, S. D., & Kucharik, C. J. (2003). Evaluating the impacts of land management and climate variability on crop production and nitrate export across the Upper Mississippi Basin. *Global Biogeochemical Cycles*, *17*(3), n/a. doi:10.1029/2001GB001808

Droogers, P., & Aerts, J. (2005). Adaptation strategies to climate change and climate variability: A comparative study between seven contrasting river basins. *Physics and Chemistry of the Earth Parts A/B/C*, *30*(6-7), 339–346. doi:10.1016/j.pce.2005.06.015

Easterling, W. E., Aggarwal, P. K., Batima, P., Brander, K. M., Erda, L., Howden, S. M., . . . Tubiello, F. N. (2007). Food, Fibre and Forest Products. In Climate Change 2007: Impacts, Adaptation and Vulnerability. Contribution of Working Group II to the Fourth Assessment Report of the Intergovernmental Panel on Climate Change (IPCC), Cambridge University Press.

FAO. (2012). *The state of food insecurity in the world 2012. Economic growth is necessary but not sufficient to accelerate reduction of hunger and malnutrition*. Rome: Food and Agriculture Organization of the United Nations.

FAO, IFAD, UNICEF, WFP, & WHO. (2017). *The State of Food Security and Nutrition in the World 2017. Building resilience for peace and food security*. Rome: FAO.

Fay, P. A., & Schultz, M. J. (2009). Germination, survival, and growth of grass and forb seedlings: Effects of soil moisture variability. *Acta Oecologica*, *35*(5), 679–684. doi:10.1016/j.actao.2009.06.007

Fischer, G., Tubiello, F. N., Van Velthuizen, H., & Wiberg, D. A. (2007). Climate change impacts on irrigation water requirements: Effects of mitigation, 1990–2080. *Technological Forecasting and Social Change*, *74*(7), 1083–1107. doi:10.1016/j.techfore.2006.05.021

Foley, J. A., DeFries, R., Asner, G. P., Barford, C., Bonan, G., Carpenter, S. R., ... Helkowski, J. H. (2005). Global consequences of land use. *Science*, *309*(5734), 570–574. doi:10.1126cience.1111772 PMID:16040698

Frey, S. D., Drijber, R., Smith, H., & Melillo, J. (2008). Microbial biomass, functional capacity, and community structure after 12 years of soil warming. *Soil Biology & Biochemistry*, *40*(11), 2904–2907. doi:10.1016/j.soilbio.2008.07.020

Funk, C., Dettinger, M. D., Michaelsen, J. C., Verdin, J. P., Brown, M. E., Barlow, M., & Hoell, A. (2008). Warming of the Indian Ocean threatens eastern and southern African food security but could be mitigated by agricultural development. *Proceedings of the National Academy of Sciences of the United States of America*, *105*(32), 11081–11086. doi:10.1073/pnas.0708196105 PMID:18685101

Garnett, T., Appleby, M. C., Balmford, A., Bateman, I. J., Benton, T. G., Bloomer, P., ... Herrero, M. (2013). Sustainable intensification in agriculture: Premises and policies. *Science*, *341*(6141), 33–34. doi:10.1126cience.1234485 PMID:23828927

Gleadow, R. M., Edwards, E. J., & Evans, J. R. (2009). Changes in nutritional value of cyanogenic Trifolium repens grown at elevated atmospheric CO2. *Journal of Chemical Ecology*, *35*(4), 476–478. doi:10.100710886-009-9617-5 PMID:19352773

Green, R. E., Cornell, S. J., Scharlemann, J. P., & Balmford, A. (2005). Farming and the fate of wild nature. *Science*, *307*(5709), 550–555. doi:10.1126cience.1106049 PMID:15618485

Gros, R. (2002). *Fonctionnement et qualité des sols soumis à des perturbations physiques et chimiques d'origines anthropiques: réponses du sol, de la flore et de la microflore bactérienne tellurique* (Doctoral dissertation). Université de Savoie.

Houghton, R. A. (2003). Revised estimates of the annual net flux of carbon to the atmosphere from changes in land use and land management 1850–2000. *Tellus. Series B, Chemical and Physical Meteorology*, *55*(2), 378–390.

Howden, S. M., Soussana, J. F., Tubiello, F. N., Chhetri, N., Dunlop, M., & Meinke, H. (2007). Adapting agriculture to climate change. *Proceedings of the National Academy of Sciences of the United States of America*, *104*(50), 19691–19696. doi:10.1073/pnas.0701890104 PMID:18077402

Huang, Z., Footitt, S., Tang, A., & Finch-Savage, W. E. (2018). Predicted global warming scenarios impact on the mother plant to alter seed dormancy and germination behaviour in Arabidopsis. *Plant, Cell & Environment*, *41*(1), 187–197. doi:10.1111/pce.13082 PMID:29044545

Hulme, M., Wigley, T. M., Barrow, E., Raper, S., Centella, A., Smith, S., & Chipanshi, A. (2000). *Using a Climate Scenario Generator for Vulnerability and Adaptation Assessments: MAGICC and SCENGEN: Version 2.4 Workbook*. Climatic Research Unit.

Johnson, J. M. F., Franzluebbers, A. J., Weyers, S. L., & Reicosky, D. C. (2007). Agricultural opportunities to mitigate greenhouse gas emissions. *Environmental Pollution*, *150*(1), 107–124. doi:10.1016/j.envpol.2007.06.030 PMID:17706849

Kandlikar, M., & Risbey, J. (2000). Agricultural impacts of climate change: If adaptation is the answer, what is the question? *Climatic Change*, *45*(3-4), 529–539. doi:10.1023/A:1005546716266

Kurukulasuriya, P., Mendelsohn, R., Hassan, R., Benhin, J., Deressa, T., Diop, M., ... Mahamadou, A. (2006). Will African agriculture survive climate change? *The World Bank Economic Review*, *20*(3), 367–388. doi:10.1093/wber/lhl004

Langkulsen, U., & Rwodzi, D. (2018). Climate Change, Air Pollution, and Human Health in Bangkok. In *Climate Change and Air Pollution* (pp. 255–271). Cham: Springer. doi:10.1007/978-3-319-61346-8_16

Le Houérou, H. N. (1992). Vegetation and land use in the Mediterranean basin by the year 2050: A prospective study. *Climatic Change and the Mediterranean*, *1*, 175–232.

Leff, B., Ramankutty, N., & Foley, J. A. (2004). Geographic distribution of major crops across the world. *Global Biogeochemical Cycles*, *18*(1), n/a. doi:10.1029/2003GB002108

Leye, E. H. M., Makoumba, D., Ndiaye, F., Bassirou, D., Maiguizo, D., & Tahir, D. (2012). Effet de la mycorhization sur la croissance et le développement de Jatropha curcas L., cultivée sous serre. *Revue des Energies Renouvelables*, *6*, 1741–1760.

Lindow-López, L., Galíndez, G., Aparicio-González, M., Sühring, S., Rojas-Aréchiga, M., Pritchard, H. W., & Ortega-Baes, P. (2018). Effects of alternating temperature on cactus seeds with a positive photoblastic response. *Journal of Arid Environments*, *148*, 74–77. doi:10.1016/j.jaridenv.2017.10.006

Lioubimtseva, E., & Henebry, G. M. (2009). Climate and environmental change in arid Central Asia: Impacts, vulnerability, and adaptations. *Journal of Arid Environments*, *73*(11), 963–977. doi:10.1016/j.jaridenv.2009.04.022

Lloret, F., Penuelas, J., & Estiarte, M. (2004). Experimental evidence of reduced diversity of seedlings due to climate modification in a Mediterranean-type community. *Global Change Biology*, *10*(2), 248–258. doi:10.1111/j.1365-2486.2004.00725.x

Loreto, F., Ciccioli, P., Cecinato, A., Brancaleoni, E., Frattoni, M., & Tricoli, D. (1996). Influence of environmental factors and air composition on the emission of [alpha]-pinene from quercus ilex leaves. *Plant Physiology*, *110*(1), 267–275. doi:10.1104/pp.110.1.267 PMID:12226182

Mahé, G., & Olivry, J. C. (1995). Variations des précipitations et des écoulements en Afrique de l'Ouest et centrale de 1951 à 1989. *Science et changements planétaires/Sécheresse, 6*(1), 109-117.

Mahé, G., Olivry, J. C., Dessouassi, R., Orange, D., Bamba, F., & Servat, E. (2000). Relations eaux de surface–eaux souterraines d'une rivière tropicale au Mali. *Comptes Rendus de l'Académie des Sciences-Series IIA-Earth and Planetary Science, 330*(10), 689–692.

McDaniel, M. D., Kaye, J. P., & Kaye, M. W. (2013). Increased temperature and precipitation had limited effects on soil extracellular enzyme activities in a post-harvest forest. *Soil Biology & Biochemistry, 56*, 90–98. doi:10.1016/j.soilbio.2012.02.026

Meddich, A., El Mokhtar, M. A., Bourzik, W., Mitsui, T., Baslam, M., & Hafidi, M. (2018). Optimizing Growth and Tolerance of Date. *Root Biology*, 239.

Mendelsohn, R., Dinar, A., & Dalfelt, A. (2000). Climate change impacts on African agriculture. Preliminary analysis prepared for the World Bank.

Misra, A. K. (2014). Climate change and challenges of water and food security. *International Journal of Sustainable Built Environment, 3*(1), 153–165. doi:10.1016/j.ijsbe.2014.04.006

Nelson, G. C., Rosegrant, M. W., Koo, J., Robertson, R., Sulser, T., Zhu, T., . . . Magalhaes, M. (2009). Climate change: Impact on agriculture and costs of adaptation (Vol. 21). Intl Food Policy Res Inst. (IFPRI).

Paeth, H., Born, K., Girmes, R., Podzun, R., & Jacob, D. (2009). Regional climate change in tropical and northern Africa due to greenhouse forcing and land use changes. *Journal of Climate, 22*(1), 114–132. doi:10.1175/2008JCLI2390.1

Parry, M., Arnell, N., McMichael, T., Nicholls, R., Martens, P., Kovats, S., ... Fischer, G. (2001). Millions at risk: Defining critical climate change threats and targets. *Global Environmental Change, 11*(3), 181–183. doi:10.1016/S0959-3780(01)00011-5

Pielke, R. A., Marland, G., Betts, R. A., Chase, T. N., Eastman, J. L., Niles, J. O., & Running, S. W. (2002). The influence of land-use change and landscape dynamics on the climate system: relevance to climate-change policy beyond the radiative effect of greenhouse gases. *Philosophical Transactions of the Royal Society of London A: Mathematical, Physical and Engineering Sciences, 360*(1797), 1705-1719.

Premanandh, J. (2011). Factors affecting food security and contribution of modern technologies in food sustainability. *Journal of the Science of Food and Agriculture, 91*(15), 2707–2714. doi:10.1002/jsfa.4666 PMID:22002569

Radhouane, L. (2013). Climate change impacts on North African countries and on some Tunisian economic sectors. *Journal of Agriculture and Environment for International Development, 107*(1), 101–113.

Ramankutty, N. (2004). Croplands in West Africa: A geographically explicit dataset for use in models. *Earth Interactions, 8*(23), 1–22. doi:10.1175/1087-3562(2004)8<1:CIWAAG>2.0.CO;2

Ramankutty, N., Foley, J. A., Norman, J., & McSweeney, K. (2002). The global distribution of cultivable lands: Current patterns and sensitivity to possible climate change. *Global Ecology and Biogeography, 11*(5), 377–392. doi:10.1046/j.1466-822x.2002.00294.x

Rosenzweig, C., & Parry, M. L. (1994). Potential impact of climate change on world food supply. *Nature*, *367*(6459), 133–138. doi:10.1038/367133a0

Rosenzweig, C., & Tubiello, F. N. (1997). Impacts of Global Climate Change on Mediterranean Agriculture: Current Methodologies and Future Directions. An Introductory Essay. *Mitigation and Adaptation Strategies for Global Change*, *1*(3), 219–232. doi:10.1007/BF00517804

Salem, A. B., Messouli, M., & Yacoubi-Khebiza, M. (2011). *Developing an oasis-based water management tool: Ecohydrologic approach and weap software for a large arid catchment in Morocco*. Academic Press.

Sarris, A. (2009). *Evolving Structure of World Agricultural Trade and Requirements for New World Trade Rules*. Paper presented at the Expert Meeting on "How to Feed the World in 2050", Rome, Italy.

Sati, V. P. (2015). Climate change and socio-ecological transformation in high mountains: An empirical study of Garhwal Himalaya. *Change and Adaptation in Socio-Ecological Systems*, *2*(1). doi:10.1515/cass-2015-0005

Schilling, J., Freier, K. P., Hertig, E., & Scheffran, J. (2012). Climate change, vulnerability and adaptation in North Africa with focus on Morocco. *Agriculture, Ecosystems & Environment*, *156*, 12–26. doi:10.1016/j.agee.2012.04.021

Schlenker, W., & Lobell, D. B. (2010). Robust negative impacts of climate change on African agriculture. *Environmental Research Letters*, *5*(1), 014010. doi:10.1088/1748-9326/5/1/014010

Schmidhuber, J., & Tubiello, F. N. (2007). Global food security under climate change. *Proceedings of the National Academy of Sciences of the United States of America*, *104*(50), 19703–19708. doi:10.1073/pnas.0701976104 PMID:18077404

Schoenholtz, S. H., Van Miegroet, H., & Burger, J. A. (2000). A review of chemical and physical properties as indicators of forest soil quality: Challenges and opportunities. *Forest Ecology and Management*, *138*(1-3), 335–356. doi:10.1016/S0378-1127(00)00423-0

Scott, L. (2008). *Climate variability and climate change: implications for chronic poverty*. Research Centre Working Paper, N 108.

Singh, J. S., & Gupta, S. R. (1977). Plant decomposition and soil respiration in terrestrial ecosystems. *Botanical Review*, *43*(4), 449–528. doi:10.1007/BF02860844

Six, J., Bossuyt, H., Degryze, S., & Denef, K. (2004). A history of research on the link between (micro) aggregates, soil biota, and soil organic matter dynamics. *Soil & Tillage Research*, *79*(1), 7–31. doi:10.1016/j.still.2004.03.008

Spash, C. L. (2007). a. Climate change: Need for new economic thought. *Economic and Political Weekly*, 483–490.

Spash, C. L. (2007). b. The economics of climate change impacts à la Stern: Novel and nuanced or rhetorically restricted? *Ecological Economics*, *63*(4), 706–713. doi:10.1016/j.ecolecon.2007.05.017

Stabler, L. B., Martin, C. A., & Brazel, A. J. (2005). Microclimates in a desert city were related to land use and vegetation index. *Urban Forestry & Urban Greening*, *3*(3-4), 137–147. doi:10.1016/j.ufug.2004.11.001

Stocker, T. F., Qin, D., Plattner, G. K., Tignor, M., Allen, S. K., Boschung, J., . . . Midgley, P. M. (2013). IPCC, 2013: Climate Change 2013: The Physical Science Basis. Contribution of Working Group I to the Fifth Assessment Report of the Intergovernmental Panel on Climate Change.

Thornton, P. K., van de Steeg, J., Notenbaert, A., & Herrero, M. (2009). The impacts of climate change on livestock and livestock systems in developing countries: A review of what we know and what we need to know. *Agricultural Systems*, *101*(3), 113–127. doi:10.1016/j.agsy.2009.05.002

Tilman, D., Fargione, J., Wolff, B., D'antonio, C., Dobson, A., Howarth, R., ... Swackhamer, D. (2001). Forecasting agriculturally driven global environmental change. *Science*, *292*(5515), 281–284. doi:10.1126cience.1057544 PMID:11303102

Verchot, L. V., Van Noordwijk, M., Kandji, S., Tomich, T., Ong, C., Albrecht, A., ... Palm, C. (2007). Climate change: Linking adaptation and mitigation through agroforestry. *Mitigation and Adaptation Strategies for Global Change*, *12*(5), 901–918. doi:10.100711027-007-9105-6

Von Braun, J. (2007). *The world food situation: new driving forces and required actions*. Washington, DC: International Food Policy Research Institute.

von Grebmer, K., von Grebmer, K., Ringler, C., Rosegrant, M. W., Olofinbiyi, T., Wiesmann, D., & Thompson, J. (2012). *Global hunger index: the challenge of hunger: Ensuring sustainable food security under land, water, and energy stresses* (Vol. 70). Intl Food Policy Res Inst.

Walck, J. L., Baskin, J. M., & Baskin, C. C. (1997). A comparative study of the seed germination biology of a narrow endemic and two geographically-widespread species of Solidago (Asteraceae). 1. Germination phenology and effect of cold stratification on germination. *Seed Science Research*, *7*(1), 47–58. doi:10.1017/S0960258500003366

Walck, J. L., Hidayati, S. N., Dixon, K. W., Thompson, K. E. N., & Poschlod, P. (2011). Climate change and plant regeneration from seed. *Global Change Biology*, *17*(6), 2145–2161. doi:10.1111/j.1365-2486.2010.02368.x

Waldrop, M., & Firestone, M. (2006). Response of microbial community composition and function to soil climate change. *Microbial Ecology*, *52*(4), 716–724. doi:10.100700248-006-9103-3 PMID:17061172

Wheeler, T., & Von Braun, J. (2013). Climate change impacts on global food security. *Science*, *341*(6145), 508–513. doi:10.1126cience.1239402 PMID:23908229

Xiao, X. (1997). *Transient climate change and potential croplands of the world in the 21st century Report*. Cambridge, MA: Massachusetts Institute of Technology.

Zhang, W., Parker, K. M., Luo, Y., Wan, S., Wallace, L. L., & Hu, S. (2005). Soil microbial responses to experimental warming and clipping in a tallgrass prairie. *Global Change Biology*, *11*(2), 266–277. doi:10.1111/j.1365-2486.2005.00902.x

Zhu, Z., Ge, T., Liu, S., Hu, Y., Ye, R., Xiao, M., ... Wu, J. (2018). Rice rhizodeposits affect organic matter priming in paddy soil: The role of N fertilization and plant growth for enzyme activities, CO2 and CH 4 emissions. *Soil Biology & Biochemistry*, *116*, 369–377. doi:10.1016/j.soilbio.2017.11.001

Ziervogel, G., & Ericksen, P. J. (2010). Adapting to climate change to sustain food security. *Wiley Interdisciplinary Reviews: Climate Change*, *1*(4), 525–540. doi:10.1002/wcc.56

Zougari-Elwedi, B., Sanaa, M., Labidi, S., & Sahraoui, A. L. H. (2012). Évaluation de l'impact de la mycorhization arbusculaire sur la nutrition minérale des plantules de palmier dattier (Phœnix dactylifera L. var. Deglet Nour). *Étude Gest. des Sols*, *19*(3), 193-202.

APPENDIX: LIST OF ABBREVIATIONS

GCM: General Circulation Models
REMO: The dynamical regional model
CRU: Climatic Research Unit
NCC: Climate action network
GDP: Gross domestic product
PAG: Percentage variation in agricultural
AVG: Average values of climate change models
CRS: Cross sectional coefficient
GHG: Greenhouse gas
IIASA: The International Institute for Applied Systems Analysis

Chapter 5
Food Safety and Climate Change:
Case of Mycotoxins

Abdellah Zinedine
Chouaib Doukkali University, Morocco

Samira El Akhdari
Ministry of National Education, Morocco

ABSTRACT

Mycotoxins are chemical compounds produced mainly by mounds of genera Aspergillus, Penicillium, and Fusarium on various grains and agricultural commodities at different stages in the field, before harvest, post-harvest, during processing, packaging, distribution, and storage. The production of mycotoxins depends on several environmental factors such as temperature and moisture. This chapter gives an overview about the major mycotoxins (e.g., aflatoxins, ochratoxin A, and Fusarium toxins), masked mycotoxins, and emerging mycotoxins. The toxicity of these mycotoxins and their negative economic impact was also discussed together with the effect of climate change on their production. A section on mycotoxins regulations by international agencies and organisms (WHO, FAO, EU, etc.) was discussed. Finally, the different strategies to reduce or eliminate the toxic effects of mycotoxins in contaminated foods and feeds by using chemical, physical, and biological/biotechnological methods or innovative approaches were explained.

INTRODUCTION

Fungi are considered ubiquitous microorganisms found in nature, their spores are able to travel across countries and continents on our planet (De Ruyck *et al*, 2015). They are known to produce active chemical compounds called mycotoxins, which are secondary metabolites that exert adverse negative effects both on human and animal health and may contaminate agricultural food products of vegetal and animal origin (Tantaoui-Elaraki *et al*., *in press*).

DOI: 10.4018/978-1-5225-7775-1.ch005

Strains of the genera *Aspergillus*, *Penicillium*, *Fusarium* and *Alternaria* are the most mycotoxins producing fungi. Mycotoxins synthesis by toxigenic species depend widely on internal parameters of fungal strains (physiological, genetic and biochemical, etc.) but also on external factors (climatic factors) such as humidity and temperature.

The genus *Aspergillus* was described by Pier Antonio Micheli in 1729. It was reported that *Aspergillus* species that produce mycotoxins are more common in the warmer, subtropical and tropical areas than in the temperate areas of the world (Wilson *et al.*, 2002). *Aspergilli* species are soil fungi or saprophytes, and several are able to produce active compounds especially aflatoxins, ochratoxins, citrinin, penicillic acid, sterigmatocystin, cyclopiazonic acid, gliotoxin, citreoviridin and other important metabolites. All these compounds are produced by a high number of fungal mycotoxin species producers belonging to the genus *Aspergillus* including *A. flavus, A. niger, A. nomius, A. ochraceus, A. parasiticus, A. candidus, A. clavatus, A. restrictus, A. tamarii, A. terreus, A. versicolor*, etc. (Frisvad & Samson, 1991).

Mycotoxins are known as low molecular weight (below 700 Da) chemical substances and are produced on various grains and agricultural commodities at different stages in the field, before harvest, post-harvest, during processing, packaging, distribution and storage (Creppy *et al.*, 2002). Modified mycotoxins called also "masked mycotoxins" can be produced by fungi or generated as part of the defense mechanism of the infected plant. They can be produced during the processing of contaminated food processing (e.g. cooking), then can be converted to the original mycotoxin by the metabolism of animals and humans (Berthiller *et al.*, 2013). The conversions of these modified mycotoxins to their free form possibly increased the bioavailability of the parent mycotoxin and induce potential risk to human health (Paris *et al.*, 2014). Some species belonging to the *Fusarium* genus are responsible for the production of another group of bioactive compounds called *emerging* or *minor* mycotoxins. This group includes enniatins (A, A1, B and B1), fusaproliferin, beauvericin, and moniliformin (Jestoi, 2008).

Some mycotoxins are known of their acute toxic properties on humans (ergot alkaloids) and animals (aflatoxins, OTA, *Fusarium* toxins), but the majority of them are of chronic toxicological effects (Zinedine & Mañes, 2009).

Several mycotoxins are correlated with toxicological effects including hepatotoxicity, teratogenicity, carcinogenicity, neurotoxicity, immunosuppressive effects as well as reproductive and developmental toxicity in humans and animals. Among these substances, aflatoxins, ochratoxin A, fumonisins, zearalenone, trichothecens, T-2, HT-2 toxins are of a great concern because of their negative impact on human and animal health (Bennett & Klich, 2003). Mycotoxins are also known with their negative economic impacts. Indeed, according to FAO, one third of all foodstuffs produced for world's population are lost from field to consumer, reaching nearly 1.3 billion metric tons each year; and that 25% of the crops in the world are damaged by mould or fungal growth (FAO, 2012).

The mycotoxin problem in public health is longstanding and all humans and animals are at risk for mycotoxin exposure. People are mainly exposed via the ingestion of contaminated foods; however, alternate routes include dermal absorption and inhalation of toxinogenic molds containing mycotoxins. The major regulated mycotoxins and the most associated producing fungi are summarized in Table 1.

The climate change is likely to alter the degree of human exposure to pollutants and the response of human populations to these exposures. The contamination of food and feed by mycotoxins, and the production of these compounds by fungi can be very sensitive to environmental factors such as temperature and humidity. Paterson and Lima (2010) showed that indirect effects of climate change may also be important; for example, changes in the distribution and activity of insect vectors may increase the exposure and vulnerability of plants to mycotoxins. It was reported that climate change could have a

Table 1. Major mycotoxins and associated fungi (AFSSA, 2009 with modifications)

	Mycotoxins	Associated Fungi
Major Legislated mycotoxins	Aflatoxins: B1, B2, G1 and G2	*Aspergillus flavus, A. parasiticus, A. nomius*
	Ochratoxin A	*Penicillium verrucosum, A. ochraceus, A. carbonarius*
	Patulin	*P. expansum, A. clavatus, Byssochlamys nivea*
	Fumonisins B1, B2 and B3	*Fusarium verticillioides, F. proliferatum*
	Trichothecens (types A and B)	*F. langsethiae, F. sporotrichioides, F. poae, F. graminearum, F. culmorumF. crookwellense, F. tricinctum, F. acuminatum*
	Zearalenone	*Fusarium graminearum, F. culmorum F. crookwellense.*
Others mycotoxins	Citrinin	*A. terreus, A. carneus, A. niveus P. verrucosum, P. citrinum, P. expansum*
	Alternaria toxins	*Alternaria alternata, Alternaria solani*
	Cyclopiazonic acid	*A. flavus, A. versicolor, A. tamari, P. camemberti*
	Sterigmatocystin	*A. nidulans, A. versicolor, A. flavus*
	Stachybotryotoxin	*Strachybotrys chartarum*

positive effect on exposure, but in most instances an adverse effect is anticipated in certain regions and it is possible that this will result in adverse health outcomes (Tirado *et al.*, 2010). In regions of some developing countries, the increase in the extent and magnitude of episodes of fungal infestations is likely to correspond to increasing mycotoxin exposures, with attendant increasing occurrence of target organ toxicity and cancer, especially in areas where fungal infestations are endemic and populations are vulnerable (Balbus *et al*, 2013).

MAJOR MYCOTOXINS

Aflatoxins

Aflatoxins (AFs) are widely distributed toxins produced by *Aspergillus* species including *A. flavus, A. parasiticus, A. nomius, A. pseudotamarii, A. bombycis, A. ochraceoroseus,* and *A. australis* (IARC, 2002). According to Creppy (2002), *A. flavus* strains produce only type B aflatoxins (B1 and B2), while the other strains produce both B and G aflatoxins (B1, B2, G1 and G2). *A. flavus* and *A. parasiticus* are responsible for the largest proportion of aflatoxins found in foodstuffs throughout the world. Of the other species, it was reported that only *A. australis*, which appears to be widespread in the southern hemisphere and is common in Australian peanut soils, may also be an important source of aflatoxins in some countries (IARC, 2002).

AFs (Figure 1) were the first mycotoxins isolated in the 1960th after hard investigations following the "Turkey-X-disease" outbreaks in the UK. AFs are well known with their heat-stability and their resistance to several food processes such as cooking, baking and extrusion (Marin *et al.*, 2013). AFs are able to contaminate a wide range of food commodities including grains (wheat, maize, barley, rice, and

Figure 1. Chemical structure of aflatoxins (B1, B2, G1, G2, M1 and M2)

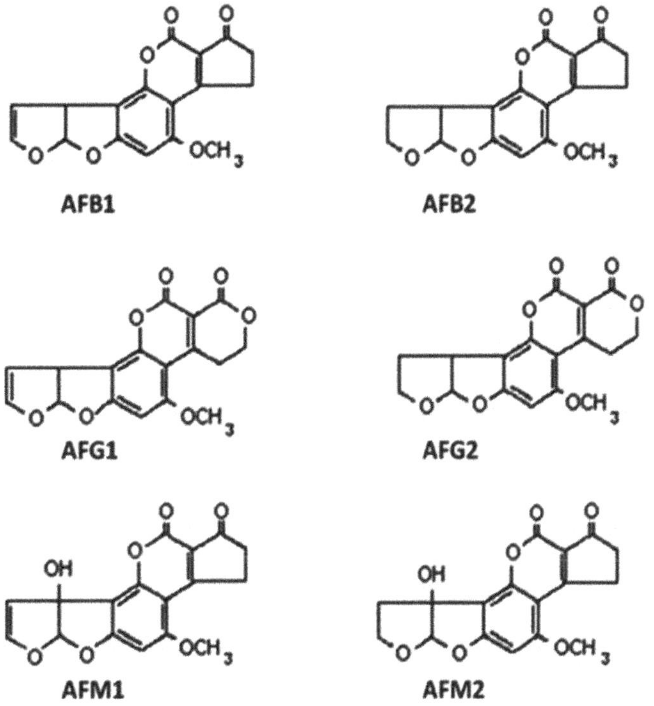

sorghum, etc.), nuts (almonds, peanuts, walnuts, and pistachios, etc.), spices (red paprika, ginger, chili powder, etc.) and animal feeds (Dragacci *et al.*, 2011). Nowadays, AFs are recognized as the most relevant mycotoxins worldwide because of their confirmed toxicity to humans and animals, their widespread occurrence and their economic losses. In term of biological activity, AFs are of great concern because of their detrimental effects on the health of humans and animals, including carcinogenic, mutagenic, teratogenic and immunosuppressive effects (Eaton & Gallagher, 1994).

Recently, AFs acute poisoning outbreaks affecting a large geographical area and causing over 123 deaths were reported in Kenya in 2004 and 2005. Epidemiological studies from this case showed a relationship between the outbreak and the local methods of harvesting, storing and preparing maize. Contamination of maize with AFs was found up to 1000 µg/Kg (Centers for Disease Control, 2004). The early symptoms of AFB1-induced hepatotoxicity in human are anorexia, malaise and low-grade fever and it can progress with vomiting abdominal pain and even may be associated to diarrhea (Sherif *et al.*, 2009).

Until now, twenty molecules of AFs are chemically known and identified in the world; however, only six aflatoxin compounds (B1, B2, G1, G2, M1 and M2) are the most significant and interesting. AFM1 and AFM2 are the main monohydroxylated derivatives respectively of AFB1 and AFB2 forming in liver of lactating mammals by means of cytochrome P450-associated enzymes (Bennett & Klich, 2003).

AFs were evaluated several times in IARC Monographs, and were confirmed as a Group 1 like carcinogenic to humans. IRAC monographs states that there is sufficient evidence in humans for the carcinogenicity of AFs. Indeed, AFs cause cancer of the liver (hepatocellular carcinoma) and there is sufficient evidence in experimental animals for the carcinogenicity of naturally occurring mixtures of AFs, and of AFB1, AFG1 and AFM1. However, there is limited evidence in experimental animals for the carcinogenicity of AFB2 and there is inadequate evidence in experimental animals for the carcinogenicity of AFG2 (IARC, 2018).

The weight of evidence for the classification of the aflatoxins as group-1 carcinogens was driven by statistically significantly increased risks for hepatocellular carcinoma (HCC) in individuals exposed to aflatoxins, as measured by aflatoxin-specific biomarkers in cohort studies in Shanghai and Taiwan provinces in China. On one hand, these studies reported statistically significant effects of exposure to AFs on the development of HCC and also confirmed that in the presence of HBV exposure, as judged by HBsAg status, there is a greater than multiplicative interaction between aflatoxin and HBV, increasing the risk for HCC. On the other, it was demonstrated that aflatoxins are able to induce a specific mutation in codon 249 of the TP53 tumor-suppressor gene (IARC, 2018).

Since milk is a major commodity for introducing AFs in the human diet, evidence of hazardous human exposure to AFM1 through dairy products has been shown by several studies. Compared to its parent molecule, AFM1 has a carcinogenicity of 2–10%. It is secreted into milk in the mammary glands of dairy cows that have consumed feeds containing AFB1. AFM1 could be detected in milk 12–24 h after the first AFB1 ingestion, reaching a high level after a few days (Zinedine *et al.* 2007a). When the intake of AFB1 has finished, AFM1 amounts decreases to undetectable levels after 72 h (Van Egmond, 1989). The amount of AFB1 present in contaminated feed is usually 1–3% but values as high as 6% were reported (Pittet, 1998). According to Yousef and Marth (1985), the residues of AFM1 remain stable when milk is heat-treated, and the level of AFM1 did not change with concentrating, drying, freezing or cold storage of contaminated milk.

Ochratoxin A

Ochratoxin A (OTA) is a mycotoxin isolated and characterized after the discovery of AFs. It was described for the first time by Van der Merwe *et al.* (1965). Chemically, OTA is known as: N-{[(3R)-5-chloro-8-hydroxy-3-methyl-1-oxo-7-isochromanyl]-carbonyl}-3-phenyl-L alanine (Figure 2). According to Abarca *et al.* (2003), OTA is a ubiquitous secondary fungal metabolite produced by wide varieties of species of the genus *Aspergillus* (e.g., *A. ochraceus, A. niger, A. carbonarius,* etc.) in temperate, tropical or warm climates (Africa, Asia), and by *Penicillium verrucosum* in cold climates (Northern Europe, Canada). OTA has been widely detected in several food commodities of plant origin (cereals, coffee beans, raisins, wine, beer, and grape juice) and commodities of animal origin such as pork and poultry meats, eggs, milk, and dairy products due to the carryover effect (Peraica *et al.*, 2014).

OTA has been implicated in a human disease of kidney referred to as Balkan endemic nephropathy, characterized by tubule interstitial nephritis and associated with high incidence of kidney, pelvis, ureter and urinary bladder tumors in some Eastern European countries (Pfohl-Leszkowicz *et al.*, 2002). In experimental and domestic animals the main target organs of OTA toxicity are the kidney and liver, but it also affects other targets as the heart, blood, lymphoid tissue, and bone marrow. Several mechanisms are involved in OTA toxicity such as the production of reactive oxygen species, the mitochondrial respiration inhibition, the protein synthesis inhibition; and particularly OTA is known to cause DNA damage (Creppy, 2002). It was described that OTA is reabsorbed by the upper gastro-intestinal tract and persists in the circulation for a long time due to binding to plasma proteins; OTA plasma half-life in humans was estimated to 35.5 days, which is extremely long and makes OTA of plasma a good biomarker of exposure (Zepnik *et al.*, 2003).

Figure 2. Chemical structure of ochratoxin A

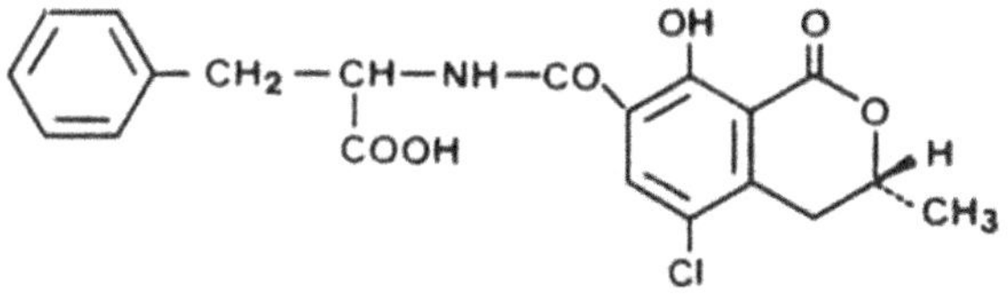

OTA has been classified as a possible human carcinogen (group 2B) by the International Agency for Research on Cancer (IARC, 1993). The Joint Committee FAO/WHO of Experts on Food Additives (JECFA) has established the provisional tolerable weekly intake (PTWI) of OTA at 100 ng/Kg of body weight (bw) corresponding to approximately 14 ng/Kg bw/day (JECFA, 2001). The European Food Safety Authority (EFSA) has proposed a new safety value of 120 ng OTA/Kg bw as a Tolerable Weekly Intake, which corresponds to a TDI of 17.1 ng/Kg bw (EFSA, 2006).

Fumonisins

Fumonisins (FUM) are a group of 15 mycotoxins produced by *Fusarium* species (e.g. *F. verticilloides, F. moniliforme*, etc.). FUM were isolated for the first time in 1988 (Gelderblom *et al.,* 1988). They contaminate various cereal grains, but are common in maize and sorghum which are the main sources of the exposure to FUM in humans. In naturally contaminated grains, the most frequent is fumonisin B1 (FB1), often accompanied by small amounts of fumonisin B2 (FB2) and fumonisin B3 (FB3). FUM are chemically similar to sphinganine which is the backbone of sphingolipids (Figure 3).

FUM are poorly absorbed by the gastro-intestinal tract and quickly eliminated from plasma with low accumulation in the kidney and liver (Soriano *et al.*, 2005). Bhat *et al.* (1997) reported a possible case of acute exposure to FB1 involved 27 villages in India, where consumption of unleavened bread made from moldy sorghum or corn, containing up to 64 mg/Kg FB1, was associated with an outbreak of human disease involving gastrointestinal symptoms (transient abdominal pain, borborygmus, and diarrhea).

Fumonisins target different organs in domestic and experimental animals: in horses they cause leukoencephalomalacia, in pig pulmonary oedema, in rats they are predominantly nephrotoxic, and in mice they are hepatotoxic and teratogenic, causing neural tube defects (Soriano *et al.*, 2005). In some world areas where maize is considered as staple food (Southern Africa, China, and Northern Italy), the high frequency of the incidence of esophageal cancer is believed to be related to exposure to FUM or their producers (Marasas *et al.*, 1988). In the early of the 1990th, a higher prevalence of neural tube defects

Figure 3. Chemical structure of fumonsin B1

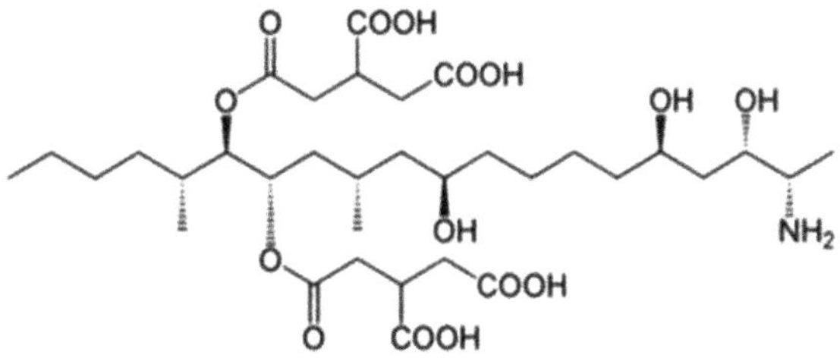

was observed in children born along the Texan and Mexican border by Mexican-American women, and it was suggested that this was caused by fumonisin exposure in the first trimester of pregnancy (Hendricks *et al.* 1999). High prevalence of these defects was also found in the Transkei region in Southern Africa, northern Iran, and several regions of China (Peraica *et al.*, 2014). FB1 was reported to be the most widespread toxin in the group of FUM and was classified by the International Agency for Research on Cancer in 2B group as a possible carcinogenic to humans (IARC, 1993). A provisional maximum for tolerable daily intake (PMTDI) is fixed for fumonisins B1, B2 and B3 alone or in combination, of 2 µg/Kg bw/day on the basis of the NOEL of 0.2 mg/Kg bw/day and a safety factor of 100 (Creppy, 2002).

Zearalenone

Zearalenone (ZEA) is a mycotoxin produced by *Fusarium* fungi in temperate and warm countries (Bennett & Klich, 2003). ZEA is chemically known as 6-[10-hydroxy-6-oxo-trans-1-undecenyl]-B-resorcyclic acid lactone (Figure 4). ZEA producing fungi contaminate cereals and derivatives, especially wheat, oats, corn, barley, millet, sorghum and rice (Zinedine *et al*, 2007b). Furthermore, the toxin has been detected in cereal products such as flour, malt, soybeans and beer. It was reported that species of the genus *Fusarium* infect cereals in the field. The ZEA derivatives (α-zearalenol, β-zearalenol, α-zearalanol, β-zearalanol and zearalanone) can also be detected in food commodities, especially in corn and derivatives. ZEA and its metabolites represent a serious hazard to animals and human health. They have been shown to bind competitively to estrogen receptors (ER-α and ER-β) because of their structural similarity to the sex hormone, 17β-estradiol, they activate the estrogen gene and cause reproductive disorders (Zinedine *et al*, 2007b). Several authors reported that ZEA is of a relatively low acute toxicity after oral administration in mice, rats and guinea pigs (SCF, 2000). ZEA produces hematologic, cytotoxic, genotoxic, immunotoxic and hepatotoxic effects. Concerning long-term toxicity, ZEA has shown adverse liver lesions with subsequent development of hepatocarcinoma. ZEA also affected some enzymatic parameters of the hepatic function in rats and in rabbits. Conková *et al.* (2001) reported the effects of ZEA on the changes in enzymatic activities of aspartate aminotransferase, alanine aminotransferase, alkaline phosphatase, γ-glutamyl transferase and total lactate dehydrogenase in rabbits, indicating the possible liver toxicity due to the chronic effects of the toxin.

ZEA was found to to induce DNA-adduct formation in vitro cultures of bovine lymphocytes (Lioi *et al.*, 2004), DNA fragmentation and micronuclei production in cultured DOK, Vero and Caco-2 cells, in Vero monkey kidney cells and in bone marrow cells of mice (Ouanes *et al.*, 2003). However, previous studies showed the potential for ZEA to stimulate growth of human breast cancer cells containing estrogen response receptors (Yu *et al.*, 2005).

For the ZEA risk assessment, JECFA (2000) established a Provisional Maximum Tolerable Daily Intake of 0.5 µg/kg b.w. This decision was based on the NOEL of 40 µg/Kg b.w. per day obtained in a 15-day study in pigs and the lowest observed effect level of 200 µg/Kg b.w. per day in this study.

Trichothecens

Trichothecens is a heterogeneous family of mycotoxins all produced by toxingenic fungi on foods and feeds. The term trichothecene is derived from trichothecin, which was the one of the first members of the family identified (Zinedine & Mañes, 2009). All trichothecenes contain a common 12,13-epoxytrichothene skeleton and an olefinic bond with various side chain substitutions (Bennett and klich, 2003).

Figure 4. Chemical structure of zearalenone

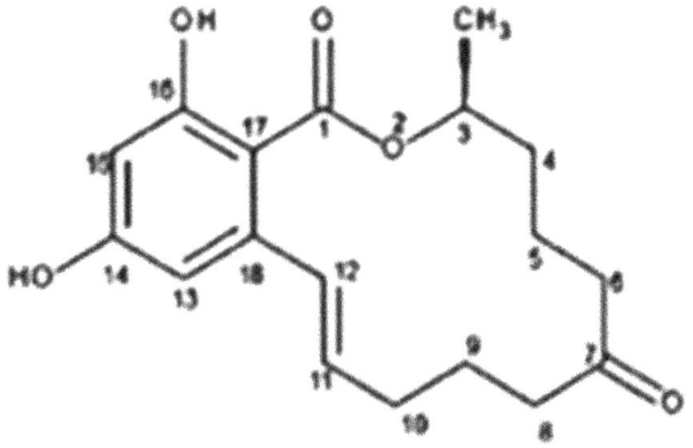

The trichothecenes constitute a family of more than sixty sesquiterpenoid metabolites produced by a number of fungal genera, including *Fusarium, Myrothecium, Phomopsis, Stachybotrys, Trichoderma and Trichothecium*. The most serious and regulated mycotoxins in this family are Deoxynivalenol (DON), T-2 and HT-2 toxins.

Deoxynivalenol

Deoxynivalenol (DON) is a mycotoxin of the type-B trichothecenes, which are epoxy-sesquiterpenoids. The toxin DON (Figure 5) often contaminates cereal grains such as wheat, barley and maize, but less found on oats, rice, rye and sorghum. According to several toxicological studies, the toxin DON shows adverse health effects after acute, short-term, or long-term administration to animals. Indeed, two characteristic toxicological effects appear after acute administration of the toxin: anorexia (decrease in feed consumption) and vomiting. For this, DON is also known under its synonym "vomitoxin" and its presence in foods can cause clinical or subclinical manifestations to humans and animals (Pestka, 2010). The toxin DON is the most prevalent in its group and was commonly described as a serious mycotoxin several cereals such as wheat, corn, barley, rye, safflower seeds, and animal mixed feeds.

T-2 and HT-2 Toxins

The toxin T-2 (Figure 6) and its derivative (HT-2) are type A trichothecenes. Chemically, they are closely related epoxy, sesquiterpenoids. Previous reports showed the occurrence of both T-2 and HT-2 toxins in cereal grains especially in wheat, maize, oats, barley, rice, beans, and soya beans and in processed cereals. According to Creppy (2002), T-2 and HT-2 toxins are produced by several *Fusarium* species such as *F. sporotrichioides, Fusarium poae, Fusarium equiseti* and *Fusarium acuminatum*. T-2 and HT-2 toxins were implicated in acute poisoning outbreaks with several symptoms that included nausea, vomiting, pharyngeal irritation, abdominal pain, diarrhea, bloody stools, dizziness and chills. According to Bennett and Klich, (2003), T-2 toxin was associated with *Alimentary Toxic Aleukia (ATA)*, a human disease that affected a large population of the former USSR during the 1940th. The symptoms include inflammation of the skin, vomiting, and damage to hematopoietic tissues.

Figure 5. Chemical structure of deoxynivalenol

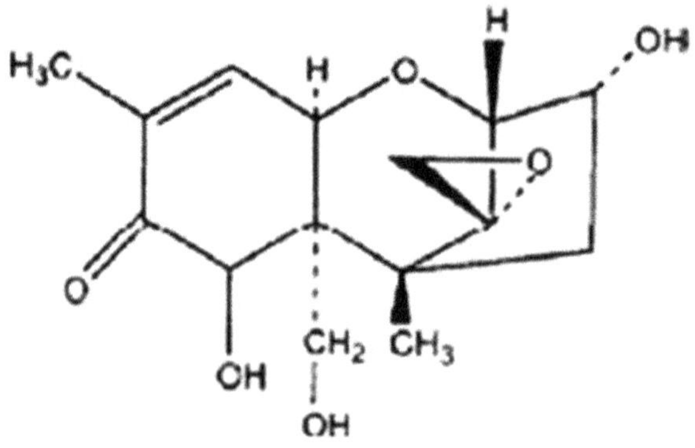

Figure 6. Chemical structure of T-2

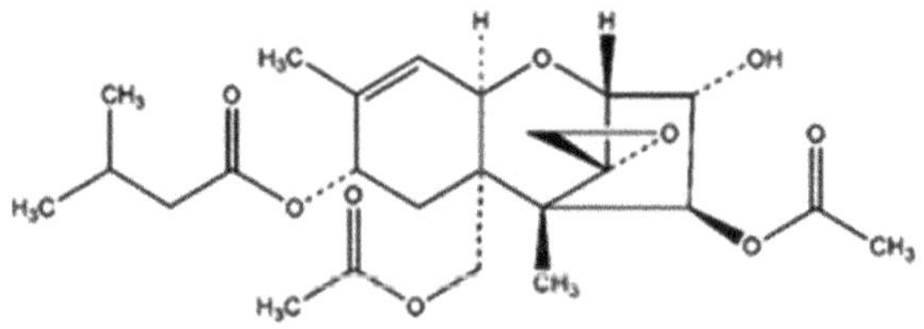

Ergot Alkaloids

The genus *Claviceps* is composed of about 36 species of fungi responsible for the production of more than 40 known ergot alkaloids (Arroyo-Manzanares et al., 2014). Although some *Aspergilli* species also produce ergot alkaloids, the most popular producer is *Claviceps purpurea.* Common growth substrates for these species include cereal grains and feeds ingredients, such as corn, barley, oats, rice, rye and sorghum. Some plants of the family of *convolvulaceae* covering the soil are also known to produce ergot alkaloids, although the consumption of these plants by humans or livestock is minimal. The most known acute effects of ergot poisoning, called "ergotism," imply powerful and very painful vasoconstrictor effects on the peripheral circulation, often leading to gangrene, to loss of limbs or even death (Schiff, 2006). Different ergot alkaloids may also produce another group of acute poisoning symptoms, including abdominal pain, seizures, vomiting, insomnia, burning skin and other hallucinations (Tudzynski & Neubauer 2014). Figure 9 below illustrates the chemical structure of ergotamine, the main mycotoxin produced by *Claviceps purpurea* and responsible for ergotism or "burning pain".

MODIFIED AND MASKED MYCOTOXINS

In addition to the known mycotoxins, scientists started to use the terms "modified mycotoxins" and "masked mycotoxin" in the mid of the 1980th. Indeed, it was reported that some cases of mycotoxicosis did not correlated with the mycotoxins detected in the suspected foods. Thus, the high toxicity could be due to the presence of unknown mycotoxins forms (Freire & Sant'Ana, 2018). The term "modified mycotoxins" was launched on the metabolites which cannot be detected during the analysis of the parent mycotoxins by official analytical methods. Nowadays, different forms of mycotoxin are still confusing, the term "masked mycotoxins" has been indicated to be used exclusively for compounds derived from mycotoxins that were formed by plants' defense mechanisms. According to Berthiller *et al.* (2013), mycotoxin-derived compounds can also be formed by other pathways, such as food processing and animal metabolism. Moreover, the term "modified mycotoxin" was commonly used for all derived-mycotoxins compounds that could be formed by food processing, produced by microorganisms, or obtained from the plants metabolism.

The first modified mycotoxin reported was AFM1, a substance formed from the hydroxylation of AFB1 and eliminated in the milk of animals that consumed AFB1 feed contaminated. The degradation of OTA after the high thermal treatment (roasting coffee) resulted in the formation of its modified forms including ochratoxin α amide, 14-(R)-ochratoxin A and 14-decarboxyochratoxin A (Bittner *et al.*, 2015). Deoxynivalenol-3-glucoside (D3G), one of the most common modified forms of DON, is formed through the plant defense mechanism, in which glycosyltransferase enzymes bind an endogenous glucose molecule to the hydroxyl group of carbon 3 of the DON molecule (Freire & Sant'Ana, 2018). In ZEA-treated barley seedlings, the metabolites ZEA-16-glucoside (ZEA-16G) and ZEA-14G were detected in the roots, although in small amounts (Paris *et al.*, 2014).

Little information is now available on the toxicity of the modified or masked mycotoxins to humans and animals. However, according to Rychlik *et al.*, (2014), there is a potential risk of release of the parent mycotoxins by hydrolysis during food processing and by the digestion and/or metabolism in humans and animals. Indeed, it was suggested that the conversions of these modified or masked mycotoxins to their free form possibly may increase the bioavailability of the parent mycotoxins and may enhance the potential risk to human health. Due to this risk, The Panel on Contaminants in the Food Chain of the European Food Safety Authority decided that during risk assessment, the modified mycotoxins must be considered with the same toxicity of the parent mycotoxins (EFSA, 2014).

EMERGING MYCOTOXINS

In addition to the production of trichothecenes, ZEA and FUM, toxigenic fungi species of *Fusarium* are also able to produce a second group of bioactive compounds already called ''minor" mycotoxins but now are known as "emerging mycotoxins" because of their widespread in cereal grain and the phenomenal levels usually detected (Jestoi, 2008). This group includes enniatins (A, A1, B and B1), beauvericin and fusaproliferin, etc. First investigations on this mycotoxin group started on the 2000th, and now several papers are worldwide published on the occurrence of these toxins in cereal grains and processed cereals in the USA, in European Nordic countries, in Italy (Uhlig, 2006; Jestoi, 2008), in Spain (Meca *et al.*, 2010) and in Morocco (Zinedine *et al.*, 2011; Mahnine *et al.*, 2011). However little information is still available in the literature about their toxicity and to assess their risk for humans.

Enniatins

The enniatines (ENs, Figure 7) were first isolated from cultures of *F. orthoceras Appl.* and *F. oxysporum*. Nowadays, more than 23 different ENs compounds (and analogues) were identified and described in the literature (Meca *et al.*, 2010). Until now, several papers reported that four ENs occurs naturally in cereal grains, these are ENA, ENA1, ENB and ENB1 (Figure7). Enniatins are produced mainly by strains of some species of *Fusarium*. But, other fungal genera including *Alternaria*, *Halosarpheia* and *Verticillium* were also described to produce ENs. These mycotoxins are of high interest because of their wide range of biological activity. This bioactivity has long been assumed to be associated with their ionophoric properties (Uhlig *et al.*, 2009). ENs inhibit the enzyme acyl-CoA:cholesterol acyl transferase (ACAT) (Tomoda *et al.*, 1992). They are also known as phytotoxins and are associated with plant diseases characterized by wilt and necrosis (Burmeister and Plattner, 1987). Other studies reported the bioactivity of ENs against Mycobacterium sp. and Plasmodium falciparum (Nilanonta *et al.*, 2000; Supothina *et al.*, 2004). Hiraga *et al.* (2005) showed that ENs have been identified as inhibitors of major drug efflux pumps in *Saccharomyces cerevisiae*.

Bauvericin

The mycotoxin beauvericin (BEA, Figure 7) is a cyclic lactone trimer containing an alternating sequence of three N-methyl L-phenylalanyl and three D-α-hydroxyisovaleryl residues. BEA was isolated for the first time from the culture of the insect-pathogenic fungus *Beauverina bassiana*. More recently, it was reported that several fungal species such as *F. bulbicola, F. denticulatum, F. lactis, F. phyllophilum, F. pseudocircinatum and F. succisae* are able to produce this mycotoxin (Moretti *et al.*, 2007).

BEA affected the electromechanical and physiological properties of isolated smooth and heart muscle preparations (Lemmens-Gruber *et al.*, 2000). BEA is a specific cholesterol acyltransferase inhibitor (Tomoda *et al.*, 1992). BEA is toxic to several human cell lines (Logrieco *et al.*, 2002) and can induce apoptosis and DNA fragmentation (Ojcius *et al.*, 1991). This mycotoxin inhibited the Ltype Ca2+ current in the NG108-15 neuronal cell line and increased the intracellular calcium by increasing the formation of cation selective channels in lipid membrane (Wu *et al.*, 2002; Kouri *et al.*, 2003). Previous studies have shown that BEA induced cell death can be prevented by administration of intracellular calcium chelator-BAPTA/AM (Jow *et al.*, 2004) in human lymphoblastic leukemia CCRF-CEM cells, indicating that the intracellular Ca2+ plays an important role in cell death signaling.

Fusaproliferin

Fusaproliferin (FUS) is a bicyclic sesterterpene (Figure 8) consisting of five isoprenic units, which was originally isolated from a pure culture of *Fusarium proliferatum* (Randazzo *et al.*, 1993). Moretti *et al.* (2007) reported that FUS could be produced by F. *antophilum, F. begoniae, F. bulbicola, F. circinatum, F. concentricum, F. succisae and F. udum*. FUS is also produced by *F. subglutinans* (Ritieni *et al.*, 1995; Meca *et al.*, 2009). FUS is produced through the isoprenoid pathway via common terpene intermediates originating from acetyl-CoA subunits. Preliminary studies indicated that FUS has been found to be toxic in the brine shrimp (*Artemia salina*) larve bioassay (Ritieni *et al.*, 1995) and mammalian cells (Logrieco *et al.*, 1996) and causes teratogenic effects on chicken embryos (Ritieni *et al.*, 1997).

Figure 7. Chemical structure of enniatins (A, A1, B, B1) and beauvericin

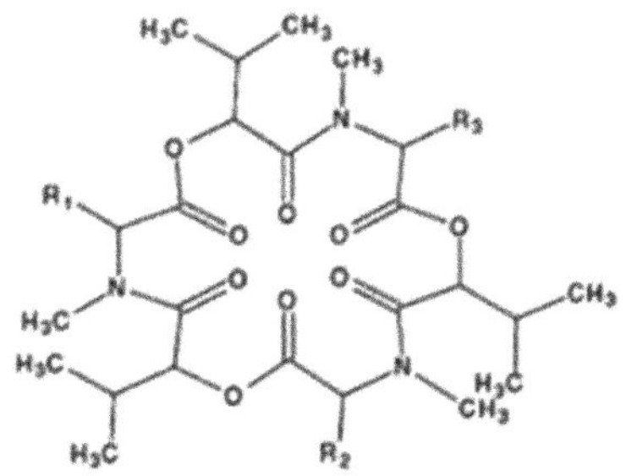

	R_1	R_2	R_3
BEA	phenylmethyl	phenylmethyl	phenylmethyl
A	*sec*-butyl	*sec*-butyl	*sec*-butyl
A_1	*sec*-butyl	*sec*-butyl	isopropyl
B	isopropyl	isopropyl	isopropyl
B_1	isopropyl	isopropyl	*sec*-butyl

Figure 8. Chemical structure of fusaproliferin

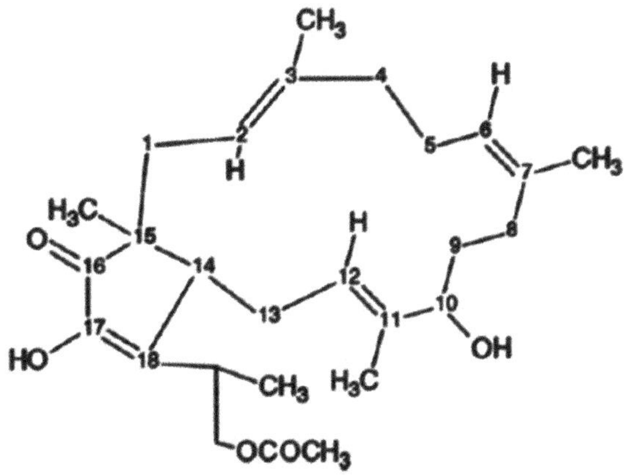

MYCOTOXINS AND ECONOMIC IMPACTS

Economic losses due to the presence of mycotoxins in food and feed are diverse and can be associated with the decrease of their quality. The food production industry is most commonly affected by mycotoxins. Major financial losses of mycotoxins are attributable to AFs because of their strict regulations and confirmed carcinogenic effects. The estimated losses in wheat and barley attributable to the *Fusarium* mycotoxins in US alone were estimated to about \$ 2.9 billion a year (Windels, 2000). In a study by Robens and Cardwell in 2003, loss due to AF contaminated corn and peanuts, as well as FB contaminated corn and DON contaminated wheat resulted in estimated loss ranging from \$ 0.5 million to over \$ 1.5 billion for the U.S. (Robens & Cardwell, 2003).

Animal Productivity Losses

Overall, most mycotoxins cause immunosuppression which can make animals more prone to disease by weakening their immune system or making them less responsive to vaccinations. In acute cases, losses are related to animal mortality. Other subclinical effects may cause many productivity losses such the decrease in the quantity and/or quality of animal by-products (milk, eggs and meat), and reduced weight gains and feed efficiency. The reduction in animal production could be due to feed refusal or diseases, increasing medical cost for toxicosis treatments, to find alternative foods, to design adequate management of contaminated supplies, to improve detection and quantification methods and to develop strategies to reduce toxin exposure. More specific reproductive effects are frequently associated with the negative effects of the estrogenic mycotoxin ZEA, these effects may cause infertility and abortion predominantly in swine. In animals, trichothecenes can cause weight loss by inhibition of protein synthesis, nutritional impairment as well as immunosuppression. In the case of dairy industry, losses involve the times farmers have to wait in order to allow animals to excrete all AFM1 form before milk collecting.

Food Unavailability

Economic loss related to unavailability of foods can be more dramatic if the affected foods are the most important ones. For example, cereals (wheat, maize and corn) are the most consumed staple food worldwide and are estimated to feed near 4 billion people all together. They are commonly reported with mycotoxin contamination; especially aflatoxins, OTA, citrinin, DON, fumonisins, fusarenon-X, nivalenol, sterigmatocystin and ZEA (Tanaka *et al.*, 2007). It is evident that loss due to mycotoxins in this widely consumed cereal may have disastrous monetary impacts not just for the producers but for all of the world population. Moreover, food recalls or food detentions due to mycotoxins are familiar to all countries that enforce regulations. Besides loss of production, food recalls of already produced cereals and grains constitute important losses. Indeed, many countries have records of recalls but not many of them make this information available.

MYCOTOXINS AND CLIMATE CHANGE

Climate change is predicted to have significant impacts on the quality and availability of staple food commodities. According to FAO guidelines, food security is determined by three key components: (a) sufficient food availability, (b) access to this food and (c) quality and use of the food in terms of both nutritional and cultural perspectives (FAO, 1996).

Several environmental factors contribute to the presence of mycotoxins in foods. Some factors are related to susceptibility of crops, plant stress, harvesting practices, and storage of grains. Intrinsic factors from the fungi are strain specific and vary according to the availability of substrates in which the mold grows. Broad environmental conditioners include climate changes that directly influence fungal contamination of foods.

The change in climate is a widely acknowledged fact and a trend of warming around the globe has been documented based on evidence taken from increased global average air and ocean temperatures, melting of snow and ice, and rising of global average sea levels (IPCC, 2007). Previous literature revisions describe how the plausible changes in temperature, rain precipitation, drought occurrence, and CO_2 increase pose a significant risk to the food supply.

Based on present available data, atmospheric concentrations of CO_2 are expected to double or triple (from 350-400 to 800-1200 ppb) in the next 25 to 50 years. Thus, different regions in Europe will be impacted by the increases in temperature of 2-5 °C coupled with elevated CO_2 and drought episodes. This will have profound impacts on pests and diseases. Similar impacts have been predicted in other areas of the world, especially parts of Asia, and Central and South America which are important producers of wheat, maize and soya beans for food and feed uses on a global basis. Miraglia *et al.* (2009) reported that recent outbreaks of aflatoxins in foods have been reported in some regions of Europe as a result of prolonged dry weather.

Temperature

Temperature is a primary determining factor that modulates fungal growth and mycotoxin production. *Fusarium* growth is more common in temperate weathers at temperatures ranging from 26–28 °C and water activity (A_w) > 0.88, while *Aspergillus* (i.e. *A. flavus*) grows better under warm temperatures. According to the strain and substrate specificities, the optimal temperature for AFs production can vary from 24 to 30 °C (Klich, 2007). Water activity is another important factor that modulates fungal growth and mycotoxin production. Recent studies evaluating the effect of A_w and temperature on AF production by *A. flavus* show that the highest AF concentrations were found in inoculated brown and polished rice under A_w ranges of 0.9-0.92 at 21 °C after 21 days of incubation.

Drought

Drought is another modulator of mycotoxin contamination that is expected to be more frequent depending on geography. For the current century, longer and more severe droughts are projected for West Africa and southern Europe, while for central Europe, central North America, Central America, northeast Brazil and southern Africa the projections of intense drought are moderated (IPCC, 2012).

Pluvial Precipitation

Pluvial precipitation is another environmental factor that plays an important role in mycotoxin food contamination. In India, sorghum grown in the rainy season (Kharif or monsoon crops) during 2006–2007 resulted with higher levels of AF compared to other years (Ratnavathi *et al.*, 2012). Severe rains, while plants are in anthesis (flowering), are associated with increase dispersion of *Fusarium* to corn ears (Parry *et al.*, 1995) which may lead to higher mycotoxin production. Similarly, the production of fumonisins has been associated with dry weather and late season rains (Munkvold & Desjardins, 1997). Unseasonable rains can also affect mycotoxin production by forcing people to harvest earlier when grains are not completely dry hence favoring mold contamination. This was the case of the maize implicated in the Kenyan outbreak of 2004, where 317 clinical cases and 125 deaths for aflatoxicosis were reported (CDC, 2004). Tantaoui-Elaraki *et al.* (*in press*) reported that stachybotryotoxicosis caused in November of 1991 the death of 216 equines in Morocco that had been fed straw moulded with *Stachybotrys atra* and that 8 strains of *S. atra* strains were isolated from the mouldy straw incriminated in this poisoning. This poisoning was explicated by the changes of the climate during autumn after unseasonable rainfall followed by a hot-climate that accelerated mould growth and mycotoxins production. Other outbreaks of aflatoxins occurred under similar conditions of unseasonable rainfall during 1981 in the Makueni District of Kenya (Ngindu *et al.*, 1982) and in western India in 1975 (Krishnamachari *et al.*, 1975). On the other hand, lack of rainfall brings other problem to consider, since it causes stress to the plants.

Due to intensifications of hydrological cycle in the planet, it's been predicted that pluvial precipitations will be reduced in subtropical regions (Solomon *et al.*, 2007) which may lead to stressed crops. The European Food Safety Authority (EFSA) has examined the potential impact of climate change in Europe and has suggested that effects will be (a) regional and (b) detrimental or advantageous depending on geographical region (Battilani et al., 2012). This suggests that in northern Europe the effects may be positive, while the Mediterranean basin may be a hot spot where many effects will be negative, with extreme changes in rainfall/drought, elevated temperatures and CO_2 impacting on food production. Effects of climate change on cereals will be significant and detrimental as ripening in southern and central Europe will occur much earlier than at present. This will influence pests and diseases with decreasing yields and increasing mycotoxins contamination. Indeed it has been suggested that climate change may be responsible for up to a 1/3 of yield variability in key staple commodities on a global basis (Ray *et al.*, 2015). This will have profound impacts on food security in different continents. Recently, several mathematical models have been elaborated to predict the potential impacts of climate change scenarios on mycotoxins and fungi. In northern Europe, Van Der Fels-Klerx *et al.*, (2012) focused a mathematical model on DON. While, another model focused on aflatoxins contamination of maize, wheat and rice grown in Europe (Battilani *et al.*, 2016).

MYCOTOXINS REGULATION

Mycotoxin-producing mold species are extremely common, and they can grow on a wide range of substrates under a wide range of environmental conditions. Mycotoxins can enter the food chain in the field, during storage, or at later points. Mycotoxin problems are exacerbated whenever shipping, handling, and storage practices are conducive to mold growth. The end result is that mycotoxins are commonly found in foods. Several authors classified mycotoxins as the most important chronic dietary risk factor, higher than synthetic contaminants, plant toxins, food additives, or pesticide residues. The economic consequences of mycotoxin contamination were well demonstrated.

Since the discovery of the AFs in 1960, regulations have been established in many countries to protect consumers from the harmful effects of mycotoxins that may contaminate foodstuffs, as well as to ensure fair practices in food trade. Various factors play a role in decision-making processes focused on setting limits for mycotoxins. These include scientific factors to assess risk (such as the availability of toxicological data), food consumption data, detailed knowledge about possibilities for sampling and analysis, the distribution of the mycotoxins over commodities and socio-economic issues. Each process for mycotoxins legislation should take into account also the situation in the other countries with which trade contacts exist.

Food security, as defined by the WHO at the World Food Summit of 1996, only exists when people have continuous access to safe and nutritious food in order to maintain a healthy and active life. Despite the efforts made by different agencies and organizations like the Food and Agriculture Organization (FAO), the World Health Organization (WHO) and country specific agencies in setting regulations to limit the amounts of mycotoxins in foods, up to day, mycotoxin regulation is not global and many countries still lack appropriate guidelines to manage these toxins (particularly in Africa and Latin America). Because of their significant toxicological impacts on both human and animal health, the focus on mycotoxins has been a high priority by the FAO and WHO and other international organisms. This has resulted in strict legislative limits for mycotoxins in many parts of the world in a wide range of foodstuffs with the strictest limits in the EU countries, USA and other countries. For example, in EU countries, several regulatory limits were adopted and revised (European Commission, 2006, 2007, 2010 and 2012), a summary of EU legislation on the aflatoxins, the most serious mycotoxins, is given in Table 2.

In contrast, if mycotoxins hazards are seriously considered and these substances strictly regulated in food and feed in developed countries, mycotoxin legislation is almost absent in some countries of the world, particularly in Africa and Latin America; and if it does exist, it serves only to control food products exported to developed countries, or it remains without any real application by official control laboratories for the determination of levels of mycotoxins in contaminated food. In these cases consumption of mycotoxins contaminated staple foods is a significant risk, especially in rural populations and sub-groups such as children and immunocompromised people.

Table 2. Maximum limits for aflatoxins set by European countries

Mycotoxins	Food Commodities	European Limits * (µg/Kg)
AFB1	Groundnuts (peanuts) and other oilseeds and processed products thereof, intended for direct human consumption	2
	Almonds, pistachios and apricot kernels, intended for direct human consumption	8
	Hazelnuts and Brazil nuts, intended for direct human consumption	5
	Dried fruit and processed products thereof, intended for direct human consumption	2
	Cereals and cereal products	2
	Spices	5
	Processed cereal-based foods and baby foods for infants and young children	0.1
	Dried figs	6
Total AFs B1+B2+G1+G2	Groundnuts (peanuts) and other oilseeds and processed products thereof, intended for direct human consumption	4
	Almonds, pistachios and apricot kernels, intended for direct human consumption	10
	Hazelnuts and Brazil nuts, intended for direct human consumption	10
	Dried fruit and processed products thereof, intended for direct human consumption	4
	Cereals and cereal products	4
	Spices	10
	Dried figs	10
Aflatoxin M1	Liquid milk	0.05
	Baby milk	0.025
	Dietary foods for special medical purposes intended specifically for infants	0.025

* European Regulations (EC, 2006, 2010 and 2012)

Zinedine and Mañes (2009) reported that fifteen countries in Africa were known to have specific mycotoxins regulations. These countries cover approximately 59% of the inhabitants of the continent. For the majority of the African countries, specific mycotoxin regulations (probably) do not exist. The fact that some countries have no specific regulatory limit for mycotoxins does not mean that the problem is ignored. Several of these countries recognize that they have problems due to mycotoxins and that regulations should be adopted as soon as possible.

STRATEGIES TO REDUCE MYCOTOXINS

Nowadays, it is well known that the good manufacturing (GMP) and good agricultural (GAP) practices are the most effective worldwide strategies recommended for the prevention of the growth of mycotoxigenic fungi and the production of mycotoxin. In developed countries, people are less exposed to mycotoxins hazards compared to those in developing countries. Indeed, the European Commission and the Codex Alimentarius Commission established several practices for the controlling of mycotoxin contamination in food and feed. The first step of each strategy is based on the control of mold growth and mycotoxins production of the prevention strategies. Common strategies for contaminated foods and feeds to reduce or eliminate the toxic effects of mycotoxins by chemical, physical, and biological/biotechnological methods are crucial to improve food safety, prevent economic losses, and reclaim contaminated products. The use of biotechnology to develop host resistance plant and to manage fungal growth and mycotoxins accumulation is the most promise approach for the prevention of mycotoxins contamination in the future. Moreover, the storage conditions are very important to control for the prevention of fungal growth and the reduction of mycotoxins biosynthesis. Biological detoxification using microorganisms (lactic acid bacteria, yeast, etc.) as biological adsorbents was widely described. Several mycotoxins, including ZEA, OTA, AFs and DON can be bind to lactic acid bacteria adsorbed by the bacterial surface (Tantaoui-Elaraki *et al.* in press). On the other, physical methods were also applied for the decontamination of mycotoxins including immersing and washing, sorting and separation, filtering and adsorption as well as irradiation. The immersing and washing is considered an effective method for the reduction of mycotoxins contamination in grains. Finally, chemical agents were also applied for their efficiency to decontaminate mycotoxin including the organic acids, bases and oxidizing agents. Although ammonia was effective to inhibit fungal growth and reduce AFs, FUM and OTA to undetectable levels in animal feeds, this process was not permitted for decontaminating food destined to human consumption (Peraica *et al.*, 2002). Some innovative strategies were also proposed to reduce mycotoxins such as the use of essential oils extracted from aromatic herbs and the use of some anti-oxidants compounds (flavonoides and polyphenols) for their inhibitory effect on fungi growth. More attention is nowadays given to the use of nanoparticles and magnetic materials, these materials appear more effective in the removal of mycotoxins and are considered promising tools for adsorption in the food industry (Sun *et al.*, 2016).

REFERENCES

Abarca, M. L., Accensi, F., Bragulat, M. R., Castella, G., & Cabañes, F. J. (2003). Aspergillus carbonarius as the main source of ochratoxin a contamination in dried vine fruits from the Spanish market. *Journal of Food Protection*, *66*(3), 504–506. doi:10.4315/0362-028X-66.3.504

AFSSA. (2009). Agence Française de Sécurité Sanitaire des Aliments. Évaluation des risques liés à la présence de mycotoxins dans les chaînes alimentaires humaine et animale. Rapport final.

Arroyo-Manzanares, N., Malysheva, S. V., Vanden Bussche, J., Vanhaecke, L., Diana Di Mavungu, J., & De Saeger, S. (2014). Holistic approach based on high resolution and multiple stage mass spectrometry to investigate ergot alkaloids in cereals. *Talanta*, *118*, 359–367. doi:10.1016/j.talanta.2013.10.002

Balbus, J. M., Boxall, A. B. A., Fenske, R. A., Mckone, T. E., & Zeisel, L. (2013). Implications of global climate change for the assessment and management of human health risks of chemicals in the natural environment. *Environmental Toxicology and Chemistry*, *32*(1), 62–78. doi:10.1002/etc.2046

Bennett, J. W., & Klich, M. (2003). Mycotoxins. *Clinical Microbiology Reviews*, *16*(3), 497–516. doi:10.1128/CMR.16.3.497-516.2003

Berthiller, F., Crews, C., Dall'Asta, C., De Saeger, S., Haesaert, G., Karlovsky, P., ... Stroka, J. (2013). Masked mycotoxins: A review. *Molecular Nutrition & Food Research*, *57*(1), 165–186. doi:10.1002/mnfr.201100764

Bhat, R. V., Shetty, P. H., Amruth, R. P., & Sudershan, R. V. (1997). A foodborne disease outbreak due to the consumption of moldy sorghum and maize containing fumonisin mycotoxins. *Journal of Toxicology. Clinical Toxicology*, *35*(3), 249–255. doi:10.3109/15563659709001208

Bittner, A., Cramer, B., Harrer, H., & Humpf, H. (2015). Structure elucidation and in vitro cytotoxicity of ochratoxin α amide, a new degradation product of ochratoxin A. *Mycotoxin Research*, *31*(2), 83–90. doi:10.100712550-014-0218-y

Bottalico, A., Visconti, A., Logrieco, A., Solfrizzo, M., & Mirocha, C. J. (1985). Occurrence of zearalenols (diastereomeric mixture) in corn stalk rot and their production by associated Fusarium species. *Applied and Environmental Microbiology*, *49*, 547–551.

Burmeister, H. R., & Plattner, R. D. (1987). Enniatin production by Fusarium tricinctum and its effects on germinating wheat seeds. *Phytopathology*, *77*(10), 1483–1487. doi:10.1094/Phyto-77-1483

Centers for Disease Control and Prevention-CDC. (2004). Outbreak of aflatoxin poisoning eastern and central provinces, Kenya, January–July 2004. *MMWR. Morbidity and Mortality Weekly Report*, *3*, 790–793.

Čonková, E., Laciaková, A., Pástorová, B., Seidel, H., & Kováč, G. (2001). The effect of zearalenone on some enzymatic parameters in rabbits. *Toxicology Letters*, *121*(3), 145–149. doi:10.1016/S0378-4274(01)00312-5

Creppy, E. E. (2002). Update of survey, regulation and toxic effects of mycotoxins in Europe. *Toxicology Letters*, *127*(1-3), 19–28. doi:10.1016/S0378-4274(01)00479-9

De Ruyck, K., De Boevre, M., Huybrechts, I., & De Saeger, S. (2015). Dietary mycotoxins, co-exposure, and carcinogenesis in humans: Short review. *Mutation Research/Reviews in Mutation Research, 766*, 32–41. doi:10.1016/j.mrrev.2015.07.003

Eaton, D. L., & Gallagher, E. P. (1994). Mechanisms of aflatoxins carcinogenesis. *Annual Review of Pharmacology and Toxicology, 34*(1), 135–172. doi:10.1146/annurev.pa.34.040194.001031

European Commission. (2006). *Commission Regulation No. 1881/2006 of December 19th setting maximum levels of certain contaminants in foodstuffs*. Official Journal of the European Union No. L364/5.

European Commission. (2010). Commission Regulation (EC) No 165/2010 of 26 February 2010 amending Regulation (EC) No 1881/2006 setting maximum levels for certain contaminants in foodstuffs as regards aflatoxins. *Official Journal of the European Union, L, 50*(8).

European Commission. (2012). Commission Regulation (EC) No 1058/2012 of 12 November 2012 amending Regulation (EC) No 1881/2006 setting maximum levels for certain contaminants in foodstuffs as regards maximum levels for aflatoxins in dried figs. *Official Journal of the European Union, L, 313*(14).

European Food Safety Authority (EFSA). (2006). Opinion of the scientific panel on contaminants in the food chain on a request from the commission related to ochratoxin A in food (Question No. EFSA-Q-2005-154). The EFSA Journal, 365.

FAO. (2012). *Statistical year book 2012*. Retrieved from http://www.fao.org /docrep/015 /i2490e/ i2490e00.htm

Freire, L., & Sant'Ana, A. S. (2018). Modified mycotoxins: An updated review on their formation, detection, occurrence, and toxic effects. *Food and Chemical Toxicology, 111*, 189–205. doi:10.1016/j.fct.2017.11.021

Frisvad, J. C., & Samson, R. A. (1991). Mycotoxins produced by species of Penicillium and Aspergillus occurring in cereals. In J. Chelkowski (Ed.), *Cereal Grain Mycotoxins, Fungi and Quality in Drying and Storage*. Amsterdam: Elsevier.

Gelderblom, W. C. A., Jaskiewicz, K., Marasas, W. F. O., Thiel, P. G., Horak, R. M., Vleggaar, R., & ... . (1988). Fumonisins-Novel mycotoxins with cancerpromoting activity produced by Fusarium moniliforme. *Applied and Environmental Microbiology, 54*, 1806–1811.

Gelderblom, W. C. A., Jaskiewicz, K., Marasas, W. F. O., Thiel, P. G., Horak, R. M., Vleggaar, R., & ... . (1988). Fumonisins – Novel mycotoxins with cancerpromoting activity produced by Fusarium moniliforme. *Applied and Environmental Microbiology, 54*, 1806–1811.

Hendricks, K. A., Simpson, J. S., & Larsen, R. D. (1999). Neural tube defect along the Texas-Mexico border, 1993-1995. *American Journal of Epidemiology, 149*(12), 119–127. doi:10.1093/oxfordjournals.aje.a009766

Hendrickse, R. G. (1997). Of sick Turkeys, Kwashiorkor, malaria, perinatal mortality, Herion addicts and food poisoning: Research on influence of aflatoxins on child health in the tropics. *Journal of Tropical Medicine and Parasitology, 91*, 87–93.

Hiraga, K., Yamamoto, S., Fukuda, H., Hamanaka, N., & Oda, K. (2005). Enniatin has a new function as an inhibitor of Pdr5p, one of the ABC transporters in Saccharomyces cerevisiae. *Biochemical and Biophysical Research Communications*, *328*(4), 1119–1125. doi:10.1016/j.bbrc.2005.01.075

IARC. (2002). *Some traditional herbal medicines, some mycotoxins, naphthalene and styrene.* IARC..

IARC. (2018). *Aflatoxins*. Available at https://monographs.iarc.fr/wp-content/uploads/2018/06/mono100F-23.pdf

International Agency for Research on Cancer (IARC). (1993). Evaluation of carcinogenic risks of chemical to humans. In Some naturally-occurring substances: Food Items and Constituent. IARC.

JECFA-Joint FAO/WHO Expert Committee on Food Additives. (2000). Zearalenone. In: JECFA (Ed.), Safety evaluation of certain food additives and contaminants. WHO.

JECFA-Joint FAO/WHO Expert Committee on Food Additives. (2001). *Safety evaluation of certain mycotoxins in food*. Prepared by the 56th Meeting of the Food Additives Series No. 47. Geneva.

Jestoi, M. (2008). Emerging *Fusarium*-mycotoxins fusaproliferin, beauvericin, enniatins, and moniliformin-a review. *Critical Reviews in Food Science and Nutrition*, *48*(1), 21–49. doi:10.1080/10408390601062021

Jow, G. M., Chou, C. J., Chen, B. F., & Tsai, J. H. (2004). Beauvericin induces cytotoxic effects in human acute lymphoblastic leukemia cells through cytochrome c release, caspase 3 activation: The causative role of calcium. *Cancer Letters*, *216*(2), 165–173. doi:10.1016/j.canlet.2004.06.005

Kouri, K., Lemmens, M., & Lemmens-Gruber, R. (2003). Beauvericin-induced channels in ventricular myocytes and liposomes. *Biochimica et Biophysica Acta*, *1609*(2), 203–210. doi:10.1016/S0005-2736(02)00689-2

Lemmens-Gruber, R., Rachoy, B., Steininger, E., Kouri, K., Saleh, P., Krska, R., ... Lemmens, M. (2000). The effect of the *Fusarium* metabolite beauvericin on electromechanical and physiological properties in isolated smooth and heart muscle preparations of guinea pigs. *Mycopathologia*, *149*(1), 5–12. doi:10.1023/A:1007293812007

Lioi, M. B., Santoro, A., Barbieri, R., Salzano, S., & Ursini, M. V. (2004). Ochratoxin and zearalenone: A comparative study on genotoxic effects and cell death induced in bovine lymphocytes. *Mutation Research*, *557*(1), 19–24. doi:10.1016/j.mrgentox.2003.09.009

Logrieco, A., Rizzo, A., Ferracane, R., & Ritieni, A. (2002). Occurrence of beauvericin and enniatins in wheat affected by Fusarium avenaceum head blight. *Applied and Environmental Microbiology*, *68*(1), 82–85. doi:10.1128/AEM.68.1.82-85.2002

Marasas, W. F., Kellerman, T. S., Gelderblom, W. C., Coetzer, J. A., Thiel, P. G., & van der Lugt, J. J. (1988). Leukoencephalomalacia in a horse induced by fumonisin B1 isolated from Fusarium moniliforme. *The Onderstepoort Journal of Veterinary Research*, *55*, 197–203.

Marasas, W. F. O., Jaskiewicz, K., Venter, F. S., & Van Schalkwyk, D. J. (1988). Fusarium moniliforme contamination of maize in esophageal cancer aereas in Transkei. *South African Medical Journal*, *74*, 110–114.

Marin, S., Ramos, A. J., Cano-Sancho, G., & Sanchis, V. (2013). Mycotoxins: Occurrence, toxicology, and exposure assessment. *Food and Chemical Toxicology, 60*, 218–237. doi:10.1016/j.fct.2013.07.047

Meca, G., Sospedra, I., Soriano, J. M., Ritieni, A., Valero, M. A., & Mañes, J. (2009). Isolation, purification and antibacterial effects of fusaproliferin produced by Fusarium subglutinans in submerged culture. *Food and Chemical Toxicology, 47*(10), 2539–2543. doi:10.1016/j.fct.2009.07.014

Meca, G., Zinedine, A., Blesa, J., Font, G., & Mañes, J. (2010). Further data on the presence of Fusarium emerging mycotoxins enniatins, fusaproliferin and beauvericin in cereals available on the Spanish markets. *Food and Chemical Toxicology, 48*(5), 1412–1416. doi:10.1016/j.fct.2010.03.010

Moretti, A., Mulè, G., Ritieni, A., & Logrieco, A. (2007). Further data on the production of beauvericin, enniatins and fusaproliferin and toxicity to Artemia salina by Fusarium species of Gibberella fujikuroi species complex. *International Journal of Food Microbiology, 18*(2), 158–163. doi:10.1016/j.ijfoodmicro.2007.07.004

Nilanonta, C., Isaka, M., Kittakoop, P., Palittapongarnpim, P., Kamchonwongpaisan, S., Pittayakhajonwut, D., ... Thebtaranonth, Y. (2000). Antimycobacterial and antiplasmodial cyclodepsipeptides from the insect pathogenic fungus Paecilomyces tenuipes BCC 1614. *Planta Medica, 66*(8), 756–758. doi:10.1055-2000-9776

Ojcius, D. M., Zychlinsky, A., Zheng, L. M., & Young, J. D. E. (1991). Ionophore-induced apoptosis: Role of DNA fragmentation and calcium fluxes. *Experimental Cell Research, 197*(1), 43–49. doi:10.1016/0014-4827(91)90477-C

Ouanes, Z., Abid, S., Ayed, I., Anane, R., Mobio, T., Creppy, E., & Bacha, H. (2003). Induction of micronuclei by zearalenone in Vero monkey kidney cells and in bone marrow cells of mice: Protective effect of vitamin E. *Mutation Research, 538*(1-2), 63–70. doi:10.1016/S1383-5718(03)00093-7

Paris, M. P. K., Schweiger, W., Hametner, C., Stückler, R., Muehlbauer, G. J., Varga, E., & ... . (2014). Zearalenone-16-O-glucoside: A new masked mycotoxin. *Journal of Agricultural and Food Chemistry, 62*(5), 1181–1189. doi:10.1021/jf405627d

Paterson, R. R., & Lima, N. (2010). Toxicology of mycotoxins. *EXS, 100*, 31–63. doi:10.1007/978-3-7643-8338-1_2

Peraica, M., Richter, D., & Rasi, C. D. (2014). Mycotoxicoses in children. *Arhiv za Higijenu Rada i Toksikologiju, 65*(4), 347–363. doi:10.2478/10004-1254-65-2014-2557

Peraica, M., Richter, D., & Rašić, D. (2014). Mycotoxicoses in children. *Arhiv za Higijenu Rada i Toksikologiju, 65*(4), 347–363. doi:10.2478/10004-1254-65-2014-2557

Pestka, J. (2010). Deoxynivalenol: Mechanisms of action, human exposure, and toxicological relevance. *Archives of Toxicology, 84*(9), 663–679. doi:10.100700204-010-0579-8

Pestka, J. J. (2010). Deoxynivalenol: Mechanism of toxicological relevance. *Archives of Toxicology, 84*, 663–679. doi:10.100700204-010-0579-8

Pfohl-Leszkowicz, A., Petkova-Bocharova, T., Chernozemsky, I. N., & Castegnaro, M. (2002). Balkan endemic nephropathy and associated urinary tract tumors: A review on etiological causes and the potential role of mycotoxins. *Food Additives and Contaminants*, *19*(3), 282–302. doi:10.1080/02652030110079815

Pittet, A. (1998). Natural occurrence of mycotoxins in foods and feeds—an updated review. *Revue de Medecine Veterinaire*, *6*, 479–492.

Randazzo, G., Fogliano, V., Ritieni, A., Mannina, L., Rossi, E., Scarallo, A., & Segre, A. L. (1993). Proliferin, a new sesterterpene from Fusarium proliferatum. *Tetrahedron*, *49*(47), 10883–10896. doi:10.1016/S0040-4020(01)80241-6

Ritieni, A., Fogliano, V., Randazzo, G., Scarallo, A., Logrieco, A., Moretti, A., ... Bottalico, A. (1995). Isolation and characterization of fusaproliferin, a new toxic metabolite from Fusarium proliferatum. *Natural Toxins*, *3*(1), 17–20. doi:10.1002/nt.2620030105

Ritieni, A., Monti, S. M., Randazzo, G., Logrieco, A., Moretti, A., Peluso, G., ... Fogliano, V. (1997). Teratogenic effects of fusaproliferin on chicken embryos. *Journal of Agricultural and Food Chemistry*, *45*(8), 3039–3043. doi:10.1021/jf960890v

Robens, J., & Cardwell, K. (2003). The costs of mycotoxin management to the USA:management of aflatoxins in the United States. *Journal of Toxicology*, *22*, 139–152.

Rychlik, M., Humpf, H. U., Marko, D., Danicke, S., Mally, A., Berthiller, F., ... Lorenz, N. (2014). Proposal of a comprehensive definition of modified and other forms of mycotoxins including "masked" mycotoxins. *Mycotoxin Research*, *30*(4), 197–205. doi:10.100712550-014-0203-5

SCF. (2000). *Opinion on Fusarium Toxins – Part 2: zearalenone (ZEA)*. Retrieved from http://ec.europa.eu/food/fs/sc/scf/out65_en.pdf

Schiff, P. L. Jr. (2006). Ergot and its alkaloids. *American Journal of Pharmaceutical Education*, *70*(5), 98. doi:10.5688/aj700598

Sherif, S. O., Salama, E. E., & Abdel-Wahhab, M. A. (2009). Mycotoxins and child health: The need for health risk assessment. *International Journal of Hygiene and Environmental Health*, *212*(4), 347–368. doi:10.1016/j.ijheh.2008.08.002

Soriano, J. M., González, L., & Catalá, A. I. (2005). Mechanism of action of sphingolipids and their metabolites in the toxicity of fumonisin B1. *Progress in Lipid Research*, *44*(6), 345–356. doi:10.1016/j.plipres.2005.09.001

Sun, X. D., Su, P., & Shan, H. (2017). Mycotoxin contamination of rice in China. *Journal of Food Science*, *82*(3), 573–584. doi:10.1111/1750-3841.13631

Supothina, S., Isaka, M., Kirtikara, K., Tanticharoen, M., & Thebtaranonth, Y. (2004). Enniatin production by the entomopathogenic fungus Verticillium hemipterigenum BCC 1449. *J. Antibiot.*, *57*(11), 732–738. doi:10.7164/antibiotics.57.732

Tantaoui-Elaraki, A., Riba, A., Oueslati, S., & Zinedine, A. (2018). Toxigenic fungi and mycotoxin occurrence and prevention in food and feed in northern Africa – a review. Special issue: "Mycotoxins in Africa. *World Mycotoxin Journal*, *11*(3), 385–400. doi:10.3920/WMJ2017.2290

Tirado, M. C., Clarke, R., Jaykus, L. A., McQuatters-Gollop, A., & Frank, J. M. (2010). Climate change and food safety: A review. *Food Research International*, *43*(7), 1745–1765. doi:10.1016/j.foodres.2010.07.003

Tomoda, H., Huang, X.-H., Cao, J., Nishida, H., Nagao, R., Okuda, S., ... Inoue, K. (1992). Inhibition of acyl-CoA: Cholesterol acyltransferase activity by cyclodepsipeptide antibiotics. *J. Antibiot.*, *45*(10), 1626–1632. doi:10.7164/antibiotics.45.1626

Tudzynski, P., & Neubauer, L. (2014). Biosynthesis and Molecular Genetics of Fungal Secondary Metabolites. Academic Press. doi:10.1007/978-1-4939-1191-2_14

Uhlig, S., Ivanova, L., Petersen, D., & Kristensen, R. (2009). Structural studies on minor enniatins from Fusarium sp. VI 03441: Novel N-methyl-threonine containing enniatins. *Toxicon*, *53*(7-8), 734–742. doi:10.1016/j.toxicon.2009.02.014

Uhlig, S., Torp, M., & Heier, B. T. (2006). Beauvericin and enniatin A, A1, B and B1 in Norwegian grain: A survey. *Food Chemistry*, *94*(2), 193–201. doi:10.1016/j.foodchem.2004.11.004

Van der Merwe, K. J., Steyne, P. S., Fourie, L. F., Scott, D. B., & Theron, J. J. (1965). Ochratoxin A, a toxic metabolite produced by Aspergillus ochraceus Wilh. *Nature*, *205*(4976), 1112–1113. doi:10.1038/2051112a0

Van Egmond, H. P. (1989). *Mycotoxins in dairy products*. London: Elsevier Applied Science.

Wilson, D. M., Mubatanhema, W., & Jurjevic, Z. (2002). Biology and ecology of mycotoxigenic aspergillus species as related to economic and health concerns. In J. W. De Vries, M. W. Trucksess, & L. S. Jackson (Eds.), *Mycotoxins and Food Safety*. New York: Kluwer Academic / Plenum Publishers. doi:10.1007/978-1-4615-0629-4_2

Wu, S. N., Chen, H., Liu, Y. C., & Chiang, H. T. (2002). Block of L-type Ca2+ current by beauvericin, a toxic cyclopeptide in the NG108-15 neuronal cell line. *Chemical Research in Toxicology*, *15*(6), 854–860. doi:10.1021/tx020003k

Yousef, A. E., & Marth, E. H. (1985). Degradation of aflatoxin M1 in milk by ultraviolet energy. *Journal of Food Protection*, *48*(8), 697–698. doi:10.4315/0362-028X-48.8.697

Yu, Z., Zhang, L., Wu, D., & Liu, F. (2005). Anti-apoptotic action of zearalenone in MCF-7 cells. *Ecotoxicology and Environmental Safety*, *62*(3), 441–446. doi:10.1016/j.ecoenv.2004.10.003

Zepnik, H., Völkel, W., & Dekant, W. (2003). Toxicokinetics of the mycotoxin ochratoxin A in F 344 rats after oral administration. *Toxicology and Applied Pharmacology*, *192*(1), 36–44. doi:10.1016/S0041-008X(03)00261-8

Zinedine, A., Gonzales-Osnaya, L., Soriano, J. M., Moltó, J. C., Idrissi, L., & Mañes, J. (2007a). Presence of aflatoxin M1 in pasteurized milk from Morocco. *International Journal of Food Microbiology*, *114*(1), 25–29. doi:10.1016/j.ijfoodmicro.2006.11.001

Zinedine, A., & Mañes, J. (2009). Occurrence and legislation of mycotoxins in food and feed from Morocco. *Food Control*, *20*(4), 334–344. doi:10.1016/j.foodcont.2008.07.002

Zinedine, A., Meca, G., Mañes, J., & Font, G. (2011). Further data on the occurrence of Fusarium emerging mycotoxins enniatins (A, A1, B, B1), fusaproliferin and beauvericin in raw cereals commercialized in Morocco. *Food Control*, *22*(1), 1–5. doi:10.1016/j.foodcont.2010.05.002

Zinedine, A., & Ruiz, M. J. (2014). Zearalenone. In *Mycotoxins and their implications in food safety*. Future Science Ltd. Retrieved from www.future-science.com

Zinedine, A., Soriano, J. M., Moltó, J. C., & Mañes, J. (2007b). Review on the toxicity, occurrence, metabolism, detoxification, regulations and intake of zearalenone: An oestrogenic mycotoxin. *Food and Chemical Toxicology*, *45*(1), 1–18. doi:10.1016/j.fct.2006.07.030

Section 3

Climate Change Leading to Emergence of Environmental Contamination

Chapter 6
Climate Change Outcomes on the Environmental Ecotoxicology

Soraia El Baz
Cadi Ayyad University, Morocco

ABSTRACT

Climate change is a daunting problem and has only recently attracted attention. This chapter presents a review on the implications of climate change on the regulation, and modelling of toxic pollutants. Also, it identifies relationships between climate fluctuations and changes in some polluants distribution (heavy metals, hydrocarbons, and pesticides). Moreover, the influence of climate change on polluant environmental behavior is explored by studying polluants response to inter-annual climate fluctuations such as precipitation and temperature. Therefore, it will be important to monitor strategies taking into account climate change and new regulatory plans should be devised in toxics polluant management.

INTRODUCTION

Climate describes the long-term behaviour of the Earth, on a time scale which starts from decadal and extends to billion years (Ciardini et al., 2016). The Earth's climate system includes the land surface, atmosphere, oceans, and ice. Many aspects of the global climate are changing rapidly, and the primary drivers of the change are human in origin (Gatto, Cabella, & Gherardi, 2016). Climate change is an increasingly urgent global problem (Stocker, 2014) and is identified as an important new challenge on the assessment of the effects of certain public and private projects on the environment (Jiricka et al., 2016). The U.N. Intergovernmental Panel on Climate Change (IPCC) has completed fifth assessments covering the evidence; impacts of climate change, and report that surface temperature is projected to rise over the 21st century. It is very likely that heat waves will occur more often and last longer, and that extreme precipitation events will become more intense and frequent in many regions. The ocean will continue to warm and acidify, and global mean sea level to rise (IPCC, 2014).

DOI: 10.4018/978-1-5225-7775-1.ch006

The effects of climate change are tangible and demand actions (Pielke Jr, Prins, Rayner, & Sarewitz, 2007; Klein, 2011). These actions can be mitigation, to prevent greenhouse gas (GHG) emissions or reduce their atmospheric concentration, or adaptation, to adjust to actual or expected climate and its effects (Field et al., 2014). Various authors (Peterson, McGuirk, Houston, Horvitz, & Wehner, 2008; Swart & Biesbroek, 2008; Haurie, Sceia, & Thénié, 2009) have looked at the consequences of extreme rainfall, including e.g. overloaded drainage systems and floods.

There is evidence that climate change, including increased climatic variability, can affect the individual organisms, the size and structure of their populations, the species composition of communities, and the structure and functioning of ecosystems (Portner & Knust, 2007). Climate change is a powerful advancing force (Oreskes, 2004; Carpenter et al., 2011), that can affect humans and ecosystems for years to come (Richardson et al., 2009; Mabey, Gulledge, Finel, & Silverthorne, 2011). It is likely that climate change will, impact the abiotic, biotic, and socioeconomic components of the landscape (Schneider et al., 2007) as a suite of stressors, or a syndrome (Mabey et al., 2011). The threats of environmental changes, survival and reproductive of individuals, and the survival of species and ecosystems are influenced by many directions: (1) habitat destruction, (2) disruption of food chains, (3) changes in disease and parasitic loads, (4) increased pollution and ultimately by (5) direct and indirect effects of climate change. The environmental changes can be detected at any place on the planet: from pole to pole (Moline et al., 2008; Robinson, 2009) and from ocean depths (Guinotte & Fabry, 2008) to the stratosphere (Wilson, Solomon, & Tang, 2007).

Environmental changes are expected to influence the distribution and toxicity of environmental chemicals, that can affect some population health (Hooper et al., 2013; Moe et al., 2013). Among the consequences of climate change is its potential to alter the environmental distribution, cycling and risks of chemical pollutants (Gouin et al., 2013; O'Driscoll, Mayer, Su, & Mathis, 2014). There is therefore now an important need for research aimed at understanding how climate change will impact the distribution and toxicity of chemical contaminants (Hooper et al., 2013).

This paper gives a short overview of the present knowledge on climate change and provides background information on the occurring and expected changes. In this paper, we analyze the likely influence of climate change on the chemical exposure. The emphasis is on expected increase in the use of pesticides, heavy metals and hydrocarbons due to the global warming, also including the consequences of change in exposure parameters conditions on those toxics polluants.

ECOTOXICOLOGY IN A CHANGING CLIMATE

The primary goal in ecotoxicology is to understand and predict effects of contaminant stress on ecological systems (Walker, 2006). One of the consequences of climate change that has recently attracted attention is its potential to alter the environmental distribution and biological effects of chemical toxicants (Hansen, Sato, & Ruedy, 2012; Huntingford, Jones, Livina, Lenton, & Cox, 2013), and changes in soil quality (Fontaine, Decker, Skagen, & Van Riper, 2009).

Climate Change Parameters and Their Impacts on Chemical Exposures

Both environmental contaminants and climate change have received more attention recently (Klerks, Xie, & Levinton, 2011; Merilä & Hendry, 2014), and studies investigating its potential interactions with contaminants in the environment (environmental fate and transport of pollutants) seem to be growing (Kallenborn, Halsall, Dellong, & Carlsson, 2012; Kimberly & Salice, 2012; Müller et al., 2012; Moe et al., 2013; Stahl et al., 2013). Changes to the abiotic and biotic components of the environment as a result of climate change may impact how we currently assess the environmental risks of chemicals (Lamon, Dalla Valle, Critto, & Marcomini, 2009; Stahl et al., 2013). Climate change is known to affect a wide range of environmental conditions, including temperature, precipitation patterns, acidification (Liu, Hofstra, & Franz, 2013), which can affect abiotic factors, such as salinity, pH and ultraviolet-radiation and have been shown to affect both uptake and toxicity of chemicals (Noyes et al., 2009). Climate change is also expected to affect the release, fate, behavior, and exposure of toxicants (Noyes et al., 2009; Moe et al., 2013), and by influencing environmental concentrations of contaminants (Gouin et al., 2013). Climate change, such as floods and droughts, have been suggested to have significant impacts on water quality around the world (Murdoch, Baron, & Miller, 2000; Worrall, Burt, & Shedden, 2003; Senhorst & Zwolsman, 2005), not only by directly changing the characteristics of the water, but also by influencing land surface processes that regulate the production, release, and transport of anthropogenic contaminants to ground and surface waters (Williamson, Dodds, Kratz, & Palmer, 2008; Campbell et al., 2009).

Temperature

Temperature is a key factor and it has long been known to cause significant alterations in the chemistry of a number of chemical pollutants that effects aquatic fauna (Schiedek, Sundelin, Readman, & Macdonald, 2007). Rise in temperature is found to increase the rate of uptake of pollutants with increase of metabolic rate and decrease in oxygen solubility. It has been reported that the upper temperature tolerance limits are decreased in the presence of certain organic chemicals for a variety of freshwater fish species (Patra, Chapman, Lim, Gehrke, & Sunderam, 2015). Estuarine systems are considered especially sensitive to climate change as they are already subject to considerable anthropogenic stress, including biological contamination and chemical pollution (Kennish, 2002). Climate change may cause numerous effects in estuarine systems, including increased water temperature, current alteration, increased salinity, increased erosion, and alteration in freshwater runoff patterns (Grabemann, Grabemann, Herbers, & Müller, 2001; Knowles & Cayan, 2002).

Factors such as temperature can greatly influence the toxicity of chemicals in a variety of taxa (Noyes et al., 2009; Holmstrup et al., 2010). In most cases, thermal stress potentiates chemical toxicity, and this appears to relate to temperature modulation of chemical uptake and to temperature-induced shifts in physiological and metabolic processes of the exposed organisms (Heugens, Hendriks, Dekker, Straalen, & Admiraal, 2001). Water temperature changes can directly influence temperature-dependent water quality parameters including dissolved oxygen, redox potentials, pH, lake stratification and mixing, and microbial activity. Changes in temperature and precipitation, along with irregular, extreme hydrologic events, can lead to changes the deterioration in water quality (Park, Duan, Kim, Mitchell, & Shibata, 2010). Therefore, increasing freshwater temperatures may cause a number of serious long-term impacts to salmon, such as increasing the toxicity of certain contaminants (Karvonen, Rintamäki, Jokela, & Valtonen, 2010; Kaushal et al., 2010). There is evidence to show that higher ambient temperature influences the response of plants to air pollutant toxicity (Heck & Dunning, 1967).

Precipitation

Hydrological impacts are expected directly through changes in precipitation, and indirectly due to changes in potential evaporation and transpiration due to atmospheric warming. These changes may have an effect on water quality and ecology (EU, 2009). Increased air temperatures due to global warming and an increased relative humidity due to heavy rainfall, could potentially lead to an increase in pathogens in food (McMichael et al., 2003). Global climate change is causing increases in the severity and frequency of droughts and extreme precipitation events, as well as regional-scale declines in air quality (e.g., increased ground-level ozone and particulate matter) in terrestrial systems (IPCC, 2007).

For example, climatic predictions in upcoming decades for southern China include lower precipitation with large year-to-year variations. The results from a four-year intensive study at a forested watershed in Chongquing province showed that acidity and the concentrations of sulfate and nitrate in soil and surface waters were generally lower in the years with lower precipitation. Results from these case studies suggest that spatially variable patterns of snow or summer precipitation associated with regional climate change across NE Asia will have significant impacts on surface water quality (Park et al., 2010).

As global warming may enhance the release of volatile contaminants from soil, secondary emissions can be also influenced by climate change (Dalla Valle, Codato, & Marcomini, 2007), and changes in wind and precipitation patterns can modifie the way chemicals are redistributed in the environment.

Salinity

The three largest symptoms of marine climate change are: (1) temperature increases, (2) fluctuations in salinity, and (3) an overall depression in pH (Brierley & Kingsford, 2009). Accompanying these symptoms, regional alterations in evaporation, sea level and precipitation rates are leading to changes in sea surface salinity, for example, saltier ocean regions trending toward increased salinity and fresher ocean regions becoming even less saline (Durack, Wijffels, & Matear, 2012; Pierce, Gleckler, Barnett, Santer, & Durack, 2012). Salinity can influence the chemical itself and its toxicity, and eventually the physiology of organisms (Schiedek et al., 2007; Noyes et al., 2009).

On the other hand there is another parametres such as atmospheric CO_2, who their concentrations predicted to continue rising until at least 2040 (Honisch et al., 2012). Consequently, high concentration of atmospheric CO_2 can cause increases in oceanic CO_2 and acidification, hypoxia, and up welling (Romero-Lankao et al., 2014). Also, the quality of the air depends on (a) how rapidly chemicals are released and (b) the reactions these substances undergo once they are released into the atmosphere. In addition, UV-radiation can render the chemical toxicants more toxic or less toxic to animals, by altering their chemical structure (Huovinen, 2001). Solar UV-B radiation (280–315 nm) provides the energy for many of the chemical transformations that occur in the atmosphere. This energy causes photolysis of a number of important atmospheric trace gases (e.g., sulfur dioxide, formaldehyde, and ozone). These processes will be altered by clouds, the elevation of the sun and attenuation by some air pollutants (Wilson et al., 2007). For example, synergistic interactions of UV-radiation and contaminants have been suggested to be a factor in population declines of apmphibiens (Blaustein, Romansic, Kiesecker, & Hatch, 2003). Figure 1 recaps impact of climate change on toxics pollutants.

Influence of Climate Change Parameters on Toxicity of Pesticides

The Environmental Protection Agency (EPA or USEPA) defines pesticide as "any substance or mixture of substances intended for preventing, destroying, repelling, or mitigating any pest". A pesticide may be a chemical substance or biological agent (such as a virus or bacteria) used against pests (Palikhe, 2007). Weather and climate affect many agricultural decisions including crop choices, water management, and crop protection (Koleva, Schneider, & Tol, 2009). Insect pests, plant pathogens and weeds represent a major constraint to crop production (Oerke, 2005), and there is agreement among scientists that warmer and wetter conditions will exacerbate agricultural weed, fungal (Kattwinkel, Kuhne, Foit, & Liess, 2011), and can have significant changes in the abundance and activity of seasonal pests (Bloomfield, Williams, Gooddy, Cape, & Guha, 2006), also requiring increased pesticide use (Kattwinkel et al., 2011). Such imbalance could drive towards an increased use of banned pesticides that would enhance the bioavailability of toxic and bioaccumulating chemicals in the environment, hence worsening the quality of the ecosystem (Macdonald, Harner, & Fyfe, 2005).

Pesticide fate and behavior will also be influenced by many indirect effects of climate change including (1) changes in cropping patterns and crop growth (Olesen et al., 2011), (2) pesticide application rates (Kattwinkel et al., 2011), and (3) soil conditions affecting fate processes such as changes in soil organic carbon content or climate induced freezing/thawing cycles (Stenrød et al., 2008).

Higher amounts of precipitation, elevated temperatures and direct exposure to sunlight may affect the amount of pesticides usage by (1) the acceleration of their dissipation and degradation, and (2) influence the chemical alteration of pesticides (Rosenzweig, Iglesias, Yang, Epstein, & Chivian, 2001). Doull (1972) also suggested that ambient temperature may affect pesticide toxicity. Temperature influences the toxicity of chemicals, by changing their degradation and volatilization rates, in addition to affecting the absorption and desorption processes and detoxification rates in exposed organisms (Mo, Yoo, Bae, & Cho, 2013). The spread of agricultural pests into higher latitudes will also serve to increase pesticide applications (Rosenzweig et al., 2001).

Figure 1. Impact of climate change on toxic polluants

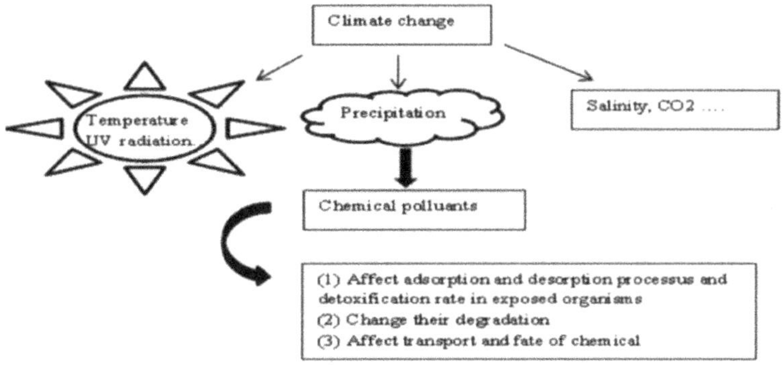

The movement of these chemicals is predicted to be altered by climate change, through increased atmospheric deposition and surface run-off (Kattwinkel et al., 2011). Volatilization from soil and vegetation is one of the main causes of pesticides in the atmosphere (Yeo, Choi, Chun, & Sunwoo, 2003) and takes place when a liquid or solid substance transfers to the gaseous phase. A humid soil after rainfall favours pesticide volatilization (Vela, Navarro, & Navarro García, 2007). After volatilization, compounds can be dispersed from areas with high concentrations and be distributed widely at low concentrations in the form of aerial inputs or wet deposition in rain (Bloomfield et al., 2006; Donald, Cessna, Sverko, & Glozier, 2007). Pesticides can be transported long distances in the atmosphere, accumulating in regions such as the Arctic, where low temperatures induce their deposition (Macdonald et al., 2005). As the climate warms, however, pesticides deposited in sinks such as water and ice are expected to revolatilize into the atmosphere (Nizzetto, Lohmann, Gioia, Dachs, & Jones, 2010). In addition, atmospheric deposition may affect the concentrations of pesticide in playa wetlands (Messing et al., 2011). Moreover, since playas contribute to recharge aquifers (Smith, 2003), in some case like in the Ogallala Aquifer, the largest aquifer in North America, the pesticides found in playas could potentially contaminate ground water (Zartman et al., 1996). Additionally, changes in water quality may change the transport and fate of chemicals by altering their toxicity within the estuarine system (DeLorenzo, Wallace, Danese, & Baird, 2009).

Physiological effects of increased salinity and temperature may modify the way how organisms respond to pesticides (Mclusky, Bryant, & Campbell, 1986). An insecticide with a positive temperature coefficient becomes more toxic with the increase in temperature, whereas, those with a negative temperature coefficient become more toxic at lower temperatures (Glunt, Blanford, & Paaijmans, 2013). However, some studies also revealed variation in the toxicity (1) within a given insecticide class (Muturi, Lampman, Costanzo, & Alto, 2011), (2) between insect species and (3) temperature range tested (Boina, Onagbola, Salyani, & Stelinski, 2009). Furthermore, the stability of insecticides can be affected by temperature. For example, carbamates persist during cold water for longer periods than in warm water, due to less hydrolysis (Aly & El-Dib, 1971).

High water temperature increases the permeability of fish tissue towards pesticide and hence, facilitates the transport of pesticides in the fish tissue (Mehta, 2017). Pesticides can be introduced into critical salmon habitat via nonpoint and point sources (Macneale, Kiffney, & Scholz, 2010; Scholz et al., 2012). Temperature stress has been shown to increase the acute and sublethal effects of some pesticides in fish (Holmstrup et al., 2010).

While commonly used pesticides and herbicides may affect lake and stream metabolism through their effects on primary producers (Seguin, Leboulanger, Rimet, Druart, & Berard, 2001), they also cause endocrine disruption in humans and wildlife. Thus, both streams and rainfall can transport these contaminants across the continent. Effects may be even more serious with exposure to multiple pesticides (Hayes et al., 2006). The factors affecting pesticidal loss rate consist of not only biological factors but also physical and chemical factors, e.g. sunlight, temperature, moisture, pH and so forth (Shui-Ming, Duo-Sen, Zong-Sheng, & Xing-Fa, 1993). Most of studies (Steffens, Larsbo, Moeys, Jarvis, & Lewan, 2013; Ahmadi, Records, & Arabi, 2014; Steffens et al., 2014) only considered the potential direct impacts of climate change on pesticides, but Kattwinkel et al. (2011) concluded that the combined effect is likely to be stronger than the direct or indirect effects of climate change (i.e. changes in land-use and insecticide use) alone. Figure 2 gives an outline of climate change effects on environmental alterations and their feasible consequences on pesticides' exposure and adsorption.

Figure 2. Climate change effects on environmental alterations and their feasible consequences on pesticides' exposure and adsorption

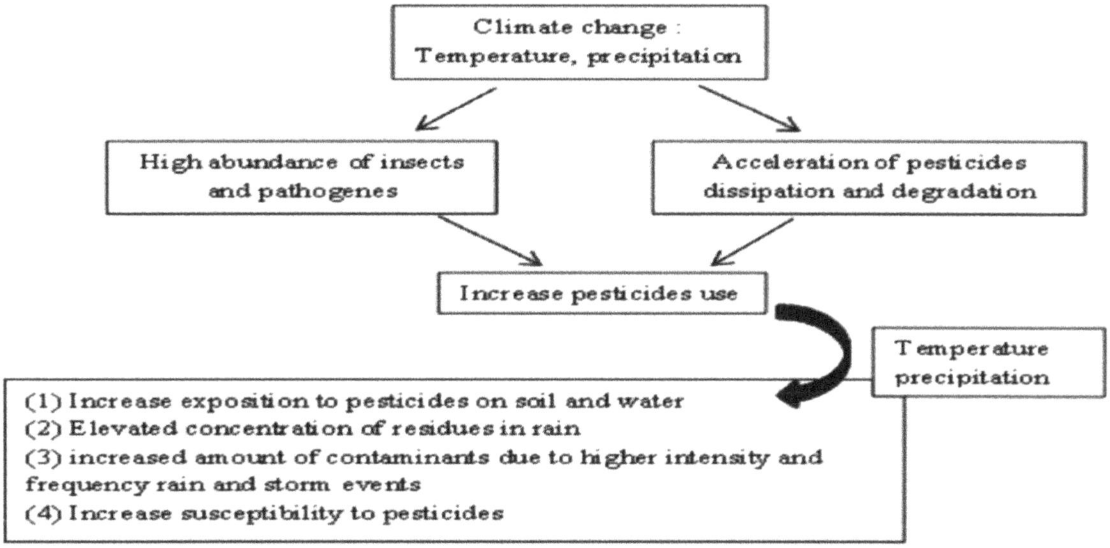

Climate Change and Toxicity of Metals

Heavy metals are defined as metals having densities greater than 5 g/cm^{-3} (Adekoya, Williams, & Ayejuyo, 2006) and with atomic weight of 40 g and above. In very small amounts, many of these metals are necessary to support life. However, in larger amounts, the heavy metals become toxic and may build up in biological systems and create a significant health hazard (WHO, 1998). Such as zinc and copper are vital micronutrients and become toxic at certain concentrations (Kuz'mina & Ushakova 2007). Climatic factors, such as rain and wind, are active to the air pollution (Breitner et al., 2011). The impacts of climate change on heavy metal concentrations have been discussed qualitatively for different marine ecosystems (Schiedek et al., 2007). The physicochemical changes alter the bioavailability of metals in the pore waters (Simpson & Batley, 2003) and the release rate (flux) of metals from the sediments (Simpson, Pryor, Mewburn, Batley, & Jolley, 2002). The speciation of the metal released from sediments is affected by the overlying water condition, like the pH, salinity, dissolved oxygen concentration, and amount of suspended solids (Simpson, Angel, & Jolley, 2004).

Temperature is an important factor that can affect the environmental chemistry of metals (uptake, removal and biotransformation rates). Toxicity may either (1) increase under high temperature, due to an increased uptake rate, or (2) decrease by stimulating detoxification and removal (Piazza et al., 2016). For example, the temperature dependence of arsenic release from flooded contaminated soils (Weber, Hofacker, Voegelin, & Kretzschmar, 2010) is one mechanism that may cause climate change to have an impact on the release of heavy metal contamination. Also, temperature can increase the inhibitory effect of metals on respiration in *Zebra mussels* (Rao & Khan, 2000).

Fritioff, Kautsky, and Greger (2005) have demonstrate that a general increase of metal uptake with increasing temperature seems likely, and consequent decrease in dissolved oxygen concentrations; thus, making the aerobic metabolism more stressful (Khan et al., 2006). The concentrations of heavy metals in bodies of water can be increased by rising temperatures as well as by acid rain, which cause their desorption from surfaces (Khan et al., 2006).

It is suggested that the concentrations of heavy metals in interstitial water of sediment decreases with decreasing temperature, because more ions are bounded to sediment colloid at high rather than low redox potentials (Förstner, 1981; Fritioff et al., 2005). This shortcoming in knowledge was also reported on metal speciation, bioavailability and toxicity (Tessier, Campbell, & Bisson, 1979). The toxicity of metals in aquatic ecosystems is enhanced by increased water temperature, through chemical interactions with dissolved organic carbon in low-pH waters (Murdoch et al., 2000).

Projected changes in air temperature and rainfall could affect river flows and, hence, the mobility and dilution of contaminants. Increased water temperatures will affect chemical reaction kinetics and, combined with deteriorations in quality of freshwater ecological status (Whitehead, Wilby, Battarbee, Kernan, & Wade, 2009). However, cool water contains more dissolved oxygen than does warm water. Thus, metal concentration in the interstitial water of the sediment may decrease with decreasing temperature, as more metals are bound to sediment colloids at high rather than low redox potentials (Fritioff et al., 2005).

The area surrounding the Keersop (Netherlands) has been contaminated with heavy metals by the atmospheric emissions of four zinc ore smelters. This heavy metal contamination, with Cd and Zn for example, has accumulated in the topsoil and leaches towards the surface water system, especially during periods with high groundwater levels and high discharge rates (Rozemeijer & Broers, 2007). Intensification of precipitation may result in larger proportions of quickflow, which could accelerate the leaching of heavy metals to surface waters. Visser et al. (2012) assessed the effects of future climate change on the hydrology and leaching of Cd and Zn in the Keersop catchment (Netherlands), and have concluded at the end of the twenty-first century, lower concentrations of Cd and Zn were projected as a result of lower discharge and lower water levels caused by higher evapotranspiration rates and an associated slowing down of groundwater flow. In addition, high hydrogen ion concentration during low pH values may replace heavy metals absorbed to the sediment particles (Chen & Lin, 2001). High temperature and low pH may thus lead to high concentrations of heavy metals.

It is difficult to assess the role of acid precipitation in relation to cadmium exposures. It has been demonstrated that the pH of soil is of great importance for the uptake of cadmium in crops such as rice or wheat (Linnman et al., 1973). Increasing salinity indeed has the potential to increase metal mobility, Two major mechanisms play a role in this process: (1) complexation capacity of salt derived anions with heavy metals, and (2) competition of salt derived cations with positively charged heavy metal species for sorption sites on the solid phase (Paalman, Van Der Weijden, & Loch, 1994). As to salinity, its decrease is known to reduce the amount of inorganic ions available for complexation of dissolved metals, thus increasing the availability of free-metal ions (Piazza et al., 2016).

The extend of mobilization depends on the type of heavy metal present, the total amount of heavy metal present and the type of salt causing the salinization. The observed inverse correlation between Cd release and clay and silt contents suggests a strong retention of Cd in the fine fractions hindering mobilization through salinization (Acosta, Jansen, Kalbitz, Faz, & Martínez-Martínez, 2011). In a temperate climate, like that of Sweden, the temperature and salinity of storm water vary with the season, salinity is very low most of the year, but salinities up to 5% have been measured during winter when NaCl is used for deicing roads. During summer, slightly elevated salinities up to 0.5% can be measured, probably due to the resuspension of sediment (Fritioff et al., 2005).

According to Wijngaard, van der Perk, van der Grift, de Nijs, and Bierkens (2017) the effects of climate change on the transport of heavy metals in catchments is multifaceted, complex, and equivocal, and requires a thorough understanding of the hydrological dynamics and pathways.

Implications of Climate Related Factors on Hydrocarbons

The petroleum industry generates a range of wastewater streams, including co-produced water and petroleum refinery effluent, which typically contain insoluble oil and grease, high levels of salinity, and a range of dissolved organics (Røe Utvik, 1999), including BTEX (i.e., benzene, toluene, ethylbenzene andxylene), phenols and polycyclic aromatic hydrocarbons have been identified as major contributors to the environmental toxicity of wastewater discharges (Younker & Walsh, 2015). Polycyclic aromatic hydrocarbons ''PAHs'' refers to compounds consisting of only carbon and hydrogen atoms. Chemically the PAHs are comprised of two or more fused aromatic rings bonded in various structural configurations (linear, cluster, or angular) (Kanaly & Harayama, 2000; Arey & Atkinson, 2003). The mode of PAHs formation can be either natural (natural petroleum, vegetative decay, rare minerals, plant synthesis, volcanic eruption…) or anthropogenic (petroleum spills, pesticides formulation, vehicles…) (Abdel-Shafy & Mansour, 2016). The environmental fate and transport of PAHs depend on the number of rings in the molecule, meteorological parameters, and the characteristics of the environment that they encounter (Lamon et al., 2009; Minai-Tehrani, Minoui, & Herfatmanesh, 2009).

Climate change has the potential of affecting the behaviour and distribution of organic pollutants. Temperature is one of the main variables that strongly influence the equilibrium and the rate of degradation reactions, partitioning, mass transfer processes and chemical composition of hydrocarbons in the environment (Rowland et al., 2000). As we know, temperature plays a significant role in controlling the nature and extent of microbial hydrocarbon metabolism (Nedwell, 1999). In counterpart, the increased volatilisation and solubility of some hydrocarbons at elevated temperature may enhance their toxicity, and may delay the onset of degradation (Coulon, Pelletier, Gourhant, & Delille, 2005). Treatments of organic pollutants such as petroleum derivatives and aromatic hydrocarbons are performed at moderate temperatures (20 to 37°C) in order to facilitate metabolic activity, diffusion, and mass transfer. A higher pollutant degradation rate is usually obtained at moderate than at lower temperatures (Leahy & Colwell, 1990; Zhou & Crawford, 1995). Temperature may increase the solubility of petroleum hydrocarbons two- to fivefold between 5 and 30°C depending upon the compound (Wolfe et al., 1998). Also, temperature plays an important role in controlling transport and sinks of some hydrocarbons at the global scale through the processes of cold condensation, global distillation, and latitudinal fractionation (Dachs et al., 2002).

Oil bioremediation in cold climates is frequently questioned (Rike, Schiewer, & Filler, 2008). However, ambient temperature close to 0°C do not completely stop oil biodegradation in sea water and even in sea ice (Delille, Bassères, Dessommes, & Rosiers, 1998) and there is no evidence that the microbial potential for degrading hydrocarbons is lower in cold regions than in warmer climates (Aislabie & Foght, 2008). On the other hand, temperature will also affect the rate at which PAHs are deposited from the atmosphere. For example, at higher temperature a greater fraction of the total PAHs will be in the vapor phase, but a low temperature will greatly increase the sorption of PAHs (Del Vento & Dachs, 2007). This phenomenon can be seen in the particle/vapor PAH distributions from winter to summer (Harrison, Smith, & Luhana, 1996).

The salt content is a main factor that affects the bioremediation process of petroleum hydrocarbons in soil (Ulrich et al., 2009). A decrease in salinity would improve the accessibility of soil organic matter to the soil microbial community (Muhammad, Müller, & Joergensen, 2008). In general, salt content is supposed to be an important factor in the bioremediation process of organic pollutants. The stress of high osmotic potential from the high salt content on living organisms resulted in the inhibition of pollutant decomposition (Hettiaratchi, Amatya, & Joshi, 2002). Organic compounds, such as PAHs, are less soluble in seawater than freshwater because of a phenomenon know as ''salting out'' resulting from the compression of seawater in the presence of high concentrations of salts (Eganhouse & Calder, 1976). Consequently, changes in environmental conditions may, in turn, modify or attenuate the influence of dispersants on the bioavailability and disposition of petroleum hydrocarbons. The negative impact of increasing salinity on hydrocarbons biodegradation is also observed in environments where halotolerant and/or slightly halophilic microorganisms tend to be dominant (Martins & Peixoto, 2012). Even in typical hypersaline environments a negative impact on hydrocarbon biodegradation induced by increasing salinity has been reported. Ward and Brock (1978) observed that the negative effect of salinity increase was pronounced on hexadecane biodegradation, more than on glutamate biodegradation.

Other climate related variables such as the frequency of storm surges, precipitations, the seasonality and the intensity of the atmospheric events, can influence the fate of PAHs in the environment. An increase of precipitations, for example, can cause an increase of PAHs deposition onto the soil. More frequent storm surges may enhance the mobilisation of chemicals stored in the soil compartment, which can be transported by land runoffs, making them available to the aquatic organisms (Dalla Valle et al., 2007). Precipitation patterns, wind, snow and ice cover and solar radiation can influence the environmental distribution of PAHs via changes in the rate and/or direction of environmental processes such as advection, dispersion, dry and wet deposition, reaction, and surface runoff (Macdonald et al., 2005; Lamon et al., 2009; Kallenborn et al., 2012).

Wet deposition is defined as the scrubbing of contaminants sorbed onto particulates out of the atmosphere by precipitation, as well as the dissolution of vapor phase contaminants into precipitation (Dickhut & Gustafson, 1995). The amounts of PAHs removed from the atmosphere by wet deposition vary depending on the phase. Also, the atmospheric concentration of the strong acids H_2SO_4 and HNO_3 has increased over large regions of the earth due to man's increasing combustion of fossil fuels as a source of energy. Alteration of atmospheric composition has caused a concurrent alteration in the composition of precipitation. This increasing acidity of precipitation can cause severe effects on the terrestrial ecosystems (Galloway & Cowling, 1978). Long-range atmospheric transport between different regions of the world could occur as a sequence of successive volatilization and condensation processes, a process called the "grasshopper effect". These small steps are strongly influenced by diurnal and seasonal variability of environmental conditions, due to its influence on air-surface partitioning (Dachs et al., 2002). The atmosphere is considered to be the major route for rapid, global distribution of PAHs (Berg, Kallenborn & Manø, 2004; Ma, Hung, & Blanchard, 2004), However, as environmental reservoirs begin to exert a greater influence on distribution patterns and processes as secondary sources for the release and, thus, remobilisation of PAHs in the environment (Kallenborn et al., 2012).

Therefore, climate change processes can influence every step along the transport of PAHs and redistribution pathways because reactivity, adsorption processes as well as accumulation are temperature dependent processes (Macdonald, Mackay, Li, & Hickie, 2003; Macdonald et al., 2005) and thus directly effected by climate change.

CONCLUSION

Over the course of this century and beyond, climate change has become a reality we must confrot in the future, with sea level rise on coastal areas will include increased flooding and inundation and increased erosion. It is possible that changes in biotic and abiotic parameters can affect the potential toxic effects. In addition, transport of chemical polluant will be also modified. So there will be an influence on human health and an ecological risk. Adaptation to climate change and limit its impact on toxic polluant, may be done through a wide range of actions structural institutional, or social.

REFERENCES

Abdel-Shafy, H. I., & Mansour, M. S. M. (2016). A review on polycyclic aromatic hydrocarbons: Source, environmental impact, effect on human health and remediation. *Egyptian Journal of Petroleum*, *25*(1), 107–123. doi:10.1016/j.ejpe.2015.03.011

Acosta, J. A., Jansen, B., Kalbitz, K., Faz, A., & Martínez-Martínez, S. (2011). Salinity increases mobility of heavy metals in soils. *Chemosphere*, *85*(8), 1318–1324. doi:10.1016/j.chemosphere.2011.07.046 PMID:21862104

Adekoya, J. A., Williams, A. B., & Ayejuyo, O. O. (2006). Distribution of heavy metals in sediments of Igbede, Ojo and Ojora rivers of Lagos, Nigeria. *The Environmentalist*, *26*(4), 277–280. doi:10.100710669-006-9077-1

Ahmadi, M., Records, R., & Arabi, M. (2014). Impact of climate change on diffuse pollutant fluxes at the watershed scale. *Hydrological Processes*, *28*(4), 1962–1972. doi:10.1002/hyp.9723

Aislabie, J., & Foght, J. (2008). Hydrocarbon-degrading bacteria in contaminated soils. In D. F. Filler, I. Snape, & D. L. Barns (Eds.), *Bioremediation of petroleum hydrocarbons in cold regions Cambridge University Press* (pp. 69–83). Cambridge. doi:10.1017/CBO9780511535956.006

Aly, O. M., & El-Dib, M. A. (1971). Studies on the persistence of some carbamate insecticides in the aquatic environment—I. Hydrolysis of sevin, baygon, pyrolan and dimetilan in waters. *Water Research*, *5*(12), 1191–1205. doi:10.1016/0043-1354(71)90084-4

Arey, J., & Atkinson, R. (2003). Photochemical reactions of PAHs in the atmosphere. In *PAHs: An Ecotoxicological Perspective* (pp. 47–63). New York: John Wiley and Sons Ltd. doi:10.1002/0470867132.ch4

Berg, T., Kallenborn, R., & Manø, S. (2004). Temporal trends in atmospheric heavy metal and organochlorine concentrations at Zeppelin, Svalbard. *Arctic, Antarctic, and Alpine Research*, *36*(3), 284–291. doi:10.1657/1523-0430(2004)036[0284:TTIAHM]2.0.CO;2

Blaustein, A. R., Romansic, J. M., Kiesecker, J. M., & Hatch, A. C. (2003). Ultraviolet radiation, toxic chemicals and amphibian population declines. *Diversity & Distributions*, *9*(2), 123–140. doi:10.1046/j.1472-4642.2003.00015.x

Bloomfield, J. P., Williams, R. J., Gooddy, D. C., Cape, J. N., & Guha, P. (2006). Impacts of climate change on the fate and behaviour of pesticides in surface and groundwater—a UK perspective. *The Science of the Total Environment*, *369*(1-3), 163–177. doi:10.1016/j.scitotenv.2006.05.019 PMID:16914182

Boina, D. R., Onagbola, E. O., Salyani, M., & Stelinski, L. L. (2009). Influence of posttreatment temperature on the toxicity of insecticides against Diaphorina citri (Hemiptera: Psyllidae). *Journal of Economic Entomology*, *102*(2), 685–691. doi:10.1603/029.102.0229 PMID:19449650

Breitner, S., Liu, L., Cyrys, J., Bruske, I., Franck, U., Schlink, U., ... Peters, A. (2011). Sub-micrometer particulate air pollution and cardiovascular mortality in Beijing, China. *The Science of the Total Environment*, *409*(24), 5196–5204. doi:10.1016/j.scitotenv.2011.08.023 PMID:21937089

Brierley, A. S., & Kingsford, M. J. (2009). Impacts of climate change on marine organisms and ecosystems. *Current Biology*, *19*(14), R602–R614. doi:10.1016/j.cub.2009.05.046 PMID:19640499

Campbell, J., Rustad, L., Boyer, E., Christopher, S., Driscoll, C., Fernandez, I., ... Ollinger, V. (2009). Consequences of climate change for biogeochemical cycling in forests of northeastern North AmericaThis article is one of a selection of papers from NE Forests 2100: A Synthesis of Climate Change Impacts on Forests of the Northeastern US and Eastern Canada. *Canadian Journal of Forest Research*, *39*(2), 264–284. doi:10.1139/X08-104

Carpenter, S. R., Cole, J. J., Pace, M. L., Batt, R., Brock, W. A., Cline, T., ... Weidel, B. (2011). Early warnings of regime shifts: A whole-ecosystem experiment. *Science*, *332*(6033), 1079–1082. doi:10.1126cience.1203672 PMID:21527677

Chen, S.-Y., & Lin, J.-G. (2001). Bioleaching of heavy metals from sediment: Significance of pH. *Chemosphere*, *44*(5), 1093–1102. doi:10.1016/S0045-6535(00)00334-9 PMID:11513396

Ciardini, V., Contessa, G. M., Falsaperla, R., Gomez-Amo, J. L., Meloni, D., Monteleone, F., ... di Sarra, A. (2016). Global and Mediterranean climate change: A short summary. *Annali dell'Istituto Superiore di Sanita*, *52*, 325–337. PMID:27698291

Coulon, F., Pelletier, E., Gourhant, L., & Delille, D. (2005). Effects of nutrient and temperature on degradation of petroleum hydrocarbons in contaminated sub-Antarctic soil. *Chemosphere*, *58*(10), 1439–1448. doi:10.1016/j.chemosphere.2004.10.007 PMID:15686763

Dachs, J., Lohmann, R., Ockenden, W. A., Mejanelle, L., Eisenreich, S. J., & Jones, K. C. (2002). Oceanic biogeochemical controls on global dynamics of persistent organic pollutants. *Environmental Science & Technology*, *36*(20), 4229–4237. doi:10.1021/es025724k PMID:12387392

Dalla Valle, M., Codato, E., & Marcomini, A. (2007). Climate change influence on POPs distribution and fate: A case study. *Chemosphere*, *67*(7), 1287–1295. doi:10.1016/j.chemosphere.2006.12.028 PMID:17258268

Del Vento, S., & Dachs, J. (2007). Atmospheric occurrence and deposition of polycyclic aromatic hydrocarbons in the northeast tropical and subtropical Atlantic Ocean. *Environmental Science & Technology*, *41*(16), 5608–5613. doi:10.1021/es0707660 PMID:17874762

Delille, D., Bassères, A., Dessommes, A., & Rosiers, C. (1998). Influence of daylight on potential biodegradation of diesel and crude oil in Antarctic seawater. *Marine Environmental Research*, *45*(3), 249–258. doi:10.1016/S0141-1136(97)00129-3

DeLorenzo, M. E., Wallace, S. C., Danese, L. E., & Baird, T. D. (2009). Temperature and salinity effects on the toxicity of common pesticides to the grass shrimp, *Palaemonetes pugio*. *Journal of Environmental Science and Health. Part. B, Pesticides, Food Contaminants, and Agricultural Wastes*, *44*(5), 455–460. doi:10.1080/03601230902935121 PMID:20183050

Dickhut, R. M., & Gustafson, K. E. (1995). Atmospheric washout of polycyclic aromatic hydrocarbons in the southern chesapeake bay region. *Environmental Science & Technology*, *29*(6), 1518–1525. doi:10.1021/es00006a013 PMID:22276872

Donald, D. B., Cessna, A. J., Sverko, E., & Glozier, N. E. (2007). Pesticides in surface drinking-water supplies of the Northern Great Plains. *Environmental Health Perspectives*, *115*(8), 1183–1191. doi:10.1289/ehp.9435 PMID:17687445

Doull, J. (1972). The effect of physical environmental factors on drug response. In W. J. Hayes (Ed.), *Essays in Toxicology* (pp. 37–63). Elsevier.

Durack, P. J., Wijffels, S. E., & Matear, R. J. (2012). Ocean salinities reveal strong global water cycle intensification during 1950 to 2000. *Science*, *336*(6080), 455–458. doi:10.1126cience.1212222 PMID:22539717

Eganhouse, R. P., & Calder, J. A. (1976). The solubility of medium molecular weight aromatic hydrocarbons and the effects of hydrocarbon co-solutes and salinity. *Geochimica et Cosmochimica Acta*, *40*(5), 555–561. doi:10.1016/0016-7037(76)90223-4

EU. (2009). *Common Implementation Strategy for the Water Framework Directive (2000/60/EC) Guidance document No. 24: River Basin Management in a Changing Climate (Technical Report — 2009–040)*. EU.

Field, C. B., Barros, V. R., Mach, K. J., Mastrandrea, M. D., van Aalst, M., Adger, W. N., & … . (2014). Technical summary. In C. B. Field, V. R. Barros, D. J. Dokken, K. J. Mach, M. D. Mastrandrea, T. E. Bilir, & ... (Eds.), *Climate Change 2014: Impacts, Adaptation and Vulnerability. Part A: Global and Sectoral Aspects. Contribution of Working Group II to the Fifth Assessment Report of the Intergovernmental Panel on Climate Change* (pp. 35–94). Cambridge, UK: Cambridge University Press.

Fontaine, J. J., Decker, K. L., Skagen, S. K., & Van Riper, C. (2009). Spatial and temporal variation in climate change: A bird's eye view. *Climatic Change*, *97*(1-2), 305–311. doi:10.100710584-009-9644-9

Förstner, U. (1981). Metal transfer between solid and aqueous phases. In *Metal Pollution in the Aquatic Environment* (pp. 197–270). Berlin: Springer Berlin Heidelberg. doi:10.1007/978-3-642-69385-4_5

Fritioff, Å., Kautsky, L., & Greger, M. (2005). Influence of temperature and salinity on heavy metal uptake by submersed plants. *Environmental Pollution*, *133*(2), 265–274. doi:10.1016/j.envpol.2004.05.036 PMID:15519457

Galloway, J. N., & Cowling, E. B. (1978). The effects of precipitation on aquatic and terrestrial ecosystems: A proposed precipitation chemistry network. *Journal of the Air Pollution Control Association, 28*(3), 229–235. doi:10.1080/00022470.1978.10470594

Gatto, M. P., Cabella, R., & Gherardi, M. (2016). Climate change: The potential impact on occupational exposure to pesticides. *Annali dell'Istituto Superiore di Sanita, 52*, 374–385. PMID:27698296

Glunt, K. D., Blanford, J. I., & Paaijmans, K. P. (2013). Chemicals, climate, and control: Increasing the effectiveness of malaria vector control tools by considering relevant temperatures. *PLoS Pathogens, 9*(10), e1003602. doi:10.1371/journal.ppat.1003602 PMID:24098111

Gouin, T., Armitage, J. M., Cousins, I. T., Muir, D. C., Ng, C. A., Reid, L., & Tao, S. (2013). Influence of global climate change on chemical fate and bioaccumulation: The role of multimedia models. *Environmental Toxicology and Chemistry, 32*(1), 20–31. doi:10.1002/etc.2044 PMID:23136071

Grabemann, H. J., Grabemann, I., Herbers, D., & Müller, A. (2001). Effects of a specific climate scenario on the hydrography and transport of conservative substances in the Weser estuary, Germany: A case study. *Climate Research, 18*, 77–87. doi:10.3354/cr018077

Guinotte, J. M., & Fabry, V. J. (2008). Ocean acidification and its potential effects on marine ecosystems. *Annals of the New York Academy of Sciences, 1134*(1), 320–342. doi:10.1196/annals.1439.013 PMID:18566099

Hansen, J., Sato, M., & Ruedy, R. (2012). Perception of climate change. *Proceedings of the National Academy of Sciences of the United States of America, 109*(37), E2415–E2423. doi:10.1073/pnas.1205276109 PMID:22869707

Harrison, R. M., Smith, D. J. T., & Luhana, L. (1996). Source apportionment of atmospheric polycyclic aromatic hydrocarbons collected from an urban location in Birmingham, U.K. *Environmental Science & Technology, 30*(3), 825–832. doi:10.1021/es950252d

Haurie, L., Sceia, A. & Thénié, J. (2009). *Inland transport and climate change.* A Literature Review UNECE.

Hayes, T. B., Case, P., Chui, S., Chung, D., Haeffele, C., Haston, K., ... Tsui, M. (2006). Pesticide mixtures, endocrine disruption, and amphibian declines: Are we underestimating the impact? *Environmental Health Perspectives, 114*(Suppl 1), 40–50. doi:10.1289/ehp.8051 PMID:16818245

Heck, W. W., & Dunning, J. A. (1967). The effects of ozone on Tobacco and Pinto Bean as conditioned by several ecological factors. *Journal of the Air Pollution Control Association, 17*(2), 112–114. doi:10.1080/00022470.1967.10468950

Hettiaratchi, J., Amatya, P., & Joshi, R. (2002). Biotreatment of flare pit waste. *Journal of Canadian Petroleum Technology, 41*, 30–36.

Heugens, E. H. W., Hendriks, A. J., Dekker, T., Straalen, N. M. V., & Admiraal, W. (2001). A review of the effects of multiple stressors on aquatic organisms and analysis of uncertainty factors for use in risk assessment. *Critical Reviews in Toxicology, 31*(3), 247–284. doi:10.1080/20014091111695 PMID:11405441

Holmstrup, M., Bindesbøl, A.-M., Oostingh, G. J., Duschl, A., Scheil, V., Köhler, H.-R., ... Spurgeon, D. J. (2010). Interactions between effects of environmental chemicals and natural stressors: A review. *The Science of the Total Environment*, *408*(18), 3746–3762. doi:10.1016/j.scitotenv.2009.10.067 PMID:19922980

Honisch, B., Ridgwell, A., Schmidt, D. N., Thomas, E., Gibbs, S. J., Sluijs, A., ... Williams, B. (2012). The geological record of ocean acidification. *Science*, *335*(6072), 1058–1063. doi:10.1126cience.1208277 PMID:22383840

Hooper, M. J., Ankley, G. T., Cristol, D. A., Maryoung, L. A., Noyes, P. D., & Pinkerton, K. E. (2013). Interactions between chemical and climate stressors: A role for mechanistic toxicology in assessing climate change risks. *Environmental Toxicology and Chemistry*, *32*(1), 32–48. doi:10.1002/etc.2043 PMID:23136056

Huntingford, C., Jones, P. D., Livina, V. N., Lenton, T. M., & Cox, P. M. (2013). No increase in global temperature variability despite changing regional patterns. *Nature*, *500*(7462), 327–330. doi:10.1038/nature12310 PMID:23883935

Huovinen, P. (2001). Resistance to trimethoprim-sulfamethoxazole. *Nephrology, Dialysis, Transplantation*, *32*, 1608–1614. PMID:11340533

IPCC. (2007). *Climate Change 2007: Synthesis report*. Cambridge, UK: Cambridge University Press.

IPCC. (2014). *Climate Change 2014: Impacts, Adaptation, and Vulnerability. Part A: Global and Sectoral Aspects. Contributions of Working Group II to the Fifth Assessment Report of the Intergovernmental Panel on Climate Change*. Cambridge, UK: Cambridge University Press.

Jiricka, A., Formayer, H., Schmidt, A., Völler, S., Leitner, M., Fischer, T. B., & Wachter, T. F. (2016). Consideration of climate change impacts and adaptation in EIA practice — Perspectives of actors in Austria and Germany. *Environmental Impact Assessment Review*, *57*, 78–88. doi:10.1016/j.eiar.2015.11.010

Kallenborn, R., Halsall, C., Dellong, M., & Carlsson, P. (2012). The influence of climate change on the global distribution and fate processes of anthropogenic persistent organic pollutants. *Journal of environmental monitoring. JEM*, *14*, 2854–2869. PMID:23014859

Kanaly, R. A., & Harayama, S. (2000). Biodegradation of high-molecular-weight polycyclic aromatic hydrocarbons by bacteria. *Journal of Bacteriology*, *182*(8), 2059–2067. doi:10.1128/JB.182.8.2059-2067.2000 PMID:10735846

Karvonen, A., Rintamäki, P., Jokela, J., & Valtonen, E. T. (2010). Increasing water temperature and disease risks in aquatic systems: Climate change increases the risk of some, but not all, diseases. *International Journal for Parasitology*, *40*(13), 1483–1488. doi:10.1016/j.ijpara.2010.04.015 PMID:20580904

Kattwinkel, M., Kuhne, J.V., Foit, K. & Liess, M. (2011) Climate change, agricultural insecticide exposure, and risk for freshwater communities. *Ecological Applications*, *21*, 2068-2081.

Kaushal, S. S., Likens, G. E., Jaworski, N. A., Pace, M. L., Sides, A. M., Seekell, D., ... Wingate, R. L. (2010). Rising stream and river temperatures in the United States. *Frontiers in Ecology and the Environment*, *8*(9), 461–466. doi:10.1890/090037

Kennish, M. J. (2002). Environmental threats and environmental future of estuaries. *Environmental Conservation*, *29*(01), 78–107. doi:10.1017/S0376892902000061

Khan, M. A., Ahmed, S. A., Catalin, B., Khodadoust, A., Ajayi, O., & Vaughn, M. (2006). Effect of temperature on heavy metal toxicity to juvenile crayfish, *Orconectes immunis* (Hagen). *Environmental Toxicology*, *21*(5), 513–520. doi:10.1002/tox.20213 PMID:16944513

Kimberly, D. A., & Salice, C. J. (2012). Understanding interactive effects of climate change and toxicants: Importance of evolutionary processes. *Integrated Environmental Assessment and Management*, *8*(2), 385–386. doi:10.1002/ieam.1286 PMID:22431356

Klein, R. (2011). Adaptation to climate change: more than technology. In Climate: Global Change and Local Adaptation (pp. 157-168). NATO.

Klerks, P. L., Xie, L., & Levinton, J. S. (2011). Quantitative genetics approaches to study evolutionary processes in ecotoxicology; a perspective from research on the evolution of resistance. *Ecotoxicology (London, England)*, *20*(3), 513–523. doi:10.100710646-011-0640-2 PMID:21516382

Knowles, N. & Cayan, D.R. (2002). Potential effects of global warming on the Sacramento/San Joaquin watershed and the San Francisco estuary. *Geophysical Research Letters, 29*, 38-31-38-34.

Koleva, N. G., Schneider, U., & Tol, T. (2009). *The impact of weather variability and climate change on pesticide applications in the US - An empirical investigation: Research unit Sustainability and Global Change*. Hamburg University.

Kuz'mina, V. V., & Ushakova, N. V. (2007). Effects of temperature, pH, and heavy metals (copper, zinc) upon proteinase activities in digestive tract mucosa of typical and facultative piscivorous fish. *Journal of Ichthyology*, *47*(6), 473–480. doi:10.1134/S0032945207060070

Lamon, L., Dalla Valle, M., Critto, A., & Marcomini, A. (2009). Introducing an integrated climate change perspective in POPs modelling, monitoring and regulation. *Environmental Pollution*, *157*(7), 1971–1980. doi:10.1016/j.envpol.2009.02.016 PMID:19272683

Leahy, J. G., & Colwell, R. R. (1990). Microbial degradation of hydrocarbons in the environment. *Microbiological Reviews*, *54*, 305–315. PMID:2215423

Linnman, L., Andersson, A., Nilsson, K. O., Lind, B., Kjellstrom, T., & Friberg, L. (1973). Cadmium uptake by wheat from sewage sludge used as a plant nutrient source. A comparative study using flameless atomic absorption and neutron activation analysis. *Archives of Environmental Health*, *27*(1), 45–47. doi:10.1080/00039896.1973.10666306 PMID:4709111

Liu, C., Hofstra, N., & Franz, E. (2013). Impacts of climate change on the microbial safety of pre-harvest leafy green vegetables as indicated by *Escherichia coli* O157 and *Salmonella spp. International Journal of Food Microbiology*, *163*(2-3), 119–128. doi:10.1016/j.ijfoodmicro.2013.02.026 PMID:23558195

Ma, J., Hung, H., & Blanchard, P. (2004). How do climate fluctuations affect persistent organic pollutant distribution in North America? Evidence from a decade of air monitoring. *Environmental Science & Technology*, *38*(9), 2538–2543. doi:10.1021/es0349610 PMID:15180048

Mabey, N., Gulledge, J., Finel, B., & Silverthorne, K. (2011). *Degrees of risk: Defining a risk management framework for climate security*. London: Third Generation Environmentalism.

Macdonald, R. W., Harner, T., & Fyfe, J. (2005). Recent climate change in the Arctic and its impact on contaminant pathways and interpretation of temporal trend data. *The Science of the Total Environment*, *342*(1-3), 5–86. doi:10.1016/j.scitotenv.2004.12.059 PMID:15866268

Macdonald, R. W., Mackay, D., Li, Y. F., & Hickie, B. (2003). How will global climate change affect risks from long-range transport of persistent organic pollutants? *Human and Ecological Risk Assessment*, *9*(3), 643–660. doi:10.1080/713609959

Macneale, K. H., Kiffney, P. M., & Scholz, N. L. (2010). Pesticides, aquatic food webs, and the conservation of Pacific salmon. *Frontiers in Ecology and the Environment*, *8*(9), 475–482. doi:10.1890/090142

Martins, L. F., & Peixoto, R. S. (2012). Biodegradation of petroleum hydrocarbons in hypersaline environments. *Brazilian Journal of Microbiology*, *43*(3), 865–872. doi:10.1590/S1517-83822012000300003 PMID:24031900

Mclusky, D. S., Bryant, V., & Campbell, R. (1986). The effect of temperature and salinity on the toxicity of heavy metals to marine and estuarine invertebrates. *Oceanography and Marine Biology - an Annual Review*, *24*, 481–520.

McMichael, A. J., Campbell Lendrum, D. H., Corvalán, C. F., Ebi, K. L., Githeko, A., Scheraga, J. D., & Woodward, A. (2003). *Climate change and human health—Risks and responses* (p. 322). Geneva: World Health Organization.

Mehta, K. (2017). Impact of temperature on contaminants toxicity in fish fauna: A review. *Indian Journal of Science and Technology*, *10*, 1–6.

Merilä, J., & Hendry, A. P. (2014). Climate change, adaptation, and phenotypic plasticity: The problem and the evidence. *Evolutionary Applications*, *7*(1), 1–14. doi:10.1111/eva.12137 PMID:24454544

Messing, P. G., Farenhorst, A., Waite, D. T., McQueen, D. A. R., Sproull, J. F., Humphries, D. A., & Thompson, L. L. (2011). Predicting wetland contamination from atmospheric deposition measurements of pesticides in the Canadian Prairie Pothole region. *Atmospheric Environment*, *45*(39), 7227–7234. doi:10.1016/j.atmosenv.2011.08.074

Minai-Tehrani, D., Minoui, S., & Herfatmanesh, A. (2009). Effect of salinity on biodegradation of polycyclic aromatic hydrocarbons (PAHs) of heavy crude oil in soil. *Bulletin of Environmental Contamination and Toxicology*, *82*(2), 179–184. doi:10.100700128-008-9548-9 PMID:18777147

Mo, H., Yoo, D., Bae, Y. J., & Cho, K. (2013). Effects of water temperature on development and heavy metal toxicity change in two midge species of *Chironomus riparius* and *C. yoshimatsui* in an era of rapid climate change. *Entomological Research*, *43*(2), 123–129. doi:10.1111/1748-5967.12014

Moe, S.J., De Schamphelaere, K., Clements, W.H., Sorensen, M.T., Van den Brink, P.J. & Liess, M. (2013). Combined and interactive effects of global climate change and toxicants on populations and communities. *Environmental Toxicology and Chemistry / Setac, 32*, 49-61.

Moline, M. A., Karnovsky, N. J., Brown, Z., Divoky, G. J., Frazer, T. K., Jacoby, C. A., ... Fraser, W. R. (2008). High latitude changes in ice dynamics and their impact on polar marine ecosystems. *Annals of the New York Academy of Sciences, 1134*(1), 267–319. doi:10.1196/annals.1439.010 PMID:18566098

Muhammad, S., Müller, T., & Joergensen, R. G. (2008). Relationships between soil biological and other soil properties in saline and alkaline arable soils from the Pakistani Punjab. *Journal of Arid Environments, 72*(4), 448–457. doi:10.1016/j.jaridenv.2007.06.016

Müller, R., Seeland, A., Jagodzinski, L. S., Diogo, J. B., Nowak, C., & Oehlmann, J. (2012). Simulated climate change conditions unveil the toxic potential of the fungicide pyrimethanil on the midge *Chironomus riparius*: A multigeneration experiment. *Ecology and Evolution, 2*(1), 196–210. doi:10.1002/ece3.71 PMID:22408736

Murdoch, P. S., Baron, J. S., & Miller, T. L. (2000). Potential effects of climate change on surface-water quality in north america. *JAWRA Journal of the American Water Resources Association, 36*(2), 347–366. doi:10.1111/j.1752-1688.2000.tb04273.x

Muturi, E. J., Lampman, R., Costanzo, K., & Alto, B. W. (2011). Effect of temperature and insecticide stress on life-history traits of *Culex restuans* and *Aedes albopictus* (Diptera: Culicidae). *Journal of Medical Entomology, 48*(2), 243–250. doi:10.1603/ME10017 PMID:21485359

Nedwell, D. B. (1999). Effect of low temperature on microbial growth: Lowered affinity for substrates limits growth at low temperature. *FEMS Microbiology Ecology, 30*(2), 101–111. doi:10.1111/j.1574-6941.1999.tb00639.x PMID:10508935

Nizzetto, L., Lohmann, R., Gioia, R., Dachs, J., & Jones, K. C. (2010). Atlantic ocean surface waters buffer declining atmospheric concentrations of persistent organic pollutants. *Environmental Science & Technology, 44*(18), 6978–6984. doi:10.1021/es101293v PMID:20726587

Noyes, P. D., McElwee, M. K., Miller, H. D., Clark, B. W., Van Tiem, L. A., Walcott, K. C., ... Levin, E. D. (2009). The toxicology of climate change: Environmental contaminants in a warming world. *Environment International, 35*(6), 971–986. doi:10.1016/j.envint.2009.02.006 PMID:19375165

O'Driscoll, K., Mayer, B., Su, J., & Mathis, M. (2014). The effects of global climate change on the cycling and processes of persistent organic pollutants (POPs) in the North Sea. *Ocean Science, 10*, 397–409. doi:10.5194/os-10-397-2014

Oerke, E. C. (2005). Crop losses to pests. *The Journal of Agricultural Science, 144*(01), 31–43. doi:10.1017/S0021859605005708

Olesen, J. E., Trnka, M., Kersebaum, K. C., Skjelvåg, A. O., Seguin, B., Peltonen-Sainio, P., ... Micale, F. (2011). Impacts and adaptation of European crop production systems to climate change. *European Journal of Agronomy, 34*(2), 96–112. doi:10.1016/j.eja.2010.11.003

Oreskes, N. (2004). The scientific consensus on climate change. *Science, 306*(5702), 1686–1686. doi:10.1126cience.1103618 PMID:15576594

Paalman, M. A. A., Van Der Weijden, C. H., & Loch, J. P. G. (1994). Sorption of cadmium on suspended matter under estuarine conditions; competition and complexation with major sea-water ions. *Water, Air, and Soil Pollution*, *73*(1), 49–60. doi:10.1007/BF00477975

Palikhe, B.R. (2007). Relationship between pesticide use and climate change for crops. Academic Press.

Park, J.-H., Duan, L., Kim, B., Mitchell, M. J., & Shibata, H. (2010). Potential effects of climate change and variability on watershed biogeochemical processes and water quality in Northeast Asia. *Environment International*, *36*(2), 212–225. doi:10.1016/j.envint.2009.10.008 PMID:19926135

Patra, R. W., Chapman, J. C., Lim, R. P., Gehrke, P. C., & Sunderam, R. M. (2015). Interactions between water temperature and contaminant toxicity to freshwater fish. *Environmental Toxicology and Chemistry*, *34*(8), 1809–1817. doi:10.1002/etc.2990 PMID:26033197

Peterson, T. C., McGuirk, M., Houston, T. G., Horvitz, A. H., & Wehner, M. F. (2008). Climate variability and change with implications for transportation. The National Research Council.

Piazza, V., Gambardella, C., Canepa, S., Costa, E., Faimali, M., & Garaventa, F. (2016). Temperature and salinity effects on cadmium toxicity on lethal and sublethal responses of *Amphibalanus amphitrite nauplii*. *Ecotoxicology and Environmental Safety*, *123*, 8–17. doi:10.1016/j.ecoenv.2015.09.023 PMID:26410667

Pielke, R. Jr, Prins, G., Rayner, S., & Sarewitz, D. (2007). Lifting the taboo on adaptation. *Nature*, *445*(7128), 597–598. doi:10.1038/445597a PMID:17287795

Pierce, D. W., Gleckler, P. J., Barnett, T. P., Santer, B. D., & Durack, P. J. (2012). The fingerprint of human-induced changes in the ocean's salinity and temperature fields. *Geophysical Research Letters*, 39.

Portner, H. O., & Knust, R. (2007). Climate change affects marine fishes through the oxygen limitation of thermal tolerance. *Science*, *315*(5808), 95–97. doi:10.1126cience.1135471 PMID:17204649

Rao, D. G. V. P., & Khan, M. A. Q. (2000). *Zebra mussels*: Enhancement of copper toxicity by high temperature and its relationship with respiration and metabolism. *Water Environment Research*, *72*(2), 175–178. doi:10.2175/106143000X137257

Richardson, K., Steffen, W., Schellnhuber, H. J., Alcamo, J., Barker, T., Kammen, D. M., . . . Waever, O. (2009) Synthesis report. Climate change: Global risks, challenges, and decisions. *Proceedings, United Nations Framework Convention on Climate Change*.

Rike, A. G., Schiewer, S., & Filler, D. M. (2008). Temperature effects on biodegradation of petroleum contaminants in cold soils. In D. L. Barnes, D. M. Filler, & I. Snape (Eds.), *Bioremediation of Petroleum Hydrocarbons in Cold Regions* (pp. 84–108). Cambridge, UK: Cambridge University Press. doi:10.1017/CBO9780511535956.007

Robinson, S. A. (2009). Introduction: Climate change biology at the ends of the Earth-International Polar year special issue. *Global Change Biology*, *15*(7), 1615–1617. doi:10.1111/j.1365-2486.2009.01977.x

Røe Utvik, T. I. (1999). Chemical characterisation of produced water from four offshore oil production platforms in the North Sea. *Chemosphere*, *39*(15), 2593–2606. doi:10.1016/S0045-6535(99)00171-X

Romero-Lankao, P., Smith, J. B., Davidson, D. J., Diffenbaugh, N. S., Kinney, P. L., Kirshen, R., … Villers Ruiz, L. (2014). North America. In Climate Change 2014: Impacts, Adaptation, and Vulnerability. Part B: Regional Aspects Contribution of Working Group II to the Fifth Assessment Report of the Intergovernmental Panel on Climate Change. Cambridge, UK: Cambridge University Press.

Rosenzweig, C., Iglesias, A., Yang, X. B. R., Epstein, P., & Chivian, E. (2001). Climate change and extreme weather events; implications for food production, plant diseases, and pests. *Global Change and Human Health*, *2*(2), 90–104. doi:10.1023/A:1015086831467

Rowland, A. P., Lindley, D. K., Hall, G. H., Rossall, M. J., Wilson, D. R., Benham, D. G., ... Daniels, R. E. (2000). Effects of beach sand properties, temperature and rainfall on the degradation rates of oil in buried oil/beach sand mixtures. *Environmental Pollution*, *109*(1), 109–118. doi:10.1016/S0269-7491(99)00224-9 PMID:15092918

Rozemeijer, J. C., & Broers, H. P. (2007). The groundwater contribution to surface water contamination in a region with intensive agricultural land use (Noord-Brabant, The Netherlands). *Environmental Pollution*, *148*(3), 695–706. doi:10.1016/j.envpol.2007.01.028 PMID:17418466

Schiedek, D., Sundelin, B., Readman, J. W., & Macdonald, R. W. (2007). Interactions between climate change and contaminants. *Marine Pollution Bulletin*, *54*(12), 1845–1856. doi:10.1016/j.marpolbul.2007.09.020 PMID:17963794

Schneider, S. H., Semenov, S., Patwardhan, A., Burton, I., Magadza, C. H., Oppenheimer, M., ... Yamin, F. (2007). Assessing key vulnerabilities and the risk from climate change. In *Climate Change 2007: Adaptation and Vulnerability Contribution of Working Group II to the Fourth Assessment Report of the Intergovernmental Panel on Climate Change* (pp. 779–810). Cambridge, UK: Cambridge University.

Scholz, N. L., Fleishman, E., Brown, L., Werner, I., Johnson, M. L., Brooks, M. L., ... Schlenk, D. (2012). A perspective on modern pesticides, pelagic fish declines, and unknown ecological resilience in highly managed ecosystems. *Bioscience*, *62*(4), 428–434. doi:10.1525/bio.2012.62.4.13

Seguin, F., Leboulanger, C., Rimet, F., Druart, J. C., & Berard, A. (2001). Effects of atrazine and nicosulfuron on phytoplankton in systems of increasing complexity. *Archives of Environmental Contamination and Toxicology*, *40*, 198–208. doi:10.1007002440010164 PMID:11243322

Senhorst, H.A. & Zwolsman, J.J. (2005) Climate change and effects on water quality: a first impression. *Water Science and Technology, 51*, 53-59.

Shui-Ming, Z., Duo-Sen, L., Zong-Sheng, W., & Xing-Fa, M. (1993). A kinetic model describing the effect of temperature on the loss rate of pesticides in soil. *Ecological Modelling*, *70*(1-2), 115–125. doi:10.1016/0304-3800(93)90075-4

Simpson, S. L., Angel, B. M., & Jolley, D. F. (2004). Metal equilibration in laboratory-contaminated (spiked) sediments used for the development of whole-sediment toxicity tests. *Chemosphere*, *54*(5), 597–609. doi:10.1016/j.chemosphere.2003.08.007 PMID:14599505

Simpson, S. L., & Batley, G. E. (2003). Disturbances to metal partitioning during toxicity testing of iron(II)-rich estuarine pore waters and whole sediments. *Environmental Toxicology and Chemistry*, *22*(2), 424–432. doi:10.1002/etc.5620220225 PMID:12558176

Simpson, S. L., Pryor, I. D., Mewburn, B. R., Batley, G. E., & Jolley, D. (2002). Considerations for capping metal-contaminated sediments in dynamic estuarine environments. *Environmental Science & Technology*, *36*(17), 3772–3778. doi:10.1021/es025632v PMID:12322750

Smith, L. M. (2003). *Playas of the Great Plains*. Austin, TX: University of Texas Press.

Stahl, R. G. Jr, Hooper, M. J., Balbus, J. M., Clements, W., Fritz, A., Gouin, T., ... Moe, S. J. (2013). The influence of global climate change on the scientific foundations and applications of Environmental Toxicology and Chemistry: Introduction to a SETAC international workshop. *Environmental Toxicology and Chemistry*, *32*(1), 13–19. doi:10.1002/etc.2037 PMID:23097130

Steffens, K., Larsbo, M., Moeys, J., Jarvis, N., & Lewan, E. (2013). Predicting pesticide leaching under climate change: Importance of model structure and parameter uncertainty. *Agriculture, Ecosystems & Environment*, *172*, 24–34. doi:10.1016/j.agee.2013.03.018

Steffens, K., Larsbo, M., Moeys, J., Kjellström, E., Jarvis, N., & Lewan, E. (2014). Modelling pesticide leaching under climate change: Parameter vs. climate input uncertainty. *Hydrology and Earth System Sciences*, *18*(2), 479–491. doi:10.5194/hess-18-479-2014

Stenrød, M., Perceval, J., Benoit, P., Almvik, M., Bolli, R. I., Eklo, O. M., ... Kværner, J. (2008). Cold climatic conditions: Effects on bioavailability and leaching of the mobile pesticide metribuzin in a silt loam soil in Norway. *Cold Regions Science and Technology*, *53*(1), 4–15. doi:10.1016/j.coldregions.2007.06.007

Stocker, T. F. (2014). *Climate change 2013: The physical science basis:Working group I contribution to the fifth assessment report of the intergovernmental panel on climate change*. Cambridge University Press.

Swart, R. & Biesbroek, R. (2008) *Adaptation of infrastructure to climate change: International inventory*. University and Research Centre Alterra/Wageningen.

Tessier, A., Campbell, P. G. C., & Bisson, M. (1979). Sequential extraction procedure for the speciation of particulate trace metals. *Analytical Chemistry*, *51*(7), 844–851. doi:10.1021/ac50043a017

Ulrich, A. C., Guigard, S. E., Foght, J. M., Semple, K. M., Pooley, K., Armstrong, J. E., & Biggar, K. W. (2009). Effect of salt on aerobic biodegradation of petroleum hydrocarbons in contaminated groundwater. *Biodegradation*, *20*(1), 27–38. doi:10.100710532-008-9196-0 PMID:18437506

Vela, N., Navarro, S., & Navarro García, G. (2007). Review. An overview on the environmental behaviour of pesticide residues in soils. *Spanish Journal of Agricultural Research*, *5*(3), 357–375. doi:10.5424jar/2007053-5344

Visser, A., Kroes, J. T. H., Van Vliet, M., Blenkinsop, S., Fowler, H. J., & Broers, H. P. (2012). Climate change impacts on the leaching of a heavy metal contamination in a small lowland catchment. *Journal of Contaminant Hydrology*, *127*, 47–64. doi:10.1016/j.jconhyd.2011.04.007 PMID:21684031

Walker, C. H. (2006). Ecotoxicity testing of chemicals with particular reference to pesticides. *Pest Management Science*, *62*(7), 571–583. doi:10.1002/ps.1218 PMID:16671067

Ward, D. M., & Brock, T. D. (1978). Hydrocarbon degradation in hypersaline environments. *Applied and Environmental Microbiology*, *35*, 353–359. PMID:16345276

Weber, F. A., Hofacker, A. F., Voegelin, A., & Kretzschmar, R. (2010). Temperature dependence and coupling of iron and arsenic reduction and release during flooding of a contaminated soil. *Environmental Science & Technology*, *44*(1), 116–122. doi:10.1021/es902100h PMID:20039741

Whitehead, P. G., Wilby, R. L., Battarbee, R. W., Kernan, M., & Wade, A. J. (2009). A review of the potential impacts of climate change on surface water quality. *Hydrological Sciences Journal*, *54*(1), 101–123. doi:10.1623/hysj.54.1.101

WHO. (1998). World Health Organization, Aluminum. In *Guidelines for Drinking-Water Quality* (2nd ed.; pp. 3–13). Geneva: World Health Organization.

Wijngaard, R. R., van der Perk, M., van der Grift, B., de Nijs, T. C. M., & Bierkens, M. F. P. (2017). The impact of climate change on metal transport in a lowland catchment. *Water, Air, and Soil Pollution*, *228*(3), 107. doi:10.100711270-017-3261-4 PMID:28260820

Williamson, C. E., Dodds, W., Kratz, T. K., & Palmer, M. A. (2008). Lakes and streams as sentinels of environmental change in terrestrial and atmospheric processes. *Frontiers in Ecology and the Environment*, *6*(5), 247–254. doi:10.1890/070140

Wilson, S.R., Solomon, K.R. & Tang, X. (2007) Changes in tropospheric composition and air quality due to stratospheric ozone depletion and climate change. *Photochemical & Photobiological Sciences, 6*, 301-310.

Wolfe, M. F., Schwartz, G. J. B., Singaram, S., Mielbrecht, E. E., Tjeerdema, R. S., & Sowby, M. L. (1998). Effects of salinity and temperature on the bioavailability of dispersed petroleum hydrocarbons to the golden-brown algae, *Isochrysis galbana*. *Archives of Environmental Contamination and Toxicology*, *35*, 268–273. doi:10.1007002449900375 PMID:9680519

Worrall, F., Burt, T., & Shedden, R. (2003). Long term records of riverine dissolved organic matter. *Biogeochemistry*, *64*(2), 165–178. doi:10.1023/A:1024924216148

Yeo, H.-G., Choi, M., Chun, M.-Y., & Sunwoo, Y. (2003). Concentration distribution of polychlorinated biphenyls and organochlorine pesticides and their relationship with temperature in rural air of Korea. *Atmospheric Environment*, *37*(27), 3831–3839. doi:10.1016/S1352-2310(03)00456-4

Younker, J. M., & Walsh, M. E. (2015). Impact of salinity and dispersed oil on adsorption of dissolved aromatic hydrocarbons by activated carbon and organoclay. *Journal of Hazardous Materials*, *299*, 562–569. doi:10.1016/j.jhazmat.2015.07.063 PMID:26259095

Zartman, R. E., Ramsey, R. H., Evans, P. W., Koenig, G., Truky, C., & Kamara, L. (1996). Outerbasin, annulus, and playa basin infiltration studies. *Texas Journal of Agriculture and Natural Resources*, *9*, 23–32.

Zhou, E., & Crawford, R. L. (1995). Effects of oxygen, nitrogen, and temperature on gasoline biodegradation in soil. *Biodegradation*, *6*(2), 127–140. doi:10.1007/BF00695343 PMID:7772939

Chapter 7
Study of the Effect of Climate Changes on the Well Water Contamination by Some Heavy Metals at a Mining Extract Region in Marrakech City, Morocco

Yassir Barkouch
University of Nantes, France

Mohyeddine El Khadiri
Independent Researcher, Morocco

Sana El Fadeli
Cadi Ayyad University, Morocco

Abdelaziz Ait Melloul
Independent Researcher, Morocco

Alain Pineau
University of Nantes, France

ABSTRACT

Seasonal variation on chemical parameters of well water at Draa Lasfar region (Marrakech, Morocco) was studied. A total of 144 samples were collected between 2012 and 2013 and were analyzed for temperature (T°), pH, total hardness (TH), chemical oxygen demand (COD), nitrates, Cd, Pb, and Zn. Significant difference between seasons was observed for these parameters. Highest temperature (28.72±3.16) was recorded during summer. COD and Zn concentration was recorded maximum during summer (167.25±31.05 mg/l, 131.4±12.0 µg/l respectively). Highest nitrates (2.67±0.75 mg/l) concentrations were recorded during spring. Highest Pb (632.14±82.54 µg/l) and Cd (1.93±0.36 µg/l) concentrations were recorded during winter. Alternating seasons can be likened to small-scale climate change. Therefore, the impacts of this change on quality of water resources include particularly the modification of parameters values. The main drawn conclusion is that a degradation trend of well water quality in the context of climate change can lead to an increase of at-risk situations related to potential health impact.

DOI: 10.4018/978-1-5225-7775-1.ch007

INTRODUCTION

Of all the natural resources, water is unarguably the most essential and precious. It is a universal solvent and as a solvent it provides the ionic balance and nutrients, which support all forms of life (Vanloon & Duffy, 2005). It is generally obtained from two principal natural sources; Surface water such as fresh water lakes, rivers, streams, and Groundwater such as borehole water and well water (Caroline & Wilfred, 2013).

In Morocco, the major source of water used to meet the domestic, agricultural and industrial needs is the ground water. The ground water is defined as water that is found underground in cracks and spaces in soil, sand and rocks. This source has two distinct functions; firstly, it is a significant source of both urban and rural population's water supply and secondly it sustains many wetland ecosystems.

Unfortunately, industrialization and human activities have partially or totally turned our environment into dumping sites for waste materials. As a result, many water resources have been rendered polluted and hazardous to man and other living systems (Bakare et al., 2003).

Water is typically referred to as polluted when it is impaired by anthropogenic contaminants and either does not support a human use, such as drinking water, and/or undergoes a marked shift in its ability to support its constituent biotic communities, such as fish.

Deposition of heavy metals in groundwater from anthropogenic activities has been implicated for an increase in heavy metal concentration above recommended levels (Maine et al., 2004; Bako et al., 2008).

The term "heavy metal" is not altogether clearly defined, but in the case of water pollution, these are metals such as arsenic, cadmium, iron, cobalt, chromium, copper, manganese, mercury, molybdenum, nickel, lead, selenium, vanadium and zinc. While heavy metals do tend to have a high atomic mass, and so are heavy in that sense, toxicity seems to be a further defining factor as to what constitutes a heavy metal and what does not.

Heavy metals are important components of agro-allied products such as pesticides, herbicides, fertilizers; manufacturing and other synthetic products such as paints and batteries. Mining activities, industrial, municipal and domestic wastes have been reported to be important sources of heavy metal pollution to the groundwater (Mathews et al., 2013).

Excessive concentration of heavy metals in the groundwater is of great concern because of their non-biodegradability. Therefore, their persistence in the environment portends health hazard plants and animals and consequently trigger ecological imbalance in the ecosystem (Ekmekyapar et al., 2012).

Another concern that high concentrations of heavy metals raise is their ability to bioaccumulate across the food chain, with members that are high up the food chain having concentration of such metals several times higher than what is obtainable in the depart point of the contamination (groundwater) (Bako et al., 2009; Megateli et al., 2009).

In Morocco, Ground water was main source of water supply in most rural communities. It had good microbiological and biological properties in general as such required minimal treatment. Actually, A variety of human activities, notably industrial and mining process have been responsible for the wider diffusion of heavy metals into this type of water (Barkouch et al., 2007).

This study was carried out to determine the spatial and seasonal variations of heavy metal deposition in groundwater in a mining area near Marrakech city in Morocco in order to assess the extent of pollution generated by the mining activity and to identify the key mechanism responsible for this contamination and its relation to this mining activity.

EXPERIMENTAL PROCEDURES

Study Area

The Draa Lasfar mine is located in northwest of the Mrabtine zone at approximately 10 Km in the west of Marrakech city (Figure 1). it's located a few hundred meters from the Tensift River, close to a rural community of about 5790 ha, which 65% are occupied by farmland. Draa Lasfar consists on deposit of pyrite mineral discovered in 1953 although their commercial exploitation did not begin until 1979. Mineral was processed by flotation after primary and secondary crushing and grinding, producing 60 Mt of products in the first two years (1979 and 1980). Industrial activity stopped in March 1981, although it restarted in 1999 due to its great resource of polymetallic components (As, Cd, Cu, Fe, Pb and Zn). During its exploitation, tailings were discharge all around the mine area posing a risk for the environment.

Sampling Methods and Sample Preparation

Water samples were collected once a month for twelve months between April 2012 and April 2013.

Water was taken from wells which are falling within 1 km radius of industrial unit of Zn and Pb extraction. Samples were taken directly from wells in sterile glass bottles of 250 milliliters capacity (Divya et al., 2011), after rinsing the bottles three times with sample water. In order to collect the samples directly from well, bottle with a string attached to neck was used. Another long clean string was tied to the end of sterile string and the bottle was lowered into the water allowed to fill up. Then the bottle was raised and stoppered. The collected samples were transported to laboratory in ice within an insulated container and analyzed within 24 hours of collection.

A total of 144 well water samples, 36 each during four different seasons of the year viz. summer (June-August), spring (March-May), autumn (September-December) and winter (January-March) were collected during the years 2012 and 2013 and analyzed for physical parameters like temperature and pH, chemical parameters like total hardness, Chemical Oxygen Demand (COD).

Temperature and pH of each sample was measured using mercury filled glass thermometer and digital pH meter respectively3. Total hardness of the samples was estimated using Total hardness test kit. Measurement of COD was made photometrically in Spectroquant NOVA 60 (Merck, Germany) after digesting the samples in preheated Thermoreactor TR 320 (Merck, Germany). Concentration of nitrate and lead in not filtered water samples was measured photometrically in Spectroquant NOVA 60 and expressed in mg/l.

Metal contents of the water samples were analyzed by AAS (Model: ECILTM AAS–4141). For the determination of heavy metals, the water samples were digested with 20 mL aqua-regia (HCl/HNO3 3:1, volume ratio) in a beaker (open beaker digestion) on a thermostatically controlled hot plate. Then 5.0 mL hydrogen peroxide was added to the sample to complete the digestion and the resulting mixture was heated again to near dryness in a fume cupboard and filtered by Whatman no. 42 filter paper and the volume was made up to 50 mL by double distilled water (Kard et al., 2008).

Estimation of zinc, lead and cadmium was carried out using Atomic Absorption Spectrophotometer (Divya et al., 2011).

RESULTS AND DISCUSSION

Results of analysis are shown in Tables 1 and 2.

pH

pH of well water was in the range of 5,78 ± 0,22 - 6,71 ± 0,24, and significant difference between seasons was not observed. pH is mainly influenced by volume of water (Divya et al., 2011), soil type(Divya et al., 2011), presence of chemicals and application of acidic fertilizers.

In the present study, pH was not within the acceptable range of pH for drinking water (6.5- 8.5) (WHO, 2006).

- Acid pH of well water during autumn may be due to dissolved carbon dioxide and organic acids such as fulvic and humic acids which are derived from decay and subsequent leaching of plant materials. During dry seasons (summer) there may be death and decay of plants due to lack of sufficient water which increases the organic acid content of water in turn causing acidity. The higher pH values during rainy season could be due to relative low photosynthesis of micro and macro vegetation resulting in production of less CO2,
- Shifting the equilibrium towards alkaline side (Kumar et al., 2010). This could be attributed to the presence of luxuriant vegetation inside most of the wells during rainy season.

Temperature

Temperature ranged from 27.5-28.1. Lowest temperature was recorded during winter and highest temperature was recorded during summer, which was in accordance with ambient temperature pattern (Kaplay et al., 2004; Agbaire et al., 2009).

Total Hardness

Total hardness was in the range of 296.28±37.14 - 423.67±27.88 mg/l, with no significant seasonal variation. Higher total hardness could be due to discharge of effluents and untreated waste (Ullah et al., 2009) from the local extracting mine industry to nearby surface water sources. Highest value of total hardness was observed during summer. It could be due to the low water level and high rate of evaporation during summer (Sisodia et al., 2006).

Chemical Oxygen Demand

COD ranged from 96.87±17.37 to 167.25±31.05 and showed significant difference between seasons. Lowest and highest values were observed during winter and summer respectively. Higher values of COD indicate the presence of oxidizable organic matter19.The entry of industrial effluents and the agricultural runoff might be responsible for increased level oxidizable organic matter (Sisodia et al., 2006). The higher COD could be due to death and decay of plants and subsequent increase in organic matter during summer (Kumar et al., 2010). The lower COD observed during winter could be due to the effect of dilution increased by rain at this season.

Nitrate

Mean nitrate concentration of well water was in the range of 1.15±0.43- 2.67±0.75 mg/l, which were within WHO guidelines (2006) for nitrate. Nitrate detected in well water samples might have originated from decaying organic matter (Subrahmanyam et al., 2001), discharge of sewage and industrial wastes and runoff from agricultural fields containing nitrate fertilizers (Kumar et al., 2010). Mean nitrate concentration was lowest during autumn and highest during spring. The highest concentration during spring might be due to application of nitrogenous fertilizers to agricultural land during rainy season and subsequent seepage through soil.

Copper

Mean copper concentration was in the range of 90.4 ± 16.47- 136,82 ± 31.91 μg/l, and was under WHO guidelines, 2006 (200 μg/l) for Cu in drinking water.

This low concentration of Cu in the groundwater is likely due to the fact that Cu is easily chemisorbed on or incorporated in clay minerals of soils (Rodriguez et al., 2009). This chemisorption can be justified by that copper is characterized by high complex constant organic matter thus it can be hypothesized that Cu is bound to labile organic matter such as lipids, proteins, and carbohydrates.

Lead

Mean lead concentration was in the range of 335.45±62.71- 632.14±82.54 μg/l, and was above WHO guidelines, 2006 (0.01 mg/l) for lead in drinking water. This mining extract zone being an industrial area is subjected to the discharge of effluent containing lead to nearby water bodies.

Analysis of waste products generated by this mine extract industry showed that significant amount of lead is generated by this industrial unit in their waste products. The effluents rich in lead are discharged to water bodies nearby and subsequently affect the groundwater quality of the area. It was shown a significant seasonal variation: spring samples showed lowest concentration and respectively winter and summer seasons showed highest concentrations. Combined effect of decreased amount of water in summer and strong leaching during winter might have contributed to higher lead concentration during these two seasons.

Zinc

Mean zinc concentration was in the range of 80.12±5.51 - 131.42 ± 7.45 μg/l, and was within the limit of 500 μg/l as prescribed by WHO guidelines, 2006. The analysis report pointed out that significant amount of zinc is generated by this industrial unit, which in turn deteriorate the ground water quality. The concentration was highest during summer when depletion of water leads to greater concentration of metals (Buragohain et al., 2010).

Cadmium

Mean cadmium concentration varied from 1.56±0.24-1.93±0.36 µg/l, and showed significant difference between seasons. Analysis of waste products had shown that this industry discharged some amount of cadmium in their waste products, deteriorating the groundwater quality. Cadmium concentration was found to be highest during winter, which might be due to leaching during this season.

CONCLUSION

Well water quality in the study zone showed seasonal variation for temperature some parameters (pH, COD, concentration of nitrate, zinc, lead and cadmium) and exceeded in most cases the limits prescribed by WHO guidelines 2006. In order to improve quality of well water and consequently to protect people and animals from the perils of well water contamination, it is crucial to initiate measures to check the pollution from industrial effluents and to establish on-site regular well water quality monitoring network stations.

REFERENCES

Agbaire, P. O., & Obi, C. E. (2009). Seasonal Variations of Some Physico-Chemical Properties Of River Ethiope Water in Abraka, Nigeria. *Journal of Applied Science & Environmental Management*, *13*(1), 55.

Bakare, A. A., Lateef, A., Amuda, O. S., & Afolabi, R. O. (2009). The Aquatic toxicity and characterization of chemical and microbiological constituents of water samples from Oba River, Odo-oba, Nigeria. *Asian J. Microbiol. Biotechnol. Environ. Exp. Sci*, *5*(1), 11.

Bako, S. P., Bhwankot, E. S., Ezealor, A. U., Chia, A. M., & Funtua, I. I. (2009). Human health implications of trace metal contents in parts of Maize (Zea mays L.) plants cultivated along highways in Nigeria's Guinea Savanna. In Soil Remediation Series: Pollution Science, Technology and Abatement. Nova Science Publishers, Inc.

Bako, S. P., Odiwo, J. E., & Ezealor, A. U. (2008). Plant diversity in relation to anthropogenic Trace Metals in soils of selected sites in Nigeria's Guinea Savanna. *International Journal of Environment and Pollution*, *33*(2/3), 185. doi:10.1504/IJEP.2008.019392

Barkouch, Y., Sedki, A., & Pineau, A. (2007). A new approach for understanding lead transfer in agricultural soil. *Water, Air, and Soil Pollution*, *186*(1-4), 3–13. doi:10.100711270-007-9450-9

Buragohain, M., Bhuyan, B., & Sarma, H. P. (2010). Seasonal variations of lead, arsenic, cadmium and aluminium contamination of groundwater in Dhemaji district, Assam, India. *Environmental Monitoring and Assessment*, *170*(1-4), 345–351. doi:10.100710661-009-1237-6 PMID:19908151

Caroline, M., & Wilfred, S. R. (2013). Studies on Leaching of Heavy Metals from E-waste. *Oriental Journal of Chemistry*, *29*(03), 1149–1154. doi:10.13005/ojc/290343

Divya, R. T., Sunil, B., & Latha, C. (2011). Physico- Chemical Analysis of Well Water at Eloor Industrial Area-Seasonal Study. *Current World Environment*, *6*(2), 259.

Ekmekyapar, F., Sabudak, T., & Seren, G. (2012). Assessment of heavy metal contamination in soil and wheat (Triticum aestivum L.) plant around the Corlu-Cerkezkoy highway in Thrace region. *Global NEST Journal*, *14*(4), 496.

Kaplay, R. D., & Patode, H. S. (2004). Groundwater pollution due to industrial effluent at Tuppa, New Nanded, Maharashtra, India. *Environ Geol.*, *46*(6-7), 871–882. doi:10.100700254-004-1095-z

Kard, S. P., Mandal, S. K., Sahu, T., & Kole, R. K. (2008). Assessment of heavy metal pollution in surface water. *International Journal of Environmental Science and Technology*, *5*(1), 119–124. doi:10.1007/BF03326004

Kumar, K. K., Hakeem, A. A., & Kumar, S. S. (2010). *Proceedings of the Twenty second Kerala Science Congress*. Kerala State Council for Science, Technology and Environment, Government of Kerala.

Maine, M. A., Sune, N. L., & Lagger, S. C. (2004). Chromium bioaccumulation: Comparison of the capacity of two floating aquatic macrophytes. *Water Research*, *38*(6), 1494–1501. doi:10.1016/j.watres.2003.12.025 PMID:15016526

Mathews, A. O. C., & Kakulu, S. (2012). Investigation of heavy metal levels in roadside agricultural soil and plant samples in Adogo, Nigeria. *Acad J Enviro Sci.*, *1*(2), 31.

Megateli, S., Semsari, S., & Couderchet, M. (2009). Toxicity and removal of heavy metals (Cadmium, Copper, and Zinc) by Lemna gibba. *Ecotoxicology and Environmental Safety*, *72*(6), 1774–1780. doi:10.1016/j.ecoenv.2009.05.004 PMID:19505721

Rodriguez, L., Ruiz, E., Alonso-Azcarate, J., & Rincon, J. (2009). Heavy metal distribution and chemical speciation in tailings and soils around a Pb–Zn mine in Spain. *Journal of Environmental Management*, *90*(2), 1106–1116. doi:10.1016/j.jenvman.2008.04.007 PMID:18572301

Sisodia, R., & Moundiotiya, C. (2006). Assessment of the water quality index of wetland Kalakho Lake, Rajasthan, India. *J Environ Hydrology.*, *14*(23), 1.

Subrahmanyam, K., & Yadaiah, Y. (2001). - Assessment of the impact of industrial effluents on water quality in Patancheru and environs, Medak district, Andhra Pradesh, India. *Hydrology Journal*, *9*(3), 297.

Ullah, R., Malik, R. N., & Qadir, A. (2009). Assessment of groundwater contamination in an industrial city, Sialkot, Pakistan. *African Journal of Environmental Science and Technology*, *3*(12), 429.

Vanloon, G. W., & Duffy, S. J. (2009). The Hydrosphere. In *Environmental Chemistry: A Global Perspective* (2nd ed.; pp. 197–211). New York: Oxford University Press.

WHO. (2006). *Guidelines for drinking water quality*. Geneva: World Health Organization.

APPENDIX

Table 1. Physical quality of wells water

parameters	seasons			
	Winter	Spring	Summer	Autumn
pH	6.71 ± 0.24	6.56 ± 0.14	6.03 ± 0.17	5.78 ± 0.21
temperature (°C)	27.5 ± 2.07	28.02 ± 3.16	28.72 ± 2.26	28.37 ± 2.68

Table 2. Chemical quality of wells water

parameters	Mean Concentration During Four Seasons			
	Winter	Spring	Summer	Autumn
Total hardness	302.56 ± 31.39	296.28 ± 37.14	423.67 ± 27.88	376.15 ± 23.75
COD	96.87 ± 17.37	117.26 ± 57.41	167.25 ± 31.09	115.68 ± 43.16
Nitrate	2.11 ± 0.47	2.67 ± 0.71	1.63 ± 0.23	1.15 ± 0.43
Copper (μg/l)	90.4 ± 16.47	102.74 ± 21.30	136,82 ± 31.91	119.52 ± 12.45
Lead (μg/l)	632.14 ± 82.54	335.45 ± 62.71	512.64 ± 60.85	433.52 ± 71.57
Zinc (μg/l)	80.12 ± 5.51	86.72 ± 6.42	131.42 ± 7.45	103.91 ± 9.25
Cadmium (μg/l)	1.93 ± 0.36	1.68 ± 0.19	1.78 ± 0.37	1.56 ± 0.24

Figure 1. Drâa Lasfar mine geographic situation in Marrakech region

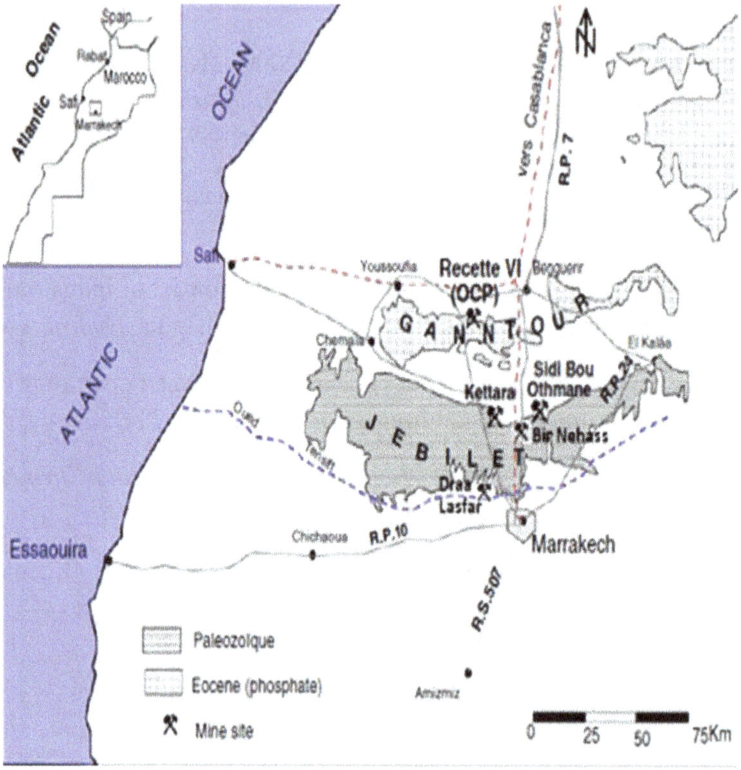

Section 4

Human Cancers Resulting From Environmental Contaminations

Chapter 8
Prostate Cancer and Environmental Exposure:
A Focus on Heavy Metals and Pesticides

Omar El Hiba
Chouaib Doukkali University, Morocco & Cadi Ayyad University, Morocco

Hicham Chatoui
Private University of Marrakech, Morocco & Cadi Ayyad University, Morocco

Mehdi Ait Laaradia
Cadi Ayyad University, Morocco

Hajar Kiai
Cadi Ayyad University, Morocco

Fatima-Zahra Azzaoui
Ibn Tofail University, Morocco

Hanane Rais
Mohammed VI University Hospital, Morocco

Malika Ait Sidi Brahim
Cadi Ayyad University, Morocco

Souad Sellami
Mohammed VI University Hospital, Morocco

Abdelali Bitar
Chouaib Doukkali University, Morocco

Ouassil El Kherchi
Hassan 1st University, Morocco

ABSTRACT

Prostate cancer (PC) is the most commonly diagnosed malignancy and the major cause of disease and mortality among men. Every year, around 1.6 million new cases are diagnosed with 366,000 death cases. Different etiologies have been associated to PC incidence including the genetic predisposition as well as the environmental influences. Recently, more interest has been given to the role of food and water contaminations with heavy metals and pesticides as direct carcinogenetic agents, particularly involved in the pathogenesis of PC. The chapter will address the relationship between heavy metals and pesticides exposures, and the development of PC with the support of epidemiological and experimental evidences.

DOI: 10.4018/978-1-5225-7775-1.ch008

INTRODUCTION

Prostate cancer (PC) is considered as the fourth leading cancer in men and women, while it is the second most common cancer in men worldwide (Arnold et al., 2015; Siegel, Naishadham, & Jemal, 2013). The incidence of PC varies largely between continents, Hence, a high incidence is observed in developed countries compared to the underdeveloped world areas (Parkin, Bray, & Devesa, 2001).

Clinically, PC is generally without apparent symptoms (asymptomatic) at the earlier phase of the disease. While in rare patients, it can be associated to some manifestations similar to those of benign prostatic hyperplasia including nocturia, hematuria and anddysuria (Mustafa et al., 2016)

Different etiologies and risk factors of PC have been elucidated including genetic and environmental exposures. Although, elderly men are preferentially affected by PC with more than 60% in people with more than 65 years old (Desantis, Siegel, & Jemal, 2014).

Several etiologies of PC have been identified including, age and ethnicity, family history, diet, obesity, sexual behavior and sexually transmitted diseases, occupation and smoking (see for review Bashir, 2015).

Beside the genetic part of PC etiologies, environmental exposures to several exogenous chemical contaminants are generally well admitted as important carcinogens. Indeed, support of this view is provided by observations of abnormal PC incidence in certain subgroups of the population, leading to suppose the implication of the non-genetic influences (environment) as a non- neglected part of the disease causes (Mullins & Loeb, 2012)

Through the present chapter, we will review the available literature on environmental exposures, especially to heavy metals and pesticides, and the relationship to PC incidence with the support of experimental as well as epidemiological studies.

PROSTATE CANCER: EPIDEMIOLOGY AND RISK FACTORS

Adenocarcinoma of prostate or prostate cancer (PC) is a masculine pathology specific of elderly people, with about 9.7% of total male cancers and approximately (Ferlay, et al. 2015).

The incidence of PC is still variable. In fact, in developed countries it reaches 15.3%, while in the underdeveloped world it accounts only for 4.3% only (table 1) (Parkin, Bray, & Devesa, 2001). In US, PC is considered as the most commonly diagnosed cancer in men and the second most common cause of death after lung cancer (Parkin, Bray, & Devesa, 2001).

PC is a particular pathology with different etiologies and risk factors, including age, ethnicity, family history, diet, obesity, sexual behavior and sexually transmitted diseases, occupation and smoking. Indeed, PC is affecting elderly men especially with 6 /10 cases diagnosed at the age of 65 years or later. Generally, the average age at the time of PC diagnosis is 66 years (Desantis, Siegel, & Jemal, 2014). While according to the race and ethnicity, studies showed that African-Americans population presents the highest rate of PC incidence in the world (Table 1). (see Bashir, 2015).

Otherwise, family history of the disease as well, is considered as important risk factor for PC. Indeed, PC incidence is increased 2 to 3 folds among male who have a first-degree relative (father, son, brother) with a positive history of PC (Stanford & Ostrander, 2001).

At the genetic level, PC results from multi-gene interaction rather than a single gene involvement. Mutations in BRCA1 and BRCA2 genes considered as important risk factors for ovarian and breast cancers,

Table 1. Prostate cancer incidence and U. S. mortality age adjusted rates by race (2007-2011) according to the surveillance, epidemiology, and end results program (SEER) of the National Cancer Institute (NCI). (modified Bashir, 2015)

Ethnicity	Incidence Rates per 100,000 Person	U. S Mortality Rates per 100,000 Persons
All Races	147.8	22.3
White	139.9	20.6
White Hispanic	120.3	19.1
White Non-Hispanic	143.3	20.7
Hispanic	121.8	18.5
Black	223.9	48.9
Asian/ Pacific Islander	79.3	10
American Indian	71.5	16.8

have also been associated to PC pathogenesis (Struewing et al., 1997). In addition other genes have been involved as well, including the Hereditary Prostate Cancer gene 1(HPC1), the androgen receptor gene, the vitamin D receptor gene and TMPRSS2-ETS gene family fusion; specifically TMPRSS2- ERG or TMPRSS2-ETV1/4 which promote cancer growth (Gallagher & Fleshner, 1998; Beuzeboc et al., 2009).

In a relatively recent study, single nucleotide polymorphism (SNPs) has been linked to PC (Eeles et al., 2008). Several SNPs have been associated to the risk of developing the disease. Indeed, individuals with TT allele pair at SNP rs10993994 were reported to be at 1,6 times higher risk of PC than those with CC allele pair. Such SNP may explain the increased PC risk of Africans compared to Americans of European descent which results from the fact that C allele is highly prevalent in the latter population (Whitaker, et al. 2010).

CLINICAL MANIFESTATION

PC is usually asymptomatic at the earlier stages. However, in some cases, it is associated with some symptoms similar to those of benign prostatic hyperplasia. The most frequent symptoms of PC are; frequent nocturia (increased nocturnal urination), hematuria, a difficulty in starting and maintaining a steady urine stream, and dysuria (see Mustafa et al., 2016).

In a study performed by the Patient Care Evaluation in the US, we showed that around 1/3 of patients diagnosed with PC manifests one or more of those symptoms, while 2/3 are still asymptomatic (Miller, Hafez, Stewart, Montie, & Wei, 2003). PC is mainly characterized by urinary dysfunction as the prostate gland surrounds the prostaticurethra. Therefore, changes within the gland, may impact on the urinary function because the vas deferens deposits seminal fluid into the prostatic urethra and secretion from prostate gland itself are included in semen content. Besides, PC may also cause problems with sexual function and performance, such as difficulty to achieve erection or painful ejaculation (Miller, Hafez, Stewart, Montie, & Wei, 2003).

Whereas, advanced prostate cancer can spread to other parts of the body, and may cause further symptoms including mainly bone pain, often in the vertebrae, pelvic, or ribs.

As well, PC in the spine can also compress the spinal cord, causing tingling leg weakness and urinary and fecal incontinence (Van Der Cruijsen-Koeter et al., 2005) and also, in some cases, feeling tired due to decreased erythrocytes density (Segal et al., 2003).

PHYSIOPATHOLOGY OF PROSTATE CANCER

The function of the prostate, as an annexed gland of the male genital system, is to contribute to formation and storage of the seminal fluid. In adult men, the average dimensions of the gland are: 3 centimeters long and an average weighs of 20 grams (Aumüller, 2012). Otherwise, the prostate contains many small glands which make 20% of the seminal fluid (Aumüller, 2012). During prostate tumors, cells undergo mutations and become cancer cells with proliferative capacity (Figure 1).

The prostate gland requires androgens (male hormones) for a normal functioning. Androgens including testosterone produced in the testis, dehydroepiandrosterone synthesized by the adrenal glands; and dihydrotestosterone converted from testosterone within the prostate.

Pathologically, PC is classified as an adenocarcinoma (glandular cancer), that begins when normal semenscreening prostate gland cells mutate into cancer cells.

The common region where prostate adenocarcinoma starts is the peripheral zone (figure 1).

Figure 1. The Human prostate gland. A: localization within the men genital system. B: anatomy of the human prostate gland, C and D: respectively normal and enlarged (hyperplasic) prostate gland and their pressure excreted on the rectum (Bibel, 2012)

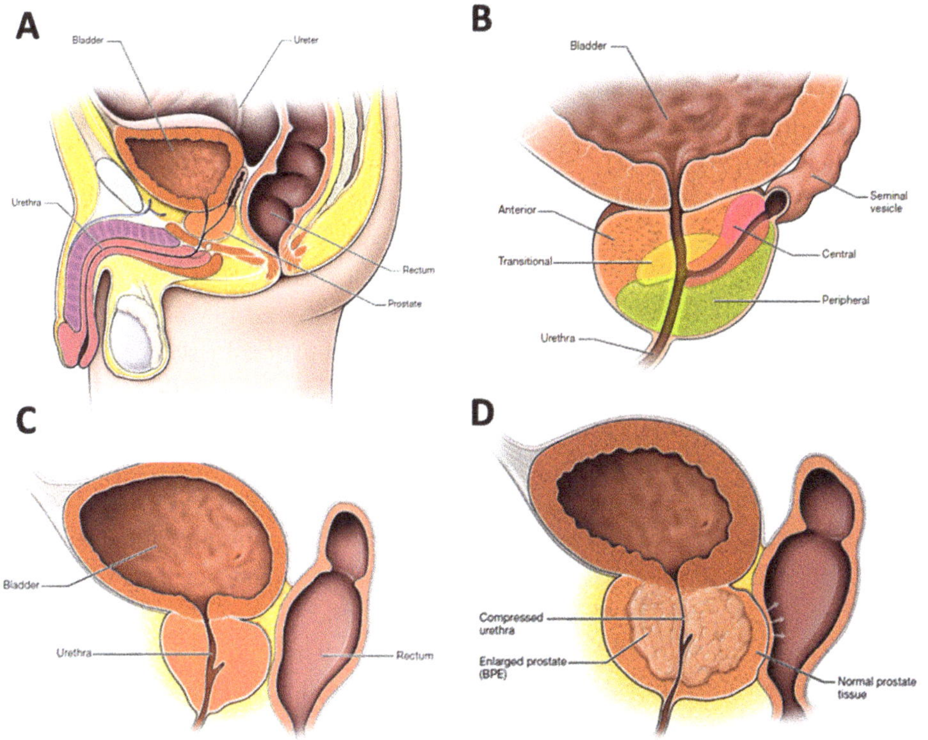

At the beginning, small clumps of tumor cells remain confined to normal prostate gland which is known as carcinoma *in situ* or prostate intraepithelial neoplasia (PIN).

It has been proposed that prostate adenocarcinoma starts with a prostatic lesion named proliferative inflammatory atrophy considered as a precursor of the prostatic intraepithelial neoplasia and therefore prostate cancer. Such atrophic areas contains proliferative epithelial cells that fail to differentiate into columnar secretory cells and occur in the peripheral region of the prostate where the cancer starts (De Marzo, Marchi, Epstein, & Nelson, 1999; Feneley, Young, Chinyama, Kirby, & Parkinson, 1996; Ruska, Sauvageot, & Epstein, 1998)

PC progress begin by multiplication of tumor cells which spread to the surrounding prostate tissue (the stroma) forming the tumor. While, the metastatic phase can be separated into three main processes: invasion, intravasation and extravasation. Firstly, during the process of invasion, a loss of cell-cell adhesion capacity occurs and allows malignant tumor cells to leave the primary tumor cell masse. Changes in cell-matrix interaction will therefore enable the malignant cells to invade the surrounding stroma. Such invasion will imply release of chemicals leading to degradation of the basement membrane and extracellular matrix, together with proteins involved in the control of motility and migration. As well, angiogenesis is solicited to elicit development of the tumor and facilitation of nutrients and waste products transports from-and-to the tumor cells (Brooks, 1996). Finally, the blood capillaries close to the tumor provide a route for tu mor cells detached to reach blood stream and therefore reaching distal organs and tissues during the metastasis: the intravasation stage (Folkman, 1984, Folkman, 1996)

ENVIRONMENTAL EXPOSURE AND PROSTATE CANCER

Involvement of Pesticides

During the last years, substantial evidences has been accumulated regarding the role of environmental and occupational exposures in the pathogenesis of PC (Belpomme, Irigaray, Ossondo, Vacque, & Martin, 2009; Mullins & Loeb, 2012).

Taken into consideration the involvement of androgens and oestrogens in the PC development (Ho, Lee, Lam, & Leung, 2011; Prezioso et al., 2007; Ricke, Wang, & Cunha, 2007), a possible association has been made between exposure to endocrine disrupting chemicals including several types of pesticides and PC incidence (Imaida & Shirai, 2000; Medjakovic et al., 2014; Mnif et al., 2011). while, it is well admitted that farming constitutes the most consistent occupational risk factor for developing PC (Blair, Zahm, Pearce, Heineman, & Fraumeni, 1992; Parent & Siemiatycki, 2001), and this may include exposures to insecticides, fertilizers, herbicides, and other chemicals (Aronson, Siemiatycki, Dewar, & Gerin, 1996: Blair, Dosemeci & Heineman, 1993: Blair & Zahm, 1991, 1995; Dosemeci, Cocco, Gómez, Stewart & Heineman, 1994; Dich & Wiklund, 1998; Morrison et al., 1993;)

A large correlation study assessed the relationship between 45 agricultural pesticides and PC incidence in a prospective cohort study of male pesticide applicators from Iowa and North Carolina. Data showed a PC standardized incidence ratio of 1.14 (95% confidence interval: 1.05, 1.24) for the agricultural health study cohort. Otherwise, the use of chlorinated pesticides as well as methyl bromide by applicators with the age of 50 years or more, was linked to PC developing risk (Alavanja, 2003).

Organochlorine (OC) pesticides are known as a class of hydrocarbon compounds and used in agriculture and pest control. This family of pesticides includes hexachlorocyclohexane (HCH) isomers, DDT and its analogs, and cyclodienes. OC are able to be accumulated in the food chain, due to their persistence in the environment, and therefore, may impact on health. In this regard, the Environmental Protection Agency in USA, banned the use of most of OC pesticides during the 1970s and 1980s (see Xu et al., 2009).

Nevertheless, substantial evidences support the view that a number of OC pesticides mimic the estrogenic or antiestrogenic pharmacological effects (Soto, et al.1995). Consequently, such reagent may interfere with the function of the endocrine system through different pathways including mimicking a hormone, blocking normal endogenous hormones, or acting as modulator of the biosynthesis of hormones (Soto et al., 1995). Hence, OC could be considered as tumor promoters via indirect effect involving hormone-mediated carcinogenesis (Hansen & Matsumura, 2001).

In this regard, some epidemiologic investigations have associated OC pesticides to divers hormone-related cancers including PC (Mills & Yang, 2003). Normally, the main source of exposure to OC pesticides is via diet (milk, fish, and meat) (Toppari et al., 1996). In addition to that, supplementary exposure sources include dust, air, and soil. A study performed between 1999 and 2004 based on the National Health and Nutrition Examination Survey data was designed to examine associations between serum concentrations of OC and PC. The study showed increased serum concentrations of β-hexachlorocyclohexane (HCH), trans-nonachlor and dieldrin and such increase was correlated to the risk developing PC (Xu et al., 2009). In addition, Hexachlorobenzene (HCB) is another category of pesticides used as a fungicide and known as endocrine disruptors (Gocmen, Peters, Cripps, Bryan, & Morris, 1989) and are bioaccumulated in the fat with a half-life estimated of 23 years in soil. HCB are potentially able to interfere with androgen hormone regulation due to similarities of chemical structures between HCB and natural hormones. Otherwise, HCB may impair the male gonadic and reproductive function in laboratory animals such as decreased fertility, and weights of both seminal vesicles and ventral prostates (Elissalde & Clark, 1979; Foster et al., 1996; Müller et al., 1978).

In vitro studies have shown that, at low concentrations (ranging from 0.5 to 5 nM), HCB with the presence of dihydrotestosterone (DHT), increases the androgen-responsive production which leads to a possible agonizing effect of HCB on the androgen receptor. While at high concentration (> 10 μM), HCB induces an opposite effect. However, HCB affinity to the androgen receptor has not been confirmed yet (Ralph, Orgebin-Crist, Lareyre, & Nelson, 2003).

In addition, using LPB-CAT transgenic mice 4 weeks age (prepuberty), prostate weight was affected following treatment with HCB, thus, at a low dose (0.5–5 nM), prostate weight increases significantly, in contrast to the high dose, where the weight decreases but not significantly (Ralph, Orgebin-Crist, Lareyre, & Nelson, 2003).

Concomitantly, HCB a the low dose, increases the frequency of prostatic dilated acini cases in the LPB-CAT mice, contrarily to the high dose which seems to dramatically reduce such frequency (Ralph, Orgebin-Crist, Lareyre, & Nelson, 2003).

In 2011, an epidemiological study has been performed on a population of Canadian farmers of British Columbia, data have shown a significant association between PC risk and high exposure to several subclasses of pesticides including DDT, Simazine and Lindane. As well, the study reports a significant increased risk for several active components like dichlone, dinoseb amine, malathion, endosulfan, 2,4-D, 2,4-DB, and carbaryl (Band et al., 2011).

Support of this view is provided by a prospective nested case–control study carried out to estimate relationship between OC exposures and metastatic PC in a population-based cohort from Norway. The study reports that metastatic PC is two times present in men with high serum concentrations of oxychlordane (Koutros et al., 2015).

Heavy Metals in Prostate Carcinogenesis

Many metals play a fundamental role in various biochemical processes in the human body, and the destabilization of these elements can influence health and be implicated in the development of various diseases including cancer (Tan and Chen, 2011). Some other metals have no physiological role in the human body and can have toxic effects that simulate or block the function of essential metals or imitate the effect of endogenous hormones, thus disrupting endocrine homeostasis (Waalkes, 2000).

Arsenic

Arsenic is a naturally occurring metalloid in the water tables of several countries. The principal sources of exposure are potable water, crops irrigated with polluted water and food prepared with contaminated water.More than 100 million people are statistically exposed to inorganicarseniclevels above 10 µg/L, the standard for drinking water quality in manycountries[International Agency for Research on Cancer (IARC), 2004].

Meat, seafood, cereals, fish and vegetables may also be a source of arsenic; however, it is found mainly in its organic, less toxic form. Industrial processes,volcanic activity and tobacco are other environmental sources of arsenic (Registry, 2007).

A causal association has been described between arsenic and prostate cancerin a number of studies of exposed human populations in the environment (Table 2). In several cases, this relationship was dose-related, suggesting an etiological role of this metalloid in prostate carcinogenesis.

The first evidence, which shows a positive relationship with increased mortality from prostate cancer, comes from four ecological monitoring studies conducted in the late 1980s and 1990s in Taiwan, USA and Australia, which are known have elevated levels of arsenic in potable water. In all these studies, prostate cancer mortality was about 6 to 10 times higher in people who had exposure to the greatest concentrations in drinking water of arsenic compared to controls (Table 2). As previously mentioned, epidemiological data may indicate an association of environmental exposure to inorganic arsenic with prostate cancer incidence and human population mortality.

Table 2. Arsenic exposure and prostate cancer epidemiological studies in humans

Researches	Localisation of Population	Source of Arsenic	Result
Chen et *al.*, 1988	Southwest Taiwan	Drinking water	Increasedmortality
Wu et *al.*, 1989	Southwest Taiwan	Drinking water	Increasedmortality
Chen and Wang, 1990	Taiwan	Drinking water	Increasedmortality
Tsai et *al.*, 1999	Southwest Taiwan	Drinking water	Increasedmortality
Lewis et *al.*, 1999	Utah, USA	Drinking water	Increasedmortality
Hindwoodet *al.*, 1999	Victoria, Australia	Local Water/Soil	Increased incidence

It is important to note that this was reported that the human prostate cells are directly sensitive to the malignant transformation induced by arsenic, strongly reinforcing the potential role of arsenic in human prostate cancer (Achanzar et *al*., 2001; BenbrahimTallaa et *al*., 2007a, b, c).

A number of in vitro investigations have demonstrated that human prostate epithelial cells are sensitive to malignant transformation induced by arsenic in vitro and guide them to an androgen-independent state (BenbrahimTallaa et al., 2007a,b,c).Thus, it is possible that Arsenic may at least partially contribute to the risk of prostate cancer through its mimicking estrogenic activity and its disruption of the endocrine system (BenbrahimTallaa et al., 2007c).

Cadmium

Cadmium is a heavy metal that is not widely distributed in its natural state and is present as impurities in various minerals such as zinc, lead and copper (Andujar and Nemery, 2009).

This metal is mostly used in the form of Ni-Cd (nickel-cadmium) in electric cells or alkaline batteries and accumulators, in paint and plastics pigments and, to a lower extent, in the cadmium plating of metal surfaces. Cadmium is also found as an automotive pollutant in the combustion of petrochemicals, phosphate fertilizers and tobacco smoke (Andujar and Nemery, 2009).

Cadmium from cigarette smoke, the environment and food accumulate in the human prostate where it induces, as in other organs, the production of Oxygen Reactive Species, alters the metabolism of steroid hormones and stimulates the growth of human prostate epithelial cells, thus causing their malignant transformation. At the prostate level, cadmium interacts with selenium to form a cadmium-selenium protein complex (Cd-Se) (Schöpfer et al., 2010).

High doses of cadmium abolish the anti-carcinogenic effects of selenium and thus increase the risk of prostate cancer (Schrauzer, 2008). This explains why smokers have an increased risk of developing prostate cancer (Coughlin et al., 1996; Plaskon et al., 2003)

Many studies have shown that increased prostate cadmium concentrations are clinically associated with prostate cancer (Feustel et al., 1982; Ogunlewe and Osegbe, 1989; Brys et al., 1997).

According to Sarafanov, cadmium has long been suspected as a risk factor in this type of cancer, although its concentrations at the prostate level are below the so-called carcinogenic concentrations in in vivo prostate cancer models (Waalkes and Rehm, 1994). Other work supports this view since men with prostate cancer have an increase in cadmium levels in systemic circulation and prostate tissue compared to healthy men (Habib et al., 1976; Ogunlewe & Osegbe, 1989; Lee et al., 2009; Golovine et al., 2010). Similarly, in vitro studies have shown that human non-tumorigenicthe epithelial cells of the human prostate are maligned after exposure to cadmium (Achanzar et al., 2001; Nakamura et al., 2002; Zhang et al., 2008). Other results show that cadmium increases cancer aggressiveness in different models (Waalkes & Rehm, 1994; Abshire et al., 1996; Haga et al., 1996) and probably in humans (Elghany et al., 1990).

Carcinogenic mechanisms of action have been conventionally proposed such as the expression of aberrant genes, the inhibition of DNA repair systems and the induction of oxidative stress (Joseph, 2009). Golovine et al., (2010) have shown that cadmium reduce the expression of XIAP (X-linked inhibitory apoptosis protein) protein in cancerous prostate cells.

Many studies have shown an association between XIAP protein expression and carcinogenesis (Uren et al., 1999; Hofer-Warbinek et al., 2000; Yang et al., 2000; Eckelman et al., 2006). An enhancement of XIAP expression is recognized in many cancers, particularly prostate cancer (Nomura et*al*., 2005).

Another mechanism of action has been proposed related to the endocrine disrupting nature of cadmium (Lacorte et *al.*, 2011). This metal exhibits deleterious activities on the reproductive system since it exerts androgenic and estrogenic effects by activating androgen and estrogen receptors after binding to the endogenous active site (Stoica et al., 2000; Martin et al., 2002; Johnson & Wang, 2003; Prins, 2008; Byrne et al., 2009). This leads to an increase in androgen receptor expression after cadmium administration. This suggests that low cadmium doses may mimic the effect of androgens in prostate epithelial cells and thus increase their proliferative activity (Lacorte et al., 2011). This same team has also demonstrated that exposure to low concentrations and short-term exposure during puberty can induce early changes in cell proliferation, apoptosis, androgen receptor expression and stromal organization of the prostate. These changes would be a contributing factor in the occurrence of prostate lesions over the life course.

Zinc

Zinc (Zn) is an essential trace mineral required for the activity of transcription factors and many enzymes (Cortesi et al., 2008). Several studies have shown that the human body contains 2 to 4 grams of (Zn), which is found in high amounts in the prostate (Rishi et al., 2003).

Ferenc and his collaborators in 1967 reported that the zinc content of prostate carcinomas was lower than that of normal prostate epithelial cells (Costello & Franklin, 2006). The concentration of zinc was higher in benign prostatic hyperplasia compared with normal prostates, while lower concentrations were found in malignant carcinomas of the prostate (Kristal et al., 1999; Gonzalez et al., 2009). The loss of zinc during prostate tumorogenesis is not clearly understood yet. Many investigators have been found to demonstrate the effects of zinc regulation in normal prostates and prostate malignant cancer. It was demonstrated that the re-introduction of physiological levels of zinc inhibits > 50% of androgen-reactive cell growth (Li et al., 2005).Zinc chloride (ZnCl2) minipumps implanted subcutaneously over a five-week period significantly slowed tumor growth in mice with PC3 tumors (Epstein et al., 2011). Similar tumour suppressive effects have also been demonstrated in other types of tumours. For example, zinc deficiency favoured esophageal tumorigenesis, which was associated with an increase in the size and tumour phase of the cancer (Wagner et al., 2009).It has also inhibited the proliferation of colon cancer cells (Gallus et al., 2007). Therefore, zinc appears to negatively affect tumor cell growth. However, the mechanism behind this action is not clear. The most widely studied zinc target is the kappa-light-chain-enhancer nuclear factor of activated B cells (NF-κB), a signalling pathway (Kolonel et al., 1988; Zhang et al., 2009).

The activation of NF-κB in prostate cancer cells is inhibited by zinc, by blocking IκB kinase (IKK) to reduce the invasive potential of prostate cancer cells. A new study shows that zinc inhibits the growth of prostate cancer cells by reducing the expression of androgen receptor proteins. This study showed that intraperitoneal administration of zinc significantly reduced tumour size in mice with TRAMP-C2 subcutaneous tumours. Analyses of xenografted tumoursand normal prostates showed reduced androgen receptor expression and increased cell death (To et al., 2018).

Iron

Iron (Fe) is the most abundant trace element in the human body. The major source is the food consumption, particularly meat and legumes.The dietary Fe is available in both heme and non-heme (ionic) forms, with the first one being the easiest to absorb. Elevated dietary Fe intake can provide a source of reactive oxygen species (ROS) and potential carcinogens, such as N-nitroso compounds (NOC), which have a negative impact on DNA and are related to several cancers, such as prostate cancer (Muzandu et al., 1997).

Nevertheless, general epidemiological studies on the association between elevated Fe concentration (total and heme) and prostate cancer incidence are limited and they are often incoherent. In three recent case-control studies, Fe levels in different biological samples (blood, hair, nails) were compared in prostate cancer patients and matched healthy controls. (Tan and Chen, 2011; Karimi et al., 2012; Qayyum & Shah, 2014). First study revealed an elevation in Fe concentration in the hair of prostate cancer patients when compared to healthy matched controls (Tan and Chen, 2011). The second study, consisting of 50 prostate cases and 50 control subjects, matched for age and ethnicity, indicated that the hair and nails of prostate cancer patients had higher levels of Fe than those of controls. The newest study among the three has confirmed an increasing level of Fe in both the blood and hair of the scalp, but not in the nails of patients with cancer, in comparison to the matched healthy control givers. (Qayyum and Shah, 2014). The findings of all of these studies suggest that elevated Fe levels in these biological samples would be a possible prostate cancer risk factor.

Ferroportin, a transmembrane transport protein that controls the Fe flow of the cell can be decreased in prostate cancer cells and its intracellular reduction is related to a reduction in prostate cancer cell differentiation and an elevated degree of disease malignancy. (Xue et al., 2015). It is thought that the reduction of ferroportin expression levels in cancer cells, by decreasing the export of intracellular Fe and causing intracellular Fe overload, stimulates ROS and DNA-related damage, thus promoting tumorigenesis and/or tumor promotion. (Xue et al., 2015). In contrast, ferroportin expression is elevated in benign prostatic hyperplasia (BPH), which confirms the possibility that ferroportin has a role to play in prostate carcinogenesis. (Xue et al., 2015).

Selenium

Selenium is an important trace element in the organism. It is present in the body in trace amounts. The selenium content of the diet depends on the concentration of selenium in the soil. It is found in abundance in Brazilian nuts, meat, fish, eggs and cereals. Cruciferous vegetables, garlic and mushrooms also contain it, but in smaller quantities. It has critical activity for the glutathione peroxidase function, which may protect DNA and the other cellular compounds from oxidative damage (Combs et *al.*, 1985).

Several pre-clinical and epidemiological research studies have indicated the benefit of selenium as a tumor suppressor, including one study in the United States that has shown lower age-specific mortality rates due to some types of cancers in states where the selenium is at higher levels on the soil (Shamberger et *al.*, 1976).

In most studies (Comstock et al., 1992) that are based on pre-diagnosed selenium concentrations in serum, there is some evidence to suggest that there is an association between lower levels of selenium and various cancer incidences.

In the recently completed double-blind, placebo-controlled cancer prevention study (Clarket al., 1957) in which 200 mg of selenium was given daily to patients with a previous history of basal cell and squamous cell carcinoma, supplementation with selenium is not protective against the development of recurring skin cancers. The primary outcome is inversely correlated both to the incidence and the death rate for prostate cancer.

In the recent trial by Clark and colleagues Clark et al., (1957), which included a total of 48 cases of prostate cancer, it was discovered that men receiving 200 mg selenium at random per day only had one third the risk of prostate cancer ($p = 0.001$) in comparison to those receiving a placebo. The challenge results of this trial are supportive of further study of the association of selenium with prostate cancer. The trial indicated that selenium supplementation at relatively higher doses than the recommended daily intake for men (70 mg) may be anticancer, but there are few in which studies have investigated the role of selenium acquired from normal food intake in the risk of prostate cancer.

Although the biological mechanism by the way selenium affects prostate cancer is not currently known, there are a several suggested mechanisms. Griffin, (1979) has suggested that selenium is an anticancer agent due to its important role as an essential component of selenium-dependent glutathione peroxidase. The hypothesis is based on the carcinogenicity and mutagenicity of malonyldialdehyde, which is believed to be developed with a selenium deficit. The regulatory role of apoptosis appears to be a key determinant of cancer incidence and the anti-tumorigenic properties of the selenium compounds have been correlated with apoptotic reactions (El-Bayoumy et al., 1992; Thompson et al., 1994).

Copper

Copper (Cu) is a mineral necessary for the proper functioning of many enzyme systems, the disruption of copper homeostasis is associated with various diseases, including cancer (Jarup, 2003). Contaminated water and food, Cd-Cu alloy battery plants, corrosion of domestic plumbing infrastructure are the main sources of copper overload. Traditional indicators of Cu homeostasis are concentrations of Cu in serum, hair and nails (Cartwright et al., 1960).

As already described for other trace elements and heavy metals, the association between high exposure to Cu and prostate cancer is still questionable, as there are few studies on this issue. 4 case-control studies compared Cu concentrations in different biological samples (hair, nails and blood) between prostate cancer patients and matched healthy controls (Tan and Chen, 2011; Karimi et al., 2012; Ozmen et al., 2006; Qayyum and Shah, 2014). In three of these studies, the concentration of Cu in the cancer group was higher on average than in the healthy group, indicating that high levels of Cu may induce an increased risk of prostate cancer. (Tan and Chen, 2011; Karimi et al., 2012; Ozmen et al., 2006). However, the most recent case-control study showed similar Cu concentrations in the patient and control groups (Qayyum and Shah, 2014).

Some studies have reported high levels of Cu (2 to 6 times) in cell lines of prostate cancer patients in vitro (Cater and Haupt, 2011; Chen et al., 2007) and in xenograft mouse models (Safi et al., 2014). Although high amounts of Cu have been found in several cancer cells, the understanding of the mechanisms underlying this phenomenon is not fully understood.In summary, the available data suggest that Cu overload may be a risk factor for prostate cancer.

Several studies have shown that exposure to metals has such toxic and carcinogenic effects on humans and animals.

CONCLUSION

According to these studies, excessive occupational and environmental exposure to metals (arsenic (As) and cadmium (Cd)) is considered a major cause of prostate cancer. Also changes in the equilibrium of optimal levels of trace elements such as copper (Cu), zinc (Zn), iron (Fe) and selenium (Se) can affect biological pathways and have been associated with many diseases, including prostate cancer.

REFERENCES

Abshire, M. K., Devor, D. E., Diwan, B. A., Shaughnessy, J. D. Jr, & Waalkes, M. P. (1996). In vitro exposure to cadmium in rat L6 myoblasts can result in both enhancement and suppression of malignant progression in vivo. *Carcinogenesis*, *17*(6), 1349–1356. doi:10.1093/carcin/17.6.1349 PMID:8681454

Achanzar, W. E., Diwan, B. A., Liu, J., Quader, S. T., Webber, M. M., & Waalkes, M. P. (2001). Cadmium-induced malignant transformation of human prostate epithelial cells. *Cancer Research*, *61*(2), 455–458. PMID:11212230

Alavanja, M. C. R. (2003). Use of Agricultural Pesticides and Prostate Cancer Risk in the Agricultural Health Study Cohort. *American Journal of Epidemiology*, *157*(9), 800–814. doi:10.1093/aje/kwg040 PMID:12727674

Andujar, P., & Nemery, B. (2009). Acute and subacute chemical pneumonitis. *Revue des Maladies Respiratoires*, *26*(8), 867–885. doi:10.1016/S0761-8425(09)73682-4 PMID:19953031

Arnold, M., Karim-Kos, H. E., Coebergh, J. W., Byrnes, G., Antilla, A., Ferlay, J., ... Soerjomataram, I. (2015). Recent trends in incidence of five common cancers in 26 European countries since 1988: Analysis of the European Cancer Observatory. *European Journal of Cancer*, *51*(9), 1164–1187. doi:10.1016/j.ejca.2013.09.002 PMID:24120180

Aronson, K. J., Siemiatycki, J., Dewar, R., & Gerin, M. (1996). Occupational risk factors for prostate cancer: Results from a case-control study in Montreal, Quebec, Canada. *American Journal of Epidemiology*, *143*(4), 363–373. doi:10.1093/oxfordjournals.aje.a008750 PMID:8633620

Aumüller, G. (2012). *Prostate gland and seminal vesicles*. Springer Science & Business Media.

Band, P. R., Abanto, Z., Bert, J., Lang, B., Fang, R., Gallagher, R. P., & Le, N. D. (2011). Prostate cancer risk and exposure to pesticides in British Columbia Farmers. *The Prostate*, *71*(2), 168–183. doi:10.1002/pros.21232 PMID:20799287

Bashir, M. N. (2015). Epidemiology of prostate cancer. *Asian Pacific Journal of Cancer Prevention*, *16*(13), 5137–5141. doi:10.7314/APJCP.2015.16.13.5137 PMID:26225642

Belpomme, D., Irigaray, P., Ossondo, M., Vacque, D., & Martin, M. (2009). Prostate cancer as an environmental disease: An ecological study in the French Caribbean islands, Martinique and Guadeloupe. *International Journal of Oncology*, *34*(4), 1037–1044. doi:10.3892/ijo_00000229 PMID:19287960

Benbrahim-Tallaa, L., Waterland, R. A., Styblo, M., Achanzar, W. E., Webber, M. M., & Waalkes, M. P. (2005). a. Molecular events associated with arsenic-induced malignant transformation of human prostatic epithelial cells: Aberrant genomic DNA methylation and K-ras oncogene activation. *Toxicology and Applied Pharmacology*, *206*(3), 288–298. doi:10.1016/j.taap.2004.11.017 PMID:16039940

Benbrahim-Tallaa, L., Webber, M. M., & Waalkes, M. P. (2005). Acquisition of androgen independence by human prostate epithelial cells during arsenic-induced malignant transformation. *Environmental Health Perspectives*, *113*(9), 1134–1139. doi:10.1289/ehp.7832 PMID:16140617

Benbrahim-Tallaa, L., Webber, M. M., & Waalkes, M. P. (2005). b. Acquisition of androgen independence by human prostate epithelial cells during arsenic-induced malignant transformation. *Environmental Health Perspectives, 113*(9), 1134–1139. doi:10.1289/ehp.7832 PMID:16140617

Benbrahim-Tallaa, L., Webber, M. M., & Waalkes, M. P. (2006). Mechanisms of acquired androgen independence during arsenic-induced malignant transformation of human prostate epithelial cells. *Environmental Health Perspectives, 115*(2), 243–247. doi:10.1289/ehp.9630 PMID:17384772

Benbrahim-Tallaa, L., Webber, M. M., & Waalkes, M. P. (2007). c. Mechanisms in acquired androgen independence during arsenic-induced malignant transformation of human prostate epithelial cells. *Environmental Health Perspectives, 115*(2), 243–247. doi:10.1289/ehp.9630 PMID:17384772

Berezovskaya, O., Schimmer, A. D., Glinskii, A. B., Pinilla, C., Hoffman, R. M., Reed, J. C., & Glinsky, G. V. (2005). Increased expression of apoptosis inhibitor protein XIAP contributes to anoikisresistance of circulating human prostate cancer metastasis precursor cells. *Cancer Research, 65*(6), 2378–2386. doi:10.1158/0008-5472.CAN-04-2649 PMID:15781653

Bibel, B. M. (2012). *The Whole Life Prostate Book: Everything that Every Man-at Every Age-Needs To Know About Maintaining Optimal Prostate Health.* Academic Press.

Blair, A., Dosemeci, M., & Heineman, E. F. (1993). Cancer and other causes of death among male and female farmers from twenty-three states. *American Journal of Industrial Medicine, 23*(5), 729–742. doi:10.1002/ajim.4700230507 PMID:8506851

Blair, A., & Zahm, S. H. (1991). Cancer among farmers. *Occupational Medicine (Philadelphia, Pa.), 6*(3), 335–354. PMID:1835166

Blair, A., & Zahm, S. H. (1995). Agricultural exposures and cancer. *Environmental Health Perspectives, 103*(suppl 8), 205–208. PMID:8741784

Blair, A., Zahm, S. H., Pearce, N. E., Heineman, E. F., & Fraumeni, J. F. (1992). Clues to cancer etiology from studies of farmers. Scandinavian Journal of Work, Environment & Health, 18(4), 209–215. Retrieved from Brooks, P. C. (1996). Cell adhesion molecules in angiogenesis. *Cancer and Metastasis Reviews, 15*(2), 187–194.

Bryś, M., Nawrocka, A. D., Miekoś, E., Zydek, C., Foksiński, M., Barecki, A., & Krajewska, W. M. (1997). Zincand cadmium analysis in human prostate neoplasms. *Biological Trace Element Research, 59*(1-3), 145–152. doi:10.1007/BF02783239 PMID:9522056

Byrne, C., Divekar, S. D., Storchan, G. B., Parodi, D. A., & Martin, M. B. (2009). Cadmium--a metallohormone. *Toxicology and Applied Pharmacology, 238*(3), 266–271. doi:10.1016/j.taap.2009.03.025 PMID:19362102

Chen, C. J., Kuo, T. L., & Wu, M. M. (1988). Arsenic and cancers. *Lancet, 1*(8582), 414–415. doi:10.1016/S0140-6736(88)91207-X PMID:2893213

Chen, C. J., & Wang, C. J. (1990). Ecological correlation between arsenic levels in well water and age adjusted mortality from malignant neoplasms. *Cancer Research, 50*, 5470–5474. PMID:2386951

Clark, L. C., Combs, G. F., Turnbull, B. W., Slate, E. H., Chalker, D. K., Chow, J., ... Taylor, J. R. (1997). Effects of selenium supplementation for cancer prevention in patients with carcinoma of the skin. A randomized controlled trial. *Journal of the American Medical Association*, *277*, 1520. PMID:8971064

Combs, G. F. Jr, & Clark, L. C. (1985). Can dietary selenium modify cancer risk? *Nutrition Reviews*, *43*(11), 325–331. doi:10.1111/j.1753-4887.1985.tb02392.x PMID:3909013

Comstock, G. W., Bush, T. L., & Helzlsouer, K. (1992). Serum retinol, beta-carotene, vitamin E, and selenium as related to subsequent cancer of specific sites. *American Journal of Epidemiology*, *135*(2), 115–121. doi:10.1093/oxfordjournals.aje.a116264 PMID:1536130

Cortesi, M., Fridman, E., Volkov, A., Shilstein, S., Chechik, R., Breskin, A., ... Raviv, G. (2008). Clinical assessment of the cancer diagnostic value of prostatic zinc: A comprehensive needle-biopsy study. *The Prostate*, *68*(9), 994–1006. doi:10.1002/pros.20766 PMID:18386292

Costello, L. C., & Franklin, R. B. (2006). The clinical relevance of the metabolism of prostate cancer; zinc and tumor suppression: Connecting the dots. *Molecular Cancer*, *5*(1), 17. doi:10.1186/1476-4598-5-17 PMID:16700911

Coughlin, S. S., Neaton, J. D., & Sengupta, A. (1996). Cigarette smoking as a predictor of death from-prostate cancer in 348,874 men screened for the Multiple Risk Factor Intervention Trial. *American Journal of Epidemiology*, *143*(10), 1002–1006. doi:10.1093/oxfordjournals.aje.a008663 PMID:8629606

De Marzo, A. M., Marchi, V. L., Epstein, J. I., & Nelson, W. G. (1999). Proliferative inflammatory atrophy of the prostate: Implications for prostatic carcinogenesis. *American Journal of Pathology*, *155*(6), 1985–1992. doi:10.1016/S0002-9440(10)65517-4 PMID:10595928

Desantis, C., Siegel, R., & Jemal, A. (2014). Cancer Treatment & Survivorship Facts and Figures 2014 & 2015. American Cancer Society.

Dich, J., & Wiklund, K. (1998). Prostate cancer in pesticide applicators in swedish agriculture. *The Prostate*, *34*(2), 100–112. doi:10.1002/(SICI)1097-0045(19980201)34:2<100::AID-PROS4>3.0.CO;2-O PMID:9465941

Dich, J., & Wiklund, K. (1998). Prostate cancer in pesticide applicators in Swedish agriculture. *The Prostate*, *34*(2), 100–112. doi:10.1002/(SICI)1097-0045(19980201)34:2<100::AID-PROS4>3.0.CO;2-O PMID:9465941

Dosemeci, M., Cocco, P., Gómez, M., Stewart, P. A., & Heineman, E. F. (1994). Effects of three features of a job-exposure matrix on risk estimates. *Epidemiology (Cambridge, Mass.)*, *5*(1), 124–127. doi:10.1097/00001648-199401000-00019 PMID:8117771

Dosemeci, M., Hoover, N., Blair, A., Figgs, W., Devesa, S., Grauman, D., ... Carolina, S. (1994). BRIEF Farming and Prostate Cancer for a grouping of three southeastern. *American Journal of Industrial Medicine*, *86*(22).

Eckelman, B. P., Salvesen, G. S., & Scott, F. L. (2006). Human inhibitor of apoptosis proteins: Why XIAP isthe black sheep of the family. *EMBO Reports*, *7*(10), 988–994. doi:10.1038j.embor.7400795 PMID:17016456

Eeles, R. A., Kote-Jarai, Z., Giles, G. G., Al Olama, A. A., Guy, M., Jugurnauth, S. K., ... Field, H. I. (2008). Multiple newly identified loci associated with prostate cancer susceptibility. *Nature Genetics*, *40*(3), 316–321. doi:10.1038/ng.90 PMID:18264097

El-Bayoumy, K., Chae, Y. H., Upadhyaya, P., Meschter, C., Cohen, L. A., & Reddy, B. S. (1992). Inhibition of 7,12- dimethylbenz(a)anthracene-induced tumors and DNA adduct formation in the mammary glands of female Sprague-Dawley rats by the synthetic organoselenium compound, 1,4- phenylenebis(methylene) selenocyanate. *Cancer Research*, *52*, 2402–2407. PMID:1568209

Elghany, N. A., Schumacher, M. C., Slattery, M. L., West, D. W., & Lee, J. S. (1990). Occupation, cadmiumexposure, and prostate cancer. *Epidemiology (Cambridge, Mass.)*, *1*(2), 107–115. doi:10.1097/00001648-199003000-00005 PMID:2073496

Elissalde, M. H., & Clark, D. E. (1979). Testosterone metabolism by hexachlorobenzene-induced hepatic microsomal enzymes. *American Journal of Veterinary Research*, *40*(12), 1762–1766.

Epstein, M. M., Kasperzyk, J. L., Andren, O., Giovannucci, E. L., Wolk, A., Hakansson, N., ... Mucci, L. A. (2011). Dietary zinc and prostate cancer survival in a Swedish cohort. *The American Journal of Clinical Nutrition*, *93*(3), 586–593.

Feneley, M. R., Young, M. P., Chinyama, C., Kirby, R. S., & Parkinson, M. C. (1996). Ki-67 expression in early prostate cancer and associated pathological lesions. *Journal of Clinical Pathology*, *49*(9), 741–748.

Feustel, A., Wennrich, R., Steiniger, D., & Klauss, P. (1982). Zinc and cadmium concentration in prostaticcarcinoma of different histological grading in comparison to normal prostate tissue andadenofibromyomatosis (BPH). *Journal of Urology and Research*, *10*(6), 301–303.

Folkman, J. (1984). Angiogenesis. In *Biology of endothelial cells* (pp. 412–428). Boston, MA: Springer. doi:10.1007/978-1-4613-2825-4_42

Folkman, J. (1996). Fighting cancer by attacking its blood supply. *Scientific American*, *275*(3), 150–154. doi:10.1038cientificamerican0996-150 PMID:8701285

Foster, W. G., Jarrell, J. F., Younglai, E. V., Wade, M. G., Arnold, D. L., & Jordan, S. (1996). An Overview of Some Reproductive Toxicology Studies Conducted At Health Canada. *Toxicology and Industrial Health*, *12*(3–4), 447–459. doi:10.1177/074823379601200316 PMID:8843561

Gallagher, R. P., & Fleshner, N. (1998). Prostate cancer: 3. Individual risk factors. *CMAJ: Canadian Medical Association Journal*, *159*(7), 807. PMID:9805030

Gallus, S., Foschi, R., Negri, E., Talamini, R., Franceschi, S., Montella, M., ... La Vecchia, C. (2007). Dietary zinc and prostate cancer risk: A case-control study from Italy. *European Urology*, *52*(4), 1052–1056. doi:10.1016/j.eururo.2007.01.094 PMID:17292532

Gocmen, A., Peters, H. A., Cripps, D. J., Bryan, G. T., & Morris, C. R. (1989). Hexachlorobenzene episode in Turkey. *Biomedical and Environmental Sciences : BES*, *2*(1), 36–43. PMID:2590490

Golovine, K., Makhov, P., Uzzo, R. G., Kutikov, A., Kaplan, D. J., Fox, E., & Kolenko, V. M. (2010). Cadmiumdown-regulates expression of XIAP at the post-transcriptional level in prostate cancer cellsthrough an NF-kappaB-independent, proteasome-mediated mechanism. *Molecular Cancer*, *9*(1), 183. doi:10.1186/1476-4598-9-183 PMID:20618956

Gonzalez, A., Peters, U., Lampe, J. W., & White, E. (2009). Zinc intake from supplements and diet and prostate cancer. *Nutrition and Cancer*, *61*(2), 206–215. doi:10.1080/01635580802419749 PMID:19235036

Griffin, A. C. (1979). Role of selenium in the chemoprevention of cancer. *Advances in Cancer Research*, *29*, 419–442. doi:10.1016/S0065-230X(08)60851-9 PMID:382777

Habib, F. K., Lee, I. R., Stitch, S. R., & Smith, P. H. (1976). Androgen levels in the plasma and prostatic tissuesof patients with benign hypertrophy and carcinoma of the prostate. *The Journal of Endocrinology*, *71*(1), 99–107. doi:10.1677/joe.0.0710099 PMID:62008

Haga, A., Nagase, H., Kito, H., & Sato, T. (1996). Enhanced invasiveness of tumour cells after host-exposure to heavy metals. *European Journal of Cancer*, *32A*(13), 2342–2347. doi:10.1016/S0959-8049(96)00349-8 PMID:9038619

Hansen, M. E., & Matsumura, F. (2001). Effects of heptachlor epoxide on components of various signal transduction pathways important in tumor promotion in mouse hepatoma cells. Determination of the most sensitive tumor promoter related effect induced by heptachlor epoxide. *Toxicology*, *160*(1–3), 139–153.

Hinwood, A. L., Jolley, D. J., & Sim, M. R. (1999). Cancer incidence and high environmental arsenic concentrations in rural populations: Results of an ecological study. *International Journal of Environmental Health Research*, *9*, 131–141.

Ho, S.-M., Lee, M.-T., Lam, H.-M., & Leung, Y.-K. (2011). Estrogens and Prostate Cancer: Etiology, Mediators, Prevention, and Management. *Endocrinology and Metabolism Clinics of North America*, *40*(3), 591–614. doi:10.1016/j.ecl.2011.05.002 PMID:21889723

Hofer-Warbinek, R., Schmid, J. A., Stehlik, C., Binder, B. R., Lipp, J., & de Martin, R. (2000). Activation ofNF-kappa B by XIAP, the X chromosome-linked inhibitor of apoptosis, in endothelial cellsinvolves TAK1. *The Journal of Biological Chemistry*, *275*(29), 22064–22068. doi:10.1074/jbc.M910346199 PMID:10807933

IARC. (2004). Some drinking-water disinfectants and contaminants, including arsenic. *IARC Monographs on the Evaluation of Carcinogenic Risks to Humans*, *84*, 269–477. PMID:15645578

Imaida, K., & Shirai, T. (2000). Endocrine disrupting chemicals and carcinogenesis--breast, testis and prostate cancer. Nihon Rinsho. *Japanese Journal of Clinical Medicine*, *58*(12), 2527–2532.

Johnson, N. P., & Wang, K. (2003). Is ovarian surgery effective for androgenic symptoms of polycystic ovarian syndrome. *Journal of Obstetrics & Gynaecology*, *23*(6), 599–606.

Joseph, P. (2009). Mechanisms of cadmium carcinogenesis. *Toxicology and Applied Pharmacology*, *238*(3), 272–279. doi:10.1016/j.taap.2009.01.011 PMID:19371617

Karimi, G., Shahar, S., Homayouni, N., Rajikan, R., Abu Bakar, N. F., & Othman, M. S. (2012). Association between trace element and heavy metal levels in hair and nail with prostate cancer. *Asian Pacific Journal of Cancer Prevention, 13*(9), 4249–4253. doi:10.7314/APJCP.2012.13.9.4249 PMID:23167323

Kolonel, L. N., Yoshizawa, C. N., & Hankin, J. H. (1988). Diet and prostatic cancer: A case-control study in Hawaii. *American Journal of Epidemiology, 127*(5), 999–1012. doi:10.1093/oxfordjournals.aje.a114903 PMID:3358418

Koutros, S., Langseth, H., Grimsrud, T. K., Barr, D. B., Vermeulen, R., Portengen, L., ... Engel, L. S. (2015). Prediagnostic Serum Organochlorine Concentrations and Metastatic Prostate Cancer: A Nested Case–Control Study in the Norwegian Janus Serum Bank Cohort. *Environmental Health Perspectives, 123*(9), 867–872. doi:10.1289/ehp.1408245 PMID:25734605

Kristal, A. R., Stanford, J. L., Cohen, J. H., Wicklund, K., & Patterson, R. E. (1999). Vitamin and mineral supplement use is associated with reduced risk of prostate cancer. *Cancer Epidemiology, Biomarkers & Prevention, 8*(10), 887–892. PMID:10548317

Lacorte, L. M., Delella, F. K., Porto Amorim, E. M., Justulin, L. A. Jr, Godinho, A. F., Almeida, A. A., ... Felisbino, S. L. (2011). Early changes induced by short-term low-dose cadmium exposure in rat ventral and dorsolateral prostates. *Microscopy Research and Technique, 74*(11), 988–997. doi:10.1002/jemt.20985 PMID:21287658

Lee, J. D., Wu, S. M., Lu, L. Y., Yang, Y. T., & Jeng, S. Y. (2009). Cadmium concentration and metallothioneinexpression in prostate cancer and benign prostatic hyperplasia of humans. *Journal of the Formosan Medical Association, 108*(7), 554–559. doi:10.1016/S0929-6646(09)60373-9 PMID:19586829

Lewis, D. R., Southwick, J. W., Ouellet-Hellstrom, R., Rench, J., & Calderon, R. L. (1999). Drinking water arsenic in Utah: A cohort mortality study. *Environmental Health Perspectives, 107*(5), 359–365. doi:10.1289/ehp.99107359 PMID:10210691

Li, X. M., Zhang, L., Li, J., Li, Y., Wang, H. L., Ji, G. Y., ... Zhao, X. J. (2005). Measurement of serum zinc improves prostate cancer detection efficiency in patients with PSA levels between 4 ng/mL and 10 ng/mL. *Asian Journal of Andrology, 7*(3), 323–328. doi:10.1111/j.1745-7262.2005.00044.x PMID:16110361

Martin, M. B., Voeller, H. J., Gelmann, E. P., Lu, J., Stoica, E. G., Hebert, E. J., ... Stoica, A. (2002). Role of cadmium in the regulation of AR gene expressionand activity. *Endocrinology, 143*(1), 263–275. doi:10.1210/endo.143.1.8581 PMID:11751618

McKinlay, R., Plant, J. A., Bell, J. N. B., & Voulvoulis, N. (2008). Endocrine disrupting pesticides: Implications for risk assessment. *Environment International, 34*(2), 168–183. doi:10.1016/j.envint.2007.07.013 PMID:17881056

Medjakovic, S., Zoechling, A., Gerster, P., Ivanova, M. M., Teng, Y., Klinge, C. M., ... Jungbauer, A. (2014). Effect of nonpersistent pesticides on estrogen receptor, androgen receptor, and aryl hydrocarbon receptor. *Environmental Toxicology, 29*(10), 1201–1216. doi:10.1002/tox.21852 PMID:23436777

Miller, D. C., Hafez, K. S., Stewart, A., Montie, J. E., & Wei, J. T. (2003). Prostate carcinoma presentation, diagnosis, and staging. *Cancer*, *98*(6), 1169–1178. doi:10.1002/cncr.11635 PMID:12973840

Mills, P. K., & Yang, R. (2003). Prostate cancer risk in California farm workers. *Journal of Occupational and Environmental Medicine*, *45*(3), 249–258. doi:10.1097/01.jom.0000058339.05741.0c PMID:12661182

Mnif, W., Hassine, A. I. H., Bouaziz, A., Bartegi, A., Thomas, O., & Roig, B. (2011). Effect of Endocrine Disruptor Pesticides: A Review. *International Journal of Environmental Research and Public Health*, *8*(6), 2265–2303. doi:10.3390/ijerph8062265 PMID:21776230

Morrison, H., Savitz, D., Semenciw, R., Hulka, B., Mao, Y., Morison, D., & Wigle, D. (1993). Farming and prostate cancer mortality. *American Journal of Epidemiology*, *137*(3), 270–280.

Morrison, H., Savitz, D., Semenciw, R., Hulka, B., Mao, Y., Morison, D., & Wigle, D. (1993). Farming and prostate cancer mortality. *American Journal of Epidemiology, 137*(3), 270-280

Mostafalou, S., & Abdollahi, M. (2013). Pesticides and human chronic diseases: Evidences, mechanisms, and perspectives. *Toxicology and Applied Pharmacology*, *268*(2), 157–177.

Müller, W. F., Hobson, W., Fuller, G. B., Knauf, W., Coulston, F., & Korte, F. (1978). Endocrine effects of chlorinated hydrocarbons in rhesus monkeys. *Ecotoxicology and Environmental Safety*, *2*(2), 161–172.

Mullins, J. K., & Loeb, S. (2012). Environmental exposures and prostate cancer. *Urologic Oncology*, *30*(2), 216–219.

Mullins, J. K., & Loeb, S. (2012, March). Environmental exposures and prostate cancer. In Urologic Oncology: Seminars and Original Investigations (Vol. 30, No. 2, pp. 216-219). Elsevier doi:10.1016/j.urolonc.2011.11.014

Mustafa, M., Salih, A., Illzam, E., Sharifa, A., Suleiman, M., & Hussain, S. (2016). Prostate Cancer: Pathophysiology, Diagnosis, and Prognosis. *IOSR Journal of Dental and Medical Sciences Ver. II*, *15*(6), 2279–2861.

Muzandu, K., Shaban, Z., Ishizuka, M., Kazusaka, A., & Fujita, S. (2005). Nitric oxide enhances catechol estrogen-induced oxidative stress in LNCaP cells. *Free Radical Research*, *39*(4), 389–398. doi:10.1080/10715760400029710 PMID:16028364

Nakamura, K., Yasunaga, Y., Ko, D., Xu, L. L., Moul, J. W., Peehl, D. M., ... Rhim, J. S. (2002). Cadmium-induced neoplastic transformation of human prostate epithelial cells. *International Journal of Oncology*, *20*(3), 543–547. PMID:11836566

Nelson, W. G., De Marzo, A. M., & Isaacs, W. B. (2003). Prostate cancer. *The New England Journal of Medicine*, *349*(4), 366–381. doi:10.1056/NEJMra021562 PMID:12878745

Nomura, T., Yamasaki, M., Nomura, Y., & Mimata, H. (2005). Expression of the inhibitors of apoptosis-proteins in cisplatin-resistant prostate cancer cells. *Oncology Reports*, *14*(4), 993–997. PMID:16142363

Ogunlewe, J. O., & Osegbe, D. N. (1989). Zinc and cadmium concentrations in indigenous blacks withnormal, hypertrophic, and malignant prostate. *Cancer, 63*(7), 1388–1392. doi:10.1002/1097-0142(19890401)63:7<1388::AID-CNCR2820630725>3.0.CO;2-M PMID:2465818

Parent, M. É., & Siemiatycki, J. (2001). Occupation and prostate cancer. *Epidemiologic Reviews, 23*(1), 138–143. doi:10.1093/oxfordjournals.epirev.a000779 PMID:11588838

Parkin, D., Bray, F., & Devesa, S. (2001). Cancer burden in the year 2000. The global picture. *European Journal of Cancer, 37*, 4–66. doi:10.1016/S0959-8049(01)00267-2 PMID:11602373

Plaskon, L. A., Penson, D. F., Vaughan, T. L., & Stanford, J. L. (2003). Cigarette smoking and risk of prostatecancer in middle-aged men. *Cancer Epidemiology, Biomarkers & Prevention, 12*(7), 604–609. PMID:12869398

Prezioso, D., Denis, L. J., Klocker, H., Sciarra, A., Reis, M., Naber, K., ... Griffiths, K. (2007). Estrogens and aspects of prostate disease. *International Journal of Urology, 14*(1), 1–16. doi:10.1111/j.1442-2042.2006.01476.x PMID:17199853

Prins, G. S. (2008). Endocrine disruptors and prostate cancer risk. *Endocrine-Related Cancer, 15*(3), 649–656. doi:10.1677/ERC-08-0043 PMID:18524946

Qayyum, M. A., & Shah, M. H. (2014). Comparative study of trace elements in blood, scalp hair and nails of prostate cancer patients in relation to healthy donors. *Biological Trace Element Research, 162*(1-3), 46–57. doi:10.100712011-014-0123-4 PMID:25231348

Ralph, J. L., Orgebin-Crist, M.-C., Lareyre, J.-J., & Nelson, C. C. (2003). Disruption of Androgen Regulation in the Prostate by the Environmental Contaminant Hexachlorobenzene. *Environmental Health Perspectives, 111*(4), 461–466. doi:10.1289/ehp.5919 PMID:12676599

Ricke, W. A., Wang, Y., & Cunha, G. R. (2007). Steroid hormones and carcinogenesis of the prostate: The role of estrogens. Differentiation. *Research in Biological Diversity, 75*(9), 871–882. PMID:17924963

Rishi, I., Baidouri, H., Abbasi, J. A., Bullard-Dillard, R., Kajdacsy-Balla, A., Pestaner, J. P., ... Bagasra, O. (2003). Prostate cancer in African American men is associated with downregulation of zinc transporters. *Applied Immunohistochemistry & Molecular Morphology, 11*(3), 253–260. doi:10.1097/00129039-200309000-00009 PMID:12966353

Ruska, K. M., Sauvageot, J., & Epstein, J. I. (1998). Histology and cellular kinetics of prostatic atrophy. The American Journal of Surgical Pathology, 22(9), 1073–1077. Retrieved from Schimmer, A.D., Dalili, S., Batey, R.A., &Riedl, S.J., (2006).*Targeting XIAP for the treatment ofmalignancy. Cell Death and Differentiation, 13*(2), 179–188.

Schöpfer, J., Drasch, G., & Schrauzer, G. N. (2010). Selenium and cadmium levels and ratios inprostates,livers, and kidneys of nonsmokers and smokers. *Biological Trace Element Research, 134*(2), 180–187. doi:10.100712011-010-8636-y PMID:20174974

Schrauzer, G. N. (2008). Interactive effects of selenium and cadmium on mammary tumor developmentand growth in MMTV-infected female mice. A model study on the roles of cadmium andselenium in human breast cancer. *Biological Trace Element Research*, *123*(1-3), 27–34. doi:10.100712011-008-8091-1 PMID:18264682

Segal, R. J., Reid, R. D., Courneya, K. S., Malone, S. C., Parliament, M. B., Scott, C. G., ... Wells, G. A. (2003). Resistance exercise in men receiving androgen deprivation therapy for prostate cancer. *Journal of Clinical Oncology*, *21*(9), 1653–1659. doi:10.1200/JCO.2003.09.534 PMID:12721238

Shamberger, R. J., Tytko, S. A., & Willis, C. E. (1976). Antioxidants and cancer. Part VI. Selenium and ageadjusted human cancer mortality. *Archives of Environmental Health*, *31*(5), 231–235. doi:10.1080/00039896.1976.10667225 PMID:973735

Siegel, R., Naishadham, D., & Jemal, A. (2013). Cancer statistics, 2013. *CA: a Cancer Journal for Clinicians*, *63*(1), 11–30. doi:10.3322/caac.21166 PMID:23335087

Soto, A. M., Sonnenschein, C., Chung, K. L., Fernandez, M. F., Olea, N., & Serrano, F. O. (1995). The E-SCREEN assay as a tool to identify estrogens: An update on estrogenic environmental pollutants. *Environmental Health Perspectives*, *103*(suppl 7), 113–122. PMID:8593856

Stanford, J. L., & Ostrander, E. (2001). Familial prostate cancer. *Epidemiologic Reviews*, *23*(1), 19–23. doi:10.1093/oxfordjournals.epirev.a000789 PMID:11588848

Stoica, A., Katzenellenbogen, B. S., & Martin, M. B. (2000). Activation of estrogen receptor-alpha by theheavy metal cadmium. *Molecular Endocrinology (Baltimore, Md.)*, *14*(4), 545–553. PMID:10770491

Struewing, J. P., Hartge, P., Wacholder, S., Baker, S. M., Berlin, M., McAdams, M., ... Tucker, M. A. (1997). The Risk of Cancer Associated with Specific Mutations of BRCA1 and BRCA2 among Ashkenazi Jews. *The New England Journal of Medicine*, *336*(20), 1401–1408. doi:10.1056/NEJM199705153362001 PMID:9145676

Tan, C., & Chen, H. (2011). Screening of prostate cancer by analyzing trace elements in hair and chemometrics. *Biological Trace Element Research*, *144*(1-3), 97–108. doi:10.100712011-011-9038-5 PMID:21452047

Thompson, H. J., Wilson, A., Lu, J., Singh, M., Jiang, C., Upadhyaya, P., ... Ip, C. (1994). Comparison of the effects of an organic and an inorganic form of selenium on a mammary carcinoma cell line. *Carcinogenesis*, *15*(2), 183–186. doi:10.1093/carcin/15.2.183 PMID:8313506

To, P.K., Do, M.H., Cho, Y.S., Kwon, S.Y., Kim, M.S., &Jung, C., (2018). Zinc Inhibits Expression of Androgen Receptor to Suppress Growth of Prostate Cancer Cells. *International Journal of Molecular Sciences*, *19*(10).

Toppari, J., Larsen, J. C., Christiansen, P., Giwercman, A., Grandjean, P., Guillette, L. J., ... Skakkebaek, N. E. (1996). Male reproductive health and environmental xenoestrogens. *Environmental Health Perspectives*, *104*(suppl 4), 741–803. PMID:8880001

Tsai, S. M., Wang, T. N., & Ko, Y. C. (1999). Mortality for certain diseases in areas with high levels of arsenic in drinking water. *Archives of Environmental Health*, *54*(3), 186–193. doi:10.1080/00039899909602258 PMID:10444040

Turner, E. L., Lane, J. A., Donovan, J. L., Davis, M. J., Metcalfe, C., Neal, D. E., ... Martin, R. M. (2011). Association of diabetes mellitus with prostate cancer: Nested case-control study (Prostate testing for cancer and Treatment study). *International Journal of Cancer*, *128*(2), 440–446. doi:10.1002/ijc.25360 PMID:20473853

Uren, A. G., Beilharz, T., O'Connell, M. J., Bugg, S. J., van Driel, R., Vaux, D. L., & Lithgow, T. (1999). Role foryeast inhibitor of apoptosis (IAP)-like proteins in cell division. *Proceedings of the National Academy of Sciences of the United States of America*, *96*(18), 10170–10175. doi:10.1073/pnas.96.18.10170 PMID:10468581

Van der Cruijsen-Koeter, I. W., Vis, A. N., Roobol, M. J., Wildhagen, M. F., de Koning, H. J., van der Kwast, T. H., & Schroder, F. (2005). Comparison of screen detected and clinically diagnosed prostate cancer in the European randomized study of screening for prostate cancer, section Rotterdam. *The Journal of Urology*, *174*(1), 121–125. doi:10.1097/01.ju.0000162061.40533.0f PMID:15947595

Waalkes, M. P. (2000). Cadmium carcinogenesis in review. *Journal of Inorganic Biochemistry*, *79*(1-4), 241–244. doi:10.1016/S0162-0134(00)00009-X PMID:10830873

Waalkes, M. P., & Rehm, S. (1994). Cadmium and prostate cancer. *Journal of Toxicology and Environmental Health*, *43*(3), 251–269. doi:10.1080/15287399409531920 PMID:7966437

Wagner, S. E., Burch, J. B., Hussey, J., Temples, T., Bolick-Aldrich, S., Mosley-Broughton, C., ... Hebert, J. R. (2009). Soil zinc content, groundwater usage, and prostate cancer incidence in South Carolina. *Cancer causes &control. CCC*, *20*(3), 345–353. PMID:18949566

Whitaker, H. C., Kote-Jarai, Z., Ross-Adams, H., Warren, A. Y., Burge, J., George, A., ... Neal, D. E. (2010). The rs10993994 Risk Allele for Prostate Cancer Results in Clinically Relevant Changes in Microseminoprotein-Beta Expression in Tissue and Urine. *PLoS One*, *5*(10), e13363. doi:10.1371/journal.pone.0013363 PMID:20967219

Wu, M. M., Kuo, T. L., Hwang, Y. H., & Chen, C. J. (1989). Dose-response relation between arsenic concentration in well water and mortality from cancers and vascular diseases. *American Journal of Epidemiology*, *130*(6), 1123–1132. doi:10.1093/oxfordjournals.aje.a115439 PMID:2589305

Xu, X., Dailey, A. B., Talbott, E. O., Ilacqua, V. A., Kearney, G., & Asal, N. R. (2009). Associations of serum concentrations of organochlorine pesticides with breast cancer and prostate cancer in US adults. *Environmental Health Perspectives*, *118*(1), 60–66. doi:10.1289/ehp.0900919 PMID:20056587

Xu, X., Dailey, A. B., Talbott, E. O., Ilacqua, V. A., Kearney, G., & Asal, N. R. (2010). Associations of serum concentrations of organochlorine pesticides with breast cancer and prostate cancer in U.S. adults. *Environmental Health Perspectives*, *118*(1), 60–66. doi:10.1289/ehp.0900919 PMID:20056587

Xue, D., Zhou, C.X., Shi, Y.B., Lu, H., &He, X.Z., (2015). Decreased expression of ferroportin in prostate cancer. *Oncology Letters*, *10*, 913-916.

Yang, Y., Fang, S., Jensen, J. P., Weissman, A. M., & Ashwell, J. D. (2000). Ubiquitin protein ligase activity ofIAPs and their degradation in proteasomes in response to apoptotic stimuli. *Science*, *288*(5467), 874–877. doi:10.1126cience.288.5467.874 PMID:10797013

Zhang, L. C., Yu, H. N., Sun, S. L., Yang, J. G., He, G. Q., Ruan, H., & Shen, S. R. (2008). Investigations of thecytotoxicity of epigallocatechin-3-gallate against PC-3 cells in the presence of Cd2+ invitro. *Toxicology In Vitro*, *22*(4), 953–960. doi:10.1016/j.tiv.2008.02.005 PMID:18359184

Zhang, Y., Coogan, P., Palmer, J. R., Strom, B. L., & Rosenberg, L. (2009). Vitamin and mineral use and risk of prostate cancer: The case-control surveillance study. *Cancer causes & control. CCC*, *20*(5), 691–698. PMID:19093214

Chapter 9
Breast Cancer With Relevance for Heavy Metals, Mycotoxines, and Pesticides

Ait Sidi Brahim Malika
Cadi Ayyad University, Morocco

Kiai Hajar
Cadi Ayyad University, Morocco

Benidire Loubna
Cadi Ayyad University, Morocco

Omar El Hiba
Chouaib Doukkali University, Morocco & Cadi Ayyad University, Morocco

Makbal Rachida
Cadi Ayyad University, Morocco

Benidire Leila
Cadi Ayyad University, Morocco

Sellami Souad
Mohammed VI University Hospital, Morocco

Rais S Hanane
Mohammed VI University Hospital, Morocco

ABSTRACT

Certain environmental contaminants such as heavy metals, pesticides, and mycotoxins are presumed to play a crucial role in the etiology of breast cancer, which is the most common tumor in women worldwide. In fact, the exposure to heavy metals poses risk in causing human cancers. Several investigations indicated strong contribution of heavy metals especially copper, arsenic, zinc, cadmium, lead, and aluminum in breast cancer. Furthermore, it has been reported that the excessive use of pesticides in agriculture in order to improve the productivity contaminates food materials and can be responsible to induce breast cancer in women. It is also noted that some fungi produce several type of mycotoxins such us zearalenone, aflatoxin, and ochratoxin that are dangerous for human health and can especially cause breast cancer. Thus, the objective of this chapter is to discuss the experimental data regarding the involvement of heavy metals, pesticides, and mycotoxins as well as the recent insights on the molecular mechanisms involved in the progress of breast cancer.

DOI: 10.4018/978-1-5225-7775-1.ch009

INTRODUCTION

Breast cancer is the most frequently diagnosed cancer among woman worldwide in terms of incidence and mortality, accounting for about 2.4 million new cases and 523,000 estimated mortalities recorded global in 2015, which corresponds to about 29% of the total incident cancer cases and 14% of all cancer deaths (Fitzmaurice et al., 2017) representing therefore a real public health problem. In Morocco, the occurrence of breast cancer, particularly in young women ranges from 8 to 25.4%, which represent the highest rates published so far (Slaoui et al., 2016). Till now, breast cancer remains highly frequent especially for women under the age of 35 in which it is aggressive (Axelrod et al., 2008). In fact, there are four stages of breast cancer: cancer in the earliest stage is appointed stage 0 (carcinoma *in situ*) and varies from stage I through IV. Stage IV is the most aggressive stage of the disease. A higher stage involves a more advanced metastatic cancer.

Depending on the expression of estrogen receptor (ER), progesterone receptor (PR) and human epidermal growth factor receptor 2 (Her2), mammary carcinomas can be classified into five molecular subtypes: Luminal A, Luminal B, HER2, Basal-like (or triple-negative) and normal-like (Tao et al., 2015). The last one was unnecessary because it is artifactual and corresponds to tumors contaminated by healthy breast tissue (Tao et al., 2015). The luminal A and luminal B subtypes account for about 65% of all breast cancers. These two subtypes are characterized by the expression of estrogen receptors (ER) and progesterone receptors (PR) (Dunnwald et al., 2007). However, there are many differences between these subtypes; mammal carcinomas of the Luminal A molecular subtype tend to express a greater amount of hormonal receptors, particularly progesterone receptors (PR) compared to luminal B (Tao et al., 2015). Due to the expression and activity of receptors hormonal mammal carcinomas luminal A and luminal B are systematically treated with endocrine treatments, including selective estrogen receptor (ER) modulators such as tamoxifen, selective negative ER regulators such as fulvestrant and inhibitors of aromatase such as letrozole which block the systemic production of the native ligand; β-estradiol (Manning, Buck, & Cook, 2016). In contrast, luminal B tumors are generally poorly studied, although they are often more proliferative, clinically more aggressive and have a lower prognosis than luminal A tumors (Tao et al., 2015). Her2 and basal-like subtypes were found to be the most aggressive breast tumors. Her2 oncogene is correlated to epidermal growth factor receptor family, overexpressed in about 20% of the breast tumors and associated with poorer short-term prognosis compared to PR and ER positive breast cancer (Tao et al., 2015). Basal-like subtypes do not express ER, PR, or Her2 and called triple negative breast cancer (Dent et al., 2007). Approximately 15% of the breast cancers fall basal-like breast cancers category and the incidence of this type is generally higher in younger women, African-American women and in patients with mutated BRCA1 (Dent et al., 2007). The precise causes of cancer are still unknown; however, several factors have been linked to its occurrence. The etiological factors such as genetics, family history, age and hormonal changes are associated to a high risk of breast cancer. Moreover, breast cancer might also be induced by external or environmental factors, like smoke, unhealthy diet, exposure to environmental contaminants, etc. There is growing evidence that exposure to environmental pollutants, including heavy metals, pesticides, persistent organic pollutants and mycotoxins is associated with increased incidence of breast cancer. The present chapter is dedicated to emphasizing the involvement of heavy metals, pesticides and mycotoxins on the incidence and progress of breast cancer.

RISK FACTORS FOR BREAST CANCER

Breast cancer is a multifactorial disease (Barnett et al., 2008). This means that several factors influence the risk of its occurrence. Nowadays, a number of risk factors for breast cancer have been identified, although, the involvement of some of them is still controverted. Female gender, age and country of birth are the strongest determinants of the disease risk. Herein, we will discuss some of the most common risk factors of breast cancer.

Age is the most important risk factor with a monotonous incidence increasing from 30 to 70 years. About 75% cases were reported in women after 50 years of their age (Farshid, 2014, pp. 899-919). However, the incidence of breast cancer in women below 40 and 35 years were found to be 6.6 and 2.4% respectively, whereas only 0.65% was reported in women below 30 years of their age (Anders et al., 2009; Farshid, 2014, pp. 899-919). Thus, the probability of developing breast cancer in women increases with age. In contrast, breast cancer in men is generally diagnosed in older age (>55 years) with more advanced stage compared to women (Brinton et al., 2008; Ly, Forman, Ferlay, Brinton, & Cook, 2013). This is due to the lack of awareness of its occurrence in men.

Another important factor in assessing breast cancer is family background. In reality, the history of 15-20% breast cancer patients showed that how family background is responsible for the cancer (Colditz, Kaphingst, Hankinson, & Rosner, 2012; Kamińska, Ciszewski, Łopacka-Szatan, Miotła, & Staroslawska, 2015). In fact, the risk increases with the number of first-degree relatives (mother, sister or daughter) having breast cancer (Collaborative Group on Hormonal Factors in Breast, 2002). For example, having a first-degree relative with breast cancer almost increases a woman's risk in doubles, whereas it increases to three-fold in the presence of two first-degree (Colditz et al., 2012). It is worth noting that women with a first-degree male relative (father, brother, son) with breast cancer also increases the risk (Haber, Ahmed, & Pekovic, 2012). It is well known that some of the familial forms of breast cancer are related to BRCA inherited mutations which are rare and could be responsible for 5 to 10% (Almutlaq et al., 2016). There are two BRCA susceptibility genes: the BRCA1 gene is located on the long arm of chromosomes 17 and encodes a protein of 1863 amino acids, while the BRCA2 gene is located on chromosome 13 and encodes an extremely large protein of 3418 amino acids (Miki et al., 1994; Tavtigian et al., 1996). With BRCA1 mutation, the estimated cumulative risk for breast cancer in women is 65% by age 70 years, while it is about 45% for BRCA2 mutation (Antoniou et al., 2003).

Similarly, men can also carry germ line mutations in the BRCA gene but with some differences comparing to women. It is estimated that cumulative risk in men by the age 70 years is 1.2% and 6.8% for BRCA1 and BRCA2 mutations, respectively. Other breast cancer susceptibility genes have been identified such as the PTEN, TP53, CDH1, STK11, CHEK2, ATM and PALB (Figure 1) (Foulkes, 2008).

Overweight and obesity can also be considered as a risk factor of breast cancer, but it depends on the menopausal status of the woman. In fact, obesity is associated with a lower risk of breast cancer at pre-menopausal period, while the risk is increased in the post-menopausal period. It is now well recognized that adipose tissue is one of the most important sources of circulating estrogen in the post-menopausal period when ovaries stop generating hormones (Cleary & Grossmann, 2009). This could explain the increase of estrogen responsive breast tumors among obese postmenopausal women. Conversely, the reason for that the obesity is a protective effect against breast cancer still poorly understood at present (Assi et al., 2013).

Figure 1. Proportions of familial risk of breast cancer explained by hereditary variants (Rudolph, Chang-Claude, & Schmidt, 2016)

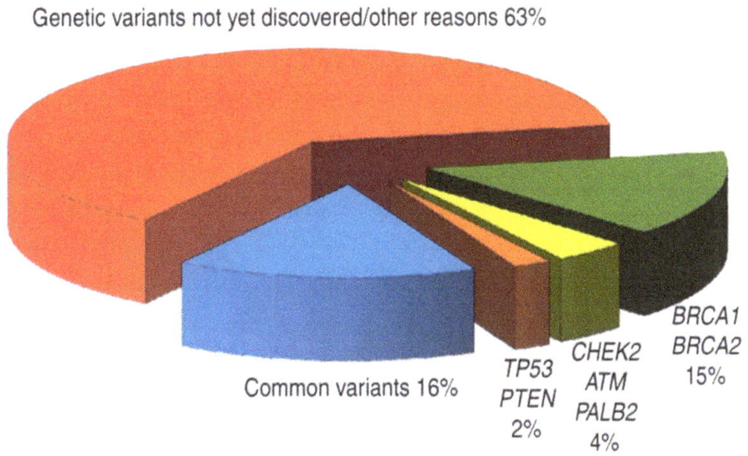

Exposure to ionizing radiation is also associated with increased risk of breast cancer (Ronckers, Erdmann, & Land, 2005). Several studies carried out on the Japanese atomic bomb survivors or on women exposed to diagnostic or therapeutic medical radiation demonstrated the existence of a deep relationship between radiation and breast cancer (Ronckers et al., 2005). Research investigations also suggested that women who are exposed to light at night are likely to develop breast cancer due to suppression of the normal nocturnal production of melatonin by the artificial light (Davis, Mirick, & Stevens, 2001). In fact, melatonin is a hormone, produced by the pineal gland, which is suggested to inhibit the growth of cancer cells. Thus, lack of melatonin by the light could increases estrogen release from the ovaries and then increases the incidence of breast cancer (Davis et al., 2001).

METALS INDUCED BREAST CANCER

Heavy metals are ubiquitous contaminants, naturally present in our environment. They can arise from natural sources such as volcanic eruption and erosion, as from anthropogenic activities such as the mining and smelting operations and agricultural and industrial productions (foundries, batteries, paintings, plastics, etc.). In human body, several metals such as copper, zinc and iron play a fundamental role in physiological functions called essential metals, while the beneficial effects of others are uncertain and called non-essential metals. Both essential and non-essential metals can, from a threshold concentration, induce toxicity. In fact, excessive exposure to heavy metals via food and water consumption, can lead to very detrimental effects on human health. The consequences resulting from intoxication by the various metals present in the environment have largely been studied. Multiple reports confirmed that long-term exposure to heavy metals induces different forms of cancer, including breast cancer (Byrne, Divekar, Storchan, Parodi & Martin, 2013). In current chapter, we will emphasis the role of environmental contaminations by heavy metals, pesticides and mycotoxins as key elements of the pathogenesis of breast cancer.

Table 1. Risk factors for women breast cancer (adapted from Gross, 2000)

a/ Major risks
Advanced age
Family history of breast cancer
Genetic predisposition: diagnosis of atypical hyperplasia or LCIS, BRCA1 or BRCA2 gene
b/Minor risks
Early menarche before age 12
Late menopause after age 55
Nulliparity
First child-birth after age 30
Estrogen replacement therapy
Daily alcohol intake
Dietary fat

Zinc

Zinc (Zn) is a crucial element involved in the activity of many enzymes, indispensable for a large number of physiological functions. It is also intimately involved in the cell growth since it is a component of DNA and RNA polymerase. Thus, it is important for the multiplication, proliferation and differentiation of mammalian cells. In fact, as cancer is characterized by uncontrolled growth, high levels of Zn could be responsible for breast tumorigenesis. The concentration of Zn can be measured in the serum, tissues and hair of breast cancer patients. Actually, there are some controversies regarding the level of Zn in breast cancer patients. Some investigations support a higher Zn level in the blood of the benign and malignant breast cancer patient than that of normal individuals (Cavallo et al., 1991; Piccinini, Borella, Bargellini, Medici, & Zoboli, 1996; Siddiqui et al.; 2006). Whereas, others seem to support the lack of significant difference in the serum level of Zn among breast cancer patients (El-Harouny, El-Morsi, Ahmed & El-Atta, 2011). Earlier studies suggest that the whole blood level of Zn was significantly lower in breast cancer patients, as compared to the healthy controls (Gupta et al., 1991; Kuo, Chen, Wu, Chen, & Lee, 2002; Saleh, Behbehani, Asfar, Khan, & Ibrahim, 2011; Tinoco-Veras et al., 2011; Yücel et al., 1994). The same result was confirmed by Adeoti, et al. (2018). As well, Wu et al. (2015) have conducted a systematic literature research to compare the Zn level in serum and hair among women with breast cancer and controls, they showed a total absence of significant difference in serum Zn levels between the two studied groups. While hair Zn levels were lower in women with breast cancer compared with those of controls. In the other hand, several studies compared the serum level of Zn as well as tissues of breast cancer patients. We can conclude from all these studies that serum level of Zn is reduced in breast cancer patients contrarily to tissues showing the opposite tendency. Lee et al. (2003) suggested that the high level of Zn in breast cancer tissues is due to the alteration of Zn homeostasis in these tissues. In fact, it is well known that breast cancer tissues are characterized by uncontrolled growth of cells which use more Zn than normal tissues. Therefore, Zn is transferred from the blood to accumulate in the interior of the tumor. Kagara et al. (2007) demonstrated that advanced breast cancers showed high expression of the ZIP10 Zn transporter implicated in the transport of large amounts of Zn into the cytosol from the serum.

Copper

Copper (Cu) is an essential micronutrient, required in minute quantities for the proper functioning of cell. The most important sources of Cu for humans are water and food especially liver and other organ meats, seafood, nuts and seeds. In the living body, Cu modulates the enzymatic activities and contributes to many biological processes including embryogenesis, growth and metabolism (Kim, Nevitt, & Thiele, 2008). Furthermore, Cu regulates the Red/Ox state in cells. However, in excessive levels, Cu binds to genetic material inducing oxidative damages (Kim et al., 2008). In fact, the alteration of Cu homeostasis causes various diseases such as cancer and chronic disorders (Jarup, 2003). It has been previously demonstrated that blood and tissues of the breast carcinoma patients contain higher concentration of Cu compared with the matching controls (Gupta et al., 1991, Kuo et al., 2002; Yücel et al., 1994). Cavallo et al. (1991) obtained a contradictory trend concerning the blood Cu level in two studied samples of breast carcinoma patients; Cu concentration was significantly lower in cases of Milan area and higher in those of Montpellier area compared to controls. Further support of the above mentioned data, an increase in Cu concentration was found in the urine samples of breast cancer patients compared to their corresponding healthy group (El-Harouny et al., 2011). A recent investigation also demonstrated that Cu is accumulated in breast cancer tissue. The accumulation was more in the parenchymal component of tumor tissue (Romaniuk et al., 2017). In addition, de Vega et al. (2017) showed that the levels of Cu in the tumor tissues are significantly higher compared to non-tumor one, as well as, a heterogeneous distribution of the investigated metal. In fact, the level of Cu in breast cancer patients could not be always higher. An investigation achieved by Saleh et al. (2011) confirmed that the level of Cu was lower in cancer patients as compared to healthy controls. In contrast, no significant differences were detected in the tissues, hair and plasma of breast cancer patients (Ionescu et al., 2006; Piccinini et al., 1996). Indeed, copper concentration is always elevated in the breast cancerous tissues, which proved that this element play an important role in the carcinogenetic process (Shah, Minhas, & Khan, 2015). It is well known that Cu has the capacity of tumor angiogenesis by activating a variety of growth factors (Skrajnowska et al., 2013). The exact mechanism by which Cu induces breast cancer is still so far from being completely understood. A study undertaken in rats with chemically induced mammary carcinogenesis has shown that excess of Cu with the existing neoplastic process accelerated the growth of mammary tumors in rats, which could be due to the inhibition of antioxidative defense in their bodies leading to a decrease of catalase activities in the serum of sick rats (Skrajnowska et al., 2013). In fact, excess of Cu level is correlated with cancer, leading to suggest that Cu might be a useful prognostic factor for breast cancer.

Cadmium

Cadmium (Cd) is a non-essential, toxic and bioaccumulating heavy metal, released into the environment from industrial and agricultural activities. This metal is widely found in soil, rocks, water and foods, especially shellfish, offal products, certain seeds, cereals, potatoes, root crops and vegetables. Chemical fertilizers, rechargeable batteries, many cosmetics and cigarettes are another important source of Cd. Normally, smokers absorb the same amount of Cd as taken through food (Jarup & Akesson, 2009). For non-smoking women, food is the most important source of cadmium intake, while for smokers, inhalation of tobacco smoke is the main source (McElroy, Shafer, Trentham-Dietz, Hampton, & Newcomb,

2006). It has been nominated by the International Agency for Research on Cancer (IARC) that Cd is a human carcinogen (IARC, 1993). It is involved in cell proliferation, differentiation, apoptosis and other cellular activities. The level of Cadmium was evaluated in tissues, urine, blood, hair and nails of breast cancer patients. Blood provides information on what the body has recently absorbed (Polkowska, 2004). Urine-Cd mainly reveals Cd accumulation in the kidney, as determined by lifetime cadmium exposure (Ju-Kun et al., 2016). Hair and nails give long-term information (Przybylowicz et al., 2012). On the other hand, there is some evidence that Cd accumulates in organs and in the blood rather than in the cutaneous appendages, which should therefore be used to predict internal Cd concentration (Vella, Malaguarnera, Lappano, Maggiolini, & Belfiore, 2017). In fact, many investigations have shown that there is a closed relationship between human breast cancer and cadmium. Previously, McElroy et al. (2006) measured cadmium levels in urine samples and found that Cd concentration is higher by two-fold in the breast cancer patients compared to controls. Similarly, Ionescu et al. (2006) reported that Cd concentration was significantly higher in cancer biopsies, as compared to matching controls. The level of Cd in the blood of patients was significantly higher, as compared to normal controls (Saleh et al., 2011). Furthermore, recent investigations revealed that there is an increased risk for breast cancer with elevated blood, urine and tissues Cd concentrations (El-Harouny et al., 2011). Furthermore, Cho et al. (2013) carried out a meta-analysis study with Western populations and found positive associations between dietary Cd intake and breast cancer risk. The same result was obtained by Julin et al. (2012) who have proved evidence of an important connection between dietary Cd intake and the risk of breast cancer in Swedish population-based cohort study. Another meta-analysis study investigating the results from 11 epidemiologic studies indicated that high concentrations of Cd were positively related to the risk of subsequent Breast cancer (Jouybari et al., 2018). It has been recently reported that cadmium induces the gene expression of cancer stem cells markers in the breast cancer cell lineage and promotes the conversion of non-cancer stem cells to cancer stem cells (Ju, Arumugam, Lee, & Song, 2017) (Figure 2). The mechanism by which Cd may induce breast cancer was investigated in an *in vitro* study evaluating the potential carcinogenetic effect of Cd. The results clearly demonstrated that Cd raises breast cancer cell proliferation by stimulating Akt, ERK1/2 and PDGFRα kinases activity likely by activating c-fos, c-jun and PDGFA by an ERα-dependent mechanism (Brama et al., 2007). Thus, Cd is suggested as anestrogenic an environmental molecule which is capable to disrupt the normal cell homeostasis triggering signals otherwise switched off. In contrast to these studies some authors found that there is no or little evidence of associations between dietary Cd and breast cancer (Adams et al., 2014).

Metals Exhibiting Estrogen-Like Effects

Arsenic (As) is universally present in the environment as a naturally occurring element. It is found in rocks, soil, air, water and food and it can be also discharged into the environment from anthropogenic sources (Martinez, Vucic, Becker-Santos, Gil, & Lam, 2011). In its organic form, it is less harmful and play trivial role in causing cancer, while its inorganic form is highly toxic. In fact, several epidemiological investigations demonstrated that As is closely associated with breast cancer development (Benderli Cihan, Sözen, & Öztürk Yıldırım, 2011; Dantzig, 2009; Joo, Kim, Jung, & Kim, 2009; Khanjani, Jafarnejad, & Tavakkoli, 2017; López-Carrillo et al., 2014). As is known as a potent endocrine disruptor which is associated with breast cancer by interfering with the estrogen signaling pathway (Davey, Bodwell,

Figure 2. Schematic illustration of the monitoring of Cd-induced CSCs based on the concurrent identification of multiple CSC markers using QD and AOTF-based cellular imaging (Ju et al., 2017)

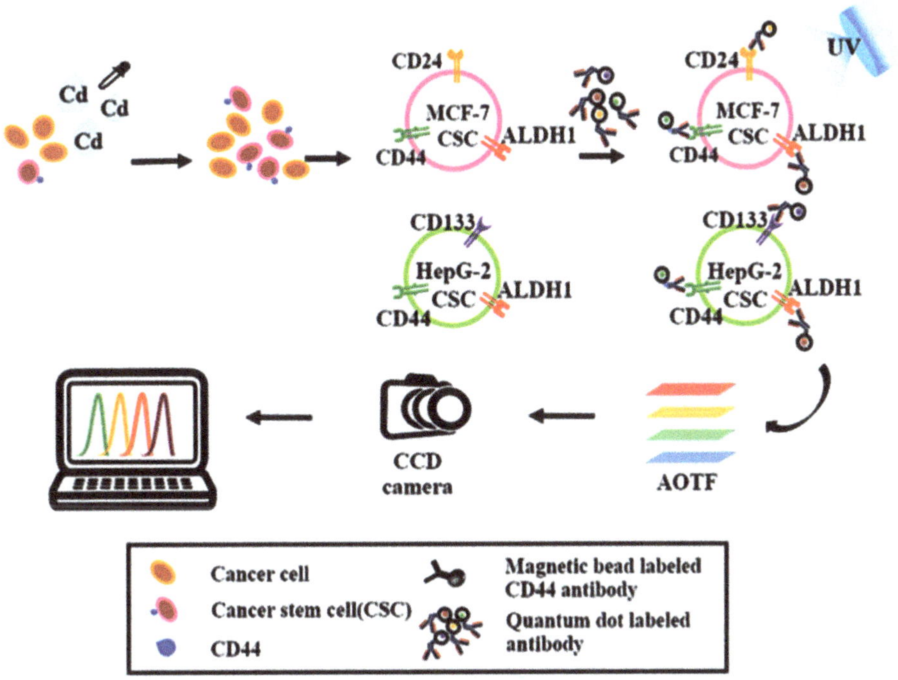

Gosse, & Hamilton, 2007). In an *in vitro* study, Xu et al. (2014) confirmed that chronic inorganic arsenic exposure changed the ER-negative human breast epithelial cells into a cancer cell phenotype through over expression of aromatase.

Aluminum (Al) is one of the most abundant metals found on Earth. It is naturally ubiquitous element generally found in food, water, some drugs and cosmetics (Darbre, Mannello, & Exley, 2013). In fact, most antiperspirants and deodorants contain Aluminum salts as ingredients because they are terribly effective at stopping perspiration (Sappino et al., 2012). Several experimental data demonstrated that extended exposure to Al salts used as antiperspirants is intimately associated with breast cancer development and progression (Darbre, 2005; Darbre et al., 2013; McGrath, 2003; Mannello, Tonti, Medda, Simone, & Darbre, 2011). It has been reported that Al induces a significant increase in estrogen receptor protein level by activating ERα and interfering with estrogen receptor gene expression in breast cancer cells (Bakir & Darbre, 2015; Darbre, 2005; Gorgogietas et al., 2018). From the above data, it can be suggested that Al is responsible for the development of breast carcinogenesis via estrogen receptor pathway.

Lead (Pb) is highly toxic heavy metal. It is found in soil, air, water, food, household dust and commercial products (Wani, Ara, & Usmani, 2015). Human beings are exposed to lead typically by ingestion, inhalation and minimally by dermal absorption (Wani et al., 2015). This element is categorized as a carcinogen according to the International Agency for Research on Cancer (IARC, 2006). It has been demonstrated that lead is associated with an increased risk of breast cancer. Several workers reported that the level of lead was high in breast cancer tumour samples as compared to control (Alatise & Schrauzer, 2010; Ben-Chioma, Obunwo, Bartimaeus & Boisa, 2017; Florea & Busselberg, 2011; Ionescu et al., 2006; Lappano, Malaguarnera, Belfiore, & Maggiolini, 2017; Mohammadi, Riyahi, Bakhtiari, & Khodabandeh, 2014; Romaniuk et al., 2017). It has been demonstrated that lead has estrogenic effects on breast cancer cells. An investigation evaluating the estrogen-like activity of metals in MCF-7 breast cancer cells confirmed that lead activates ERα, suggesting that this metal may represent a new class of nonsteroidal environmental estrogens (Martin et al., 2003).

PESTICIDES ASSOCIATED WITH BREAST CANCER

According to Food and Agriculture Organization (FAO); pesticides is any substance or mixture destined for preventing, destroying or controlling any pest (insects, bacteria, fungi, weeds, etc) able to transfer human or animal disease or affecting the production, processing, storage and transport of food. They can also be used as a plant growth regulator or agent for thinning fruit or preventing the premature fall of fruit and protecting them after harvesting from damage during storage and transport.

Based on the target organism, pesticides can be divided into Algaecides (algae), Avicides (birds), Bactericides (bacteria), Fungicides (fungi), Insecticides (insects), Miticides or Acaricides (mites), Molluscicides (snails), Nematicides (nematodes), Rodenticide (rodents) and Herbicides (undesired plants).

The main pesticides used global, reported to be associated with breast cancer are Di-Chloro Di-Phenyl Trichloro Ethane (DDT/DDE), Dieldrin, Heptachlor, Triazine, Chlordane, Malathion, Hexachlorobenzene (HCB) and Hexachlorocyclohexane (HCH).

Di-Chloro Di-Phenyl Trichloro Ethane (DDT/DDE)

DDT was one of the most widely used insecticides in the globe over the past century. It was banned in developed countries in the '70s due to its adverse environmental effects and human health hazards. DDT includes *p,p'*-DDT (1,1,1-trichloro-2,2-bis(*p*-chlorophenyl)ethane), *o,p'*-DDT [1,1,1-trichloro-2-(*p*-chlorophenyl)-2-(*o*-chloro-phenyl)ethane], *p,p'*-DDE [1,1'-dichloro-2,2'-bis(*p*-chloro-phenyl)ethylene] and *p,p'*-DDD or *p,p'*-TDE (1,1-dichloro-2,2-bis(p-chlorophenyl) ethane). DDT and DDE possess persistent and lipophilic characteristics which led to their bioaccumulation through the food chain. Owing to continued use of DDT in developing countries to eradicate malaria and some agricultural pests and the long environmental half-life of these compounds. DDE and its metabolites remain present in the environment and in the population nowadays (Woodruff & Morello-Frosch, 2011).

Several epidemiological studies have been conducted to investigate the relationship between DDT exposure and breast cancer risk. Charlier et al. (2003) demonstrated that DDT and its metabolites are present with significantly higher concentration in women with breast cancer compared to the control group. Also, Boada et al. (2012) reported that an organochlorine pesticide mixture, counting aldrin, p,p'-DDE and DDD, might have a relevant role in breast cancer development. A nested case control study, to analyze the relationship between age of exposure to DDT and subsequent breast cancer occurrence later in their life, by Cohn et al. (2007) showed that women who were exposed to DDT and its metabolites prior to age 14 years exhibited a 5 fold increase in risk of breast cancer than those women who had same quantity of exposure after 14 years of their life; risk was greatest for women who were exposed even earlier, by age 4 years. Supportive evidence comes from Niehoff et al. (2016) work, which established that young girls up to age 18 years who were exposed to plane spraying of DDT or fogger truck were at a non-significant increased risk of breast cancer for premenopausal breast cancer. Likewise, the Long Island breast cancer study project demonstrated increased odds of estrogen receptor/progesterone receptor-positive (ER+PR+) breast cancer amid women who reported chasing after a fogger truck in their youth (White et al., 2013). Cohn. et al. (2015) investigated the relationship between in utero exposure to DDT and risk of breast cancer and noticed a significant association in young women. Besides, a case control study to evaluate the relationship between DDT and DDE levels, lactation history and breast cancer occurrence, by Romieu et al. (2000) showed that a long period of lactation was associated with a slightly decreased risk of breast cancer independently of serum DDE levels, although the high levels of exposure to DDE may increase women's risk of breast cancer, in particular amid postmenopausal women. Furthermore, Demers et al. (2000) studied the relationship between blood DDE levels with staging and grading of tumor, a positive correlation was observed between increased doses of DDE and its metabolites with aggressive tumors with lymph node involvement showing that DDT/DDE might worsen malignancy of mammary glands, if not initiate it. Although, some epidemiologic works, demonstrated no consistent association between serum or adipose tissue DDE levels and breast cancer (Aronson et al., 2000; Gammon et al., 2002; Ibarluzea et al., 2004; Laden et al., 2001a; McCready et al., 2004; Muscat et al., 2003; Raaschou-Nielsen et al., 2005). A combined analysis of five case-control studies (1400 cases, 1642 controls) carried out in the northeastern United States revealed no association between breast cancer risk and p,p'-DDE (Laden et al., 2001b). A meta-analysis of 22 studies performed up to the year 2001 reported similar results (Lopez-Cervantes, Torres-Sanchez, Tobias, & Lopez-Carrillo, 2004).

DDT and its metabolites are not mutagenic; however, they exhibit estrogenic activity (Andersen et al., 1999; Snedeker, 2001, pp. 35-47). Uppala et al. (2005) demonstrated that administering DDT with DMBA (7,12-Dimethylbenz(a)anthracene) may provoke cellular and chromosomal alterations in the rat mammary gland. Li et al. (2008) showed that both p,p′-DDE and p,p′-DDT revealed agonist activity toward ER-alpha, however DDE acted like an antagonist to both androgen and progesterone receptors, and p,p′-DDT had no effect on the progesterone receptor.

Triazine

Triazine group are the most widely used herbicides worldwide. This group includes atrazine, cyanazine, propazine, and simazine. Amongst all triazines, atrazine is of specific concern since it is one of the most commonly used herbicides with 76 million pounds of it applied each year as it is cheap and efficient. Excessive use of atrazine as the main herbicide for several years has resulted into contamination of groundwater and soil with it; consequently, its potential health effects have been the subject of considerable studies. Exposure to atrazine and other triazine herbicides via diet is very low. The main non-occupational way of exposure is throughout contamination of drinking water supplies. Such contamination is frequent, however based on oversight done by Environmental Protection Agency [EPA], (2010); it is generally at levels that are very low from a population risk outlook. The NHANES III study reported absence of atrazine metabolites in the urine of more than 4,000 samples collected between 1999 and 2004 (Centers for Disease Control and Prevention [CDC], 2009). EPA revealed that the exposure of atrazine via groundwater is periodic. For example, in March no detectable atrazine (<2 ppb) was found in samples of drinking water supply in Ohio, however a high level (36 ppb) was observed in mid-April, which return below the detection limit by mid-May. Similar trends have been observed in other water supplies (EPA, 2010). An ecologic study in Kentucky demonstrated a weak relationship between contamination of drinking water by triazine herbicides and breast cancer incidence rates (Kettles, Browning, Prince, &

Horstman, 1997). In addition, atrazine exposure from groundwater was not associated with breast cancer incidence in a Wisconsin study (McElroy et al., 2007). Similarly, several studies investigating spatial patterns of atrazine use and breast cancer incidence have not reported an association (Hopenhayn-Rich, Stump, & Browning, 2002; Mills & Yang, 2006; Muir et al., 2004; Sathiakumar, MacLennan, Mandel, & Delzell, 2011).

Atrazine is a known endocrine disruptor. It does not have a direct estrogenic activity but can indirectly modulate sex hormone levels (decrease prolactin and luteinizing hormone levels) and thus contribute to an increase in mammary gland tumor (Fan et al., 2007; Higley, Newsted, Zhang, Giesy, & Hecker, 2010; Tinfo et al., 2011). Atrazine leading to dramatic damage to reproductive structures in frogs, fish and other wildlife (Hayes et al., 2003), nevertheless correlational human studies are missing. Atrazine also provoke increased aromatase enzyme activity ensuing in increased levels of estrogen that is directly associated with breast cancer (Fan et al., 2007). Ueda et al. (2005) observed a significant speeding up in tumor cell proliferation when experimental rats with existing breast cancer were exposed to atrazine. Raynor et al. (2005) deduced that in utero exposure of atrazine delay development of mammary glands, a known risk factor for breast neoplasm. Other studies in animals found that atrazine administered to Sprague Dawley female rats affects neuroendocrine pathways to speed up reproductive senescence and induce mammary tumors not noticed in mice or other rat strains (Davis et al., 2011; Hovey et al., 2011; Raynor, Enoch, & Fenton, 2005). The hormonal manifestations of reproductive aging in humans are very different from those of Sprague Dawley rats; consequently, this mechanism is not considered to be pertinent to humans (International Agency for Research on Cancer [IARC], 1999). Similar deductions were drawn concerning cyanazine, from a 2-year bioassay in Sprague Dawley rats (Bogdanffy et al., 2000). No epidemiologic studies have investigated the effects of timing of exposure to atrazine. There are contradictory data from animal studies regarding if low-dose atrazine exposures in utero may assist to developmental abnormalities of mammary tissue in offspring (IARC, 1999; EPA, 2009). Jointly, these results point out that maternal atrazine exposure has no long-term effects on mammary gland development in female offspring beyond a transitory response to high doses.

Atrazine was banned in European Union in 2005 due to the persistent groundwater contamination caused by it and investigations demonstrating its carcinogenic potential for mammary gland, prostate and also its association with ecological disturbance. However, it is still widely used in United States as EPA is reticent to ban atrazine as it considers atrazine's risk benefit ratio in favor of its use as an herbicide.

Dieldrin and Aldrin

Dieldrin and aldrin are persistent organochlorines. Aldrin decomposes to dieldrin in the human body and in the environment. They were commonly used as insecticides to control crops damage, although concerns about harmful effects to the environment and human health led EPA to ban dieldrin and aldrin in 1987 (CDC, 2009; Agency for Toxic Substances and Disease Registry [ATSDR], 2011). Since these last persists in soil and is a water contaminant, exposure may arise by consuming contaminated food (ATSDR, 2011; Snedeker, 2001, pp. 35-47).

Few epidemiologic studies regarding exposure to dieldrin and subsequent risk of breast cancer were reported in the literature. A prospective cohort study carried out by Copenhagen City Heart Study for studies showed a dose-related increase in risk of breast cancer. The result from the study revealed that blood dieldrin levels are directly proportional to tumor grading and staging. This study also deduced that dieldrin is associated with increased incidence, incidence of aggressive tumor and higher mortality in breast cancer. This work as well revealed an association between serum dieldrin levels and reduced survival of incident cases among women with mutant p53 breast cancers (Hoyer, Gerdes, Jorgensen, Rank, & Hartvig, 2002a). A subsequent case-control study in Long Island demonstrated that a high risk of breast cancer was associated with exposure to dieldrin (Gammon et al., 2002). Another case-control study undertaken by Ibarluzea et al. (2004) in Spain showed that twofold increases in risk for breast cancer amongst postmenopausal women with perceptible adipose levels of the pesticides aldrin and lindane. Slender women with a BMI under 28.6 kg/m^2 and in the highest quartile of total xenoestrogen body load had a 3.4-fold increased risk of postmenopausal breast cancer. Conversely of cited studies earlier, a nested case-control study by Raaschou-Nielsen et al. (2005) reported no association between dieldrin levels and estrogen negative or positive breast cancer.

Dieldrin has xenoestrogenic potential and also disrupts androgenic pathways. Andersen et al. (2002) showed that addition of dieldrin into MCF-7 (Michigan Cancer Foundation-7) human breast cancer cell in vitro accelerated their growth and proliferation. The gavage of rats, prenatally and neonatally with environmentally relevant doses of dieldrin induced an increased incidence of breast cancer among them, likely mechanism of it can be dieldrin mediated changes in cellular expression of BNDF and cell signal receptors Hero in breast tissue (Cameron & Foster, 2009)

Heptachlor

Heptachlor is a renowned carcinogenic insecticide that was widely used before ‘80s. It was banned in 1988 but still omnipresent in the environment. Exposure may arise through contaminated water, food, soil or air. Heptachlor has persistent (long half-life in environment for up to 14 years) and lipophilic properties (accumulated in adipose tissues counting mammary gland).

Researches concerning heptachlor exposure and breast cancer in humans and animals are scarce. Results of studies in animals demonstrated that heptachlor caused DNA damage contributing to the development of mammary cancer (Cassidy, Natarajan, & Vaughan, 2005). Also, rats exposure to heptachlor caused liver cancer. This influenced the liver’s capacity to process hormones, ensuing in increased estrogen levels and thus increase the risk of breast cancer (Dich, Zahm, Hanberg, & Adami, 1997).

Allen et al. (1997) posit an association between heptachlor use and breast cancer incidence in Hawaiian population. Another study undertaken by Classidy et al. (2005) to determine the levels of heptachlor epoxide (HE), oxychlordane, and DDE in adipose tissue in breast biopsies in a series of 34 women evaluated for breast abnormality, merely HE exhibited positive association with prevalence of breast cancer in the biopsies. HE is a xenoestrogenic compound, combined with HE’s capacity to interact with nitric oxide provokes an inverted-U increase in intracellular oxidants inducing DNA damage and subsequent malignancy. Heptachlor also activates kinase signaling pathways ensuing in accelerated proliferation of cancer cells (Cassidy, 2010, pp. 1363-1364).

Other Pesticides

A registry-based case-control study assessed by Mills and Yang (2005) in farm labor union members in California, revealed an association between chlordane, malathion, 2,4-D and increased risk of breast cancer, particularly in young women or who had early onset of cancer. Another large prospective cohort study carried out by Lawrence et al. (2005) to investigate the relationship between pesticides and occurrence of breast cancer amid women whose husbands work in agricultural fields, showed an increased incidence of breast cancer in women exposed to 2, 4, 5-triclorophenoxypropionic acid. A population-based case-control study demonstrated a relationship between plasma β-hexachlorocyclohexane, p,p¢-DDE, oxychlordane, and transnonachlor levels and aggressive breast cancers (Demers et al., 2000). A nested case-control study in Denmark exhibited inverse associations with adipose concentrations of β-hexachlorocyclohexane, oxychlordane, and transnonachlor (Raaschou-Nielsen et al., 2005), and inaccurate associations were noticed in a case-control study in Long Island (Muscat et al., 2003). Other case-control studies revealed no relationship with β-hexachlorocyclohexane (Aronson et al., 2000; Zheng et al., 1999).

Four case-control studies found no relationship between hexachlorobenzene (HCB) and breast cancer (Aronson et al., 2000; Dorgan et al., 1999; Lopez-Carrillo et al., 2002), whereas four others proposed positive relationship (Charlier et al., 2003; Liljegren, Hardell, Lindstrom, Dahl, & Magnuson, 1998; Muscat et al., 2003). A nested case-control study revealed a positive association, but merely amid women sampled less than 3 years before breast cancer diagnosis (Moysich et al., 1998). In contrary, an inverse association was found in another nested case-control study for ER-negative breast cancer amid those in the highest quartile of HCB adipose levels (Raaschou-Nielsen et al., 2005).

Mussalo-Rauhamaa et al. (1990) undertaken a study in Finland and observed that residues of β- HCH were found more often in breast cancer patients. After attuned for age and parity, β-HCH was found as a significant risk factor for breast cancer. An Egyptian study carried out by Soliman et al. (2003) reported no association between β-HCH and breast cancer risk. In India the average dietary intake of HCH (115 ng/person) was higher than those reported in developed countries (Kannan, Tanabe, Ramesh, Subramanian, & Tasukawa, 1992). The effect of HCH exposure in spray men and general population in India were investigated by Joshi et al. (1996). The mean residue amount of α, β, and γ isomers and the total β, β, and β isomers in spray men were twice those of the general population. A significant association was noticed between their length of exposure and the levels of HCH isomers in the blood of spray men. Bhatnagar et al. (2004), showed that the mean levels (± SE) of α, β, and γ-HCH in serum samples in India were found to be 4.49 (±0.73), 35.06 (±3.50), and 1.69 (±0.15), respectively. In another study by Rusieki et al. (2005), the mean concentrations of β-HCH amid the breast cancer cases were 14.45 ppb and 2876.15 ppb for lipid unadjusted and adjusted respectively. It is reported that the average Indian dietary intake of HCH is hundred-fold that of the US and UK (Pesticide Action Network UK [PAN UK], 1993).

Persistent Organic Polluants

Persistent organic pollutants (POPs) are organic compounds that degrade slowly in the environment and bioaccumulate through the food chain. They can cause endocrine, immune and reproductive system dysfunctions and malignancy counting breast cancer. They include polychlorinated biphenyls, polybrominated diphenyl ethers, dioxins, DDT, furans, etc.

Polychlorinated Biphenyls (PCBs)

PCBs are chlorinated organic biphenyl compounds with 209 congeners. They are highly stable and lipophilic chemicals broadly spread in the environment. PCBs were used as flame retardants, paints, and plasticizers, and broadly used in the electrical industry. Notwithstanding a ban on their manufacture in the United States in the 1970 s, PCBs persist in the environment, biomagnify across food chains and thus accumulate in adipose tissues in humans (Park et al., 2007; Shields, 2006, pp. 830-839). Individuals are exposed to PCBs via the ingestion of fish from contaminated waters, and through inhalation and dermal contact (Carpenter, 2006, pp. 1-23). For several years, PCBs are concerned as etiologic agents for cancers, including breast cancer.

A number of epidemiologic studies demonstrated an association between levels of serum or plasma PCBs and breast cancer (Demers et al., 2000; Hoyer, Jorgensen, Grandjean, & Hartvig, 2000; Lucena, Allam, Costabeber, Villarejo, & Navajas, 2001; Millikan et al., 2000; Woolcott et al., 2001) while others have not (Gammon et al., 2002; Laden et al., 2001a; Lopez-Carrillo et al., 2002; Pavuk et al., 2003; Wolff, Zeleniuch-Jacquotte, Dubin, & Toniolo, 2000). In some works, associations were only noticeable in subgroups (e.g., amid African-American but not white women; Millikan et al., 2000). Holford et al. (2000) reported that some congeners were positively associated with breast cancer while others were inversely related. Likewise, Demers et al. (2002) noticed positive association between mono-*ortho*-substituted PCBs (105, 158, 156) and breast cancer risk. A hospital-based case control study examining 43 patients operated for invasive breast cancer (cases) and 35 patients operated for benign breast disease (controls) revealed positive but inaccurate association between postmenopausal breast cancer and breast adipose tissue levels of coplanar PCB congeners (Liljegren et al., 1998). Retrospective case control study by Aronson et al. (2000) showed associations with breast adipose tissue levels of PCB congeners, in particular congeners 105 and 118 among premenopausal women and with congeners 170 and 180 amid postmenopausal women.

Numerous studies investigated the potential role of genetic polymorphisms in breast cancer risk related to PCBs. Li, et al. (2005) reported that women with CYP1A1-m2 genetic variant, also known as the exon 7 variant (present in 10 to 15% of white women and in larger proportion of African American women) are far more susceptible to PCB induced malignant changes in breast tissues. Hoyer et al. (2002) revealed a threefold increase in risk of breast cancer associated with total serum PCB level among women with variants of the p53 suppressor gene. A nested case control study done by Helzlsouer et al. (1999) did not show modifying effects of the GSTM1, GSTT1, GSTP1, COMT, and CYP17 genotypes on the association between PCB levels and breast cancer risk.

Three studies have related PCB exposure to breast cancer recurrence or survival. Muscat et al. (2003) reported that high PCB levels in breast tissue were related with an increased risk of breast cancer recurrence. Hoyer et al. (2001) noticed a strong association between high PCB levels and mortality among women with estrogen receptorpositive (ER1) tumors. Demers et al. (2002) revealed that higher levels of PCB 153 were associated with more aggressive breast cancer. A study carried out by Sijin Liue et al. (2010) demonstrated that PCBs enhance metastatic potential of breast cancer cells by activating Rho-associated Kinase (ROCK). Likewise, other studies indicated that PCBs were involved in promoting breast cancer progression, and some particular PCB congeners might contribute to high-grade tumors and overall poor prognosis in breast cancer patients (Demers et al., 2000; Snedeker, 2001, pp. 35-47; Woolcott et al., 2001).

Polybrominated Diphenyl Ethers (PBDE)

Polybrominated diphenyl ethers (PBDEs) are mainly used as flame retardant to reduce the inflammability of combustible materials to prevent and reduce fire hazards. PBDEs are widely added to building materials, plastics, textiles, furniture, electronics appliances and polyurethane foams. They comprise 209 possible congeners with 1-10 bromine atoms.

With rife use and persistent properties, PBDEs have become ubiquitous contaminants and are found in air, water, soil, dust, animal products, breast milk, human tissue and organs (Hurley et al., 2011; Ma, Qiu, Zhang, Duan, & Zhu, 2012; Wang, Jiang, Lam, & Li, 2007). PBDEs are lipophilic and persistent compounds which can bioconcentrate and biomagnify through the food chain (De Wit, Herzke, & Vorkamp, 2010). Human exposure is chiefly by food and dust ingestion (Domingo, 2012, pp. 238-249; Huwe et al., 2008). PBDEs are structurally and functionally alike to PCBs and were widely used after PCBs were banned. PBDEs are endocrine disrupting substances related primarily with thyroid disruption. Some data also associated them to breast cancer.

A study undertaken by Li et al. (2012) in china deduced that PBDEs, particularly PBDE-209 improve proliferation of tumor cell lines in a dose dependent manner by altering cell growth cycle inducing S phase between G2 and M phase. PBDE-209 is also known to partly inhibit the cell apoptosis in breast cancer cells (MCF-7) and also suppresses Gö6976- and PD98059- induced apoptosis in all cell lines. Furthermore, some PBDE congeners were reported as agonists of estrogen receptor α (ERα) and ERβ (Meerts et al., 2001). A few in vitro studies proposed that PBDE exposure affected the proliferation and apoptosis of breast cancer cell (Karpeta, Maniecka, & Gregoraszczuk, 2016; Li et al. 2012). PBDEs have estrogenic forcefulness; low doses could induce MCF-7 cell proliferation in vitro (Li et al., 2012; Llabjani, Trevisan, Jones, Shore, & Martin, 2011). Given the estrogen-like effects of PBDE, exposure can be associated to breast cancer development. Thus, PBDE exposure was found a suspected risk factor for breast cancer (McDonald, 2002, pp. 745-755). Epidemiological studies of the effect of PBDE exposure in breast cancer risk are rare and incompatible. A recent case-control study done by He et al. (2018) in China found out that the sum of PBDE and most individual PBDE congeners were positively associated with breast cancer risk. Another case-control study carried out by Holmes et al. (2014) in Alaska Native showed significant association between BDE-47 level and breast cancer. In contrary, a case–control study also in the U.S based on hospital data did not reveal a significant relationship between adipose PBDE exposure and risk of breast cancer (Hurley et al., 2011). Hence, more epidemiologic studies are required to supply evidence or indexes for investigating the relationship between PBDEs and risk of breast cancer.

Dioxins

Dioxins are a group of highly toxic, lipophilic, and persistent by-products of industrial processes and incineration. The dioxin-like chemicals comprise diverse furans and coplanar PCBs, however the congener 2,3,7,8-tetrachlorodibenzo-*para*-dioxin (TCDD) is considered as the most potent of the dioxins and dioxin-like compounds and has been a major focus of concerns about carcinogenicity (CDC, 2009). TCDD has a half-life of 7–9 years in humans (Pirkle et al., 1989). It has been classified by IARC (1997; Baan et al., 2009) as a human carcinogen (Group 1) and by EPA as carcinogenic to humans (EPA, 2000), while the classification of dioxins as "known human carcinogens" by IARC and EPA remains contentious (National Research Council [NRC], 2006).

Evidence implicating dioxins and dioxin-like chemicals as human carcinogens has first and foremost been based on elevated incidence of some cancers amongst residents of Seveso, Italy, who sustained high levels of exposure from a major 1976 industrial accident.

Few epidemiological studies have examined the relation between dioxin exposure and breast cancer risk, and there have been conflicting results. Amongst the cohort of residents exposed to dioxin subsequent the Seveso accident (Italy, 1976), no increased breast cancer risk was noticed amongst women living in the areas most exposed to TCDD (Bertazzi, Pesatori, Consonni, Tironi, Landi, & Zocchetti, 1993; Pesatori,

Consonni, Rubagotti, Grillo, & Bertazzi, 2009). However, high serum TCDD levels measured in Seveso women that lived in the two most contaminated zones were related with increased breast cancer risk in a first analysis undertaken in 2002 (Warner et al., 2002). Nevertheless, this was not long-established with longer follow up of the same population (Warner, Mocarelli, Samuels, Needham, Brambilla, & Eskenazi, 2011). Studies done amongst cohorts of workers exposed to dioxins through herbicides production revealed higher breast cancer mortality in these populations (Brody, Moysich, Humblet, Attfield, Beehler, & Rudel, 2007; Manuwald, Velasco-Garrido, Berger, Manz, & Baur, 2012). Other studies also showed high breast cancer mortality (Revich et al., 2001) and increased breast cancer risk (Fabre & Gardiès, 2008) in cohorts of community residents exposed to dioxins released by industrial amenities. Nonetheless in a French case-control study, a decrease in breast cancer risk was noticed amongst older women living near a municipal solid-waste incinerator and most exposed to dioxins (Viel et al., 2008).

Several TCDD-related cancer bioassays (Kociba et al., 1978; National Toxicology Program [NTP], 2004) have shown induction of numerous types of cancer in both rats and mice. In few studies, mammary gland tumors in Sprague Dawley rats were significantly reduced at the highest doses of TCDD (Kociba et al., 1978; NTP, 2006). Other studies have also reported that early-life exposure to TCDD may alter mammary gland development (Vorderstrasse, Fenton, Bohn, Cundiff, & Lawrence, 2004; Wang et al., 2011).

Dioxins are not mutagenic and do not bind to the estrogen receptor, though one study revealed that TCDD can provoke oxidative stress and subsequent DNA strand breaks in MCF7 breast cancer cells (Lin et al., 2007).

Dioxins mode of action as a putative hepatocarcinogen needs binding and activation of the AhR that induces a cascade of downstream effects on gene expression for genes implicated in an assortment of biological processes. Two in vivo works investigating whether early-life exposure to dioxins may increase the incidence of carcinogen-initiated mammary tumors did not supply evidence of such an effect (Desaulniers et al., 2004; Wang et al., 2011). It has also been hypothesized that via interactions with other factors, early-life exposure to dioxins may change mammary gland development and finally tumorigenesis. La Merrill et al. (2010) investigated the combined effect of TCDD exposure and a high-fat diet in mothers and offspring of mice and observed that this combined exposure increased mammary cancer incidence in the offspring by two-fold after oral administration of a standard cancer inducing agent.

Though 3-methylcholanthrene has been demonstrated to stimulate estrogen receptor alpha in numerous different ER response assays (Shipley & Waxman, 2006), TCDD and other dioxin analogs provoked tissue-specific inhibition of estrogen-induced genes and pathways (Safe & Wormke, 2003; Safe, 2005, pp. 139-144). Indeed, many structural analogs of chlorinated dioxins have been suggested as tamoxifen-like antiestrogens for treatment of ER negative breast cancer (Zhang et al., 2009).

Mycotoxins and Breast Cancer

The term mycotoxin is derived from the Greek prefix "*mycos*" alludes to mold or fungus and the Latin word "*toxicum*" meaning poison. These compounds are considered as the most significant food contaminants in terms of impact on public health and the economy of many countries (Pitt, Basilico, Abarca, & Lopez, 2000). The main fungal toxigenic species belong to the genera *Aspergillus*, *Fusarium*, *Penicillium Alternaria* and *Claviceps*, because of their ubiquitous nature (Misihairabgwi, Ezekiel, Sulyok, Shephard, & Krska, 2017; Sidhu, 2002, pp.705-711). Mycotoxin contamination of food products occurs when a combination of environmental conditions in the field, as well as poorly controlled processes of harvesting, storage and processing are brought together (Tola & Kebede, 2016; Bennett & Klich, 2003).

In general, mycotoxins enter the body through the consumption of contaminated foodstuffs. The primary route of exposure consists of the intake of foods of plant origin and the secondary route can be indirectly by ingestion of food products of animal origin (milk and meat); where mycotoxins are present as the result of feeding the animals with mycotoxins contaminated feeds (Maciorowski, Herrera, Jones, Pillai, & Ricke, 2007). Some others routes of exposure can happen through inhaling spores transported in aerosol-bound dust and dermal contact of mycotoxins (Jarvis & Miller, 2005).

The impact of such toxins on human health has attracted worldwide attention. Acute or chronic exposure to mycotoxin contamination can exert serious health problems in consumer populations. The mycotoxicosis effects are diverse and extremely variable from one mycotoxin to another; it may be carcinogenic, mutagenic, teratogenic, hepatotoxic, nephrotoxic, neurotoxic, estrogenic and cytotoxic or may induce immunosuppression in humans (Misihairabgwi et al., 2017; De Ruyck, Boevre, Huybrechts, & De Saeger, 2015; Liu & Wu, 2010 ; Johanning, Gareis, Nielsen, Dietrich, & Märtlbauer, 2002).

Aflatoxin

Isolated from mold-contaminated peanut and identified for the first time in the early '60s by several research teams after the death of more than 100,000 turkeys and other poultry (turkey X disease) in England (Dollear & Goldblatt, 1969; Blount, 1961, pp. 52-55; Sargeant, Sheridan, O'kelly, & Carnaghan, 1961). The term aflatoxin (AF) has been assigned to these toxins in reference to *Aspergillus flavus* (*A. flavus*) which is the first fungus identified as being responsible for the production of these toxins. Currently, aflatoxins are known to be produced by various *Aspergillus spp.* and other fungi, most commonly by *Aspergillus flvus*, *Aspergillus parasiticus*, *Aspergillus nomius*, *Aspergillus pseudotamarii*, and *Aspergillus bombycis* (Becker-Algeri et al., 2016; Zain, 2011, pp. 129-144). Some of them produce more than one AF. For example, *A. flavus* typically produces AFB_1 and AFB_2, whereas *A. parasiticus* produces aflatoxins B_1, B_2, G_1, and G_2 and therefore these toxins can often be found in foods as mixtures (Amaike & Keller, 2011; Wilson & King, 1995).

Since the identification of aflatoxins, numerous reports of *in vitro* and *in vivo* studies have provided considerable evidence on their mutagenic, genotoxic and carcinogenic activity. An array of rodents, human or animal-derived cells and microbes have been used to examine the mutagenicity effects of AFB_1. For example, studies in bacterial and human cells containing a shuttle vector as a mutation target revealed an increased mutation frequency with G:C to T:A transversion being the predominant mutation type following AFB_1 exposure compared to untreated control (Trottier, Waithe, & Anderson,1992; Foster, Eisenstadt, & Miller, 1983). Other researchers have reported a high frequency of mutations affecting the tumour suppressor p53 gene in codon 249 associated with GC—TA transversions in liver tumours obtained from aflatoxin B_1-exposed individuals (Hsu et al., 1991; Bressac, Kew, Wands, & Ozturk, 1991; Vautier, et al., 1999).

The DNA-toxin interaction is the key point in the development of the carcinogenesis process. It has been clearly demonstrated in experimental animal studies that AFB_1 reacts with DNA and proteins. AFB1 after being metabolized, via the cytochrome P450-catalyzedoxidation, to aflatoxin B_1-8,9-epoxide will covalently binds to guanine at their N7 nitrogen and form the major adduct 8,9-dihydro-8-(N7-guanyl)-9-hydroxy aflatoxin B_1(AFB1-N7-GUA) (Kew, 2013; Wang & Groopman, 1999). This epoxide can also covalently binds to the nucleophilic atoms in nucleic acids (RNA and DNA) or proteins such as histones and albumin. The AFB1-N7-GUA formed may induce both transversion and transition mutations which then results in inactivation of tumor suppressor genes.

Today, aflatoxins are conceded as a major risk factor for hepatocellular carcinomas (5-28%) worldwide (Theumer et al., 2018; Gursoy-Yuzugullu, Yuzugullu, Yilmaz, & Ozturk, 2011), but little is known about their potential responsibility in breast cancer. Eldridge et al. (1992) noticed that the treatment with AFB_1 can lead to the transformation of normal human mammary epithelial cells to cancerous one. In addition, the examination of breast tumorous tissues obtained from cancer patients in United Kingdom showed a higher AFB_1-DNA adducts levels than did normal tissue (Harrison, Carvajal, & Garner, 1993). Lately, the effect of AFB1 on cell growth and cell cycle progression in a human breast cancer cell line, MCF-7 was evaluated by Yip et al. (2017). The obtained results revealed that AFB1 was cytotoxic and affected MAPK pathways (Yip et al., 2017). Interestingly, breast cancer resistance proteins (Bcrp1/BCRP), mentioned to be significantly expressed in MCF-7 and BT20 cells (Faneyte et al., 2002), resulted capable to bind AFB1 *in vitro* and *in vivo* works, to block AFB1 accumulation in tissues expressing these proteins and to provoke a higher secretion in milk (Herwaarden et al., 2005).

Ochratoxin

In 1965, a group of South African researchers discovered new mycotoxins through the screen of fungal secondary metabolites (Van der Merwe, Steyn, & Fourie, 1965; Van der Merwe, Steyn, Fourie, Scott, & Theron, 1965). Ochratoxin A (OTA) was first isolated from *Aspergillus ochraceus* in South Africa and soon after from a commercial corn sample in the United States (Shotwell, Hesseltine, & Goulden, 1969). It has been recognized as a toxic metabolite and considered as one of the most frequently diagnosed mycotoxicosis in human. Indeed, OTA was reportedly detected in 60% of a healthy Moroccan study population of blood donors (Filali et al., 2002).

There are three types of ochratoxins that occur in nature (A, B, and C), mainly produced by some *Aspergillus* and *Penicillium* species (Heussner & Bingle, 2015). The major species implicated in the production of ochratoxin family includes *A. ochraceus*, *A. carbonarius*, *A. niger* and *P. verrucosum* (Zain, 2010, pp. 129-144; Bui-Klimke & Wu, 2015; Wang et al., 2016).

According to IARC, ochratoxin A has been classified as possibly carcinogenic to humans (Group 2B) due to evidence of OTA-mediated carcinogenicity in laboratory (Kujawa, 1994, pp. 351).

The mechanisms by which this toxin induces carcinogenesis are still unknown but two general hypotheses have been suggested. The first propose a direct genotoxic mechanisms, which OTA being metabolized into an electrophilic species may led to the formation of DNA adducts by binding to some nucleotide bases (Pfohl-Leszkowicz & Manderville, 2011; De Ruyck et al., 2015). Kuroda et al. (2015) reported that p53 tumor suppressor protein was upregulated during OTA treatment, and investigated the extent to which the p53 protein inhibits progression of OTA induced DNA damage. The second hypotheses suggest that OTA cause oxidative damage to DNA, leading to mutagenesis and potential carcinogenesis (De Ruyck et al., 2015).

There is also evidence that contamination with OTA can cause other cancers such as mammary gland (breast) tumors in experimental animals such as rats and mice. A significant increased incidence of fibroadenomas of the mammary glands was observed in females Fischer F344 rats treated with the highest dose of OTA (28/50) compared to controls (17/50) (Son, Kamino, Lee, & Kang, 2003). The incidence of multiple fibroadenomas is even more significant (14/50 treated animals versus 4/50 controls). No mammary gland tumors in male rats have been observed (Son et al., 2003). Furthermore, ochratoxin has been identified as a substrate for BCRP (Schrickx, Lektarau, & Fink-Gremmels, 2006). A rodent study confirmed the involvement of ochratoxin A in the progress of breast cancer through genetic damage and hypothesized that ochratoxin A could induce breast cancer. This could be explained by the obtained data that showed an increase in breast cancer cases when a number of people in a population migrate from one geographic area to another (Dunnick, Elwell, Huff, & Barrett, 1995). Indeed, no adequate epidemiological studies have been conducted to fully assess the extent to which OTA levels may or may not be associated with cancer or other adverse health effects in humans.

Zearalenone

The first description of the clinical symptoms associated with this toxin was in 1927 by Buxton, although the toxin itself was not identified. Later, the active component produced by fungi infecting maize grains was isolated from cultures of *Gibberella zeae* (*Fusarium graminearum*) by Stob et al. (1962) and named zearalenone (Urry, Wehrmeister, Hodge & Hidy, 1966).

Zearalenone (ZEA) is produced by *Fusarium* species, including *F. graminearum*, *F. culmorum*, *F. oxysporum*, *F. tricinctum* and others under favorable climatic conditions (Dänicke & Winkler, 2015; Kuciel-Lisieska et al., 2008).

The presence of zearalenone is a serious problem and the consumption of high levels of this toxin proved to be dangerous for humans and animals. Indeed, ZEA is structurally similar to 17 β-estradiol, and its high affinity for estrogen receptors (ERs) explains the health issues of liver, hypothalamus, uterus, ovaries, and breast of different species including human (De Ruyck et al., 2015; Yazar & Omurtag, 2008; Zinedine, Soriano, Molto, & Manes, 2007).

ZEA is rapidly metabolized in animals and humans by enterocytes and hepatocytes after oral administration. As an endogenous sex hormone, zearalenone and its active metabolites can occupy and stimulate estrogenic receptors (ER) (Mbundi et al., 2014). Some epidemiological studies suggest a correlation between ZEA exposure and the occurrence of central precocious puberty in young girls (Pazaiti, Kontos, & Fentiman, 2012) and endometrial adenocarcinoma in women (Gajecki et al., 2004; Tomaszewski, Miturski, Semczuk, Kotarski, & Jakowicki, 1998). Zearalenone and its metabolites are suspected to be responsible for premature thelarche (premature breast development in girls under 8 years of age) in Puerto Rico (Sáenz de Rodríguez, 1984, pp. 1741-1742; Comas, 1982, pp. 1299-1300) and south-eastern Hungary (Szuets, Mesterhazy, Falkay, & Bartok, 1997) where high levels of this oestrogenic mycotoxins have been detected in serum. In New Jersey, a study of 163 girls (9-10 years) showed that ZEN and its derivatives were detectable in urine in 78.5% of the girls. These girls tended to be shorter and less likely to have reached the onset of breast development (Bandera et al., 2011). Other studies suggest the possibility that ZEA could be involved in carcinogenesis in humans. Indeed, the examination of ZEA concentrations in endometrial tissues showed that women with adenocarcinoma had high concentrations of toxin in comparison with women without endometrial changes where the mycoestrogen was not detectable (Tomaszewski et al., 1998). In Poland, the data analyse of 200 patients with neoplastic lesions in the mammary glands reported a higher concentration of ZEA in patients with benign and malignant tumours (Kuciel-Lisieska et al., 2008). Belhassen et al. (2015) assessed urine levels of ZEA and its metabolites in 69 women with breast cancer and in 41 controls and noticed a positive correlation between increased risk of breast cancer and ZEA concentrations. However, some other researchers suggest that the presence of exogenous myco-oestrogen in plasma of patients with breast cancer does not indicate causal relationship since no significant differences between the groups in the levels of zearalenone and its congeners was found (Pillay et al., 2002).

On the other hand, *in vitro* studies showed that ZEA, may play an important role in increasing the risk of estrogen-dependent tumors in particular breast tumor. Indeed, a number of studies have reported that treatment with ZEA enhance the proliferation and the viability of human breast cancer cell lines MCF-7, through the estrogen mediated pathways and by increasing the protein biosynthesis as well as the lipid metabolism (Nittoli et al., 2018; Yip, Wan, Wong, Korach, & El-Nezami, 2017; Parandin, Rassouli, Sisakhtnezhad, & Shahri, 2015).

CONCLUSION

On the basis of the above data, it is concluded that exposure to environmental contaminant especially heavy metals, pesticides and mycotoxins is associated with an increased risk of breast cancer. In fact, breast cancer is associated with the increase of some heavy metals such as Zn, copper, cadmium, arsenic, aluminium and lead. It is worth quote to mention that the concentration of these elements is high in the serum, hair, urine and tissues of breast cancer patients, suggesting that heavy metals play an important role in the breast carcinogenesis. In addition, exposure to pesticide increased occurrence and aggressiveness of breast cancer. Furthermore, the presences of mycotoxins in food constitute a constant health risk for human population. Carcinogenic effects and related pathway of aflatoxins are well known. However, for ZEA and OTA there is a need for more *in vitro* and *in vivo* research to clarify the human biological mechanisms in which these two mycotoxins are involved.

The excessive urbanization and industrialization may lead to a high exposure to environment contaminants. Thus, extensive efforts by researchers and government should be diverted to address the issue.

REFERENCES

Adams, S. V., Quraishi, S. M., Shafer, M. M., Passarelli, M. N., Freney, E. P., Chlebowski, R. T., ... Newcomb, P. A. (2014). Dietary cadmium exposure and risk of breast, endometrial, and ovarian cancer in the Women's Health Initiative. *Environmental Health Perspectives*, *122*(6), 594–600. doi:10.1289/ehp.1307054

Adeoti, M. L., Oguntola, A. S., Akanni, E. O., Agodirin, O. S., & Oyeyemi, G. M. (2015). Trace elements; copper, zinc and selenium, in breast cancer afflicted female patients in LAUTECH Osogbo, Nigeria. *Indian Journal of Cancer*, *52*(1), 106–109. doi:10.4103/0019-509X.175573

Agency for Toxic Substances and Disease Registry. (2011). *Toxicological profile for aldrin/dieldrin*. Retrieved from http://www.atsdr.cdc.gov/toxprofiles/tp.asp?id=317&tid=56

Alatise, O. I., & Schrauzer, G. N. (2010). Lead exposure: A contributing cause of the current breast cancer epidemic in Nigerian Women. *Biological Trace Element Research*, *136*(2), 127–139. doi:10.100712011-010-8608-2

Allen, R. H., Gottlieb, M., Clute, E., Pongsiri, M. J., Sherman, J., & Obrams, G. I. (1997). Breast cancer and pesticides in Hawaii: The need for further study. *Environmental Health Perspectives*, *105*, 679–683.

Almutlaq, B. A., Almuazzi, R. F., Almuhayfir, A. A., Alfouzan, A. M., Alshammari, B. T., AlAnzi, H. S., & Ahmed, H. G. (2017). Breast cancer in Saudi Arabia and its possible risk factors. *Journal of Cancer Policy*, *12*, 83–89. doi:10.1016/j.jcpo.2017.03.004

Amaike, S., & Keller, N. P. (2011). Aspergillus flavus. *Annual Review of Phytopathology*, *49*(1), 107–133. doi:10.1146/annurev-phyto-072910-095221

Andersen, H. R., Andersson, A. M., Arnold, S. F., Autrup, H., Barfoed, M., Beresford, N. A., ... Grandjean, P. (1999). Comparison of short-term estrogenicity tests for identification of hormone-disrupting chemicals. *Environmental Health Perspectives*, *107*(suppl 1), 89–108. doi:10.1289/ehp.99107s189

Andersen, H. R., Vinggaard, A. M., Rasmussen, T. H., Gjermandsen, I. M., & Bonefeld-Jørgensen, E. C. (2002). Effects of Currently Used Pesticides in Assays for Estrogenicity, Androgenicity, and Aromatase Activity in Vitro. *Toxicology and Applied Pharmacology*, *179*(1), 1–12. doi:10.1006/taap.2001.9347

Antoniou, A., Pharoah, P. D. P., Narod, S., Risch, H. A., Eyfjord, J. E., Hopper, J. L., ... Easton, D. F. (2003). Average risks of breast and ovarian cancer associated with BRCA1 or BRCA2 mutations detected in case Series unselected for family history: A combined analysis of 22 studies. *American Journal of Human Genetics*, *72*(5), 1117–1130. doi:10.1086/375033

Aronson, K., Miller, A., Woolcott, C., Sterns, E., McCready, D., Lickley, L., ... Weber, J. P. (2000). Breast adipose tissue concentrations of polychlorinated biphenyls and other organochlorines and breast cancer risk. *Cancer Epidemiology, Biomarkers & Prevention*, *9*, 55–63.

Arvis, B. B., & Miller, J. D. (2005). Mycotoxins as harmful indoor air contaminants. *Applied Microbiology and Biotechnology*, *66*, 367-372.

Assi, H. A., Khoury, K. E., Dbouk, H., Khalil, L. E., Mouhieddine, T. H., & El Saghir, N. S. (2013). Epidemiology and prognosis of breast cancer in young women. *Journal of Thoracic Disease*, *5*, S2–S8.

Axelrod, D., Smith, J., Kornreich, D., Grinstead, E., Singh, B., Cangiarella, J., & Guth, A. A. (2008). Breast Cancer in Young Women. *Journal of the American College of Surgeons*, *206*(6), 1193–1203. doi:10.1016/j.jamcollsurg.2007.12.026

Baan, R., Grosse, Y., Straif, K., Secretan, B., El Ghissassi, F., Bouvard, V., ... Cogliano, V. (2009). A review of human carcinogens—Part F: Chemical agents and related occupations. *The Lancet Oncology*, *10*(12), 1143–1144. doi:10.1016/S1470-2045(09)70358-4

Bakir, A., & Darbre, P. D. (2015). Effect of aluminium on migration of oestrogen unresponsive MD-AMB-231 human breast cancer cells in culture. *Journal of Inorganic Biochemistry*, *152*, 180–185. doi:10.1016/j.jinorgbio.2015.09.001

Bandera, E. V., Chandran, U., Buckley, B., Lin, Y., Isukapalli, S., Marshall, I., ... Zarbl, H. (2011). Urinary mycoestrogens, body size and breast development in New Jersey girls. *The Science of the Total Environment*, *409*(24), 5221–5227. doi:10.1016/j.scitotenv.2011.09.029

Barnett, G. C., Shah, M., Redman, K., Easton, D. F., Ponder, B. A., & Pharoah, P. D. (2008). Risk factors for the incidence of breast cancer: Do they affect survival from the disease? *Journal of Clinical Oncology*, *26*(20), 3310–3336. doi:10.1200/JCO.2006.10.3168

Becker-Algeri, T. A., Castagnaro, D., de Bortoli, K., de Souza, C., Drunkler, D. A., & Badiale-Furlong, E. (2016). Mycotoxins in bovine milk and dairy products: A review. *Journal of Food Science*, *81*(3), R544–R552. doi:10.1111/1750-3841.13204

Belhassen, H., Jiménez-Díaz, I., Arrebola, J. P., Ghali, R., Ghorbel, H., Olea, N., & Hedili, A. (2015). Zearalenone and its metabolites in urine and breast cancer risk: A case-control study in Tunisia. *Chemosphere*, *128*, 1–6. doi:10.1016/j.chemosphere.2014.12.055

Ben-Chioma, A. E., Obunwo, C. C., Bartimaeus, E. S., & Boisa, N. (2017). Evaluation of Heavy Metal Levels in the Serum of Breast Cancer Patients in Port Harcourt, Nigeria. *International Journal of Contemporary Medical Research.*, *4*, 2393–915X.

Benderli Cihan, Y., Sözen, S., Yıldırım, O., Benderli Cihan, Y., Sözen, S., & Öztürk Yıldırım, S. (2011). nTrace elements and heavy metals in hair of stage III breast cancer patients. *Biological Trace Element Research*, *144*, 360-379.

Bennett, J., & Klich, M. (2003). Mycotoxins. *Clinical Microbiology Reviews*, *16*(3), 497–516. doi:10.1128/CMR.16.3.497-516.2003

Bertazzi, P., Pesatori, A., Consonni, D., Tironi, A., Landi, M., & Zocchetti, C. (1993). Cancer incidence in a population accidentally exposed to 2,3,7,8-tetrachlorodibenzo-para-dioxin. *Epidemiology (Cambridge, Mass.)*, *4*(5), 398–406. doi:10.1097/00001648-199309000-00004

Bhatnagar, V. K., Kashyap, R., Zaidi, S. S. A., Kulkarni, P. K., & Saiyed, H. N. (2004). Levels of DDT, HCH, and HCB residues in human blood in Ahmedabad. *Bulletin of Environmental Contamination and Toxicology*, *72*(2), 261–265. doi:10.100700128-003-9049-9

Blount, W. P. (1961). Turkey "X" disease. *Turkeys*, *9*, 52–55.

Boada, L. D., Zumbado, M., Henríquez-Hernández, L. A., Almeida-González, M., Alvarez-León, E. E., Serra-Majem, L., & Luzardo, O. P. (2012). Complex organochlorine pesticide mixtures as determinant factor for breast cancer risk: A population-based case-control study in the Canary Islands (Spain). *Environmental Health*, *11*(1), 28. doi:10.1186/1476-069X-11-28

Bogdanffy, M. S., O'Connor, J. C., Hansen, J. F., Gaddamidi, V., Van Pelt, C. S., Green, J. W., & Cook, J. C. (2000). Chronic toxicity and oncogenicity bioassay in rats with the chlorostriazine herbicide cyanazine. *Journal of Toxicology and Environmental Health*, *60*(8), 567–586. doi:10.1080/00984100050082102

Boorman, G. (1989). Toxicology and carcinogenesis studies of ochratoxin A F344/N rats (gavage studies). *National Toxicology Program*, *358*, 1-65.

Brama, M., Gnessi, L., Basciani, S., Cerulli, N., Politi, L., Spera, G., ... Migliaccio, S. (2007). Cadmium induces mitogenic signaling in breast cancer cell by an ERalpha-dependent mechanism. *Molecular and Cellular Endocrinology*, *264*(1-2), 102–108. doi:10.1016/j.mce.2006.10.013

Bressac, B., Kew, M., Wands, J., & Ozturk, M. (1991). Selective G to T mutations of p53 gene in hepatocellular carcinoma from southern Africa. *Nature*, *350*(6317), 429–431. doi:10.1038/350429a0

Brinton, L. A., Richesson, D. A., Gierach, G. L., Lacey, J. V., Park, Y., Hollenbeck, A. R., & Schatzkin, A. (2008). Prospective evaluation of risk factors for male breast cancer. *Journal of the National Cancer Institute*, *100*(20), 1477–1481. doi:10.1093/jnci/djn329

Brody, J. G., Moysich, K. B., Humblet, O., Attfield, K. R., Beehler, G. P., & Rudel, R. A. (2007). Environmental pollutants and breast cancer: Epidemiologic studies. *Cancer*, *109*(S12), 2667–2711. doi:10.1002/cncr.22655

Bui-Klimke, T. R., & Wu, F. (2015). Ochratoxin A and human health risk: A review of the evidence. *Critical Reviews in Food Science and Nutrition*, *55*(13), 1860–1869. doi:10.1080/10408398.2012.724480

Byrne, C., Divekar, S. D., Storchan, G. B., Parodi, D. A., & Martin, M. B. (2013). Metals and Breast Cancer. *Journal of Mammary Gland Biology and Neoplasia*, *18*(1), 63–73. doi:10.100710911-013-9273-9

Cameron, H. L., & Foster, W. G. (2009). Developmental and Lactational Exposure to Dieldrin Alters Mammary Tumorigenesis in Her2/Neu Transgenic Mice. *PLoS One*, *4*(1), 4303. doi:10.1371/journal.pone.0004303

Carpenter, D. O. (2006). Polychlorinated biphenyls (PCBs): Routes of exposure and effects on human health. *Reviews on Environmental Health*, *21*(1), 1–23. doi:10.1515/REVEH.2006.21.1.1

Cassidy, R. (2010). Cancer and Chlordane-Treated Homes: A Pinch of Prevention Is Worth a Pound of Cure. *Leukemia & Lymphoma*, *51*(7), 1363–1364. doi:10.3109/10428194.2010.483304

Cassidy, R. A., Natarajan, S., & Vaughan, G. M. (2005). The link between the insecticide heptachlor epoxide, estradiol, and breast cancer. *Breast Cancer Research and Treatment*, *90*(1), 55–64. doi:10.100710549-004-2755-0

Cavallo, F., Gerber, M., Marubini, E., Richardson, S., Barbieri, A., Costa, A., ... Pujol, H. (1991). Zinc and copper in breast cancer. A joint study in northern Italy and southern France. *Cancer*, *67*(3), 738–745. doi:10.1002/1097-0142(19910201)67:3<738::AID-CNCR2820670335>3.0.CO;2-#

Centers for Disease Control and Prevention. (2009). *National report on human exposure to environmental chemicals*. Retrieved from http://www.cdc.gov/exposurereport/

Charlier, C., Albert, A., Herman, P., Hamoir, E., Gaspard, U., Meurisse, M., & Plomteux, G. (2003). Breast Cancer and Serum Organochlorine Residues. *Occupational and Environmental Medicine*, *60*(5), 348–351. doi:10.1136/oem.60.5.348

Cho, Y. A., Kim, J., Woo, H. D., & Kang, M. (2013). Dietary Cadmium Intake and the Risk of Cancer: A Meta-Analysis. *PLoS One*, *8*(9), e75087. doi:10.1371/journal.pone.0075087

Clark, H. A., & Snedeker, S. M. (2006). Ochratoxin A: Its cancer risk and potential for exposure. *Journal of Toxicology and Environmental Health. Part B, Critical Reviews*, *9*(3), 265–296. doi:10.1080/15287390500195570

Cleary, M. P., & Grossmann, M. E. (2009). Obesity and Breast Cancer: The Estrogen Connection. *Endocrinology*, 150, 2537–2542.

Cohn, B. A., La Merrill, M., Krigbaum, N. Y., Yeh, G., Park, J. S., Zimmermann, L., & Cirillo, P. M. (2015). DDT exposure in utero and breast cancer. *The Journal of Clinical Endocrinology and Metabolism*, *100*(8), 2865–2872. doi:10.1210/jc.2015-1841

Cohn, B. A., Wolff, M. S., Cirillo, P. M., & Sholtz, R. I. (2007). DDT and breast cancer in young women: New data on the significance of age at exposure. *Environmental Health Perspectives*, *115*, 1406–1414.

Colditz, G. A., Kaphingst, K. A., Hankinson, S. E., & Rosner, B. (2012). Family history and risk of breast cancer: Nurses' health study. *Breast Cancer Research and Treatment*, *133*(3), 1097–1104. doi:10.100710549-012-1985-9

Collaborative Group on Hormonal Factors in Breast Cancer. (2002). Breast cancer and breast feeding: Collaborative reanalysis of individual data from 47 epidemiological studies in 30 countries, including 50302 women with breast cancer and 96973 women without the disease. *Lancet*, *360*(9328), 187–195. doi:10.1016/S0140-6736(02)09454-0

Comas, A. P. (1982). Precocious sexual development in Puerto Rico. *Lancet*, *319*(8284), 1299–1300. doi:10.1016/S0140-6736(82)92857-4

Dänicke, S., & Winkler, J. (2015). Invited review: Diagnosis of zearalenone (ZEN) exposure of farm animals and transfer of its residues into edible tissues (carry over). *Food and Chemical Toxicology*, *84*, 225–249. doi:10.1016/j.fct.2015.08.009

Dantzig, P. I. (2009). Breast cancer, dermatofibromas and arsenic. *Indian Journal of Dermatology*, *54*(1), 23–25. doi:10.4103/0019-5154.48981

Darbre, P. D. (2005). Aluminium, antiperspirants and breast cancer. *Journal of Inorganic Biochemistry*, *99*(9), 1912–1919. doi:10.1016/j.jinorgbio.2005.06.001

Darbre, P. D., Mannello, F., & Exley, C. (2013). Aluminium and breast cancer: Sources of exposure, tissue measurements and mechanisms of toxicological actions on breast biology. J. *Journal of Inorganic Biochemistry*, *128*, 257–261. doi:10.1016/j.jinorgbio.2013.07.005

Davey, J. C., Bodwell, J. E., Gosse, J. A., & Hamilton, J. W. (2007). Arsenic as an endocrine disruptor: Effects of arsenic on estrogen receptor-mediated gene expression *in vivo* and in cell culture. *Toxicological Sciences*, *98*(1), 75–86. doi:10.1093/toxsci/kfm013

Davis, L. K., Murr, A. S., Best, D. S., Fraites, M. J., Zorrilla, L. M., Narotsky, M. G., ... Cooper, R. L. (2011). The effects of prenatal exposure to atrazine on pubertal and postnatal reproductive indices in the female rat. *Reproductive Toxicology (Elmsford, N.Y.)*, *32*(1), 43–51. doi:10.1016/j.reprotox.2011.04.004

Davis, S., Mirick, D. K., & Stevens, R. G. (2001). Night shift work, light at night, and risk of breast cancer. *Journal of the National Cancer Institute*, *93*(20), 1557–1562. doi:10.1093/jnci/93.20.1557

De Ruyck, K., De Boevre, M., Huybrechts, I., & De Saeger, S. (2015). Dietary mycotoxins, co-exposure, and carcinogenesis in humans: Short review. *Mutation Research/Reviews in Mutation Research*, *766*, 32–41. doi:10.1016/j.mrrev.2015.07.003

De Vega, G. R., Fernández-Sánchez, M. L., Pisonero, J., Eiró, N., Vizoso, F., & Sanz-Medel, A. (2017). Quantitative Bioimaging of Ca, Fe, Cu and Zn in Breast Cancer Tissues by LA-ICP-MS. *Journal of Analytical Atomic Spectrometry*, *32*(3), 671–677. doi:10.1039/C6JA00390G

De Wit, C. A., Herzke, D., & Vorkamp, K. (2010). Brominated flame retardants in the Arctic environment trends and new candidates. *The Science of the Total Environment*, *408*(15), 2885–2918. doi:10.1016/j.scitotenv.2009.08.037

Demers, A., Ayotte, P., Brisson, J., Dodin, S., Robert, J., & Dewailly, E. (2000). Risk and aggressiveness of breast cancer in relation to plasma organochlorine concentration. *Cancer Epidemiology, Biomarkers & Prevention*, *9*, 161–166.

Demers, A., Ayotte, P., Brisson, J., Dodin, S., Robert, J., & Dewailly, E. (2002). Plasma concentrations of polychlorinated biphenyls and the risk of breast cancer: A congener-specific analysis. *American Journal of Epidemiology*, *155*(7), 629–635. doi:10.1093/aje/155.7.629

Dent, R., Trudeau, M., Pritchard, K. I., Hanna, W. M., Kahn, H. K., Sawka, C. A., ... Narod, S. A. (2007). Triple-negative breast cancer: Clinical features and patterns of recurrence. *Clinical Cancer Research*, *13*(15), 4429–4434. doi:10.1158/1078-0432.CCR-06-3045

Desaulniers, D., Leingartner, K., Musicki, B., Cole, J., Li, M., Charbonneau, M., & Tsang, B. K. (2004). Lack of effects of postnatal exposure to a mixture of aryl hydrocarbon-receptor agonists on the development of methylnitrosourea-induced mammary tumors in Sprague- Dawley rats. *Journal of Toxicology and Environmental Health. Part A.*, *67*(18), 1457–1475. doi:10.1080/15287390490483818

Dich, J., Zahm, S., Hanberg, A., Adami, H., & Adami, H.-O. (1997). Pesticides and cancer. *Cancer Causes & Control*, *8*(3), 420–443. doi:10.1023/A:1018413522959

Dollear, F. G., & Goldblatt, L. A. (1969). *Aflatoxin, scientific background, control and implications.* New York: Academic Press.

Domingo, J. L. (2012). Polybrominated diphenyl ethers in food and human dietary exposure: A review of the recent scientific literature. *Food and Chemical Toxicology*, *50*(2), 238–249. doi:10.1016/j.fct.2011.11.004

Dorgan, J. F., Brock, J. W., Rothman, N., Needham, L. L., Miller, R., Stephenson, H. E. Jr, ... Taylor, P. R. (1999). Serum organochlorine pesticides and PCBs and breast cancer risk: Results from a prospective analysis (USA). *Cancer Causes & Control*, *10*(1), 1–11. doi:10.1023/A:1008824131727

Dunnwald, L. K., Rossing, M. A., & Li, C. I. (2007). Hormone receptor status, tumor characteristics, and prognosis: A prospective cohort of breast cancer patients. *Breast Cancer Research*, *9*(1), R6. doi:10.1186/bcr1639

El-Harouny, M. A., El-Morsi, D. A., Ahmed, B. R. A., & El-Atta, H. M. A. (2011). Chronic Toxicity of Some Heavy Metals and Breast Cancer in Egyptian Females. *Journal of Clinical Toxicology*, *1*, 109.

Eldridge, S. R., Gould, M. N., & Butterworth, B. E. (1992). Genotoxicity of environmental agents in human mammary epithelial cells. *Cancer Research*, *52*, 5617–5621.

Environmental Protection Agency. (2000). *Exposure and human health reassessment of 2,3,7,8-tetrachlorodibenzo-p-dioxin (TCDD) and related compounds. Part II: Health assessment for 2,3,7,8-tetrachlorodibenzo- p-dioxin (TCDD) and related compounds*. Retrieved from http://www.epa.gov/ncea/pdfs/dioxin/nasreview/pdfs/part2/dioxin_pt2_ch08_dec2003.pdf

Environmental Protection Agency. (2009). *Atrazine science reevaluation: Potential health impacts document.* Retrieved from http://www. epa.gov/pesticides/reregistration/atrazine/atrazine_update.htm

Environmental Protection Agency. (2010). *Atrazine updates.* Retrieved from http://www.epa.gov/opp00001/reregistration/atrazine/ atrazine_update.htm#cancer

Eze, U. A., Routledge, M. N., Okonofua, F. E., Huntriss, J., & Gong, Y. Y. (2018). Mycotoxin exposure and adverse reproductive health outcomes in Africa: A review. *World Mycotoxin Journal*, 1–20.

Fabre, I., & Gardiès, C. (2008). L'accès à l'information scientifique numérique: Organisation des savoirs et enjeu de pouvoir dans une communauté scientifique. *Science and Society*, *75*, 85–99.

Fan, W., Yanase, T., Morinaga, H., Gondo, S., Okabe, T., Nomura, M., ... Nawata, H. (2007). Atrazine-induced aromatase expression is SF-1 dependent: Implications for endocrine disruption in wildlife and reproductive cancers in humans. *Environmental Health Perspectives*, *115*(5), 720–727. doi:10.1289/ehp.9758

Faneyte, I. F., Kristel, P. M., Maliepaard, M., Scheffer, G. L., Scheper, R. J., Schellens, J. H., & van de Vijver, M. J. (2002). Expression of the breast cancer resistance protein in breast cancer. *Clinical Cancer Research*, *8*, 1068–1074.

Farshid, G. (2014). The Normal Breast and Risk Factors for Breast Cancer. Pathobiology of Human Disease: A Dynamic Encyclopedia of Disease Mechanisms, 899-919.

Filali, A., Betbeder, A. M., Baudrimont, I., Benayada, A., Soulaymani, R., & Creppy, E. E. (2002). Ochratoxin A in human plasma in Morocco: A preliminary survey. *Human and Experimental Toxicology*, *21*(5), 241–245. doi:10.1191/0960327102ht249oa

Fitzmaurice, C., Allen, C., Barber, R. M., Barregard, L., Bhutta, Z. A., Brenner, H., ... Naghavi, M. (2017). Global, Regional, and National Cancer Incidence, Mortality, Years of Life Lost, Years Lived With Disability, and Disability-Adjusted Life-Years for 29 Cancer Groups, 1990 to 2016: A Systematic Analysis for the Global Burden of Disease Study. *JAMA Oncology*, *3*, 524. doi:10.1001/jamaoncol.2016.5688

Florea, A. M., & Büsselberg, D. (2011). Metals and Breast Cancer: Risk Factors or Healing Agents? *Journal of Toxicology*.

Foster, P. L., Eisenstadt, E., & Miller, J. H. (1983). Base substitution mutations induced by metabolically activated aflatoxin B1. *Proceedings of the National Academy of Sciences of the United States of America*, *80*(9), 2695–2698. doi:10.1073/pnas.80.9.2695

Foulkes, W. D. (2008). Inherited susceptibility to common cancers. *The New England Journal of Medicine*, *359*(20), 2143–2153. doi:10.1056/NEJMra0802968

Gadzała-Kopciuch, R., Cendrowski, K., Cesarz, A., Kiełbasa, P., & Buszewski, B. (2011). Determination of zearalenone and its metabolites in endometrial cancer by coupled separation techniques. *Analytical and Bioanalytical Chemistry*, *401*(7), 2069–2078. doi:10.100700216-011-5206-x

Gajecki, M., Przybyłowicz, M., Zielonka, L., Zwierzchowski, W., Obremski, K., Skorska-Wyszyńska, E., ... Jakimiuk, E. (2004). Preliminary results of monitoring research on zearalenone presence in blood of women with neoplastic lesions in reproductive system. *Polish Journal of Veterinary Sciences*, *7*, 153–156.

Gammon, M. D., Wolff, M. S., Neugut, A. I., Eng, S. M., Teitelbaum, S. L., Britton, J. A., ... Santella, R. M. (2002). Environmental toxins and breast cancer on Long Island. II. Organochlorine compound levels in blood. *Cancer Epidemiology, Biomarkers & Prevention*, *11*, 686–697.

Gorgogietas, V. A., Tsialtas, I., Sotiriou, N., Laschou, V. C., Karra, A. G., Leonidas, D. D., ... Psarra, A. G. (2018). Potential interference of aluminum chlorohydrate with estrogen receptor signaling in breast cancer cells. *Journal of Molecular Biochemistry*, *7*, 1–13.

Gross, R. E. (2000). Breast cancer: Risk factors, screening, and prevention. *Journal of Seminars in Oncology Nursing*, *16*(3), 176–184. doi:10.1053onc.2000.8110

Gupta, S. K., Shukla, V. K., Vaidya, M. P., Roy, S. K., & Gupta, S. (1991). Serum trace elements and Cu/Zn ratio in breast cancer patients. *Journal of Surgical Oncology*, *46*(3), 178–181. doi:10.1002/jso.2930460311

Gursoy-Yuzugullu, O., Yuzugullu, H., Yilmaz, M., & Ozturk, M. (2011). Aflatoxin genotoxicity is associated with a defective DNA damage response bypassing p53 activation. *Liver International*, *31*(4), 561–571. doi:10.1111/j.1478-3231.2011.02474.x

Haber, G., Ahmed, N. U., & Pekovic, V. (2012). Family History of Cancer and Its Association With Breast Cancer Risk Perception and Repeat Mammography. *American Journal of Public Health*, *102*(12), 2322–2329. doi:10.2105/AJPH.2012.300786

Harrison, J. C., Carvajal, M., & Garner, R. C. (1993). Does aflatoxin exposure in the United Kingdom constitute a cancer risk? *Environmental Health Perspectives*, *99*, 99–105. doi:10.1289/ehp.939999

Hayes, T., Haston, K., Tsui, M., Honag, A., Haeffele, C., & Vonk, A. (2003). Atrazine-Induced Hermaphroditism at 0.1 ppb in American Leopard Frogs (Ranapipiens): Laboratory and Field Evidence. *Environmental Health Perspectives*, *111*(4), 568–575. doi:10.1289/ehp.5932

He, Y., Peng, L., Zhang, W., Liu, C., Yang, Q., Zheng, S., ... Wu, K. (2018). Adipose tissue levels of polybrominated diphenyl ethers and breast cancer risk in Chinese women: A case-control study. *Environmental Research*, *167*, 160–168. doi:10.1016/j.envres.2018.07.009

Helzlsouer, K., Alberg, A., Huang, H., Hoffman, S., Strickland, P., Brock, J., ... Yager, J. (1999). Serum concentrations of organochlorine compounds and the subsequent development of breast cancer. *Cancer Epidemiology, Biomarkers & Prevention*, *8*, 525–532.

Hennig, A., Fink-Gremmels, J., & Leistner, L. (1991). Mutagenicity and effects of ochratoxin A on the frequency of sister chromatid exchange after metabolic activation. *IARC Scientific Publications*, 255–260.

Herwaarden, A. E. V., Wagenaar, E., Karnekamp, B., Merino, G., Jonker, J. W., & Schinkel, A. H. (2005). Breast cancer resistance protein (Bcrp1/Abcg2) reduces systemic exposure of the dietary carcinogens aflatoxin B1, IQ and Trp-P-1 but also mediates their secretion into breast milk. *Carcinogenesis*, *27*(1), 123–130. doi:10.1093/carcin/bgi176

Heussner, A. H., & Bingle, L. E. (2015). Comparative ochratoxin toxicity: A review of the available data. *Toxins*, *7*(10), 4253–4282. doi:10.3390/toxins7104253

Higley, E. B., Newsted, J. L., Zhang, X., Giesy, J. P., & Hecker, M. (2010). Assessment of chemical effects on aromatase activity using the H295R cell line. *Environmental Science and Pollution Research International*, *17*(5), 1137–1148. doi:10.100711356-009-0285-3

Holford, T., Zheng, T., Mayne, S., Zahm, S., Tessari, J., & Boyle, P. (2000). Joint effects of nine polychlorinated biphenyl (PCB) congeners on breast cancer risk. *International Journal of Epidemiology*, *29*(6), 975–982. doi:10.1093/ije/29.6.975

Holmes, A. K., Koller, K. R., Kieszak, S. M., Sjodin, A., Calafat, A. M., Sacco, F. D., ... Rubin, C. H. (2014). Case-control study of breast cancer and exposure to synthetic environmental chemicals among Alaska Native women. *International Journal of Circumpolar Health*, *73*(1), 25760. doi:10.3402/ijch.v73.25760

Hopenhayn-Rich, C., Stump, M. L., & Browning, S. R. (2002). Regional assessment of atrazine exposure and incidence of breast and ovarian cancers in Kentucky. *Archives of Environmental Contamination and Toxicology*, *42*, 127–136. doi:10.1007002440010300

Hovey, R. C., Coder, P. S., Wolf, J. C., Sielken, R. L., Tisdel, M. O., & Breckenridge, C. B. (2011). Quantitative assessment of mammary gland development in female Long Evans rats following in utero exposure to atrazine. *Toxicological Sciences*, *119*(2), 380–390. doi:10.1093/toxsci/kfq337

Hoyer, A., Gerdes, A., Jorgensen, T., Rank, F., & Hartvig, H. (2002a). Organochlorines, p53 mutations in relation to breast cancer risk and survival. *Breast Cancer Research and Treatment*, *71*(1), 59–65. doi:10.1023/A:1013340327099

Hoyer, A., Jorgensen, T., Grandjean, P., & Hartvig, H. (2000). Repeated measurements of organochlorine exposure and breast cancer risk (Denmark). *Cancer Causes & Control*, *11*(2), 177–184. doi:10.1023/A:1008926219539

Hoyer, A. P., Grandjean, P., Jorgensen, T., Brock, J. W., & Hartvig, H. B. (1998). Organochlorine exposure and risk of breast cancer. *Lancet*, *352*(9143), 1816–1820. doi:10.1016/S0140-6736(98)04504-8

Hoyer, A. P., Jørgensen, T., Rank, F., & Grandjea, P. (2001). Organo-Chlorine Exposures Influence on Breast Cancer Risk and Survival According to Estrogen Receptor Status: A Danish Cohort-Nested Case-Control Study. *Biomedical Central Cancer*, *1*, 8.

Hoyer, P., Grandjean, P., Jorgensen, T., Brock, J. W., & Hartvig, H. B. (2002b). Dieldrin as a Risk Factor for Breast Cancer and for Increased Mortality Once Breast Cancer Is Detected. *Journal of Clinical Epidemiology*, *53*, 323–330.

Hsu, I. C., Metcalf, R. A., Sun, T., Welsh, J. A., Wang, N. J., & Harris, C. C. (1991). Mutational hot spot in the p53 gene in human hepatocellular carcinomas. *Nature, 350*(6317), 427–428. doi:10.1038/350427a0

Hurley, S., Reynolds, P., Goldberg, D., Nelson, D. O., Jeffrey, S. S., & Petreas, M. (2011). Adipose levels of polybrominated diphenyl ethers and risk of breast cancer. *Breast Cancer Research and Treatment, 129*(2), 505–511. doi:10.100710549-011-1481-7

Huwe, J. K., Hakk, H., Smith, D. J., Diliberto, J. J., Richardson, V., Stapleton, H. M., & Birnbaum, L. S. (2008). Comparative absorption and bioaccumulation of polybrominated diphenyl ethers following ingestion via dust and oil in male rats. *Environmental Science & Technology, 42*(7), 2694–2700. doi:10.1021/es702644k

Ibarluzea, J., Fernandez, M., Santa-Marina, L., Olea-Serrano, M., Rivas, A., Aurrekoetxea, J., ... Olea, N. (2004). Breast cancer risk and the combined effect of environmental estrogens. *Cancer Causes & Control, 15*(6), 591–600. doi:10.1023/B:CACO.0000036167.51236.86

International Agency for Research on Cancer. (1993). IARC Scientific Publications: Vol. 58. *Be, Cd, Hg, and Exposures in the Glass Manufacturing Industry, IARC Monographs on the Evaluation of Carcinogenic Risks of Chemicals to Humans*. Lyon, France: Author.

International Agency for Research on Cancer. (1997). *Polychlorinated dibenzo-para-dioxins and polychlorinated dibenzofurans. IARC monographs on the evaluation of carcinogenic risks to humans* (Vol. 69). Lyon, France: IARC.

International Agency for Research on Cancer. (1999). *Some chemicals that cause tumours of the kidney or urinary bladder in rodents and some other substances: Summary of data reported and evaluation. IARC monographs on the evaluation of carcinogenic risks to humans* (Vol. 73). Lyon, France: IARC.

International Agency for Research on Cancer (IARC). (2006). Inorganic and Organic Lead Compounds. Lyon, France: Author.

Ionescu, J. G., Novotny, J., Stejskal, V., Latsch, A., Blaurock-Busch, E., & Eisenmann-Klein, M. (2006). Increased levels of transition metals in breast cancer tissue. *Neuroendocrinology Letters, 27*, 36–39.

Jarup, L. (2003). Hazards of heavy metal contamination. *British Medical Bulletin, 68*(1), 167–182. doi:10.1093/bmb/ldg032

Jarup, L., & Akesson, A. (2009). Current status of cadmium as an environmental health problem. *Toxicology and Applied Pharmacology, 238*(3), 201–208. doi:10.1016/j.taap.2009.04.020

Johanning, E., Gareis, M., Nielsen, K., Dietrich, R., & Märtlbauer, E. (2002). Airborne mycotoxin sampling and screening analysis. *Proceedings of the 9th International Conference on Indoor Air Quality and Climate (Indoor Air).*

Joo, N. S., Kim, S. M., Jung, Y. S., & Kim, K. M. (2009). Hair iron and other minerals' level in breast cancer patients. *Biological Trace Element Research, 129*(1-3), 28–35. doi:10.100712011-008-8281-x

Joshi, P. L., Bhattacharya, M., Yadava, R. L., Chand, B., Narasimham, M. V., Nigam, D. K., & Jain, C. L. (1996). A community-based study on the effect of hexachlorocyclohexane (HCH) exposure in spraymen and general population. *The Journal of Communicable Diseases*, *28*, 189–198.

Jouybari, L., Saei Ghare Naz, M., Sanagoo, A., Kiani, F., Sayehmiri, F., Sayehmiri, K., & Hasanpour Dehkordi, A. (2018). Toxic elements as biomarkers for breast cancer: A meta-analysis study. *Cancer Management and Research*, *10*, 69–79. doi:10.2147/CMAR.S151324

Ju, H., Arumugam, P., Lee, J., & Song, J. M. (2017). Impact of Environmental Pollutant Cadmium on the Establishment of a Cancer Stem Cell Population in Breast and Hepatic Cancer. *ACS Omega*, *2*(2), 563–572. doi:10.1021/acsomega.6b00181

Ju-Kun, S., Yuan, D. B., Rao, H. F., Chen, T. F., Luan, B. S., Xu, X. M., ... Zhu, J. G. (2016). Association between Cd exposure and risk of prostate Cancer: A PRISMA-compliant systematic review and meta-analysis. *Medicine; Analytical Reviews of General Medicine, Neurology, Psychiatry, Dermatology, and Pediatries*, *95*(6), e2708. doi:10.1097/MD.0000000000002708

Julin, B., Wolk, A., Johansson, J. E., Andersson, S. O., Andren, O., & Akesson, A. (2012). Dietary cadmium exposure and prostate cancer incidence: A population-based prospective cohort study. *British Journal of Cancer*, *107*(5), 895–900. doi:10.1038/bjc.2012.311

Kagara, N., Tanaka, N., Noguchi, S., & Hirano, T. (2007). Zinc and its transporter ZIP10 are involved in invasive behavior of breast cancer cells. *Cancer Science*, *98*(5), 692–697. doi:10.1111/j.1349-7006.2007.00446.x

Kamińska, M., Ciszewski, T., Łopacka-Szatan, K., Miotła, P., & Staroslawska, E. (2015). Breast cancer risk factors. *Przeglad Menopauzalny Journal*, *14*, 196–202.

Kannan, K., Tanabe, S., Ramesh, A., Subramanian, A., & Tasukawa, R. (1992). Persistent organochlorine residues in food stuffs from India and their implications on human dietary exposure. *Journal of Agricultural and Food Chemistry*, *40*(3), 518–524. doi:10.1021/jf00015a032

Karpeta, A., Maniecka, A., & Gregoraszczuk, E. L. (2016). Different mechanisms of action of 2, 2′, 4, 4′-tetrabromodiphenyl ether (BDE-47) and its metabolites (5-OH-BDE-47 and 6-OH-BDE-47) on cell proliferation in OVCAR-3 ovarian cancer cells and MCF-7 breast cancer cells. *Journal of Applied Toxicology*, *36*(12), 1558–1567. doi:10.1002/jat.3316

Kettles, M., Browning, S., Prince, T., & Horstman, S. (1997). Triazine herbicide exposure and breast cancer incidence: An ecologic study of Kentucky counties. *Environmental Health Perspectives*, *105*(11), 1222–1227. doi:10.1289/ehp.971051222

Kew, M. C. (2013). Aflatoxins as a cause of hepatocellular carcinoma. *Journal of Gastrointestinal and Liver Diseases; JGLD*, *22*, 305–310.

Khanjani, N., Jafarnejad, A. B., & Tavakkoli, L. (2017). Arsenic and breast cancer: A systematic review of epidemiologic studies. *Reviews on Environmental Health*, *32*(3), 267–277. doi:10.1515/reveh-2016-0068

Kim, B. E., Nevitt, T., & Thiele, D. J. (2008). Mechanisms for copper acquisition, distribution and regulation. *Nature Chemical Biology*, *4*(3), 176–185. doi:10.1038/nchembio.72

Kociba, R. J., Keyes, D. G., Beyer, J. E., Carreon, R. M., Wade, C. E., Dittenber, D. A., ... Humiston, C. G. (1978). Results of a two-year chronic toxicity and oncogenicity study of 2,3,7,8-tetrachlorodibenzo-p-dioxin in rats. *Toxicology and Applied Pharmacology*, *46*(2), 279–303. doi:10.1016/0041-008X(78)90075-3

Kuciel-Lisieska, G. R. A. Ż. Y. N. A., Obremski, K., Stelmachów, J. E. R. Z. Y., Gajecka, M., Zielonka, Ł. U. K. A. S. Z., Jakimiuk, E., & Gajecki, M. (2008). Presence of zearalenone in blood plasma in women with neoplastic lesions in the mammary gland. *Bulletin of the Veterinary Institute in Pulawy*, *52*, 671–674.

Kujawa, M. (1994). Some Naturally Occurring Substances: Food Items and Constituents, Heterocyclic Aromatic Amines and Mycotoxins. Food/Nahrung, 38, 351-351.

Kumar, R., Razab, S., Prabhu, K., Ray, S., & Prakash, B. (2017). Serum butyrylcholinesterase and zinc in breast cancer. *Journal of Cancer Research and Therapeutics*, *13*(2), 367–370. doi:10.4103/0973-1482.165869

Kuo, H. W., Chen, S. F., Wu, C. C., Chen, D. R., & Lee, J. H. (2002). Serum and tissue trace elements in patients with breast cancer in Taiwan. *Biological Trace Element Research*, *89*(1), 1–11. doi:10.1385/BTER:89:1:1

Kuroda, K., Hibi, D., Ishii, Y., Yokoo, Y., Takasu, S., Kijima, A., ... Sakai, H. (2015). Role of p53 in the progression from ochratoxin A-Induced DNA damage to gene mutations in the kidneys of mice. *Toxicological Sciences*, *144*(1), 65–76. doi:10.1093/toxsci/kfu267

La Merrill, M., Harper, R., Birnbaum, L. S., Cardiff, R. D., & Threadgill, D. W. (2010). Maternal dioxin exposure combined with a diet high in fat increases mammary cancer incidence in mice. *Environmental Health Perspectives*, *118*(5), 596–601. doi:10.1289/ehp.0901047

Laden, F., Collman, G., Iwamoto, K., Alberg, A., Berkowitz, G., Freudenheim, J., ... Hunter, D. J. (2001a). 1,1-Dichloro-2,2-bis(p-chlorophenyl)ethylene and polychlorinated biphenyls and breast cancer: Combined analysis of five U.S. studies. *Journal of the National Cancer Institute*, *93*(10), 768–776. doi:10.1093/jnci/93.10.768

Laden, F., Hankinson, S., Wolff, M., Colditz, G., Willett, W., Speizer, F., & Hunter, D. (2001b). Plasma organochlorine levels and the risk of breast cancer: An extended follow-up in the Nurses' Health Study. *International Journal of Cancer*, *91*(4), 568–574. doi:10.1002/1097-0215(200002)9999:9999<::AID-IJC1081>3.0.CO;2-W

Lappano, R., Malaguarnera, R., Belfiore, A., & Maggiolini, M. (2017). Recent advances on the stimulatory effects of metals in breast cancer. *Molecular and Cellular Endocrinology*, *5*, 49–56. doi:10.1016/j.mce.2016.10.017

Lawrence, E., Hill, D., Hoppin, J., Lubin, J., Lynch, C., Pierce, J., ... Alavanja, M. (2005). 'Pesticide Use and Breast Cancer Risk among Farmers' Wives in the Agricultural Health Study. *American Journal of Epidemiology*, *161*(2), 121–135. doi:10.1093/aje/kwi022

Lee, R., Woo, W., Kummer, A., Duminy, H., & Xu, Z. (2003). Zinc accumulation in *N*-methyl-*N*-nitrosourea-induced rat mammary tumors is accompanied by an altered expression of ZnT-1 and metallothionein. *Experimental Biology and Medicine*, *228*, 689–696.

Li, J., Li, N., Ma, M., Giesy, J. P., & Wang, Z. (2008). In vitro profiling of the endocrine disrupting potency of organochlorine pesticides. *Toxicology Letters*, *183*, 65–71.

Li, Y., Millikan, R. C., Bell, D. A., Cui, L., Tse, C. K., Newman, B., & Conway, K. (2005). Polychlorinated biphenyls, cytochrome P450 1A1 (CYP1A1) polymorphisms, and breast cancer risk among African American women and white women in North Carolina: A population-based case-control study. *Breast Cancer Research*, *7*(1), 12–18. doi:10.1186/bcr941

Li, Z. H., Liu, X. Y., Wang, N., Chen, J. S., Chen, Y. H., Huang, J. T., ... Chen, D. J. (2012). Effects of decabrominated diphenyl ether (PBDE-209) in regulation of growth and apoptosis of breast, ovarian and cervical cancer cells. *Environmental Health Perspectives*, *120*(4), 541–546. doi:10.1289/ehp.1104051

Liljegren, G., Hardell, L., Lindstrom, G., Dahl, P., & Magnuson, A. (1998). Case-control study on breast cancer and adipose tissue concentrations of congener specific polychlorinated biphenyls, DDE and hexachlorobenzene. *European Journal of Cancer Prevention*, *7*, 135–140.

Lin, P. H., Lin, C. H., Huang, C. C., Chuang, M. C., & Lin, P. (2007). 2,3,7,8-Tetrachlorodibenzo- p-dioxin (TCDD) induces oxidative stress, DNA strand breaks, and poly(ADPribose) polymerase-1 activation in human breast carcinoma cell lines. *Toxicology Letters*, *172*(3), 146–158. doi:10.1016/j.toxlet.2007.06.003

Liu, J., Wu, S., Shen, H., Cui, J., Wang, Y., Xing, L., ... Zhang, X. (2015). Ochratoxin A induces DNA damage and G2 phase arrest in human esophageal epithelium Het-1A cells in vitro. *The Journal of Toxicological Sciences*, *40*(5), 657–665. doi:10.2131/jts.40.657

Liu, S., Li, S., & Du, Y. (2010). Polychlorinated Biphenyls (PCBs) Enhance Metastatic Properties of Breast Cancer Cells by Activating Rho-Associated Kinase (ROCK). *PLoS One*, *5*(6), 11272. doi:10.1371/journal.pone.0011272

Liu, Y., & Wu, F. (2010). Global burden of aflatoxin-induced hepatocellular carcinoma: A risk assessment. *Environmental Health Perspectives*, *118*(6), 818–824. doi:10.1289/ehp.0901388

Llabjani, V., Trevisan, J., Jones, K. C., Shore, R. F., & Martin, F. L. (2011). Derivation by infrared spectroscopy with multivariate analysis of bimodal contaminant-induced dose-response effects in MCF-7 cells. *Environmental Science & Technology*, *45*(14), 6129–6135. doi:10.1021/es200383a

López-Carrillo, L., Hernández-Ramírez, R. U., Gandolfi, A. J., Ornelas Aguirre, J. M., Torres-Sánchez, L., & Cebrian, M. E. (2014). Arsenic methylation capacity is associated with breast cancer in northern Mexico. *Toxicology and Applied Pharmacology*, *280*(1), 53–59. doi:10.1016/j.taap.2014.07.013

Lopez-Carrillo, L., Lopez-Cervantes, M., Torres-Sanchez, L., Blair, A., Cebrain, M., & Garcia, R. (2002). Serum levels of betahexachlorocyclohexane, hexachlorobenzene and polychlorinated biphenyls and breast cancer in Mexican women. *European Journal of Cancer Prevention*, *11*(2), 129–135. doi:10.1097/00008469-200204000-00004

Lopez-Cervantes, M., Torres-Sanchez, L., Tobias, A., & Lopez-Carrillo, L. (2004). Dichlorodiphenyldichloroethane burden and breast cancer risk: A meta-analysis of the epidemiologic evidence. *Environmental Health Perspectives*, *112*(2), 207–214. doi:10.1289/ehp.112-1241830

Lucena, R., Allam, M., Costabeber, I., Villarejo, M., & Navajas, R. (2001). Breast cancer risk factors: PCB congeners. *European Journal of Cancer Prevention*, *10*(1), 117–119. doi:10.1097/00008469-200102000-00018

Ly, D., Forman, D., Ferlay, J., Brinton, L. A., & Cook, M. B. (2013). An international comparison of male and female breast cancer incidence rates. *International Journal of Cancer*, *132*(8), 1918–1926. doi:10.1002/ijc.27841

Ma, J., Qiu, X., Zhang, J., Duan, X., & Zhu, T. (2012). State of polybrominated diphenyl ethers in China: An overview. *Chemosphere*, *88*(7), 769–778. doi:10.1016/j.chemosphere.2012.03.093

Maciorowski, K. G., Herrera, P., Jones, F. T., Pillai, S. D., & Ricke, S. C. (2007). Effects on poultry and livestock of feed contamination with bacteria and fungi. *Animal Feed Science and Technology*, *133*(1-2), 109–136. doi:10.1016/j.anifeedsci.2006.08.006

Mannello, F., Tonti, G. A., Medda, V., Simone, P., & Darbre, P. D. (2011). Analysis of aluminium content and iron homeostasis in nipple aspirate fluids from healthy women and breast cancer-affected patients. *Journal of Applied Toxicology*, *31*(3), 262–269. doi:10.1002/jat.1641

Manning, H. C., Buck, J. R., & Cook, R. S. (2016). Mouse models of breast cancer: platforms for discovering precision imaging diagnostics and future cancer medicine. *Journal of Nuclear Medicine*, *57*, 60S–68S.

Manolova, Y., Manolov, G., Parvanova, L., Petkova-Bocharova, T., Castegnaro, M., & Chernozemsky, I. N. (1990). Induction of characteristic chromosomal aberrations, particularly X-trisomy, in cultured human lymphocytes treated by ochratoxin A, a mycotoxin implicated in Balkan endemic nephropathy. *Mutation Research. Fundamental and Molecular Mechanisms of Mutagenesis*, *231*(2), 143–149. doi:10.1016/0027-5107(90)90021-U

Mantle, P., Kulinskaya, E., & Nestler, S. (2005). Renal tumourigenesis in male rats in response to chronic dietary ochratoxin A. *Food Additives and Contaminants*, *22*(sup1), 58–64. doi:10.1080/02652030500358431

Manuwald, U., Velasco-Garrido, M., Berger, J., Manz, A., & Baur, X. (2012). Mortality study of chemical workers exposed to dioxins: Follow-up 23 years after chemical plant closure. *Occupational and Environmental Medicine*, *69*(9), 636–642. doi:10.1136/oemed-2012-100682

Martin, M. B., Reiter, R., Pham, T., Avellanet, Y. R., Camara, J., Lahm, M., ... Stoica, A. (2003). Estrogen like activity of metals in MCF-7 breast cancer cells. *The Journal of Endocrinology*, *144*(6), 2425–2436. doi:10.1210/en.2002-221054

Mbundi, L., Gallar-Ayala, H., Khan, M. R., Barber, J. L., Losada, S., & Busquets, R. (2014). Advances in the analysis of challenging food contaminants: Nanoparticles, bisphenols, mycotoxins, and brominated flame retardants. In J. C. Fishbein & J. M. Heilman (Eds.), *Advances in Molecular Toxicology* (Vol. 8, pp. 35–105). Elsevier.

McCready, D., Aronson, K., Chu, W., Fan, W., Vesprini, D., & Narod, S. (2004). Breast tissue organochlorine levels and metabolic genotypes in relation to breast cancer risk Canada. *Cancer Causes & Control*, *15*(4), 399–418. doi:10.1023/B:CACO.0000027505.32564.c2

McDonald, T. A. A. (2002). Perspective on the potential health risks of PBDEs. *Chemosphere*, *46*(5), 745–755. doi:10.1016/S0045-6535(01)00239-9

McElroy, J. A., Gangnon, R. E., Newcomb, P. A., Kanarek, M. S., Anderson, H. A., Brook, J. V., ... Remington, P. L. (2007). Risk of breast cancer for women living in rural reas from adult exposure to atrazine from well water in Wisconsin. *Journal of Exposure Science & Environmental Epidemiology*, *17*(2), 207–214. doi:10.1038j.jes.7500511

McElroy, J. A., Shafer, M. M., Trentham-Dietz, A., Hampton, J. M., & Newcomb, P. A. (2006). Cadmium exposure and breast cancer risk. *Journal of the National Cancer Institute*, *98*(12), 869–873. doi:10.1093/jnci/djj233

McGrath, K. G. (2003). An earlier age of breast cancer diagnosis related to more frequent use of anti-perspirants/deodorants and underarm shaving. *European Journal of Cancer Prevention*, *12*(6), 479–485. doi:10.1097/00008469-200312000-00006

Meerts, I. A., Letcher, R. J., Hoving, S., Marsh, G., Bergman, A., Lemmen, J. G., ... Brouwer, A. (2001). *In vitro* estrogenicity of polybrominated diphenyl ethers, hydroxylated PDBEs, and polybrominated bisphenol A compounds. *Environmental Health Perspectives*, *109*(4), 399–407. doi:10.1289/ehp.01109399

Miki, Y., Swensen, J., Shattuck-Eidens, D., Futreal, P. A., Harshman, K., Tavtigian, S., ... Skolnick, M. H. (1994). A strong candidate for the breast and ovarian cancer susceptibility gene BRCA1. *Science*, *266*(5182), 66–71. doi:10.1126cience.7545954

Millikan, R., DeVoto, E., Duell, E., Tse, C., Savitz, D., Beach, J., ... Newman, B. (2000). Dichlorodiphenyldichloroethene, polychlorinated biphenyls, and breast cancer among African-American and white women in North Carolina. *Cancer Epidemiology, Biomarkers & Prevention*, *9*, 1233–1240.

Mills, P. K., & Yang, R. (2005). Breast Cancer Risk in Hispanic Agricultural Workers in California. *International Journal of Occupational and Environmental Health*, *11*(2), 123–131. doi:10.1179/oeh.2005.11.2.123

Mills, P. K., & Yang, R. (2006). Regression analysis of pesticide use and breast cancer incidence in California Latinas. *Journal of Environmental Health*, *68*, 15–22.

Misihairabgwi, J. M., Ezekiel, C. N., Sulyok, M., Shephard, G. S., & Krska, R. (2017). Mycotoxin contamination of foods in Southern Africa: A 10-year review (2007–2016). *Critical Reviews in Food Science and Nutrition*, 1–16. doi:10.1080/10408398.2017.1357003

Mohammadi, M., Riyahi Bakhtiari, A., & Khodabandeh, S. (2014). Concentration of Cd, Pb, Hg, and Se in different parts of human breast cancer tissues. *Journal of Toxicology*, *2014*, 1–5. doi:10.1155/2014/413870

Mori, H., Kawai, K., Ohbayashi, F., Kuniyasu, T., Yamazaki, M., Hamasaki, T., & Williams, G. M. (1984). Genotoxicity of a variety of mycotoxins in the hepatocyte primary culture/DNA repair test using rat and mouse hepatocytes. *Cancer Research*, *44*, 2918–2923.

Moysich, K., Ambrosone, C., Vena, J., Shields, P., Mendola, P., Kostyniak, P., ... Freudenheim, J. (1998). Environmental organochlorine exposure and postmenopausal breast cancer risk. *Cancer Epidemiology, Biomarkers & Prevention*, *7*, 181–188.

Muir, K., Rattanamongkolgul, S., Smallman-Raynor, M., Thomas, M., Downer, S., & Jenkinson, C. (2004). Breast cancer incidence and its possible spatial association with pesticide application in two counties of England. *Public Health, 118*(7), 513–520. doi:10.1016/j.puhe.2003.12.019

Mulware, S. J. (2013). Article Comparative Trace Elemental Analysis in Cancerous and Noncancerous Human Tissues Using PIXE. *Journal of Biophysics.*

Muscat, J., Britton, J., Djordjevic, M., Citron, M., Kemeny, M., Busch-Devereaux, E., ... Stellman, S. (2003). Adipose concentrations of organochlorine compounds and breast cancer recurrence in Long Island, New York. *Cancer Epidemiology, Biomarkers & Prevention, 12*, 1474–1478.

Mussalo-Rauhamaa, H., Hasanen, E., Pyysalo, H., Antervo, K., Kauppila, R., & Pantzar, P. (1990). Occurrence of beta-hexachlorocyclohexane in breast cancer patients. *Cancer, 66*(10), 2124–2128. doi:10.1002/1097-0142(19901115)66:10<2124::AID-CNCR2820661014>3.0.CO;2-A

National Research Council. (2006). *Health risks from dioxin and related compounds: Evaluation of the EPA reassessment.* Washington, DC: The National Academies Press.

National Toxicology Program. (2004). *Toxicology and carcinogenesis studies of 2,3,7,8-tetrachlorodibenzo-p-dioxin (TCDD) (CAS No. 1746-01-6) in female Harlan Sprague-Dawley rats (gavage studies). TR-521.* Research Triangle Park, NC: NIEHS.

National Toxicology Program. (2006). *Toxicology and carcinogenesis studies of 2,3,7,8-tetrachlorodibenzo-p-dioxin (TCDD) (CAS No. 1746–01–6) in female Harlan Sprague-Dawley rats (gavage studies).* TR-521. Research Triangle Park, NC: NIEHS. Retrieved from http://ntp.niehs.nih.gov/files/521_web.pdf

Niehoff, N. M., Nichols, H. B., White, A. J., Parks, C. G., D'Aloisio, A. A., & Sandler, D. P. (2016). Childhood and adolescent pesticide exposure and breast cancer risk. *Epidemiology (Cambridge, Mass.), 27*(3), 326–333. doi:10.1097/EDE.0000000000000451

Nittoli, A. C., Costantini, S., Sorice, A., Capone, F., Ciarcia, R., Marzocco, S., ... Severino, L. (2018). Effects of α-zearalenol on the metabolome of two breast cancer cell lines by 1H-NMR approach. *Metabolomics, 14*(3), 33. doi:10.100711306-018-1330-3

Parandin, R., Rassouli, M. B., Sisakhtnezhad, S., & Shahri, N. M. (2015). In vitro evaluation of the effects of zearalenone and α-zearalenol on MCF-7 and MDA-MB-468 cell lines of human breast cancer. *Razavi International Journal of Medicine, 3.*

Park, J. S., Linderholm, L., Charles, M. J., Athanasiadou, M., Petrik, J., Kocan, A., ... Hertz-Picciotto, I. (2007). Polychlorinated biphenyls their hydroxylated metabolites (OH-PCBS) in pregnant women from eastern Slovakia. *Environmental Health Perspectives, 115*(1), 20–27. doi:10.1289/ehp.8913

Pazaiti, A., Kontos, M., & Fentiman, I. S. (2012). ZEN and the art of breast health maintenance. *International Journal of Clinical Practice, 66*(1), 28–36. doi:10.1111/j.1742-1241.2011.02805.x

Pesatori, A. C., Consonni, D., Rubagotti, M., Grillo, P., & Bertazzi, P. A. (2009). Cancer incidence in the population exposed to dioxin after the "Seveso accident": Twenty years of follow-up. *Environmental Health, 8*(1), 39. doi:10.1186/1476-069X-8-39

Pesticide Action Network UK. (1993). High residues in India. In *A history of HCH contamination*. Academic Press. Retrieved from http://www.pan-uk.org/ pestnews/pn21/pn21p15c.htm

Pfohl-Leszkowicz, A., & Manderville, R. A. (2011). An update on direct genotoxicity as a molecular mechanism of ochratoxin A carcinogenicity. *Chemical Research in Toxicology*, *25*(2), 252–262. doi:10.1021/tx200430f

Piccinini, L., Borella, P., Bargellini, A., Medici, C. I., & Zoboli, A. (1996). A case-control study on selenium, zinc, and copper in plasma and hair of subjects affected by breast and lung cancer. *Biological Trace Element Research*, *51*(1), 23–30. doi:10.1007/BF02790144

Pillay, D., Chuturgoon, A. A., Nevines, E., Manickum, T., Deppe, W., & Dutton, M. F. (2002). The quantitative analysis of zearalenone and its derivatives in plasma of patients with breast and cervical cancer. *Clinical Chemistry and Laboratory Medicine*, *40*(9), 946–951. doi:10.1515/CCLM.2002.166

Pirkle, J., Wolfe, W., Patterson, D., Needham, L., Michalek, J., Miner, J., ... Philips, D. (1989). Estimates of the half-life of 2,3,7,8-tetrachlorodibenzo-*p*-dioxin in Vietnam veterans of Operation Ranch Hand. *Journal of Toxicology and Environmental Health*, *27*(2), 165–171. doi:10.1080/15287398909531288

Pitt, J. I., Basilico, J. C., Abarca, M. L., & Lopez, C. (2000). Mycotoxins and toxigenic fungi. *Medical Mycology*, *38*(s1), 41–46. doi:10.1080/mmy.38.s1.41.46

Polkowska, Z., Kozłowska, K., Namieśnik, J., & Przyjazny, A. (2004). Biological fluids as a source of information on the exposure of man to environmental chemical agents. *Critical Reviews in Analytical Chemistry*, *34*(2), 105–119. doi:10.1080/10408340490475911

Przybylowicz, A., Chesy, P., Herman, M., Parczewski, A., Walas, S., & Piekoszewski, W. (2012). Examination of distribution of trace elements in hair, fingernails and toenails as alternative biological materials. Application of chemometric methods. *Central European Journal of Chemistry*, *10*, 1590–1599.

Raaschou-Nielsen, O., Pavuk, M., Leblanc, A., Dumas, P., Weber, J., Olsen, A., ... Olsen, J. (2005). Adipose organochlorine concentrations and risk of breast cancer among postmenopausal Danish women. *Cancer Epidemiology, Biomarkers & Prevention*, *14*, 67–74.

Raynor, J. L., Enoch, R. R., & Fenton, S. E. (2005). Adverse Effects of Prenatal Exposure to Atrazine during a Critical Period of Mammary Gland Growth. *Toxicological Sciences*, *87*(1), 255–266. doi:10.1093/toxsci/kfi213

Revich, E., Aksel, E., Ushakova, T., Ivanova, J., Zhuchenko, N., Klyuev, N., ... Sotskov, Y. (2001). Dioxin exposure and public health in Chapaevsk, Russia. *Chemosphere*, *43*(4-7), 951–966. doi:10.1016/S0045-6535(00)00456-2

Romaniuk, A., Lyndin, M., Sikora, V., Lyndina, Y., Romaniuk, S., & Sikora, K. (2017). Heavy metals effect on breast cancer progression. *Journal of Occupational Medicine and Toxicology*, *28*, 12-32.

Romieu, I., Hernandez, M., Lazcano-Ponce, E., Weber, J. P., & Dewailly, E. (2000). Breast Cancer, Lactation History, and Serum Organochlorines. *American Journal of Epidemiology*, *152*(4), 363–370. doi:10.1093/aje/152.4.363

Ronckers, C. M., Erdmann, C. A., & Land, C. E. (2005). Radiation and breast cancer: A review of current evidence. *Breast Cancer Research*, *7*(1), 21–32. doi:10.1186/bcr970

Rudolph, A., Chang-Claude, J., & Schmidt, M. K. (2016). Gene–environment interaction and risk of breast cancer. *British Journal of Cancer*, *114*(2), 125–133. doi:10.1038/bjc.2015.439

Rusiecki, J. A., Holford, T. R., Zahm, S. H., & Zheng, T. (2004). Polychlorinated biphenyls and breast cancer risk by combined estrogen and progesterone receptor status. *European Journal of Epidemiology*, *19*(8), 793–801. doi:10.1023/B:EJEP.0000036580.05471.31

Sáenz de Rodríguez, C. A. (1984). Environmental hormone contamination in Puerto Rico. *The New England Journal of Medicine*, *310*(26), 1741–1742. doi:10.1056/NEJM198406283102612

Safe, S. (2005). Clinical correlates of environmental endocrine disruptors. *Trends in Endocrinology and Metabolism*, *16*(4), 139–144. doi:10.1016/j.tem.2005.03.004

Safe, S., & Wormke, M. (2003). Inhibitory aryl hydrocarbon receptor-estrogen receptor alpha cross-talk and mechanisms of action. *Chemical Research in Toxicology*, *16*(7), 807–816. doi:10.1021/tx034036r

Saleh, F., Behbehani, A., Asfar, S., Khan, I., & Ibrahim, G. (2011). Abnormal blood levels of trace elements and metals, DNA damage, and breast cancer in the state of Kuwait. *Biological Trace Element Research*, *141*(1-3), 96–109. doi:10.100712011-010-8724-z

Sappino, A. P., Buser, R., Lesne, L., Gimelli, S., Bena, F., Belin, D., & Mandriota, S. J. (2012). Aluminium chloride promotes anchorage-independent growth in human mammary epithelial cells. *Journal of Applied Toxicology*, *32*(3), 233–243. doi:10.1002/jat.1793

Sargeant, K., Sheridan, A., O'kelly, J., & Carnaghan, R. B. A. (1961). Toxicity associated with certain samples of groundnuts. *Nature*, *192*(4807), 1096–1097. doi:10.1038/1921096a0

Sathiakumar, N., MacLennan, P. A., Mandel, J., & Delzell, E. (2011). A review of epidemiologic studies of triazine herbicides and cancer. *Critical Reviews in Toxicology*, *41*(sup1), 1–34. doi:10.3109/10408444.2011.554793

Schwartz, G. G. (2002). Hypothesis: Does ochratoxin A cause testicular cancer? *Cancer Causes & Control*, *13*(1), 91–100. doi:10.1023/A:1013973715289

Shah, S. I. A., Minhas, U., & Khan, H. A. (2015). Trace minerals and heavy metals: Implications in prostate cancer. *World Cancer Research Journal*, *2*, e615.

Shields, P. G. (2006). Understanding population and individual risk assessment: The case of polychlorinated biphenyls. *Cancer Epidemiology, Biomarkers & Prevention*, *15*(5), 830–839. doi:10.1158/1055-9965.EPI-06-0222

Shipley, J. M., & Waxman, D. J. (2006). Aryl hydrocarbon receptor-independent activation of estrogen receptor-dependent transcription by 3-methylcholanthrene. *Toxicology and Applied Pharmacology*, *213*(2), 87–97. doi:10.1016/j.taap.2005.09.011

Shotwell, O. L., Hesseltine, C. W., & Goulden, M. L. (1969). Ochratoxin A: Occurrence as natural contaminant of a corn sample. *Applied Microbiology, 17*, 765.

Siddiqui, M. K., Jyoti, Singh, S., Mehrotra, P., Singh, K., & Sarangi, R. (2006). Comparison of some trace elements concentration in blood, tumor free breast and tumor tissues of women with benign and malignant breast lesions: An Indian study. *Environment International, 32*(5), 630–637. doi:10.1016/j.envint.2006.02.002

Sidhu, G. S. (2002). Mycotoxin genetics and gene clusters. *European Journal of Plant Pathology, 108*(7), 705–711. doi:10.1023/A:1020613413483

Silva, M. P., Soave, D. F., Ribeiro-Silva, A., & Poletti, M. E. (2012). Trace elements as tumor biomarkers and prognostic factors in breast cancer: a study through energy dispersive x-ray fluorescence. *BMC Research Notes, 6*.

Skrajnowska, D., Bobrowska-Korczak, B., Tokarz, A., Bialek, S., Jezierska, E., & Makowska, J. (2013). Copper and resveratrol attenuates serum catalase, glutathione peroxidase, and element values in rats with DMBA-induced mammary carcinogenesis. *Biological Trace Element Research, 156*(1-3), 271–278. doi:10.100712011-013-9854-x

Slaoui, M., Mouh, F. Z., Ghanname, I., Razine, R., El Mzibri, M., & Amrani, M. (2016). Outcome of Breast Cancer in Moroccan Young Women Correlated to Clinic-Pathological Features, Risk Factors and Treatment: A Comparative Study of 716 Cases in a Single Institution. *PLoS One, 11*(10), e0164841. doi:10.1371/journal.pone.0164841

Snedeker, S. M. (2001). Pesticides and breast cancer risk: A review of DDT, DDE, and dieldrin. *Environmental Health Perspectives, 109*, 35–47.

Soliman, A. S., Wang, X., Digiovanni, J., Eissa, S., Morad, M., Vulimiri, S., ... Bondy, M. L. (2003). Serum organochlorine levels and history of lactation in Egypt. *Environmental Research, 92*(2), 110–117. doi:10.1016/S0013-9351(02)00056-7

Son, W. C., Kamino, K., Lee, Y. S., & Kang, K. S. (2003). Strain-specific mammary proliferative lesion development following lifetime oral administration of ochratoxin A in DA and Lewis rats. *International Journal of Cancer, 105*(3), 305–311. doi:10.1002/ijc.11069

Stob, M., Baldwin, R. S., Tuite, J., Andrews, F. N., & Gillette, K. G. (1962). Isolation of an anabolic, uterotrophic compound from corn infected with *Gibberella zeae. Nature, 196*(4861), 1318. doi:10.1038/1961318a0

Strumylaite, L., Bogusevicius, A., Abdrachmanovas, O., Baranauskiene, D., Kregzdyte, R., Pranys, D., & Poskiene, L. (2011). Cadmium concentration in biological media of breast cancer patients. *Breast Cancer Research and Treatment, 125*(2), 511–517. doi:10.100710549-010-1007-8

Szuets, P., Mesterhazy, A., Falkay, G. Y., & Bartok, T. (1997). Early telarche symptoms in children and their relations to zearalenon contamination in foodstuffs. *Cereal Research Communications*, 429–436.

Tao, Z., Shi, A., Lu, C., Song, T., Zhang, Z., & Zhao, J. (2015). Breast Cancer: Epidemiology and Etiology. *Cell Biochemistry and Biophysics, 72*(2), 333–338. doi:10.100712013-014-0459-6

Tavtigian, S. V., Simard, J., Rommens, J., Couch, F., Shattuck-Eidens, D., Neuhausen, S., ... Goldgar, D. E. (1996). The complete BRCA2 gene and mutations in chromosome 13q-linked kindreds. *Nature Genetics*, *12*(3), 333–337. doi:10.1038/ng0396-333

Theumer, M. G., Henneb, Y., Khoury, L., Snini, S. P., Tadrist, S., Canlet, C., ... Audebert, M. (2018). Genotoxicity of aflatoxins and their precursors in human cells. *Toxicology Letters*, *287*, 100–107. doi:10.1016/j.toxlet.2018.02.007

Tinfo, N. S., Hotchkiss, M. G., Buckalew, A. R., Zorrilla, L. M., Cooper, R. L., & Laws, S. C. (2011). Understanding the effects of atrazine on steroidogenesis in rat granulosa and H295R adrenal cortical carcinoma cells. *Reproductive Toxicology (Elmsford, N.Y.)*, *31*(2), 184–193. doi:10.1016/j.reprotox.2010.11.005

Tinoco-Veras, C. M., Bezerra Sousa, M. S., da Silva, B. B., Franciscato Cozzolino, S. M., Viana Pires, L., Coelho Pimentel, J. A., ... do Nascimento-Marreiro, D. (2011). Analysis of plasma and erythrocyte zinc levels in premenopausal women with breast cancer. *Nutrición Hospitalaria*, *26*, 293–297.

Tola, M., & Kebede, B. (2016). Occurrence, importance and control of mycotoxins: A review. *Cogent Food & Agriculture*, *2*(1), 1191103. doi:10.1080/23311932.2016.1191103

Tomaszewski, J., Miturski, R., Semczuk, A., Kotarski, J., & Jakowicki, J. (1998). Tissue zearalenone concentration in normal, hyperplastic and neoplastic human endometrium. *Ginekologia Polska*, *69*, 363–366.

Trottier, Y., Waithe, W. I., & Anderson, A. (1992). Kinds of mutations induced by aflatoxin B1 in a shuttle vector replicating in human cells transiently expressing cytochrome P450IA2 cDNA. *Molecular Carcinogenesis*, *6*(2), 140–147. doi:10.1002/mc.2940060209

Ueda, M., Imai, T., Takizawa, T., Onodera, H., Mitsumori, K., Matsui, T., & Hirose, M. (2005). Possible Enhancing Effects of Atrazine on Growth of 7, 12-Dimethylbenz(a) Anthracene-Induced Mammary Tumors in Ovariectomized Sprague-Dawley Rats. *Cancer Science*, *96*(1), 19–25. doi:10.1111/j.1349-7006.2005.00008.x

Uppala, P. T., Roy, S. K., Tousson, A., Barnes, S., Uppala, G. R., & Eastmond, D. A. (2005). Induction of cell proliferation, micronuclei and hyperdiploidy/polyploidy in the mammary cells of DDT- and DMBA-treated pubertal rats. *Environmental and Molecular Mutagenesis*, *46*(1), 43–52. doi:10.1002/em.20131

Urry, W. H., Wehrmeister, H. L., Hodge, E. B., & Hidy, P. H. (1966). The structure of zearalenone. *Tetrahedron Letters*, *7*(27), 3109–3114. doi:10.1016/S0040-4039(01)99923-X

Van der Merwe, K. J., Steyn, P. S., & Fourie, L. (1965). Mycotoxins. II. The constitution of ochratoxins A, B, and C, metabolites of *Aspergillus ochraceus* Wilh. *Journal of the Chemical Society, Perkin Transactions 1: Organic and Bio-Organic Chemistry*, 7083–7088.

Van der Merwe, K. J., Steyn, P. S., Fourie, L., Scott, D. B., & Theron, J. J. (1965). Ochratoxin A, a toxic metabolite produced by *Aspergillus ochraceus* Wilh. *Nature*, *205*(4976), 1112–1113. doi:10.1038/2051112a0

Vautier, G., Bomford, A. B., Portmann, B. C., Metivier, E., Williams, R., & Ryder, S. D. (1999). p53 mutations in British patients with hepatocellular carcinoma: Clustering in genetic hemochromatosis. *Gastroenterology*, *117*(1), 154–160. doi:10.1016/S0016-5085(99)70562-7

Vella, V., Malaguarnera, R., Lappano, R., Maggiolini, M., & Belfiore, A. (2017). Recent views of heavy metals as possible risk factors and potential preventive and therapeutic agents in prostate cancer. *Molecular and Cellular Endocrinology*, *5*, 57–72. doi:10.1016/j.mce.2016.10.020

Viel, T., Dransart, E., Nemati, F., Henry, E., Thézé, B., Decaudin, D., ... Tavitian, B. (2008). In vivo tumor targeting by the B-subunit of shiga toxin. *Molecular Imaging*, *7*(6), 239–247. doi:10.2310/7290.2008.00022

Vorderstrasse, B. A., Fenton, S. E., Bohn, A. A., Cundiff, J. A., & Lawrence, B. P. (2004). A novel effect of dioxin: Exposure during pregnancy severely impairs mammary gland differentiation. *Toxicological Sciences*, *78*(2), 248–257. doi:10.1093/toxsci/kfh062

Wang, J. S., & Groopman, J. D. (1999). DNA damage by mycotoxins. *Mutation Research. Fundamental and Molecular Mechanisms of Mutagenesis*, *424*(1-2), 167–181. doi:10.1016/S0027-5107(99)00017-2

Wang, T., Gavin, H. M., Arlt, V. M., Lawrence, B. P., Fenton, S. E., Medina, D., & Vorderstrasse, B. A. (2011). Aryl hydrocarbon receptor activation during pregnancy, and in adult nulliparous mice, delays the subsequent development of DMBA-induced mammary tumors. *International Journal of Cancer*, *128*(7), 1509–1523. doi:10.1002/ijc.25493

Wang, Y., Jiang, G., Lam, P. K., & Li, A. (2007). Polybrominated diphenyl ether in the East Asian environment: A critical review. *Environment International*, *33*(7), 963–973. doi:10.1016/j.envint.2007.03.016

Wang, Y., Wang, L., Liu, F., Wang, Q., Selvaraj, J. N., Xing, F., ... Liu, Y. (2016). Ochratoxin A producing fungi, biosynthetic pathway and regulatory mechanisms. *Toxins*, *8*(3), 83. doi:10.3390/toxins8030083

Wani, A. L., Ara, A., & Usmani, J. A. (2015). Lead toxicity: A review. *Interdisciplinary Toxicology*, *8*(2), 55–64. doi:10.1515/intox-2015-0009

Warner, M., Eskenazi, B., Mocarelli, P., Gerthoux, P. M., Samuels, S., Needham, L., ... Brambilla, P. (2002). Serum dioxin concentrations and breast cancer risk in the Seveso Women's Health Study. *Environmental Health Perspectives*, *110*(7), 625–628. doi:10.1289/ehp.02110625

Warner, M., Mocarelli, P., Samuels, S., Needham, L. L., Brambilla, P., & Eskenazi, B. (2011). Dioxin exposure and cancer risk in the Seveso Women's Health Study. *Environmental Health Perspectives*, *119*(12), 1700–1705. doi:10.1289/ehp.1103720

White, A. J., Teitelbaum, S. L., Wolff, M. S., Stellman, S. D., Neugut, A. I., & Gammon, M. D. (2013). Exposure to fogger trucks and breast cancer incidence in the Long Island Breast Cancer Study Project: A case-control study. *Environmental Health*, *12*(1), 24. doi:10.1186/1476-069X-12-24

Wilson, D. M., & King, J. K. (1995). Production of aflatoxins B_1, B_2, G_1, and G_2 in pure and mixed cultures of *Aspergillus parasiticus* and *Aspergillus flavus*. *Food Additives and Contaminants*, *12*(3), 521–525. doi:10.1080/02652039509374339

Wolff, M., Zeleniuch-Jacquotte, A., Dubin, N., & Toniolo, P. (2000). Risk of breast cancer and organochlorine exposure. *Cancer Epidemiology, Biomarkers & Prevention*, *9*, 271–277.

Woodruff, T., & Morello-Frosch, R. (2011). Communicating about chemical body burden, with Tracey Woodruff and Rachel Morello-Frosch. *Environmental Health Perspectives*, 119.

Woolcott, C. G., Aronson, K. J., Hanna, W. M., SenGupta, S. K., McCready, D. R., Sterns, E., & Miller, A. (2001). Organochlorines and breast cancer risk by receptor status, tumor size, and grade (Canada). *Cancer Causes & Control*, *12*(5), 395–404. doi:10.1023/A:1011289905751

Wu, X., Tang, J., & Xie, M. (2015). Serum and hair zinc levels in breast cancer: A meta-analysis. *Scientific Reports*, *5*(1), 12249. doi:10.1038rep12249

Xu, Y., Tokar, E. J., & Waalkes, M. P. (2014). Arsenic-induced cancer cell phenotype in human breast epithelia is estrogen receptor-independent but involves aromatase activation. *Archives of Toxicology*, *88*(2), 263–274. doi:10.100700204-013-1131-4

Yazar, S., & Omurtag, G. Z. (2008). Fumonisins, trichothecenes and zearalenone in cereals. *International Journal of Molecular Sciences*, *9*(11), 2062–2090. doi:10.3390/ijms9112062

Yip, K. Y., Wan, M. L. Y., Wong, A. S. T., Korach, K. S., & El-Nezami, H. (2017). Combined low-dose zearalenone and aflatoxin B_1 on cell growth and cell-cycle progression in breast cancer MCF-7 cells. *Toxicology Letters*, *281*, 139–151. doi:10.1016/j.toxlet.2017.09.022

Yücel, I., Arpaci, F., Ozet, A., Döner, B., Karayilanoğlu, T., Sayar, A., & Berk, O. (1994). Serum copper and zinc levels and copper/zinc ratio in patients with breast cancer. *Biological Trace Element Research*, *40*(1), 31–38. doi:10.1007/BF02916818

Zain, M. E. (2011). Impact of mycotoxins on humans and animals. *Journal of Saudi Chemical Society*, *15*(2), 129–144. doi:10.1016/j.jscs.2010.06.006

Zhang, S., Lei, P., Liu, X., Li, X., Walker, K., Kotha, L., ... Safe, S. (2009). The aryl hydrocarbon receptor as a target for estrogen receptor-negative breast cancer chemotherapy. *Endocrine-Related Cancer*, *16*(3), 835–844. doi:10.1677/ERC-09-0054

Zheng, T., Holford, T., Mayne, S., Owens, P., Ward, B., Carter, D., ... Tessari, J. (1999). Beta-benzene hexachloride in breast adipose tissue and risk of breast carcinoma. *Cancer*, *85*(10), 2212–2218. doi:10.1002/(SICI)1097-0142(19990515)85:10<2212::AID-CNCR16>3.0.CO;2-F

Zinedine, A., Soriano, J. M., Molto, J. C., & Manes, J. (2007). Review on the toxicity, occurrence, metabolism, detoxification, regulations and intake of zearalenone: An oestrogenic mycotoxin. *Food and Chemical Toxicology*, *45*(1), 1–18. doi:10.1016/j.fct.2006.07.030

Chapter 10
Childhood Leukemia and Environmental Risk Factors

Mounir Tilaoui
Sultan Moulay Slimane University, Morocco

Abdelamjid Zyad
Sultan Moulay Slimane University, Morocco

Hicham Chatoui
Private University of Marrakech, Morocco & Cadi Ayyad University, Morocco

Tarik Rakkaa
Private University of Marrakech, Morocco

ABSTRACT

Leukemia is a cancer that starts in blood forming cells which occurs in several forms of chronic or acute diseases. It is the most common cause of pediatric malignancy, accounting for approximately 25% of all cancers occurring before age 20. It represents 32% of all cancer cases occurring among children younger than 15 years of age, with an annual incidence rate of 43 cases per million. In the last decade of the 20th century, the occurrence of childhood leukemia has shown a rise. This disease, like most cancers, has a multifactorial etiological causal mechanism and a heterogeneous biological composition involving the interaction between different aspects originating from the environment as well as human genetics. This chapter discusses, through the current published literature, the relationship between cancer, particularly childhood leukemia, and environmental exposures to heavy metals, pesticides, and mycotoxins.

INTRODUCTION

Cancer is the second most common cause of death worldwide following cardiovascular diseases (Siegel, Miller, & Jemal, 2017). Globally in 2018, there were documented 9.5 million deaths because of cancer excluding non-melanoma skin cancer (and 18.1 million new cases of cancer excluding non-melanoma skin cancer (Bray et al., 2018).

DOI: 10.4018/978-1-5225-7775-1.ch010

Cancer include large group of diseases that can impair any organs. It is a multifactorial disease because the combined effect of external and genetic factors acting sequentially and concurrently. Several factors both inside and outside the body contributes to the promote cancer. Overwhelming evidence shows that the predominant contributor to many form of cancer is associated to the environment (Boffetta & Nyberg, 2003).These environmental factors risks including all non-genetic factors like lifestyle choices such as cigarette smoking, excessive alcohol consumption, excessive sunlight exposure, diet, and infectious agents. These factors environment is linked to cause the majority of cancers in human. Other factors include indoor air pollution, drinking water contamination and soil (Boffetta & Nyberg, 2003). The association with cancer and many environmental chemicals have been well studies through occupational groups that have elevated exposures to these chemicals compounds than the general population.

Different environmental exposures are associated to specific types of cancer. For example, exposure to benzidine, a chemical found in certain dyes, is linked with bladder cancer (Sun et al., 2018). Whereas exposure to asbestos is associated primarily to lung cancer (Richardson, Keil, Cole, & Dement, 2018). Also, environmental risk factors are influences in our surroundings, such as certain chemicals, that increase the risk of having diseases such as leukemias especially childhood leukemia (Fucic, Guszak, & Mantovani, 2017). Cancers Childhood remains one of the principal causes of death for children 1 to 14 years old worldwide. Annually, 30% to 40% of all pediatric cancers is leukemia and their origin is still understood (Zachek et al., 2015). However, Several studies reported that environmental hazards are implicated in the certain childhood cancers etiology (Zachek et al., 2015).

There are more than 12 types of leukemia, with the four mainly ones are chronic lymphocytic leukemia (CLL), acute myeloid leukemia (AML), acute lymphocytic leukemia (ALL) and chronic myeloid leukemia (CML). The differences between themes are related to the kind of blood cells that are affected and the multiple rates of progressions. ALL is the most leukemia common in the population pediatric; it about 75% of all pediatric leukemia cases and is five times more common than AML (Greenlee, Murray, Bolden, & Wingo, 2000). After AML and ALL, CML makes up most of the other leukemia childhood cases. Despite the advances in of childhood leukemia diseases, it is still unclear exactly what leads to this disease. Epidemiologic reports of acute leukemia in childhood have analyzed a number of potentially risk factors (e.g., environment, genetic, or infectious) in an effort to determine the etiology of the disease. The two most important environmental risk factors significantly linked with either ALL or AML are exposure to hydrocarbons and ionizing radiation; other environmental risk factors such as cigarette smoking, electromagnetic fields, alcohol and drugs have been weakly or inconsistently associated with either form of childhood leukemia (Belson, Kingsley, & Holmes, 2006). Therefore, this chapter book discusses concerns and evidences linking exposure to environmental factors such as heavy metals, pesticides and mycotoxins and risk of cancer with the focus on childhood leukemia. Through recognizing and understanding the discussed risk factors, in this chapter, of cancer and childhood leukemia then perhaps we can decrease the occurrence of this disease and help to reduce harmful exposure of childhood (Jin, Xu, An, & Wang, 2016).

HEAVY METALS

Heavy Metals in the Environment

Throughout the development of technology, numerous inorganic chemicals were discovered and studied without regard to any possible harmful effects, and undoubtedly the toxicity of many of these substances was experienced by man long before the effect was related to cause cancer (Duffus, 1997). Although such toxic substances have always been present in the environment, they are not often found concentrated in a harmful form in nature.

There are numerous examples of the use or misuse of toxic metals during the industrial era, and not unexpectedly, those engaged in the metal processing industries were often the first victims. The specific problems of the toxicity of metals in industry have been well documented elsewhere and recent reports have established that metals known to be toxic have been accumulated in the environment which may present a substantial hazard to man (Sritharan et al., 2018). In particular the so-called heavy metals, cadmium, lead and mercury have become widely distributed in the environment within this century through industrial and agricultural applications (Lane, Canty, & More, 2015).

Effects of Heavy Metals on Public Health

Heavy metals are naturally present but in very small quantities in soil, water and air but the problem of these metals are not biodegradable (Azimi, Azari, Rezakazemi, & Ansarpour, 2017). They are present in the environment by different ways: mining and mineral exploitation, the use of fossil fuels, solid waste incineration, forest fires, cigarette smoke, etc (Jan et al., 2015). Heavy metals can be divided into two categories: essential and nonessential (Tuteja & Gill, 2016). Essential heavy metals play important roles in cell functions such as copper and iron which plays a key functional role in cell growth and proliferation. Thus, deficiencies or excesses of these tow essential metals lead to impaired cellular functions and eventually cell death. Also, it was shown that high levels of copper in drinking water can cause kidney and liver damage with possible development of cirrhosis in childhood (National Research Council, 2000). However, nonessential heavy metals are highly toxic for living organisms, such as cadmium (Cd), lead (Pb), and mercury (Hg), their toxicity are known even at very low concentration levels. Toxicity of heavy metals is due mainly to their affinity to sulfur, once added to proteins the soluble salts of heavy metals such as Pb, Hg and Cd form insoluble precipitates (Florkin, 1956) which are less bioavailable and toxic (Florkin, 1956). Once absorbed, these heavy metals are generally takes a long time to be removed and presents most of them a biological half-life (15 to 30 years) very long in the human body (Sugita, 1978). Also, it is well documented that metal toxicity increase with smokers because of the variety of toxic heavy metals in cigarette tobacco (Caruso, O'Connor, Stephens, Cummings, & Fong, 2013). Furthermore, several studies showed that exposure to metals in the air is capable of causing a numerous effects of human health, ranging from pulmonary inflammation and cardiovascular to damage of vital organs and cancers diseases (Lodovici & Bigagli, 2011).

Heavy Metals and Cancer

The relation between exposure to heavy metals, either environmental or occupational, and cancer has been widely investigated (Waisberg, Joseph, Hale, & Beyersmann, 2003). The mechanisms which heavy metal induced carcinogenic effect appear to be multiples and are far from being deeply understood (Hartwig, 2013). Almost all heavy metals are associated with various diseases and cancers (Hartwig, 2013). Heavy metal induces reactive oxygen species (ROS) is a well-known mechanism to induce damages through oxidative stress (Nieboer, Tsuji, Martin, & Liberda, 2013). Oxidative stress, in turn, may damage DNA and producing mutations that initiate cancer. Also, Heavy metal disrupt the apoptosis process leading to various diseases like cancer (Baig et al., 2016). Heavy metals affect intra and extra cellular environments promoting pathogens microbes' growth such as viruses, bacteria and fungi which may increase a person's risk for certain types of cancers (David, Krishna, & Sangeetha, 2016).

Arsenic (As), cadmium (Cd), chromium (Cr), and nickel (Ni) are classified in category one as carcinogenic heavy metals to humans based to the International agency for Research on Cancer. Many reports have found that exposures to these metals were linked to tumor suppressor, gene expression disruptions, damage repair processes and metabolism enzymatic activities (Bánfalvi, 2011). Epidemiological studies have shown that chronic exposure to arsenic increase incidence of kidneys, lung, skin, bladder, liver and colon cancers (Stevens, Graham, Walker, Tchounwou, & Rogers, 2010).

Cadmium is classified as a human carcinogen that can cause cancers of the kidney, respiratory tract (lungs, nasopharynx), liver, stomach, hematopoietic system, pancreas and prostate (Huff, Lunn, Waalkes, Tomatis, & Infante, 2007). The International Agency for Research on Cancer through epidemiologic and experimental studies about nickel-linked to cancer concluded that nickel compounds were carcinogenic to humans (e.g., lung and nasal cancers) (IARC, 1990). There are numerous epidemiological studies of cancer pulmonary induced in humans after exposition to hexavalent Cr compounds in the work area. Amonge them, one a study done by Gibb et al 2000 showed the association between lung cancer of 2357 workers and exposition to hexavalent Cr (Gibb, Lees, Pinsky, & Rooney, 2000). The risk of heavy metal exposure is also connected with the contamination source. As example, Grimsrud & Andersen 2012, showed an elevated risk of occupational pathology and cancer in individuals working in industrial areas using heavy metal (Grimsrud & Andersen, 2012, 2012).

Childhood Leukemia and Heavy Metals

Young children are particularly exposed to the lead derivatives present in the environment and particularly sensitive to their toxicity. Indeed, they explore the environment with hands and mouth, which can lead to ingestion of dust and paint scales contaminated with lead. In addition, lead is absorbed more by children than by adults (the proportion of ingested lead passing into the blood is nearly 50% in children against 10% in adults). Lead can have a variety of acute and chronic toxic effects, but is primarily characterized by chronic neurotoxicity. Indeed, several recent epidemiological studies have observed an effect on the cognitive development of the child at blood lead levels of less than 100 eg / L.

Inorganic lead compounds, classified since 1987 by the International Agency for Research on Cancer (IARC) in group 2B (potentially carcinogenic to humans) have recently been reclassified to group 2A (probably carcinogenic to humans) (IARC, 2006). Children may be exposed to lead from various environmental sources: old paint in buildings, contaminated water from pipelines, industrial emissions atmospheres, polluted soils, contaminated food (in particular by ceramic enameled poor quality), cosmetics and traditional remedies, professional or leisure activities of parents and various objects. It has recently been shown that - even at low doses - it has a cytotoxic effect on central nervous system stem cells. Also, mercury can affect cell division and intracellular processes and possibly leading to cause leukemia, mononucleosis, Hodgkin's disease (Rice, Walker Jr, Wu, Gillette, & Blough, 2014). On the animal model, it can induce blood alterations (Massanyi et al., 2007), apoptosis and chromosomal aberrations (Silva-Pereira et al., 2005).

Leukemias are cancers that are characterized by a proliferation of immature lymphocytes in the marrow. There are two types: Acute Lymphoid Leukemias (ALL) and Acute Myeloid Leukemias (AML) more dangerous than ALL. Currently 80% of leukemias are curable (O'Leary, Krailo, Anderson, & Reaman, 2008), but, treated individuals could develop sequelae that have an impact on their quality of life (Leplège, Ecosse, Verdier, & Perneger, 1998). It is well documented that young children are very sensitive and more vulnerable than adults to environmental pollution risks. These happen due to children have no control over the environment in the postnatal and prenatal periods, such as the food they eat, the quality of the air they breathe, the water they drink, and also their exposure to disease-transmitting vectors (Perlroth & Branco, 2017).

Contamination occurs through exposition between the physical environment agent (parenteral, breast milk, oral, in utero) and the child development period which was exposed (neonatal, puberty, fetal and childhood) (Perlroth & Branco, 2017). It is important to know the role of adverse substance magnitude, the susceptibility of child's individual, the introduction pathways of the pathological agents and the exposition frequency (Perlroth & Branco, 2017). As children breathe more air and tend to be more physically active than adults, inhaling toxic gases may compromise their pulmonary function, elevating the incidence of acute respiratory infections (Bener, Kamal, & Shanks, 2007). These substances include particularly dust, smoke, aerosols), pollutants of public health concern, like NO_2, O_3, CO, SO_2; mercury, lead and volatile organic substances; biomass burning residue as well as allergens, polycyclic aromatic hydrocarbons halogens and ionizing radiation (Perlroth & Branco, 2017). maternal exposure during pregnancy to chemicals resulting from mining, agricultural and industrial activities might be linked with the possible increased occurrence of behavioral disorders, birth defects, asthma, neurological, endocrine, prematurity, and hematopoietic cancers of their children (Perlroth & Branco, 2017). As example, chronic nickel poisoning promotes the development of leukemias childhood (Y. Yang et al., 2008).

Childhood cancers are reported to be increasing in several industrialized countries (Terracini, 2011). In France for example, according to INSERM survey, nearly 1,500 new cases of cancer occur each year in children before the age of 6 years. The environment is probably involved in a number of these cancers (INSERM Collective Expertise Centre, 2000). Lead poisoning still affected 85,000 children in France in 1999 (Pichery et al., 2011). In a cohort study of about 1,800 children from Nancy and Poitiers cities, that more women were exposed to urban air pollution during pregnancy, the lower the immune cells number was in their newborns since their intra-uterine life (Vrijheid et al., 2011). In the United States, leukemia and brain tumors in young children have increased from 30% to 40% in twenty-five years (Belson et al., 2006).

PESTICIDES AND CANCER

Pesticides commonly used in agriculture, include plant growth regulators insecticides, herbicides, fungicides, molluscicides, algaecides etc, used to protect plantations from diseases, harmful micro-organisms and vermin (Lyons & Watterson, 2010). Chemicals used in pesticides are intended to be toxic to living organisms, thus, they have been linked with a range of adverse health effects, including neurological, dermatological, respiratory diseases (Nicolopoulou-Stamati, Maipas, Kotampasi, Stamatis, & Hens, 2016). Also, some chemicals pesticides have been linked to cancer (Table 1) in adults and children through laboratory assay and epidemiological studies (Sabarwal, Kumar, & Singh, 2018). This part of this chapter highlights the role that pesticides are suspected to play in some cancers.

Pesticide Components and Carcinogenicity

Finding evidence of carcinogenicity in humans is difficult as investigations need very large numbers of people followed for decades, with detailed information about specific pesticide exposure including how much pesticide and length of exposure time. Animal assays can give some indication of possible carcinogenicity of pesticides, but their results are not always valid to humans. As example, the IARC classified atrazine herbicide as a possible carcinogen to human (Group 2B) on the basis of rat studies. Nevertheless, this compound did not show carcinogenic effect to humans, the chemical was classified in Group 3 (unclassifiable) (IARC, 1999) and later epidemiological investigations showed no relationship between atrazine and cancer (Rusiecki et al., 2004; Sathiakumar, MacLennan, Mandel, & Delzell, 2011). In 2015, IARC upgraded the glyphosate herbicide from a possible (Group 2B) to probable human carcinogen (Group 2A) based on strong mechanistic effect evidence (Guyton et al., 2015). Ethylene oxide is classed within Group 1 (IARC, 2012) as an ingredient in five fumigant products. In 2015, IARC classified lindane insecticide as Group 1 due to epidemiological research which showed significant

Table 1. Pesticides and cancers

Type of cancer	Comments
Brain	Various studies have raised the hypothesis of the role of pesticides in the occurrence of brain tumors (Samanic et al., 2008). Nevertheless, it is not possible to conclude definitively an association between the roles of pesticides in the occurrence of brain tumors.
Thyroid	Exposure to pesticides increases the risk of thyroid cancers in rodents (Zeng et al., 2017). Nevertheless, it is not possible to confirm or invalidate in humans the results obtained in animals assays.
Lung	Exposure to certain pesticides is a debated issue as to its relationship with the risk of lung cancer(Alavanja, Dosemeci, et al., 2004) .
Breast	Organochlorine pesticides in particular DDE (1,1-dichloro-2,2-bis (p-chlorophenyl) ethylene) and DDT are possible factors in promoting breast cancer (He, Zuo, Wang, & Zhao, 2017).
Ovary	Data are insufficient on this cancer and few chemical agents have been investigated to conclude the association between pesticide and ovary cancer (Shah, Bhat, Sharma, Banerjee, & Guleria, 2018).
Testicule	Many studies have put forward a possible role of the link between pesticides on the occurrence of testicular cancer (Giannandrea & Fargnoli, 2017) but have not, to date, identified evident risk factors.
Prostate	Exposure to certain pesticides, especially professional employees of plants production, could be responsible for an increased prostate cancer risk (Lemarchand et al., 2016). Studies reported that herbicides do not increase the prostate cancer risk (Davoren & Schiestl, 2018).

increases in non-Hodgkin lymphoma risk with increasing occupational exposure to lindane (Loomis et al., 2015). A side from these clear exceptions, no specific pesticide has been conclusively linked to a specific cancer, and suggested links do not group by class or type of pesticide.

The IARC has also classified the "spraying and application of non-arsenical insecticides" as a probable cause of cancer (IARC, 1991) . However, only six specific pesticides – captafol, ethylene dibromide, glyphosate, malathion, diazinon and dichlorodiphenyltrichloroethane (DDT) are classed within this category. While there was limited evidence of carcinogenicity in humans found, there was strong mechanistic evidence for the carcinogenicity of glyphosate, malathion and diazinon with all three agents inducing DNA and/or chromosomal damage in human and animal cells in vitro. Several pesticides have been classed as possible causes of cancer (Group 2B). Chlorothalonil, sodium ortho-phenylphenate, phenoxy herbicides and dichlorvos are licensed for agricultural use. Para-dichlorobenzene pesticide is not used in agricultural, but is used in urinal cakes and mothballs. In March 2015, IARC classified the parathion and tetrachlorvinphos insecticides as possibly carcinogenic to humans (Group 2B) based on evidence that these pesticides are carcinogenic in animals (Guyton et al., 2015). Also, the herbicide 2,4-dichlorophenoxyacetic acid (2,4-D) was classified as Group 2B based on animals assays through promoting oxidative stress (Loomis et al., 2015).

Epidemiologic investigations, although sometimes contradictory, have linked phenoxy acid herbicides with malignant lymphoma and soft tissue sarcoma (STS); organochlorine insecticides are linked with STS, non-Hodgkin's lymphoma (NHL), leukemia, lung and breast cancers; organophosphorous compounds are linked with leukemia and NHL, atriazine herbicides with ovarian cancer (Coggon, Ntani, Harris, Jayakody, & Palmer, 2015; Jayakody, Harris, & Coggon, 2015).

Epidemiological Studies

The first study on the role of pesticides in cancer induction is based on the observation of differences in mortality between socio-professional categories and farmers for certain cancer pathologies: cancers of connective tissues, the stomach the brain the skin the lips and the hematopoietic (Baldi, Mohammed-Brahim, Brochard, Dartigues, & Salamon, 1998). Other cancers may occur, including those regulated by the sex hormones (prostate, breast, testicle, endometrium...). Epidemiological studies investigating the link between breast cancer and exposure to environmental pesticides are relatively abundant, but have focused mainly on organochlorine insecticides such as DDT (dichloro-diphenyl-trichloroethane). In 1993, Wolff et al demonstrated that the risk of breast cancer was increasing in women with high serum levels of DDE compared with women with low concentrations of DDE (dichloro-diphenyl-dichloroethene). However, in 2002 Calle and his collaborators concluded that there was insufficient evidence in favor of a relationship between concentrations of pesticides organochlorine measured in blood and adipose tissue and the breast cancer risk. Furthermore, Lopez-Cervantes et al (2004) confirmed the same results based on a meta-analysis of DDE tissue levels (López-Cervantes, Torres-Sánchez, Tobías, & López-Carrillo, 2004). Other studies have showed conflicting results with 4 studies demonstrating no link between organochlorine levels and breast cancer (Itoh et al., 2009; Iwasaki et al., 2008; McCready et al., 2004; Soliman et al., 2003) and 4 other investigations demonstrating positive associations (Cassidy, Natarajan, & Vaughan, 2005; Charlier et al., 2003; Cohn, Wolff, Cirillo, & Sholtz, 2007; Ibarluzea et al., 2004). Although there are some studies on non-occupational exposures (breast cancer and childhood tumors), data on the link between pesticides and cancers comes mainly from epidemiological investigations in farmers. A study based on Meta-analyzes data showed lower incidence of cancer among farmers, especially lung, esophageal and bladder cancers; these cancers for which the role of smoking is particularly marked. Thus, the restricted exposure to the risk factor "tobacco" may mask the potential impact of pesticides in the occurrence of these cancers. In fact, the arsenic compounds used in arboriculture and viticulture may play a possible role in lung cancer occurrence. Although the risk of cancer is lower due to the lower-incidence of smoking-linked cancers in the agricultural population, some tumors are regularly found such as lips, prostate, brain and hematological tumors (Blair & Freeman, 2009).

Childhood Leukemia and Pesticide Exposure

Childhood leukemia, the most common childhood malignancy, and its main sub-types, acute lymphoblastic leukemia and acute myeloid leukemia, occur mainly in children under five years of age, suggesting a role for parental exposures before birth or for the child's exposure in early childhood in their etiology. Research shows that infants and children are more sensitive to pesticides than adults. This is probably due to the speed of their growth and development. The mode of exposure is also involved. When playing on a pesticide-treated lawn, children enter more than adults in contact with these products (Roberts & Karr, 2012). Numerous case-control studies evoked the possible association between pesticides and the risk factors for childhood leukemia (Belson et al., 2006; Van Maele-Fabry, Gamet-Payrastre, & Lison, 2018). Nevertheless, in developed countries and due to the infrequency of childhood leukemia with an annual incidence rate of 30-50 per million for lymphoblastic leukemia and 4-8 per million for acute myeloid leukemia, individual reports rarely have sufficient statistical analysis to detect an effect, particularly for leukemias subtypes.

Increases childhood cancer incidence gives evidence that environmental exposure, including pesticides exposure, plays a role in some childhood cancer. In industrialized countries, one child out of 500 develops a cancer before the age of fifteen. Moreover, childhood cancers were increased by 1% per year in some European countries. Several studies have linked parental and/or a child's pesticide exposures to higher risks of childhood cancer, these including leukaemia, brain, lymphomas cancers, Ewing's sarcoma and Wilms' tumour (Zahm & Ward, 1998). Furthermore, a study reported that children living in country with moderate to high levels of agricultural activity were diagnosed with various cancers (Lyons & Watterson, 2010). In 2007, researchers reported the association between pesticide exposure and childhood cancer, and that maternal pesticide exposure during pregnancy was most consistently linked with childhood cancer. Similarly, in 2009 other research concluded that prenatal maternal exposure was most strongly associated with increased risk of cancer shildhood (Lyons & Watterson, 2010). A meta-analysis study finds that children's exposure to pesticides in and/or around the home increase the risk of developing some cancers childhood, and that cancer risks were linked to the type of pesticide applied and the area where it was used (McLean et al., 2009). In another study, scientist showed that household and garden pesticide use may increase the risk of childhood leukemia (Lyons & Watterson, 2010).

Pesticides can increase susceptibility to some cancers types by decreasing the immune system's surveillance toward cancer cells. Children and infants, the aged and the chronically illness are at highest risk from chemically-induced immune dysfunction and suppression (Wigle, Turner, & Krewski, 2009).

Children are also exposed to pesticides even before birth (in utero) by transplacental route of compounds to which the mother is exposed. Childs may also be in touch with persistent pesticides that may be bio-accumulated by breast milk. Therefore, it is necessary to diminish the exposure of the pregnant woman to pesticides. A study published in the Environmental Health Perspectives shows that infant born to mothers living in households with pesticide use during pregnancy had much higher risk of getting cancer, especially non Hodgkin lymphoma (NHL) or acute leukemia (AL) (Rudant et al., 2007). Similarly, a meta-analyzes study showed an increased risk of hematological malignancies (multiple myeloma), (from 9 to 39%) among exposed professionals compared to the general population. Excess risk of Hodgkin's disease and leukemia, has been diagnosed in farmers, but the findings still inconclusive. Also, the present investigations on melanomas in agricultural settings do not, in many cases, distinguish the role of UV radiation and pesticide exposure. Only the Agricultural Health Study demonstrates the role of pesticides exposure in the melanomas occurrence, taking into consideration other known risk factors for cancers such as brain tumors and testicular cancer. . A link between the professional uses of pesticides by parents has also been reported. However, no pesticide has been designed specifically. In children, the exposure of mother and/or father to pesticides in the workplace increases brain tumors risk by 30 to 53%, particularly if the exposure happened in the prenatal period (Van Maele-Fabry, Hoet, & Lison, 2013). An increased risk of leukemia was also reported when the mother is exposed to pesticides (herbicides or insecticides) during pregnancy period. This increase is about 54% and 60% with residential and occupational exposures to pesticides, respectively. (M. C. Turner, Wigle, & Krewski, 2010; Van Maele-Fabry, Lantin, Hoet, & Lison, 2010).

Malignant Haemopathies and Pesticides

Haematological malignancies are the most studied cancer linked to pesticide. A meta-analyzes performed on Haematological cancers converge towards a significant risk elevation, ranging from 3 to 34%, + 23% for multiple myeloma, + 25% for Hodgkin's disease and + 34% for non-Hodgkin lymphoma (Alavanja, Hoppin, & Kamel, 2004). The association between malignant haemopathies and pesticides was documented in the 1980s in Sweden, mostly for non-Hodgkin's malignant lymphomas, with exposure to chlorophenols and phenoxy herbicides. Later, investigations conducted in New Zealand, the USA, Australia, and Canada found this association but with a lower level of risk. The geographical difference between the studies suggested various explanations such as the contamination of these herbicides by dibenzofurans and dioxins in the Scandinavian countries as well as differences in the modalities exposure to these herbicides. In addition, published meta-analyses report the link between home pesticide exposure and the risk of lymphoblastic leukemia (ALL) and acute myeloid leukemia (AML). From these meta-analyses scientist reported increased odds ratios for ALL with exposure to herbicides and insecticides during pregnancy period. For exposures after birth, they reported no link with herbicides and childhood malignancy but there was with insecticides exposure (Bailey et al., 2015).

Mechanisms of Action of Pesticides in Hematological Diseases

Hematopoiesis is a set of processes that contribute to the daily production of millions of blood cells as different in their morphology as in their functions. Hematopoiesis starts in the yolk sac after 6 to 8 weeks of gestation, then in the fetal liver, spleen and bone marrow, the only hematopoietic compartment in adult humans. In adults, hematopoiesis results from progressive and multiple differentiations from a small number of stem cells capable of self-renewal. The first step leads to the formation of progenitors at the origin of the precursors and mature cells that will join the blood compartment. Pesticide association with the immune system and hematopoiesis have been given deep attention in the last two decades because to their large use in industries, agriculture, and domestic purposes. Many studies showed that chronic exposure to pesticides is linked with the risk of hematopoietic malignancies (Merhi et al., 2007; Van Maele-Fabry et al., 2018) and possibly lead to different diseases. Several studies provide scientific evidences on the plausibility of an association between pesticides and health, especially with hematopoietic pathologies, bibliographic studies has focused on in vivo and/or in vitro investigations of the impact of pesticides on gene expression, DNA damage and oxidative stress in hematological diseases.

Monocrotophos an organophosphate pesticide, has been shown to induce apoptosis in hUCBSCs, up-regulate the expression of ASK1/pJNK1/2, Bad, P21, Bax, Wnt/GSKb and activate caspase-9, 3 (Kashyap et al., 2013). Malathion, increase the level of the pro-apoptotic protein Bax and decreased the levels of the anti-apoptotic proteins p-Akt (Venkatesan, Park, Ji, Yeo, & Kim, 2017).

On relation with the immune system and hematopoisis, Dichlorvos (organophosphate insecticide) Causes microglial activation with the induction of NADPH oxidase and pro-inflammatory cytokines (TNF-α, IL-1β and IL-6) (Lewis, Gehman, Baer, & Jackson, 2013). Cypermethrin and mancozeb, decrease granulocyte-macrophage colony formation (Mandarapu & Prakhya, 2015) and damage the erythropoietic (colony-forming unit-erythroid and burst-forming unit-erythroid) progenitors (Mandarapu & Prakhya, 2015). In addition, Alphamethrin, Cypermethrin, Profenofos and Chloropyrofos (commonly used as agricultural pesticides), decrease the expression of IL-2, IL-3,IL-5 and the fibroblastic colony in mature erythrocyte formation, colony forming unit granulocytes and colony forming units' granulocyte erythrocyte monocyte macrophage (Chatterjee et al., 2013).

Concerning the effect of pesticides on DNA damage and oxidative stress, pirimiphosmethyl and chlorpyrifos was shown to increased DNA damage of horticultural farmers occupationally exposed to those pesticide mixtures(Atherton et al., 2009) also to induce oxidative stress accompanied with DNA damage at a time when blood AChE activity was lower (Singh, Kaur, & Budhiraja, 2013). Furthermore, Malathione, carbamate and lindane demonstrate lipid peroxidation, with a simultaneous increase in the activities of SOD, CAT, GPx, S-glutathione transferase and a decrease in GSH level hospitalized in patients exposed to those pesticides (Sunkireddy, Sriramoju, Roy, Kanwar, & Kanwar, 2018).

In conclusion, several epidemiological data suggest a link between exposure to certain pesticides and hematopoietic pathologies, in particular lymphomas and leukemias. This relationship was augmented with the analysis of the pesticides at the cellular and molecular level. Pesticides such as DDT, lindane, chlordane, certain organophosphorus products, permethrin and triazines affect both hematopoiesis and immunity, characteristics that give them a possible role in the hematopoietic pathologies etiology (Mokarizadeh, Faryabi, Rezvanfar, & Abdollahi, 2015). Nevertheless, further investigations particularly using hematopoietic models seem important to better understand the molecular events involved.

The association between pesticides and cancer risk is a difficult question to document considering the many products used and their evolution according to the period of use and the types of cultures. Nevertheless, pesticide exposures have been particularly in certain types of cancers. Most studies, however, are highly inaccurate (difficulties with retrospective assessment of exposures, lack of prospective data) and still not precise to elaborate conclusions about the increased risk from exposure to pesticides.

MYCOTOXINS

Mycotoxins are secondary metabolites with low molecular weight (not exceeding <1000 Da), produced by several fungus species that can develop on the plant in the field or during storage (Alshannaq & Yu, 2017). From approximately 200 recognized filamentous fungi, the most toxigenic species belong to the genera *Alternaria*, *Penicillium*, *Fusarium* and *Aspergillus*. Alternaria and Fusarium usually demonstrate a high mycotoxicological risk at both freshly harvested products on drying or in pre-harvest level, while Penicillium and Aspergillus toxigenic species present a higher risk for feed products and stored food (Afssa & AFSSA, 2009).

Mycotoxins are endowed with potential toxicity to humans and animals. More than 300 secondary metabolites have been characterized however only about 30 strains have demonstrated high toxicity. These toxins are found as natural contaminants of many foods of plant origin, including nuts grains almonds and also fruits (Afssa & AFSSA, 2009). Mycotoxins may be classified into nitrogen metabolites, cyclopeptides polyketoids and terpenes according on their structure and biological origin (Bhat, Rai, & Karim, 2010). Also, they can be classified depending to their high toxic effects.

Mycotoxins and the Risk to the Consumer

The toxicity of mycotoxins may be linked directly or indirectly to organisms consuming contaminated food (Afssa & AFSSA, 2009). Chronic effects are the most dangerous because of the persistence of metabolites toxins. Certain toxins exhibit a hepatotoxic effect (aflatoxins), others are immuno / haematotoxic (trichothecenes, patulin, fumonisins), dermonecrosis (trichothecenes), estrogenic (zearalenone), nephrotoxic (ochratoxin A) and neurotoxic (tremorgenous toxins). Furthermore, studies from published papers report that certain mycotoxins are known or suspected to show a carcinogenic effect (Afssa & AFSSA, 2009; Saad-Hussein et al., 2016). Indeed, these metabolites toxins can also be responsible for cancers occurrence (Zain, 2011).

Mycotoxins and Cancers

Humans are exposed daily to varieties of mycotoxins, occupational exposure by inhalation and consumption of contaminated products. Ochratoxin A, zearalenone, aflatoxins (AF), trichothecenes and fumonisins are of greatest concern among other mycotoxins as they are greatly toxic and potently carcinogenic (Bullerman, 2003). These mycotoxins are responsible for immunotoxicity, hepatotoxicity, nephrotoxicity, reprotoxicity, and carcinogenic effects (Christian, 2013) after long-term exposure at low doses via food or inhalation of dusts.

Fumonisins are produced by the fungus *Fusarium verticillioides* and *Fusarium proliferatum* are know with their high rate risk of human esophageal through disrupting sphingolipid synthesis, kidney and liver cancers (Wilson, 2012). In addition, investigations showed that Ochratoxin A is associated with upper urinary tract cancers after metabolism into derivatives of quinine (Pfohl-Leszkowicz, 2009). Ochratoxin A, another mycotoxin, has been documented to promote skin cancer and induce cell proliferation in mice through activating AP-1 (Kumar et al., 2012).

Aflatoxin consumption increases the risk of developing cancer of liver by more than 10-fold (Miller & Marasas, 2002). Furthermore, epidemiological studies carried out since 1970 in China Philippines, Thailand, Mozambique, Uganda, Kenya and Swaziland have established the role of Aflatoxins in hepatic carcinogenesis (Wu & Santella, 2012). Aflatoxins are documented as hepatocarcinogens, and causing lung cancer after metabolism epoxidation, origin of DNA adduct. Aflatoxin, particularly AFB1, becomes carcinogenic compound after metabolism in the body through P450 cytochromes, and also by prostaglandin synthetases and lipo-oxygenases (Wen, Mu, & Deng, 2016). It is essentially converted to various metabolites: aflatoxicol, aflatoxicol H1, AFP1, M1, AFM1, AFB1, AFQ1 and AFB2. These metabolites compounds are less toxic compared to AFB1 with the exception of epoxide (AFB1-8,9-exo-epoxide) (Bbosa, Kitya, Odda, & Ogwal-Okeng, 2013). Nevertheless, at the same exposure dose, the incidence of hepatocarcinoma is elevated after AFQ1 exposure compared to AFB1 metabolite. Similarly, findings from two studies of patient came from Mozambique and Qidong Province, China reported mutations in the p53 gene at the third base of codon 249 in hepatocarcinoma. These two areas are known for the consumption of contaminated food by fungi such as *Aspergillus parasiticus* or *Aspergillus flavis* which are aflatoxin B1 metabolites producers. Epidemiological studies have demonstrated the association between mutation of codon 249 and countries whose food is contaminated with aflatoxin (El-Din et al., 2010). It is a well established that hepatic carcinogen plays a crucial role in the genesis of hepatocarcinoma (HCC) and can acts synergistically to increase the occurrence of hepatitis B virus (El-Din et al., 2010). The same findings were observed on monkeys in which p53 gene is similar to that of humans (M. Yang et al., 1997).

Mycotoxins and Leukemia

The fetus is especially susceptible to the effects of environmental toxins due to immaturity and rapidly developing organs (Grigg, 2004). Many studies have related immunological disorders and childhood cancer, particularly leukemia to mycotoxins (PAULO, 2016; Wang et al., 2007). Among the mycotoxins which can cause leukemia in childhood; ochratoxin A, aflatoxin and fusarium. Ochratoxin A can induce leukemic lymphocyte line and apoptosis effect in human peripheral human lymphocytes (Kit 225 cells) (Assaf, Azouri, & Pallardy, 2004). Aflatoxin also known to be associated with childhood leukemia risk (Christopher Paul Wild & Kleinjans, 2003). Fusarium genus mycotoxins with their toxin products, in particular, trichothecenes, fumonisins, and zearalenones are possible to induce leukemia through Inhibition of nucleic acids and proteins synthesis, arresting cell division and disturbing mitochondrial function, as well as, destabilizing cell membranes (Escrivá, Font, Manyes, & Berrada, 2017). In another study, Klara R. Klein et al, 2005 demonstrated that mycotoxin-producing fungi from house were associated with four leukemic patients. The results from this study demonstrate that Gliotoxin and Patulin mycotoxins are cytotoxic to PBMC (Klein, Woodward, Waller, Lechowicz, & Rosenthal, 2005). Similarly, Stanzani et al, 2005 also concluded the same results regarding Gliotoxin, they showed that mycotoxin is most cytotoxic toward the antigen presenting cells, which impairs T cell activation (Stanzani et al., 2005). In Addition, Aleksandrowicz et al, observed an increase in the presence of toxins produced by mycotoxins such as *A. flavus* in comparison with non-leukemia-associated houses (Aleksandrowicz, Czachor, Schiffer, & Smyk, 1970).Wray et al. (1975) published multiples of investigations reporting the association between residential clusters of leukemia and fungal growth indoor. However, another study showed that Aflatoxin produced by *Aspergillus parasiticus* isolated from houses was not linked with leukemia clusters (Wray & O'Steen, 1975). Thus, we have to keep in mind that multiple leukemia cases in the same house in unrelated individuals was an extremely rare phenomenon, which required further study.

Mycotoxins and Childhood Leukemia

Aflatoxins metabolites such as AFB1, AFB1, and AFB1-albumin adducts have been identified in cord blood of babies after maternal exposure during pregnancy period. Several investigations have reported increase levels of aflatoxins in cord blood of mothers living in contaminated areas, which confirms that aflatoxins metabolite may traverse human placenta (Abdulrazzaq, Osman, & Ibrahim, 2002; Christopher P. Wild, Miller, & Groopman, 2015). Also, aflatoxin-albumin adducts have been located both in cord blood of mothers and maternal (Groopman et al., 2014). In young children and infants, AF-alb adducts have been detected in blood (P. C. Turner et al., 2007). It is well established that Aflatoxins, such as AFB1, like most carcinogens compounds requires metabolic activation to play its role in cellular toxicity. It is metabolized by 3A (3A4, 3A5, 3A7) and CYP1A2 enzymes into several metabolites, which may react with cellular macromolecules like proteins and DNA (Partanen et al., 2010).

REFERENCES

Abdulrazzaq, Y. M., Osman, N., & Ibrahim, A. (2002). Fetal exposure to aflatoxins in the United Arab Emirates. *Annals of Tropical Paediatrics*, *22*(1), 3–9. doi:10.1179/027249302125000094 PMID:11926047

AFSSA. (2009). *Évaluation des risques liés à la présence de mycotoxines dans les chaînes alimentaires humaine et animale.* Rapport final.

Alavanja, M. C. R., Dosemeci, M., Samanic, C., Lubin, J., Lynch, C. F., Knott, C., ... Blair, A. (2004). Pesticides and lung cancer risk in the agricultural health study cohort. *American Journal of Epidemiology*, *160*(9), 876–885. doi:10.1093/aje/kwh290 PMID:15496540

Alavanja, M. C. R., Hoppin, J. A., & Kamel, F. (2004). Health effects of chronic pesticide exposure: Cancer and neurotoxicity. *Annual Review of Public Health*, *25*(1), 155–197. doi:10.1146/annurev.publhealth.25.101802.123020 PMID:15015917

Aleksandrowicz, J., Czachor, M., Schiffer, Z., & Smyk, B. (1970). Mycotoxins and their role in oncogenesis, with special reference to blood diseases. *Haematologica Latina*, *13*(2), 115–124. PMID:5504037

Alshannaq, A., & Yu, J.-H. (2017). Occurrence, toxicity, and analysis of major mycotoxins in food. *International Journal of Environmental Research and Public Health*, *14*(6), 632. doi:10.3390/ijerph14060632 PMID:28608841

Assaf, H., Azouri, H., & Pallardy, M. (2004). Ochratoxin A induces apoptosis in human lymphocytes through down regulation of Bcl-xL. *Toxicological Sciences: An Official Journal of the Society of Toxicology*, *79*(2), 335–344. doi:10.1093/toxsci/kfh123 PMID:15056805

Atherton, K. M., Williams, F. M., Egea González, F. J., Glass, R., Rushton, S., Blain, P. G., & Mutch, E. (2009). DNA damage in horticultural farmers: A pilot study showing an association with organophosphate pesticide exposure. *Biomarkers: Biochemical Indicators of Exposure, Response, and Susceptibility to Chemicals*, *14*(7), 443–451. doi:10.3109/13547500903137265 PMID:19863182

Azimi, A., Azari, A., Rezakazemi, M., & Ansarpour, M. (2017). Removal of heavy metals from industrial wastewaters: A review. *ChemBioEng Reviews*, *4*(1), 37–59. doi:10.1002/cben.201600010

Baig, S., Seevasant, I., Mohamad, J., Mukheem, A., Huri, H. Z., & Kamarul, T. (2016). Potential of apoptotic pathway-targeted cancer therapeutic research: Where do we stand? *Cell Death & Disease*, *7*(1), e2058. doi:10.1038/cddis.2015.275 PMID:26775709

Bailey, H. D., Infante-Rivard, C., Metayer, C., Clavel, J., Lightfoot, T., Kaatsch, P., ... Schüz, J. (2015). Home pesticide exposures and risk of childhood leukemia: Findings from the childhood leukemia international consortium. *International Journal of Cancer, 137*(11), 2644–2663. doi:10.1002/ijc.29631 PMID:26061779

Baldi, I., Mohammed-Brahim, B., Brochard, P., Dartigues, J. F., & Salamon, R. (1998). Delayed health effects of pesticides: review of current epidemiological knowledge. *Revue d'Epidemiologie et de Sante Publique, 46*(2), 134–142. PMID:9592856

Bánfalvi, G. (2011). Heavy Metals, Trace Elements and Their Cellular Effects. In G. Banfalvi (Ed.), *Cellular Effects of Heavy Metals* (pp. 3–28). Dordrecht: Springer Netherlands; doi:10.1007/978-94-007-0428-2_1

Bbosa, G. S., Kitya, D., Odda, J., & Ogwal-Okeng, J. (2013). Aflatoxins metabolism, effects on epigenetic mechanisms and their role in carcinogenesis. *Health, 05*(10), 14–34. doi:10.4236/health.2013.510A1003

Belson, M., Kingsley, B., & Holmes, A. (2006). Risk factors for acute leukemia in children: A review. *Environmental Health Perspectives, 115*(1), 138–145. doi:10.1289/ehp.9023 PMID:17366834

Bener, A., Kamal, M., & Shanks, N. J. (2007). Impact of asthma and air pollution on school attendance of primary school children: Are they at increased risk of school absenteeism? *The Journal of Asthma, 44*(4), 249–252. doi:10.1080/02770900701246725 PMID:17530520

Bhat, R., Rai, R. V., & Karim, A. A. (2010). Mycotoxins in food and feed: Present status and future concerns. *Comprehensive Reviews in Food Science and Food Safety, 9*(1), 57–81. doi:10.1111/j.1541-4337.2009.00094.x

Blair, A., & Freeman, L. B. (2009). Epidemiologic Studies of Cancer in Agricultural Populations: Observations and Future Directions. *Journal of Agromedicine, 14*(2), 125–131. doi:10.1080/10599240902779436 PMID:19437268

Boffetta, P., & Nyberg, F. (2003). Contribution of environmental factors to cancer risk. *British Medical Bulletin, 68*(1), 71–94. doi:10.1093/bmp/ldg023 PMID:14757710

Bray, F., Ferlay, J., Soerjomataram, I., Siegel, R. L., Torre, L. A., & Jemal, A. (2018). Global Cancer Statistics 2018: GLOBOCAN Estimates of Incidence and Mortality Worldwide for 36 *CA: A Cancer Journal for Clinicians*, 185. PMID:30207593

Bullerman, L. B. (2003). MYCOTOXINS I Classifications. In B. Caballero (Ed.), *Encyclopedia of Food Sciences and Nutrition* (2nd ed.; pp. 4080–4089). Oxford, UK: Academic Press; doi:10.1016/B0-12-227055-X/00821-X

Caruso, R. V., O'Connor, R. J., Stephens, W. E., Cummings, K. M., & Fong, G. T. (2013). Toxic metal concentrations in cigarettes obtained from US smokers in 2009: Results from the International Tobacco Control (ITC) United States survey cohort. *International Journal of Environmental Research and Public Health, 11*(1), 202–217. doi:10.3390/ijerph110100202 PMID:24452255

Cassidy, R. A., Natarajan, S., & Vaughan, G. M. (2005). The link between the insecticide heptachlor epoxide, estradiol, and breast cancer. *Breast Cancer Research and Treatment, 90*(1), 55–64. doi:10.100710549-004-2755-0 PMID:15770527

Charlier, C., Albert, A., Herman, P., Hamoir, E., Gaspard, U., Meurisse, M., & Plomteux, G. (2003). Breast cancer and serum organochlorine residues. *Occupational and Environmental Medicine, 60*(5), 348–351. doi:10.1136/oem.60.5.348 PMID:12709520

Chatterjee, S., Basak, P., Chaklader, M., Das, P., Pereira, J. A., Chaudhuri, S., & Law, S. (2013). Pesticide induced marrow toxicity and effects on marrow cell population and on hematopoietic stroma. *Experimental and Toxicologic Pathology: Official Journal of the Gesellschaft Fur Toxikologische Pathologie, 65*(3), 287–295. doi:10.1016/j.etp.2011.09.002 PMID:22005501

Christian, R. (2013). *Mycologie médicale*. Lavoisier.

Coggon, D., Ntani, G., Harris, E. C., Jayakody, N., & Palmer, K. T. (2015). Soft tissue sarcoma, non-Hodgkin's lymphoma and chronic lymphocytic leukaemia in workers exposed to phenoxy herbicides: extended follow-up of a UK cohort. *Occup Environ Med*.

Cohn, B. A., Wolff, M. S., Cirillo, P. M., & Sholtz, R. I. (2007). DDT and Breast Cancer in Young Women: New Data on the Significance of Age at Exposure. *Environmental Health Perspectives, 115*(10), 1406–1414. doi:10.1289/ehp.10260 PMID:17938728

David, M., Krishna, P. M., & Sangeetha, J. (2016). Elucidation of impact of heavy metal pollution on soil bacterial growth and extracellular polymeric substances flexibility. *3 Biotech, 6*(2). doi:10.100713205-016-0475-x

Davoren, M. J., & Schiestl, R. H. (2018). Glyphosate-based herbicides and cancer risk: A post-IARC decision review of potential mechanisms, policy and avenues of research. *Carcinogenesis, 39*(10), 1207–1215. doi:10.1093/carcin/bgy105 PMID:30060078

Duffus, J. H. (1997). *Carcinogenicity of Inorganic Substances: Risks from Occupational Exposure*. Woodhead Publishing.

El-Din, H. G., Ghafar, N. A., Saad, N. E., Aziz, M., Rasheed, D., & Hassan, E. M. (2010). Relationship between codon 249 mutation in exon 7 of p53 gene and diagnosis of hepatocellular carcinoma. *Archives of Medical Science : AMS, 6*(3), 348–355. doi:10.5114/aoms.2010.14254 PMID:22371770

Escrivá, L., Font, G., Manyes, L., & Berrada, H. (2017). Studies on the Presence of Mycotoxins in Biological Samples: An Overview. *Toxins, 9*(8). doi:10.3390/toxins9080251 PMID:28820481

Florkin, M. (1956). *Aspects biochimiques communs aux êtres vivants: introduction à la biochimie générale des organismes*. Masson.

Fucic, A., Guszak, V., & Mantovani, A. (2017). Transplacental exposure to environmental carcinogens: Association with childhood cancer risks and the role of modulating factors. *Reproductive Toxicology (Elmsford, N.Y.), 72*, 182–190. doi:10.1016/j.reprotox.2017.06.044 PMID:28624605

Fumonisin Toxicosis. (2012). InWilson, D. A. (Ed.), *Clinical Veterinary Advisor* (pp. 214–215). St. Louis, MO: W.B. Saunders; doi:10.1016/B978-1-4160-9979-6.00147-1

Giannandrea, F., & Fargnoli, S. (2017). Environmental Factors Affecting Growth and Occurrence of Testicular Cancer in Childhood: An Overview of the Current Epidemiological Evidence. *Children, 4*(1), 1. doi:10.3390/children4010001 PMID:28067779

Gibb, H. J., Lees, P. S., Pinsky, P. F., & Rooney, B. C. (2000). Lung cancer among workers in chromium chemical production. *American Journal of Industrial Medicine*, *38*(2), 115–126. doi:10.1002/1097-0274(200008)38:2<115::AID-AJIM1>3.0.CO;2-Y PMID:10893504

Greenlee, R. T., Murray, T., Bolden, S., & Wingo, P. A. (2000). Cancer statistics, 2000. *CA: a Cancer Journal for Clinicians*, *50*(1), 7–33. doi:10.3322/canjclin.50.1.7 PMID:10735013

Grigg, J. (2004). Environmental toxins; their impact on children's health. *Archives of Disease in Childhood*, *89*(3), 244–250. doi:10.1136/adc.2002.022202 PMID:14977703

Grimsrud, T. K., & Andersen, A. (2012). Unrecognized risks of nickel-related respiratory cancer among Canadian electrolysis workers. *Scandinavian Journal of Work, Environment & Health*, *38*(6), 503–515. doi:10.5271jweh.3274 PMID:22258056

Groopman, J. D., Egner, P. A., Schulze, K. J., Wu, L. S.-F., Merrill, R., Mehra, S., … Christian, P. (2014). Aflatoxin exposure during the first 1000 days of life in rural South Asia assessed by aflatoxin B1-lysine albumin biomarkers. *Food and Chemical Toxicology: An International Journal, 74*, 184-189. doi:10.1016/j.fct.2014.09.016

Guyton, K. Z., Loomis, D., Grosse, Y., El Ghissassi, F., Benbrahim-Tallaa, L., Guha, N., ... Straif, K. (2015). Carcinogenicity of tetrachlorvinphos, parathion, malathion, diazinon, and glyphosate. *The Lancet Oncology*, *16*(5), 490–491. doi:10.1016/S1470-2045(15)70134-8 PMID:25801782

Hartwig, A. (2013). Cadmium and cancer. *Metal Ions in Life Sciences*, *11*, 491–507. doi:10.1007/978-94-007-5179-8_15 PMID:23430782

He, T.-T., Zuo, A.-J., Wang, J.-G., & Zhao, P. (2017). Organochlorine pesticides accumulation and breast cancer: A hospital-based case–control study. *Tumour Biology*, *39*(5), 1010428317699114. doi:10.1177/1010428317699114 PMID:28459199

Huff, J., Lunn, R. M., Waalkes, M. P., Tomatis, L., & Infante, P. F. (2007). Cadmium-induced cancers in animals and in humans. *International Journal of Occupational and Environmental Health*, *13*(2), 202–212. doi:10.1179/oeh.2007.13.2.202 PMID:17718178

IARC. (1990). *IARC Monographs on the Evaluation of Carcinogenic Risks to Humans: Chromium, Nickel and Welding*. International Agency for Research on Cancer.

IARC. (1991). Occupational exposures in insecticide application, and some pesticides. *IARC Monographs on the Evaluation of Carcinogenic Risks to Humans*, 53. PMID:1688189

IARC. (1999). Some chemicals that cause tumours of the kidney or urinary bladder in rodents and some other substances. *IARC Monographs on the Evaluation of Carcinogenic Risks to Humans*, *73*, 131–182. PMID:10804954

IARC. (2006). *Inorganic and organic lead compounds*. IARC.

IARC. (2012). review of human carcinogens, Part C: Arsenic, Metals, Fibres, and Dusts. *IARC Monographs, Lyon, France*, *100*, 196–211.

Ibarluzea, J. M., Fernández, M. F., Santa-Marina, L., Olea-Serrano, M. F., Rivas, A. M., Aurrekoetxea, J. J., ... Olea, N. (2004). Breast Cancer Risk and the Combined Effect of Environmental Estrogens. *Cancer Causes & Control, 15*(6), 591–600. doi:10.1023/B:CACO.0000036167.51236.86 PMID:15280638

INSERM Collective Expertise Centre. (2000). Cancer: A methodological approach for studying the link between cancer and the environment. In *INSERM Collective Expert Reports*. Paris: Institut national de la santé et de la recherche médicale. Retrieved from http://www.ncbi.nlm.nih.gov/books/NBK7114/

Itoh, H., Iwasaki, M., Hanaoka, T., Kasuga, Y., Yokoyama, S., Onuma, H., ... Tsugane, S. (2009). Serum organochlorines and breast cancer risk in Japanese women: A case–control study. *Cancer Causes & Control, 20*(5), 567–580. doi:10.100710552-008-9265-z PMID:19031103

Iwasaki, M., Inoue, M., Sasazuki, S., Kurahashi, N., Itoh, H., Usuda, M., & Tsugane, S. (2008). Plasma organochlorine levels and subsequent risk of breast cancer among Japanese women: A nested case–control study. *The Science of the Total Environment, 402*(2-3), 176–183. doi:10.1016/j.scitotenv.2008.05.009 PMID:18555519

Jan, A. T., Azam, M., Siddiqui, K., Ali, A., Choi, I., & Haq, Q. M. R. (2015). Heavy metals and human health: Mechanistic insight into toxicity and counter defense system of antioxidants. *International Journal of Molecular Sciences, 16*(12), 29592–29630. doi:10.3390/ijms161226183 PMID:26690422

Jayakody, N., Harris, E. C., & Coggon, D. (2015). Phenoxy herbicides, soft-tissue sarcoma and non-Hodgkin lymphoma: A systematic review of evidence from cohort and case–control studies. *British Medical Bulletin, 114*(1), 75–94. doi:10.1093/bmb/ldv008 PMID:25790819

Jin, M. W., Xu, S. M., An, Q., & Wang, P. (2016). A review of risk factors for childhood leukemia. *European Review for Medical and Pharmacological Sciences, 20*(18), 3760–3764. PMID:27735044

Kashyap, M. P., Singh, A. K., Kumar, V., Yadav, D. K., Khan, F., Jahan, S., ... Pant, A. B. (2013). Pkb/Akt1 Mediates Wnt/GSK3β/β-Catenin Signaling-Induced Apoptosis in Human Cord Blood Stem Cells Exposed to Organophosphate Pesticide Monocrotophos. *Stem Cells and Development, 22*(2), 224–238. doi:10.1089cd.2012.0220 PMID:22897592

Klein, K. R., Woodward, C. S., Waller, E. K., Lechowicz, M. J., & Rosenthal, H. (2005). Effects of Mycotoxins on Mononuclear Cells (MNCs) in Normal Blood, T-Cell Leukemia and Lymphoma Cell Lines. *Blood, 106*(11), 4828–4828.

Kumar, R., Alam, S., Chaudhari, B. P., Dwivedi, P. D., Jain, S. K., Ansari, K. M., & Das, M. (2012). Ochratoxin A-induced cell proliferation and tumor promotion in mouse skin by activating the expression of cyclin-D1 and cyclooxygenase-2 through nuclear factor-kappa B and activator protein-1. *Carcinogenesis, 34*(3), 647–657. doi:10.1093/carcin/bgs368 PMID:23172667

Lane, E. A., Canty, M. J., & More, S. J. (2015). Cadmium exposure and consequence for the health and productivity of farmed ruminants. *Research in Veterinary Science, 101*, 132–139. doi:10.1016/j.rvsc.2015.06.004 PMID:26267103

Lemarchand, C., Tual, S., Boulanger, M., Levêque-Morlais, N., Perrier, S., Clin, B., ... Baldi, I. (2016). Prostate cancer risk among French farmers in the AGRICAN cohort. *Scandinavian Journal of Work, Environment & Health*, *42*(2), 144–152. doi:10.5271jweh.3552 PMID:26932763

Leplège, A., Ecosse, E., Verdier, A., & Perneger, T. V. (1998). The French SF-36 Health Survey: Translation, cultural adaptation and preliminary psychometric evaluation. *Journal of Clinical Epidemiology*, *51*(11), 1013–1023. doi:10.1016/S0895-4356(98)00093-6 PMID:9817119

Lewis, J. A., Gehman, E. A., Baer, C. E., & Jackson, D. A. (2013). Alterations in gene expression in Caenorhabditis elegans associated with organophosphate pesticide intoxication and recovery. *BMC Genomics*, *14*(1), 291. doi:10.1186/1471-2164-14-291 PMID:23631360

Lodovici, M., & Bigagli, E. (2011). *Oxidative Stress and Air Pollution Exposure* [Research article]. doi:10.1155/2011/487074

Loomis, D., Guyton, K., Grosse, Y., El Ghissasi, F., Bouvard, V., Benbrahim-Tallaa, L., ... Straif, K. (2015). Carcinogenicity of lindane, DDT, and 2, 4-dichlorophenoxyacetic acid. *The Lancet Oncology*, *16*(8), 891–892. doi:10.1016/S1470-2045(15)00081-9 PMID:26111929

López-Cervantes, M., Torres-Sánchez, L., Tobías, A., & López-Carrillo, L. (2004). Dichlorodiphenyldichloroethane burden and breast cancer risk: A meta-analysis of the epidemiologic evidence. *Environmental Health Perspectives*, *112*(2), 207–214. doi:10.1289/ehp.112-1241830 PMID:14754575

Lyons, G., & Watterson, A. (2010). *Review of the Role Pesticides Play in Some Cancers: Children, Farmers and Pesticide Users at Risk*. CHEMTrust Report.

Mandarapu, R., & Prakhya, B. M. (2015). In vitro myelotoxic effects of cypermethrin and mancozeb on human hematopoietic progenitor cells. *Journal of Immunotoxicology*, *12*(1), 48–55. doi:10.3109/1547691X.2014.880535 PMID:24499300

Massanyi, P., Lukac, N., Slivkova, J., Kovacik, J., Makarevich, A. V., Chrenek, P., ... Stawarz, R. (2007). Mercury-induced alterations in rat kidneys and testes in vivo. *Journal of Environmental Science and Health Part A*, *42*(7), 865–870. doi:10.1080/10934520701370410 PMID:17558766

McCready, D., Aronson, K. J., Chu, W., Fan, W., Vesprini, D., & Narod, S. A. (2004). Breast tissue organochlorine levels and metabolic genotypes in relation to breast cancer risk Canada. *Cancer Causes & Control*, *15*(4), 399–418. doi:10.1023/B:CACO.0000027505.32564.c2 PMID:15141140

McLean, D., Mannetje, A., Dryson, E., Walls, C., McKenzie, F., & Maule, M. (2009). Leukaemia and occupation: A New Zealand Cancer Registry-based case–control Study. *International Journal of Epidemiology*, *38*(2), 594–606. doi:10.1093/ije/dyn220 PMID:18953052

Merhi, M., Raynal, H., Cahuzac, E., Vinson, F., Cravedi, J. P., & Gamet-Payrastre, L. (2007). Occupational exposure to pesticides and risk of hematopoietic cancers: Meta-analysis of case–control studies. *Cancer Causes & Control*, *18*(10), 1209–1226. doi:10.100710552-007-9061-1 PMID:17874193

Miller, D., & Marasas, W. (2002). Ecology of mycotoxins in maize and groundnuts. *LEISA-LEUSDEN*, *17*, 23–24.

Mokarizadeh, A., Faryabi, M. R., Rezvanfar, M. A., & Abdollahi, M. (2015). A comprehensive review of pesticides and the immune dysregulation: Mechanisms, evidence and consequences. *Toxicology Mechanisms and Methods*, *25*(4), 258–278. doi:10.3109/15376516.2015.1020182 PMID:25757504

National Research Council. (2000). *Health Effects of Excess Copper*. Author.

Nicolopoulou-Stamati, P., Maipas, S., Kotampasi, C., Stamatis, P., & Hens, L. (2016). Chemical Pesticides and Human Health: The Urgent Need for a New Concept in Agriculture. *Frontiers in Public Health*, *4*. doi:10.3389/fpubh.2016.00148 PMID:27486573

Nieboer, E., Tsuji, L. J. S., Martin, I. D., & Liberda, E. N. (2013). Human biomonitoring issues related to lead exposure. *Environmental Science. Processes & Impacts*, *15*(10), 1824–1829. doi:10.1039/c3em00270e PMID:23970117

O'Leary, M., Krailo, M., Anderson, J. R., & Reaman, G. H. (2008). Progress in Childhood Cancer: 50 Years of Research Collaboration, A Report from the Children's Oncology Group. *Seminars in Oncology*, *35*(5), 484–493. doi:10.1053/j.seminoncol.2008.07.008 PMID:18929147

Partanen, H. A., El-Nezami, H. S., Leppänen, J. M., Myllynen, P. K., Woodhouse, H. J., & Vähäkangas, K. H. (2010). Aflatoxin B1 transfer and metabolism in human placenta. *Toxicological Sciences: An Official Journal of the Society of Toxicology*, *113*(1), 216–225. doi:10.1093/toxsci/kfp257 PMID:19875679

Paulo & Sergio. (2016). *Mycotoxins: Biochemical Approach*. Editora Albatroz.

Perlroth, N. H., & Branco, C. W. C. (2017). Current knowledge of environmental exposure in children during the sensitive developmental periods. *Jornal de Pediatria*, *93*(1), 17–27. doi:10.1016/j.jped.2016.07.002 PMID:27821252

Pfohl-Leszkowicz, A. (2009). Mycotoxines: Facteur de risque de cancers. *Journal africain du cancer/ African. Journal of Cancer*, *1*(1), 42–55.

Pichery, C., Bellanger, M., Zmirou-Navier, D., Glorennec, P., Hartemann, P., & Grandjean, P. (2011). Childhood lead exposure in France: Benefit estimation and partial cost-benefit analysis of lead hazard control. *Environmental Health*, *10*(1), 44. doi:10.1186/1476-069X-10-44 PMID:21599937

Rice, K. M., Walker, E. M. Jr, Wu, M., Gillette, C., & Blough, E. R. (2014). Environmental mercury and its toxic effects. *Journal of Preventive Medicine and Public Health*, *47*(2), 74–83. doi:10.3961/jpmph.2014.47.2.74 PMID:24744824

Richardson, D. B., Keil, A. P., Cole, S. R., & Dement, J. (2018). Asbestos standards: Impact of currently uncounted chrysotile asbestos fibers on lifetime lung cancer risk. *American Journal of Industrial Medicine*, *61*(5), 383–390. doi:10.1002/ajim.22836 PMID:29573442

Roberts, J. R., & Karr, C. J. (2012). Pesticide Exposure in Children. *Pediatrics*, *130*(6), e1765–e1788. doi:10.1542/peds.2012-2758 PMID:23184105

Rudant, J., Menegaux, F., Leverger, G., Baruchel, A., Nelken, B., Bertrand, Y., ... Clavel, J. (2007). Household Exposure to Pesticides and Risk of Childhood Hematopoietic Malignancies: The ESCALE Study (SFCE). *Environmental Health Perspectives*, *115*(12), 1787–1793. doi:10.1289/ehp.10596 PMID:18087601

Rusiecki, J. A., De Roos, A., Lee, W. J., Dosemeci, M., Lubin, J. H., Hoppin, J. A., ... Alavanja, M. C. (2004). Cancer incidence among pesticide applicators exposed to atrazine in the Agricultural Health Study. *Journal of the National Cancer Institute*, *96*(18), 1375–1382. doi:10.1093/jnci/djh264 PMID:15367570

Saad-Hussein, A., Shahy, E. M., Shaheen, W., Taha, M. M., Mahdy-Abdallah, H., Ibrahim, K. S., ... El-Shamy, K. A. (2016). Comparative Hepatotoxicity of Aflatoxin B1 among Workers Exposed to Different Organic Dust with Emphasis on Polymorphism Role of Glutathione S-Transferase Gene. *Open Access Macedonian Journal of Medical Sciences*, *4*(2), 312. doi:10.3889/oamjms.2016.051 PMID:27335608

Sabarwal, A., Kumar, K., & Singh, R. P. (2018). Hazardous Effects of Chemical Pesticides on Human Health-Cancer and Other Associated Disorders. *Environmental Toxicology and Pharmacology*, *63*, 103–114. doi:10.1016/j.etap.2018.08.018 PMID:30199797

Samanic, C. M., De Roos, A. J., Stewart, P. A., Rajaraman, P., Waters, M. A., & Inskip, P. D. (2008). Occupational exposure to pesticides and risk of adult brain tumors. *American Journal of Epidemiology*, *167*(8), 976–985. doi:10.1093/aje/kwm401 PMID:18299277

Sathiakumar, N., MacLennan, P. A., Mandel, J., & Delzell, E. (2011). A review of epidemiologic studies of triazine herbicides and cancer. *Critical Reviews in Toxicology, 41*(sup1), 1–34.

Shah, H. K., Bhat, M. A., Sharma, T., Banerjee, B. D., & Guleria, K. (2018). Delineating Potential Transcriptomic Association with Organochlorine Pesticides in the Etiology of Epithelial Ovarian Cancer. *The Open Biochemistry Journal*, *12*(1), 16–28. doi:10.2174/1874091X01812010016 PMID:29576811

Siegel, R. L., Miller, K. D., & Jemal, A. (2017). Cancer statistics, 2017. *CA: a Cancer Journal for Clinicians*, *67*(1), 7–30. doi:10.3322/caac.21387 PMID:28055103

Silva-Pereira, L. C., Cardoso, P. C. dos S., Leite, D. S., Bahia, M. de O., Bastos, W. R., Smith, M. A. C., & Burbano, R. R. (2005). Cytotoxicity and genotoxicity of low doses of mercury chloride and methylmercury chloride on human lymphocytes in vitro. *Brazilian Journal of Medical and Biological Research*, *38*(6), 901–907. doi:10.1590/S0100-879X2005000600012 PMID:15933784

Singh, S., Kaur, S., & Budhiraja, R. D. (2013). Chlorpyrifos-induced oxidative stress in rat's brain and protective effect of grape seed extract. *The Journal of Phytopharmacology*, *2*(3), 26–33.

Soliman, A. S., Wang, X., DiGiovanni, J., Eissa, S., Morad, M., Vulimiri, S., ... Bondy, M. L. (2003). Serum organochlorine levels and history of lactation in Egypt. *Environmental Research*, *92*(2), 110–117. doi:10.1016/S0013-9351(02)00056-7 PMID:12854690

Sritharan, J., MacLeod, J., Harris, S., Cole, D. C., Harris, A., Tjepkema, M., ... Demers, P. A. (2018). Prostate cancer surveillance by occupation and industry: The Canadian Census Health and Environment Cohort (CanCHEC). *Cancer Medicine*, *7*(4), 1468–1478. doi:10.1002/cam4.1358 PMID:29493883

Stanzani, M., Orciuolo, E., Lewis, R., Kontoyiannis, D. P., Martins, S. L. R., St John, L. S., & Komanduri, K. V. (2005). Aspergillus fumigatus suppresses the human cellular immune response via gliotoxin-mediated apoptosis of monocytes. *Blood*, *105*(6), 2258–2265. doi:10.1182/blood-2004-09-3421 PMID:15546954

Stevens, J. J., Graham, B., Walker, A. M., Tchounwou, P. B., & Rogers, C. (2010). The effects of arsenic trioxide on DNA synthesis and genotoxicity in human colon cancer cells. *International Journal of Environmental Research and Public Health*, *7*(5), 2018–2032. doi:10.3390/ijerph7052018 PMID:20623008

Sugita, M. (1978). The biological half-time of heavy metals. *International Archives of Occupational and Environmental Health*, *41*(1), 25–40. doi:10.1007/BF00377797 PMID:627414

Sun, X., Zhang, T., Deng, Q., Zhou, Q., Sun, X., Li, E., ... Zhong, C. (2018). Benzidine Induces Epithelial–Mesenchymal Transition of Human Bladder Cancer Cells through Activation of ERK5 Pathway. *Molecules and Cells*, *41*(3), 188. PMID:29463068

Sunkireddy, P., Sriramoju, B., Roy, K., Kanwar, R. K., & Kanwar, J. R. (2018). Oxidative Stress in Ocular Disorders: Exploring the Link to Pesticide Exposure and Potential for Using Nanotechnology for Antioxidant Delivery. In Drug Delivery for the Retina and Posterior Segment Disease (pp. 151–167). Springer.

Terracini, B. (2011). Epidemiology of childhood cancer. *Environmental Health*, *10*(Suppl 1), S8. doi:10.1186/1476-069X-10-S1-S8 PMID:21489218

Turner, M. C., Wigle, D. T., & Krewski, D. (2010). Residential Pesticides and Childhood Leukemia: A Systematic Review and Meta-Analysis. *Environmental Health Perspectives*, *118*(1), 33–41. doi:10.1289/ehp.0900966 PMID:20056585

Turner, P. C., Collinson, A. C., Cheung, Y. B., Gong, Y., Hall, A. J., Prentice, A. M., & Wild, C. P. (2007). Aflatoxin exposure in utero causes growth faltering in Gambian infants. *International Journal of Epidemiology*, *36*(5), 1119–1125. doi:10.1093/ije/dym122 PMID:17576701

Tuteja, N., & Gill, S. S. (2016). *Abiotic Stress Response in Plants*. John Wiley & Sons. doi:10.1002/9783527694570

Van Maele-Fabry, G., Gamet-Payrastre, L., & Lison, D. (2018). Household exposure to pesticides and risk of leukemia in children and adolescents: Updated systematic review and meta-analysis. *International Journal of Hygiene and Environmental Health*. PMID:30268646

Van Maele-Fabry, G., Hoet, P., & Lison, D. (2013). Parental occupational exposure to pesticides as risk factor for brain tumors in children and young adults: A systematic review and meta-analysis. *Environment International*, *56*, 19–31. doi:10.1016/j.envint.2013.02.011 PMID:23567326

Van Maele-Fabry, G., Lantin, A.-C., Hoet, P., & Lison, D. (2010). Childhood leukaemia and parental occupational exposure to pesticides: A systematic review and meta-analysis. *Cancer Causes & Control: CCC*, *21*(6), 787–809. doi:10.100710552-010-9516-7 PMID:20467891

Venkatesan, R., Park, Y. U., Ji, E., Yeo, E.-J., & Kim, S. Y. (2017). Malathion increases apoptotic cell death by inducing lysosomal membrane permeabilization in N2a neuroblastoma cells: A model for neurodegeneration in Alzheimer's disease. *Cell Death & Disease*, *3*, 17007. doi:10.1038/cddiscovery.2017.7 PMID:28487766

Vrijheid, M., Casas, M., Bergström, A., Carmichael, A., Cordier, S., Eggesbø, M., ... Fernández-Somoano, A. (2011). European birth cohorts for environmental health research. *Environmental Health Perspectives*, *120*(1), 29–37. doi:10.1289/ehp.1103823 PMID:21878421

Waisberg, M., Joseph, P., Hale, B., & Beyersmann, D. (2003). Molecular and cellular mechanisms of cadmium carcinogenesis. *Toxicology*, *192*(2), 95–117. doi:10.1016/S0300-483X(03)00305-6 PMID:14580780

Wang, F., David, E., Lonial, S., Arellano, M., Khoury, H. J., Langston, A., ... Waller, N. (2007). Patulin, a Mycotoxin, Has Potent Cytotoxic Activity on Actue Lymphoblastic Leukemia and Induce Apoptosis through Intrinsic and Extrinsic Pathways. *Blood*, *110*(11), 4334–4334.

Wen, J., Mu, P., & Deng, Y. (2016). Mycotoxins: Cytotoxicity and biotransformation in animal cells. *Toxicology Research*, *5*(2), 377–387. doi:10.1039/C5TX00293A PMID:30090353

Wigle, D. T., Turner, M. C., & Krewski, D. (2009). A Systematic Review and Meta-analysis of Childhood Leukemia and Parental Occupational Pesticide Exposure. *Environmental Health Perspectives*, *117*(10), 1505–1513. doi:10.1289/ehp.0900582 PMID:20019898

Wild, C. P., & Kleinjans, J. (2003). Children and increased susceptibility to environmental carcinogens: Evidence or empathy? *Cancer Epidemiology, Biomarkers & Prevention*, *12*(12), 1389–1394. PMID:14693727

Wild, C. P., Miller, J. D., & Groopman, J. D. (2015). *Effects of aflatoxins and fumonisins on child growth*. International Agency for Research on Cancer. Retrieved from https://www.ncbi.nlm.nih.gov/books/NBK350566/

Wray, B. B., & O'Steen, K. G. (1975). Mycotoxin-producing fungi from house associated with leukemia. *Archives of Environmental Health*, *30*(12), 571–573. doi:10.1080/00039896.1975.10666781 PMID:1239242

Wu, H. C., & Santella, R. (2012). The role of aflatoxins in hepatocellular carcinoma. *Hepatitis Monthly*, *12*(10).

Yang, M., Zhou, H., Kong, R. Y., Fong, W. F., Ren, L. Q., Liao, X. H., ... Yang, S. (1997). Mutations at codon 249 of p53 gene in human hepatocellular carcinomas from Tongan, China. *Mutation Research*, *381*(1), 25–29. doi:10.1016/S0027-5107(97)00142-5 PMID:9403027

Yang, Y., Jin, X. M., Yan, C. H., Tian, Y., Tang, J. Y., & Shen, X. M. (2008). Urinary level of nickel and acute leukaemia in Chinese children. *Toxicology and Industrial Health*, *24*(9), 603–610. doi:10.1177/0748233708100091 PMID:19106127

Zachek, C. M., Miller, M. D., Hsu, C., Schiffman, J. D., Sallan, S., Metayer, C., & Dahl, G. V. (2015). Children's cancer and environmental exposures: Professional attitudes and practices. *Journal of Pediatric Hematology/Oncology*, *37*(7), 491–497. doi:10.1097/MPH.0000000000000416 PMID:26334434

Zahm, S. H., & Ward, M. H. (1998). Pesticides and childhood cancer. *Environmental Health Perspectives*, *106*(Suppl 3), 893. PMID:9646054

Zain, M. E. (2011). Impact of mycotoxins on humans and animals. *Journal of Saudi Chemical Society*, *15*(2), 129–144. doi:10.1016/j.jscs.2010.06.006

Zeng, F., Lerro, C., Lavoué, J., Huang, H., Siemiatycki, J., Zhao, N., ... Zhang, Y. (2017). Occupational Exposure to Pesticides and Other Biocides and Risk of Thyroid Cancer. *Occupational and Environmental Medicine*, *74*(7), 502–510. doi:10.1136/oemed-2016-103931 PMID:28202579

Section 5

Neurological and Neurophsychiatric Disorders Associated to Environmental Contaminations

Chapter 11
Metal Toxicity and Brain-Liver Axis:
The Good, the Bad, and the Neurodegenerated

Tiziano Balzano
Príncipe Felipe Research Center, Spain

Omar El Hiba
Chouaib Doukkali University, Morocco & Cadi Ayyad University, Morocco

ABSTRACT

The liver is the main detoxifier organ of the body. When normal liver function is compromised, other systems in the body can be affected, including the brain. Hepatocerebral disorder is the term used to describe some neuropsychiatric conditions that result from liver failure and characterized by the accumulation of these toxic metals in brain. Examples of such disorders are Wilson's disease (WD), an autosomal recessive disorder that is characterized by the deposition of copper in liver and brain tissues and acquired (non-Wilsonian) hepatocerebral degeneration (AHCD), a complication that occurs most frequently in patients with hepatic coma or that suffered multiple episodes of severe HE. AHCD is characterized by accumulation in brain of manganese. This chapter will focus on the crucial importance of relationship between liver and brain functioning and on the effects produced when this relationship is compromised. Specifically, the chapter will discuss on the physiopathology of WD and AHCD and on the role that toxic metals play on neurological symptoms in such disorders.

LIVER-BRAIN AXIS: FROM BABYLONIANS TO OUR DAYS

The liver is the main detoxifier organ of the body. When normal liver function is compromised, also other systems in the body can be affected, including the brain. This makes the relationship between the liver and the brain very important and the normal brain functioning depends on normal liver functioning.

DOI: 10.4018/978-1-5225-7775-1.ch011

This relationship was already known more than 4000 years ago, when the Babylonians attributed powers of augury and divination to the liver, designating it by the term used also for 'soul' or 'mood' (Davidson & Summerskill, 1956).

Nearly 2500 years ago, Hippocrates (460-377 B.C.) was the first to make an association between hepatic disease and impaired brain function. In fact, the Father of Medicine observed that "Those who are mad on account of phlegm are quiet, but those on account of bile are vociferous, vicious, and do not keep quiet" (Summerskill, Davidson, Sherlock, & Steiner, 1956). This was the beginning of humoral medicine, developed by Hippocrates and consisting in four humors: 1) the blood with its warm and wet properties corresponded to air; 2) warm and dry yellow bile corresponds to fire; 3) cold and dry black bile to soil and 4) the phlegm with its cold and wet properties corresponds to water. (Nam, 2014). According to this theory, the human healthy depended on the balance and harmony of these four elements.

This belief dominated medical practice for many centuries, from Greek to Romans. In fact, Galen (A.D. 131-200) a Greek physician and Celsus, a Greek philosopher (c. A.D. 30) also attributed the influence of liver function on personality, behaviour, and physical infirmity (Summerskill et al., 1956).

With the fall of the Roman Empire, also the status of the liver declined from that of a 'vital organ' to a 'great, bile producer'. Only in the 18th century, the relationship between hepatic function and neuropsychiatric disorder was emphasized again and new definition of 'hepatic coma' was elaborated to include an immense variety of clinical features. This definition was created by Gianbattista Morgagni, a professor of Theoretical Medicine in Padua. Morgagni described an initial phase observed in his patients where they appear stupid, forgetful and suffered from a 'suppression of the senses', later followed by violent delirium, progressing to terminal coma (Morgagni, 1820).

At the end of the 20th century, the interdependence of liver and brain functioning was finally demonstrated by Marcel Nencki and Ivan Pavlov. They described the physiological consequences of a portacaval shunt (PCS) in dogs, a surgical procedure still used today in rats to reproduce hepatic encephalopathy and first described by Eck in 1879 ("Eck's fistula") (Starzl, PORTER, & FRANCAVILLA, 1983). Such dogs developed neurobehavioral changes in 10 days to 6 weeks postsurgery; symptoms included aggression, irritability, ataxia, convulsions, and coma (Hahn, Massen, Nencki, & Pawlow, 1893).

Today, the main role of liver as detoxifier organ and its interconnection with brain functioning is well studied. The liver cleanses the blood of substances, which if not eliminated, could penetrate the blood-brain-barrier and damage the brain cells. These substances which include many metals such as copper, manganese, and iron are responsible of some conditions known as hepatocerebral disorders. Examples of such disorders are: Wilson's disease (WD), Acquired (non-Wilsonian) hepatocerebral degeneration (AHCD) and Hepatic encephalopathy (HE).

ACQUIRED HEPATOCEREBRAL DEGENERATION: STORY AND PHYSIOPATHOLOGY OF A RARE NEUROPSYCHIATRIC SYNDROME

Acquired hepatocerebral degeneration (AHCD) is a brain disorder that occurs in people with liver damage. Its symptomatology is often confused with that observed in patients with hepatic encephalopathy or Wilson's disease. AHCD was first described by Van Woerkom in 1914 but only 50 years after, Victor and coworkers realized the first and complete anatomo-pathological description of disease (Victor, Adams, & Cole, 1965). With their work, Victor et al., fundamentally distinguished AHCD from Wilson's disease. Neuropathological findings included patchy cortical laminar neuronal loss, neuronal drop-out in the cerebellum and basal ganglia, proliferation of Alzheimer type II glia and cytoplasmic glycogen granules in basal ganglia (Burgos, Bermejo, Calleja, Vaquero, & Abreu, 2009).

Such alterations were responsible to produce in the 27 patients included in the study the typical neurological impairments observed in people suffering of severe hepatic diseases and recurrent episode of coma. The impairments described were progressive dementia, dysarthria, involuntary movements (including tremor, asterixis, and choreoathetosis), ataxia of limb and gait and mild pyramidal tract signs (Victor et al., 1965).

The physiopathological origin of AHCD is multifactorial although is mainly described in patients with severe hepatic disease due to surgically induced or spontaneous porto-systemic shunts (Spencer & Forno, 2000). Other causes include hepatic parenchymal diseases such as cirrhosis, chronic and acute hepatitis, and hemochromatosis or cholestatic diseases such as primary sclerosing cholangitis and primary biliary cirrhosis. AHCD has also been described in patients without a history of liver disease such as Manganese-exposed workers (welders, miners) (Y. C. Shin et al., 2007), patients receiving total parenteral nutrition (Fitzgerald et al., 1999; Santos, Batoreu, Mateus, dos Santos, & Aschner, 2014) or patients with portosystemic shunts due to Schistosomiasis (Okinaka et al., 1962), congenital hepatic fibrosis (Lewis, MacQuillan, Bamford, & Howdle, 2000) and portal vein thrombosis (Nolte et al., 1998).

Independently of the presence of liver disease, manganese was demonstrated to play a central role in the pathogenesis of AHCD.

SOURCE AND PHYSIOLOGICAL ROLE: THE GOOD SIDE OF MANGANESE

Manganese is the twelfth most abundant element and the fifth most abundant metal on the Earth's surface, forming approximately 0.1% of the earth's crust (Williams et al., 2012). Manganese does not exist in nature as an elemental form, but it is a component of over 100 numerous complex minerals including pyroluosite, rhodochrosite, rhodanite, braunite, pyrochroite and manganite (Keen & Zidenberg-Cherr, 2003). Chemical forms of manganese in their natural deposits include oxides, sulfides, carbonates and silicates (Keen & Zidenberg-Cherr, 2003).

Manganese (Mn) ion is a normal dietary component of food and water. According to Food Composition Databases from United States Department of Agriculture (USDA) the biggest source of the element are teas (Table 1) containing 133 mg of the element (per 100g of the food). Some types of spices, such as ginger, turmeric, cinnamon, curry, bay leaf, tarragon, thyme, savory and marjoram also represent a good source of Mn, ranging from 5,43 mg to 33,30 mg of the element per 100 g of food. Instead, the content of Mn in meat, fish and poultry is low, ranging from 0,02 mg to 1,09 mg per 100 g of food. Despite this, the absorption of Mn contained in meat products is higher than that contained in teas. In fact, tannin found in tea binds manganese and prevents its absorption from the gastrointestinal tract (Keen & Zidenberg-Cherr, 2003). Thus, the levels of Manganese present in a food do not always correlate with the absorbed levels.

The average daily consumption of Mn ranges from 2.0 to 8.8 mg Mn/day (Marques dos Santos, Aschner, & Marreilha dos Santos, 2017). Excess dietary of manganese is rapidly cleared by the liver before reaching the systemic circulation. In patients with cirrhosis and portosystemic shunting, manganese can bypass the liver and accumulates in the brain, while serum manganese levels may be normal or increased (Pomier-Layrargues, Spahr, & Butterworth, 1995).

Just ~5% of ingested Mn is absorbed into the gastrointestinal tract. Physiologically, the Mn is involved in many important functions in the body. It was demonstrated to be necessary for the maintenance of proper amounts of mucopolysaccharides, essential to the formation of bone, cartilage and connective tissue (Leach & Muenster, 1962). Furthermore, Manganese is an essential cofactor of cellular processes promoting the activity of many Mn-dependent enzymes such as superoxide dismutase (Stallings, Metzger, Pattridge, Fee, & Ludwig, 1991), arginase (Hellerman & Perkins, 1935), glutamine synthetase (Wedler, Denman, & Roby, 1982) and neurotransmitter synthetic enzymes (Michael Aschner, Guilarte, Schneider, & Zheng, 2007).

Mn is found in all mammalian tissues, with concentrations ranging from 0.3 to 2.9 μg Mn/g tissue of wet-weight (ww). The highest Mn concentrations in humans were found in liver (1.2–2.1 mg/Kg = 21.8–38.2 μmol/L), kidneys (0.6–0.9 mg/Kg = 10.9–16.4 μmol/L), endocrine glands, and intestines (Marques dos Santos et al., 2017). The brain also represents an important target for manganese, mainly in pathological conditions. Manganese (Mn) accumulation in the brain is detected as symmetrical hyperintensity signal in the basal ganglia (especially in globus pallidum and striatum) on T1-weighted MR images without an abnormal signal on T2-weighted images (Figure 1).

BRAIN ACCUMULATION AND PATHOLOGICAL ROLE: THE BAD SIDE OF MANGANESE

The metal was first considered to be neurotoxic more than 150 years ago, when five workers employed in grinding black oxide of manganese developed paraplegia, festinating gait, muscular weakness, tremor of the extremities, whispering speech and salivation (COUPER, 1837). The clinical description of "manganese poisoning" was further characterized by Rodier in 1955 describing the symptoms observed in Moroccan miners. The author described subjects with a rapidly progressive neurobehavioral syndrome characterized by Parkinsonism, dystonia, emotional lability, gait impairment, and psychosis (Rodier, 1955). Since that time, many cases of manganism, have been reported particularly in miners, smelters, and workers involved in the alloy industry (O'Neal & Zheng, 2015). Cumulative evidence has established that Mn exposure induces similar but not identical signs and symptoms to Parkinson's disease. Studies performed in human and nonhuman primates, suggest that degeneration of nigrostriatal dopaminergic neurons and their terminals in the striatum, which are selectively lesioned in Parkinson's disease, remain intact after Mn intoxication (Guilarte, 2010). This finding is consistent with the fact that L-dopa therapy does not provide a benefit to the Mn-induced movement abnormalities as it does in PD (Lu, Huang, Chu, & Calne, 1994).

Recent investigation of Mn neurotoxicity was also extended to the field of adult neurogenesis. The study revealed that young adult mice daily exposed to dietary Mn presented a reduced numbers of local Pvalb (+) GABAergic interneurons in the subgranular zone of the dentate gyrus in the hippocampus (Kikuchihara et al., 2015).

Mn transport within the CNS is mediated by several transporter proteins as a free ion or a non-specific protein-bound species (M. Aschner & Gannon, 1994). In the 3+ oxidation state Mn binds to the transferrin receptor (TfR) and, in a similar manner to iron (Fe), is transported by a TfR-mediated mechanism (M. Aschner & Aschner, 1991). Transport of divalent manganese (Mn^{2+}) is mediated by different channels/transporters such as divalent metal transporter 1 (DMT1) (Wolff et al., 2018), bicarbonate ion symporters ZIP8 and ZIP14 (Jenkitkasemwong, Wang, Mackenzie, & Knutson, 2012), various calcium channels such as voltage-gated calcium channels (VGCs) and store-operated calcium channels (SOCs) (Costa & Aschner, 2014), park9/ATP13A2 (Tan et al., 2011), the magnesium transporter HIP14 (Goytain, Hines, & Quamme, 2008) and the transient receptor potential melastatin 7 (TRPM7) (Costa & Aschner, 2014).

In the central nervous system, the main cellular type involved in detoxifying the excess of manganese are the astrocytes, showing a high affinity for this element. In fact, it was demonstrated that astrocytes have the ability to concentrate Mn at levels 50-fold higher than neurons and preferentially in the mitochondria (M. Aschner, Gannon, & Kimelberg, 1992).

Astrocytes are key regulators in brain function being involved in many functions such as physical and metabolic support for neurons, blood-brain-barrier formation, detoxification, clearance of neurotransmitters and regulation of energy metabolism. An elevated exposure to Mn may alter astrocytes function leading to further neuronal synaptic dysfunction and activation of an excitotoxic state. Manganese was shown to alter astrocytes function producing an oxidative stress state that creates a mitochondrial dysfunction and energy insufficiency. In addition, it was demonstrated that Mn alters glucose metabolism in astrocytes by inhibition of glutamine synthetase, an astrocyte-specific enzyme. Glucose metabolism is crucial for the normal brain functioning and for the biosynthesis of neurotransmitters such as Glutamate, aspartate and GABA.

Mn was also shown to upregulate the expression of genes encoding proinflammatory chemokines such as CCL7, CXCL14, CXCL2, CXCL3 and CXCL6 and some proinflammatory interleukins such as IL12A and IL7 (Sengupta et al., 2007). Thus Mn-mediated disturbances in astrocytes function would be expected to cause neuroinflammation and neuronal demise explaining the appearance of the most of neurological impairments observed in patients affected by manganism (Sidoryk-Wegrzynowicz & Aschner, 2013).

THE ROLE OF MANGANESE IN AHCD: BAD NEWS FOR THE BRAIN

Manganese plays a crucial role in the pathogenesis of AHCD. As previously described it can produce neurodegenerative changes in basal ganglia, cerebral cortex and cerebellum. These neurodegenerative alterations are often accompanied by the presence of Alzheimer type II astrocytes in the region where the lesion is observed (Figure 2). Alzheimer type II astrocytes have enlarged, pale nuclei with margination of chromatin and prominent nucleoli (Norenberg, 1987). Manganese stimulates stellation of cultured rat cortical astrocytes involving the cytoskeletal system, a potential contributor toward transformation to the Alzheimer type II form (Liao & Chen, 2001) and suggesting an important role for astrocytes in the pathophysiology of manganese neurotoxicity in AHCD.

The relationship between manganese accumulation and brain neurotoxixity was demonstrated by the presence of hyperintensity signal in T1-weighted magnetic resonance imaging (MRI) localized in the basal ganglia structures, mainly in globus pallidus (Nelson, Golnick, Korn, & Angle, 1993). These hyperintensity signals occur as a consequence of the accumulation of manganese (Figure 1). The correlation between the MRI-associated T1 relaxation time and brain manganese concentration was directly established in rats (Gallez et al., 2001) and in nonhuman primates (Shinotoh et al., 1995). Thus, MRI may represent a very useful tool to diagnose manganese neurotoxicity in patients with AHCD and at same time quantify the levels of the element in brain of such patients.

As previously reported, many membrane importers are involved in Mn transportation inside brain cells. One of these is DMT1, a transporter present in the outer mitochondrial membrane which was demonstrated to be highly expressed in basal ganglia including the globus pallidus and striatum (Huang, Ong, & Connor, 2004). This could explain the preferentially accumulation of manganese in basal ganglia and the hyperintensity signals in these regions by MRI.

Other proteins involved in manganese efflux are SLC30A10 and ATP13A2. Alteration related with the function or expression of these proteins could also contribute to the appearance of parkinsonism in AHCD. SLC30A10 is mainly expressed in the basal ganglia and liver, and mutations in the *SLC30A10* gene are associated with parkinsonism, hepatic cirrhosis, dystonia, hypermanganesemia and polycythemia, (Lechpammer et al., 2014; H.-W. Shin & Park, 2017). *ATP13A2* mutations cause Kufor-Rakeb syndrome, characterized by early onset parkinsonism, spasticity, supranuclear gaze palsy, and dementia (H.-W. Shin & Park, 2017). Based on these findings, membrane transporters of manganese may play key roles in the regional specificity and neurologic manifestations of manganese accumulation in the brain (H.-W. Shin & Park, 2017).

Another main player participating in the induction of neurological alterations observed in patients with AHCD is the neuroinflammation. In AHCD, the principal contributors to start a brain inflamed state are the astrocytes. The astrocytes are indispensable for cellular homeostatic maintenance regulating, among other things, extracellular glutamate levels and neuronal synapses. Astrocytes showed to have to 50-fold more affinity for Manganese than other cells of SNC (M. Aschner et al., 1992) and excessive Mn can alter the physiological function of astrocytes and converting them into an inflammatory machine activating NFκB pathway, production of NOS2 and several cytokines, such as tumor necrosis factor-α (TNFα) (Sarkar et al., 2018).

Studies realized during two last decades demonstrate a clear relationship between the presence and grade of neuroinflammation with the progression in major neurodegenerative diseases including Alzheimer's disease (AD) (Heneka et al., 2015), Parkinson's disease (PD) (Tansey & Goldberg, 2010) and amyotrophic lateral sclerosis (ALS) (Philips & Robberecht, 2011).

Neuroinflammation in AD, PD, and ALS is typified instead by a reactive morphology of glial cells, including both astrocytes and microglia, accompanied by low to moderate levels of inflammatory mediators in the parenchyma (Ransohoff, 2016). This reaction largely mediates neurodegenerative effects observed in these diseases. Neuroinflammation-induced neuronal loss was also observed in cerebellum of patients with different grades of chronic liver disease (Balzano, Forteza, Borreda, et al., 2018; Balzano, Forteza, Molina, et al., 2018). Thus, taken together, these results suggest that chronic neuroinflammation may also contribute to neurodegeneration in AHCD.

WILSON'S DISEASE: EPIDEMIOLOGY AND CLINICAL MANIFESTATIONS

Wilson's disease (WD) is a monogenetic autosomal-recessive disorder caused by mutation in a particular gene; ATP7B encoding a copper transporting P-type ATPase resulting in impairment of biliary excretion of copper (Cu) (Bull, Thomas, Rommens, Forbes, & Cox, 1993; Petrukhin et al., 1993; Tanzi et al., 1993). Dysfunction of the ATP7B gene results in excessive accumulation of Cu in liver hepatocytes progressively and leads to compromising liver function as well as hepatic storage capacity of the ion. When liver storage capacity is exceeded, Cu can be deposited in other compartments and organ tissues such as brain and eye provoking profound injuries (Huster, 2010; Pfeiffer, 2007; H. Scheinberg & I. Sternlieb, 1984).

The history of WD begun with Samuel Alexander Kinnier Wilson who was the first to recognize the disease process in 1912, he provided the first detailed clinical as well as the pathological descriptions of WD (Compston, 2009). In 1993 the gene responsible for WD has been identified to be localized in the chromosome 13, simultaneously by three separate laboratories (Bull et al., 1993; Frydman et al., 1985; Tanzi et al., 1993; Yamaguchi, Heiny, & Gitlin, 1993). Till now, more than 500 ATP7B mutations have been identified as missense mutations (small deletions or insertions of the gene) or splice junction mutations. In addition, other types of mutations have been also recognized such as whole exon deletions, promoter region mutations, multiple mutations and monogenic disomy (Table 2) (Coffey et al., 2013; Møller, Ott, Lund, & Horn, 2005).

WD is a rare pathology with low incidence. WD is generally present in1:30,000 individuals of all populations worldwide with high frequency in case consanguinity (Table 2) (Weiss, 1993). Clinically, WD patients show different symptoms including liver disease, neurological and neuropsychiatric disorders and ophthalmologic manifestations, osteoarthritis, renal tubular dysfunction, and cardiomyopathy (Bandmann, Weiss, & Kaler, 2015; Lorincz, 2010; Wu, Wang, Pu, Qiao, & Jiang, 2015).

PHYSIOPATHOLOGY OF WILSON'S DISEASE

Cu is known as one of the essential elements for cellular function. However, free copper is extremely toxic and can trigger profound and irreversible cellular damage. The organism has a particular system that can bind Cu ions to ensure safe transport to their targets and safe biliary excretion.

The ATP7B protein is contained in the trans-Golgi network of hepatocytes. The protein ensures incorporation of six copper molecules into apoceruloplasmin to form ceruloplasmin (Hung et al., 1997). In normal case, Cu 2+ is absorbed in the small intestine by the enterocyte uptake using the human copper transporter 1 (hCTR1), therefore, Cu2+ is released into blood circulation by the transporter ATP7A at the basolateral part of duodenal epithelia. Cu2+ is then transported the liver through portal circulation, where the excessive amounts of Cu2+ should be removed.

In the liver, ATP7B can be distributed to cytoplasmic vesicles in order to remove excess Cu across the hepatocyte apical membrane into the bile canaliculus to reach the biliary circulation (Forbes & Cox, 2000; La Fontaine et al., 2001). Such vital process is impaired by mutations in ATP7B gene. However, other mutations of different gens could be behind several diseases including mutations in the AP1S1 gene implicated in MEDNIK (mental retardation, enteropathy, deafness, neuropathy, ichthyosis, and keratodermia) syndrome (mutation in the *AP1S1* gene encoding for AP-1 protein subunit involved in the intracellular protein), the acetyl CoA transporter SLC33A1, and the cytosolic copper chaperone CCS. As well, mutations in the transporter of Mn; SLC30A10 is associated to hepatic cirrhosis arising from Mn accumulation which presents analogy to Wilson's disease. At the level of the CNS, two proteins seem to be involved in Cu transport such as ATP7A at the blood–CSF barrier and ATP7B within choroid plexus epithelia. Such mutations underlie either excessive (Wilson disease) or deficiency (Menkens disease) brain Cu (Figure 3) (Bandmann et al., 2015). It is well established that the cellular damages observed in WD results from a direct excessive Cu toxicity, thus, recent finding support a possible loss of a particular protein: X-linked inhibitor of apoptosis (XIAP), triggered by elevated Cu level. Such inhibition of this protein, results in acceleration of caspase 3–initiated apoptosis and cell death (Mufti et al., 2006).

In almost cases, the main onset age of WD is variable between 5 and 35 years old and may appear later even in 80 years old (Ala, Borjigin, Rochwarger, & Schilsky, 2005; Ferenci et al., 2007). The clinical features of WD include a large spectrum of manifestations starting with hepatic, then neurologic and finally psychiatric symptoms (Table 3) (European Association for Study of Liver, 2012; Roberts, Schilsky, & American Association for Study of Liver Diseases, 2008).

CONCLUSION

The liver is the main detoxifier organ of the body cleaning the blood of substances, which if not eliminated, could penetrate the blood-brain-barrier and damage the brain cells. These substances which include many metals such as copper and manganese are responsible of some conditions known as hepatocerebral disorders. Acquired hepatocerebral degeneration is a rare neurological syndrome that mainly occurs in patients with chronic liver disease, though it was also observed in people without a history of liver disease. Many studies demonstrated a direct contribution of manganese to the neurological dysfunctions present in patients with AHCD. Manganese is a normal dietary component of food and water that plays an essential role during the formation of bone, cartilage and connective tissue and participating as cofactor for many enzymes involved in different cellular processes. Despite its physiological functions, accumulation of Mn into the brain was related with the neurological alterations in AHCD, such as progressive dementia, dysarthria, involuntary movements, ataxia and mild pyramidal tract signs. A clear evidence of the involvement of manganese in these neurological manifestations is the localized presence of hyperintensity signals in T1-weighted magnetic resonance imaging in the basal ganglia structures. The neuroinflammation and more specifically the astrocytes would mediate the effects of manganese on the brain. Increased levels of manganese in the brain alter the metabolism and the physiological function of these star-shape cells, converting them into Alzheimer type 2 astrocytes. These cells are the responsible to recruit the whole inflammatory machine promoting the production of different cytokines which acting synergistically with the resulting oxidative and nitrosative stress cause neurodegeneration in AHCD.

Wilson's disease is a monogenetic autosomal-recessive disorder caused by mutation of ATP7B gene that produces an impairment of biliary excretion of copper and resulting in its excessive accumulation. Copper accumulation produces different symptoms including many neurological and neuropsychiatric alterations such as movement disorders, speech disturbances, autonomic dysfunction, gait and balance disturbances and alterations of personality and consciousness, most of them produced by the neurodegenerative effects caused by an accumulation of copper into the brain.

As conclusion, the liver and the heavy metals represent the good and the bads of this story. When the normal liver function is compromised, the heavy metals take the upper hand on many organs, including the brain. Brain accumulation of heavy metals produces neurodegeneration that is the main responsible of the impairments observed in patients with Acquired hepato-cerebral degeneration and Wilson's disease.

REFERENCES

Ala, A., Borjigin, J., Rochwarger, A., & Schilsky, M. (2005). Wilson disease in septuagenarian siblings: Raising the bar for diagnosis. *Hepatology (Baltimore, Md.)*, *41*(3), 668–670. doi:10.1002/hep.20601 PMID:15723329

Aschner, M., & Aschner, J. L. (1991). Manganese neurotoxicity: Cellular effects and blood-brain barrier transport. *Neuroscience and Biobehavioral Reviews*, *15*(3), 333–340. doi:10.1016/S0149-7634(05)80026-0 PMID:1956602

Aschner, M., & Gannon, M. (1994). Manganese (Mn) transport across the rat blood-brain barrier: Saturable and transferrin-dependent transport mechanisms. *Brain Research Bulletin*, *33*(3), 345–349. doi:10.1016/0361-9230(94)90204-6 PMID:8293318

Aschner, M., Gannon, M., & Kimelberg, H. K. (1992). Manganese uptake and efflux in cultured rat astrocytes. *Journal of Neurochemistry*, *58*(2), 730–735. doi:10.1111/j.1471-4159.1992.tb09778.x PMID:1729413

Aschner, M., Guilarte, T. R., Schneider, J. S., & Zheng, W. (2007). Manganese: Recent advances in understanding its transport and neurotoxicity. *Toxicology and Applied Pharmacology*, *221*(2), 131–147. doi:10.1016/j.taap.2007.03.001 PMID:17466353

Balzano, T., Forteza, J., Borreda, I., Molina, P., Giner, J., & Leone, P. … Felipo, V. (2018). Histological Features of Cerebellar Neuropathology in Patients With Alcoholic and Nonalcoholic Steatohepatitis. *Journal of Neuropathology and Experimental Neurology*. doi:10.1093/jnen/nly061

Balzano, T., Forteza, J., Molina, P., Giner, J., Monzó, A., Sancho-Jiménez, J., ... Felipo, V. (2018). The Cerebellum of Patients with Steatohepatitis Shows Lymphocyte Infiltration, Microglial Activation and Loss of Purkinje and Granular Neurons. *Scientific Reports*, *8*(1), 3004. doi:10.103841598-018-21399-6 PMID:29445232

Bandmann, O., Weiss, K. H., & Kaler, S. G. (2015). Wilson's disease and other neurological copper disorders. *Lancet Neurology*, *14*(1), 103–113. doi:10.1016/S1474-4422(14)70190-5 PMID:25496901

Bull, P. C., Thomas, G. R., Rommens, J. M., Forbes, J. R., & Cox, D. W. (1993). The Wilson disease gene is a putative copper transporting P-type ATPase similar to the Menkes gene. *Nature Genetics*, *5*(4), 327–337. doi:10.1038/ng1293-327 PMID:8298639

Burgos, A., Bermejo, P. E., Calleja, J. L., Vaquero, A., & Abreu, L. E. (2009). Acquired chronic hepatocerebral degeneration due to cirrhosis from non-alcoholic steatohepatitis. *Revista Espanola De Enfermedades Digestivas: Organo Oficial De La Sociedad Espanola De Patologia Digestiva*, *101*(11), 806–811. PMID:20001159

Coffey, A. J., Durkie, M., Hague, S., McLay, K., Emmerson, J., Lo, C., ... Bandmann, O. (2013). A genetic study of Wilson's disease in the United Kingdom. *Brain*, *136*(Pt 5), 1476–1487. doi:10.1093/brain/awt035 PMID:23518715

Compston, A. (2009). Progressive lenticular degeneration: a familial nervous disease associated with cirrhosis of the liver, by S. A. Kinnier Wilson, (From the National Hospital, and the Laboratory of the National Hospital, Queen Square, London). *Brain*, *132*(Pt 8), 1997–2001. doi:10.1093/brain/awp193 PMID:19634211

Costa, L. G., & Aschner, M. (2014). *Manganese in Health and Disease*. Royal Society of Chemistry. doi:10.1039/9781782622383

Couper, J. (1837). On the effects of black oxide of manganese when inhaled into the lungs. *Br Ann Med Pharmacol, 1*, 41–42.

Davidson, E. A., & Summerskill, W. H. J. (1956). Psychiatric Aspects of Liver Disease. *Postgraduate Medical Journal, 32*(372), 487–494. doi:10.1136/pgmj.32.372.487 PMID:13389012

European Association for Study of Liver. (2012). EASL Clinical Practice Guidelines: Wilson's disease. *Journal of Hepatology, 56*(3), 671–685. doi:10.1016/j.jhep.2011.11.007 PMID:22340672

Ferenci, P., Członkowska, A., Merle, U., Ferenc, S., Gromadzka, G., Yurdaydin, C., ... Stremmel, W. (2007). Late-onset Wilson's disease. *Gastroenterology, 132*(4), 1294–1298. doi:10.1053/j.gastro.2007.02.057 PMID:17433323

Fitzgerald, K., Mikalunas, V., Rubin, H., McCarthey, R., Vanagunas, A., & Craig, R. M. (1999). Hypermanganesemia in patients receiving total parenteral nutrition. *JPEN. Journal of Parenteral and Enteral Nutrition, 23*(6), 333–336. doi:10.1177/0148607199023006333 PMID:10574481

Forbes, J. R., & Cox, D. W. (2000). Copper-dependent trafficking of Wilson disease mutant ATP7B proteins. *Human Molecular Genetics, 9*(13), 1927–1935. doi:10.1093/hmg/9.13.1927 PMID:10942420

Frydman, M., Bonné-Tamir, B., Farrer, L. A., Conneally, P. M., Magazanik, A., Ashbel, S., & Goldwitch, Z. (1985). Assignment of the gene for Wilson disease to chromosome 13: Linkage to the esterase D locus. *Proceedings of the National Academy of Sciences of the United States of America, 82*(6), 1819–1821. doi:10.1073/pnas.82.6.1819 PMID:3856863

Gallez, B., Demeure, R., Baudelet, C., Abdelouahab, N., Beghein, N., Jordan, B., ... Roels, H. A. (2001). Non invasive quantification of manganese deposits in the rat brain by local measurement of NMR proton T1 relaxation times. *Neurotoxicology, 22*(3), 387–392. doi:10.1016/S0161-813X(01)00020-1 PMID:11456339

Ginat, D. T., & Meyers, S. P. (2012). Intracranial lesions with high signal intensity on T1-weighted MR images: differential diagnosis. *Radiographics: A Review Publication of the Radiological Society of North America. Inc, 32*(2), 499–516. doi:10.1148/rg.322105761

Goytain, A., Hines, R. M., & Quamme, G. A. (2008). Huntingtin-interacting Proteins, HIP14 and HIP14L, Mediate Dual Functions, Palmitoyl Acyltransferase and Mg2+ Transport. *The Journal of Biological Chemistry, 283*(48), 33365–33374. doi:10.1074/jbc.M801469200 PMID:18794299

Guilarte, T. R. (2010). Manganese and Parkinson's Disease: A Critical Review and New Findings. *Environmental Health Perspectives, 118*(8), 1071–1080. doi:10.1289/ehp.0901748 PMID:20403794

Hahn, M., Massen, O., Nencki, M., & Pawlow, J. (1893). Die Eck'sche Fistel zwischen der unteren Hohlvene und der Pfortader und ihre Folgen für den Organismus. *Archiv für experimentelle Pathologie und Pharmakologie, 32*(3–4), 161–210. doi:10.1007/BF01995065

Hellerman, L., & Perkins, M. E. (1935). Activation of Enzymes Iii. the Role of Metal Ions in the Activation of Arginase. the Hydrolysis of Arginine Induced by Certain Metal Ions with Urease. *The Journal of Biological Chemistry, 112*(1), 175–194.

Heneka, M. T., Carson, M. J., El Khoury, J., Landreth, G. E., Brosseron, F., Feinstein, D. L., ... Kummer, M. P. (2015). Neuroinflammation in Alzheimer's disease. *Lancet Neurology*, *14*(4), 388–405. doi:10.1016/S1474-4422(15)70016-5 PMID:25792098

Huang, E., Ong, W. Y., & Connor, J. R. (2004). Distribution of divalent metal transporter-1 in the monkey basal ganglia. *Neuroscience*, *128*(3), 487–496. doi:10.1016/j.neuroscience.2004.06.055 PMID:15381278

Hung, I. H., Suzuki, M., Yamaguchi, Y., Yuan, D. S., Klausner, R. D., & Gitlin, J. D. (1997). Biochemical characterization of the Wilson disease protein and functional expression in the yeast Saccharomyces cerevisiae. *The Journal of Biological Chemistry*, *272*(34), 21461–21466. doi:10.1074/jbc.272.34.21461 PMID:9261163

Huster, D. (2010). Wilson disease. *Best Practice & Research. Clinical Gastroenterology*, *24*(5), 531–539. doi:10.1016/j.bpg.2010.07.014 PMID:20955957

Jenkitkasemwong, S., Wang, C.-Y., Mackenzie, B., & Knutson, M. D. (2012). Physiologic implications of metal-ion transport by ZIP14 and ZIP8. *Biometals : An International Journal on the Role of Metal Ions in Biology. Biochemical Medicine*, *25*(4), 643–655. doi:10.100710534-012-9526-x

Keen, C. L., & Zidenberg-Cherr, S. (2003). MANGANESE. In B. Caballero (Ed.), *Encyclopedia of Food Sciences and Nutrition* (2nd ed.; pp. 3686–3691). Oxford, UK: Academic Press; doi:10.1016/B0-12-227055-X/00732-X

Kikuchihara, Y., Abe, H., Tanaka, T., Kato, M., Wang, L., Ikarashi, Y., ... Shibutani, M. (2015). Relationship between brain accumulation of manganese and aberration of hippocampal adult neurogenesis after oral exposure to manganese chloride in mice. *Toxicology*, *331*, 24–34. doi:10.1016/j.tox.2015.02.005 PMID:25698507

La Fontaine, S., Theophilos, M. B., Firth, S. D., Gould, R., Parton, R. G., & Mercer, J. F. (2001). Effect of the toxic milk mutation (tx) on the function and intracellular localization of Wnd, the murine homologue of the Wilson copper ATPase. *Human Molecular Genetics*, *10*(4), 361–370. doi:10.1093/hmg/10.4.361 PMID:11157799

Leach, R. M. Jr, & Muenster, A. M. (1962). Studies on the role of manganese in bone formation. I. Effect upon the mucopolysaccharide content of chick bone. *The Journal of Nutrition*, *78*(1), 51–56. doi:10.1093/jn/78.1.51 PMID:14463214

Lechpammer, M., Clegg, M. S., Muzar, Z., Huebner, P. A., Jin, L.-W., & Gospe, S. M. Jr. (2014). Pathology of inherited manganese transporter deficiency. *Annals of Neurology*, *75*(4), 608–612. doi:10.1002/ana.24131 PMID:24599576

Lewis, M. B., MacQuillan, G., Bamford, J. M., & Howdle, P. D. (2000). Delayed myelopathic presentation of the acquired hepatocerebral degeneration syndrome. *Neurology*, *54*(4), 1011. doi:10.1212/WNL.54.4.1011 PMID:10691011

Liao, S. L., & Chen, C. J. (2001). Manganese stimulates stellation of cultured rat cortical astrocytes. *Neuroreport*, *12*(18), 3877–3881. doi:10.1097/00001756-200112210-00004 PMID:11742202

Lorincz, M. T. (2010). Neurologic Wilson's disease. *Annals of the New York Academy of Sciences, 1184*(1), 173–187. doi:10.1111/j.1749-6632.2009.05109.x PMID:20146697

Lu, C. S., Huang, C. C., Chu, N. S., & Calne, D. B. (1994). Levodopa failure in chronic manganism. *Neurology, 44*(9), 1600–1602. doi:10.1212/WNL.44.9.1600 PMID:7936281

Marques dos Santos, D., Aschner, M., & Marreilha dos Santos, A. P. (2017). Manganese and Neurodegeneration. In A. R. White, M. Aschner, L. G. Costa, & A. I. Bush (Eds.), *Biometals in Neurodegenerative Diseases* (pp. 117–151). Academic Press. doi:10.1016/B978-0-12-804562-6.00008-7

Møller, L. B., Ott, P., Lund, C., & Horn, N. (2005). Homozygosity for a gross partial gene deletion of the C-terminal end of ATP7B in a Wilson patient with hepatic and no neurological manifestations. *American Journal of Medical Genetics. Part A, 138*(4), 340–343. doi:10.1002/ajmg.a.30977 PMID:16222684

Morgagni, G. B. (1820). *De sedibus et causis morborum per anatomen indagatis.* Typographia Remondiniana. Retrieved from http://archive.org/details/desedibusetcaus06adelgoog

Mufti, A. R., Burstein, E., Csomos, R. A., Graf, P. C. F., Wilkinson, J. C., Dick, R. D., ... Duckett, C. S. (2006). XIAP Is a copper binding protein deregulated in Wilson's disease and other copper toxicosis disorders. *Molecular Cell, 21*(6), 775–785. doi:10.1016/j.molcel.2006.01.033 PMID:16543147

Nam, J. K. (2014). Medieval European medicine and Asian spices. *Ui Sahak, 23*(2), 319–342. doi:10.13081/kjmh.2014.23.319 PMID:25223223

Nelson, K., Golnick, J., Korn, T., & Angle, C. (1993). Manganese encephalopathy: Utility of early magnetic resonance imaging. *British Journal of Industrial Medicine, 50*(6), 510–513. PMID:8329316

Nephron. (2018). In *Wikipedia*. Retrieved from https://en.wikipedia.org/w/index.php?title=Alzheimer_type_II_astrocyte&oldid=838733504

Nolte, W., Wiltfang, J., Schindler, C. G., Unterberg, K., Finkenstaedt, M., Niedmann, P. D., ... Ramadori, G. (1998). Bright basal ganglia in T1-weighted magnetic resonance images are frequent in patients with portal vein thrombosis without liver cirrhosis and not suggestive of hepatic encephalopathy. *Journal of Hepatology, 29*(3), 443–449. doi:10.1016/S0168-8278(98)80063-9 PMID:9764992

Norenberg, M. D. (1987). The role of astrocytes in hepatic encephalopathy. *Neurochemical Pathology, 6*(1–2), 13–33. doi:10.1007/BF02833599 PMID:3306480

O'Neal, S. L., & Zheng, W. (2015). Manganese Toxicity Upon Overexposure: A Decade in Review. *Current Environmental Health Reports, 2*(3), 315–328. doi:10.100740572-015-0056-x PMID:26231508

Okinaka, S., Yoshikawa, M., Mozai, T., Ogihara, K., Watanabe, H., Terao, T., ... Hirai, S. (1962). Studies on hepatocerebral disease. IV. Portal-systemic encephalopathy in chronic schistosomiasis japonica. *Archives of Neurology, 7*(1), 1–9. doi:10.1001/archneur.1962.04210010007001 PMID:14481571

Petrukhin, K., Fischer, S. G., Pirastu, M., Tanzi, R. E., Chernov, I., Devoto, M., ... Russo, J. J. (1993). Mapping, cloning and genetic characterization of the region containing the Wilson disease gene. *Nature Genetics, 5*(4), 338–343. doi:10.1038/ng1293-338 PMID:8298640

Pfeiffer, R. F. (2007). Wilson's Disease. *Seminars in Neurology*, *27*(2), 123–132. doi:10.1055-2007-971173 PMID:17390257

Philips, T., & Robberecht, W. (2011). Neuroinflammation in amyotrophic lateral sclerosis: Role of glial activation in motor neuron disease. *Lancet Neurology*, *10*(3), 253–263. doi:10.1016/S1474-4422(11)70015-1 PMID:21349440

Pomier-Layrargues, G., Spahr, L., & Butterworth, R. F. (1995). Increased manganese concentrations in pallidum of cirrhotic patients. *Lancet (London, England)*, *345*(8951), 735. doi:10.1016/S0140-6736(95)90909-5 PMID:7885158

Ransohoff, R. M. (2016). How neuroinflammation contributes to neurodegeneration. *Science*, *353*(6301), 777–783. doi:10.1126cience.aag2590 PMID:27540165

Roberts, E. A., & Schilsky, M. L. (2008). Diagnosis and treatment of Wilson disease: An update. *Hepatology (Baltimore, Md.)*, *47*(6), 2089–2111. doi:10.1002/hep.22261 PMID:18506894

Rodier, J. (1955). Manganese poisoning in Moroccan miners. *British Journal of Industrial Medicine*, *12*(1), 21–35. PMID:14351643

Santos, D., Batoreu, C., Mateus, L., dos Santos, A. M., & Aschner, M. (2014). Manganese in Human Parenteral Nutrition: Considerations for Toxicity and Biomonitoring. *Neurotoxicology*, *43*, 36–45. doi:10.1016/j.neuro.2013.10.003 PMID:24184781

Sarkar, S., Malovic, E., Harischandra, D. S., Ngwa, H. A., Ghosh, A., Hogan, C., ... Kanthasamy, A. (2018). Manganese exposure induces neuroinflammation by impairing mitochondrial dynamics in astrocytes. *Neurotoxicology*, *64*, 204–218. doi:10.1016/j.neuro.2017.05.009 PMID:28539244

Sengupta, A., Mense, S. M., Lan, C., Zhou, M., Mauro, R. E., Kellerman, L., ... Zhang, L. (2007). Gene Expression Profiling of Human Primary Astrocytes Exposed to Manganese Chloride Indicates Selective Effects on Several Functions of the Cells. *Neurotoxicology*, *28*(3), 478–489. doi:10.1016/j.neuro.2006.10.005 PMID:17175027

Shin, H.-W., & Park, H. K. (2017). Recent Updates on Acquired Hepatocerebral Degeneration. *Tremor and Other Hyperkinetic Movements (New York, N.Y.)*, *7*, 463. doi:10.7916/D8TB1K44 PMID:28975044

Shin, Y. C., Kim, E., Cheong, H.-K., Cho, S., Sakong, J., Kim, K. S., ... Kim, Y. (2007). High signal intensity on magnetic resonance imaging as a predictor of neurobehavioral performance of workers exposed to manganese. *Neurotoxicology*, *28*(2), 257–262. doi:10.1016/j.neuro.2006.03.014 PMID:16647136

Shinotoh, H., Snow, B. J., Hewitt, K. A., Pate, B. D., Doudet, D., Nugent, R., ... Calne, D. B. (1995). MRI and PET studies of manganese-intoxicated monkeys. *Neurology*, *45*(6), 1199–1204. doi:10.1212/WNL.45.6.1199 PMID:7783889

Sidoryk-Wegrzynowicz, M., & Aschner, M. (2013). Role of astrocytes in manganese mediated neurotoxicity. *BMC Pharmacology & Toxicology*, *14*(1), 23. doi:10.1186/2050-6511-14-23 PMID:23594835

Spencer, D. C., & Forno, L. S. (2000). February 2000: Dementia with motor dysfunction in a patient with liver disease. *Brain Pathology (Zurich, Switzerland)*, *10*(2), 315–316, 319. doi:10.1111/j.1750-3639.2000.tb00205.x PMID:10764052

Stallings, W. C., Metzger, A. L., Pattridge, K. A., Fee, J. A., & Ludwig, M. L. (1991). Structure-function relationships in iron and manganese superoxide dismutases. *Free Radical Research Communications*, *12-13*(Pt 1), 259–268. doi:10.3109/10715769109145794 PMID:2071033

Starzl, T. E., Porter, K. A., Francavilla, A., & McDermott, W. V. Jr. (1983). THE ECK FISTULA IN ANIMALS AND HUMANS. *Current Problems in Surgery*, *20*(11), 687–752. doi:10.1016/S0011-3840(83)80010-0 PMID:6357642

Summerskill, W. H., Davidson, E. A., Sherlock, S., & Steiner, R. E. (1956). The neuropsychiatric syndrome associated with hepatic cirrhosis and an extensive portal collateral circulation. *The Quarterly Journal of Medicine*, *25*(98), 245–266. PMID:13323252

Tan, J., Zhang, T., Jiang, L., Chi, J., Hu, D., Pan, Q., … Zhang, Z. (2011). Regulation of intracellular manganese hemeostasis by kufor-rakeb syndrome associated ATP13A2. *Journal of Biological Chemistry*. doi:10.1074/jbc.M111.233874

Tansey, M. G., & Goldberg, M. S. (2010). Neuroinflammation in Parkinson's disease: Its role in neuronal death and implications for therapeutic intervention. *Neurobiology of Disease*, *37*(3), 510–518. doi:10.1016/j.nbd.2009.11.004 PMID:19913097

Tanzi, R. E., Petrukhin, K., Chernov, I., Pellequer, J. L., Wasco, W., Ross, B., ... Brzustowicz, L. M. (1993). The Wilson disease gene is a copper transporting ATPase with homology to the Menkes disease gene. *Nature Genetics*, *5*(4), 344–350. doi:10.1038/ng1293-344 PMID:8298641

Victor, M., Adams, R. D., & Cole, M. (1965). The acquired (non-Wilsonian) type of chronic hepatocerebral degeneration. *Medicine*, *44*(5), 345–396. doi:10.1097/00005792-196509000-00001 PMID:5318075

Wedler, F. C., Denman, R. B., & Roby, W. G. (1982). Glutamine synthetase from ovine brain is a manganese(II) enzyme. *Biochemistry*, *21*(25), 6389–6396. doi:10.1021/bi00268a011 PMID:6129892

Weiss, K. H. (1993). Wilson Disease. In M. P. Adam, H. H. Ardinger, R. A. Pagon, S. E. Wallace, L. J. Bean, K. Stephens, & A. Amemiya (Eds.), *GeneReviews®*. Seattle, WA: University of Washington. Retrieved from http://www.ncbi.nlm.nih.gov/books/NBK1512/

Williams, M., Todd, G. D., Roney, N., Crawford, J., Coles, C., McClure, P. R., … Citra, M. (2012). *Relevance to public health*. Agency for Toxic Substances and Disease Registry (US). Retrieved from https://www.ncbi.nlm.nih.gov/books/NBK158871/

Wilson's disease (a volume in the major problems in internal medicine series) By I. H. Scheinberg and I. Sternlieb Philadelphia, W. B. Saunders, 1984 192 pp, illustrated, $35.00. (1984). *Annals of Neurology*, *16*(5), 626–626. doi:10.1002/ana.410160531

Wolff, N. A., Garrick, M. D., Zhao, L., Garrick, L. M., Ghio, A. J., & Thévenod, F. (2018). A role for divalent metal transporter (DMT1) in mitochondrial uptake of iron and manganese. *Scientific Reports*, *8*(1), 211. doi:10.103841598-017-18584-4 PMID:29317744

Wu, F., Wang, J., Pu, C., Qiao, L., & Jiang, C. (2015). Wilson's disease: A comprehensive review of the molecular mechanisms. *International Journal of Molecular Sciences*, *16*(3), 6419–6431. doi:10.3390/ijms16036419 PMID:25803104

Yamaguchi, Y., Heiny, M. E., & Gitlin, J. D. (1993). Isolation and characterization of a human liver cDNA as a candidate gene for Wilson disease. *Biochemical and Biophysical Research Communications*, *197*(1), 271–277. doi:10.1006/bbrc.1993.2471 PMID:8250934

KEY TERMS AND DEFINITIONS

Ataxia: Loss of muscle control and/or voluntary motor coordination.

Basal Ganglia: Group of structures (including caudate, putamen, and globus pallidus) mainly involved in voluntary and involuntary motor control.

Cerebellum: Located at the base of the brain, is the area involved in coordination, posture, and balance.

Dysarthria: Speech disorder caused by altered tone and/or incoordination of the muscles used in speech.

Dystonia: State of abnormal muscle tone resulting in muscular spasm, contractions, and abnormal posture.

Neurodegeneration: Progressive loss of the number and function of neurons.

Neuroinflammation: Inflammatory response within the central nervous system.

APPENDIX

Figure 1. Example of symmetrical high signal intensity in basal ganglia on T1-weighted MR images related with an accumulation of manganese. Image obtained by (Ginat & Meyers, 2012)

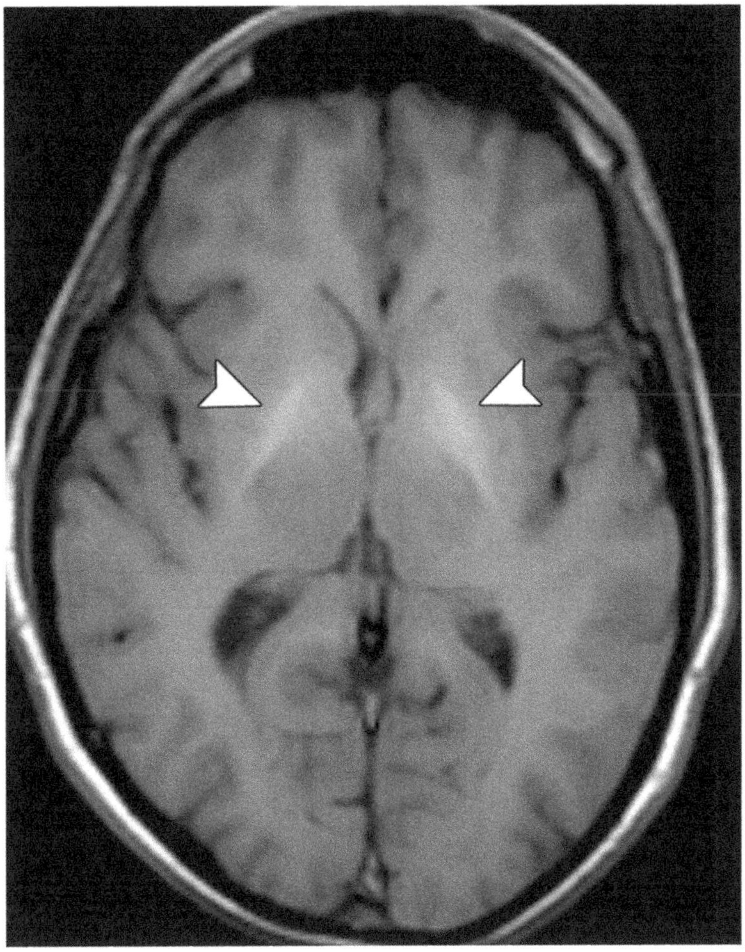

Figure 2. Two astrocytes exhibiting enlarged, pale nuclei with margination of chromatin and prominent nucleoli characteristic of Alzheimer type II astrocytes (arrow) are shown. Hematoxylin and eosin. Image obtained by (Nephron, 2018)

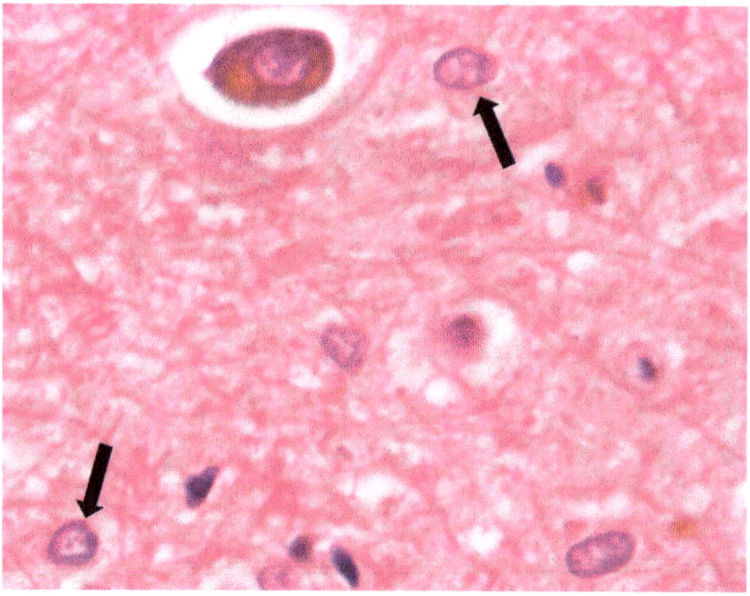

Figure 3. Normal Cu metabolism and pathomechanism of Cu-related disorders. Modified according to (Bandmann et al., 2015)

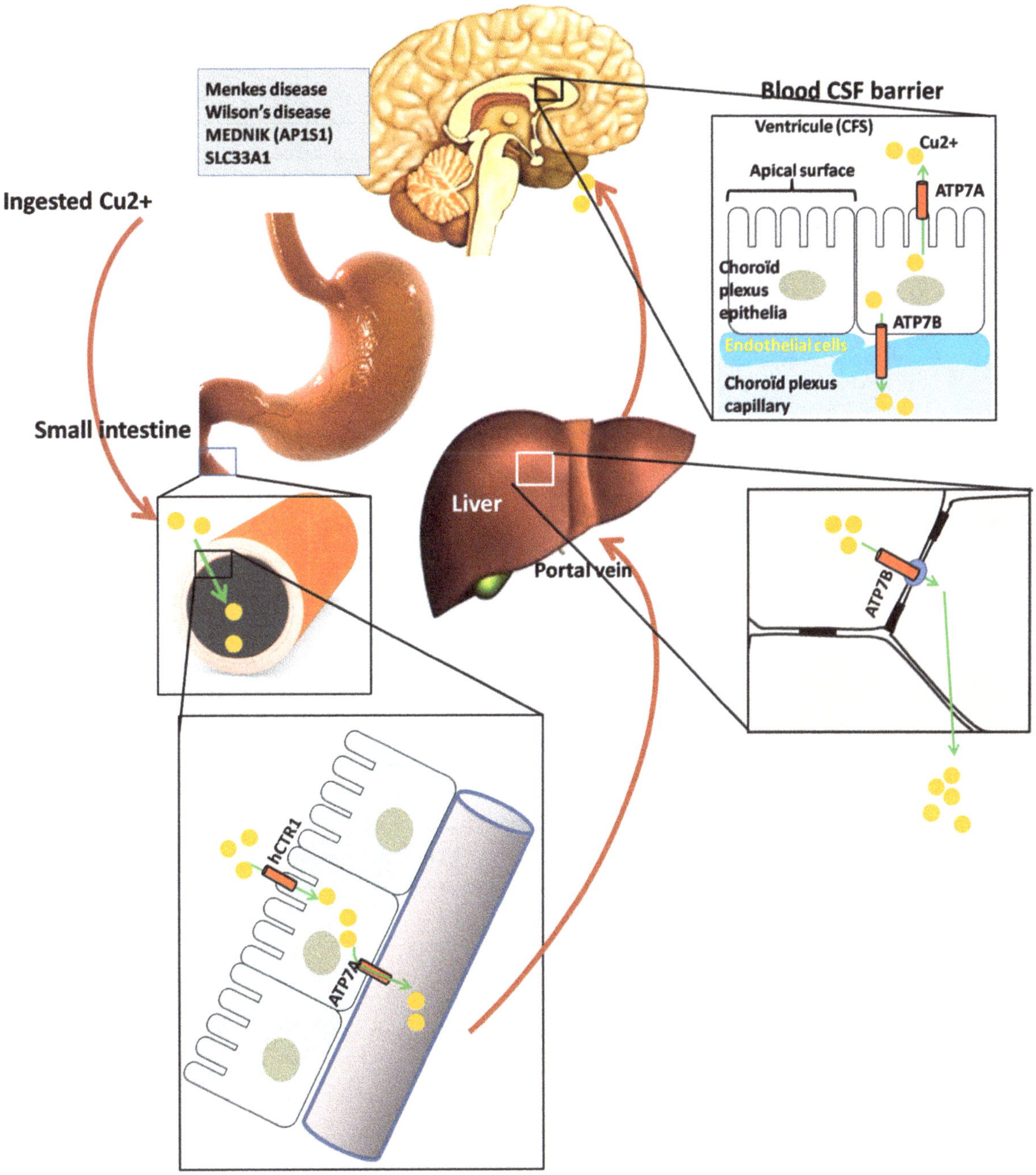

Table 1. Mean concentrations of Manganese (milligrams per 100 grams of food) in some types of typical foods. The values shown in the table were obtained from the Food Composition Databases from United States Department of Agriculture (USDA). NDB_No = Nutrient Database number. ('USDA Food Composition Databases', n.d.)

NDB_No	Description	Manganese, Mn(mg) Value Per
14366	Beverages, tea, instant, unsweetened, powder	133,0
	Spices, ginger, ground	
02021	Spices, turmeric, ground	33,30
02043	Spices, cinnamon, ground	19,80
02010	Spices, curry powder	17,47
02015	Spices, bay leaf	8,30
02004	Spices, tarragon, dried	8,17
02041	Spices, thyme, dried	7,97
02042	Nuts, butternuts, dried	7,87
12084	Snacks, rice cakes, brown rice, buckwheat	6,56
19052	Nuts, hazelnuts or filberts	6,18
12120	Spices, savory, ground	6,17
02039	Spices, marjoram, drie	6,10
02023	Fish, trout, mixed species, cooked, dry heat	5,43
15219	Chicken, broilers or fryers, leg, meat only, cooked, fried	1,09
05081	Beef, round, top round, steak, separable lean and fat, trimmed to 1/8" fat,	0,03
13902	prime,	0,02
	cooked, broiled	

Table 2. Different types of ATP7B mutations commonly seen in different world populations. Modified from (Weiss, 1993)

	DNA Nucleotide Change	Protein Aminoacide Change	Exon	Frequency	Other Common Mutations
East Asia	2333G→T	Arg778Leu	8	30–50%	2871delC
Europe	207C→A	His1069Gln	14	35–45%	2299insG 1934T→G
India	813C→A	Cys271Stop	2	~20%	3305T→C 2975C→T
Middle East	4196A→G	Gln1399Arg	21	~30%	

Table 3. Clinical manifestations of WD

Type of Clinical Features	Symptoms
Hepatic	Asymptomatic elevation aminotransferases Acute hepatitis (jaundice, abdomen pain) Acute liver failure (coagulopathy, jaundice, encephalopathy) Liver cirrhosis symptoms (compensated or decompensated) (fatigue, spider naevi, portal hypertension, splenomegaly, bleeding)
Neurologic	Movement disorders (tremor, dystonia, ataxia, ballism, chorea, parkinsonian syndrome) Speech disturbances: dysarthria (extrapyramidal, dystonic, cerebellar, mixed, unclassified) Dysphagia Autonomic dysfunction (eg, salivation, electrocardiographic abnormalities, orthostatic hypotension) Gait and balance disturbances (due to involuntary movements, cerebellar ataxia, impairment of postural reflexes
Psychiatric	Cognitive impairment (neurodegeneration due to brain copper accumulation, hepatic encephalopathy) Personality disorders (abnormal, antisocial behavior, irritability, disinhibition) Mood disorders (bipolar disorders, depression, suicidal attempts) Psychosis and other psychiatric alterations (, rarely psychosis, anorexia, sleep disturbances)
Ophthalmologic	Kayser-Fleischer ring; sunflower cataract
Other	Renal (nephrolithiasis, aminoaciduria) Bone (osteoporosis, joint pain) Heart (cardiac arrhythmia) Skin (hyperpigmentation of lower legs) Hematopoietic system (thrombocytopenia, hemolytic anemia, leukopenia)

Chapter 12
Environmental Aspects of Alzheimer's and Parkinson's Diseases Neuropathologies:
A Focus on Heavy Metals and Pesticides

Nadia Zouhairi
Cadi Ayyad University, Morocco

Omar El Hiba
Chouaib Doukkali University, Morocco & Cadi Ayyad University, Morocco

Hasna Lahouaoui
Cadi Ayyad University, Morocco

Hind Benammi
Cadi Ayyad University, Morocco

Hicham Chatoui
Private University of Marrakech, Morocco & Cadi Ayyad University, Morocco

Abdeljalil El Got
Hassan 1st University, Morocco

Abdelmohcine Aimrane
Cadi Ayyad University, Morocco

Abdelali Bitar
Chouaib Doukkali University, Morocco

Kholoud Kahime
Cadi Ayyad University, Morocco

Ahmed Draoui
Cadi Ayyad University, Morocco

Ouassil El Kherchi
Hassan 1st University, Morocco

Wafa Ait Hmyed
Cadi Ayyad University, Morocco

ABSTRACT

This chapter presents a literature review on the effect of environmental changes factors exposure in the etiology of Alzheimer's and Parkinson's diseases. The use of pesticides is more intense and somehow erratic as it aims to face climate change consequences like drought and water scarcity. The rural population is getting to be more vulnerable to have these neurodegenerative diseases. However, intense food production and economic models mean also the use of heavy metals in many stages as well during the production and the consumption processes and practices. Evidence from experimental studies shows that such heavy metals may also be a factor for the occurrence of Parkinson's and Alzheimer's diseases. At least, the environmental lifestyle and, likely, genetic factors, individually and collectively, play a significant role in the etiology of the diseases.

DOI: 10.4018/978-1-5225-7775-1.ch012

INTRODUCTION

In human history, adaptation to environmental challenges has been always a key criterion for development and survival in challenging periods. This adaptation implies natural and cognitive responses to the environmental challenges through science and technology improving human health and survival rates.

Since the industrial age, the quality of life as a concept has gained much importance in people's life. In the last decades, health has been considered as one of the major standards to measure human well-being and life quality by the major international rating organizations. Therefore, human health is becoming a priority for the policymakers in the majority of national, regional and international organizations. Climate change is an example where this understanding is becoming urgent. Humans are obliged as never before to adapt to their changing environment, due to the raising concerns and scientific evidence of health issues caused directly or indirectly by climate change.

Although the human body has the capability to avoid many of the climate change negative effects, the costs remain obviously hazardous spreading worldwide to a great extent. For that reason, and from an ethical concern, it is crucial to mobilize the efforts in order to spare the risk of affection from future generations. This awareness may furthermore be fundamental for short-term protection and long-term alleviation of health consequences.

Food production is usually correlated with pesticides and synthetic fertilizers use. These products to increase crops yields to meet the population needs in these countries are usually a response to climate change consequences such as the lack of rainfall and drought. In such situations, farmers are also using sewage, which is harmful to population health.

Many countries in the world such the United States have seen an increased prevalence of neurological diseases and deficits (Steenland, MacNeil, Vega, &Levey, 2009). The onsets of diseases like Alzheimer (AD) and Parkinson (PD) disease is occurring at earlier ages across the population. Environmental factors are supposed to be responsible for both the onset and severity of these diseases. However, there is a gap in the understanding of this role, especially in relation to genetics, aging, and other factors. (Bronstein et al., 2009). Despite the fact that changes in neurological health are likely due to the aging of a large portion of the population, learning disabilities that affect children also are on the rise. There is also evidence that environmental factors may be involved including changes in climate that may exacerbate factors affecting the rates and severity (Altevogt, Hanson, & Leshner, 2008)of neurological conditions. Neurological conditions related to AD and PD generally carry high costs in terms of quality of life for both the patient and the caregiver and increases healthcare stresses on the financial charges and the workforce. The presence of these factors may affect an extended portion of a population and have significant impacts on productivity.

PARKINSON AND ALZHEIMER DISEASES AS NEURODEGENERATIVE DISORDERS: ETIOLOGY AND HISTORY

Health care improvements have contributed to people living longer leading to an increasing life expectancy in the last decades. However, the cases of AD and PD increased as well. Parkinson's diseases are among the most common neurodegenerative disorder after AD, and overall incidence rates for all age groups have been reported to range from 1.5 to 22 per 100,000 person-years. Other reports estimate the prevalence range of PD from 167 to 5703 per 100,000 with preference in men (Wirdefeldt, Adami, Cole,

Trichopoulos, & Mandel, 2011)and these diseases are increasingly raising big international concerns in last decades due to their high incidence. Etiology of AD and PD still little understood and genetic and environmental factors may contribute to the onset and/or development of these illnesses, which highlights the need to broaden the research scope by identifying the environmental risk factors that predispose to the development of AD and PD. It is known that the etiology of neurodegenerative diseases is multifactorial, and there is evidence that external factors like lifestyle and chemical exposures are linked with the risks of onset of these diseases (Campdelacreu, 2014).The vast majority of AD and PD cases are observed among elderly persons. As the exposure to risk factors occur many years before the diagnosis, the assessment of these chronic exposures are difficult to perform in retrospective studies to associate them with the onset/development of AD and PD.

Pre-and post-natal exposures to environmental factors are also proved to predispose to the onset of AD and PD in later life. Neurotoxic, pesticides and metal-based nanoparticles have been involved in AD by increasing beta-amyloid (Ab) peptide and the phosphorylation of Tau protein (P-Tau), causing senile/amyloid plaques and neurofibrillary tangles (NFTs) specific of AD. Lead, manganese, solvents and some pesticides have been linked to some forms PD like the mitochondrial dysfunction, alterations in metal homeostasis and aggregation of proteins such as a-synuclein (a-syn), which is the main constituent of Lewy bodies (LB), a crucial factor in PD pathogenesis. There is evidence that environmental pollutants may increase Ab, P-Tau, a-syn and neuronal death have been reported, including the oxidative stress mainly involved in the increase of Ab and a-syn, and the reduced activity/protein levels of Ab degrading enzyme (IDE) such as neprilysin or insulin IDE. Furthermore, epigenetic mechanisms by maternal nutrient supplementation and exposure to heavy metals and pesticides have been addressed to cause phenotypic diversity and susceptibility to neurodegenerative diseases in the offspring.

AD and PD are correlated with the amyloid fibrils formation in the brain from b-amyloid and a-synuclein proteins. Oligomericfibrillization may intermediates (protofibrils), rather than the fibrils themselves, are pathogenic. However, the mechanisms responsible for neuronal death remain unknown. Mutant amyloid proteins linked to familial AD and PD form morphologically indistinguishable annular protofibrils similar to a class of pore-forming bacterial toxins. This suggests that abnormal membrane permeabilization might be the cause of cell disorder and death in amyloid diseases.

Since the 1980s, interest in the relationship between pesticides and PD was increased after the discovery of the dangerous effect of exposure to1-methyl-4-phenyl-1, 2, 3, 6,-tetrahydropyridine (MPTP), a substance structurally similar to the herbicide paraquat to humans, as it resulted in chronic Parkinsonism and degeneration of dopaminergic neurons in humans (Langston & Ballard, 1983). There is more and more evidence of a biological plausibility about an association between pesticide exposure and PD. In addition to MPTP and paraquat, animal studies have revealed that pesticides with related properties such as rotenone, maneb, dieldrin, heptachlor and atrazine are responsible fora-synuclein accumulation and to dopaminergic cell degeneration and apoptosis (Betarbet et al., 2000; Caudle, Richardson, Wang, & Miller, 2005; Cicchetti et al., 2005; Costa, Giordano, Guizzetti, & Vitalone, 2008; Dinis-Oliveira et al., 2006; Filipov, Stewart, Carr, & Sistrunk, 2007; Hatcher, Pennell, & Miller, 2008; Kanthasamy, Kitazawa, Kanthasamy, & Anantharam, 2005; Meredith, Halliday, & Totterdell, 2004; Moretto & Colosio, 2011). Exposure to a combination of agents affecting dopaminergic systems at multiple endpoints was also pointed as responsible for neuropathological changes beyond those caused by each individual agent (Thiruchelvam, Richfield, Goodman, Baggs, & Cory-Slechta, 2002).

Proposed mechanisms to highlight the role of pesticides in PD concern c oxidative stress, interference with dopamine transporters, mitochondrial dysfunction, promotion of a-synuclein fibrillation, and

Figure 1. Comparison of PD prevalence in different populations (for review see Kasten, Chade, & Tanner, 2007)

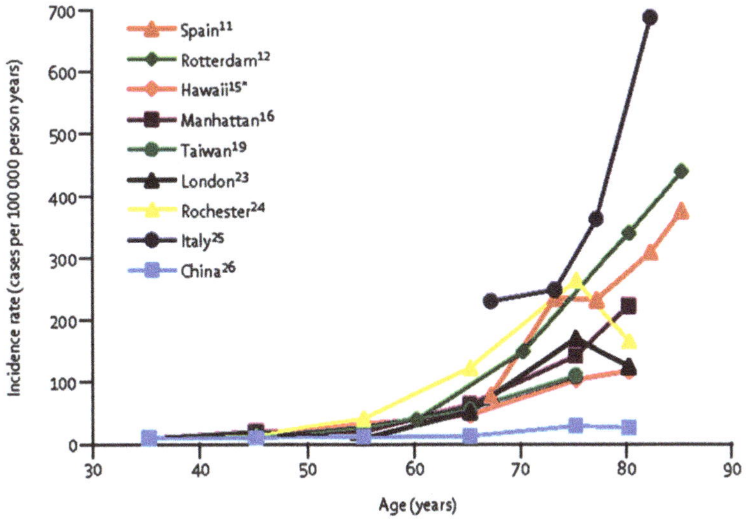

inflammation (Brown, Rumsby, Capleton, Rushton, & Levy, 2006). Furthermore, Berry, La Vecchia, & Nicotera, (2010)suggested the involvement of selected mechanisms in neuronal activities at different endpoints, leading synaptic, dendritic, and axonal damage. However, little toxicological data exists for many specific pesticides (Brown et al., 2006; Costa et al., 2008).

Familial susceptibility to PD was suggested to be mediated by genetic variability in enzymes involved in the detoxification of neurotoxins (Barbeau et al., 1985). Therefore polymorphic variations in xenobiotic metabolizing enzymes, such as glutathione transferase and paraoxonase, are supposed to influence PD occurrence risk by affecting the antioxidative defensive potential against pesticides and other putative neurotoxins (Franco, Li, Rodriguez-Rocha, Burns, & Panayiotidis, 2010; Lopez et al., 2007; Manthripragada, Costello, Cockburn, Bronstein, & Ritz, 2010).In the regard, the present chapter will discuss the possible involvement of environmental contamination with heavy metals and pesticides, in the pathogenesis of Parkinson's and Alzheimer diseases.

Clinical Features and Epidemiology of Parkinson's Disease

PD is considered as the common movement disorder and represents the second most common degenerative disease of the central nervous system. Generally it is admitted that the prevalence of PD is generally ranged from 1 to 2 / 1000 in unselected populations with high variability worldwide depending on the ethnicity and socio-economical status (Figure 1) (von Campenhausen et al., 2005), as well, the disease may affect 1% of the population older than -(de Lau & Breteler, 2006). While it is rare before the age of 50 and may reach a prevalence of 4% in the highest age groups (de Rijk et al., 1995, 2000).

Pathologically, PD is known by depletion of dopaminergic DA cell bodies in the substantianigra pars compacta with subsequent loss of DA in the nigrostriatal system (Barlow, Coryslechta, Richfield, & Thiruchelvam, 2007).The cardinal features of the syndrome are related to motor dysfunction including resting tremor, rigidity, akinesia (or bradykinesia), and postural instability. The motor symptoms appear when at least 60% of dopaminergic neurons are lost and approximately 80–85% of dopamine content in the striatum is depleted (Figure 2) (Jankovic, 2008; Wirdefeldt et al., 2011). PD pathology is not restricted to the DA system, however progressively involving noradrenergic and serotoninergic neurons within the locus ceruleus and Raphe nucleus. Degeneration in these and other structures induces non-motor symptoms but associated with the disease such as (including) autonomic sexual and sleepy disease which may lead to cognitive, psychiatric, autonomic and sensory disturbances (DeMaagd & Philip, 2015; Emre et al., 2004; Lim, Fox, & Lang, 2009). Neuronal death mechanisms in PD pathogenesis may include defective protein handling, mitochondrial dysfunction, oxidative stress, and inflammation. Furthermore, environmental factors might trigger a cascade of dopaminergic nigral neuron degeneration in genetically susceptible individuals (Hatcher et al., 2008; Manning-BoĞ & Langston, 2007). In addition, pesticides, and factors related to pesticide exposure have been pointed as potential risk factors for PD (Kontakos & Stokes, 1999; Li et al., 2005; Anumeet Priyadarshi, Khuder, Schaub, & Priyadarshi, 2001).As an environmental cause, pesticide exposure is said to be linked to over 90% of PD is (Hatcher et al., 2008).

Figure 2. 18F-dopa PET scans of basal ganglia showing innervations of the striatum in different stages of PD: mild (early), moderate and severe, compared to healthy control, The color scale represents the maximum of 18F-dopa uptake in red and minimum uptake in dark blue (see Lang & Obeso, 2004)

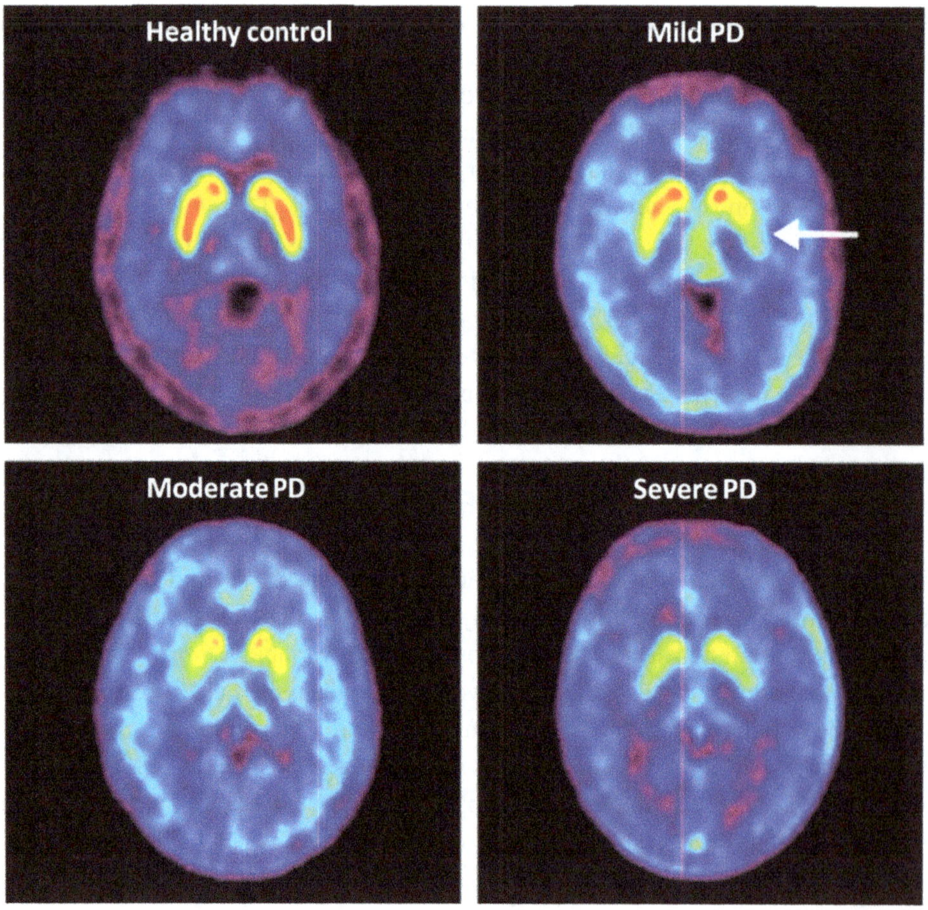

Physiopathology of Parkinson Disease: Involvement of Environmental Toxins

Pesticides Exposure

The etiology of PD is not fully understood. It is possible to involve both environmental and genetic factors (Schapira, 2009; Warner & Schapira, 2003). Several genes involved in PD pathogenesis have been identified. Alpha-synuclein and parkin are implicated in familial PD, autosomal dominant and recessive PD, respectively. In addition to other genes, including Omi/ HtrA2, PINK-1, DJ-1 and dardarin, their function has been the subject of intense studies unveiling nested gene pathways. Nevertheless, as 90% of PD cases are sporadic, genetic factors can't be the unique cause, which suggests that PD have a multifactorial etiology(Goldman, 2014). Findings from Epidemiological research suggests that the exposure to environmental toxicants, mainly pesticides, metals, and solvents may increase the risk of developing PD (Gao & Hong, 2011; Hatcher et al., 2008). A study of male twins has confirmed that genetic heritability is not a basis of sporadic PD with onset over the age of 50 (Tanner, Ottman, Goldman, Ellenberg, Chan, Mayeux, & Langston, 1999).Although it has been suggested that PD may be caused by a single toxic environmental exposure (Langston, Ballard, Tetrud, & Irwin, 1983),findings have suggested that the main PD cases may result from interactions between genetic and environmental exposures to certain toxins (Greenamyre, Betarbet, & Sherer, 2003);these findings led to the hypothesis that PD may be initiated or precipitated by environmental or endogenous toxins in genetically-predisposed individuals (Corrigan, Wienburg, Shore, Daniel, & Mann, 2000; Matsubara et al., 1995). Many epidemiological studies related the increased risk for developing PD to pesticides exposure including paraquat and rotenone and other environmental factors (table 1) (i.e. farming, well-water drinking, and rural living) (Brown et al., 2006; Dick, 2006; Le Couteur, McLean, Taylor, Woodham, & Board, 1999; Li et al., 2005; Priyadarshi, Khuder, Schaub, & Shrivastava, 2000).

The hypothesis that pesticide exposures may be related to PD development was prompted by the discovery that intravenous injection of the pre-toxin 1-methyl-4pheny 1-1, 2, 3, 6-tetrahydropyridine (MPTP). This substance was proven as a chemical contaminant of synthetic heroin that causes an acute parkinsonian syndrome in humans and in mice similar to idiopathic PD (Langston et al., 1983; Duvoisin, 1986; Freyaldenhoven, Ali, & Hart, 1995).MPTP was also proven to selectively damaged dopaminergic neurons in the substantianigra (Langston, Forno, Rebert, & Irwin, 1984; Langston & Ballard 1983). Once in the brain, MPTP is transformed to 1-methyl-4-phenyl-4-phenylpyridinium ion (MPP+) by monoamine oxidase B (Fahn, 1988). MPP+ is then stocked into the dopaminergic neurons by the high-affinity DA uptake system (Frederickson, 1989). Biochemically, MPTP administration causes damage to the nigrostriatal system, with subsequent DA depletion similar to that seen in PD (Hallman, Lange, Olson, Strömberg, & Jonsson, 1985),and therefore, it is possible that compounds similar to MPTP may cause PD. Since then, environmental factors with similar toxicological profiles have received attention as potential risk factors for PD.

Paraquat (1,1'-dimethyl-4, 4'-bipyridinium dichloride) is a highly toxic quaternary nitrogen largely used herbicide. Because of its low cost, rapid action, it is widely used in agricultural and other settings. Due to its structural resemblance with MPP (the active metabolite of MPTP), it was thought to be toxic to dopaminergic neurons and thus might be related to PD.

Paraquat toxicity is due to its ability to redox cycle, accepting an electron from an appropriate donor with subsequent reduction in oxygen to produce superoxide anion while also regenerating the parent compound (Bus & Gibson, 1984).

Table 1. Literature summary of epidemiological studies of PD related to pesticides exposure (Rotenone and Paraquat)

Study	Design	Method of assessing pesticide use	No. enrolled (cases, controls)	Finding	OR (95% CI) or p-value	Exposed *n*
			Paraquat			
Sanchez-Ramos et al. 1987	Case report	Medicalhistory	1 case	Symptoms "comparable to PD" in a 32-yearold farmer after 15 years of paraquat use	NA	1 case
Hertzman et al. 1990	Case–control	Specific question re: paraquat use	57 cases 122 controls	Association with PD	p = 0.0* [a]	4 cases 0 controls
Semchuk et al. 1992	Case–control	General questions re: pesticide use	130 cases 260 cases	One case with early-onset PD (< age 40 years) reported using paraquat (ages 26–31 years)	NA	1 case
Hertzman et al. 1994	Case–control	Specific question re: paraquat use	127 cases 245 controls	No association with PD	1.11 (0.32–3.87)	6 cases 5 controls
Firestone et al. 2010	Case–control	Checklist [c]	404 incident cases 526 controls	Added subjects to 2005 interim population No association with PD in reanalysis	0.9 (0.14–5.43)	2 cases 3controls
			Rotenone			
Kamel et al. 2007	Case–control	Specific question re: rotenone use in a supplementary questionnaire	83 prevalent cases 79,557 controls 78 incident cases 55,931 controls	Association of PD with ever use Could not determine	1.7 (0.6–4.7)	4 prevalent 671 controls 1 incident 565 controls
Tanner et al. 2009	Case–control	Specific question re: occupational rotenone use	519 cases 511 controls	No association with PD	0.82 (0.05–13.34)	1 case 1 control

In vivo research has shown that repeated administration of paraquat in mice or rats produces weak or moderate decreases in the number of dopaminergic neurons of the substantianigra, striatal levels of dopamine and its metabolites, expression and activity of tyrosine hydroxylase, and immunoreactivity of the dopamine transporter (McCormack et al., 2002; Thiruchelvam, Richfield, Baggs, Tank, & Cory-Slechta, 2000).These results suggest that chronic poisoning of rodent with paraquat may increasingly set off cellular mechanism similar to those operating in PD. There is evidence that paraquat causes selective degeneration of dopaminergic neurons in the substantianigra pars compacta (SNpc), reproducing a characteristic feature of PD (Fei, McCormack, Di Monte, & Ethell, 2008). In addition, regions of high paraquat use overlap with regions of high PD incidence (Lanska, 1997; Ritz & Yu, 2000; Shimizu et al., 2003).Because of its use in agriculture, paraquat exposure may increase the risk of developing PD in the human population. Liou, Tsai, Chen, Jeng, Chang, Chen, & Chen, (1997)reported the possible link between paraquat and PD observed in PD patients in Taiwan, in which the pesticide use was associated with an increased risk of developing PD, being higher for those individuals using paraquat(Liou et al., 1997). Also, Tanner, Kamel, Ross, Hoppin, Goldman, Korell, M., … Langston, (2011) reported acorrelation between PD and the use of oxidative pesticides, including paraquat (Tanner et al., 2011). Other epidemiological studies have also linked the exposure to paraquat with PD (Ascherio et al., 2006; Hertzman, Wiens, Bowering, Snow, & Calne, 1990; Kamel et al., 2006; Wang et al., 2011).

Paraquat is conducted to dopaminergic terminals by the dopamine transport and organic cation transporter (Rappold et al., 2011), and causes cellular toxicity (figure 3) by oxidative stress through the cellular redox cycling generating superoxide radical by the oxidation of NADPH, which in turn impairs the restoration of GSH levels and thus the activity of several antioxidant systems (Berry et al., 2010; Franco et al., 2010). Observations from repeated administrations of paraquat to adult mice and rats (5–10 mg/Kg/ week/at least 3 weeks, i.p.) increases ROS levels in the striatal homogenate, causes a dose-dependent decrease in dopaminergic neurons from the substantianigra, and decreases thestriatal dopamine nerve terminal density, and neurobehavioral syndrome characterized by reduced ambulatory activity (Brooks, Chadwick, Gelbard, Cory-Slechta, & Federoff, 1999; Kuter, Nowak, Gołembiowska,& Ossowska, 2010; McCormack et al., 2002). Paraquat belongs to a class of redox cycling compounds capable of inducing mitochondrial damage, increases reactive oxygen species(ROS) production and oxidative stress (Castello, Drechsel, & Patel, 2007; Jenner, 2003)which is considered to be a key indicator (foundation) of neurotoxicity. Experiments indicate that paraquat targets mitochondria of brain cells and the damage to the latter cells is the main source of ROS. In-vitro assays paraquat have been proved to disturb the mitochondria isolated from mouse brains and from differentiated human neuroblastoma cells (McCarthy, Somayajulu, Sikorska, Borowy-Borowski, & Pandey, 2004).Also, Fukushima, Yamada, Isobe, Shiwaku, & Yamane,(1993)showed that paraquat may generate ROS by accepting electrons from purified Complex I of the mitochondrial respiratory chain (Fukushima, et al., 1993), with the subse-

Figure 3. Photomicrographs of frontal sections through midbrain substantianigracompacta (SNpc) stained with anti-Tyrosine hydroxylase in saline control and Paraquat treated rat at 12 months of age with statistical analysis of TH immunoreactivity. Note a drastic loss of TH-immunopositive neurons in Paraquat treated group (modified according to Peng, Peng, Stevenson, Doctrow, & Andersen, 2007)

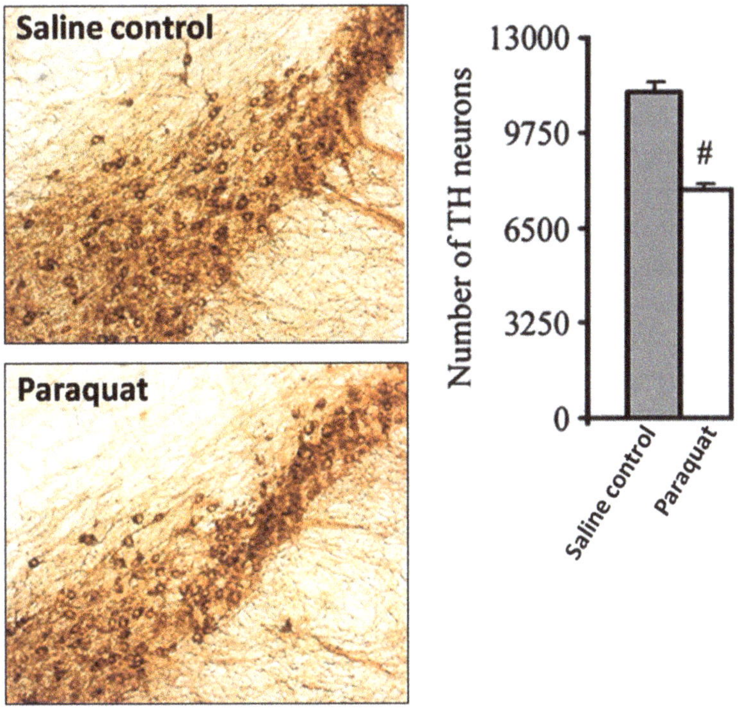

quent inhibition of this mitochondrial complex; suggesting the importance of the interaction of paraquat with mitochondria in its toxicity. Many neurodegenerative diseases have been linked to mitochondrial dysfunction (Lin & Beal, 2006; Schon & Manfredi, 2003). Abnormal mitochondrial electron transfer mechanism has been detected in Alzheimer, Parkinson, and Huntington disease (Franco et al., 2010; Hanagasi, Ayribas, Baysal, &Emre, 2005).

Heavy Metals Exposure

An increased risk of developing PD has been associated with exposure to heavy metals such as iron, Pb and Mn and their combinations. Heavy metals may accumulate in the substantianigra and cause oxidative stress. As a heavy metal, iron is an essential element transported into the brain through the transferrin receptor and divalent metal transporter1 (DMT1) (Zheng, & Monnot (2012). Postmortem and in vitro studies confirmed links between iron accumulation and the cardinal pathological features of PD, the loss of dopamine neurons in the substantianigra (SN) pars compacta (Dexter, Wells, Agid, Agid, Lees, Jenner, & Marsden, 1987) and the presence of a-synuclein-rich Lewy bodies (Hashimoto, Hsu, Xia, Takeda, Sisk, Sundsmo, & Masliah, 1999). Iron has also been linked to risk of PD in some epidemiological studies without strong evidence (Abbott et al., 2012; Logroscino, Gao, Chen, Wing, & Ascherio, 2008; Miyake et al., 2011; Rybicki, Johnson, Uman, & Gorell, 1993).Nevertheless, iron and its deregulated homeostasis have been suggested to play a role in the pathogenesis of PD because of its prooxidants characteristics that may lead to ROS generation via Fenton and Haber–Weiss reactions (Sian-Hülsmann, Mandel, Youdim, & Riederer, 2011; Stohs & Bagchi, 1995). The substantianigra has the highest levels of iron in the human brain, which may be caused by the presence of neuromelanin in pigmented SNpc dopaminergic neurons known for their high chelating capacity for metals, and iron in particular. However, this may increase their vulnerability since iron may react with ROS produced from dopamine metabolism and promote the further generation of highly toxic radicals (Zecca et al., 2002; Zecca, Youdim, Riederer, Connor, & Crichton, 2004). Dexter et al.,(1987) found that free ferrous iron ($Fe2+$) acts as a catalyst in a reaction with hydrogen peroxide (Fenton reaction), producing highly toxic reactive oxygen species and ferric iron ($Fe3+$) that lead to oxidative-stress-related damage to cellular components including proteins, lipids, and DNA. To relieve potential toxicity, excess ferrous iron within deep brain nuclei is accumulated and transported into the core of apoferritin molecules where it is oxidized and safely stored in high concentrations in the ferric state (Haacke et al., 2005). As ferritin (iron-containing apoferritin) is highly paramagnetic and indeed one of the only non-haem iron molecules present in a sufficient quantity to cause local field inhomogeneities (Haacke et al., 2005; Schenck & Zimmerman, 2004).

The epidemiological proof of Pb association with PD is more accurate as the accumulative lifetime exposure can be estimated through Pb concentration in bone that has a half-life of years to decades. Initially, Kuhn, Winkel, Woitalla, Meves, Przuntek, & Muller, (1998) reported that 7 out of 9 postal workers exposed to lead sulfate batteries for up to 30 years developed parkinsonian symptoms, concluding that Pb intoxication may be a cause of these symptoms (Kuhn et al., 1998).

As well, experimental studies showed that acute and chronic, in vivo and in vitro, exposures to lead may trigger a potent neurotoxic effect on the midbrain SNpcdopaminergic neurons. NourEddine, Miloud, & Abdelkader, (2005) have shown a clear dose-dependent inhibition of 3H-DA uptake in presence of lead which was associated with a deficit in locomotor behavior (Figure 4), leading to suggest a possible involvement of such heavy metal in the pathogenesis of PD (NourEddine, et al., 2005).

Figure 4. (A): Effect of lead acetate on ^{3}H-DA uptake rates in synaptosomes isolated from the brain in vitro of controls and Pb treated (amounts selected of lead acetate from 10-3 M to10-5M). (B): Effect of lead acetate on locomotor activity in controls (C) and lead-exposed rats (Pb) (modified NourEddine, Miloud, & Abdelkader, 2005)

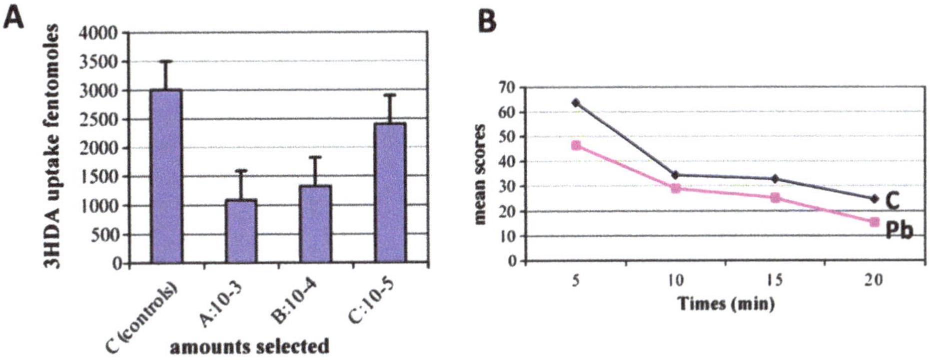

As a heavy metal, manganese is an essential element with important physiological functions for cellular homeostasis. The epidemiologic evidence does not provide sufficient proof for an association between Mn exposure and PD risk (Mortimer, Borenstein, & Nelson, 2012; Wirdefeldt et al., 2011). Only one case-control study (144 cases vs. 464 controls) in a population from the metropolitan Detroit area reported an increase of PD risk when the exposure to Mn was over 20 years (OR = 10.63, 95% CI, 1.07–105.99) (Gorell, Peterson, Rybicki, & Johnson, 2004). However, occupational or environmental exposures to Mn have been associated with a neurological syndrome that includes cognitive deficits, neuropsychological abnormalities and Parkinsonism (Guilarte, 2013).

Clinical Features and Epidemiology of Alzheimer Disease

AD is the most common and age-related neurodegenerative disorder and affects approximately 47 million people worldwide (Prince et al., 2015) with heterogeneity of prevalence rate between nations, depending on race and the socioeconomic status (Figure 5). Almost 50% of cases of AD are found in individuals older than 85 years. Few cases are due to genetic mutations and the majority is occurring without known familial genetic mutation (Ceccom et al., 2012). Typical symptoms of AD patients are early memory loss, personality and behavioral changes and deficits in sensory and motor functions. The brain of AD patients presents synaptic and neuronal loss resulting in brain atrophy, affecting regions including the entorhinal cortex, hippocampus (figure 6), basal forebrain and amygdala (Bertram, Lill, & Tanzi, 2010; Ittner & Götz, 2011). Clinical symptoms of AD are characterized by a progressive deterioration of higher mental functions and cognitive functions, ending in severe dementia (Nunomura, Moreira, Castellani, Lee, Zhu, Smith, & Perry, 2012). The neuropathology is characterized by the accumulation of two types of fibrous material in the brain: the Aß-amyloid extracellular peptide deposited in senile plaques and intraneuronal neurofibrillary tangles (NFT) composed mainly of abnormal and hyperphosphorylated forms of the microtubule-bound tau protein. A large number of reactive astrocytes is also observed as-signs of neuroinflammation, metal dyshomeostasis, and oxidative stress (Nunomura et al., 2012; Reddy, 2006). The formation of NTFs and the resulting degeneration first appear in the hippocampus, especially pyramidal neurons in CA1 and layer II of the entorhinal cortex and the neocortex (Kepp, 2012).

Figure 5. Age-specific prevalence of Alzheimer's disease (per 100 populations) across continents and countries (Qiu, Kivipelto, & von Strauss, 2009)

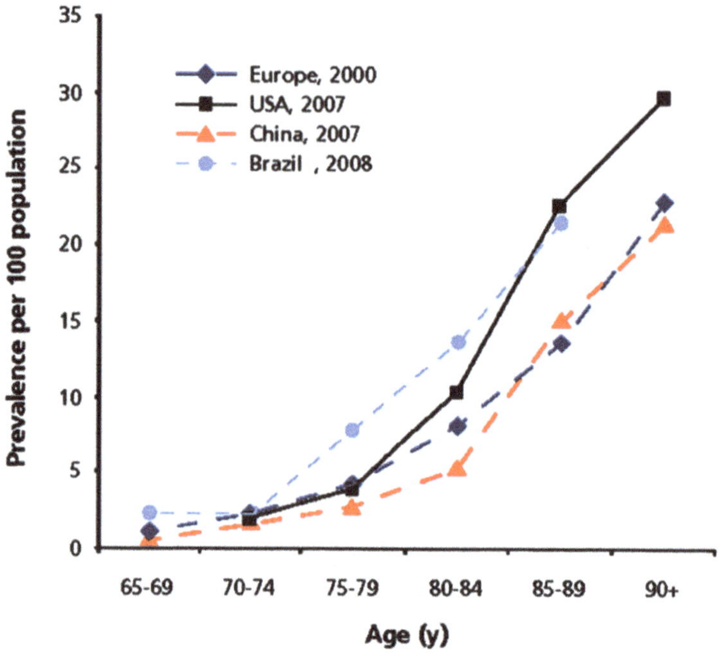

Figure 6. T1-weighted MRI image of a patient with Alzheimer's disease, the arrow pointing at right hippocampal atrophy (Bonifacio & Zamboni, 2016)

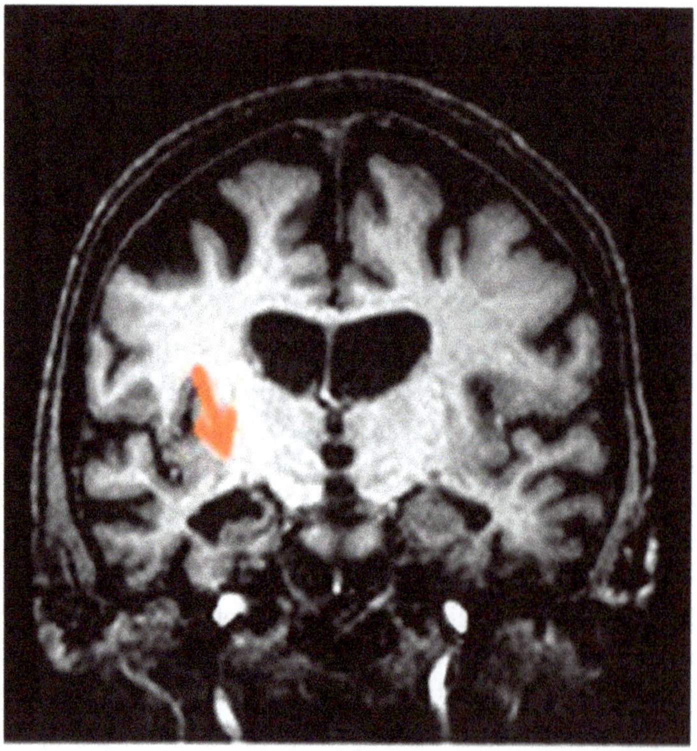

Environmental factors such as diet (fat-rich), heavy metals, biogenic metals, and pesticides have been involved in AD development due to their ability to disrupt metabolic pathways involved in the homeostasis of (amyloid betaAb). However, factors such as lifestyle (antioxidants and exercise) can prevent AD development (Paillard, Rolland, & de Souto Barreto, 2015).

Many of these environmental factors are oxidative agents acting through different mechanisms. The brain is particularly vulnerable to oxidative stress due to its high glucose-based metabolic rate, low levels of antioxidants, high levels of polyunsaturated fatty acids, and high enzymatic activities related to transition metals that catalyze the formation of free radicals (Halliwell, 1992).

Etiology of Alzheimer Disease

The etiology of AD still unclear. AD may be caused by a multifactorial process involving both a genetic predisposition and exposure to environmental factors modulated by the biological aging process (Gauvreau, 1987).

AD is a progressive neurodegenerative disorder characterized by extracellular Aß plaques and intracellular neurofibrillary tangles (NFTs) with high synaptic defects (Correia et al., 2012). In 2010, around 36 million people were affected by AD in the world with projections to elevate to 66 million by 2030 and 115 million by 2050 (Honjo, Black, & Verhoeff, 2012).

Oxidative stress and inflammatory mechanisms are the main factors responsible for AD progression and targeting these mechanisms may be the best way in its treatment. Oxidative stress is an early feature of AD pathogenesis that contributes to Aß generation and NFTs formation (Cai, Zhao, & Ratka, 2011; Kalra, Kumar, Majeed, & Prakash, 2016). The inflammatory reactions involving cytokines such as IL-1, IL-6, NF-κß, COX-2, and TNF-a, contribute dramatically to AD progression (Bronzuoli, Iacomino, Steardo, & Scuderi, 2016). Moreover, neuroinflammatory cytokines have been detected at high levels in post-mortem brains of AD patients. Findings reported that aspirin and non-steroidal anti-inflammatory drug treatment (NSAID) could reduce the incidence of AD (Hung et al., 2016; Su et al., 2016).

AD is the fourth leading cause of death in the elderly (McDowell, Hill, Lindsay, Helliwell,Costa, Beattie, ... & Parboosingh, 1994). AD is a neurodegenerative disease defined by specific neuropathological and neurochemical features. The main lesions observed in the brain of AD cases are senile plaques (Duyckaerts et al., 1986) and neurofibrillary tangles (NFT) (Ball, 1977). Some studies identified a major decrease of choline acetyltransferase and acetylcholinesterase (AchE) in AD cases (Davies& Maloney, 1976; Perry, Perry, Blessed, & Tomlinson, 1977). Others reported oxidative stress by free radicals (FR) as a risk factor for AD (Frölich & Riederer, 1995).Ikeda, Toda, Wang, & Nakazawa, (1994) observed that FR are highly toxic and can initiate lipid peroxidation specifically because of the increase of NFT.

However, the genetic components account is known for only few cases of AD (Van Broeckhoven, 1995), suggesting that other genes or risk factors, such as environmental exposure, may be implicated in the etiology AD. Epidemiological studies revealed a higher prevalence of AD in rural environments than in urban settings (Jean et al., 1996; Shibayama, Kasahara, & Kobayashi, 1986). For many decades, pesticides have been used more intensively in rural than in urban areas and their use has been increasing with the growth of productivity and the specialization of cultures (Brundtland, 1989). Many pesticides have been known to have neurotoxic properties (Corriveau & Lajoie, 1987). Toxicological studies showed that some pesticide families (carbamates, organophosphates, organochlorines, bipyridyls) may be responsible for serious damage to the nervous system (Blain, 1990).

The organophosphorus compounds like parathion and fenitrothion and carbamates such as aldicarb are known inhibitors of AchE (Bardin, 1994; Fukuto, 1990). Bipyridyls such as paraquat can generate FR that is able to cross the blood-brain barrier and inflict irreversible damage to neurons (Corasaniti, Defilippo, Rodinò, Nappi, & Nisticò, 1991; Elroy-Stein, Bernstein, & Groner, 1986). Organochlorines such as DDT and dieldrin can disturb the functioning of the mitochondrial system and produce FR (Kimbrough, Gaines, & Linder, 1971). Thus, exposure to pesticides may contribute to the cholinergic system deficiency and the production of FR observed in AD. Epidemiological studies in the workplace have identified some damage to the central nervous system and neuropsychological disorders due to pesticide exposure (Boyd, Weiler, & Porter, 1990).Others have shown significant relationships between exposure to pesticides and neurodegenerative diseases such as PD(Barbeau, Roy, Cloutier, Plasse, & Paris, 1987; Fleming, Mann, Bean, Briggle, & Sanchez-Ramos, 1994) and AD (Cannas, Costa, Tacconi, Pinna, & Fiaschi, 1992; McDowell, Hill, Lindsay, Helliwell, Costa, Beattie, ... &Parboosingh, 1994). The Canadian Study of Health and Aging has estimated an odds ratio (OR) of 2.17 (95% CI, 1.18-3.99) associated to AD with respect to occupational exposure to pesticides (Pasture & Onkia, 1994)

However, the scope of this result is limited because the study did not long-term environmental exposure to pesticides. In addition, occupational exposure to pesticides was established from answers provided by contacts to a simple binary question on whether the subjects were exposed or not exposed. This does not permit the quantification of exposure. Finally, the Canadian study of Health and Aging (1994) (Pasture & Onkia, 1994) did not consider potential confounding factors such as age, sex, education level, occupation, and genetic factors that might influence the relationship between pesticides and AD.

Physiopathology of Alzheimer Disease Involving Environmental Neurotoxicity

Role of Heavy Metals

As neurodegenerative disorders progenitor, heavy metals play a role in the precipitation of abnormal ß-amyloid protein and hyperphosphorylated tau, which are the main hallmarks of AD.

Environmental heavy metals affect brain development. There is evidence of a link between heavy metals and neurodegenerative disorders such as AD and PD(Brown, Lockwood, & Sonawane, 2005; Nallagouni & Pratap, 2017). Aluminum is known as one of the heavy metals causing neurodegenerative diseases. It upregulates the expression of pathogenic genes that are implicated in AD such as APP, IL-1ß, IL-6, TNF-a, AChE and MAO (Cao, Wang, Xiu, Zhang, & Li, 2017; Lukiw, Percy, & Kruck, 2005). Further studies have suggested that the interaction of metal ions such as iron and aluminum with Ab, and also copper and zinc, could play a crucial role in the release of reactive oxygen species (ROS) that contribute to neurodegenerative damage (House et al., 2004; Khan, Dobson, & Exley, 2006; Smith, Richey Harris, Sayre, Beckman, & Perry, 1997).

Other heavy metals like cadmium and fluoride enhance neurotoxicity by several mechanisms including free radical generation and lipid peroxidation in the hippocampus (Blaylock, 2004; B. Wang & Du, 2013). Transition metals such as Cu or Fe, together with Zn, cause oxidative stress as well as to the aggregation and precipitation of amyloid- peptides in the AD brain (Bush, 2003; Bush et al., 1994). Influence of Transgenic Metallothionein-1 on Gliosis, CA1 Neuronal Loss, and Brain Metal Levels of theTg2576 Mouse Model of AD.

Transition metals, particularly iron (Fe) zin (Zn) and copper (Cu) have been studied in AD etiology. These metals bind to amyloid ßpeptide (Aß) accelerate (Aß) aggregation and consequently promote neurotoxic plaque formation. Fe and Cu are also likely to be responsible for causing oxidative stress and neuroinflammatory changes in the AD brain (Choo, Alukaidey, White, & Grubman, 2013; Praticò, 2008; Sayre, Perry, & Smith, 2008). In addition, cognitive decline in AD has been associated with the interference of the processing and function of the amyloid-ß precursor protein (APP) and the phosphorylation and aggregation of the microtubule-associated protein (MAP), tau both of which are associated with altered metal homeostasis (Crouch et al., 2009).

Lead (Pb) is a heavy metal characterized by its neurological toxic effects, even if a direct effect on AD development has not been reported. Pb affects cognitive abilities, intelligence, memory, speed processing and motor functions in children (Mason, Harp, & Han, 2014). However, studies in the elderly are still limited. A cohort study reported that bone Pb levels were associated with poor cognitive performance scores in old workers, suggesting that past Pb exposure can lead to late cognitive deterioration (Dorsey, Lee, Bolla, Weaver, Lee, Lee, … Schwartz, 2006).

However, a recent study reported no evidence on the role of serum Pb levels in AD (Park, Lee, Park, & Joung, 2014). Experimental studies are showing more solid evidence compared to epidemiological studies related to the link between Pb exposure with AD. Mercury (Hg) is known as a heavy metal with a high potential to cause neurotoxicity. Studies about Minamata and Iraq disasters helped to understand the neurotoxicity of this metal. There is a large consensus that Hg disrupts the brain development and produces cognitive and motor disabilities (Johansson et al., 2007), and in adults, Hg exposure causes memory loss and cognitive alterations (Chang et al., 2008; Wojcik, Godfrey, Christie, & Haley, 2006). An early study suggested a relation between Hg exposure and AD. Authors reported increased levels of Hg (in microsomes) and bromide (in the whole brain), and lower levels of rubidium (in the whole brain, microsomes, and nuclei), selenium (Se; in microsomes) and zinc (Zn; in nuclei) in AD brains compared with controls (Wenstrup, Ehman, & Markesbery, 1990). Cadmium is another toxic heavy metal associated with neurological alterations including memory loss and mental retardation (Wang & Du, 2013). Another study found an increase of plasma levels of various metals including Cadmium, aluminum, mercury, and selenium in 24 AD patients compared with 28 healthy volunteers (Basun, Forssell, Wetterberg, & Winblad, 1991).

Aluminum (Al) is the most used neurotoxic metal, widely bioavailable to humans and involved in the etiology of neurodegenerative diseases such as AD. It is known to accumulate at higher concentrations in brain regions that are selectively affected in AD, including the entorhinal cortex (an area that shows the earliest pathological changes in AD), hippocampus, and the amygdala (McDermott, Smith, Iqbal, & Wisniewski,1979; Walton, 2009b, 2009a). Pyramidal cells in the hippocampus and cortex, basal forebrain cholinergic neurons, and upper brainstem catecholaminergic neurons are the most affected neuronal populations in AD and also, the most susceptible to Al-induced neurofibrillary degeneration (Bertholf & Gambetti, 1987;Kowall, Pendlebury, Kessler, Perl, & Beal, 1989; Matyja, 2000; Walton, 2006).

The incident of Al pollution in Cornwall, UK (1998) after a discharge of 20 tonnes of aluminum sulfate into the drinking water, gave evidence of Al potential neurotoxicity. Another brain pathological characteristic found in AD patients were observed in subjects exposed to Al in this region (Exley & Esiri, 2006), as well as damage in cerebral functions (Altmann et al., 1999). A Chinese study reported that high concentrations of iron and copper in the soil might be associated with the high AD annual mortality. However, no association between Al levels in soil and the mortality caused by AD (Shen, Yu, Zhang, Xie, & Jiang, 2014)was reported.

Role of Pesticides

Epidemiological studies revealed a higher prevalence of AD in rural environments compared to urban settings (Jean et al., 1996; Hiroto Shibayama et al., 1986). Pesticides have been used more intensively in rural than in urban areas and their use has been increasing with the growth of productivity and the specialization of cultures (Brundtland, 1989). Many pesticides have shown neurotoxic properties (Corriveau & Lajoie, 1987). Toxicological studies revealed that some pesticide families (carbamates, organophosphates, organochlorines, bipyridyls) can cause serious damage to the nervous system (Blain, 1990).

The organophosphorus compounds such as parathion and fenitrothion and carbamates including aldicarb are known inhibitors of AchE (Bardin, 1994; Fukuto, 1990). Bipyridyls such as paraquat can generate free radicals (FR) that is able to cross the blood-brain barrier and cause irreversible damage to neurons (Corasaniti et al.,1991; Elroy-Stein et al., 1986). Organochlorines such as DDT and dieldrin can disturb the functioning of the mitochondrial system and produce FR(Fleming et al., 1994; Kimbrough et al., 1971).Thus, exposure to pesticides may contribute to the cholinergic system deficiency and the production of FR observed in AD. Epidemiological studies in the workplace have identified some damage to the central nervous system and neuropsychological disorders due to pesticide exposure (Boyd et al., 1990; Stephens, Spurgeon, Calvert, Beach, Levy, Harrington, & Berry, 1995). Others have shown significant correlation between exposure to pesticides and neurodegenerative diseases such as PD(Butterfield, Valanis, Spencer, Lindeman, & Nutt, 1993; Fleming et al., 1994) and AD (Cannas et al., 1992; McDowell, Hill, Lindsay, Helliwell, Costa, Beattie, ... &Parboosingh, 1994). Studies in the workplace have identified some damage to the central nervous system and neuropsychological disorders due to pesticide exposure (Boyd et al., 1990; Stephens et al., 1995).

Paraquat is a commonly used herbicide that has been supposed to be involved in AD development. A recent study showed that treatment of wild-type and β-amyloid precursor protein (APP) transgenic mice with PQ produced a significant increase in β-amyloid(Aβ) levels in transgenic mice that was associated with mitochondrial oxidative damage in cerebral cortex causing mitochondrial dysfunction which induces an impairment of learning and memory (Chen, Yoo, Na, Liu, & Ran, 2012). This result demonstrates that mitochondrial damage is a key mechanism underlying cognitive impairment and elevated amyloido genesis induced by paraquat, suggesting that pro-oxidant xenobiotics like PQ can cause AD.

CONCLUSION

A variety of environmental and occupational exposures have been implicated in the etiology of PD and AD. These include exposure to well water, pesticides, herbicides, rural living, and certain heavy metals. Chronic low-level exposure to pesticide and heavy metals may potentiate the neurodegeneration caused by aging (Bolin, Basha, Cox, Zawia, Maloney, Lahiri, & Cardozo-Pelaez, 2006; Peng et al., 2007), these pollutants present similar mechanisms of toxicity, which lead to a generalized mechanism based on the generation of oxidative stress causing common hallmarks of both neurodegenerative disorders.

More studies are needed to clarify the mechanisms involved in the early lifetime exposure to neurotoxic pesticide residues and other possible drinking water contaminants effects on AD and PD disorders in humans. This knowledge would be crucial for improving the life quality of elderly people and would significantly help the health authorities to develop global prevention programs against neurodegenerative diseases.

REFERENCES

Abbott, R. D., Ross, G. W., Tanner, C. M., Andersen, J. K., Masaki, K. H., Rodriguez, B. L., ... Petrovitch, H. (2012). Late-life hemoglobin and the incidence of Parkinson's disease. *Neurobiology of Aging*, *33*(5), 914–920. doi:10.1016/j.neurobiolaging.2010.06.023 PMID:20709430

Altevogt, B. M., Hanson, S. L., & Leshner, A. I. (2008). Autism and the Environment: Challenges and Opportunities for Research. *Pediatrics*, *121*(6), 1225–1229. doi:10.1542/peds.2007-3000 PMID:18519493

Altmann, P., Cunningham, J., Dhanesha, U., Ballard, M., Thompson, J., & Marsh, F. (1999). Disturbance of cerebral function in people exposed to drinking water contaminated with aluminium sulphate: Retrospective study of the Camelford water incident. *BMJ (Clinical Research Ed.)*, *319*(7213), 807–811. doi:10.1136/bmj.319.7213.807 PMID:10496822

Ascherio, A., Chen, H., Weisskopf, M. G., O'Reilly, E., McCullough, M. L., Calle, E. E., ... Thun, M. J. (2006). Pesticide exposure and risk for Parkinson's disease. *Annals of Neurology*, *60*(2), 197–203. doi:10.1002/ana.20904 PMID:16802290

Ball, M. J. (1977). Neuronal loss, neurofibrillary tangles and granulovacuolar degeneration in the hippocampus with ageing and dementia. *Acta Neuropathologica*, *37*(2), 111–118. doi:10.1007/BF00692056 PMID:848276

Barbeau, A., Cloutier, T., Roy, M., Plasse, L., Paris, S., & Poirier, J. (1985). Ecogenetics of Parkinson's disease: 4-hydroxylation of debrisoquine. *Lancet (London, England)*, *2*(8466), 1213–1216. Retrieved from http://www.ncbi.nlm.nih.gov/pubmed/2866293

Barbeau, A., Roy, M., Cloutier, T., Plasse, L., & Paris, S. (1987). Environmental and genetic factors in the etiology of Parkinson's disease. *Advances in Neurology*, *45*, 299–306. Retrieved from http://www.ncbi.nlm.nih.gov/pubmed/3493629 PMID:3493629

Bardin, P. G. (1994). Organophosphate and Carbamate Poisoning. *Archives of Internal Medicine*, *154*(13), 1433. doi:10.1001/archinte.1994.00420130020005 PMID:8017998

Barlow, B., Coryslechta, D., Richfield, E., & Thiruchelvam, M. (2007). The gestational environment and Parkinson's disease: Evidence for neurodevelopmental origins of a neurodegenerative disorder. *Reproductive Toxicology (Elmsford, N.Y.)*, *23*(3), 457–470. doi:10.1016/j.reprotox.2007.01.007 PMID:17350799

Basun, H., Forssell, L. G., Wetterberg, L., & Winblad, B. (1991). Metals and trace elements in plasma and cerebrospinal fluid in normal aging and Alzheimer's disease. *Journal of Neural Transmission. Parkinson's Disease and Dementia Section*, *3*(4), 231–258. Retrieved from http://www.ncbi.nlm.nih.gov/pubmed/1772577 PMID:1772577

Berry, C., La Vecchia, C., & Nicotera, P. (2010). Paraquat and Parkinson's disease. *Cell Death and Differentiation*, *17*(7), 1115–1125. doi:10.1038/cdd.2009.217 PMID:20094060

Bertholf, R. L., & Gambetti, P. (1987). Aluminum and Alzheimer's disease: Perspectives for a cytoskeletal mechanism. *CRC Critical Reviews in Clinical Laboratory Sciences*, *25*(3), 195–210. doi:10.3109/10408368709105882 PMID:3315463

Bertram, L., Lill, C. M., & Tanzi, R. E. (2010). The Genetics of Alzheimer Disease: Back to the Future. *Neuron, 68*(2), 270–281. doi:10.1016/j.neuron.2010.10.013 PMID:20955934

Betarbet, R., Sherer, T. B., MacKenzie, G., Garcia-Osuna, M., Panov, A. V., & Greenamyre, J. T. (2000). Chronic systemic pesticide exposure reproduces features of Parkinson's disease. *Nature Neuroscience, 3*(12), 1301–1306. doi:10.1038/81834 PMID:11100151

Blain, P. G. (1990). Aspects of pesticide toxicology. *Adverse Drug Reactions and Acute Poisoning Reviews, 9*(1), 37–68. Retrieved from http://www.ncbi.nlm.nih.gov/pubmed/2190453 PMID:2190453

Blaylock, R. L. (2004). *Excitotoxicity : a possible central mechanism in fluoride neurotoxicity compelling evidence indicates that fluoride produces injury to the central nervous system (CNS) by several mechanisms.* Academic Press.

Bolin, C. M., Basha, R., Cox, D., Zawia, N. H., Maloney, B., Lahiri, D. K., & Cardozo-Pelaez, F. (2006). Exposure to lead and the developmental origin of oxidative DNA damage in the aging brain. *The FASEB Journal, 20*(6), 788–790. doi:10.1096/fj.05-5091fje PMID:16484331

Bonifacio, G., & Zamboni, G. (2016). Brain imaging in dementia. *Postgraduate Medical Journal, 92*(1088), 333–340. doi:10.1136/postgradmedj-2015-133759 PMID:26933232

Boyd, C. A., Weiler, M. H., & Porter, W. P. (1990). Behavioral and neurochemical changes associated with chronic exposure to low-level concentration of pesticide mixtures. *Journal of Toxicology and Environmental Health, 30*(3), 209–221. doi:10.1080/15287399009531424 PMID:2366258

Bronstein, J., Carvey, P., Chen, H., Cory-Slechta, D., DiMonte, D., Duda, J., ... Weisskopf, M. (2009). Meeting Report: Consensus Statement—Parkinson's Disease and the Environment: Collaborative on Health and the Environment and Parkinson's Action Network (CHE PAN) Conference 26–28 June 2007. *Environmental Health Perspectives, 117*(1), 117–121. doi:10.1289/ehp.11702 PMID:19165397

Bronzuoli, M. R., Iacomino, A., Steardo, L., & Scuderi, C. (2016). Targeting neuroinflammation in Alzheimer's disease. *Journal of Inflammation Research, 9*, 199–208. doi:10.2147/JIR.S86958 PMID:27843334

Brooks, A., Chadwick, C., Gelbard, H., Cory-Slechta, D., & Federoff, H. (1999). Paraquat elicited neurobehavioral syndrome caused by dopaminergic neuron loss. *Brain Research, 823*(1–2), 1–10. doi:10.1016/S0006-8993(98)01192-5 PMID:10095006

Brown, R. C., Lockwood, A. H., & Sonawane, B. R. (2005). Neurodegenerative Diseases: An Overview of Environmental Risk Factors. *Neurodegenerative Diseases, 113*(9), 1250–1256. doi:10.1289/ehp.7567 PMID:16140637

Brown, T. P., Rumsby, P. C., Capleton, A. C., Rushton, L., & Levy, L. S. (2006). Pesticides and Parkinson's Disease—Is There a Link? *Environmental Health Perspectives, 114*(2), 156–164. doi:10.1289/ehp.8095 PMID:16451848

Brundtland, G. H. (1989). *Notre avenir à tous*. Academic Press.

Bus, J. S., & Gibson, J. E. (1984). Paraquat: Model for oxidant-initiated toxicity. *Environmental Health Perspectives*, *55*, 37–46. doi:10.1289/ehp.845537 PMID:6329674

Bush, A., Pettingell, W., Multhaup, G., d Paradis, M., Vonsattel, J., Gusella, J., ... Tanzi, R. (1994). Rapid induction of Alzheimer A beta amyloid formation by zinc. *Science*, *265*(5177), 1464–1467. doi:10.1126cience.8073293 PMID:8073293

Bush, A. I. (2003). The metallobiology of Alzheimer's disease. *Trends in Neurosciences*, *26*(4), 207–214. doi:10.1016/S0166-2236(03)00067-5 PMID:12689772

Butterfield, P. G., Valanis, B. G., Spencer, P. S., Lindeman, C. A., & Nutt, J. G. (1993). Environmental antecedents of young-onset Parkinson's disease. *Neurology*, *43*(6), 1150–1158. doi:10.1212/WNL.43.6.1150 PMID:8170560

Cai, Z., Zhao, B., & Ratka, A. (2011). Oxidative Stress and β-Amyloid Protein in Alzheimer's Disease. *Neuromolecular Medicine*, *13*(4), 223–250. doi:10.100712017-011-8155-9 PMID:21901428

Campdelacreu, J. (2014). Parkinson's disease and Alzheimer disease: Environmental risk factors. [English Edition]. *Neurologia (Barcelona, Spain)*, *29*(9), 541–549. doi:10.1016/j.nrl.2012.04.001 PMID:22703631

Cannas, A., Costa, B., Tacconi, P., Pinna, L., & Fiaschi, A. (1992). Dementia of Alzheimer type (DAT) in a man chronically exposed to pesticides. *Acta Neurologica*, *14*(3), 220–223. Retrieved from http://www.ncbi.nlm.nih.gov/pubmed/1442221 PMID:1442221

Cao, Z., Wang, F., Xiu, C., Zhang, J., & Li, Y. (2017). Hypericum perforatum extract attenuates behavioral, biochemical, and neurochemical abnormalities in Aluminum chloride-induced Alzheimer's disease rats. *Biomedicine and Pharmacotherapy*, *91*, 931–937. doi:10.1016/j.biopha.2017.05.022 PMID:28514831

Castello, P. R., Drechsel, D. A., & Patel, M. (2007). Mitochondria Are a Major Source of Paraquat-induced Reactive Oxygen Species Production in the Brain. *The Journal of Biological Chemistry*, *282*(19), 14186–14193. doi:10.1074/jbc.M700827200 PMID:17389593

Caudle, W. M., Richardson, J. R., Wang, M., & Miller, G. W. (2005). Perinatal heptachlor exposure increases expression of presynaptic dopaminergic markers in mouse striatum. *Neurotoxicology*, *26*(4), 721–728. doi:10.1016/j.neuro.2004.09.003 PMID:16112329

Ceccom, J., Coslédan, F., Halley, H., Francès, B., Lassalle, J. M., & Meunier, B. (2012). Copper Chelator Induced Efficient Episodic Memory Recovery in a Non-Transgenic Alzheimer's Mouse Model. *PLoS One*, *7*(8), e43105. doi:10.1371/journal.pone.0043105 PMID:22927947

Chang, J.-W., Pai, M.-C., Chen, H.-L., Guo, H.-R., Su, H.-J., & Lee, C.-C. (2008). Cognitive function and blood methylmercury in adults living near a deserted chloralkali factory. *Environmental Research*, *108*(3), 334–339. doi:10.1016/j.envres.2008.06.006 PMID:18675410

Chen, L., Yoo, S. E., Na, R., Liu, Y., & Ran, Q. (2012). Cognitive impairment and increased Aβ levels induced by paraquat exposure are attenuated by enhanced removal of mitochondrial H 2O 2. *Neurobiology of Aging*, *33*(2), 15–26. doi:10.1016/j.neurobiolaging.2011.01.008 PMID:21429624

Choo, X. Y., Alukaidey, L., White, A. R., & Grubman, A. (2013). Neuroinflammation and Copper in Alzheimer's Disease. *International Journal of Alzheimer's Disease, 2013*, 1–12. doi:10.1155/2013/145345 PMID:24369524

Cicchetti, F., Lapointe, N., Roberge-Tremblay, A., Saint-Pierre, M., Jimenez, L., Ficke, B. W., & Gross, R. E. (2005). Systemic exposure to paraquat and maneb models early Parkinson's disease in young adult rats. *Neurobiology of Disease*, *20*(2), 360–371. doi:10.1016/j.nbd.2005.03.018 PMID:16242641

Corasaniti, M. T., Defilippo, R., Rodinò, P., Nappi, G., & Nisticò, G. (1991). Evidence that paraquat is able to cross the blood-brain barrier to a different extent in rats of various age. *Functional Neurology*, *6*(4), 385–391. Retrieved from http://www.ncbi.nlm.nih.gov/pubmed/1810839 PMID:1810839

Correia, S. C., Santos, R. X., Perry, G., Zhu, X., Moreira, P. I., & Smith, M. A. (2012). Mitochondrial Importance in Alzheimer's, Huntington's and Parkinson's Diseases (Vol. 724). New York, NY: Springer US. doi:10.1007/978-1-4614-0653-2_16

Corrigan, F. M., Wienburg, C. L., Shore, R. F., Daniel, S. E., & Mann, D. (2000). Organochlorine insecticides in substantia nigra in Parkinson's disease. *Journal of Toxicology and Environmental Health. Part A.*, *59*(4), 229–234. doi:10.1080/009841000156907 PMID:10706031

Corriveau, L., & Lajoie, P. (1987). *Les Pesticides les plus utilisés en agriculture au Québec et la santé: recueil des données toxicologiques et épidémiologiques*. Academic Press.

Costa, L. G., Giordano, G., Guizzetti, M., & Vitalone, A. (2008). Neurotoxicity of pesticides: A brief review. *Frontiers in Bioscience*, *13*(13), 1240. doi:10.2741/2758 PMID:17981626

Crouch, P. J., Hung, L. W., Adlard, P. A., Cortes, M., Lal, V., Filiz, G., … Barnham, K. J. (2009). Increasing Cu bioavailability inhibits A oligomers and tau phosphorylation. *Proceedings of the National Academy of Sciences*, *106*(2), 381–386. 10.1073/pnas.0809057106

Davies, P., & Maloney, A. J. F. (1976). Selective loss of central cholinergic neurons in alzheimer's disease. *Lancet*, *308*(8000), 1403. doi:10.1016/S0140-6736(76)91936-X PMID:63862

de Lau, L. M., & Breteler, M. M. (2006). Epidemiology of Parkinson's disease. *Lancet Neurology*, *5*(6), 525–535. doi:10.1016/S1474-4422(06)70471-9 PMID:16713924

de Rijk, M. C., Breteler, M. M., Graveland, G. A., Ott, A., Grobbee, D. E., van der Meché, F. G., & Hofman, A. (1995). Prevalence of Parkinson's disease in the elderly: The Rotterdam Study. *Neurology*, *45*(12), 2143–2146. doi:10.1212/WNL.45.12.2143 PMID:8848182

de Rijk, M. C., Launer, L. J., Berger, K., Breteler, M. M., Dartigues, J. F., Baldereschi, M., … Hofman, A. (2000). Prevalence of Parkinson's disease in Europe: A collaborative study of population-based cohorts. Neurologic Diseases in the Elderly Research Group. *Neurology, 54*(11), S21-3. Retrieved from http://www.ncbi.nlm.nih.gov/pubmed/10854357

DeMaagd, G., & Philip, A. (2015). Parkinson's Disease and Its Management: Part 1: Disease Entity, Risk Factors, Pathophysiology, Clinical Presentation, and Diagnosis. *P & T : A Peer-Reviewed Journal for Formulary Management, 40*(8), 504–532. Retrieved from http://www.ncbi.nlm.nih.gov/pubmed/26236139

Dexter, D. T., Wells, F. R., Agid, F., Agid, Y., Lees, A. J., Jenner, P., & Marsden, C. D. (1987). Increased nigral iron content in postmortem parkinsonian brain. *Lancet*, *330*(8569), 1219–1220. doi:10.1016/S0140-6736(87)91361-4 PMID:2890848

Dick, F. D. (2006). Parkinson's disease and pesticide exposures. *British Medical Bulletin*, *79–80*(1), 219–231. doi:10.1093/bmb/ldl018 PMID:17242039

Dinis-Oliveira, R. J., Remião, F., Carmo, H., Duarte, J. A., Navarro, A. S., Bastos, M. L., & Carvalho, F. (2006). Paraquat exposure as an etiological factor of Parkinson's disease. *Neurotoxicology*, *27*(6), 1110–1122. doi:10.1016/j.neuro.2006.05.012 PMID:16815551

Dorsey, C. D., Lee, B.-K., Bolla, K. I., Weaver, V. M., Lee, S.-S., Lee, G.-S., ... Schwartz, B. S. (2006). Comparison of Patella Lead With Blood Lead and Tibia Lead and Their Associations With Neurobehavioral Test Scores. *Journal of Occupational and Environmental Medicine*, *48*(5), 489–496. doi:10.1097/01.jom.0000199678.86629.3b PMID:16688005

Duvoisin, R. C. (1986). Etiology of Parkinson's disease: Current concepts. *Clinical Neuropharmacology*, *9*(Suppl 1), S3–S21. doi:10.1097/00002826-198609001-00003 PMID:2423234

Duyckaerts, C., Hauw, J.-J., Bastenaire, F., Piette, F., Poulain, C., Rainsard, V., ... Berthaux, P. (1986). Laminar distribution of neocortical senile plaques in senile dementia of the alzheimer type. *Acta Neuropathologica*, *70*(3–4), 249–256. doi:10.1007/BF00686079 PMID:3766125

Elroy-Stein, O., Bernstein, Y., & Groner, Y. (1986). Overproduction of human Cu/Zn-superoxide dismutase in transfected cells: Extenuation of paraquat-mediated cytotoxicity and enhancement of lipid peroxidation. *The EMBO Journal*, *5*(3), 615–622. doi:10.1002/j.1460-2075.1986.tb04255.x PMID:3011416

Emre, M., Aarsland, D., Albanese, A., Byrne, E. J., Deuschl, G., De Deyn, P. P., ... Lane, R. (2004). Rivastigmine for Dementia Associated with Parkinson's Disease. *The New England Journal of Medicine*, *351*(24), 2509–2518. doi:10.1056/NEJMoa041470 PMID:15590953

Exley, C., & Esiri, M. M. (2006). Severe cerebral congophilic angiopathy coincident with increased brain aluminium in a resident of Camelford, Cornwall, UK. *Journal of Neurology, Neurosurgery, and Psychiatry*, *77*(7), 877–879. doi:10.1136/jnnp.2005.086553 PMID:16627535

Fahn, S. (1988). Parkinsonism. In J. B. Wyngaarden & L. H. Smith Jr., (Eds.), *Cecil's textbook of medicine* (pp. 2143–2147). Philadelphia: Saunders.

Fei, Q., McCormack, A. L., Di Monte, D. A., & Ethell, D. W. (2008). Paraquat Neurotoxicity Is Mediated by a Bak-dependent Mechanism. *The Journal of Biological Chemistry*, *283*(6), 3357–3364. doi:10.1074/jbc.M708451200 PMID:18056701

Filipov, N. M., Stewart, M. A., Carr, R. L., & Sistrunk, S. C. (2007). Dopaminergic toxicity of the herbicide atrazine in rat striatal slices. *Toxicology*, *232*(1–2), 68–78. doi:10.1016/j.tox.2006.12.007 PMID:17218051

Fleming, L., Mann, J. B., Bean, J., Briggle, T., & Sanchez-Ramos, J. R. (1994). Parkinson's disease and brain levels of organochlorine pesticides. *Annals of Neurology*, *36*(1), 100–103. doi:10.1002/ana.410360119 PMID:7517654

Franco, R., Li, S., Rodriguez-Rocha, H., Burns, M., & Panayiotidis, M. I. (2010). Molecular mechanisms of pesticide-induced neurotoxicity: Relevance to Parkinson's disease. *Chemico-Biological Interactions*, *188*(2), 289–300. doi:10.1016/j.cbi.2010.06.003 PMID:20542017

Frederickson, C. J. (1989). Neurobiology of Zinc and Zinc-Containing Neurons. *International Review of Neurobiology*, *31*, 145–238. doi:10.1016/S0074-7742(08)60279-2 PMID:2689380

Freyaldenhoven, T. E., Ali, S. F., & Hart, R. W. (1995). MPTP- and MPP+-induced effects on body temperature exhibit age- and strain-dependence in mice. *Brain Research*, *688*(1–2), 161–170. doi:10.1016/0006-8993(95)00529-Y PMID:8542303

Frölich, L., & Riederer, P. (1995). Free radical mechanisms in dementia of Alzheimer type and the potential for antioxidative treatment. *Arzneimittel-Forschung*, *45*(3A), 443–446. Retrieved from http://www.ncbi.nlm.nih.gov/pubmed/7763342 PMID:7763342

Fukushima, T., Yamada, K., Isobe, A., Shiwaku, K., & Yamane, Y. (1993). Mechanism of cytotoxicity of paraquat. *Experimental and Toxicologic Pathology*, *45*(5–6), 345–349. doi:10.1016/S0940-2993(11)80424-0 PMID:8312721

Fukuto, T. R. (1990). Mechanism of action of organophosphorus and carbamate insecticides. *Environmental Health Perspectives*, *87*, 245–254. doi:10.1289/ehp.9087245 PMID:2176588

Gao, H.-M., & Hong, J.-S. (2011). Gene-environment interactions: Key to unraveling the mystery of Parkinson's disease. *Progress in Neurobiology*, *94*(1), 1–19. doi:10.1016/j.pneurobio.2011.03.005 PMID:21439347

Gauvreau, D. (1987). Le paradigme de la maladie d'Alzheimer. *Interface*, *8*, 16–21.

Goldman, S. M. (2014). Environmental Toxins and Parkinson's Disease. *Annual Review of Pharmacology and Toxicology*, *54*(1), 141–164. doi:10.1146/annurev-pharmtox-011613-135937 PMID:24050700

Gorell, J. M., Peterson, E. L., Rybicki, B. A., & Johnson, C. C. (2004). Multiple risk factors for Parkinson's disease. *Journal of the Neurological Sciences*, *217*(2), 169–174. doi:10.1016/j.jns.2003.09.014 PMID:14706220

Greenamyre, J. T., Betarbet, R., & Sherer, T. B. (2003). The rotenone model of Parkinson's disease: Genes, environment and mitochondria. *Parkinsonism & Related Disorders*, *9*, 59–64. doi:10.1016/S1353-8020(03)00023-3 PMID:12915069

Guilarte, T. R. (2013). Manganese neurotoxicity: New perspectives from behavioral, neuroimaging, and neuropathological studies in humans and non-human primates. *Frontiers in Aging Neuroscience*, *5*. doi:10.3389/fnagi.2013.00023 PMID:23805100

Haacke, E. M., Cheng, N. Y. C., House, M. J., Liu, Q., Neelavalli, J., Ogg, R. J., ... Obenaus, A. (2005). Imaging iron stores in the brain using magnetic resonance imaging. *Magnetic Resonance Imaging*, *23*(1), 1–25. doi:10.1016/j.mri.2004.10.001 PMID:15733784

Halliwell, B. (1992). Reactive Oxygen Species and the Central Nervous System. *Journal of Neurochemistry*, *59*(5), 1609–1623. doi:10.1111/j.1471-4159.1992.tb10990.x PMID:1402908

Hallman, H., Lange, L., Olson, L., Strömberg, I., & Jonsson, G. (1985). Neurochemical and Histochemical Characterization of Neurotoxic Effects of 1-Methyl-4-Phenyl-1,2,3,6-Tetrahydropyridine on Brain Catecholamine Neurones in the Mouse. *Journal of Neurochemistry*, *44*(1), 117–127. doi:10.1111/j.1471-4159.1985.tb07120.x PMID:3880581

Hanagasi, H. A., Ayribas, D., Baysal, K., & Emre, M. (2005). Mitochondrial complex i, ii/iii, and iv activities in familial and sporadic parkinson's disease. *The International Journal of Neuroscience*, *115*(4), 479–493. doi:10.1080/00207450590523017 PMID:15809215

Hashimoto, M., Hsu, L. J., Xia, Y., Takeda, A., Sisk, A., Sundsmo, M., & Masliah, E. (1999). Oxidative stress induces amyloid–like aggregate formation of NACP/α–synuclein in vitro. *Neuroreport*, *10*(4), 717–721. doi:10.1097/00001756-199903170-00011 PMID:10208537

Hatcher, J., Pennell, K., & Miller, G. (2008). Parkinson's disease and pesticides: A toxicological perspective. *Trends in Pharmacological Sciences*, *29*(6), 322–329. doi:10.1016/j.tips.2008.03.007 PMID:18453001

Hertzman, C., Wiens, M., Bowering, D., Snow, B., & Calne, D. (1990). Parkinson's disease: A case-control study of occupational and environmental risk factors. *American Journal of Industrial Medicine*, *17*(3), 349–355. doi:10.1002/ajim.4700170307 PMID:2305814

Honjo, K., Black, S. E., & Verhoeff, N. P. L. G. (2012). Alzheimer's Disease, Cerebrovascular Disease, and the β-amyloid Cascade. *The Canadian Journal of Neurological Sciences*, *39*(06), 712–728. doi:10.1017/S0317167100015547 PMID:23227576

House, E., Collingwood, J., Khan, A., Korchazkina, O., Berthon, G., & Exley, C. (2004). Aluminium, iron, zinc and copper influence the in vitro formation of amyloid fibrils of Aβ42 in a manner which may have consequences for metal chelation therapy in Alzheimer's disease. *Journal of Alzheimer's Disease*, *6*(3), 291–301. doi:10.3233/JAD-2004-6310 PMID:15201484

Hung, A. S. M., Liang, Y., Chow, T. C. H., Tang, H. C., Wu, S. L. Y., Wai, M. S. M., & Yew, D. T. (2016). Mutated tau, amyloid and neuroinflammation in Alzheimer disease—A brief review. *Progress in Histochemistry and Cytochemistry*, *51*(1), 1–8. doi:10.1016/j.proghi.2016.01.001 PMID:26851150

Ikeda, Y., Toda, S., Wang, M., & Nakazawa, S. (1994). Early changes of blood-brain barrier and superoxide scavenging activity in rat cryogenic brain injury. In *Brain Edema IX* (pp. 136–138). Vienna: Springer. doi:10.1007/978-3-7091-9334-1_36

Ittner, L. M., & Götz, J. (2011). Amyloid-β and tau — a toxic pas de deux in Alzheimer's disease. *Nature Reviews. Neuroscience*, *12*(2), 67–72. doi:10.1038/nrn2967 PMID:21193853

Jankovic, J. (2008). Parkinson's disease: Clinical features and diagnosis. *Journal of Neurology, Neurosurgery, and Psychiatry*, *79*(4), 368–376. doi:10.1136/jnnp.2007.131045 PMID:18344392

Jean, H., Emard, J.-F., Thouez, J.-P., Houde, L., Robitaille, Y., Mathieu, J., ... Gauvreau, D. (1996). Alzheimer's disease: Preliminary study of spatial distribution at birth place. *Social Science & Medicine*, *42*(6), 871–878. doi:10.1016/0277-9536(95)00185-9 PMID:8778999

Jenner, P. (2003). Oxidative stress in Parkinson's disease. *Annals of Neurology*, *53*(S3), S26–S38. doi:10.1002/ana.10483 PMID:12666096

Johansson, C., Castoldi, A. F., Onishchenko, N., Manzo, L., Vahter, M., & Ceccatelli, S. (2007). Neurobehavioural and molecular changes induced by methylmercury exposure during development. *Neurotoxicity Research*, *11*(3–4), 241–260. doi:10.1007/BF03033570 PMID:17449462

Kalra, J., Kumar, P., Majeed, A. B. A., & Prakash, A. (2016). Modulation of LOX and COX pathways via inhibition of amyloidogenesis contributes to mitoprotection against β-amyloid oligomer-induced toxicity in an animal model of Alzheimer's disease in rats. *Pharmacology, Biochemistry, and Behavior*, *146–147*, 1–12. doi:10.1016/j.pbb.2016.04.002 PMID:27106205

Kamel, F., Tanner, C., Umbach, D., Hoppin, J., Alavanja, M., Blair, A., ... Sandler, D. (2006). Pesticide Exposure and Self-reported Parkinson's Disease in the Agricultural Health Study. *American Journal of Epidemiology*, *165*(4), 364–374. doi:10.1093/aje/kwk024 PMID:17116648

Kanthasamy, A. G., Kitazawa, M., Kanthasamy, A., & Anantharam, V. (2005). Dieldrin-Induced Neurotoxicity: Relevance to Parkinson's Disease Pathogenesis. *Neurotoxicology*, *26*(4), 701–719. doi:10.1016/j.neuro.2004.07.010 PMID:16112328

Kasten, M., Chade, A., & Tanner, C. M. (2007). Epidemiology of Parkinson's disease. *Handbook of Clinical Neurology*, *83*, 129–151. doi:10.1016/S0072-9752(07)83006-5 PMID:18808913

Kepp, K. P. (2012). Bioinorganic Chemistry of Alzheimer's Disease. *Chemical Reviews*, *112*(10), 5193–5239. doi:10.1021/cr300009x PMID:22793492

Khan, A., Dobson, J. P., & Exley, C. (2006). Redox cycling of iron by Aβ42. *Free Radical Biology & Medicine*, *40*(4), 557–569. doi:10.1016/j.freeradbiomed.2005.09.013 PMID:16458186

Kimbrough, R. D., Gaines, T. B., & Linder, R. E. (1971). The Ultrastructure of Livers of Rats Fed DDT and Dieldrin. *Archives of Environmental Health: An International Journal*, *22*(4), 460–467. doi:10.1080/00039896.1971.10665879 PMID:5545497

Kontakos, N., & Stokes, J. (1999). Monograph Series on Aging-related Diseases: XII. Parkinson s Disease Recent Developments and New Directions. *Chronic Diseases and Injuries in Canada*, *20*(2), 58–76. PMID:10455038

Kowall, N. W., Pendlebury, W. W., Kessler, J. B., Perl, D. P., & Beal, M. F. (1989). Aluminum-induced neurofibrillary degeneration affects a subset of neurons in rabbit cerebral cortex, basal forebrain and upper brainstem. *Neuroscience*, *29*(2), 329–337. doi:10.1016/0306-4522(89)90060-2 PMID:2725861

Kuhn, W., Winkel, R., Woitalla, D., Meves, S., Przuntek, H., & Muller, T. (1998). High prevalence of parkinsonism after occupational exposure to lead-sulfate batteries. *Neurology*, *50*(6), 1885–1886. doi:10.1212/WNL.50.6.1885 PMID:9633752

Kuter, K., Nowak, P., Gołembiowska, K., & Ossowska, K. (2010). Increased Reactive Oxygen Species Production in the Brain After Repeated Low-Dose Pesticide Paraquat Exposure in Rats. A Comparison with Peripheral Tissues. *Neurochemical Research*, *35*(8), 1121–1130. doi:10.100711064-010-0163-x PMID:20369291

Lang, A. E., & Obeso, J. A. (2004). Challenges in Parkinson's disease: Restoration of the nigrostriatal dopamine system is not enough. *Lancet Neurology*, *3*(5), 309–316. doi:10.1016/S1474-4422(04)00740-9 PMID:15099546

Langston, J. W., Ballard, P., Tetrud, J. W., & Irwin, I. (1983). Chronic Parkinsonism in humans due to a product of meperidine-analog synthesis. *Science*, *219*(4587), 979–980. doi:10.1126cience.6823561 PMID:6823561

Langston, J. W., Ballard, P., Tetrud, J. W., & Irwin, I. (1983). Chronic Parkinsonism in humans due to a product of meperidine-analog synthesis. *Science*, *219*(4587), 979–980. doi:10.1126/science.6823561 PubMed

Langston, J. W., & Ballard, P. A. Jr. (1983). Parkinson's disease in a chemist working with l-methyl-4-phenyl-l, 2, 5, 6-tetrahydropyridine. *The New England Journal of Medicine*, *309*(5). PMID:6602944

Langston, J. W., Forno, L. S., Rebert, C. S., & Irwin, I. (1984). Selective nigral toxicity after systemic administration of 1-methyl-4-phenyl-1, 2, 5, 6-tetrahydropyrine (MPTP) in the squirrel monkey. *Brain Research*, *292*(2), 390–394.

Lanska, D. J. (1997). The geographic distribution of Parkinson's disease mortality in the United States. *Journal of the Neurological Sciences*, *150*(1), 63–70. doi:10.1016/S0022-510X(97)05371-9 PMID:9260859

Le Couteur, D. G., McLean, A. J., Taylor, M. C., Woodham, B. L., & Board, P. G. (1999). Pesticides and Parkinson's disease. *Biomedicine and Pharmacotherapy*, *53*(3), 122–130. doi:10.1016/S0753-3322(99)80077-8 PMID:10349500

Li, X., Yin, J., Cheng, C., Sun, J., Li, Z., & Wu, Y. (2005). Paraquat induces selective dopaminergic nigrostriatal degeneration in aging C57BL/6 mice. *Chinese Medical Journal*, *118*(16), 1357–1361. Retrieved from http://www.ncbi.nlm.nih.gov/pubmed/16157030 PMID:16157030

Lim, S.-Y., Fox, S. H., & Lang, A. E. (2009). Overview of the Extranigral Aspects of Parkinson Disease. *Archives of Neurology*, *66*(2). doi:10.1001/archneurol.2008.561 PMID:19204152

Lin, M. T., & Beal, M. F. (2006). Mitochondrial dysfunction and oxidative stress in neurodegenerative diseases. *Nature*, *443*(7113), 787–795. doi:10.1038/nature05292 PMID:17051205

Liou, H. H., Tsai, M. C., Chen, C. J., Jeng, J. S., Chang, Y. C., Chen, S. Y., & Chen, R. C. (1997). Environmental risk factors and Parkinson's disease: A case-control study in Taiwan. *Neurology*, *48*(6), 1583–1588. doi:10.1212/WNL.48.6.1583 PMID:9191770

Logroscino, G., Gao, X., Chen, H., Wing, A., & Ascherio, A. (2008). Dietary Iron Intake and Risk of Parkinson's Disease. *American Journal of Epidemiology*, *168*(12), 1381–1388. doi:10.1093/aje/kwn273 PMID:18945687

Lopez, O., Hernandez, A., Rodrigo, L., Gil, F., Pena, G., Serrano, J., ... Pla, A. (2007). Changes in antioxidant enzymes in humans with long-term exposure to pesticides. *Toxicology Letters*, *171*(3), 146–153. doi:10.1016/j.toxlet.2007.05.004 PMID:17590542

Lukiw, W. J., Percy, M. E., & Kruck, T. P. (2005). Nanomolar aluminum induces pro-inflammatory and pro-apoptotic gene expression in human brain cells in primary culture. *Journal of Inorganic Biochemistry, 99*(9), 1895–1898. doi:10.1016/j.jinorgbio.2005.04.021 PMID:15961160

Manning-BoĞ, A. B., & Langston, J. W. (2007). Model fusion: The next phase in developing animal models for parkinson's disease. *Neurotoxicity Research, 11*(3–4), 219–240. doi:10.1007/BF03033569 PMID:17449461

Manthripragada, A. D., Costello, S., Cockburn, M. G., Bronstein, J. M., & Ritz, B. (2010). Paraoxonase 1, Agricultural Organophosphate Exposure, and Parkinson Disease. *Epidemiology (Cambridge, Mass.), 21*(1), 87–94. doi:10.1097/EDE.0b013e3181c15ec6 PMID:19907334

Mason, L. H., Harp, J. P., & Han, D. Y. (2014). Pb Neurotoxicity: Neuropsychological Effects of Lead Toxicity. *BioMed Research International, 2014*, 1–8. doi:10.1155/2014/840547 PMID:24516855

Matsubara, K., Kobayashi, S., Kobayashi, Y., Yamashita, K., Koide, H., Hatta, M., ... Kimura, K. (1995). beta-Carbolinium cations, endogenous MPPplus analogs, in the lumbar cerebrospinal fluid of patients with Parkinson's disease. *Neurology, 45*(12), 2240–2245. doi:10.1212/WNL.45.12.2240 PMID:8848200

Matyja, E. (2000). Aluminum enhances glutamate-mediated neurotoxicity in organotypic cultures of rat hippocampus. *Folia Neuropathologica, 38*(2), 47–53. Retrieved from http://www.ncbi.nlm.nih.gov/pubmed/11057035 PMID:11057035

McCarthy, S., Somayajulu, M., Sikorska, M., Borowy-Borowski, H., & Pandey, S. (2004). Paraquat induces oxidative stress and neuronal cell death; neuroprotection by water-soluble Coenzyme Q10. *Toxicology and Applied Pharmacology, 201*(1), 21–31. doi:10.1016/j.taap.2004.04.019 PMID:15519605

McCormack, A. L., Thiruchelvam, M., Manning-Bog, A. B., Thiffault, C., Langston, J. W., Cory-Slechta, D. A., & Di Monte, D. A. (2002). Environmental Risk Factors and Parkinson's Disease: Selective Degeneration of Nigral Dopaminergic Neurons Caused by the Herbicide Paraquat. *Neurobiology of Disease, 10*(2), 119–127. doi:10.1006/nbdi.2002.0507 PMID:12127150

McDermott, J. R., Smith, A. I., Iqbal, K., & Wisniewski, M.D., H. M. (. (1979). Brain aluminum in aging and Alzheimer disease. *Neurology, 29*(6), 809–809. doi:10.1212/WNL.29.6.809 PMID:572003

McDowell, I., Hill, G., Lindsay, J., Helliwell, B., Costa, L., Beattie, L., ... Parboosingh, J. (1994). The Canadian Study of Health and Aging: Risk factors for Alzheimer's disease in Canada. *Neurology, 44*(11), 2073–2080. doi:10.1212/WNL.44.11.2073 PMID:7969962

Meredith, G. E., Halliday, G. M., & Totterdell, S. (2004). A critical review of the development and importance of proteinaceous aggregates in animal models of Parkinson's disease: New insights into Lewy body formation. *Parkinsonism & Related Disorders, 10*(4), 191–202. doi:10.1016/j.parkreldis.2004.01.001 PMID:15120093

Miyake, Y., Tanaka, K., Fukushima, W., Sasaki, S., Kiyohara, C., Tsuboi, Y., ... Nagai, M. (2011). Dietary intake of metals and risk of Parkinson's disease: A case-control study in Japan. *Journal of the Neurological Sciences, 306*(1–2), 98–102. doi:10.1016/j.jns.2011.03.035 PMID:21497832

Moretto, A., & Colosio, C. (2011). Biochemical and toxicological evidence of neurological effects of pesticides: The example of Parkinson's disease. *Neurotoxicology*, *32*(4), 383–391. doi:10.1016/j.neuro.2011.03.004 PMID:21402100

Mortimer, J. A., Borenstein, A. R., & Nelson, L. M. (2012). Associations of welding and manganese exposure with Parkinson disease: Review and meta-analysis. *Neurology*, *79*(11), 1174–1180. doi:10.1212/WNL.0b013e3182698ced PMID:22965675

Nallagouni, C. S. R., & Pratap, K., R. (2017). Aluminum and fluoride impacts cortex, hippocampus and dentate gyrus structure in rats: Protective role of resveratrol. *International Journal of Applied Biology and Pharmaceutical Technology*, *8*, 89–97.

NourEddine, D., Miloud, S., & Abdelkader, A. (2005). Effect of lead exposure on dopaminergic transmission in the rat brain. *Toxicology*, *207*(3), 363–368. doi:10.1016/j.tox.2004.10.016 PMID:15664264

Nunomura, A., Moreira, P. I., Castellani, R. J., Lee, H., Zhu, X., Smith, M. A., & Perry, G. (2012). Oxidative Damage to RNA in Aging and Neurodegenerative Disorders. *Neurotoxicity Research*, *22*(3), 231–248. doi:10.100712640-012-9331-x PMID:22669748

Paillard, T., Rolland, Y., & de Souto Barreto, P. (2015). Protective Effects of Physical Exercise in Alzheimer's Disease and Parkinson's Disease: A Narrative Review. *Journal of Clinical Neurology (Seoul, Korea)*, *11*(3), 212. doi:10.3988/jcn.2015.11.3.212 PMID:26174783

Park, J. H., Lee, D. W., Park, K. S., & Joung, H. (2014). Serum trace metal levels in Alzheimer's disease and normal control groups. *American Journal of Alzheimer's Disease and Other Dementias*, *29*(1), 76–83. doi:10.1177/1533317513506778 PMID:24164932

Pasture, O., & Onkia, O. (1994). Canadian Study of Health and Aging: Study methods and prevalence of dementia. *Canadian Medical Association Journal*, *150*(6), 899–913. PMID:8131123

Peng, J., Peng, L., Stevenson, F. F., Doctrow, S. R., & Andersen, J. K. (2007). Iron and Paraquat as Synergistic Environmental Risk Factors in Sporadic Parkinson's Disease Accelerate Age-Related Neurodegeneration. *The Journal of Neuroscience*, *27*(26), 6914–6922. doi:10.1523/JNEUROSCI.1569-07.2007 PMID:17596439

Perry, E., Perry, R., Blessed, G., & Tomlinson, B. (1977). Necropsy evidence of central cholinergic deficits in senile dementia. *Lancet*, *309*(8004), 189. doi:10.1016/S0140-6736(77)91780-9 PMID:64712

Praticò, D. (2008). Oxidative stress hypothesis in Alzheimer's disease: A reappraisal. *Trends in Pharmacological Sciences*, *29*(12), 609–615. doi:10.1016/j.tips.2008.09.001 PMID:18838179

Prince, M., Wimo, A., Guerchet, M., Ali, G. C., Wu, Y. T., & Prina, M. (2015). *World Alzheimer Report 2015. The Global Impact of Dementia. Alzheimer's Disease International. Alzheimer's Disease International*. London: ADI.

Priyadarshi, A., Khuder, S. A., Schaub, E. A., & Priyadarshi, S. S. (2001). Environmental Risk Factors and Parkinson's Disease: A Metaanalysis. *Environmental Research*, *86*(2), 122–127. doi:10.1006/enrs.2001.4264 PMID:11437458

Priyadarshi, A., Khuder, S. A., Schaub, E. A., & Shrivastava, S. (2000). A meta-analysis of Parkinson's disease and exposure to pesticides. *Neurotoxicology*, *21*(4), 435–440. Retrieved from http://www.ncbi.nlm.nih.gov/pubmed/11022853 PMID:11022853

Qiu, C., Kivipelto, M., & von Strauss, E. (2009). Epidemiology of Alzheimer's disease: Occurrence, determinants, and strategies toward intervention. *Dialogues in Clinical Neuroscience*, *11*(2), 111–128. Retrieved from http://www.ncbi.nlm.nih.gov/pubmed/19585947 PMID:19585947

Rappold, P. M., Cui, M., Chesser, A. S., Tibbett, J., Grima, J. C., & Duan, L., … Tieu, K. (2011). Paraquat neurotoxicity is mediated by the dopamine transporter and organic cation transporter-3. *Proceedings of the National Academy of Sciences*, *108*(51), 20766–20771. 10.1073/pnas.1115141108

Reddy, P. H. (2006). Amyloid precursor protein-mediated free radicals and oxidative damage: Implications for the development and progression of Alzheimer's disease. *Journal of Neurochemistry*, *96*(1), 1–13. doi:10.1111/j.1471-4159.2005.03530.x PMID:16305625

Ritz, B., & Yu, F. (2000). Parkinson's disease mortality and pesticide exposure in California 1984–1994. *International Journal of Epidemiology*, *29*(2), 323–329. doi:10.1093/ije/29.2.323 PMID:10817132

Rybicki, B. A., Johnson, C. C., Uman, J., & Gorell, J. M. (1993). Parkinson's disease mortality and the industrial use of heavy metals in Michigan. *Movement Disorders*, *8*(1), 87–92. doi:10.1002/mds.870080116 PMID:8419812

Sayre, L. M., Perry, G., & Smith, M. A. (2008). Oxidative Stress and Neurotoxicity. *Chemical Research in Toxicology*, *21*(1), 172–188. doi:10.1021/tx700210j PMID:18052107

Schapira, A. H. V. (2009). Neurobiology and treatment of Parkinson's disease. *Trends in Pharmacological Sciences*, *30*(1), 41–47. doi:10.1016/j.tips.2008.10.005 PMID:19042040

Schenck, J. F., & Zimmerman, E. A. (2004). High-field magnetic resonance imaging of brain iron: Birth of a biomarker? *NMR in Biomedicine*, *17*(7), 433–445. doi:10.1002/nbm.922 PMID:15523705

Schon, E. A., & Manfredi, G. (2003). Neuronal degeneration and mitochondrial dysfunction. *The Journal of Clinical Investigation*, *111*(3), 303–312. doi:10.1172/JCI200317741 PMID:12569152

Shen, X.-L., Yu, J.-H., Zhang, D.-F., Xie, J.-X., & Jiang, H. (2014). Positive Relationship between Mortality from Alzheimer's Disease and Soil Metal Concentration in Mainland China. *Journal of Alzheimer's Disease*, *42*(3), 893–900. doi:10.3233/JAD-140153 PMID:25024310

Shibayama, H., Kasahara, Y., & Kobayashi, H. (1986). Prevalence of dementia in a Japanese elderly population. *Acta Psychiatrica Scandinavica*, *74*(2), 144–151. doi:10.1111/j.1600-0447.1986.tb10598.x PMID:3490741

Shibayama, H., Kobayashi, H., Iwase, S., Nakagawa, M., Marui, Y., Kayukawa, Y., … Takeuchi, T. (1986). Unusual Cases of Presenile Dementia with Fahr's Syndrome. *Psychiatry and Clinical Neurosciences*, *40*(1), 85–100. doi:10.1111/j.1440-1819.1986.tb01615.x PMID:3773356

Shimizu, K., Matsubara, K., Ohtaki, K., Fujimaru, S., Saito, O., & Shiono, H. (2003). Paraquat induces long-lasting dopamine overflow through the excitotoxic pathway in the striatum of freely moving rats. *Brain Research*, *976*(2), 243–252. doi:10.1016/S0006-8993(03)02750-1 PMID:12763259

Sian-Hülsmann, J., Mandel, S., Youdim, M. B. H., & Riederer, P. (2011). The relevance of iron in the pathogenesis of Parkinson's disease. *Journal of Neurochemistry*, *118*(6), 939–957. doi:10.1111/j.1471-4159.2010.07132.x PMID:21138437

Smith, M. A., Richey Harris, P. L., Sayre, L. M., Beckman, J. S., & Perry, G. (1997). Widespread Peroxynitrite-Mediated Damage in Alzheimer's Disease. *The Journal of Neuroscience*, *17*(8), 2653–2657. doi:10.1523/JNEUROSCI.17-08-02653.1997 PMID:9092586

Steenland, K., MacNeil, J., Vega, I., & Levey, A. (2009). Recent trends in Alzheimer disease mortality in the United States, 1999 to 2004. *Alzheimer Disease and Associated Disorders*, *23*(2), 165–170. doi:10.1097/WAD.0b013e3181902c3e PMID:19484918

Stephens, R., Spurgeon, A., Calvert, I. A., Beach, J., Levy, L. S., Harrington, J. M., & Berry, H. (1995). Neuropsychological effects of long-term exposure to organophosphates in sheep dip. *Lancet*, *345*(8958), 1135–1139. doi:10.1016/S0140-6736(95)90976-1 PMID:7723544

Stohs, S. J., & Bagchi, D. (1995). Oxidative mechanisms in the toxicity of metal ions. *Free Radical Biology & Medicine*, *18*(2), 321–336. doi:10.1016/0891-5849(94)00159-H PMID:7744317

Su, L., Blamire, A. M., Watson, R., He, J., Hayes, L., & O'Brien, J. T. (2016). Whole-brain patterns of 1H-magnetic resonance spectroscopy imaging in Alzheimer's disease and dementia with Lewy bodies. *Translational Psychiatry*, *6*(8), e877–e877. doi:10.1038/tp.2016.140 PMID:27576166

Tanner, C. M., Kamel, F., Ross, G. W., Hoppin, J. A., Goldman, S. M., Korell, M., ... Langston, J. W. (2011). Rotenone, Paraquat, and Parkinson's Disease. *Environmental Health Perspectives*, *119*(6), 866–872. doi:10.1289/ehp.1002839 PMID:21269927

Tanner, C. M., Ottman, R., Goldman, S. M., Ellenberg, J., Chan, P., Mayeux, R., & Langston, J. W. (1999). Parkinson disease in twins: An etiologic study. *Journal of the American Medical Association*, *281*(4), 341–346. doi:10.1001/jama.281.4.341 PMID:9929087

Thiruchelvam, M., Richfield, E. K., Baggs, R. B., Tank, A. W., & Cory-Slechta, D. A. (2000). The nigrostriatal dopaminergic system as a preferential target of repeated exposures to combined paraquat and maneb: Implications for Parkinson's disease. *The Journal of Neuroscience*, *20*(24), 9207–9214. doi:10.1523/JNEUROSCI.20-24-09207.2000 PMID:11124998

Thiruchelvam, M., Richfield, E. K., Goodman, B. M., Baggs, R. B., & Cory-Slechta, D. A. (2002). Developmental Exposure to the Pesticides Paraquat and Maneb and the Parkinson's Disease Phenotype. *Neurotoxicology*, *23*(4–5), 621–633. doi:10.1016/S0161-813X(02)00092-X PMID:12428734

Van Broeckhoven, C. L. (1995). Molecular genetics of Alzheimer disease: Identification of genes and gene mutations. *European Neurology*, *35*(1), 8–19. doi:10.1159/000117083 PMID:7737252 von Campenhausen, S., Bornschein, B., Wick, R., Bötzel, K., Sampaio, C., Poewe, W., ... Dodel, R. (2005). Prevalence and incidence of Parkinson's disease in Europe. *European Neuropsychopharmacology : The Journal of the European College of Neuropsychopharmacology*, *15*(4), 473–490. doi:10.1016/j.euroneuro.2005.04.007 PMID:15963700

Walton, J. R. (2006). Aluminum in hippocampal neurons from humans with Alzheimer's disease. *Neurotoxicology*, *27*(3), 385–394. doi:10.1016/j.neuro.2005.11.007 PMID:16458972

Walton, J. R. (2009a). Brain lesions comprised of aluminum-rich cells that lack microtubules may be associated with the cognitive deficit of Alzheimer's disease. *Neurotoxicology*, *30*(6), 1059–1069. doi:10.1016/j.neuro.2009.06.010 PMID:19591863

Walton, J. R. (2009b). Functional impairment in aged rats chronically exposed to human range dietary aluminum equivalents. *Neurotoxicology*, *30*(2), 182–193. doi:10.1016/j.neuro.2008.11.012 PMID:19109991

Wang, A., Costello, S., Cockburn, M., Zhang, X., Bronstein, J., & Ritz, B. (2011). Parkinson's disease risk from ambient exposure to pesticides. *European Journal of Epidemiology*, *26*(7), 547–555. doi:10.100710654-011-9574-5 PMID:21505849

Wang, B., & Du, Y. (2013). Cadmium and Its Neurotoxic Effects. *Oxidative Medicine and Cellular Longevity*, *2013*, 1–12. doi:10.1155/2013/898034 PMID:23997854

Warner, T. T., & Schapira, A. H. V. (2003). Genetic and environmental factors in the cause of Parkinson's disease. *Annals of Neurology*, *53*(S3), S16–S25. doi:10.1002/ana.10487 PMID:12666095

Wenstrup, D., Ehman, W. D., & Markesbery, W. R. (1990). Trace element imbalances in isolated subcellular fractions of Alzheimer's disease brains. *Brain Research*, *533*(1), 125–131. doi:10.1016/0006-8993(90)91804-P PMID:2085723

William Langston, J., Forno, L. S., Rebert, C. S., & Irwin, I. (1984). Selective nigral toxicity after systemic administration of 1-methyl-4-phenyl-1,2,5,6-tetrahydropyrine (MPTP) in the squirrel monkey. *Brain Research*, *292*(2), 390–394. doi:10.1016/0006-8993(84)90777-7 PMID:6607092

Wirdefeldt, K., Adami, H.-O., Cole, P., Trichopoulos, D., & Mandel, J. (2011). Epidemiology and etiology of Parkinson's disease: A review of the evidence. *European Journal of Epidemiology*, *26*(S1), 1–58. doi:10.100710654-011-9581-6 PMID:21626386

Wojcik, D. P., Godfrey, M. E., Christie, D., & Haley, B. E. (2006). Mercury toxicity presenting as chronic fatigue, memory impairment and depression: Diagnosis, treatment, susceptibility, and outcomes in a New Zealand general practice setting (1994-2006). *Neuroendocrinology Letters*, *27*(4), 415–423. Retrieved from http://www.ncbi.nlm.nih.gov/pubmed/16891999 PMID:16891999

Zecca, L., Fariello, R., Riederer, P., Sulzer, D., Gatti, A., & Tampellini, D. (2002). The absolute concentration of nigral neuromelanin, assayed by a new sensitive method, increases throughout the life and is dramatically decreased in Parkinson's disease. *FEBS Letters*, *510*(3), 216–220. doi:10.1016/S0014-5793(01)03269-0 PMID:11801257

Zecca, L., Youdim, M. B. H., Riederer, P., Connor, J. R., & Crichton, R. R. (2004). Iron, brain ageing and neurodegenerative disorders. *Nature Reviews. Neuroscience*, *5*(11), 863–873. doi:10.1038/nrn1537 PMID:15496864

Zheng, W., & Monnot, A. D. (2012). Regulation of brain iron and copper homeostasis by brain barrier systems: Implication in neurodegenerative diseases. *Pharmacology & Therapeutics*, *133*(2), 177–188. doi:10.1016/j.pharmthera.2011.10.006 PMID:22115751

Chapter 13
Neuropathogenesis of Multiple Sclerosis and Huntington's Disease:
An Overview of Environment Patterns

Omar El Hiba
Chouaib Doukkali University, Morocco & Cadi Ayyad University, Morocco

Nadia Zouhairi
Cadi Ayyad University, Morocco

Hicham Chatoui
Private University of Marrakech, Morocco & Cadi Ayyad University, Morocco

Tiziano Balzano
Príncipe Felipe Research Center, Spain

Hind Benammi
Cadi Ayyad University, Morocco

Faical Isbaine
Emory University, USA

Abdelali Bitar
Chouaib Doukkali University, Morocco

Hamid Rguibi Idrissi
Mohamed V University, Morocco

Ismail El Moudden
Mohammed V University, Morocco

Fatima Zahra Marhoume
Cadi Ayyad University, Morocco

Halima Gamrani
Cadi Ayyad University, Morocco

ABSTRACT

Substantial epidemiological studies have established a link between environmental factors and prevalence of neurodegenerative diseases. Increasingly, some studies report a direct cause-and-effect relation between environmental agents such as heavy metals and pesticides, and some neurodegenerative disorders including multiple sclerosis (MS) and Huntington's disease (HD). Interestingly, high blood level of heavy metals and pesticides has been shown in patients with MS and HD. Those agents could be involved directly or indirectly in the pathogenesis of MS and HD. The underlying mechanisms may imply an immune breaking of self-tolerance or neurodegeneration onset of several neurotransmission systems. The chapter will discuss the role of different metals and pesticides in the onset and progress of MS and HD with an overview of the possible underlying pathomechanisms.

DOI: 10.4018/978-1-5225-7775-1.ch013

INTRODUCTION

Multiple sclerosis (MS) and Huntington disease (HD) are two different neurodegenerative disorders, however, in spite of being pathologically distinguished, quite similarities between the two disorders exist including several cognitive functions that have been shown to be altered similarly including: reading, interpretation of metaphors and proverbs, sequential finger tapping (motor speed), motor praxis, Trails errors, and orientation (Caine, Bamford, Schiffer, Shoulson & Levy, 1986).

MS is known as a neurological disease with chronic autoimmune and neuroinflammatory features (Calabresi, 2004; Fauci, Braunwald, Kasper & Hauser, 2008). Pathologicallay, MS is primarily characterized by attack of the myelinated axons of the central nervous system (CNS), leading to a progressive destruction of the myelin and consequently axons (Weinshenker, 1996).

Neurologically, MS patients show episodes of reversible neurological deficits, generally followed by progressive and irreversible neurological deterioration. Statistics in U.S, affirm that 250,000 to 350,000 patients are suffering from MS with more incidence in men compared to women (Singh, Mehrotra & Agarwal, 1999), while 50% of those patients need help walking within 15 years after the onset of the disease (Navikas & Link, 1996). While HD, firstly named as Huntington chorea, is a neurodegenerative disorder generally of a genetic origin with onset in middle age. HD is characterized by abnormal choreatic movements, as well as several behavioral and psychiatric disturbances and even dementia (Bruyn, 1968). Its old appellation (H chorea) was maintained during several decades until 1990s, when light have been shed on the extensive non-motor symptoms of the disease, hence, H chorea was changed to H disease (Hunington's disease collaborative research group, 1993). Increased body of evidence support the involvement of environmental factors in the pathogenesis of MS and HD. Indeed, numerous environmental exposures such as heavy metals, organic solvents, ultraviolet radiation, infection, and diet have been investigated as possible etiologic factors which could interact directly with responsible genes and my precipitate symptoms onset and progress of the disease. (Aminzadeh & Etminan, 2007; Ascherio & Munger, 2007; Ascherio, Munger & Simon, 2010; Irvine, Schiefer & Hader, 1988; Ketelaar, Hofstra & Hayden, 2012; Lauer & Firnhaber, 1985; Landtblom, Flodin, Söderfeldt, Wolfson & Axelson, 1996; Marrie, 2004; Moszczynski, 2011).

In the present chapter, we will discuss the role of environmental exposures, with a focus on heavy metals and pesticides, both as a risk factors for MS and HD pathogenesis and progress with the underlying pathomechanisms in order to delineate the role of the environmental part as piece of puzzle of neurodegenerative disorders.

CLINICAL FEATURES AND PHYSIOPATHOLOGY OF MS

Symptomatology of MS

The history of MS dates back to the 14th century when anatomoclinical methods made the first time a correlation between the clinical features of the disease and the post-mortem neuropathological characteristics (Lublin, 2005). The recognition of MS as a distinct disease was quite a feat for the time. Since, many diseases in the early 19th century that would now be categorized as either neurological or psychiatric would have been classified as "nervous disorders" without separation between individual conditions (Murray, 2009). Such an attempt at the classification of neurological diseases had not been

Figure 1. Classification of the different forms of MS. A: Relapsing-remitting MS, B: secondary progressive MS, C: primary progressive MS, D: progressive-relapsing MS

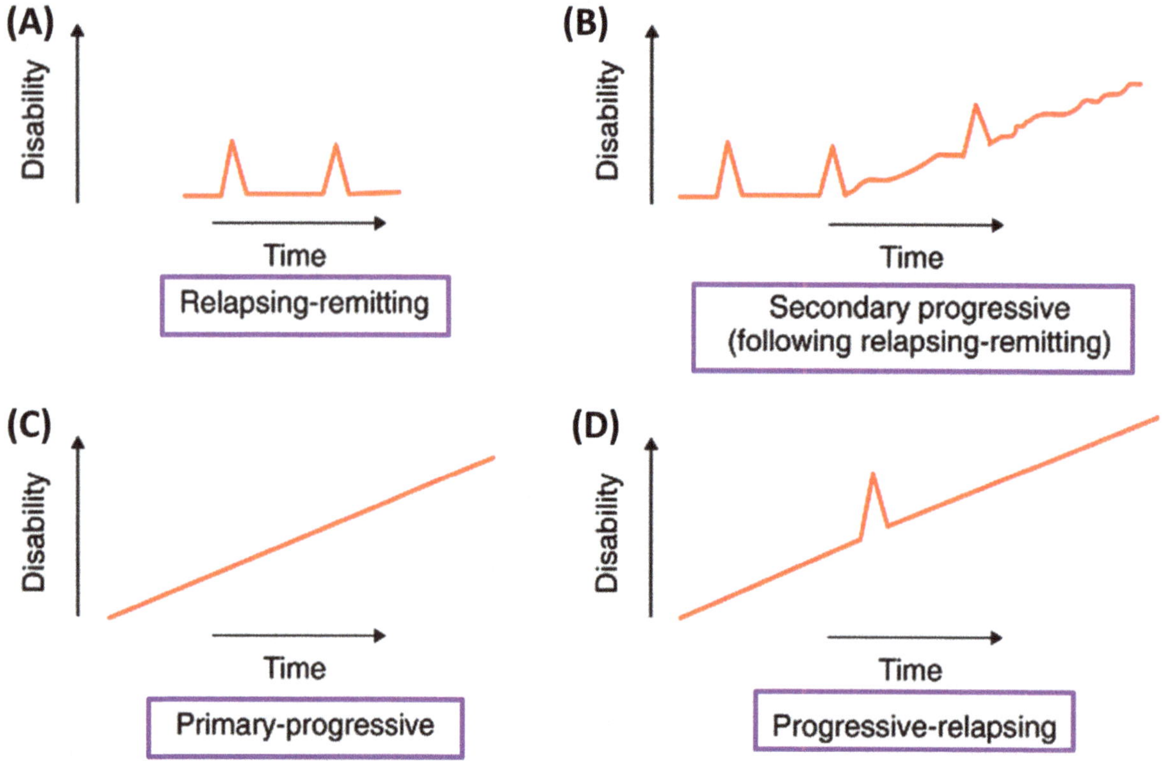

undertaken prior to Charcot (Tollis, 1996). Only a small group of illnesses such as epilepsy, paraplegia, and neurosyphilis were differentiated at the time (Murray, 2009).

On the basis of the neurological features of MS, neurologists agreed about grouping patients into 4 major categories depending on the development of the disease (Figure 1) (Fauci, Braunwald, Kasper & Hauser, 2008).

Physiopathology of MS

Pathology Basis of MS

The hallmark of MS is the presence of plaques of inflammatory demyelination, CNS injury and axonal loss (Figure 2) (Frohman, Racke & Raine, 2006). Plaques are present in the brain as well as in the spinal cord, preferentially within the white matter surrounding the ventricles, the optic nerve and tracts, the corpus callosum, cerebellar peduncles, the brainstem, and also in the gray matter (Compston & Coles, 2008).

At the beginning of relapsing remitting stage, a pronounced demyelination appears with axonal loss and reactive gliosis (Van Der Valk & De Groot, 2000), while patients show focal inflammatory plaques containing demyelinated axons, reduced number of oligodendrocytes, astrocyte proliferation with subsequent gliosis, transected axons, and perivenular as well as parenchymal infiltrates of lymphocytes and macrophages. At the advanced stages, a diffuse gray and white matter atrophy is usually noticed which

Figure 2. Normal and destroyed myelin of Schwann cells and oligodendrocytes in MS

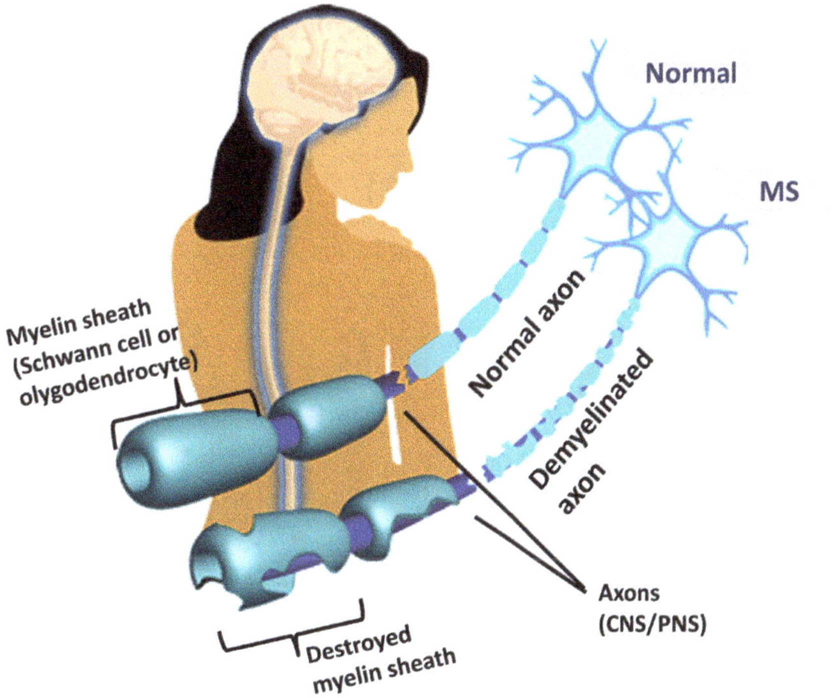

is featured by low-grade inflammation and microglial activation at the plaque borders concomitant with diffuse injury of the normal-appearing white matter outside the plaque (Figure 3) (Mahad, Trapp & Lassmann, 2015).

In the burnt-out stage the presence of iron in the brain in the absence of additional oxidative burst from microglia does not lead to further neurodegeneration, although patients may become affected by concomitant diseases such as stroke or Alzheimer's disease (figure 3).

Involvement of Heavy Metals in the Pathogenesis of MS

Beside the genetic etiology of MS, the influence of environment is well sustained by cumulative evidences that environmental agents can drive a T-cell-mediated chronic inflammatory response to myelin proteins in genetically predisposed individuals (Napier et al., 2016).

Role of Lead

The theory of Pb as an etiology of MS is very old and its originator was Putnam (1887a, b) who discovered the presence of Pb in the urine of a small number of patients with MS and all of whom had been exposed to Pb during occupation.

Several years later, Cone et al. (1934), confirmed the presence of Pb in CSF, urine, and faeces of 6 patients with the disease.

A recent study performed by Aliomrani et al., (2016), in 69 patients with relapsing-remitting MS compared to 74 age/gender-matched healthy individuals resident in Tehran as most polluted city in Iran,

Figure 3. Kinetic evolution of MS patterns pathology in cortical, white matter and deep grey matter demyelination. Acute MS (A), Relapsing-remitting MS (B), Secondary Progressive MS (C) and Primary Progressive MS (D). Green color refers to white matter demyelination, yellow color refers to white matter remyelination, the blue refers to deep gray matter lesion and the red one refers to cortical lesion (modified Haider et al., 2014).

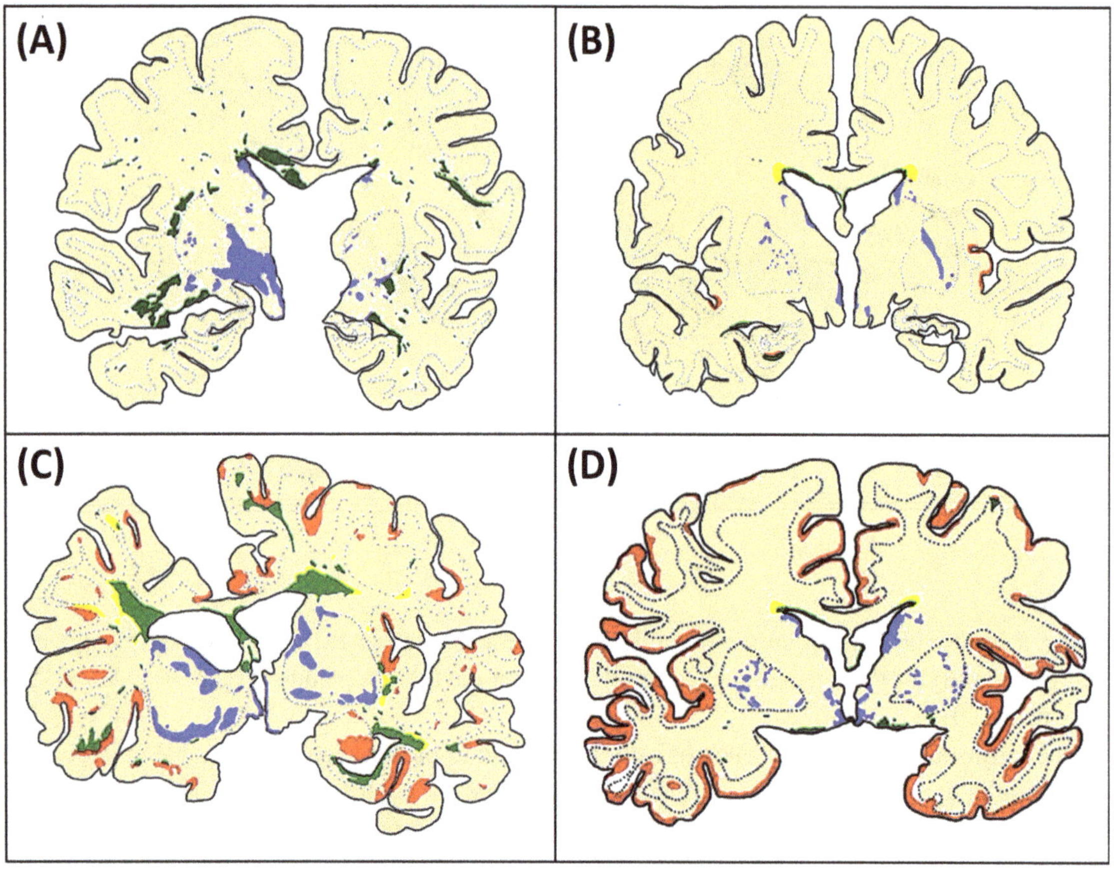

authors measured the level of Pb in blood of males and females of the studied population, they showed an increased Pb level in MS male patients particularly, while female MS patients didn't showed any difference compared to healthy controls (Aliomrani et al., 2016).

In addition, a low-to-moderate level of Pb exposure is able to trigger several functional alterations in T-lymphocytes and macrophages, and consequently increased hypersensitivity together with impaired cytokine production, which increases risk of inflammation-associated to tissue damage (Dietert & Piepenbrink, 2006).

Support of the view that Pb is mainly involved in the pathophysiology of MS, is provided by this interesting investigation published in Sciences, 1970 (Sauer, Zook & Garner, 1970) where authors investigated Pb poisoning in four primates, they showed by histological examination, a focal, symmetrical demyelination of subcortical white matter, which has been associated with degenerative and proliferative vascular lesions. In addition, wet liver and kidney tissue contained lead in concentrations of 110 and 120 PPM, respectively.

Role of Mercury

Mercury as well, can be considered as one of the heavy metals that can be associated to MS. Indeed, a comparison between circulating blood Pb in MS subjects with removed dental amalgams to MS subjects with amalgams, showed decreased erythrocytes density, hemoglobin and hematocrit levels in MS patients. Other parameters were changed including thyroxine levels which were lower in the MS amalgam subjects with significantly lower total T Lymphocytes and T-8 (CD8) suppressor cells number. Similarly, MS amalgam patients exhibited significantly increased blood urea nitrogen and lower serum IgG, while in hair, mercury level was largely increased in MS patients compared controls (non-MS) (Siblerud & Kienholz, 1994).

Indeed, in humans and animals, cell accumulation of mercury as well as Pb has been widely associated with the development of autoantibodies against different neuronal components such as: cytoskeletal proteins, neurofilaments, and myelin basic protein (El-Fawal, Waterman, De Feo & Shamy, 1999; Stejskal & Stejskal, 1999; Hansson et al., 2005). Whereas, excessive exposure to those metals has been shown to be highly neurotoxic, especially for motor neurons (Callaghan, Feldman, Gruis & Feldman, 2011). Otherwise, in non-predisposed animals, exposure to inorganic mercury is able to trigger prior autoimmune disease (Via et al., 2003), while in genetically susceptible animals, the metal, even at low doses, can precipitate autoimmune diseases leading to impaired cytokine production (Hansson et al., 2005).

Role Cadmium

Cadmium is one of the most important neurotoxic metals that have been associated to the pathogenesis of MS. Indeed, Aliomrani et al. (2016), have assessed blood cadmium levels in 69 patients with relapsing-remitting MS resident in Tehran; they have found an increased Cd level in MS patients particularly men while compared to controls. (Aliomrani et al., 2016).

Different routes by which Cd can reach the body including respiratory and digestive tracts. Once in the body, Cd can bind to metallothionein forming a complex: Cd-MT. However, the precise underlying mechanisms of Cd neurotoxicity is not yet fully understood. It has been suggested that Cd indirectly enhances the free radical generation and participates in oxidative stress via Fenton reaction (Jomova & Valko, 2011).

Role of Pesticides

Likewise, pesticides are also another environment agent able de drive MS. Several findings support such view, indeed, a correlation between exposure levels to pesticides and MS prevalence in a Spanish population, indicates that the prevalence and risk of MS is greater in populations living in areas of high pesticide use (indicating a higher exposure potential) relative to those of low pesticide use (Table 1 and 2). Otherwise, females showed an increased risk for MS in areas of high pesticide exposure compared to males (Parrón Requena, Hernández & Alarcón, 2011).

The sex differences in the prevalence of MS between men and women may involve the physiological differences between both sexes. Indeed, taken into account the involvement of heavy metals in the pathogenesis of MS, it has been shown that, estrogen causes enhanced copper absorption and extends its half-life (Johnson, Milne & Lykken, 1992). During adolescence, females have a high Zn requirement while, Cu excessively absorbed Cu interacts with Zn metabolism by reducing its absorption. Otherwise, the brain production of melatonin which is required for Zn absorption is highly reduced adolescence.

Table 1. Representation of the MS prevalence (100,000 inhabitants), sex differences, odds ratio (OR) and 95% confidence interval (95% CI) for a number of MS in the population living in high exposure areas relative to that resided in low exposure areas (according to Parrón, Requena, Hernández & Alarcón, 2011)

	Prevalence rate		OR	95% CI	P value
	High Exposure	Low Exposure			
Total population	41.62	33.88	1.23	1.05–1.43	0.007
Males	28.21	26.89	1.05	0.81-1.35	0.007
Females	55.08	40.79	1.35	1.11-1.64	0.001

Table 2. Representation of differences in the mean age for the medical conditions at the time of the first hospital discharge after diagnosis (according to the Andalusian Minimum Dataset) between high exposure and low exposure areas in the study population and after stratifying by gender. (according to Parrón, Requena, Hernández & Alarcón, 2011)

Exposure	Mean Age (Total Population)	P Value	Mean Age (Males)	P Value	Mean Age (Females)	P Value
High dose **Low dose**	39.44 42.78	<0.05	41.33 44.16	<0.10	38.47 41.84	<0.05

Consequently, a gradual depletion of Zn occurs. Otherwise, Zn is necessary for the zinc fingers of the protein TPO-1 involved in myelin synthesis in the oligodendrocyte (Krueger et al., 1997).

In addition, most of the damage in MS is due to peroxinitrite which results from excessive superoxide, due to deficient CuZn SOD affinity for Zn (Kesel et al., 1999). Also, Cu inhibits the synthesis of uric acid by Xanthine oxidase (Konstantinova, Russanova & Russanov, 1991), explaining the abnormally low levels of uric acid in the CSF during MS

Physiopathology of HD

Pathology Basis of HD

HD is a rare neuropsychiatric disorder genetically caused by dominant autosomal mutations. Clinically HD is characterized by movement disorders and cognitive decline. The cardinal features of the disease are motor defects including chorea and loss of coordination. Whereas, other psychiatric symptoms are also common to HD such as depression, psychosis, and obsessive–compulsive disorder (Rosenblatt. 2007). The prevalence of HD is 4-10 cases/100,000 in populations of Western European origin. Neurologically, HD patients manifest a general shrinkage of the brain and degeneration of the striatum especially efferent medium spiny neurons (Reiner et al., 1988). Although, specific thinning of the brain cortex was also observed in patients with HD (Rosas et al., 2002). Cortical degeneration extends from posterior to the anterior parts during the disease progression, which may be behind the heterogeneity the clinical features of HD. Genetically, HD is caused by a mutations in the gene encoding for huntingtin (HTT gene) which is an ubiquitously expressed protein of 350 kDa (figure 4) (Huntington's Disease Collaborative Research Group, 1993) with several functions including early embryonic development and neurogenesis, us knockout mice show embryonic lethality around day 8.5, before the emergence of the nervous system

Figure 4. Localization of the HD gene in band 4p16.3 of chromosome 4 (Adapted from Gusella, MacDonald, Ambrose & Duyao, 1994)

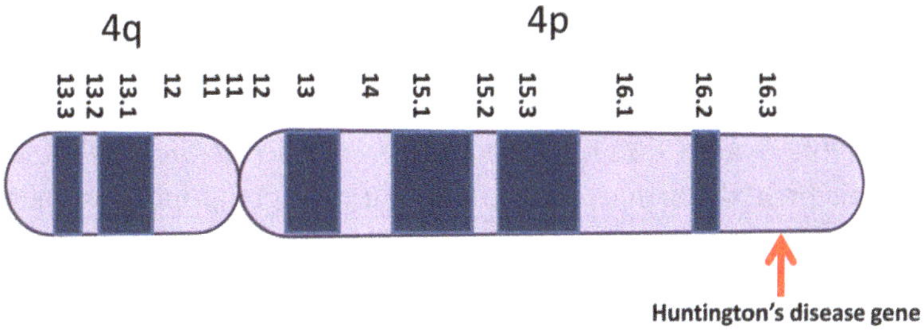

(Nasir et al. 1995; Zeitlin, Liu, Chapman, Papaioannou & Efstratiadis, 1995). Huntingtin is a scaffolding protein which interacts with b-tubulin and binds to microtubules (Hoffner, Kahlem & Djian. 2002). It may be considered also as a transcriptional factor; us the nuclear localization confers huntingtin a role in transcriptional regulation (Kegel et al. 2002) as well as synaptic connectivity (Di Figlia et al. 1995)

The mutant gene of HTT contains a polyglutamine tract encoded by uninterrupted CAG trinucleotide repeats (more then 36) while the wild type up to 35 CAG repeats (Rubinsztein et al.,1996). A particular aspect of HD is phenomenon of anticipation. In successive generations, the onset age of the disease become earlier resulting from cumulative mutations of the HTT gene due to the unstable nature of the CAG repeats which increase in size, particularly when passed through the male germline (Trottier, Biancalana & Mandel, 1994).

The different pathomechanisms suggested for HD are summarized in the Figure 5.

Figure 5. Schematic representation of the different pathophysiological mechanisms of HD

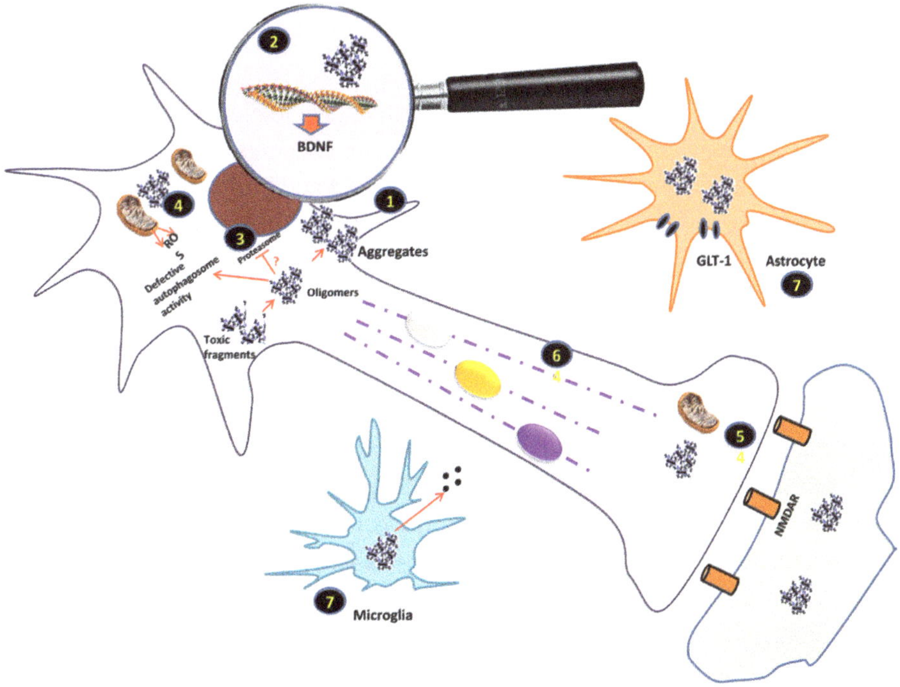

1. The mutant HTT protein (polyglutamine expansion) is cleaved in fragments with toxicity ptotential arising from the oligomeric soluble parts or from the insoluble inclusions (aggregates) that scavenge essential proteins (Jimenez-Sanchez, Licitra, Underwood & Rubinsztein, 2017).
2. Transcription of essential genes such as BDNF and chromatin modifications is also altered during HD.
3. During HD, a possible dysfunction of proteasome could be another pathomechanism as well as impaired autophagic breakdown due to cargo sequestration or autophagosome transport leading to alteration of proteins homeostasis.
4. Defective Ca2+ homeostasis leads to a mitochondrial dysfunction especially of protein import, impaired mitochondrial dynamics (increased fragmentation), oxidative stress onset and decreased ATP genesis.
5. Synaptic plasticity could be also altered during HD;a possible excitotoxicity could arise from alteration of postsynaptic NMDAR, from glutamate uptake, or from impaired synaptic vesicle trafficking.
6. Impairment of synaptic transport of organelle (mitochondria, autophagosomes, synaptic vesicles) and neurotransmitters
7. Neuroglia dysfunction: beyond neurons, astrocytes and microglia are involved as well. Secretion of pro-inflammatory cytokines and chemokines originating from microglia and alteration of astrocytic glutamate uptake.

BDNF (Brain-derivedneurotrophicfactor), ROS (reactive oxygen species), NMDAR (N-methyl-D-aspartate receptor). GLT-1 (glutamate transporter 1). (modified Jimenez-Sanchez, Licitra, Underwood & Rubinsztein, 2017)

Macroscopical Noticeable Brain Lesions in HD Patients

Gross examination of postmortem brains of HD shows a remarkable and characteristic bilateral atrophy of the striatum (Vonsattel & DiFiglia, 1998). Chronologically, brain atrophy and neurodegeneration adopt a topographical distribution: the tail and body of the caudate nucleus show more degeneration than the head in the very early stages of the degenerative process. Degeneration in the caudate nucleus and the putamen occurs earlier in the tail and progress to the head and body (Vonsattel & DiFiglia, 1998). Other brain areas are also touched such as the cerebral cortex and the centrum semi-ovale which show features of sever atrophy.

Involvement of Heavy Metals in the Pathogenesis of HD

It well established that not all HD patients with similar polyQ tract length develop the disease at the same age (Rosenblatt et al., 2001). Such observation leads to suppose the effect of environment on the progress of the disease. Whereas, the lack of correlation between phenotype and genotype in some cases of HD, especially monozygotic twins, may provide clues to the role of environment on the onset/progress of monogenic diseases (Ketelaar, Hofstra & Hayden, 2012) as well as other polyglutamine disorders such as spinocerebellar ataxias (Anderson et al., 2002; Durr, 2010).

In a relatively recent report, correlation between substance abuse and HD showed that individuals with histories of alcohol or drug abuse had a significantly lower age of onset of the disease (Byars et al., 2012).

Involvement of heavy metals in basal ganglia disorders is very well established, hence, such association may involve those metals in the pathogenesis of HD as an environmental rather than genetic cause of the disease.

Due to its particularity as a highly metabolizing organ and the detoxification function against reactive oxygen species, the brain solicits the presence of high concentrations of the biologically essential heavy metal ions (Fe, Mn, Cu, Zn) which are important cofactors for enzymes that regulate these processes. However, accumulation of metal ion leads to oxidative stress onset, and cellular damages (mitochondrial dysfunction, protein aggregation, and apoptosis).

Similarities between metals (especially heavy metals) neurotoxicity and neurodegenerative diseases (Gaeta & Hider, 2005; Jomova, Vondrakova, Lawson & Valko, 2010; Molina-Holgado, Hider, Gaeta, Williams & Francis, 2007) support the view of a possible involvement of those metals in the pathophysiology of HD as a neurodegenerative disorder.

Role of Manganese (Mn)

An over view of literature shows that Mn exposure at high levels is associated to neurotoxicity with preference to some brain regions such as the globus pallidus, striatum, substantia nigra and the subthalamic nucleus (Olanow, 2004; Pal, Samii & Calne, 1999).

A comparison of striatal proteomes of wild-type and HD (YAC128Q) mice by two-dimensional differential gel electrophoresis (DIGE) identified 16 proteins whose expression was altered on exposure to Mn in case of HD mice (Wegrzynowicz, Holt, Friedman, & Bowman, 2012). Most of these proteins were associated with cytoskeletal mobility, glutamatergic neurotransmission and energy metabolism. Such finding lead to suggest that exposure to Mn induces toxicity in HD mice by altering the interaction of the mutant huntingtin (128Q) with these proteins, thus affecting various cellular processes.

In addition, Mn is associated with alterations in integrity of DAergic striatal neurons and DA neurochemistry, including decreased DA transport function and/or striatal DA levels (Aschner, Guilarte, Schneider & Zheng, 2007; Perl & Olanow, 2007). The medium spiny neurons (MSN) are the target of the dopaminergic innervation of the striatum, comprising more than 90% of striatal neurons (Deutch, Colbran & Winder, 2007). MSN have radially projecting dendrites that are densely studded with spines, synapsing with dopamine and glutamate axons and providing the site of integration of several key inputs and outputs of the striatum (Day et al., 2006). A relevant study showed following quantitative morphometric analyses of Golgi-impregnated striatal sections from mice exposed to single or three Mn injections, a progressive spine degeneration and dendritic damage of medium spiny neurons (MSNs) (Milatovic, Zaja-Milatovic, Gupta, Yu & Aschner, 2009). Consequently, alterations in dendritic length and dendritic spine number may destabilize the structural basis of synaptic communication and thus compromise MSN function.

Whereas, an increase in brain Mn levels is also known to cause motor dysfunction in humans, non-human primates and rodents (Aschner, 2000; Autissier et al., 1982; Calne, Chu, Huang, Lu & Olanow, 1994; Gianutsos & Murray, 1982; Guilarte et al., 2006; Gwiazda, Lee, Sheridan & Smith, 2002; Kontur & Fechter, 1988; Milatovic, Zaja-Milatovic, Gupta, Yu & Aschner, 2009; Olanow et al., 1996; Pal, Samii & Calne, 1999; Olanow, 2004; Reaney, Bench, & Smith, 2006; Struve, McManus, Wong, & Dorman, 2007; Vezer et al., 2005; Vidal et al., 2005). Since the basal ganglia, including striatum are a common target for both HD neuropathology and Mn accumulation, this provides the opportunity to observe a disease-toxicant interaction.

Role of Copper

Previous findings support an accumulation of copper in the striatum of post-mortem HD human brain (Dexter et al., 1991). Preferentially, such accumulation occurs within basal ganglia. Support of this view is provided by observation in patients with Wilson's disease; a genetic disorder in which copper builds up in the body involving the brain and liver axis, in which 40% of cases present with neurologic symptoms related to copper-mediated striatal degeneration (Kitzberger, Madl & Ferenci, 2005). Such data suggest a possible involvement of copper induced disorders of basal ganglia and the neuropathogenesis of HD. *In vitro* studies have shown that Cu is able to accelerate the fibrillation and aggregation of purified recombinant Htt exon1 with polyQ tract (Hands, Mason, Sajjad, Giorgini & Wyttenbach, 2010), whereas, it has been also reported that the fragment containing the first 171 amino acids of human wild-type Htt and its glutamine-expanded mutant form can directly interact with copper (Fox et al. 2007). An experimental study in transgenic mice of HD; CAG140 knock-in with a CAG expansion within one of the endogenous full-length huntingtin alleles and R6/2 expressing human huntingtin exon-1 containing a glutamine coding CAG representing respectively the preclinical and the late stage of HD, Cu levels was increased significantly by 26% and 51% in the striatum and the cortex respectively (Figure 6) (Fox et al. 2007).

Pesticides and HD

Pesticides as well, have been linked to the pathogenesis of HD as environmental agents. Indeed, the complex II inhibitor, 3-nitropropionic acid (3-NP), on the aggregation profile of mutant huntingtin has been studied. In fact, exposure to 3-NP, a fungal-derived neurotoxin, has been reported to cause HD-like symptoms in Drosophila, rats and mice (Beal et al., 1993; Ludolph, He, Spencer, Hammerstad & Sabri, 1991; McConoughey et al., 2010). 3-NP is a mitochondrial complex II inhibitor and is increasingly being used to generate HD animal models (McConoughey et al., 2010; Túnez et al., 2006).

*Figure 6. Copper level in brains of 12-week-old R6/2 HD mice (late-stage HD) as measured by ICP spectroscopy. p-values: *<0.05 (modified Fox et al. 2007)*

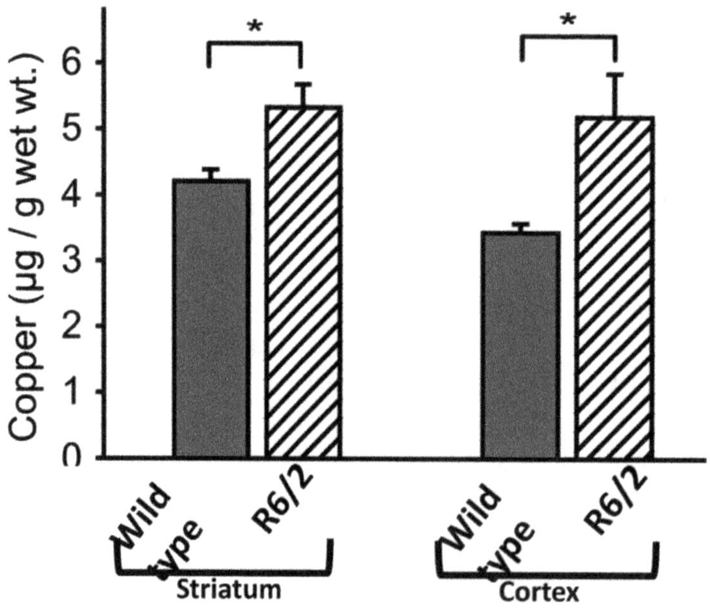

Beal et al. (1993), have assessed intrastriatal administration of 3-NP in rat, they have shown that 3-NP produced dose-dependent striatal lesions. Whereas, neurochemical and histological evaluation showed that markers of both spiny projection neurons were affected. While subacute systemic administration of 3-NP produced age-dependent bilateral striatal lesions with a similar neurochemical profile. As well, chronic administration showed growth-related proliferative changes in dendrites of spiny neurons similar to changes in Huntington's disease (HD) (Beal et al., 1993).

Other researchers have studied the effect of mitochondrial complex I inhibitors, by means of 1-methyl-4-phenyl-1,2,3,6-tetrahydropyridine and rotenone, and complex II inhibitor using 3- nitropropionic acid, on the aggregation of the mutant huntingtin (mthtt) protein whose misfolding and aggregation is known to result in cellular abnormalities which characterize HD. Data showed that all these inhibitors accelerate the aggregation of mthtt in vitro with differences in the amount of aggregates formed depending on the agent used (Deshmukh, Chaudhary & Roy, 2012).

CONCLUSION

Neurodegenerative diseases including MS and HD are often highly debilitating to both mental and physical functioning. MS and HD have been considered for long time as an exclusive genetic CNS disorders. However, recent data of experimental and clinical investigations seem to revolutionize our comprehension of those pathologies, since, heavy metals and pesticides seem to be a new key factors that underlie, in some cases, the earlier onset and/or the aggravation of the disease progression.

REFERENCES

Aliomrani, M., Sahraian, M. A., Shirkhanloo, H., Sharifzadeh, M., Khoshayand, M. R., & Ghahremani, M. H. (2016). Blood Concentrations of Cadmium and Lead in Multiple Sclerosis Patients from Iran. *Iranian Journal of Pharmaceutical Research*, *15*(4), 825–833. PMID:28243279

Aminzadeh, K. K., & Etminan, M. (2007). Dental amalgam and multiple sclerosis: A systematic review and meta-analysis. *Journal of Public Health Dentistry*, *67*(1), 64–66. doi:10.1111/j.1752-7325.2007.00011.x PMID:17436982

Anderson, J. H., Christova, P. S., Xie, T. D., Schott, K. S., Ward, K., & Gomez, C. M. (2002). Spinocerebellar ataxia in monozygotic twins. *Archives of Neurology*, *59*, 1945–1951.

Ascherio, A., & Munger, K. L. (2007). Environmental risk factors for multiple sclerosis. Part II: Noninfectious factors. *Annals of Neurology*, *61*(6), 504–513. doi:10.1002/ana.21141 PMID:17492755

Ascherio, A., Munger, K. L., & Simon, K. C. (2010). Vitamin D and multiple sclerosis. *Lancet Neurology*, *9*(6), 599–612. doi:10.1016/S1474-4422(10)70086-7 PMID:20494325

Aschner, M. (2000). Manganese: Brain transport and emerging research needs. *Environmental Health Perspectives*, *103*, 429–432. PMID:10852840

Aschner, M., Guilarte, T. R., Schneider, J. C., & Zheng, W. (2007). Manganese: Recent advances in understanding its transport and neurotoxiicty. *Toxicology and Applied Pharmacology*, *221*(2), 131–147. doi:10.1016/j.taap.2007.03.001 PMID:17466353

Autissier, N., Rochette, L., Dumas, P., Beley, A., Loireau, A., & Bralet, J. (1982). Dopamine and norepinephrine turnover in various regions of the rat brain after chronic manganese chloride administration. *Toxicology*, *24*(2), 175–182. doi:10.1016/0300-483X(82)90055-5 PMID:7135412

Beal, M. F., Brouillet, E., Jenkins, B. G., Ferrante, R. J., Kowall, N. W., Miller, J. M., ... Hyman, B. T. (1993). Neurochemical and histologic characterization of striatal excitotoxic lesions produced by the mitochondrial toxin 3-nitropropionic acid. *The Journal of Neuroscience*, *13*(10), 4181–4192. doi:10.1523/JNEUROSCI.13-10-04181.1993 PMID:7692009

Bruyn, G. W. (1968). Huntington's chorea: historical, clinical and laboratory synopsis. In Handbook of Clinical Neurology (pp. 298-378). Amsterdam: Elsevier.

Byars, J. A., Beglinger, L. J., Moser, D. J., Gonzalez-Alegre, P., & Nopoulos, P. (2012). Substance abuse may be a risk factor for earlier onset of Huntington disease. *Journal of Neurology*, *259*(9), 1824–1831. doi:10.100700415-012-6415-8 PMID:22274789

Caine, E.D., Bamford, KA, Schiffer, RB., Shoulson, I., & Levy, S. (1986). A controlled neuropsychological comparison of Huntington's disease and multiple sclerosis. *Archives of Neurology, 43*, 249 54.

Calabresi, P. A. (2004). Diagnosis management of multiple sclerosis. *American Family Physician Journal*, *70*, 1935–1944. PMID:15571060

Callaghan, B., Feldman, D., Gruis, K., & Feldman, E. (2011). The association of exposure to lead, mercury, and selenium and the development of amyotrophic lateral sclerosis and the epigenetic implications. *Neurodegenerative Diseases*, *8*(1-2), 1–8. doi:10.1159/000315405 PMID:20689252

Calne, D. B., Chu, N. S., Huang, C. C., Lu, C. S., & Olanow, W. (1994). Manganism and idiopathic parkinsonism: Similarities and differences. *Neurology*, *44*(9), 1583–1586. doi:10.1212/WNL.44.9.1583 PMID:7936278

Compston, A., & Coles, A. (2008). Multiple sclerosis. *Lancet*, *372*(9648), 1502–1517. doi:10.1016/S0140-6736(08)61620-7 PMID:18970977

Cone, W., Russell, C., & Harwood, R. U. (1934). lead as a possible cause of multiple sclerosis. *Archives of Neurology and Psychiatry*, *31*(2), 236–263. doi:10.1001/archneurpsyc.1934.02250020024002

Day, M., Wang, Z., Ding, J., An, X., Ingham, C. A., Shering, A. F., ... Surmeier, D. J. (2006). Selective elimination of glutamatergic synapses on striatopallidal neurons in Parkinson disease models. *Nature Neuroscience*, *9*(2), 251–259. doi:10.1038/nn1632 PMID:16415865

Deshmukh, R. S., Chaudhary, R. K., & Roy, I. (2012). Effect of Pesticides on the Aggregation of Mutant Huntingtin Protein. *Molecular Neurobiology*, *45*(3), 405–414. doi:10.100712035-012-8252-2 PMID:22415443

Deutch, A.Y., Colbran, R.J., & Winder, D.J. (2007). Striatal plasticity and medium spiny neuron dendritic remodeling in Parkinsonism. *Parkinsonism Related Disorders, 13*, S251 S258.

Dexter, D. T., Carayon, A., Javoy-Agid, F., Agid, Y., Wells, F. R., Daniel, S. E., ... Marsden, C. D. (1991). Alterations in the levels of iron, ferritin and other trace metals in Parkinson's disease and other neurodegenerative diseases affecting the basal ganglia. *Brain, 114*(Pt 4), 1953–1975. doi:10.1093/brain/114.4.1953 PMID:1832073

Dietert, R. R., & Piepenbrink, M. S. (2006). Lead and immune function. *Critical Reviews in Toxicology, 36*(4), 359–385. doi:10.1080/10408440500534297 PMID:16809103

DiFiglia, M., Sapp, E., Chase, K., Schwarz, C., Meloni, A., Young, C., ... Aronin, N. (1995). Huntingtin is a cytoplasmic protein associated with vesicles in human and rat brain neurons. *Neuron, 14*(5), 1075–1081. doi:10.1016/0896-6273(95)90346-1 PMID:7748555

Durr, A. (2010). Autosomal dominant cerebellar ataxias: Polyglutamine expansions and beyond. *Lancet Neurology, 9*, 885–894.

El-Fawal, H. A., Waterman, S. J., De Feo, A., & Shamy, M. Y. (1999). Neuroimmunotoxicology: Humoral assessment of neurotoxicity and autoimmune mechanisms. *Environmental Health Perspectives, 107*(Suppl 5), 767–775. PMID:10502543

Fauci, A. S., Braunwald, E., Kasper, D. L., & Hauser, S. L. (Eds.). (2008). *Harrison's Principles of Internal Medicine*. New York: McGraw-Hill Medical.

Fox, J. H., Kama, J. A., Lieberman, G., Chopra, R., Dorsey, K., Chopra, V., ... Hersch, S. (2007). Mechanisms of copper ion mediated Huntington's disease progression. *PLoS One, 2*(3), e334. doi:10.1371/journal.pone.0000334 PMID:17396163

Frohman, E. M., Racke, M. K., & Raine, C. S. (2006). Multiple sclerosis--the plaque and its pathogenesis. *The New England Journal of Medicine, 354*(9), 942–955. doi:10.1056/NEJMra052130 PMID:16510748

Gaeta, A., & Hider, R. C. (2005). The crucial role of metal ions in neurodegeneration: The basis for a promising therapeutic strategy. *British Journal of Pharmacology, 146*(8), 1041–1059. doi:10.1038j.bjp.0706416 PMID:16205720

Gianutsos, G., & Murray, M. T. (1982). Alterations in brain dopamine and GABA following inorganic or organic manganese administration. *Neurotoxicology, 3*, 75–81. PMID:6891761

Guilarte, T. R., Chen, M. K., McGlothan, J. L., Verina, T., Wong, D. F., Zhou, Y., ... Schneider, J. S. (2006). Dopamine system dysfunction and subtle motor deficits in manganese-exposed non-human primates. *Experimental Neurology, 202*(2), 381–390. doi:10.1016/j.expneurol.2006.06.015 PMID:16925997

Gusella, J. F., MacDonald, M. E., Ambrose, C. M., & Duyao, M. P. (1994). Molecular Genetics of Huntington's Disease. *Archives of Neurology, 50*(11), 1157–1163. doi:10.1001/archneur.1993.00540110037003 PMID:8215974

Gwiazda, R. H., Lee, D., Sheridan, J., & Smith, D. R. (2002). Low cumulative manganese exposure affects striatal GABA but not dopamine. *Neurotoxicology, 23*(1), 69–76. doi:10.1016/S0161-813X(02)00002-5 PMID:12164549

Haider, L., Simeonidou, C., Steinberger, G., Hametner, S., Grigoriadis, N., Deretzi, G., ... Frischer, J. M. (2014). Multiple sclerosis deep grey matter: The relation between demyelination, neurodegeneration, inflammation and iron. *Journal of Neurology, Neurosurgery, and Psychiatry*, *85*(12), 1386–1395. doi:10.1136/jnnp-2014-307712 PMID:24899728

Hands, S. L., Mason, R., Sajjad, M. U., Giorgini, F., & Wyttenbach, A. (2010). Metallothioneins and copper metabolism are candidate therapeutic targets in Huntington's disease. *Biochemical Society Transactions*, *38*(2), 552–558. doi:10.1042/BST0380552 PMID:20298220

Hansson, M., Djerbi, M., Rabbani, H., Mellstedt, H., Gharibdoost, F., Hassan, M., ... Abedi-Valugerdi, M. (2005). Exposure to mercuric chloride during the induction phase and after the onset of collagen-induced arthritis enhances immune/autoimmune responses and exacerbates the disease in DBA/1 mice. *Immunology*, *114*(3), 428–437. doi:10.1111/j.1365-2567.2005.02105.x PMID:15720444

Hansson, M., Djerbi, M., Rabbani, H., Mellstedt, H., Gharibdoost, F., Hassan, M., ... Abedi-Valugerdi, M. (2005). Exposure to mercuric chloride during the induction phase and after the onset of collagen-induced arthritis enhances immune/autoimmune responses and exacerbates the disease in DBA/1 mice. *Immunology*, *114*(3), 428–437. doi:10.1111/j.1365-2567.2005.02105.x PMID:15720444

Herndon, E. S., Hladik, C. L., Shang, P., Burns, D. K., Raisanen, J., & White, C. L. III. (2009). Neuroanatomic profile of polyglutamine immunoreactivity in Huntington disease brains. *Journal of Neuropathology and Experimental Neurology*, *68*(3), 250–261. doi:10.1097/NEN.0b013e318198d320 PMID:19225411

Hoffner, G.1., Kahlem, P., & Djian, P. (2002). Perinuclear localization of huntingtin as a consequence of its binding to microtubules through an interaction with beta-tubulin: Relevance to Huntington's disease. *Journal of Cell Science*, *115*(Pt 5), 941–948. PMID:11870213

Irvine, D. G., Schiefer, H. B., & Hader, W. J. (1988). Geotoxicology of multiple sclerosis: The Henribourg, Saskatchewan, cluster focus. II. The soil. *The Science of the Total Environment*, *77*(2-3), 175–188. doi:10.1016/0048-9697(88)90054-X PMID:3241961

Jimenez-Sanchez, M., Licitra, F., Underwood, B. R., & Rubinsztein, D. C. (2017). Huntington's Disease: Mechanisms of Pathogenesis and Therapeutic Strategies. *Cold Spring Harbor Perspectives in Medicine*, *7*(7), a024240. doi:10.1101/cshperspect.a024240 PMID:27940602

Johnson, P. E., Milne, D. B., & Lykken, G. I. (1992). Effect of age and sex on copper absorption, biological half life and status in humans. *The American Journal of Clinical Nutrition*, *56*(5), 917–925. doi:10.1093/ajcn/56.5.917 PMID:1329483

Jomova, K., & Valko, M. (2011). Advances in metal-induced oxidative stress and human disease. *Toxicology*, *283*(2-3), 65–87. doi:10.1016/j.tox.2011.03.001 PMID:21414382

Jomova, K., Vondrakova, D., Lawson, M., & Valko, M. (2010). Metals, oxidative stress and neurodegenerative disorders. *Molecular and Cellular Biochemistry*, *345*(1-2), 91–104. doi:10.100711010-010-0563-x PMID:20730621

Kegel, K. B., Meloni, A. R., Yi, Y., Kim, Y. J., Doyle, E., Cuiffo, B. G., ... DiFiglia, M. (2002). Huntingtin is present in the nucleus, interacts with the transcriptional corepressor C-terminal binding protein, and represses transcription. *The Journal of Biological Chemistry, 277*(9), 7466–7476. doi:10.1074/jbc.M103946200 PMID:11739372

Kesel, A. J., Sonnenbichler, I., Polborn, K., Gürtler, L., Klinkert, W. E., Modolell, M., ... Oberthür, W. (1999). A new antioxidative vitamin B6 analogue modulates pathophysiological cell proliferation and damage. *Bioorganic & Medicinal Chemistry, 7*(2), 359–367. doi:10.1016/S0968-0896(98)00229-6 PMID:10218829

Ketelaar, M., Hofstra, R., & Hayden, M. (2012). What monozygotic twins discordant for phenotype illustrate about mechanisms influencing genetic forms of neurodegeneration. *Clinical Genetics, 81*(4), 325–333. doi:10.1111/j.1399-0004.2011.01795.x PMID:21981075

Kitzberger, R., Madl, C., & Ferenci, P. (2005). Wilson disease. *Metabolic Brain Disease, 20*(4), 295–302. doi:10.100711011-005-7910-8 PMID:16382340

Konstantinova, S. G., Russanova, I. E., & Russanov, E. M. (1991). Do the copper complexes of histamine, histidine and two H2antagonists react with O2? *Free Radical Research Communications, 12-13*(Pt 1), 215–220. doi:10.3109/10715769109145789 PMID:1676975

Kontur, P. J., & Fechter, L. D. (1988). Brain regional manganese levels and monoamine metabolism in manganese-treated neonatal rats. *Neurotoxicology and Teratology, 10*(4), 295–303. doi:10.1016/0892-0362(88)90031-1 PMID:3226372

Krueger, W. H., Gonye, G. E., Madison, D. L., Murray, K. E., Kumar, M., Spoerel, N., & Pfeiffer, S. E. (1997). TPO1, a member of a novel protein family, is developmentally regulated in cultured oligodendrocytes. *Journal of Neurochemistry, 69*(4), 1343–1355. doi:10.1046/j.1471-4159.1997.69041343.x PMID:9326262

Landtblom, A. M., Flodin, U., Söderfeldt, B., Wolfson, C., & Axelson, O. (1996). Organic solvents and multiple sclerosis: A synthesis of the current evidence. *Epidemiology (Cambridge, Mass.), 7*(4), 429–433. doi:10.1097/00001648-199607000-00015 PMID:8793371

Lassmann, H., & van Horssen, J. (2011). The molecular basis of neurodegeneration in multiple sclerosis. *FEBS Letters, 585*(23), 3715–3723. doi:10.1016/j.febslet.2011.08.004 PMID:21854776

Lauer, K., & Firnhaber, W. (1985). Epidemiological investigations into multiple sclerosis in Southern Hesse. III. The possible influence of occupation on the risk of disease. *Acta Neurologica Scandinavica, 72*(4), 397–402. doi:10.1111/j.1600-0404.1985.tb00890.x PMID:4082904

Lublin, F. (2005). History of modern multiple sclerosis therapy. *Journal of Neurology, 252*(S3), iii3–iii9. doi:10.100700415-005-2010-6 PMID:16170498

Ludolph, A. C., He, F., Spencer, P. S., Hammerstad, J., & Sabri, M. (1991). 3- Nitropropionic acid exogenous animal neurotoxin and possible human striatal toxin. *The Canadian Journal of Neurological Sciences, 18*(04), 492–498. doi:10.1017/S0317167100032212 PMID:1782616

MacDonald, M.Huntington's Disease Collaborative Research Group. (1993). A novel gene containing a trinucleotide repeat that is expanded and unstable on Huntington's disease chromosomes. *Cell*, *72*(6), 971–983. doi:10.1016/0092-8674(93)90585-E PMID:8458085

Mahad, D. H., Trapp, B. D., & Lassmann, H. (2015). Pathological mechanisms in progressive multiple sclerosis. *Lancet Neurology*, *14*(2), 183–193. doi:10.1016/S1474-4422(14)70256-X PMID:25772897

Marrie, R. A. (2004). Environmental risk factors in multiple sclerosis aetiology. *Lancet Neurology*, *3*(12), 709–718. doi:10.1016/S1474-4422(04)00933-0 PMID:15556803

McConoughey, S. J., Basso, M., Niatsetskaya, Z. V., Sleiman, S. F., Smirnova, N. A., Langley, B. C., ... Ratan, R. R. (2010). Inhibition of transglutaminase 2 mitigates transcriptional dysregulation in models of Huntington disease. *EMBO Molecular Medicine*, *2*(9), 349–370. doi:10.1002/emmm.201000084 PMID:20665636

Milatovic, D., Zaja-Milatovic, S., Gupta, R. C., Yu, Y., & Aschner, M. (2009). Oxidative damage and neurodegeneration in manganese-induced neurotoxicity. *Toxicology and Applied Pharmacology*, *240*(2), 219–225. doi:10.1016/j.taap.2009.07.004 PMID:19607852

Molina-Holgado, F., Hider, R. C., Gaeta, A., Williams, R., & Francis, P. (2007). Metals ions and neurodegeneration. *Biometals*, *20*(3-4), 639–654. doi:10.100710534-006-9033-z PMID:17294125

Moszczynski, P. (1981). Organic solvents and T lymphocytes. Lancet, 1(8217), 438. Murray, T.J. (2009). The history of multiple sclerosis: The changing frame of the disease over the centuries. *Journal of the Neurological Sciences*, *277*(Suppl 1), S3–S8.

Murray, T. J. (2009). The history of multiple sclerosis: The changing frame of the disease over the centuries. *Journal of the Neurological Sciences*, *277*(Suppl 1), S3–S8. doi:10.1016/S0022-510X(09)70003-6 PMID:19200863

Napier, M. D., Poole, C., Satten, G. A., Ashley-Koch, A., Marrie, R. A., & Williamson, D. M. (2016). Heavy metals, organic solvents, and multiple sclerosis: An exploratory look at gene-environment interactions. *Archives of Environmental & Occupational Health*, *71*(1), 26–34. doi:10.1080/19338244.2014.937381 PMID:25137520

Nasir, J., Floresco, S. B., O'Kusky, J. R., Borowski, A., Richman, J. M., Zeisler, J., ... Hayden, M. R. (1995). Targeted disruption of the HD gene results in embryonic lethality and behavioral and morphological changes in heterozygotes. *Cell*, *81*, 811–823. doi:10.1016/0092-8674(95)90542-1 PMID:7774020

Navikas, V., & Link, H. (1996). Review: Cytokines and the pathogenesis of multiple sclerosis. *Journal of Neuroscience Research*, *45*(4), 322–333. doi:10.1002/(SICI)1097-4547(19960815)45:4<322::AID-JNR1>3.0.CO;2-B PMID:8872892

Olanow, C. W. (2004). Manganese Induced Parkinsonism and Parkinson's Disease. *Annals of the New York Academy of Sciences*, *1012*(1), 209–223. doi:10.1196/annals.1306.018 PMID:15105268

Olanow, C. W. (2004). Manganese-Induced Parkinsonism and Parkinson's Disease. *Annals of the New York Academy of Sciences*, *1012*(1), 209–223. doi:10.1196/annals.1306.018 PMID:15105268

Olanow, C. W., Good, P. F., Shinotoh, H., Hewitt, K. A., Vingerhoets, F., Snow, B. J., ... Perl, D. P. (1996). Manganese intoxication in the rhesus monkey: A clinical, imaging, pathologic, and biochemical study. *Neurology*, *46*(2), 492–498. doi:10.1212/WNL.46.2.492 PMID:8614520

Pal, P. K., Samii, A., & Calne, D. B. (1999). Manganese neurotoxicity: A review of clinical features, imaging and pathology. *Neurotoxicology*, *20*, 227–238. PMID:10385886

Pal, P. K., Samii, A., & Calne, D. B. (1999). Manganese neurotoxicity: A review of clinical features, imaging and pathology. *Neurotoxicology*, *20*, 227–238. PMID:10385886

Parrón, T., Requena, M., Hernández, A. F., & Alarcón, R. (2011). Association between environmental exposure to pesticides and neurodegenerative diseases. *Toxicology and Applied Pharmacology*, *256*(3), 379–385. doi:10.1016/j.taap.2011.05.006 PMID:21601587

Perl, D. P., & Olanow, C. W. (2007). The neuropathology of manganese-induced Parkinsonism. *Journal of Neuropathology and Experimental Neurology*, *66*(8), 675–682. doi:10.1097/nen.0b013e31812503cf PMID:17882011

Putnam, J. J. (1887a). On the Frequency with Which Lead Is Found in the Urine and on Certain Points in the Symptomatology of Chronic Lead Poisoning. *Boston Medical and Surgical Journal*, *117*(4), 73–76. doi:10.1056/NEJM188707281170401

Putnam, J. J. (1887b). On the Frequency with Which Lead Is Found in the Urine, and on Certain Points in the Symptomatology of Chronic Lead Poisoning. *Boston Medical and Surgical Journal*, *117*(5), 97–99. doi:10.1056/NEJM188708041170501

Reaney, S. H., Bench, G., & Smith, D. R. (2006). Brain accumulation and toxicity of Mn(II) and Mn(III) exposures. *Toxicological Sciences*, *93*(1), 114–124. doi:10.1093/toxsci/kfl028 PMID:16740617

Reiner, A., Albin, R. L., Anderson, K. D., D'Amato, C. J., Penney, J. B., & Young, A. B. (1988). Differential loss of striatal projection neurons in Huntington disease. *Proceedings of the National Academy of Sciences of the United States of America*, *85*(15), 5733–5737. doi:10.1073/pnas.85.15.5733 PMID:2456581

Riise, T., Kirkeleit, J., Aarseth, J. H., Farbu, E., & Midgard, R. (2011). Risk of MS is not associated with exposure to crude oil, but increases with low level of education. *Multiple Sclerosis*, *17*(7), 780–787. doi:10.1177/1352458510397686 PMID:21343231

Rosas, H. D., Liu, A. K., Hersch, S., Glessner, M., Ferrante, R. J., Salat, D. H., ... Fischl, B. (2002). Regional and progressive thinning of the cortical ribbon in Huntington's disease. *Neurology*, *58*(5), 695–701. doi:10.1212/WNL.58.5.695 PMID:11889230

Rosenblatt, A. (2007). Neuropsychiatry of Huntington's disease. *Dialogues in Clinical Neuroscience*, *9*, 191–197. PMID:17726917

Rosenblatt, A., Brinkman, R. R., Liang, K. Y., Almqvist, E. W., Margolis, R. L., Huang, C. Y., ... Ross, C. A. (2001). Familial influence on age of onset among siblings with Huntington disease. *American Journal of Medical Genetics*, *105*(5), 399–403. doi:10.1002/ajmg.1400 PMID:11449389

Rubinsztein, D. C., Leggo, J., Coles, R., Almqvist, E., Biancalana, V., Cassiman, J. J., ... Hayden, M. R. (1996). Phenotypic characterization of individuals with 30-40 CAG repeats in the Huntington disease (HD) gene reveals HD cases with 36 repeats and apparently normal elderly individuals with 36-39 repeats. *American Journal of Human Genetics*, *59*, 16–22. PMID:8659522

Sauer, R. M., Zook, B. C., & Garner, F. M. (1970). Demyelinating encephalomyelopathy associated with lead poisoning in nonhuman primates. *Science*, *169*(3950), 1091–1093. doi:10.1126cience.169.3950.1091 PMID:5449319

Siblerud, R. L., & Kienholz, E. (1994). Evidence that mercury from silver dental fillings may be an etiological factor in multiple sclerosis. *The Science of the Total Environment*, *42*(3), 191–205. doi:10.1016/0048-9697(94)90327-1 PMID:8191275

Singh, V. K., Mehrotra, S., & Agarwal, S. S. (1999). The paradigm of Th1 and Th2 cytokines: Its relevance to autoimmunity and allergy. *Immunologic Research*, *20*(3), 147–161. doi:10.1007/BF02786470 PMID:10580639

Squitieri, F., Frati, L., Ciarmiello, A., Lastoria, S., & Quarrell, O. (2006). Juvenile Huntington's disease: Does a dosage-effect pathogenic mechanism differ from the classical adult disease? *Mechanisms of Ageing and Development*, *127*(2), 208–212. doi:10.1016/j.mad.2005.09.012 PMID:16274727

Stejskal, J., & Stejskal, V. D. (1999). The role of metals in autoimmunity and the link to neuroendocrinology. *Neuroendocrinology Letters*, *20*(6), 351–364. PMID:11458198

Struve, M. F., McManus, B. E., Wong, B. A., & Dorman, D. C. (2007). Basal ganglia neurotransmitter concentrations in rhesus monkeys following subchronic manganese sulfate inhalation. *American Journal of Industrial Medicine*, *50*(10), 772–778. doi:10.1002/ajim.20489 PMID:17620281

Tollis, D. (1996). Who was Charcot? *Nursing Times*, *92*, 56–57. PMID:8710677

Trottier, Y., Biancalana, V., & Mandel, J. L. (1994). Instability of CAG repeats in Huntington's disease: Relation to parental transmission and age of onset. *Journal of Medical Genetics*, *31*(5), 377–382. doi:10.1136/jmg.31.5.377 PMID:8064815

Túnez, I., Drucker-Colín, R., Jimena, I., Medina, F. J., Muñoz Mdel, C., Peña, J., & Montilla, P. (2006). Transcranial magnetic stimulation attenuates cell loss and oxidative damage in the striatum induced in the 3-nitropropionic model of Huntington's disease. *Journal of Neurochemistry*, *97*(3), 619–630. doi:10.1111/j.1471-4159.2006.03724.x PMID:16524377

Van Der Valk, P., & De Groot, C. J. (2000). Staging of multiple sclerosis (MS) lesions: Pathology of the time frame of MS. *Neuropathology and Applied Neurobiology*, *26*(1), 2–10. doi:10.1046/j.1365-2990.2000.00217.x PMID:10736062

Vezér, T., Papp, A., Hoyk, Z., Varga, C., Náray, M., & Nagymajtényi, L. (2005). Behavioral and neurotoxicological effects of subchronic manganese exposure in rats. *Environmental Toxicology and Pharmacology*, *19*(3), 797–810. doi:10.1016/j.etap.2004.12.046 PMID:21783557

Via, C. S., Nguyen, P., Niculescu, F., Papadimitriou, J., Hoover, D., & Silbergeld, E. K. (2003). Low-dose exposure to inorganic mercury accelerates disease and mortality in acquired murine lupus. *Environmental Health Perspectives, 111*(10), 1273–1277. doi:10.1289/ehp.6064 PMID:12896845

Vidal, L., Alfonso, M., Campos, F., Faro, L. R., Cervantes, R. C., & Durán, R. (2005). Effects of manganese on extracellular levels of dopamine in rat striatum: An analysis in vivo by brain microdialysis. *Neurochemical Research, 30*(9), 1147–1154. doi:10.100711064-005-7775-6 PMID:16292508

Vonsattel, J. P., & DiFiglia, M. (1998). Huntington disease. *Journal of Neuropathology and Experimental Neurology, 57*(5), 369–384. doi:10.1097/00005072-199805000-00001 PMID:9596408

Vonsattel, J. P., Keller, C., & Del Pilar Amaya, M. (2008). Neuropathology of Huntington's disease. *Handbook of Clinical Neurology, 89*, 599–618. doi:10.1016/S0072-9752(07)01256-0 PMID:18631782

Wegrzynowicz, M., Holt, H. K., Friedman, D. B., & Bowman, A. B. (2012). Changes in the striatal proteome of YAC128Q mice exhibit geneenvironment interactions between mutant huntingtin and manganese. *Journal of Proteome Research, 1*(2), 1118–1132. doi:10.1021/pr200839d PMID:22191580

Weinshenker, B. C. (1996). Epidemiology of multiple sclerosis. *Neurologic Clinics, 142*, 1–308.

Zeitlin, S., Liu, J. P., Chapman, D. L., Papaioannou, V. E., & Efstratiadis, A. (1995). *Increased apoptosis and early embryonic lethality in mice nullizygous for the HD gene*. Academic Press.

Chapter 14
Amyotrophic Lateral Sclerosis Disease and Environmental Risk Factors:
Role of Heavy Metals and Pesticides

Abdeljalil Elgot
Hassan 1st University, Morocco

Hasna Lahouaoui
Cadi Ayyad University, Morocco

Ouassil El Kherchi
Hassan 1st University, Morocco

Nadia Zouhairi
Cadi Ayyad University, Morocco

ABSTRACT

Amyotrophic lateral sclerosis (ALS) is a progressive neurodegenerative disorder that affects central and peripheral motor neuron cells leading to a severe muscle weakness. Until now, no efficient cures exist and those existing are limited. The ALS etiology remains obscure, although the relationship between genetic background and environmental insults including pesticides and heavy metals is well documented. These latter may play a major role in the onset of the ALS neurodegenerative process. Pesticides are known to have many benefits to mankind in the agricultural and industrial areas, but their toxicities in humans have always been a debatable issue. The pathophysiological mechanisms involve, among others, inflammation processes, oxidative stress, and mitochondrial function impairments. The aim of this chapter was to examine the association between the risk of amyotrophic lateral sclerosis (ALS) and exposure to pesticides and heavy metals.

DOI: 10.4018/978-1-5225-7775-1.ch014

INTRODUCTION

Amyotrophic lateral sclerosis (ALS) is a fast progressive degenerative brain disease. About 90% of patients diagnosed with ALS have been classified as having sporadic sickness (Kiernan et al., 2011), only 10% of ALS is classified familial. Even though numerous environmental risk factors have been considered, the real and direct causes of ALS still unknown. Therefore, the association between ALS and exposure to many neurotoxic substances, including pesticides, solvents and heavy metals, has been studied in several epidemiologic studies (Ahmed & Wicklund, 2011; Weisskopf et al., 2009).

Several studies suggest that pesticides are proven to be significant risk factors for the onset of ALS and other neurodegenerative brain diseases, such as Parkinson's and Alzheimer's (Parrón, Requena, Hernández & Alarcón, 2011). While the pathophysiological mechanisms contributing to risk of ALS related with pesticides exposure are unidentified. Several pesticides are considered as potential neurotoxins, which affect the brain functioning in various ways (Keifer & Firestone, 2007).

Epidemiology of Amyotrophic Lateral Sclerosis

Recent studies showed that the crude incidence of ALS in Europe and North America countries, especially among white people, is quite stable, about 2-3 patients/ 100,000 people/year (Cronin, Hardiman & Traynor, 2007; Alonso, Logroscino, Jick & Hernan, 2009). Even though the incidence of ALS in Asia continent is generally reduced, it has been shown that the highest incidences (50-100 times more than the world average) have been reported in Asia, especially in Guam and the surrounding Pacific islands between 1940 and 1960 (Plato et al., 2003; Okamoto et al., 2009b; Spencer, Palmer & Ludolph, 2005; Waring et al., 2004; Steele, 2005; Yoshida et al., 1998). Moreover, populations of Hispanic or African root exhibit lower incidences of ALS (Cronin, Hardiman & Traynor, 2007). Based on the relatively short life span of the ALS patients, the incidence of the disease is almost similar to the mortality rate. (Marin, Couratier, Preux & Logroscino, 2011a).

According to new ALS epidemiological studies, both the incidence and prevalence are higher in men compared to women, the ratio being approximately 1.5:1 (McCombe & Henderson, 2010). Concerning the age of the ALS onset, data showed that it's positively correlated with the male sex (Gordon et al., 2011a). A large amount of ALS in male's sex has been assigned to respiratory, arms flailing, classic and pure lower motor neuron phenotypes (Chio, Calvo, Moglia, Mazzini & Mora, 2011). Recently, this male preponderance which suggests that changing the exposure to environmental risk factors for ALS, may partly explain this gender difference. (Gordon et al., 2011a).

ALS is an age-related neurodegenerative disease, sharing some similarities with Alzheimer's disease; however, unlike Alzheimer's disease, ALS is not strictly an aging associated condition (Brody & Grant, 2001). It has been shown that the incidence of ALS rises with age, and peaks are reached at about seventy years of age, and then declines quickly thereafter (Logroscino et al., 2008). Despite the fact that ALS is a rare neurodegenerative disease, it has been reported that the proportion of global mortality related to ALS has increased in recent years, so, women suffer more than men (Gordon et al., 2011a).

Genetics of Amyotrophic Lateral Sclerosis

The proportion of familial ALS is approximately 10% of cases, most of them are autosomal dominant inheritance pattern. The vast majority of cases (90%) are sporadic with no clear genetic basis. It has been shown that over 20 genes mutations have been associated with ALS, including superoxide dismutase 1, and TAR DNA binding protein (Abel, Powell, Andersen & Al-Chalabi, 2012). The genes involved in ALS have been implicated in many others cellular pathways, leading to suppose that ALS is a multi-risk factor disease. The pathogenic mechanistic exhibiting motor neuron degeneration include among others, RNA toxicity, neurons excitotoxicity, axonal transport impairment, oxidative stress, and mitochondrial dysfunction (Figure 1) (Ferraiuolo, Kirby, Grierson, Sendtner & Shaw, 2011; Yu & Pamphlett, 2017).

Overview on Pesticides

The Environmental Protection Agency (EPA), the government establishment that legalizes the use of pesticides in the United states, defines the pesticides as a substance or combination of substances intended for preventing, abolishing, repelling or mitigating any pest. Nevertheless, it's often misunderstood to

Figure 1. Model of ALS pathogenesis. Environmental factors especially the presence of genetic susceptibilities to toxicants, would damage the mitochondria and trigger the decompensating process in motor neurons. Any attempt to compensate via excitatory transmission in the surviving neurons would increase their own metabolic load and adversely affect this delicate balance. These ineffective compensatory leads then initiate a chain reaction of mitochondrial crisis and neuronal apoptosis, leading to ALS.
(Figure adapted from Yu & Pamphlett, 2017)

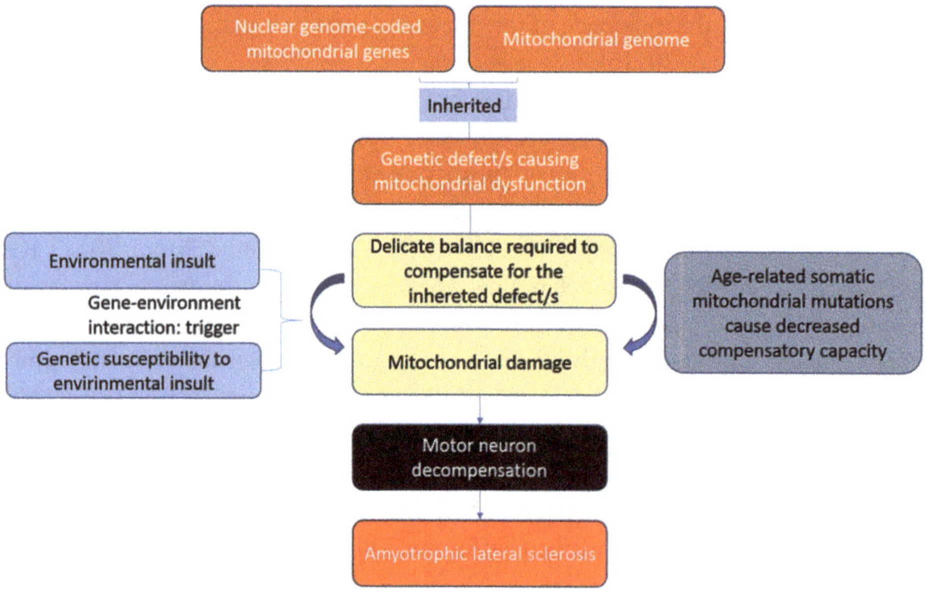

refer only to insecticides, the term pesticide also applies to herbicides, fungicides and rodenticides (US Environmental, 2007).

The main families of pesticides largely used in agriculture and their toxicities are summarized in the following table.

Pesticides Use in the World

Globally, about 9000 species of insects and mites, 50000 species of plant pathogens, 8000 species of weeds cause damage to crops (Zhang, Jiang & Ou, 2011) and 85 to 170 species of rodents are pests agricultural or urban (Stenseth et al., 2003).

Table 1. Classification of pesticides used by farmers and their toxicological class

Active Ingredient	WHO Toxicity Class *	Classification by Main Group
Primiphos methyl	II	Organophosphate
Diazinon	II	Organophosphate
Acephate	II	Organophosphate
Monocrotophos	Hh	Organophosphate
Malathion	III	Organophosphate
Profenofos	II	Organophosphate
Trichlorfon	II	Organophosphate
Dimethoate	II	Organophosphate
Dimethoate	II	Organophosphate
Fenvalerate	II	Pyrethroid
Lambda-cyhalothrin	II	Pyrethroid
Azoxystrobine	U	Fungicide and bactericides
Difenoconazole	II	Fungicide and bactericides
Propamocarb hydroxide	U	Fungicide and bactericides
Benomyl	U	Fungicide and bactericides
Dazomet	II	Fungicide and bactericides
Tebuconazole	II	Fungicide and bactericides
Hymexazol	III	Fungicide and bactericides
Tolclofos-methyl Maneb	U	Fungicide and bactericides
Cuprous oxide	U	Fungicide and bactericides
Fenhexamid Mancozeb	U	Fungicide and bactericide
Copper hydroxide	II	Fungicide and bactericides
Copper oxychloride	II	Fungicide and bactericides
Metam-sodium	II	Nematicides
Benomyl	U	Nematicides
Triazophos	Hh	Nematicides
Carbosulfan	II	Nematicides
Diamidafos	NC	Nematicides
Fenamiphos	Hh	Nematicides
Thionazin	NC	Nematicides
Fensulfothion	NC	Nematicides
Dichlofenthion	NC	Nematicides
Pendimethalin	II	Herbicides
Glyphosate	III	Herbicides
Oxadiazon	U	Herbicides
Metolachor	III	Herbicides
Sethoxydim	III	Herbicides

Table showing the classification of pesticides used by farmers and their toxicological class.

Hh: Highly hazardous; II: Moderately hazardous; III: Slightly hazardous; U: Unlikely to pose an acute hazard in normal use; NC: Not classified.

(Table adapted from Jallow et al., 2017).

Globally, 4.6 million tons of pesticides are used each year and 500 substances are used extensively (Zhang, Jiang & Ou, 2011).

Herbicides account for the majority of global pesticide sales (in 2005, 48% of sales in millions of US dollars); they are followed by insecticides (25% of sales in 2005), bactericidal fungicides (24% of sales in 2005) and other pesticides including rodenticides and molluscicides (3% of sales in 2005) (Zhang, Jiang & Ou, 2011). Europe, which accounts only for 8% of agricultural land, is the largest pesticide user in the world (Birch, Begg & Squire, 2011). It is followed by Asia (Zhang, Jiang & Ou, 2011). In Europe, France is the country that consumes the most pesticides, followed by Germany (Zhang, Jiang & Ou, 2011).

Toxicology of Pesticides

Many recent studies claim that lack of selectivity of the pesticides toward their target causes most of the harmful effects to the environment. Humans absorb pesticides through different pathways especially food, feed water, air or through the skin. Having various franchises barriers, the toxic has reached the sites of metabolism or is stored. The dose (or concentration) that causes a particular effect in humans is called effect concentration 50% or EC50 and is usually used to estimate the toxicity of a chemical. If this effect is death, it is called median lethal dose (LD50 or LC50). The highest no-effect dose (HNED) is the dose immediately below that causing the effect in the same experimental challenge (Severn and Ballard, 1990). In cases where the rates of excretion or metabolization of the molecule are low where it is strongly bound to other constituents, its final concentration in the body will be greater than in the medium of this organism (Madhun & Freed, 1990). This is called bioaccumulation (Cooper, 1991). And it is obvious that a pesticide that adapts to this process, with equal exposure and toxicity, is more dangerous for another product that does not accumulate.

INVOLVEMENT OF PESTICIDES ON THE ONSET OF ALS

The etiology of ALS still unknown but there is presumed to be an interaction between genes and environmental factors (Al-Chalabi & Hardiman, 2013). Pesticides have an important risk factors for ALS and other neurodegenerative diseases such as Parkinson's and Alzheimer's (Parrón, Requena, Hernández & Alarcón, 2011).

Pesticides, like heavy metals, have been widely documented (Cassereau et al., 2017, Krewski et al., 2017) and the association between these environmental factors exposure and ALS risk comes from human studies including case reports (Fonseca, Resende, Silva & Camargo, 1993; Franchignoni, Mora, Giordano, Volanti & Chio, 2013) as well as epidemiologic studies, mainly of occupational exposure (Bozzoni et al., 2016; Couratier et al., 2016; Ingre, Roos, Piehl, Kamel & Fang, 2015).

Evidences showed an elevated risk associated with pesticide exposure as assessed though a simple indicator of rural residence (Mandrioli, Faglioni, Merelli & Sola, 2003) or occupational pesticide exposure (Bonvicini, Marcello, Mandrioli, Pietrini & Vinceti, 2010), but not when pesticide content in cerebrospinal fluid was used as biomarker of exposure (Vinceti et al., 2017). Also, it was shown that this association between pesticides and the initiation of ALS was only found among men; moreover, the increased ALS risk was demonstrated in rural residence and farmer's exposure to particular pesticides (Malek, Barchowsky, Bowser, Youk & Talbott, 2012).

Environmental and biomonitoring studies showed that residential proximity to agricultural fields increased exposure to various pesticide categories (Bradman et al., 2011; Lu, Fenske, Simcox & Kalman, 2000). Overall, investigators found that residence near agricultural land characterized by crop types for which pesticides are used was associated with higher risk of ALS (Vinceti et al., 2017).

Neurotoxic pesticides are divided in: Organochlorine pesticides (OCPs), a group of compounds used widely from the 1940 to the 1970s which include hexachlorobenzene (HCB) and dichlorodiphenyltrichloroethane (DDT) (Richardson et al., 2014). DDT contains about 75% of the p,p′-isomer (p,p′-DDT), though the o,p′-isomer (o,p′-DDT) is also present in significant amounts (Venier & Hites, 2014). In mammals, after being absorbed by the oral and respiratory route, p,p′DDT is broken down into a few metabolites and may alter neuronal electrical excitability and increase susceptibility of nerve cells to stimulation (Guimaraes et al., 2007; Rinkevich et al., 2015).

Most previous studies reported that ALS occurring after exposure to a dithiocarbamate (Hoogenraad, 1988), organochlorine insecticides (Fonseca et al., 1993), pyrethroids (Doi et al., 2006), a mixture of an OC and a pyrethroid (Pall et al., 1987), or the fumigant methyl bromide (Shaw, 2010). An excess of ALS cases was shown among workers exposed to 2,4-D (Burns, Beard & Cartmill, 2001). Less direct evidence has implicated organophosphate insecticides.

Pesticides differ in chemical structure and vary widely in neurotoxicity, so evaluating pesticides as a group may obscure associations with specific pesticides. It was shown that ALS is primarily related to use of OCs (Kamel et al., 2012).

Furthermore, biological evidence on the relation between pesticides and ALS offers some biological plausibility for an association, through mechanisms such as altered protein synthesis, oxidative stress, inflammation, and mitochondrial dysfunction (Bjorling-Poulsen, Andersen & Grandjean, 2008; Choi, Polcher & Joas, 2016; Sanchez-Santed, Colomina & Herrero, 2016; Vinceti, Bottecchi, Fan, Finkelstein & Mandrioli, 2012; Yu & Pamphlett, 2017) (See figure 1).

Other studies have reported that ALS is due to polymorphisms in the gene for paraoxonase 1, which detoxifies organothiophosphates (Costa et al., 2005). However, a recent study found no relationship of ALS to coding polymorphisms in paraoxonase 1 (Wills et al., 2009). In addition, some reports have shown a possible association between OCPs and ALS (Fonseca et al., 1993; Kanavouras et al., 2011). Laboratory studies demonstrated the involvement of OCPs and PCBs in ALS etiopathogenesis, showing the ability of these compounds to adversely affect RNA processing, protein metabolism, glutamate uptake, calcium homeostasis, oxidative stress, axonal transport, neurotransmitters release, and motor function (Baltazar et al., 2014; Selvakumar et al., 2013; Zufiria et al., 2016)

Finally, exposure to pesticides has been reported to be associated with ALS risk in some investigations (Bonvicini, Marcello, Mandrioli, Pietrini & Vinceti, 2010; Morahan & Pamphlett, 2006), but others have found no relationship (Gunnarsso, Bodin, Söderfeldt & Axelson, 1992; Savettieri et al., 1991; Weisskopf et al., 2009). One possible explanation for these inconsistencies may be low statistical power since the number of cases available for study is typically limited in the case of rare diseases such as ALS.

HEAVY METALS-BASED PESTICIDES

Because pesticides are generally not sufficiently purified during the processes of production, for some economic purposes, they mostly contain various impurities, including heavy metals. Moreover, heavy metals frequently are a portion of the active composition of pesticides. An excess of this heavy metals in lands is often bring about by the use of fertilizers, and the concentration of this metals in the ground can raise by repeated and unreasonable pesticide applications. In the light of this, many countries have fixed sever limits and restrictions regarding the soil addition of heavy metal because their long-term effects remain obscure (Barth & L'Hermite, 1987; Hãini, 1990; Celardin, Chatenoux, & D'Ersu, 1990)

Agricultural pesticides and abnormal exposure to heavy metals could be a risk factor for ALS because these candidates are widespread in the environmental media such as air, water, wind drift, multiple dietary sources and volcanic eruptions. Oxidative stress may have a role in the appearance of ALS which may be one mechanism by which metals cause neurotoxicity, this oxidative stress is present in the sera and spinal fluid of ALS patients (Pedersen et al., 1998; Simpson, Henry, Henkel, Smith & Appel, 2004).

The pathophysiology of agricultural pesticides and heavy metals is not yet known but their human neurotoxicity is well established and documented (Albanito et al., 2008; Mutter &Yeter, 2008; Roscoe et al., 2002; Thakur, Rao, Rajwanshi, Parwana & Kumar, 2008) (Figure 2).

Amyotrophic Lateral Sclerosis Toxicity of Heavy Metals

Lead

Lead exposure remains the most environmental risk factor of ALS. The link between this heavy metal and the pathology backs to 1850, when Aran first described motor neuron disease in the context of lead exposure. 3 of his 11 patients presenting a progressive muscular atrophy had been exposed to lead, 2 of them having suffered from out-right lead poisoning (Campbell, Williams & Barltrop, 1970). After, other studies reported 4 additional cases of ALS associated with chronic lead poisoning (Conradi, Ronnevi &Vesterberg, 1976). Since the second half of the twentieth century, these associations have been examined in much greater detail.

Figure 2. Graphic representation showing the effect of different environmental factors including heavy metals, drugs and pesticides on neurons and mitochondria inducing neuronal apoptosis
(Figure adapted from Modgil et al., 2014)

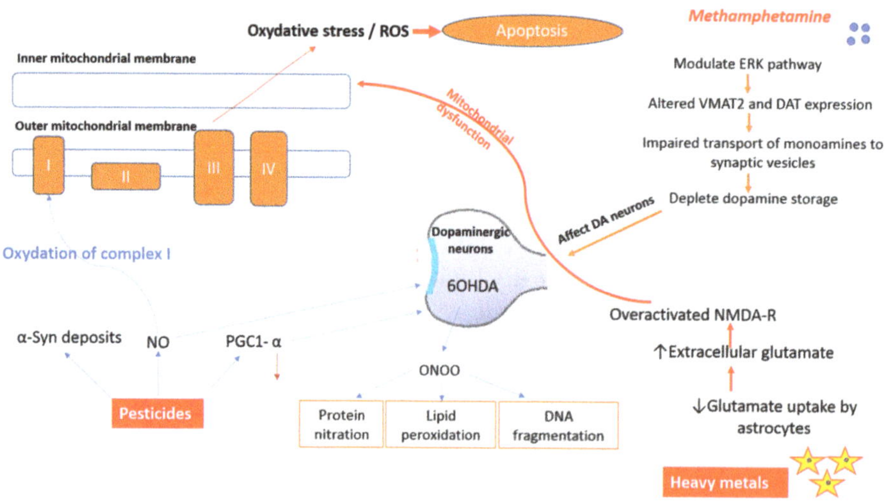

Some studies examined lead levels within different body compartments of ALS patients compared to controls. The results showed an abnormal turnover, tissue distribution, detoxification and excretion of lead. The abnormal lead level may be due to a subsequent lead uptake by neurons manifesting by the toxic neurodegeneration (Conradi, Ronnevi & Vesterberg, 1978). Several results demonstrated an increased level of lead in the cerebrospinal fluid and in spinal ventral horn tissue of ALS patients (Conradi, Ronnevi &Vesterberg, 1976; Kurlander & Patten, 1979). Also, it has been shown that lead may be associated with greater survival in ALS patients and human mutant superoxide dismutase 1 G93A transgenic mice (Campbell, Williams & Barltrop, 1970; Kamel et al., 2005). Boille´e et al (2006) showed that different cell types contribute to different stages of ALS onset and development. This was confirmed by the observation that greater tibia and blood lead levels increase survival of ALS patients (Kamel et al., 2008).

Epigenetic changes are also involved in the pathogenesis of ALS comes from studies of genes known to cause familial ALS. Two of these genes, FUS and ELP3, encode proteins can cause alterations in histone acetylation. In the case of ALS, FUS gene interacts with the CREB binding protein decreasing this protein's HAT activity, this process leads to a downregulation of CREB target genes that may cause the pathogenesis of ALS (Urdinguio, Sanchez-Mut & Esteller, 2009). Similarly, ELP3 alter the expression of HSP70 through its HAT activity (Han et al., 2008). Interestingly, overexpression of HSP70 has been shown to decrease the amount of insoluble SOD1 in a SOD1 mouse model of familial ALS as well as prevent neuronal cell death in a mammalian cell model (Koyama et al., 2006; Patel, Payne Smith, de Belleroche & Latchman, 2005). These findings confirm that epigenetic changes are involved in the pathogenesis of ALS and open up a whole way for potential experiments and perhaps future treatments of this devastating condition.

Mercury

The neurotoxicity of mercury is well documented especially on the motor neurons. In human as well as in animals, mercury accumulates in motor neurons after its injection into blood (Su, Wakabayashi, Kakita, Ikuta &Takahashi, 1998).

Mercury, especially methylmercury (MeHg) intoxication is a real concern. However, the link between mercury and neurodegenerative diseases still not well documented in humans or animals (Praline et al., 2007; Zumstein & Regli, 1982). Epidemiological studies have shown a correlation between mercury exposure and the ALS development (Praline et al., 2007; Provinciali & Giovagnoli, 1990).

In a mouse model of ALS with an overexpression of SOD1 (Superoxide Dismutase 1) gene, the chronic exposure to methylmercury (MeHg) causes early onset of hind limb weakness (Johnson & Atchison, 2008). This result suggests that if an individual has an underlying genetic polymorphism for ALS, exposure to the heavy metal like mercury could hasten the onset of ALS.

In human, a case report from 1996 showed the appearance of ALS in a 38-year-old woman 3.5 years after an accidental injection of mercury into her hand, also, the same study confirmed that after 3 years of an acute exposure to the mercury, the symptoms were similar to those in ALS patients (Schwarz, Husstedt, Bertram & Kuchelmeister, 1992). The neurotoxicity of mercury causes spasticity, fasciculation, extremity weakness and ataxia, the same syndromes showed in ALS diagnosis.

However, a cause and effect relationship between exposure to MeHg and ALS has never been specifically demonstrated. A unique dose of HgCl2 in mice causes deposition of mercury in localized spinal motor neurons, brainstem motor nuclei and cerebral cortex (Arvidson, 1992; Chuu, Liu & Lin-Shiau, 2007). However, those results disagree with Pamphlett and Wely (1998) work, they found a similar concentration of the inorganic mercury in both, the upper and lower motoneurons of ALS patients and controls.

Furthermore, it was shown that mercury was not associated with an increased risk of ALS in two studies (Gresham, Molgaard, Golbeck & Smith, 1986; McGuire et al., 1986) but was associated to an elevated risk in another study (Pierce-Ruhland & Patten, 1981). A cohort study, after 18 years of follow-up, amongst 83 Japanese mercury miners poisoned by or exposed to mercury vapor failed to identify any cases of ALS in 18 years of follow-up. No cases were reported among the 65 controls either.

Overall, there is no clear evidence to suggest that mercury exposure is associated with ALS, the literature on this subject still limited.

Selenium

Selenium (Se) as a metal ion was well documented neurotoxic in humans (Yang, Wang, Zhou & Sun, 1983), individuals are exposed to Se especially during diet, whilst drinking water and occupational environments (Vinceti, Rovesti, Bergomi & Vivoli, 2000). The inorganic form of Se is considerably more toxic than the others forms.

Epidemiologic studies found a causal relation between the Se exposure and ALS (Kilness & Hochberg, 1977; Vinceti et al., 1996). The first investigation reported in 1977 cluster of four ALS cases in a "sparsely populated county", with a population of around 4000, located in west central South Dakota (Kilness & Hochberg, 1977). Investigators described that the area was affected by selenosis and they suggested that the high environmental Se can cause the ALS.

Further support to an association between Se and ALS comes from Yang et al work (Yang, Wang, Zhou & Sun, 1983). They documented in Se-contaminated areas of China the development of neurological signs and symptoms which estimate that Se may adversely affect motor function in humans. However, these observations are of interest since they confirm that the nervous system is a target organ in Se human intoxication and that motor abnormalities are among the neurological effects induced by excess Se exposure.

Finally, a number of case-control studies have examined the body tissue levels of Se in ALS patients compared with controls (Bergomi et al., 2002; Vinceti, Maraldi, Bergomi & Malagoli, 2009). Also, investigations focusing on spinal cord Se content and have shown an excess Se concentration in patients compared with controls (Markesbery et al., 1995; Mitchell, East, Harris & Pentland, 1991).

Then, Selenium at very low concentrations is considered an essential element for the body but chronic exposure and/or high doses can show toxic effects (Vinceti, Maraldi, Bergomi & Malagoli, 2009). The inorganic and less frequently organic Se forms have been shown to interfere with several pathogenetic mechanisms related to neurotoxicity, such as inhibition of prostaglandin D synthase in the brain, inhibition of squalene monooxygenase, increase in dopamine and its metabolites, inhibition of succinic dehydrogenase, acetylcholine esterase and Na+/K+ ATPase, and induction of seizures (Liu, Fu & Lin-Shiau, 1989; Matsumura, Takahata & Hayaishi, 1991).

CONCLUSION

In conclusion, the influence of pesticides and heavy metals on ALS is not yet fully understood. Moreover, exposure to pesticides exhibits a continuous health hazard, especially in the agricultural working environment. Most pesticides show a high degree of intoxication because they are used to kill some organisms and then create some health problems. In the light of this, identifying the preventable exposures that enhance risk of ALS is a crucial step in reducing the incidence of this degenerative disease. For this reason, it is very important to develop health education programs based on knowledge, skills and practices to minimize exposure to pesticides.

REFERENCES

Abel, O., Powell, J. F., Andersen, P. M., & Al-Chalabi, A. (2012). ALSoD: A user-friendly online bioinformatics tool for amyotrophic lateral sclerosis genetics. *Human Mutation*, *33*(9), 1345–1351. doi:10.1002/humu.22157 PMID:22753137

Ahmed, A., & Wicklund, M. P. (2011). Amyotrophic lateral sclerosis: What role does environment play? *Neurologic Clinics*, *29*(3), 689–711. doi:10.1016/j.ncl.2011.06.001 PMID:21803219

Al-Chalabi, A., & Hardiman, O. (2013). The epidemiology of ALS: A conspiracy of genes, environment and time. *Nature Reviews. Neurology*, *9*(11), 617–628. doi:10.1038/nrneurol.2013.203 PMID:24126629

Albanito, L., Lappano, R., Madeo, A., Chimento, A., Prossnitz, E. R., Cappello, A. R., ... Maggiolini, M. (2008). G protein-coupled receptor 30 and estrogen receptor-alpha are involved in the proliferative effects induced by atrazine in ovarian cancer cells. *Environmental Health Perspectives*, *116*(12), 1648–1655. doi:10.1289/ehp.11297 PMID:19079715

Alonso, A., Logroscino, G., Jick, S. S., & Hernan, M. A. (2009). Incidence and lifetime risk of motor neuron disease in the United Kingdom: A population-based study. *European Journal of Neurology*, *16*(6), 745–751. doi:10.1111/j.1468-1331.2009.02586.x PMID:19475756

Arvidson, B. (1992). Inorganic mercury is transported from muscular nerve terminals to spinal and brainstem motoneurons. *Muscle & Nerve*, *15*(10), 1089–1094. doi:10.1002/mus.880151006 PMID:1383815

Baltazar, M. T., Dinis-Oliveira, R. J., de Lourdes Bastos, M., Tsatsakis, A. M., Duarte, J. A., & Carvalho, F. (2014). Pesticides exposure as etiological factors of Parkinson's disease and other neurodegenerative diseases–a mechanistic approach. *Toxicology Letters*, *230*(2), 85–103. doi:10.1016/j.toxlet.2014.01.039 PMID:24503016

Barth, H., & L'Hermite, P. (1987). *Scientific Basis for Soil Protection in the European Community, Elsevier*. Amsterdam, The Netherlands: Academic Press. doi:10.1007/978-94-009-3451-1

Bergomi, M., Vinceti, M., Nacci, G., Pietrini, V., Bratter, P., Alber, D., ... Vivoli, G. (2002). Environmental exposure to trace elements and risk of amyotrophic lateral sclerosis: A population-based case-control study. *Environmental Research*, *89*(2), 116–123. doi:10.1006/enrs.2002.4361 PMID:12123644

Bing, Y., & Pamphlett, R. (2017). Environmental insults: Critical triggers for amyotrophic lateral sclerosis. *Translational Neurodegeneration*, *6*(1), 15. doi:10.118640035-017-0087-3 PMID:28638596

Birch, A. N. E., Begg, G. S., & Squire, G. R. (2011). How agro-ecological research helps to address food security issues under new IPM and pesticide reduction policies for global crop production systems. *Journal of Experimental Botany*, *62*, 3251–3261. doi:10.1093/jxb/err064 PMID:21669880

Bjorling-Poulsen, M., Andersen, H. R., & Grandjean, P. (2008). Potential developmental neurotoxicity of pesticides used in Europe. *Environmental Health*, *7*(1), 50. doi:10.1186/1476-069X-7-50 PMID:18945337

Boille'e, S., Yamanaka, K., Lobsiger, C. S., Copeland, N. G., Jenkins, N. A., Kassiotis, G., ... Cleveland, D. W. (2006). Onset and progression in inherited ALS determined by motor neurons and microglia. *Science*, *312*(5778), 1389–1392. doi:10.1126cience.1123511 PMID:16741123

Bonvicini, F., Marcello, N., Mandrioli, J., Pietrini, V., &Vinceti, M. (2010). Exposure to pesticides and risk of amyotrophic lateral sclerosis: a population-based case control study. *Annalidell'istituto superior di sanita*, *46*, 284–7.doi: . doi:10.4415/ANN_10_03_10

Bozzoni, V., Pansarasa, O., Diamanti, L., Nosari, G., Cereda, C., & Ceroni, M. (2016). Amyotrophic lateral sclerosis and environmental factors. *Functional Neurology*, *31*, 7–19. doi:10.11138/FNeur/2016.31.1.007 PMID:27027889

Bradman, A., Castorina, R., Barr, D. B., Chevrier, J., Harnly, M. E., Eisen, E. A., ... Eskenazi, B. (2011). Determinants of organophosphorus pesticide urinary metabolite levels in young children living in an agricultural community. *International Journal of Environmental Research and Public Health*, *8*(4), 1061–1083. doi:10.3390/ijerph8041061 PMID:21695029

Brody, J. A., & Grant, M. D. (2001). Age-associated diseases and conditions: Implications for decreasing late life morbidity. *Aging (Milan, Italy)*, *13*, 64–67. PMID:11405387

Burns, C. J., Beard, K. K., & Cartmill, J. B. (2001). Mortality in chemical workers potentially exposed to 2,4- dichlorophenoxyacetic acid (2,4-D) 1945-94: An update. *Occupational and Environmental Medicine*, *58*(1), 24–30. doi:10.1136/oem.58.1.24 PMID:11119631

Campbell, A. M., Williams, E. R., & Barltrop, D. (1970). Motor neurone disease and exposure to lead. *Journal of Neurology, Neurosurgery, and Psychiatry*, *33*(6), 877–885. doi:10.1136/jnnp.33.6.877 PMID:5497881

Campbell, A. M., Williams, E. R., & Barltrop, D. (1970). Motor neurone disease and exposure to lead. *Journal of Neurology, Neurosurgery, and Psychiatry*, *33*(6), 877–885. doi:10.1136/jnnp.33.6.877 PMID:5497881

Cassereau, J., Ferre, M., Chevrollier, A., Codron, P., Verny, C., Homedan, C., ... Reynier, P. (2017). (in press). Neurotoxicity of insecticides. *Current Medicinal Chemistry*, *24*(27). doi:10.2174/09298673246 66170526122654 PMID:28552054

Celardin, F., Chatenoux, L., & D'Ersu, P. H. (1990). Metaux lourds dans les sols du basin genovois: 6tat de la question (II). *Archs. Sci. Geneve*, *42*, 493–498.

Chio, A., Calvo, A., Moglia, C., Mazzini, L., & Mora, G. (2011). Phenotypic heterogeneity of amyotrophic lateral sclerosis: A population based study. *Journal of Neurology, Neurosurgery, and Psychiatry*, *82*(7), 740–746. doi:10.1136/jnnp.2010.235952 PMID:21402743

Choi, J., Polche,r A.,& Joas, A. (2016). Systematic literature review on Parkinson's disease and childhood leukaemia and mode of actions for pesticides. *European Food Safety Authority Supporting Publications*, *13*, EN–955. doi:10.2903p.efsa.2016.EN-955

Chuu, J. J., Liu, S. H., & Lin-Shiau, S. Y. (2007). Differential neurotoxic effects of methylmercury and mercuric sulfide in rats. *Toxicology Letters*, *169*(2), 109–120. doi:10.1016/j.toxlet.2006.12.006 PMID:17292570

Conradi, S., Ronnevi, L.O., & Vesteberg, O. (1976). Abnormal tissue distribution of lead in amyotrophic lateral sclerosis. *Journal of the Neurological Sciences*, *29*, 259–265.doi: (76)90175-1. doi:10.1016/0022-510X

Conradi, S., Ronnevi, L. O., & Vesterberg, O. (1978). Lead concentration in skeletal muscle in amyotrophic lateral sclerosis patients and control subjects. *Journal of Neurology, Neurosurgery, and Psychiatry*, *41*(11), 1001–1004. doi:10.1136/jnnp.41.11.1001 PMID:81872

Cooper, K. (1991). Effects of pesticides on wildlife. In W. J. Hayes (Ed.), *Handbook of Pesticide Toxicology* (pp. 463–496). San Diego, CA: Academic Press.

Costa, L. G., Vitalone, A., Cole, T. B., & Furlong, C. E. (2005). Modulation of paraoxonase (PON1) activity. *Biochemical Pharmacology*, *69*(4), 541–550. doi:10.1016/j.bcp.2004.08.027 PMID:15670573

Couratier, P., Corcia, P., Lautrette, G., Nicol, M., Preux, P. M., & Marin, B. (2016). Epidemiology of amyotrophic lateral sclerosis: A review of literature. *Revue Neurologique*, *172*(1), 37–45. doi:10.1016/j.neurol.2015.11.002 PMID:26727307

Cronin, S., Hardiman, O., & Traynor, B. J. (2007). Ethnic variation in the incidence of ALS: A systematic review. *Neurology*, *68*(13), 1002–1007. doi:10.1212/01.wnl.0000258551.96893.6f PMID:17389304

Doi, H., Kikuchi, H., Murai, H., Kawano, Y., Shigeto, H., Ohyagi, Y., & Kira, J. (2006). Motor neuron disorder simulating ALS induced by chronic inhalation of pyrethroid insecticides. *Neurology*, *67*(10), 1894–1895. doi:10.1212/01.wnl.0000244489.65670.9f PMID:17130437

US Environmental. (2007), *What is a pesticide?*. epa.gov.

Ferraiuolo, L., Kirby, J., Grierson, A. J., Sendtner, M., & Shaw, P. J. (2011). Molecular pathways of motor neuron injury in amyotrophic lateral sclerosis. *Nature Reviews. Neurology*, *7*(11), 616–630. doi:10.1038/nrneurol.2011.152 PMID:22051914

Fonseca, R. G., Resende, L. A., Silva, M. D., & Camargo, A. (1993). Chronic motor neuron disease possibly related to intoxication with organochlorine insecticides. *Acta Neurologica Scandinavica*, *88*(1), 56–58. doi:10.1111/j.1600-0404.1993.tb04187.x PMID:7690513

Franchignoni, F., Mora, G., Giordano, A., Volanti, P., & Chio, A. (2013). Evidence of multidimensionality in the ALSFRS-R scale: A critical appraisal on its measurement properties using Rasch analysis. *Journal of Neurology, Neurosurgery, and Psychiatry*, *84*(12), 1340–1345. doi:10.1136/jnnp-2012-304701 PMID:23516308

Gordon, P. H., Artaud, F., Aouba, A., Laurent, F., Meininger, V., & Elbaz, A. (2011a). Changing mortality for motor neuron disease in France (1968-2007): An age-period-cohort analysis. *European Journal of Epidemiology*, *26*(9), 729–737. doi:10.100710654-011-9595-0 PMID:21674216

Gresham, L. S., Molgaard, C. A., Golbeck, A. L., & Smith, R. (1986). Amyotrophic lateral sclerosis and occupational heavy metal exposure: A case control study. *Neuroepidemiology*, *5*(1), 29–38. doi:10.1159/000110810 PMID:3748267

Guimaraes, R. M., Asmus, C. I., & Meyer, A. (2007). DDT reintroduction for malaria control: The cost-benefit debate for public health. *Cadernos de Saude Publica*, *23*(12), 2835–2844. doi:10.1590/S0102-311X2007001200004 PMID:18157325

Gunnarsson, L. G., Bodin, L., Söderfeldt, B., & Axelson, O. (1992). A case-control study of motor neurone disease: Its relation to heritability, and occupational exposures, particularly to solvents. *Occupational and Environmental Medicine*, *49*(11), 791–798. doi:10.1136/oem.49.11.791 PMID:1463680

Han, Q., Lu, J., Duan, J., Su, D., Hou, X., Li, F., ... Huang, B. (2008). Gcn5- and elp3-induced histone H3 acetylation regulates hsp70 gene transcription in yeast. *The Biochemical Journal*, *409*(3), 779–788. doi:10.1042/BJ20070578 PMID:17910533

Häni, H.Häni. (1990). The Analysis of Inorganic and Organic Pollutants in Soil with Special Regard to Their Bioavailability. *International Journal of Environmental Analytical Chemistry*, *39*(2), 197–208. doi:10.1080/03067319008027697

Hoogenraad, T. (1988). Dithiocarbamates and Parkinson's disease. *Lancet*, *1*(8588), 767. doi:10.1016/S0140-6736(88)91573-5 PMID:2895292

Ingre, C., Roos, P. M., Piehl, F., Kamel, F., & Fang, F. (2015). Risk factors for amyotrophic lateral sclerosis. *Clinical Epidemiology*, *7*, 181–193. doi:10.2147/CLEP.S37505 PMID:25709501

Jallow, M., Awadh, D. J., Albaho, M. S., Devi, V. Y., & Thomas, B. M. (2017). Pesticide Knowledge and Safety Practices among Farm Workers in Kuwait: Results of a Survey. *International Journal of Environmental Research and Public Health*, *14*(4), 340. doi:10.3390/ijerph14040340 PMID:28338612

Johnson, F. O., & Atchison, W. D. (2008). Potential enhanced development of paralytic phenotype in SOD1-G93a mice by methylmercury. Environmental etiologies of neurological disorders: scientific, translational and policy implications. *25th International Neurotoxicology Conference.*

Kamel, F., Umbach, D. M., Bedlack, R. S., Richards, M., Watson, M., Alavanja, M. C., ... Sandler, D. P. (2012). Pesticide exposure and amyotrophic lateral sclerosis. *Neurotoxicology*, *33*(3), 457–462. doi:10.1016/j.neuro.2012.04.001 PMID:22521219

Kamel, F., Umbach, D. M., Hu, H., Munsat, T. L., Shefner, J. M., Taylor, J. A., & Sandler, D. P. (2005). Lead exposure as a risk factor for amyotrophic lateral sclerosis. *Neurodegenerative Diseases*, *2*(3-4), 195–201. doi:10.1159/000089625 PMID:16909025

Kamel, F., Umbach, D. M., Stallone, L., Richards, M., Hu, H., & Sandler, D. P. (2008). Association of lead exposure with survival in amyotrophic lateral sclerosis. *Environmental Health Perspectives*, *116*(7), 943–947. doi:10.1289/ehp.11193 PMID:18629318

Kanavouras, K., Tzatzarakis, M. N., Mastorodemos, V., Plaitakis, A., & Tsatsakis, A. M. (2011). A case report of motor neuron disease in a patient showing significant level of DDTs, HCHs and organophosphate metabolites in hair as well as levels of hexane and toluene in blood. *Toxicology and Applied Pharmacology*, *256*(3), 399–404. doi:10.1016/j.taap.2011.07.022 PMID:21851830

Keifer, M. C., & Firestone, J. (2007). Neurotoxicity of pesticides. *Journal of Agromedicine*, *12*(1), 17–25. doi:10.1300/J096v12n01_03 PMID:18032333

Kiernan, M. C., Vucic, S., Cheah, B. C., Turner, M. R., Eisen, A., Hardiman, O., ... Zoing, M. C. (2011). Amyotrophic lateral sclerosis. *Lancet*, *377*(9769), 942–955. doi:10.1016/S0140-6736(10)61156-7 PMID:21296405

Kilness, A. W., & Hochberg, F. H. (1977). Amyotrophic lateral sclerosis in a high selenium environment. *Journal of the American Medical Association*, *237*(26), 2843–2844. doi:10.1001/jama.1977.03270530051023 PMID:577250

Koyama, S., Arawaka, S., Chang-Hong, R., Wada, M., Kawanami, T., Kurita, K., ... Kato, T. (2006). Alteration of familial ALS-linked mutant SOD1 solubility with disease progression: Its modulation by the proteasome and hsp70. *Biochemical and Biophysical Research Communications*, *343*(3), 719–730. doi:10.1016/j.bbrc.2006.02.170 PMID:16563356

Krewski, D., Barakat-Haddad, C., Donnan, J., Martino, R., Pringsheim, T., Tremlett, H., ... Cashman, N. R. (2017). (in press). Determinants of neurological disease: Synthesis of systematic reviews. *Neurotoxicology*, *2017*. doi:10.1016/j.neuro.2017.04.002 PMID:28410962

Kurlander, H. M., & Patten, B. M. (1979). Metals in spinal cord tissue of patients dying of motor neuron disease. *Annals of Neurology*, *6*(1), 21–24. doi:10.1002/ana.410060105 PMID:507754

Liu, S. H., Fu, W. M., & Lin-Shiau, S. Y. (1989). Effects of sodium selenite on neuromuscular junction of the mouse phrenic nerve-diaphragm preparation. *Neuropharmacology*, *28*(7), 733–739. doi:10.1016/0028-3908(89)90159-7 PMID:2474766

Logroscino, G., Traynor, B. J., Hardiman, O., Chio, A., Couratier, P., Mitchell, J. D., ... Beghi, E. (2008). Descriptive epidemiology of amyotrophic lateral sclerosis: New evidence and unsolved issues. *Journal of Neurology, Neurosurgery, and Psychiatry*, *79*(1), 6–11. doi:10.1136/jnnp.2006.104828 PMID:18079297

Lu, C., Fenske, R. A., Simcox, N. J., & Kalman, D. (2000). Pesticide exposure of children in an agricultural community: Evidence of household proximity to farmland and take home exposure pathways. *Environmental Research*, *84*(3), 290–302. doi:10.1006/enrs.2000.4076 PMID:11097803

Madhun, Y. A., & Freed, V. H. (1990). Impact of pesticides on the environment. In Pesticides in the soil environment. Soil Science Society of America.

Malek, A. M., Barchowsky, A., Bowser, R., Youk, A., & Talbott, E. O. (2012). Pesticide exposure as a risk factor for amyotrophic lateral sclerosis: a meta-analysis of epidemiological studies: pesticide exposure as a risk factor for ALS. *Environmental Research*, *117*, 112–119. doi:10.1016/j.envres.2012.06.007 PMID:22819005

Mandrioli, J., Faglioni, P., Merelli, E., & Sola, P. (2003). The epidemiology of ALS in Modena, Italy. *Neurology*, *60*(4), 683–689. doi:10.1212/01.WNL.0000048208.54755.78 PMID:12601112

Marin, B., Couratier, P., Preux, P. M., & Logroscino, G. (2011). Can mortality data be used to estimate amyotrophic lateral sclerosis incidence? *Neuroepidemiology*, *36*(1), 29–38. doi:10.1159/000321930 PMID:21088432

Markesbery, W. R., Ehmann, W. D., Candy, J. M., Ince, P. G., Shaw, P. J., Tandon, L., & Deibel, M. A. (1995). Neutron activation analysis of trace elements in motor neuron disease spinal cord. *Neurodegeneration*, *4*(4), 383–390. doi:10.1006/neur.1995.0046 PMID:8846231

Matsumura, H., Takahata, R., & Hayaishi, O. (1991). Inhibition of sleep in rats by inorganic selenium compounds, inhibitors of prostaglandin D synthase. *Proceedings of the national avademy of sciences of the united states of America*, *88*, 9046-50.10.1073/pnas.88.20.9046

McCombe, P. A., & Henderson, R. D. (2010). Effects of gender in amyotrophic lateral sclerosis. *Gender Medicine*, *7*(6), 557–570. doi:10.1016/j.genm.2010.11.010 PMID:21195356

McGuire, V., Longstreth, W. T., Nelson, L. M., Koepsell, T. D., Checkoway, H., Morgan, M. S., & van Belle, G. (1997). Occupational exposures and amyotrophic lateral sclerosis. A population based case-control study. *American Journal of Epidemiology*, *145*(12), 1076–1088. doi:10.1093/oxfordjournals.aje.a009070 PMID:9199537

Mitchell, J. D., East, B. W., Harris, I. A., & Pentland, B. (1991). Manganese, selenium and other trace elements in spinal cord, liver and bone in motor neurone disease. *European Neurology*, *31*(1), 7–11. doi:10.1159/000116626 PMID:2015841

Modgil, S., Lahiri, D. K., Sharma, V. L., & Anand, A. (2014). Role of early life exposure and environment on neurodegeneration: Implications on brain disorders. *Translational Neurodegeneration*, *29*, 3–9. doi:10.1186/2047-9158-3-9 PMID:24847438

Morahan, J. M., & Pamphlett, R. (2006). Amyotrophic lateral sclerosis and exposure to environmental toxins: An Australian case-control study. *Neuroepidemiolog*, *27*(3), 130–135. doi:10.1159/000095552 PMID:16946624

Mutter, J., & Yeter, D. (2008). Kawasaki's disease, acrodynia, and mercury. *Current Medicinal Chemistry*, *15*(28), 3000–3010. doi:10.2174/092986708786848712 PMID:19075648

Okamoto, K., Kihira, T., Kondo, T., Kobashi, G., Washio, M., Sasaki, S., ... Nagai, M. (2009b). Lifestyle Factors and Risk of Amyotrophic Lateral Sclerosis: A Case-Control Study in Japan. *Annals of Epidemiology, 19*(6), 359–364. doi:10.1016/j.annepidem.2009.01.015 PMID:19342254

Pall, H. A., Williams, A. C., Waring, R., & Elias, E. (1987). Motorneurone disease as manifestation of pesticide toxicity. *Lancet, ii*(8560), 685. doi:10.1016/S0140-6736(87)92468-8 PMID:2887964

Pamphlett, R., & Waley, P. (1998). Mercury in human spinal motor neurons. *Acta Neuropathologica, 96*(5), 515–519. doi:10.1007004010050927 PMID:9829816

Parrón, T., Requena, M., Hernández, A. F., & Alarcón, R. (2011). Association between environmental exposure to pesticides and neurodegenerative diseases. *Toxicology and Applied Pharmacology, 256*(3), 379–385. doi:10.1016/j.taap.2011.05.006 PMID:21601587

Parrón, T., Requena, M., Hernández, A. F., & Alarcón, R. (2011). Association between environmental exposure to pesticides and neurodegenerative diseases. *Toxicology and Applied Pharmacology, 256*(3), 379–385. doi:10.1016/j.taap.2011.05.006 PMID:21601587

Patel, Y. J., Payne, S. M. D., de Belleroche, J., & Latchman, D. S. (2005). Hsp27 and Hsp70 administered in combination have a potent protective effect against FALS-associated SOD1- mutant-induced cell death in mammalian neuronal cells. *Brain Research. Molecular Brain Research, 134*(2), 256–274. doi:10.1016/j.molbrainres.2004.10.028 PMID:15836922

Pedersen, W. A., Fu, W., Keller, J. N., Markesbery, W. R., Appel, S., Smith, R. G., ... Mattson, M. P. (1998). Protein modification by the lipid peroxidation product 4-hydroxynonenal in the spinal cords of amyotrophic lateral sclerosis patients. *Annals of Neurology, 44*(5), 819–824. doi:10.1002/ana.410440518 PMID:9818940

Pierce-Ruhland, R., & Patten, B. M. (1981). Repeat study of antecedent events in motor neuron disease. *Annals of Clinical Research, 13*, 102–107. PMID:7235610

Plato, C. C., Garruto, R. M., Galasko, D., Craig, U., Plato, M., Gamst, A., ... Wiederholt, W. (2003). Amyotrophic Lateral Sclerosis and Parkinsonism-Dementia Complex of Guam: Changing Incidence Rates during the Past 60 Years. *American Journal of Epidemiology, 157*(2), 149–157. doi:10.1093/aje/kwf175 PMID:12522022

Praline, J., Guennoc, A., Limousin, N., Hallak, H., de Toffol, B., & Corcia, P. (2007). Case report; ALS and mercury intoxication: A relationship? *Clinical Neurology and Neurosurgery, 109*(10), 880–883. doi:10.1016/j.clineuro.2007.07.008 PMID:17719172

Provinciali, L., & Giovagnoli, A. R. (1990). Antecedent events in amyotrophic lateral sclerosis: Do they influence clinical onset and progression? *Neuroepidemiology, 9*(5), 255–262. doi:10.1159/000110782 PMID:2087250

Rinkevich, F. D., Du, Y., Tolinski, J., Ueda, A., Wu, C. F., Zhorov, B. S., & Dong, K. (2015). Distinct roles of the DmNav and DSC1 channels in the action of DDT and pyrethroids. *Neurotoxicology, 47*, 99–106. doi:10.1016/j.neuro.2015.02.001 PMID:25687544

Roscoe, R. J., Ball, W., Curran, J. J., DeLaurier, C., Falken, M. C., Fitchett, R., ... Whittaker, S. G. (2002). Adult blood lead epidemiology and surveillance—United States, 1998–2001. *MMWR. Surveillance Summaries, 13*, 1–10. PMID:12528812

Sanchez-Santed, F., Colomina, M. T., & Herrero, H. E. (2016). Organophosphate pesticide exposure and neurodegeneration. *Cortex, 74*, 417–426. doi:10.1016/j.cortex.2015.10.003 PMID:26687930

Savettieri, G., Salemi, G., Arcara, A., Cassata, M., Castiglione, M. G., & Fierro, B. (1991). A case-control study of amyotrophic lateral sclerosis. *Neuroepidemiology, 10*(5-6), 242–245. doi:10.1159/000110279 PMID:1798425

Schwarz, S., Husstedt, I., Bertram, H. P., & Kuchelmeister, K. (1996). Amyotrophic lateral sclerosis after accidental injection of mercury. *Journal of Neurology, Neurosurgery, and Psychiatry, 60*(6), 698. doi:10.1136/jnnp.60.6.698 PMID:8648348

Selvakumar, K., Bavithra, S., Ganesh, L., Krishnamoorthy, G., Venkataraman, P., & Arunakaran, J. (2013). Polychlorinated biphenyls induced oxidative stress mediated neurodegeneration in hippocampus and behavioral changes of adult rats: Anxiolyticlike effects of quercetin. *Toxicology Letters, 222*(1), 45–54. doi:10.1016/j.toxlet.2013.06.237 PMID:23831945

Severn, D. J., & Ballard, G. (1990). Risk/benefit and regulations. In Pesticides in the soil environment. Soil Science Society of America Book.

Shaw, I. (2010). Motor neurone disease - a methyl bromide exposure cluster points to a causal mechanism. *Human and Experimental Toxicology, 29*(3), 241–242. doi:10.1177/0960327109359462 PMID:20071473

Simpson, E. P., Henry, Y. K., Henkel, J. S., Smith, R. G., & Appel, S. H. (2004). Increased lipid peroxidation in sera of ALS patients: A potential biomarker of disease burden. *Neurology, 62*(10), 1758–1765. doi:10.1212/WNL.62.10.1758 PMID:15159474

Spencer, P. S., Palmer, V. S., & Ludolph, A. C. (2005). On the decline and etiology of high-incidence motor system disease in West Papua (southwest New Guinea). *Movement Disorders, 12*(S12), S119–S126. doi:10.1002/mds.20552 PMID:16092101

Steele, J. C. (2005). Parkinsonism-dementia complex of Guam. *Movement Disorders, 12*(S12), S99–S107. doi:10.1002/mds.20547 PMID:16092098

Stenseth, N. C., Leirs, H., Skonhoft, A., Davis, S. A., Pech, R. P., Andreassen, H. P., ... Wan, X. R. (2003). Mice, rats, and people: The bio-economics of agricultural rodent pests. *Frontiers in Ecology and the Environment, 1*(7), 367–375. doi:10.1890/1540-9295(2003)001[0367:MRAPTB]2.0.CO;2

Su, M., Wakabayashi, K., Kakita, A., Ikuta, F., & Takahashi, H. (1998). Selective involvement of large motor neurons in the spinal cord of rats treated with methylmercury. *Journal of the Neurological Sciences, 156*(1), 12–17. doi:10.1016/S0022-510X(98)00030-6 PMID:9559981

Thakur, J. S., Rao, B. T., Rajwanshi, A., Parwana, H. K., & Kumar, R. (2008). Epidemiological study of high cancer among rural agricultural community of Punjab in Northern India. *International Journal of Environmental Research and Public Health, 5*(5), 399–407. doi:10.3390/ijerph5050399 PMID:19151435

Urdinguio, R. G., Sanchez-Mut, J. V., & Esteller, M. (2009). Epigenetic mechanisms in neurological diseases: Genes, syndromes, and therapies. *Lancet Neurology*, *8*(11), 1056–1072. doi:10.1016/S1474-4422(09)70262-5 PMID:19833297

Venier, M., & Hites, R. A. (2014). DDT and HCH, two discontinued organochlorine insecticides in the Great Lakes region: Isomer trends and sources. *Environment International*, *69*, 159–165. doi:10.1016/j.envint.2014.03.028 PMID:24858704

Vinceti, M., Bottecchi, I., Fan, A., Finkelstein, Y., & Mandrioli, J. (2012). Are environmental exposures to selenium, heavy metals, and pesticides risk factors for amyotrophic lateral sclerosis? *Reviews on Environmental Health*, *27*(1), 19–41. doi:10.1515/reveh-2012-0002 PMID:22755265

Vinceti, M., Guidetti, D., Pinotti, M., Rovesti, S., Merlin, M., Vescovi, L., ... Vivoli, G. (1996). Amyotrophic lateral sclerosis after long-term exposure to drinking water with high selenium content. *Epidemiology (Cambridge, Mass.)*, *7*(5), 529–532. doi:10.1097/00001648-199609000-00014 PMID:8862986

Vinceti, M., Maraldi, T., Bergomi, M., & Malagoli, C. (2009). Risk of chronic low-dose selenium overexposure in humans: Insights from epidemiology and biochemistry. *Reviews on Environmental Health*, *24*(3), 231–248. doi:10.1515/REVEH.2009.24.3.231 PMID:19891121

Vinceti, M., Rovesti, S., Bergomi, M.,& Vivoli, G. (2000). The epidemiology of selenium and human cancer. *Tumori Journal, 86*, 105-18. doi: . doi:10.1177/030089160008600201

Vinceti, M., Violi, F., Tzatzarakis, M., Mandrioli, J., Malagoli, C., Hatch, E. E., ... Tsatsakis, A. (2017). Pesticides, polychlorinated biphenyls and polycyclic aromatic hydrocarbons in cerebrospinal fluid of amyotrophic lateral sclerosis patients: A case-control study. *Environmental Research*, *155*, 261–267. doi:10.1016/j.envres.2017.02.025 PMID:28242563

Waring, S. C., Esteban-Santillan, C., Reed, D. M., Craig, U., Labarthe, D. R., Petersen, R. C., & Kurland, L. T. (2004). Incidence of Amyotrophic Lateral Sclerosis and of the Parkinsonism-Dementia Complex of Guam, 1950-1989. *Neuroepidemiology*, *23*(4), 192–200. doi:10.1159/000078505 PMID:15272222

Weisskopf, M. G., Morozova, N., O'Reilly, E. J., McCullough, M. L., Calle, E. E., Thun, M. J., & Ascherio, A. (2009). Prospective study of chemical exposures and amyotrophic lateral sclerosis. *Journal of Neurology, Neurosurgery, and Psychiatry*, *80*(5), 558–561. doi:10.1136/jnnp.2008.156976 PMID:19372290

Weisskopf, M. G., Morozova, N., O'Reilly, E. J., McCullough, M. L., Calle, E. E., Thun, M. J., & Ascherio, A. (2009). Prospective study of chemical exposures and amyotrophic lateral sclerosis. *Journal of Neurology, Neurosurgery, and Psychiatry*, *80*(5), 558–561. doi:10.1136/jnnp.2008.156976 PMID:19372290

Wills, A. M., Cronin, S., Slowik, A., Kasperaviciute, A., Van Es, A. M., Morahan, J. M., ... Brown, R. H. (2009). A large-scale international meta-analysis of paraoxonase gene polymorphisms in sporadic ALS. *Neurology*, *73*(1), 16–24. doi:10.1212/WNL.0b013e3181a18674 PMID:19321847

Yang, G. Q., Wang, S. Z., Zhou, R. H., & Sun, S. Z. (1983). Endemic selenium intoxication of humans in China. *The American Journal of Clinical Nutrition*, *37*(5), 872–881. doi:10.1093/ajcn/37.5.872 PMID:6846228

Yoshida, S., Uebayashi, Y., Kihira, T., Kohmoto, J., Wakayama, I., Taguchi, S., & Yase, Y. (1998). Epidemiology of motor neuron disease in the Kii Peninsula of Japan, 1989–1993: Active or disappearing focus? *Journal of the Neurological Sciences*, *155*(2), 146–155. doi:10.1016/S0022-510X(97)00300-6 PMID:9562259

Zhang, W., Jiang, F., & Ou, J. (2011). Global pesticide consumption and pollution: With China as a focus. *Proceedings of the International Academy of Ecology and Environmental Sciences*, *1*(2), 125–144.

Zufiria, M., Gil-Bea, F. J., Fernandez-Torron, R., Poza, J. J., Munoz-Blanco, J. L., Rojas-García, R., ... López de Munain, A. (2016). ALS: A bucket of genes, environment, metabolism and unknown ingredients. *Progress in Neurobiology*, *142*, 104–129. doi:10.1016/j.pneurobio.2016.05.004 PMID:27236050

Zumstein, V., & Regli, F. (1982). Amyotrophic lateral sclerosis: Epidemiologic study. *Revue swisse de médecine praxis*, *14*, 1424–7.

Chapter 15

Depression and Anxiety Emerging From Heavy Metals: What Relationship?

Hasna Lahouaoui
Cadi Ayyad University, Morocco

Hind Benammi
Cadi Ayyad University, Morocco

Abdelmohcine Aimrane
Cadi Ayyad University, Morocco

Moulay Abdelmonaim El Hidan
Ibn Zohr University, Morocco

Youssef Khamsi
University of Arizona, USA

Ahmed Draoui
Cadi Ayyad University, Morocco

Nadia Zouhairi
Cadi Ayyad University, Morocco

Hassan Alahyane
Cadi Ayyad University, Morocco

Abdellah Bouazza
Cadi Ayyad University, Morocco

ABSTRACT

Depression and anxiety are among the most serious disorders spread-out all over the world. In most cases, patients with depression present features of anxiety. Interestingly and inversely, patients with anxiety also present depression. Thus, both disorders may occur together, with one meeting criteria of the other. The extent of the two disorders has been shown through the high rates of their prevalence. They are, furthermore, associated with significant morbidity which shows how important is to identify and treat both illnesses. However, several epidemiological studies have reported such illnesses to be intensified with the influence of environmental factors such as the toxic effect of heavy metals. Furthermore, the influence of climate change exacerbates the negative effect of these elements. Biological and preclinical investigations have reconciled the mechanism of action by which heavy metals set off emotional disorders. Though its potential harms are important, more studies are needed to understand heavy metals' influence on the evoked pathways.

DOI: 10.4018/978-1-5225-7775-1.ch015

INTRODUCTION

Depression is perceived to be a common and serious medical illness among adults. The nature of this disorder has changed considerably in the last twenty years. It is now seen as an acute and self-limiting illness as opposed to a chronic, and lifelong one (J. Johnson, Weissman, & Klerman, 1992). The term depression is depicted as a wide spectrum of symptoms, including depressed mood, sleep disorders, anhedonia, fatigue and loss of energy, lack of concentration, low self-esteem, negative thinking, and suicidality (Cryan, Markou, & Lucki, 2002 ; Wong & Licinio, 2004). Epidemiological studies have shown that depression and anxiety are common throughout the life cycle (Beekman et al., 1998 ; Regier, 1988). Other studies, which were carried out on children, adolescents, and adults have demonstrated the fact that anxiety and depression are closely related(Oliver et al., 2018 ; Werner-Seidler, Perry, Calear, Newby, & Christensen, 2017 ; Yap, Martin, & Jorm, 2018).

Anxiety is defined as a high impact disorder that causes various difficult problems such as excessive worries, motor tension, and fatigue (Nutt, 2005). Some of the different subtypes that anxiety has include general anxiety disorder, panic disorder, social anxiety, agoraphobia, posttraumatic stress disorder, and obsessive-compulsive disorder (Clement & Chapouthier, 1998 ; Nutt, 2005).

Interestingly, patients with depression often have anxiety disorders and those with anxiety disorders have depression as well. Since both disorders may co-occur together, it is very difficult to differentiate between them (Ayis, Ayerbe, Ashworth, & DA Wolfe, 2018 ; Graeff, Guimarães, De Andrade, & Deakin, 1996).Furthermore, epidemiological evidences have revealed that depression is higher in industrialized countries than in non-developed ones (Jaga & Dharmani, 2007). Such a fact may be related to a plausible high exposure to heavy metals in the contaminated areas, which are specifically industrialized zones.

A plethora of studies, indeed, has declared the existence of a relationship between exposure to heavy metals and mood disorders(Berk et al., 2014 ; Dean, Lam, Scarr, & Duce, 2019) where the most important are anxiety and depression (Berk et al., 2014 ; Theorell et al., 2015). To the epidemiological studies, the high level of affection in the industrialized areas was a strong indicator on the link between exposure to heavy metals and mood disorders (Theorell et al., 2015). However, preclinical investigation have presented more than one explanation to the mechanisms underlying the intoxication with heavy metals (Benammi, El Hiba, Romane, & Gamrani, 2014 ; del Blanco & Barco, 2018 ; Yang et al., 2018).

As preclinical and epidemiological studies have advanced knowledge on the effects of intoxication with heavy metals, recent investigation have revealed the existence of a link between the risk of intoxication and the climate change(Ackah et al., 2014 ; del Blanco & Barco, 2018). In fact, the global warming is a strong factor exacerbating the negative effects of heavy metals on human's health. The actual increase of temperature rise up the amount of released metals in the air as well as in water. Therefore, the amount of intoxicated species become higher, including vegetables, fishes, and many other alimentary products (Ackah et al., 2014 ; Bosch, O'Neill, Sigge, Kerwath, & Hoffman, 2016 ; Kohzadi, Shahmoradi, Ghaderi, Loqmani, & Maleki, 2018). As a result, the ways of human being's infection turn out to be more important. Between this and that, the risk of infection remain a real threat not only to the ecosystem and human's wellbeing, but also for the future generations.

For this reason and in response to the acute need to reconcile the situation, as a starting point to move the needle for sustainable solutions, this chapter represent a synthesis and an analytical view to the point in response to some questions that may come to readers' mind:

What are the common ways of exposure to the most dangerous heavy metals? What are the links between exposure to heavy metals and mood disorders? How could exposure to heavy metals promote mood disorders, specifically depression and anxiety?

Subtypes of Depression

Many studies have shown that depression is characterized by numerous depressive subtypes, including Melancholic Depression, Hostile Depression, Anxious Depression, Atypical Depression and Double Depression (Rush & Weissenburger, 1994).

- **Melancholic Depression:** Is characterized by different symptoms such as: late insomnia, weight loss, lack of reactivity of mood or loss of pleasure in almost all activities, distinct quality of depressed mood and finally marked psychomotor retardation(Rush & Weissenburger, 1994).
- **Hostile Depression:** First introduced by John Overall (Overall, 1976), is characterized by somatic symptoms, anxiety, high levels of hostility, and suspiciousness.
- **Anxious Depression:** Is depicted as a Major Depressive Disorder (MDD) with either comorbid Panic Disorder (Abramson, Keshavan, & Sitaram, 1989), or comorbid anxiety disorder. It is understandable that patients with comorbid anxiety take longer to recover from depression than those with comorbid PD (Clayton et al., 1991).The latter (MDD paired with comorbid anxiety disorder) is determined by the presence of any concomitant anxiety disorder (Hiller, Zaudig, & von Bose, 1989).
- **Atypical Depression:** Describes a number of non-endogenous, mood-reactive types of depression (Rosenbaum, Pollock, Otto, & Pollack, 1995). According to The Diagnostic and Statistical Manual of Mental Disorders--Fourth Edition (DSMIV), atypical depression is determined by a lack of melancholia's diagnosis, mood reactivity, overeating, hypersomnia and oversensitivity to rejection.
- **Double Depression:** A concept introduced by Keller and Shapiro (1982), constitutes of a major depressive episode during the course of at least 2 years of minor depression or dysthymia (Keller & Shapiro, 1982).

Subtypes of Anxiety

Anxiety, as well, is clinically subdivided into several types, where the most commonly defined are:

- **Generalized Anxiety Disorder (GAD):** Is characterized by an excessive worrying that is persistent and not restricted to particular circumstances. Patients with GAD have physical anxiety symptoms like tachycardia and key psychological symptoms. Restlessness, fatigue, difficulty in concentrating, and disturbed sleep are all examples of tachycardia and key psychological symptoms (Alvaro, Roberts, & Harris, 2014).
- **Social Anxiety Disorder (SAD):** Also known as social phobia, is one of the most common psychiatric disorders, with a lifetime prevalence of 12%. According to (Leichsenring & Leweke, 2006), patients show fear as well as avoidance inmost social situations.

- **Separation Anxiety Disorder (SAD):** Confers risk for the development of psychopathology during young adulthood (ages 19-30). The central phenomenology of SAD focuses on a child`s reluctance to be separated from major attachment figures because of his or her fear that something awful may happen to the attachment figure (Lewinsohn, Holm-Denoma, Small, Seeley, & Joiner, 2008).

EPIDEMIOLOGY OF DEPRESSION AND ANXIETY

Significant depressive symptoms were shown among community-dwelling older adults ranging from approximately 8% to 16% (Berkman et al., 1986 ; Dan G Blazer, Burchett, Service, & George, 1991). Studies suggest that in the U.S., depression is more frequent among Mexican Americans than among non-Hispanic Caucasians and African Americans. Nevertheless, African Americans are generally seen by psychiatrists to have fewer depressive symptoms and are much less likely to be treated with antidepressant medications (Dan G Blazer & Ph, 2000 ; Teresi et al., 2002). It was also shown that depression symptoms are more frequent among the eldest age group. This higher frequency is explained by factors associated with aging such as, a higher proportion of women, more physical disability, more cognitive impairment, and lower socioeconomic status (Dan G. Blazer, Hybels, Simonsick, & Hanlon, 2000). In the United States, the estimate of incidence (new cases over a year) of major depression from the ECA overall was 3 per 1000, with a peak in subjects in their fifties (Eaton et al., 1989). The incidence of major depression in elderly people was 0.15%, similar to the rate in younger age groups (Eaton et al., 1989 ; Forsell & Winblad, 1999). In a study from Sweden, the incidence of major depression was 12 per 1000 person-years in men and 30 per 1000 person-years in women between the ages of 70 and 85. The incidence increased from 17 per 1000 person-years between the ages of 70 and 79 to 44 per 1000 person-years between 79 and 85 (D. G. Blazer, 2003).

The strongest comorbidity of depression is with the anxiety disorders of one form or another. Comorbid anxiety disorders are present in 50% of subjects with major depression (Fava et al., 1997). Comorbidity is associated with greater severity of depressive symptoms, and lower treatment response rates (Young, Mufson, & Davies, 2006), alongside the greater social and occupational impairment (Ansseau et al., 2008). The association with depression is particularly strong for post-traumatic stress disorder, generalized anxiety disorder, obsessive-compulsive disorder, and social phobia. The Zurich study notes higher prevalence of depression and panic disorder, reporting that 12% of their sample displayed comorbidity (Volirath & Angst, 1989).

PATHOPHYSIOLOGY OF HEAVY METALS

Hazards evoked by heavy metals and insecticides are quite evident, and these substances are present in many objects commonly used by human beings. Vegetables and fruits are usually treated by pesticides and groundwater, soil and plants are infected in many regions (Kim, Kabir, & Jahan, 2017). Heavy metals are also used for multiple purposes. They are employed, for instance, in medicals and pharmaceuticals as the case are for mercury (Hg) (Homme et al., 2014), in manufactories (Wu et al., 2018), or within some object used just to have pleasure like tobacco and hookah water used in coffee houses (Yousefinejad, Mansouri, Ramezani, Mohammadzadeh, & Akhlaghi, 2018). Studies have well documented the consequences and the potentialities of exposures of harmful metals and pesticides. In this review, we will put the focus on the most prominent studies in this subject to understand the picture of the leading pathophysiological processes that lead to evoked depression and anxiety illnesses.

Heavy Metals

The negative effects of heavy metals on the biological functions of human body have been proven worldwide. Indeed, several preclinical, epidemiological, and biological studies have affirmed the correlativity between heavy metal contamination and emotional illnesses such as depression and anxiety (Benammi et al., 2014 ; Komulainen & Tuomisto, 1980 ; Wu et al., 2018). The US National Health and Nutrition Examination Survey have conducted an investigation on a group of 5,560 Americans aged 20–80 years. The results have proven the reliance between the burden of heavy metals in tested urine and the presence of depression symptoms. The conclusion being the greater the concentration of heavy metals, the higher prevalence of the developed depressive disorder symptoms (Shiue, 2015). Similarly, various studies that have diagnosed patients with depression and panic attacks disorders showed an excess of some metals concentration such as cadmium (Cd), lead (Pb) (Jurczak et al., 2018), and mercury (Hg) (Bratel, Haraldson, & Ottosson, 1997 ; Homme et al., 2014). Quite a lot of investigation has been undertaken in the purpose of human body's burden of these chemicals and their potential effects. It is difficult to present all the possible related heavy metals and cases, but these are the most prominent and will present good explanatory examples.

Cadmium (Cd)

Cadmium is a malleable metal found naturally in the crust of the earth. It has the intent to be combined with other elements such as chlorine, oxygen, and sulfur, and exist most frequently in the form of cadmium sulfide (CdS2). Cadmium also is a widespread industrial and environmental pollutant, released through anthropogenic processes such as metal mining and refining, production of plastics, organic phosphate fertilization, and from cigarette smoke (Ashley J. Malin, 2018 ; Nordberg, 2004). Thus, the rate of absorption has been revealed to range from 10% to 50%, depending on the route of exposure (Ashley J. Malin, 2018). As a consequence of the potential frequent exposition, many developmental, intellectual as well as behavioral complications have been revealed to begin even as soon as prenatal exposure (Orisakwe, 2014). Furthermore, adverse neurodevelopmental outcomes among children and adolescents have been linked to the amount of exposure and the cumulative body burden (Ashley J. Malin, 2018).

To identify Cd^{2+} exposure several biological studies use a variety of biomarkers. Once exposure is identified, body burden confirmation of Cd^{2+} is assessed and correlates most often with emotional problems at different age ranges. Prenatal exposure within 281 women have revealed that Cd^{2+} at means of 0.22 μg/L in umbilical cord blood increase the risk of emotional problems to 1.53-fold greater within descending boys at the age of 7–8 years for every doubling of cord blood Cd^{2+} concentration (Sioen et al., 2013). Even within adults, the confirmation of significant loads of Cd^{2+} is revealed to cause depressive disorder. A study conducted on 2,892 people aged 20–39 showed that there was a dependence between the Cd^{2+} load to a geometric mean level of 0.31 μg/l and the prevalence of depressive symptoms. Furthermore, the authors suggested that decreasing the amount of Cd^{2+} exposure may be reflected by a reduction in frequency of depression appearance (Scinicariello & Buser, 2015). However, the induced emotional symptoms may depend on the level of the body burden. In 198 healthy postmenopausal women from Poland, as a case in point, an exposure to this metal that does not exceed acceptable limits did not confirm the relation between the level of selected Cd^{2+} and the severity of anxiety and depression (Jurczak et al., 2018). In sum, Cd^{2+} is a toxic metal that may provoke anxiety and depression when the body burden reached is beyond a certain level.

Lead (Pb)

Lead (Pb) is one of the oldest toxicants known worldwide that damages many organs including the central nervous system which can lead to provoked anxiety (Benammi et al., 2014 ; Leret, San Millán, & Antonio, 2003). The neurological alterations related to Pb continue to be major health problem worldwide. In fact, the exposure is widespread because of the broad use as a major component in many commercial products such as lead-based paint (Mathee et al., 2009). Moreover, Pb related alterations will continue to happen even within people who have been exposed to the metal a long time ago, mainly because of its prolonged half life in the brain and the bones (Maret, 2017) where it may last respectively 1 year to decades (Lidsky & Schneider, 2003). Another reason why this metal is highly forbidden is its high reactivity. A fact that, again, may lead to cognitive impairments (Mansouri, Muñoz-Fambuena, & Cauli, 2018).

Unlike various other toxic metals, Pb is able to substitute to Ca^{2+} in many fundamental cellular signaling molecules such as the neuronal complex Ca^{2+}-calmodulin that have a crucial role in the regulation of monoamine neuro-transmissions. Thus, its menacing effects are multiple and may even lead to neuronal death (Lidsky & Schneider, 2003). This affirmation may be regarded as an explanatory statement of the magnitude of lead's revealed intoxication in humans, especially within pregnant women and children. Indeed, children with a blood lead level of 70µg/dl have been either severely compromised cognitively or seriously mentally handicapped (Cory-Schlecta, D. A., Schaumburg, 2000). At lower doses, a blood lead level of 15µg/dl, Sciarillo and colleagues have observed in 4 to 5 year-old children an increased incidence of a range of behavioral troubles including depression (Sciarillo, Alexander, & Farrell, 1992).

Sadly, lead has also been revealed actually to provoke more lasting effects. There is a strong correlation between maternal and umbilical cord blood lead levels (Gardella, 2001), a matter of fact which explains the transfer of lead from mother to fetus. Mechanically, the lead burdens the bones for decades is supposed to be mobilized in blood from infected women to their children during pregnancy, the period of high resorption (M. A. Johnson, 2001). Even with low placental blood lead levels (e.g. beyond 10µg/dl), this mobilization is revealed to induce psychological emotional and cognitive deleterious effects within infants at different ages (Freire et al., 2018 ; Lidsky & Schneider, 2003).

Mercury (Hg)

Mercury is another heavy metal commonly used in human population and has been known to cause several health complications including anxiety and depression (Siblerud, Motl, & Kienholz, 1994). According to the World Health Organization, the major causes of human exposure to Hg have been related to outgassing of mercury from dental amalgam, ingestion of contaminated fish, or occupational exposure (World Health Organization, 1991). Acute exposures has been revealed to induce serious health complaints (Bose-O'Reilly, McCarty, Steckling, & Lettmeier, 2010). But, several studies have also linked the chronic low dose Hg exposure to many physiological complications. For instance, several genes are susceptible to Hg toxicity (Bose-O'Reilly et al., 2010 ; Onwuzuligbo et al., 2018). In general, Hg toxicity leads to a personality change, known as Erethism. This change displays excessive timidity, shyness, loss of self-confidence, anxiety and depression which are key characteristics, as stated before, to Hg poisoning (Bernhoft, 2012 ; Onwuzuligbo et al., 2018). The extent of Hg effect is widespread, mainly because of it being used in medical fields such as dental amalgams (Bernhoft, 2012 ; Homme et al., 2014). In fact, the body burden has been stated to cross considerable levels (Homme et al., 2014) although the toxicity differs with the chemical form of Hg, the dose, and the rate of exposure (Bernhoft, 2012).

The evidence linking the elemental mercury to depression and anxiety has been detailed for many decades. It has been stated that when Hg is introduced it goes through the blood stream (Hursh & Clarkson, 1976) and crosses the blood brain barrier where it is ionized (Thomas W. Clarkson, 1972). This, correlation between the mercury burden and psychiatric disorders trends have been further revealed and supported by many studies. As an illustrative instance, an investigation on a group of 50 consecutive patients with dental amalgam restoration has revealed 70% of cases, versus 14% within control, with psychiatric disorders including anxiety and depression symptoms (Bratel et al., 1997). This correlation and affirmation, again, has been repeatedly long-established in many recent studies across different populations. To provide further insight, Nelson and Bauman (2003) have described some Hg-associated symptoms in children which includes anxiety and depression. Furthermore, reports regarding neurological symptoms in teenagers exposed to Hg are in both of the two indexed (Bose-O'Reilly et al., 2010 ; Onwuzuligbo et al., 2018).

Knowing heavy metal exposures can lead to major psychological issues (including depression and anxiety), serious efforts must be done worldwide to reduce the sources of human exposure to such dangerous metals. These heavy metals have been presented in the literature as being effectively and most commonly the most harmful. For these reasons, we emphasize on the mechanisms of action by which they commonly these metals induce fatality within the CNS, specifically among the regulatory systems of evoked emotions.

Plausible Mechanisms of Action by Which Heavy Metals Cause Illness

The mechanisms by which a broad range of heavy metals set off anxiety and depression has been well documented. It is revealed that the observed symptoms in most cases are linked to the alteration of the functionality within monoamines pathways (Komulainen & Tuomisto, 1980). These pathways generate from the dorsal raphe nucleus (5-HT pathways) to the amygdale and from the median raphe nucleus (mainly dopamine pathways) to the hippocampus. The two systems regulate successively evoked anxiety and depression (Graeff et al., 1996) (figure 1). These sets of processes are complicated and needs a fundamental understanding of the subject, mainly because of the variability of the ways of action, although the influenced pathways tend to overlap.

First, it has been stated that heavy metals tend to accumulate in many regions of the CNS, specifically, and these metals gather with greater extent in the hippocampus compared to other parts of the brain (Stoltenburg-Didinger, 1994). This is also preponderant with divalent metals such as lead (Pb) and Cadmium (Cd) (Leret et al., 2003). The infected organs play prominently crucial roles in the regulation and control of many other parts of the central nervous system (Ashley J. Malin, 2018). However, in our case we will focus on how the CNS's functional processes in the modulation of anxiety and depression are blocked and/or imbalanced. These processes involve links between the hippocampus and the amygdale, and the main involved pathways, which the links are specifically made by a set of monoamine neurotransmitters, mainly dopamine and serotonin 5- hydroxytriptamine (5-HT) (Graeff et al., 1996).

Figure 1. Schematic representation of the role serotonin and dopamine ascending pathways implicated in the modulation of depression and anxiety. DRN: dorsal raphe nucleus; MRN: median raphe nucleus (modified according to Graeff et al., 1996).

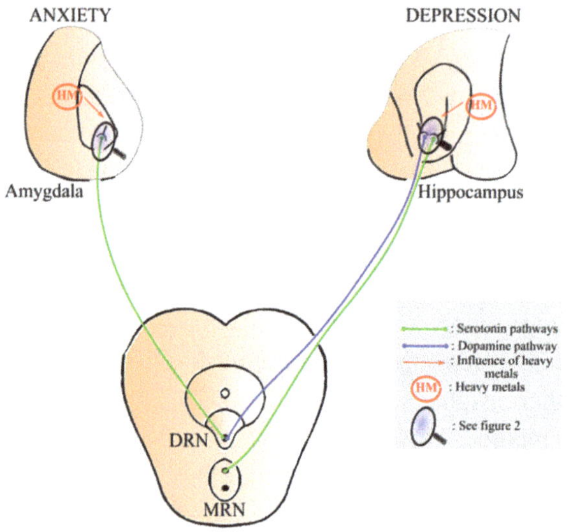

Involvement of 5-HT Pathways

The emotion regulatory serotonin pathways are ascending pathways that rise up from the dorsal raphe nucleus (DRN) to the medial forebrain bundle, and innervate the amygdala and the frontal cortex (Blanchard, R. J., Blanchard, D. C., & Hori, 1989). The change of neuronal interconnections that could follow an alteration within the synthesis or the modulation processes of these neurotransmitters has been demonstrated to cause several psychological harms (Njung'e & Handley, 1991). As a supportive example, in an investigation on hippocampal and hypothalamic neurotransmitters' regulatory processes, perinatal exposure to Pb and Cd has been shown to modulate the regulatory properties in the brain (Leret et al., 2003). Such a modulation may promote an increase in the amount of serotonin delivered in the synapses, an action that has been shown to trigger to anxiety and depression (Graeff et al., 1996). Even though the underlying process is not yet fully understood, two main explanatory suggestions may be proposed (figure 2). From one side, Bondy (1988) has stated that Pb may impede the metabolism of thryptophan through an action on tryptophan-pyrolase; a tryptophane metabolite that contains a heme where Pb trend to substitute an atom of iron (Fe) decreasing the functional level of the molecule and, consequently, the amount of disposed tryptophan become higher (Bondy, 1988). From the other side, Pb and Cd have been shown to affect, in different ways, brain calmodulin and modify its regulatory properties. Indeed, a complex Ca^{++}-calmodulin is implicated in the activation of tyrosine hydroxylase (TH) via a reaction of phosphrylation. Following an action of Pb, TH is excessively activated and consequently could activate tryptophane-hydroxylase (TrpH) that facilitate the synthesis of 5-HT(Vig, PJ and Nath, 1991), a fact that is stated to be anxiolytic (Tanya, Chutima, Andre F, Michel, & Michael, 2018). This being said, the associated pathways may be influenced as well, either by direct effect of some heavy metals on the related regulatory patterns or following the prescribed calmodulin system disturbance.

Figure 2. schematic synthesis of the plausible mechanisms of heavy metals' action on the nervous system. DR: dopamine receptor; 5-HTR: 5-hydroxytryptamine receptor; 5-HT$_{1A}$: 5-HT 1A subtype receptor; 5-HTUR: 5-HT uptake system receptor.

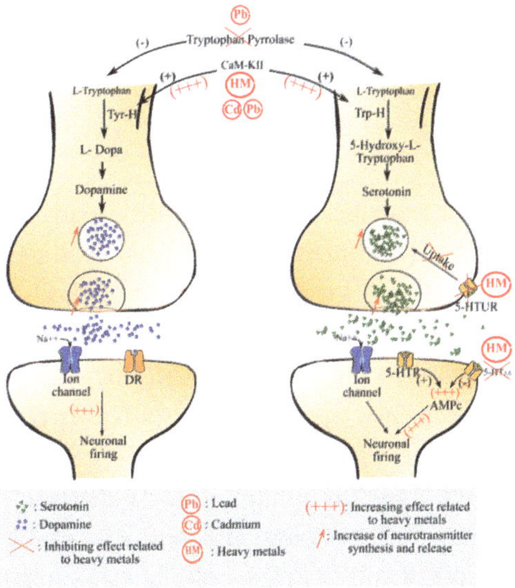

Besides calmodulin functionality and tryptophan metabolism, the regulation of the release and the uptake of 5-HT are central courses by which evoked emotions are modulated. Given clinical information presume that antidepressants are prescribed to improve many types of anxiety, like panic and obsessive compulsive disorder (de Vries, Roest, Burgerhof, & de Jonge, 2018), a fact which refers to an imbalance that occurs within released 5-HT and 5-HT uptake system (Blier, Ph, Montigny, & Ph, 1999). Moreover, depression and anxiety often occur together in the same patient (Ayis et al., 2018), a fact which refers to an eventual disturbance within the implicated ascending pathways that facilitates active escape or avoidance behaviors (Graeff et al., 1996). An overview on some electrophysiological, microdialysis, and drug administration based studies has led us to propose an explanatory interpretation of the general occurring process. The release of 5-HT is under an indirect predominant control of 5-HT receptors. This control resides in the reduction of serotonin release via an action on 5-HT1A subtype receptor, which lead to 5-HT neurons firing reduction (Blier et al., 1999). As 5-HT seems to inhibit aversions generated in the dorsal periacuedectural gray (DPAG), the brain structure related to anxiety and depression, initial aggravation of anxiety and depression happens (Graeff et al., 1996). The same pathophysiological process may be set off following the action of some metals like Cu^{2+}, Pb^{2+} and Cd^{2+}, since they have been experimentally demonstrated to have a potent inhibitor effect on hypothalamic 5-HT uptake system (Komulainen & Tuomisto, 1980).

Involvement of Dopamine Pathways

The mechanism by which heavy metals alter dopamine (DA) pathways is not yet well elucidated. However, many studies have pointed out the link between heavy metals intoxication and dopamine uptake and release modulation. For instance, Cd has been demonstrated to inhibit the Na+/ K+-ATP-ase activity and the uptake of DA in brain synaptosomes (Hobson, Milhouse, Rajanna, Natural, & Sciences, 1986). This process may include the ascending pathways from the median raph nucleus that command prevention of depression. Moreover, some studies on the presence of significant loads of Pb2+ has proven an excessive related activation of TH, as mentioned above (Graeff et al., 1996), which consequently may also increases the levels of DA in the hippocampus.

General Oxidative Stress

The most cytotoxic effect of heavy metals occurs in their fully oxidized states, as for Hg^{2+}, Pb^{2+}, Cd^{2+}, and other toxic metals such as arsenic and nickel. These fully oxidized forms of heavy metals have a strong affinity toward proteins' functional groups, commonly called nucleophiles, like selenol (SeH) and thiols (-SH) (Jan et al., 2015). The role of nucleophiles is fundamental in vital cellular processes. For instance, cysteine is a thiol found in the active sites of enzymes, cofactors, receptors, cytokines, ion channels, transport proteins, and transcription factors (Álvarez et al., 2017). Therefore, a binding of these elements to nucleophiles, either specifically across the central nervous system or generally in many organs, could generate a plethora of interacting effects that affect membrane potential, cellular excitability, cell signaling, neurotransmitter release, and gene expression (Jan et al., 2015). Though this oxidative process is general and may take place in any cellular system in the body, it has plausible effects on the central nervous system, which lead to considerable emotional troubles among which anxiety and depression are prominent.

CONCLUSION

A variety of environmental factors promotes depression and anxiety disorders. Prominently, exposure to heavy metals increase largely the risk of affection. The studies conducted in the last decades advanced knowledge on the pathophysiological processes of toxicity with these elements. Interestingly, heavy metals influence negatively on the biological system either via specific actions in precise sites or through reactions that could occur in all cells. The resulting damages are important and tend to be incurable. Therefore, every protection must be taken to avoid the intoxication and to alleviate climate change influence on the expansion heavy metals' distribution.

REFERENCES

Abramson, L., Keshavan, M. S., & Sitaram, N. (1989). Elevated sleep and waking heart rates in anxious depressions. *Biological Psychiatry*, *26*(5), 496–499. doi:10.1016/0006-3223(89)90070-X PMID:2790092

Ackah, M., Anim, A. K., Gyamfi, E. T., Zakaria, N., Hanson, J., Tulasi, D., ... Osei, J. (2014). Uptake of heavy metals by some edible vegetables irrigated using wastewater: A preliminary study in Accra, Ghana. *Environmental Monitoring and Assessment*, *186*(1), 621–634. doi:10.100710661-013-3403-0 PMID:24114276

Álvarez, L., Bianco, C. L., Toscano, J. P., Lin, J., Akaike, T., & Fukuto, J. M. (2017). Chemical Biology of Hydropersulfides and Related Species: Possible Roles in Cellular Protection and Redox Signaling. *Antioxidants & Redox Signalling*, *27*(10), 622–633. doi:10.1089/ars.2017.7081 PMID:28398141

Alvaro, P. K., Roberts, R. M., & Harris, J. K. (2014). The independent relationships between insomnia, depression, subtypes of anxiety, and chronotype during adolescence. *Sleep Medicine*, *15*(8), 934–941. doi:10.1016/j.sleep.2014.03.019 PMID:24958244

Ansseau, M., Fischler, B., Dierick, M., Albert, A., Leyman, S., & Mignon, A. (2008). Socioeconomic correlates of generalized anxiety disorder and major depression in primary care: The GADIS II study (Generalized Anxiety and Depression Impact Survey II). *Depression and Anxiety*, *25*(6), 506–513. doi:10.1002/da.20306 PMID:17595015

Ashley, J., & Malin, R. O. W. (2018). *The Developmental Neurotoxicity of Cadmium. Handbook of Developmental Neurotoxicology (2nd ed.)*. Elsevier Inc.; doi:10.1007/978-94-009-5594-3

Ayis, S. A., Ayerbe, L., Ashworth, M., & Wolfe, D. A. (2017, September). C. (2018). Evaluation of the Hospital Anxiety and Depression Scale (HADS) in screening stroke patients for symptoms: Item Response Theory (IRT) analysis. *Journal of Affective Disorders*, *228*, 33–40. doi:10.1016/j.jad.2017.11.037 PMID:29202444

Beekman, A. T. F., Bremmer, M. A., Deeg, D. J. H., Van Balkom, A. J. L. M., Smit, J. H., De Beurs, E., ... Van Tilburg, W. (1998). Anxiety disorders in later life: A report from the longitudinal aging study Amsterdam. *International Journal of Geriatric Psychiatry*, *13*(10), 717–726. doi:10.1002/(SICI)1099-1166(1998100)13:10<717::AID-GPS857>3.0.CO;2-M PMID:9818308

Benammi, H., El Hiba, O., Romane, A., & Gamrani, H. (2014). A blunted anxiolytic like effect of curcumin against acute lead induced anxiety in rat: Involvement of serotonin. *Acta Histochemica*, *116*(5), 920–925. doi:10.1016/j.acthis.2014.03.002 PMID:24721902

Berk, M., Williams, L. J., Andreazza, A. C., Pasco, J. A., Dodd, S., Jacka, F. N., ... Magalhaes, P. V. S. (2014). Pop, heavy metal and the blues: Secondary analysis of persistent organic pollutants (POP), heavy metals and depressive symptoms in the NHANES National Epidemiological Survey. *BMJ Open*, *4*(7), e005142–e005142. doi:10.1136/bmjopen-2014-005142 PMID:25037643

Berkman, L. F., Berkman, C. S., Kasl, S., Freeman, D. H. Jr, Leo, L., Ostfeld, A. M., ... Brody, J. A. (1986). Depressive Symptoms in Relation to Physical Health and Functioning in the Elderly. *American Journal of Epidemiology*, *124*(3), 372–388. doi:10.1093/oxfordjournals.aje.a114408 PMID:3740038

Bernhoft, R. A. (2012). Mercury toxicity and treatment: A review of the literature. *Journal of Environmental and Public Health*, *2012*, 1–10. doi:10.1155/2012/460508 PMID:22235210

Blanchard, R. J., Blanchard, D. C., & Hori, K. (1989). *An ethoexperimental approach to the study of defense*. Academic Press.

Blazer, D. G. (2003). Depression in late life: Review and commentary. *The Journals of Gerontology Series: Biological Sciences and Medical Sciences*, *58*(3), M249–M265. doi:10.1093/gerona/58.3.M249 PMID:12634292

Blazer, D. G., Burchett, B., Service, C., & George, L. K. (1991). The association of age and depression among the elderly: An epidemiologic explanation. *The Journals of Gerontology. Series A, Biological Sciences and Medical Sciences*, *46*(6), M210–M215. PMID:1834726

Blazer, D. G., Hybels, C. F., Simonsick, E. M., & Hanlon, J. T. (2000). Marked differences in antidepressant use by race in an elderly community sample: 1986-1996. *The American Journal of Psychiatry*, *157*(7), 1089–1094. doi:10.1176/appi.ajp.157.7.1089 PMID:10873916

Blazer, D. G., & Ph, D. (2000). Reviews and Overviews Psychiatry and the Oldest Old. *Am J PsychiatryAm J Psychiatry*, *15712*(December), 1915–1924. doi:10.1176/appi.ajp.157.12.1915

Blier, P., Ph, D., De Montigny, C., & Ph, D. (1999). Serotonin and Drug-Induced Therapeutic Responses in Major Depression, Obsessive – Compulsive and Panic Disorders. *Neuropsychopharmacology*, *21*(99), 15–20. PMID:10432494

Bondy, S. C. (1988). *The neurotoxicity of organic and inorganic lead*. Academic Press.

Bosch, A. C., O'Neill, B., Sigge, G. O., Kerwath, S. E., & Hoffman, L. C. (2016). Heavy metals in marine fish meat and consumer health: A review. *Journal of the Science of Food and Agriculture*, *96*(1), 32–48. doi:10.1002/jsfa.7360 PMID:26238481

Bose-O'Reilly, S., McCarty, K. M., Steckling, N., & Lettmeier, B. (2010). Mercury exposure and children's health. *Current Problems in Pediatric and Adolescent Health Care*, *40*(8), 186–215. doi:10.1016/j.cppeds.2010.07.002 PMID:20816346

Bratel, J., Haraldson, T., & Ottosson, J. O. (1997). Potential side effects of dental amalgam restorations. (II). No relation between mercury levels in the body and mental disorders. *European Journal of Oral Sciences*, *105*(3), 244–250. doi:10.1111/j.1600-0722.1997.tb00207.x PMID:9249191

Clarkson, T. W. (1972). The pharmacology of mercury. *Journal of the American Medical Association*, (12), 375–406. doi:10.1001/jama.1922.02640250025010 PMID:4556945

Clayton, P. J., Grove, W. M., Coryell, W., Keller, M., Hirschfeld, R., & Fawcett, J. (1991). *Follow-Up and Family Study of Anxious Depression*. Academic Press.

Clement, Y., & Chapouthier, G. (1998). Article. *Biological Bases of Anxiety*, *22*(5), 623–633.

Cory-Schlecta, D. A., & Schaumburg, H. H. (2000). Lead, inorganic. Oxford University Press.

Cryan, J. F., Markou, A., & Lucki, I. (2002). Assessing antidepressant activity in rodents: Recent developments and future needs. *Trends in Pharmacological Sciences*, *23*(5), 238–245. doi:10.1016/S0165-6147(02)02017-5 PMID:12008002

de Vries, Y. A., Roest, A. M., Burgerhof, J. G. M., & de Jonge, P. (2018). Initial severity and antidepressant efficacy for anxiety disorders, obsessive-compulsive disorder, and posttraumatic stress disorder: An individual patient data meta-analysis. *Depression and Anxiety*, 1–8. doi:10.1002/da.22737 PMID:29659102

Dean, B., Lam, L. Q., Scarr, E., & Duce, J. A. (2019). Cortical biometals: Changed levels in suicide and with mood disorders. *Journal of Affective Disorders*, *243*, 539–544. doi:10.1016/j.jad.2018.09.026 PMID:30292148 del Blanco, B., & Barco, A. (2018). Impact of environmental conditions and chemicals on the neuronal epigenome. *Current Opinion in Chemical Biology*, *45*, 157–165. doi:10.1016/j.cbpa.2018.06.003 PMID:29913328

Eaton, W. W., Kramer, M., Anthony, J. C., Dryman, A., Shapiro, S., & Locke, B. Z. (1989). The incidence of specific DIS/DSM-III mental disorders: Data from the NIMH Epidemiologic Catchment Area Program. *Acta Psychiatrica Scandinavica*, *79*(2), 163–178. doi:10.1111/j.1600-0447.1989.tb08584.x PMID:2784251

Fava, M., Uebelacker, L. A., Alpert, J. E., Nierenberg, A. A., Pava, J. A., & Rosenbaum, J. F. (1997). Major depressive subtypes and treatment response. *Biological Psychiatry*, *42*(7), 568–576. doi:10.1016/S0006-3223(96)00440-4 PMID:9376453

Forsell, Y., & Winblad, B. (1999). Incidence of major depression in a very elderly population. *International Journal of Geriatric Psychiatry*, *14*(5), 368–372. doi:10.1002/(SICI)1099-1166(199905)14:5<368::AID-GPS919>3.0.CO;2-Y PMID:10389040

Freire, C., Amaya, E., Gil, F., Fernández, M. F., Murcia, M., Llop, S., ... Olea, N. (2017, November). Prenatal co-exposure to neurotoxic metals and neurodevelopment in preschool children: The Environment and Childhood (INMA) Project. *The Science of the Total Environment*, *621*, 340–351. doi:10.1016/j.scitotenv.2017.11.273 PMID:29190557

Gardella, C. (2001). Lead exposure in pregnancy: A review of the literature and argument for routine prenatal screening. *Obstetrical & Gynecological Survey*, *56*(4), 231–238. doi:10.1097/00006254-200104000-00024 PMID:11285436

Graeff, F. G., Guimarães, F. S., De Andrade, T. G. C. S., & Deakin, J. F. W. (1996). Role of 5-HT in stress, anxiety, and depression. *Pharmacology, Biochemistry, and Behavior*, *54*(I), 129–141. doi:10.1016/0091-3057(95)02135-3 PMID:8728550

Hiller, W., Zaudig, M., & von Bose, M. (1989). The overlap between depression and anxiety on different levels of psychopathology. *Journal of Affective Disorders, 16*(2-3), 223-31. Retrieved from http://www.ncbi.nlm.nih.gov/pubmed/2522121

Hobson, M., Milhouse, M., Rajanna, B., Natural, D., & Sciences, A. (1986). *~ S Effects of Cadmium on the Uptake of Dopamine and Norepinephrine in Rat Brain Synaptosomes*. Academic Press.

Homme, K., Kern, J. K., Geier, D. A., Bjørklund, G., King, P. G., Homme, K. G., … Geier, M. R. (2014). *Amalgams and Chronic Illness, Fatigue, Depression, Anxiety, and Suicide*. Academic Press.

Hursh, J. B., Clarkson, T. W., Cherian, M. G., Vostal, J. J., & Mallie, R. V. (1976). Clearance of mercury vapor inhaled by human subjects. *Archives of Environmental Health*, *31*(6), 302–309. doi:10.1080/00039896.1976.10667240 PMID:999343

Jaga, K., & Dharmani, C. (2007). The interrelation between organophosphate toxicity and the epidemiology of depression and suicide. *Reviews on Environmental Health*, *22*(1), 57–73. doi:10.1515/REVEH.2007.22.1.57 PMID:17508698

Jan, A. T., Azam, M., Siddiqui, K., Ali, A., Choi, I., & Haq, Q. M. R. (2015). Heavy metals and human health: Mechanistic insight into toxicity and counter defense system of antioxidants. *International Journal of Molecular Sciences*, *16*(12), 29592–29630. doi:10.3390/ijms161226183 PMID:26690422

Johnson, J., Weissman, M. M., & Klerman, G. L. (1992). Service utilization and social morbidity associated with depressive symptoms in the community. *Journal of the American Medical Association*, *267*(11), 1478–1483. doi:10.1001/jama.1992.03480110054033 PMID:1538538

Johnson, M. A. (2001). High calcium intake blunts pregnancy-induced increases in maternal blood lead. *Nutrition Reviews, 59*(5), 152-156. Retrieved from http://www.scopus.com/inward/record.url?eid=2-s2.0-0034989948&partnerID=tZOtx3y1

Jurczak, A., Brodowska, A., Szkup, M., Prokopowicz, A., Karakiewicz, B., Łój, B., ... Grochans, E. (2018). Influence of Pb and Cd levels in whole blood of postmenopausal women on the incidence of anxiety and depressive symptoms. *Annals of Agricultural and Environmental Medicine*, *25*(2), 219–223. doi:10.26444/aaem/85929 PMID:29936823

Keller, B., & Shapiro, W. (1982). *Double Episodes.* Academic Press.

Kim, K. H., Kabir, E., & Jahan, S. A. (2017). Exposure to pesticides and the associated human health effects. *The Science of the Total Environment*, *575*, 525–535. doi:10.1016/j.scitotenv.2016.09.009 PMID:27614863

Kohzadi, S., Shahmoradi, B., Ghaderi, E., Loqmani, H., & Maleki, A. (2018). Concentration, Source, and Potential Human Health Risk of Heavy Metals in the Commonly Consumed Medicinal Plants. *Biological Trace Element Research.* doi:10.100712011-018-1357-3 PMID:29700699

Komulainen, H., & Tuomisto, J. (1980). Effect of Heavy Metals on Dopamine, Noradrenaline and Serotonin Uptake and Release in Rat Brain Synaptosomes. *Acta Pharmacologica et Toxicologica*, *48*(3), 199–204. doi:10.1111/j.1600-0773.1981.tb01609.x

Leichsenring, F., & Leweke, F. (2006). Social Anxiety Disorder. *The New England Journal of Medicine*, *355*(10), 1029–1036. doi:10.1056/NEJMcp060145 PMID:16957148

Leret, M. L., San Millán, J. A., & Antonio, M. T. (2003). Perinatal exposure to lead and cadmium affects anxiety-like behaviour. *Toxicology*, *186*(1-2), 125–130. doi:10.1016/S0300-483X(02)00728-X PMID:12604176

Lewinsohn, P. M., Holm-Denoma, J. M., Small, J. W., Seeley, J. R., & Joiner, T. E. Jr. (2008). Separation anxiety disorder in childhood as a risk factor for future mental illness. *Journal of the American Academy of Child and Adolescent Psychiatry*, *47*(5), 548–555. doi:10.1097/CHI.0b013e31816765e7 PMID:18356763

Lidsky, T. I., & Schneider, J. S. (2003). Lead neurotoxicity in children: Basic mechanisms and clinical correlates. *Brain*, *126*(1), 5–19. doi:10.1093/brain/awg014 PMID:12477693

Mansouri, M. T., Muñoz-Fambuena, I., & Cauli, O. (2018). Cognitive impairment associated with chronic lead exposure in adults. *Neurology, Psychiatry & Brain Research*, *30*(March), 5–8. doi:10.1016/j.npbr.2018.04.001

Maret, W. (2017). The bioinorganic chemistry of lead in the context of its toxicity. *Metal Ions in Biological Systems*. doi:10.1515/9783110434330-001

Mathee, A., Singh, E., Mogotsi, M., Timothy, G., Maduka, B., Olivier, J., & Ing, D. (2009). Lead-based paint on playground equipment in public children's parks in Johannesburg, Tshwane and Ekurhuleni. *South African Medical Journal*, *99*(11), 819–821. PMID:20218484

Nelson, K. B., & Bauman, M. L. (2003). Thimerosal and Autism? *Pediatrics*, *111*(3), 674–679. doi:10.1542/peds.111.3.674 PMID:12612255

Njung'e, K., & Handley, S. L. (1991). Effects of 5???HT uptake inhibitors, agonists and antagonists on the burying of harmless objects by mice; a putative test for anxiolytic agents. *British Journal of Pharmacology*, *104*(1), 105–112. doi:10.1111/j.1476-5381.1991.tb12392.x PMID:1686200

Nordberg, G. F. (2004). Cadmium and health in the 21st century--historical remarks and trends for the future. *Biometals: An International Journal on the Role of Metal Ions in Biology, Biochemistry, and Medicine, 17*(5), 485-489. doi:10.1023/B:BIOM.0000045726.75367.85

Nutt, B. D. J. (2005). *Overview of Diagnosis and Drug Treatments of Anxiety Disorders*. Academic Press.

Oliver, A. M., Wright, K. D., Kakadekar, A., Pharis, S., Pockett, C., Bradley, T. J., ... Erlandson, M. C. (2018). Health anxiety and associated constructs in children and adolescents with congenital heart disease: A CHAMPS cohort study. *Journal of Health Psychology*. doi:10.1177/1359105318755263 PMID:29402140

Onwuzuligbo, O., Hendricks, A. R., Hassler, J., Domanski, K., Goto, C., & Wolf, M. T. F. (2018). Mercury Intoxication as a Rare Cause of Membranous Nephropathy in a Child. *American Journal of Kidney Diseases*, *1-5*. doi:10.1053/j.ajkd.2018.05.013 PMID:30041877

Orisakwe, O. E. (2014). The role of lead and cadmium in psychiatry. *North American Journal of Medical Sciences*, *6*(8), 370–376. doi:10.4103/1947-2714.139283 PMID:25210669

Overall, J. E. (1976). The Brief Psychiatric Rating Scale in psychopharmacologic research. *ECDEU Assessment Manual for Psychopharmocol*, *7*, 166–169.

Regier, D. (1988). One-Month Prevalence of Mental Disorders in the United States. *Archives of General Psychiatry*, *45*(11), 977. doi:10.1001/archpsyc.1988.01800350011002 PMID:3263101

Rosenbaum, J. F., Pollock, R. A., Otto, M. W., & Pollack, M. H. (1995). Integrated treatment of panic disorder. *Bulletin of the Menninger Clinic, 59*(2 Suppl A), A4-26. Retrieved from http://www.ncbi.nlm.nih.gov/pubmed/7795571

Rush, A. J., & Weissenburger, J. E. (1994). Melancholic symptom features and DSM-IV. *The American Journal of Psychiatry*, *151*(4), 489–498. doi:10.1176/ajp.151.4.489 PMID:8147445

Sciarillo, W. G., Alexander, G., & Farrell, K. P. (1992). Lead exposure and child behavior. *American Journal of Public Health*, *82*(10), 1356–1360. doi:10.2105/AJPH.82.10.1356 PMID:1415859

Scinicariello, F., & Buser, M. C. (2015). Blood cadmium and depressive symptoms in young adults (aged 20-39 years). *Psychological Medicine*, *45*(4), 807–815. doi:10.1017/S0033291714001883 PMID:25115444

Shiue, I. (2015). Urinary heavy metals, phthalates and polyaromatic hydrocarbons independent of health events are associated with adult depression: USA NHANES, 2011–2012. *Environmental Science and Pollution Research International*, *22*(21), 17095–17103. doi:10.100711356-015-4944-2 PMID:26126689

Siblerud, R. L., Motl, J., & Kienholz, E. (1994). Psychometric evidence that mercury from silver dental fillings may be an etiological factor in depression, excessive anger, and anxiety. *Psychological Reports*, *74*(36), 67–80. doi:10.2466/pr0.1994.74.1.67 PMID:8153237

Sioen, I., Den Hond, E., Nelen, V., Van de Mieroop, E., Croes, K., Van Larebeke, N., ... Schoeters, G. (2013). Prenatal exposure to environmental contaminants and behavioural problems at age 7-8years. *Environment International*, *59*, 225–231. doi:10.1016/j.envint.2013.06.014 PMID:23845936

Stoltenburg-Didinger, G. (1994). *Neuropathology of the hippocampus and its susceptibility to neurotoxic insult.* Retrieved from https://sci-hub.tw/https://europepmc.org/abstract/med/7854577

Tanya, S., Chutima, R., Andre, F. C., Michel, B., & Michael, M. (2018). Anxiety disorders: Sex differences in serotonin and tryptophan metabolism. *Current Topics in Medicinal Chemistry*, 1–46. doi:10.20944/preprints201808.0487.v1 PMID:30430940

Teresi, J. A., Abrams, R., Holmes, D., Ramirez, M., Shapiro, C., & Eimicke, J. P. (2002). Influence of cognitive impairment, illness, gender, and African-American status on psychiatric ratings and staff recognition of depression. *The American Journal of Geriatric Psychiatry*, *10*(5), 506–514. doi:10.1097/00019442-200209000-00003 PMID:12213684

Theorell, T., Hammarström, A., Aronsson, G., Träskman Bendz, L., Grape, T., Hogstedt, C., ... Hall, C. (2015). A systematic review including meta-analysis of work environment and depressive symptoms. *BMC Public Health*, *15*(1), 738. doi:10.118612889-015-1954-4 PMID:26232123

Vig, P.J., & Nath, R. (1991). *In vivo effects of cadmium on calmodulin and calmodulin regulated enzymes in rat brain*. Academic Press.

Volirath, M., & Angst, J. (1989). Outcome of panic and depression in a seven???year follow???up: Results of the Zurich study. *Acta Psychiatrica Scandinavica*, *80*(6), 591–596. doi:10.1111/j.1600-0447.1989.tb03031.x PMID:2618782

Werner-Seidler, A., Perry, Y., Calear, A. L., Newby, J. M., & Christensen, H. (2017). School-based depression and anxiety prevention programs for young people: A systematic review and meta-analysis. *Clinical Psychology Review*, *51*, 30–47. doi:10.1016/j.cpr.2016.10.005 PMID:27821267

Wong, M. L., & Licinio, J. (2004). From monoamines to genomic targets: A paradigm shift for drug discovery in depression. *Nature Reviews. Drug Discovery*, *3*(2), 136–151. doi:10.1038/nrd1303 PMID:15040578

World Health Organization. (1991). *Inorganic mercury: environ- mental health criteria 118*. Geneva, Switzerland: World Health Organization.

Wu, W., Wu, P., Yang, F., Sun, D., Zhang, D.-X., & Zhou, Y.-K. (2018). Assessment of heavy metal pollution and human health risks in urban soils around an electronics manufacturing facility. *The Science of the Total Environment*, *630*, 53–61. doi:10.1016/j.scitotenv.2018.02.183 PMID:29475113

Yang, Y.-W., Liou, S.-H., Hsueh, Y.-M., Lyu, W.-S., Liu, C.-S., Liu, H.-J., ... Chung, C.-J. (2018). Risk of Alzheimer's disease with metal concentrations in whole blood and urine: A case–control study using propensity score matching. *Toxicology and Applied Pharmacology*, *356*, 8–14. doi:10.1016/j.taap.2018.07.015 PMID:30025849

Yap, M. B. H., Martin, P. D., & Jorm, A. F. (2018). Online parenting guidelines to prevent adolescent depression and anxiety: Evaluating user characteristics and usefulness. *Early Intervention in Psychiatry*, *12*(5), 951–958. doi:10.1111/eip.12503 PMID:29076246

Young, J. F., Mufson, L., & Davies, M. (2006). Impact of comorbid anxiety in an effectiveness study of interpersonal psychotherapy for depressed adolescents. *Journal of the American Academy of Child and Adolescent Psychiatry*, *45*(8), 904–912. doi:10.1097/01.chi.0000222791.23927.5f PMID:16865032

Yousefinejad, V., Mansouri, B., Ramezani, Z., Mohammadzadeh, N., & Akhlaghi, M. (2018). Evaluation of heavy metals in tobacco and hookah water used in coffee houses in Sanandaj city in 2017. *Majallah-i Ilmi-i Danishgah-i Ulum-i Pizishki-i Kurdistan*, *22*(6), 96–106.

Chapter 16
Epilepsy and Stroke Emerging From Climate Change-Related Neurotoxicity:
Involvement of Food and Water Contaminations

Hind Benammi
Cadi Ayyad University, Morocco

Omar El Hiba
Chouaib Doukkali University, Morocco & Cadi Ayyad University, Morocco

Abdelmohcine Aimrane
Cadi Ayyad University, Morocco

Nadia Zouhairi
Cadi Ayyad University, Morocco

Hicham Chatoui
Private University of Marrakech, (UPM), Morocco & Cady Ayyad University, Morocco

Hasna Lahouaoui
Cadi Ayyad University, Morocco

Kholoud Kahime
Cadi Ayyad University, Morocco

Morad Belkouri
Centre Hospitalier Universitaire, Morocco

Rokia Ghchime
Mohamed V University, Morocco

Ahmed Draoui
Cadi Ayyad University, Morocco

Halima Gamrani
Cadi Ayyad University, Morocco

ABSTRACT

Climate change has an important impact on the environment. As it degrades the quality of water, soil, and area, it also spreads the distribution of many toxic elements, specifically heavy metals and pesticides. The impact of climate change on contamination with heavy metals and pesticides has been well investigated and discussed. The influence of these elements on human health is obviously exacerbated following their extended distribution. Moreover, a wide range of health problems have been associated to such intoxication, among which impairment and dysfunction of the nervous system are prominent. In this chapter, the authors will shed light on two most common neurological diseases such as epilepsy and stroke affecting people worldwide arising from food and water contaminations, mainly with heavy metals and pesticides.

DOI: 10.4018/978-1-5225-7775-1.ch016

INTRODUCTION

Climate change endangers human health and affects all sectors of society, both domestically and globally. The impacts of climate change on the environment are obviously discernable through degradation of the quality of water, soil and area. Several challenges are facing scientists today on that issue. One of the most prominent questions preoccupying scientists, organizations and governments is the impacts of climate change on the risk of contamination with heavy metals. Thus, the related effects have been discussed qualitatively for the marine ecosystems (Schiedek, Sundelin, Readman, & Macdonald, 2007). In fact, the risk of contamination may increasewith the increasing risk of flooding caused by climate change. Incontaminated lands, flooding promotes remobilization of contaminants in sedimentation water,in marine areas and freshwater environment. Furthermore, human management of the system, such as industrialization and progresses to an exposure to toxic metals, have an important influence on the quality of water table, which consequently leads to the contamination of soil and food. Moreover, climate change has a powerful effect on the fate and the behavior of pesticides in the environment by modifying the fundamental partitioning mechanisms between environmental compartments, therefore affecting the use of pesticides (Figure 1) (Noyes et al., 2009). A lower pesticide residue on crops, due to climate change, increases the susceptibility to pests and diseases. That is to say, farmers in the future may need to spray more often during the growing season(Jackson et al., 2011). Higher parasitic or pathological pressure will also increase application frequencies and volumes. As a result, the residue levels detected could double for some products (Seeland, Oehlmann, & Müller, 2012).Consequently, the risk of depositing pesticide residues in food increases as well as the exposure of consumer to pesticides residues at the end of the food supply chain(Noyes et al., 2009).In this regard, food safety issues related to increased exposure to pesticide residues, due to climate change, are increasingly being addressed (Delcour, Spanoghe, & Uyttendaele, 2015).

Heavy metals and pesticides are toxics that may aggregate in the biological system setting off significant health hazards (Rajan, Nadiah, Firdaus, Appukutty, & Rama-, 2012; Tago, Andersson, & Treich, 2014) Overexposure to heavy metals and pesticides can cause myriad of problems in our bodies, autoimmune diseases, infertility, dementia, epilepsy, stroke and more (Tago et al., 2014). In this chapter we will put the focus on neurological diseases, especially epilepsy and stroke. Epilepsy is a common medical and social disorder or groupof disorders with unique characteristics (Day, Wu, Strauss, Shavelle, & Reynolds, 2005). Epilepsy is usually defined as neurological disease characterized by the recurrence of excessive paroxysmal and hypersynchronous discharges of more or less extensive neuronal brain population called epileptic seizures (Devinsky, Vezzani, Jette, Curtis, & Perucca, 2018; Stafstrom & Carmant, 2015). Epilepsy occurs in approximately 0.7% of the population at any time. More than two-thirds of seizure problems begin in childhood, with a second peak of onset in the elderly. Usually, epilepsy does not significantly alter life expectancy, but the quality of life may be seriously compromised when seizures are not satisfactorily managed(Ann & Gus, 2008).The second disease, Stroke, also called brain attack, is a brain injury caused by a sudden interruption in the blood supply of the brain (Gund, Jagtap, Ingale, & Patil, 2013). World Health Organization (WHO) has defined stroke as ''rapidly developed clinical signs of focal disturbance of cerebral function, lasting more than 24 hours or leading to death, with no apparent cause other than vascular origin.'' (WHO, 2014). Worldwide burden of stroke, estimated that stroke is the second most common cause of death worldwide (Lim, Lee, Lee, & Ha, 2018).

Figure 1. Illustration of the environmental factors that influence pesticide fate after application (modified (Delcour et al., 2015))

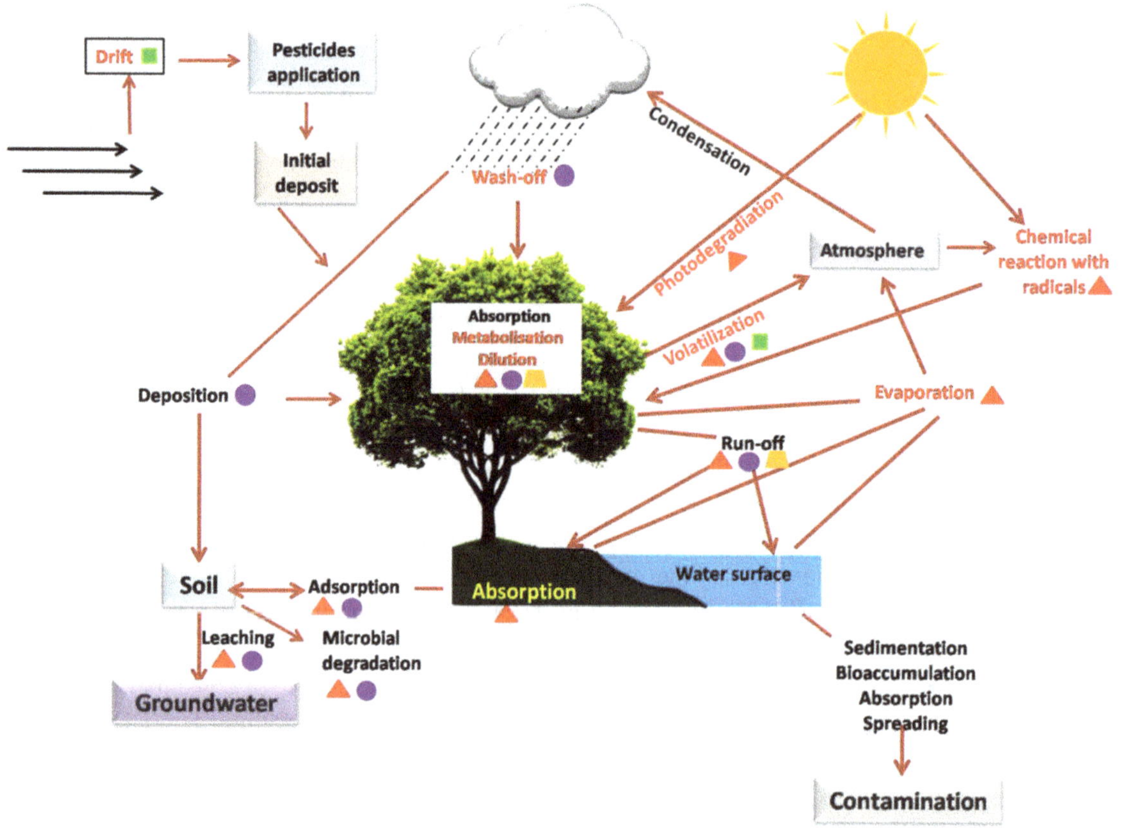

Given the severity and the complication of these diseases, we will discuss in this chapter the implication of environmental contamination (water, soil, food and area) by heavy metals, pesticides as well as their exacerbation by climate change, therefore causing diseases in humans.

THE INCIDENCE AND PREVALENCE OF EPILEPSY

Epilepsy is a chronic affection of the nervous system characterized by the recurrence of excessive paroxysmal and hypersynchronous discharges of a more or less extensive neuronal brain population, called epileptic seizures, which can be recorded by electroencephalogram (EEG)(Devinsky et al., 2018; Stafstrom & Carmant, 2015). Interestingly, epidemiological studies have classified epilepsy as one common neurological disease worldwide that needs to be handled with the utmost care, since it affects roughly 50 millons of individuals with no age discrimination. It, therefore, leads to a plethora of social, behavioural, health and economic problems to the patients and their families(Ngugi, Bottomley, Kleinschmidt, Sander, & Newton, 2010). The incidence of epilepsy is also exacerbated with poverty. In fact, countries with low- to medium-income are affected with a prevalence two to three times higher than industrialized ones (Hennion & Hennion, 2017).

Though a lot of studies have declared the high influence of epilepsy on populations' health and wellbeing, precise evaluation remains challenging because of the difficulties faced when defining and classifying epilepsy (passive or active), the significant variations within region in the same country, the difficulties in obtaining epidemiological surveys with valid statistics, the scarcity of rigorous work, the scarcity of information concerning mortality and the absence of study protocols and uniform investigation questionnaires(Bhalla, 2014).Nevertheless, medical diagnostics appeal to use a variety of tools to identify and classify the type/epilepsy syndrome and etiology. Basically, the most pertinent are electroencephalogram (EEG), magnetic resonance imaging (MRI), positron emission tomography (PET), single photon emission computed tomography (SPECT), magneto encephalogram (MEG), in addition with neuropsychiatric testing.

Electroencephalography is one among the pertinent devise in physical checkup for some electroclinical syndromes. Thus, it is very helpful for examination of patient with epilepsy, as it affords vital clues related to epileptiform discharges in epilepsy and other conditions(Owolabi et al., 2018). To confirm the type of the seizure and to estimate the amount of cortex representing the epileptogenic zone, Video-EEG monitoring may be the most adequate device to use, especially when neuroimaging do not provide precise results, and surgery must be done (Rama K. Maganti, MD; Paul Rutecki, 2013).

Furthermore, computerized tomography (CT) scans are useful in emergent conditions, although the focal lesions are detected in only 30 per cent of patients (Guerreiro, 2016). Functional imaging, such as positron emission tomography (PET), is used to demonstrate regional differences in metabolic activity. Moreover, single photon emission CT (SPECT) serves to analyze regional differences in blood flow during a seizure (ictal SPECT) and between seizures (interictal SPECT)(S. Kim & Mountz, 2011).MEG and functional MRI are additional important tools that help providing further information to localize the zones with epileptogenic lesion and to identify the areas surrounding the eloquent cortex(Shin, Jewells, Hadar, Fisher, & Hinn, 2014). Altogether, these devices are of particular assistance with cases in which focal seizures are suspected, but the brain MRI is negative.

Etiology of Epilepsy

The factors that may increase the risk of epilepsy are multiple and vary worldwide across regions(Mac et al., 2007). The most pertinent among them are the following:

- **Brain Tumors:** either benign or malignant, are major factors setting of epilepsy. Interestingly, brain tumors cause an epilepsy incidence of nearly 30% (Aive Liigant, Sulev Haldre, Andre Oun, Ülla Linnamägi, Anu Saar, Toomas Asser, 2001).
- **Brain Injury:** That constitutes more than 20% of all symptomatic epilepsy cases(Lowenstein, 2009).
- **Neurocutaneous Syndromes:** Like tuberous sclerosis complex characterized by the emergence of benign tumors in collective organ systems, including the brain,
- **Stroke:** Which is one of the important risk factor of epilepsy. Several studies have shown that 2-4% of stroke cases have epilepsy in their lifetime(Tom Skyhøj Olsen, MD, 2001).
- **Bacterial/Viral Brain Infection:** Which is a risk factor more prevalent in the developing countries and linked to several pathologies like *Toxocariasis,* predominantly seen in children (aged <5 years),which is caused by the larvae of *Toxocara canis* that have a tendency to lodge in the brain(Nicoletti, 2010).

- **Bacterial Brain Abscesses:** Which may be due to contiguous suppurating process, blood-bourne infection or complications of brain surgery. Commonly, epilepsy begins within the 3 years following abscess in these cases(Chuang et al., 2010),
- **Toxoplasmosis**: An opportunistic infection among immunocompromised individuals. In these patients, toxoplasmosis is a frequent etiological factor for acute seizures and epilepsy, as are progressive multifocal leukoencephalopathy and other acute cerebral infections (Kellinghaus et al., 2008). The developmental cortical malformations are the first etiology of intractable epilepsy and widespread in children(Fujiwara & Shigematsu, 2004).
- **Neurodegenerative Disorders:** Such as Alzheimer disease, are more and more increasingly recognized as a cause of epilepsy oncoming(Hans Forstl, MD; Alistair Burns, MD; Raymond Levy, PhD; Nigel Cairns, PhD; Philip Luthert, MRCPath; Peter Lantos, 1992). Furthermore, there is a significant relationship between epilepsy and chronic use of alcohol(Samokhvalov, Irving, & Mohapatra, 2010)as well aspathologies of inflammatory origin such as autoimmune encephalitis (Bien & Schramm, 2009).

Pathophysiology of Epileptic Disorders

Epilepsy emerges from an imbalance between the excitatory and the inhibitory networks. In fact, neurons have electrical activity related to ionic movements through specific membrane channels. The fluxes of calcium (Ca $^{++}$) and sodium (Na $^{+}$) produce a depolarization of the cell making it hyperexcitable. However, the fluxes of potassium (k $^{+}$) and chlorine (Cl^{-}) cause a hyperpolarization making the cellless excitable. The opening of these channels is either voltage-dependent or linked to a neurotransmitter (Safini, Bouskraoui, Pédiatrie, Mohammed, & Marrakech, 2009).

As an illustration, the paroxysmal depolarization shift (PDS), responsible for all types of epileptic seizures (Gregory L Holmes & Yezekiel Ben-Ari, 2003)is acellular event where rapid repetitive action potentials are not accompanied by the usual refractory period. This generates a prolonged membraned epolarization (Figure 2); which is more prolonged than the usual occurrence of the normal excitatory postsynaptic potentials [EPSPs] (Thomas, 2012).GABA-ergic neurons and after the occurrence of epileptogenic attacks (e.g. status epilepticus, cerebrovascular accident) followed by the reorganization of neuronal circuits, favor ahypersynchrony of neuronal populations (e.g. aberrant axon connections of granular cells teeth of the dentate gyrus), a fact known as germination of mossy fibers(Thomas, 2012).

Therefore, the deficiency of signals mediated with GABAergic neurons and the increase in glutamatergic transmission that have been labeled in many types of epilepsy(not all), represent the basis of drug therapy for the disease(Kovac & Walker, 2013). The glutamate-GABA epilepsy hypothesis is based on Dale's principle which states that the neuropharmacological profile of each neuron is defined by a single neurotransmitter. Sincethis principle has been proven to be wrong, it is obviously clear that the neuropharmacological basis of epilepsy is much more complex and multifaceted (Mazarati, 2009).

Involvement of Heavy Metals

Toxic metals, counting "heavy metals", are individual metals and metal compounds that deleteriously affect people's health. In very small amounts, many of these metals are oligo-elements indispensable to body function. However, in larger amounts, they become toxic and may build up in biological systems causing significant health hazard (Rajan et al., 2012).The influence of climate change on heavy metal

Figure 2. Paroxysmal depolarization shift (PDS). When a PDS occurs as an abnormally prolonged run of action potentials during sustained membrane depolarization in a single neuron, as shown in the upper trace in B, the event is detectable only with microelectrodes; increased glutamate concentration is associated with influx of cations initially, followed by increased GABA concentration with efflux of potassium. When PDSs in a large number of neurons are synchronized for less than 200 ms, as shown in A, these electrical potentials may summate as a spike-wave complex that is recorded with macroelectrodes, as shown in the lower trace in B. When sustained repetitive firing of PDSs in a large number of neurons becomes synchronized for many seconds or longer, an electrographic seizure occurs, as shown in C. (Adapted with permission from Holmes GL, Ben-Ari Y. Seizing hold of seizures. Nature Med 2003; 9:994-6).

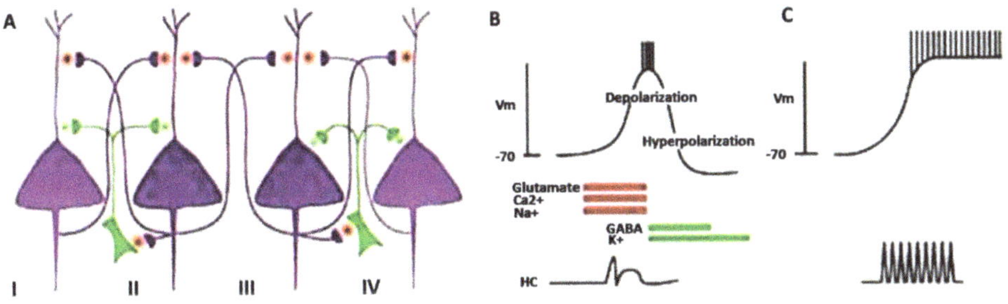

contamination have been discussed for water, soils (foods) and area (Schiedek et al., 2007; Sobolev & Begonia, 2008; Visser et al., 2012). Climate change trend to exacerbate the intoxication with metals on the population, where the most noticeable effects are observed among children and elderly, more vulnerable to toxicity since their immune systems are either underdeveloped or age-compromised(Acosta, Cano, Arocena, Debela, & Martínez-martínez, 2009).

The toxic amounts that could promote poisoning are variable according to the metal accumulated. An exposure to a small amount of arsenic or lead may induce a lot of damage, whereas several other metals serves for medical uses such as barium, nickel, and cadmium have (Zander, 2012). Moreover, the surrounding living area remain the source of varying levels of contamination because of the increasing amount of toxic waste, contaminated work environment as well as contaminated food and water supplies, and legacy materials like lead pipes, (WHO, 2010). Furthermore, teeth amalgam is one common way through which people experience heavy metal toxicity in a great extent. For instance, dental appliance made of mercury, gold crowns, and other metals may "leak" into the body and aggregate in different organs, mainly the brain (Zander, 2012). Altogether, human increasing tendencies for using compounds containing heavy metals and leaving in contaminated areas build up the burden of heavy metals in the body. Therefore, the threat within the central nervous system increases and give rise for many mental illnesses such as epilepsy (Rajan et al., 2012), a disease on which we will put the focus in this part.

Meany toxic elements of heavy metals can induce epileptic-like seizures. For instance, overloaded iron, variably existing as ferrous or ferric ion, mediates the generation of hydroxyl radical and affects the oxidative stress level of cortical neurons (Zou et al., 2017), leading to epileptogenesis (Mori et *al.,* 1990). The capacity of metal ions to induce epilepsy has long been recognized. Focal intracerebral applications of metallic substances, such as alumina cream iron or haem(Zhang & Pardridge, 2001; Zhao, Xi, Liu, Keep, & Hua, 2016). Furthermore, zinc chloride (Ganesh, Janakiraman, & Meenakshi, 2011)(Reid et *al.,* 2017) and copper chloride(Kheradmand et al., 2014)induce epileptiform discharges and sometimes spontaneous seizures in animals. Likewise, exposure to Pb can cause seizures, mental retardation, and behavioral disorders. The threat that heavy metals pose to human and animal health is aggravated by their low environmental mobility, even under high precipitations, and their long-termpersistence in the environment (Sobolev & Begonia, 2008).

Altogether, these heavy metals are absorbed by the body and stored in the nerve cells, the brain and the neuroganglions. They are the opponents of many trace elements (e.g. manganese, selenium, magnesium, zinc, etc), a fact which lower the nerve stimulus threshold (Meinardi, Scott, Reis, & Sander, 2001). Magnesium is essential in neurotransmitters metabolism and in modulating neurotransmitters receptors function. Moreover, ionized magnesium is important in seizure control, and is a NMDA antagonist (Nechifor, 2011) that may be a contributing factor in some epilepsies (Martynyuk, Glushakov, Sumners, & Laipis, 2005). The activation of NMDA receptor by glutamate results in calcium influx, which is pro-epileptogenic (Frye et al., 2016). Low ionized magnesium or altered balance between ionized magnesium and ionized calcium may accelerate the occurrence of seizures(Bjorn Schelter, Jens Timmer, 2008). Patients with epilepsy have been shown to have significantly lower means of ionized magnesium levels and an increased level of ionized calcium to ionized magnesium ratio (Sinert et al., 2007). Furthermore, low zinc levels have been associated with seizures in children(Ganesh & Janakiraman, 2008; Mollah et al., 2008). Although the role of zinc in epilepsy is not fully clear, it has been stated to act as an anticonvulsant that decreases seizure susceptibility(Cole, Robbins, & Ju, 2000). In fact, Zinc co-localizes with glutamate, inhibiting the reuptake of synaptic GABA, and eventually increasing the cortical inhibitory tone (Cohen-kfir, Lee, Eskandari, & Nelson, 2005). Thus, zinc deficiency, which may results from heavy metals competitive effect, could increase the relative excitatory-to-inhibitory balance (Frye et al., 2016).

Moreover, heavy metals may have toxic effects on the brain by reducing mitochondrial function (Belyaeva, Korotkov, & Saris, 2011; Rose, Wynne, Frye, Melnyk, & James, 2015), causing apoptosis (Pal, Pal, Das, & Sil, 2012), and increasing levels of reactive oxygen species(Chen Li, Liu Jian, Zhang Qiao Jun, Feng Jian Jun, Gui Zhen Hua, Ali Umar, Wang Yong, Fan Ling Ling, Hou Chen, 2011; Furieri et al., 2011).Although the mechanism(s) by which heavy metals cause epilepsy are not clear, both mitochondrial dysfunction(Rossignol & Frye, 2011) and high levels of reactive oxygen species(Specchio, Fusco, Claps, & Vigevano, 2010; Waldbaum & Patel, 2010)are additional factors that have been linked to epilepsy(Frye et al., 2016).

Involvement of Pesticides

Most agricultural productions depend on the use of pesticides to maintain high crop yields (Palikhe, 2007). In modern farming, using these chemicals has been considered as a key factor of agricultural industry success. However, most of the pesticides applied to agricultural lands may affect non-target organisms and contaminate soil and water media (Palikhe, 2002; Green Facts, 2007). Obviously, the concern about the risk of general population contamination with insecticides, via residues in food supplies, have increased in recent years. (Palikhe, 2003).Pesticide transfer to food potentially represents the most important exposure pathway for humans (Palikhe, 2007).

Previous literature reports sustain the potential effects on health of pesticides and first focused on the risks of acute intoxication among people with direct exposure(Ferreira et al., 2013). Interestingly, the main concern of scientific investigations has progressively shifted, thanks to the availability of longitudinal data, to studying the risks related to chronic intoxication and environmental contamination (Bhanti

& Taneja, 2007). Furthermore, the concern has more recently been given to studying the association of diseases, such as cancer, with chronic exposure to pesticides instead of investigating their instant effects (Weichenthal et al., 2014). Nevertheless, the possible results of exposure to pesticides are classified to short term effect (e.g. nausea, vomiting, diarrhea, abdominal pain, headaches, etc) and long-term effect (e.g. diabetes, cancer, genetic disorders, depression, neurological deficits, and death) (Tago et al., 2014). As a supportive example, chronic neurologic sequelae could result from acute organophosphates pesticides poisoning, possibly as a consequence of the onset of neurodegenerative processes by excitotoxic responses(Jett, 2011).

Epilepsy can result from chronic exposure to pesticides. In fact, the exposure to nerve agents and organophosphorus pesticides can occur through dermal contact, ingestion, or inhalation (Pierluigi Cocco, 2016). Such chemicals bind irreversibly to the enzyme acetylcholinesterase, thereby preventing clearance of acetylcholine from peripheral and central nerve terminals (War, 2008). The resultant cholinergic overstimulation manifests clinically as excessive secretions, weakness/paralysis, convulsions, and often eventual respiratory failure (Sevim, Aktekin, Dogu, Ozturk, & Ertas, 2002). In addition to these peripheral symptoms, overexcitation of cholinergic neurons in the brain leads to development of status epilepticus, a state of prolonged seizure activity. Individuals who survive in acute toxic exposure are subject to the long-term consequences of central nervous system damage caused by status epilepticus(Beavers, Parker, Flinchum, Weakley-jones, & Jortani, 2014). Additionally, in cases of pesticides poisoning, survivors often experience peripheral polyneuropathies and general multifaceted health problems (Mccarren & Jr, 2016). Other study show that some kind of pesticides like Dichlorodiphenyltrichloroethane (DDT), aldrin, and dieldrin are also known to induce seizures in mammals by affecting various neuronal pathways(Tattersall, 2009). Instead the resemblance of their effects, the targets and the mechanisms of action of these chemicals may be variables. For instance, DDT induce excitatory effects on CNS through increasing the amount of free ammonia, a fact that may reduce the level of GABA (Matin, Jaffery, & Siddiqui, 1981). However, aldrin tends to reduce CNS excitability threshold for seizure, a fact mediated by noradrenergic pathways. (Arora, Batra, Sharma, Banerjee, & Gupta, 2013)

Stroke Diseases

Stroke, also called brain attack, is a brain injury caused by a sudden interruption in the blood supply of the brain(Adams et al., 2007). This lack of brain blood supplies may results from one of these two reasons, either a sudden interruption of the blood supply to part of the brain, or following a rupture within a blood vessel in the which invades the surrounding areas (Gund et al., 2013). Approximately two million brain cells are estimated to die every minute during a stroke, thus increasing the risks of brain damage, disability, and death. Depending on the type of stork, the affected area in the brain as well as the size of the induced damage, the level of disability varies from patient to patient, (American Heart Association, 2003).World Health Organization (WHO) has defined stroke as "rapidly developed clinical signs of focal disturbance of cerebral function, lasting more than 24 hours or leading to death, with no apparent cause other than vascular origin." According to a study conducted by the same organization, storks accounts for 9.6% of all deaths causes. Furthermore, it has been considered as the principal cause of disability in adults(World & Report, 2003).

Epidemiology and Etiologies of Stroke Diseases

Stroke is estimated as the second most common cause of death worldwide(Lim et al., 2018). The public health burden of stroke is large in every country, and in many countries the burden continues to grow. Worldwide, in 2005, global stroke incidence was estimated at 16 million cases (Sahathevan, Brodtmann, & Donnan, 2012; Strong, Mathers, & Bonita, 2007) and 16.9 million new strokes occurred in 2010, representing an increase of 68%since 1990 (Feigin et al., 2014) which therefore was forecasted to reach23 million by 2030 (Strong et al., 2007). Obviously, incidence and mortality related to stroke have been remarked to decline in high-income countries while they increase in middle- and low-income countries of the world (Feigin, Lawes, Bennett, Barker-collo, & Parag, 2009; Johnston, Mendis, & Mathers, 2009). Indeed, the observed alleviation of stroke incidence in high-income countries may be the result of better healthcare standards and intervention strategies. However, at least one study demonstrates that these survival rates translate to an increased number of cognitively impaired stroke survivors ((Svetlana Ukraintseva, PhD; Frank Sloan, PhD; Konstantin Arbeev, PhD; Anatoly Yashin, 2015).

Risk factors for stroke include advanced age, hypertension(Tu, 2010), diabetes, high cholesterol, and atrial fibrillation. Recently, sleep and sleep disorders have been associated with stroke as risk factors (Patyar, Pharm, & Patyar, 2015). As well as smoking, stress, alcohol consumption, overweight, poor nutrition, and physical inactivity(AUDHILD HJALMARSEN, 2008; Carlson, Neelon, Carlson, Hartman, & Dogra, 2005; Corfield & Meadows, 2006).

Apart from these fundamental risk factors, environmental toxic substances including heavy metals and pesticides may affect novel pathways of risk such as inflammation and oxidative stress (Lukaszewicz-hussain, 2010).

Pathophysiology of Stroke Diseases

The brain is the central information-processing organ of the body responsible of the control of multiple complex functions. Without the blood supply, brain cells start to die (cerebral infarction) leading to brain damage. Normal cerebral blood flow (CBF) is approximately 50-60 ml/100g/min. The reduction in CBF below 20 ml/100g/min results in an electrical silence and less than 10 ml/100g/min causes irreversible neuronal injury (Patyar et al., 2015). Lack of blood circulation to the brain deprives neurons of necessary glucose and oxygen. Neurons are the impulse transmitters; hence they require a constant supply of energy(Koji Abe, 2008). Acute strokes have been classified into two major categories: ischemic and hemorrhagic strokes (Figure 3). Over 85% of acute stroke cases are resulted from disruption of the blood supply to the brain due to thrombosis or embolism (ischemia), leading to dysfunction of the brain tissue(Bath & Lees, 2007)Less frequently the remaining stroke cases are hemorrhagic, involving either intracerebral or subarachnoid hemorrhage(Escudero Augusto, Marqués Álvarez, & Taboada Costa, 2008). Ischemic stroke may manifest in the form of thrombotic stroke (large vessel and small vessel types); embolic stroke (with/without knowing cardiac and/or arterial factor); systemic hypoperfusion (Watershed or Border Zone stroke); or venous thrombosis (Mir, Al-Baradie, & Alhussainawi, 2014).

- **Hemorrhagic Stroke:** Constitutes only 10-15% of all strokes(Escudero Augusto et al., 2008), and results from blood vessel rupture in the brain. Its harmful effect resulted in (a) hypoxia due to disrupted vascular supply; (b) irritant effect of releasing blood on brain parenchyma and vasculature; and (c) raised ICP (intracranial pressure) due to continued bleeding, which may further restrict ce-

Figure 3. The two types of Stroke: hemorrhagic and ischemic
(Gund et al., 2013)

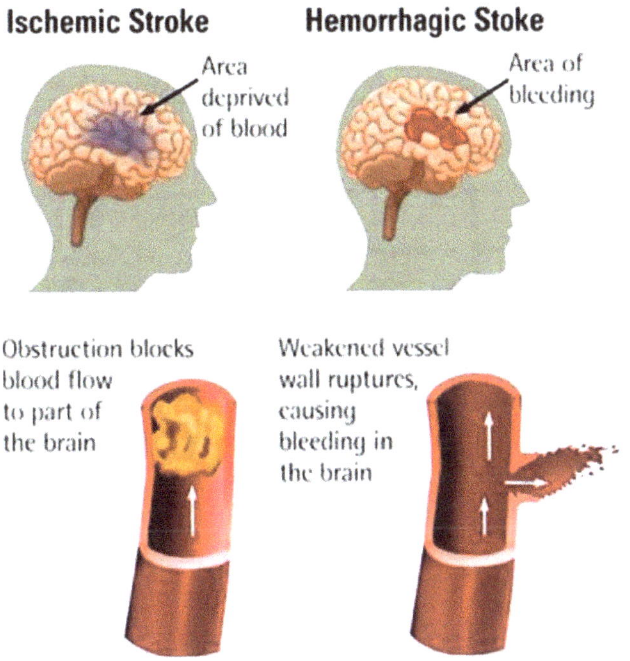

rebral blood flow. In this respect, hemorrhagic strokes are more dangerous than ischemic strokes. Intra-cerebral hemorrhage generally occurs in small arteries or arterioles and is commonly due to hypertension, trauma, bleeding disorders, amyloid angiopathy, drug use like amphetamines or cocaine, and vascular malformations, while sub-arachnoid hemorrhage results from rupture of aneurysms from the base of the brain and bleeding from vascular malformations near the pial surface)

- **Ischemic Brain Damage:** Ischemic stroke is usually caused by blockade of blood circulation by an embolus or by in situ thrombus (Savitz & Fisher, 2007), and it represents 87% of the total stroke (Roger et al., 2012). During ischemic brain injury, the neuronal damage is immediately initiated at the site of ischemic core. According to the ischemic duration, the necrotic lesions of neuronal death expand. Even with a successful reperfusion within few minutes after ischemic onset, the neuronal death can occur after several days representing delayed neuronal death (Figure 4). The major cell types affected by ischemic insults are neurons, but also astrocytes and CECs can be also damaged by sustained duration of ischemia(J. Kim, Byun, Chung, Chung, & Bae, 2013).
- **Molecular Mechanisms of Ischemic Damage:** Ischemic insult in brain induces a complicated array of pathological mechanisms. Ischemic cascade, which ultimately results in irreversible neuronal injury and brain infarction(Lo, Dalkara, & Moskowitz, 2003) . The most typical phenomenon of ischemic neuronal death is an excessive excitation (Macdonald, Xiong, & Jackson, 2006). There are several types of ion channels that regulate the membrane potential for neuronal excitation, and ATP is critical to maintain the function of these ion channels. Upon ischemic insult, ATP depletion immediately results in depolarization of neuronal membrane and excessive influx of calcium and sodium into neuronal cytosol(Lo et al., 2003). Also, dysregulated release of neurotransmitters such as glutamate occurs by reversed operation of transporters, and the excessive activation of

Figure 4. The different types ofstroke and cascade of events following ischemia (modified (Mir et al., 2014))

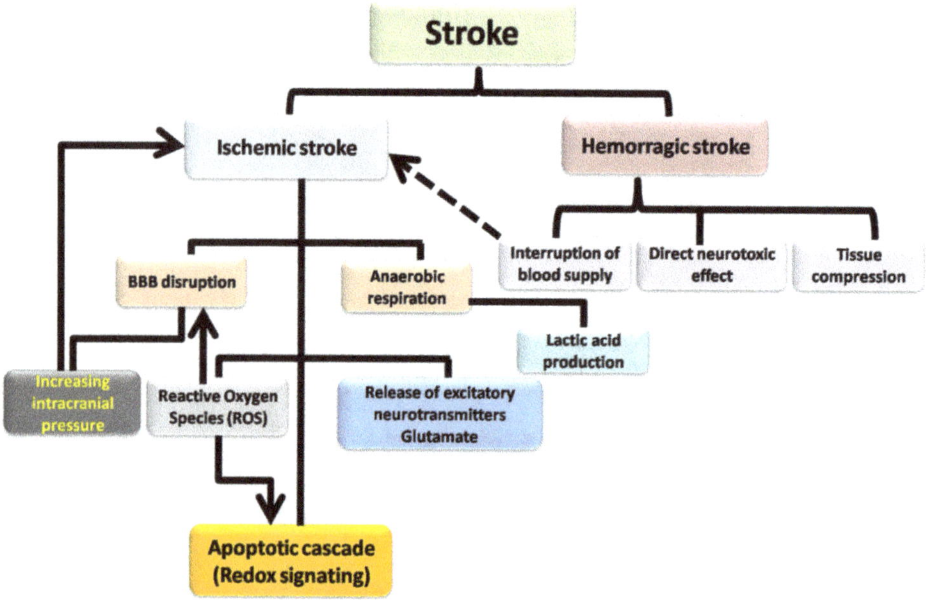

glutamate receptors further increase calcium influx into neuronal cytosol. Following calcium ion elevation, cytoskeletal degradation and enzymatic activation occurs, resulting in further activation of cell death pathways including mitochondrial perturbation (Graham & Chen, 2001). Along with the exciting stimulation, radical generation and suicidal apoptotic pathways are also involved in neuronal cell death (Figure 5) (S. Kim & Mountz, 2011).

Involvement of Heavy Metals

The general responsible factors of abnormal metal accumulation are believed to be either genetic or environmental. The most considered environmental contaminant are especially contaminated water and food, malnutrition, occupational exposure as well as medical procedures (Mitra et al., 2014). The extensive use and production of heavy metals in industries result in ubiquitous pollution in soil, water and air, which contributes to human exposure. High levels of some heavy metals can be found in leafy vegetables, fruits, seeds, grains, shellfish, and organ meats (Chen, 2017). Furthermore, intoxication with such metals can obviously be detected in many diseases, including thrombosis, hypertension, coronary heart disease, myocardial infarction, cardiac arrhythmias, increased carotid intima-media thickness, carotid artery obstruction, generalized atherosclerosis, and cerebrovascular accidents (Houston, 2011). However, the relevance of the association between heavy metals and the risk of stroke remains inconclusive (T. Lee, Tseng, Chen, & Lin, 2009). Therefore, a better understanding of their implication on the pathophysiological process of this disease is needed.

The prevalence of ischemic stroke increases following exposure to several kinds of xenobiotics, counting heavy metals. This suggests the possible toxicological contribution of these compounds to the onset and the aggravation of stroke.(J. Kim et al., 2013).Moreover, the permeability of the Blood-Brain-

Figure 5. Ischemic cascade leading to cerebral damage: Ischemic stroke leads to hypo-perfusion of a brain area that initiates a complex series of events. Excitotoxicity, oxidative stress, microvascular injury, blood-brain barrier dysfunction and post-ischemic inflammation lead ultimately to cell death of neurons, glia and CECs (cerebral endothelial cells). The degree and duration of ischemia determine the extent of cerebral damage
(Mir et al., 2014).

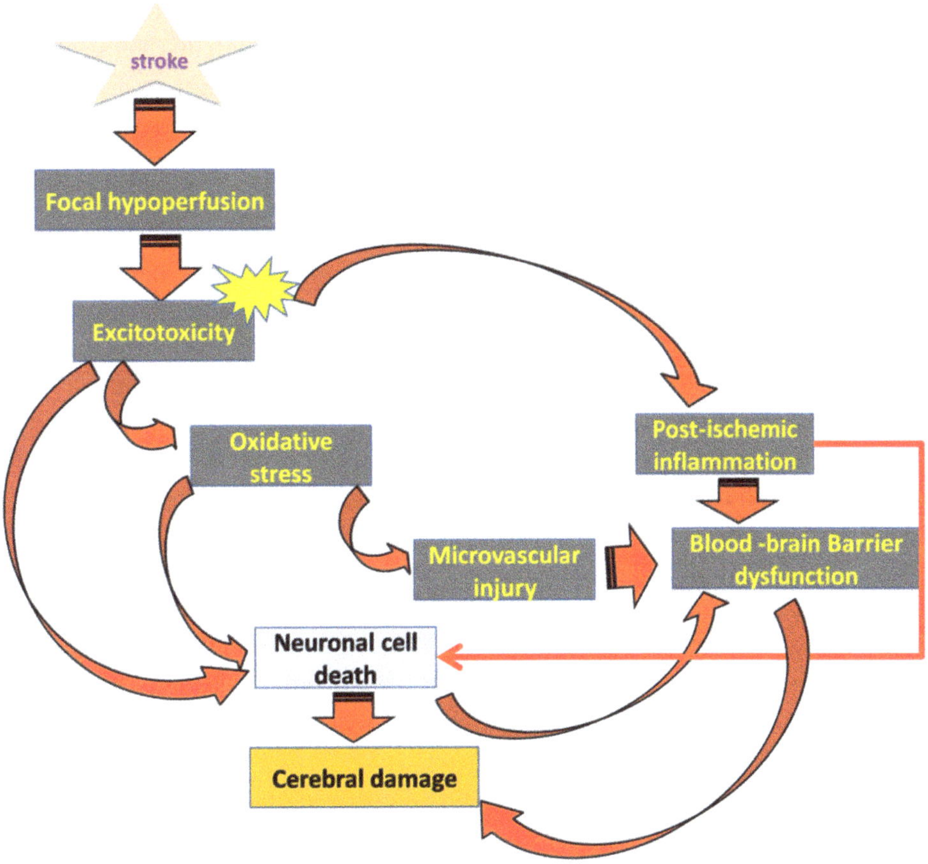

Barrier(BBB)is emphasized to be an important mediator for determination of the clinical outcome following ischemic stroke, such as edema, inflammation, and epileptogenesis (Dietrich, Alonso, & Busto, 1993). Even though the studies investigating the effect of heavy metals on BBB integrity are limited, several reports have given insights for heavy metal induced BBB disruption(Zheng, Aschner, & Ghersi-egea, 2003). Acute lead poisoning results in brain swelling with micro vascular damage with Tight Junctions opening (Struzyiqska & Gadamski, 1997), suggesting that BBB is one of the main targets of lead-associated toxicity. Notably, lead showed a higher affinity to CECs, so that it accumulates in high concentration in CECs than in other brain cells(Toews, Kolber, Hayward, Krigman, & Morell, 1978), and irreversibly potentiates cytokine and glutamate-mediated decrease in BBB resistance (Dyatlov, Platoshin, Lawrence, & Carpenter, 1998). As well, in vivo exposure to cadmium in drinking water increased the BBB permeability and the concentration of malondialdehyde in brain microvessels of rats, whereas the activities of anti-oxidant enzymes were significantly decreased(Arti Shukla, 1996). In CECs, cadmium

(Cd) induced activation of apoptotic signaling(Jung et al., 2008), and stimulated expression of adhesion molecules which is an indicator for BBB injury(J. Kim et al., 2013). Similarly, the CNS damage following mercury (Hg) exposure has been largely focused on BBB transport of methyl-mercury (MeHg). While MeHg easily crosses BBB due to its high lipophilicity (Clarkson, 1992). resulting in encephalopathy. Inorganic mercury has been considered as less neurotoxic, but BBB breakdown was induced by inorganic mercury. Chronic arsenic exposure is also associated with the increased prevalence of micro vascular diseases including neurological diseases (Chiou et al., 2005). Another heavy metal has been also associated to BBB damage such as Arsenic; in fact, alteration of vascular permeability by arsenic was reported in rats (Chen, Tsai, Wang, Yu, & Chang, 2004)Further studies on heavy metals-induced BBB permeability damage are summarized in Table 1.

Acute metal toxicity is implicated in stroke and certain conditions that result from hereditary defects in the regulation of metal homeostasis. For example, the dysregulation of metal ions due to the acute release of free iron (Fe) following hemorrhagic stroke causes massive neuronal injury (Gao et al., 2014; Li & Zhang, 2012) Neurotoxicity from acute increase in the level of zinc (Zn) and other transition metals may as well play a critical role after ischemic focal brain injury(Galasso & Dyck, 2007; Mitra et al., 2014).

Moreover, intoxication with Hghas importantly been widely revealed. Therefore, it has been recently recommended to evaluate Hg toxicity in each patient showing hypertension, coronary heart disease, cerebrovascular disease, cerebrovascular accident (stroke), or other vascular diseases(Chen et *al.,* 2018). Even though the previous studies have been conclusive on the risk of developing stroke, neither following direct intoxication with mercury nor in relation with fish increased intake(Torres, Rai, & Hardiek, 2000), the link between them both has been obviously revealed. For instance, in a population-based cohort, Hg levels and relative content of fatty acids were determined in erythrocyte membranes in the population consuming one meal per week as fish (Skoczyńska et al., 2010).Another study suggested, as well, that the risk for stroke between sexes differs with increasing fish intake, EPA (eicosapentaenoic acid),and DHA (DocosaHexaenoic Acid)consumption, nevertheless it has not shown any association between stroke risk and mercury at these lower levels of one meal of fish per week (Houston, 2011). Nevertheless, to elucidate the link between the intoxication with Hg and stroke, it is necessary to point out some of the plausible involved mechanism from the literature.

Interestingly, several mechanisms have been stated to underlay stroke following intoxication with mercury (Skoczyńska et al., 2010). The increase, either in brain or pulse pressures(In, Intermittently, To, & Mercury, 2000). the increased thrombotic risk related to increased platelet aggregation(Lin, Huang, & Huang, 1996), increase in Factor VIII, thrombin, platelet factor 423 and reduction in protein C (Wierzbicki & Praz, 2002),as well as endothelial dysfunction from reducedNO bioavailability may account for much of the observed elevation in CVA (cerebro-vascular accidents) risk with Hg.One recent study showed that mercury increases thrombotic risk by enhancement of procoagulant activity in erythrocytes by protein thiol depletion–mediated phosphatidyl serine exposure and microvesical generation(Lundh et al., 2007).

Generally, the mechanism of heavy metals induced stroke may imply different pathways. Hence, the intracellular and extracellular levels of metals are tightly regulated through a complex network(Mitra et al., 2014). Therefore, excessive concentration of non-sequestered metal salts could cause cellular toxicity and pathological damage(Balamurugan, Egli, Zhang, & Georgiev, 2004). In addition to altering the membrane potential, particularly in neurons where metal ions can bind to proteins/enzymes and nucleic acid and affect their activity, which may cause cytotoxicity(Kondoh et al., 2002). In addition, the major cause of oxidative toxicity from transition metals is the generation of ROS; the most pervasive oxidant

Table 1. Summary of experimental studies on BBB permeability impairment by by heavy metals

Heavy Metal		System	Concentration	Exposure Duration	Toxicity	Ref
Pb	*in vivo*	Rats of 5 days age	highest non-lethal dosage (1 mg Pb/g body weight/day)	2 days	Hemorrhagic encephalopathy, abnormal morphology in capillaries and microvessels, swollen and vacuolated ECs	(Toews et al., 1978)
	in vivo	Mice between 10~40 days of age	2.5 and 5 μg/g (subcutaneous injection)	five injections between 2 and 10 days after birth	Potentiating BBB disruption by lipopolysaccharide, interleukin-6 or glutamate measured by transendothelial electrical resistance	(Dyatlov et al., 1998)
Cd	*in vivo*	Rats of 21 days age	10 ppm	90 days	Decrease in microvessel antioxidant potential	(Arti Shukla, 1996)
	in vivo	Cerebral endothelial cells (bEnd.3)	10, 30 and 50 μM	upto 24 hr	Remarkable decrease in cell viability	(Jung et al., 2008)
Hg	*in vivo*	Cerebral endothelial cells (bEnd.3)	1, 3 and 10 μM	upto 24 hr	Stimulation of the expression of ICAM-1	(Baik, Moon, & Jung, 2004)
	in vivo	Adult cats	6×10-5 M solution of mercuric chloride (i.p)	30 min	BBB damage examined by Evans blue	(Peterson & Cardoso, 1983)
	in vivo	Adult rats	1.92, 3.85 and 7.70 μM of sodium arsenite (i.p)	upto 60 min	Vascular dysfunction (vascular leakage) determined by Evans blue	(Chen et al., 2004)

(Kim at *al.*, 2013)

in cells (Figure 6)(Chang, Hsu, Liu, Su, & Yen, 2013). In addition, some many heavy metals, such as Cd and Pb, are highly toxic. and could cause membrane depolarization by blocking calcium-ion influx and cell death (Pentyala, Ruggeri, Veerraju, Yu, & Bhatia, 2010).

Involvement of Pesticides

A body of evidence sustains the concept of environmental exposures to pesticides as a potent cause for cardiovascular diseases, including stroke(D. Lee et al., 2012). In addition, epidemiological data provide some evidence to support associations between environmental exposures and well established stroke risk factors including diabetes, metabolic syndrome, weight gain, and other cardiovascular conditions (Barmeir, 2007; E. B. Montgomery & Gale, 2008). The Agricultural Health Study (AHS) provides a unique opportunity to examine associations between a range of agricultural exposures and stroke mortality in a large cohort of pesticide applicators, consisting mainly of farmers(Taylor et al., 2013).

There is some evidence that pesticides may be associated with stroke and consequently stroke mortality, through inflammatory processes producing tissue damage leading to various disease states, coagulation disturbances, and through effects on the cardiac system(Anderson & Thundiyil, 2012; M. P. Montgomery, Kamel, Saldana, Alavanja, & Sandler, 2008; Tore, Bergdahl, Nilsson, & Ja, 2007).The direct association between pesticides exposure and stroke diseases is still controversial. Indeed, a large cohort study on agricultural exposures to pesticides in association to stroke incidence in the elderly. demonstrated that

Figure 6. Molecular basis for metal neurotoxicity and its potential as a therapeutic target. Studies suggest a charge-based dyshomeostasis of metals in neurons affected by degenerative diseases. Typically, trivalent metals increase in late-stage AD, whereas divalent metal ions increase in early AD. The increase in metal ions could reversibly inhibit DNA repair enzymes, inducing genomic damage. Metal chelation therapy should address these challenges based on recent molecular understanding of the phenomenon (Mitra et al., 2014).

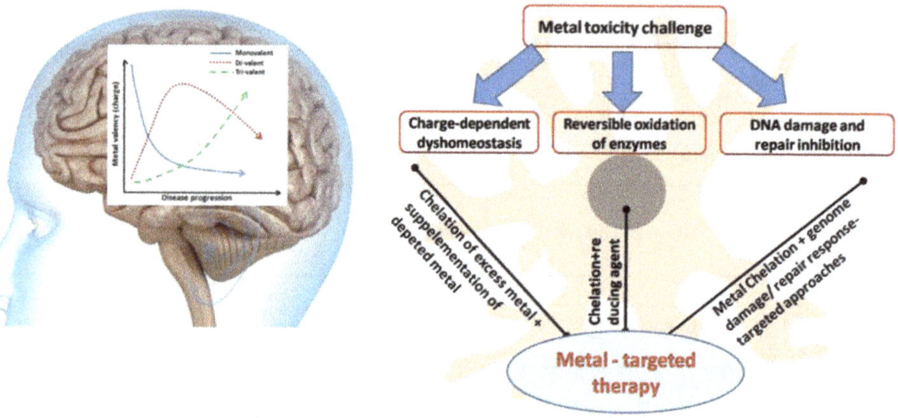

polychlorinated biphenyl pesticides which were trans-nonachlor and octachlorodibenzo-p-dioxin were significantly associated with increased risk for stroke incidence(D. Lee et al., 2012). While, others reported increased stroke-related mortality among individuals occupationally exposed to pesticides(Charles et al., 2010). A 70% increased rate of stroke mortality was observed among those determined to have the highest level of exposure to pesticides compared to those with no exposure(Charles et al., 2010). Similarly, a study of aerial pesticide applicators reported an almost two-fold elevation in stroke-related mortality as compared to flight instructors in the United States(Taylor et al., 2013)However, others seem to deny such association(Taylor et al., 2013).

The association of pesticides with atherosclerosis have also been studied by (Min et al., 2011) who reported that organochlorine (OC) pesticides are a potent risk factor for peripheral arterial diseases. However, (Goncharov, Haase, Santiago-rivera, & Morse, 2008), postulated that polychlorinated biphenyls (PCBs) are directly involved in increased synthesis of cholesterol and triglycerides, substances known to be major risk factors for cardiovascular diseases including stroke (Goncharov et al., 2008; Min et al., 2011). Such finding was sustained by a previous study showing that increased serum level of pesticides was associated with increased risk of developing ischemic stroke (Lim et al., 2018).

CONCLUSION

The damage emerging from climate change's influence on the intoxication with heavy metals and pesticides have been clearly elucidated. In this chapter, we argued that epilepsy and stroke are among the most prominent illustrations of the handicaps resulting from climate change. Therefore, most serious global mobilizations must be organized on this issue to alleviate such a hazardous effect.

REFERENCES

Abe. (2008). Neuroprotective therapy for ischemic stroke with free radical scavenger and gene-stem cell therapy. *Clin Neurol*, 896–898.

Acosta, J. A., Cano, A. F., Arocena, J. M., Debela, F., & Martínez-martínez, S. (2009). Distribution of metals in soil particle size fractions and its implication to risk assessment of playgrounds in Murcia City (Spain). *Geoderma*, *149*(1–2), 101–109. doi:10.1016/j.geoderma.2008.11.034

Adams, H. P., Zoppo, G., Alberts, M. J., Bhatt, D. L., Brass, L., Furlan, A., ... Wijdicks, E. F. M. (2007). AHA / ASA Guideline Guidelines for the Early Management of Adults A Guideline From the American Heart Association. *AME*, *115*, e478–e534. doi:10.1161/CIRCULATIONAHA.107.181486

Anderson, J. O., & Thundiyil, J. G. (2012). Clearing the Air : A Review of the Effects of Particulate Matter Air Pollution on Human Health. *J.Med, Toxicol*, 166–175. doi:10.100713181-011-0203-1

Ann, J., & Gus, A. B. (2008). Quality of Life Trajectories in Epilepsy: A review of the literature. *Epilepsy & Behavior*, *12*(4), 557–571. doi:10.1016/j.yebeh.2007.11.013 PMID:18158270

Arora, S. K., Batra, P., Sharma, T., Banerjee, B. D., & Gupta, S. (2013). Role of Organochlorine Pesticides in Children with Idiopathic Seizures. *ISRN Pediatrics*, 5. PMID:24368944

Arti Shukla, G. S. S. and R. S. (1996). brain barrier permeability and its possible correlation with decreased microvessel antioxidant potential in rat. *Human et Experimental Toxicolgy*, (January).

Audhild Hjalmarsen, D. L. H. (2008). Severe nocturnal hypoxaemia in geriatric. Oxford University Press. doi:10.1093/ageing/afn110

Baik, E. J., Moon, K., & Jung, Y. (2004). Cadmium stimulates the expression of ICAM-1 via NF-. *Biochemical and Biophysical Research Communications*, *320*(3), 887–892. doi:10.1016/j.bbrc.2004.05.218 PMID:15240131

Balamurugan, K., Egli, D., Zhang, B., & Georgiev, O. (2004). Metal-responsive transcription factor (MTF-1) and heavy metal stress response in Drosophila and mammalian cells : A functional comparison. *Biological Chemistry*, *385*(July), 597–603. PMID:15318808

Bar-meir, E. (2007). Guidelines for Treating Cardiac Manifestations of Organophosphates Poisoning with Special Emphasis on Long QT and Torsades De Pointes. *Informa Healthcare*, 279–285. doi:10.1080/10408440601177855

Bath, P., & Lees, K. R. (2007). Acute Stroke. Textbook of Clinical Trials, 173, 179–214. doi:10.1002/9780470010167.ch13

Beavers, C. T., Parker, J. J., Flinchum, D. A., Weakley-jones, B. A., & Jortani, S. A. (2014). Pesticide-Induced Quadriplegia in a 55-Year-Old Woman. *Forensic Med Pathol*, *35*(4), 239–241. doi:10.1097/PAF.0000000000000108 PMID:25354225

Belyaeva, E. A., Korotkov, S. M., & Saris, N. (2011). Journal of Trace Elements in Medicine and Biology In vitro modulation of heavy metal-induced rat liver mitochondria dysfunction : A comparison of copper and mercury with cadmium. *Journal of Trace Elements in Medicine and Biology*, *25*, S63–S73. doi:10.1016/j.jtemb.2010.10.007 PMID:21146384

Bhalla, D. (2014). Michel Dumas and Pierre-Marie Preux: Promoting tropical neurology silently -The gist of their contributions. *Neuroepidemiology*, *42*(3), 139–143. doi:10.1159/000357354 PMID:24481209

Bhanti, M., & Taneja, A. (2007). Contamination of vegetables of different seasons with organophosphorous pesticides and related health risk assessment in northern India. *Chemosphere*, *69*(1), 63–68. doi:10.1016/j.chemosphere.2007.04.071 PMID:17568651

Bien, C. G., & Schramm, J. (2009). Treatment of Rasmussen encephalitis half a century after its initial description : Promising prospects and a dilemma. *Epilepsy Research*, *86*(2-3), 101–112. doi:10.1016/j.eplepsyres.2009.06.001 PMID:19615863

Carlson, B. W., Neelon, V. J., Carlson, J. R., Hartman, M., & Dogra, S. (2005). Cerebrovascular Disease and Patterns of Cerebral Oxygenation During Sleep in Elders. *Biological Research for Nursing*, 307–317. PMID:19144652

Chang, K., Hsu, C., Liu, S., Su, C., Yen, C., Lee, M.-J., ... Huang, C.-F. (2013). Cadmium Induces Apoptosis in Pancreatic b -Cells through a Mitochondria-Dependent Pathway : The Role of Oxidative Stress-Mediated c-Jun N-Terminal Kinase Activation. *PLoS One*, *8*(2), e54374. doi:10.1371/journal.pone.0054374 PMID:23405080

Charles, L. E., Burchfiel, C. M., Fekedulegn, D., Gu, J. K., Petrovitch, H., Sanderson, W. T., ... Ross, G. W. (2010). Occupational exposure to pesticides, metals, and solvents: The impact on mortality rates in the Honolulu Heart Program. *Work (Reading, Mass.)*, *37*(2), 205–215. doi:10.3233/WOR-2010-1071 PMID:20938081

Chen, S.-C., Tsai, M.-H., Wang, H.-J., Yu, H.-S., & Chang, L. W. (2004). Vascular permeability alterations induced by arsenic. *Human and Experimental Toxicology*, *23*(1), 1–7. doi:10.1191/0960327104ht407oa PMID:15027809

Chiou, J., Wang, S., Chen, C., Deng, C., Lin, W., & Tai, T. (2005). Arsenic ingestion and increased microvascular disease risk : Observations from the south- western arseniasis-endemic area in Taiwan. *International Journal of Epidemiology*, *34*(May), 936–943. doi:10.1093/ije/dyi108 PMID:15911542

Chuang, M. J., Chang, W. N., Chang, H. W., Lin, W. C., Tsai, N. W., Hsieh, M. J., ... Lu, C. H. (2010). Predictors and long-term outcome of seizures after bacterial brain abscess. *Journal of Neurology, Neurosurgery, and Psychiatry*, *81*(8), 913–917. doi:10.1136/jnnp.2009.195073 PMID:20682720

Clarkson, T. W. (1992). Mercury : Major Issues in Environmental Health. *Environmental Health Perspectives*, *100*, 31–38. doi:10.1289/ehp.9310031 PMID:8354179

Cohen-kfir, E., Lee, W., Eskandari, S., & Nelson, N. (2005). Zinc inhibition of γ -aminobutyric acid transporter 4 (GAT4) reveals a link between excitatory and inhibitory neurotransmission. *Proceedings of the National Academy of Sciences of the United States of America*, *102*(17), 17. doi:10.1073/pnas.0501431102 PMID:15829583

Cole, T. B., Robbins, C. A., & Ju, H. (2000). Seizures and neuronal damage in mice lacking vesicular zinc. *Science B.V*, *39*, 153–169. PMID:10759303

Corfield, D. R., & Meadows, G. E. (2006). Control of cerebral blood flow during sleep and the effects of hypoxia. *Advances in Experimental Medicine and Biology*, *588*, 65–73. doi:10.1007/978-0-387-34817-9_7 PMID:17089880

Day, S. M., Wu, Y. W., Strauss, D. J., Shavelle, R. M., & Reynolds, R. J. (2005). Causes of death in remote symptomatic epilepsy. *Neurology*, *65*(2), 216–222. doi:10.1212/01.wnl.0000169018.44950.68 PMID:16043789

Delcour, I., Spanoghe, P., & Uyttendaele, M. (2015). Literature review : Impact of climate change on pesticide use. *Food Research International Journal*, *68*, 7–15. doi:10.1016/j.foodres.2014.09.030

Devinsky, O., Vezzani, A., Jette, N., Curtis, M. De, & Perucca, P. (2018). Epilepsy. *Nature Reviews*, *3*. doi:10.1038/nrdp.2018.24

Dietrich, W. D., Alonso, O., & Busto, R. (1993). Moderate Hyperglycemia Worsens Acute Blood-Brain Barrier Injury After Forebrain Ischemia in Rats. *Stroke*, 24. PMID:8418533

Dyatlov, V. A., Platoshin, A. V., Lawrence, D. A., & Carpenter, D. O. (1998). Lead potentiates cytokine- and glutamate-mediated increases in permeability of the blood-brain barrier. *Neurotoxicology*, *19*(2), 283–291. Retrieved from http://www.ncbi.nlm.nih.gov/pubmed/9553965 PMID:9553965

Escudero Augusto, D., Marqués Álvarez, L., & Taboada Costa, F. (2008). Actualización en hemorragia cerebral espontánea. *Medicina Intensiva*, *32*(6), 282–295. doi:10.1016/S0210-5691(08)70956-2 PMID:18601836

Feigin, V. L., Forouzanfar, M. H., Krishnamurthi, R., Mensah, G. A., Connor, M., Bennett, D. A., ... Murray, C. (2014). Global and regional burden of stroke during 1990 – 2010 : Fi ndings from the Global Burden of Disease Study 2010. *Lancet*, *383*(9913), 245–255. doi:10.1016/S0140-6736(13)61953-4 PMID:24449944

Feigin, V. L., Lawes, C. M. M., Bennett, D. A., Barker-collo, S. L., & Parag, V. (2009). Worldwide stroke incidence and early case fatality reported in 56 population-based studies : A systematic review. *Lancet Neurology*, *8*(4), 355–369. doi:10.1016/S1474-4422(09)70025-0 PMID:19233729

Ferreira, D., Nobre, A., Silva, M. L., Tulha, J., Faria-oliveira, F., Ferreira, C., & Lucas, C. (2013). encodes a xylose/H + symporter from the highly related yeast species Debaryomyces fabryi and Debaryomyces hansenii. *FEMS Yeast Research*, *13*(7), 585–596. doi:10.1111/1567-1364.12061 PMID:23809840

Forstl, & Burns, Levy, Cairns, Luthert, & Lantos. (1992). Neurologic Signs in Alzheimer ' s. *Archives of Neurology*, 49.

Frye, R. E., Casanova, M. F., Fatemi, S. H., Folsom, T. D., Reutiman, T. J., Brown, G. L., ... Zimmerman, A. W. (2016). Neuropathological Mechanisms of Seizures in Autism Spectrum Disorder. *Frontiers in Neuroscience*, *10*(May), 1–9. doi:10.3389/fnins.2016.00192 PMID:27242398

Fujiwara, T., & Shigematsu, H. (2004). Etiologic factors and clinical features of symptomatic epilepsy : Focus on pediatric cases. *Psychiatry and Clinical Neurosciences*, *58*(3), 9–12. doi:10.1111/j.1440-1819.2004.01244_3.x

Furieri, L. B., Galán, M., Avendaño, M. S., García-redondo, A. B., Aguado, A., Martínez, S., ... Salaices, M. (2011). Endothelial dysfunction of rat coronary arteries after exposure to low concentrations of mercury is dependent on reactive oxygen species. *British Journal of Pharmacology*, *162*(8), 1819–1831. doi:10.1111/j.1476-5381.2011.01203.x PMID:21232032

Galasso, S. L., & Dyck, R. H. (2007). The Role of Zinc in Cerebral Ischemia. *Molecular Medicine (Cambridge, Mass.)*, *13*(9), 380–387. doi:10.2119/2007 PMID:17622314

Ganesh, R., & Janakiraman, L. (2008). Serum Zinc Levels in Children With Simple Febrile Seizure. *Clinical Pediatrics*, *47*(2), 164–166. doi:10.1177/0009922807306165 PMID:17873242

Ganesh, R., Janakiraman, L., & Meenakshi, B. (2011). Serum zinc levels are low in children with simple febrile seizures compared with those in children with epileptic seizures and controls. *Annals of Tropical Paediatrics*, *345–349*. doi:10.1179/1465328111Y.0000000032 PMID:22041469

Gao, C., Du, H., Hua, Y., Keep, R. F., Strahle, J., & Xi, G. (2014). Role of red blood cell lysis and iron in hydrocephalus after intraventricular hemorrhage. *Journal of Cerebral Blood Flow and Metabolism*, *34*(6), 1070–1075. doi:10.1038/jcbfm.2014.56 PMID:24667910

Goncharov, A., Haase, R. F., Santiago-rivera, A., Morse, G., McCaffrey, R. J., Rej, R., & Carpenter, D. O. (2008). High serum PCBs are associated with elevation of serum lipids and cardiovascular disease in a Native American population $. *Environmental Research*, *106*(2), 226–239. doi:10.1016/j.envres.2007.10.006 PMID:18054906

Graham, S. H., & Chen, J. (2001). Programmed Cell Death in Cerebral Ischemia. *Journal of Cerebral Blood Flow and Metabolism*, *21*(2), 99–109. doi:10.1097/00004647-200102000-00001 PMID:11176275

Guerreiro, C. A. M. (2016). Epilepsy: Is there hope? *The Indian Journal of Medical Research*, *144*(5), 657–660. doi:10.4103/ijmr.IJMR_1051_16 PMID:28361817

Gund, G.-B. M., Jagtap, P. N., Ingale, V. B., & Patil, R. Y. (2013). Stroke : A Brain Attack. *IOSR Journal Of Pharmacy*, *3*(8), 1–23. doi:10.9790/3013-038201-23 PMID:27024833

Hennion, S., & Hennion, S. (2017). *Neuropsychologie et neuroimagerie des troubles de cognition sociale dans l ' épilepsie du lobe temporal*. Academic Press.

Houston, M. C. (2011). Role of Mercury Toxicity in Hypertension, Cardiovascular Disease, and Stroke. *Journal of Clinical Hypertension*, *13*(8), 621–627. doi:10.1111/j.1751-7176.2011.00489.x PMID:21806773

In, R. F., Intermittently, M., To, E., & Mercury, E. (2000). *Renal function in miners intermittently exposed to elemental mercury vapour.* Academic Press.

Jackson, L. E., Wheeler, S. M., Hollander, A. D., Orlove, B. S., Six, J., Sumner, D. A., ... Tomich, T. P. (2011). Case study on potential agricultural responses to climate change in a California landscape. *Climatic Change*, *109*(S1), 407–427. doi:10.100710584-011-0306-3

Jett, D. A. (2011). Neurotoxic Pesticides and Neurologic Effects. *Neurologic Clinics*, *29*(3), 667–677. doi:10.1016/j.ncl.2011.06.002 PMID:21803217

Johnston, S. C., Mendis, S., & Mathers, C. D. (2009). Global variation in stroke burden and mortality : Estimates from monitoring, surveillance, and modelling. *Lancet Neurology*, *8*(4), 345–354. doi:10.1016/S1474-4422(09)70023-7 PMID:19233730

Jung, Y., Jeong, E., Kyung, E., Kim, Y., Sohn, S., Hwan, S., ... Moon, C. (2008). Cadmium induces apoptotic cell death through p38 MAPK in brain microvessel endothelial cells. *European Journal of Pharmacology*, *578*(1), 11–18. doi:10.1016/j.ejphar.2007.08.049 PMID:17961543

Kellinghaus, C., Engbring, C., Kovac, S., Möddel, G., Boesebeck, F., Fischera, M., ... Husstedt, I. W. (2008). Frequency of seizures and epilepsy in neurological HIV-infected patients. *Seizure*, *17*(1), 27–33. doi:10.1016/j.seizure.2007.05.017 PMID:17618132

Kheradmand, A., Yarali, B., Zare, A., Pourpak, Z., Shams, S., & Comparison, A. (2014). Comparison of Serum Zinc and Copper levels in Children and adolescents with Intractable and Controlled Epilepsy. *Iranian Journal of Child Neurology*, *8*(3), 49–54. PMID:25143774

Kim, J., Byun, H., Chung, E., Chung, H., & Bae, O. (2013). Loss of Integrity : Impairment of the Blood-brain Barrier in Heavy Metal-associated Ischemic Stroke. *Toxicological Research*, *29*(3), 157–164. doi:10.5487/TR.2013.29.3.157 PMID:24386515

Kim, S., & Mountz, J. M. (2011). SPECT Imaging of Epilepsy : An Overview and Comparison with F-18 FDG PET. *International Journal OfMolecular Imaging*, *2011*. doi:10.1155/2011/813028 PMID:21785722

Kondoh, M., Araragi, S., Sato, K., Higashimoto, M., Takiguchi, M., & Sato, M. (2002). Cadmium induces apoptosis partly via caspase-9 activation in HL-60 cells. *Toxicology*, *170*(1-2), 111–117. doi:10.1016/S0300-483X(01)00536-4 PMID:11750088

Kovac, S., & Walker, M. C. (2013). Neuropeptides Neuropeptides in epilepsy. *Neuropeptides*, *47*(6), 467–475. doi:10.1016/j.npep.2013.10.015 PMID:24210141

Lee, D., Lind, P. M. Jr, Salihovic, D. R. J., Bavel, S., Van, B., & Lind, L. (2012). Background exposure to persistent organic pollutants predicts stroke in the elderly. *Environment International*, *47*, 115–120. doi:10.1016/j.envint.2012.06.009 PMID:22809777

Lee, T., Tseng, M., Chen, C., & Lin, J. (2009). NeuroToxicology Association of high body lead store with severe intracranial carotid atherosclerosis. *Neurotoxicology*, *30*(6), 876–880. doi:10.1016/j.neuro.2009.07.004 PMID:19616024

Li, C., Jian, L., Zhang, Q. J., Feng, J. J., & Gui, Z. H. (2011). Alterations of emotion, cognition and firing activity of the basolateral nucleus of the amygdala after partial bilateral lesions of the nigrostriatal pathway in rats. *Brain Research Bulletin*, *85*(6), 329–338. doi:10.1016/j.brainresbull.2011.05.009 PMID:21624440

Li, Y. V, & Zhang, J. H. (2012). Metal Ions in Stroke Pathophysiology. *Metal Ion in Stroke*, 1–12. doi:10.1007/978-1-4419-9663-3

Liigant, A., Haldre, S., Oun, A., Linnamägi, Ü., Saar, A., & Asser, T. A.-E. K. (2001). Seizure Disorders in Patients with Brain Tumors. *European Neurology, 45*, 46–51.

Lim, J., Lee, S., Lee, S., & Ha, S. (2018). Serum persistent organic pollutants levels and stroke risk. *Environmental Pollution, 233*, 855–861. doi:10.1016/j.envpol.2017.12.031 PMID:29248762

Lin, T., Huang, Y., & Huang, S. (1996). Lipid Peroxidation in Liver of Rats Administrated with Methyl Mercuric Chloride. *Biological Trace Element Research, 33*, 54. PMID:8862759

Lo, E. H., Dalkara, T., & Moskowitz, M. A. (2003). Mechanisms, challenges and opportunities in stroke. *Nature Reviews. Neuroscience, 4*. doi:10.1038/nrn1106 PMID:12728267

Lowenstein, D. H. (2009). Posttraumatic Epilepsy: Epilepsy after head injury : An overview. *Epilepsia, 50*, 4–9. doi:10.1111/j.1528-1167.2008.02004.x PMID:19187288

Lukaszewicz-hussain, A. (2010). Role of oxidative stress in organophosphate insecticide toxicity – Short review. *Pesticide Biochemistry and Physiology, 98*(2), 145–150. doi:10.1016/j.pestbp.2010.07.006

Lundh, T., Wennberg, M., Bergdahl, I. A., Stegmayr, B., Vessby, B., Skerfving, S., & Stro, U. (2007). And risk of stroke in northern Sweden. *British Journal of Nutrition*, 1038–1045. doi:10.1017/S0007114507756519

Mac, T. L., Tran, D., Quet, F., Odermatt, P., Preux, P., & Tan, C. T. (2007). Epidemiology, aetiology, and clinical management of epilepsy in Asia : A systematic review. *Lancet Neurology, 6*(6), 533–543. doi:10.1016/S1474-4422(07)70127-8 PMID:17509488

Macdonald, J. F., Xiong, Z., & Jackson, M. F. (2006). Paradox of Ca 2 C signaling, cell death and stroke. *Neurosciences, 29*(2). doi:10.1016/j.tins.2005.12.001

Martynyuk, A. E., Glushakov, A. V., Sumners, C., Laipis, P. J., Dennis, D. M., & Seubert, C. N. (2005). Impaired glutamatergic synaptic transmission in the PKU brain. *Molecular Genetics and Metabolism, 86*, 34–42. doi:10.1016/j.ymgme.2005.06.014 PMID:16153867

Matin, M. A., Jaffery, F. N., & Siddiqui, R. A. (1981). A Possible Neurochemical Basis of the Central Stimulatory Effects of pp ' DDT. *Neurochemistry, 36*(3), 1000–1005. doi:10.1111/j.1471-4159.1981.tb01692.x PMID:7205252

Mccarren, H. S., & Jr, J. H. M. (2016). Anticonvulsant discovery through animal models of status epilepticus induced by organophosphorus nerve agents and pesticides. *Annals of the New York Academy of Sciences*, 1–7. doi:10.1111/nyas.13092 PMID:27258770

Meinardi, H., Scott, R. A., Reis, R., & Sander, J. W. A. S. (2001). The Treatment Gap in Epilepsy : The Current Situation and Ways Forward. *Epilepsia, 42*(1), 136–149. doi:10.1046/j.1528-1157.2001.32800.x PMID:11207798

Min, J., Cho, J., Lee, K., Park, J., Park, S., Yong, J., & Min, K. (2011). Potential role for organochlorine pesticides in the prevalence of peripheral arterial diseases in obese persons : Results from the National Health and Nutrition Examination Survey 1999 – 2004. *Atherosclerosis*, *218*(1), 200–206. doi:10.1016/j.atherosclerosis.2011.04.044 PMID:21620405

Mir, M. A., Al-Baradie, R. S., & Alhussainawi, M. D. (2014). *Pathophysiology of Strokes. Recent Advances in Stroke Therapeutics*. Retrieved from https://www.researchgate.net/publication/273061843_Pathophysiology_of_Strokes?enrichId=rgreq-b11151049e7ecb01203e75e2f76d2998-XXX&enrichSource=Y292ZXJQYWdlOzI3MzA2MTg0MztBUzoyMTA4MjY5NTE2OTYzODdAMTQyNzI3NjQ3MDY1NQ%3D%3D&el=1_x_2&_esc=publicationCoverPdf

Mitra, J., Vasquez, V., Hegde, P. M., Boldogh, I., Mitra, S., Kent, T. A., & Kosagi, S. (2014). Neurological Research and Therapy Stroke : Therapeutic Potential. *Neurol Res Ther*, *1*(2), 1–8. doi:10.14437/NRTOA-1-107.Revisiting PMID:25717476

Mollah, M. A. H., Rakshit, S. C., Anwar, K. S., Arslan, M. I., Saha, N., Ahmed, S., ... Hassan, T. (2008). Zinc concentration in serum and cerebrospinal fluid simultaneously decrease in children with febrile seizure : Findings from a prospective study in Bangladesh. *Acta Paediatrica (Oslo, Norway)*, *97*(1), 1707–1711. doi:10.1111/j.1651-2227.2008.01001.x PMID:18795906

Montgomery, E. B. Jr, & Gale, J. T. (2008). Mechanisms of action of deep brain stimulation (DBS). *Neuroscience and Biobehavioral Reviews*, *32*(3), 388–407. doi:10.1016/j.neubiorev.2007.06.003 PMID:17706780

Montgomery, M. P., Kamel, F., Saldana, T. M., Alavanja, M. C. R., & Sandler, D. P. (2008). Original Contribution Incident Diabetes and Pesticide Exposure among Licensed Pesticide Applicators : Agricultural Health Study, 1993 – 2003. *American Journal of Epidemiology*, *167*(10), 1235–1246. doi:10.1093/aje/kwn028 PMID:18343878

Most, P., Remppis, A., Pleger, S. T., Löffler, E., Ehlermann, P., Bernotat, J., ... Koch, W. J. (2003). Transgenic Overexpression of the Ca2+-binding Protein S100A1 in the Heart Leads to Increased in Vivo Myocardial Contractile Performance. *The Journal of Biological Chemistry*, *278*(36), 33809–33817. doi:10.1074/jbc.M301788200 PMID:12777394

Nechifor, R., & Vink, M. (2011). Magnesium in the central nervous system. University of Adelaide Press.

Ngugi, A. K., Bottomley, C., Kleinschmidt, I., Sander, J. W., & Newton, C. R. (2010). Estimation of the burden of active and life-time epilepsy : A meta-analytic approach. *Epilepsia*, *51*(5), 883–890. doi:10.1111/j.1528-1167.2009.02481.x PMID:20067507

Nicoletti, A. (2010). Toxocariasis. *Épilepsies, 22*(2).

Noyes, P. D., Mcelwee, M. K., Miller, H. D., Clark, B. W., Tiem, L. A., Van, & Walcott, K. C. (2009). The toxicology of climate change : Environmental contaminants in a warming world. *Environment International*, *35*(6), 971–986. doi:10.1016/j.envint.2009.02.006 PMID:19375165

Owolabi, L., Sale, S., Owolabi, S., Nalado, A., Umar, M., & Taura, A. (2018). Electroencephalography Abnormalities in Generalized Epilepsy and Their Predictors: A Multicenter Experience. *Annals of African Medicine*, *17*(2), 64. doi:10.4103/aam.aam_2_17 PMID:29536959

Pal, P. B., Pal, S., Das, J., & Sil, P. C. (2012). Modulation of mercury-induced mitochondria-dependent apoptosis by glycine in hepatocytes. *Amino Acids*, *42*(5), 1669–1683. doi:10.100700726-011-0869-3 PMID:21373768

Palikhe, B. R. (2007). *Relationship between pesticide use and climate change for crops*. Academic Press.

Patyar, S., Pharm, M. S., & Patyar, R. R. (2015). Correlation between Sleep Duration and Risk of Stroke. *Journal of Stroke and Cerebrovascular Diseases*, 1–7. doi:10.1016/j.jstrokecerebrovasdis.2014.12.038 PMID:25817615

Pentyala, S., Ruggeri, J., Veerraju, A., Yu, Z., & Bhatia, A. (2010). Microsomal Ca 2 + flux modulation as an indicator of heavy metal toxicity. *Indian Journal of Experimental Biology*, *48*(July), 737–743. PMID:20929057

Peterson, E. W., & Cardoso, E. R. (1983). The blood-brain barrier following experimental subarachnoid hemorrhage. Part 2: Response to mercuric chloride infusion. *Journal of Neurosurgery*, *58*(3), 345–351. doi:10.3171/jns.1983.58.3.0345 PMID:6219189

Rajan, S., Nadiah, N., Firdaus, M., & Appukutty, M., & Rama-, K. (2012). Effects of climate changes on dissolved heavy metal concentrations among recreational park tributaries in Pahang, Malaysia. *Biomedical Research*, *23*(1), 23–30.

Rama K. Maganti, MD; Paul Rutecki, M. (2013). EEG and Epilepsy Monitoring. *American Academy of Neurology*, (June), 598–622.

Roger, V. L., Go, A. S., Lloyd-Jones, D. M., Benjamin, E. J., Berry, J. D., & Borden, W. B., & American Heart Association Statistics Committee and Stroke Statistics Subcommittee. (2012). Heart disease and stroke statistics--2012 update: A report from the American Heart Association. *Circulation*, *125*(1), e2–e220. doi:10.1161/CIR.0b013e31823ac046 PMID:22179539

Rose, S., Wynne, R., Frye, R. E., Melnyk, S., & James, S. J. (2015). Increased Susceptibility to Ethylmercury-Induced Mitochondrial Dysfunction in a Subset of Autism Lymphoblastoid Cell Lines. *Journal of Toxicology*, *2015*, 13. doi:10.1155/2015/573701 PMID:25688267

Rossignol, D. A., & Frye, R. E. (2011). Mitochondrial dysfunction in autism spectrum disorders : A systematic review and meta-analysis. *Molecular Psychiatry*, 1–25. doi:10.1038/mp.2010.136 PMID:21263444

Safini, F., Bouskraoui, M., Pédiatrie, S. De, Mohammed, A. C. H. U., & Marrakech, V. I. (2009). *Epidémiologie de l ' épilepsie à l ' hôpital de jour pédiatrique*. Academic Press.

Sahathevan, R., Brodtmann, A., & Donnan, G. A. (2012). Systematic Review Dementia, stroke, and vascular risk factors; a review. *International Journal of Stroke*, *7*, 61–73. doi:10.1111/j.1747-4949.2011.00731.x PMID:22188853

Samokhvalov, A. V., Irving, H., Mohapatra, S., & Rehm, J. (2010). Alcohol consumption, unprovoked seizures, and epilepsy : A systematic review and meta-analysis. *Epilepsia*, *51*(7), 1177–1184. doi:10.1111/j.1528-1167.2009.02426.x PMID:20074233

Savitz, S. I., & Fisher, M. (2007). Future of Neuroprotection for Acute Stroke : In the Aftermath of the SAINT Trials. *Annals of Neurology*, *61*(5), 396–402. doi:10.1002/ana.21127 PMID:17420989

Schelter, B. (2008). Seizure Prediction in Epilepsy. Academic Press.

Schiedek, D., Sundelin, B., Readman, J. W., & Macdonald, R. W. (2007). Interactions between climate change and contaminants. *Marine Pollution Bulletin*, *54*(12), 1845–1856. doi:10.1016/j.marpolbul.2007.09.020 PMID:17963794

Seeland, A., Oehlmann, J., & Müller, R. (2012). Aquatic ecotoxicity of the fungicide pyrimethanil : Effect pro fi le under optimal and thermal stress conditions. *Environmental Pollution*, *168*, 161–169. doi:10.1016/j.envpol.2012.04.020 PMID:22622013

Sevim, S., Aktekin, M., Dogu, O., Ozturk, H., & Ertas, M. (2002). Late Onset Polyneuropathy due to Organophosphate (DDVP) Intoxication. *The Canadian Journal of Neurological Sciences*, 75–78. PMID:12619790

Shin, H. W., Jewells, V., Hadar, E., Fisher, T., & Hinn, A. (2014). Review of Epilepsy - Etiology, Diagnostic Evaluation and Treatment. *International Journal of Neurorehabilitation*, *1*(3), 1–8. doi:10.4172/2376-0281.1000129

Sinert, R., Zehtabchi, S., Desai, S., Peacock, P., Altura, B. T., & Altura, B. M. (2007). Serum ionized magnesium and calcium levels in adult patients with seizures. *Scand J Clin Lab Invest*, 317–326. doi:10.1080/00365510601051441

Skoczyńska, A., Jedrejko, M., Martynowicz, H., Poreba, R., Affelska-Jercha, A., Steinmetz-Beck, A., … Jedrychowska, I. (2010). The cardiovascular risk in chemical factory workers exposed to mercury vapor. *Medycyna Pracy*, *61*(4), 381–391. Retrieved from http://www.ncbi.nlm.nih.gov/pubmed/20865852

Skyhøj Olsen. (2001). Post-stroke epilepsy. *Medicina*, *78*(2), 86–90. doi:10.1016/j.neuint.2017.02.002 PMID:29659357

Sobolev, D., & Begonia, M. F. T. (2008). Effects of Heavy Metal Contamination upon Soil Microbes : Lead-induced Changes in General and Denitrifying Microbial Communities as Evidenced by Molecular Markers. *International Journal of Environmental Research and Public Health*, *5*(5), 450–456. doi:10.3390/ijerph5050450 PMID:19151442

Specchio, N., Fusco, L., Claps, D., & Vigevano, F. (2010). Epileptic encephalopathy in children possibly related to immune-mediated pathogenesis. *Brain & Development*, *32*(1), 51–56. doi:10.1016/j.braindev.2009.09.017 PMID:19850427

Stafstrom, C. E., & Carmant, L. (2015). *Seizures and Epilepsy : An Overview for Neuroscientists*. Cold Spring Harbor Laboratory Press.

Strong, K., Mathers, C., & Bonita, R. (2007). Preventing stroke : Saving lives around the world. *Lancet Neurology*, *6*(February), 182–187. doi:10.1016/S1474-4422(07)70031-5 PMID:17239805

Struzyiqska, L., & Gadamski, R. (1997). Lead-lnduced Abnormalities in Blood-Brain Barrier Permeability in Experimental Chronic Toxicity. *Molecular and Chemical Neuropathology*, *31*(3), 207–224. doi:10.1007/BF02815125 PMID:9336764

Tago, D., Andersson, H., & Treich, N. (2014). *Preference Measurement in Health Article information : Users who downloaded this article also downloaded*. Preference Measurement in Health.

Tattersall, J. (2009). Biomedical Sciences Department, Dstl Porton Down, Salisbury, Wiltshire, SP4 0JQ, UK. *Frontiers in Bioscience*, *14*, 3688–3711. doi:10.2741/3481

Taylor, P., Rinsky, J. L., Hoppin, J. A., Blair, A., He, K., & Freeman, L. E. B. (2013). Article. *Journal of Toxicology and Environmental Health, Part A: Current Issues*, 37–41. doi:10.1080/15287394.2013.819308

Thomas, R. H. (2012). seizures and Epilepsy: Pathophysiology and principles of Diagnosis. *Epilepsy*, *1*, 28–42. doi:10.1016/j.engfracmech.2017.04.040

Toews, A. D., Kolber, A., Hayward, J., Krigman, M. R., & Morell, P. (1978). Experimental animals Isolation of capillaries and microvessels. *North-Holland Biomedical Press*, *147*, 131–138.

Tore, K., Bergdahl, I. A., Nilsson, T., & Ja, B. (2007). Occupational exposure to particulate air pollution and mortality due to ischaemic heart disease and cerebrovascular disease. *Occupational and Environmental Medicine*. doi:10.1136/oem.2006.029488 PMID:17303673

Torres, A. D., Rai, A. N., & Hardiek, M. L. (2000). Mercury Intoxication and Arterial Hypertension : Report of Two Patients and Review of the Literature. *Pediatrics*, *105*(3), e34. doi:10.1542/peds.105.3.e34 PMID:10699136

Tu, J. V. (2010). Reducing the global burden of stroke : INTERSTROKE Public health research funding : independence is important The Diet and Health Research Industry Club (DRINC) is. *Lancet*, *376*(9735), 74–75. doi:10.1016/S0140-6736(10)60975-0 PMID:20561674

Ukraintseva, Sloan, Arbeev, & Yashin. (2015). *Increasing Rates of Dementia at Time of Declining Mortality From Stroke*. American Heart Association. doi:10.1161/01.STR.0000217971.88034.e9

Visser, A., Kroes, J., Van Vliet, M. T. H., Blenkinsop, S., Fowler, H. J., & Peter, H. (2012). Climate change impacts on the leaching of a heavy metal contamination in a small lowland catchment. *Journal of Contaminant Hydrology*, *127*(1–4), 47–64. doi:10.1016/j.jconhyd.2011.04.007 PMID:21684031

Waldbaum, S., & Patel, M. (2010). Mitochondria, oxidative stress, and temporal lobe epilepsy. *Epilepsy Research*, *88*(1), 23–45. doi:10.1016/j.eplepsyres.2009.09.020 PMID:19850449

War, G. (2008). Challenges to 'Acetylcholinesterase inhibitors and Gulf War illnesses'. *Proceedings of the National Academy of Sciences of the United States of America*, 33.

Weichenthal, S., Moase, C., Chan, P., Environmental, S., Perspectives, H., August, N., ... Chan, P. (2014). A Review of Pesticide Exposure and Cancer Incidence in the Agricultural Health Study Cohort All use subject to JSTOR Terms and Conditions Review A Review of Pesticide Exposure and Cancer Agricultural Health Study Cohort in the. *Environmental Health Perspectives*, *118*(8), 1117–1125. doi:10.1289/ehp.0901731 PMID:20444670

Wierzbicki, R., & Praz, M. (2002). Disorders in Blood Coagulation in Humans Occupationally Exposed to Mercuric Vapors. *The Journal of Trace Elements in Experimental Medicine, 29*, 21–29. doi:10.1002/jtra.1055

Zhang, Y., & Pardridge, W. M. (2001). Rapid transferrin ef ¯ ux from brain to blood across the blood ± brain barrier. *Journal of Neurochemistry*, *76*(5), 1597–1600. doi:10.1046/j.1471-4159.2001.00222.x PMID:11238745

Zhao, F., Xi, G., Liu, W., Keep, R. F., & Hua, Y. (2016). Minocycline Attenuates Iron-Induced Brain Injury Materials and Methods. *Acta Neurochirurgica*, (S121). doi:10.1007/978-3-319-18497-5

Zheng, W., Aschner, M., & Ghersi-egea, J. (2003). Brain barrier systems : A new frontier in metal neurotoxicological research. *Toxicology and Applied Pharmacology*, *192*(1), 1–11. doi:10.1016/S0041-008X(03)00251-5 PMID:14554098

Zou, X., Jiang, S., Wu, Z., Shi, Y., Cai, S., Zhu, R., & Chen, L. (2017). Effectiveness of deferoxamine on ferric chloride-induced epilepsy in rats. *Brain Research*, *1658*, 25–30. doi:10.1016/j.brainres.2017.01.001 PMID:28063856

Section 6

Climate Change Leading to Vector Born Disease Emergence

Chapter 17
Vector Borne Diseases and Climate Change

Kholoud Kahime
Cadi Ayyad University, Morocco

Moulay Abdelmonaim El Hidan
Ibn Zohr University, Morocco

Denis Sereno
https://orcid.org/0000-0002-0034-9291
University of Montpellier, France

Bounoua Lahouari
NASA, USA

Ahmed Karmaoui
https://orcid.org/0000-0003-3881-4029
Southern Center for Culture and Sciences, Morocco

Abdellaziz Ait Mansour
Cadi Ayyad University, Morocco

Nadia Zouhairi
Cadi Ayyad University, Morocco

Hind Benammi
Cadi Ayyad University, Morocco

Khalid Elkalay
Cadi Ayyad University, Morocco

ABSTRACT

The incidence of emergence diseases including vector borne diseases, water diseases, and some physiologic impairment is considered sensitive to climate. Malaria, leishmaniasis, dengue, and viral encephalitis are among those diseases most influenced by climate. Variation in the incidence of vector borne diseases is associated with extreme weather events and annual changes in weather conditions. Africa in general and Morocco in particular are designated as an area of significant impact by numerous the Intergovernmental Panel on Climate Change (IPCC) reports and notably susceptible to such drastic climate-related health consequences. Climatic parameter change would directly affect disease transmission by acting on the vector's geographic range, activity, or reproduction and by reduction the period of pathogen incubation. This chapter will discuss the increasing risk of some vector-borne diseases in hazard-prone localities. It further identifies the severe challenges both of health adaptation to climate change by highlighting Moroccan adaptive capacity to such crises.

DOI: 10.4018/978-1-5225-7775-1.ch017

INTRODUCTION

Climate change is affecting ecosystems and may have direct or indirect effects on human and animal health. This change affects the distribution and abundance as well as the spatial dynamics of vector species and reservoirs, which in turn disrupts ecosystem composition, vector and reservoir reproduction cycles. Climate change also acts on viruses, bacteria or parasites pathogens, forcing a selection of populations better adapted to environmental conditions (Bounoua *et al.*, 2013).

Climate-related increases in sea surface temperature and sea level can lead to higher incidence of water-borne infectious and toxin-related illnesses, such as cholera and shellfish poisoning (Patz *et al.*, 1996). Focusing on Moroccan outlook, the main diseases considered as health problems and may be aggravate by climate change, such malaria, bilharzia, typhoid, leishmaniasis, dengue and cholera, especially among the most vulnerable groups. Human migration and damage to health infrastructures from the projected increase in climate variability could indirectly contribute to disease transmission (Patz *et al.*, 1996).

Climatic parameters influence the apparition, emergence and reemergence of infectious diseases, in addition to multiple human, biological, and ecological determinants. Changing environmental conditions caused by climate change, often have a negative impact on the life cycle of vector/reservoir populations, by modifying their behavior and areas inhabited by migrating host species and, consequently, on epidemiology of disease. Some vector and reservoir species may disappear, others may become more abundant. In this optic, climate change is expected to affect the distribution and prevalence of vector-borne diseases such as Leishmaniasis and Malaria, as well as waterborne diseases such as Schistosomiasis or cholera. These neglected diseases still ravage lives covertly in the world(WHO, 2012a).

In Morocco, infectious diseases such Leishmaniasis, Malaria and Schistosomiasis are still a public health problem that may be more complicated by climate change. Despite the domestic program to fight against these parasitic diseases, currently, the kingdom is aware to all socioeconomic problems that may be link to health population damage.

Moroccan Ministry of Health (MMH) (MMH, 2014), declared about 2086 cases of Malaria imported between 2005 and 2014. The risk of autochthonous malaria resumption is important in Morocco because of the possible presence of gametocytes carriers in the last malaria focus(Faraj *et al.*, 2008).Leishmaniasis shows significant increase in the number of recorded cases during the last couple of years (Kahime *et al.*, 2014). 43163 cases of leishmaniasis – 41867 cases of Cutaneous Leishmaniasis against 1296 cases of Visceral Leishmaniasis – were recorded between 2005 and 2014 by MMH(MMH, 2014). While, for schistosomiasis, 39 cases were reported in Morocco between 2005 and 2014(MMH, 2014).

For all these diseases, the emergence/reemergence, the outbreaks remain sensitive to climate factors, as well as to the local disease control and monitoring that influence disease trends. In the other hand, Factors such as social economics, health-seeking behavior, geographical location, and population growth will determine the vulnerability of populations to climate change (Githeko *et al.*, 2000).

In this chapter, we aims to highly some sensitive diseases to climate change in Morocco.

OVERVIEW OF SOME SENSITIVE PARASITIC DISEASES TO CLIMATE CHANGE IN MOROCCO CONTEXT

Evidence for the past and current health experiences in Morocco, shows that parasitic diseases pose a public health challenge. Their spread is very closely tied to risk factors (climatic, ecological, and socioeconomic) and other factors, such as urbanization, agricultural practices, beyond health authorities' traditional sphere of influence. Climate change is imposing a burden that is unevenly distributed across populations. The most vulnerable populations are in worse positions to effectively face these challenges to the extent that they have a limited ability to adapt to changing conditions. Therefore, a particular case of vulnerability in this context is the emergence of vector-borne-diseases, such, Malaria, Schistosomiasis and Leishmaniasis. In view of life threatening nature of most of these diseases, there is need for understanding having eco-epidemiological approach in respect of each disease with emphasis on habitats of reservoirs of infection, micro niche of arthropod/insect vectors, showing the trend of occurrence cases (Figure 1. a: leishmaniasis, b: Schistosomiasis, c: Malaria) and understanding life cycle of each disease.

Figure 1. Annual evolution of parasitic diseases in Morocco between 2005-2014, (a) Leishmaniasis forms, (b) Schistosomiasis (including some imported cases), (c) Malaria (Imported cases)

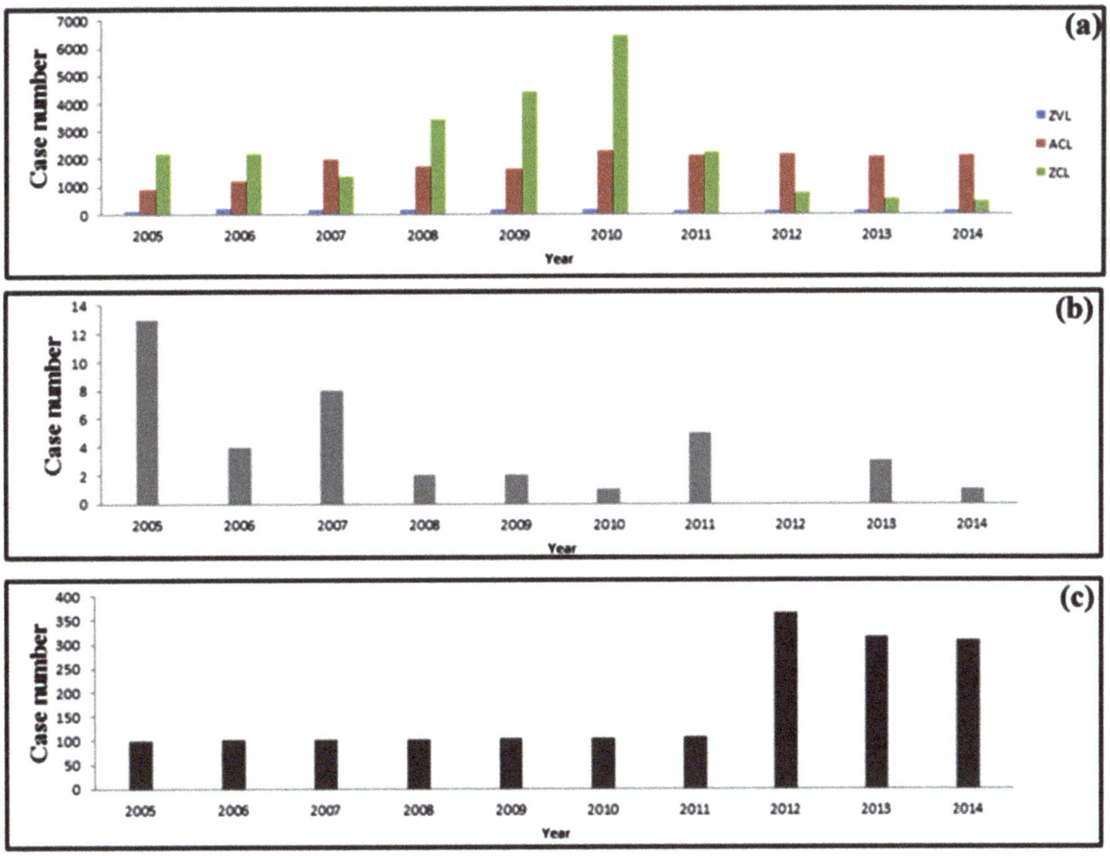

Malaria

Malaria is a leading cause of infectious disease and death worldwide with 3.3 billion people at risk for contracting the disease. In 2010, an estimated 219 million (range 154 million to 289 million) became infected with malaria, of which an estimated 660,000 people died (WHO, 2012b).

According to While *et al.* (2014), malaria, a parasite infection of red blood cells, still kills roughly 2000 people per day, most of whom are children in Africa. Malaria is a mosquito-borne infectious disease caused by parasitic protozoans of the genus *Plasmodium* (*P. vivax, P. malariae, P. ovale and P. falciparum*) and is transmitted by female mosquito vectors of the Anopheles species. The cycle of Plasmodium is carried out in several stages which spread between mosquitoes and humans.

Malaria raged in Morocco for centuries and was an endemic disease in the majority of provinces. In 1960, a domestic program to fight the disease was launched. It allowed controlling the situation after 40 years of bitter struggle. By 1999, malaria was occurring as sporadic cases of Plasmodium vivax in some residual foci in the north. The epidemiological assessment undertaken by the Moroccan Ministry of Health showed a shift towards the elimination of indigenous cases and the last indigenous case was reported in 2004 (Figure 1c.). Morocco is now WHO (World Health Organization) certified Malaria free but imported cases are reported constantly.

In Morocco, *Anopheles labranchiae* is the main vector of malaria. In common with leishmaniasis, it is the female (vector) that bites when it needs a blood meal before laying eggs. The vectorial capacity of *A. labranchiae* is considerably high during the summer which corresponds to the rice cultivation period in northern Morocco (Faraj *et al.*, 2018). The North center area at high risk of transmission MMH (MMH, 2004; Trari *et al.*, 2011).

The risk of malaria resurgence in Morocco is higher because of the possible presence of parasite in human and invertebrate hosts in the last malaria foci, numerous imported cases and the presence of the main vector of Malaria in Morocco. This risk may be increasing by climate change. In this context, Malaria is one of the few climate-sensitive health outcomes that has been modeled by more than one research group and can therefore facilitate a more thorough assessment of possible climate change effects using a multimodal intercomparison (Caminade *et al.*, 2014 a).

Sensitivity of malaria incidence to climate factors in Africa is generally consistent with literature (Caminade*et al.*, 2014b). Arab *et al.* (2014) study shows, a statistically significant correspondence between malaria rates and the climate variables considered. The two most important climate factors are found to be average annual temperature and total annual precipitation.

Schistosomiasis

According to the World Health Organization (WHO), more than 206 million people are infected with Schistosoma (of whom 120 million have symptoms and around 20 million show severe disease sequelae) (WHO, 1995).

Schistosomiasis is an indirectly (snail)-transmitted disease whose distribution is particularly sensitive to environmental changes, including changes of human origin. Transmission of the parasite is highly focal, with heterogeneity reflecting numerous human and snail host as well as environmental factors (Fenwick *et al.*, 2006).

Schistosomiasis is a disease caused by parasitic worms belonging to the class of Trematodes and genus *Schistosoma* (*S. haematobium, S. mansoni, S. japonicum and S. intercalatum*). The parasite develops successively in two hosts: mollusk and human. This parasite is commonly found in ponds, streams and irrigation canals, housed in freshwater mollusks, and infest humans through the skin via contact with contaminated water.

In Morocco, the disease had spread in the oases in the south and along the southern side of the Atlas. The majority of cases have been filed in south Morocco(WHO, 2012c). In recent years, many *S. haematobium* foci proved unstable and some even disappeared (Fig 1b.). However, creating large water supply for irrigation may lead to the onset of new foci.

Despite the existence of monitoring programs and adequate health services, cases of Malaria, Leishmaniasis and Schistosomiasis are still recorded in Morocco. As shown in Figure 1a,leishmaniasis is the most reported disease and all cases are indigenous. In contrast, all Malaria cases are imported. For Schistosomiasis, the number of indigenous cases is rarely associated with the risk of introduction of new species, which is linked to travelers or immigrants from countries where these species are endemic.

These diseases constitute a real social problem in Morocco because they disproportionately affect the poor, particularly those in vulnerable housing and environmental conditions. Loss of income and healthcare costs exacerbate the economic situation of already-disadvantaged households (Kahime *et al.*, 2016a). Urban population growth, responsible for the spatial expansion of cities, creates enormous need for community facilities and equipment related to decent housing, access to clean water and sanitation and environmental preservation. The vulnerability of populations to these disease risks varies across spatial and temporal scales in response to environmental change, economic development dynamics, social capital, demographics and population structure. Understanding the interactions between human health and the environment is quite complex. In fact, the very term environment can be confusing because it is used in various acceptances. The ecosystem approach to human health is distinguished from traditional approaches by integrating all determinants of health, namely economic and environmental factors and community needs.

Leishmaniasis

Leishmaniasis is a complex disease caused by Leishmania species and transmitted by a Phlebotomine sand fly (Diptera: Psychodidae). Two forms are known, Cutaneous and Visceral Leishmaniasis. The main reservoirs are dogs for Zoonotic Visceral Leishmaniasis (ZVL), rodents for Zoonotic Cutaneous Leishmaniasis (ZCL) and human for Anthroponotic Cutaneous Leishmaniasis (ACL)(Kahime *et al.*, 2014, Kahime*et al.*, 2016b).

Actually, three parasite species co-exist in Morocco. *Leishmaniainfantum*, causes mainly ZVL and transmitted by species of the subgenus Larroussius. It is widespread in the whole country and is more frequent in its northern part(Kahime *et al.*, 2014, Kahime*et al.*, 2015); *L. infantum* can cause Cutaneous leishmaniasis (CL) as well even if it is a rare condition with a few sporadic cases in the North of the country (especially in SidiKacem province) with little epidemiological data available(Kahime *et al.*, 2016b). L. major causes ZCL with *Phlebotomuspapatasi* as the vector in pre-Saharan area(Bounoua *et al.*, 2013, Kahime*et al.*, 2014); and *L. tropica*, causative agent of ACL is widespread in northern and central areas, especially in semi-arid regions with *P. sergenti* as the vector(Kahime *et al.*, 2014, Boussaa *et al.*, 2016).

Leishmaniases are significantly affected by the change and instability of the climate (WHO, 2015). The propagation of leishmaniasis forms in Morocco may be attributed to a low efficiency of the leishmaniasis control program (Kahime *et al.*, 2018). Nevertheless, bioclimatic, soil, vegetation, and climate change, as well as other socioeconomic factors, may also contribute to the dynamic and geographic expansions of leishmaniaisis in this country (Kahime *et al.*, 2018).

POTENTIELS IMPACTS OF CLIMATE CHANGE TO INFECTIOUS DISEASES: IN MOROCCO CONTEXTE

Anthropogenic climate change may directly affect the behavior and geographical distribution of mosquito vectors and the lifecycle of the parasite, and thus change the incidence of the disease. Indirectly, climate change could also have various effects by influencing environmental factors such as climate variables, soil PH and the availability of breeding sites(Ulrik *et al.*, 2014).

As the climate has an important influence on environmental and socio-economic components, the juxtaposition of many of socio-economic factors e.g., poverty, economic degradation over time may convincingly increase as result to climate change. This change could, depending on the region, have adverse impacts on water resources, agriculture, biodiversity, ecosystems, food and health security.

A wide range of infectious diseases may change their geographic range, seasonality and incidence due to climate change, but there is still limited and uncertain scientific evidence about the extent of health vulnerability to climate change. For example, over the past decade, the epidemiological situation of CL has changed significantly. It is acquiring an increasingly epidemic status with geographic expansion to previously free areas and the emergence of new foci in several provinces of Morocco. 24 804 cases of zoonotic *L. major* CL and 16 852 cases of *L. tropica* anthroponotic CL were recorded between 2004 and 2013 in Morocco (Kahime *et al.*, 2016b). ZCL acquires an increasingly epidemic status with geographic expansion into previously free areas in several provinces of South-East Morocco (Kahime *et al.*, 2016a). Whereas, ACL has an extensive geographical distribution (Kahime*et al.*, 2016b).

The leishmaniasis lifecycle, parasite-reservoir-vector, evolves depending on geographic regions and is sensitive to environmental changes that can affect the parasite, the reservoir and the vector, as well as their dynamic interaction and territorial expansion (Bounoua *et al.*, 2013). Seasonal patterns and the correlation between the vector density and the number of cases have been documented (Bounoua *et al.*, 2013). Weather and climate variables also play an important role in leishmaniasis incidence as they can constrain or exacerbate favorable conditions for the disease, such as an acceleration of the development of the parasite or synergistic changes in reservoir and vector populations that cause an explosion in the vector population. For example, increase in precipitation may increase the vegetation foliage, and thus the number and quality of breeding sites for both rodents and sandflies(Bounoua *et al.*, 2013).

Distribution and seasonal activity of malaria are also sensitive to climate factors and impact the local capacity to control the disease. Malaria is one of the few climate-sensitive diseases that has been modeled by many research groups, which facilitates a thorough assessment of possible climate change effects using a multi-model inter-comparison (Caminade *et al.*, 2014). The distribution and population dynamics of malaria are probably more governed by abiotic than biotic factors. Of the possible abiotic influences on the transmission cycle of malaria, temperature and rainfall are the most important; therefore the situation could be exacerbated by climate change. Although this linkage is still uncertain, an increase in the incidence of malaria has been identified as a potential impact of climate change in South America

(Lieshout *et al.*, 2004) and Africa (Ebi *et al.*, 2005). Climatic factors that feed into the phenomenon could have a direct bearing on the number of malaria cases. Actually, a number of studies have reported the association between malaria cases, rainfall and temperature (Bhattarai*et al.*, 2007). For example, a study carried out in Ethiopia revealed an association of malaria with rainfall and minimum temperature, the strength of which varied with altitude (Teklehaimanot *et al.*, 2004). In South Africa, variations in annual cases of malaria were shown to be related to rainfall and temperature patterns (Craig *et al.*, 2004).

According to McCreesh and Booth (McCreesh*et al.*, 2013), climate change will inevitably influence both the distribution of Schistosoma sp. and incidence of schistosomiasis in areas where it is currently endemic. Global warming is expected to be accompanied by perturbations in the global hydrologic cycle(Allen *et al.*, 2002), precipitation levels and pronounced changes in water availability(Allen *et al.*, 2002; Olveda*et al.*, 2016). However, only a few attempts have been made to predict changes in the spatial distribution of Schistosomiasis due to global warming (Ulrik*et al.*, 2014). Thus, the nature and extent of climate change on the transmission of Schistosomiasis remains poorly understood(Morgan *et al.*, 2001). The lack of research in this field probably explains why climate change implications for Schistosomiasis control and elimination have been largely ignored (McCrees*et al.*, 2015).

Snails of the genus *Bulinus* serve as the intermediate hosts of *S. haematobium* in Africa and the Eastern Mediterranean (WHO, 2012c). These aquatic snails breed in waters (barrages, irrigation canals...), contaminated with infected faeces or urine from infected person or by infected wastewater, used also for irrigation. Water resources development schemes in certain areas, particularly irrigation schemes, can contribute to the introduction and spread of Schistosomiasis. Consequently, intermediate hosts can determine the geographical area of Schistosomiasis risk. In Morocco, the disease foci, especially in BeniMellal, El Kelaa and Nador Provinces were linked to hydrologic programs for irrigation (barrages and irrigation canals and wells) in the 1970s (WHO, 2012c).

Potential impact of climate change on the transmission of Schistosomiasis and other diseases must be juxtaposed to profound ecological, demographic, and socioeconomic changes. Also, it should be mentioned that, in Morocco, there is no study that addresses the relationship between malaria or Schistosomiasis and climate change.

MOROCCO'S ADAPTATION POTENTIAL RELATED TO HEALTH-CLIMATE ISSU

The impact of Climate Change on Morocco is drastic. A structured but adaptable guidance framework is needed to mitigate and respond to myriad health threats posed throughout the kingdom. Specifically, this must manage environmental and socio-economic aspects including access to education, nutrition, financial stability, etc. Of particular pertinence to Morocco are: legal and political attempts to manage health and environmental issues, health and environmentally-related human rights, local climate-health policies and programs, social and human science approaches to climate-health links, policy maker interaction with climate and health specialists, and civil society's unfettered access to health policy decision makers.

A national plan in regard to climate change and health must clearly be closely informed by the UNFCC (United Nations Framework Convention on Climate Change). In addition, it must develop projects and responses focusing on: regional equity in terms of financial and human resource distribution, national-level infrastructure sufficiently robust to ensure urban and rural parity in healthcare access and distribution, appropriate management and implementation of good health governance, healthcare reform, obligatory medical assistance and consistent financial investment to support all the preceding elements.

REFERENCES

Allen, M. R., & Ingram, W. J. (2002). Constraints on future changes in climate and the hydrologic cycle. *Nature*, *419*(6903), 224–232. doi:10.1038/nature01092 PMID:12226677

Arab, A., Jackson, M. C., & Kongoli, C. (2014). Modelling the effects of weather and climate on malaria distributions in West Africa. *Malaria Journal*, *13*(1), 126. doi:10.1186/1475-2875-13-126 PMID:24678602

Bhattarai, A., Ali, A. S., Kachur, S. P., Mårtensson, A., Abbas, A. K., Khatib, R., ... Molteni, F. (2007). Impact of artemisinin-based combination therapy and insecticide-treated nets on malaria burden in Zanzibar. *PLoS Medicine*, *4*(11), e309. doi:10.1371/journal.pmed.0040309 PMID:17988171

Bounoua, L., Kahime, K., Houti, L., Blakey, T., Ebi, K. L., Zhang, P., ... Boumezzough, A. (2013). Linking climate to incidence of zoonotic cutaneous leishmaniasis (*L. major*) in Pre-Saharan North Africa. *International Journal of Environmental Research and Public Health*, *10*(8), 3172–3191. doi:10.3390/ijerph10083172 PMID:23912199

Boussaa, S., Kahime, K., Samy, A. M., Bensalem, A., & Boumezzough, A. (2016). Species composition of sand flies and bionomics of Phlebotomuspapatasi and P. sergenti (Diptera: Psychodidae) in cutaneous leishmaniasis endemic foci, Morocco. *Parasites & Vectors*, *9*(1), 60. doi:10.118613071-016-1343-6 PMID:26830689

Caminade, C., Kovats, S., Rocklov, J., Tompkins, A. M., Morse, A. P., Colón-González, F. J., ... Lloyd, S. J. (2014). Impact of climate change on global malaria distribution. *Proceedings of the National Academy of Sciences of the United States of America*, *111*(9), 3286–3291. doi:10.1073/pnas.1302089111 PMID:24596427

Caminade, C., Kovats, S., Rocklov, J., Tompkins, A. M., Morse, A. P., Colón-González, F. J., ... Lloyd, S. J. (2014). Impact of climate change on global malaria distribution. [PubMed]. *Proceedings of the National Academy of Sciences of the United States of America*, *111*(9), 3286–3291.

Craig, M. H., Kleinschmidt, I., Nawn, J. B., Le Sueur, D., & Sharp, B. L. (2004). Exploring 30 years of malaria case data in KwaZulu-Natal, South Africa: Part I. The impact of climatic factors. *Tropical Medicine & International Health*, *9*(12), 1247–1257. doi:10.1111/j.1365-3156.2004.01340.x PMID:15598256

Ebi, K. L., Hartman, J., Chan, N., Mcconnell, J., Schlesinger, M., & Weyant, J. (2005). Climate suitability for stable malaria transmission in Zimbabwe under different climate change scenarios. *Climatic Change*, *73*(3), 375–393. doi:10.100710584-005-6875-2

Faraj, C., Ouahabi, S., Adlaoui, E., Boccolini, D., Romi, R., & El Aouad, R. (2008). Risque de reemergence du paludisme au Maroc étude de la capacité vectorielle d'*Anopheleslabranchiae*dans une zone rizicole au nord du pays. *Parasite (Paris, France)*, *15*(4), 605–610. doi:10.1051/parasite/2008154605 PMID:19202769

Fenwick, A. (2006). Waterborne infectious diseases—could they be consigned to history? *Science*, *313*(5790), 1077–1081. doi:10.1126cience.1127184 PMID:16931751

Githeko, A. K., Lindsay, S. W., Confalonieri, U. E., & Patz, J. A. (2000). Climate change and vector-borne diseases: A regional analysis. *Bulletin of the World Health Organization*, *78*, 1136–1147. PMID:11019462

Kahime, K., Bounoua, B., Messouli, M., Boussaa, S., & Boumezzough, A. (2016a). Evaluation of Eco-Adaptation Strategies of Health to Climate change: Case of Zoonotic Cutaneous Leishmaniasis (ZCL) as Vulnerability Indicator in Pre-Saharan region of Morocco. In *Environmental Change and Human Security in the Middle East and Africa. GECS2012 Conference Proceedings*. Springer.

Kahime, K., Boussaa, S., Laamrani, A., Nhammi, H., & Boumezzough, A. (2016b). Epidemiological study on acute cutaneous leishmaniasis in Morocco. *Journal ofAcute Disease*, *5*(1), 41–45. doi:10.1016/j.joad.2015.08.004

Kahime, K., Boussaa, S., Ouanaimi, F., Boumezzough, A. (2015). Entomological Assessment of Human Visceral Leishmaniasis Risk in area with sporadic cases, Morocco. *ActaTropica*, *48*, 58-65. doi: tropica.2015.04.006 doi:10.1016/j.acta

Kahime, K., Denis, S., Lahouari, B., El Hidan, M., & Souad, B. (2018). Management of Leishmaniases in the Era of Climate Change in Morocco. *International Journal of Environmental Research and Public Health*, *15*(7), 1542. doi:10.3390/ijerph15071542 PMID:30037049

McCreesh, N., & Booth, M. (2013). Challenges in predicting the effects of climate change on Schistosoma mansoni and Schistosoma haematobium transmission potential. *Trends in Parasitology*, *29*(11), 548–555. doi:10.1016/j.pt.2013.08.007 PMID:24064438

McCreesh, N., Nikulin, G., & Booth, M. (2015). Predicting the effects of climate change on Schistosomamansoni transmission in eastern Africa. *Parasites & Vectors*, *8*(1), 4. doi:10.118613071-014-0617-0 PMID:25558917

Morgan, J. A. T., Dejong, R. J., Snyder, S. D., Mkoji, G. M., & Loker, E. S. (2001). Schistosomamansoni and Biomphalaria: Past history and future trends. *Parasitology*, *123*(7), 211–228. doi:10.1017/S0031182001007703 PMID:11769285

Moroccan Ministry of Health. (2004). Direction de l'épidémiologie et de lutte contre les maladies (1966–2004). Données épidémiologiques des maladies sous surveillance. Bulletin épidémiologique, ministère de la Santé, Maroc.

Moroccan Ministry of Health. (2014). *A report on progress of control programs against parasitic diseases*. Directorate of Epidemiology and Disease Control, Ministry of Health, Rabat, Morocco. Available from: http://www.sante.gov.ma/departements/delm/index-delm.htm

Olveda, R. M., Tallo, V., Olveda, D. U., Inobaya, M. T., Chau, T. N., & Ross, A. G. (2016). National survey data for zoonotic schistosomiasis in the Philippines grossly underestimates the true burden of disease within endemic zones: Implications for future control. *International Journal of Infectious Diseases*, *45*, 13–17. doi:10.1016/j.ijid.2016.01.011 PMID:26820760

Patz, J. A., Epstein, P. R., Burke, T. A., & Balbus, J. M. (1996). Global climate change and emerging infectious diseases. *Journal of the American Medical Association*, *275*(3), 217–223. doi:10.1001/jama.1996.03530270057032 PMID:8604175

Pedersen, U. B., Stendel, M., Midzi, N., Mduluza, T., Soko, W., Stensgaard, A. S., ... Kristensen, T. K. (2014). Modelling climate change impact on the spatial distribution of fresh water snails hosting trematodes in Zimbabwe. *Parasites & Vectors*, *7*(1), 536.

Teklehaimanot, H. D., Schwartz, J., Teklehaimanot, A., & Lipsitch, M. (2004). Alert threshold algorithms and malaria epidemic detection. *Emerging Infectious Diseases*, *10*(7), 1220–1226. doi:10.3201/eid1007.030722 PMID:15324541

Trari, B., & Carnevale, P. (2011). De la préélimination à l'élimination du paludisme au Maroc. Quels risques pour l'avenir? *Bulletin de la Société de Pathologie Exotique*, *104*(4), 291–295. doi:10.100713149-011-0156-2 PMID:21638201

Van Lieshout, M., Kovats, R. S., Livermore, M. T. J., & Martens, P. (2004). Climate change and malaria: Analysis of the SRES climate and socio-economic scenarios. *Global Environmental Change*, *14*(1), 87–99. doi:10.1016/j.gloenvcha.2003.10.009

White, N.J., Pukrittayakamee, S., Hien, T.T., Faiz, M.A., Mokuolu, O.A., & Dondorp, A. M. (2014). Malaria. *Lancet, 383*(9918), 723-35. doi: . doi:10.1016/S0140-6736(13)60024-0

World Health Organization. (1995). Social and Economic Research Reports TAR/SER/PRS, Identification of high-risk communities for control of schistosomiasis in Africa: A multicommunity study. WHO.

World Health Organization. (2012b). *World malaria report*. Geneva: World Health Organization. Retrieved from http://www.who.int/malaria/publications/world_malaria_report_2012/report/en

World Health Organization. (2012c). *Élimination de la schistosomiaseau Maroc Une réalité et un succèsaprès trois décennies de lutte*. WHO.

World Health Organization. (2015). *Leishmaniasis Fact Sheet No 375*. Available online: http://www.who.int/mediacentre/factsheets/fs375/en/

World Health Organization (WHO). (2012 a). *Accelerating work to overcome the global impact of neglected tropical diseases: a roadmap for implementation*. Geneva, Switzerland: WHO.

Chapter 18
Climatic Factors Impacting Leishmaniasis Risk in a Global View: Case of Morocco

Kholoud Kahime
Cadi Ayyad University, Morocco

Moulay Abdelmonaim El Hidan
Ibn Zohr University, Morocco

Denis Sereno
https://orcid.org/0000-0002-0034-9291
University of Montpellier, France

Bounoua Lahouari
NASA, USA

Abdelkrim Ben Salem
Cadi Ayyad University, Morocco

Abdellaziz Ait Mansour
Cadi Ayyad University, Morocco

Mohammed Messouli
Cadi Ayyad University, Morocco

Mohammed Yacoubi Khebiza
Cadi Ayyad University, Morocco

ABSTRACT

Leishmaniasis is a complex disease comprised of multiple organisms in association. Each of these organisms responds differently to external factors. The environmental and socio-economic associations contribute to the dynamics and emergence of leishmaniasis across the globe. Ecological dynamics of the vector-parasite-host system of leishmaniasis influenced directly and indirectly both human and animal health. The transitions and rapid climate and socio-economic changes caused a transition of emergence and re-emergence of leishmaniasis outbreaks. The pattern of changes is influenced by the distribution and abundance as well as the spatial dynamics of vector and reservoirs species, which in turn disrupts ecosystem structure of vector and parasite. In Morocco, leishmaniasis are endemic and constitute a major public health threat. The observations showed significant variations in its spatial distribution and forms through Morocco with increase in the number of recorded cases during the last couple of years. Here, the authors discuss disease change related to climate and socio-economic influence.

DOI: 10.4018/978-1-5225-7775-1.ch018

INTRODUCTION

Environmental changes would directly affect the transmission patterns of vector-borne diseases (Patz et al., 2000) through their impacts at different spatial and temporal ecological scales (i.e. ecozones, biomes and ecotopes) (Ready 2008). Changes in climate are expected to affect the occurrence, the spatio-temporal distribution and the lifecycle of several vector borne diseases such as leishmaniases.

Leishmaniases are among the second most apparent resurging vector-borne disease after malaria; Leishmaniasis cause death to 20 000 to 40 000 deaths occur annually, and Malaria to 367 000 to 755 000 across the world (WHO 2015; Avar et al. 2012; WHO 2010). Leishmaniases occurs in 88 countries and four continents: Africa, North and South America, Asia and Europe with around 350 Million persons are at risk, 14 Million people infected worldwide and about 2 Million new cases recorded per year (WHO 2007, 2010). Thus, in some countries, leishmaniasis became an emergency health problem (e.g. India, Bangladesh, Sudan, Ethiopia, Brazil, Afghanistan, Algeria, Colombia, Iran, Syria, Costa Rica and Peru (Alvar et al., 2012). In Morocco, they present a serious public health problem. Three *Leishmania* species representing two leishmaniasis co-occur (Figure 1): *Leishmania infantum* (visceral leishaniasis), *Leishmania major* and *Leishmania tropica* (cutaneous leishmaniasis) were warned. In 2001-2010, the Ministry of Health reported about 38500, and 1300 cases of CL, and VL, respectively (MMH, 2011).

Depending on the origin of the infection, leishmaniasis in Morocco can be grouped into three forms: 1) Zoonotic Visceral Leishmaniasis (ZVL) caused by *L. infantum* and maintained in long term by dogs and bites of three vector species; *P. ariasi, P. perniciosus,* and possibly *P. longicuspis*, 2) Zoonotic Cutaneous Leishmaniasis (ZCL) caused by *L. major*, where the parasite is maintained among small mammals and human by the bites of *P. papatasi*, and 3) Anthroponotic Cutaneous Leishmaniasis (ACL) caused by *L. tropica*, human is the main host-reservoir in Morocco (Kahime et al., 2014, Kahime et al., 2015a), while zoonotic foci have been reported from Israel, Kenya and Egypt (Sang et al., 1992; Shehata et al., 2009). It transmitted among human hosts by the bites of *P. sergenti* (e.g.; Nejjar et al., 1998; Rioux et al., 1982; Rhajaoui, 2009; Pratlong et al., 1991; Kahime et al., 2015b).

Dynamics of leishmaniasis foci can be addressed by an eco-epidemiological risk approach, characterized by three principal components: temporal, spatial and demographic components. Each component has an ensemble of influencing factors. This study presents an overview of Environmental and socio-economic factors influencing leishmaniasis trend in Morocco.

The assumption is to demonstrate a linking between changing in leishmaniasis incidence over time (temporal component), across different areas of Morocco (spatial component) and the increase in population (demographic component) which may contribute to increases the leishmaniasis risk.

Spatiotemporal components analysis, in parallel, is useful to highlight the variations of leishmaniasis in Morocco. Many studies in Morocco examined sparely the spatial and temporal leishmaniasis spread (Rioux, 2001; Rispail et al., 2002; Rioux and de La Rocque, 2003, Boussaa et al., 2005; Kahime et al., 2015b; Kahime et al. in press) however; more studies are needed. To date, also, little is known about the probable linking between environmental and socio-economic conditions on the evolution of leishmaniasis transmission across Morocco.

Figure 1. Components of the leishmaniasis transmission cycles reported in Morocco. The diagram show the association of several organisms in the dynamics of three different leishmaniasis forms occurred across the country

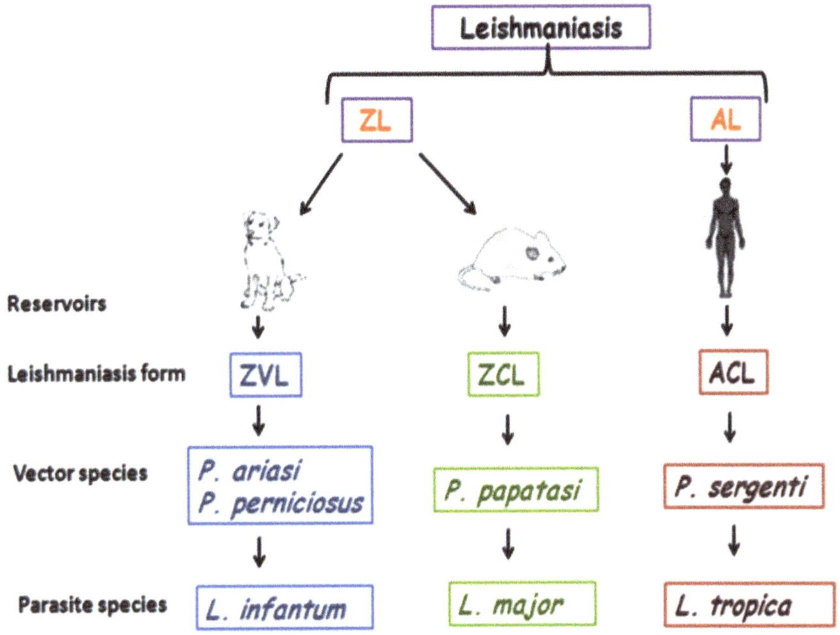

MATERIAL AND METHODS

Study Area and Population

Morocco is localized in northwest of the African continent between the Atlantic and the Mediterranean between latitudes 21N to 36N and longitudes 1W to 17W, suffers from a lack of water during all seasons [Bennani et al., 2015]. Morocco's climate is Mediterranean and mainly characterized by hot and dry summer and a temperate to mild winter in the coastal strip, cool to cold in the country's interior, on the chains of the Atlas, in the Rif and the highlands of the Eastern [Esper et al., 2007].

Epidemiological Data

Epidemiological data were obtained from the national medical services registers and reports. These epidemiological data were recorded after active or passive screening (leishmaniasis is a certifiable disease in Morocco). We used clinical and epidemiological data provided by the Moroccan Directorate of Epidemiology and Fight Against Diseases.

GIS Data Base and Statistic Analysis

Mapping the spatial distributions of infectious diseases like leishmaniasis is key to both investigating their epidemiology and identifying populations at risk. In this study, digital maps were produced by using ArcMap GIS Version 10. The output is two maps.

Investigation and Vulnerability Assessing

Data collection was achieved through: i) individual interviews (in the field), ii) electronic questionnaires (specifically addressed to health personal) and iii) literature search (Health Ministry reports).

We were able to reach and obtain detailed information on 100 people. Before handing over the questionnaire. The survey questionnaires focused on three main themes:

- Identification of the respondent in terms of age, gender, education, and profession
- Issues related to Leishmaniasis: notion and risk factors (climate change, vulnerability, risk factors, intra-domestic investigation)
- Adaptation strategies as solutions

Concept and Use of Vulnerability in Leishmaniasis Context

Vulnerability is conceptualized in different ways by different disciplines. Turner et al. (2003) recognize that vulnerability is not only determined by exposure to hazards, but also depends on the sensitivity and resilience of the system experiencing the risk to such a hazard. The vulnerability of a region to climate change is described by three elements: exposure, sensitivity and adaptive capacity (IPCC 2001), where i) exposure can be interpreted as a direct threat (stress factor), for example: temperature, precipitation, or extreme weather events; ii) sensitivity describes the conditions of human environment that can increase the risk, improve the danger, or trigger an impact; and iii) the adaptive capacity is the ability to adapt to measures that avoid the potential harmful impacts.

This concept of vulnerability is also useful for assessing risks to human societies from vector-borne diseases to the extent these diseases are influenced by climate. According to the World Health Organization (WHO, 2010), leishmaniasis are significantly associated with socio-economic indicators (or sensitivity factors) and climate change (or factors of exposition). Accordingly, we have considerate leishmaniasis as an indicator of vulnerability of population infected.

RESULTS AND DISCUSSION

Changes in Moroccan Leishmaniasis Status

In Morocco, recent data about leishmaniasis prevalence and distribution revealed considerable shifts in the CL disease dynamics toward North and VL toward the Center and South of the country. These changes in the potential distribution of leishmaniasis across the country are resulted from direct and indirect responses to local changes in climate and socio-economic factors associated with different forms of leishmaniasis.

Moreover, these dynamic changes in disease occurrences may be highlight as change of spatiotemporal components and illustrate as indirect result of demographic components.

Spatial Component

Leishmaniasis forms are increasingly widespread in terms of spatial propagation (Figure 2). During the early 90s, the nosogeographical form of each entity of leishmaniasis has been related to its own territory. *L. major* was mainly located in the South-east, *L. tropica*, to the west in the region of Azilal, Marrakesh and Essaouira, and *L. infantum* confined to the northern province of Taounate. In 1997, we observed the wide spatial distribution of these species across the country with no longer availability of such geographic-based patterns. We noticed a migration of *L. major* (ZCL) from the South-east to North-east, *L. tropica* (ACL) was increasingly spread over new foci in all direction; also the overlapping foci, notably with *L. infantum* (ZVL) become more common (Rajhaoui, 2009; kahime et al., 2014). Moreover, in southern Morocco, *L. infantum* infections are sporadic, however their appearance was more frequent recently (Kahime et al., 2015a, Zarrouk et al., 2015).

Figure 2. Geographical distribution of leishmaniasis foci in Morocco: a- before and b- after 1997, with chronological order. L. infantum in A) Taounate (Rioux et al., 1996), B) Sidi Kacem province (Rhajaoui et al., 2007) and C) Chefchaoun, D) Taza, E) Fes, F) Moulay yacoub, G) Meknes, H) Sefrou, I) Al Houceima (Amro et al. (2013). L. tropica in 1) Azilal (Marty et al., 1989 2) Essaouira, 3) Agadir, 4) Guelmim (Pratlong et al., 1991), 5) Taza (Guessous-Idrissi et al., 1997), 6) Moulay Yacoub (Rhajaoui et al., 2004), 7) Chichaoua (Guernaoui et al., 2005), 8) Taroudant, 9) Beni Mellal, 10) Boulemane (Rhajaoui et al., 2007), 11) Al Haouz (Ramaoui et al., 2008), 12) Sefrou (Hmamouch et al.,2014). L. major in I) Tata (Rioux et al., 1986), II) Ouarzazate and III) Errachidia (Rhajaoui et al., 2007). And the Mucocutaneous leishmaniasis (MCL) in Taza and Taounate (Iguermia et al., 2011)

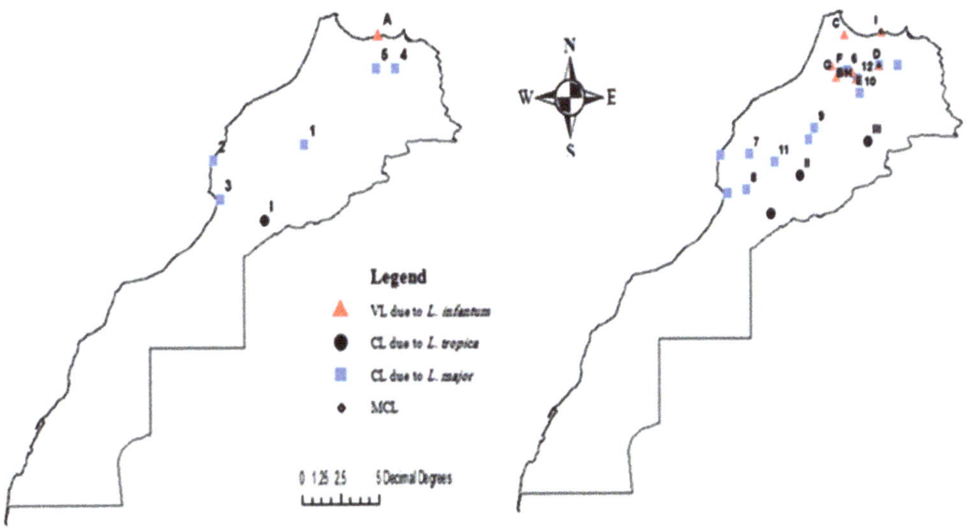

Temporal Component

According to Moroccan ministry of health report (MMH, 2011), a rapid and continuous significant increase in the number of cases infected with *L. major* from 1990 to 2010 was noted (Figure 3). The same pattern was also observed among cases infected with ACL; however, we reported rapid jumps in the number of ACL cases between 2001 and 2010. In 1990-2000, ZCL (*L. major*) form showed important variations; however, no major changes in the number of occurrences of ACL (*L. tropica*) or VL (*L. infantum*). Beginning of 2001, the total number of CL cases had experienced an alarming increase (784 cases) in 2000; (2029 cases) in 2001; (3361 cases) in 2006 and (8707 cases) in 2010. While, ZVL has experienced an increase from 105 cases in 1999 to 163 cases in 2008 (MMH, 2011).

It is worth noting that ACL increased the most (16 times) followed by CL due to *L major* and VL (2 times) between the periods of 1991-2000 and 2001-2010. An increase in VL occurrence was relatively smaller with 154 cases declared in 2004 versus only 76 in 1997 (Tamimy, 2011). Despite monitoring actions against leishmaniasis, all leishmaniasis foci reported in 1997 are still active and recorded a higher number of yearly incidences.

There is ample evidence that these disease geographical and temporal variations may be linked to environmental and climatic change (Yates et al., 2002; Bounoua et al., 2013) but trend of demography can influence.

Probable Contribution of Demographic Component

Population in Morocco increase exponentially from 1960 to 2011; population increased from 11,626,000 to 32, 187000 inhabitants in 1960, and 2011, respectively (MMH, 2012). The same population reached about 34 million inhabitants in 2014 (HCP, 2014). According to High Commission for Planning of Morocco, 2014, the urbanization rate never ceases to progress (from 29.2 in 1960 to 59.6 in 2014). This increase in urban population is the result of three major factors: natural movement (like travel), migration (labour and others), and promotion of rural centers to urban status (improving living conditions).

Figure 3. Leishmaniasis cases (Y axis) reported between 1991-2000 and 2001-2010 (MMH, 2011)

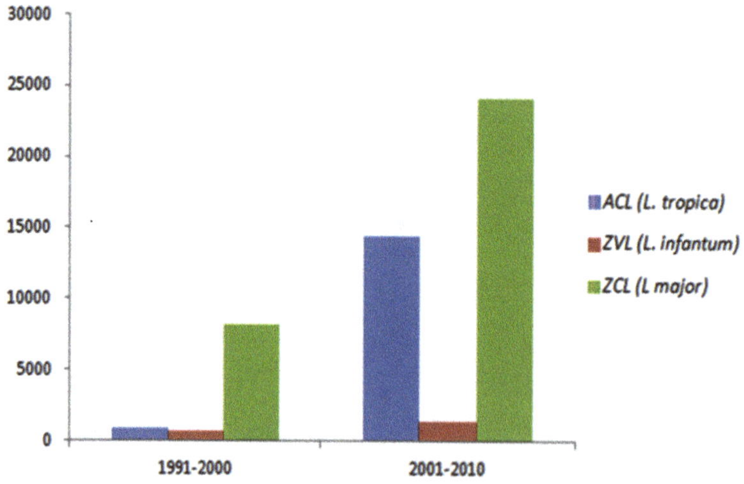

Most of Morocco's large cities have expanded quickly because of the mass emigration from rural areas to the city. These changes raised the possibility for the presence of several unplanned geographic pockets developed as extensions from local villages, where poor facilities, low hygiene, and low sanitation were most common in these populations (Boussaa et al. 2007). The rapid increase in the dynamics of leishmaniases across urban areas is probably the result of the partial urbanization (WHO, 2002; Desjeux, 2004). Recently, we noted the increase of urban ACL foci in Morocco (Boussaa et al. 2007). Immigration and active transportation allowed introduction of infection by a previously known parasite circulated in the country or by new forms of leishmaniasis which are rare or never recorded from across the country, for example, mucocutaneous leishmaniasis was reported for the first time in Taza and Taounate provinces (Iguermia et al., 2011).

CLIMATE CHANGE CONTRIBUTIONS

Climate change impacts on human health span the trajectory of time—past, present, and future (Semenza, 2014). The key finding from the Working Group II, Fifth Assessment Report (AR5) of the Intergovernmental Panel on Climate Change (IPCC) states that health impacts due to climate change have already occurred in the past, are currently occurring and will continue to occur (IPCC, 2015).

Infectious vector-borne diseases, such as leishmaniasis, are affected by environmental changes (e.g.; Semenza, 2009, Yates et al., 2002; Bounoua et al., 2013). Although, at different periods and in different regions, epidemics, disease emergence and reemergence, have increased during the last decade. In fact despite the efforts deployed by the health department and the rigorous control of the disease, all known forms of leishmaniasis still constitute a major threat to public health in Morocco. It is therefore important to better understand the risk factors associated with these trends, and to establish realistic and effective strategies aimed at fighting the disease and manage the man-made environmental changes in order to improve the health of the burdened populations.

For Leishmaniasis, the vector's activity, including its reproduction, constitutes an important transmission factor. This can control the vector's density and therefore the biting rate and number of cases of the disease. All three components of the Leishmaniasis complex - parasite-reservoir-vectors - are sensitive to climate, thus, affect their dynamic interaction and geographic expansions (WHO, 2010; Boudrissa et al., 2012; Bounoua et al., 2013). For example, increases in precipitation and vegetation foliage are favorable habitats to both rodents and sand-fly (Yates et al., 2002, Bounoua et al., 2013). On the other hand, droughts and above-normal temperatures stress vegetation and negatively affect sandflies' reproductive activity (Bounoua et al., 2013).

In the Mediterranean regions, several studies have revealed associations between the incidence of leishmaniasis and bio-climatological factors (Rispail 2002; Toumi et al., 2012). ZCL due to *L. major* seems most clearly related to climate and vegetation (Samy et al., 2014). In Morocco, a recent study relating climate to incidence of CL, postulated that surface climate characteristics, such as temperatures, precipitation and vegetation may influence the dynamic of ZCL in the pre-Saharan regions (Bounoua et al., 2013). They further suggested that the number of ZCL occurrence is related to increase in minimum temperatures which allowed sandfly larvae to survive winters and created conditions suitable for endemicity that did not previously exist (Bounoua et al., 2013). Rapid environmental changes caused by climate changes, are characterized by changes in vegetation cover and density, reduction in surface water, urban growth, as well as conditions that creates new, favorable or unfavorable habitats for survival of leishmaniasis vectors and reservoirs (Kahime et al., in press).

Climate change may also affect the modes of transmission via socio-economic changes; which in turn affect the number of human contact with elements of the transmission cycle (Ready, 2008). In general, the potential transmission rate may be sensitive to temperature change, however the incidence may not be altered if the host population is immune or adapted to infection (Sutherst, 2004).

All these changes make it necessary to identify and study all relationships between environmental changes and leishmaniasis as a vector borne disease and create the need to assess the risk of the socio-economic dimension.

Socio-Economic Issues and Overview of Vulnerability Assessment

Independent of climate change, socio-economic factors will also alter the status of public health. These, in turn, will interact with climate change, leading to different risks under different development pathways. Many projections of the health risks to climate change make limited socio-economic assumptions, sometimes incorporating demographic change and economic growth (Ebi, 2014).

Similar to infectious disease risks, vulnerability of populations varies across spatial and temporal scales as a response to changes in economic development, social capital, demographic population structure (such as the proportion of elderly, children, males and females in a population), urbanization style, trade and travel patterns, prevalence of pre-existing medical conditions and genetic (Steinbruner et al., 2012; Balbus et al., 2009; Suk & Steinbruner, 2011). Lieshmaniasis can be an indicator of vulnerability which is related to poverty and climate change.

According to Steinbruner et al. (2012), a vulnerability indicator must consider who is exposed at present and in the future to a particular hazard induced by changes in the mean climate and its change; the susceptibility of exposed people and communities and their capacity to prevent, prepare for, cope with, and recover from impacts.

Rapid urbanization combined with poverty leads to concentration of people with insufficient infrastructure for safe storage and distribution of clean water and drainage of wastewater. Although, in Morocco foci, in many instances, abandoned containers and used tires provide ideal breeding sites for the disease's vectors. Poverty often contributes to ecological factors that increase risk, such as poor housing conditions (cracked mud walls that provide daytime resting places for sandflies, damp earthen floors that prolong sandfly survival, open areas in walls that enable sandfly entry) and sleeping outside or on the ground which increases exposure (Ranjan et al. 2005; Bern et al. 2005). Generally, adaptation to changes reduces vulnerability and increases resilience to impacts (Smit and Pilifosova, 2001). Low social capital is found as a factor in shaping the vulnerability of population groups who are excluded from access to resources.

In Figure 4, we adapt the Vulnerability Scoping Diagram of Polsky et al. (2007) to the situation in Morocco, with "leishmaniasis" representing the risk and "population" the exposed unit. Through this diagram, we aim to show all the factors of sensitivity and exposure which exacerbate an epidemic situation. According to this adapted diagram, Moroccan leishmaniasis can be considered as a 'degraded' ecosystem providing favorable conditions to disease induction.

From the perspective of risk assessments, we propose taking into account changes in climate, urban growth and urbanization pattern, disease cycle, disease incidence, activities, movements and lifestyle of the population and all interactions for strengthening of citizens' capacity especially, in marginalized

Figure 4. Vulnerability Scoping Diagram, already adapted and completed here with the Risk is "leishmaniasis" and the unit exposed as "population"
(Polsky et al. 2007)

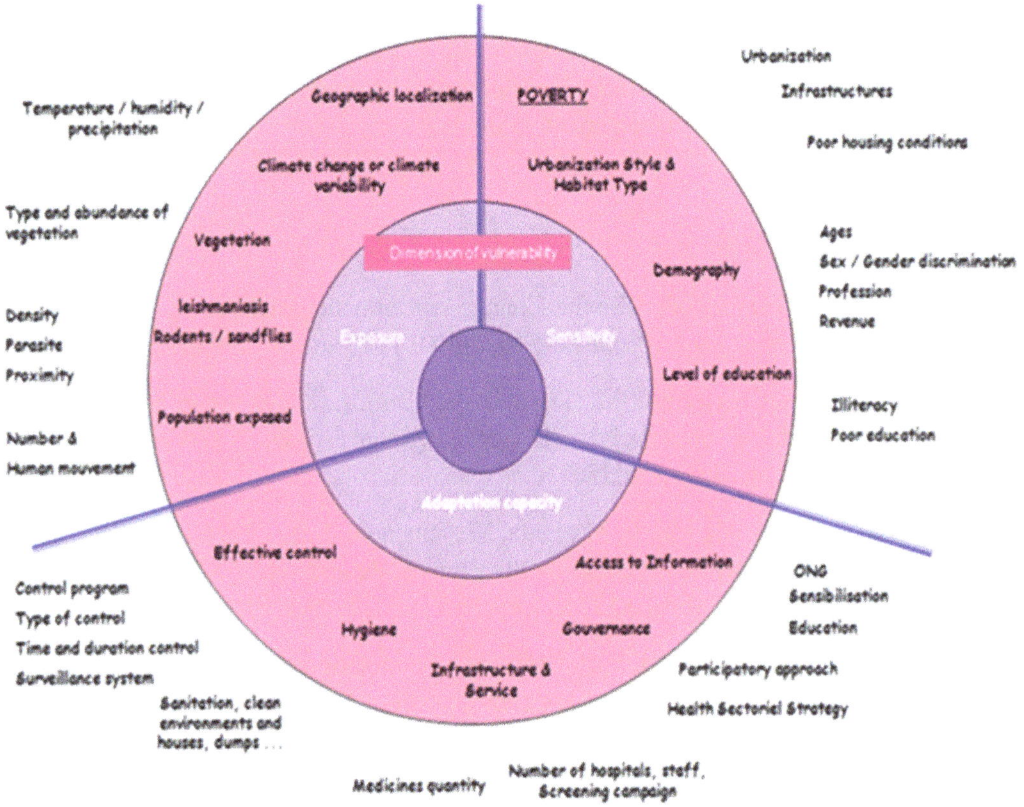

rural and peri-rural areas, in order to reduce their exposure. This will lead to better management of the health risk associated with this vector-borne disease and the ensuing problems across the social ladder. This approach will also fight lack of information via an enhancement of awareness, education, and integration of leishmaniasis prevention measures in the school system of affected regions. Improving the urban infrastructure, by rehabilitation and restructuring of old buildings with suitable drainage and waste disposal systems and an environmental management which consists in removing the habitats of the elements of the disease complex and reducing its incidence (reduction of vector and reservoirs densities).

CONCLUSION

Evidence suggests that annual and decadal climate variability have a direct influence on the epidemiology of vector-borne diseases (Githeko et al., 2000), the exact consequences are still not well comprehended. Warmer and wetter conditions can facilitate or not the multiplication of vectors, and the spread of the disease depends on a wider range of environmental and societal factors that must be identified in order to develop reasonable control strategies.

Climate and social changes may affect leishmaniasis evolution and territorial expansion either by acting on the vector, the parasite, the reservoirs and human host. It is therefore important to improve our knowledge on how change in climate affects the cycle of these vector-borne diseases in order to strengthen health our care systems.

This requires a seamless collaboration between health, education, agriculture, sociology and environment sectors to better observe, characterize, understand and predict risk factors and implement an effective and sustainable integrated management strategy.

REFERENCES

Altezer. (2013). Predictive Framework Climate Change and Infectious Diseases: From Evidence to a Predictive Framework. *Science*, *341*(6145), 514–519. doi:10.1126cience.1239401 PMID:23908230

Alvar, J., Velez, I. D., Bern, C., Herrero, M., Desjeux, P., Cano, J., ... den Boer, M. W. H. O. (2012). Leishmaniasis Control Team. Leishmaniasis worldwide and global estimates of its incidence. *PLoS One*, *7*(5), e35671. doi:10.1371/journal.pone.0035671

Amro, A., Hamdi, S., Lemrani, M., Idrissi, M., Hida, M., Rhajaoui, M., & ... . (2013). Moroccan *Leishmaniainfantum*: Genetic Diversity and Population Structure as Revealed by Multi-Locus Microsatellite Typing. *PLoS One*, *8*(10), e77778. doi:10.1371/journal.pone.0077778 PMID:24147078

Balbus, J. M., & Malina, C. (2009). Identifying vulnerable subpopulations for climate change health effects in the United States. *Journal of Occupational and Environmental Medicine*, *51*(1), 33–37. doi:10.1097/JOM.0b013e318193e12e PMID:19136871

Bennani, A., Buret, J., & Senhaji, F. (2001). *Initial national communication to the UN framework convention on climate change.* Kingdom of Morocco, Ministry of Land Management, Urban Planning, Housing and Environment. Available from: http://unfccc. int/resource/docs/natc/mornc1fres.pdf (in French)

Bern, C., Hightower, A. W., Chowdhury, R., Ali, M., Amann, J., Wagatsuma, Y., ... Maguire, J. H. (2005). Risk factors for kala-azar in Bangladesh. *Emerging Infectious Diseases*, *11*(5), 655–662. doi:10.3201/eid1105.040718 PMID:15890115

Boelaert, M., Meheus, F., Sanchez, A., Singh, S. P., Vanlerberghe, V., Picado, A., ... Sundar, S. (2009). The poorest of the poor: A poverty appraisal of households affected by visceral leishmaniasis in Bihar, India. *Tropical Medicine & International Health*, *14*(6), 639–644. doi:10.1111/j.1365-3156.2009.02279.x PMID:19392741

Born, K., Fink, A. H., & Paeth, H. (2009). *Dry and Wet Periods in the Northwestern Maghreb for Present Day and Future Climate Conditions.* Retrieved from http://www.impetus.unikoeln.de/fileadmin/content/veroeffentlichungen/publikationsliste/1415_Born_full_article.pdf

Boudrissa, A., Cherif, K., Kherrachi, I., Benbetka, S., Bouiba, L., Boubidi, S. C., ... Harrat, Z. (2012). Extension de Leishmania major au nord de l'Algérie. *Bulletin de la Société de Pathologie Exotique*, *105*(1), 30–35. doi:10.100713149-011-0199-4 PMID:22170408

Bounoua, L., Kahime, K., Houti, L., Blakey, T., Ebi, K. L., Zhang, P., ... Boumezzough, A. (2013). Linking climate to incidence of zoonotic cutaneous leishmaniasis (*L. major*) in Pre-Saharan North Africa. *International Journal of Environmental Research and Public Health, 10*(8), 3172–3191. doi:10.3390/ijerph10083172 PMID:23912199

Githeko, Lindsay, Confalonieri, & Patz. (2000). Climate change and vector-borne diseases: A regional analysis. *Bulletin of the World Health Organization, 78*(9). PMID:11019462

Desjeux, P. (2004). – Leishmaniasis: Current situation and new perspectives. *Comparative Immunology, Microbiology and Infectious Diseases, 27*(5), 305–318. doi:10.1016/j.cimid.2004.03.004 PMID:15225981

Esper, J., Franck, D., B¨untgen, U., Verstege, A., Luterbacher, J., & Xoplaki, E. (2007). Long-term drought severity variations in Morocco. *Geophysical Research Letters, 34*(17), L17702. doi:10.1029/2007GL030844

Feliciangeli, M. D., & Rabinovich, J. E. (1998). Abundance of *Lutzomyiaovallesi*but not *Lu. gomezi*(Diptera: Psychodidae) correlated with cutaneous leishmaniasis incidence in north-central Venezuela. *Medical and Veterinary Entomology, 12*(2), 121–131. doi:10.1046/j.1365-2915.1998.00072.x PMID:9622364

Gubler. (1998). Resurgent vector borne diseases as a global health problem. *EmergInfec Dis, 4*, 442-450.

Guernaoui, S., Boumezzough, A., Pesson, B., & Pichon, G. (2005). Entomological investigations in Chichaoua: An emerging epidemic focus of cutaneous leishmaniasis in Morocco. *Journal of Medical Entomology, 42*(4), 697–70. doi:10.1093/jmedent/42.4.697 PMID:16119562

Guessous-Idrissi, N., Chiheb, S., Hamdani, A., Riyad, M., Bichichi, M., Hamdani, S., & Krimech, A. (1997). Cutaneous leishmaniasis: An emerging epidemic focus of *Leishmaniatropica* in north Morocco. *Transactions of the Royal Society of Tropical Medicine and Hygiene, 91*(6), 660–663. doi:10.1016/S0035-9203(97)90511-3 PMID:9509172

Haines, A. (1998). Global warming and vector-borne diseases. *Lancet, 351*(9117), 1737–1738. doi:10.1016/S0140-6736(05)77778-3 PMID:9734919

High Planning Commission of Morocco (HCP). (2014). *Report of General Census of Population and Housing of 2004*. Retrieved from http://www.hcp.ma/Recensement-General-de-la-Population-et-de-l-Habitat-2004_a92.html

Hmamouch, A., Amarir, F., Fellah, H., Karzaz, M., Bekhti, K., & Rhajaoui, M. (2014). Coexistence of *Leishmaniatropica* and *Leishmaniainfantum*in Sefrou province, Morocco. *Acta Tropica, 130*, 94–99. doi:10.1016/j.actatropica.2013.10.012 PMID:24161534

Iguermia, S., Harmouche, T., Mikou, O., Amarti, A., & Mernissi, F. Z. (2011). Mucocutaneousleishmaniasis in Morocco, evidence of the parasite's ecological evolution. *Medecine et Maladies Infectieuses, 41*, 47–48. doi:10.1016/j.medmal.2010.07.004 PMID:20947275

Intergovernmental Panel on Climate Change (IPCC). (2014). IPCC Fifth Assessment Report - *The Synthesis Report*. Author.

Kahime, K., Bounoua, B., Messouli, M., & Boussaa, S. (in press). Evaluation of Eco-Adaptation Strategies of Health to Climate change: Case of Zoonotic Cutaneous Leishmaniasis (ZCL) as Vulnerability Indicator in Pre-Saharan region of Morocco. In *Environmental Change and Human Security in the Middle East and Africa. GECS2012 Conference Proceedings*. Springer.

Kahime, K., Boussaa, S., Bounoua, L., Fouad, O., Messouli, M., & Boumezzough, A. (2014). Leishmaniasis in Morocco: Diseases & vectors. *Asian Pacific Journal of Tropical Disease*, *4*, 530–534. doi:10.1016/S2222-1808(14)60671-X

Kahime, K., Boussaa, S., Idrissi, A. L.-E., Nhammi, H., & Boumezzough, A. (in press). Epidemiological study on acute cutaneous leishmaniasis in Morocco. *Journal of Acute Disease*. doi:10.1016/j.joad.2015.08.004

Kahime, K., Boussaa, S., Ouanaimi, F., & Boumezzough, A. (2015). Species composition of phlebotomine sand fly fauna in an area with sporadic cases of *Leishmania infantum* human visceral leishmaniasis, Morocco. *Acta Tropica*, *148*, 58–65. doi:10.1016/j.actatropica.2015.04.006 PMID:25930188

Kristie, L. (2014). Health in the New Scenarios for Climate Change Research. *International Journal of Environmental Research and Public Health*.

Mahjour, J., Akalay, O., & Saddiki, A. (1992). Les leishmanioses au Maroc de l'analyse écoépidémiologie à la prévention. *DEPS Bul. Epidém.*, *7*.

Marthy, P., LeFichoux, Y., Pratlong, F., Rioux, J. A., Rostain, G., & Lacour, J. P. (1989). Cutaneous leishmaniasis due to *Leishmaniatropica*in a young Moroccan child observed in Nice, France. *Transactions of the Royal Society of Tropical Medicine and Hygiene*, *83*(4), 510. doi:10.1016/0035-9203(89)90268-X PMID:2617601

Moroccan Ministry of Health. (2011). *A report on progress of control programs against parasitic diseases.* Directorate of Epidemiology and Disease Control, Ministry of Health, Rabat, Morocco. Available from: http://www.sante.gov.ma/departements/delm/index-delm.htm

Moroccan Ministry of Health. (2012). *A report on State of Health of the Moroccan population*. Author.

Patz, J. A., McGeehin, M. A., Bernard, S. M., Ebi, K. L., Epstein, P. R., Grambsch, A., ... Trtanj, J. (2000). The potential health impacts of climate variability and change for the United States: Executive summary of the report of the health sector of the U.S. National Assessment. *Environmental Health Perspectives*, *108*(4), 367–376. doi:10.1289/ehp.00108367 PMID:10753097

Pelling, M., & High, C. (2005). Understanding adaptation: What can social capital offer assessments of adaptive capacity? *Global Environmental Change*, *15*(4), 308–319. doi:10.1016/j.gloenvcha.2005.02.001

Polsky, P., Neff, R., & Yarnal, B. (2007). Building comparable global change vulnerability assessments: The vulnerability scoping diagram. *Global Environmental Change*, *17*(3-4), 472–485. doi:10.1016/j.gloenvcha.2007.01.005

Pratlong, F., Rioux, J. A., Dereure, J., Mahjour, J., Gallego, M., Guilvard, E., ... Saddiki, A. (1991). *Leishmania tropica*au Maroc. IV. Diversité isozymiqueintrafocale. *Annales de Parasitologie Humaine et Comparee*, *66*(3), 100–104. doi:10.1051/parasite/1991663100 PMID:1776781

Ramaoui, K., Guernaoui, S., & Boumezzough, A. (2008). Entomological and epidemiological study of a new focus of cutaneous leishmaniasis in Morocco. *Parasitology Research*, *103*(4), 859–863. doi:10.100700436-008-1068-3 PMID:18592272

Ranjan, A. (2005). Risk factors for Indian kala-azar. *The American Journal of Tropical Medicine and Hygiene*, *73*(1), 74–78. doi:10.4269/ajtmh.2005.73.74 PMID:16014837

Ready, P.D. (2008). Leishmaniasis emergence and climate change. *Rev Sci Tech Off intEpiz, 27*, 399-412.

Reiter, P. (2001). Climate change and mosquito-borne disease. *Environmental Health Perspectives*, *109*, 141–161. PMID:11250812

Rhajaoui, M., Abedelmajeed, N., Fellah, H., Azmi, K., Amrir, F., Al-jawabreh, A., & ... . (2007). Molecular typing reveals emergence of a new clinic-epidemiologic profile of cutaneous leishmanisis in Morocco. *Emerging Infectious Diseases*, *3*(9), 1358–1360. doi:10.3201/eid1309.060946 PMID:18252108

Rhajaoui, M., Fellah, H., Pratlong, F., Dedet, J. P., & Lyagoubi, M. (2004). Leishmaniasis due to *Leishmaniatropica*MON-102 in a new Moroccan focus. *Transactions of the Royal Society of Tropical Medicine and Hygiene*, *98*(5), 299–301. doi:10.1016/S0035-9203(03)00071-3 PMID:15109554

Rioux, J. A. (2001). Trente ans de coopération franco–marocaine sur les leishmanioses: Dépistage et analyse des foyers. Facteurs de risque. Changements climatiques et dynamique nosogéographique. *Association des Anciens Elèves de l'Institut Pasteur*, *168*, 90–101.

Rioux, J. A., Akalay, O., Périères, J., Dereure, J., & Mahjour, J. (1997). Evolution of Eco-epidemiological of 'leishmaniasis risk' at Atlantic Moroccan Sahara. Heuristic interest of the relationship 'sandflies bioclimates'. *Ecologia Mediterranea*, *23*, 73–92.

Rioux, J. A., & de La Rocque, S. (2003). Climats, leishmanioses et trypanosomiases. Changements climatiques, maladies infectieuses et allergiques. *Annales de l'Institut Pasteur*, *16*, 41–62.

Rioux, J. A., Guilvard, E., Dereure, J., Lanotte, G., Denial, M., & Pratlong, F. (1986) Infestation naturelle de Phlebotomuspapatasi(Scopoli, 1786) par Leishmania major MON-25. A propos de 28 souches isolées dans un foyer du Sud Marocain. In:Leishmania. Taxinomie et Phylogenèse. Applications écoépidémiologiques. (Coll. Int. CNRS/INSERM, 1984). IMEEE, 471–480.

Rioux, J. A., Mahjour, J., Gallego, M., Dereure, J., Périères, J., Laamrani, A., & ... . (1996). Leishmaniose cutanée humaine à *Leishmania infantum*MON-24 au Maroc. *Bulletin de la Société Française de Parasitologie*, *14*, 179–183.

Rioux, J. A., Rispail, P., Lanotte, G., & Lepart, J. (1984). Relations Phlébotomes-bioclimats en écologie des leishmanioses Corollaires épidémiologiques. L'exemple du Maroc. *Bull Soc Bot Fr, 131*, 549–557.

Rispail, P., Dereure, J., & Jarry, D. (2002). Risk zones of human leishmaniases in the western Mediterranean basin. Correlations between vectors sand flies, bioclimatology and phytosociology. *MemInstOswaldo Cruz Rio de Janeiro, 97*, 477-483.

Royaume du Maroc. (2009). Ministère de l'Aménagement du territoire de l'habitat et de l'environnement. Second National Communication à la Convention Cadre des Nations Unis sur les Changements Climatiques.

Sang, D. K., Njeru, W. K., & Ashford, R. W. (1992). A possible animal reservoir for Leishmania 331 tropica s.l. in Kenya. *Annals of Tropical Medicine and Parasitology, 86*(3), 311–312. doi:10.1080/00034983.1992.11812670 PMID:1449281

Semenza, J. C., & Menne, B. (2009). Climate change and infectious diseases in Europe. *The Lancet Infectious Diseases, 9*(6), 365–375. doi:10.1016/S1473-3099(09)70104-5 PMID:19467476

Shehata, M. G., Samy, A. M., Doha, S. A., Fahmy, A. R., Kaldas, R. M., & … . (2009). First report of 335 Leishmania tropica from a classical focus of L. major in North-Sinai, Egypt. Am J Trop 336. *Medecine et Hygiene, 81*, 213–218. PMID:19635872

Smit, B., & Pilifosova, O. (2001). Adaptation to climate change in the context of sustainable development and equity. In Climate Change 2001: Impacts, Adaptation and Vulnerability. Cambridge University Press.

Steinbruner, J. D., Stern, P. C., & Husbands, J. L. (2012). Climate and Social Stress: Implications for Security Analysis. National Research Council.

Suk, J. E., & Semenza, J. C. (2011). Future infectious disease threats to Europe. *American Journal of Public Health, 101*(11), 2068–2079. doi:10.2105/AJPH.2011.300181 PMID:21940915

Sutherst, R. B. (2004). Global Change and Human Vulnerability to Vector-Borne Diseases. *Clinical Microbiology Reviews*, 136–173.

Sutherst, R. W. (1998). Implications of global change and climate variability for vector-borne diseases: Generic approaches to impact assessments. *International Journal for Parasitology, 28*(6), 935–945. doi:10.1016/S0020-7519(98)00056-3 PMID:9673872

Tamimy, H. (2011). *The infantile visceral leishmaniasis (About 73 cases)*. PhD in Medicine. Sidi Mohammed Ben Abdellah University, Faculty of Medicine and Pharmacy. Available from: http://scolarite.fmp-usmba.ac.ma/cdim/mediatheque/e_theses/89-11.pdf

The United Nations Framework Convention on Climate Change Portal. (n.d.). Retrieved from http://unfccc.int

Toumi, A., Chlif, S., Bettaieb, J., Alaya, N. B., Boukthir, A., Ahmadi, Z. E., & Salah, A. B. (2012). Temporal dynamics and impact of climate factors on the cases of zoonotic cutaneous Leishmaniasis in central Tunisia. *PLoS Neglected Tropical Diseases, 6*(5), e1633. doi:10.1371/journal.pntd.0001633 PMID:22563513

World Health Organization (WHO). (2007). *Report of the Sixtieth Worldwide Assembly on Health*. Geneva, Switzerland: WHO.

World Health Organization (WHO). (2015). *Media centre, Malaria Fact sheet N°94*. Retrieved from http://www.who.int/mediacentre/factsheets/fs094/en/

World Health Organization (WHO). (2010). *Control of the Leishmaniasis: Report of a Meeting of the WHO Expert Committee on the Control of Leishmaniases*. World Health Organization Technical Report Series No. 949. WHO.

Yates, T. L., Mills, J. N., Parmenter, C. A., Ksiazek, T. G., Parmenter, R. R., Vande Castle, J. R., ... Peters, C. J. (2002). The ecology and evolutionary history of an emergent disease: Hantavirus pulmonary syndrome. *Bioscience*, *52*(11), 989–998. doi:10.1641/0006-3568(2002)052[0989:TEAEHO]2.0.CO;2

Chapter 19
Climate Impact on Seasonal Distribution of Zoonotic Cutaneous Leishmaniasis in Southern Morocco

Ahmed Karmaoui
https://orcid.org/0000-0003-3881-4029
Southern Center for Culture and Sciences, Morocco

Siham Zerouali
Southern Center for Culture and Sciences, Morocco

Kholoud Kahime
Cadi Ayyad University, Morocco

ABSTRACT

As in most arid and pre-Saharan North Africa regions, the local population is especially rural, socio-economically vulnerable, and exposed to vector-borne disease like cutaneous leishmaniasis. Information on geographical and seasonal distribution of this disease can be helpful in the control of this disease. In this context, four sites were used to identify the seasonality of zoonotic cutaneous leishmaniasis (ZCL) in Middle Draa Valley (MDV). The seasonal occurrence of this disease was correlated with some climatic and hydrologic variables. The findings show that the most part of patients (86%) were from rural areas with a bit high rate of affected people is female gender (54%) and children less than 9 years (51%). The distribution of cases shows a prevalence of affected in the upstream area than in downstream where the climate is more and more arid. The highest number of cases were detected in the winter season (from December to March).

DOI: 10.4018/978-1-5225-7775-1.ch019

INTRODUCTION

There is a wide bibliographic data about leishmaniasis infection through the world, mainly in Mediterranean region and central Asia. Favorable bio-physical conditions (climate and environment) affect parasite-vector-reservoir increasing the transmission risk of leishmaniasis (Ready, 2008; Cardenas *et al.*, 2006). In the pre-Sahara of Morocco, Zoonotic cutaneous leishmaniasis (ZCL) is the dominant type of leishmaniasis. It is caused by *L. major* and transmitted from the rodent *Meriones shawi* via the vector *Phlebotomus papatasi*. The parasite is therefore transmitted to humans. In this region, the lesions appear generally in autumn, if left untreated, it leaves persistent scars.

Many researches explored the impacts of climate variables such as rainfall, temperature, and humidity on zoonotic ZCL infection (Toumi *et al.,* 2012; Bounoua *et al.,* 2013). According to Shirzadi *et al.,* 2015, the relative humidity variable is the most significant factor; and the infection increases in area with higher temperature, and lower relative humidity. However, few studies have dealt with the subject of leishmaniasis on the scale of Morocco and the pre-Sahara (Bounoua *et al.,* 2013; Karmaoui, 2018), and no studies have been carried out especially on the seasonality of infection of the ZCL in the Middle Draa Valley (MDV) from upstream to downstream at sub-catchment scale. In the Moroccan pre-Sahara, due to the impact of drought and flood, the government-built dams in upstream basins (Karmaoui *et al.,* 2015a), for example, Hassan Edakhil in Tafilalet and Mansour Eddahbi in Draa basin. These dams change the watercourses from the permanent rhythm to the seasonal rhythm, which leads to favorable climatic and biological conditions for the growth and proliferation of vertebrates as reservoirs (hosts supporting the leishmaniasis transmission) and also the insect vectors. While the reservoir *Psammomys obesus* is frequent in the arid region (Mili *et al.,* 2012) its food preference *Salicornia* and *Atriplex* lives in sandy and salty soil (Fichet-Calvet *et al.,* 2003).

Observation along the oases shows the densification of rodent galleries in the outskirts of palm groves along Wadis or vegetation and abundant especially halophilous plants. The changes in vegetation and agricultural production are linked to the water availability, which impacts the cycle of the insect vectors. The drought impacts the water quality and quantity and then influences the vector borne disease. In addition to parasite-vector-reservoir presence, the incidence of cutaneous leishmaniasis depends on several factors like climate, altitude, urbanization, socio-economical conditions. The leishmaniasis vectors are sensitive to wind, daylight, and winter (temperate regions) (Abonnenc, 1972).

In this paper, the study was conducted in the middle part of the Basin of the Draa, the Middle Draa Valley (MDV). The MDV has known to be among the most endemic areas for ZCL for more than a decade. It is an oasean zone, in which the human population is located largely in the province of Zagora, one of the poorest regions of Morocco. In this context, this article sets out the following objectives:

- Determine the distribution and physical and anthropogenic characteristics of this disease in four sites from the MDV;
- Explore the seasonal transmission rate of this disease in the four sites and correlate it with the main climatic parameters;
- Examine the transmission following the upstream-downstream situation of the sub-basin.

This paper used real seasonal data of ZCL cases, climatic, hydrologic, and biologic variables in 2005 and also from upstream to downstream sites. Pearson correlation coefficient was used to depict the association between ZCL incidence and the risk factors like air temperature, relative humidity, rainfall, wind direction, and speed.

MATERIAL AND METHODS

Study Area

This paper involved four sites from upstream to downstream, Agdez, Tamezmoute, Zagora, and Tagounite (Figure 1). These sites are located in the Middle Draa Valley (South East of Morocco), currently situated in the Draa-Tafilalet region. The average altitude of this valley varies between 1800m in the upstream Saghro zone and ~ 400m in the downstream Iriki (Iriqui) National Park. Topographically, the MDV consists of three levels, low-lying mountains, plateaus where the six palm groves are located, along Oued Draa (Figure 1), and a plateau dominated by the Saharan erg-reg system. The palm groves from upstream to downstream are, Mezguita in the center of which is the urban center of Agdez, the palm grove of Tinzouline (Tamezmoute), then in the middle of the palm plantations of Ternata and Fezouata whose center is occupied by The urban center of Zagora, the largest city of this province, downstream there is the palm grove of Ktaoua where is the rural center of Tagounite. For the hydrographic network, this valley is crossed by Oued Draa which is fed upstream by the Mansour Eddahbi dam (Figure 1).

The climate is arid and the vegetation is concentrated mainly in the 6 palm groves, from upstream to downstream, Mezguita, Tinzouline, Ternata, Fezouata, Ktaoua and M'Hamid (Karmaoui *et al.*, 2014)

Figure 1. Localization of the Middle Draa Valley in Morocco including the four sites with leishmaniasis data, Agdez (a), Tamezmoute (b) Zagora (c), and Tagonite (d)
(Karmaoui & Zerouali, 2018)

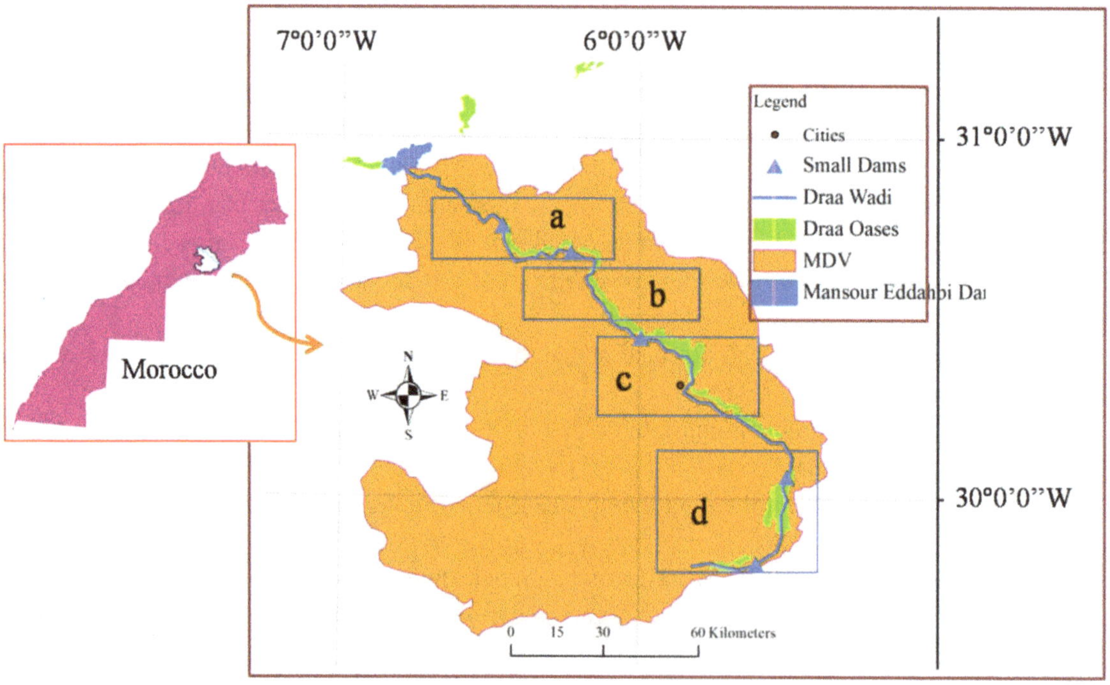

Methodology

Data on ZCL cases in Morocco in 2005 were studied, age, seasonality, period with high incidence of cutaneous leishmaniasis in general at Moroccan scale and for ZCL at MDV scale (Zagora Province).

The use of 2005 is justified by the fact that available seasonal and upstream-downstream data in the MDV is complete for this year. The year of 2005 was the beginning (starting point) of a growing trend in the number of cases registered in the MDV. The major limitation of this study is that it is based on a single year, it may not reflect reality. For this reason, the ZCL occurrence of 2005 trend was compared with the available data of 2017 (climatic and ZCL cases). The methodology is based on three steps:

In the first place, quantitative and qualitative data on the transmission of ZCL cases recorded in the four sites of the Draa Valley (Agdez, Zagora, Tamezmoute, and Tagonite) were used. In this study, data on the gender, the age of leishmaniasis patients were considered for a purpose of comparison between the four sites. Physical characteristics, including geographic parameters such as altitude, latitude, longitude, and biotope were also compared. All of these data were correlated with the air temperature, relative humidity, rainfall, wind direction and speed data for the 2005 transmission period. The data on population was provided by the archives of the national census of the population of 2004.

Secondly, the seasonal transmission was discussed using monthly climatic data. The collected and compiled data were correlated with the change in the number of ZCL cases recorded at the four sites during the 12 months of the year.

Finally, the upstream-downstream infection aspect of the hydraulic basin was studied. In this stage, data on agricultural production, the number of cases of leishmaniasis, downstream upstream, and urbanization were approached from upstream to downstream.

The Distribution and Characteristics of the Infection

Table 1 depicts the physical, anthropogenic, leishmaniasis patients and climatic characteristic of the study area. An overall number of 155 cases of ZCL were recorded during 2005 (Table 1). About 86% of the patients were installed in rural areas and 21% from the urban area and the females are the most affected by the disease (54%). This is in accordance with Mili *et al.*, (2012), which reported that the ZCL disperses to cover a large rural area in Tunisia with a sex ratio F/M = 1.36. Regarding the age factor, the 00-09years is the most threatened followed by 10-19, 20-40, and 41>. According to Kelly-Hope & Thomson (2008), the most vulnerable age to the vector-borne diseases is the age of children, and the most vulnerable people are the nomad sand poor rural people. The distribution of cases shows a high number of affected in the upstream area than in downstream where the climate is more and more arid. The situation could be explained follows: the upstream area is more populated and more rural it means that more cases are in the upstream (not because of upstream, but because of more people and more villagers which are in contact with rodents and disease vector).

The yearly rainfall is generally lower and disperses from the north to south (up to downstream). The High Atlas Mountains in the northern part of Draa basin intercept some precipitation. The contrast in precipitation from upstream to downstream was reported by Middleton and Thomas (1997) in this region and by Karmaoui *et al.* (2015a). Additionally, Born et al., (2008) found the same contrast also in temperature for present and future projection following the topography.

Table 1. Physical, anthropogenic, number of houses, large families, leishmaniasis infection and climatic characteristic of the Middle Draa Valley (MDV), and the ZCL cases

Characteristics	Unit/Type	Agdez	Tamezmoute	Zagora	Tagonite	Total
Centers	Type	Urban & rural	Rural	Urban and rural	Rural	MDV
Latitude	N	30°42'16	30°39'18"	30°20'26"	29°58'52"	
Longitude	W	6°27'10	6°08'36"	5°50'26"	5°34'46"	
Altitude	m	939	855	733	609	
Land area	Km^2	2456,3	1132,31	2965,65	7017,3	13571,56
Population	Nb	35465	23566	82728	36474	178233
Popul. growth rate	%	3,1	1,4	2,9	0,5	-
Houses	Nb	4544	2891	10050	4519	22004
Size of the family	Nb	7,8	8,15	8,23	8	-
Total Cases	Nb	90	18	42	5	155
Age of patients (years)	0-09	51	6	20	1	78
	10-19	12	6	15	2	35
	20-40	19	5	6	2	32
	41>	6	1	1	0	8
Urbanization	Urban	11	0	10	0	21 (14%)
	Rural	78	18	32	5	133 (86%)
Gender of patients	Male (M)	44 (48%)	4 (22%)	19 (45%)	4 (80%)	73
	Female (F)	46 (52%)	14 (78%)	23 (55%)	1 (20%)	84
Yearly rainfall 2005-2006	(mm)	49.5	73	58	35	215

Seasonal Infection and Climate

When gathering the relative humidity, air temperature, number of cases of Zoonotic cutaneous leishmaniasis, and the wind speed in 2005 in the whole Middle Draa Valley, we obtained the change from January to December of 2005 (Figure 2). The most cases were recorded (detected) in the winter season (wet season). The incidence peaks exactly from December to March. In the same period, the seasonal distribution pattern of zoonotic cutaneous leishmaniasis cases registered in the Balkh Province (Afghanistan), September to December (Faulde *et al.*, 2006); (2008) from October to January in Central Tunisia (Chelbi *et al.*, 2007). Boudrissa, (2005) find in El-Hodna (Wilaya de M'Sila, Algeria) that the peak incidence starts from October to January during the period 1995-2000.

After Figure 2, a positive association between relative humidity and ZCL cases is apparent, and a negative association between both air temperature and wind speed with ZCL cases.

The use of Pearson correlation coefficient (at a=0.05) revealed a positive strong correlation (Table 2) between relative humidity and ZCL cases (r=0.735). This result is in contrast with the result of Bellali *et al.*, (2017) which found no association between the ZCL and the humidity in central Tunisia. However, Toumi et al., (2012), found that ZCL occurrence increase with the increase in humidity during the sum-

Figure 2. rh_2, relative humidity at 2m above ground (%); at_2, air temperature at 2m above ground (°C); ZCL Cases, number of cases of Zoonotic cutaneous leishmaniasis represented by the left axis; and wv_2, wind speed at 2m above ground (m/s) represented by the right axis, in 2005 in the whole Middle Draa Valley (MDV).
Data from IMPETUS project and Moroccan health ministry archives,

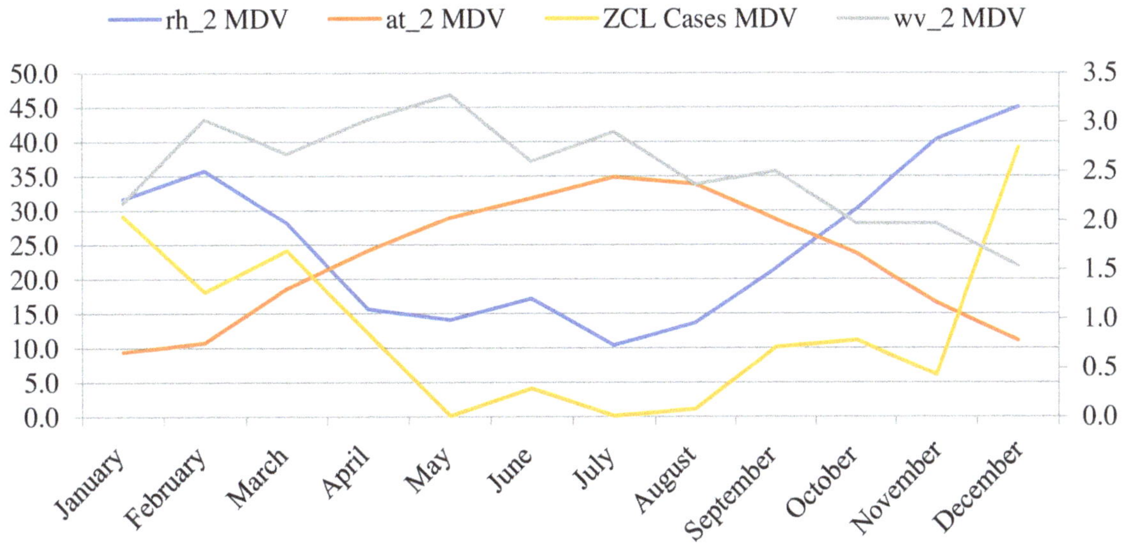

Table 2. Correlation between relative humidity, air temperature, wind speed, and the number of cases of Zoonotic cutaneous leishmaniasis in 2005 in the whole Middle Draa Valley

	Relative Humidity	Air Temperature	Wind Speed	Precipitation
ZCL cases	0.735	-0.831	-0.527	0.128
Calculated	3.428	-4.724	-1.961	0.408
From table	2.228	-1.812	-1.812	2.228
Number	12	12	12	12
Liberty degree	10	10	10	10
Correlation	Strong	Very Strong	Medium	Very Low

mer. Shirzadi *et al.*, 2015; Roger *et al.*, (2013); and Sing (1999) recorded that the cutaneous leishmaniasis declined with decreased relative humidity.

A negative very strong correlation between air temperature (r=-0.831) and ZCL was recorded, and a negative middle correlation between wind speed (r=-0.572) and ZCL cases (Table 3). However, Bellali *et al.*, (2017) did not record a relationship between this disease and the wind speed. The middle or no association of wind speed and the ZCL cases can be explained by the moderate sensitivity of the disease vector toward the wind because the sand fly behavior depends mainly on microhabitats, and thus it is really very difficult to compare wind and prevalence of sand flies.

The coefficient shows a very low correlation between the number of ZCL cases in the MDV with precipitation (mm) at 1.5m above ground (r=0.735). These findings are not consistent with Shirzadi *et al.*, 2015; Roger et al. (2013); and Sing (1999) which recorded that the cutaneous leishmaniasis declined with decreased rainfall. According to Bellali *et al.*, (2017), in central Tunisia, high rainfall would occasion floods that decreasing the chenopods abundance and then the host (rodents). This is true also in Iran, the rainfall (seasonal) affect the potential repartition of ZCL reservoirs (Gholamrezaei *et al.*, 2016).

Avoiding the unrepresentativeness of the data in a single-year, the data of 2005 were compared with the data of climatic and the ZCL cases of 2017. The same trend was recorded between data of 2005 and 2017 as showed in Figure 3.

Regarding the correlation between the ZCL cases and the wind direction, Figure 4 depicts the values of the Wind direction degree (of 360°) at 3m above ground in Bou-Skour (upstream zone of the Middle Draa Valley) and the Normalized values of ZCL cases.

The Pearson correlation (Table 3) between the normalized number of ZCL and the direction is strong (r=0.653, with a=0.05).

Figure 3. a, the seasonal trend of the ZCL cases both in 2005 and 2017. b, rh (2017), relative humidity (%); at (2017), air temperature (°C); ZCL Cases (2017), number of cases of Zoonotic cutaneous leishmaniasis represented by the left axis; and wv (2017), wind speed (m/s) represented by the right axis, in the whole Middle Draa Valley (MDV).

Data from the Moroccan health ministry archives, and Ouarzazate Watershed Agency and Regional Office of Agricultural Development of Ouarzazate

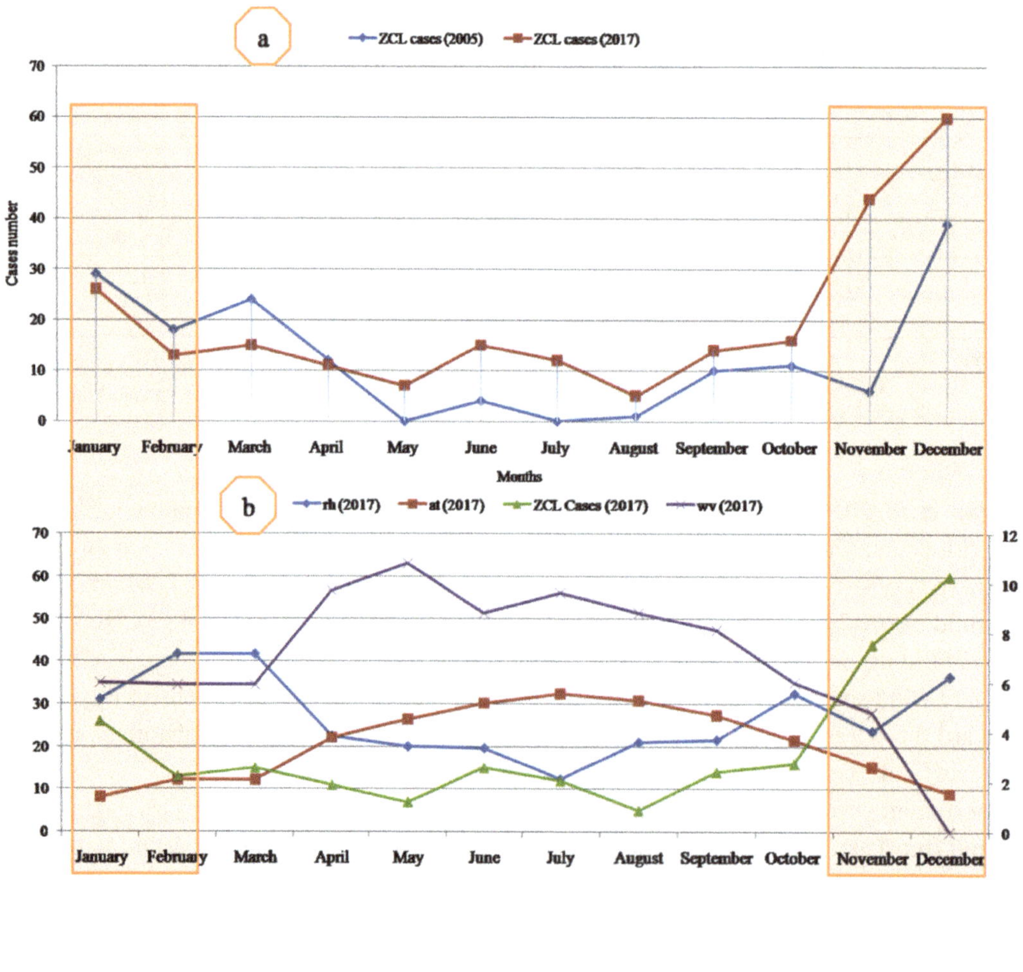

Figure 4. The normalized values of the ZCL cases (left axis) and the wind direction (wd) expressed by degrees (right axis) in Middle Draa Valley

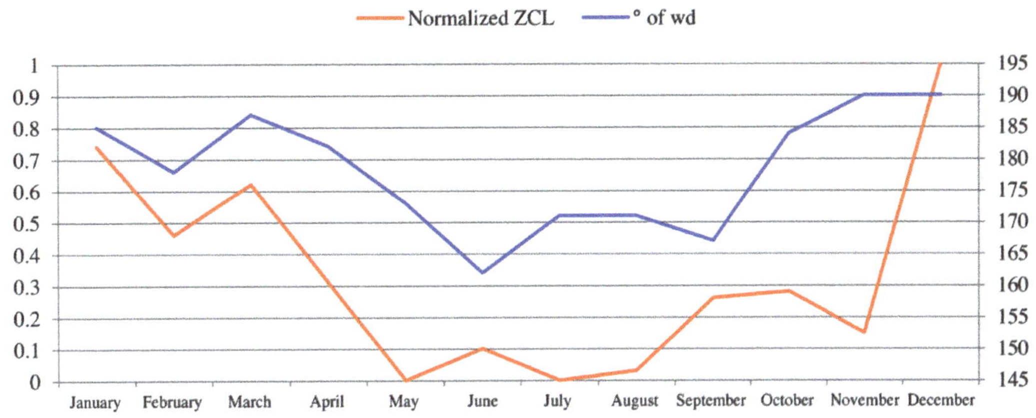

The Table 3 shows that the high and very high number of ZCL is seen mainly when the direction of the wind in the upstream zone is south-south-west (from high Atlas to Sahara direction). However, Bellali *et al.*, (2017) did not identify a significant relationship between this disease and wind direction.

Aspect Infection Upstream-Downstream of the Hydraulic Basin

For the ZCL cases, Agdez recorded the high number of ZCL cases and Ktaoua the low number of cases. A decreasing trend has been identified also from upstream (Mezguita) to downstream (Ktaoua) regard-

Table 3. Wind direction (wd) ° of 360° at 3m above ground in Bou-Skour (upstream zone of the Middle Draa Valley). Normalized values of ZCL cases, where, 0.8-1 indicates very high number; 0.8-0.6, high; 0.6-0.4, average; 0.4-0.2, low, and 0.2-0, very low number, where the normalized number of ZCL and the direction are strongly associated (r=0.653, with a=0.05).

Months	° of wd	Direction	Normalized ZCL	Number of ZCL
January	185	SSW	0,74	high
February	178	SSE	0,46	average
March	187	SSW	0,62	high
April	182	SSW	0,31	low
May	173	SSE	0,00	very low
June	162	SSE	0,10	very low
July	171	SSE	0,00	very low
August	171	SSE	0,03	very low
September	167	SSE	0,26	low
October	184	SSW	0,28	low
November	190	SSW	0,15	very low
December	190	SSW	1,00	very high

ing the percentage of crop area, the water pumping for irrigation, the outflow repartition and the runoff (Figure 5).

The surface water (Draa Wadi) supplies the underground aquifers of downstream palm groves. This oasean region depends on groundwater. The negative trend of vegetation cover from upstream to downstream (Figure 5) is evidently due to the gradual decrease of these resources (water surface and groundwater). These findings are consistent with Hnichi (1989) that recorded that the average annual rainfall differs from 150mm in upstream and 54mm in downstream of this region. Additionally, the construction of Mansour Eddahbi Dam in upstream decreased the water quantity in downstream of this valley and prevented the river to reach the ocean (Busche 2008). This decrease in surface water caused a parallel decrease in the level of the groundwater as illustrated in Figure 5. After the installation of this dam, the downstream is becoming dry. Facing this change (Dam construction) the ecosystem is transforming which modifies the species in this place (Souchon & Nicolas 2011).

In this context, the state built dams to regulate flow and also for water storage for periods of drought to maintain agriculture in the change in flow or releases for irrigation causes a change in temperatures and local humidity, which influences vegetation and crop and livestock production. In this dry year, the water stored was decreased which affected the amount of water directed to the irrigation mainly in the downstream.

From upstream to downstream, the correlation between the normalized values of rainfall (r=0.742) and the runoff (r=0.697) with the normalized number of ZCL show a strong correlation. A medium correlation was recorded between the ZCL cases and the dam outflow for irrigation (r=0. 484) and the cropping area (r=0.534). These results are consistent with the work of Bounoua et al., (2013) recording that a rise in rainfall and vegetation density would sustain local rodents and insect vector activity which increase the ZCL incidence. In addition, gradually as cropping area increase as the time spent on the farm increased. However, a low correlation was seen between the number of ZCL and the groundwater pumped for irrigation (r=0.279) following the direction from upstream to downstream (Table 4).

In addition to biophysical changes from upstream to downstream, the socio-economical change consequently was observed. After the general census of population (2004 and 2014), the growth rate depicts a negative rate in downstream. In downstream, a very low activity (20.5%) was recorded and

Figure 5. Trend of the used variables in 2005 from upstream (Mezguita or Agdez) to downstream (Ktaoua or Tagounite)

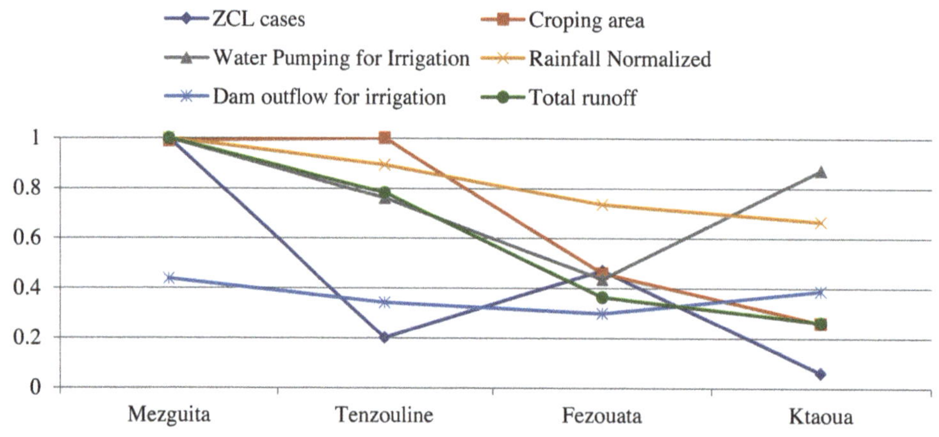

Table 4. Pearson correlation coefficient at a=0.05 (N=4) from upstream to downstream of the Middle Draa Valley in 2005. Data of the percentage of area cultivated in 2005 (survey from IMPETUS project archives)

	Rainfall	Runoff	Groundwater Pumped	Dam Outflow	Cropping Area
ZCL cases	0.742	0.697	0.279	0.484	0.534
Correlation	Strong	Strong	low	Medium	Medium

38.3% of rural households accesses drinking and water sewerage system is absent (HCEFLCD 2007). In response to this situation, the local population migrates towards big cities in the north and the west (Karmaoui *et al.,* 2015a).

In the downstream area (Ktaoua and M'Hamid), the drought is more frequent (Karmaoui *et al.,* 2015a) and the population growth rate is negative (MZ, 2016). This forces the local population to migrate toward the most upstream, which increase the exposure to the sand fly vectors.

The last years of drought have resulted in the loss of a number of bred animals, due to lack of forage. The decline is quite pronounced because of insufficient rainfall and, consequently, the low availability of forage in the region (Schulz & Judex, 2008). Droughts and demographic pressure in the area are impacting this resource. Groundwater has been declining in the past 30 years, for the six palm groves of the MDV, mainly in downstream (Tagounite and M'Hamid). The rapid and continuous decrease in the groundwater resources level of the M'Hamid aquifer (Downstream of the study area) is due to the growth of water exploitation and the degradation of the favorable conditions of the recharge of the aquifers. Deterioration of the vegetation cover limits the possibility of infiltration and consequently, the supply of the water table is reduced (Zainabi, 2003).

Figure 6 gives a spatial overview of the different elements along a downstream upstream direction. The watercourse combined with a decreasing slope towards the south causes important modifications based on the availability of the water resource. This availability of water and climatic and social conditions can cause water-related infections such as leishmaniasis. Following this decreasing trend of water (surface and underground) and vegetation density, a decrease in ZCL cases have been identified. This leads us to say that a change in climate and land use /land cover can interact to cause an incidence of this disease.

CONCLUSION

The most part of the MDV population is rural (Karmaoui *et al.,* 2015b & 2015d). In this area, risk factors related to ZCL infection are several. Additionally, the majority uses the traditional medicine. In summer, almost all rural population sleep outdoor and do not use a bed net. Data included the socioeconomic characteristics (age, gender, rural population, and family size). In water basins like Draa basin, the state constructs dams to regulate flows and to store water for use in periods of drought in order to maintain agriculture. Change in flow or releases for irrigation causes a change in temperature and local humidity. This influences the density of the vegetation and thus the plant and animal production. The change will undoubtedly affect the density of natural reservoirs and vectors (insect).

The results of this paper depict that the major part of affected people in rural areas with the major rate of affected are females. Regarding the age, the 00-09 is the most threatened. The distribution of cases

Figure 6. The contrast observed from the upstream (Agdez or Mezguita) concentrating water, energy, and biomass to downstream (Tagounite or Ktaoua), showing also climatic and leishmaniasis trend

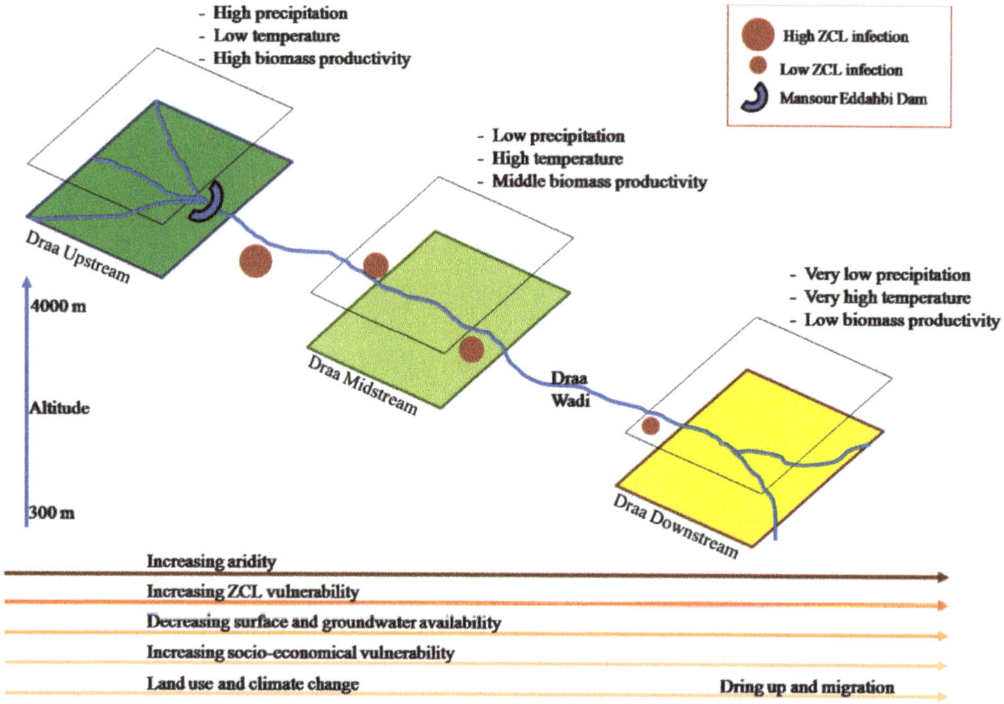

shows a high number of affected in the upstream area than in downstream where the climate is more and more arid. The most cases were detected in the winter season. Statistically, a positive strong correlation between ZCL cases and relative humidity, a negative very strong correlation with air temperature, and a negative middle correlation wind speed, a strong association with the wind direction. From upstream to downstream, the correlation between the normalized values of rainfall and the runoff with the normalized number of ZCL show a strong correlation. A medium correlation was recorded between the ZCL cases and the dam outflow for irrigation and the cropping area.

Large numbers of leishmaniasis patients were found speared in the MDV. The local population (mainly in the rural area) can use different methods to prevent sand flies like the use of bed net, clean the environment, use insecticide and rodenticide with precaution. The identified correlations between climatic, biologic and hydrologic risk factors offer an important knowledge to develop local and regional plans to control the spread of the disease.

REFERENCES

Abedi-Astaneh, F., Akhavan, A. A., Shirzadi, M. R., Rassi, Y., Yaghoobi-Ershadi, M. R., Hanafi-Bojd, A. A., ... Raufi, H. (2015). Species diversity of sand flies and ecological niche model of Phlebotomus papatasi in central Iran. *Acta Tropica*, *149*, 246–253. doi:10.1016/j.actatropica.2015.05.030 PMID:26071647

Akhavan, A. A., Yaghoobi-Ershadi, M. R., Mehdipour, D., Abdoli, H., Farzinnia, B., Mohebali, M., & Hajjaran, H. (2003). Epidemic outbreak of cutaneous leishmaniasis due to Leishmania major in Ghanavat rural district, Qom Province, Central Iran. *Iranian Journal of Public Health*, *32*(4), 35–41.

Bellali, H., Talmoudi, K., Alaya, N. B., Mahfoudhi, M., Ennigrou, S., & Chahed, M. K. (2017). Effect of temperature, rainfall and relative density of rodent reservoir hosts on zoonotic cutaneous leishmaniasis incidence in Central Tunisia. *Asian Pacific Journal of Tropical Disease*, *7*(2), 88–96. doi:10.12980/apjtd.7.2017D6-330

Boudrissa, A. (2005). *Étude éco-épidémiologique de la leishmaniose cutanée dans la wilaya de M'Sila* (thèse de Magister). Algérie.

Bounoua, L., Kahime, K., Houti, L., Blakey, T., Ebi, K., Zhang, P., ... Boumezzough, A. (2013). Linking Climate To Incidence Of Zoonotic Cutaneous Leishmaniasis (L. Major) In Pre-Saharan North Africa. *International Journal of Environmental Research and Public Health*, *10*(8), 3172–3191. doi:10.3390/ijerph10083172 PMID:23912199

Boussaa, S., Guernaoui, S., Pesson, B., & Boumezzough, A. (2005). Seasonal fluctuations of phlebotomine sand fly populations (Diptera: Psychodidae) in the urban area of Marrakech, Morocco. *Acta Tropica*, *95*(2), 86–91. doi:10.1016/j.actatropica.2005.05.002 PMID:15985259

Busche, H. (2008). Hydrology of the Drâa Basin. In IMPETUS Atlas Morocco. University of Bonn.

Cardenas, R., Sandoval, C. M., Rodríguez-Morales, A. J., & Franco-Paredes, C. (2006). Impact of climate variability in the occurrence of leishmaniasis in northeastern Colombia. *The American Journal of Tropical Medicine and Hygiene*, *75*(2), 273–277. doi:10.4269/ajtmh.2006.75.273 PMID:16896132

Chelbi, I., Derbali, M., Al-Ahmadi, Z., Zaafouri, B., El Fahem, A., & Zhioua, E. (2007). Phenology of Phlebotomus papatasi (Diptera: Psychodidae) relative to the seasonal prevalence of zoonotic cutaneous leishmaniasis in central Tunisia. *Journal of Medical Entomology*, *44*(2), 385–388. doi:10.1093/jmedent/44.2.385 PMID:17427714

Doha, S. A., & Samy, A. M. (2010). Bionomics of phlebotomine sand flies (Diptera: Psychodidae) in the province of Al-Baha, Saudi Arabia. *Memorias do Instituto Oswaldo Cruz*, *105*(7), 850–856. doi:10.1590/S0074-02762010000700002 PMID:21120352

Dokhan, M. R., Kenawy, M. A., Doha, S. A., El-Hosary, S. S., Shaibi, T., & Annajar, B. B. (2016). Entomological studies of phlebotomine sand flies (Diptera: Psychodidae) in relation to cutaneous leishmaniasis transmission in Al Rabta, North West of Libya. *Acta Tropica*, *154*, 95–101. doi:10.1016/j.actatropica.2015.11.004 PMID:26589378

Faulde, M., Schrader, J., Heyl, G., & Amirih, M. (2008). Differences in transmission seasons as an epidemiological tool for characterization of anthroponotic and zoonotic cutaneous leishmaniasis in northern Afghanistan. *Acta Tropica*, *105*(2), 131–138. doi:10.1016/j.actatropica.2007.10.011 PMID:18068684

Faulde, M. K., Heyl, G., & Amirih, M. L. (2006). Zoonotic cutaneous leishmaniasis, Afghanistan. *Emerging Infectious Diseases*, *12*(10), 1623–1624. doi:10.3201/eid1210.060076 PMID:17176597

Fichet-Calvet, E., Jommaa, I., Ben Ismail, R., & Ashford, R. W. (2003). *Leishmania major* infection in the fat sand rat *Psammomys obesus* in Tunisia: Interaction of host and parasite populations. *Annals of Tropical Medicine and Parasitology, 97*(6), 593–603. doi:10.1179/000349803225001517 PMID:14511558

Gholamrezaei, M., Mohebali, M., Hanafi-Bojd, A. A., Sedaghat, M. M., & Shirzadi, M. R. (2016). Ecological niche modeling of main reservoir hosts of zoonotic cutaneous leishmaniasis in Iran. *Acta Tropica, 160*, 44–52. doi:10.1016/j.actatropica.2016.04.014 PMID:27150212

HCEFLCD. (2007). *Etude d'aménagement du parc National d'Iriqui*. Maroc: HCEFLCD-DREF SO.

Hnichi, M. (1989). Un parc national en perspective. *Le Matin du Maghreb, 12*, 142.

Karmaoui, A. (2018). The Cutaneous Leishmaniasis Vulnerability index (CLVI). *Acta Ecologica Sinica, 38*(4), 288–295. doi:10.1016/j.chnaes.2018.01.001

Karmaoui A, Ifaadassan I, Babqiqi A, Messouli M & Yacoubi Khebiza M. (2016c). Analysis of the water supply-demand relationship in the Draa valley basin, Morocco, under climate change and socio-economic scenarios. *Journal of Scientific Research & Reports*. DOI: doi:10.9734/JSRR/2016/21536

Karmaoui, A., Ifaadassan, I., Messouli, M., & Khebiza, M. Y. (2015b). Sustainability of the Moroccan oasean system (Case study: Middle Draa Valley). *Global J Technol Optim, 6*(01), 170. doi:10.4172/2229-8711.1000170

Karmaoui, A., Ifaadassan, I., Messouli, M., & Yacoubi Khebiza, M. (2015d). Monetarization of Ecosystem Services of Oasean Biome (Case study: Provisioning services of Middle Draa Valley Oases). *British Journal of Applied Science and Technology, 11*(6), 1–18. doi:10.9734/BJAST/2015/19689

Karmaoui, A., Messouli, M., & Yacoubi, M. (2015a). Vulnerability of Ecosystem Services to Climate Change and Anthropogenic Impacts in South East of Morocco (Case study: Drying up of Iriki lake). *The International Journal of Climate Change: Impacts and Responses., 7*(3), 1835–7156.

Karmaoui A, Messouli M, Yacoubi Khebiza M & Ifaadassan I. (2014). Environmental Vulnerability to Climate Change and Anthropogenic Impacts in Dryland, (Pilot Study: Middle Draa Valley, South Morocco). *J Earth Sci Clim Change, S11*, 2. doi:. doi:10.4172/2157-7617.S11-002

Karmaoui, A., & Zerouali, S. (2018). Modeling the vulnerability to Zoonotic Cutaneous Leishmaniasis at the local scale. *Asian Journal of Applied Sciences*.

Kelly-Hope, L., & Thomson, M. C. (2008). Climate and infectious diseases. In M. C. Thomson, R. Garcia-Herrera, & M. Beniston (Eds.), *Seasonal forecasts, climate change and human health* (p. 31). New York: Springer. doi:10.1007/978-1-4020-6877-5_3

Lahouiti, K., Maniar, S., & Bekhti, K. (2013). Seasonal fluctuations of phlebotomines sand fly populations (Diptera: Psychodidae) in the Moulay Yacoub province, centre Morocco: Effect of ecological factors. *African Journal of Environmental Science and Technology, 7*(11), 1028–1031.

Mili, A. F., Saghrouni, F., BenSaid, Z., Saadi-BenAoun, Y., Guizani, I., & BenSaid, M. (2012). Retrospective Analysis of Leishmaniasis in Central Tunisia: An Update on Emerging Epidemiological Trends. In Current Topics in Tropical Medicine. InTech.

Moroccan Ministry of Health (Direction de la Planification et des Ressources Financière, Division de la Planification et des Etudes, Service des Etudes et de l'Information Sanitaire). (2016). Santé En Chiffre. Rabat: Moroccan Ministry of Health.

Orshan, L., Szekely, D., Khalfa, Z., & Bitton, S. (2010). Distribution and seasonality of Phlebotomus sand flies in cutaneous leishmaniasis foci, Judean Desert, Israel. *Journal of Medical Entomology*, *47*(3), 319–328. doi:10.1093/jmedent/47.3.319 PMID:20496578

Ready, P. D. (2008). Leishmaniasis emergence and climate change. *Revue Scientifique et Technique (International Office of Epizootics)*, *27*, 399–412. doi:10.20506/rst.27.2.1803 PMID:18819668

Roger, A., Nacher, M., Hanf, M., Drogoul, A. S., Adenis, A., Basurko, C., & Dufour, J. (2013). Climate and Leishmaniasis in French Guiana. *The American Journal of Tropical Medicine and Hygiene*, *89*(3), 564–569. doi:10.4269/ajtmh.12-0771 PMID:23939706

Shirzadi, M. R., Mollalo, A., & Yaghoobi-Ershadi, M. R. (2015). Dynamic relations between incidence of zoonotic cutaneous leishmaniasis and climatic factors in Golestan Province. *Iranian Journal of Arthropod-Borne Diseases*, *9*(2), 148. PMID:26623427

Singh, K. V. (1999). Studies on the role of climatological factors in the distribution of *Phlebotomine* sandflies (Diptera: Psychodidae) in semi-arid areas of Rajasthan, India. *Journal of Arid Environments*, *42*(1), 43–48. doi:10.1006/jare.1999.0499

Soltani, Z., Fakoorziba, M. R., Moemenbellah-Fard, M. D., Kalantari, M., Akbarpour, M., Faramarzi, H., & Azizi, K. (2015). Phlebotomus papatasi (Diptera: Psychodidae) as the Vector of Leishmania Major in Kharameh District, Southern Iran. *Journal of Health Sciences and Surveillance System, 3*(4), 160-164.

Souchon, Y., & Nicolas, V. (2011). *Barrages et seuils: principaux impacts environnementaux: Dynamique Physique.* ONEMA. Retrieved from www.irstea.fr/sites/default/files/

Talbi, F. Z., El Ouali Lalami, A., Janati Idrissi, A., Sebti, F., & Faraj, C. (2015). Leishmaniasis in central Morocco: Seasonal fluctuations of phlebotomine sand fly in Aichoun locality, from Sefrou province. *Pathology Research International*. PMID:25741448

Toumi, A., Chlif, S., Bettaieb, J., Alaya, N. B., Boukthir, A., Ahmadi, Z. E., & Salah, A. B. (2012). Temporal dynamics and impact of climate factors on the incidence of zoonotic cutaneous leishmaniasis in central Tunisia. *PLoS Neglected Tropical Diseases*, *6*(5), e1633. doi:10.1371/journal.pntd.0001633 PMID:22563513

Wasserberg, G., Yarom, I., & Warburg, A. (2003). Seasonal abundance patterns of the sandfly Phlebotomus papatasi in climatically distinct foci of cutaneous leishmaniasis in Israeli deserts. *Medical and Veterinary Entomology*, *17*(4), 452–456. doi:10.1111/j.1365-2915.2003.00461.x PMID:14651661

Yaghoobi-Ershadi, M. R., Akhavan, A. A., Zahraei-Ramazani, A. R., Jalali-Zand, A. R., & Piazak, N. (2005). Bionomics of Phlebotomus papatasi (Diptera: Psychodidae) in an endemic focus of zoonotic cutaneous leishmaniasis in central Iran. *Journal of Vector Ecology*, *30*(1), 115. PMID:16007964

Yaghoobi Ershadi, M. R., Akhavan, A. A., Zahraei Ramazani, A. V., Abai, M. R., Ebrahimi, B., Vafaei Nezhad, R., ... Jafari, R. (2003). *Epidemiological study in a new focus of cutaneous leishmaniasis in the Islamic Republic of Iran*. Academic Press.

Yaghoobi-Ershadi, M. R., Hakimiparizi, M., Zahraei-Ramazani, A. R., Abdoli, H., Akhavan, A. A., Aghasi, M., ... Ranjbar, A. A. (2010). Sand fly surveillance within an emerging epidemic focus of cutaneous leishmaniasis in southeastern Iran. *Journal of Arthropod-Borne Diseases*, *4*(1), 17. PMID:22808384

Yaghoobi-Ershadi, M. R., Marvi-Moghadam, N., Jafari, R., Akhavan, A. A., Solimani, H., Zahrai-Ramazani, A. R., ... Dehghan-Dehnavi, A. R. (2015). Some epidemiological aspects of cutaneous leishmaniasis in a new focus, central Iran. *Dermatology Research and Practice*. PMID:26483838

Chapter 20

Biology, Epidemiology, and Public Health Significance of Malaria Disease Linked to Climate Changes

Moulay Abdelmonaim El Hidan
Ibn Zohr University, Morocco

Kholoud Kahime
Cadi Ayyad University, Morocco

Aimrane Abdelmohcine
Cadi Ayyad University, Morocco

Abdellatif Abbaoui
Cadi Ayyad University, Morocco

Mohamed Echchakery
Cadi Ayyad University, Morocco

Ahmed Draoui
Cadi Ayyad University, Morocco

Hicham Chatoui
Private University of Marrakech, Morocco & Cadi Ayyad University, Morocco

Mohamed Ait El Mokhtar
https://orcid.org/0000-0001-8096-9927
Cadi Ayyad University, Morocco

Zakaria Tazart
Cadi Ayyad University, Morocco

Aaziz Kebbou
Cadi Ayyad University, Morocco

Hamid Rguibi Idrissi
University Mohammed V, Morocco

ABSTRACT

Malaria is a mosquito-borne infectious disease caused by obligate intraerythrocytic protozoa of the genus Plasmodium. As one of the most devastating global health issues, malaria is a sensitive disease to weather and climate conditions, in such a way the ongoing trends of increasing temperature and more variable weather could lead to malaria transmission spreading. Spatial and temporal variations in precipitation, temperature, and humidity that are projected to take place under different climate change scenarios will impact the biology and ecology of malaria vectors and subsequently the risk of disease transmission. Here, the authors review how climate and climate change may impact malaria transmission. They contrast ecological and behavioral characteristics of malaria vectors and parasites and how weather, climate, climate change, and socioeconomic factors may have very different impacts on their spatiotemporal occurrence and abundance and the resulting malaria risk.

DOI: 10.4018/978-1-5225-7775-1.ch020

INTRODUCTION

Malaria is one of the most devastating global health issues. World-wide an estimated 216 million people contracted malaria in 2016, resulting in 445,000 deaths (WHO, 2017). It is the most deadly and widespread tropical mosquito-borne parasitic disease. In 2014, ninety-seven countries and territories throughout Africa, Asia, and Latin America had malaria transmission while an estimated 1.2 billion people were at high risk (Dasgupta, 2018).

Malaria is a vector-borne infectious disease caused by obligate intra-erythrocytic protozoa of the genus *Plasmodium*. Humans can be infected with one (or more) of the following four species: *P. falciparum*, *P. vivax*, *P. ovale*, and *P. malariae*. Plasmodia are primarily transmitted by the bite of an infected female Anopheles mosquito, but infections can also occur through exposure to infected blood products (transfusion malaria) and by congenital transmission.

Climate and environmental conditions play an important role in the dynamics and distribution of malaria (Martens et al., 1997; Kim et al., 2012). Climatological research over the past two decades makes clear that Earth's climate will change in response to atmospheric greenhouse gas accumulation. The unusually rapid temperature rise (0·5ºC) since the mid- 1970s is substantially attributable to this anthropogenic increase in greenhouse gases (Watson & Albritton, 2001; Trenberth, 2001). The Intergovernmental Panel on Climate Change (IPCC), drawing on the published results of leading modeling groups around the world, forecasts an increase in world average temperature by 2100 within the range 1·4–5·8ºC (Griggs & Noguer, 2002). Various effects of this recent warming on non-human systems are apparent (Easterling et al., 2000; Parmesan & Yohe, 2003). Thus, the increase of temperature will be greater at higher latitudes and over land. Global average annual rainfall will increase, although many mid latitude and lower latitude land regions will become drier, whereas elsewhere precipitation events (and flooding) could become more severe. Climate variability is expected to increase in a warmer world. A fundamental global environmental change, affecting physical systems and ecosystems, will affect vector-borne disease and particularly Malaria transmission in many ways. Hence, spatial and temporal variations in precipitation, temperature and humidity those are projected to take place under different climate change scenarios will impact the biology and ecology of Malaria vectors and subsequently the risk of disease transmission. Even though rainfall is important in providing suitable environments for mosquitoes to breed (Craig, Snow, & le Sueur, 1999; Grover-Kopec et al., 2005; Thomson et al., 2005). In addition, temperature is the second main driver of Malaria vector and parasite life history traits that both determine transmission intensity, counting mosquito development rate, biting rate, and development rate and survival of the parasite within the mosquito (Mordecai et al., 2010). In this chapter, authors will review important characteristics of Malaria transmission and how they may respond to changes in weather and climate.

OVERVIEW OF MALARIA DISEASE

Life Cycle and Morphology

Plasmodium has a complex, multistage life cycle occurring within two living beings, the vector mosquitoes and the vertebrate hosts. Malaria parasite present different shapes and structures and protein complements depending on their stages of development. The surface proteins and metabolic pathways keep changing during these different stages, which help the parasite to evade the immune clearance, while also creating problems for the development of drugs and vaccines.

Schematically, when the infected anopheline mosquito takes a blood meal, sporozoites are inoculated into the bloodstream. Within an hour sporozoites pass in hepatocytes and start to divide intensely for several days without causing symptoms into exoerythrocytic merozoites (tissue schizogony). This is the so-called "pre-erythrocyte phase". In some cases (*P. vivax* or *P. ovale*), the parasite can persist in the liver in a latent form, and cause recurrence of malaria months or years after the first malaria attack. During the second phase the symptoms appear: thousands of merozoites formed leave the hepatocytes, invade erythrocytes and develop into early trophozoites, which are ring shaped, vacuolated and uninucleated. Once the parasite begins to divide, the trophozoites are called schizonts, consisting of many daughter merozoites (blood schizogony). Eventually, the infected erythrocytes are lysed by the merozoites, which consequently attack other erythrocytes, beginning a new cycle of schizogony. The time of each cycle in *P. falciparum* is about 48 hours. In non-immune humans, the infection is amplified about 20-fold each cycle. After several cycles, some of the merozoites transform into gametocytes, the sexual stage of malaria, which cause no symptoms, but are infective for mosquitoes.

The second part of the parasite's reproductive cycle takes place in the mosquito: when an anopheles bites a sick person, the gametocytes begin another, different cycle of growth and multiplication in the mosquito known as the sporogonic cycle. Inside the mosquito's gut, the infected human blood cell bursts, releasing the gametocytes that mature into sex cells called gametes. Male and female gametes then fuse forming zygotes. The zygotes develop into active elongated ookinetes, which burrow into the mosquito mid-gut wall to form oocysts. The oocysts grow and divide producing thousands of sporozoites; after 8-15 days, the oocyst ruptures releasing the sporozoites inside the mosquito. The sporozoites travel within the mosquito body eventually invading the salivary glands. The human plasmodium cycle begins again when the female mosquito takes a blood meal, injecting the sporozoites from its salivary glands into the human bloodstream.

Signs and Symptoms of Malaria

Malaria may be considered as uncomplicated or severe (complicated) disease, symptoms start when the parasite develops in red blood cells. Thus, clinical symptoms of malaria are primarily due to schizont rupture and destruction of erythrocytes. Malaria can have a gradual or a severe and sudden course with nonspecific symptoms. The presentation of malaria generally is similar to those of common viral infections; this could lead to a delay in diagnosis (Murphy and Oldfield, 1996). First signs of illness include fever (>92% of cases), chills (79%), headaches (70%), and diaphoresis (64%) (Genton & D'Acremont, 2001). Other common symptoms comprising dizziness, malaise, myalgia, abdominal pain, nausea, vomiting, mild diarrhea, and dry cough. Physical signs include fever, tachycardia, jaundice, pallor, orthostatic hypotension, hepatomegaly, and splenomegaly. The beginning of symptoms takes place 10-15 days after infection, and differs in severity depending on the infecting species. In general, all severe forms and deaths from malaria are induced by *P. falciparum*. Occasionally, *P. vivax* or *P. ovale* cause grave complications, devastating deterioration, and even death (Svenson et al., 1996). In order to unify future clinical and epidemiological studies, the World Health Organization (WHO) established criteria for severe malaria, in 1990 (WHO, 1990). The main complications of grave malaria comprise cerebral malaria, pulmonary edema, acute renal failure, severe anemia, and/or bleeding. The most common metabolic complications are acidosis and hypoglycemia. Any of these complications can develop rapidly and progress to death within hours or days (WHO, 2000).

Risk factors for severe malaria and death comprise age (superior to 65 years), female sex (particularly when accompanied by pregnancy), non-immune status, co-existing medical conditions, no antimalarial prophylaxis, delay in treatment, and severity of the illness at admission (coma, acute renal failure, shock, pulmonary edema, coagulation dis- orders) (Bruneel et al., 2003; Schwartz et al., 2001). In hyperendemic areas of malaria, severe malaria is mainly observed in young children (1 month to 5 years of age). However, in developed countries, most grave malaria cases occur in non-immune travelers returning from endemic areas. Severe malaria accounts for around 5% of imported malaria cases (range 1–38%) (Genton & D'Acremont, 2001).The rate of fatality case in returning travelers with falciparum malaria fluctuates from 0.6% to 3.8%, and for severe malaria it might surpass 20% (Kain et al., 1998).

Epidemiology of Malaria Worldwide

Malaria is a life-threatening emergency and a common public health problem in many regions of the world, in term of morbidity and mortality, comprising more than 200 million cases and 655.000 deaths every year. Malaria infection is threatening human population in 91 countries in the world as stated by the World Health Organization (WHO) Malaria Report 2017. Thus, according to current estimates of the WHO, around 216 million malaria cases occurred in 2016, 81% of which were reported in the African Region, followed by South East Asia (13%) and Eastern Mediterranean Region (5%)(Table 1). The total number of malaria deaths was estimated to be 445.000 in 2016; 91% of whom reported in the African Region, 6% in South-East Asia and 3% in Eastern Mediterranean Region (Figure 1).

The number of people at risk of malaria infection had augmented from 0.8 billion in 1900 to 3.3 billion in 2010, as a result of the rise of human population inhabiting endemic regions of malaria (WHO, 2011; Hay et al., 2004). Conversely, a reduction in malaria cases was observed from 244 million to 216 million between 2005 and 2010; moreover, a noticeable decrease in malaria mortality rates (26%) was also shown between 2000 and 2010.

Table 1. Estimated number of Malaria by WHO region, 2010-2016

	Number of Deaths						
	2010	**2011**	**2012**	**2013**	**2014**	**2015**	**2016**
African	538 000	484 000	445 000	430 000	423 000	409 000	407 000
Eastern Mediterranean	7 200	7 100	7 700	7 800	7 800	7 600	8 200
European	0	0	0	0	0	0	0
Americas	830	790	630	620	420	450	650
South-East Asia	41 700	34 000	29 000	22 000	25 000	26 000	27 000
Western Pacific	3 800	3 300	4 000	4 300	2 900	2 600	3 300
World	591 000	529 000	487 000	465 000	459 000	446 000	445 000

(WHO, 2017)

Figure 1. Proportion of estimated malaria deaths attributable to the 15 countries with nearly 80% of malaria deaths globally in 2016
(WHO, 2017)

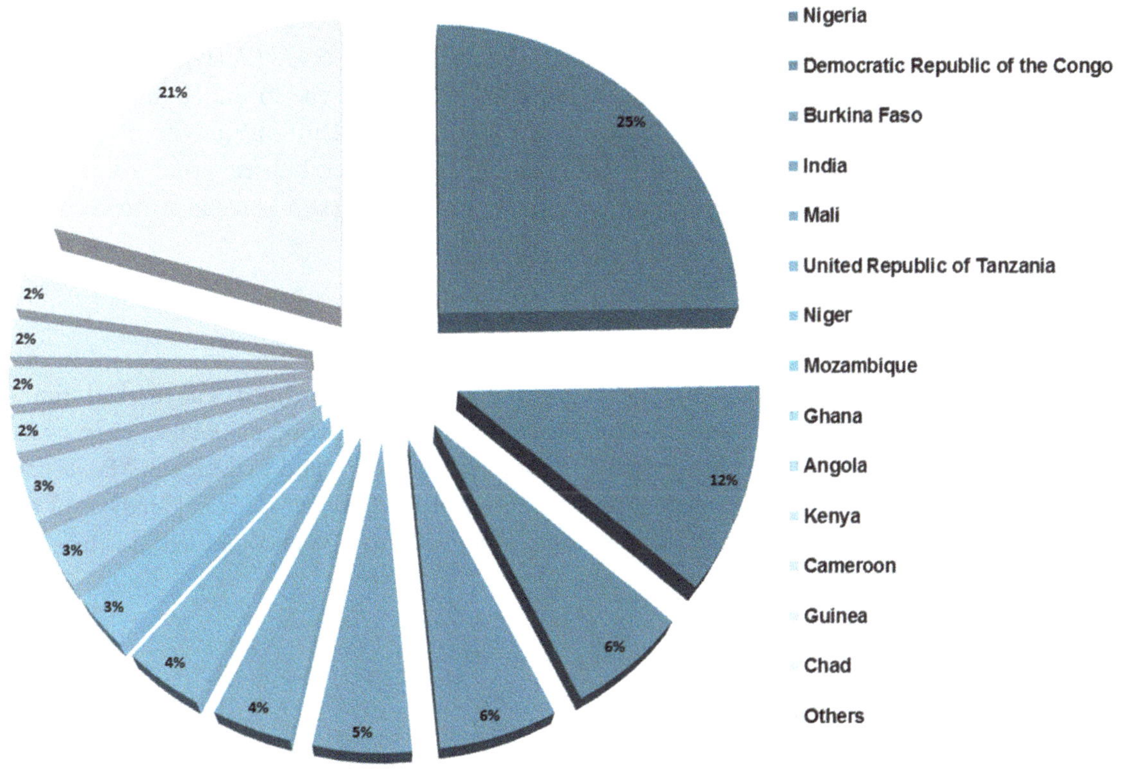

Plasmodium Species Distribution

Five Plasmodium species are responsible of malaria in humans: *Plasmodium falciparum*, *P. vivax*, *P. malariae*, *P. ovale* and *P. knowlesi*. Prevalence of malaria cases and deaths differs according to seasons, and the specie of *Plasmodium* involved. The current *Plasmodium* species distribution in the world is as follows:

Plasmodium Falciparum

P. falciparum is one of the most widespread Malaria parasite species in nearly all Malaria endemic countries, with 2.37 billion people at risk of transmission worldwide. About 26% of population at risk of *P. falciparum* transmission is located in the African Region and 62% in South East Asian and Western Pacific regions (Guerra et al., 2008). Thus, Culleton et al., (2008) were conducted a study between 1998 and 2006 in nine different African countries showed that out of 2.588 samples, 1.737 were positive for *Plasmodium* species and 1.711 (98,5%) were positive for *P. falciparum* considering both mono and mixed infection (Culleton et al., 2008). In Asia, *P. falciparum* and *P. vivax* are the two prevalent species. In South America *P. vivax* is the predominant species, followed by *P. falciparum* (25.7%) (Arevalo et al., 2012)

Plasmodium Vivax

Geographic distribution of *P. vivax* varies widely among the World regions Health Organization (WHO) regions. High endemicity of *P. vivax* occurs in the South-East Asian, Western Pacific regions and in most of the Americas where high prevalence of infection by this specie is observed (WHO, 2015). In Africa, the geographic extent of *P. vivax* transmission is difficult to estimate due to the High *P. falciparum* endemicity in much of that continent coupled with high prevalence of Duffy negativity overshadowed collection of *P. vivax*-specific data as a priority (Shanks, 2012). However, Indirect proof from returning travelers suggests that *P. vivax* is present at low endemicity in almost all sub-Saharan African countries (Guerra et al., 2010; Howes et al., 2015).

Plasmodium Ovale

Plasmodium ovale was the last of the malaria parasites of humans to be described. The marked stippling of the infected erythrocyte and its tertian periodicity led early investigators to consider it as a variant form of *Plasmodium vivax*. Therefore, the real extent of *P. ovale* malaria still hard to evaluate. Many reports have been made on the presence of *P. ovale* throughout the world. However, a critical analysis of these reports by Lysenko and Bejaev indicated that *Plasmodium ovale* may be encountered in sub-Saharan Africa and the islands of the western Pacific.

Plasmodium Malariae

In general, the distribution of *P. malariae* overlaps with that of *P. falciparum*. In areas of endemicity in Africa, infections of *P. malariae* are mixed with *P. falciparum* infections. In many cases, the occurrence of *P. malariae* infections is imperceptible unless the use of PCR techniques. *Plasmodium malariae* is spread in sub-Saharan Africa, in Southeast Asia, in Indonesia, in many islands in western Pacific and in areas of the Amazon Basin of South America. In the recent past, *P. malariae* was prevalent in Europe and in southern parts of the United States.

P. knowlesi

Plasmodium knowlesi infection is localized only in the South East Asia Region and interest both monkeys, where it was first reported, and humans. Forest areas are the reservoirs of *P. knowlesi*, that was first reported in humans in 1965 in a man who had worked in the jungle of Pahang, Peninsular Malaysia.

Malaria Vectors

Malaria is transmitted to humans through the bite of an infectious Anopheles mosquito. Of more than 512 species of Anopheles, only about seventy of these species have the capacity to transmit Plasmodium parasite to human hosts (Snow & Gilles, 2002) and 41 are considered here to be dominant vector species/species complexes (DVS). DVS are Anopheles species capable of transmitting malaria at a level of major concern to public health (Hay et al., 2010; Sinka et al., 2010), They are characterized by their propensity for humans feeding, longevity, abundance and elevate vectorial capacity (Takken & Scott, 2003)

Anopheles distribution was assessed through several studies in Africa (Moffett, Shackelford & Sarkar, 2007; Moffett et al., 2009), Americas (Levine, Peterson & Benedict, 2004; Foley et al., 2008), Europe (Kuhn, Campbell-Lendrum, & Davies, 2002) Central and South East Asia (Manguin et al., 200).

Africa

As stated above, African Malaria endemic regions present the highest entomological inoculation rates and the highest malaria prevalence worldwide (Guerra et al., 2008); due to the wide spread presence of the most effective and efficient DVS of human malaria, known as the *Anopheles gambiae* complex. There are four principal species belonging to *An. gambiae* complex: *An. gambiae*, *An. arabiensis*, *An. merus* and *An. melas*. In addition to gambiae complex there are three highly anthropophilic DVS spread in African region: *An. funestus* subgroup of which *An. funestus* is another highly effective vector, *An. moucheti* a restricted, but highly anthropophilic vector and the more widespread *An. nili* complex proved to be highly efficient in malaria transmission and equally difficult to control (Who, 2009).

Americas

Nine DVS were found in the Americas (Sinka et al., 2010). Of this nine DVS, six species covering areas from southern North America, through Central America and into South America, including the northern reaches of Argentina. The distribution of the two remaining species (*An. freeborni* and *An. Quadrimaculatus*) is restricted to the North America. As stated above, in South America, *An. darlingi* is considered to be the most important of the DVS where it is found (Manguin et al., 2008). However there is increasing evidence of the importance of other species, comprising members of the *An. albitarsis* complex (e.g. *An. marajoara*), that may have a higher influence in malaria transmission than previously thought (Conn et al., 2002).

Asian-Pacific Region

The Asian-Pacific region presents a high vector species diversity. Many species complexes and suspected species complexes could occur sympatrically and exhibiting a high level of behavioral plasticity.

Of 41 DVS documented globally (Hay et al., 2010), 19 occur in the Asian-Pacific region and of these, at least ten are now considered as species complexes (Harbach, 2004). Distribution of these DVS vary according to the country, in fact, some of them are predominant in the Arabian Peninsula (e.g. *An. stephensi* and the *An. culicifacies* complex), others in the Indian subcontinent, China and Korea (e.g. *An. lesteri*), in the Solomon Islands and Vanuatu (e.g. *the An. farauti* complex) and finally in Queensland and the Northern Territory of Australia (e.g. the *An. farauti* complex). Myanmar appeared to contain the greatest number of DVS, of wich *An. aconitus*, *An. annularis*, *An. barbirostris* complex, *An. culicifacies* complex, *An. dirus* complex, *An. maculatus* group, *An. minimus* complex, *An. sinensis* complex, *An. stephensi*, *An. subpictus* complex and, in some coastal site, *An. sundaicus* complex (Sinka et al., 2011).

MALARIA EMERGENCE AND/ OR RE-EMERGENCE AND CLIMATE CHANGE

How Climate Changes Affect the Emergence and Re-Emergence of Malaria?

Climate change is considered as a statistically significant change in either the mean state of the climate or in its variability, lasting for a long period -- typically decades or longer -- that could be a consequence of natural internal processes, external forcing, or anthropogenic changes in the composition of the atmosphere or in land use.

Malaria is considered as one of the most sensitive vector-borne disease to climate change (Rogers & Randolph, 2000; Kim et al., 2012). Manifestations of climate change are generally summarized as a rise in mean temperatures, mainly at higher latitudes; variations in precipitation, (Heavy rainfall in some regions and drought prone in others); and extreme weather events (severe storms, extreme heat events, heavy rainfall events, etc.) (Pachauri et al., 2014). All These climatic disturbances could lead to the emergence and re-emergence of Malaria in several ways as it was summarized in the review of Ogen and Lindsay (2016) (Figure 2):

- Spread of Anopheles and Plasmodium as climate warms in temperate zones and becomes more suitable for these species.
- Higher probability and frequency of introduction and endemic establishment of Malaria in non-malarious countries by a combination of increasing temperatures accompanied by a high Malaria vector and parasite survival and so, in their abundance and finally growing rates of import of plasmodium pathogens due to enlarged climate change-related human migration.

Figure 2. Factors influencing Anopheles spread and Malaria transmission

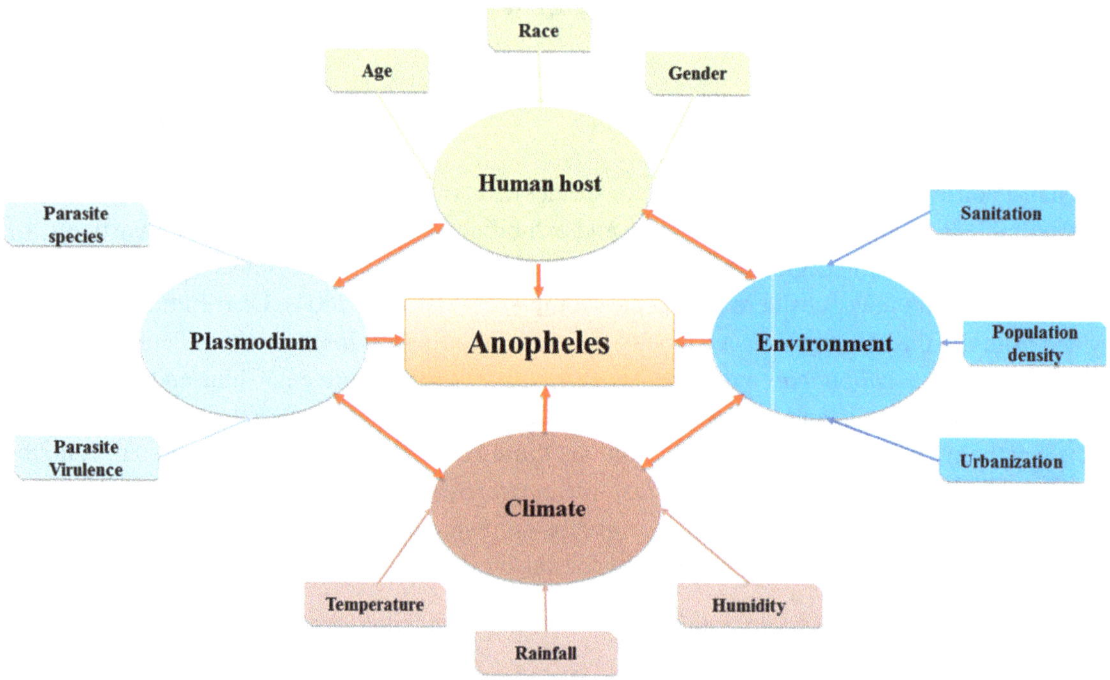

- Re-emergence of Malaria in endemic areas associated with increasing temperatures and weather and climate variability.
- Emergence and fixation of novel genotypes of *Plasmodium sp.* as a consequence of climate change-driven changes to animal host and vector dynamics.

What Are the Impacts of Weather, Climate and Climate Change on Anophelese and Plasmodium Transmission?

Direct Effects of Temperature and Humidity on Anophelese Mortality Rates

Temperature directly impacts Anopheles mortality rates. Because mosquitoes are ectotherms, each life stage is dependent on temperature in the developmental and mortality rates. In the laboratory, mortality rates of Anopheles larval and adult forms (Bayoh & Lindsay, 2004; Christiansen-Jucht et al., 2014) vary according to vector species, for example optimal temperature for *Anopheles gambiae* survival is about 27°C. Interestingly, the temperature-dependencies are not the same among the stages, leading to nonlinearities in population responses to temperature. Mortality rates of larval and pupal Anopheles depend to the temperature of the water in which they occur, which is relatively directly influenced by diurnal changes in ambient temperatures (Paaijmans et al., 2010). Juvenile stage mortality rate data present greater variability to temperature and stayed low over a smaller temperature range than the adult mortality rates. Moreover, adult mosquito mortality rates did not change much across the temperature range, except at the extremely warm temperatures. Daily adult survival is generally high for all temperatures comprised between 18 and 34°C. Additionally, some studies had reported that adult abundance decrease at cooler temperatures. Also, the juvenile stage temperature sensitivities impact recruitment into the adult stage, which made the abundance of adults more temperature-dependent. Consequently, the ability for anopheles to find refuge from temperature extremes, and the capacity of the habitat and/or environment of the vector to provide such refuges, will determine how well vector populations survive at the limits of the geographic regions where temperature conditions are within optimal conditions (Eisen et al., 2014). Urban areas may provide refuges for adult mosquitoes that permit survival of populations outside their expected geographic ranges (Lima et al., 2016).

Humidity is known to be one of the most important drivers of mosquito dynamics and malaria risk. In fact, generally all stages of anopheles are exterminated by desiccation (Benedict et al., 2010), and anopheles could only subsist in environments that (i) support sources of water for rehydration for adult Anopheles; and (ii) afford still-water aquatic environments for undeveloped stages of Anopheles.

Effects of Temperature and Humidity on Activity

Beside the effect of temperature and humidity on Anopheles mortality rate, they have also strong impacts on the rate of mosquito biting rates, which determines malaria transmission. In fact, together high and low temperatures stop Anopheles meal-seeking activity (Rowley and Graham, 1968). Thus, diurnal duration of vector activity, the seasons when vectors are active and even the geographic regions where the climate is suitable for Anopheles activity, will be affected by temperature and humidity. Moreover, humidity plays a key role in the duration spent in host and/or meal-seeking activity. Actually, Anopheles will spend more energy in drier environments for rehydration and they will use mostly water or nectar (Benoit and Denlinger, 2010). In this case, time spent in seeking host meals will be reduced and as a consequence will reduced the probability that vectors will survive to successfully feed on a host. High (100%) humidity does not impact Anopheles activity, but heavy rainfall could inhibit it.

Effects of Temperature on Interstadial Development

Among the most important determinants of adult mosquito abundance and distribution is the success of anopheles interstatdial development, which is closely related to environmental factors, such as hydrology, temperature, light/ shade, pH, salinity, nutrient availability (Chase & Knight, 2003; Stresman, 2010). Of all of these factors, temperature seems to be the most important factor that affects interstadial development of Malaria vector. Indeed, temperature influences all the main processes such as the proportion of larval development and survivorship, pupation rates, larval-to-adult survivorship and larval-to adult development time (Paaijmans et al., 2008; Bayoh & Lindsay, 2004).

Aquatic environments -in which the immature stages occur- temperature is impacted by different elements, such as local climate, water depth and movement, land cover type, habitat geometry and size, presence of vegetation, water properties and turbidity (Paaijmans et al., 2008).

The influence of temperature on interstadial development has been studied in a number of different species of mosquitoes (Diptera: Culicidae), including *Anopheles quadrimaculatus* Say (Huffaker, 1944), *Aedes aegypti* Linnaeus (Bar-Zeev, 1958; Tun-Lin et al., 2000), Culex and Anopheles species (Shelton, 1973), *Toxorhynchites brevipaplis* Theobald (Trips, 1972) and *Wyeomyia smithii* Coquillett (Bradshaw, 1980). Therefore, Studies on the temperature of larval habitats of *An. gambiae* and *An. funestus* show that one of the principal reasons for higher productivity of these species in habitats marked by the presence of agricultural crops or swamp margins is increased temperature as compared to shaded dense papyrus swamps (Munga et al., 2006; Imbahale et al., 2011). Another study had reported a reduction of *An. gambiae* density in shaded water channels (by napier grass, Pennisetum purpureum) in comparison with the unshaded water channels in Western Kenya highlands. In this work, the shading decreased anopheline larvae by more than 75%, apparently due to ~ 3 degrees C reduced water temperature. High water temperature pools (30-33 degrees C) were reported as the most productive habitats for *An. gambiae* in Gambia. Larvae are not generally able to survive temperatures over 40 degrees °C as documented by Muirhead-Thomson (1951). Recent detailed study on the longevity and mortality of *An. gambiae* under a wide range of temperatures (Bayoh & Lindsay, 2004) concluded that under extremely cold (10–12°C) or hot (38–40°C) temperatures all larvae died within a few days. While the low temperature range is rarely experienced in larval habitats of *An. gambiae*, the higher temperatures are frequently encountered in most tropical regions. In nature, however, such high temperatures occur for no more than a few hours and larvae may survive these short periods. Accordingly, within the limits of a lower development threshold and an upper lethal temperature, the aquatic stages of mosquitoes develop faster as temperature increases (Brust, 1967; Hagstrum & Workman; 1971; Lyimo & Takken, 1993). The shortening of aquatic life is important since it will increase adult turnover, with consequences for increased vector biting rates and disease transmission (Garett-Jones, 1964).

Effects of Weather and Climate on Reproduction

The mosquito reproduction is successful only if larval habitats remain stable for duration equivalent to the development of immature stages (Barros et al., 2011). Thus, reproduction rates of Anopheles are highly impacted by the occurrence of standing- water larval development habitat in and/or on which eggs are placed and where all transitions from egg to adult occurs. Reproductive frequency in any specific habitat varies according to the quantity of rainfall. If there is less rainfall, the drying of mosquito breeding sites would, in most cases, decrease the contact of local communities with infective mosquitoes. However, extended periods of dryness could also lead to rivers drying out, creating small pools of water along the old riverbed. In Sri Lanka, such conditions provide ideal breeding sites for *An. culicifacies* and have resulted in major epidemics of malaria in the past (Wijesundera, 1988).

Increased rainfall could possibly lead to a rise of breeding sites and consequently enhance the transmission of parasites. Earlier, several regional epidemics of malaria in the Far East have happened after heavy rains in regions where rivers have flooded and generated suitable conditions for an increase in transmission (Gill, 1928). Though, excessive rainfall might also decrease transmission intensity in certain cases. Anopheles species which breed in slow-moving streams or in pools on dried-up riverbeds may be flushed out by increased stream flow. During the breeding of *An. farauti* in Papua New Guinea, for example, heavy rain can breach sandbars, flush out immature stages and consequently reduce the number of subsequent adult mosquitoes (Charlwood et al., 1985).

Effects of Temperature on Malaria Transmission

One of the most influential parameters in the spatiotemporal occurrence of Malaria is the extrinsic incubation period (EIP) of malaria, also called the period of sporogony. The EIP describes the time it takes for parasites to develop in the mosquito from point of ingestion via an infected blood meal, through to the point at which sporozoites enter the salivary glands and the mosquito becomes infectious. Smith and McKenzie (2004) had reported that the importance of the EIP in Malaria transmission lies in its interaction with adult mosquito survival rate as an exponential term, meaning that even very small changes in EIP can have a large effect on the number of mosquitoes living long enough to be able to transmit parasites. Changes in EIP potentially have much greater impact than equivalent changes in traits such as vector competence or vector density. The speed with which EIP occurs depends on temperature (Ogden & Lindsay, 2016): the warmer the temperature, the faster EIP occurs. If temperature conditions are too low, the EIP is longer than the average life expectancy of the Anopheles and, therefore, pathogen transmission cycles cannot occur, even if vector populations are abundant. Also, at high temperatures, mosquitoes mortality rates may increase faster than the shortening of EIP, resulting in upper temperature limits for pathogen transmission (Reisen, Lothrop & Hardy, 1995; Danforth, Reisen & Barker; 2015).

CONCLUSION

Climate change would, rise malaria prevalence in tropical countries currently disposed to the disease. Either, increase of temperatures and global travel may lead to reintroduce or enhance transmission of malaria in tropical and temperate countries that have both eliminated or controlled transmission. Such regions could be disposed to epidemics, as surveillance of malaria control could not be as strong as while malaria was a main public health concern in these countries. It is in this context that surveillance and preparedness need to be emphasized, not compromised. This is especially true in developing countries that have to balance competing interests for scarce resources, some of which need not necessarily be health related.

FUNDING INFORMATION

This research received no specific grant from any funding agency in the public, commercial, or not-for-profit sectors.

REFERENCES

Arevalo-Herrera, M., Quiñones, M. L., Guerra, C., Céspedes, N., Giron, S., Ahumada, M., ... Padilla, J. C. (2012). Malaria in selected non-Amazonian countries of Latin America. *Acta Tropica, 121*(3), 303–314. doi:10.1016/j.actatropica.2011.06.008

Bar-Zeev, M. (1958). The effect of temperature on the growth rate and survival of the immature stages of Aedes aegypti (L.). *Bulletin of Entomological Research, 49*(1), 157–163. doi:10.1017/S0007485300053499

Barros, F. S. M., Arruda, M. E., Gurgel, H. C., & Honorio, N. A. (2011). Spatial clustering and longitudinal variation of Anopheles darlingi (Diptera: Culicidae) larvae in a river of the Amazon: the importance of the forest fringe and of obstructions to flow in frontier malaria. *Bulletin of Entomological Research, 101*(6), 643–658. doi:10.1017/S0007485311000265

Bayoh, M. N., & Lindsay, S. W. (2004). Temperature-related duration of aquatic stages of the Afrotropical malaria vector mosquito Anopheles gambiae in the laboratory. *Medical and Veterinary Entomology, 18*(2), 174–179. doi:10.1111/j.0269-283X.2004.00495.x

Benedict, M. Q., Sandve, S. R., Wilkins, E. E., & Roberts, J. M. (2010). Relationship of larval desiccation to Anopheles gambiae Giles and An. arabiensis Patton survival. *Journal of Vector Ecology, 35*(1), 116–123. doi:10.1111/j.1948-7134.2010.00067.x

Benoit, J. B., & Denlinger, D. L. (2010). Meeting the challenges of on-host and off-host water balance in blood-feeding arthropods. *Journal of Insect Physiology, 56*(10), 1366–1376. doi:10.1016/j.jinsphys.2010.02.014

Bradshaw, W. E. (1980). Thermoperiodism and the thermal environment of the pitcher-plant mosquito, Wyeomyia smithii. *Oecologia, 46*(1), 13–17. doi:10.1007/BF00346959

Bruneel, F., Hocqueloux, L., Alberti, C., Wolff, M., Chevret, S., Bédos, J. P., ... Vachon, F. (2003). The clinical spectrum of severe imported falciparum malaria in the intensive care unit: Report of 188 cases in adults. *American Journal of Respiratory and Critical Care Medicine, 167*(5), 684–689. doi:10.1164/rccm.200206-631OC

Brust, R. A. (1967). Weight and development time of different stadia of mosquitoes reared at various constant temperatures. *Canadian Entomologist, 99*(9), 986–993. doi:10.4039/Ent99986-9

Charlwood, J. D., Birley, M. H., Dagoro, H., Paru, R., & Holmes, P. R. (1985). Assessing survival rates of Anopheles farauti (diptera: Culicidae) from Papua New Guinea. *Journal of Animal Ecology, 54*(3), 1003–1016. doi:10.2307/4393

Chase, J. M., & Knight, T. M. (2003). Drought-induced mosquito outbreaks in wetlands. *Ecology Letters, 6*(11), 1017–1024. doi:10.1046/j.1461-0248.2003.00533.x

Christiansen-Jucht, C., Parham, P. E., Saddler, A., Koella, J. C., & Basáñez, M. G. (2014). Temperature during larval development and adult maintenance influences the survival of Anopheles gambiae ss. *Parasites & Vectors*, *7*(1), 489. doi:10.118613071-014-0489-3

Conn, J. E., Wilkerson, R. C., Segura, M. N. O., de Souza, R. T., Schlichting, C. D., Wirtz, R. A., & Póvoa, M. M. (2002). Emergence of a new neotropical malaria vector facilitated by human migration and changes in land use. *The American Journal of Tropical Medicine and Hygiene*, *66*(1), 18–22. doi:10.4269/ajtmh.2002.66.18

Craig, M. H., Snow, R. W., & le Sueur, D. (1999). A climate-based distribution model of malaria transmission in sub-Saharan Africa. *Parasitology Today (Personal Ed.)*, *15*(3), 105–111. doi:10.1016/S0169-4758(99)01396-4

Culleton, R. L., Mita, T., Ndounga, M., Unger, H., Cravo, P. V., Paganotti, G. M., ... Karema, C. (2008). Failure to detect Plasmodium vivax in West and Central Africa by PCR species typing. *Malaria Journal*, *7*(1), 174. doi:10.1186/1475-2875-7-174

Danforth, M. E., Reisen, W. K., & Barker, C. M. (2015). Extrinsic incubation rate is not accelerated in recent California strains of West Nile virus in Culex tarsalis (Diptera: Culicidae). *Journal of Medical Entomology*, *52*(5), 1083–1089. doi:10.1093/jme/tjv082

Dasgupta, S. (2018). Burden of climate change on malaria mortality. *International Journal of Hygiene and Environmental Health*, *221*(5), 782–791. doi:10.1016/j.ijheh.2018.04.003

Easterling, D. R., Meehl, G. A., Parmesan, C., Changnon, S. A., Karl, T. R., & Mearns, L. O. (2000). Climate extremes: observations, modeling, and impacts. *Science, 289*(5487), 2068-2074.

Eisen, L., Monaghan, A. J., Lozano-Fuentes, S., Steinhoff, D. F., Hayden, M. H., & Bieringer, P. E. (2014). The impact of temperature on the bionomics of Aedes (Stegomyia) aegypti, with special reference to the cool geographic range margins. *Journal of Medical Entomology*, *51*(3), 496–516. doi:10.1603/ME13214

Foley, D. H., Klein, T. A., Kim, H. C., Wilkerson, R. C., & Rueda, L. M. (2008). *Malaria risk assessment for the Republic of Korea based on models of mosquito distribution*. Walter Reed Army Inst Of Research.

Foley, D. H., Weitzman, A. L., Miller, S. E., Faran, M. E., Rueda, L. M., & Wilkerson, R. C. (2008). The value of georeferenced collection records for predicting patterns of mosquito species richness and endemism in the Neotropics. *Ecological Entomology*, *33*(1), 12–23.

Garrett-Jones, C. (1964). Prognosis for interruption of malaria transmission through assessment of the mosquito's vectorial capacity. *Nature*, *204*(4964), 1173–1175. doi:10.1038/2041173a0

Genton, B., & D'Acremont, V. (2001). *Clinical features of malaria in returning travelers and migrants. Travelers' malaria*. Hamilton, Canada: Decker.

Gill, C. A. (1928). *The Genesis of Epidemics and the Natural History of Disease. An Introduction to the Science of Epidemiology based upon the Study of Epidemics of Malaria, Influenza, and Plague. Academic Press.*

Griggs, D. J., & Noguer, M. (2002). Climate change 2001: The scientific basis. Contribution of working group I to the third assessment report of the intergovernmental panel on climate change. *Weather*, *57*(8), 267–269. doi:10.1256/004316502320517344

Grover-Kopec, E., Kawano, M., Klaver, R. W., Blumenthal, B., Ceccato, P., & Connor, S. J. (2005). An online operational rainfall-monitoring resource for epidemic malaria early warning systems in Africa. *Malaria Journal*, *4*(1), 6. doi:10.1186/1475-2875-4-6

Guerra, C. A., Gikandi, P. W., Tatem, A. J., Noor, A. M., Smith, D. L., Hay, S. I., & Snow, R. W. (2008). The limits and intensity of Plasmodium falciparum transmission: Emplications for malaria control and elimination worldwide. *PLoS Medicine*, *5*(2), 300–312. doi:10.1371/journal.pmed.0050038

Guerra, C. A., Howes, R. E., Patil, A. P., Gething, P. W., Van Boeckel, T. P., Temperley, W. H., ... Baird, J. K. (2010). The international limits and population at risk of Plasmodium vivax transmission in 2009. *PLoS Neglected Tropical Diseases*, *4*(8), e774. doi:10.1371/journal.pntd.0000774

Hagstrum, D. W., & Workman, E. B. (1971). Interaction of temperature and feeding rate in determining the rate of development of larval Culex tarsalis (Diptera, Culicidae). *Annals of the Entomological Society of America*, *64*(3), 668–671. doi:10.1093/aesa/64.3.668

Harbach, R. E. (2004). The classification of genus Anopheles (Diptera: Culicidae): a working hypothesis of phylogenetic relationships. *Bulletin of Entomological Research*, *94*(6), 537–553. doi:10.1079/BER2004321

Hay, S. I., Guerra, C. A., Tatem, A. J., Noor, A. M., & Snow, R. W. (2004). The global distribution and population at risk of malaria: Past, present, and future. *The Lancet Infectious Diseases*, *4*(6), 327–336. doi:10.1016/S1473-3099(04)01043-6

Hay, S. I., Sinka, M. E., Okara, R. M., Kabaria, C. W., Mbithi, P. M., Tago, C. C., ... Temperley, W. H. (2010). Developing global maps of the dominant Anopheles vectors of human malaria. *PLoS Medicine*, *7*(2), e1000209. doi:10.1371/journal.pmed.1000209

Howes, R. E., Reiner, R. C. Jr, Battle, K. E., Longbottom, J., Mappin, B., Ordanovich, D., ... Smith, D. L. (2015). Plasmodium vivax transmission in Africa. *PLoS Neglected Tropical Diseases*, *9*(11), e0004222. doi:10.1371/journal.pntd.0004222

Huffaker, C. B. (1944). The temperature relations of the immature stages of the malarial mosquito, Anopheles quadrimaculatus Say, with a comparison of the developmental power of constant and variable temperatures in insect metabolism. *Annals of the Entomological Society of America*, *37*(1), 1–27. doi:10.1093/aesa/37.1.1

Imbahale, S. S., Mweresa, C. K., Takken, W., & Mukabana, W. R. (2011). Development of environmental tools for anopheline larval control. *Parasites & Vectors*, *4*(1), 130. doi:10.1186/1756-3305-4-130

Kain, K. C., Harrington, M. A., Tennyson, S., & Keystone, J. S. (1998). Imported malaria: Prospective analysis of problems in diagnosis and management. *Clinical Infectious Diseases*, *27*(1), 142–149. doi:10.1086/514616

Kim, Y. M., Park, J. W., & Cheong, H. K. (2012). Estimated effect of climatic variables on the transmission of Plasmodium vivax malaria in the Republic of Korea. *Environmental Health Perspectives*, *120*(9), 1314–1319. doi:10.1289/ehp.1104577

Kuhn, K. G., Campbell-Lendrum, D. H., & Davies, C. R. (2002). A continental risk map for malaria mosquito (Diptera: Culicidae) vectors in Europe. *Journal of Medical Entomology*, *39*(4), 621–630. doi:10.1603/0022-2585-39.4.621

Levine, R. S., Peterson, A. T., & Benedict, M. Q. (2004). Distribution of members of Anopheles quadrimaculatus Say sl (Diptera: Culicidae) and implications for their roles in malaria transmission in the United States. *Journal of Medical Entomology*, *41*(4), 607–613. doi:10.1603/0022-2585-41.4.607

Levine, R. S., Peterson, A. T., & Benedict, M. Q. (2004). Geographic and ecologic distributions of the Anopheles gambiae complex predicted using a genetic algorithm. *The American Journal of Tropical Medicine and Hygiene*, *70*(2), 105–109. doi:10.4269/ajtmh.2004.70.105

Lima, A., Lovin, D. D., Hickner, P. V., & Severson, D. W. (2016). Evidence for an overwintering population of Aedes aegypti in Capitol Hill neighborhood, Washington, DC. *The American Journal of Tropical Medicine and Hygiene*, *94*(1), 231–235. doi:10.4269/ajtmh.15-0351

Lyimo, E. O., & Takken, W. (1993). Effects of adult body size on fecundity and the pre-gravid rate of Anopheles gambiae females in Tanzania. *Medical and Veterinary Entomology*, *7*(4), 328–332. doi:10.1111/j.1365-2915.1993.tb00700.x

Lysenko, A. J., & Beljaev, A. E. (1969). An analysis of the geographical distribution of Plasmodium ovale. *Bulletin of the World Health Organization*, *40*(3), 383.

Manguin, S., Carnevale, P., Mouchet, J., Coosemans, M., Julvez, J., Richard-Lenoble, D., & Sircoulon, J. (2008). *Biodiversity of malaria in the world*. Academic Press.

Manguin, S., Garros, C., Dusfour, I., Harbach, R. E., & Coosemans, M. (2008). Bionomics, taxonomy, and distribution of the major malaria vector taxa of Anopheles subgenus Cellia in Southeast Asia: An updated review. *Infection, Genetics and Evolution*, *8*(4), 489–503. doi:10.1016/j.meegid.2007.11.004

Martens, W. J., Jetten, T. H., & Focks, D. A. (1997). Sensitivity of malaria, schistosomiasis and dengue to global warming. *Climatic Change*, *35*(2), 145–156. doi:10.1023/A:1005365413932

Moffett, A., Shackelford, N., & Sarkar, S. (2007). Malaria in Africa: Vector species' niche models and relative risk maps. *PLoS One*, *2*(9), e824. doi:10.1371/journal.pone.0000824

Moffett, A., Strutz, S., Guda, N., González, C., Ferro, M. C., Sánchez-Cordero, V., & Sarkar, S. (2009). A global public database of disease vector and reservoir distributions. *PLoS Neglected Tropical Diseases*, *3*(3), e378. doi:10.1371/journal.pntd.0000378

Mordecai, E. A., Paaijmans, K. P., Johnson, L. R., Balzer, C., Ben-Horin, T., de Moor, E., ... Lafferty, K. D. (2013). Optimal temperature for malaria transmission is dramatically lower than previously predicted. *Ecology Letters*, *16*(1), 22–30. doi:10.1111/ele.12015

Muirhead-Thomson, R. C. (1951). Mosquito Behaviour in Relation to Malaria Transmission and Control in the Tropics. London: Edward Arnold.

Munga, S., Minakawa, N., Zhou, G., Barrack, O. O. J., Githeko, A. K., & Yan, G. (2006). Effects of larval competitors and predators on oviposition site selection of Anopheles gambiae sensu stricto. *Journal of Medical Entomology*, *43*(2), 221–224. doi:10.1093/jmedent/43.2.221

Murphy, G. S., & Oldfield, E. C. III. (1996). Falciparum malaria. *Infectious Disease Clinics of North America*, *10*(4), 747–775. doi:10.1016/S0891-5520(05)70325-1

Ogden, N. H., & Lindsay, L. R. (2016). Effects of climate and climate change on vectors and vector-borne diseases: Ticks are different. *Trends in Parasitology*, *32*(8), 646–656. doi:10.1016/j.pt.2016.04.015

Paaijmans, K. P., Imbahale, S. S., Thomas, M. B., & Takken, W. (2010). Relevant microclimate for determining the development rate of malaria mosquitoes and possible implications of climate change. *Malaria Journal*, *9*(1), 196. doi:10.1186/1475-2875-9-196

Paaijmans, K. P., Jacobs, A. F. G., Takken, W., Heusinkveld, B. G., Githeko, A. K., Dicke, M., & Holtslag, A. A. M. (2008). Observations and model estimates of diurnal water temperature dynamics in mosquito breeding sites in western Kenya. *Hydrological Processes: An International Journal*, *22*(24), 4789–4801. doi:10.1002/hyp.7099

Pachauri, R. K., Allen, M. R., Barros, V. R., Broome, J., Cramer, W., Christ, R., & Dubash, N. K. (2014). *Climate change 2014: synthesis report. Contribution of Working Groups I, II and III to the fifth assessment report of the Intergovernmental Panel on Climate Change*. IPCC.

Parmesan, C., & Yohe, G. (2003). A globally coherent fingerprint of climate change impacts across natural systems. *Nature*, *421*(6918), 37–42. doi:10.1038/nature01286

Reisen, W. K., Lothrop, H. D., & Hardy, J. L. (1995). Bionomics of Culex tarsalis (Diptera: Culicidae) in relation to arbovirus transmission in southeastern California. *Journal of Medical Entomology*, *32*(3), 316–327. doi:10.1093/jmedent/32.3.316

Rogers, D. J., & Randolph, S. E. (1985). Population ecology of tsetse. *Annual Review of Entomology*, *30*(1), 197–216. doi:10.1146/annurev.en.30.010185.001213

Rowley, W. A., & Graham, C. L. (1968). The effect of temperature and relative humidity on the flight performance of female Aedes aegypti. *Journal of Insect Physiology*, *14*(9), 1251–1257. doi:10.1016/0022-1910(68)90018-8

Schwartz, E., Sadetzki, S., Murad, H., & Raveh, D. (2001). Age as a risk factor for severe Plasmodium falciparum malaria in nonimmune patients. *Clinical Infectious Diseases*, *33*(10), 1774–1777. doi:10.1086/322522

Shanks, G. D. (2012). Control and elimination of Plasmodium vivax. *Advances in Parasitology*, *80*, 301–341. doi:10.1016/B978-0-12-397900-1.00006-2

Shelton, R. M. (1973). The effect of temperatures on development of eight mosquito species. *Mosquito News*, *33*(1), 1–12.

Sinka, M. E., Bangs, M. J., Manguin, S., Chareonviriyaphap, T., Patil, A. P., Temperley, W. H., ... Hay, S. I. (2011). The dominant Anopheles vectors of human malaria in the Asia-Pacific region: Occurrence data, distribution maps and bionomic précis. *Parasites & Vectors*, *4*(1), 89. doi:10.1186/1756-3305-4-89

Sinka, M. E., Bangs, M. J., Manguin, S., Coetzee, M., Mbogo, C. M., Hemingway, J., ... Okara, R. M. (2010). The dominant Anopheles vectors of human malaria in Africa, Europe and the Middle East: Occurrence data, distribution maps and bionomic précis. *Parasites & Vectors, 3*(1), 117. doi:10.1186/1756-3305-3-117

Sinka, M. E., Rubio-Palis, Y., Manguin, S., Patil, A. P., Temperley, W. H., Gething, P. W., ... Hay, S. I. (2010). The dominant Anopheles vectors of human malaria in the Americas: Occurrence data, distribution maps and bionomic précis. *Parasites & Vectors, 3*(1), 72. doi:10.1186/1756-3305-3-72

Smith, D. L., & McKenzie, F. E. (2004). Statics and dynamics of malaria infection in Anopheles mosquitoes. *Malaria Journal, 3*(1), 13. doi:10.1186/1475-2875-3-13

Snow, R. W., & Gilles, H. M. (2002). The epidemiology of malaria. *Essential Malariology, 4.*

Stresman, G. H. (2010). Beyond temperature and precipitation: Ecological risk factors that modify malaria transmission. *Acta Tropica, 116*(3), 167–172. doi:10.1016/j.actatropica.2010.08.005

Svenson, J. E., MacLean, J. D., Gyorkos, T. W., & Keystone, J. (1995). Imported malaria: Clinical presentation and examination of symptomatic travelers. *Archives of Internal Medicine, 155*(8), 861–868. doi:10.1001/archinte.1995.00430080109013

Takken, W., & Scott, T. W. (Eds.). (2003). *Ecological aspects for application of genetically modified mosquitoes* (Vol. 2). Springer Science & Business Media.

Thomson, M. C., Mason, S. J., Phindela, T., & Connor, S. J. (2005). Use of rainfall and sea surface temperature monitoring for malaria early warning in Botswana. *The American Journal of Tropical Medicine and Hygiene, 73*(1), 214–221. doi:10.4269/ajtmh.2005.73.214

Trenberth, K. E. (2001). Climate variability and global warming. *Science, 293*(5527), 48–49. doi:10.1126cience.293.5527.48

Trpis, M. (1972). Development and predatory behavior of Toxorhynchites brevipalpis (Diptera: Culicidae) in relation to temperature. *Environmental Entomology, 1*(5), 537–546. doi:10.1093/ee/1.5.537

Tun-Lin, W., Burkot, T. R., & Kay, B. H. (2000). Effects of temperature and larval diet on development rates and survival of the dengue vector Aedes aegypti in north Queensland, Australia. *Medical and Veterinary Entomology, 14*(1), 31–37. doi:10.1046/j.1365-2915.2000.00207.x

Watson, R. T., & Albritton, D. L. (Eds.). (2001). *Climate change 2001: Synthesis report: Third assessment report of the Intergovernmental Panel on Climate Change.* Cambridge University Press.

WHO. (2009). *Global Malaria Action Plan World Health Organization.* Geneva, Switzerland: World Health Organization.

WHO. (2017). *World Malaria report 2017.* Geneva: World Health Organization. Available http//www.who.int/malaria/publication/world-malaria-report-2017/en/

Wijesundera, M. D. S. (1988). Malaria outbreaks in new foci in Sri Lanka. *Parasitology Today (Personal Ed.), 4*(5), 147–150. doi:10.1016/0169-4758(88)90193-7

World Health Organization. (1990). Severe and complicated malaria. *Transactions of the Royal Society of Tropical Medicine and Hygiene*, *84*(suppl 2), S1–S65. doi:10.1016/0035-9203(90)90363-J

World Health Organization. (2000). Severe falciparum malaria. *Transactions of the Royal Society of Tropical Medicine and Hygiene*, *94*(suppl 1), S1–S90. doi:10.1016/S0035-9203(00)90300-6

World Health Organization. (2015). *World Malaria Report 2015*. Geneva, Switzerland: World Health Organization.

World Malaria Report. (2011). Retrieved from http://www.who.int/malaria/world_malaria_report_2011/97892415 64403_eng.pdf

Section 7

Vector Born Diseases Related to Climate Change

Chapter 21
Waterborne Diseases Arising From Climate Change:
An Overview on the Possible Link

Soraia El Baz
Cadi Ayyad University, Morocco

Kholoud Kahime
Cadi Ayyad University, Morocco

ABSTRACT

As a result of increased frequency and intensity of heat waves, increased floods and droughts, change in climate will affect biological, physical, and chemical components of water through different paths thus enhancing the risk of waterborne diseases. Identifying the role of weather in waterborne infection is a priority public health research issue as climate change is predicted to increase the frequency of extreme precipitation and temperature events. This chapter provides evidence that precipitation and temperature can affect directly or indirectly water quality and consequently affect the health human. This chapter also highlights the complex relationship between precipitation or temperature and transmission of waterborne disease such as diarrheal disease, gastroenteritis, cryptosporidiosis, giardiasis, and cholera.

INTRODUCTION

It has been known for thousands of years, that climate has wide ranging impacts on human health. In recent years there has been growing concern about the possible effects of climatic changes including their effect on health (Gosling *et al.* 2009). Climate change is expected to affect health by several mechanisms, (1) directly, as for instance drowning or trauma in extreme weather events, or (2) indirectly, by environmental degradation, scarcity of resources,, high rates of infectious disease, weak infrastructure, and overpopulation (Patz *et al.* 2005),increasing the exposure of human populations to risk factors (Funari *et al.* 2012). Vulnerable populations include the elderly, children, urban populations, and the poor (Ebi & Paulson 2010). Climate change is expected to affect the frequency, intensity and duration of extreme water related weather events such as excessive rainfall, storm surges, floods, and drought (Semenza &

DOI: 10.4018/978-1-5225-7775-1.ch021

Menne 2009). Changes in the global water cycle also are expected, with increasing contrast in precipitation between wet and dry regions and between wet and dry seasons(Levy *et al.* 2016).

Altered pressure and temperature patterns, caused by global warming, may also shift the distribution of when and where extreme water-related events usually occur (Cann *et al.* 2013). A considerable amount of research is being conducted to map and assess risks, vulnerabilities and the impact of climate change in waterborne disease (Semenza *et al.* 2012). Some work like, Cann *et al.* (2013) have identified waterborne outbreaks potentially linked to an extreme water-related weather event and assessed the different types of extreme weather events impact the occurrence of waterborne disease. The impact of climate change on increasing of total and dissolved chemical and organic products as pesticides, heavy metals and hydrocarbons, and among the consequences on water quality is an increase pathogens (Hunter 2003; van Vliet & Zwolsman 2008). Heat waves also promotes harmful cyanobacteria blooms (Jöhnk *et al.* 2008). Water quality degradation is controlled by many factors (1) chemical (2) physical (3) biological and also socioeconomic factors (Figure 1). Most of these factors are often correlated with extreme meteorological events like heavy rainstorms or droughts. Short-term peaks in pathogen concentration including bacteria, virus and parasitic protozoa may increase disease risks considerably, and may result in outbreaks of waterborne diseases when associated with insufficient treatment at the water treatment works (WHO 2004).

The purpose of this chapter was to analyze the relationship between waterborne diseases (such as diarrheal disease, gastroenteritis, cryptosporidiosis, giardiasis and cholera) and environmental factors such as temperature, heavy rainfall, flooding, and drought. We chose these meteorological conditions because of the strong evidence that these factors are increasing due to climate change, and because of prior publications suggesting associations with waterborne diseases. Rainfall intensity and temperature are assumed to be a key determining factor in the fate and transport of pathogenic microorganisms.

Figure 1. Different factors impacting water quality

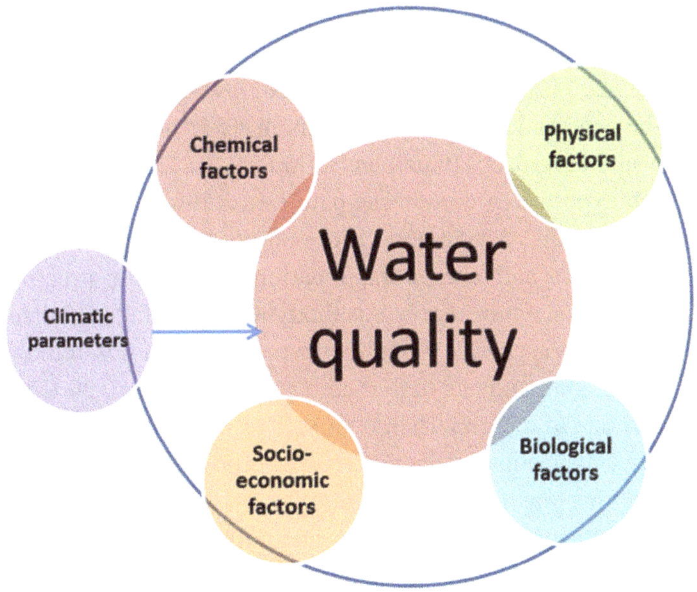

THE IMPACT OF CLIMATE CHANGE ON HUMAN HEALTH

Climatic factors have an impact on water (surface water, rivers, and oceans), but temperature and precipitations are the key factors that control water parameters (like light, turbidity, heat, dissolved oxygen...). A short-term or long-term change in one or both of these factors can affect water quality, so we study the effect of climate change on water quality, to get closer to the direct or indirect impact of temperature and precipitation on human health via waters.

Precipitation and Water Quality

Surface waters are one of the major sources for human and animal usage and their quality depends on natural or anthropogenic factors in nature. Surface water bodies are more vulnerable to fecal contamination than ground water reservoirs. In the case of heavy precipitation the microbial loads of running waters may suddenly increase substantially and reach reservoir bodies very quickly (Kistemann *et al.* 2001). Consequently, climate change can increase microbiological risks primarily in surface water sources, because surface water shows greater and more rapid variations in quality compared to ground water (Delpla *et al.* 2009).

In recent years, most of reservoirs, particularly deep-water reservoirs, have faced eutrophication and seasonal water shortage (Yu & Wang 2011). Many studies reported significant effects of extreme rainfall events on the physical and chemical characteristics of reservoirs (Girmay *et al.* 2009; O'Neil *et al.* 2012; Wang *et al.* 2012; Wu *et al.* 2012; Liu *et al.* 2014). Rainfall events cause disturbances to water bodies because they can influencing the thermal structure and they can also changes in the environment and in the hydrological conditions of reservoirs (Huang *et al.* 2014). Moreover, the particulate pollutants carried by runoff are brought into reservoirs, then they cause serious exogenous pollution, which can promote the overproduction of toxic algae (Zhang *et al.* 2006).

On the other hand, heavy precipitation events increase the likelihood of water supply contamination due to the risk of sewer overflows (Moors *et al.* 2013). Particularly, where water infrastructure is old the vulnerability (the risk of contamination) of the drinking water increases after heavy precipitation events (Semenza & Nichols 2007; Delpla *et al.* 2009).Also, low rainfall may contribute to waterborne infections by increasing the percentage of sewage effluent in rivers or by increasing risk of groundwater contamination when the water table drops (Guzman Herrador *et al.* 2015). Then, extreme rainfall events may also increasing human exposure to waterborne pathogens (Fewtrell *et al.* 2011).

It is generally acknowledged that during periods of contribution from rainfall-induced runoff, concentrations of microbial pathogens and faecal contamination indicators in streams and rivers are higher (Ashbolt & Roser 2003). In these circumstances it is thought that rainfall mobilizes and transports non-point source microbial particles via runoff, though in some cases it is possible that the increased flow also leads to re-suspension of contaminants in the sediment (Nagels *et al.* 2002).

Surface runoff, which is one of the important components of the hydrologic cycle, is the important water source of many rivers, the primary driver of river hydrodynamics as well as the carrier of a variety of pollutants into receiving water bodies (Kalkhoff *et al.* 2016). Many studies have demonstrated the increase of concentrations and/or loadings of various pollutants including total suspended solids, nutrients, and microorganisms (e.g., *E. coli* and pathogens) during storm events and extensive rainfalls (Hamilton & Luffman 2009; McCarthy *et al.* 2012; Chen & Chang 2014). Heavy rainfalls, especially after a long dry period which allows more pollutants accumulated in runoff pathways and available for transport by runoff, could wash-off a lot of pollutants into rivers (Bae 2013).

In 2001 it was estimated that 20% of the world's population lacked access to safe drinking water, and that more than five million people died annually from illnesses associated with unsafe drinking water or inadequate sanitation (Hunter *et al.* 2001). Additionally, precipitation and other hydro-meteorological parameters affect directly or indirectly water quantity (Ackerman & Weisberg 2003; Olyphant *et al.* 2003). High concentrations of fecal indicator-bacteria have been linked to GI-related health risks (Wade *et al.* 2010), specifically *E. coli* concentrations in freshwater (Pruss 1998). Heavy precipitation and subsequent storm water runoff can flush pathogens and other microorganisms directly into nearby surface water, resulting in increased concentrations of bacteria, and increased risk of waterborne disease (Curriero *et al.* 2001; Schuster *et al.* 2005). It's reported that precipitation is positively correlated with *E. coli* concentrations in recreational water (Byappanahalli *et al.* 2010; Nevers & Whitman 2011).

Rain and spring thaw events influence the horizontal and vertical movement of pathogens at the soil level (McGechan and Lewis 2002),by increasing water flow, therefore they could increase the speed at which the pathogen enters the source water and drinking water supply as well as pathogen load(Unc & Goss 2003). For example, for typical Norwegian lakes, during autumn and spring circulations an increase in faecal indicator bacteria is observed at deep water intake (Hem 2008). Rainfall intensity is considered the primary force driving the inflow of sediment into a certain area, while the occurrence and duration of rainfall events have significant effects on the inflow water temperature (Qin *et al.* 2010). The stratification of productive reservoirs could lead to long periods of anaerobic conditions at the bottom of the reservoir and a significant deterioration of water quality is envisaged (Fan & Kao 2008).

In addition to precipitation and river discharge, other physical mechanisms such as tides, wind-driven transport and river plumes may be responsible for the concentration or dispersal of pathogens in estuaries and the coastal ocean (Olyphant 2005; Pettigrew *et al.* 2005; Warrick *et al.* 2007). Consequently, the pollutants can be transported from their sources to new locations, resulting in reduced water quality far from the source of the pollution (Washburn *et al.* 2003). Also, river plumes can be the important pathways for the transport of faecal pollution and the accompanying pathogens (Schiff & Bay 2003; Gersberg *et al.* 2004; Nagvenkar & Ramaiah 2009).

Temperature and Water Quality

Climate change has both a direct and indirect effect on the quality of surface waters. It is well known that all physico-chemical vary with temperature, and frequently increasing endothermic reactions. For these reasons, temperature must be as the main factor affecting almost all biological and physico-chemical reactions (Whitehead *et al.* 2009). Consequently, several transformations like dissolution, solubilisation, complexation, degradation, evaporation, etc. related to water will be favored by water temperature increase (Delpla *et al.* 2009). Humans are exposed to a wide range of climate sensitive pathogens (such as bacteria, viruses, parasites, and algae) through drinking water consumption, and recreational water use. At elevated temperatures, some pathogens proliferate, whereas other pathogens show faster die off or inactivation (Schijven & de Roda Husman 2005; Semenza *et al.* 2012).

For low river flow rates, the main effect on water quality is affected by (1) a temperature increase, a concentration increase, (2) dissolved substances in water (3) a concentration decrease of dissolved oxygen (Prathumratana *et al.* 2008; van Vliet & Zwolsman 2008). Changing trends in average temperature is anticipated to have an impact on vector-borne disease incidence and distribution (Patz *et al.* 1996).

Also, climate-related increases in sea surface temperature and sea level can lead to higher incidence of waterborne infectious and toxin-related illnesses (Bezirtzoglou *et al.* 2011). Because of high temperature, some population and certain regions are more vulnerable to infectious disease due to their lack of the ability to effectively respond to the stresses imposed by this factor (Wei *et al.* 2012). In fact, many works suggest the increasing risk of parasite development with temperature (Zhou *et al.* 2008; McCreesh *et al.* 2015).In general, pathogen growth and reproduction are temperature dependent (Thomas *et al.* 2006). Water temperature is known to influence the potential for *Legionella* to colonize water systems (Parker *et al.* 1983). In freshwaters, free-living parasites are of a concern to swimmers such as *Naegleria fowleri*. This parasite causes primary amoebic meningoencephalitis, generally in children or young adults, and is acquired through the parasite entering the nasal passages. Water temperature is a significant factor in the occurrence and distribution of this organism (Kilvington & Beeching 1995).

Changes to temperature can accentuate the wildfires, encourage invasive species, or increase forest mortality, resulting in both short-term impacts on water quality (Stanford *et al.* 2014). Higher temperatures can facilitate the growth of algae and toxic cyanobacteria (Paerl & Paul 2012), and decrease survival of enteric pathogens in the aquatic environment (Azevedo *et al.* 2008; Carratala *et al.* 2013). In a warmer climate cyanobacteria could be more abundant than phytoplankton (Arheimer *et al.* 2005).High water temperature and nutrient concentration promote the appearance of massive cyanobacteria bloom in many waterbodies (Hunter 2003). Like *Microcystis* which can produce microcystin, could become invasive with climate warming, over diatoms and green algae (Jöhnk *et al.* 2008). Moreover, an earlier annual warming in temperate countries permits an earlier and more important growth of alga (Wiedner *et al.* 2007).

On the other hand, waterborne pathogens could be spread within the freshwater after a contamination by animal or human waste due to heavy rainfall discharge in combined sewer systems. When the flow exceeds the combined sewer systems capacity, the sewers overflow directly into surface water body (Charron *et al.* 2004). Moreover, higher water temperatures will probably lead to a pathogen survival increase in the environment (Hunter 2003). Additionally, water quality can be altered by solar irradiation increase and especially characteristics of natural organic matter in freshwaters systems both by warming and UVB radiation (increasing photolysis) (Soh *et al.* 2008). Phototransformation can promote the formation of UV transformation products from organic micropollutants such as pharmaceuticals (Canonica *et al.* 2008).

IMPACTS OF CLIMATE CHANGE ON WATERBORNE DISEASES

The transmission of infectious diseases is determined by many factors (1) social, (2) economic and ecological conditions, (3) access to health care, and (3) intrinsic human immunity (Jones *et al.* 2008). The relation between climate change and infectious diseases can be explain by (a) higher proliferation and reproduction rates at higher temperatures, (b) extended transmission season, (c) changes in ecological balances, and (d) climate-related migration of vectors, reservoir hosts, or human populations (Lafferty 2009).

As Putignani and Menichella (2010), Parasite, host and environmental indexes acting as key factors for the global burden of water disease (Figure 2).

Several international studies have shown an association between heavy rainfall and elevated concentrations of pathogens in surface waters (Kistemann *et al.* 2002) as well as with waterborne outbreaks (Nichols *et al.* 2009), through the transport of bacteria, viruses, or small parasites into water systems or wellheads (Patz *et al.* 2000; Rose *et al.* 2000). For quite sometimes, scientists have endeavored to predict

Figure 2. Diagram of factors leading to water-diseases apparition and spread

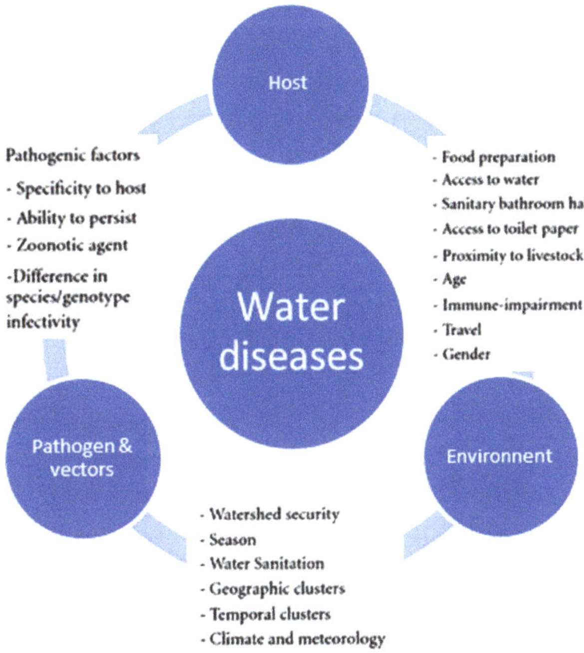

large-scale responses of infectious diseases to climate change (Altizer *et al.* 2013), as many components of the transmission cycles of vectorborne diseases are inextricably tied to climate (Harvell *et al.* 2002; Altizer *et al.* 2013).

A study of the relationship between rainfall and waterborne diseases in the USA used 548 reported outbreaks between 1948 and 1994 and precipitation data of the National Climatic Data Centre (Curriero *et al.* 2001). They found that 51 percent of waterborne outbreaks occurring in the USA were preceded by precipitation above the 90th percentile, and 68% by events above the 80th percentile, with outbreaks due to surface water contamination showing the strongest association with extreme precipitation during the month of the outbreak. Thomas *et al.* (2006) reported that there is a association between extreme rainfall and snowmelt with 92 Canadian waterborne disease outbreaks between 1975 and 2001. Additionally, Rose *et al.* (2000) observed that between 1971 and 2004, 20 to 40 percent of outbreaks occurring in the USA were associated with precipitation above the 90th percentile. There have been 111 outbreaks of disease in England and Wales that were thought to be transmitted through the consumption of drinking water over the 20th century. While outbreaks in the nineteenth century were caused by typhoid, paratyphoid and cholera and in the first half of the 20th by typhoid and paratyphoid, these outbreaks were uncommon in the second half of the 20th century and in the last 20 years outbreaks have been dominated by the more newly recognized pathogens cryptosporidium and campylobacter (Galbraith *et al.* 1987; Furtado *et al.* 1998; Smith *et al.* 2006). Many studies from elsewhere in the world suggests that waterborne disease outbreaks are preceded by extreme precipitation events (Curriero *et al.* 2001) and that the seasonal contamination of surface water may explain some of the variability in the occurrence of many waterborne diseases (Patz *et al.* 2008).

Among waterborne diseases, we chose to study gastroenteritis, diarrheal disease, cryptosporidiosis, giardiasis and cholera in order to assess their relationship with climate change.

Gastroenteritis

Gastrointestinal infection (or infectious gastroenteritis) is a medical condition associated with inflammation of the gastrointestinal (GI) tract that involves both the stomach and the small intestine.Gastrointestinal infection is caused by bacteria, viruses, and parasites (Singh and Fleurat 2010).Changes in the environmental conditions, such as humidity(Chew *et al.* 1998), temperature cycles (Checkley *et al.* 2000), rain patterns and winds, are associated with seasonality of infectious diseases.The incidence of GI infections and the rate of outbreaks are seasonal and related to environmental parameters, particularly temperature (Ghazani *et al.* 2018). For example, In Scotland, the peak of GI infection was in summer (Eze *et al.* 2013), caused by particularly *Campylobacter* and *Salmonella*. But, in Fukuoka, Japan, GI infection was in winter and spring (Onozuka and Hashizume), caused by *Norovirus* and *Rotavirus* respectively (Onozuka & Hashizume 2011).

Changes in humidity have been reported to facilitate viral persistence for gastrointestinal infectious agents such as rotaviruses, increasing the risk of transmission through contaminated surfaces (Dowell 2001). Weather factors such as temperature, humidity, and rainfall have been suggested as important factors in the spread and seasonality of infectious gastroenteritis (Onozuka & Hashizume 2011). As water consumption is higher during the hot seasons, this may increase the transmission of waterborne GI pathogens (Zhou *et al.* 2013). It is reported that rainfall may increase the rate of gastroenteritis: (1) heavy rainfall can contribute to contaminate water supplies with either human or animal fecal pathogens through flushing them from sewers and leaching them into water supply bodies (rivers and lakes), and (2) during low rainfall and drought periods, the concentration of fecal pathogens may increase (Jofre *et al.* 2010).

The largest reported outbreak of *Escherichia coli* O157:H7, which was the significant pathogen for the Walkerton outbreak of waterborne gastroenteritis and was linked to contaminated well water. Unusually heavy rainfall, which was preceded by a drought, coincided with this major outbreak (Patz *et al.* 2000). Investigative on the Walkerton outbreak, further stated that the onset of the majority of *E. coli* O157:H7 and Campylobacter cases occurred after 12 May, with a peak in illness between 17 and 19 May. These suggest that heavy rainfall in mid-May was responsible for the significant contamination of the water distribution system, resulting in the majority of the illnesses (Auld *et al.* 2004).

Climate change study investigating the relationship of gastroenteritis with increasing temperature in Australia showed that the number of cases was predicted to rise substantially over the coming century (Bambrick *et al.* 2008), due to increases in cases caused by *Salmonella* and other bacteria. Cases of gastroenteritis caused by enteric pathogens such as *Clostridium, Vibrio, Aeromonas spp.* appear to peak in summer (Bambrick *et al.* 2008). Previous studies have documented a positive association between salmonellosis and increasing temperature (Lal *et al.* 2013) this positive association has been reported from Europe, Canada, the United States, Australia and NZ (Lal *et al.* 2016). A number of cases of gastroenteritis in Australia caused by *Clostridium perfringens, Vibrio parahaemolyticus, Aeromonas spp., Giardia, Norovirus* and *Rotavirus* have been shown to peak in summer (Dear *et al.* 2007).

El Niño Southern Oscillation (ENSO) is the most prominent source of inter-annual global climate variability which affects weather conditions throughout the world (Shaman and Lipsitch 2013). In addition to local weather factors, several studies reported that ENSO play important roles in the transmission of infectious diseases (Onozuka 2014). After analyzed monthly data for cases of infectious gastroenteritis in Fukuoka, Japan from 2000 to 2012 using cross-wavelet coherency analysis to assess the pattern of associations between indices for the Indian Ocean Dipole (IOD) and ENSO, Onozuka (2014)have reported that there is a significant association between infectious gastroenteritis cases and ENSO. Additionally, an increase in the cases of infectious gastroenteritis to more than 200% was reported in Lima, Peru as a result of the ENSO, when the ambient temperature increased to >5°C above the expected values for the winter season (Checkley *et al.* 2000).

Diarrheal Disease

Climate change has the potential to severely impact human health worldwide, including diarrheal diseases, which are highly affected by environmental drivers (Eisenberg *et al.* 2007; Patz *et al.* 2014). Some authors like (Campbell-Lendrum and Woodruff 2006; Kolstad & Johansson 2011) have reviewed associations between temperature and all-cause diarrhea, also others authors examined the relation meteorological variables and specific diarrheal pathogens, extreme weather events and waterborne diseases, and climatic influences on pathogens in the environment (Levy *et al.* 2009; Jagai *et al.* 2012; Cann *et al.* 2013; Sterk *et al.* 2013). Diarrheal disease remains among the top five causes of death in low- and middle-income countries, particularly among children under 5 years of age (Boschi-Pinto *et al.* 2008). High levels of water volume were associated with infectious gastrointestinal (GI) illness in Canada (Harper *et al.* 2011), as well in Bangladesh (Hashizume *et al.* 2007), and in Taiwan (Chen *et al.* 2012).

Several studies have shown relationships with wet weather, and raw water quality parameters (Kistemann *et al.* 2002; Signor *et al.* 2005; Astrom *et al.* 2007). Particularly, Heavy precipitation has been linked to the majority of observed drinking water-related outbreaks of gastrointestinal diseases in developed nations worldwide (Curriero *et al.* 2001; Thomas *et al.* 2006; Nichols *et al.* 2009). Also, previous studies have reported a delayed onset of diarrheal disease following heavy rainfall events (Aramini *et al.* 2000; Schwartz *et al.* 2000; Egorov *et al.* 2003). One explanation for the observed lag could be that the incubation period of waterborne pathogens ranges from one day, for pathogens (for *Shigella*, *Salmonella*, and *Rotavirus*), to two weeks for pathogens (for *Cryptosporidium* and *E. coli*) (Haley *et al.* 2009; Jagai *et al.* 2009). In general, cases of GI peak within seven days of exposure to contaminated water (Eisenberg *et al.* 1998; Naumova *et al.* 2003). Health risks associated with exposure to contaminated recreational water (e.g. *E. coli*) include skin, eye, ear, and upper respiratory irritations and infections, as well as GI illness (Naumova *et al.* 2003; Wade *et al.* 2008). Consequently, Swimmers may be directly impacted by poor recreational water quality; elderly non-swimmers may be exposed to pathogens via drinking water as a result of increased turbidity following extreme events (Schwartz *et al.* 2000). In USA, extreme weather events can cause low water pressure following a freezing weather can impact water supply systems, and led to increased rates of acute gastrointestinal illness (Gargano *et al.* 2015). According to Bush *et al.* (2014) who evaluated the association between extreme precipitation and GI illness-related hospital admissions in Chennai, India, from 2004 to 2007. They concluded that the Hospital admissions related to GI illness were positively associated with extreme precipitation in Chennai, with positive cumulative risk ratios for a 15-day period following an extreme event in all age groups. Projected changes in precipitation and extreme weather events suggest that climate change will have important implications for human health in India, where health disparities already exist.

On the other hand, after a period of heavy rainfall in USA, an runoff can compromised the efficiency of the drinking water treatment plant and caused 403000 cases of intestinal illness and 54 deaths occurred (MacKenzie *et al.* 1995; Hoxie *et al.* 1997).

Moreover, the World Health Organization (WHO) reported that warming by 1°C was associated with a 5% increase in diarrhea (WHO 2004). Additionally, hot weather might cause a higher water consumption and less conscientious hygiene practices, which favors diarrhea transmission (Black & Lanata 1995).

Cryptosporidiosis and Giardiasis

Worldwide cryptosporidiosis is widespread in many developed and developing countries (Hunter 2003). People become infected with *Cryptosporidium* through animal and human transmission (Casemore *et al.* 1997), including via swimming pools (Rose 1997), or by infected food and drink (Casemore *et al.* 1997). Giardia is another protozoan enteropathogen, according to Hoque *et al.* (2004)approximately 200 million people are infected with the parasite (*Giardia*) globally. Studies of outbreaks have demonstrated the importance of weather events (heavy precipitation and high river flows) in transmission through public and private water supplies (Curriero *et al.* 2001).

One of the main routes of transmission of *Cryptosporidium* and *Giardia* to humans is through water (Rose 1997). Heavy precipitation onto saturated soil could plausibly facilitate the transfer of *Cryptosporidium* oocysts in animal manure into surface water (Lake *et al.* 2005; Thurston-Enriquez *et al.* 2005). This factor has been found to be positively correlated with the concentration of *Cryptosporidium* oocysts and/or *Giardia* cysts in surface and runoff water (Miller *et al.* 2007), and to waterborne disease outbreaks of giardiasis (Thomas *et al.* 2006). Another factor like temperature also was positively associated with cryptosporidiosis rates (Lake *et al.* 2008).

There is strong evidence that the concentration of *Cryptosporidium* oocysts increases in drinking water and surface water sources after heavy rainfall due to runoff (Muchiri *et al.* 2009). Further evidence suggests that seasonal changes in temperature and precipitation affect the incidence of cryptosporidiosis around the world (Jagai *et al.* 2009). It has been shown that long-term climate change can affect seasonal patterns of infectious diseases by lengthening the transmission cycle (Harvell *et al.* 2002).

Once the *Cryptosporidium* has been washed into rivers the cool and moist conditions are ideal for its survival. However, the oocysts may not simply remain suspended in the water column but can pass into river sediments and survive for extended periods (Robertson *et al.* 1992). The mechanisms for oocyst mobilization, sedimentation and resuspension were coinciding with heavy rainfall events (Casemore *et al.* 1997). Young *et al.* (2015) have revealed that the average odds of identifying *Cryptosporidium* oocysts and *Giardia* cysts in fresh surface waters were increased between 2-3 times during and after extreme weather events, compared to normal conditions. The concentrations of these protozoans under these conditions were also higher

Other studies have attributed specific waterborne disease outbreaks and contamination incidents to adverse consequences that have resulted from preceding rainfall events, e.g. the waterborne disease outbreak of giardiasis in Montana was related to rainfall(Weniger *et al.* 1983), as was the largest reported waterborne disease outbreak ever documented, which occurred in Milwaukee, Wis, in 1993. The Milwaukee outbreak, which resulted in the deaths of 54 people and more than 403,000 ill, was related to heavy rainfall and associated runoff (Mac Kenzie *et al.* 1994; Hoxie *et al.* 1997).

A recent large outbreak of waterborne giardiasis occurred in Bergen, Norway during 2004 and affected over 1,500 individuals (Robertson *et al.* 2006). The Bergen outbreak was most probably caused by heavy rainfall which caused sewage contamination of the lake used as drinking water source, coupled with deficiencies in the water treatment.

Land use (agriculture and urbanization) has also a direct effect on the relationship between precipitation, run-off and water quality (Mallin *et al.* 2009). Changes in land use can result in greater erosion, more run-off and increased turbidity in rivers (Cebecauer & Hofierka 2008). For example, Impermeable surfaces of urbanized areas (Burns *et al.* 2005) increase run-off and the transport of pollutants into streams, rivers and the ocean. The strong relationship between land use and *Giardia* and *Cryptosporidium* contamination of water may be related to agricultural land use (Hunter & Thompson 2005).

Cholera

During a cholera epidemic in 1854 in Florence, Italy, Pacini (1854) first described the comma-shaped gram-negative Vibrio, the "comma bacillus," responsible for cholera, which was subsequently named *Vibrio cholerae* by Robert Koch. Outbreaks of Cholera were linked to extreme precipitation and temperature in the Lake Victoria Basin (Olago *et al.* 2007), Bangladesh (Pascual *et al.* 2000), and Peru caused 30,000 cases and 114 deaths in the first 7 days (Checkley *et al.* 2000).

The presence of *Vibrio spp.* is mainly detected in warmer waters. Many studies have demonstrated that the abundance of V*ibrio spp.* are largely correlated with high water temperature (e.g. sea, coastal and brackish waters) (Lutz *et al.* 2013; Rashid *et al.* 2013; Vezzulli *et al.* 2013). The warmer temperatures enhance the persistence of V*ibrio spp.* by promoting biofilm formation and colonization of environmental (Stauder *et al.* 2010). An increase in temperature in Asia and South America threats water quality with regard to waterborne diseases particularly cholera disease (Hunter 2003).

Vibrio cholerae and other bacteria thrive in warm waters of moderate salinity (Lipp *et al.* 2001) and are closely associated with aquatic invertebrates (Kaneko & Colwell 1975). Therefore, with a changing climate, the geographic range of these pathogens may also change, potentially resulting in increased exposure and risk of infection for humans.

Some analysis of in situ hydrographic and meteorological data sets, and cholera case data for Bangladesh, have demonstrate evidence that cholera cases occur following a rise in ocean surface temperatures(Colwell 1996). Regional correlation maps have shown a significant association of temperature with the recent cholera data for Dhaka, Bangladesh (Pascual *et al.* 2000). The effect of more extreme ENSO and global warming conditions of drought and flood, might impact the cholera transmission. Cases of cholera have been imported to Europe from Kenya (Houghton 1997), where the recent spreading epidemic has been linked to the El Niño phenomenon, which originates in the pacific ocean. Other factors such as human migration and damage to health infrastructures also could indirectly contribute to disease transmission.

The copepods (zooplankton), which feed on algae, can serve as reservoirs for *V. cholerae* and other enteric pathogens (Colwell 1996); warm sea surface temperature are generally associated with El Niño, and promote copepods bloom. In Peru, El Niño phenomenon in 1991 may have been the trigger that resulted in the resurgence of cholera. Warm waters along the coast, coupled with plankton blooms driven by rains of this phenomenon, may have helped to amplify the population of *V. cholerae* already in the environment (Colwell 1996; Mourino-Perez 1998). Both Wandiga *et al.* (2010) and Cash *et al.* (2014) have concluded that ENSO driven climate variation has been linked to increases in the transmission cycle of vector-borne diseases and fecal-oral disease. The focus of *V. cholerae* may thrive or may disappear as a result of the changes of temperature, pH, salinity, and composition of plant and animal life (Shope 1991).

Figure 3. Diagram showing how climate impacts human population health, directly by acting on natural component and indirectly by impacting population and their socioeconomic factors

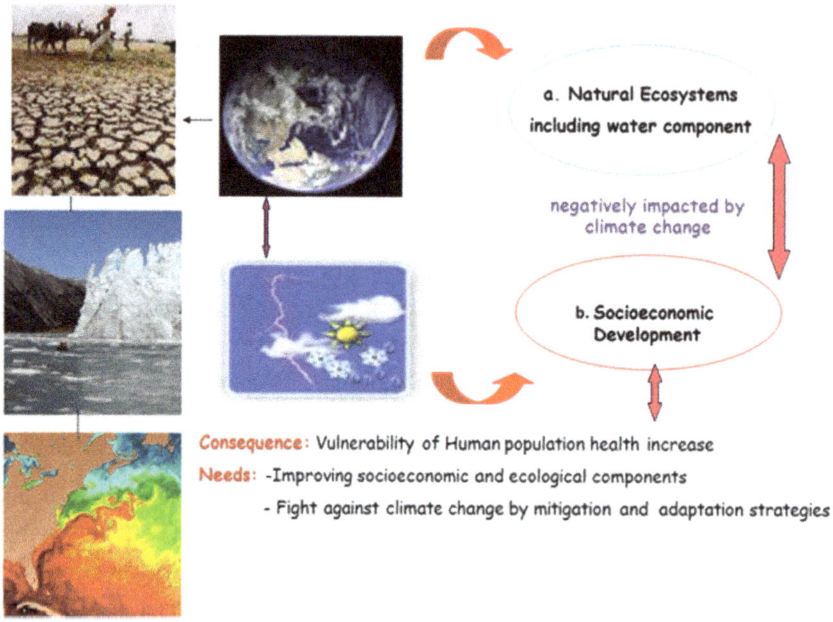

CONCLUSION

The World Health Organization (WHO) have emphasized the need for strengthening partnerships between health and climate experts, to improve scientific evidence of the linkages between health and climate drivers (WHO 2012). Furthermore, extreme weather events can damage electrical, communication and transportation infrastructure, leaving water supply systems and operations vulnerable to other water quality impacts (Stanford *et al.* 2014). Also, Direct impacts to water quality from extreme weather are relatively simple to identify, but their indirect impacts can be overlooked, especially when they occur even years, after the onset of the particular event (Stanford *et al.* 2014). According to our review, we can summarize the factors that are simultaneously involved in waterborne disease outbreaks as follows: (1) a source of contamination (humans or animals); (2) fate and transport of the contaminant from source; (3) inadequate treatment; and (4) detection and reporting of the outbreak. Therefore, understanding the contextual interactions between climate change and diseases transmission and spread in global view combined is the first challenge in understanding the impact of long-term climate change on waterborne disease (Figure 3.).

REFERENCES

Ackerman, D., & Weisberg, S. B. (2003). Relationship between rainfall and beach bacterial concentrations on Santa Monica bay beaches. *Journal of Water and Health*, *1*(2), 85–89. doi:10.2166/wh.2003.0010 PMID:15382737

Altizer, S., Ostfeld, R. S., Johnson, P. T., Kutz, S., & Harvell, C. D. (2013). Climate change and infectious diseases: From evidence to a predictive framework. *Science, 341*(6145), 514–519. doi:10.1126cience.1239401 PMID:23908230

Aramini, J., McLean, M., Wilson, J., Holt, J., Copes, R., Allen, B. and Sears, W. (2000) Drinking water quality and health-care utilization for gastrointestinal illness in greater Vancouver. *Releve des maladies transmissibles au Canada, 26*, 211-214.

Arheimer, B., Andreasson, J., Fogelberg, S., Johnsson, H., Pers, C. B., & Persson, K. (2005). Climate change impact on water quality: Model results from southern Sweden. *Ambio, 34*(7), 559–566. doi:10.1579/0044-7447-34.7.559 PMID:16435746

Ashbolt, N. J., & Roser, D. J. (2003). Interpretation and management implications of event and baseflow pathogen data (CDROM). In M. J. Pfeffer, D. J. V. Abs, & K. N. Brooks (Eds.), *Watershed Management for Water Supply Systems*. New York: American Water Resources Association.

Astrom, J., Petterson, S., Bergstedt, O., Pettersson, T. J., & Stenstrom, T. A. (2007). Evaluation of the microbial risk reduction due to selective closure of the raw water intake before drinking water treatment. *Journal of Water and Health, 5*(S1), 81–97. doi:10.2166/wh.2007.139 PMID:17890838

Auld, H., MacIver, D., & Klaassen, J. (2004). Heavy rainfall and waterborne disease outbreaks: The Walkerton example. *Journal of Toxicology and Environmental Health. Part A., 67*(20-22), 1879–1887. doi:10.1080/15287390490493475 PMID:15371222

Azevedo, N. F., Almeida, C., Fernandes, I., Cerqueira, L., Dias, S., Keevil, C. W., & Vieira, M. J. (2008). Survival of gastric and enterohepatic *Helicobacter spp.* in water: Implications for transmission. *Applied and Environmental Microbiology, 74*(6), 1805–1811. doi:10.1128/AEM.02241-07 PMID:18245254

Bae, H.-K. (2013). Changes of river's water quality responded to rainfall events. *Environ Ecol Res, 1*, 21–25.

Bambrick, H., Dear, K., Woodruff, R., Hanigan, I., & McMichael, A. (2008) The impacts of climate change on three health outcomes: Temperature-related mortality and hospitalisations, salmonellosis and other bacterial gastroenteritis, and population at risk from dengue. Garnaut Climate Change Review.

Bezirtzoglou, C., Dekas, K., & Charvalos, E. (2011). Climate changes, environment and infection: Facts, scenarios and growing awareness from the public health community within Europe. *Anaerobe, 17*(6), 337–340. doi:10.1016/j.anaerobe.2011.05.016 PMID:21664978

Black, R. E., & Lanata, C. F. (1995). *Epidemiology of diarrhoeal diseases in developing countries. Infections of the gastrointestinal tract*. New York: Raven Press.

Boschi-Pinto, C., Velebit, L., & Shibuya, K. (2008). Estimating child mortality due to diarrhoea in developing countries. *Bulletin of the World Health Organization, 86*(9), 710–717. doi:10.2471/BLT.07.050054 PMID:18797647

Burns, D., Vitvar, T., McDonnell, J., Hassett, J., Duncan, J., & Kendall, C. (2005). Effects of suburban development on runoff generation in the Croton River basin, New York, USA. *Journal of Hydrology (Amsterdam), 311*(1-4), 266–281. doi:10.1016/j.jhydrol.2005.01.022

Bush, K. F., O'Neill, M. S., Li, S., Mukherjee, B., Hu, H., Ghosh, S., & Balakrishnan, K. (2014). Associations between extreme precipitation and gastrointestinal-related hospital admissions in Chennai, India. *Environmental Health Perspectives*, *122*, 249–254. PMID:24345350

Byappanahalli, M. N., Whitman, R. L., Shively, D. A., & Nevers, M. B. (2010). Linking non-culturable (qPCR) and culturable enterococci densities with hydrometeorological conditions. *The Science of the Total Environment*, *408*(16), 3096–3101. doi:10.1016/j.scitotenv.2010.04.051 PMID:20546850

Campbell-Lendrum, D., & Woodruff, R. (2006). Comparative risk assessment of the burden of disease from climate change. *Environmental Health Perspectives*, *114*(12), 1935–1941. doi:10.1289/ehp.8432 PMID:17185288

Cann, K. F., Thomas, D. R., Salmon, R. L., Wyn-Jones, A. P., & Kay, D. (2013). Extreme water-related weather events and waterborne disease. *Epidemiology and Infection*, *141*(04), 671–686. doi:10.1017/S0950268812001653 PMID:22877498

Canonica, S., Meunier, L., & von Gunten, U. (2008). Phototransformation of selected pharmaceuticals during UV treatment of drinking water. *Water Research*, *42*(1-2), 121–128. doi:10.1016/j.watres.2007.07.026 PMID:17709124

Carratala, A., Rusinol, M., Rodriguez-Manzano, J., Guerrero-Latorre, L., Sommer, R., & Girones, R. (2013). Environmental effectors on the inactivation of human adenoviruses in water. *Food and Environmental Virology*, *5*(4), 203–214. doi:10.100712560-013-9123-3 PMID:23955425

Casemore, D. P., Wright, S. E., & Coop, R. L. (1997). Cryptosporidiosis: Human and Animal Epidemiology. In R. Fayer (Ed.), *Cryptosporidium and Cryptosporidiosis*. Boca Raton, FL: CRC Press.

Cash, B. A., Rodo, X., Emch, M., Yunus, M., Faruque, A. S., & Pascual, M. (2014). Cholera and shigellosis: Different epidemiology but similar responses to climate variability. *PLoS One*, *9*(9), e107223. doi:10.1371/journal.pone.0107223 PMID:25229494

Cebecauer, T., & Hofierka, J. (2008). The consequences of land-cover changes on soil erosion distribution in Slovakia. *Geomorphology*, *98*(3-4), 187–198. doi:10.1016/j.geomorph.2006.12.035

Charron, D., Thomas, M., Waltner-Toews, D., Aramini, J., Edge, T., Kent, R., ... Wilson, J. (2004). Vulnerability of waterborne diseases to climate change in Canada: A review. *Journal of Toxicology and Environmental Health. Part A.*, *67*(20-22), 1667–1677. doi:10.1080/15287390490492313 PMID:15371208

Checkley, W., Epstein, L. D., Gilman, R. H., Figueroa, D., Cama, R. I., Patz, J. A., & Black, R. E. (2000). Effect of El Nino and ambient temperature on hospital admissions for diarrhoeal diseases in Peruvian children. *Lancet (London, England)*, *355*, 442–450. PMID:10841124

Chen, H. J., & Chang, H. (2014). Response of discharge, TSS, and *E. coli* to rainfall events in urban, suburban, and rural watersheds. *Environmental Science. Processes & Impacts*, *16*(10), 2313–2324. doi:10.1039/C4EM00327F PMID:25096028

Chen, M. J., Lin, C. Y., Wu, Y. T., Wu, P. C., Lung, S. C., & Su, H. J. (2012). Effects of extreme precipitation to the distribution of infectious diseases in Taiwan, 1994-2008. *PLoS One*, *7*(6), e34651. doi:10.1371/journal.pone.0034651 PMID:22737206

Chew, F. T., Doraisingham, S., Ling, A. E., Kumarasinghe, G., & Lee, B. W. (1998). Seasonal trends of viral respiratory tract infections in the tropics. *Epidemiology and Infection, 121*(1), 121–128. doi:10.1017/S0950268898008905 PMID:9747763

Colwell, R. R. (1996). Global climate and infectious disease: The cholera paradigm. *Science, 274*(5295), 2025–2031. doi:10.1126cience.274.5295.2025 PMID:8953025

Curriero, F. C., Patz, J. A., Rose, J. B., & Lele, S. (2001). The association between extreme precipitation and waterborne disease outbreaks in the United States, 1948-1994. *American Journal of Public Health, 91*(8), 1194–1199. doi:10.2105/AJPH.91.8.1194 PMID:11499103

Dear, K., Woodruff, R., Hanigan, I. and McMichael, A. (2007) The impacts of climate change on three health outcomes: temperature-related mortality and hospitalisations, salmonellosis and other bacterial gastroenteritis, and population at risk from dengue. *Garnaut Climate Change Review*, 1-47.

Delpla, I., Jung, A. V., Baures, E., Clement, M., & Thomas, O. (2009). Impacts of climate change on surface water quality in relation to drinking water production. *Environment International, 35*(8), 1225–1233. doi:10.1016/j.envint.2009.07.001 PMID:19640587

Dowell, S. F. (2001). Seasonal variation in host susceptibility and cycles of certain infectious diseases. *Emerging Infectious Diseases, 7*(3), 369–374. doi:10.3201/eid0703.017301 PMID:11384511

Ebi, K. L., & Paulson, J. A. (2010). Climate change and child health in the United States. *Current Problems in Pediatric and Adolescent Health Care, 40*(1), 2–18. doi:10.1016/j.cppeds.2009.12.001 PMID:20114000

Egorov, A. I., Naumova, E. N., Tereschenko, A. A., Kislitsin, V. A., & Ford, T. E. (2003). Daily variations in effluent water turbidity and diarrhoeal illness in a Russian city. *International Journal of Environmental Health Research, 13*(1), 81–94. doi:10.1080/0960312021000071567 PMID:12745350

Eisenberg, J. N., Desai, M. A., Levy, K., Bates, S. J., Liang, S., Naumoff, K., & Scott, J. C. (2007). Environmental determinants of infectious disease: A framework for tracking causal links and guiding public health research. *Environmental Health Perspectives, 115*(8), 1216–1223. doi:10.1289/ehp.9806 PMID:17687450

Eisenberg, J. N., Seto, E. Y., Colford, J. M. Jr, Olivieri, A., & Spear, R. C. (1998). An analysis of the Milwaukee cryptosporidiosis outbreak based on a dynamic model of the infection process. *Epidemiology (Cambridge, Mass.), 9*(3), 255–263. doi:10.1097/00001648-199805000-00008 PMID:9583416

Eze, J. I., Scott, E. M., Pollock, K. G., Stidson, R., Miller, C. A., & Lee, D. (2013). The association of weather and bathing water quality on the incidence of gastrointestinal illness in the west of Scotland. *Epidemiology and Infection, 142*(06), 1289–1299. doi:10.1017/S0950268813002148 PMID:24007797

Fan, C. W., & Kao, S. J. (2008). Effects of climate events driven hydrodynamics on dissolved oxygen in a subtropical deep reservoir in Taiwan. *The Science of the Total Environment, 393*(2-3), 326–332. doi:10.1016/j.scitotenv.2007.12.018 PMID:18243280

Fewtrell, L., Kay, D., Watkins, J., Davies, C., & Francis, C. (2011). The microbiology of urban UK floodwaters and a quantitative microbial risk assessment of flooding and gastrointestinal illness. *Journal of Flood Risk Management, 4*(2), 77–87. doi:10.1111/j.1753-318X.2011.01092.x

Funari, E., Manganelli, M., & Sinisi, L. (2012). Impact of climate change on waterborne diseases. *Annali dell'Istituto Superiore di Sanita*, *48*(4), 473–487. doi:10.4415/ANN_12_04_13 PMID:23247142

Furtado, C., Adak, G. K., Stuart, J. M., Wall, P. G., Evans, H. S., & Casemore, D. P. (1998). Outbreaks of waterborne infectious intestinal disease in England and Wales, 1992-5. *Epidemiology and Infection*, *121*(1), 109–119. doi:10.1017/S0950268898001083 PMID:9747762

Galbraith, N. S., Barrett, N. J., & Stanwell-smith, R. (1987). Water and disease after croydon: A review of water-borne and water-associated disease in the UK 1937–86. *Water and Environment Journal: the Journal / the Chartered Institution of Water and Environmental Management*, *1*(1), 7–21. doi:10.1111/j.1747-6593.1987.tb01184.x

Gargano, J. W., Freeland, A. L., Morrison, M. A., Stevens, K., Zajac, L., Wolkon, A., ... Brunkard, J. M. (2015). Acute gastrointestinal illness following a prolonged community-wide water emergency. *Epidemiology and Infection*, *143*(13), 2766–2776. doi:10.1017/S0950268814003501 PMID:25608522

Gersberg, R. M., Daft, D., & Yorkey, D. (2004). Temporal pattern of toxicity in runoff from the Tijuana River Watershed. *Water Research*, *38*(3), 559–568. doi:10.1016/j.watres.2003.11.002 PMID:14723924

Ghazani, M., FitzGerald, G., Hu, W., Toloo, G. S., & Xu, Z. (2018). Temperature variability and gastrointestinal infections: A review of impacts and future perspectives. *International Journal of Environmental Research and Public Health*, *15*(4), 766. doi:10.3390/ijerph15040766 PMID:29659519

Girmay, G., Singh, B. R., Nyssen, J., & Borrosen, T. (2009). Runoff and sediment-associated nutrient losses under different land uses in Tigray, Northern Ethiopia. *Journal of Hydrology (Amsterdam)*, *376*(1-2), 70–80. doi:10.1016/j.jhydrol.2009.07.066

Gosling, S. N., Lowe, J. A., McGregor, G. R., Pelling, M., & Malamud, B. D. (2009). Associations between elevated atmospheric temperature and human mortality: A critical review of the literature. *Climatic Change*, *92*(3-4), 299–341. doi:10.100710584-008-9441-x

Guzman Herrador, B. R., de Blasio, B. F., MacDonald, E., Nichols, G., Sudre, B., Vold, L., ... Nygård, K. (2015). Analytical studies assessing the association between extreme precipitation or temperature and drinking water-related waterborne infections: A review. *Environmental Health*, *14*(1), 29. doi:10.118612940-015-0014-y PMID:25885050

Haley, B. J., Cole, D. J., & Lipp, E. K. (2009). Distribution, diversity, and seasonality of waterborne *Salmonellae* in a rural watershed. *Applied and Environmental Microbiology*, *75*(5), 1248–1255. doi:10.1128/AEM.01648-08 PMID:19124594

Hamilton, J. L., & Luffman, I. (2009). Precipitation, pathogens, and turbidity trends in the little river, tennessee. *Physical Geography*, *30*(3), 236–248. doi:10.2747/0272-3646.30.3.236

Harper, S. L., Edge, V. L., Schuster-Wallace, C. J., Berke, O., & McEwen, S. A. (2011). Weather, water quality and infectious gastrointestinal illness in two Inuit communities in Nunatsiavut, Canada: Potential implications for climate change. *EcoHealth*, *8*(1), 93–108. doi:10.100710393-011-0690-1 PMID:21785890

Harvell, C. D., Mitchell, C. E., Ward, J. R., Altizer, S., Dobson, A. P., Ostfeld, R. S., & Samuel, M. D. (2002). Climate warming and disease risks for terrestrial and marine biota. *Science*, *296*(5576), 2158–2162. doi:10.1126cience.1063699 PMID:12077394

Hashizume, M., Armstrong, B., Hajat, S., Wagatsuma, Y., Faruque, A. S., Hayashi, T., & Sack, D. A. (2007). Association between climate variability and hospital visits for non-cholera diarrhoea in Bangladesh: Effects and vulnerable groups. *International Journal of Epidemiology*, *36*(5), 1030–1037. doi:10.1093/ije/dym148 PMID:17664224

Hem, L. (2008). Modeller for vurdering av vannkilden som hygienisk barriere (Models for evaluation of the hygienic barrier effect of water sources). *Vann*, *1*, 19–34.

Hoque, E., Hope, V., Scragg, R., Baker, M., & Shrestha, R. (2004). A descriptive epidemiology of giardiasis in New Zealand and gaps in surveillance data. *The New Zealand Medical Journal*, *117*, U1149. PMID:15570332

Houghton, M. (1997). *Global warming: the complete briefing* (2nd ed.). Cambridge, UK: Cambridge University Press. doi:10.1017/CBO9781139164252

Hoxie, N. J., Davis, J. P., Vergeront, J. M., Nashold, R. D., & Blair, K. A. (1997). Cryptosporidiosis-associated mortality following a massive waterborne outbreak in Milwaukee, Wisconsin. *American Journal of Public Health*, *87*(12), 2032–2035. doi:10.2105/AJPH.87.12.2032 PMID:9431298

Huang, T., Li, X., Rijnaarts, H., Grotenhuis, T., Ma, W., Sun, X., & Xu, J. (2014). Effects of storm runoff on the thermal regime and water quality of a deep, stratified reservoir in a temperate monsoon zone, in Northwest China. *The Science of the Total Environment*, *485-486*, 820–827. doi:10.1016/j.scitotenv.2014.01.008 PMID:24485908

Hunter, P., Colford, J., LeChevallier, M., Binder, S., & Berger, P. (2001). Waterborne diseases. *Emerg Infect Dis, 7*, 544 – 545.

Hunter, P.R. (2003). Climate change and waterborne and vector-borne disease. *J Appl Microbiol, 94*, 37S-46S.

Hunter, P. R., & Thompson, R. C. (2005). The zoonotic transmission of *Giardia* and *Cryptosporidium. International Journal for Parasitology*, *35*(11-12), 1181–1190. doi:10.1016/j.ijpara.2005.07.009 PMID:16159658

Jagai, J. S., Castronovo, D. A., Monchak, J., & Naumova, E. N. (2009). Seasonality of cryptosporidiosis: A meta-analysis approach. *Environmental Research*, *109*(4), 465–478. doi:10.1016/j.envres.2009.02.008 PMID:19328462

Jagai, J. S., Sarkar, R., Castronovo, D., Kattula, D., McEntee, J., Ward, H., ... Naumova, E. N. (2012). Seasonality of rotavirus in South Asia: A meta-analysis approach assessing associations with temperature, precipitation, and vegetation index. *PLoS One*, *7*(5), e38168. doi:10.1371/journal.pone.0038168 PMID:22693594

Jofre, J., Blanch, A. R., & Lucena, F. (2010) Waterborne infectious disease outbreaks associated with water scarcity and rainfall events. In Water Scarcity in the Mediterranean: Perspectives Under Global Change (pp. 147-159). Berlin: Springer Berlin Heidelberg.

Jöhnk, K. D., Huisman, J., Sharples, J., Sommeijer, B., Visser, P. M., & Stroom, J. M. (2008). Summer heatwaves promote blooms of harmful cyanobacteria. *Global Change Biology*, *14*(3), 495–512. doi:10.1111/j.1365-2486.2007.01510.x

Jones, K. E., Patel, N. G., Levy, M. A., Storeygard, A., Balk, D., Gittleman, J. L., & Daszak, P. (2008). Global trends in emerging infectious diseases. *Nature*, *451*(7181), 990–993. doi:10.1038/nature06536 PMID:18288193

Kalkhoff, S. J., Hubbard, L. E., Tomer, M. D., & James, D. E. (2016). Effect of variable annual precipitation and nutrient input on nitrogen and phosphorus transport from two Midwestern agricultural watersheds. *The Science of the Total Environment*, *559*, 53–62. doi:10.1016/j.scitotenv.2016.03.127 PMID:27054493

Kaneko, T., & Colwell, R. R. (1975). Incidence of *Vibrio parahaemolyticus* in Chesapeake Bay. *Applied Microbiology*, *30*, 251–257. PMID:1164012

Kilvington, S., & Beeching, J. (1995). Identification and epidemiological typing of *Naegleria fowleri* with DNA probes. *Applied and Environmental Microbiology*, *61*, 2071–2078. PMID:7793928

Kistemann, T., Claßen, T., Koch, C., Dangendorf, F., Fischeder, R., Gebel, J., ... Exner, M. (2002). Microbial load of drinking water reservoir tributaries during extreme rainfall and runoff. *Applied and Environmental Microbiology*, *68*(5), 2188–2197. doi:10.1128/AEM.68.5.2188-2197.2002 PMID:11976088

Kistemann, T., Dangendorf, F., & Exner, M. (2001). A Geographical Information System (CIS) as a tool for microbial risk assessment in catchment areas of drinking water reservoirs. *International Journal of Hygiene and Environmental Health*, *203*(3), 225–233. doi:10.1078/S1438-4639(04)70033-4 PMID:11279819

Kolstad, E. W., & Johansson, K. A. (2011). Uncertainties associated with quantifying climate change impacts on human health: A case study for diarrhea. *Environmental Health Perspectives*, *119*(3), 299–305. doi:10.1289/ehp.1002060 PMID:20929684

Lafferty, K. D. (2009). The ecology of climate change and infectious diseases. *Ecology*, *90*(4), 888–900. doi:10.1890/08-0079.1 PMID:19449681

Lake, I. R., Bentham, G., Kovats, R. S., & Nichols, G. L. (2005). Effects of weather and river flow on cryptosporidiosis. *Journal of Water and Health*, *3*(4), 469–474. doi:10.2166/wh.2005.048 PMID:16459850

Lake, I. R., Pearce, J., & Savill, M. (2008). The seasonality of human cryptosporidiosis in New Zealand. *Epidemiology and Infection*, *136*(10), 1383–1387. doi:10.1017/S0950268807009922 PMID:18053274

Lal, A., Hales, S., Kirk, M., Baker, M. G., & French, N. P. (2016). Spatial and temporal variation in the association between temperature and salmonellosis in NZ. *Australian and New Zealand Journal of Public Health*, *40*(2), 165–169. doi:10.1111/1753-6405.12413 PMID:26260292

Lal, A., Ikeda, T., French, N., Baker, M. G., & Hales, S. (2013). Climate variability, weather and enteric disease incidence in new zealand: Time series analysis. *PLoS One*, *8*(12), e83484. doi:10.1371/journal.pone.0083484 PMID:24376707

Levy, K., Hubbard, A. E., & Eisenberg, J. N. (2009). Seasonality of rotavirus disease in the tropics: A systematic review and meta-analysis. *International Journal of Epidemiology*, *38*(6), 1487–1496. doi:10.1093/ije/dyn260 PMID:19056806

Levy, K., Woster, A. P., Goldstein, R. S., & Carlton, E. J. (2016). Untangling the impacts of climate change on waterborne diseases: A systematic review of relationships between diarrheal diseases and temperature, rainfall, flooding, and drought. *Environmental Science & Technology*, *50*(10), 4905–4922. doi:10.1021/acs.est.5b06186 PMID:27058059

Lipp, E. K., Farrah, S. A., & Rose, J. B. (2001). Assessment and impact of microbial fecal pollution and human enteric pathogens in a coastal community. *Marine Pollution Bulletin*, *42*(4), 286–293. doi:10.1016/S0025-326X(00)00152-1 PMID:11381749

Liu, R., Wang, J., Shi, J., Chen, Y., Sun, C., Zhang, P., & Shen, Z. (2014). Runoff characteristics and nutrient loss mechanism from plain farmland under simulated rainfall conditions. *The Science of the Total Environment*, *468-469*, 1069–1077. doi:10.1016/j.scitotenv.2013.09.035 PMID:24095969

Lutz, C., Erken, M., Noorian, P., Sun, S., & McDougald, D. (2013). Environmental reservoirs and mechanisms of persistence of *Vibrio cholerae*. *Frontiers in Microbiology*, *4*, 375. doi:10.3389/fmicb.2013.00375 PMID:24379807

Mac Kenzie, W. R., Hoxie, N. J., Proctor, M. E., Gradus, M. S., Blair, K. A., Peterson, D. E., ... Davis, J. P. (1994). A massive outbreak in Milwaukee of cryptosporidium infection transmitted through the public water supply. *The New England Journal of Medicine*, *331*(3), 161–167. doi:10.1056/NEJM199407213310304 PMID:7818640

MacKenzie, W. R., Schell, W. L., Blair, K. A., Addiss, D. G., Peterson, D. E., Hoxie, N. J., ... Davis, J. P. (1995). Massive outbreak of waterborne cryptosporidium infection in Milwaukee, Wisconsin: Recurrence of illness and risk of secondary transmission. *Nephrology, Dialysis, Transplantation*, *21*, 57–62. PMID:7578760

Mallin, M. A., Johnson, V. L., & Ensign, S. H. (2009). Comparative impacts of stormwater runoff on water quality of an urban, a suburban, and a rural stream. *Environmental Monitoring and Assessment*, *159*(1-4), 475–491. doi:10.100710661-008-0644-4 PMID:19067207

McCarthy, D. T., Hathaway, J. M., Hunt, W. F., & Deletic, A. (2012). Intra-event variability of *Escherichia coli* and total suspended solids in urban stormwater runoff. *Water Research*, *46*(20), 6661–6670. doi:10.1016/j.watres.2012.01.006 PMID:22321764

McCreesh, N., Nikulin, G., & Booth, M. (2015). Predicting the effects of climate change on *Schistosoma mansoni* transmission in eastern Africa. *Parasites & Vectors*, *8*(1), 4. doi:10.118613071-014-0617-0 PMID:25558917

McGechan, M. B., & Lewis, D. R. (2002). SW—Soil and Water: Transport of particulate and colloid-sorbed contaminants through Soil, Part 1: General Principles. *Biosystems Engineering*, *83*(3), 255–273. doi:10.1006/bioe.2002.0125

Miller, W. A., Lewis, D. J., Lennox, M., Pereira, M. G., Tate, K. W., Conrad, P. A., & Atwill, E. R. (2007). Climate and on-farm risk factors associated with *Giardia duodenalis* cysts in storm runoff from California coastal dairies. *Applied and Environmental Microbiology*, *73*(21), 6972–6979. doi:10.1128/AEM.00100-07 PMID:17873066

Moors, E., Singh, T., Siderius, C., Balakrishnan, S., & Mishra, A. (2013). Climate change and waterborne diarrhoea in northern India: Impacts and adaptation strategies. *The Science of the Total Environment*, *468-469*, S139–S151. doi:10.1016/j.scitotenv.2013.07.021 PMID:23972324

Mourino-Perez, R. R. (1998). Oceanography and the seventh cholera pandemic. *Epidemiology (Cambridge, Mass.)*, *9*, 355–357. doi:10.1097/00001648-199805000-00024 PMID:9583431

Muchiri, J. M., Ascolillo, L., Mugambi, M., Mutwiri, T., Ward, H. D., Naumova, E. N., ... Griffiths, J. K. (2009). Seasonality of *Cryptosporidium* oocyst detection in surface waters of Meru, Kenya as determined by two isolation methods followed by PCR. *Journal of Water and Health*, *7*(1), 67–75. doi:10.2166/wh.2009.109 PMID:18957776

Nagels, J.W., Davies-Colley, R.J., Donnison, A.M., & Muirhead, R.W. (2002) Faecal contamination over flood events in a pastoral agricultural stream in New Zealand. *Water Science and Technology: A Journal of the International Association on Water Pollution Research, 45*, 45-52.

Nagvenkar, G. S., & Ramaiah, N. (2009). Abundance of sewage-pollution indicator and human pathogenic bacteria in a tropical estuarine complex. *Environmental Monitoring and Assessment*, *155*(1-4), 245–256. doi:10.100710661-008-0432-1 PMID:18633722

Naumova, E. N., Egorov, A. I., Morris, R. D., & Griffiths, J. K. (2003). The elderly and waterborne cryptosporidium infection: Gastroenteritis hospitalizations before and during the 1993 Milwaukee outbreak. *Emerging Infectious Diseases*, *9*(4), 418–425. doi:10.3201/eid0904.020260 PMID:12702220

Nevers, M. B., & Whitman, R. L. (2011). Efficacy of monitoring and empirical predictive modeling at improving public health protection at Chicago beaches. *Water Research*, *45*(4), 1659–1668. doi:10.1016/j.watres.2010.12.010 PMID:21195447

Nichols, G., Lane, C., Asgari, N., Verlander, N. Q., & Charlett, A. (2009). Rainfall and outbreaks of drinking water related disease and in England and Wales. *Journal of Water and Health*, *7*(1), 1–8. doi:10.2166/wh.2009.143 PMID:18957770

O'Neil, J. M., Davis, T. W., Burford, M. A., & Gobler, C. J. (2012). The rise of harmful cyanobacteria blooms: The potential roles of eutrophication and climate change. *Harmful Algae*, *14*, 313–334. doi:10.1016/j.hal.2011.10.027

Olago, D., Marshall, M., Wandiga, S. O., Opondo, M., Yanda, P. Z., Kanalawe, R., ... Achola, P. (2007). Climatic, socio-economic, and health factors affecting human vulnerability to cholera in the Lake Victoria basin, East Africa. *Ambio*, *36*(4), 350–358. doi:10.1579/0044-7447(2007)36[350:CSAHFA]2.0.CO;2 PMID:17626474

Olyphant, G. A. (2005). Statistical basis for predicting the need for bacterially induced beach closures: Emergence of a paradigm? *Water Research*, *39*(20), 4953–4960. doi:10.1016/j.watres.2005.09.031 PMID:16290180

Olyphant, G. A., Thomas, J., Whitman, R. L., & Harper, D. (2003). Characterization and statistical modeling of bacterial (*Escherichia coli*) outflows from watersheds that discharge into southern Lake Michigan. *Environmental Monitoring and Assessment*, *81*(1/3), 289–300. doi:10.1023/A:1021345512203 PMID:12620022

Onozuka, D. (2014). Effect of non-stationary climate on infectious gastroenteritis transmission in Japan. *Scientific Reports*, *4*(1), 5157. doi:10.1038rep05157 PMID:24889802

Onozuka, D., & Hashizume, M. (2011). Weather variability and paediatric infectious gastroenteritis. *Epidemiology and Infection*, *139*(09), 1369–1378. doi:10.1017/S0950268810002451 PMID:21044404

Pacini, F. (1854). Observazioni microscopiche e deduzioni patologiche sul cholera asiatico. *Gazzetta Medica Italiana*, *6*, 405–412.

Paerl, H. W., & Paul, V. J. (2012). Climate change: Links to global expansion of harmful cyanobacteria. *Water Research*, *46*(5), 1349–1363. doi:10.1016/j.watres.2011.08.002 PMID:21893330

Parker, B. C., Ford, M. A., Gruft, H., & Falkinham, J. O. III. (1983). Epidemiology of infection by non-tuberculous mycobacteria. IV. Preferential aerosolization of *Mycobacterium* intracellulare from natural waters. *The American Review of Respiratory Disease*, *128*, 652–656. PMID:6354024

Pascual, M., Rodo, X., Ellner, S. P., Colwell, R., & Bouma, M. J. (2000). Cholera dynamics and El Nino-Southern Oscillation. *Science*, *289*(5485), 1766–1769. doi:10.1126cience.289.5485.1766 PMID:10976073

Patz, J. A., Campbell-Lendrum, D., Holloway, T., & Foley, J. A. (2005). Impact of regional climate change on human health. *Nature*, *438*(7066), 310–317. doi:10.1038/nature04188 PMID:16292302

Patz, J. A., Epstein, P. R., Burke, T. A., & Balbus, J. M. (1996). Global climate change and emerging infectious diseases. *Journal of the American Medical Association*, *275*(3), 217–223. doi:10.1001/jama.1996.03530270057032 PMID:8604175

Patz, J. A., Frumkin, H., Holloway, T., Vimont, D. J., & Haines, A. (2014). Climate change: Challenges and opportunities for global health. *Journal of the American Medical Association*, *312*(15), 1565–1580. doi:10.1001/jama.2014.13186 PMID:25244362

Patz, J. A., McGeehin, M. A., Bernard, S. M., Ebi, K. L., Epstein, P. R., Grambsch, A., ... Trtanj, J. (2000). The potential health impacts of climate variability and change for the United States: Executive summary of the report of the health sector of the U.S. National Assessment. *Environmental Health Perspectives*, *108*(4), 367–376. doi:10.1289/ehp.00108367 PMID:10753097

Patz, J. A., Vavrus, S. J., Uejio, C. K., & McLellan, S. L. (2008). Climate change and waterborne disease risk in the Great Lakes region of the U.S. *American Journal of Preventive Medicine*, *35*(5), 451–458. doi:10.1016/j.amepre.2008.08.026 PMID:18929971

Pettigrew, N. R., Churchill, J. H., Janzen, C. D., Mangum, L. J., Signell, R. P., Thomas, A. C., ... Xue, H. (2005). The kinematic and hydrographic structure of the Gulf of Maine Coastal Current. *Deep-sea Research. Part II, Topical Studies in Oceanography*, *52*(19-21), 2369–2391. doi:10.1016/j.dsr2.2005.06.033

Prathumratana, L., Sthiannopkao, S., & Kim, K. W. (2008). The relationship of climatic and hydrological parameters to surface water quality in the lower Mekong River. *Environment International*, *34*(6), 860–866. doi:10.1016/j.envint.2007.10.011 PMID:18068783

Pruss, A. (1998). Review of epidemiological studies on health effects from exposure to recreational water. *International Journal of Epidemiology*, *27*(1), 1–9. doi:10.1093/ije/27.1.1 PMID:9563686

Putignani, L., & Menichella, D. (2010). Global distribution, public health and clinical impact of the protozoan pathogen Cryptosporidium. *Interdisciplinary Perspectives on Infectious Diseases*. PMID:20706669

Qin, H. P., Khu, S. T., & Yu, X. Y. (2010). Spatial variations of storm runoff pollution and their correlation with land-use in a rapidly urbanizing catchment in China. *The Science of the Total Environment*, *408*(20), 4613–4623. doi:10.1016/j.scitotenv.2010.07.021 PMID:20667581

Rashid, A., Haley, B. J., Rajabov, M., Ahmadova, S., Gurbanov, S., Colwell, R. R., & Huq, A. (2013). Detection of *Vibrio cholerae* in environmental waters including drinking water reservoirs of Azerbaijan. *Environmental Microbiology Reports*, *5*(1), 30–38. doi:10.1111/j.1758-2229.2012.00369.x PMID:23757128

Robertson, L. J., Campbell, A. T., & Smith, H. V. (1992). Survival of *Cryptosporidium parvum* oocysts under various environmental pressures. *Applied and Environmental Microbiology*, *58*, 3494–3500. PMID:1482175

Robertson, L. J., Hermansen, L., & Gjerde, B. K. (2006). Occurrence of *Cryptosporidium* Oocysts and Giardia Cysts in Sewage in Norway. *Applied and Environmental Microbiology*, *72*(8), 5297–5303. doi:10.1128/AEM.00464-06 PMID:16885279

Rose, J. B. (1997). Environmental ecology of *Cryptosporidium* and public health implications. *Annual Review of Public Health*, *18*(1), 135–161. doi:10.1146/annurev.publhealth.18.1.135 PMID:9143715

Rose, J. B., Daeschner, S., Easterling, D. R., Curriero, F. C., Lele, S., & Patz, J. A. (2000). Climate and waterborne disease outbreaks. *Journal - American Water Works Association*, *9*(9), 77–87. doi:10.1002/j.1551-8833.2000.tb09006.x

Schiff, K., & Bay, S. (2003). Impacts of stormwater discharges on the nearshore benthic environment of Santa Monica Bay. *Marine Environmental Research*, *56*(1-2), 225–243. doi:10.1016/S0141-1136(02)00332-X PMID:12648957

Schijven, J.F., & de Roda Husman, A.M. (2005) Effect of climate changes on waterborne disease in The Netherlands. *Water Science and Technology: A Journal of the International Association on Water Pollution Research, 51*, 79-87.

Schuster, C.J., Ellis, A.G., Robertson, W.J., Charron, D.F., Aramini, J.J., Marshall, B.J., & Medeiros, D.T. (2005). Infectious disease outbreaks related to drinking water in Canada, 1974-2001. *Revue canadienne de sante publique, 96*, 254-258.

Schwartz, J., Levin, R., & Goldstein, R. (2000). Drinking water turbidity and gastrointestinal illness in the elderly of Philadelphia. *Journal of Epidemiology and Community Health*, *54*(1), 45–51. doi:10.1136/jech.54.1.45 PMID:10692962

Semenza, J. C., Höuser, C., Herbst, S., Rechenburg, A., Suk, J. E., Frechen, T., & Kistemann, T. (2012). Knowledge mapping for climate change and food- and waterborne diseases. *Critical Reviews in Environmental Science and Technology*, *42*(4), 378–411. doi:10.1080/10643389.2010.518520 PMID:24771989

Semenza, J. C., & Menne, B. (2009). Climate change and infectious diseases in Europe. *The Lancet Infectious Diseases*, *9*(6), 365–375. doi:10.1016/S1473-3099(09)70104-5 PMID:19467476

Semenza, J.C., & Nichols, G. (2007) Cryptosporidiosis surveillance and water-borne outbreaks in Europe. *Euro Surveillance: Bulletin Europeen sur les maladies transmissibles, 12*, E13-14.

Shaman, J., & Lipsitch, M. (2013). The El Nino-Southern Oscillation (ENSO)-pandemic influenza connection: Coincident or causal? *Proceedings of the National Academy of Sciences of the United States of America*, *110*(Suppl 1), 3689–3691. doi:10.1073/pnas.1107485109 PMID:22308322

Shope, R. (1991). Global climate change and infectious diseases. *Environmental Health Perspectives*, *96*, 171–174. doi:10.1289/ehp.9196171 PMID:1820262

Signor, R. S., Roser, D. J., Ashbolt, N. J., & Ball, J. E. (2005). Quantifying the impact of runoff events on microbiological contaminant concentrations entering surface drinking source waters. *Journal of Water and Health*, *3*(4), 453–468. doi:10.2166/wh.2005.052 PMID:16459849

Singh, A., & Fleurat, M. (2010). Pediatric emergency medicine practice acute gastroenteritis-an update. *Pediatric Emergency Medicine Practice*, *7*, 1–12.

Smith, A., Reacher, M., Smerdon, W., Adak, G. K., Nichols, G., & Chalmers, R. M. (2006). Outbreaks of waterborne infectious intestinal disease in England and Wales, 1992–2003. *Epidemiology and Infection*, *134*(06), 1141–1149. doi:10.1017/S0950268806006406 PMID:16690002

Soh, Y. C., Roddick, F., & van Leeuwen, J. (2008). The future of water in Australia: The potential effects of climate change and ozone depletion on Australian water quality, quantity and treatability. *The Environmentalist*, *28*(2), 158–165. doi:10.100710669-007-9123-7

Stanford, B. D., Wright, B., Routt, J. C., & Khan, S. J. (2014). *Water quality impacts of extreme weather-related events*. Alexandria, VA: Academic Press.

Stauder, M., Vezzulli, L., Pezzati, E., Repetto, B., & Pruzzo, C. (2010). Temperature affects *Vibrio cholerae* O1 El Tor persistence in the aquatic environment via an enhanced expression of GbpA and MSHA adhesins. *Environmental Microbiology Reports*, *2*(1), 140–144. doi:10.1111/j.1758-2229.2009.00121.x PMID:23766009

Sterk, A., Schijven, J., de Nijs, T., & de Roda Husman, A. M. (2013). Direct and indirect effects of climate change on the risk of infection by water-transmitted pathogens. *Environmental Science & Technology*, *47*(22), 12648–12660. doi:10.1021/es403549s PMID:24125400

Thomas, K. M., Charron, D. F., Waltner-Toews, D., Schuster, C., Maarouf, A. R., & Holt, J. D. (2006). A role of high impact weather events in waterborne disease outbreaks in Canada, 1975 - 2001. *International Journal of Environmental Health Research, 16*(3), 167–180. doi:10.1080/09603120600641326 PMID:16611562

Thurston-Enriquez, J. A., Gilley, J. E., & Eghball, B. (2005). Microbial quality of runoff following land application of cattle manure and swine slurry. *Journal of Water and Health, 3*(2), 157–171. doi:10.2166/wh.2005.0015 PMID:16075941

Unc, A., & Goss, M. J. (2003). Movement of faecal bacteria through the Vadose zone. *Water, Air, and Soil Pollution, 149*(1/4), 327–337. doi:10.1023/A:1025693109248

van Vliet, M. T. H., & Zwolsman, J. J. G. (2008). Impact of summer droughts on the water quality of the Meuse river. *Journal of Hydrology (Amsterdam), 353*(1-2), 1–17. doi:10.1016/j.jhydrol.2008.01.001

Vezzulli, L., Colwell, R. R., & Pruzzo, C. (2013). Ocean warming and spread of pathogenic vibrios in the aquatic environment. *Microbial Ecology, 65*(4), 817–825. doi:10.100700248-012-0163-2 PMID:23280498

Wade, T. J., Calderon, R. L., Brenner, K. P., Sams, E., Beach, M., Haugland, R., ... Dufour, A. P. (2008). High sensitivity of children to swimming-associated gastrointestinal illness: Results using a rapid assay of recreational water quality. *Epidemiology (Cambridge, Mass.), 19*(3), 375–383. doi:10.1097/EDE.0b013e318169cc87 PMID:18379427

Wade, T.J., Sams, E., Brenner, K.P., Haugland, R., Chern, E., Beach, M., … Dufour, A.P. (2010). Rapidly measured indicators of recreational water quality and swimming-associated illness at marine beaches: A prospective cohort study. *Environmental Health: A Global Access Science Source, 9*, 66.

Wandiga, S. O., Opondo, M., Olago, D., Githeko, A., Githui, F., Marshall, M., ... Achola, P. (2010). Vulnerability to epidemic malaria in the highlands of Lake Victoria basin: The role of climate change/variability, hydrology and socio-economic factors. *Climatic Change, 99*(3-4), 473–497. doi:10.100710584-009-9670-7

Wang, S., Qian, X., Han, B.-P., Luo, L.-C., & Hamilton, D. P. (2012). Effects of local climate and hydrological conditions on the thermal regime of a reservoir at Tropic of Cancer, in southern China. *Water Research, 46*(8), 2591–2604. doi:10.1016/j.watres.2012.02.014 PMID:22391017

Warrick, J. A., DiGiacomo, P. M., Weisberg, S. B., Nezlin, N. P., Mengel, M., Jones, B. H., ... Farnsworth, K. L. (2007). River plume patterns and dynamics within the Southern California Bight. *Continental Shelf Research, 27*(19), 2427–2448. doi:10.1016/j.csr.2007.06.015

Washburn, L., McClure, K. A., Jones, B. H., & Bay, S. M. (2003). Spatial scales and evolution of stormwater plumes in Santa Monica Bay. *Marine Environmental Research, 56*(1-2), 103–125. doi:10.1016/S0141-1136(02)00327-6 PMID:12648952

Wei, T., Yang, S., Moore, J. C., Shi, P., Cui, X., Duan, Q., ... Dong, W. (2012). Developed and developing world responsibilities for historical climate change and CO(2) mitigation. *Proceedings of the National Academy of Sciences of the United States of America, 109*(32), 12911–12915. doi:10.1073/pnas.1203282109 PMID:22826257

Weniger, B. G., Blaser, M. J., Gedrose, J., Lippy, E. C., & Juranek, D. D. (1983). An outbreak of waterborne giardiasis associated with heavy water runoff due to warm weather and volcanic ashfall. *American Journal of Public Health*, *73*(8), 868–872. doi:10.2105/AJPH.73.8.868 PMID:6869640

Whitehead, P. G., Wilby, R. L., Battarbee, R. W., Kernan, M., & Wade, A. J. (2009). A review of the potential impacts of climate change on surface water quality. *Hydrological Sciences Journal*, *54*(1), 101–123. doi:10.1623/hysj.54.1.101

WHO. (2004). *Guidelines for Drinking Water Quality* (3rd ed.). Geneva: World Health Organization.

WHO. (2012). *Atlas of health and climate*. Joint publication World Health Organization and meteorogical world organization. Available at http://www.who.int/globalchange/publications/atlas/en/

Wiedner, C., Rücker, J., Brüggemann, R., & Nixdorf, B. (2007). Climate change affects timing and size of populations of an invasive *Cyanobacterium* in temperate regions. *Oecologia*, *152*(3), 473–484. doi:10.100700442-007-0683-5 PMID:17375336

Wu, L., Long, T.-y., Liu, X., & Guo, J. (2012). Impacts of climate and land-use changes on the migration of non-point source nitrogen and phosphorus during rainfall-runoff in the Jialing River Watershed, China. *Journal of Hydrology (Amsterdam)*, *475*, 26–41. doi:10.1016/j.jhydrol.2012.08.022

Young, I., Smith, B. A., & Fazil, A. (2015). A systematic review and meta-analysis of the effects of extreme weather events and other weather-related variables on *Cryptosporidium* and *Giardia* in fresh surface waters. *Journal of Water and Health*, *13*(1), 1–17. doi:10.2166/wh.2014.079 PMID:25719461

Yu, Z.-z., & Wang, L.-l. (2011). Factors influencing thermal structure in a tributary bay of Three Gorges Reservoir. *Journal of Hydrodynamics. Ser B*, *23*, 407–415.

Zhang, Q., Xu, C.-y., Becker, S., & Jiang, T. (2006). Sediment and runoff changes in the Yangtze River basin during past 50 years. *Journal of Hydrology (Amsterdam)*, *331*(3-4), 511–523. doi:10.1016/j.jhydrol.2006.05.036

Zhou, X., Zhou, Y., Chen, R., Ma, W., Deng, H., & Kan, H. (2013). High temperature as a risk factor for infectious diarrhea in Shanghai, China. *Journal of Epidemiology*, *23*(6), 418–423. doi:10.2188/jea.JE20130012 PMID:23994865

Zhou, X. N., Yang, G. J., Yang, K., Wang, X. H., Hong, Q. B., Sun, L. P., ... Utzinger, J. (2008). Potential impact of climate change on schistosomiasis transmission in China. *The American Journal of Tropical Medicine and Hygiene*, *78*(2), 188–194. doi:10.4269/ajtmh.2008.78.188 PMID:18256410

Chapter 22
Climate Change and Fecal Peril: Possible Impacts and Emerging Trends

Ahmed Khadra
Cadi Ayyad University, Morocco

ABSTRACT

Fecal peril caused by intestinal parasites is commonly reported to be causing health problems in the world. Furthermore, global climate change is inevitable. The purpose of this chapter is to examine the health effects of climate change. Water shortage contribute to increase the pressure on regional water resources and force a greater number of people to use urban wastewater as an alternative for irrigation. Therefore, unsafe management and inappropriate wastewater use in urban agriculture is likely to be responsible of exacerbating the transmission of infectious diseases, including those caused by intestinal protozoa and helminths parasitic worms. It should be taken into account that waterborne diseases are influencedby climate change. The frequency and severity of intertwined extreme weather events driven by climate change are occurring worldwide and likely to cause epidemics of waterborne gastroenteritis. The association found between both rainfall, river flooding, and the majority of waterborne disease outbreaks was frequently proved to be preceded by climatic change events.

INTRODUCTION

At the global level, there has been a growing shortage of freshwater reserves, mainly those of good quality, as a result of increasing human consumption and, in some regions, decreases in the annual rainfall or annual rainfall consisting mostly of heavy rain, which is poorly absorbed by the soil (Karl and Knight, 1996). Additionally, Climate change will likely affect the water resources due to the expected changes in precipitation and evapotranspiration and the spatio-temporal distribution of these essential water balance components (Kumar et al., 2017). Water scarcity is considered a key threat for the twenty-first century. FAO defines dry lands as areas where water shortage constrains the length of the growing season below 179 days (FAO 2000); this includes regions classified climatically as arid, semi-arid and dry sub-humid. Currently, 36% of the world population is living in regions where water is a limited resource (Safriel et al., 2005).

DOI: 10.4018/978-1-5225-7775-1.ch022

Population growth associated with urbanization is increasing the pressure on regional water resources. In 2014, the United Nations estimated that 3.9 billion of the world's population resided in urban areas, with 2.8 billion living in developing countries. The continuing population growth and urbanization are projected to add 3.1 billion people to the urban world population by 2050, with a 2.25 billion-population increase in Asian and African cities (WUP. 2014). Hydrometeorological and climate extremes and change can exert profound stresses on urban environments. This is especially critical in regions where climate change is expected to exacerbate water stress and increase precipitation variability, thus reducing the amount of water available for both irrigation and the environment (Hanjra et al., 2012). When projecting the current trend of global annual water usage, it will rise to 6.9 trillion cubic meters by 2030, being 40% more than can be provided by available water supplies (Gilbert, 2010). Water quality degradation is quickly joining water scarcity in most countries of these regions. Growing demands for water to produce food, supply industries, and support human populations have led to competition for scarce freshwater supplies. A whole range of possibilities of mechanisms that can be used to reduce the pressure on freshwater resources for irrigation use.

They are based on the mobilization of conventional and unconventional resources. It is based under three pillars:

- Management of water demand and water valuation.
- Mobilization of unconventional resources including promotion of projects desalination of seawater and the reuse of wastewater.
- Protection of water resources, the natural environment and adaptation to climatic changes.

Wastewater reusing in agriculture is an ancient practice that has been generally implemented worldwide (Gupta et al., 2009; Blumenthal et al. 2001; Ensink et al. 2002; Sharma et al. 2007; WHO 2006). Rough estimates indicate that at least 20 million hectares in 50 countries are irrigated with raw or partially treated wastewater (Scott et al. 2004; Hussain et al. 2001). The major objectives of wastewater irrigation are that it provides a reliable source of water supply to farmers and has the beneficial aspects of adding valuable plant nutrients and organic matter to the soil (Liu et al. 2005b; Horswell et al. 2003). The discharge of wastewater treated or non-treated, into the receiving environment, such as lakes, rivers, and the coastal marine environments can cause severe degradation of these waters. The degradation is often related to the presence of organic and inorganic nutrients, which can cause problems as eutrophication and algal blooms. Additionally, reuse of wastewater and faeces in agriculture and aquaculture might contribute to the transmission of intestinal parasites (Pham-Duc et al., 2013). It is conceivable that increasing volumes of wastewater might exacerbate the spread of intestinal parasites, enteric bacteria and viruses (WHO, 2006; Strande et al., 2014). Without adequate treatment and protective measures on farms and in markets, workers and the public are exposed to pathogens and hazardous contaminants that pose risks to human health (Fuhrimann et al., 2016). Diarrhoeal disease, helminth infections and intestinal protozoa are considered the greatest risks to human health transmitted by consumption of wastewater-irrigated products or through contact with contaminated soils and irrigation water (Abaidoo et al., 2010; Pham-Duc., 2013).

Urban agriculture production systems can provide 41% of a city's total food supply and up to 90% of its demand for perishable vegetables (Drechsel and Dongus, 2016). Due to water scarcity and cost, urban farmers generally use untreated wastewater irrigation, since more than 80% of the sewage generated in LMICs is discharged untreated into the environment; (Mateo-Sagasta et al., 2013) this is both easily accessible and has fertilizing capacity. Approximately 50% of the world's population depend on polluted water sources, including for irrigation of agricultural lands; of which 20 million hectares of arable land worldwide, giving rise to 10% of the world population' s food production, with one billion consumers, is reported to be irrigated with wastewaters (Mateo-Sagasta et al., 2013). It has been shown that wastewater used for this form of agriculture expose humans and animals to various enteric diseases caused by pathogenic bacteria, protozoa, and helminths (Dickin et al., 2016). Furthermore, ingestion of faecally contaminated water and/or food by microorganisms is one of the major reasons for the higher number of gastro-intestinal and waterborne diseases in developing countries (Liu et al., 2016). In 2015, globally 1.87 million children under 5 years old worldwide died from diarrhea, representing 19% of total child deaths, with 78% (1.46 million) occurring in Africa and South-East Asia (Liu et al., 2016). With such shortage of water-related with climate change led to extensive use of wastewater in irrigated agriculture, contamination of soils and crops is expected. In this chapter, the water shortage and reuse wastewater in agriculture use of wastewater, as well as the related negative impacts are discussed. In addition, the influence of climate change on common pathogens contained in wastewater are studied to evaluate the relationship between climate variability and intestinal pathogens.

WATER SHORTAGE AND REUSE WASTEWATER IN AGRICULTURE

Many authors and studies point out that water scarcity1 is the main issue and a major challenge that currently affects a large part of the global population (Hadipour et al., 2016; Lyu et al., 2016 Pintilie et al., 2016; Riemenschneider et al., 2016; Ternes et al., 2007). Jāhān (2015) argues that more than 40% of the global population is facing water scarcity nowadays. Factors such as climate change, rapid urbanization and population growth related to increasing anthropogenic activities are putting immense pressure on available water resources and impact even more the global availability of water (Bixio et al., 2006; Elmeddahi et al., 2016; Hadipour et al., 2016; Luo et al., 2015; Pintilie et al., 2016; Gao et al., 2017). It is estimated that by 2025, 60% of the total population will be affected by water scarcity (Qadir et al., 2007), whereas by the same year, around 2 billion people will be affected by absolute water scarcity (Riemenschneider et al., 2016; Watkins, 2006). Among those factors the most important is climate change has been concerned especially for parts located in arid and semi-arid regions (Herrington et al., 1997; Frederick et al., 1999; Lettenmaier et al., 1999; Beuhler, 2003; Merritt et al., 2003; Ojo et al., 2003). Also, Howard and Calow (2016) present the global-scale assessment to date of the impact of climate change on water scarcity. Patterns of climate change from 21 Global Climate Models (GCMs) using two index: the Water Crowding Index (WCI) and the Water Stress Index (WSI) to calculate exposure to increases and decreases in global water scarcity due to climate change. 1.6 (WCI) and 2.4 (WSI) billion people are estimated to be currently living within watersheds exposed to water scarcity. Using the WCI, by 2050 under the A1B scenario, 0.5 to 3.1 billion people are exposed to an increase in water scarcity due to climate change. Climate change has generally accelerated the hydrological processes to make Drought set in quicker and become more intense, with many consequences. Nobre et al (2016) investigate the consequences and impacts of the 2013/14 and 2014/15 droughts and related water crisis in Sao Paulo located in southeastern Brazil. Since the austral summer of 2014, this area has been experiencing one of the most severe droughts in decades. This rainfall deficiency has generated water shortages and a water crisis By January 2015; main reservoirs had reached storage levels of only 5% of their 1.3 billion m3 capacity (Nobre et al., 2016). In Morocco, the majority of the country is characterized by arid and semi-arid climate which occupied 75% of the total surface. The scarcity of water in Morocco will worsen as more probabilities of dry-year cycles are high and rainfall and water inflows are likely to decrease as a result of climate change, water resources mobilized per capita will decrease by 15% by 2030 and by 20% by 2040 (FAO / World Bank / National Meteorological Directorate / INRA). The scarcity of water in Morocco will worsen as more. Furthermore, the projections of regional models of climate change in Morocco based on the 1961-1990 rainfall series (FAO / World Bank / National Meteorological Directorate / INRA) indicate that by 2050, rainfall declines could reach 26% in certain regions.

Kishiwa et al (2017) investigate Assessment of impacts of climate change on current and future surface water availability in the upper Pangani River Basin, Tanzania by a multitier modeling technique. The results show the impact of future shortage will be more severe in irrigation where 71.12% of its future demand will be unmet. Future water demands of Hydropower and Livestock will be unmet by 27.47 and 1.41% respectively. Saudi Arabia is another example of a country with demonstrated impacts on natural water resources due to increasing demands on groundwater by the agricultural sector (Bushnak, 2002). Large decreases in groundwater levels (up to 200 m in some places) have been observed due to over-extraction. Water resources management in arid and semi-arid areas calls for solutions able to provide responses to the decrease of available resources as effect of, among others, climate change (Vergine et al., 2015) and to ensure the sustainability of water uses, mainly in agriculture (D'Agostino et al., 2014). Especially, the agricultural sector is currently the largest user of water and wastewater globally, accounting for approximately 70% of water use on average (Winpenny et al., 2010). Estimates show that at least 20 million ha are irrigated with wastewater, and about 200 million farmers are involved (Jimenez & Asano, 2008).

The volume of wastewater generated by domestic and industrial sources has increased with population, urbanization, improved living conditions, and economic development. In developing countries like for Morocco, the growing urbanization in many resulted in a significant increase in generated amounts of urban wastewater. Indeed, the annual volume of urban wastewater discharges has reached 500 million m3 in the year 2000. These discharges will rapidly continue to increase to reach 900 million m3 by 2020. Wastewater reusing in agriculture allows the irrigation of 7000 hectares, mainly continental cities (Dssouli, El Halouani & Berrichi, 2006). The reuse of wastewater could provide a key solution to address sustainable water resources management in agriculture due to the potentially large volumes of water that can be used. The reuse of wastewater for irrigation is devised in tow forms treated and untreated, depending on economic contexts, although with the majority in the untreated form in developing countries (Scott et al. 2010). An estimated 90% of wastewater in developing countries undergoes no treatment (Corcoran et al., 2010). Estimated median levels of treated wastewater to be about 35% in Asia, 14% in Latin America and the Caribbean and 0% in sub-Saharan Africa (WHO, 2000). For diverse reasons many developing countries are still unable to build wastewater treatment plants. The construction, operation and maintenance costs of wastewater treatment plants are high in addition to restricted budgets and lack of funding, result in inadequate operation of wastewater treatment plants in developing countries (Paraskevas et al., 2002; Scott, Faruqui and Raschid-Sally, 2004). For example in Kumasi, Ghana, only 8% of the wastewater undergoes some form of treatment (Keraita et al., 2002), while the remaining raw sewage flows to wetlands linked to small streams or is discharged via stormwater drains and gutters into surface streams, along which, irrigated vegetable production is practiced (Keraita et al., 2002). A study Hanoi, Vietnam, about 80 percent of vegetable production is from urban and peri-urban areas irrigated with diluted wastewater (Lai et al., 2002). The use of household sewage, and human and animal excreta in agriculture and aquaculture has a long tradition in Vietnam (Shuval et al., 1986).

Urban agriculture production systems can provide 41% of a city's total food supply and up to 90% of its demand for perishable vegetables (Drechsel and Dongus, 2016). Urban farmers generally use untreated wastewater irrigation, since more than 80% of the sewage generated in developing countries is discharged untreated into the environment; (Mateo-Sagasta et al., 2013) this is both easily accessible and has fertilizing capacity. Approximately 50% of the world' s population depend on polluted water sources, including for irrigation of agricultural lands; of which 20 million hectares of arable land worldwide, giving rise to 10% of the world population's food production, with one billion consumers, is reported to be irrigated with wastewaters (Mateo-Sagasta et al., 2013).

Ensink et al (2009) showed in India that the direct use of untreated wastewater for agriculture, particularly vegetable production, was common in most cities. The main reasons for this use were the absence of alternative water sources, the reliability of the wastewater supply, the nutrient value and the proximity to urban markets. It was estimated that 26% of the total domestic vegetable production of Pakistan was cultivated with wastewater. These unconventional resources can usually contain significant concentrations of organic and inorganic nutrients such as nitrogen and phosphate. There is many authors have noted the positive correlation between the metabolic activity of soil microorganisms and sewage effluent used in irrigation (Meli et al., 2002; Ramirez-Fuentes et al., 2002).

There are concerns and unknowns, however, about the impact of the quality of the recycled water, both on the crop itself and on the end users of the crops. Water quality issues that can create real or perceived problems in agriculture include nutrient and sodium concentrations, heavy metals, and the presence of contaminants such as human and animal pathogens, pharmaceuticals and endocrine disruptors. The organic contaminants are subdivided into two types; Priority organic micro-pollutants including polycyclic aromatic hydrocarbons, polychlorinated biphenyls (PCBs), phthalates, phenol, pesticides and other pollutants (Wang et al., 2017) and emerging organic micro-pollutants including pharmaceuticals and personal care products and endocrine disrupting compounds (Chen et al., 2011). Wastewater not only contains organic and non-organic contaminants but it can also contain the high concentration of pathogenic protozoa and helminths which expose humans and animals to various enteric diseases with significant health implications (WHO, 2006; Dickin et al., 2016). Furthermore, ingestion of faecal contaminated water and/or food by microorganisms is one of the major reasons for the higher number of gastro-intestinal and waterborne diseases in developing countries (Liu et al., 2016). In 2015, globally 1.87 million children under 5 years old worldwide died from diarrhea, representing 19% of total child deaths, with 78% (1.46 million) occurring in Africa and South-East Asia (Liu et al., 2016).

The process of treatment of wastewater reduces water pollution and generates large volumes of sludge (municipal solid waste). Sludge is high in organic matter and nutrients (Hernandez et al., 2017), making it a valuable amendment in restoring degraded soil (Jimenez and Alvarez, 2005). Applications of sludge on agricultural lands improve fertility and crop productivity (Bittencourt et al., 2013) and it may also increase the water holding capacity of soil (Almendro-Candel et al., 2014). However, municipal solid waste contains pathogens, many of these microorganisms such as intestinal parasites (Ascaris, Trichuris, Cryptosporidium, and Giardia) are trapped in, or adsorbed to particulates and are concentrated in the sludge. For example, during the separation of solid material from wastewater in both primary (settling) and secondary treatments, numerous pathogens remain associated with particulates thus concentrating them in the sludge (Godfree & Farrell, 2005: Smith et al., 2008). Due to the massive amounts of sewage sludge used by agriculture or deposited in landfills (Gale, 2004) and the environmental robustness of these pathogens (Hutchinson et al., 2005) environmental contamination derived from landfill leachates when water from rainfall contacts the waste and sewage sludge presents serious public and veterinary threats. To overcome this risk, vegetables which have been grown on land to which treated sewage sludge has been applied in accordance with the UK guidance as set out in the Safe Sludge Matrix (http://www.adas.co.uk/matrix) that a harvest interval is observed between application of the treated sludge and the harvesting of crops. In the case of vegetable crops (which are not eaten raw, e.g. potatoes), the harvest interval is 12 months. For ready-to-eat crops, such as salad crops and carrots this is extended to 30 months. Gale (2004) quantified the exposures to humans from consumption of root crops for seven pathogens known to be present in raw sewage, namely salmonellas, *Cryptosporidium parvum*, *Listeria monocytogenes*, campylobacters, *Escherichia coli* O157, Giardia, and enteroviruses and translated these pathogen exposures into risks of infection in humans using available dose–response data (Figure 1). The results confirmed the risks to humans from consumption of vegetable crops are remote. Furthermore, the harvest intervals stipulated by the Safe Sludge Matrix compensate for potential lapses in the operational efficiency of sludge treatment. Significance and Impact of the Study: The models demonstrate the huge potential impact of decay in the soil over the 12/30-month intervals specified by the Matrix, although lack of knowledge on the exact nature of soil decay processes is a source of uncertainty.

EFFECT OF CLIMATE CHANGE ON COMMON PATHOGENS CONTAINED IN WASTEWATER

Protozoan Parasites

Enteric protozoan pathogens are unicellular eukaryotes, which are obligate parasites. Outside of an infected host, they persist as dormant stages known as cysts or oocysts. There are several protozoan pathogens, which have been isolated from wastewater and recycled water sources (Gennaccaro et al., 2003). The most common waterborne diseases detected are cryptosporidiosis, giardiasis and Amoebiasis with a total of 30,000 cases reported every year in the United States alone (Yoder et al., 2012a, 2012b; Razakandrainibe et al., 2104). Cryptosporidium was identified as a cause of human infection in 1976 (White, 2010). The two enteric protozoa mainly infect humans, but also several domestic pets livestock and wild animals and induce zoonotic illnesses (Haas et al., 1999) (Chalmers et al., 2011; Giangaspero, Berrilli & Brandonisio, 2007)

In general, the factors contribute to the success of infection of two parasites include the large numbers of cysts/oocysts are excreted by infected individuals into the environment. Also, Oocysts and cysts are environmentally hardy resistants and can survive for many months in temperate and moist conditions. And the infection can be initiated by a very small number of oocysts (low infectious dose (between 10 and 25 cysts) (Fayer et al., 2000; Caccio et al., 2003). Other hands, the studies from human volunteer studies a median infectious dose for *Cryptosporidium hominis* ranged from 10 to 83 oocysts and for *Cryptosporidium parvum* from below 10 to over 1000 oocysts can initiate infection (Chappell et al. 2006).

Cryptosporidiosis is a highly prevalent and extremely widespread disease documented by over 1000 reports in humans in 95 countries on all continents except Antarctica (Fayer, Speer and Dubey, 1997). Considering that cryptosporidiosis is primarily spread by ingestion of contaminated water, was ranked fifth among the 24 most important food-borne parasites in a global ranking by a joint FAO Food and WHO expert committee in 2012 Agriculture Organization, and can be spread by close proximity to infected humans and animals, the importance of this genus in human and animal health has long been underestimated. Cryptosporidium is related to a wide range of symptoms, especially severe and chronic in the case of extremely weak patients and is even able to cause death (Ottoson et al., 2006; Reinoso, Becares & Smith, 2008). During the early 1980s, cryptosporidiosis was recognized as the major cause of chronic diarrhoea in patients with AIDS and as a cause of diarrhoea in children and was linked with childhood malnutrition and premature death in low-resource settings (Current et al., 1983; Sallon et al., 1988). The *Cryptosporidium parvum* is propagated through the anthropozonotic cycles (Graczyk, Fayer & Cranfield, 1997), and *Cryptosporidium hominis* cycles among people (Morgan-Ryan et al., 2002). The oocysts can remain viable for almost a year in the environment (Tamburrini & Pozio, 1999), e.g., animal liquid waste (Hutchinson et al., 2005a, b). An infected person can shed up to 3.10^9 oocysts (Okhuysen et al., 1999) or up to 5.10^8 cysts over the course of infection (Medema & Schijven, 2001). The other common protozoa in wastewaters Giardia such as *Giardia duodenalis, Giardia lamblia*, and *Giardia intestinalis*), a flagellate protozoan parasite, and the etiological agent of giardiasis, is one of the most prevalent and widespread intestinal parasite in humans and several vertebrate animal species worldwide (Feng & Xiao, 2011). As a parasite, Giardia has a broad host range, however, the adverse consequences of infection and its pathogenic potential are best recognised in humans (Thompson, 2004) *Giardia intestinalis* also has several hosts, including a number of warm-blooded animals (Feachem et al., 1983). It causes an estimated $2.8.10^8$ human cases per annum (Lane & Lloyd, 2002). In Asia, Africa and Latin America, about 200 million people have symptomatic giardiasis with some 500000 new cases reported each year (WHO, 1996). The cyst is the infective stage and represents the resting stage of the organism. Its rigid outer wall protects the parasite against changes in environmental temperature, dehydration and chlorination, all of which would destroy the trophozoite (Thompson, 2004; Olson et al., 2004; Thompson & Monis, 2012). Its simple life cycle involving an environmentally resistant cyst provides greater opportunities for the parasite to be transmitted directly from one infected individual to another, or indirectly through contamination of the environment or food (Feng & Xiao, 2011).

Amoebiasis caused by the intestinal parasite Entamoeba histolytica WHO reported that Entamoeba histolytica affects approximately 500 million people worldwide, resulting in symptomatic diseases in 50 millions and mortality in 100,000 persons (Lozano et al., 2012). About 80-90% of infections are asymptomatic and are likely due to the non-pathogenic species *Entamoeba dispar* or *Entamoeba moshkovskii*, therefore the worldwide incidence of *Entamoeba histolytica* is, more likely, estimated to be 5 million cases annually, with global mortality still at 100.000 persons per annum. Approximately 4 to 10% of the carriers of this amoeba infection develop clinical symptoms within a year and amoebic dysentery is considered as the third leading cause of death from parasitic disease worldwide after Malaria and Schistosomiasis (Mortimer & Chadee, 2010; Ghasemi, Rahdar & Rostami, 2015). Transmission by the fecal-oral route renders most of these protozoa capable of infecting humans through various means of fecal contamination of land and rivers by feces of both human and animal origin (Plutzer & Karanis, 2016; Lanata, 2003). It is an important health problem, especially in developing countries. Pham Duc et al. (2011) studied the risk factors for *Entamoeba histolytica* infection in an agricultural community in Hanam province, Vietnam. They suggest that in settings where human and animal excreta and Nhue River water are intensively used in agriculture, socio-economic and personal hygiene factors determine infection with *Entamoeba* histolytica, rather than exposure to human and animal excreta in agricultural activities. The rate of infection by E. histolytica differs among countries, socio-economic and sanitary conditions and populations (Al-Harthi & Jamjoom, 2007).

Thirteen species of Cryptospridium and three of widespread Giardia, including *Cryptosporidium parvum*, *Cryptosporidium hominis* and *Giardia intestinalis*, are responsible for the majority of human infections (Rimhanen-Finne et al., 2004). Cryptosporidium and Giardia cysts have been isolated from wastewater and digested sludge (Caccio et al., 2003, Quintero-Betancourt et al., 2003; Rimhanen-Finne et al., 2004; Payment et al., 2001). Caccio et al. (2003) reported that Giardia cysts are present throughout the year and more common in wastewater compared to Cryptosporidium oocysts that have a seasonal distribution.

In the group of parasitic protozoa, several other species may be transmitted to humans through the drinking water route and cause infections: *Toxoplasma gondii, Balantidium coli, Cyclospora cayetanensis*, Microsporidia, Isospora, Naegleria spp. Sarcocystis spp., *Balantidium coli*, Acanthamoeba spp. and *Blastocystis hominis* (Plutzer & Karanis, 2006; Efstratiou et al 2017).

They are present in water in resistant forms (cysts and oocysts) that protect them from environmental stress. Protozoa dispersion forms are much more resistant to disinfection processes than bacteria. The resistance to disinfectants of Giardia cysts, combined with their viability for 1–3 months in water, making them one of the main vectors for contagion.

The raw sewage is almost all loaded: from 10^5 to 10^6 Oocysts of Cryptosporidium and from 10^6 to 10^7 Giardia cysts per 100 liters of water. Therefore, contamination of surface water and food products with runoff from land treated with animal manure is also of concern (Slifko, Smith & Rose, 2000; Fayer et al., 2000). Giardia may cause several gastrointestinal disorders and episodes of malnutrition arising from deficient absorption of fats (Vera et al., 2006; Gomez-Couso et al., 2005). Kato et al. (2004) investigate the environmental parameters that affect the inactivation of *Cryptosporidium parvum* oocysts in soil under field conditions. At 120 days after placement, 5–30% of oocysts remained viable and potentially infective after their deposition indicates, however, that they continued to persist and if they had been naturally deposited, they could be transported by runoff from rain events to surface waters. When the United States Environmental Protection Agency (USEPA) promulgated the Surface Water Treatment Rule for treatment of drinking water, its goal was to ensure that the population consuming water would not be subject to a risk of greater than one infection of giardiasis per 10 000 persons year^{-1} (Regli et al. 1991); the acceptable risk to the individual consumer of infection by Giardia through drinking water is thus 10^{-4} per person per year. In a population of 60 000 000 persons (e.g in the UK) this would translate into 6000 Giardia infections per year through drinking water alone.

Also, land application of wastewater and sewage sludge facilitates circulation of Cryptosporidium and Giardia in the environment. An epidemiological study of over 22000 infants and children in Africa and Asia found that Cryptosporidium was one of the four pathogens responsible for most of the severe diarrhoea (Kotloff et al. 2013) and was considered the second greatest cause of diarrhoea and death in children after rotavirus (Striepen, 2013). Infection from all three of these protozoan pathogens can occur after consumption of food or water contaminated with the (oo)cysts or through person to person contact (Carey et al., 2004). For instance, Amoros et al. (2010) demonstrated 47.5 oocysts L-1 in water used for irrigation of lettuce in Spain and in Mexico, 98% of irrigation water samples contained Cryptosporidium, Giardia or both parasites (Figure 2).

Helminths Parasites

Helminths are flat (Platyhelminthes) as Schistosoma spp., Taenia spp. or round (roundworms) as Ascaris spp., Trichuris spp., Capillaria spp., Hookworm as *Ancylostoma duodenale* or *Necator americanus* and Toxocara spp. multicellular worms. Like protozoa, they are mostly parasitic organisms. Toxocara canis, commonly known as dog roundworm, can affect humans and cause human toxocariasis (Keffala Harerimana & Vasel, 2012). Hookworm eggs causing human hookworm occur in the stool of infected person; infection is through skin penetration by infective larvae. Schistosoma spp. is a parasite responsible for schistosomiasis, Schistosoma eggs released into the external environment through feces or urine of infected people by transcutaneous; furthermore, Schistosoma can be observed in other mammals as mice and wistar rats (Sene 1994; Wang et al. 2016)

Taenia spp. is a tapeworm of the class of Cestoda, responsible for Taeniasis disease; the usual hosts for this parasite are cattle and swine, from which humans become infected with the adult tapeworm (Parkhouse & Harrison 2014). Ascaris spp. is the most common intestinal parasites in the world; it is responsible for the human Ascariasis. The natural habitats of adult larvae is the small intestine (Stephenson, 2009), the infected individuals excrete up to 200,000 eggs per day (Bethony et al. 2006; Karkashan et al. 2015). Capillaria is responsible for Capillariasis disease, and whipworm (Trichuris trichiur) are responsible for Trichuriasis disease, which is among the common human parasitic infection (Pullan et al., 2014).

A great variety of helminth eggs may be found in wastewaters. The most important group is the nematodes, known as intestinal worms, especially those belonging to the genera Ascaris, Ancylostoma, Necator, Trichuris, and others. Their main epidemiological characteristics are long persistence in the environment, a minimal infective dose, limited immune response and the ability to remain viable in soils over long periods of time (Vera et al., 2006; Amahmid, Asmama & Bouhoum, 2002). It is estimated that over 1.5 billion people are infected with at least one species of soil-transmitted helminths (STHs) world-wide (WHO, 2015), with the majority of these infections caused by the roundworms (*Ascaris lumbricoides* and *Strongyloides stercoralis*), whipworms (*Trichuris trichiura*) and hookworms (Necator ameri-canus or *Ancylostoma duodenale*) (Strunz et al., 2014).

In 2010, an estimated 439 million people were reported to be infected with hookworm, 819 million with *Ascaris lumbricoides*, and 465 million with *Trichuris trichiura*, representing a morbidity rate of 1.8 and 1.6 respectively (Bethony et al., 2006). And after five years the people infected increase, Ascariasis is reported reached 771.7–891.6 million people, while 429.6–508.0 million people have trichurias is and 406.3–480.2 million are infected with hookworm (Pullan et al., 2014). Most of these infections occur in tropical and subtropical regions of the world where poverty results in poor sanitary conditions (Stolk et al., 2016). STHs infections are mostly caused by exposure to faecally contaminated water, soil or contaminated food (Keraita & Amoah, 2011).

The low helminths, their high survival capacity in the external environment and their abundant emission in the stool, make them pathogens of particular concern.

Schistosomes are endemic in at least 74 countries of Africa, Asia and Latin America where approximately 200 million people are infected and another 600 million are at risk (Chitsulo et al., 2000). Schistosomiasis or bilharzia is a chronic disease caused by an infestation of parasitic worms called Schistosoma. In 2011, at least 243 million people required treatment for the disease, but only 28.1 million people were actually treated for it (World Health Organization (WHO), (Drudge-coates Turner, 2013)

Blaszkowska et al. (2013) reported also that, the helminth eggs were found in 15.7% of the samples taken from the area around school sport fields, 7.7% from park playgrounds and 1.4% from fenced sandpits. Etewa and Naglaa (2016) studied soil-transmitted helminths. The study was carried out in Sharkyia governorate of Egypt, Analysis of prevalence data on helminths showed a close relationship between the physicochemical characters of the soil and the distribution of helminths but the climate has the upper hand upon the characters of the soil. Among the examined sites helminths were much more numerous in the soil of rural areas especially in the spring and summer seasons, while the contamination of canal banks by helminths was the worst (80%).

A part of the fertilizing element's contents, wastewater is an important vehicle of biological agents (Khallayoun et al., 2009) that can be transmitted by direct contact, or indirectly through consumption of crops irrigated with wastewater (Khallayoune, 2009). On the other hand, although the consumption of fresh vegetables plays an important role in the transmission of parasitic contaminations (Anuar & Ramachandran, 1977), their recovery in vegetables used as the source of contamination, may be helpful in indicating the incidence of intestinal parasites among a community.

In developing countries, intestinal parasites are very common and the use of sewage contaminated water for irrigation of vegetables is a common practice in developing countries including Morocco. In this country, some authors in their studies about crops irrigated by RWW reported a values of 2 eggs/kg (Rhallabi et al., 1990), around 4 eggs/100 g in crops got from market (Idrissa et al., 2010), 32 eggs/kg in *alfalfa* (El Amiri et al., 2009), and loads varying between 10.5 and 4.65 eggs/100 g have been noted by Dssouli (2006). Hajjami et al (2010) had reported Helminth eggs loads in mint, coriander, *alfalfa* and cereals have respective values of 4 eggs/100 g, 2.2 eggs/100 g, 2 and 0.3 eggs/100 g of fresh weight. Also, Amahmid and Bouhoum (2000) and Bouhoum and Amahmid (2000) detected that the incidence of parasitic diseases in consumers of sewage irrigated crops was higher than that of the control population. In Turkey, Ulukanligil et al. (2001) detected soil-transmitted helminths (mainly A. lumbricoides) in 14% of fresh vegetables, in 84% of soil samples where vegetables are cultivated and in 61% of irrigation water. Daryani et al. (2008) reported the detection of intestinal parasites in 29% of native garden vegetables consumed in Ardabil city, Iran. Abougrain et al. (2010) examined 126 samples of four different types of fresh salad vegetables from wholesale and retail markets in Tripoli, of which 58% were positive for helminth eggs and Giardia spp. Cysts. These include 14% of tomato, 42% of cucumber, 96% of lettuce and 100% of cress samples.

In Mexico, farmer workers and their children that work in fields irrigated with untreated sewage effluent have been found to have a greater prevalence of roundworm infection than the general population (Peasey et al., 2000). Public and commercial concern does exist regarding pathogens through the use of recycled water and biosolids on cereal crops (Crute et al., 2004). An earlier study conducted by Bryan (1977) reported 3 epidemics of Ascariasis in Germany, associated with food contaminated by wastewater. The use of untreated wastewater for irrigation can be contaminated vegetables with enteric bacterial, viral and parasitic pathogens (Abougrain et al., 2010).

RISK TO FAECAL PERIL IN HUMAN WITH EXPOSURES TO CLIMATE CHANGE IMPACTS

Climate change is certainly one of the more durable aspects of anthropogenic disruptions to natural resources (Beniston 2010). The increase in temperature, even if restricted to 1.5°C, is expected to result in significant changes in climatic factors precipitation patterns. Influence of precipitation patterns, quantity, intensity, frequency, and duration and, and will be, more frequent and severe floods, droughts, storms, and heat waves will impact local hydrology and consequently groundwater subsequently affect environmental conditions that predispose developing countries to the transmission of waterborne disease (Harper et al., 2011). Major causes of diarrhoea linked to contaminated water supplies are: cholera, cryptosporidium, *E.coli*, giardia, shigella, typhoid, and viruses such as hepatitis A. Of cases of enteric illness that are reported, often the source is not identified and may be any of travel, waterborne, food-borne, or person-to-person transmission.

Especially, in developing countries the high morbidity caused by enteric protozoa in human infections such as Giardia and Cryptosporidium in developing countries has been attributed to factors including the multiple exposure routes reflecting living, sources of organisms and cycling through domestic animals, resistance of the (oo)cysts against common water disinfectants, infection and the unavailability of effective water treatment and irrigation with wastewater in agricultural secteur have sustained conditions for their transmission (Slifko et al., 2000; Bakheit et al., 2008; Escobedo et al., 2010; Mor Siobhan & Tzipori, 2008;). However, majority of developing countries, where gastrointestinal infections are suspected to be grossly under-reported, it is at least partly due to their endemic characteristics (Ashbolt, 2004). Also, the tow protozoa Cryptosporidium and Giardia has been shown to be universally distributed but under detected due to poorly planned and implemented monitoring (Ongerth, 2013).

There seems to be a generally positive association between increases in temperature and rainfall and transmission of Giardia and Cryptosporidium parasites. For example, a correlation between environmental temperatures and cryptosporidiosis in tropical and temperate regions has been reported (Lal et al., 2013) and also between heavy rainfall and the likelihood of detecting Giardia and Cryptosporidium in surface and river water (Hunter, 2003; Sterk et al., 2013). Precipitation influences the dispersion and transport of oocysts. Kistemann et al. (2002) showed an increase in 15 to 38 times the number of oocysts in the environment after heavy rain or snowmelt. Spring peaks in cryptosporidiosis incidence may be related to contamination of water supplies through heavy rainfall events. A study in North West England showed that in areas with marked seasonal patterns, cryptosporidiosis was associated with increased rainfall (Naumova et al., 2005). With leaching agricultural soils, oocysts reach the waters of surfaces and are transported over a long distance. Oocysts can be dispersed over a distance of 160 km in 7 days. In this context, these parasites are most often found in river water (Olson et al., 1999) as recreational waters. Thus, it is estimated that 60% of North American surface waters contain oocysts of Cryptosporidium. More recently, a study that we carried out in Haute-Normandie we showed that 70% of the karst aquifers used to production of drinking water were contaminated by oocysts of Cryptosporidium (Khaldi et al., 2011). For the last 35 years, waterborne disease outbreaks caused by enteric protozoa (Giardia & Cryptosporidium) have been shown to be universally present although widely undetected in raw water samples throughout the USA (Ongerth, 2017). Giardia infection has been recognized as the number one intestinal parasitic disease affecting humans in the USA (Kappus et al., 1994). Outbreaks due to both organisms have been heavily associated with consumption of surface water or groundwater that has been contaminated (Adam et al., 2016; Hlavsa et al., 2011, 2015). Cryptosporidiosis is the most frequently reported gastrointestinal illness in outbreaks associated with treated recreational water in the US (Yoder & Beach, 2007; Hlavsa et al., 2015). Epidemics of waterborne gastroenteritis such as outbreaks of cryptosporidiosis, giardia, and other infections have been shown to be associated with severe weather events, including heavy rainfall and river flooding which are on the rise worldwide, also in countries with a regulated public water supply (Gertler et al., 2015; Atherton, Newman & Casemore, 1995; Curriero et al., 2001; Lisle & Rose, 1995, Rose et al., 2000). An association between drinking water turbidity and gastrointestinal illness has been reported (Schwartz & Levin 1996). All this contributes to the occurrence of waterborne epidemics. More than 50 epidemic episodes of water-borne cryptosporidiosis have been reported, mainly in the United States, Canada, the United Kingdom, Japan and New Zealand. The largest outbreak occurred in 1993 in Milwaukee, United States, where an estimated 403000 people were infected (Mac Kenzie et al., 1994). Many of the waterborne outbreaks for which a cause could be identified were secondary to a malfunction in the water treatment procedure. Both *Cryptosporidium hominis* and *Cryptosporidium parvum* have been implicated, but the majority of the epidemics are due to C. hominis, and the genotype Ib of the gene encoding gp60 is most often involved. In France, 4 epidemics have been reported since 1998. In one case, the epidemic was due to the contamination of a surface resource during the flood of a river secondary to a storm. Two outbreaks were due to abnormalities in the water treatment plants (Dalle et al., 2003). A study of waterborne disease outbreaks in the United States has shown that about half were significantly associated with extreme rainfall (Curriero et al., 2001). Outbreak locations from an Environmental Protection Agency database were assigned to watersheds. Also, the severity of the Milwaukee outbreak has been attributed to the heavy spring rainfall and runoff from melting snow. Similarly, in the Delaware River, an association was found between the amount of rainfall and the number of Cryptosporidium oocysts (Patz et al., 2000). In 2012, European Union countries reported 18 cases of cholera. The United Kingdom reported 12 cases, France reported four cases, and Austria and Sweden all reported one case each. In the same year, 9591 cases of Cryptosporidium were reported in several countries (United Kingdom, Ireland, Belgium and Germany) largely due to *Cryptosporidium parvum* gp60 subtype IIaA15GR1. A total of 16,368 cases of giardiasis were reported by 23 EU and countries part of the European Economic Area such as Bulgaria, Estonia, Sweden and Finland, in 2012 (Mac Kenzie et al., 1994)

Climate and weather conditions also exert a range of more effects, for example as drought may affect water-storage and led to wastewater reuse use and irrigation practices, and human exposure to infection. Helminths parasites commonly detected in wastewaters that are of significant health risk in reused waters include the round worm, the hookworm, and the whipworm. These helminths have a simple life cycle with no intermediate hosts and are capable of causing infection via the faecal-oral route (Toze, 1997).

Additionaly, helminthiasis are belonging to waterborne infectious diseases, among this group the most affected by climate changes is trematodes. For exemple, Climate change may affect the transmission of *fasciola hepatica* by two ways, directly by affecting the parasite, but also indirectly, by affecting the success and distribution of the intermediate host. In Europe the major intermediate host, *Lymnaea truncatula*, has its northern distribution limited by temperature, and is recognised as a rather stenoecious species (Caron et al., 2014); higher temperatures may enable northward spread of this intermediate host, and hence northward spread of this parasite. In the trematodes themselves, temperature has a direct and pronounced effect on crucial lifecycle stages, including the production of cercariae in the snail host, which is a key component of the parasite's success (Mas–Coma et al., 2009). An increase of only a few degrees in environmental temperatures may result unmarked increases in cercarial emergence; within the range of temperatures in which the snails and parasites can live, an increase in temperature is almost invariably coupled with an increase in cercarial output.

Several nematodes are described as soil-transmitted helminths. These nematodes are very common in tropical or semi-tropical countries and are good indicators of poor basic infrastructure, such as effective sewage disposal. Although having very different life cycles, transmission of both Ascaris and Trichuris is dependent on ingestion of the robust eggs that are not immediately infective but embryonate in the environment. Embryonation is a temperature dependent function, and in areas where temperatures are rising lifecycle speed may increase; indeed, accelerated transmission of Ascaris due to increases in global temperatures and ecosystem changes has been suggested (Kim et al., 2012).

CONCLUSION

Fecal peril is a global concern for worldwide public health. Since fecal pollution of water exposes humans to intestinal parasites and water are still a major cause of severe illness and mortality. Multiple factors produce outbreaks of parasitic diseases. Inappropriate management and reuse of wastewater might exacerbate the risk of human infections. In the use of sewage or untreated wastewater in agriculture poses a number of health risks. Predominant among these is the risk of intestinal helminths and protozoa infection. The global water scarcity increases wastewater use for crop irrigation. Giardiasis and cryptosporidiosis are the most common waterborne protozoan diseases. Outbreaks due to both organisms have been heavily associated with consumption of surface water or groundwater that has been contaminated. Furthermore, adverse climatic events driven by climate change climate change resulted to the change in temperature, including heavy rainfall and river flooding, are on the rise world- wide, and are likely to cause epidemics of waterborne gastroenteritis and altered the distribution of diseases. Areas that were previously non-endemic for certain infections are now at risk (Gertler et al., 2015). Floods and droughts are each associated with an increased risk of diarrhoeal diseases. The suggestion is plausible, however, since heavy rainfall can wash contaminants into water supplies, while drought conditions can reduce the availability of fresh water leading to an increase in hygiene-related diseases.

REFERENCES

Abaidoo, R. C., Keraita, B., Drechsel, P., & Dissanayake, P. (2014). Soil and Crop Contamination Through Wastewater Irrigation and Options for Risk Reduction in Developing Countries. Academic Press. doi:10.1007/978-3-642-05076-3

Abaidoo, R. C., Keraita, B., Drechsel, P., Dissanayake, P., & Maxwell, A. S. (2010). Soil and crop contamination through wastewater irrigation and options for risk reduction in developing countries. In P. Dion (Ed.), *Soil biology and agriculture in the tropics* (pp. 275–297). Berlin: Springer-Verlag. doi:10.1007/978-3-642-05076-3_13

Abougrain, A. K., Nahaisi, M. H., Madi, N. S., Saied, M. M., & Ghenghesh, K. S. (2010). Parasitological contamination in salad vegetables in Tripoli-Libya. *Food Control*, *21*(5), 760–762. doi:10.1016/j.foodcont.2009.11.005

Adam, E., Yoder, J., Gould, L.H., Hlavsa, M.C., & Gargano, J.W. (2016). Giardiasis outbreaks in the United States, 1971–2011. *Epidemiol. Infect.*, 1-12.

Al-Harthi, S., & Sci, M. J.-W. J. M. (2007). *Diagnosis and differentiation of Entamoeba infection in Makkah Al Mukarramah using microscopy and stool antigen detection kits*. Retrieved from https://pdfs.semanticscholar.org/1f1f/e5c2911c0a5e684b9ab952e42571537ca6bc.pdf

Almendro-Candel, M., & Navarro-Pedreno, J., Vidal, M.M., Gomez, I., & Melendez-Pastor, I. (2014). *Use of municipal solid waste compost to reclaim limestone quarries mine spoils as soil amendments: Effects on Cd and Ni*. Elsevier. Retrieved from https://www.sciencedirect.com/science/article/pii/S0375674213002215

Amahmid, O., Asmama, S., & Bouhoum, K. (2002). Urban wastewater treatment in stabilization ponds: Occurrence and removal of pathogens. *Urban Water, 4*(3), 255–262. doi:10.1016/S1462-0758(01)00071-1

Amahmid, O., & Bouhoum, K. (2000). Health effect of urban wastewater reuse in a peri-urban area in Morocco. *Environmental Management and Health, 11*(3), 263–269. doi:10.1108/09566160010372004

Amoros, I., Alonso, J. L., & Cuesta, G. (2010). Cryptosporidium oocysts and Giardia cysts on salad products irrigated with contaminated water. *Journal of Food Protection, 73*(6), 1138–1140. doi:10.4315/0362-028X-73.6.1138 PMID:20537274

Anuar, A. K., & Ramachandran, C. P. (1977). A study on the prevalence of soil transmitted helminthes among lettuce leaves sold in local markets in Penang, Malaysia. *The Medical Journal of Malaysia, 31*, 262–265. PMID:562969

Ashbolt, N.J. (2004). Microbial contamination of drinking water and disease out- comes in developing regions. *Toxicology, 198*(1), 229-238.

Atherton, F., Newman, C. P. S., & Casemore, D. P. (1995). *An outbreak of waterborne cryptosporidiosis associated with a public water supply in the UK.* Academic Press.

Bakheit, M. A., Torra, D., Palomino, L. A., Thekisoe, O. M. M., Mbati, P. A., Ongerth, J., & Karanis, P. (2008). *Sensitive and specific detection of Cryptosporidium species in PCR-negative samples by loop-mediated isothermal DNA amplification and confirmation of generated LAMP products by sequencing.* Academic Press. doi:10.1016/j.vetpar.2008.09.012

Bethony, J., Brooker, S., Albonico, M., Lancet, S. G.-T., Diemert, D., & Hotez, P. J. (2006). Soil-transmitted helminth infections: Ascariasis, trichuriasis, and hookworm. *Lancet, 367*(9521), 1521–1532. doi:10.1016/S0140-6736(06)68653-4 PMID:16679166

Beuhler, M. (2003). Potential impacts of global warming on water resources in southern California. *Water Science and Technology, 47*(7-8), 165–168. doi:10.2166/wst.2003.0685 PMID:12793676

Bittencourt, S., Serrat, B.M., Andreoli, C.V., Moura, E.N., Togny, F.L., & Silva, L.A.T.P. (2013). Sewage sludge subjected to mixed: effect on total solids, pH and viable helminth eggs. *Rev. Acad. Ciencias Agrarias eAmbientais PUCPR Impresso, 11*, S191-S200. (in Portuguese)

Blaszkowska, J., Wojcik, A., Kurnatowski, P., & Szwabe, K. (2013). Biologicalinteractions between soil saprotrophic fungi and Ascaris suum eggs. *Vector Parasitology., 196*(3-4), 401–408. doi:10.1016/j.vetpar.2013.02.029 PMID:23534981

Blumenthal, U. J., Cifuentes, E., Bennett, S., Quigley, M., & Ruiz-Palacios, G. (2001). The risk of enteric infections associated with wastewater reuse; the effect of season and degree of storage of wastewater. *Transactions of the Royal Society of Tropical Medicine and Hygiene, 95*(2), 131–137. doi:10.1016/S0035-9203(01)90136-1 PMID:11355541

Bouhoum, K., & Amahmid, O. (2000). Health effect of wastewater reuse in agriculture. *WHO Water Series, 105*, 241–247. PMID:10842821

Bridge, J. W., Oliver, D. M., Chadwick, D., Godfray, H. C. J., Heathwaite, A. L., Kay, D., … Porter, J. (2010). *Engaging with the water sector for public health benefits : waterborne pathogens and diseases in developed countries*. Academic Press. doi:10.2471/BLT.09.072512

Bryan, F. L. (1977). Diseases transmitted by foods contaminated by wastewater. *Journal of Food Protection*, *40*(1), 45–56. doi:10.4315/0362-028X-40.1.45

Bushnak, A. A. (2002). Future strategy for water resources management in Saudi Arabia. *Proceeding of A Future Vision for the Saudi Economy Symposium*, 37.

Caccio, S. M., De Giacomo, M., Aulicino, F. A., Pozio, E., & Superiore, I. (2003). *Giardia Cysts in Wastewater Treatment Plants in Italy*. Academic Press. doi:10.1128/AEM.69.6.3393

Cacciò, S. M., Thompson, R. C., McLauchlin, J., & Smith, H. V. (2005). Unravelling Cryptosporidium and Giardia epidemiology. *Trends in Parasitology*, *21*(9), 430–437. doi:10.1016/j.pt.2005.06.013 PMID:16046184

Caron, Y., Martens, K., Lempereur, L., Saegerman, C., & Losson, B. (2014). New insight in lymnaeid snails (Mollusca, Gastropoda) as intermediate hosts of Fasciola hepatica (Trematoda, Digenea) in Belgium and Luxembourg. *Parasites & Vectors*, *7*(1), 66. doi:10.1186/1756-3305-7-66 PMID:24524623

Chalmers, R. M., Smith, R. P., Hadfield, S. J., Elwin, K., & Giles, M. (2011). Zoonotic linkage and variation in Cryptosporidium parvum from patients in the United Kingdom. *Parasitology Research*, *108*(5), 1321–1325. doi:10.100700436-010-2199-x PMID:21193928

Chappell, C. L., Okhuysen, P. C., Langer-Curry, R., Widmer, G., Akiyoshi, D. E., Tanriverdi, S., & Tzipori, S. (2006). Cryptosporidium hominis: Experimental challenge of healthy adults. *The American Journal of Tropical Medicine and Hygiene*, *75*(5), 851–857. doi:10.4269/ajtmh.2006.75.851 PMID:17123976

Chauret, C., Springthorpe, S., & Sattar, S. (1999). Fate of Cryptosporidium oocysts, Giardia cysts, and microbial indicators during wastewater treatment and anaerobic sludge digestion. *Canadian Journal of Microbiology*, *45*(3), 257–262. doi:10.1139/w99-001 PMID:10408099

Chen, F., Ying, G., Kong, L., & Wang, L. (n.d.). *Distribution and accumulation of endocrine-disrupting chemicals and pharmaceuticals in wastewater irrigated soils in Hebei, China*. Elsevier. Retrieved from https://www.sciencedirect.com/science/article/pii/S0269749111001680

Chitsulo, L., Engels, D., Montresor, L., & Savioli, L. (2000). The global status of schistosomiasis and its control. *Acta Tropica*, *77*(1), 41–51. doi:10.1016/S0001-706X(00)00122-4 PMID:10996119

Corcoran, E., Nellemann, C., Baker, E., Bos, R., Osborn, D., & Savelli, H. (2010). *Sick Water The central role of wastewater management in sustainable development. A Rapid Response Assessment.* UNEP.

Crute, K., Toze, S., Pritchard, D., & Penney, N. (2004). Proposed research to determine risk from microbial pathogens in grains due to land application of biosolids. *Proceeding of Biosolids Specialty II Conference*.

Current, W. L., Reese, N. C., Ernst, J. V., Bailey, W. S., Heyman, M. B., & Weinstein, W. M. (1983). Human Cryptosporidiosis in Immunocompetent and Immunodeficient Persons. *The New England Journal of Medicine*, *308*(21), 1252–1257. doi:10.1056/NEJM198305263082102 PMID:6843609

Curriero, F. C., Patz, J. A., Rose, J. B., & Lele, S. (2001). The Association Between Extreme Precipitation and Waterborne Disease Outbreaks in the United States, 1948–1994. *American Journal of Public Health, 91*(8), 1194–1199. doi:10.2105/AJPH.91.8.1194 PMID:11499103

D'Agostino, D. R., Scardigno, A., Lamaddalena, N., & El Chami, D. (2014). Sensitivity analysis of coupled hydro-economic models: Quantifying climate change uncertainty for decision making. *Water Resources Management, 28*(12), 4303–4318. doi:10.100711269-014-0748-2

Dalle, F., Roz, P., Dautin, G., Di-Palma, M., Kohli, E., Sire-Bidault, C., ... Bonnin, A. (2003). Molecular characterization of isolates of waterborne Cryptosporidium spp. collected during an outbreak of gastroenteritis in South Burgundy, France. Microbiol. *Journal of Clinical Microbiology, 41*(6), 2690–2693. doi:10.1128/JCM.41.6.2690-2693.2003 PMID:12791906

Daryani, A., Ettehad, G. H., Sharif, M., Ghorbani, L., & Ziaei, H. (2008). Prevalence of intestinal parasites in vegetables consumed in Ardabil. Iran. *Food Control, 19*(8), 790–794. doi:10.1016/j.foodcont.2007.08.004

Dickin, S. K., Schuster-Wallace, C. J., Qadir, M., & Pizzacalla, K. (2016). A Review of Health Risks and Pathways for Exposure to Wastewater Use in Agriculture. *Environmental Health Perspectives, 124*(7). doi:10.1289/ehp.1509995 PMID:26824464

Dongus, S. (2016). *Dynamics and sustainability of urban agriculture : Examples from sub-Saharan Africa*. Academic Press. doi:10.100711625-009-0097-x

Drudge-Coates, L., & Turner, B. (2013). *Parasitic waterborne disease*. Academic Press.

Dssouli, K. (2006). *Treatment and reuse of wastewater in agriculture Marocoriental: Study of helminth parasites* (PhD in Social Sciences). Faculty of Sciences, Oujda.

Dssouli, K., El Halouani, H., & Berrichi, A. (2006). The health impact of the reuse of raw sewage from the city of Oujda in agriculture: A study of the parasite load in helminth eggs in some vegetable crops. *Biology & Health, 6*, 1.

Efstratiou, A., Ongerth, J. E., & Karanis, P. (2017). Waterborne transmission of protozoan parasites : Review of worldwide outbreaks - An update 2011 e 2016. *Water Research, 114*, 14–22. doi:10.1016/j.watres.2017.01.036 PMID:28214721

El Amiri, B., Sibaoueih, M., Hadarbach, D., Cohen, N., & Chriyaa, A. (2009). Impact of wastewater use on cattle in the common Mzamza, Province of Settat. *Laboratory Technology, 17*.

Ensink, J. H. J., Blumenthal, U. J., & Brooker, S. (2009). *Europe PMC Funders Group Wastewater quality and the risk of intestinal nematode infection in sewage farming families in Hyderabad, India*. Academic Press.

Ensink, J. H. J., Mahmood, T., van der Hoek, W., Raschid-Sally, L., & Amerasinghe, F. P. (2004). A nation-wide assessment of wastewater use in Pakistan: An obscure activity or a vitally important one? *Water Policy, 6*(3), 197–206. doi:10.2166/wp.2004.0013

Escobedo, A. A., Almirall, P., Robertson, J., Franco, R. M. B., Hanevik, K., Mørch, K., & Cimerman, S. (2010). *Giardiasis : The Ever-Present Threat of a Neglected Disease, (Ci)*. Academic Press.

Etewa, S. E., & Naglaa, S. A. A. (2016). Geohelminths distribution as affected by soil properties, physicochemical factors and climate in Sharkyia governorate. *Journal of Parasitic Diseases: Official Organ of the Indian Society for Parasitology, 40*(2), 496–504. doi:10.100712639-014-0532-5 PMID:27413327

FAO (Food Agriculture Organization). (2000). *The Energy and Agriculture Nexus*. Environment and Natural Resources Working Paper No. 4. Rome: FAO.

FAO/World Health Organization (WHO) expert committee in 2012. (n.d.). Retrieved from http://www.who.int/foodsafety/micro/jemra/meetings/sep12/en/

Fayer, R., Speer, C. A., & Dubey, J. P. (1997). The general biology of Cryptosporidium. In R. Fayer (Ed.), *Cryptosporidium and Cryptosporidiosis* (pp. 1–41). Boca Raton, FL: Taylor & Francis.

Fayer, R., Trout, J. M., Graczyk, T. K., & Lewis, E. J. (2000). Prevalence of Cryptosporidium, Eimeria and Giardia infections in post-weaned and adult cattle on three Maryland farms. *Veterinary Parasitology, 93*(2), 103–112. doi:10.1016/S0304-4017(00)00356-3 PMID:11035228

Feachem, R. G., Bradley, D. J., Garelick, H., & Mara, D. D. (1983). *Sanitation and disease health aspects of excreta and wastewater management*. Chichester, UK: Wiley.

Feng, Y., & Xiao, L. (2011). Zoonotic Potential and Molecular Epidemiology of Giardia Species and Giardiasis. *Clinical Microbiology Reviews, 24*(1), 110–140. doi:10.1128/CMR.00033-10 PMID:21233509

Frederick, K. D., & Schwarz, G. E. (1999). Socioeconomic impacts of climate change on U.S. water supplies. *Journal of the American Water Resources Association, 35*(6), 1563–1583. doi:10.1111/j.1752-1688.1999.tb04238.x

Fuhrimann, S., & Medlicott, K. (2017). *Sanitation safety planning as a tool for achieving safely managed sanitation systems and safe use of wastewater*. Academic Press. doi:10.4103/2224-3151.213790

Fuhrimann, S., Pham-duc, P., Cissé, G., Thuy Tram, N., Ha Thu, H., Do Trung, D., & … . (2016). Microbial contamination along the main open wastewater and storm water channel of Hanoi, Vietnam, and potential health risks for urban farmers. *The Science of the Total Environment, 566–7*, 1014–1022. doi:10.1016/j.scitotenv.2016.05.080 PMID:27325013

Gao, T., Wang, X. C., Chen, R., Ngo, H. H., & Guo, W. (2015). Disability adjusted life year (DALY): A useful tool for quantitative assessment of environmental pollution. *The Science of the Total Environment, 511*, 268–287. doi:10.1016/j.scitotenv.2014.11.048 PMID:25549348

Gertler, M., Dürr, M., Renner, P., Poppert, S., Askar, M., Breidenbach, J., ... Chalmers, R. (2015). Outbreak of Crypto- sporidium hominis following river flooding in the city of Halle (Saale), Germany, August 2013. *BMC Infectious Diseases, 15*(1), 1. doi:10.118612879-015-0807-1 PMID:25879490

Ghasemi, E., Rahdar, M., & Rostami, M. (2015). *Prevalence of Entamoeba histolytica / dispar in drinking water in the city of Shush, Khuzestan Province in 2011*. Academic Press.

Giangaspero, A., Berrilli, F., & Brandonisio, O. B.-P. (2007). *Giardia and Cryptosporidium and public health: the epidemiological scenario from the Italian perspective.* Springer. Retrieved from https://link.springer.com/article/10.1007/s00436-007-0598-4

Godfree, A., & Farrell, J. (2005). Processes for managing pathogens. *Journal of Environmental Quality*, *34*(1), 105–113. doi:10.2134/jeq2005.0105 PMID:15647539

Gomez-Couso, H., Mendez-Hermida, F., Castro-Hermida, J., & Ares-Mazas, E. (2005). Giardia in shellfish-farming areas: Detection in mussels, river water and wastewaters. *Veterinary Parasitology*, *133*(1), 13–18. doi:10.1016/j.vetpar.2005.04.043 PMID:15982819

Graczyk, T., Fayer, R., & Cranfield, M. R. (1997). Zoonotic transmission of Cryptosporidium parvum: Implications for water-borne cryptosporidiosis. *Parasitol. Today, 13*, 348–351. Retrieved from https://www.cell.com/partod/abstract/S0169-4758(97)01076-4

Graczyk, T. K., Fayer, R., & Cranfield, M. R. (1997a). Zoonotic potential of Cryptosporidium parvum: Implications for waterborne cryptosporidiosis. *Parasitology Today (Personal Ed.)*, *13*(9), 348–351. doi:10.1016/S0169-4758(97)01076-4 PMID:15275048

Gupta, N., Khan, D. K., & Santra, S. C. (2009). An assessment of heavy metal contamination in vegetables grown in wastewater-irrigated areas of Titagarh, West Bengal, India. *Bulletin of Environmental Contamination and Toxicology*, *80*(2), 115–118. doi:10.100700128-007-9327-z PMID:18165915

Gupta, N., Khan, D. K., & Santra, S. C. (2009). Prevalence of intestinal helminth eggs on vegetables grown in wastewater-irrigated areas of Titagarh West Bengal, India. *Food Control*, *20*(10), 942–945. doi:10.1016/j.foodcont.2009.02.003

Hadipour, A., Rajaee, T., Hadipour, V., & Seidirad, S. (2016). Multi-criteria decision-making model for wastewater reuse application: A case study from Iran. *Desalination and Water Treatment*, *57*(30), 13857–13864. doi:10.1080/19443994.2015.1060905

Hajjami, K., Ennaji, M. M., Fouad, S., Oubrim, N., & Cohen, N. (2013). Wastewater Reuse for Irrigation in Morocco: Helminth Eggs Contamination's Level of Irrigated Crops and Sanitary Risk (A Case Study of Settat and Soualem Regions). *Journal of Bacteriology & Parasitology*, *4*, 163. doi:10.4172/2155-9597.1000163

Hanjra, M. A., Blackwell, J., Carr, G., Zhang, F., & Jackson, T. M. (2012). Wastewater irrigation and environmental health: Implications for water governance and public policy. *International Journal of Hygiene and Environmental Health*, *215*(3), 255–269. doi:10.1016/j.ijheh.2011.10.003 PMID:22093903

Harper, S., Edge, V., Schuster-Wallace, C., Berke, O., & McEwen, S. (2011). Weather, water quality and infectious gastrointestinal illness in two Inuit communities in Nunatsiavut, Canada: Potential implications for climate change. *EcoHealth*, *2011*, 1–16. PMID:21785890

Herrington, R., Johnson, B., & Hunter, F. (1997). Canada Country Study: Climate Impacts and Adaptation, Volume 3: Responding to Global Climate Change in the Prairies. Environment Canada.

Howard, G., Calow, R., Macdonald, A., & Bartram, J. (2016). *Climate Change and Water and Sanitation : Likely Impacts and Emerging Trends for Action*. Academic Press. . doi:10.1146/annurev-environ-110615-085856

Hunter, P. R. (2003). Climate change and waterborne and vector-borne disease. *Journal of Applied Microbiology*, *94*(s1), 37–46. doi:10.1046/j.1365-2672.94.s1.5.x PMID:12675935

Hutchinson, M. L., Walters, L. D., Moore, T., Thomas, D. J. I., & Avery, S. M. (2005a). Fate of pathogens present in livestock wastes spread onto fescue lots. *Applied and Environmental Microbiology, 71*(2), 691–696. doi:10.1128/AEM.71.2.691-696.2005 PMID:15691918

Hutchinson, M. L., Walters, L. D., Moore, T., Thomas, D. J. I., & Avery, S. M. (2005b). Decline of zoonotic agents in liquid livestock wastes stored in batches on-farm. *Letters in Applied Microbiology, 99*(1), 58–65. doi:10.1111/j.1365-2672.2005.02585.x

Idrisa, S., Belghyti, D., El Kharrim, K., & Yoro, C. (2010). *Parasitic contamination of the mint and turnip irrigated by untreated wastewater in Sidi Yahia Gharb*. Balwois, Ohrid, Republic of Macedonia.

Jāhān, S. (2015). *Human Development Report 2015 - Work for Human Development*. New York: United Nations Development Programme.

Jimenez, B., & Asano, T. (2008). *Water Reuse: An International Survey. Contrasts, Issues and Needs around the World*. London: IWA Publishing.

Jimenez, R., & Alvarez, A. M. (2005). Control de la degradaci_on de suelos. Universidad Aut_onoma de Madrid, Madrid.

Kappus, K., Lundgren, R. G., Jr., Juranek, D. D., Roberts, J. M., & Spencer, H. C. (1994). *Intestinal parasitism in the United States: update on a continuing problem*. ASTMH. Retrieved from http://ajtmh.org/content/journals/10.4269/ajtmh.1994.50.705

Karkashan, A., Khallaf, B., Morris, J., Thurbon, N., Rouch, D., Smith, S. R., & Deighton, M. (2015). Comparison of methodologies for enumerating and detecting the viability of Ascaris eggs in sewage sludge by standard incubation-microscopy, the BacLight Live/Dead viability assay and other vital dyes. *Water Research, 68*, 533–544. doi:10.1016/j.watres.2014.10.003 PMID:25462759

Karl, T. R., Knight, R. W., Easterling, D. R., & Quayle, R. G. (1996). Indices of climate change for the United States. *Bulletin of the American Meteorological Society, 77*(2), 279–303. doi:10.1175/1520-0477(1996)077<0279:IOCCFT>2.0.CO;2

Kato, S., Jenkins, M., Fogarty, E., & Bowman, D. (2004). *Cryptosporidium parvum oocyst inactivation in field soil and its relation to soil characteristics : analyses using the geographic information systems*. Academic Press. doi:10.1016/j.scitotenv.2003.08.027

Keffala, C., Harerimana, C., & Vasel, J. L. (2012). Oeufs d'helminthes dans les eaux us ées et les boues de station d' épuration: Enjeux sanitaires et int eret du traitement par lagunage. *Environ Risque Sante, 11*, 511–520. doi:10.1684/ers.2012.0573

Keraita, B., & Amoah, P. (2011). Fecal Exposure Pathways in Accra: A Literature Reviewwith Specific Focus on IWMI's Work on Wastewater Irrigated Agriculture (IWMI). Literature Review Report. Centre for Global SafeWater, Emory University.

Keraita, B., Drechsel, P., Huibers, F., & Raschid-Sally, L., (2002). Wastewater use in informal irrigation in urban and periurban areas of Kumasi, Ghana. *UA Magazine - Wastewater Reuse in Urban Agriculture, 8*.

Khaldi, S., Ratajczak, M., Gargala, G., Fournier, M., Berthe, T., Favennec, L., & Dupont, J. P. (2011). Intensive exploitation of a karst aquifer leads to Cryptosporidium water supply contamination. *Water Research*, *45*(9), 2906–2914. doi:10.1016/j.watres.2011.03.010 PMID:21477840

Khalifa, A., Hadi, M., Sahli, N., & Mohamed, M. (2010). Parasitological contamination in salad vegetables in Tripoli-Libya. *Food Control*, *21*(5), 760–762. doi:10.1016/j.foodcont.2009.11.005

Khallayoun, K., Ziad, H., Lhadi, E. K., & Cabaret, J. (2009). Reuse of untreated wastewater in agriculture in Morocco: Potential contamination of crops and soil helminth eggs. In Collection Environment/Health aspects Waste Water/Treatment and reuse of wastewater: impact on health and the environment. IAV Edition.

Khallayoune, K. (2009). *Aspects of the sanitary wastewater reuse in agriculture, Treatment and reuse of wastewater: impact on health and the environment*. IAV Edition.

Kim, M.-K., Pyo, K.-H., Hwang, Y.-S., Park, K. H., Hwang, I. G., Chai, J.-Y., & Shin, E.-H. (2012, August 13). temperature on embryonation of Ascaris suum eggs in an environmental chamber. *Korean Journal of Parasitology*, *50*(3), 239–242. doi:10.3347/kjp.2012.50.3.239 PMID:22949753

Kishiwa, P., Nobert, J., Kongo, V., & Ndomba, P. (2018). Assessment of impacts of climate change on surface water availability using coupled SWAT and WEAP models : case of upper Pangani River Basin, Tanzania. Academic Press.

Kistemann, T., Classen, T., Koch, C., Dangendorf, F., Fischeder, R., Gebel, J., ... Exner, M. (2002). Microbial load of drinking water reservoir tributaries during extreme rainfall and runoff. *Applied and Environmental Microbiology*, *68*(5), 2188–2197. doi:10.1128/AEM.68.5.2188-2197.2002 PMID:11976088

Kotloff, K.L., Nataro, J.P., Blackwelder, W.C., Nasrin, D., Farag, T.H., Panchalingam, S., … Faruque, A.S. (2013). Burden and aetiology of diarrhoeal disease in infants and young children in developing countries (the Global Enteric Multicenter Study, GEMS): a prospective, case-control study. *Lancet*, *382*(9888), 209-222.

Kumar, N., Tischbeinb, B., Kuschec, J., Laux, P., Bege, M. K., & Bogardi, J. J. (2017). Impact of climate change on water resources of upper Kharun catchment in Chhattisgarh, India. *Journal of Hydrology: Regional Studies*, *13*, 189–207.

Lai, T. (2002). *Perspectives of peri-urban vegetable production in Hanoi*. Paper prepared for the Action Planning Workshop of the CGIAR Strategic Initiative for Urban and Peri-urban Agriculture (SIUPA), Lima, Peru.

Lal, A., Baker, M. G., Hales, S., & French, N. P. (2013). Potential effects of global environmental changes on cryptosporidiosis and giardiasis transmission. *Trends in Parasitology*, *29*(2), 83–90. doi:10.1016/j.pt.2012.10.005 PMID:23219188

Lanata, C.F. (2003). Studies of food hygiene and diarrhoeal disease. *Int. J. Environ. Health Res.*, *13*(Suppl. 1), S175-S183.

Lane, S., & Lloyd, D. (2002). Current Trends in Research into the Waterborne Parasite Giardia. *Critical Reviews in Microbiology*, *28*(2), 123–147. doi:10.1080/1040-840291046713 PMID:12109771

Lettenmaier, D. P., Wood, A. W., Palmer, R. N., Wood, E. F., & Stakhiv, E. Z. (1999). Water resources implications of global warming: A U.S. regional perspective. *Climatic Change*, *43*(3), 537–579. doi:10.1023/A:1005448007910

Lisle, J. T., & Rose, J. B. (1995). Cryptosporidium contamination of water in the USA and UK: A mini-review. *Aqua*, *44*(3), 103–117.

Liu, L., Oza, S., Hogan, D., Chu, Y., Perin, J., & Zhu, J. (2016). *Global, regional, and national causes of under-5 mortality in 2000–15: an updated systematic analysis with implications for the Sustainable Development*. Elsevier. Retrieved from https://www.sciencedirect.com/science/article/pii/S0140673616315938

Lozano, R., Naghavi, M., Foreman, K., & Lim, S. (n.d.). *Global and regional mortality from 235 causes of death for 20 age groups in 1990 and 2010: a systematic analysis for the Global Burden of Disease Study 2010*. Elsevier. Retrieved from https://www.sciencedirect.com/science/article/pii/S0140673612617280

Luo, T., Young, R., & Reig, P. (2015). *Aqueduct Projected Stress Country Rankings*. Retrieved from http://www.wri.org/blog/2015/08/ranking-world's-most-water-stressed-countries-2040

Mac Kenzie, W. R., Hoxie, N. J., Proctor, M. E., Gradus, M. S., Blair, K. A., Peterson, D. E., ... Davis, J. P. (1994). A massive outbreak in Milwaukee of Cryptosporidium infection transmitted through the public water supply. *The New England Journal of Medicine*, *331*(3), 161–167. doi:10.1056/NEJM199407213310304 PMID:7818640

Mas–Coma, S., Valero, M. A., & Bargues, M. D. (2009). Climate change effects in trematodiases, with emphasis on zoonotic fasciolasis and schistosomiasis. *Veterinary Parasitology*, *163*(4), 264–280. doi:10.1016/j.vetpar.2009.03.024 PMID:19375233

Mateo-Sagasta, J., Medlicott, J., Qadir, M., Raschid-Sally, L., Drechsel, P., & Liebe, J. (2013). *Proceedings of the UN-Water project on the safe use of wastewater in agriculture*. UN-Water Decade Programme on Capacity Development Proceedings Series No. 11.

Mateo-Sagasta, J., Medlicott, K., Qadir, M., Rashid-Sally, L., Dreschel, P., & Liebe, J. (2013). *Proceedings of the UN-Water project on the Safe Use of Wastewater in Agriculture*. Retrieved from http://collections.unu.edu/eserv/UNU:2661/ proceedings-no-11_WEB.pdf

Medema, G. J., & Schijven, J. F. (2001). Modeling the sewage discharge and dispersion of Cryptosporidium and Giardia in surface water. *Water Research*, *18*(18), 4307–4316. doi:10.1016/S0043-1354(01)00161-0 PMID:11763032

Meli, S., Maurizio, M., Belligno, A., Bufo, S. A., Mazzatura, A., & Scopa, A. (2002). Influence of irrigation with lagooned urban wastewater on chemical and microbial soil parameters in a citrus orchard under Mediterranean condition. *The Science of the Total Environment*, *285*(1-3), 69–77. doi:10.1016/S0048-9697(01)00896-8 PMID:11874049

Merritt, W. S., Alila, Y., Barton, M., Taylor, B., & Cohen, S. J. (2003). Exploring Impacts of Climate Change on the Hydrology of the Okanagan Basin. Proc. of the Canadian Water Resources Association.

Mor Siobhan, M., & Tzipori, S. (2008). Cryptosporidiosis in children in sub-Saharan Africa: A lingering challenge. *Clinical Infectious Diseases*, *47*(7), 915–921. doi:10.1086/591539 PMID:18715159

Morgan-Ryan, U. M., Fall, A., Ward, L. A., Hijjawi, N., Sulaiman, N., Fayer, R., ... Xiao, L. (2002). Cryptosporidium hominis n. sp. (Apicomplexa: Cryptosporidiidae) from Homo sapiens. *The Journal of Eukaryotic Microbiology*, *49*(6), 433–440. doi:10.1111/j.1550-7408.2002.tb00224.x PMID:12503676

Mortimer, L., & Chadee, K. (2010). The immunopathogenesis of Entamoeba histolytica. *Experimental Parasitology*, *126*(3), 366–380. doi:10.1016/j.exppara.2010.03.005 PMID:20303955

Naumova, E. N., Christodouleas, J., Hunter, P. R., & Syed, Q. (2005). Effect of precipitation on seasonal variability in cryptosporidiosis recorded by the North West England surveillance system in 1990–1999. *Journal of Water and Health*, *3*(2), 185–196. doi:10.2166/wh.2005.0017 PMID:16075943

Nobre, C. A., Marengo, J. A., Seluchi, M. E., Cuartas, L. A., & Alves, L. M. (2016). Some Characteristics and Impacts of the Drought and Water Crisis in Southeastern Brazil during 2014 and 2015. Academic Press. doi:10.4236/jwarp.2016.82022

Ojo, O., Oni, F., & Ogunkunle, O. (2003). Implications of climatic variability and climate change on water resources availability and water resources management in West Africa. *IAHS-AISH Publication*, *280*, 37–47.

Okhuysen, P. C., Chappell, C. L., Crabb, J. H., Sterling, C. R., & Dupont, H. L. (1999). Virulence of three distinct Cryptosporidium parvum isolates for healthy adults. *The Journal of Infectious Diseases*, *180*(4), 1275–1281. doi:10.1086/315033 PMID:10479158

Olson, M., O'Handley, R., Ralston, B., McAllister, T., & Thompson, R. (2004). Update on Cryptosporidium and Giardia infections in cattle. *Trends in Parasitology*, *20*, 151-198. DOI: . pt.2004.01.015 doi:10.1016/j

Olson, M. E., Goh, J., Phillips, M., Guselle, N., & McAllister, T. A. (1999). Giardia cyst and Cryptosporidium oocyst survivaln in water, soil, and cattle feces. *Journal of Environmental Quality*, *28*(6), 1991–1996. doi:10.2134/jeq1999.00472425002800060040x

Ongerth, J. E. (2013). The concentration of Cryptosporidium and Giardia in water--the role and importance of recovery efficiency. *Water Research*, *47*(7), 2479–2488. doi:10.1016/j.watres.2013.02.015 PMID:23490099

Ongerth, J. E. (2017). *Cryptosporidium* and *Giardia* in Water: Reassessment of Occurrence and Significance. *Journal of Environmental Engineering*, *143*(3), 04016084. doi:10.1061/(ASCE)EE.1943-7870.0001161

Ottoson, J., Hansen, A., Bjorlenius, B., Norder, H., & Stenstrom, T. (2006). Removal of virus, parasitic protozoa and microbial indicators in conventional and membrane processes in a wastewater pilot plant. *Water Research*, *40*(7), 1449–1457. doi:10.1016/j.watres.2006.01.039 PMID:16533517

Paraskevas, P. A., Giokas, D. L., & Lekkas, T. D. (2002). Wastewater management in coastal urban areas: The case of Greece. *Water Science and Technology*, *46*(8), 177–186. doi:10.2166/wst.2002.0174 PMID:12420981

Parkhouse, R. M. E., & Harrison, L. J. S. (2014). Helminth-Cestode: Taenia saginata and Taenia solium. *Encycl Food Saf*, *2*, 70–77. doi:10.1016/B978-0-12-378612-8.00148-7

Patz, J. A., Graczyk, T. K., Geller, N., & Vittor, A. Y. (2000). Effects of environmental change on emerging parasitic diseases. *International Journal for Parasitology*, *30*(12), 1395–1405. doi:10.1016/S0020-7519(00)00141-7 PMID:11113264

Payment, P., Berte, A., Prevost, M., Menard, B., & Barbeau, B. (2000). Occurence of pathogenic micro-organisims in the Saint Lawrence River (Canada) and comparison of health risks for populations using it as their source of drinking water. *Canadian Journal of Microbiology*, *46*(6), 565–576. doi:10.1139/w00-022 PMID:10913979

Peasey, A., Blumenthal, U., Mara, D., & Ruiz-Palacios, G. (2000). *A review of policy and standards for wastewater reuse in agriculture: a Latin American perspective*. WELL Study. Retrieved from http://www.iboro.ac.uk/well/

Pham Duc, P., Nguyen-Viet, H., Hattendorf, J., Zinsstag, J., Dac Cam, P., & Odermatt, P. (2011). Risk factors for Entamoeba histolytica infection in an agricultural community in Hanam province, Vietnam. *Parasites & Vectors*, *4*(1), 102. doi:10.1186/1756-3305-4-102 PMID:21663665

Pham-Duc, P., Nguyen-Viet, H., Hattendorf, J., Zinsstag, J., Phung-Dac, C., Zurbrügg, C., & Odermatt, P. (2013). Ascaris lumbricoides and Trichuris trichiura infections associated with wastewater and human excreta use in agriculture in Vietnam. *Parasitol Int.*, *62*, 172–80. doi:. Parint.2012.12.007. doi:10.1016/j

Pintilie, L., Torres, C. M., Teodosiu, C., & Castells, F. (2016). Urban wastewater reclamation for industrial reuse: An LCA case study. *Journal of Cleaner Production*, *139*, 1–14. doi:10.1016/j.jclepro.2016.07.209

Plutzer, J., & Karanis, P. (2016). Neglected waterborne parasitic protozoa and their detection in water. *Water Res.*, *101*, 318-332.

Pullan, R. L., Smith, J. L., Jasrasaria, R., & Brooker, S. J. (2014). Global numbers of infection and disease burden of soil-transmitted helminth infections in 2010. *Parasites & Vectors*, *7*(1), 37. doi:10.1186/1756-3305-7-37 PMID:24447578

Qadir, M., Sharma, B. R., Bruggeman, A., Choukr-Allah, R., & Karajeh, F. (2007). Non-conventional water resources and opportunities for water augmentation to achieve food security in water scarce countries. *Agricultural Water Management*, *87*(1), 2–22. doi:10.1016/j.agwat.2006.03.018

Qishlaqi, A., Moore, F., & Forghani, G. (2008). *Impact of untreated wastewater irrigation on soils and crops in Shiraz suburban area*. SW Iran.

Quintero-Betancourt, W., Peele, E. R., & Rose, J. B. (2002). Cryptosporidium parvum and Cyclospora cayetanensis: A review of laboratory methods for detection of these waterborne parasites. *Journal of Microbiological Methods*, *49*(3), 209–224. doi:10.1016/S0167-7012(02)00007-6 PMID:11869786

Ramirez-Fuentes, E., Lucho-Constantino, C., Escamilla-Silva, E., & Dendooven, L. (2002). Characteristics, and carbon and nitrogen dynamics in soil irrigated with wastewater for different lengths of time. *Bioresource Technology*, *85*(2), 179–187. doi:10.1016/S0960-8524(02)00035-4 PMID:12227543

Razakandrainibe, R., Le, L., Gargala, G., & Favennec, L. (2014). Giardiose et cryptosporidiose : deux parasitoses à transmission hydrique. *RFL Rev. Francoph. des Lab.*, *459*, 51–56.

Reinoso, R., Becares, E., & Smith, H. V. (2008). Effect of various environmental factors on the viability of Cryptosporidium parvum oocysts. *Journal of Applied Microbiology, 104*(4), 980–986. doi:10.1111/j.1365-2672.2007.03620.x PMID:17973913

Rhallabi, N.., Moundib, R., Maaroufi, M., Marghich, M., & Khallayoune, K.H. (1990). Effects of irrigation with raw sewage and purified by the soil, crop yield of tomato and hygienic quality of the harvest. *Vet Inst Agron acts, 10*, 57-66.

Rijsberman, F. R. (2006). Water scarcity: Fact or fiction? *Agricultural Water Management, 80*(1–3), 5–22. doi:10.1016/j.agwat.2005.07.001

Rimhanen-Finne, R., Vuorinen, A., Marmo, S., Malmberg, S., & Hänninen, M. L. (2004). Comparative analysis of Cryptosporidium, Giardia and indicator bacteria during sewage sludge hygienization in various composting processes. *Letters in Applied Microbiology, 38*(4), 301–305. doi:10.1111/j.1472-765X.2004.01487.x PMID:15214729

Rose, J. B., Daeschner, S., Easterling, D. R., Curriero, F. C., Lele, S., & Patz, J. A. (2000). Climate and waterborne disease outbreaks. *Journal - American Water Works Association, 92*(9), 77–87. doi:10.1002/j.1551-8833.2000.tb09006.x

Sallon, S., Deckelbaum, R. J., Schmid, I. I., Harlap, S., Baras, M., & Spira, D. T. (1988). *Cryptosporidium*, malnutrition, and chronic diarrhea in children. *American Journal of Diseases of Children, 142*, 312–315. PMID:3344720

Schwartz, J., & Levin, R. (1999). Drinking water turbidity and health. *Epidemiology (Cambridge, Mass.), 10*(1), 86–89. doi:10.1097/00001648-199901000-00017 PMID:9888287

Scott, C. A., Drechsel, P., Raschid-Sally, L., Bahri, A., Mara, D., Redwood, M., & ... . (2010). Wastewater irrigation and health: challenges and outlook for mitigating risks in low-income countries. In P. Drechsel, C. A. Scott, L. Raschid-Sally, M. Redwood, & A. Bahri (Eds.), *Wastewater Irrigation and Health: Assessing and Mitigating Risk in Low-Income Countries* (pp. 381–394). London, UK: Earthscan.

Scott, C. A., Faruqui, N. I., & Raschid-Sally, L. (2004). *Wastewater use in irrigated agriculture: confronting the livelihood and environmental realities*. Wallingford: Cabi Publishing. doi:10.1079/9780851998237.0000

Sene, M. (1994). *Etude de la Schistosomose intestinale à Schistosoma mansoni chez les rongeurs sauvages à Richard-Toil (Sénégal): Suivi de L'infestation naturelle et transmission expérimentale* (Dissertation). Université Cheikh Anta Diop de Dakar.

Shuval, H. I., Adin, A., Fattal, B., Rawitz, E., & Yekutiel, P. (1986). *Wastewater Irrigation in Developing Countries: Health Effects and Technical Solutions*. World Bank Technical Paper Number 51, The World Bank.

Slifko, T. R., Smith, H. V., & Rose, J. B. (2000). Emerging parasite zoonoses associated with water and food. *International Journal for Parasitology, 30*(12-13), 1379–1393. doi:10.1016/S0020-7519(00)00128-4 PMID:11113263

Smith, J. E. Jr, Surampalli, R. Y., Reimers, R. S., Tyagi, R. D., & Lohani, B. N. (2008). Disinfection Processes and Stability Refinements to Biosolids Treatment Technologies. *Practice Periodical of Hazardous, Toxic, and Radioactive Waste Management, 12*(1), 10–17. doi:10.1061/(ASCE)1090-025X(2008)12:1(10)

Stephenson, L. S. (2009). Nutritional and economic implications of soiltransmitted helminths with special reference to Ascariasis. In F. Lifshitz (Ed.), *Clinical disorders in pediatric gastroenterology and nutrition* (pp. 390–409). New York: Marcel Dekker.

Sterk, A., Schijven, J., de Nijs, T., & de Roda Husman, A. M. (2013). Direct and indirect effects of climate change on the risk of infection by water-transmitted pathogens. *Environmental Science & Technology, 47*(22), 12648–12660. doi:10.1021/es403549s PMID:24125400

Stolk, W. A., Kulik, M. C., le Rutte, E. A., Jacobson, J., Richardus, J. H., & de Vlas, S. J. (2016). Between-country inequalities in the neglected tropical disease burden in 1990 and 2010, with projections for 2020. *PLoS Neglected Tropical Diseases, 10*(5), e0004560. doi:10.1371/journal.pntd.0004560

Stolk, W. A., Kulik, M. C., le Rutte, E. A., Jacobson, J., Richardus, J. H., de Vlas, S. J., & Houweling, T. A. J. (2016). Between-country inequalities in the neglected tropical disease burden in 1990and 2010, with projections for 2020. *PLoS Neglected Tropical Diseases, 10*(5), e0004560. doi:10.1371/journal.pntd.0004560 PMID:27171193

Strande, L., Ronteltap, M., & Brdjanovic, D. (2014). *Faecal sludge management: systems approach for implementation and operation* (1st ed.). London: IWA Publishing. doi:10.26530/OAPEN_578132

Striepen, B. (2013). Time to tackle cryptosporidiosis. *Nature, 503*, 189–191.

Strunz, E. C., Addiss, D. G., Stocks, M. E., Ogden, S., Utzinger, J., & Freeman, M. C. (2014). Water, sanitation, hygiene, and soil-transmitted helminth infection: Asystematic review and meta-analysis. *PLoS Medicine, 11*(3), e1001620. doi:10.1371/journal.pmed.1001620 PMID:24667810

Szostakowska, B., Kruminis-Lozowska, W., Racewicz, M., Knight, R., Tamang, L., Myjak, P., & Graczyk, T. K. (2004). Cryptosporidium parvum and Giardia lamblia recovered from flies on a cattle farm and in a landfill. *Applied and Environmental Microbiology, 70*(6), 3742–3744. doi:10.1128/AEM.70.6.3742-3744.2004 PMID:15184182

Tamburrini, A., & Pozio, E. (1999). Long-term survival of Cryptosporidium parvum oocysts in sea water and in experimentally infected mussels (Mytilus galloprovincialis). *International Journal for Parasitology, 29*(5), 711–715. doi:10.1016/S0020-7519(99)00033-8 PMID:10404265

Ternes, T. A., Bonerz, M., Herrmann, N., Teiser, B., & Andersen, H. R. (2007). Irrigation of treated wastewater in Braunschweig, Germany: An option to remove pharmaceuticals and musk fragrances. *Chemosphere, 66*(5), 894–904. doi:10.1016/j.chemosphere.2006.06.035 PMID:16872661

Thompson, R., & Monis, P. (2012). Giardia – from genome to proteome. *Advances in Parasitology, 78*, 57–95. doi:10.1016/B978-0-12-394303-3.00003-7 PMID:22520441

Thompson, R. C. A. (2004). The zoonotic significance and molecular epidemiology of Giardia and giardiasis. *Veterinary Parasitology, 126*(1-2), 15–35. doi:10.1016/j.vetpar.2004.09.008 PMID:15567577

Toz, S. (1999). PCR and the detection of microbial pathogens in water and wastewater. *Water Research, 33*(17), 3545–3556. doi:10.1016/S0043-1354(99)00071-8

Ulukanligil, M., Seyrek, A., Aslan, G., Ozbilge, H., & Atay, S. (2001). Environmental pollution with soil-transmitted helminthes in Sanliurfa, Turkey. *Memorias do Instituto Oswaldo Cruz, 96*(7), 903–909. doi:10.1590/S0074-02762001000700004 PMID:11685253

Vera, L., Martel, G., Gutierrez, J., & Marquez, M. (2006). Gesti on sostenible del agua residual en entornos rurales. Capitulo 3: Evaluaci ´on de los sistemas de depuraci ´on natural. Proyecto depuranat. Instituto Tecnol ´ogico de Canarias. Departamento de agua, Direcci ´on de investigaci ´on y desarrollo tecnol ´ogico.

Vergine, P., Saliba, R., Salerno, C., Laera, G., Berardi, G., & Pollice, A. (2015). Fate of the fecal indicator Escherichia coli in irrigation with partially treated wastewater. *Water Research, 85*, 66–73. doi:10.1016/j.watres.2015.08.001 PMID:26302216

Wang, T., Guo, X., Hong, Y., Han, H., Cao, X., & Gene, Y. H.. (n.d.). *Comparison of apoptosis between adult worms of Schistosoma japonicum from susceptible (BALB/c mice) and less-susceptible (Wistar rats) hosts.* Elsevier. Retrieved from https://www.sciencedirect.com/science/article/pii/S0378111916305911

Watkins, K. (2006). *Human Development Report 2006 - Beyond scarcity: Power, poverty and the global water crisis*. New York: United Nations Development Programme. doi:10.1016/S1352-0237(02)00387-8

Weniger, B. G., Blaser, M. J., Gedrose, J., Lippy, E. C., & Juranek, D. D. (1983). An outbreak of waterborne giardiasis associated with heavy water runoff due to warm weather and volcanic ashfall. *American Journal of Public Health, 73*(8), 868–872. doi:10.2105/AJPH.73.8.868 PMID:6869640

White, A. C., Robinson, P., Okhuysen, P. C., Lewis, D. E., Shahab, I., Lahoti, S., ... Chappell, C. L. (2000). Interferon-gamma expression in jejunal biopsies in experimental human cryptosporidiosis correlates with prior sensitization and control of oocyst excretion. *The Journal of Infectious Diseases, 181*(2), 701–709. doi:10.1086/315261 PMID:10669358

WHO. (1996). *The World Health Report 1996. Fighting Disease Fostering Development*. Geneva: World Health Organization.

Winpenny, J., Heinz, I., Koo-Oshima, S., Salgot, M., Collado, J., & Hernandex, F. (2010). *The Wealth of Waste: The Economics of Wastewater Use in Agriculture.* Rome, Italy: Food and Agriculture Organization of the United Nations. Available http://www.fao.org/docrep/012/i1629e/i1629e.pdf

World Health Organization (WHO). (2006). *Guidelines for the Safe Use of Wastewater Excreta and Greywater* (Vol. 4). Geneva: World Health Organization.

World Health Organization (WHO). (2015). Investing To Overcome The Global ImpactOf Neglected Tropical Diseases Third Who Report On Neglected TropicalDiseases. WHO Document Production Services (WHO/HTM/NTD/2015.1).

Yangin, C., Yilmaz, S., Altinbas, M., & Ozturk, I. (2002). A new process for the combined treatment of municipal wastewaters and land leachates in coastal areas. *Water Science and Technology, 46*(8), 111–118. doi:10.2166/wst.2002.0156 PMID:12420973

Yoder, J.S., Gargano, J.W., Wallace, R.M., & Beach, M.J. (2012a). Giardiasis surveillancee United States, 2009e2010. *MMWR Surveill. Summ., 61*(5), 13-23.

Yoder, J.S., Wallace, R.M., Collier, S.A., Beach, M.J., & Hlavsa, M.C. (2012b). Centers for Disease Control and Prevention (CDC), Cryptosporidiosis surveillanced United States, 2009-2010. *MMWR Surveill. Summ., 61*(5), 1-12.

Section 8

Scorpionic and Ophidian Envenomation Arising From Climate Change

Chapter 23
Climate Change, Scorpion Ecology, and Envenomation: What Are the Links?

Moulay Abdelmonaim El Hidan
Ibn Zohr University, Morocco

Kholoud Kahime
Cadi Ayyad University, Morocco

Mehdi Ait Laaradia
Cadi Ayyad University, Morocco

Btissam Bouimeja
Cadi Ayyad University, Morocco & Sultan Moulay Slimane University, Morocco

Fadoua Aabadi
Cadi Ayyad University, Morocco

Abdellaziz Ait Mansour
Cadi Ayyad University, Morocco

Oulaid Touloun
Sultan Moulay Slimane University, Morocco

Abderrahman Chait
Cadi Ayyad University, Morocco

ABSTRACT

Every year, more than 1 million cases of scorpion envenomation are reported worldwide. Scorpions are thermophilic organisms. They are sensitive to weather and climate conditions, in such a way the ongoing trends of increasing temperature and more variable weather could lead to scorpionism spreading. There has been considerable debate as to whether global envenomation will be impacted by climate change which has focused on snake and spider envenomation risk. This debate didn't give enough interest to scorpion stings and its burden risks, in spite their widespread potential effects in many regions. Here, the authors review how climate and climate change may impact scorpion activity as well as scorpion envenomation. They contrast ecological and behavioral characteristics of these arthropods, and how weather, climate, climate change, and socioeconomic factors may have very different impacts on the spatiotemporal occurrence and abundance of scorpions, and the resulting scorpion envenomation.

DOI: 10.4018/978-1-5225-7775-1.ch023

INTRODUCTION

Scorpion envenoming is one of the leading causes of deaths in many parts of the world. With an estimated almost two and half billion people at risk of scorpion envenomation. Additionally, it is expected that the real extent of the scorpion envenoming is more elevated than the hospital-based approximation, as confirmed by several studies (El hidan et al., 2015). Scorpion stings it's a result of human and scorpions interaction. In fact, from an ecological point of view, scorpion envenomations mostly take place in scorpions' natural habitats invaded, disrupted, or destructed by humans. Scorpions are generally nocturnal animals and mostly active in warm season, thus, scorpion stings occur in summer, mainly during the night and at home. The risk also exists in city downtowns, even if it is definitely higher in rural areas (Celis et al., 2007).

Scorpions live in dry and hot areas but some species can be adapted to manmade environment. As an ectothermic organism, scorpions' distribution, foraging patterns, activity and life history strategies vary according to weather fluctuations. Under a warming climate, an increased envenomation incidence has been noticed, supporting climate change effects on scorpion stings. In parallel to the modification of scorpion distribution, climate change could increase the occurrence of scorpion stings and aggravate poor population vulnerability exposed to scorpions because of the absence of suitable shelter in the face of extreme weather events. Underlining the importance of socioeconomic aspects to appreciate changes in venomous scorpions' stings models related to weather, climate change, and variability, according raising the need for understanding the interaction of both factors given the observations and expectations: of more frequent extreme weather events with climate change.

Several aspects of climate change and socioeconomic factors may drive changes in scorpion envenomation distribution. In this chapter, authors will focus on how climate and climate change may impact scorpion activity as well as scorpion envenomation. They contrast ecological and behavioral characteristics of these arthropods, and how weather, climate, climate change and socioeconomic factors may have very different impacts on the spatiotemporal occurrence and abundance of scorpions, and the resulting scorpion envenomation.

SCORPION ENVENOMATION: AN OVERVIEW

Epidemiology of Scorpion Envenomation

Scorpionism, the pathology of scorpion envenomation, is a problem that occurs across five continents. Because of its frequency and severity, it represents a serious concern for public health on a worrying scale in many countries of North Africa, the Middle East, India and South America (Lourenço, 1988). Epidemiological studies report that more than 5,000 people worldwide die each year (Stockmann and Ythier, 2010). According to Chippaux (2009), the incidence of scorpion envenomation is essentially limited to four highly endemic regions: Mexico, South America to the East of the Andes, North Africa, Near and Middle East.

In North Africa, *A. australis*, *Leiurus quinquestriatus* and *Androctonus mauritanicus* (last endemic in Morocco) are considered the most dangerous. However, *Androctonus aeneas*, *Buthus occitanus* and *Hottentota gentili*, especially in the Saharan zone (Goyffon and Guette, 2005), can also cause serious, sometimes fatal envenomations. Thus the main North African countries affected by this scourge are Algeria Morocco and Tunisia with an incidence that exceeds 29,900 cases / year.

In the Near- and Middle-East, scorpion stings, mainly due to species belonging to *Androctonus, Buthus* and *Leiurus* such as *L. quinquestriatus*, *A. crassicauda* and *B. occitanus* and *P. liosoma*. The annual incidence of scorpion envenomation in this region varies according to the countries from 16 scorpion stings per 100,000 inhabitants in Oman to 90 scorpion stings per 100,000 inhabitants in Saudi Arabia. While the annual mortality is about 0.0013 death to .01 per 100,000 inhabitants recorded respectively in Saudi Arabia and Turkey (Ozkan et al., 2008; El-Aminn and Berair, 1995).

In Asia, Iran and India are the main countries affected by scorpion envenomation. The total annual incidence is about 140 per 100,000 inhabitants. The most dangerous species are *Hemiscorpius lepturus*, *Androctonus crassicauda* and *Hottentota tamulus* (Pipelzadeh et al., 2007).

In Latin America, although there is a wide geographical variations in scorpion envenomation incidence. Scorpion stings incidence is high and could reach 1350 scorpion stings per 100,000 inhabitants in some states of Mexico. Scorpion species such as *Centruroides limpidus*, *C. sufussus*, *C. gracilisTityus pachyurus*, *Tityus serrulatus*, *T. bahiensis*, *T. stigmurus* and *T. brazilae* are responsible for most of the severe scorpion envenomations (Otero et al., 2004).

Physiopathology of Scorpion Envenomation

Many toxins have been identified in scorpion venoms, most of which are small peptide toxins that target ion channels present in mammals and insects (Quintero-Hernández et al., 2013).

The toxins that have the greatest medical importance are α-toxins, specifically bind to voltage-gated sodium channels in mammals. Once attached to the site, the toxin inhibits channel inactivation resulting in prolonged depolarization and, consequently, neuronal excitation (Quintero-Hernández et al., 2013).

Neuronal excitation stimulates autonomous centers, both sympathetic and parasympathetic, resulting in autonomic excitation (Quintero-Hernández et al., 2013). In addition, α-toxins cause massive endogenous release of catecholamines, adrenaline and noradrenaline, as well as other vasoactive peptide hormones, such as neuropeptide Y and endothelin-1 (Abroug et al., 2003). In comparison with the sympathetic effects, the parasympathetic effects are less severe; they occur immediately after the bite and may cause respiratory failure. In contrast, the combination of sympathetic excitation and release of catecholamines in plasma induces the majority of severe systemic effects, including myocardial injury, pulmonary edema, and cardiogenic shock. The mechanisms of cardiac dysfunction and pulmonary edema after scorpion envenomation are complex, but they appear to result from a combination of catecholamine-mediated myocarditis and myocardial ischemia (coronary vasoconstriction) and probably direct toxins on the myocardium (Bahloul et al., 2013).

Despite the diversity of scorpions, systemic scorpion envenomation is characterized by relatively similar neurotoxic excitation syndromes, regardless of the species involved, although some differences exist. In fact, envenomation by species belonging to the genera Centruroides and Parabuthus is mainly associated with neuromuscular toxicity (Boyer et al., 2009), whereas severe envenomations by *Androctonus*, *Buthus* and *Mesobuthus* species are linked to cardiovascular toxicity, which results from hyperstimulation of autonomic centers and the release of catecholamines (Bouaziz et al., 2008, Bawaskar and Bawskar, 2011).

Most scorpion stings cause localized pain, while only about 10% of stings, even the most dangerous scorpions, result in severe systemic envenomation. Edema, erythema, paresthesia, muscle fibrillation, and numbness may occur at the site of the sting. It is often difficult to identify this site, despite significant local pain. The majority of cases of severe envenomation occur in children. Systemic envenomation is characterized by neuromuscular abnormalities resulting from effects on the somatic and cranial nerves, both by cholinergic and adrenergic excitation of the autonomic nervous system, which induces cardiac disorders and pulmonary edema. The Most of the clinical manifestations observed in the organs are caused by neuronal excitation and release of neurotransmitters (Bouaziz et al., 2008; Maia et al., 1994; Boyer et al., 2009; Bahloul et al., 2010).

SENSITIVITY OF SCORPION TO WEATHER, CLIMATE, AND CLIMATE CHANGE

Effects of Temperature and Humidity on Defensive Behaviors

Scorpions generally use their defense behaviors when they are being threatened by Human. However, the kind of defensive behavior used by the scorpion may vary depending on the type of threat and environmental conditions. In fact, temperature profoundly modulates several physiological processes counting, but not limited to, digestion, oxygen consumption (Bobka et al. 1981; Zhang & Ji, 2004), growth (Angilletta et al., 2004), and movement performance in ectotherm animals such as scorpion (Forsman, 1999). The last process is associated to temperature mainly due to well known thermal dependencies of muscular contraction and relaxation (Bennett 1984). The influence of temperature on movement performance is ecologically important for scorpions, as it impacts both hunting ability and predator avoidance via such activities as sprinting and stinging (Waldschmidt & Tracy, 1983; Bauwens et al., 1995).

Sprint speed and defensive stinging, in particular, can be critical for a scorpion attempting escape from a potential predatoras well as for prey incapacitation (Rein 1993; Van Berkum et al., 1986; Hertz et al., 1988). For this reason, preferred body temperatures (Tp) is often strongly correlated with the optimal temperature (To) for sprinting and stinging capacity (Miller 1982; Bauwens et al., 1995; Forsman, 1999), providing evidence that scorpionas an ectothermic organism typically selects body temperatures that maximize their movement capabilities. Moreover, Scorpions had significantly higher movement performance at warmer temperatures, with males significantly faster than females. Additionally, sting latency was longer and sting rate lower when scorpions were cooler (Carlson and Rowe, 2009).

If a prey organism's ability to survive encounters with and escape from predators is dependent on locomotor performance, then maintaining Tp will favor survival in the face of predation; indeed, this has been demonstrated in wild populations (Christian & Tracy, 1981).

Some reports suggest that humidity may indeed affect movement performance in scorpions (Hadley, 1974; Warburg & Polis, 1990). In fact, desiccated scorpions exhibit limited functionality of the limbs (Sensenig & Shultz, 2004). This is likely due to the use of hydraulics in some locomotor activities (Sensenig & Shultz, 2004) and the decreased effectiveness of hydrostatic systems with fluid loss (Anderson & Prestwich, 1975). Additionally, dehydration may reduce the oxygen-transporting abilities of the hemolymph, further inhibiting movement performance (Gefen & Ar 2005). Thus, at high levels of desiccation, scorpions do have difficulty moving (Sensenig & Shultz, 2004).

Several studies have shown that ectotherms subject to high temperatures, but within the normal range they experience in the field, suffer denaturation of cellular proteins (Hofmann & Somero, 1995); replacing these proteins requires energy that must be diverted from other tasks including, perhaps, maintaining the structures required for limb extension and retraction. Moreover scorpions, like many ectotherms, exhibit temperature-sensitive metabolic rates (Lighton et al., 2001); the energetic demands of higher body temperatures (Tb) would reduce glycogen stores (Sinha & Kanungo, 1967), further squeezing the pipeline that fuels locomotion. Whether desiccated scorpions are buffered from protein damage at high Tb is unknown. Dehydration, however, might actually protect scorpions from depleting their carbohydrate reserves, as desiccation significantly reduces their metabolic rates (Gefen, 2008).

All of this serves to highlight the potential importance of thermoregulation and water balance for movement-based defensive behaviors. Indeed, the primary defensive mechanisms of scorpions, sprinting (Shaffer & Formanowicz, 1996; 2000) and stinging, fall into this category.

Effects of Climatic Factors on Scorpion Surface Activity and Abundance

On Surface Activity

Surface activity of scorpions depends on biotic and abiotic factors. The most relevant abiotic factors for scorpion activity are pluvial precipitation and surface ambient temperature. Several studies have reported the temperature influence on scorpion activity (Toren 1973; Polis 1980; Bradley 1988; Benton, 1992). Thus, ecological reports indicate that higher temperatures favor scorpion surface activity. Polis (1980) showed that members of S. mesaensis responded to variation in temperature, even within seasons. Moreover, it has been shown that nocturnal air temperature was positively correlated with activity levels in adults of Paruroctonus utahensis, and it was observed that members of this species retreat into their burrows when the temperature falls below 6–8°C (Bradley, 1988).

Other studies had exhibited that the combination of temperature decrease and precipitation result in a reduction of scorpion surface activity (Polis, 1980; Bradley, 1988; Calderon-Aranda et al., 1996; Araujo et al., 2010; Nime et al., 2013). However, other data assume that precipitation alone can stimulate scorpion surface activity (Araujo et al., 2010).

Beside Temperature and precipitations, effect of cloud cover on scorpion activity was reported in the study of Nime et al. (2013). Authors suggest that cloud could, presumably cover significantly reduced moonlight penetration through the atmosphere. An effect of cloud cover has been previously demonstrated in other animal groups, for example, on migration of bats (Cryan & Brown, 2007), the activity of rodents (Vickery & Bider, 1981), and light trap captures of lepidopterans (Yela & Holyoak, 1997).

Moreover, biotic seasonal factors such as prey abundance and breeding season for scorpions can increase their surface activity (Polis, 1980; Polis and Farley, 1980; Bradley, 1988). Therefore, the influence of biotic and abiotic factors can modify scorpion behavior, which can result in stings that threaten human health (Chowell et al., 2006).

On Scorpion Abundance

Scorpion abundance is generally low in winter, intermediate in autumn, starting to grow in the spring, and has its peak in the summer. This standard is common for various species of scorpions (Polis and Farley, 1979), and he points out that these standards refer to natural environments rather than urban areas.

It has been pointed out in several studies that scorpion populations appear in higher abundance in cold and dry months. Thus, higher appearance was observed for *Microtityus jaumei* in Sierra de Canasta, Cuba in dry months (Cala-Riquelme and Colombo, 2011). But no significant correlation has been found between the distribution of scorpions during the months and rainfall; however, it is possible to observe that in the summer rainy months (January to March) there was a record peak which is very clear for T.

bahiensis. Corroborating with this pattern of *T. bahiensiss* Chowell et al. (2006) confirmed in his study that the peaks of scorpions' occurrence correlate well with rainfall.

Effects of Climatic Factors on Scorpion Reproduction

Reproductive investment concerns both the total amount of investment (energy or resources) by an individual into reproduction and the way that investment is apportioned among offspring (Wilbur, 1977). Reproduction rates of scorpion species could be influenced by differences in the environment experienced by scorpion population, such as differences in: resource abundance or utilization, competition for resources, degree of variability (or stressfulness), abiotic factors such as temperature or precipitation, or predation risk (Stearns, 1976). Lourenço (1995) demonstrates that, *Tityus spp*, females require only a single insemination to produce multiple broods (one to four) during the favorable wet season. As soon as the dry season begins, inseminated females develop reproductive diapause and gestation is blocked until the return of the following wet season, six months later. A diapause is hormonally triggered and manifests itself depending on the abiotic pressure caused by the environment. The occurrence of diapauses is mainly related to humidity, and, thus, also related to rainfall. This reproductive strategy is characteristic of the non-opportunist species (Lourenço, 1995). However, Polis (1990a) considers these strategies as being characteristic of opportunistic species.

CLIMATE AS ONE OF MANY INTERACTING DETERMINANTS OF SCORPION ENVENOMATION

Over the past decades, human forces have driven enormous changes in the natural systems, through the extension of its own distribution because of urban growth, rising exploitation of natural resources, and transformation of natural habitats into agroecosystems (Lawler et al., 2013, Venter et al., 2016).

The first consequence of this strong human impact on biodiversity is the loss of species of medicinal, commercial and evolutionary value. As species disappear we lose both known and unknown benefits they provide. The second outcome is to lead to a rising contact amid man and wild species (Conover, 2002). Especially, for poisonous species such as scorpions, snakes and spiders which is gradually becoming an important public health problem in several countries (Kasturiratne et al., 2008; Ediriweera et al., 2016). Moreover, climate change had modified the geographic distributions of poisonous species (e.g. scorpions: El Hidan et al., 2018; spiders: Saupe et al., 2011; snakes: Nori et al, 2013; Yañes-Arenas et al., 2015) (Figure 1).

In view of poisonous animals, envenomations with scorpions are amid the most alarming from an epidemiological point of view (Day et al., 2004; Chippaux & Goyffon, 2008). It has been estimated that about 1.5 million cases of envenomation by scorpion stings occur annually worldwide with a 3% risk of a fatal outcome (Chippaux & Goyffon, 2008). However, in spite of their large disease burden, widespread occurrence and high sensitivity to climatic factors, scorpion envenomation is among the most little studied of the diseases associated with climate change. In fact, the temporal incidence of scorpion stings shows a strong seasonal pattern that correlates to climatological variables. An increase in scorpion activity in the warmer months has been observed in the state of Guanajuato, Mexico, where the dangerous scorpion species *Centruroides infamatus infamatus* inhabits (Dehesa-Davila & Possani, 1994), and in Argentina, where an increase in the scorpion sting incidence caused by *Tityus trivittatus* is observed

Figure 1. Interaction between scorpion envenomation, climate, environment and socioeconomic factors

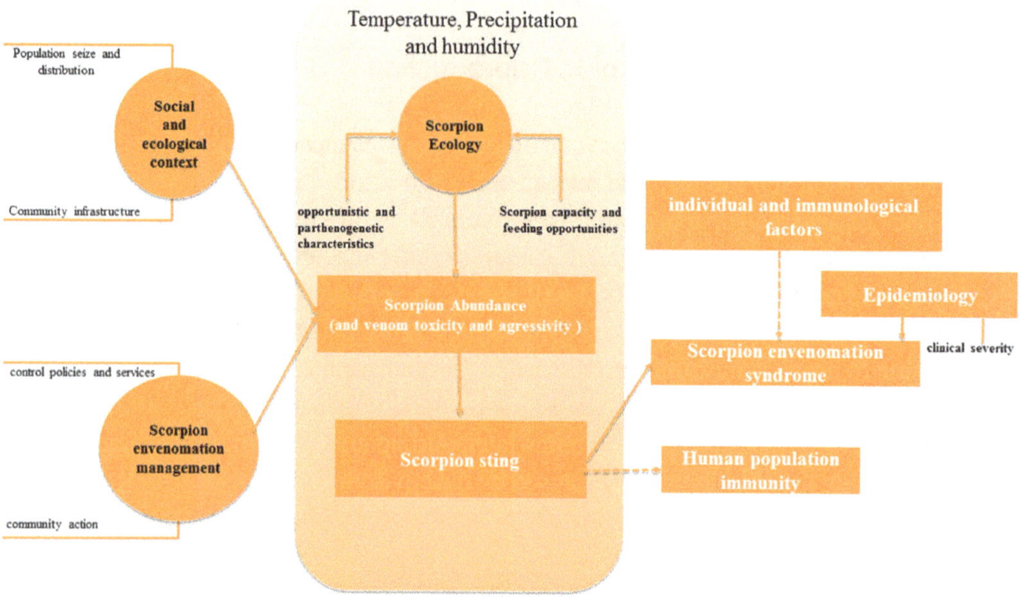

in the warmer months of October through April (De Roodt et al., 2003). Moreover, the influence of climatic factors in scorpion sting incidence has previously been studied in Mexico, showing a positive correlation between the increase in minimum temperature and scorpion sting incidence (Chowell et al., 2005).The impact of rainfall is more ambiguous. Mazzoti and Bravo-Becherelle (1963) report a higher scorpion sting incidence during the rainy season. They explain this increase in scorpion activity to the flooding from the rainfall of the burrows where scorpions live. On the contrary, Dehesa-Davila (1989) associates the beginning of the rainy season with a decrease in the number of scorpion stings.

Other factor that could influence scorpion stings is the scorpion aptitude to occupy modified environment by human activity. In fact, scorpions might be divided into two big ecological groups, «equilibrium» and «opportunistic», according to the nature of environment they occupy and their life history strategies (Polis, 1990). Equilibrium species inhabit stable, natural environments, while opportunistic species, for instance some *Androctonus* in North-Africa and *Tityus spp.* in Brazil, can easily invade disturbed habitats(Lourenço, 1991; 1994).Alterations in the natural area, environmental conditions become less and less adequate to afford the ecological necessities of the equilibrium species. Their populations, consequently, start a process of regression until, in some cases, they even disappear. This has already taken place in the Caribbean, where the island ecosystems are principally fragile. Numerous species of the oriental Amazonia and Brazilian coastal Atlantic forest may, as well, have been eradicated. Impacts of environmental changes lead to a selection against equilibrium species, which are unable of readaptation to new environmental conditions, and the encouragement of opportunistic species. This act on both species possessing very toxic venom and those possessing innocuous venom. In the first case, the phenomenon is noticed by most people, as the species are dangerous to human beings, as in the case of some *Androctonus* species in North Africa such as, *Androctonus liouvillei*, *Androctonus mauritanicus* and *A. australis*, *Centruroides* species in Mexico (e.g. *Centruroides suffusus* Pocock, the famous Durango scorpion), *Tityus serrulatus* (Lutz and Mello) in Brazil. In the second case, though, the phenomenon has been noticedjust by specialists concerned in the study of scorpions in general. Examples include innocuous buthid scorpions, such as *Isometrus maculatus* (de Geer) which lives in most tropical coastal zones of the world, and bothriurid scorpions, such as *Bothriurus bonariensis* (Koch). The latter occupies urban areas in meridional South America, being common in the region of Buenos Aires, Argentina. These two species do not present any menace to human beings. The positive selection of poisonous opportunistic species is also directly related to human activity. In fact, in a short period of time, the three principal factors required to convert the environment into an important center of scorpionism appear to be: (i) extension of the human population; (ii) fast increase of the toxic opportunistic scorpion species which soon inhabit all the niches left empty by the decrease or fading of the equilibrium species (in many cases the opportunistic species changes its behavior and begins to live inside human dwellings); (iii) overlapping of a large human population with a large population of noxious scorpions greatly increases the probability of incidents of scorpionism (Figure 1). This situation is characteristic of certain regions of Brazil, Mexico and North Africa.

Socioeconomic status is another factor that affects scorpion stings. Thus, according to the literature, the incidence of scorpion stings prevails amid populations with a low socioeconomic level. Income level, schooling, and residence location are linked to the living conditions and can be determined by indicators known as Gini index and Human Development Index (HDI). It is observed that in Brazil, the Northeast presented the highest frequencies of scorpionism (48%) and the second in number of deaths (43.5%) almost the same (44%) of the Southeast region. The Northeast, also, presented the highest annual average incidence of scorpionism (34.3/100,000 inhabitants) and the worst Gini index of Brazil: 0.501 (Reckziegel and Pinto, 2014). Additionally, other factors like the construction work in urban areas (with accumulation of materials like brick, wood, and clay), public and residential illumination, regular trash collection (that leads to the accumulation of cockroaches), piped water, and the sewage system impact the socioeconomic status of the population and may affect the frequency of scorpion envenomation (Stutz et al., 2008; Barbosa et al., 2014). Moreover, previous study in Minas Gerais, which showed that the highest incidence rates were noted in old and populous slum areas characterized by the lack of basic sanitation and other specific geographic aspects (Nunes, Bevilacqua and Jardim, 2000). The populations of Campina Grande follow a standard of inequality that is also seen in other big urban centers and is characterized by living conditions that are worse in regions localized more distant to the urban center (Queiroga et al., 2012).

The complexity of these interactions means that the effect of climate change, and the nature and extent of interaction with non-climate factors, varies markedly by location. The effects of climate on scorpion stings may be obscured, for example where the scorpion species are relatively buffered against weather and climate owing to living entirely inside houses (such as *Androctnus sp*).

COMBINING EFFECTS OF WEATHER AND CLIMATE ON SCORPION ENVENOMATION: SCENARIOS FOR IMPACTS OF CLIMATE CHANGE

Via long-term changes over decades, it is likely that climate warming will increase the basic reproductive number of scorpions in temperate regions by reducing the direct effects of cold on mortality rates and lengthening seasons for scorpion activity. This will likely facilitate the spread of the geographic distribution of scorpion species. Key limitations to spread would be: (i) occurrence of suitable habitat, particularly habitats that are important as refuge for scorpions; (ii) the occurrence of preys at suitable densities; (iii) mechanisms for dispersion (anthropic agents, via roads or rail); and (iv) preventive efforts by public and animal health organisations. While the extent of impacts will differ from one scorpion species to another, overall effects of a warming climate on changing geographic ranges may be similar for all scorpion species.

A warming climate would likely change the seasonality of scorpion arrows activity. Earlier springs would be expected to advance the seasonal of activity of scorpions where they are already endemic. A warming climate may also render temperate zones more susceptible to introduction and endemic establishment of tropical and/or subtropical scorpion species introduced by travelers or trade products by allowing the greater survival of exotic scorpion species.

Changes in long-term rainfall patterns are anticipated to modulate any warming-associated range changes: overall wetter conditions will likely favor scorpion reproduction. Long-term drought will likely reduce prey density and inhibit the reproduction of scorpion. Moreover, the study of Shachak and Brand (1983) had postulated that two alternative strategies are available for *S. maurus palmatus* to overcome prey unpredictability; dispersal during the rainy season and inactivity during the dry seasons.

CONCLUSION

It is likely that climate change will increase the impact of scorpion envenomation on human around the world by direct effects on the biology of the scorpions, or indirectly via socioeconomic mechanisms affecting the sensitivity of humans and animals to this disease. To prepare for these risks, we need to predict the potential impacts of climate change. Here, we have identified key features that might affect how these organisms respond to a changing climate. Scorpion species will likely change their geographic ranges over the coming decades in response to climate warming and changes in rainfall patterns. These characteristics need to be considered in preparatory planning for changing scorpionism risks with climate change.

Funding Information

This research received no specific grant from any funding agency in the public, commercial, or not-for-profit sectors.

REFERENCES

Abroug, F., Nouira, S., El Atrous, S., Besbes, L., Boukef, R., Boussarsar, M., ... Dellagi, K. (2003). A canine study of immunotherapy in scorpion envenomation. *Intensive Care Medicine*, *29*(12), 2266–2276. doi:10.100700134-003-1947-0 PMID:12955186

Anderson, J. F., & Prestwich, K. N. (1975). The fluid pressure pumps of spiders (Chelicerata, Araneae). *Zeitschrift für Morphologie der Tiere*, *81*(4), 257–277. doi:10.1007/BF00298488

Angilletta, M. J. Jr, Steury, T. D., & Sears, M. W. (2004). Temperature, growth rate, and body size in ectotherms: Fitting pieces of a life-history puzzle. *Integrative and Comparative Biology*, *44*(6), 498–509. doi:10.1093/icb/44.6.498 PMID:21676736

Araújo, A. S., Lobato, Z. I. P., Chávez-Olórtegui, C., & Velarde, D. T. (2010). Brazilian IgY-Bothrops antivenom: Studies on the development of a process in chicken egg yolk. *Toxicon*, *55*(4), 739–744. doi:10.1016/j.toxicon.2009.11.004 PMID:19925817

Bahloul, M., Chaari, A., Dammak, H., Samet, M., Chtara, K., Chelly, H., ... Bouaziz, M. (2013). Pulmonary edema following scorpion envenomation: Mechanisms, clinical manifestations, diagnosis and treatment. *International Journal of Cardiology*, *162*(2), 86–91. doi:10.1016/j.ijcard.2011.10.013 PMID:22075406

Barbosa, A., Silva, J. A., Cardoso, M. F. E. C., Meneses, J. N. C., Cunha, M. C. M., Haddad, J. P. A., ... Magalhães, D. F. (2014). Distribuição espacial de acidentes escorpiônicos em Belo Horizonte, Minas Gerais, 2005 a 2009 (Spatial distribution of scorpion sting accidents in Belo Horizonte, Minas Gerais, 2005 to 2009). *Arquivo Brasileiro de Medicina Veterinária e Zootecnia*, *66*(3), 721–730. doi:10.1590/16784162-7116

Bauwens, D., Garland, T. Jr, Castilla, A. M., & Van Damme, R. (1995). Evolution of sprint speed in lacertid lizards: Morphological, physiological, and behavioral covariation. *Evolution; International Journal of Organic Evolution*, *49*(5), 848–863. PMID:28564867

Bauwens, D., Garland, T. Jr, Castilla, A. M., & Van Damme, R. (1995). Evolution of sprint speed in lacertid lizards: Morphological, physiological, and behavioral covariation. *Evolution; International Journal of Organic Evolution*, *49*(5), 848–863. PMID:28564867

Bawaskar, H. S., & Bawaskar, P. H. (2011). Efficacy and safety of scorpion antivenom plus prazosin compared with prazosin alone for venomous scorpion (Mesobuthus tamulus) sting: Randomised open label clinical trial. *BMJ (Clinical Research Ed.)*, *342*(3), c7136. doi:10.1136/bmj.c7136 PMID:21209062

Bennett, A. F. (1984). Thermal dependence of muscle function. *American Journal of Physiology. Regulatory, Integrative and Comparative Physiology*, *247*(2), R217–R229. doi:10.1152/ajpregu.1984.247.2.R217 PMID:6380314

Bobka, M. S., Jaeger, R. G., & McNaught, D. C. (1981). Temperature dependent assimilation efficiencies of two species of terrestrial salamanders. *Copeia*, *1981*(2), 417–421. doi:10.2307/1444232

Bouaziz, M., Bahloul, M., Kallel, H., Samet, M., Ksibi, H., Dammak, H., ... Rekik, N. (2008). Epidemiological, clinical characteristics and outcome of severe scorpion envenomation in South Tunisia: Multivariate analysis of 951 cases. *Toxicon*, *52*(8), 918–926. doi:10.1016/j.toxicon.2008.09.004 PMID:18930073

Boyer, L. V., Theodorou, A. A., Berg, R. A., Mallie, J., Chávez-Méndez, A., García-Ubbelohde, W., ... Alagón, A. (2009). Antivenom for critically ill children with neurotoxicity from scorpion stings. *The New England Journal of Medicine*, *360*(20), 2090–2098. doi:10.1056/NEJMoa0808455 PMID:19439743

Bradley, R. A. (1988). The influence of weather and biotic factors on the behaviour of the scorpion (Paruroctonus utahensis). *Journal of Animal Ecology*, *57*(2), 533–551. doi:10.2307/4923

Cala-Riquelme, F., & Colombo, M. (2011). Ecology of the scorpion, Microtityus jaumei in Sierra de Canasta, Cuba. *Journal of Insect Science*, *11*(1). PMID:21870972

Calderón-Aranda, E. S., Dehesa-Dávila, M., Chavez-Haro, A., & Possani, L. D. (1996). Scorpion stings and their treatment in Mexico. *Envenomings and their treatments*, 311-320.

Carlson, B. E., & Rowe, M. P. (2009). Temperature and desiccation effects on the antipredator behavior of Centruroides vittatus (Scorpiones: Buthidae). *The Journal of Arachnology*, *37*(3), 321–330. doi:10.1636/Hi09-06.1

Celis, A., Gaxiola-Robles, R., Sevilla-Godínez, E., Valerio, M. D. J. O., & Armas, J. (2007). Tendencia de la mortalidad por picaduras de alacrán en México, 1979-2003. *Revista Panamericana de Salud Pública*, *21*(6), 373–380. doi:10.1590/S1020-49892007000500005 PMID:17761049

Chippaux, J. P. (2009). Incidence mondiale et prise en charge des envenimations ophidiennes et scorpioniques. *Médecine/Sciences*, *25*(2), 197-200.

Chippaux, J. P., & Goyffon, M. (2008). Epidemiology of scorpionism: A global appraisal. *Acta Tropica*, *107*(2), 71–79. doi:10.1016/j.actatropica.2008.05.021 PMID:18579104

Chowell, G., Díaz-Dueñas, P., Bustos-Saldaña, R., Mireles, A. A., & Fet, V. (2006). Epidemiological and clinical characteristics of scorpionism in Colima, Mexico (2000–2001). *Toxicon*, *47*(7), 753–758. doi:10.1016/j.toxicon.2006.02.004 PMID:16574179

Chowell, G., Hyman, J. M., Díaz-Dueñas, P., & Hengartner, N. W. (2005). Predicting scorpion sting incidence in an endemic region using climatological variables. *International Journal of Environmental Health Research*, *15*(6), 425–435. doi:10.1080/09603120500392475 PMID:16506436

Christian, K. A., & Tracy, C. R. (1981). The effect of the thermal environment on the ability of hatchling Galapagos land iguanas to avoid predation during dispersal. *Oecologia*, *49*(2), 218–223. doi:10.1007/BF00349191 PMID:28309312

Conover, M. R. (2001). *Resolving human-wildlife conflicts: the science of wildlife damage management*. CRC press. doi:10.1201/9781420032581

Cryan, P. M., & Brown, A. C. (2007). Migration of bats past a remote island offers clues toward the problem of bat fatalities at wind turbines. *Biological Conservation*, *139*(1-2), 1–11. doi:10.1016/j.biocon.2007.05.019

Day, J. F., Edman, J. D., Kunz, S. E., & Wikel, S. K. (2004). Direct injury: phobias, psychoses, annoyance, allergies, toxins, venoms and myiasis. In *Medical Entomology* (pp. 99–149). Dordrecht: Springer. doi:10.1007/978-94-007-1009-2_4

de Roodt, A. R., Lago, N. R., Salomón, O. D., Laskowicz, R. D., de Román, L. E. N., López, R. A., ... Vega, V. D. V. (2009). A new venomous scorpion responsible for severe envenomation in Argentina: Tityus confluens. *Toxicon*, *53*(1), 1–8. doi:10.1016/j.toxicon.2008.10.003 PMID:18983868

Dehesa-Dávila, M. (1989). Epidemiological characteristics of scorpion sting in Leon, Guanajuato, Mexico. *Toxicon*, *27*(3), 281–286. doi:10.1016/0041-0101(89)90175-X PMID:2728020

Dehesa-Dávila, M., & Possani, L. D. (1994). Scorpionism and serotherapy in Mexico. *Toxicon*, *32*(9), 1015–1018. doi:10.1016/0041-0101(94)90383-2 PMID:7801335

Ediriweera, D. S., Kasturiratne, A., Pathmeswaran, A., Gunawardena, N. K., Wijayawickrama, B. A., Jayamanne, S. F., ... Lalloo, D. G. (2016). Mapping the risk of snakebite in Sri Lanka-a national survey with geospatial analysis. *PLoS Neglected Tropical Diseases*, *10*(7), e0004813. doi:10.1371/journal.pntd.0004813 PMID:27391023

El Aminn, E. O., & Berair, R. (1995). Piqûres de scorpion chez l'enfant. Expérience saoudienne. *Archives de Pediatrie*, *2*(8), 766–773. doi:10.1016/0929-693X(96)81248-6 PMID:7550843

El Hidan, M. A., Touloun, O., Bouazza, A., Laaradia, M. A., & Boumezzough, A. (2018). Androctonus genus species in arid regions: Ecological niche models, geographical distributions, and envenomation risk. *Veterinary World*, *11*(3), 286–292. doi:10.14202/vetworld.2018.286-292 PMID:29657418

El Hidan, M. A., Touloun, O., & Boumezzough, A. (2015). An epidemiological study on scorpion envenomation in the Zagora oases (Morocco). *Journal of Coastal Life Medicine*, *3*(9), 704–707. doi:10.12980/jclm.3.2015j5-86

Forsman, A. (1999). Variation in thermal sensitivity of performance among colour morphs of a pygmy grasshopper. *Journal of Evolutionary Biology*, *12*(5), 869–878. doi:10.1046/j.1420-9101.1999.00084.x

Freire-Maia, L., Campos, J. A., & Amaral, C. F. S. (1994). Approaches to the treatment of scorpion envenoming. *Toxicon*, *32*(9), 1009–1014. doi:10.1016/0041-0101(94)90382-4 PMID:7801334

Gefen, E. (2008). Sexual dimorphism in desiccation responses of the sand scorpion Smeringurus mesaensis (Vaejovidae). *Journal of Insect Physiology*, *54*(5), 798–805. doi:10.1016/j.jinsphys.2008.02.004 PMID:18374353

Gefen, E., & Ar, A. (2005). The effect of desiccation on water management and compartmentalisation in scorpions: The hepatopancreas as a water reservoir. *The Journal of Experimental Biology*, *208*(10), 1887–1894. doi:10.1242/jeb.01568 PMID:15879069

Goyffon, M., & Guette, C. (2005). Scorpions dangereux du Niger. *Bulletin de la Société de Pathologie Exotique*, *98*(4), 293–295. PMID:16402579

Hadley, N. F. (1970). Water relations of the desert scorpion, Hadrurus arizonensis. *The Journal of Experimental Biology*, *53*(3), 547–558. PMID:5487163

Hertz, P. E., Huey, R. B., & Garland, T. Jr. (1988). Time budgets, thermoregulation, and maximal locomotor performance: Are reptiles olympians or boy scouts? *American Zoologist*, *28*(3), 927–938. doi:10.1093/icb/28.3.927

Hofmann, G., & Somero, G. (1995). Evidence for protein damage at environmental temperatures: Seasonal changes in levels of ubiquitin conjugates and hsp70 in the intertidal mussel Mytilus trossulus. *The Journal of Experimental Biology*, *198*(7), 1509–1518. PMID:9319406

Larkin, N. K., & Harrison, D. E. (2005). Global seasonal temperature and precipitation anomalies during El Niño autumn and winter. *Geophysical Research Letters*, *32*(16), L16705. doi:10.1029/2005GL022860

Lawler, J. J., Ruesch, A. S., Olden, J. D., & McRae, B. H. (2013). Projected climate-driven faunal movement routes. *Ecology Letters*, *16*(8), 1014–1022. doi:10.1111/ele.12132 PMID:23782906

Lighton, J., Brownell, P., Joos, B., & Turner, R. (2001). Low metabolic rate in scorpions: Implications for population biomass and cannibalism. *The Journal of Experimental Biology, 204*(3), 607–613. PMID:11171311

Lourenço, W. (1991). Biogéographie évolutive, écologies et les stratégies biodémographiques chez les scorpions néotropicaux. *Compte rendu des séances de la société de biogéographie, 67*(4), 171-190.

Lourenço, W. R. (1988). Peut-on parler d'une biogéographie du scorpionisme. *CR Seances Soc Biogéogr, 64*(4), 137–143.

Lourenço, W. R. (1994). Diversity and endemism in tropical versus temperate scorpion communities. *Compte rendu des séances de la société de biogéographie, 70*(3), 155-160.

Lourenço, W. R., & Cuellar, O. (1995). Scorpions, scorpionism, life history strategies and parthenogenesis. *The Journal of Venomous Animals and Toxins, 1*(2), 51–62. doi:10.1590/S0104-79301995000200002

Lourengo, W. (1991). Biogéographie évolutive, écologies et les stratégies biodémographiques chez les scorpions néotropicaux. *Compte rendu des séances de la société de biogéographie, 67*(4), 171-190.

Mazzotti, L., & Bravo-Becherelle, M. A. (1961). Scorpionism in the Mexican republic. *Revista del Instituto de Salubridad y Enfermedades Tropicales, 21*, 3. PMID:14471889

Miller, K. (1982). Effect of temperature on sprint performance in the frog Xenopus laevis and the salamander Necturus maculosus. *Copeia, 1982*(3), 695–698. doi:10.2307/1444671

Nime, M. F., Casanoves, F., Vrech, D. E., & Mattoni, C. I. (2013). Relationship between environmental variables and surface activity of scorpions in the Arid Chaco ecoregion of Argentina. *Invertebrate Biology, 132*(2), 145–155. doi:10.1111/ivb.12019

Nori, J., Carrasco, P. A., & Leynaud, G. C. (2014). Venomous snakes and climate change: Ophidism as a dynamic problem. *Climatic Change, 122*(1-2), 67–80. doi:10.100710584-013-1019-6

Nunes, C. S., Bevilacqua, P. D., & Jardim, C. C. G. (2000). Aspectos demográficos e espaciais dos acidentes escorpiônicos no Distrito Sanitário Noroeste, Município de Belo Horizonte, Minas Gerais, 1993 a 1996. *Cadernos de Saude Publica, 16*(1), 213–223. doi:10.1590/S0102-311X2000000100022 PMID:10738166

Otero, R., Navio, E., Céspedes, F. A., Núñez, M. J., Lozano, L., Moscoso, E. R., ... Rodas, J. H. (2004). Scorpion envenoming in two regions of Colombia: Clinical, epidemiological and therapeutic aspects. *Transactions of the Royal Society of Tropical Medicine and Hygiene, 98*(12), 742–750. doi:10.1016/j.trstmh.2003.12.018 PMID:15485705

Ozkan, O., Uzun, R., Adiguzel, S., Cesaretli, Y., & Ertek, M. (2008). Evaluation of scorpion sting incidence in Turkey. *The Journal of Venomous Animals and Toxins Including Tropical Diseases, 14*(1), 128–140. doi:10.1590/S1678-91992008000100010

Pipelzadeh, M. H., Jalali, A., Taraz, M., Pourabbas, R., & Zaremirakabadi, A. (2007). An epidemiological and a clinical study on scorpionism by the Iranian scorpion Hemiscorpius lepturus. *Toxicon, 50*(7), 984–992. doi:10.1016/j.toxicon.2007.07.018 PMID:17854855

Polis, G. A. (1980). Seasonal patterns and age-specific variation in the surface activity of a population of desert scorpions in relation to environmental factors. *Journal of Animal Ecology*, *49*(1), 1–18. doi:10.2307/4275

Polis, G. A. (1990). *The biology of scorpions* (No. 595.46 B5). Academic Press.

Polis, G. A., & Farley, R. D. (1980). Population biology of a desert scorpion: Survivorship, microhabitat, and the evolution of life history strategy. *Ecology*, *61*(3), 620–629. doi:10.2307/1937428

Queiroga, R. P. F. D., Sá, L. D. D., Nogueira, J. D. A., Lima, E. R. V. D., Silva, A. C. O., Pinheiro, P. G. O. D., & Braga, J. U. (2012). Distribuição espacial da tuberculose ea relação com condições de vida na área urbana do município de Campina Grande-2004 a 2007. *Revista Brasileira de Epidemiologia*, *15*(1), 222–232. doi:10.1590/S1415-790X2012000100020 PMID:22450507

Quintero-Hernández, V., Jiménez-Vargas, J. M., Gurrola, G. B., Valdivia, H. H., & Possani, L. D. (2013). Scorpion venom components that affect ion-channels function. *Toxicon*, *76*, 328–342. doi:10.1016/j.toxicon.2013.07.012 PMID:23891887

Reckziegel, G. C., & Pinto, V. L. (2014). Scorpionism in Brazil in the years 2000 to 2012. *The Journal of Venomous Animals and Toxins Including Tropical Diseases*, *20*(1), 46. doi:10.1186/1678-9199-20-46 PMID:25873937

Rein, J. O. (1993). Sting use in two species of Parabuthus scorpions (Buthidae). *The Journal of Arachnology*, 60–63.

Saupe, E. E., Papes, M., Selden, P. A., & Vetter, R. S. (2011). Tracking a medically important spider: Climate change, ecological niche modeling, and the brown recluse (Loxosceles reclusa). *PLoS One*, *6*(3), e17731. doi:10.1371/journal.pone.0017731 PMID:21464985

Sensenig, A. T., & Shultz, J. W. (2004). Elastic energy storage in the pedipalpal joints of scorpions and sun-spiders (Arachnida, Scorpiones, Solifugae). *The Journal of Arachnology*, *32*(1), 1–10. doi:10.1636/S02-73

Shachak, M., & Brand, S. (1983). The relationship between sit and wait foraging strategy and dispersal in the desert scorpion, Scorpio maurus palmatus. *Oecologia*, *60*(3), 371–377. doi:10.1007/BF00376854 PMID:28310698

Shaffer, L. R., Daniel, R., & Formanowicz, J. R. (1996). A cost of viviparity and parental care in scorpions: Reduced sprint speed and behavioural compensation. *Animal Behaviour*, *51*(5), 1017–1024. doi:10.1006/anbe.1996.0104

Shaffer, L. R., & Formanowicz, D. R. Jr. (2000). Sprint speeds of juvenile scorpions: Among family differences and parent offspring correlations. *Journal of Insect Behavior*, *13*(1), 45–54. doi:10.1023/A:1007711508847

Sinha, R. C., & Kanungo, M. S. (1967). Effect of starvation on the scorpion Palamnaeus bengalensis. *Physiological Zoology*, *40*(4), 386–390. doi:10.1086/physzool.40.4.30158456

Stearns, S. C. (1976). Life-history tactics: A review of the ideas. *The Quarterly Review of Biology*, *51*(1), 3–47. doi:10.1086/409052 PMID:778893

Stockmann, R., & Ythier, E. (2010). *Scorpions of the world.* N.A.P. Editions, Verrières-le- Buisson.

Stutz, W. H., Bendeck, O., Macedo, E. M., Camargo, J. C. C., Oliveira, F. S., & Bonito, R. F. (2008). Bioensaio Visando Controle de Escorpionídeos (Tityus serrulatus), através do uso de Bendiocarb. *Deltamethrin e Lambda-cyhalothrin, 2003.*

Toren, T. J. (1973). *Biology of the California Coast Range scorpion Vaejovis gertschi striatus Hjelle* (Unpublished Masters thesis). California State University at San Francisco.

Van Berkum, F. H., Huey, R. B., & Adams, B. A. (1986). Physiological consequences of thermoregulation in a tropical lizard (Ameiva festiva). *Physiological Zoology, 59*(4), 464–472. doi:10.1086/physzool.59.4.30158599

Venter, O., Sanderson, E. W., Magrach, A., Allan, J. R., Beher, J., Jones, K. R., ... Levy, M. A. (2016). Sixteen years of change in the global terrestrial human footprint and implications for biodiversity conservation. *Nature Communications, 7*(1), 12558. doi:10.1038/ncomms12558 PMID:27552116

Vickery, W. L., & Bider, J. R. (1981). The influence of weather on rodent activity. *Journal of Mammalogy, 62*(1), 140–145. doi:10.2307/1380484

Waldschmidt, S., & Tracy, C. R. (1983). Interactions between a lizard and its thermal environment: Implications for sprint performance and space utilization in the lizard Uta stansburiana. *Ecology, 64*(3), 476–484. doi:10.2307/1939967

Warburg, M. R., & Polis, G. A. (1990). Behavioral responses, rhythms, and activity patterns. *The Biology of Scorpions, 224*, 246.

Wilbur, H. M. (1977). Propagule size, number, and dispersion pattern in Ambystoma and Asclepias. *American Naturalist, 111*(977), 43–68. doi:10.1086/283137

Wright, S. J., & Calderón, O. (2006). Seasonal, El Nino and longer term changes in flower and seed production in a moist tropical forest. *Ecology Letters, 9*(1), 35–44. PMID:16958866

Yañez-Arenas, C., Peterson, A. T., Rodríguez-Medina, K., & Barve, N. (2016). Mapping current and future potential snakebite risk in the new world. *Climatic Change, 134*(4), 697–711. doi:10.100710584-015-1544-6

Yela, J. L., & Holyoak, M. (1997). Effects of moonlight and meteorological factors on light and bait trap catches of noctuid moths (Lepidoptera: Noctuidae). *Environmental Entomology, 26*(6), 1283–1290. doi:10.1093/ee/26.6.1283

Zhang, Y. P., & Ji, X. (2004). The thermal dependence of food assimilation and locomotor performance in southern grass lizards, Takydromus sexlineatus (Lacertidae). *Journal of Thermal Biology, 29*(1), 45–53. doi:10.1016/j.jtherbio.2003.10.007

Chapter 24
Climate Change Effects on Venomous Snakes:
Distribution and Snakebite Epidemiology

Abdellah Bouazza
Cadi Ayyad University, Morocco

Moulay Abdelmonaim El Hidan
Ibn Zohr University, Morocco

Abdelmohcine Aimrane
Cadi Ayyad University, Morocco

Kholoud Kahime
Cadi Ayyad University, Morocco

Aziza Lansari
Cadi Ayyad University, Morocco

Mehdi Ait Laaradia
Cadi Ayyad University, Morocco

Hasna Lahouaoui
Cadi Ayyad University, Morocco

Abdellatif Moukrim
Abdelmalek Essaadi University, Morocco

ABSTRACT

The purpose of this chapter is to examine the evidence of a relationship between climatic changes and snake species distribution in relation with the snakebites risk increment against human populations. The global climatic change is a key factor leading to snake species behavioral changes mainly because of the rise of temperature. The variety of venomous snakes and their related potency toward human being have been well documented. Thus, this may serve as a basic knowledge for any preventive act in the face of snake toxins and their caused physiopathological and clinical effects. In addition, several studies have shown that global warming have caused a change in snake habitat and distribution, thus leading to an increase of overlapped human and snake populations living territories which raise up the risk of envenomation. Globally, more than 20,000 deaths occur every year with a high tendency to increase. Thus, consideration of human risk of envenomation may be fundamental to the effective intervention in epidemiological and clinical scales.

DOI: 10.4018/978-1-5225-7775-1.ch024

INTRODUCTION

Numerous animal species have the potential to bite and cause envenoming in humans, but snakes have a considerable epidemiological importance. Worldwide, up to five million people are bitten by snakes every year, within which venomous snakes cause considerable morbidity and mortality (World Health Organization, WHO 2018). However, snakebite received little attention from national and international health authorities (Gutiérrez et *al.*, 2006; Williams et *al.*, 2010), and is now categorized as a neglected tropical disease (WHO 2018; White, 2018). At least 421,000 envenoming's and 20,000 deaths occur each year due to snakebite globally, but these numbers may be as high as1.8millionenvenomings and 94,000 deaths (Kasturiratne et *al.*, 2008). In fact, morbidity and mortality resulting from snake bites remain unclear and often go unrecorded in many undeveloped regions of the world.

In practical terms, the true global incidence of current snakebite risk predictions and its associated mortality are difficult to estimate (Kasturiratne et *al.*, 2008). This is due to the fact that information on epidemiological indicators, incidence, mortality, and morbidity of snakebites is not comprehensive both within particular countries and globally for several reasons. For example, few reliable incidence data are available from the rural tropics where snakebites are higher but reliable data are mostly limited to a few developed countries where bites are rare (WHO, 2018). Thus, information on the total number of envenomed individuals cannot be obtained, and actually tracing such people within health institutions is practically impossible. Also, the only data available is on patients dealt with by toxicology centers or hospitalized at particular institutions (Chippaux, 2008).

Snakebites incidence depends on several factors such as climate, ecological parameters, distribution of venomous snakes, human population density and economic activities (Chippaux, 2008). However, treatment of snakebite envenoming is depending on the species responsible for the bite, and the only approved and accepted treatment is the use of antivenoms. Hence, antivenoms should be carefully distributed based on the distribution of the problem (Gutiérrez, 2012; Hansson et *al.*, 2013). In this way, knowledge regarding the distributional patterns (and potential changes) of venomous snakes becomes essential for rapid identification of dangerous species, and so the suitable treatment of snakebite accidents. Therefore, increasing the knowledge of distributional ranges, activities, behavior and suitable habitats of venomous snakes is necessary in treating snakebite envenoming (Chippaux, 2008; Valenta, 2010). In addition, it would be important to evaluate if climate change can produce changes in snake's species' suitable climate spaces, which in turn could be important when addressing the problem (Nori et *al.*, 2014; Yañez-Arenas et *al.*, 2016). In this work the authors try to identify effects of climate change on venomous snake distribution and snakebites epidemiology.

DISTRIBUTION AND EPIDEMIOLOGY OF VENOMOUS SNAKES IN THE WORLD

Distribution

Venomous snakes occur throughout a major part of the world, including many oceans (Vitt and Caldwell, 2013). However, there are few places without wild snakes. In general, they involve cold regions with an extreme climate, or islands that snakes could not reach by migrating through evolution or via the intervention of humans (Figure 1).

Figure 1. Global distribution of venomous snakes (according to Vitt and Caldwell 2013). The light gray sections indicate the approximated home ranges of marine snakes. The white areas are territories estimated to be free of venomous snakes.

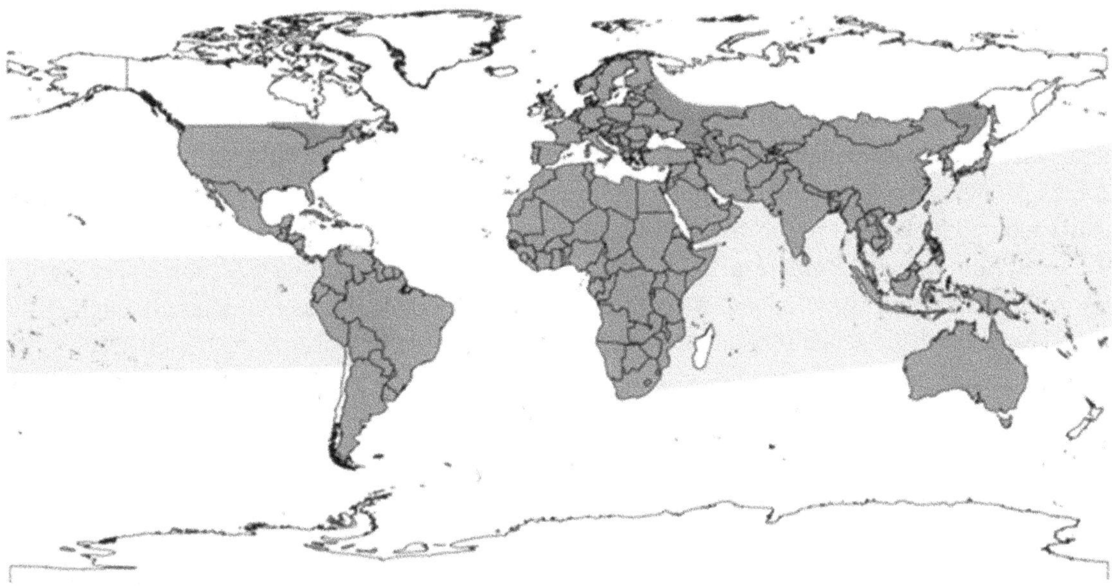

Snakes do not inhabit Antarctica because cold climate is inhospitable for reptiles whose cold-blooded bodies need heat from the surroundings to function (Pough, 1980). This is the case in many other areas, especially Greenland, Alaska, Iceland, Ireland and Newfoundland. While in Chile, snakes spreading through history was blocked from the east by the Andes Mountains which set up a physical barrier (Vitt & Caldwell, 2013).

Within other areas with no extreme weather, islands and archipelagos are the most distinguished for their characteristic of not being at all hospitable for terrestrial venomous snakes, a matter of fact which could be related to the isolation by water or to their volcanic nature, with exception of some specific islands (Underwood, 1979). In fact, venomous snakes cannot be found neither on the Canary, the Cape Verde and the Balearic archipelagos, nor in Corsica, Sardinia, Malta, or in Crete and other Greek islands. In Galápagos Islands, New Zealand and Madagascar no terrestrial venomous snakes are found furthermore. While, in Central America, they does not occur throughout any of the Caribbean islands except of Tobago, Trinidad, Saint Lucia and Martinique where venomous snakes do reside (Vitt & Caldwell, 2013).

Sea snakes occur in Indian and Pacific oceans from the east coast of Africa to the Gulf of Panama. Most species are found in the Indo-Malayan Archipelago, China seas, Indonesia, and the Australian region. However, no sea snakes can be found in Atlantic Ocean or the Mediterranean Sea (Chippaux, 1998).

In terms of altitude, most snakes inhabit localities below 2,500 m, with some exceptions manifested in some notable cases. For example, the common adder (*Vipera berus*) can ascend to 3,000 m in the Alps and Scandinavian mountains (Vitt & Caldwell, 2013) and Mountain Viper (*Vipera monticola*) can found at 3,900 m in the High Atlas of Morocco (Bons & Geniez, 1996). The dusky rattlesnake (*Crotalus triseriatus*) of Central America that is able to live as high up as 4,400 m above sea level (Campbell and Lamar, 2004), and finally the Himalayan pit viper (*Gloydius hima layanus*) that could bear extreme conditions in glacial zones up to 5,000 m (Shi et *al.*, 2017).

Although venomous snakes have adapted to living in a variety of land and water habitats, some species prefer specific types of landscapes (Shine and Shetty, 2001). For instance, members of the same *Viperinae* subfamilies, represent a typical illustration of how different they are from one another in relation to their preferred home ground. In fact, they were found in both temperate and tropical environments, and in a wide variety of habitats including terrestrial, arboreal, grasslands, dry steppe, mountains, forests, savannas, and mountainous areas, etc (Phelps, 2010). As an example, the Gabon and Rhinoceros vipers avoid dense forested enclaves with restricted sunlight, while, *Atheris*, another species of the *Viperinae* genera, has specialized in hunting and dwelling in trees and bushes (Chippaux, 2006). *Crotalinae* members, showing however usually similar patterns in habitats requirements, are found in savannahs, sparse woodlandthickets in the sub-mountains and mountains of America and Asia (Fenker et *al.*, 2014). Thus, some *Bothrops* and *Trimeresurus* species occur in forests areas, as well as, some species of the *Agkistrodon* genus prefer proximity to water (Vitt & Caldwell, 2013). These examples all together show how snakes distribution depends on a set of preferences that could be linked to restricted species' ways of life such as diet and shelter.

Speciation inside the same family, as it is supposed to be via the historical adaptation to surrounding factors' influence, does not only create divergences in terms of terrestrial habitats life form specification, it also leads to branching off species developed to live in aquatic area. The *Elapidae* family is venomous snakes found in the tropics and subtropics around the world with terrestrial forms in Asia, Australia, Africa, North America and South America, and marine forms in the Pacific and Indian Oceans (Vitt & Caldwell, 2013). As well, some terrestriallapids occur in damp or dense tropical forests, as do a number of Asian cobra species of *Naja* and *Ophiophagus* genera, as well as coral snakes of the *Micrurus* genus. Other African cobras and mambas mainly live areas of grass savannas or shrublands. Some elapids are arboreal (e.g. African mambas), while many others are more or less specialized burrowers (e.g.*Ogmodon, Parapistocalamus, Simoselaps, Toxicocalamus, Vermicella*) in humid or arid environments. Some sea snakes that once again belong to the group of elapid snakes (*Elapidae*)are partially marine, they are inhabiting coral reefs where they feed, but come onto beaches and rocks to rest and lay eggs, while other sea snakes are fully marine (Spalding et *al.*, 2001).

Others venomous snakes have different degrees of envenoming potentials toward human that has been revealed within snake families. The *Atractaspididae* (burrowing asps) is a family of venomous snakes found in Africa and the Middle East. All burrowing asps are venomous but few are dangerous or lethal to humans (Warrell, 2010), where most species live underground; some utilizing existing animal tunnels and others pushing through loose sand or leaf litter (Vitt and Caldwell, 2013). *Colubridae* (colubrids), which have been diagnosed as non-venomous or to have venom that is not dangerous to humans, includes few groups, such as genus Boiga, which can produce medically significant bites. There is five species including the Boomslang (*Dispholidus typus*) and Twig Snake (*Thelotornis capensis*) from Africa (Kuch & Mebs, 2002) and, also, Red-necked Keelback (*Rhabdophis subminiatus*) from Asia (Weinstein, 2017), about, have caused human fatalities.

Epidemiology

The most important challenge of health system authorities in many countries regarding snakebites issue is to collect accurate information because of lack of infrastructure and resources. In fact, accurate data about venomous snake distribution and snakebite epidemiology may help health authorities to better understand therapeutic requirements, and thus facilitate the management and prioritization of scarce health care resources for prevention and treatment of this health problem (Chippaux, 2005).Inspire of this, information on epidemiological incidence, mortality, and morbidity of snake bites remain, until now, not comprehensive, either within particular countries or globally (i.e. the high number of untreated and unreported cases world-wide).Consequently, epidemiological studies from most countries are often elaborated so as to estimate, as largely as possible, the amount of produced envenomings thanks to evident data from health institutions, consultation centers, along with surveys on families from randomized areas (Chippaux, 1998; Chippaux, 2011; Kasturiratne et *al.*, 2005; Kasturiratne et *al.*, 2008).

Some authors estimate that at least 421,000 envenomings and 20,000deaths occur worldwide from snakebite annually (Kasturiratne et *al.*, 2008). As these estimates may not be entirely correct because of undeclared snakebites and its potential caused deaths in many regions where patients resort to traditional methods rather than seeking help from health centers, these figures may be as high as 1,841,000 envenoming and 94,000deaths (Kasturiratne et *al.*, 2008). Moreover, on the basis of modeling estimations, the real total number of snakebites is pretended to be two to three times the number of evident envenomation. In fact, Kasturiratne et *al.*, (2008) suggested that approximately 1,200,000 to5,500,000 snakebites may occur globally, with higher exacerbations in the underdeveloped countries.

Though estimations could present an alternative to approach the real snakebites incidence, it may not be with a high accuracy because of the challenge faced by scientists in reaching information. Nevertheless, some studies had reported, in the base of estimations, that a clear majority of the burden of snake bites is located specifically in tropical countries of South and Southeast Asia, sub-Saharan Africa, and in Central and South America (Chippaux, 1998; Kasturiratne et *al.*, 2008). Giving these, collected data show lower snakebites incidence in sub-Saharan Africa which is a fact that may relates to the lack of incidence declaration and prospective studies within these regions (Kasturiratne et *al.*, 2008). However, venomous snakes and snakebites are numerically lowest in Europe and North America where populations are more urban, but highest in tropical countries (Kasturiratne et *al.*, 2008). Most of the population living in tropical regions and rural areas comes to overlap snake habitats, thus bringing humans and snakes into direct conflict. Consequently, this makes snakebite much more of an occupational disease primarily affecting underage agricultural workers who usually perform their activities without any kind of protection against it (Chippaux, 1992; Cruz et *al.*, 2009). In this context, and to prevent the possible substantial risks related to snake and human habitats overlaps, it is of high interest to represent a recall of the main venomous snakes classification, to upgrade knowledge about the distributional patterns, and further, to take into account the potential changes and expansions in the range of venomous snakes habitats distribution. This matter of fact has become more and more important, notably during the last decades, under the influence of climate change.

Main Venomous Snake in the World

Venomous snakes constitute, from a medical perspective, the most important venomous animal taxa considering their high potential mortality and morbidity rates in comparison with all other groups combined. There are approximately 2900 snake species worldwide (Uetz et *al.*, 2018) where roughly 375 are venomous. Even though a small number of snake's families represent potential dangerous harm to humans, the evidence of the caused hazards and their widespread extents all over the world justify the particular emphasis we put on their classification.

All the venomous snakes are parts of the *Colubroidea* superfamily. They are divided into 4 families listed below:

Colubridae (Colubrid Snakes)

Colubrid snakes (Colubridae) include over 1922 snake species (Uetz et *al.*, 2018) distributed all over the world except Antarctica and oceanic islands (Warrell, 1993; Asmundsson et *al.*, 2001; Vitt and Caldwell, 2013). The majority of colubrid snakes are most often considered as harmless to humans mainly because of the small size and the position of their fangs at the back of the mouth (Weinstein and Kardong, 1994). However, this general anatomic characteristic does not always outweigh their potential harm to human. In fact, some snakes of this family can produce toxic oral secretions that some authors argue it may constitute venom, a fact that is supported by apparent possible DNA coding for toxins (Fry et *al.*, 2006). Thus, some relatively dangerous species of this family are able to induce severe, sometimes lethal systemic envenoming usually associated with blood coagulation and a bleeding tendency. For instance, Boomslang (*Dispholidustypus*) and vine snakes (*Thelotorniscapensis*) are southern African arboreal snakes that have caused a number of fatalities associated with coagulopathy. As well, Dumeril's Diadem Snake (*Phalotrislemniscatus*), Montpellier Snake (*Malpolonmonspessulanus*) and Yamakagashi families (*Rhabdophissubminiatus* and *R.tigrinus*) represent other significant illustrative examples; even though they were thought to be harmless in the beginning, several severe, even fatal bites have proven their ability to cause major envenoming and coagulopathy (Minton, 1990).

Elapidae (Cobras, Mambas, Kraits, Coral Snakes)

Elapidae (360+ species) are mainly found in tropical and sub-tropical regions of The Americas, Africa, Asia, and Australia (Warrell, 1993; Enwere et *al.*, 2000). They are, without exception, venomous, having well-developed venom glands and paired anterior placed proteroglyphous fangs. Besides, the size of these extremely venomous snakes varies from few centimeters to more than 5 meters long. Thus, some pertinent examples of Elapidae are presented bellow to reconcile the variability of their characteristics. We mainly underline on the cobras (*Naja spp.*; Figure 3), the mambas (*Dendroaspis spp.*), the kraits (*Bungarus spp.*), and the coral snakes (*Micrurus spp.* and *Micruroides spp.*).

- **Cobras:** Represent one of the most major medical important snakes widely distributed. This group of elapid snakes induces mortality and morbidity in roughly thousands to tens of thousands of humans every year (Hung et *al.*, 2003). There are two kind of cobra envenomations (Watt et *al.*, 1988; Hung et *al.*, 2003): (*i*) mainly local envenomation with necrosis and mild to moderate neurotoxicity, and (*ii*) mainly neurotoxic envenomation without major local effects. The former group comprises several species in Asia and Africa characterized by spitting venom and inducing grave venom ophthalmia.
- **Mambas (Dendroaspis):** Is one of the most dangerous snakes in African mambas within which some species have a high lethality potential. Their venom produces complex neurotoxicity, leading to both muscle fasciculation and paralysis, but commonly few local effects (Tan et *al.*, 1991).
- **Krait Snakes (Bungarus):** Are known to cause lethal envenomation especially in Asia. Envenomation caused by this group is characterized by a painless bite and later development of progressive severe paralysis, habitually linked with abdominal pain and, at least for some species, myotoxicity as well (Faiz et *al.*, 2010).
- **Coral Snakes (Micrurus, etc.):** Comprise the most venomous species in the Americas, especially in South and Central America (Roodt et *al.*, 2013). Thus, they may induce grave paralysis and/

or myolys is with minimal local effects. However, the coral snakes cause sometimes fatal consequences, even though they are infrequent and represented by few species in their range (Walter et *al.*, 2010).

Atractaspididae (Burrowing Vipers)

Atractaspididae (mole vipers, burrowing vipers): is a family subdivided into many subfamilies and comprise over 50 species that have a distribution limited to Africa and the Middle East (Vitt and Caldwell, 2013). The most common characteristics over all taxa are their small size and often, the adoption of "venomous glands". They are also known for the variety of their fangs' feature.

Some species may be fangless while others, as well as the *Atractaspis aterrima*, have long fangs. Sarafatoxin is the main toxic toxins present in venomous Atractaspididae species (Underwood and Kochva, 1993). It is a toxin similar to human endothelins and can induces severe or lethal cardiac effects. However, local tissue injury and sometimes necrosis is a far more common consequence in the humans' envenomation (Warrell et *al.*, 1976). Thus, save for Atractaspis sp., which is the only taxon capable of fatal envenomation to human, this family's species cause rarely death within human being.

Viperidae (Vipers, Adders, Pitvipers, and Mocassins)

Viperidae (vipers, pit vipers) have a wide distribution over the Americas, Europe, Africa, and Asia (Malina et *al.*, 2008; Walter et *al.*, 2009; Alirol et *al.*, 2010). They are characterized by long hollow fangs in the front of their mouth, a triangular-shaped head and vertical pupils. There are four subfamilies with more than 340 species of Viperidae (Uetz et *al.*, 2018). Among these subfamilies Viperinae and Crotalinae comprise some of the most dangerous snakes to human beings.

Viperinae

This subfamily comprises mainly classic vipers found specifically throughout the "old world" and is linked to a large portion of the human snakebite envenomation. The most eminent species belong to the carpet vipers (genus *Echis*), Russell vipers (genus *Daboia*) and African adders (genus *Bitis*).

Crotalinae

"Pit vipers" is the common name of snakes ranged amongst this subfamily, a name which refers to heat-sensing glands (pits) located in either side of the head. Therefore, this characteristic provides them with rangefinder capabilities and rises up their hazard potential. Moreover, Crotalinae vipers are widely distributed in both, the "Old World" and the "New World" (Vitt and Caldwell, 2013) with a greater diversity in the Americas, as a result the related causality of envenomation is much important (Juckett & Hancox, 2002).

EFFECT OF CLIMATE CHANGE ON VENOMOUS SNAKES

Snakes responses to global warming include short-term effects on populations (e.g., changes in activities, behavior, habitat use and abundance), but also long-term effects (e.g., shifts in species distribution).

Activities and Behavior

As ectotherms, snakes depend directly on the external thermal conditions for their activities; they regulate their body temperature by taking advantage of the sun and warm surfaces in the environment for heat gain, and shade, retreats, water, and cool surfaces for heat loss (Bogert, 1949). Thus, this show how important is the natural environment equilibrium for snakes to keep their body temperature and vital functions constant. However, climate change is causing an average temperature increment which affects several aspects of organisms' biology, especially in ectotherms (Kearney et *al.*, 2009; Paaijmans et *al.*, 2013). Many species are showing alterations in phenology, distribution, morphology, and population dynamics in response to those changes (Clusella-Trullas et *al.*, 2009; Broennimann et *al.*, 2014; Walther et *al.*, 2002). Snakes should adapt to climate change by two natural compensatory ways: having enough time and dispersal, species may shift to more favorable thermal environments, or they may adjust to new environments by behavioral plasticity, physiological plasticity, or adaptation (Nori et *al.*, 2014; Yañez-Arenas et *al.*, 2015). Otherwise, a failure to adjust or adapt to any potential demographic culmination may lead to collapse and extinction (Yañez-Arenas et *al.*, 2015).

The evidence of global warming influence on snakes can be clearly seen in the rising air temperature in nearly all parts of the world (Rahmstorf et *al.*, 2007; Foufopoulos et *al.*, 2010). The warmer the global temperature, the more important will be the potentialities that lead to reactions chains of other environmental changes. Increasing the air temperature indeed, affects the oceans, weather patterns, snow and ice, and plants and animals (Stenseth et *al.*, 2002). Faced with these, snakes show variation in activity in a way that has frequently been linked to weather conditions such as temperature, rainfall, etc, most obviously in temperate zones where snake populations activity isoften controlled by ambient temperature (Lueth, 1941). Actually, snakes may shift between nocturnal and diurnalactivity depending on seasonal temperature regimes. In another hand, the activity of snake species inhabiting tropical zone are unlikely to be constrained by temperature (Shine & Madsen, 1996) instead of other factors such as relative humidity or moisture (Henderson & Hoevers, 1977; Daltry et *al.*, 1998), which may be more important in this respect.

In cold climate regions, some studies draw conclusions concerning the effect of a rising temperature on snake-population's dynamics (López-Alcaide et *al.*, 2011). The high temperature increases activity, survival and/or breeding success of some snakes' species. There may be several causes for this. To illustrate, when the temperature of a given year is higher, snakes have more available time for foraging and basking (Sinervo and Adolph, 1994), which might diminish their mortality rate for reasons: less torpor (Bennett, 1980), which diminishes predation risk (Goode and Duvall, 1989), and improved feeding ability (Greenwald, 1974), which increases immune capacity (French et *al.*, 2007) and the quantity of resources to survive the winter (Nayaet *al.*, 2008).In temperate regions, an increase in environmental temperature could also enlarge the snake's available time for feeding (Wang et *al.*, 2002), body growth (Lindell, 1997), breeding success (Chamaillé-Jammes et *al.*, 2006), and survival(Altwegg et *al.*, 2005), which could lead to increased population sizes.

Tropical snakes however, have been hypothesized to exhibit a polymodal activity pattern in response to wet-dry cycles (Brown and Shine, 2002). Studies on African tropical snakes have shown a strong relationship between rainfall and activity, with peaks occurring at the onset of the wet seasons that continue to increase throughout wet months (e.g., Akani et *al.*, 20013). This general pattern has been confirmed in both savannah species, such as *Naja nigricollis* (Luiselli, 2001) and *Crotaphopeltis hotamboeia* (Eniang et *al.*,2013), and mainly within forest-dwelling species such as *Najamelanoleuca* (Luiselli, 2001), *Bitis gabonica* and *Bitis nasicornis* (Luiselli, 2006; Akani et *al.*, 2013). Moreover, the seasonal incidences of humans getting bitten by snakes (a proxy of snake activity intensity) showed thatin agricultural landscapes snakebites occur more frequently in the wet season. This pattern observed in Bangladesh, Mali, Burkina Faso, Nigeria, Ivory Coast, Cameroon, Ghana and Benin (see Chippaux, 2006; Rahma et *al.*, 2010).

Weather conditions deeply influence the phenology of several snakes, and temperature and rainfall regimes influence snake activity patterns in tropical regions (e.g. in West Africa and southeastern Brazil; see Akani et *al.*, 2013). Also, in the reproduction season, snakes become particularly active (mating and find suitable sites for oviposition). In fact, these attributes (weather and activity) are functionally linked, which influences snake envenoming (Salomão et *al.*, 1995).

Distributional Patterns

As the climate-change was demonstrated to have an influence upon snakes' behavior, empirical data show more evident effects on distributional patterns shift. In fact, studies on behavioral adaptations are valuable contribution to describe the influence of global changes, but they do not capture the full complexity of the phenomenon as they are able neither to take into account the ecological factors nor to predict the potential risks of related snakebites. Yet, increasing the knowledge of suitable snakes' ranges of distribution and taking advantage of it is crucial in treating snakebite envenoming (Yañez-Arenas et *al.*, 2016).

As mentioned, the average temperature increment and the weather change in respect with species tendency to specific climate are key factors in defining distributional patterns. Accordingly, two scenarios in distributional changes can be graded: expanded distributional potential and decreased distributional potential. Though the patterns seem to be different, each of the two situations could increase the probabilities of human confrontation to snake species of medical importance, increasing thus the incidence and the risk of snakebites. As a result, many studies on the new mapping of venomous snakes' species distribution have developed models of maps predictions in terms of snakebite risk and incidence rather than species richness (Yañez-Arenas et al., 2016; Needleman et al., 2018).

Several studies have forecasted venomous snakes' distribution and snakebite risks *in silico* with good performance. They have been elaborated by the use of tools such as geographic information system (GIS), ecological niche models (ENM), species distributional models (SDMs) and others (Yañez-Arenas et *al.*, 2016; Needleman et *al.*, 2018; El Hidan et *al.*, 2018). These tools are programmed to finding significant links and establishing correlations between current occurrences of venomous snake species and specific environmental variables. Consequently, those models can be projected into hypothetic climate conditions to generate hypothesis about species responses in different climatic conditions. Thus, the developed models are useful to provide information for both, assessing appropriate climate spaces for current conditions, and predicting future potential distributional ranges of snakes. So, it is possible to evaluate the potentialities of global climatic changes in producing changes in species' suitable climate areas, which in turn could be a valuable asset when facing the problem. Furthermore, the models are developed to address epidemiological relative issues. As a matter of fact, localization of regions with high vulnerability is mainly based on the collection of evident incidence data so that the performance of the models in forecasting the possible scenarios could assimilate closely the reality.

An overview over these models applications, via assets of empirical evidences and predictive studies, show that the geographic distributions of many species have been demonstrated to begun to shift, a pattern that have the tendency to continue (e.g. Parmesan & Yohe, 2003). Thus, in terms of snakebites risks, ENM techniques have been applied across the America to predict current and future potential displacement of suitable areas, to produce snakebite risk maps and to evaluate the human rural population at risk. For instance, Nori et al. (2014) anticipated a moderate north to south displacement of *Bothrops alternatus, B. ammodytoides, B. diporus, Crotalus durissus terrificus* and *Micrurus pyrrhocryptus*, which are five species with the most harmful effect on the Argentina populations. In a study that models the current and the future snakebite risk expansion across North and Latin America with larger selection of taxa (90 taxa) within the most medically important snakes, the current projections showed that the snakebite risk area covered almost29,708,000 km^2 with ~ 1,353,337,000 rural populations exposed. The future projection of the same study shows that these numbers trend to develop so that the total risk area and the rural population at risk will spread out in the year 2050 to reach ~31,685,000 and ~1,365,397,000, respectively. Furthermore, the expansion of snakebite risk is forecasted to expand toward the north in Canada and toward the south in Argentina and Chile, with increases of the risk ranged 0.83– 30.85% in area and 2.74–18.38% for rural population exposed. Otherwise, empirical studies in Europe show a northward shift of species in Britain (Hickling et al. 2006; Le Galliard et al., 2012) and Spain (Moreno-Rueda et al. 2012) putting the light on the influence of the global warming on species distribution. In Africa however, unless some preliminary investigations based on a small selection of taxa, few studies have been developed to model the snakebite risk, a fact which could be related to lack of empirical data and resources. For instance, Molesworth et *al.*, (2003) have done a preliminary investigation through GIS approach to risk mapping of one viper specie incidences, *Echis ocellatus*, in northern Ghana and Nigeria. Even though it was highly interesting in identifying the association between environmental variables and marked seasonal and geographical variation in snakebite incidences, deep further studies are required to enable larger and more reliable forecast of potential geographical shifts and snakebite risk across Africa.

CONCLUSION

This review has presented evidence suggesting a relationship between the climate changes and the increase of the incidences caused by snakebites. The global warming is revealed to be the principal cause that set off snakes behavioral and habitat alterations, and thus the overlap between venomous snakes and human being habitats. On behalf of the given evidences, efforts must be done worldwide to face the situation and to reduce its extent.

REFERENCES

Akani, G. C., Ebere, N., Franco, D., Eniang, E. A., Petrozzi, F., Politano, E., & Luiselli, L. (2013). Correlation between annual activity patterns of venomous snakes and rural people in the Niger Delta, southern Nigeria. *The Journal of Venomous Animals and Toxins Including Tropical Diseases, 19*(1), 2. doi:10.1186/1678-9199-19-2 PMID:23849681

Alirol, E., Sharma, S. K., & Bawasar, H. S. (2010). Snake bite in South Asia: A review. *PLoS Neglected Tropical Diseases, 4*(1), e603. doi:10.1371/journal.pntd.0000603 PMID:20126271

Altwegg, R., Dummermuth, S., Anholt, B. R., & Flatt, T. (2005). Winter weather affects asp viper Viperaaspis population dynamics through susceptible juveniles. *Oikos*, *110*(1), 55–66. doi:10.1111/j.0030-1299.2001.13723.x

Auerbach, P. S. (2005). The Venomous Reptiles of the Western Hemisphere. *Wilderness & Environmental Medicine*, *16*(3), e13. doi:10.1580/1080-6032(2005)16[e13a:BR]2.0.CO;2

Bennett, A. F. (1980). The metabolic foundations of vertebrate behavior. *Bioscience*, *30*(7), 452–456. doi:10.2307/1307946

Bogert, C. M. (1949). Thermoregulation in reptiles, a factor in evolution. *Evolution; International Journal of Organic Evolution*, *3*(3), 195–211. doi:10.1111/j.1558-5646.1949.tb00021.x PMID:18138377

Bons, J., & Geniez, P. (1996). *Amphibiens et reptiles du Maroc (Sahara occidental compris). Atlas biogéographique*. AsociaciónHerpetologica Espanola.

Broennimann, O., Ursenbacher, S., Meyer, A., Golay, P., Monney, J. C., Schmocker, H., ... Dubey, S. (2014). Influence of climate on the presence of colour polymorphism in two montane reptile species. *Biology Letters*, *10*(11), 20140638. doi:10.1098/rsbl.2014.0638 PMID:25392313

Brown, G. P., & Shine, R. (2002). Influence of weather conditions on activity of tropical snakes. *Austral Ecology*, *27*(6), 596–605. doi:10.1046/j.1442-9993.2002.01218.x

Chamaille-Jammes, S., Massot, M., Aragon, P., & Clobert, J. (2006). Global warming and positive fitness response in mountain populations of common lizards Lacerta vivipara. *Global Change Biology*, *12*(2), 392–402. doi:10.1111/j.1365-2486.2005.01088.x

Chippaux, J. P. (1992). Les morsures de serpents en Afrique intertropicale. *Cahiers d'études et de recherches francophones/Santé, 2*(4), 221-234.

Chippaux, J. P. (1998). Snake-bites: Appraisal of the global situation. *Bulletin of the World Health Organization*, *76*(5), 515. PMID:9868843

Chippaux, J. P. (2006). *Les serpents d'Afriqueoccidentaleetcentrale*. IRD éditions.

Chippaux, J. P. (2008). Estimating the global burden of snakebite can help to improve management. *PLoS Medicine*, *5*(11), e221. doi:10.1371/journal.pmed.0050221 PMID:18986211

Chippaux, J. P. (2011). Estimate of the burden of snakebites in sub-Saharan Africa: A meta-analytic approach. *Toxicon*, *57*(4), 586–599. doi:10.1016/j.toxicon.2010.12.022 PMID:21223975

Clusella-Trullas, S., Wyk, J. H., & Spotila, J. R. (2009). Thermal benefits of melanism in cordylid lizards: A theoretical and field test. *Ecology*, *90*(8), 2297–2312. doi:10.1890/08-1502.1 PMID:19739391

Cruz, L. S., Vargas, R., & Lopes, A. A. (2009). Snakebite envenomation and death in the developing world. *Ethnicity & Disease*, *19*(1), 42. PMID:19484874

Daltry, J. C., Ross, T., Thorpe, R. S., & Wüster, W. (1998). Evidence that humidity influences snake activity patterns: A field study of the Malayan pit viper Calloselasmarhodostoma. *Ecography*, *21*(1), 25–34. doi:10.1111/j.1600-0587.1998.tb00391.x

El Hidan, M. A., Touloun, O., Bouazza, A., Laaradia, M. A., & Boumezzough, A. (2018). Androctonus genus species in arid regions: Ecological niche models, geographical distributions, and envenomation risk. *Veterinary World*, *11*(3), 286–292. doi:10.14202/vetworld.2018.286-292 PMID:29657418

Eniang, E. A., Akani, G. C., Rugiero, L., Vignoli, L., & Luiselli, L. (2013). Ecological data of Nigerian Crotaphopeltishotamboeia (Colubridae) populations. *The Herpetological Journal*, *23*(1), 5–9.

Faiz, M. A., Ghose, A., Ahsan, M. F., Rahman, M. R., Amin, M. R., Hassan, M. M. U., ... Theakston, R. D. G. (2010). The greater black krait (Bungarusniger), a newly recognized cause of neuro-myotoxic snake bite envenoming in Bangladesh. *Brain*, *133*(11), 3181–3193. doi:10.1093/brain/awq265 PMID:20855420

Fenker, J., Tedeschi, L. G., Pyron, R. A., & Nogueira, C. D. C. (2014). Phylogenetic diversity, habitat loss and conservation in South American pitvipers (Crotalinae: Bothrops and Bothrocophias). *Diversity & Distributions*, *20*(10), 1108–1119. doi:10.1111/ddi.12217

Foufopoulos, J., Kilpatrick, A. M., & Ives, A. R. (2010). Climate change and elevated extinction rates of reptiles from Mediterranean islands. *American Naturalist*, *177*(1), 119–129. doi:10.1086/657624 PMID:21091198

French, S. S., Johnston, G. I. H., & Moore, M. C. (2007). Immune activity suppresses reproduction in food-limited female tree lizards Urosaurusornatus. *Functional Ecology*, *21*(6), 1115–1122. doi:10.1111/j.1365-2435.2007.01311.x

Fry, B. G., Vidal, N., Norman, J. A., Vonk, F. J., Scheib, H., Ramjan, S. F., ... Kochva, E. (2006). Early evolution of the venom system in lizards and snakes. *Nature*, *439*(7076), 584–588. doi:10.1038/nature04328 PMID:16292255

Goode, M. J., & Duvall, D. (1989). Body temperature and defensive behaviour of free-ranging prairie rattlesnakes, Crotalusviridisviridis. *Animal Behaviour*, *38*(2), 360–362. doi:10.1016/S0003-3472(89)80100-9

Greenwald, O. E. (1974). Thermal dependence of striking and prey capture by gopher snakes. *Copeia*, *1974*(1), 141–148. doi:10.2307/1443016

Gutiérrez, J. M. (2012). Snakebite envenoming: a public health perspective. In *Public health-methodology, environmental and systems issues*. InTech.

Gutiérrez, J. M., Theakston, R. D. G., & Warrell, D. A. (2006). Confronting the neglected problem of snake bite envenoming: The need for a global partnership. *PLoS Medicine*, *3*(6), e150. doi:10.1371/journal.pmed.0030150 PMID:16729843

Hansson, E., Sasa, M., Mattisson, K., Robles, A., & Gutiérrez, J. M. (2013). Using geographical information systems to identify populations in need of improved accessibility to antivenom treatment for snakebite envenoming in Costa Rica. *PLoS Neglected Tropical Diseases*, *7*(1), e2009. doi:10.1371/journal.pntd.0002009 PMID:23383352

Henderson, R. W., & Hoevers, L. G. (1977). The seasonal incidence of snakes at a locality in northern Belize. *Copeia*, *1977*(2), 349–355. doi:10.2307/1443914

Hickling, R., Roy, D. B., Hill, J. K., Fox, R., & Thomas, C. D. (2006). The distributions of a wide range of taxonomic groups are expanding polewards. *Global Change Biology*, *12*(3), 450–455. doi:10.1111/j.1365-2486.2006.01116.x

Hung, D. Z., Liau, M. Y., & Lin-Shiau, S. Y. (2003). The clinical significance of venom detection in patients of cobra snakebite. *Toxicon*, *41*(4), 409–415. doi:10.1016/S0041-0101(02)00336-7 PMID:12657310

Juckett, G., & Hancox, J. G. (2002). Venomous snakebites in the United States: Management review and update. *American Family Physician*, *65*(7), 1367–1378. PMID:11996419

Kasturiratne, A., Pathmeswaran, A., Fonseka, M. M. D., Lalloo, D. G., Brooker, S., & De Silva, H. J. (2005). Estimates of disease burden due to land-snake bite in Sri Lankan hospitals. *The Southeast Asian Journal of Tropical Medicine and Public Health*, *36*(3), 733. PMID:16124448

Kasturiratne, A., Wickremasinghe, A. R., de Silva, N., Gunawardena, N. K., Pathmeswaran, A., Premaratna, R., ... de Silva, H. J. (2008). The global burden of snakebite: A literature analysis and modelling based on regional estimates of envenoming and deaths. *PLoS Medicine*, *5*(11), e218. doi:10.1371/journal.pmed.0050218 PMID:18986210

Kearney, M., Shine, R., & Porter, W. P. (2009). The potential for behavioral thermoregulation to buffer "cold-blooded" animals against climate warming. *Proceedings of the National Academy of Sciences of the United States of America*, *106*(10), 3835–3840. doi:10.1073/pnas.0808913106 PMID:19234117

Kuch, U., & Mebs, D. (2002). Envenomations by colubrid snakes in Africa, Europe, and the Middle East. *Journal of Toxicology. Toxin Reviews*, *21*(1-2), 159–179. doi:10.1081/TXR-120004745

Le Galliard, J. F., Massot, M., Baron, J. P., &Clobert, J. (2012). Ecological effects of climate change on European reptiles. *Wildlife conservation in a changing climate*, 179-203.

Lindell, L. E. (1997). Annual variation in growth rate and body condition of adders, Viperaberus: Effects of food availability and weather. *Canadian Journal of Zoology*, *75*(2), 261–270. doi:10.1139/z97-033

López-Alcaide, S., & Macip-Ríos, R. (2011). Effects of climate change in amphibians and reptiles. In *Biodiversity loss in a changing planet*. InTech. doi:10.5772/24663

Lueth, F. X. (1941). Effects of temperature on snakes. *Copeia*, *1941*(3), 125–132. doi:10.2307/1437733

Luiselli, L. (2001). The ghost of a recent invasion in the reduced feeding rates of spitting cobras during the dry season in a rainforest region of tropical Africa? *Acta Oecologica*, *22*(5-6), 311–314. doi:10.1016/S1146-609X(01)01113-4

Luiselli, L. (2006). Site occupancy and density of sympatric Gaboon viper (Bitisgabonica) and nose-horned viper (Bitisnasicornis). *Journal of Tropical Ecology*, *22*(5), 555–564. doi:10.1017/S0266467406003397

Malina, T., Krecsak, L., & Warrell, D. A. (2008). Neurotoxicity and hypertension following European adder (*Viperaberusberus*) bites in Hungary: Case report and review. *QJM*, *101*(10), 801–806. doi:10.1093/qjmed/hcn079 PMID:18647797

Minton, S. A. (1990). Venomous bites by nonvenomous snakes: An annotated bibliography of colubrid envenomation. *Wilderness & Environmental Medicine*, *1*(2), 119–127. doi:10.1580/0953-9859-1.2.119

Molesworth, A. M., Harrison, R., David, R., Theakston, G., & Lalloo, D. G. (2003). Geographic information system mapping of snakebite incidence in northern Ghana and Nigeria using environmental indicators: A preliminary study. *Transactions of the Royal Society of Tropical Medicine and Hygiene*, *97*(2), 188–192. doi:10.1016/S0035-9203(03)90115-5 PMID:14584375

Moreno-Rueda, G., Pleguezuelos, J. M., Pizarro, M., & Montori, A. (2012). Northward shifts of the distributions of Spanish reptiles in association with climate change. *Conservation Biology*, *26*(2), 278–283. doi:10.1111/j.1523-1739.2011.01793.x PMID:22098311

Naya, D. E., Veloso, C., & Bozinovic, F. (2008). Physiological flexibility in the Andean lizard Liolaemusbellii: Seasonal changes in energy acquisition, storage and expenditure. *Journal of Comparative Physiology. B, Biochemical, Systemic, and Environmental Physiology*, *178*(8), 1007–1015. doi:10.100700360-008-0292-6 PMID:18626649

Needleman, R. K., Neylan, I. P., & Erickson, T. (2018). Potential Environmental and Ecological Effects of Global Climate Change on Venomous Terrestrial Species in the Wilderness. *Wilderness & Environmental Medicine*, *29*(2), 226–238. doi:10.1016/j.wem.2017.11.004 PMID:29395962

Nori, J., Carrasco, P. A., & Leynaud, G. C. (2014). Venomous snakes and climate change: Ophidism as a dynamic problem. *Climatic Change*, *122*(1-2), 67–80. doi:10.100710584-013-1019-6

Paaijmans, K. P., Heinig, R. L., Seliga, R. A., Blanford, J. I., Blanford, S., Murdock, C. C., & Thomas, M. B. (2013). Temperature variation makes ectotherms more sensitive to climate change. *Global Change Biology*, *19*(8), 2373–2380. doi:10.1111/gcb.12240 PMID:23630036

Parmesan, C., & Yohe, G. (2003). A globally coherent fingerprint of climate change impacts across natural systems. *Nature*, *421*(6918), 37–42. doi:10.1038/nature01286 PMID:12511946

Phelps, T. (2010). *Old World Vipers, a natural history of the azemiopinae and viperinae*. Edition Chimaira.

Pough, F. H. (1980). The advantages of ectothermy for tetrapods. *American Naturalist*, *115*(1), 92–112. doi:10.1086/283547

Rahman, R., Faiz, M. A., Selim, S., Rahman, B., Basher, A., Jones, A., ... Milton, A. H. (2010). Annual incidence of snake bite in rural Bangladesh. *PLoS Neglected Tropical Diseases*, *4*(10), e860. doi:10.1371/journal.pntd.0000860 PMID:21049056

Rahmstorf, S., Cazenave, A., Church, J. A., Hansen, J. E., Keeling, R. F., Parker, D. E., & Somerville, R. C. (2007). Recent climate observations compared to projections. *Science*, *316*(5825), 709–709. doi:10.1126cience.1136843 PMID:17272686

Roodt, A. R. D., De Titto, E., Dolab, J. A., & Chippaux, J. P. (2013). Envenoming by coral snakes (Micrurus) in Argentina, during the period between 1979-2003. *Revista do Instituto de Medicina Tropical de São Paulo*, *55*(1), 13–18. doi:10.1590/S0036-46652013000100003 PMID:23328720

Salomão, M. D., Santos, S. M. A., & Puorto, G. (1995). Activity pattern of Crotalusdurissus (Viperidae, Crotalinae): Feeding, reproduction and snakebite. *Studies on Neotropical Fauna and Environment*, *30*(2), 101–106. doi:10.1080/01650529509360946

Shi, J., Wang, G., Fang, Y., Ding, L., Huang, S., Hou, M., ... Li, P. (2017). A new moth-preying alpine pit viper species from Qinghai-Tibetan Plateau (Viperidae, Crotalinae). *Amphibia-Reptilia, 38*(4), 517–532. doi:10.1163/15685381-00003134

Shine, R., & Madsen, T. (1996). Is thermoregulation unimportant for most reptiles? An example using water pythons (Liasisfuscus) in tropical Australia. *Physiological Zoology, 69*(2), 252–269. doi:10.1086/physzool.69.2.30164182

Shine, R., & Shetty, S. (2001). Moving in two worlds: Aquatic and terrestrial locomotion in sea snakes (Laticaudacolubrina, Laticaudidae). *Journal of Evolutionary Biology, 14*(2), 338–346. doi:10.1046/j.1420-9101.2001.00265.x

Sinervo, B., & Adolph, S. C. (1994). Growth plasticity and thermal opportunity in Sceloporus lizards. *Ecology, 75*(3), 776–790. doi:10.2307/1941734

Spalding, M., Spalding, M. D., Ravilious, C., & Green, E. P. (2001). *World atlas of coral reefs.* Univ of California Press.

Stenseth, N. C., Mysterud, A., Ottersen, G., Hurrell, J. W., Chan, K. S., & Lima, M. (2002). Ecological effects of climate fluctuations. *Science, 297*(5585), 1292–1296. doi:10.1126cience.1071281 PMID:12193777

Tan, N. H., Arunmozhiarasi, A., & Ponnudurai, G. (1991). A comparative study of the biological properties of Dendroaspis (mamba) snake venoms. *Comparative Biochemistry and Physiology. C. Comparative Pharmacology and Toxicology, 99*(3), 463–466. doi:10.1016/0742-8413(91)90272-U PMID:1685421

Uetz, P., Freed, P., & Hošek, J. (2018). *The Reptile Database.* Retrieved from http://www.reptile-database.org

Underwood, G. (1979). Classification and distribution of venomous snakes in the world. In *Snake venoms* (pp. 15–40). Berlin: Springer. doi:10.1007/978-3-642-66913-2_2

Underwood, G., & Kochva, E. (1993). On the affinities of the burrowing asps Atractaspis (Serpentes: Atractaspididae). *Zoological Journal of the Linnean Society, 107*(1), 3–64. doi:10.1111/j.1096-3642.1993.tb01252.x

Valenta, J., Stach, Z., & Svítek, M. (2010). Acute pancreatitis after viperid snake cerastescerastes envenoming: A case report. *Prague Medical Report, 111*(1), 69–75. PMID:20359440

Vitt, L. J., & Caldwell, J. P. (2013). *Herpetology: an introductory biology of amphibians and reptiles.* Academic Press.

Walter, F. G., Stolz, U., Shirazi, F., & McNally, J. (2009). Epidemiology of severe and fatal rattlesnake bites published in the American Association of Poison Control Center's Annual Reports. *Clinical Toxicology (Philadelphia, PA), 47*(7), 663–669. doi:10.1080/15563650903113701 PMID:19640239

Walter, F. G., Stolz, U., Shirazi, F., & Mcnally, J. (2010). Temporal analyses of coral snakebite severity published in the American Association of Poison Control Centers' Annual Reports from 1983 through 2007. *Clinical Toxicology, 48*(1), 72–78. doi:10.3109/15563650903430944 PMID:20070180

Walther, G. R., Post, E., Convey, P., Menzel, A., Parmesan, C., Beebee, T. J., ... Bairlein, F. (2002). Ecological responses to recent climate change. *Nature*, *416*(6879), 389–395. doi:10.1038/416389a PMID:11919621

Wang, T., Zaar, M., Arvedsen, S., Vedel-Smith, C., & Overgaard, J. (2002). Effects of temperature on the metabolic response to feeding in Python molurus. *Comparative Biochemistry and Physiology. Part A, Molecular & Integrative Physiology*, *133*(3), 519–527. doi:10.1016/S1095-6433(02)00250-7 PMID:12443910

Warrell, D. A. (1993). Venomous bites and stings in the tropical world. *The Medical Journal of Australia*, *159*, 773–779. PMID:8264466

Warrell, D. A. (2010). Snake bite. *Lancet*, *375*(9708), 77–88. doi:10.1016/S0140-6736(09)61754-2 PMID:20109866

Warrell, D. A., Ormerod, L. D., & Davidson, N. M. (1976). Bites by the night adder (*Causus maculatus*) and burrowing vipers (genus Atractaspis) in Nigeria. *The American Journal of Tropical Medicine and Hygiene*, *25*(3), 517–524. doi:10.4269/ajtmh.1976.25.517 PMID:945703

Watt, G., Padre, L., Tuazon, M. L., Theakston, R. D. G., & Laughlin, L. (1988). Bites by the Philippine cobra (Najanajaphilippinensis): Prominent neurotoxicity with minimal local signs. *The American Journal of Tropical Medicine and Hygiene*, *39*(3), 306–311. doi:10.4269/ajtmh.1988.39.306 PMID:3177741

Weinstein, S. A. (2017). Non-Front-Fanged Colubroid Snakes. In *Critical Care Toxicology* (pp. 1–41). Cham: Springer.

Weinstein, S. A., & Kardong, K. V. (1994). Properties of Duvernoy's secretions from opisthoglyphous and aglyphous colubrid snakes. *Toxicon*, *32*(10), 1161–1185. doi:10.1016/0041-0101(94)90347-6 PMID:7846688

White, J. (2018). Snakebite as a neglected disease. *ToxicologieAnalytiqueet Clinique*, *30*(3), 161.

WHO. (2018). *Neglected tropical diseases: snakebite.* Retrieved from http://www.who.int/neglected_diseases/diseases/snakebites/en/index.html

Williams, D., Gutiérrez, J. M., Harrison, R., Warrell, D. A., White, J., Winkel, K. D., & Gopalakrishnakone, P. (2010). The Global Snake Bite Initiative: An antidote for snake bite. *Lancet*, *375*(9708), 89–91. doi:10.1016/S0140-6736(09)61159-4 PMID:20109867

Yañez-Arenas, C., Peterson, A. T., Rodríguez-Medina, K., & Barve, N. (2016). Mapping current and future potential snakebite risk in the new world. *Climatic Change*, *134*(4), 697–711. doi:10.100710584-015-1544-6

Compilation of References

Abahussain, A. A., Abdu, A. S., Al-Zubari, W. K., El-Deen, N. A., & Abdul-Raheem, M. (2002). Desertification in the Arab Region: Analysis of current status and trends. *Journal of Arid Environments*, *51*(4), 521–545. doi:10.1016/S0140-1963(02)90975-4

Abaidoo, R. C., Keraita, B., Drechsel, P., & Dissanayake, P. (2014). Soil and Crop Contamination Through Wastewater Irrigation and Options for Risk Reduction in Developing Countries. Academic Press. doi:10.1007/978-3-642-05076-3

Abaidoo, R. C., Keraita, B., Drechsel, P., Dissanayake, P., & Maxwell, A. S. (2010). Soil and crop contamination through wastewater irrigation and options for risk reduction in developing countries. In P. Dion (Ed.), *Soil biology and agriculture in the tropics* (pp. 275–297). Berlin: Springer-Verlag. doi:10.1007/978-3-642-05076-3_13

Abarca, M. L., Accensi, F., Bragulat, M. R., Castella, G., & Cabañes, F. J. (2003). Aspergillus carbonarius as the main source of ochratoxin a contamination in dried vine fruits from the Spanish market. *Journal of Food Protection*, *66*(3), 504–506. doi:10.4315/0362-028X-66.3.504

Abbott, R. D., Ross, G. W., Tanner, C. M., Andersen, J. K., Masaki, K. H., Rodriguez, B. L., ... Petrovitch, H. (2012). Late-life hemoglobin and the incidence of Parkinson's disease. *Neurobiology of Aging*, *33*(5), 914–920. doi:10.1016/j.neurobiolaging.2010.06.023 PMID:20709430

Abdel-Shafy, H. I., & Mansour, M. S. M. (2016). A review on polycyclic aromatic hydrocarbons: Source, environmental impact, effect on human health and remediation. *Egyptian Journal of Petroleum*, *25*(1), 107–123. doi:10.1016/j.ejpe.2015.03.011

Abdulrazzaq, Y. M., Osman, N., & Ibrahim, A. (2002). Fetal exposure to aflatoxins in the United Arab Emirates. *Annals of Tropical Paediatrics*, *22*(1), 3–9. doi:10.1179/027249302125000094 PMID:11926047

Abe. (2008). Neuroprotective therapy for ischemic stroke with free radical scavenger and gene-stem cell therapy. *Clin Neurol*, 896–898.

Abedi-Astaneh, F., Akhavan, A. A., Shirzadi, M. R., Rassi, Y., Yaghoobi-Ershadi, M. R., Hanafi-Bojd, A. A., ... Raufi, H. (2015). Species diversity of sand flies and ecological niche model of Phlebotomus papatasi in central Iran. *Acta Tropica*, *149*, 246–253. doi:10.1016/j.actatropica.2015.05.030 PMID:26071647

Abel, O., Powell, J. F., Andersen, P. M., & Al-Chalabi, A. (2012). ALSoD: A user-friendly online bioinformatics tool for amyotrophic lateral sclerosis genetics. *Human Mutation*, *33*(9), 1345–1351. doi:10.1002/humu.22157 PMID:22753137

Abougrain, A. K., Nahaisi, M. H., Madi, N. S., Saied, M. M., & Ghenghesh, K. S. (2010). Parasitological contamination in salad vegetables in Tripoli-Libya. *Food Control*, *21*(5), 760–762. doi:10.1016/j.foodcont.2009.11.005

Abramson, L., Keshavan, M. S., & Sitaram, N. (1989). Elevated sleep and waking heart rates in anxious depressions. *Biological Psychiatry*, *26*(5), 496–499. doi:10.1016/0006-3223(89)90070-X PMID:2790092

Abroug, F., Nouira, S., El Atrous, S., Besbes, L., Boukef, R., Boussarsar, M., ... Dellagi, K. (2003). A canine study of immunotherapy in scorpion envenomation. *Intensive Care Medicine*, *29*(12), 2266–2276. doi:10.100700134-003-1947-0 PMID:12955186

Abshire, M. K., Devor, D. E., Diwan, B. A., Shaughnessy, J. D. Jr, & Waalkes, M. P. (1996). In vitro exposure to cadmium in rat L6 myoblasts can result in both enhancement and suppression of malignant progression in vivo. *Carcinogenesis*, *17*(6), 1349–1356. doi:10.1093/carcin/17.6.1349 PMID:8681454

Achanzar, W. E., Diwan, B. A., Liu, J., Quader, S. T., Webber, M. M., & Waalkes, M. P. (2001). Cadmium-induced malignant transformation of human prostate epithelial cells. *Cancer Research*, *61*(2), 455–458. PMID:11212230

Ackah, M., Anim, A. K., Gyamfi, E. T., Zakaria, N., Hanson, J., Tulasi, D., ... Osei, J. (2014). Uptake of heavy metals by some edible vegetables irrigated using wastewater: A preliminary study in Accra, Ghana. *Environmental Monitoring and Assessment*, *186*(1), 621–634. doi:10.100710661-013-3403-0 PMID:24114276

Ackerman, D., & Weisberg, S. B. (2003). Relationship between rainfall and beach bacterial concentrations on Santa Monica bay beaches. *Journal of Water and Health*, *1*(2), 85–89. doi:10.2166/wh.2003.0010 PMID:15382737

Acosta, J. A., Cano, A. F., Arocena, J. M., Debela, F., & Martínez-martínez, S. (2009). Distribution of metals in soil particle size fractions and its implication to risk assessment of playgrounds in Murcia City (Spain). *Geoderma*, *149*(1–2), 101–109. doi:10.1016/j.geoderma.2008.11.034

Acosta, J. A., Jansen, B., Kalbitz, K., Faz, A., & Martínez-Martínez, S. (2011). Salinity increases mobility of heavy metals in soils. *Chemosphere*, *85*(8), 1318–1324. doi:10.1016/j.chemosphere.2011.07.046 PMID:21862104

Adam, E., Yoder, J., Gould, L.H., Hlavsa, M.C., & Gargano, J.W. (2016). Giardiasis outbreaks in the United States, 1971–2011. *Epidemiol. Infect.*, 1-12.

Adams, H. P., Zoppo, G., Alberts, M. J., Bhatt, D. L., Brass, L., Furlan, A., ... Wijdicks, E. F. M. (2007). AHA / ASA Guideline Guidelines for the Early Management of Adults A Guideline From the American Heart Association. *AME*, *115*, e478–e534. doi:10.1161/CIRCULATIONAHA.107.181486

Adams, S. V., Quraishi, S. M., Shafer, M. M., Passarelli, M. N., Freney, E. P., Chlebowski, R. T., ... Newcomb, P. A. (2014). Dietary cadmium exposure and risk of breast, endometrial, and ovarian cancer in the Women's Health Initiative. *Environmental Health Perspectives*, *122*(6), 594–600. doi:10.1289/ehp.1307054

Adekoya, J. A., Williams, A. B., & Ayejuyo, O. O. (2006). Distribution of heavy metals in sediments of Igbede, Ojo and Ojora rivers of Lagos, Nigeria. *The Environmentalist*, *26*(4), 277–280. doi:10.100710669-006-9077-1

Adeoti, M. L., Oguntola, A. S., Akanni, E. O., Agodirin, O. S., & Oyeyemi, G. M. (2015). Trace elements; copper, zinc and selenium, in breast cancer afflicted female patients in LAUTECH Osogbo, Nigeria. *Indian Journal of Cancer*, *52*(1), 106–109. doi:10.4103/0019-509X.175573

Adidi, A. (2009). *La question de l'intégration sociale dans les programmes de lutte contre l'habitat insalubre au Maroc.* Retrieved from Ministère de la Réforme de l'Administration et de la Fonction Publique: http://www.mmsp.gov.ma

AFSSA. (2009). Agence Française de Sécurité Sanitaire des Aliments. Évaluation des risques liés à la présence de mycotoxins dans les chaînes alimentaires humaine et animale. Rapport final.

AFSSA. (2009). *Évaluation des risques liés à la présence de mycotoxines dans les chaînes alimentaires humaine et animale.* Rapport final.

Agbaire, P. O., & Obi, C. E. (2009). Seasonal Variations of Some Physico-Chemical Properties Of River Ethiope Water in Abraka, Nigeria. *Journal of Applied Science & Environmental Management*, *13*(1), 55.

Agency for Toxic Substances and Disease Registry. (2011). *Toxicological profile for aldrin/dieldrin*. Retrieved from http://www.atsdr.cdc.gov/toxprofiles/tp.asp?id=317&tid=56

Aghamohammadi, N., & Isahak, M. (2018). Climate Change and Air Pollution in Malaysia. In *Climate Change and Air Pollution* (pp. 241–254). Cham: Springer. doi:10.1007/978-3-319-61346-8_15

Ahamed, T. N., Rao, K. G., & Murthy, J. S. R. (2000). GIS-based fuzzy membership model for crop-land suitability analysis. *Agricultural Systems*, *63*(2), 75–95. doi:10.1016/S0308-521X(99)00036-0

Ahmadi, M., Records, R., & Arabi, M. (2014). Impact of climate change on diffuse pollutant fluxes at the watershed scale. *Hydrological Processes*, *28*(4), 1962–1972. doi:10.1002/hyp.9723

Ahmed, A., & Wicklund, M. P. (2011). Amyotrophic lateral sclerosis: What role does environment play? *Neurologic Clinics*, *29*(3), 689–711. doi:10.1016/j.ncl.2011.06.001 PMID:21803219

Ahmed, S., Diffenbaugh, N., & Hertel, T. (2009). Climate volatility deepens poverty vulnerability in developing countries. *Environmental Research Letters*, *4*(3), 034004. doi:10.1088/1748-9326/4/3/034004

Ahsan, M. N., & Warner, J. (2014). The socioeconomic vulnerability index: Apragmatic approach for assessing climate change led risks –A case study in thesouth-western coastal Bangladesh. *International Journal of Disaster Risk Reduction*, *8*, 32–49. doi:10.1016/j.ijdrr.2013.12.009

Aislabie, J., & Foght, J. (2008). Hydrocarbon-degrading bacteria in contaminated soils. In D. F. Filler, I. Snape, & D. L. Barns (Eds.), *Bioremediation of petroleum hydrocarbons in cold regions Cambridge University Press* (pp. 69–83). Cambridge. doi:10.1017/CBO9780511535956.006

Akani, G. C., Ebere, N., Franco, D., Eniang, E. A., Petrozzi, F., Politano, E., & Luiselli, L. (2013). Correlation between annual activity patterns of venomous snakes and rural people in the Niger Delta, southern Nigeria. *The Journal of Venomous Animals and Toxins Including Tropical Diseases*, *19*(1), 2. doi:10.1186/1678-9199-19-2 PMID:23849681

Akhavan, A. A., Yaghoobi-Ershadi, M. R., Mehdipour, D., Abdoli, H., Farzinnia, B., Mohebali, M., & Hajjaran, H. (2003). Epidemic outbreak of cutaneous leishmaniasis due to Leishmania major in Ghanavat rural district, Qom Province, Central Iran. *Iranian Journal of Public Health*, *32*(4), 35–41.

Ala, A., Borjigin, J., Rochwarger, A., & Schilsky, M. (2005). Wilson disease in septuagenarian siblings: Raising the bar for diagnosis. *Hepatology (Baltimore, Md.)*, *41*(3), 668–670. doi:10.1002/hep.20601 PMID:15723329

Alatise, O. I., & Schrauzer, G. N. (2010). Lead exposure: A contributing cause of the current breast cancer epidemic in Nigerian Women. *Biological Trace Element Research*, *136*(2), 127–139. doi:10.100712011-010-8608-2

Alavanja, M. C. R. (2003). Use of Agricultural Pesticides and Prostate Cancer Risk in the Agricultural Health Study Cohort. *American Journal of Epidemiology*, *157*(9), 800–814. doi:10.1093/aje/kwg040 PMID:12727674

Alavanja, M. C. R., Dosemeci, M., Samanic, C., Lubin, J., Lynch, C. F., Knott, C., ... Blair, A. (2004). Pesticides and lung cancer risk in the agricultural health study cohort. *American Journal of Epidemiology*, *160*(9), 876–885. doi:10.1093/aje/kwh290 PMID:15496540

Alavanja, M. C. R., Hoppin, J. A., & Kamel, F. (2004). Health effects of chronic pesticide exposure: Cancer and neurotoxicity. *Annual Review of Public Health*, *25*(1), 155–197. doi:10.1146/annurev.publhealth.25.101802.123020 PMID:15015917

Albanito, L., Lappano, R., Madeo, A., Chimento, A., Prossnitz, E. R., Cappello, A. R., ... Maggiolini, M. (2008). G protein-coupled receptor 30 and estrogen receptor-alpha are involved in the proliferative effects induced by atrazine in ovarian cancer cells. *Environmental Health Perspectives*, *116*(12), 1648–1655. doi:10.1289/ehp.11297 PMID:19079715

Alcamo, J., Dronin, N., Endejan, M., Golubev, G., & Kirilenko, A. (2007). A new assessment of climate change impacts on food production shortfalls and water availability in Russia. *Global Environmental Change*, *17*(3-4), 429–444. doi:10.1016/j.gloenvcha.2006.12.006

Al-Chalabi, A., & Hardiman, O. (2013). The epidemiology of ALS: A conspiracy of genes, environment and time. *Nature Reviews. Neurology*, *9*(11), 617–628. doi:10.1038/nrneurol.2013.203 PMID:24126629

Aleksandrowicz, J., Czachor, M., Schiffer, Z., & Smyk, B. (1970). Mycotoxins and their role in oncogenesis, with special reference to blood diseases. *Haematologica Latina*, *13*(2), 115–124. PMID:5504037

Al-Harthi, S., & Sci, M. J.-W. J. M. (2007). *Diagnosis and differentiation of Entamoeba infection in Makkah Al Mukarramah using microscopy and stool antigen detection kits.* Retrieved from https://pdfs.semanticscholar.org/1f1f/e5c2911c0a5e684b9ab952e42571537ca6bc.pdf

Aliomrani, M., Sahraian, M. A., Shirkhanloo, H., Sharifzadeh, M., Khoshayand, M. R., & Ghahremani, M. H. (2016). Blood Concentrations of Cadmium and Lead in Multiple Sclerosis Patients from Iran. *Iranian Journal of Pharmaceutical Research*, *15*(4), 825–833. PMID:28243279

Alirol, E., Sharma, S. K., & Bawasar, H. S. (2010). Snake bite in South Asia: A review. *PLoS Neglected Tropical Diseases*, *4*(1), e603. doi:10.1371/journal.pntd.0000603 PMID:20126271

Al-Karaki, G., McMichael, B. Z. A. K. J., & Zak, J. (2004). Field response of wheat to arbuscular mycorrhizal fungi and drought stress. *Mycorrhiza*, *14*(4), 263–269. doi:10.100700572-003-0265-2 PMID:12942358

Allen, M. R., & Ingram, W. J. (2002). Constraints on future changes in climate and the hydrologic cycle. *Nature*, *419*(6903), 224–232. doi:10.1038/nature01092 PMID:12226677

Allen, R. H., Gottlieb, M., Clute, E., Pongsiri, M. J., Sherman, J., & Obrams, G. I. (1997). Breast cancer and pesticides in Hawaii: The need for further study. *Environmental Health Perspectives*, *105*, 679–683.

Almendro-Candel, M., & Navarro-Pedreno, J., Vidal, M.M., Gomez, I., & Melendez-Pastor, I. (2014). *Use of municipal solid waste compost to reclaim limestone quarries mine spoils as soil amendments: Effects on Cd and Ni.* Elsevier. Retrieved from https://www.sciencedirect.com/science/article/pii/S0375674213002215

Almutlaq, B. A., Almuazzi, R. F., Almuhayfir, A. A., Alfouzan, A. M., Alshammari, B. T., AlAnzi, H. S., & Ahmed, H. G. (2017). Breast cancer in Saudi Arabia and its possible risk factors. *Journal of Cancer Policy*, *12*, 83–89. doi:10.1016/j.jcpo.2017.03.004

Alonso, A., Logroscino, G., Jick, S. S., & Hernan, M. A. (2009). Incidence and lifetime risk of motor neuron disease in the United Kingdom: A population-based study. *European Journal of Neurology*, *16*(6), 745–751. doi:10.1111/j.1468-1331.2009.02586.x PMID:19475756

Alshannaq, A., & Yu, J.-H. (2017). Occurrence, toxicity, and analysis of major mycotoxins in food. *International Journal of Environmental Research and Public Health*, *14*(6), 632. doi:10.3390/ijerph14060632 PMID:28608841

Altevogt, B. M., Hanson, S. L., & Leshner, A. I. (2008). Autism and the Environment: Challenges and Opportunities for Research. *Pediatrics*, *121*(6), 1225–1229. doi:10.1542/peds.2007-3000 PMID:18519493

Altezer. (2013). Predictive Framework Climate Change and Infectious Diseases: From Evidence to a Predictive Framework. *Science*, *341*(6145), 514–519. doi:10.1126cience.1239401 PMID:23908230

Altmann, P., Cunningham, J., Dhanesha, U., Ballard, M., Thompson, J., & Marsh, F. (1999). Disturbance of cerebral function in people exposed to drinking water contaminated with aluminium sulphate: Retrospective study of the Camelford water incident. *BMJ (Clinical Research Ed.)*, *319*(7213), 807–811. doi:10.1136/bmj.319.7213.807 PMID:10496822

Altwegg, R., Dummermuth, S., Anholt, B. R., & Flatt, T. (2005). Winter weather affects asp viper Viperaaspis population dynamics through susceptible juveniles. *Oikos*, *110*(1), 55–66. doi:10.1111/j.0030-1299.2001.13723.x

Álvarez, L., Bianco, C. L., Toscano, J. P., Lin, J., Akaike, T., & Fukuto, J. M. (2017). Chemical Biology of Hydropersulfides and Related Species: Possible Roles in Cellular Protection and Redox Signaling. *Antioxidants & Redox Signalling*, *27*(10), 622–633. doi:10.1089/ars.2017.7081 PMID:28398141

Alvar, J., Velez, I. D., Bern, C., Herrero, M., Desjeux, P., Cano, J., ... den Boer, M. W. H. O. (2012). Leishmaniasis Control Team. Leishmaniasis worldwide and global estimates of its incidence. *PLoS One*, *7*(5), e35671. doi:10.1371/journal.pone.0035671

Alvaro, P. K., Roberts, R. M., & Harris, J. K. (2014). The independent relationships between insomnia, depression, subtypes of anxiety, and chronotype during adolescence. *Sleep Medicine*, *15*(8), 934–941. doi:10.1016/j.sleep.2014.03.019 PMID:24958244

Aly, O. M., & El-Dib, M. A. (1971). Studies on the persistence of some carbamate insecticides in the aquatic environment—I. Hydrolysis of sevin, baygon, pyrolan and dimetilan in waters. *Water Research*, *5*(12), 1191–1205. doi:10.1016/0043-1354(71)90084-4

Amahmid, O., Asmama, S., & Bouhoum, K. (2002). Urban wastewater treatment in stabilization ponds: Occurrence and removal of pathogens. *Urban Water*, *4*(3), 255–262. doi:10.1016/S1462-0758(01)00071-1

Amahmid, O., & Bouhoum, K. (2000). Health effect of urban wastewater reuse in a peri-urban area in Morocco. *Environmental Management and Health*, *11*(3), 263–269. doi:10.1108/09566160010372004

Amaike, S., & Keller, N. P. (2011). Aspergillus flavus. *Annual Review of Phytopathology*, *49*(1), 107–133. doi:10.1146/annurev-phyto-072910-095221

Aminzadeh, K. K., & Etminan, M. (2007). Dental amalgam and multiple sclerosis: A systematic review and meta-analysis. *Journal of Public Health Dentistry*, *67*(1), 64–66. doi:10.1111/j.1752-7325.2007.00011.x PMID:17436982

Amoros, I., Alonso, J. L., & Cuesta, G. (2010). Cryptosporidium oocysts and Giardia cysts on salad products irrigated with contaminated water. *Journal of Food Protection*, *73*(6), 1138–1140. doi:10.4315/0362-028X-73.6.1138 PMID:20537274

Amro, A., Hamdi, S., Lemrani, M., Idrissi, M., Hida, M., Rhajaoui, M., & ... . (2013). Moroccan *Leishmaniainfantum*: Genetic Diversity and Population Structure as Revealed by Multi-Locus Microsatellite Typing. *PLoS One*, *8*(10), e77778. doi:10.1371/journal.pone.0077778 PMID:24147078

Andersen, H. R., Andersson, A. M., Arnold, S. F., Autrup, H., Barfoed, M., Beresford, N. A., ... Grandjean, P. (1999). Comparison of short-term estrogenicity tests for identification of hormone-disrupting chemicals. *Environmental Health Perspectives*, *107*(suppl 1), 89–108. doi:10.1289/ehp.99107s189

Andersen, H. R., Vinggaard, A. M., Rasmussen, T. H., Gjermandsen, I. M., & Bonefeld-Jørgensen, E. C. (2002). Effects of Currently Used Pesticides in Assays for Estrogenicity, Androgenicity, and Aromatase Activity in Vitro. *Toxicology and Applied Pharmacology*, *179*(1), 1–12. doi:10.1006/taap.2001.9347

Anderson, J. O., & Thundiyil, J. G. (2012). Clearing the Air : A Review of the Effects of Particulate Matter Air Pollution on Human Health. *J.Med, Toxicol,* 166–175. doi:10.100713181-011-0203-1

Anderson, J. F., & Prestwich, K. N. (1975). The fluid pressure pumps of spiders (Chelicerata, Araneae). *Zeitschrift für Morphologie der Tiere*, *81*(4), 257–277. doi:10.1007/BF00298488

Anderson, J. H., Christova, P. S., Xie, T. D., Schott, K. S., Ward, K., & Gomez, C. M. (2002). Spinocerebellar ataxia in monozygotic twins. *Archives of Neurology*, *59*, 1945–1951.

Andujar, P., & Nemery, B. (2009). Acute and subacute chemical pneumonitis. *Revue des Maladies Respiratoires*, *26*(8), 867–885. doi:10.1016/S0761-8425(09)73682-4 PMID:19953031

Angilletta, M. J. Jr, Steury, T. D., & Sears, M. W. (2004). Temperature, growth rate, and body size in ectotherms: Fitting pieces of a life-history puzzle. *Integrative and Comparative Biology*, *44*(6), 498–509. doi:10.1093/icb/44.6.498 PMID:21676736

Ann, J., & Gus, A. B. (2008). Quality of Life Trajectories in Epilepsy: A review of the literature. *Epilepsy & Behavior*, *12*(4), 557–571. doi:10.1016/j.yebeh.2007.11.013 PMID:18158270

Ansseau, M., Fischler, B., Dierick, M., Albert, A., Leyman, S., & Mignon, A. (2008). Socioeconomic correlates of generalized anxiety disorder and major depression in primary care: The GADIS II study (Generalized Anxiety and Depression Impact Survey II). *Depression and Anxiety*, *25*(6), 506–513. doi:10.1002/da.20306 PMID:17595015

Antoniou, A., Pharoah, P. D. P., Narod, S., Risch, H. A., Eyfjord, J. E., Hopper, J. L., ... Easton, D. F. (2003). Average risks of breast and ovarian cancer associated with BRCA1 or BRCA2 mutations detected in case Series unselected for family history: A combined analysis of 22 studies. *American Journal of Human Genetics*, *72*(5), 1117–1130. doi:10.1086/375033

Anuar, A. K., & Ramachandran, C. P. (1977). A study on the prevalence of soil transmitted helminthes among lettuce leaves sold in local markets in Penang, Malaysia. *The Medical Journal of Malaysia*, *31*, 262–265. PMID:562969

Arab, A., Jackson, M. C., & Kongoli, C. (2014). Modelling the effects of weather and climate on malaria distributions in West Africa. *Malaria Journal*, *13*(1), 126. doi:10.1186/1475-2875-13-126 PMID:24678602

Aramini, J., McLean, M., Wilson, J., Holt, J., Copes, R., Allen, B. and Sears, W. (2000) Drinking water quality and health-care utilization for gastrointestinal illness in greater Vancouver. *Releve des maladies transmissibles au Canada, 26*, 211-214.

Araújo, A. S., Lobato, Z. I. P., Chávez-Olórtegui, C., & Velarde, D. T. (2010). Brazilian IgY-Bothrops antivenom: Studies on the development of a process in chicken egg yolk. *Toxicon*, *55*(4), 739–744. doi:10.1016/j.toxicon.2009.11.004 PMID:19925817

Arevalo-Herrera, M., Quiñones, M. L., Guerra, C., Céspedes, N., Giron, S., Ahumada, M., ... Padilla, J. C. (2012). Malaria in selected non-Amazonian countries of Latin America. *Acta Tropica*, *121*(3), 303–314. doi:10.1016/j.actatropica.2011.06.008

Arey, J., & Atkinson, R. (2003). Photochemical reactions of PAHs in the atmosphere. In *PAHs: An Ecotoxicological Perspective* (pp. 47–63). New York: John Wiley and Sons Ltd. doi:10.1002/0470867132.ch4

Arheimer, B., Andreasson, J., Fogelberg, S., Johnsson, H., Pers, C. B., & Persson, K. (2005). Climate change impact on water quality: Model results from southern Sweden. *Ambio*, *34*(7), 559–566. doi:10.1579/0044-7447-34.7.559 PMID:16435746

Arnell, N. W., Livermore, M. J., Kovats, S., Levy, P. E., Nicholls, R., Parry, M. L., & Gaffin, S. R. (2004). Climate and socio-economic scenarios for global-scale climate change impacts assessments: Characterising the SRES storylines. *Global Environmental Change*, *14*(1), 3–20. doi:10.1016/j.gloenvcha.2003.10.004

Arnell, W. N. (2004). Climate change and global water resources: SRES emissions and socio-economic scenarios. *Global Environmental Change*, *14*(1), 31–52. doi:10.1016/j.gloenvcha.2003.10.006

Arnold, M., Karim-Kos, H. E., Coebergh, J. W., Byrnes, G., Antilla, A., Ferlay, J., ... Soerjomataram, I. (2015). Recent trends in incidence of five common cancers in 26 European countries since 1988: Analysis of the European Cancer Observatory. *European Journal of Cancer*, *51*(9), 1164–1187. doi:10.1016/j.ejca.2013.09.002 PMID:24120180

Aronson, K. J., Siemiatycki, J., Dewar, R., & Gerin, M. (1996). Occupational risk factors for prostate cancer: Results from a case-control study in Montreal, Quebec, Canada. *American Journal of Epidemiology*, *143*(4), 363–373. doi:10.1093/oxfordjournals.aje.a008750 PMID:8633620

Aronson, K., Miller, A., Woolcott, C., Sterns, E., McCready, D., Lickley, L., ... Weber, J. P. (2000). Breast adipose tissue concentrations of polychlorinated biphenyls and other organochlorines and breast cancer risk. *Cancer Epidemiology, Biomarkers & Prevention*, *9*, 55–63.

Arora, S. K., Batra, P., Sharma, T., Banerjee, B. D., & Gupta, S. (2013). Role of Organochlorine Pesticides in Children with Idiopathic Seizures. *ISRN Pediatrics*, 5. PMID:24368944

Arroyo-Manzanares, N., Malysheva, S. V., Vanden Bussche, J., Vanhaecke, L., Diana Di Mavungu, J., & De Saeger, S. (2014). Holistic approach based on high resolution and multiple stage mass spectrometry to investigate ergot alkaloids in cereals. *Talanta*, *118*, 359–367. doi:10.1016/j.talanta.2013.10.002

Arti Shukla, G. S. S. and R. S. (1996). brain barrier permeability and its possible correlation with decreased microvessel antioxidant potential in rat. *Human et Experimental Toxicolgy*, (January).

Arvidson, B. (1992). Inorganic mercury is transported from muscular nerve terminals to spinal and brainstem motoneurons. *Muscle & Nerve*, *15*(10), 1089–1094. doi:10.1002/mus.880151006 PMID:1383815

Arvis, B. B., & Miller, J. D. (2005). Mycotoxins as harmful indoor air contaminants. *Applied Microbiology and Biotechnology*, *66*, 367-372.

Ascherio, A., Chen, H., Weisskopf, M. G., O'Reilly, E., McCullough, M. L., Calle, E. E., ... Thun, M. J. (2006). Pesticide exposure and risk for Parkinson's disease. *Annals of Neurology*, *60*(2), 197–203. doi:10.1002/ana.20904 PMID:16802290

Ascherio, A., & Munger, K. L. (2007). Environmental risk factors for multiple sclerosis. Part II: Noninfectious factors. *Annals of Neurology*, *61*(6), 504–513. doi:10.1002/ana.21141 PMID:17492755

Ascherio, A., Munger, K. L., & Simon, K. C. (2010). Vitamin D and multiple sclerosis. *Lancet Neurology*, *9*(6), 599–612. doi:10.1016/S1474-4422(10)70086-7 PMID:20494325

Aschner, M. (2000). Manganese: Brain transport and emerging research needs. *Environmental Health Perspectives*, *103*, 429–432. PMID:10852840

Aschner, M., & Aschner, J. L. (1991). Manganese neurotoxicity: Cellular effects and blood-brain barrier transport. *Neuroscience and Biobehavioral Reviews*, *15*(3), 333–340. doi:10.1016/S0149-7634(05)80026-0 PMID:1956602

Aschner, M., & Gannon, M. (1994). Manganese (Mn) transport across the rat blood-brain barrier: Saturable and transferrin-dependent transport mechanisms. *Brain Research Bulletin*, *33*(3), 345–349. doi:10.1016/0361-9230(94)90204-6 PMID:8293318

Aschner, M., Gannon, M., & Kimelberg, H. K. (1992). Manganese uptake and efflux in cultured rat astrocytes. *Journal of Neurochemistry*, *58*(2), 730–735. doi:10.1111/j.1471-4159.1992.tb09778.x PMID:1729413

Aschner, M., Guilarte, T. R., Schneider, J. S., & Zheng, W. (2007). Manganese: Recent advances in understanding its transport and neurotoxicity. *Toxicology and Applied Pharmacology*, *221*(2), 131–147. doi:10.1016/j.taap.2007.03.001 PMID:17466353

Ashbolt, N.J. (2004). Microbial contamination of drinking water and disease out- comes in developing regions. *Toxicology*, *198*(1), 229-238.

Ashbolt, N. J. (2004). Microbial contamination of drinking water and disease outcomes in developing regions. *Toxicology, 198*(1-3), 229–238. doi:10.1016/j.tox.2004.01.030 PMID:15138046

Ashbolt, N. J., & Roser, D. J. (2003). Interpretation and management implications of event and baseflow pathogen data (CDROM). In M. J. Pfeffer, D. J. V. Abs, & K. N. Brooks (Eds.), *Watershed Management for Water Supply Systems*. New York: American Water Resources Association.

Ashley, J., & Malin, R. O. W. (2018). *The Developmental Neurotoxicity of Cadmium. Handbook of Developmental Neurotoxicology (2nd ed.)*. Elsevier Inc.; doi:10.1007/978-94-009-5594-3

Assaf, H., Azouri, H., & Pallardy, M. (2004). Ochratoxin A induces apoptosis in human lymphocytes through down regulation of Bcl-xL. *Toxicological Sciences: An Official Journal of the Society of Toxicology, 79*(2), 335–344. doi:10.1093/toxsci/kfh123 PMID:15056805

Assi, H. A., Khoury, K. E., Dbouk, H., Khalil, L. E., Mouhieddine, T. H., & El Saghir, N. S. (2013). Epidemiology and prognosis of breast cancer in young women. *Journal of Thoracic Disease, 5*, S2–S8.

Astrom, J., Petterson, S., Bergstedt, O., Pettersson, T. J., & Stenstrom, T. A. (2007). Evaluation of the microbial risk reduction due to selective closure of the raw water intake before drinking water treatment. *Journal of Water and Health, 5*(S1), 81–97. doi:10.2166/wh.2007.139 PMID:17890838

Atherton, F., Newman, C. P. S., & Casemore, D. P. (1995). *An outbreak of waterborne cryptosporidiosis associated with a public water supply in the UK*. Academic Press.

Atherton, K. M., Williams, F. M., Egea González, F. J., Glass, R., Rushton, S., Blain, P. G., & Mutch, E. (2009). DNA damage in horticultural farmers: A pilot study showing an association with organophosphate pesticide exposure. *Biomarkers: Biochemical Indicators of Exposure, Response, and Susceptibility to Chemicals, 14*(7), 443–451. doi:10.3109/13547500903137265 PMID:19863182

Audhild Hjalmarsen, D. L. H. (2008). Severe nocturnal hypoxaemia in geriatric. Oxford University Press. doi:10.1093/ageing/afn110

Auerbach, P. S. (2005). The Venomous Reptiles of the Western Hemisphere. *Wilderness & Environmental Medicine, 16*(3), e13. doi:10.1580/1080-6032(2005)16[e13a:BR]2.0.CO;2

Auld, H., MacIver, D., & Klaassen, J. (2004). Heavy rainfall and waterborne disease outbreaks: The Walkerton example. *Journal of Toxicology and Environmental Health. Part A., 67*(20-22), 1879–1887. doi:10.1080/15287390490493475 PMID:15371222

Aumüller, G. (2012). *Prostate gland and seminal vesicles*. Springer Science & Business Media.

Autissier, N., Rochette, L., Dumas, P., Beley, A., Loireau, A., & Bralet, J. (1982). Dopamine and norepinephrine turnover in various regions of the rat brain after chronic manganese chloride administration. *Toxicology, 24*(2), 175–182. doi:10.1016/0300-483X(82)90055-5 PMID:7135412

Axelrod, D., Smith, J., Kornreich, D., Grinstead, E., Singh, B., Cangiarella, J., & Guth, A. A. (2008). Breast Cancer in Young Women. *Journal of the American College of Surgeons, 206*(6), 1193–1203. doi:10.1016/j.jamcollsurg.2007.12.026

Ayis, S. A., Ayerbe, L., Ashworth, M., & Wolfe, D. A. (2017, September). C. (2018). Evaluation of the Hospital Anxiety and Depression Scale (HADS) in screening stroke patients for symptoms: Item Response Theory (IRT) analysis. *Journal of Affective Disorders, 228*, 33–40. doi:10.1016/j.jad.2017.11.037 PMID:29202444

Azevedo, N. F., Almeida, C., Fernandes, I., Cerqueira, L., Dias, S., Keevil, C. W., & Vieira, M. J. (2008). Survival of gastric and enterohepatic *Helicobacter spp.* in water: Implications for transmission. *Applied and Environmental Microbiology*, *74*(6), 1805–1811. doi:10.1128/AEM.02241-07 PMID:18245254

Azimi, A., Azari, A., Rezakazemi, M., & Ansarpour, M. (2017). Removal of heavy metals from industrial wastewaters: A review. *ChemBioEng Reviews*, *4*(1), 37–59. doi:10.1002/cben.201600010

Baan, R., Grosse, Y., Straif, K., Secretan, B., El Ghissassi, F., Bouvard, V., ... Cogliano, V. (2009). A review of human carcinogens—Part F: Chemical agents and related occupations. *The Lancet Oncology*, *10*(12), 1143–1144. doi:10.1016/S1470-2045(09)70358-4

Bae, H.-K. (2013). Changes of river's water quality responded to rainfall events. *Environ Ecol Res*, *1*, 21–25.

Bahloul, M., Chaari, A., Dammak, H., Samet, M., Chtara, K., Chelly, H., ... Bouaziz, M. (2013). Pulmonary edema following scorpion envenomation: Mechanisms, clinical manifestations, diagnosis and treatment. *International Journal of Cardiology*, *162*(2), 86–91. doi:10.1016/j.ijcard.2011.10.013 PMID:22075406

Baig, S., Seevasant, I., Mohamad, J., Mukheem, A., Huri, H. Z., & Kamarul, T. (2016). Potential of apoptotic pathway-targeted cancer therapeutic research: Where do we stand? *Cell Death & Disease*, *7*(1), e2058. doi:10.1038/cddis.2015.275 PMID:26775709

Baik, E. J., Moon, K., & Jung, Y. (2004). Cadmium stimulates the expression of ICAM-1 via NF-. *Biochemical and Biophysical Research Communications*, *320*(3), 887–892. doi:10.1016/j.bbrc.2004.05.218 PMID:15240131

Bailey, H. D., Infante-Rivard, C., Metayer, C., Clavel, J., Lightfoot, T., Kaatsch, P., ... Schüz, J. (2015). Home pesticide exposures and risk of childhood leukemia: Findings from the childhood leukemia international consortium. *International Journal of Cancer*, *137*(11), 2644–2663. doi:10.1002/ijc.29631 PMID:26061779

Bakare, A. A., Lateef, A., Amuda, O. S., & Afolabi, R. O. (2009). The Aquatic toxicity and characterization of chemical and microbiological constituents of water samples from Oba River, Odo-oba, Nigeria. *Asian J. Microbiol. Biotechnol. Environ. Exp. Sci*, *5*(1), 11.

Baker, D. N., Kanekal, S. G., Pulkkinen, T. I., & Blake, J. B. (1999). Equinoctial and Solstitial Averages of Magnetospheric Relativistic Electrons: A Strong Semiannual Modulation. *Geophysical Research Letters*, *26*(20), 3193–3196. doi:10.1029/1999GL003638

Bakheit, M. A., Torra, D., Palomino, L. A., Thekisoe, O. M. M., Mbati, P. A., Ongerth, J., & Karanis, P. (2008). *Sensitive and specific detection of Cryptosporidium species in PCR-negative samples by loop-mediated isothermal DNA amplification and confirmation of generated LAMP products by sequencing*. Academic Press. doi:10.1016/j.vetpar.2008.09.012

Bakir, A., & Darbre, P. D. (2015). Effect of aluminium on migration of oestrogen unresponsive MDAMB-231 human breast cancer cells in culture. *Journal of Inorganic Biochemistry*, *152*, 180–185. doi:10.1016/j.jinorgbio.2015.09.001

Bako, S. P., Bhwankot, E. S., Ezealor, A. U., Chia, A. M., & Funtua, I. I. (2009). Human health implications of trace metal contents in parts of Maize (Zea mays L.) plants cultivated along highways in Nigeria's Guinea Savanna. In Soil Remediation Series: Pollution Science, Technology and Abatement. Nova Science Publishers, Inc.

Bako, S. P., Odiwo, J. E., & Ezealor, A. U. (2008). Plant diversity in relation to anthropogenic Trace Metals in soils of selected sites in Nigeria's Guinea Savanna. *International Journal of Environment and Pollution*, *33*(2/3), 185. doi:10.1504/IJEP.2008.019392

Balamurugan, K., Egli, D., Zhang, B., & Georgiev, O. (2004). Metal-responsive transcription factor (MTF-1) and heavy metal stress response in Drosophila and mammalian cells : A functional comparison. *Biological Chemistry*, *385*(July), 597–603. PMID:15318808

Balbus, J. M., Boxall, A. B. A., Fenske, R. A., Mckone, T. E., & Zeisel, L. (2013). Implications of global climate change for the assessment and management of human health risks of chemicals in the natural environment. *Environmental Toxicology and Chemistry*, *32*(1), 62–78. doi:10.1002/etc.2046

Balbus, J. M., & Malina, C. (2009). Identifying vulnerable subpopulations for climate change health effects in the United States. *Journal of Occupational and Environmental Medicine*, *51*(1), 33–37. doi:10.1097/JOM.0b013e318193e12e PMID:19136871

Baldi, I., Mohammed-Brahim, B., Brochard, P., Dartigues, J. F., & Salamon, R. (1998). Delayed health effects of pesticides: review of current epidemiological knowledge. *Revue d'Epidemiologie et de Sante Publique*, *46*(2), 134–142. PMID:9592856

Ball, M. J. (1977). Neuronal loss, neurofibrillary tangles and granulovacuolar degeneration in the hippocampus with ageing and dementia. *Acta Neuropathologica*, *37*(2), 111–118. doi:10.1007/BF00692056 PMID:848276

Baltazar, M. T., Dinis-Oliveira, R. J., de Lourdes Bastos, M., Tsatsakis, A. M., Duarte, J. A., & Carvalho, F. (2014). Pesticides exposure as etiological factors of Parkinson's disease and other neurodegenerative diseases–a mechanistic approach. *Toxicology Letters*, *230*(2), 85–103. doi:10.1016/j.toxlet.2014.01.039 PMID:24503016

Balzano, T., Forteza, J., Borreda, I., Molina, P., Giner, J., & Leone, P. ... Felipo, V. (2018). Histological Features of Cerebellar Neuropathology in Patients With Alcoholic and Nonalcoholic Steatohepatitis. *Journal of Neuropathology and Experimental Neurology*. doi:10.1093/jnen/nly061

Balzano, T., Forteza, J., Molina, P., Giner, J., Monzó, A., Sancho-Jiménez, J., ... Felipo, V. (2018). The Cerebellum of Patients with Steatohepatitis Shows Lymphocyte Infiltration, Microglial Activation and Loss of Purkinje and Granular Neurons. *Scientific Reports*, *8*(1), 3004. doi:10.103841598-018-21399-6 PMID:29445232

Bambrick, H., Dear, K., Woodruff, R., Hanigan, I., & McMichael, A. (2008) The impacts of climate change on three health outcomes: Temperature-related mortality and hospitalisations, salmonellosis and other bacterial gastroenteritis, and population at risk from dengue. Garnaut Climate Change Review.

Ban, K. (2012). *A message from the UN Secretary General for the opening session of the 39th Session of the Committee on World Food Security, Rome, 15 to 20 October 2012*. Retrieved from www.un-foodsecurity.org/node/1356)

Bandera, E. V., Chandran, U., Buckley, B., Lin, Y., Isukapalli, S., Marshall, I., ... Zarbl, H. (2011). Urinary mycoestrogens, body size and breast development in New Jersey girls. *The Science of the Total Environment*, *409*(24), 5221–5227. doi:10.1016/j.scitotenv.2011.09.029

Bandmann, O., Weiss, K. H., & Kaler, S. G. (2015). Wilson's disease and other neurological copper disorders. *Lancet Neurology*, *14*(1), 103–113. doi:10.1016/S1474-4422(14)70190-5 PMID:25496901

Band, P. R., Abanto, Z., Bert, J., Lang, B., Fang, R., Gallagher, R. P., & Le, N. D. (2011). Prostate cancer risk and exposure to pesticides in British Columbia Farmers. *The Prostate*, *71*(2), 168–183. doi:10.1002/pros.21232 PMID:20799287

Bánfalvi, G. (2011). Heavy Metals, Trace Elements and Their Cellular Effects. In G. Banfalvi (Ed.), *Cellular Effects of Heavy Metals* (pp. 3–28). Dordrecht: Springer Netherlands; doi:10.1007/978-94-007-0428-2_1

Barbeau, A., Cloutier, T., Roy, M., Plasse, L., Paris, S., & Poirier, J. (1985). Ecogenetics of Parkinson's disease: 4-hydroxylation of debrisoquine. *Lancet (London, England), 2*(8466), 1213–1216. Retrieved from http://www.ncbi.nlm.nih.gov/pubmed/2866293

Barbeau, A., Roy, M., Cloutier, T., Plasse, L., & Paris, S. (1987). Environmental and genetic factors in the etiology of Parkinson's disease. *Advances in Neurology, 45*, 299–306. Retrieved from http://www.ncbi.nlm.nih.gov/pubmed/3493629 PMID:3493629

Barbosa, A., Silva, J. A., Cardoso, M. F. E. C., Meneses, J. N. C., Cunha, M. C. M., Haddad, J. P. A., ... Magalhães, D. F. (2014). Distribuição espacial de acidentes escorpiônicos em Belo Horizonte, Minas Gerais, 2005 a 2009 (Spatial distribution of scorpion sting accidents in Belo Horizonte, Minas Gerais, 2005 to 2009). *Arquivo Brasileiro de Medicina Veterinária e Zootecnia, 66*(3), 721–730. doi:10.1590/16784162-7116

Bardin, P. G. (1994). Organophosphate and Carbamate Poisoning. *Archives of Internal Medicine, 154*(13), 1433. doi:10.1001/archinte.1994.00420130020005 PMID:8017998

Barkouch, Y., Sedki, A., & Pineau, A. (2007). A new approach for understanding lead transfer in agricultural soil. *Water, Air, and Soil Pollution, 186*(1-4), 3–13. doi:10.100711270-007-9450-9

Barlow, B., Coryslechta, D., Richfield, E., & Thiruchelvam, M. (2007). The gestational environment and Parkinson's disease: Evidence for neurodevelopmental origins of a neurodegenerative disorder. *Reproductive Toxicology (Elmsford, N.Y.), 23*(3), 457–470. doi:10.1016/j.reprotox.2007.01.007 PMID:17350799

Bar-meir, E. (2007). Guidelines for Treating Cardiac Manifestations of Organophosphates Poisoning with Special Emphasis on Long QT and Torsades De Pointes. *Informa Healthcare*, 279–285. doi:10.1080/10408440601177855

Barnett, G. C., Shah, M., Redman, K., Easton, D. F., Ponder, B. A., & Pharoah, P. D. (2008). Risk factors for the incidence of breast cancer: Do they affect survival from the disease? *Journal of Clinical Oncology, 26*(20), 3310–3336. doi:10.1200/JCO.2006.10.3168

Barnett, T. P., Adam, J. C., & Lettenmaier, D. P. (2005). Potential impacts of a warming climate on water availability in snow-dominated regions. *Nature, 438*(7066), 303–309. doi:10.1038/nature04141 PMID:16292301

Barros, F. S. M., Arruda, M. E., Gurgel, H. C., & Honorio, N. A. (2011). Spatial clustering and longitudinal variation of Anopheles darlingi (Diptera: Culicidae) larvae in a river of the Amazon: the importance of the forest fringe and of obstructions to flow in frontier malaria. *Bulletin of Entomological Research, 101*(6), 643–658. doi:10.1017/S0007485311000265

Barth, H., & L'Hermite, P. (1987). *Scientific Basis for Soil Protection in the European Community, Elsevier*. Amsterdam, The Netherlands: Academic Press. doi:10.1007/978-94-009-3451-1

Bar-Zeev, M. (1958). The effect of temperature on the growth rate and survival of the immature stages of Aedes aegypti (L.). *Bulletin of Entomological Research, 49*(1), 157–163. doi:10.1017/S0007485300053499

Bashir, M. N. (2015). Epidemiology of prostate cancer. *Asian Pacific Journal of Cancer Prevention, 16*(13), 5137–5141. doi:10.7314/APJCP.2015.16.13.5137 PMID:26225642

Basun, H., Forssell, L. G., Wetterberg, L., & Winblad, B. (1991). Metals and trace elements in plasma and cerebrospinal fluid in normal aging and Alzheimer's disease. *Journal of Neural Transmission. Parkinson's Disease and Dementia Section, 3*(4), 231–258. Retrieved from http://www.ncbi.nlm.nih.gov/pubmed/1772577 PMID:1772577

Bath, P., & Lees, K. R. (2007). Acute Stroke. Textbook of Clinical Trials, 173, 179–214. doi:10.1002/9780470010167.ch13

Batterman, S., Eisenberg, J., Hardin, R., Kruk, M. E., Lemos, M. C., Michalak, A. M., ... Wilson, M. L. (2009). Sustainable control of water-related infectious diseases: A review and proposal for interdisciplinary health-based systems research. *Environmental Health Perspectives*, *117*(7), 1023–1032. doi:10.1289/ehp.0800423 PMID:19654908

Bauer, E., Claussen, M., Brovkin, V., & Huenerbein, A. (2003). Assessing climate forcings of the Earth system for the past millennium. *Geophysical Research Letters*, *30*(6). doi:10.1029/2002GL016639

Bauwens, D., Garland, T. Jr, Castilla, A. M., & Van Damme, R. (1995). Evolution of sprint speed in lacertid lizards: Morphological, physiological, and behavioral covariation. *Evolution; International Journal of Organic Evolution*, *49*(5), 848–863. PMID:28564867

Bawaskar, H. S., & Bawaskar, P. H. (2011). Efficacy and safety of scorpion antivenom plus prazosin compared with prazosin alone for venomous scorpion (Mesobuthus tamulus) sting: Randomised open label clinical trial. *BMJ (Clinical Research Ed.)*, *342*(3), c7136. doi:10.1136/bmj.c7136 PMID:21209062

Bayoh, M. N., & Lindsay, S. W. (2004). Temperature-related duration of aquatic stages of the Afrotropical malaria vector mosquito Anopheles gambiae in the laboratory. *Medical and Veterinary Entomology*, *18*(2), 174–179. doi:10.1111/j.0269-283X.2004.00495.x

Bbosa, G. S., Kitya, D., Odda, J., & Ogwal-Okeng, J. (2013). Aflatoxins metabolism, effects on epigenetic mechanisms and their role in carcinogenesis. *Health*, *05*(10), 14–34. doi:10.4236/health.2013.510A1003

Beal, M. F., Brouillet, E., Jenkins, B. G., Ferrante, R. J., Kowall, N. W., Miller, J. M., ... Hyman, B. T. (1993). Neurochemical and histologic characterization of striatal excitotoxic lesions produced by the mitochondrial toxin 3-nitropropionic acid. *The Journal of Neuroscience*, *13*(10), 4181–4192. doi:10.1523/JNEUROSCI.13-10-04181.1993 PMID:7692009

Beavers, C. T., Parker, J. J., Flinchum, D. A., Weakley-jones, B. A., & Jortani, S. A. (2014). Pesticide-Induced Quadriplegia in a 55-Year-Old Woman. *Forensic Med Pathol*, *35*(4), 239–241. doi:10.1097/PAF.0000000000000108 PMID:25354225

Becker-Algeri, T. A., Castagnaro, D., de Bortoli, K., de Souza, C., Drunkler, D. A., & Badiale-Furlong, E. (2016). Mycotoxins in bovine milk and dairy products: A review. *Journal of Food Science*, *81*(3), R544–R552. doi:10.1111/1750-3841.13204

Becker, S. L., Vogt, J., Knopp, S., Panning, M., Warhurst, D. C., Polman, K., ... Utzinger, J. (2013). Persistent digestive disorders in the tropics: Causative infectious pathogens and reference diagnostic tests. *BMC Infectious Diseases*, *13*(1), 37. doi:10.1186/1471-2334-13-37 PMID:23347408

Beekman, A. T. F., Bremmer, M. A., Deeg, D. J. H., Van Balkom, A. J. L. M., Smit, J. H., De Beurs, E., ... Van Tilburg, W. (1998). Anxiety disorders in later life: A report from the longitudinal aging study Amsterdam. *International Journal of Geriatric Psychiatry*, *13*(10), 717–726. doi:10.1002/(SICI)1099-1166(1998100)13:10<717::AID-GPS857>3.0.CO;2-M PMID:9818308

Belhassen, H., Jiménez-Díaz, I., Arrebola, J. P., Ghali, R., Ghorbel, H., Olea, N., & Hedili, A. (2015). Zearalenone and its metabolites in urine and breast cancer risk: A case-control study in Tunisia. *Chemosphere*, *128*, 1–6. doi:10.1016/j.chemosphere.2014.12.055

Bellali, H., Talmoudi, K., Alaya, N. B., Mahfoudhi, M., Ennigrou, S., & Chahed, M. K. (2017). Effect of temperature, rainfall and relative density of rodent reservoir hosts on zoonotic cutaneous leishmaniasis incidence in Central Tunisia. *Asian Pacific Journal of Tropical Disease*, *7*(2), 88–96. doi:10.12980/apjtd.7.2017D6-330

Belpomme, D., Irigaray, P., Ossondo, M., Vacque, D., & Martin, M. (2009). Prostate cancer as an environmental disease: An ecological study in the French Caribbean islands, Martinique and Guadeloupe. *International Journal of Oncology*, *34*(4), 1037–1044. doi:10.3892/ijo_00000229 PMID:19287960

Belson, M., Kingsley, B., & Holmes, A. (2006). Risk factors for acute leukemia in children: A review. *Environmental Health Perspectives*, *115*(1), 138–145. doi:10.1289/ehp.9023 PMID:17366834

Belyaeva, E. A., Korotkov, S. M., & Saris, N. (2011). Journal of Trace Elements in Medicine and Biology In vitro modulation of heavy metal-induced rat liver mitochondria dysfunction : A comparison of copper and mercury with cadmium. *Journal of Trace Elements in Medicine and Biology*, *25*, S63–S73. doi:10.1016/j.jtemb.2010.10.007 PMID:21146384

Benabid, A. (1982). *Bref aperçu sur la zonation altitudinale de la végétation climacique du Maroc*. Marseille: Centre National de Documentation Pedagogique.

Benammi, H., El Hiba, O., Romane, A., & Gamrani, H. (2014). A blunted anxiolytic like effect of curcumin against acute lead induced anxiety in rat: Involvement of serotonin. *Acta Histochemica*, *116*(5), 920–925. doi:10.1016/j.acthis.2014.03.002 PMID:24721902

Benbrahim-Tallaa, L., Waterland, R. A., Styblo, M., Achanzar, W. E., Webber, M. M., & Waalkes, M. P. (2005). a. Molecular events associated with arsenic-induced malignant transformation of human prostatic epithelial cells: Aberrant genomic DNA methylation and K-ras oncogene activation. *Toxicology and Applied Pharmacology*, *206*(3), 288–298. doi:10.1016/j.taap.2004.11.017 PMID:16039940

Benbrahim-Tallaa, L., Webber, M. M., & Waalkes, M. P. (2005). Acquisition of androgen independence by human prostate epithelial cells during arsenic-induced malignant transformation. *Environmental Health Perspectives*, *113*(9), 1134–1139. doi:10.1289/ehp.7832 PMID:16140617

Benbrahim-Tallaa, L., Webber, M. M., & Waalkes, M. P. (2006). Mechanisms of acquired androgen independence during arsenic-induced malignant transformation of human prostate epithelial cells. *Environmental Health Perspectives*, *115*(2), 243–247. doi:10.1289/ehp.9630 PMID:17384772

Ben-Chioma, A. E., Obunwo, C. C., Bartimaeus, E. S., & Boisa, N. (2017). Evaluation of Heavy Metal Levels in the Serum of Breast Cancer Patients in Port Harcourt, Nigeria. *International Journal of Contemporary Medical Research.*, *4*, 2393–915X.

Benderli Cihan, Y., Sözen, S., Yıldırım, O., Benderli Cihan, Y., Sözen, S., & Öztürk Yıldırım, S. (2011). nTrace elements and heavy metals in hair of stage III breast cancer patients. *Biological Trace Element Research*, *144*, 360-379.

Benedict, M. Q., Sandve, S. R., Wilkins, E. E., & Roberts, J. M. (2010). Relationship of larval desiccation to Anopheles gambiae Giles and An. arabiensis Patton survival. *Journal of Vector Ecology*, *35*(1), 116–123. doi:10.1111/j.1948-7134.2010.00067.x

Bener, A., Kamal, M., & Shanks, N. J. (2007). Impact of asthma and air pollution on school attendance of primary school children: Are they at increased risk of school absenteeism? *The Journal of Asthma*, *44*(4), 249–252. doi:10.1080/02770900701246725 PMID:17530520

Bennani, A., Buret, J., & Senhaji, F. (2001). *Initial national communication to the UN framework convention on climate change*. Kingdom of Morocco, Ministry of Land Management, Urban Planning, Housing and Environment. Available from: http://unfccc. int/resource/docs/natc/mornc1fres.pdf (in French)

Bennett, A. F. (1980). The metabolic foundations of vertebrate behavior. *Bioscience*, *30*(7), 452–456. doi:10.2307/1307946

Bennett, A. F. (1984). Thermal dependence of muscle function. *American Journal of Physiology. Regulatory, Integrative and Comparative Physiology*, *247*(2), R217–R229. doi:10.1152/ajpregu.1984.247.2.R217 PMID:6380314

Bennett, J. W., & Klich, M. (2003). Mycotoxins. *Clinical Microbiology Reviews*, *16*(3), 497–516. doi:10.1128/CMR.16.3.497-516.2003

Benoit, J. B., & Denlinger, D. L. (2010). Meeting the challenges of on-host and off-host water balance in blood-feeding arthropods. *Journal of Insect Physiology*, *56*(10), 1366–1376. doi:10.1016/j.jinsphys.2010.02.014

Berezovskaya, O., Schimmer, A. D., Glinskii, A. B., Pinilla, C., Hoffman, R. M., Reed, J. C., & Glinsky, G. V. (2005). Increased expression of apoptosis inhibitor protein XIAP contributes to anoikisresistance of circulating human prostate cancer metastasis precursor cells. *Cancer Research*, *65*(6), 2378–2386. doi:10.1158/0008-5472.CAN-04-2649 PMID:15781653

Bergomi, M., Vinceti, M., Nacci, G., Pietrini, V., Bratter, P., Alber, D., ... Vivoli, G. (2002). Environmental exposure to trace elements and risk of amyotrophic lateral sclerosis: A population-based case-control study. *Environmental Research*, *89*(2), 116–123. doi:10.1006/enrs.2002.4361 PMID:12123644

Berg, T., Kallenborn, R., & Manø, S. (2004). Temporal trends in atmospheric heavy metal and organochlorine concentrations at Zeppelin, Svalbard. *Arctic, Antarctic, and Alpine Research*, *36*(3), 284–291. doi:10.1657/1523-0430(2004)036[0284:TTIAHM]2.0.CO;2

Berk, M., Williams, L. J., Andreazza, A. C., Pasco, J. A., Dodd, S., Jacka, F. N., ... Magalhaes, P. V. S. (2014). Pop, heavy metal and the blues: Secondary analysis of persistent organic pollutants (POP), heavy metals and depressive symptoms in the NHANES National Epidemiological Survey. *BMJ Open*, *4*(7), e005142–e005142. doi:10.1136/bmjopen-2014-005142 PMID:25037643

Berkman, L. F., Berkman, C. S., Kasl, S., Freeman, D. H. Jr, Leo, L., Ostfeld, A. M., ... Brody, J. A. (1986). Depressive Symptoms in Relation to Physical Health and Functioning in the Elderly. *American Journal of Epidemiology*, *124*(3), 372–388. doi:10.1093/oxfordjournals.aje.a114408 PMID:3740038

Bern, C., Hightower, A. W., Chowdhury, R., Ali, M., Amann, J., Wagatsuma, Y., ... Maguire, J. H. (2005). Risk factors for kala-azar in Bangladesh. *Emerging Infectious Diseases*, *11*(5), 655–662. doi:10.3201/eid1105.040718 PMID:15890115

Bernhoft, R. A. (2012). Mercury toxicity and treatment: A review of the literature. *Journal of Environmental and Public Health*, *2012*, 1–10. doi:10.1155/2012/460508 PMID:22235210

Berry, C., La Vecchia, C., & Nicotera, P. (2010). Paraquat and Parkinson's disease. *Cell Death and Differentiation*, *17*(7), 1115–1125. doi:10.1038/cdd.2009.217 PMID:20094060

Bertazzi, P., Pesatori, A., Consonni, D., Tironi, A., Landi, M., & Zocchetti, C. (1993). Cancer incidence in a population accidentally exposed to 2,3,7,8-tetrachlorodibenzo-para-dioxin. *Epidemiology (Cambridge, Mass.)*, *4*(5), 398–406. doi:10.1097/00001648-199309000-00004

Berthiller, F., Crews, C., Dall'Asta, C., De Saeger, S., Haesaert, G., Karlovsky, P., ... Stroka, J. (2013). Masked mycotoxins: A review. *Molecular Nutrition & Food Research*, *57*(1), 165–186. doi:10.1002/mnfr.201100764

Bertholf, R. L., & Gambetti, P. (1987). Aluminum and Alzheimer's disease: Perspectives for a cytoskeletal mechanism. *CRC Critical Reviews in Clinical Laboratory Sciences*, *25*(3), 195–210. doi:10.3109/10408368709105882 PMID:3315463

Bertram, L., Lill, C. M., & Tanzi, R. E. (2010). The Genetics of Alzheimer Disease: Back to the Future. *Neuron*, *68*(2), 270–281. doi:10.1016/j.neuron.2010.10.013 PMID:20955934

Betarbet, R., Sherer, T. B., MacKenzie, G., Garcia-Osuna, M., Panov, A. V., & Greenamyre, J. T. (2000). Chronic systemic pesticide exposure reproduces features of Parkinson's disease. *Nature Neuroscience*, *3*(12), 1301–1306. doi:10.1038/81834 PMID:11100151

Bethony, J., Brooker, S., Albonico, M., Lancet, S. G.-T., Diemert, D., & Hotez, P. J. (2006). Soil-transmitted helminth infections: Ascariasis, trichuriasis, and hookworm. *Lancet*, *367*(9521), 1521–1532. doi:10.1016/S0140-6736(06)68653-4 PMID:16679166

Beuhler, M. (2003). Potential impacts of global warming on water resources in southern California. *Water Science and Technology*, *47*(7-8), 165–168. doi:10.2166/wst.2003.0685 PMID:12793676

Bezirtzoglou, C., Dekas, K., & Charvalos, E. (2011). Climate changes, environment and infection: Facts, scenarios and growing awareness from the public health community within Europe. *Anaerobe*, *17*(6), 337–340. doi:10.1016/j.anaerobe.2011.05.016 PMID:21664978

Bhalla, D. (2014). Michel Dumas and Pierre-Marie Preux: Promoting tropical neurology silently -The gist of their contributions. *Neuroepidemiology*, *42*(3), 139–143. doi:10.1159/000357354 PMID:24481209

Bhanti, M., & Taneja, A. (2007). Contamination of vegetables of different seasons with organophosphorous pesticides and related health risk assessment in northern India. *Chemosphere*, *69*(1), 63–68. doi:10.1016/j.chemosphere.2007.04.071 PMID:17568651

Bhatnagar, V. K., Kashyap, R., Zaidi, S. S. A., Kulkarni, P. K., & Saiyed, H. N. (2004). Levels of DDT, HCH, and HCB residues in human blood in Ahmedabad. *Bulletin of Environmental Contamination and Toxicology*, *72*(2), 261–265. doi:10.100700128-003-9049-9

Bhat, R. V., Shetty, P. H., Amruth, R. P., & Sudershan, R. V. (1997). A foodborne disease outbreak due to the consumption of moldy sorghum and maize containing fumonisin mycotoxins. *Journal of Toxicology. Clinical Toxicology*, *35*(3), 249–255. doi:10.3109/15563659709001208

Bhat, R., Rai, R. V., & Karim, A. A. (2010). Mycotoxins in food and feed: Present status and future concerns. *Comprehensive Reviews in Food Science and Food Safety*, *9*(1), 57–81. doi:10.1111/j.1541-4337.2009.00094.x

Bhattarai, A., Ali, A. S., Kachur, S. P., Mårtensson, A., Abbas, A. K., Khatib, R., ... Molteni, F. (2007). Impact of artemisinin-based combination therapy and insecticide-treated nets on malaria burden in Zanzibar. *PLoS Medicine*, *4*(11), e309. doi:10.1371/journal.pmed.0040309 PMID:17988171

Bibel, B. M. (2012). *The Whole Life Prostate Book: Everything that Every Man-at Every Age-Needs To Know About Maintaining Optimal Prostate Health.* Academic Press.

Bien, C. G., & Schramm, J. (2009). Treatment of Rasmussen encephalitis half a century after its initial description : Promising prospects and a dilemma. *Epilepsy Research*, *86*(2-3), 101–112. doi:10.1016/j.eplepsyres.2009.06.001 PMID:19615863

Bindi, M., & Moriondo, M. (2005). Impact of a 2°C global temperature rise on the Mediterranean region: Agriculture analysis assessment In C. Giannakopoulos, M, Bindi, M. Moriondo, P. Le Sager, & T. Tin (Eds.), Climate change impacts in the Mediterranean resulting from a 2°C global temperature rise (pp. 54-66). WWF Report.

Bing, Y., & Pamphlett, R. (2017). Environmental insults: Critical triggers for amyotrophic lateral sclerosis. *Translational Neurodegeneration*, *6*(1), 15. doi:10.118640035-017-0087-3 PMID:28638596

Birch, A. N. E., Begg, G. S., & Squire, G. R. (2011). How agro-ecological research helps to address food security issues under new IPM and pesticide reduction policies for global crop production systems. *Journal of Experimental Botany*, *62*, 3251–3261. doi:10.1093/jxb/err064 PMID:21669880

Bittencourt, S., Serrat, B.M., Andreoli, C.V., Moura, E.N., Togny, F.L., & Silva, L.A.T.P. (2013). Sewage sludge subjected to mixed: effect on total solids, pH and viable helminth eggs. *Rev. Acad. Ciencias Agrarias eAmbientais PUCPR Impresso, 11*, S191-S200. (in Portuguese)

Bittner, A., Cramer, B., Harrer, H., & Humpf, H. (2015). Structure elucidation and in vitro cytotoxicity of ochratoxin α amide, a new degradation product of ochratoxin A. *Mycotoxin Research, 31*(2), 83–90. doi:10.100712550-014-0218-y

Bjorling-Poulsen, M., Andersen, H. R., & Grandjean, P. (2008). Potential developmental neurotoxicity of pesticides used in Europe. *Environmental Health, 7*(1), 50. doi:10.1186/1476-069X-7-50 PMID:18945337

Black, R. E., & Lanata, C. F. (1995). *Epidemiology of diarrhoeal diseases in developing countries. Infections of the gastrointestinal tract.* New York: Raven Press.

Blain, P. G. (1990). Aspects of pesticide toxicology. *Adverse Drug Reactions and Acute Poisoning Reviews, 9*(1), 37–68. Retrieved from http://www.ncbi.nlm.nih.gov/pubmed/2190453 PMID:2190453

Blair, A., Dosemeci, M., & Heineman, E. F. (1993). Cancer and other causes of death among male and female farmers from twenty-three states. *American Journal of Industrial Medicine, 23*(5), 729–742. doi:10.1002/ajim.4700230507 PMID:8506851

Blair, A., & Freeman, L. B. (2009). Epidemiologic Studies of Cancer in Agricultural Populations: Observations and Future Directions. *Journal of Agromedicine, 14*(2), 125–131. doi:10.1080/10599240902779436 PMID:19437268

Blair, A., & Zahm, S. H. (1991). Cancer among farmers. *Occupational Medicine (Philadelphia, Pa.), 6*(3), 335–354. PMID:1835166

Blair, A., & Zahm, S. H. (1995). Agricultural exposures and cancer. *Environmental Health Perspectives, 103*(suppl 8), 205–208. PMID:8741784

Blair, A., Zahm, S. H., Pearce, N. E., Heineman, E. F., & Fraumeni, J. F. (1992). Clues to cancer etiology from studies of farmers. Scandinavian Journal of Work, Environment & Health, 18(4), 209–215. Retrieved from Brooks, P. C. (1996). Cell adhesion molecules in angiogenesis. *Cancer and Metastasis Reviews, 15*(2), 187–194.

Blanchard, R. J., Blanchard, D. C., & Hori, K. (1989). *An ethoexperimental approach to the study of defense*. Academic Press.

Blaszkowska, J., Wojcik, A., Kurnatowski, P., & Szwabe, K. (2013). Biologicalinteractions between soil saprotrophic fungi and Ascaris suum eggs. *Vector Parasitology., 196*(3-4), 401–408. doi:10.1016/j.vetpar.2013.02.029 PMID:23534981

Blaustein, A. R., Romansic, J. M., Kiesecker, J. M., & Hatch, A. C. (2003). Ultraviolet radiation, toxic chemicals and amphibian population declines. *Diversity & Distributions, 9*(2), 123–140. doi:10.1046/j.1472-4642.2003.00015.x

Blaylock, R. L. (2004). *Excitotoxicity : a possible central mechanism in fluoride neurotoxicity compelling evidence indicates that fluoride produces injury to the central nervous system (CNS) by several mechanisms.* Academic Press.

Blazer, D. G. (2003). Depression in late life: Review and commentary. *The Journals of Gerontology Series: Biological Sciences and Medical Sciences, 58*(3), M249–M265. doi:10.1093/gerona/58.3.M249 PMID:12634292

Blazer, D. G., Burchett, B., Service, C., & George, L. K. (1991). The association of age and depression among the elderly: An epidemiologic explanation. *The Journals of Gerontology. Series A, Biological Sciences and Medical Sciences, 46*(6), M210–M215. PMID:1834726

Blazer, D. G., Hybels, C. F., Simonsick, E. M., & Hanlon, J. T. (2000). Marked differences in antidepressant use by race in an elderly community sample: 1986-1996. *The American Journal of Psychiatry*, *157*(7), 1089–1094. doi:10.1176/appi.ajp.157.7.1089 PMID:10873916

Blazer, D. G., & Ph, D. (2000). Reviews and Overviews Psychiatry and the Oldest Old. *Am J PsychiatryAm J Psychiatry*, *15712*(December), 1915–1924. doi:10.1176/appi.ajp.157.12.1915

Blier, P., Ph, D., De Montigny, C., & Ph, D. (1999). Serotonin and Drug-Induced Therapeutic Responses in Major Depression, Obsessive – Compulsive and Panic Disorders. *Neuropsychopharmacology*, *21*(99), 15–20. PMID:10432494

Bloomfield, J. P., Williams, R. J., Gooddy, D. C., Cape, J. N., & Guha, P. (2006). Impacts of climate change on the fate and behaviour of pesticides in surface and groundwater—a UK perspective. *The Science of the Total Environment*, *369*(1-3), 163–177. doi:10.1016/j.scitotenv.2006.05.019 PMID:16914182

Blount, W. P. (1961). Turkey "X" disease. *Turkeys*, *9*, 52–55.

Blumenthal, U. J., & Peasey, A. (2002). *Critical review of epidemiological evidence of the health effects of wastewater and excreta use in agriculture*. Repéré à http://www.who.int/water_sanitation_health/wastewater/whocriticalrev.pdf

Blumenthal, U. J., Cifuentes, E., Bennett, S., Quigley, M., & Ruiz-Palacios, G. (2001). The risk of enteric infections associated with wastewater reuse; the effect of season and degree of storage of wastewater. *Transactions of the Royal Society of Tropical Medicine and Hygiene*, *95*(2), 131–137. doi:10.1016/S0035-9203(01)90136-1 PMID:11355541

Boada, L. D., Zumbado, M., Henríquez-Hernández, L. A., Almeida-González, M., Alvarez-León, E. E., Serra-Majem, L., & Luzardo, O. P. (2012). Complex organochlorine pesticide mixtures as determinant factor for breast cancer risk: A population-based case-control study in the Canary Islands (Spain). *Environmental Health*, *11*(1), 28. doi:10.1186/1476-069X-11-28

Bobka, M. S., Jaeger, R. G., & McNaught, D. C. (1981). Temperature dependent assimilation efficiencies of two species of terrestrial salamanders. *Copeia*, *1981*(2), 417–421. doi:10.2307/1444232

Boelaert, M., Meheus, F., Sanchez, A., Singh, S. P., Vanlerberghe, V., Picado, A., ... Sundar, S. (2009). The poorest of the poor: A poverty appraisal of households affected by visceral leishmaniasis in Bihar, India. *Tropical Medicine & International Health*, *14*(6), 639–644. doi:10.1111/j.1365-3156.2009.02279.x PMID:19392741

Boffetta, P., & Nyberg, F. (2003). Contribution of environmental factors to cancer risk. *British Medical Bulletin*, *68*(1), 71–94. doi:10.1093/bmp/ldg023 PMID:14757710

Bogdanffy, M. S., O'Connor, J. C., Hansen, J. F., Gaddamidi, V., Van Pelt, C. S., Green, J. W., & Cook, J. C. (2000). Chronic toxicity and oncogenicity bioassay in rats with the chlorostriazine herbicide cyanazine. *Journal of Toxicology and Environmental Health*, *60*(8), 567–586. doi:10.1080/00984100050082102

Bogert, C. M. (1949). Thermoregulation in reptiles, a factor in evolution. *Evolution; International Journal of Organic Evolution*, *3*(3), 195–211. doi:10.1111/j.1558-5646.1949.tb00021.x PMID:18138377

Boille'e, S., Yamanaka, K., Lobsiger, C. S., Copeland, N. G., Jenkins, N. A., Kassiotis, G., ... Cleveland, D. W. (2006). Onset and progression in inherited ALS determined by motor neurons and microglia. *Science*, *312*(5778), 1389–1392. doi:10.1126cience.1123511 PMID:16741123

Boina, D. R., Onagbola, E. O., Salyani, M., & Stelinski, L. L. (2009). Influence of posttreatment temperature on the toxicity of insecticides against Diaphorina citri (Hemiptera: Psyllidae). *Journal of Economic Entomology*, *102*(2), 685–691. doi:10.1603/029.102.0229 PMID:19449650

Bolin, C. M., Basha, R., Cox, D., Zawia, N. H., Maloney, B., Lahiri, D. K., & Cardozo-Pelaez, F. (2006). Exposure to lead and the developmental origin of oxidative DNA damage in the aging brain. *The FASEB Journal*, *20*(6), 788–790. doi:10.1096/fj.05-5091fje PMID:16484331

Bond, N. R., Burrows, R. M., Kennard, M. J., & Bunn, S. E. (2019). *Water Scarcity as a Driver of Multiple Stressor Effects. Multiple Stressors in River Ecosystems. (S.l.)*. Elsevier Inc.; doi:10.1016/B978-0-12-811713-2.00006-6

Bondy, S. C. (1988). *The neurotoxicity of organic and inorganic lead.* Academic Press.

Bonifacio, G., & Zamboni, G. (2016). Brain imaging in dementia. *Postgraduate Medical Journal*, *92*(1088), 333–340. doi:10.1136/postgradmedj-2015-133759 PMID:26933232

Bonner, F. T. (2008). Storage of seeds. *Woody plant seed manual*, 85-96.

Bons, J., & Geniez, P. (1996). *Amphibiens et reptiles du Maroc (Sahara occidental compris). Atlas biogéographique.* AsociaciónHerpetologica Espanola.

Bonvicini, F., Marcello, N., Mandrioli, J., Pietrini, V., &Vinceti, M. (2010). Exposure to pesticides and risk of amyotrophic lateral sclerosis: a population-based case control study. *Annalidell'istituto superior di sanita*, *46*, 284–7.doi: . doi:10.4415/ANN_10_03_10

Boorman, G. (1989). Toxicology and carcinogenesis studies of ochratoxin A F344/N rats (gavage studies). *National Toxicology Program*, *358*, 1-65.

Born, K., Fink, A. H., & Paeth, H. (2009). *Dry and Wet Periods in the Northwestern Maghreb for Present Day and Future Climate Conditions.* Retrieved from http://www.impetus.unikoeln.de/fileadmin/content/veroeffentlichungen/publikationsliste/1415_Born_full_article.pdf

Bosch, A. C., O'Neill, B., Sigge, G. O., Kerwath, S. E., & Hoffman, L. C. (2016). Heavy metals in marine fish meat and consumer health: A review. *Journal of the Science of Food and Agriculture*, *96*(1), 32–48. doi:10.1002/jsfa.7360 PMID:26238481

Boschi-Pinto, C., Velebit, L., & Shibuya, K. (2008). Estimating child mortality due to diarrhoea in developing countries. *Bulletin of the World Health Organization*, *86*(9), 710–717. doi:10.2471/BLT.07.050054 PMID:18797647

Bose-O'Reilly, S., McCarty, K. M., Steckling, N., & Lettmeier, B. (2010). Mercury exposure and children's health. *Current Problems in Pediatric and Adolescent Health Care*, *40*(8), 186–215. doi:10.1016/j.cppeds.2010.07.002 PMID:20816346

Bot, A., Nachtergaele, F., & Young, A. (2000). *Land resource potential and constraints at regional and country levels (No. 90).* Food & Agriculture Org.

Bottalico, A., Visconti, A., Logrieco, A., Solfrizzo, M., & Mirocha, C. J. (1985). Occurrence of zearalenols (diastereomeric mixture) in corn stalk rot and their production by associated Fusarium species. *Applied and Environmental Microbiology*, *49*, 547–551.

Bouaziz, M., Bahloul, M., Kallel, H., Samet, M., Ksibi, H., Dammak, H., ... Rekik, N. (2008). Epidemiological, clinical characteristics and outcome of severe scorpion envenomation in South Tunisia: Multivariate analysis of 951 cases. *Toxicon*, *52*(8), 918–926. doi:10.1016/j.toxicon.2008.09.004 PMID:18930073

Boudrissa, A. (2005). *Étude éco-épidémiologique de la leishmaniose cutanée dans la wilaya de M'Sila* (thèse de Magister). Algérie.

Boudrissa, A., Cherif, K., Kherrachi, I., Benbetka, S., Bouiba, L., Boubidi, S. C., ... Harrat, Z. (2012). Extension de Leishmania major au nord de l'Algérie. *Bulletin de la Société de Pathologie Exotique*, *105*(1), 30–35. doi:10.100713149-011-0199-4 PMID:22170408

Bouhoum, K., & Amahmid, O. (2000). Health effect of wastewater reuse in agriculture. *WHO Water Series*, *105*, 241–247. PMID:10842821

Bounoua, L., Kahime, K., Houti, L., Blakey, T., Ebi, K. L., Zhang, P., ... Boumezzough, A. (2013). Linking climate to incidence of zoonotic cutaneous leishmaniasis (*L. major*) in Pre-Saharan North Africa. *International Journal of Environmental Research and Public Health*, *10*(8), 3172–3191. doi:10.3390/ijerph10083172 PMID:23912199

Boussaa, S., Guernaoui, S., Pesson, B., & Boumezzough, A. (2005). Seasonal fluctuations of phlebotomine sand fly populations (Diptera: Psychodidae) in the urban area of Marrakech, Morocco. *Acta Tropica*, *95*(2), 86–91. doi:10.1016/j.actatropica.2005.05.002 PMID:15985259

Boussaa, S., Kahime, K., Samy, A. M., Bensalem, A., & Boumezzough, A. (2016). Species composition of sand flies and bionomics of Phlebotomuspapatasi and P. sergenti (Diptera: Psychodidae) in cutaneous leishmaniasis endemic foci, Morocco. *Parasites & Vectors*, *9*(1), 60. doi:10.118613071-016-1343-6 PMID:26830689

Bouziani, M. (2006). L'eau dans tous ses états. Ed. Dar El Gharb.

Boyd, C. A., Weiler, M. H., & Porter, W. P. (1990). Behavioral and neurochemical changes associated with chronic exposure to low-level concentration of pesticide mixtures. *Journal of Toxicology and Environmental Health*, *30*(3), 209–221. doi:10.1080/15287399009531424 PMID:2366258

Boyer, L. V., Theodorou, A. A., Berg, R. A., Mallie, J., Chávez-Méndez, A., García-Ubbelohde, W., ... Alagón, A. (2009). Antivenom for critically ill children with neurotoxicity from scorpion stings. *The New England Journal of Medicine*, *360*(20), 2090–2098. doi:10.1056/NEJMoa0808455 PMID:19439743

Bozzoni, V., Pansarasa, O., Diamanti, L., Nosari, G., Cereda, C., & Ceroni, M. (2016). Amyotrophic lateral sclerosis and environmental factors. *Functional Neurology*, *31*, 7–19. doi:10.11138/FNeur/2016.31.1.007 PMID:27027889

Bradley, R. A. (1988). The influence of weather and biotic factors on the behaviour of the scorpion (Paruroctonus utahensis). *Journal of Animal Ecology*, *57*(2), 533–551. doi:10.2307/4923

Bradman, A., Castorina, R., Barr, D. B., Chevrier, J., Harnly, M. E., Eisen, E. A., ... Eskenazi, B. (2011). Determinants of organophosphorus pesticide urinary metabolite levels in young children living in an agricultural community. *International Journal of Environmental Research and Public Health*, *8*(4), 1061–1083. doi:10.3390/ijerph8041061 PMID:21695029

Bradshaw, W. E. (1980). Thermoperiodism and the thermal environment of the pitcher-plant mosquito, Wyeomyia smithii. *Oecologia*, *46*(1), 13–17. doi:10.1007/BF00346959

Brama, M., Gnessi, L., Basciani, S., Cerulli, N., Politi, L., Spera, G., ... Migliaccio, S. (2007). Cadmium induces mitogenic signaling in breast cancer cell by an ERalpha-dependent mechanism. *Molecular and Cellular Endocrinology*, *264*(1-2), 102–108. doi:10.1016/j.mce.2006.10.013

Bratel, J., Haraldson, T., & Ottosson, J. O. (1997). Potential side effects of dental amalgam restorations. (II). No relation between mercury levels in the body and mental disorders. *European Journal of Oral Sciences*, *105*(3), 244–250. doi:10.1111/j.1600-0722.1997.tb00207.x PMID:9249191

Bray, F., Ferlay, J., Soerjomataram, I., Siegel, R. L., Torre, L. A., & Jemal, A. (2018). Global Cancer Statistics 2018: GLOBOCAN Estimates of Incidence and Mortality Worldwide for 36 *CA: A Cancer Journal for Clinicians*, 185. PMID:30207593

Breitner, S., Liu, L., Cyrys, J., Bruske, I., Franck, U., Schlink, U., ... Peters, A. (2011). Sub-micrometer particulate air pollution and cardiovascular mortality in Beijing, China. *The Science of the Total Environment*, *409*(24), 5196–5204. doi:10.1016/j.scitotenv.2011.08.023 PMID:21937089

Bressac, B., Kew, M., Wands, J., & Ozturk, M. (1991). Selective G to T mutations of p53 gene in hepatocellular carcinoma from southern Africa. *Nature*, *350*(6317), 429–431. doi:10.1038/350429a0

Bricquet, J., Bamba, F., Mahe, G., Toure, M., & Olivry, J. (1997). Evolution récente des ressources en eau de l'Afrique atlantique. *Revue des sciences de l'eau/Journal of Water Science*, *10*(3), 321-337.

Bridge, J. W., Oliver, D. M., Chadwick, D., Godfray, H. C. J., Heathwaite, A. L., Kay, D., … Porter, J. (2010). *Engaging with the water sector for public health benefits : waterborne pathogens and diseases in developed countries*. Academic Press. doi:10.2471/BLT.09.072512

Brierley, A. S., & Kingsford, M. J. (2009). Impacts of climate change on marine organisms and ecosystems. *Current Biology*, *19*(14), R602–R614. doi:10.1016/j.cub.2009.05.046 PMID:19640499

Brinton, L. A., Richesson, D. A., Gierach, G. L., Lacey, J. V., Park, Y., Hollenbeck, A. R., & Schatzkin, A. (2008). Prospective evaluation of risk factors for male breast cancer. *Journal of the National Cancer Institute*, *100*(20), 1477–1481. doi:10.1093/jnci/djn329

Brody, J. A., & Grant, M. D. (2001). Age-associated diseases and conditions: Implications for decreasing late life morbidity. *Aging (Milan, Italy)*, *13*, 64–67. PMID:11405387

Brody, J. G., Moysich, K. B., Humblet, O., Attfield, K. R., Beehler, G. P., & Rudel, R. A. (2007). Environmental pollutants and breast cancer: Epidemiologic studies. *Cancer*, *109*(S12), 2667–2711. doi:10.1002/cncr.22655

Broennimann, O., Ursenbacher, S., Meyer, A., Golay, P., Monney, J. C., Schmocker, H., ... Dubey, S. (2014). Influence of climate on the presence of colour polymorphism in two montane reptile species. *Biology Letters*, *10*(11), 20140638. doi:10.1098/rsbl.2014.0638 PMID:25392313

Bronstein, J., Carvey, P., Chen, H., Cory-Slechta, D., DiMonte, D., Duda, J., ... Weisskopf, M. (2009). Meeting Report: Consensus Statement—Parkinson's Disease and the Environment: Collaborative on Health and the Environment and Parkinson's Action Network (CHE PAN) Conference 26–28 June 2007. *Environmental Health Perspectives*, *117*(1), 117–121. doi:10.1289/ehp.11702 PMID:19165397

Bronzuoli, M. R., Iacomino, A., Steardo, L., & Scuderi, C. (2016). Targeting neuroinflammation in Alzheimer's disease. *Journal of Inflammation Research*, *9*, 199–208. doi:10.2147/JIR.S86958 PMID:27843334

Brooks, A., Chadwick, C., Gelbard, H., Cory-Slechta, D., & Federoff, H. (1999). Paraquat elicited neurobehavioral syndrome caused by dopaminergic neuron loss. *Brain Research*, *823*(1–2), 1–10. doi:10.1016/S0006-8993(98)01192-5 PMID:10095006

Brown, C., & Lall, U. (2006). Water and economic development: The role of variability and a framework for resilience. *Natural Resources Forum*, *30*(4), 306–317. doi:10.1111/j.1477-8947.2006.00118.x

Brown, G. P., & Shine, R. (2002). Influence of weather conditions on activity of tropical snakes. *Austral Ecology*, *27*(6), 596–605. doi:10.1046/j.1442-9993.2002.01218.x

Brown, M. E., & Funk, C. C. (2008). Food security under climate change. *Science*, *319*(5863), 580–581. doi:10.1126cience.1154102 PMID:18239116

Brown, R. C., Lockwood, A. H., & Sonawane, B. R. (2005). Neurodegenerative Diseases: An Overview of Environmental Risk Factors. *Neurodegenerative Diseases*, *113*(9), 1250–1256. doi:10.1289/ehp.7567 PMID:16140637

Brown, T. P., Rumsby, P. C., Capleton, A. C., Rushton, L., & Levy, L. S. (2006). Pesticides and Parkinson's Disease—Is There a Link? *Environmental Health Perspectives*, *114*(2), 156–164. doi:10.1289/ehp.8095 PMID:16451848

Bruinsma, J. (2003). *World Agriculture: Towards 2015/2030—an FAO perspective.* London: Earthscan.

Brundtland, G. H. (1989). *Notre avenir à tous*. Academic Press.

Bruneel, F., Hocqueloux, L., Alberti, C., Wolff, M., Chevret, S., Bédos, J. P., ... Vachon, F. (2003). The clinical spectrum of severe imported falciparum malaria in the intensive care unit: Report of 188 cases in adults. *American Journal of Respiratory and Critical Care Medicine*, *167*(5), 684–689. doi:10.1164/rccm.200206-631OC

Brust, R. A. (1967). Weight and development time of different stadia of mosquitoes reared at various constant temperatures. *Canadian Entomologist*, *99*(9), 986–993. doi:10.4039/Ent99986-9

Bruyn, G. W. (1968). Huntington's chorea: historical, clinical and laboratory synopsis. In Handbook of Clinical Neurology (pp. 298-378). Amsterdam: Elsevier.

Bryan, F. L. (1977). Diseases transmitted by foods contaminated by wastewater. *Journal of Food Protection*, *40*(1), 45–56. doi:10.4315/0362-028X-40.1.45

Bryś, M., Nawrocka, A. D., Miekoś, E., Zydek, C., Foksiński, M., Barecki, A., & Krajewska, W. M. (1997). Zincand cadmium analysis in human prostate neoplasms. *Biological Trace Element Research*, *59*(1-3), 145–152. doi:10.1007/BF02783239 PMID:9522056

Bui-Klimke, T. R., & Wu, F. (2015). Ochratoxin A and human health risk: A review of the evidence. *Critical Reviews in Food Science and Nutrition*, *55*(13), 1860–1869. doi:10.1080/10408398.2012.724480

Bullerman, L. B. (2003). MYCOTOXINS | Classifications. In B. Caballero (Ed.), *Encyclopedia of Food Sciences and Nutrition* (2nd ed.; pp. 4080–4089). Oxford, UK: Academic Press; doi:10.1016/B0-12-227055-X/00821-X

Bull, P. C., Thomas, G. R., Rommens, J. M., Forbes, J. R., & Cox, D. W. (1993). The Wilson disease gene is a putative copper transporting P-type ATPase similar to the Menkes gene. *Nature Genetics*, *5*(4), 327–337. doi:10.1038/ng1293-327 PMID:8298639

Bulluck, L.R., III, Brosius, M., Evanylo, G.K., & Ristaino, J. B. (2002). Organic and synthetic fertility amendments influence soil microbial, physical and chemical properties on organic and conventional farms. *Applied Soil Ecology, 19*(2), 147-160.

Buragohain, M., Bhuyan, B., & Sarma, H. P. (2010). Seasonal variations of lead, arsenic, cadmium and aluminium contamination of groundwater in Dhemaji district, Assam, India. *Environmental Monitoring and Assessment*, *170*(1-4), 345–351. doi:10.100710661-009-1237-6 PMID:19908151

Burgos, A., Bermejo, P. E., Calleja, J. L., Vaquero, A., & Abreu, L. E. (2009). Acquired chronic hepatocerebral degeneration due to cirrhosis from non-alcoholic steatohepatitis. *Revista Espanola De Enfermedades Digestivas: Organo Oficial De La Sociedad Espanola De Patologia Digestiva*, *101*(11), 806–811. PMID:20001159

Burmeister, H. R., & Plattner, R. D. (1987). Enniatin production by Fusarium tricinctum and its effects on germinating wheat seeds. *Phytopathology*, *77*(10), 1483–1487. doi:10.1094/Phyto-77-1483

Burns, C. J., Beard, K. K., & Cartmill, J. B. (2001). Mortality in chemical workers potentially exposed to 2,4- dichlorophenoxyacetic acid (2,4-D) 1945-94: An update. *Occupational and Environmental Medicine*, *58*(1), 24–30. doi:10.1136/oem.58.1.24 PMID:11119631

Burns, D., Vitvar, T., McDonnell, J., Hassett, J., Duncan, J., & Kendall, C. (2005). Effects of suburban development on runoff generation in the Croton River basin, New York, USA. *Journal of Hydrology (Amsterdam), 311*(1-4), 266–281. doi:10.1016/j.jhydrol.2005.01.022

Busche, H. (2008). Hydrology of the Drâa Basin. In IMPETUS Atlas Morocco. University of Bonn.

Bush, A. I. (2003). The metallobiology of Alzheimer's disease. *Trends in Neurosciences, 26*(4), 207–214. doi:10.1016/S0166-2236(03)00067-5 PMID:12689772

Bush, A., Pettingell, W., Multhaup, G., d Paradis, M., Vonsattel, J., Gusella, J., ... Tanzi, R. (1994). Rapid induction of Alzheimer A beta amyloid formation by zinc. *Science, 265*(5177), 1464–1467. doi:10.1126cience.8073293 PMID:8073293

Bush, K. F., O'Neill, M. S., Li, S., Mukherjee, B., Hu, H., Ghosh, S., & Balakrishnan, K. (2014). Associations between extreme precipitation and gastrointestinal-related hospital admissions in Chennai, India. *Environmental Health Perspectives, 122*, 249–254. PMID:24345350

Bushnak, A. A. (2002). Future strategy for water resources management in Saudi Arabia. *Proceeding of A Future Vision for the Saudi Economy Symposium*, 37.

Bus, J. S., & Gibson, J. E. (1984). Paraquat: Model for oxidant-initiated toxicity. *Environmental Health Perspectives, 55*, 37–46. doi:10.1289/ehp.845537 PMID:6329674

Butterfield, P. G., Valanis, B. G., Spencer, P. S., Lindeman, C. A., & Nutt, J. G. (1993). Environmental antecedents of young-onset Parkinson's disease. *Neurology, 43*(6), 1150–1158. doi:10.1212/WNL.43.6.1150 PMID:8170560

Byappanahalli, M. N., Whitman, R. L., Shively, D. A., & Nevers, M. B. (2010). Linking non-culturable (qPCR) and culturable enterococci densities with hydrometeorological conditions. *The Science of the Total Environment, 408*(16), 3096–3101. doi:10.1016/j.scitotenv.2010.04.051 PMID:20546850

Byars, J. A., Beglinger, L. J., Moser, D. J., Gonzalez-Alegre, P., & Nopoulos, P. (2012). Substance abuse may be a risk factor for earlier onset of Huntington disease. *Journal of Neurology, 259*(9), 1824–1831. doi:10.100700415-012-6415-8 PMID:22274789

Byrne, C., Divekar, S. D., Storchan, G. B., Parodi, D. A., & Martin, M. B. (2009). Cadmium--a metallohormone. *Toxicology and Applied Pharmacology, 238*(3), 266–271. doi:10.1016/j.taap.2009.03.025 PMID:19362102

Byrne, C., Divekar, S. D., Storchan, G. B., Parodi, D. A., & Martin, M. B. (2013). Metals and Breast Cancer. *Journal of Mammary Gland Biology and Neoplasia, 18*(1), 63–73. doi:10.100710911-013-9273-9

Caccio, S. M., De Giacomo, M., Aulicino, F. A., Pozio, E., & Superiore, I. (2003). *Giardia Cysts in Wastewater Treatment Plants in Italy*. Academic Press. doi:10.1128/AEM.69.6.3393

Cacciò, S. M., Thompson, R. C., McLauchlin, J., & Smith, H. V. (2005). Unravelling Cryptosporidium and Giardia epidemiology. *Trends in Parasitology, 21*(9), 430–437. doi:10.1016/j.pt.2005.06.013 PMID:16046184

Caine, E.D., Bamford, KA, Schiffer, RB., Shoulson, I., & Levy, S. (1986). A controlled neuropsychological comparison of Huntington's disease and multiple sclerosis. *Archives of Neurology, 43*, 249 54.

Cairncross, E., Dalvie, A., Euripidou, R., Irlam, J., & Naidoo, R. N. (2018). Climate Change, Air Pollution and Health in South Africa. In *Climate Change and Air Pollution* (pp. 327–347). Cham: Springer. doi:10.1007/978-3-319-61346-8_20

Cai, X., Zhang, X., & Wang, D. (2010). Land availability for biofuel production. *Environmental Science & Technology, 45*(1), 334–339. doi:10.1021/es103338e PMID:21142000

Cai, Z., Xing, G., Yan, X., Xu, H., Tsuruta, H., Yagi, K., & Minami, K. (1997). Methane and nitrous oxide emissions from rice paddy fields as affected by nitrogen fertilisers and water management. *Plant and Soil*, *196*(1), 7–14. doi:10.1023/A:1004263405020

Cai, Z., Zhao, B., & Ratka, A. (2011). Oxidative Stress and β-Amyloid Protein in Alzheimer's Disease. *Neuromolecular Medicine*, *13*(4), 223–250. doi:10.100712017-011-8155-9 PMID:21901428

Calabresi, P. A. (2004). Diagnosis management of multiple sclerosis. *American Family Physician Journal*, *70*, 1935–1944. PMID:15571060

Cala-Riquelme, F., & Colombo, M. (2011). Ecology of the scorpion, Microtityus jaumei in Sierra de Canasta, Cuba. *Journal of Insect Science*, *11*(1). PMID:21870972

Calderón-Aranda, E. S., Dehesa-Dávila, M., Chavez-Haro, A., & Possani, L. D. (1996). Scorpion stings and their treatment in Mexico. *Envenomings and their treatments*, 311-320.

Callaghan, B., Feldman, D., Gruis, K., & Feldman, E. (2011). The association of exposure to lead, mercury, and selenium and the development of amyotrophic lateral sclerosis and the epigenetic implications. *Neurodegenerative Diseases*, *8*(1-2), 1–8. doi:10.1159/000315405 PMID:20689252

Calne, D. B., Chu, N. S., Huang, C. C., Lu, C. S., & Olanow, W. (1994). Manganism and idiopathic parkinsonism: Similarities and differences. *Neurology*, *44*(9), 1583–1586. doi:10.1212/WNL.44.9.1583 PMID:7936278

Cameron, H. L., & Foster, W. G. (2009). Developmental and Lactational Exposure to Dieldrin Alters Mammary Tumorigenesis in Her2/Neu Transgenic Mice. *PLoS One*, *4*(1), 4303. doi:10.1371/journal.pone.0004303

Caminade, C., Kovats, S., Rocklov, J., Tompkins, A. M., Morse, A. P., Colón-González, F. J., ... Lloyd, S. J. (2014). Impact of climate change on global malaria distribution. *Proceedings of the National Academy of Sciences of the United States of America*, *111*(9), 3286–3291. doi:10.1073/pnas.1302089111 PMID:24596427

Campbell, A. M., Williams, E. R., & Barltrop, D. (1970). Motor neurone disease and exposure to lead. *Journal of Neurology, Neurosurgery, and Psychiatry*, *33*(6), 877–885. doi:10.1136/jnnp.33.6.877 PMID:5497881

Campbell, J., Rustad, L., Boyer, E., Christopher, S., Driscoll, C., Fernandez, I., ... Ollinger, V. (2009). Consequences of climate change for biogeochemical cycling in forests of northeastern North AmericaThis article is one of a selection of papers from NE Forests 2100: A Synthesis of Climate Change Impacts on Forests of the Northeastern US and Eastern Canada. *Canadian Journal of Forest Research*, *39*(2), 264–284. doi:10.1139/X08-104

Campbell-Lendrum, D., & Woodruff, R. (2006). Comparative risk assessment of the burden of disease from climate change. *Environmental Health Perspectives*, *114*(12), 1935–1941. doi:10.1289/ehp.8432 PMID:17185288

Campdelacreu, J. (2014). Parkinson's disease and Alzheimer disease: Environmental risk factors. [English Edition]. *Neurologia (Barcelona, Spain)*, *29*(9), 541–549. doi:10.1016/j.nrl.2012.04.001 PMID:22703631

Cannas, A., Costa, B., Tacconi, P., Pinna, L., & Fiaschi, A. (1992). Dementia of Alzheimer type (DAT) in a man chronically exposed to pesticides. *Acta Neurologica*, *14*(3), 220–223. Retrieved from http://www.ncbi.nlm.nih.gov/pubmed/1442221 PMID:1442221

Cann, K. F., Thomas, D. R., Salmon, R. L., Wyn-Jones, A. P., & Kay, D. (2013). Extreme water-related weather events and waterborne disease. *Epidemiology and Infection*, *141*(4), 671–686. doi:10.1017/S0950268812001653 PMID:22877498

Canonica, S., Meunier, L., & von Gunten, U. (2008). Phototransformation of selected pharmaceuticals during UV treatment of drinking water. *Water Research*, *42*(1-2), 121–128. doi:10.1016/j.watres.2007.07.026 PMID:17709124

Cao, Z., Wang, F., Xiu, C., Zhang, J., & Li, Y. (2017). Hypericum perforatum extract attenuates behavioral, biochemical, and neurochemical abnormalities in Aluminum chloride-induced Alzheimer's disease rats. *Biomedicine and Pharmacotherapy*, *91*, 931–937. doi:10.1016/j.biopha.2017.05.022 PMID:28514831

Cardenas, R., Sandoval, C. M., Rodríguez-Morales, A. J., & Franco-Paredes, C. (2006). Impact of climate variability in the occurrence of leishmaniasis in northeastern Colombia. *The American Journal of Tropical Medicine and Hygiene*, *75*(2), 273–277. doi:10.4269/ajtmh.2006.75.273 PMID:16896132

Carlson, B. E., & Rowe, M. P. (2009). Temperature and desiccation effects on the antipredator behavior of Centruroides vittatus (Scorpiones: Buthidae). *The Journal of Arachnology*, *37*(3), 321–330. doi:10.1636/Hi09-06.1

Carlson, B. W., Neelon, V. J., Carlson, J. R., Hartman, M., & Dogra, S. (2005). Cerebrovascular Disease and Patterns of Cerebral Oxygenation During Sleep in Elders. *Biological Research for Nursing*, 307–317. PMID:19144652

Carmichael, W. W., Jones, C. L. A., Mahmood, N. A., Theiss, W. C., & Krogh, P. (1985). Algal toxins and water-based diseases. *Critical Reviews in Environmental Control*, *15*(3), 275–313. doi:10.1080/10643388509381734

Caroline, M., & Wilfred, S. R. (2013). Studies on Leaching of Heavy Metals from E-waste. *Oriental Journal of Chemistry*, *29*(03), 1149–1154. doi:10.13005/ojc/290343

Caron, Y., Martens, K., Lempereur, L., Saegerman, C., & Losson, B. (2014). New insight in lymnaeid snails (Mollusca, Gastropoda) as intermediate hosts of Fasciola hepatica (Trematoda, Digenea) in Belgium and Luxembourg. *Parasites & Vectors*, *7*(1), 66. doi:10.1186/1756-3305-7-66 PMID:24524623

Carpenter, D. O. (2006). Polychlorinated biphenyls (PCBs): Routes of exposure and effects on human health. *Reviews on Environmental Health*, *21*(1), 1–23. doi:10.1515/REVEH.2006.21.1.1

Carpenter, S. R., Cole, J. J., Pace, M. L., Batt, R., Brock, W. A., Cline, T., ... Weidel, B. (2011). Early warnings of regime shifts: A whole-ecosystem experiment. *Science*, *332*(6033), 1079–1082. doi:10.1126cience.1203672 PMID:21527677

Carratala, A., Rusinol, M., Rodriguez-Manzano, J., Guerrero-Latorre, L., Sommer, R., & Girones, R. (2013). Environmental effectors on the inactivation of human adenoviruses in water. *Food and Environmental Virology*, *5*(4), 203–214. doi:10.100712560-013-9123-3 PMID:23955425

Carslaw, K. S., Harrison, R. G., & Kirkby, J. (2002). Cosmic Rays, Clouds, and Climate. *Science*, *298*(5599), 1732–1737. doi:10.1126cience.1076964 PMID:12459578

Caruso, R. V., O'Connor, R. J., Stephens, W. E., Cummings, K. M., & Fong, G. T. (2013). Toxic metal concentrations in cigarettes obtained from US smokers in 2009: Results from the International Tobacco Control (ITC) United States survey cohort. *International Journal of Environmental Research and Public Health*, *11*(1), 202–217. doi:10.3390/ijerph110100202 PMID:24452255

Casemore, D. P., Wright, S. E., & Coop, R. L. (1997). Cryptosporidiosis: Human and Animal Epidemiology. In R. Fayer (Ed.), *Cryptosporidium and Cryptosporidiosis*. Boca Raton, FL: CRC Press.

Cash, B. A., Rodo, X., Emch, M., Yunus, M., Faruque, A. S., & Pascual, M. (2014). Cholera and shigellosis: Different epidemiology but similar responses to climate variability. *PLoS One*, *9*(9), e107223. doi:10.1371/journal.pone.0107223 PMID:25229494

Cassereau, J., Ferre, M., Chevrollier, A., Codron, P., Verny, C., Homedan, C., ... Reynier, P. (2017). (in press). Neurotoxicity of insecticides. *Current Medicinal Chemistry*, *24*(27). doi:10.2174/0929867324666170526122654 PMID:28552054

Cassidy, R. (2010). Cancer and Chlordane-Treated Homes: A Pinch of Prevention Is Worth a Pound of Cure. *Leukemia & Lymphoma*, *51*(7), 1363–1364. doi:10.3109/10428194.2010.483304

Cassidy, R. A., Natarajan, S., & Vaughan, G. M. (2005). The link between the insecticide heptachlor epoxide, estradiol, and breast cancer. *Breast Cancer Research and Treatment*, *90*(1), 55–64. doi:10.100710549-004-2755-0

Cassman, K. G., & Wood, S. (2005). Cultivated systems. In R. Hassan & ... (Eds.), *Ecosystems and Human Well-being: Current State and Trends* (Vol. 1, pp. 745–794). Washington, D. C.: Island Press.

Castello, P. R., Drechsel, D. A., & Patel, M. (2007). Mitochondria Are a Major Source of Paraquat-induced Reactive Oxygen Species Production in the Brain. *The Journal of Biological Chemistry*, *282*(19), 14186–14193. doi:10.1074/jbc.M700827200 PMID:17389593

Caudle, W. M., Richardson, J. R., Wang, M., & Miller, G. W. (2005). Perinatal heptachlor exposure increases expression of presynaptic dopaminergic markers in mouse striatum. *Neurotoxicology*, *26*(4), 721–728. doi:10.1016/j.neuro.2004.09.003 PMID:16112329

Cavallo, F., Gerber, M., Marubini, E., Richardson, S., Barbieri, A., Costa, A., ... Pujol, H. (1991). Zinc and copper in breast cancer. A joint study in northern Italy and southern France. *Cancer*, *67*(3), 738–745. doi:10.1002/1097-0142(19910201)67:3<738::AID-CNCR2820670335>3.0.CO;2-#

CCR, L. C. (2011). *Rapport sur la régionalisation avancée.* Royaume du Maroc, La Commission Consultative de la Régionalisation (CCR).

Cebecauer, T., & Hofierka, J. (2008). The consequences of land-cover changes on soil erosion distribution in Slovakia. *Geomorphology*, *98*(3-4), 187–198. doi:10.1016/j.geomorph.2006.12.035

Ceccom, J., Coslédan, F., Halley, H., Francès, B., Lassalle, J. M., & Meunier, B. (2012). Copper Chelator Induced Efficient Episodic Memory Recovery in a Non-Transgenic Alzheimer's Mouse Model. *PLoS One*, *7*(8), e43105. doi:10.1371/journal.pone.0043105 PMID:22927947

Celardin, F., Chatenoux, L., & D'Ersu, P. H. (1990). Metaux lourds dans les sols du basin genovois: 6tat de la question (II). *Archs. Sci. Geneve*, *42*, 493–498.

Celis, A., Gaxiola-Robles, R., Sevilla-Godínez, E., Valerio, M. D. J. O., & Armas, J. (2007). Tendencia de la mortalidad por picaduras de alacrán en México, 1979-2003. *Revista Panamericana de Salud Pública*, *21*(6), 373–380. doi:10.1590/S1020-49892007000500005 PMID:17761049

Centers for Disease Control and Prevention. (2009). *National report on human exposure to environmental chemicals.* Retrieved from http://www.cdc.gov/exposurereport/

Centers for Disease Control and Prevention-CDC. (2004). Outbreak of aflatoxin poisoning eastern and central provinces, Kenya, January–July 2004. *MMWR. Morbidity and Mortality Weekly Report*, *3*, 790–793.

Chalmers, R. M., Smith, R. P., Hadfield, S. J., Elwin, K., & Giles, M. (2011). Zoonotic linkage and variation in Cryptosporidium parvum from patients in the United Kingdom. *Parasitology Research*, *108*(5), 1321–1325. doi:10.100700436-010-2199-x PMID:21193928

Chamaille-Jammes, S., Massot, M., Aragon, P., & Clobert, J. (2006). Global warming and positive fitness response in mountain populations of common lizards Lacerta vivipara. *Global Change Biology*, *12*(2), 392–402. doi:10.1111/j.1365-2486.2005.01088.x

Chang, J.-W., Pai, M.-C., Chen, H.-L., Guo, H.-R., Su, H.-J., & Lee, C.-C. (2008). Cognitive function and blood methylmercury in adults living near a deserted chloralkali factory. *Environmental Research*, *108*(3), 334–339. doi:10.1016/j.envres.2008.06.006 PMID:18675410

Chang, K., Hsu, C., Liu, S., Su, C., Yen, C., Lee, M.-J., ... Huang, C.-F. (2013). Cadmium Induces Apoptosis in Pancreatic b -Cells through a Mitochondria-Dependent Pathway : The Role of Oxidative Stress-Mediated c-Jun N-Terminal Kinase Activation. *PLoS One*, *8*(2), e54374. doi:10.1371/journal.pone.0054374 PMID:23405080

Chappell, C. L., Okhuysen, P. C., Langer-Curry, R., Widmer, G., Akiyoshi, D. E., Tanriverdi, S., & Tzipori, S. (2006). Cryptosporidium hominis: Experimental challenge of healthy adults. *The American Journal of Tropical Medicine and Hygiene*, *75*(5), 851–857. doi:10.4269/ajtmh.2006.75.851 PMID:17123976

Charles, L. E., Burchfiel, C. M., Fekedulegn, D., Gu, J. K., Petrovitch, H., Sanderson, W. T., ... Ross, G. W. (2010). Occupational exposure to pesticides, metals, and solvents: The impact on mortality rates in the Honolulu Heart Program. *Work (Reading, Mass.)*, *37*(2), 205–215. doi:10.3233/WOR-2010-1071 PMID:20938081

Charlier, C., Albert, A., Herman, P., Hamoir, E., Gaspard, U., Meurisse, M., & Plomteux, G. (2003). Breast Cancer and Serum Organochlorine Residues. *Occupational and Environmental Medicine*, *60*(5), 348–351. doi:10.1136/oem.60.5.348

Charlwood, J. D., Birley, M. H., Dagoro, H., Paru, R., & Holmes, P. R. (1985). Assessing survival rates of Anopheles farauti (diptera: Culicidae) from Papua New Guinea. *Journal of Animal Ecology*, *54*(3), 1003–1016. doi:10.2307/4393

Charron, D., Thomas, M., Waltner-Toews, D., Aramini, J., Edge, T., Kent, R., ... Wilson, J. (2004). Vulnerability of waterborne diseases to climate change in Canada: A review. *Journal of Toxicology and Environmental Health. Part A.*, *67*(20-22), 1667–1677. doi:10.1080/15287390490492313 PMID:15371208

Chase, J. M., & Knight, T. M. (2003). Drought-induced mosquito outbreaks in wetlands. *Ecology Letters*, *6*(11), 1017–1024. doi:10.1046/j.1461-0248.2003.00533.x

Chatterjee, S., Basak, P., Chaklader, M., Das, P., Pereira, J. A., Chaudhuri, S., & Law, S. (2013). Pesticide induced marrow toxicity and effects on marrow cell population and on hematopoietic stroma. *Experimental and Toxicologic Pathology: Official Journal of the Gesellschaft Fur Toxikologische Pathologie*, *65*(3), 287–295. doi:10.1016/j.etp.2011.09.002 PMID:22005501

Chauret, C., Springthorpe, S., & Sattar, S. (1999). Fate of Cryptosporidium oocysts, Giardia cysts, and microbial indicators during wastewater treatment and anaerobic sludge digestion. *Canadian Journal of Microbiology*, *45*(3), 257–262. doi:10.1139/w99-001 PMID:10408099

Checkley, W., Epstein, L. D., Gilman, R. H., Figueroa, D., Cama, R. I., Patz, J. A., & Black, R. E. (2000). Effect of El Nino and ambient temperature on hospital admissions for diarrhoeal diseases in Peruvian children. *Lancet (London, England)*, *355*, 442–450. PMID:10841124

Chelbi, I., Derbali, M., Al-Ahmadi, Z., Zaafouri, B., El Fahem, A., & Zhioua, E. (2007). Phenology of Phlebotomus papatasi (Diptera: Psychodidae) relative to the seasonal prevalence of zoonotic cutaneous leishmaniasis in central Tunisia. *Journal of Medical Entomology*, *44*(2), 385–388. doi:10.1093/jmedent/44.2.385 PMID:17427714

Chen, F., Ying, G., Kong, L., & Wang, L. (n.d.). *Distribution and accumulation of endocrine-disrupting chemicals and pharmaceuticals in wastewater irrigated soils in Hebei, China*. Elsevier. Retrieved from https://www.sciencedirect.com/science/article/pii/S0269749111001680

Chen, C. J., Kuo, T. L., & Wu, M. M. (1988). Arsenic and cancers. *Lancet*, *1*(8582), 414–415. doi:10.1016/S0140-6736(88)91207-X PMID:2893213

Chen, C. J., & Wang, C. J. (1990). Ecological correlation between arsenic levels in well water and age adjusted mortality from malignant neoplasms. *Cancer Research*, *50*, 5470–5474. PMID:2386951

Chen, H. J., & Chang, H. (2014). Response of discharge, TSS, and *E. coli* to rainfall events in urban, suburban, and rural watersheds. *Environmental Science. Processes & Impacts*, *16*(10), 2313–2324. doi:10.1039/C4EM00327F PMID:25096028

Chen, L., Yoo, S. E., Na, R., Liu, Y., & Ran, Q. (2012). Cognitive impairment and increased Aβ levels induced by paraquat exposure are attenuated by enhanced removal of mitochondrial H 2O 2. *Neurobiology of Aging*, *33*(2), 15–26. doi:10.1016/j.neurobiolaging.2011.01.008 PMID:21429624

Chen, M. J., Lin, C. Y., Wu, Y. T., Wu, P. C., Lung, S. C., & Su, H. J. (2012). Effects of extreme precipitation to the distribution of infectious diseases in Taiwan, 1994-2008. *PLoS One*, *7*(6), e34651. doi:10.1371/journal.pone.0034651 PMID:22737206

Chen, S.-C., Tsai, M.-H., Wang, H.-J., Yu, H.-S., & Chang, L. W. (2004). Vascular permeability alterations induced by arsenic. *Human and Experimental Toxicology*, *23*(1), 1–7. doi:10.1191/0960327104ht407oa PMID:15027809

Chen, S.-Y., & Lin, J.-G. (2001). Bioleaching of heavy metals from sediment: Significance of pH. *Chemosphere*, *44*(5), 1093–1102. doi:10.1016/S0045-6535(00)00334-9 PMID:11513396

Chew, F. T., Doraisingham, S., Ling, A. E., Kumarasinghe, G., & Lee, B. W. (1998). Seasonal trends of viral respiratory tract infections in the tropics. *Epidemiology and Infection*, *121*(1), 121–128. doi:10.1017/S0950268898008905 PMID:9747763

Chiarelli, D. D., Rosa, L., Rulli, M. C., & D'Odorico, P. (2018). The water-land-food nexus of natural rubber production. *Journal of Cleaner Production*, *172*, 1739–1747. doi:10.1016/j.jclepro.2017.12.021

Chio, A., Calvo, A., Moglia, C., Mazzini, L., & Mora, G. (2011). Phenotypic heterogeneity of amyotrophic lateral sclerosis: A population based study. *Journal of Neurology, Neurosurgery, and Psychiatry*, *82*(7), 740–746. doi:10.1136/jnnp.2010.235952 PMID:21402743

Chiou, J., Wang, S., Chen, C., Deng, C., Lin, W., & Tai, T. (2005). Arsenic ingestion and increased microvascular disease risk : Observations from the south- western arseniasis-endemic area in Taiwan. *International Journal of Epidemiology*, *34*(May), 936–943. doi:10.1093/ije/dyi108 PMID:15911542

Chippaux, J. P. (1992). Les morsures de serpents en Afrique intertropicale. *Cahiers d'études et de recherches francophones/Santé*, *2*(4), 221-234.

Chippaux, J. P. (2006). *Les serpents d'Afriqueoccidentaleetcentrale*. IRD éditions.

Chippaux, J. P. (2009). Incidence mondiale et prise en charge des envenimations ophidiennes et scorpioniques. *Médecine/Sciences*, *25*(2), 197-200.

Chippaux, J. P. (1998). Snake-bites: Appraisal of the global situation. *Bulletin of the World Health Organization*, *76*(5), 515. PMID:9868843

Chippaux, J. P. (2008). Estimating the global burden of snakebite can help to improve management. *PLoS Medicine*, *5*(11), e221. doi:10.1371/journal.pmed.0050221 PMID:18986211

Chippaux, J. P. (2011). Estimate of the burden of snakebites in sub-Saharan Africa: A meta-analytic approach. *Toxicon*, *57*(4), 586–599. doi:10.1016/j.toxicon.2010.12.022 PMID:21223975

Chippaux, J. P., & Goyffon, M. (2008). Epidemiology of scorpionism: A global appraisal. *Acta Tropica*, *107*(2), 71–79. doi:10.1016/j.actatropica.2008.05.021 PMID:18579104

Chitsulo, L., Engels, D., Montresor, L., & Savioli, L. (2000). The global status of schistosomiasis and its control. *Acta Tropica*, *77*(1), 41–51. doi:10.1016/S0001-706X(00)00122-4 PMID:10996119

Choi, J., Polche,r A.,& Joas, A. (2016). Systematic literature review on Parkinson's disease and childhood leukaemia and mode of actions for pesticides. *European Food Safety Authority Supporting Publications*, *13*, EN–955. doi:10.2903p.efsa.2016.EN-955

Choo, X. Y., Alukaidey, L., White, A. R., & Grubman, A. (2013). Neuroinflammation and Copper in Alzheimer's Disease. *International Journal of Alzheimer's Disease*, *2013*, 1–12. doi:10.1155/2013/145345 PMID:24369524

Choudhary, D. (2018). *An Assignment on Climate Change and Agriculture Agmet-602, Fundamentals of Climatology (3+0).* Agrometeorologist, Dept. of Agricultural Meteorology B.A.C.A., AAU, ANAND.

Chowell, G., Díaz-Dueñas, P., Bustos-Saldaña, R., Mireles, A. A., & Fet, V. (2006). Epidemiological and clinical characteristics of scorpionism in Colima, Mexico (2000–2001). *Toxicon*, *47*(7), 753–758. doi:10.1016/j.toxicon.2006.02.004 PMID:16574179

Chowell, G., Hyman, J. M., Díaz-Dueñas, P., & Hengartner, N. W. (2005). Predicting scorpion sting incidence in an endemic region using climatological variables. *International Journal of Environmental Health Research*, *15*(6), 425–435. doi:10.1080/09603120500392475 PMID:16506436

Cho, Y. A., Kim, J., Woo, H. D., & Kang, M. (2013). Dietary Cadmium Intake and the Risk of Cancer: A Meta-Analysis. *PLoS One*, *8*(9), e75087. doi:10.1371/journal.pone.0075087

Christian, K. A., & Tracy, C. R. (1981). The effect of the thermal environment on the ability of hatchling Galapagos land iguanas to avoid predation during dispersal. *Oecologia*, *49*(2), 218–223. doi:10.1007/BF00349191 PMID:28309312

Christian, R. (2013). *Mycologie médicale*. Lavoisier.

Christiansen-Jucht, C., Parham, P. E., Saddler, A., Koella, J. C., & Basáñez, M. G. (2014). Temperature during larval development and adult maintenance influences the survival of Anopheles gambiae ss. *Parasites & Vectors*, *7*(1), 489. doi:10.118613071-014-0489-3

Chuang, M. J., Chang, W. N., Chang, H. W., Lin, W. C., Tsai, N. W., Hsieh, M. J., ... Lu, C. H. (2010). Predictors and long-term outcome of seizures after bacterial brain abscess. *Journal of Neurology, Neurosurgery, and Psychiatry*, *81*(8), 913–917. doi:10.1136/jnnp.2009.195073 PMID:20682720

Chuu, J. J., Liu, S. H., & Lin-Shiau, S. Y. (2007). Differential neurotoxic effects of methylmercury and mercuric sulfide in rats. *Toxicology Letters*, *169*(2), 109–120. doi:10.1016/j.toxlet.2006.12.006 PMID:17292570

Ciardini, V., Contessa, G. M., Falsaperla, R., Gomez-Amo, J. L., Meloni, D., Monteleone, F., ... di Sarra, A. (2016). Global and Mediterranean climate change: A short summary. *Annali dell'Istituto Superiore di Sanita*, *52*, 325–337. PMID:27698291

Cicchetti, F., Lapointe, N., Roberge-Tremblay, A., Saint-Pierre, M., Jimenez, L., Ficke, B. W., & Gross, R. E. (2005). Systemic exposure to paraquat and maneb models early Parkinson's disease in young adult rats. *Neurobiology of Disease*, *20*(2), 360–371. doi:10.1016/j.nbd.2005.03.018 PMID:16242641

Clark, H. A., & Snedeker, S. M. (2006). Ochratoxin A: Its cancer risk and potential for exposure. *Journal of Toxicology and Environmental Health. Part B, Critical Reviews*, *9*(3), 265–296. doi:10.1080/15287390500195570

Clark, L. C., Combs, G. F., Turnbull, B. W., Slate, E. H., Chalker, D. K., Chow, J., ... Taylor, J. R. (1997). Effects of selenium supplementation for cancer prevention in patients with carcinoma of the skin. A randomized controlled trial. *Journal of the American Medical Association*, *277*, 1520. PMID:8971064

Clarkson, T. W. (1972). The pharmacology of mercury. *Journal of the American Medical Association*, (12), 375–406. doi:10.1001/jama.1922.02640250025010 PMID:4556945

Clarkson, T. W. (1992). Mercury : Major Issues in Environmental Health. *Environmental Health Perspectives*, *100*, 31–38. doi:10.1289/ehp.9310031 PMID:8354179

Clayton, P. J., Grove, W. M., Coryell, W., Keller, M., Hirschfeld, R., & Fawcett, J. (1991). *Follow-Up and Family Study of Anxious Depression*. Academic Press.

Cleary, M. P., & Grossmann, M. E. (2009). Obesity and Breast Cancer: The Estrogen Connection. *Endocrinology*, 150, 2537–2542.

Clement, Y., & Chapouthier, G. (1998). Article. *Biological Bases of Anxiety*, *22*(5), 623–633.

Cline, W. R. (2007). *World agriculture faces serious decline from global warming*. Washington, DC: Centre for Global Development.

Clusella-Trullas, S., Wyk, J. H., & Spotila, J. R. (2009). Thermal benefits of melanism in cordylid lizards: A theoretical and field test. *Ecology*, *90*(8), 2297–2312. doi:10.1890/08-1502.1 PMID:19739391

Coffey, A. J., Durkie, M., Hague, S., McLay, K., Emmerson, J., Lo, C., ... Bandmann, O. (2013). A genetic study of Wilson's disease in the United Kingdom. *Brain*, *136*(Pt 5), 1476–1487. doi:10.1093/brain/awt035 PMID:23518715

Coggon, D., Ntani, G., Harris, E. C., Jayakody, N., & Palmer, K. T. (2015). Soft tissue sarcoma, non-Hodgkin's lymphoma and chronic lymphocytic leukaemia in workers exposed to phenoxy herbicides: extended follow-up of a UK cohort. *Occup Environ Med.*

Cohen, M. J., Tirado, C., Aberman, N. L., & Thompson, B. (2008). Impact of climate change and bioenergy on nutrition. Rome: Food and Agricultural Organisations of the United Nations (FAO) and International Food Policy Research Institute (IFPRI).

Cohen-kfir, E., Lee, W., Eskandari, S., & Nelson, N. (2005). Zinc inhibition of γ -aminobutyric acid transporter 4 (GAT4) reveals a link between excitatory and inhibitory neurotransmission. *Proceedings of the National Academy of Sciences of the United States of America*, *102*(17), 17. doi:10.1073/pnas.0501431102 PMID:15829583

Cohn, B. A., La Merrill, M., Krigbaum, N. Y., Yeh, G., Park, J. S., Zimmermann, L., & Cirillo, P. M. (2015). DDT exposure in utero and breast cancer. *The Journal of Clinical Endocrinology and Metabolism*, *100*(8), 2865–2872. doi:10.1210/jc.2015-1841

Cohn, B. A., Wolff, M. S., Cirillo, P. M., & Sholtz, R. I. (2007). DDT and Breast Cancer in Young Women: New Data on the Significance of Age at Exposure. *Environmental Health Perspectives*, *115*(10), 1406–1414. doi:10.1289/ehp.10260 PMID:17938728

Cohn, B. A., Wolff, M. S., Cirillo, P. M., & Sholtz, R. I. (2007). DDT and breast cancer in young women: New data on the significance of age at exposure. *Environmental Health Perspectives*, *115*, 1406–1414.

Colditz, G. A., Kaphingst, K. A., Hankinson, S. E., & Rosner, B. (2012). Family history and risk of breast cancer: Nurses' health study. *Breast Cancer Research and Treatment*, *133*(3), 1097–1104. doi:10.100710549-012-1985-9

Cole, T. B., Robbins, C. A., & Ju, H. (2000). Seizures and neuronal damage in mice lacking vesicular zinc. *Science B.V*, *39*, 153–169. PMID:10759303

Collaborative Group on Hormonal Factors in Breast Cancer. (2002). Breast cancer and breast feeding: Collaborative reanalysis of individual data from 47 epidemiological studies in 30 countries, including 50302 women with breast cancer and 96973 women without the disease. *Lancet*, *360*(9328), 187–195. doi:10.1016/S0140-6736(02)09454-0

Colwell, R. R. (1996). Global climate and infectious disease: The cholera paradigm. *Science, 274*(5295), 2025–2031. doi:10.1126cience.274.5295.2025 PMID:8953025

Comas, A. P. (1982). Precocious sexual development in Puerto Rico. *Lancet, 319*(8284), 1299–1300. doi:10.1016/S0140-6736(82)92857-4

Combs, G. F. Jr, & Clark, L. C. (1985). Can dietary selenium modify cancer risk? *Nutrition Reviews, 43*(11), 325–331. doi:10.1111/j.1753-4887.1985.tb02392.x PMID:3909013

Compston, A. (2009). Progressive lenticular degeneration: a familial nervous disease associated with cirrhosis of the liver, by S. A. Kinnier Wilson, (From the National Hospital, and the Laboratory of the National Hospital, Queen Square, London). *Brain, 132*(Pt 8), 1997–2001. doi:10.1093/brain/awp193 PMID:19634211

Compston, A., & Coles, A. (2008). Multiple sclerosis. *Lancet, 372*(9648), 1502–1517. doi:10.1016/S0140-6736(08)61620-7 PMID:18970977

Comstock, G. W., Bush, T. L., & Helzlsouer, K. (1992). Serum retinol, beta-carotene, vitamin E, and selenium as related to subsequent cancer of specific sites. *American Journal of Epidemiology, 135*(2), 115–121. doi:10.1093/oxfordjournals.aje.a116264 PMID:1536130

Cone, W., Russell, C., & Harwood, R. U. (1934). lead as a possible cause of multiple sclerosis. *Archives of Neurology and Psychiatry, 31*(2), 236–263. doi:10.1001/archneurpsyc.1934.02250020024002

Čonková, E., Laciaková, A., Pástorová, B., Scidel, H., & Kováč, G. (2001). The effect of zearalenone on some enzymatic parameters in rabbits. *Toxicology Letters, 121*(3), 145–149. doi:10.1016/S0378-4274(01)00312-5

Conn, J. E., Wilkerson, R. C., Segura, M. N. O., de Souza, R. T., Schlichting, C. D., Wirtz, R. A., & Póvoa, M. M. (2002). Emergence of a new neotropical malaria vector facilitated by human migration and changes in land use. *The American Journal of Tropical Medicine and Hygiene, 66*(1), 18–22. doi:10.4269/ajtmh.2002.66.18

Conover, M. R. (2001). *Resolving human-wildlife conflicts: the science of wildlife damage management.* CRC press. doi:10.1201/9781420032581

Conradi, S., Ronnevi, L.O., & Vesteberg, O. (1976). Abnormal tissue distribution of lead in amyotrophic lateral sclerosis. *Journal of the Neurological Sciences, 29*, 259–265.doi: (76)90175-1. doi:10.1016/0022-510X

Conradi, S., Ronnevi, L. O., & Vesterberg, O. (1978). Lead concentration in skeletal muscle in amyotrophic lateral sclerosis patients and control subjects. *Journal of Neurology, Neurosurgery, and Psychiatry, 41*(11), 1001–1004. doi:10.1136/jnnp.41.11.1001 PMID:81872

Cooper, K. (1991). Effects of pesticides on wildlife. In W. J. Hayes (Ed.), *Handbook of Pesticide Toxicology* (pp. 463–496). San Diego, CA: Academic Press.

Corasaniti, M. T., Defilippo, R., Rodinò, P., Nappi, G., & Nisticò, G. (1991). Evidence that paraquat is able to cross the blood-brain barrier to a different extent in rats of various age. *Functional Neurology, 6*(4), 385–391. Retrieved from http://www.ncbi.nlm.nih.gov/pubmed/1810839 PMID:1810839

Corcoran, E., Nellemann, C., Baker, E., Bos, R., Osborn, D., & Savelli, H. (2010). *Sick Water The central role of waste-water management in sustainable development. A Rapid Response Assessment.* UNEP.

Corfield, D. R., & Meadows, G. E. (2006). Control of cerebral blood flow during sleep and the effects of hypoxia. *Advances in Experimental Medicine and Biology, 588*, 65–73. doi:10.1007/978-0-387-34817-9_7 PMID:17089880

Correia, S. C., Santos, R. X., Perry, G., Zhu, X., Moreira, P. I., & Smith, M. A. (2012). Mitochondrial Importance in Alzheimer's, Huntington's and Parkinson's Diseases (Vol. 724). New York, NY: Springer US. doi:10.1007/978-1-4614-0653-2_16

Corrigan, F. M., Wienburg, C. L., Shore, R. F., Daniel, S. E., & Mann, D. (2000). Organochlorine insecticides in substantia nigra in Parkinson's disease. *Journal of Toxicology and Environmental Health. Part A.*, *59*(4), 229–234. doi:10.1080/009841000156907 PMID:10706031

Corriveau, L., & Lajoie, P. (1987). *Les Pesticides les plus utilisés en agriculture au Québec et la santé: recueil des données toxicologiques et épidémiologiques.* Academic Press.

Cortesi, M., Fridman, E., Volkov, A., Shilstein, S., Chechik, R., Breskin, A., ... Raviv, G. (2008). Clinical assessment of the cancer diagnostic value of prostatic zinc: A comprehensive needle-biopsy study. *The Prostate*, *68*(9), 994–1006. doi:10.1002/pros.20766 PMID:18386292

Cory-Schlecta, D. A., & Schaumburg, H. H. (2000). Lead, inorganic. Oxford University Press.

Costa, C. R. X., Pivetta, K. F. L., de Souza, G. R. B., Mazzini-Guedes, R. B., Pereira, S. T. S., & Nogueira, M. R. (2018). Effects of Temperature, Light and Seed Moisture Content on Germination of Euterpe precatoria Palm. *American Journal of Plant Sciences*, *9*(01), 98–106. doi:10.4236/ajps.2018.91009

Costa, L. G., & Aschner, M. (2014). *Manganese in Health and Disease.* Royal Society of Chemistry. doi:10.1039/9781782622383

Costa, L. G., Giordano, G., Guizzetti, M., & Vitalone, A. (2008). Neurotoxicity of pesticides: A brief review. *Frontiers in Bioscience*, *13*(13), 1240. doi:10.2741/2758 PMID:17981626

Costa, L. G., Vitalone, A., Cole, T. B., & Furlong, C. E. (2005). Modulation of paraoxonase (PON1) activity. *Biochemical Pharmacology*, *69*(4), 541–550. doi:10.1016/j.bcp.2004.08.027 PMID:15670573

Costello, L. C., & Franklin, R. B. (2006). The clinical relevance of the metabolism of prostate cancer; zinc and tumor suppression: Connecting the dots. *Molecular Cancer*, *5*(1), 17. doi:10.1186/1476-4598-5-17 PMID:16700911

Coughlin, S. S., Neaton, J. D., & Sengupta, A. (1996). Cigarette smoking as a predictor of death fromprostate cancer in 348,874 men screened for the Multiple Risk Factor Intervention Trial. *American Journal of Epidemiology*, *143*(10), 1002–1006. doi:10.1093/oxfordjournals.aje.a008663 PMID:8629606

Coulon, F., Pelletier, E., Gourhant, L., & Delille, D. (2005). Effects of nutrient and temperature on degradation of petroleum hydrocarbons in contaminated sub-Antarctic soil. *Chemosphere*, *58*(10), 1439–1448. doi:10.1016/j.chemosphere.2004.10.007 PMID:15686763

Couper, J. (1837). On the effects of black oxide of manganese when inhaled into the lungs. *Br Ann Med Pharmacol*, *1*, 41–42.

Couratier, P., Corcia, P., Lautrette, G., Nicol, M., Preux, P. M., & Marin, B. (2016). Epidemiology of amyotrophic lateral sclerosis: A review of literature. *Revue Neurologique*, *172*(1), 37–45. doi:10.1016/j.neurol.2015.11.002 PMID:26727307

Craig, M. H., Kleinschmidt, I., Nawn, J. B., Le Sueur, D., & Sharp, B. L. (2004). Exploring 30 years of malaria case data in KwaZulu-Natal, South Africa: Part I. The impact of climatic factors. *Tropical Medicine & International Health*, *9*(12), 1247–1257. doi:10.1111/j.1365-3156.2004.01340.x PMID:15598256

Craig, M. H., Snow, R. W., & le Sueur, D. (1999). A climate-based distribution model of malaria transmission in sub-Saharan Africa. *Parasitology Today (Personal Ed.)*, *15*(3), 105–111. doi:10.1016/S0169-4758(99)01396-4

Cramer, W. P., & Solomon, A. M. (1993). Climatic classification and future global redistribution of agricultural land. *Climate Research*, *3*, 97–110. doi:10.3354/cr003097

Creppy, E. E. (2002). Update of survey, regulation and toxic effects of mycotoxins in Europe. *Toxicology Letters*, *127*(1-3), 19–28. doi:10.1016/S0378-4274(01)00479-9

Cronin, S., Hardiman, O., & Traynor, B. J. (2007). Ethnic variation in the incidence of ALS: A systematic review. *Neurology*, *68*(13), 1002–1007. doi:10.1212/01.wnl.0000258551.96893.6f PMID:17389304

Crouch, P. J., Hung, L. W., Adlard, P. A., Cortes, M., Lal, V., Filiz, G., … Barnham, K. J. (2009). Increasing Cu bioavailability inhibits A oligomers and tau phosphorylation. *Proceedings of the National Academy of Sciences*, *106*(2), 381–386. 10.1073/pnas.0809057106

Crute, K., Toze, S., Pritchard, D., & Penney, N. (2004). Proposed research to determine risk from microbial pathogens in grains due to land application of biosolids. *Proceeding of Biosolids Specialty II Conference*.

Cruz, L. S., Vargas, R., & Lopes, A. A. (2009). Snakebite envenomation and death in the developing world. *Ethnicity & Disease*, *19*(1), 42. PMID:19484874

Cryan, J. F., Markou, A., & Lucki, I. (2002). Assessing antidepressant activity in rodents: Recent developments and future needs. *Trends in Pharmacological Sciences*, *23*(5), 238–245. doi:10.1016/S0165-6147(02)02017-5 PMID:12008002

Cryan, P. M., & Brown, A. C. (2007). Migration of bats past a remote island offers clues toward the problem of bat fatalities at wind turbines. *Biological Conservation*, *139*(1-2), 1–11. doi:10.1016/j.biocon.2007.05.019

Culleton, R. L., Mita, T., Ndounga, M., Unger, H., Cravo, P. V., Paganotti, G. M., ... Karema, C. (2008). Failure to detect Plasmodium vivax in West and Central Africa by PCR species typing. *Malaria Journal*, *7*(1), 174. doi:10.1186/1475-2875-7-174

Current, W. L., Reese, N. C., Ernst, J. V., Bailey, W. S., Heyman, M. B., & Weinstein, W. M. (1983). Human Cryptosporidiosis in Immunocompetent and Immunodeficient Persons. *The New England Journal of Medicine*, *308*(21), 1252–1257. doi:10.1056/NEJM198305263082102 PMID:6843609

Curriero, F. C., Patz, J. A., Rose, J. B., & Lele, S. (2001). The association between extreme precipitation and waterborne disease outbreaks in the United States, 1948-1994. *American Journal of Public Health*, *91*(8), 1194–1199. doi:10.2105/AJPH.91.8.1194 PMID:11499103

D'Agostino, D. R., Scardigno, A., Lamaddalena, N., & El Chami, D. (2014). Sensitivity analysis of coupled hydro-economic models: Quantifying climate change uncertainty for decision making. *Water Resources Management*, *28*(12), 4303–4318. doi:10.100711269-014-0748-2

D'Aleo, S. (2016). Solar Changes and the Climate. In D. J. Easterbrook (Ed.), *Evidence-Based Climate Science: Data Opposing CO_2 Emissions as the Primary Source of Global Warming* (pp. 263–282). Elsevier. doi:10.1016/B978-0-12-804588-6.00015-X

Dachs, J., Lohmann, R., Ockenden, W. A., Mejanelle, L., Eisenreich, S. J., & Jones, K. C. (2002). Oceanic biogeochemical controls on global dynamics of persistent organic pollutants. *Environmental Science & Technology*, *36*(20), 4229–4237. doi:10.1021/es025724k PMID:12387392

Dalgleish, H. J., Koons, D. N., & Adler, P. B. (2010). Can life-history traits predict the response of forb populations to changes in climate variability? *Journal of Ecology*, *98*(1), 209–217. doi:10.1111/j.1365-2745.2009.01585.x

Dalla Valle, M., Codato, E., & Marcomini, A. (2007). Climate change influence on POPs distribution and fate: A case study. *Chemosphere*, *67*(7), 1287–1295. doi:10.1016/j.chemosphere.2006.12.028 PMID:17258268

Dalle, F., Roz, P., Dautin, G., Di-Palma, M., Kohli, E., Sire-Bidault, C., ... Bonnin, A. (2003). Molecular characterization of isolates of waterborne Cryptosporidium spp. collected during an outbreak of gastroenteritis in South Burgundy, France. Microbiol. *Journal of Clinical Microbiology*, *41*(6), 2690–2693. doi:10.1128/JCM.41.6.2690-2693.2003 PMID:12791906

Daltry, J. C., Ross, T., Thorpe, R. S., & Wüster, W. (1998). Evidence that humidity influences snake activity patterns: A field study of the Malayan pit viper Calloselasmarhodostoma. *Ecography*, *21*(1), 25–34. doi:10.1111/j.1600-0587.1998.tb00391.x

Danforth, M. E., Reisen, W. K., & Barker, C. M. (2015). Extrinsic incubation rate is not accelerated in recent California strains of West Nile virus in Culex tarsalis (Diptera: Culicidae). *Journal of Medical Entomology*, *52*(5), 1083–1089. doi:10.1093/jme/tjv082

Dänicke, S., & Winkler, J. (2015). Invited review: Diagnosis of zearalenone (ZEN) exposure of farm animals and transfer of its residues into edible tissues (carry over). *Food and Chemical Toxicology*, *84*, 225–249. doi:10.1016/j.fct.2015.08.009

Dantzig, P. I. (2009). Breast cancer, dermatofibromas and arsenic. *Indian Journal of Dermatology*, *54*(1), 23–25. doi:10.4103/0019-5154.48981

Darbre, P. D. (2005). Aluminium, antiperspirants and breast cancer. *Journal of Inorganic Biochemistry*, *99*(9), 1912–1919. doi:10.1016/j.jinorgbio.2005.06.001

Darbre, P. D., Mannello, F., & Exley, C. (2013). Aluminium and breast cancer: Sources of exposure, tissue measurements and mechanisms of toxicological actions on breast biology. J. *Journal of Inorganic Biochemistry*, *128*, 257–261. doi:10.1016/j.jinorgbio.2013.07.005

Daryani, A., Ettehad, G. H., Sharif, M., Ghorbani, L., & Ziaei, H. (2008). Prevalence of intestinal parasites in vegetables consumed in Ardabil. Iran. *Food Control*, *19*(8), 790–794. doi:10.1016/j.foodcont.2007.08.004

Dasgupta, S. (2018). Burden of climate change on malaria mortality. *International Journal of Hygiene and Environmental Health*, *221*(5), 782–791. doi:10.1016/j.ijheh.2018.04.003

Davey, J. C., Bodwell, J. E., Gosse, J. A., & Hamilton, J. W. (2007). Arsenic as an endocrine disruptor: Effects of arsenic on estrogen receptor-mediated gene expression *in vivo* and in cell culture. *Toxicological Sciences*, *98*(1), 75–86. doi:10.1093/toxsci/kfm013

David, M., Krishna, P. M., & Sangeetha, J. (2016). Elucidation of impact of heavy metal pollution on soil bacterial growth and extracellular polymeric substances flexibility. *3 Biotech*, *6*(2). doi:10.100713205-016-0475-x

Davidson, E. A., & Summerskill, W. H. J. (1956). Psychiatric Aspects of Liver Disease. *Postgraduate Medical Journal*, *32*(372), 487–494. doi:10.1136/pgmj.32.372.487 PMID:13389012

Davies, P., & Maloney, A. J. F. (1976). Selective loss of central cholinergic neurons in alzheimer's disease. *Lancet*, *308*(8000), 1403. doi:10.1016/S0140-6736(76)91936-X PMID:63862

Davis, L. K., Murr, A. S., Best, D. S., Fraites, M. J., Zorrilla, L. M., Narotsky, M. G., ... Cooper, R. L. (2011). The effects of prenatal exposure to atrazine on pubertal and postnatal reproductive indices in the female rat. *Reproductive Toxicology (Elmsford, N.Y.)*, *32*(1), 43–51. doi:10.1016/j.reprotox.2011.04.004

Davis, S., Mirick, D. K., & Stevens, R. G. (2001). Night shift work, light at night, and risk of breast cancer. *Journal of the National Cancer Institute*, *93*(20), 1557–1562. doi:10.1093/jnci/93.20.1557

Davoren, M. J., & Schiestl, R. H. (2018). Glyphosate-based herbicides and cancer risk: A post-IARC decision review of potential mechanisms, policy and avenues of research. *Carcinogenesis*, *39*(10), 1207–1215. doi:10.1093/carcin/bgy105 PMID:30060078

Day, J. F., Edman, J. D., Kunz, S. E., & Wikel, S. K. (2004). Direct injury: phobias, psychoses, annoyance, allergies, toxins, venoms and myiasis. In *Medical Entomology* (pp. 99–149). Dordrecht: Springer. doi:10.1007/978-94-007-1009-2_4

Day, M., Wang, Z., Ding, J., An, X., Ingham, C. A., Shering, A. F., ... Surmeier, D. J. (2006). Selective elimination of glutamatergic synapses on striatopallidal neurons in Parkinson disease models. *Nature Neuroscience*, *9*(2), 251–259. doi:10.1038/nn1632 PMID:16415865

Day, S. M., Wu, Y. W., Strauss, D. J., Shavelle, R. M., & Reynolds, R. J. (2005). Causes of death in remote symptomatic epilepsy. *Neurology*, *65*(2), 216–222. doi:10.1212/01.wnl.0000169018.44950.68 PMID:16043789

de Fraiture, C., Smakhtin, V., Bossio, D., McCornick, P., Hoanh, C., Noble, A., ... Turral, H. (2008). *Facing Climate Change by Securing Water for Food, Livelihoods and Ecosystems*. Colombo, Sri Lanka: International Water Management Institute.

de Lau, L. M., & Breteler, M. M. (2006). Epidemiology of Parkinson's disease. *Lancet Neurology*, *5*(6), 525–535. doi:10.1016/S1474-4422(06)70471-9 PMID:16713924

De Marzo, A. M., Marchi, V. L., Epstein, J. I., & Nelson, W. G. (1999). Proliferative inflammatory atrophy of the prostate: Implications for prostatic carcinogenesis. *American Journal of Pathology*, *155*(6), 1985–1992. doi:10.1016/S0002-9440(10)65517-4 PMID:10595928

de Rijk, M. C., Launer, L. J., Berger, K., Breteler, M. M., Dartigues, J. F., Baldereschi, M., … Hofman, A. (2000). Prevalence of Parkinson's disease in Europe: A collaborative study of population-based cohorts. Neurologic Diseases in the Elderly Research Group. *Neurology*, *54*(11), S21-3. Retrieved from http://www.ncbi.nlm.nih.gov/pubmed/10854357

de Rijk, M. C., Breteler, M. M., Graveland, G. A., Ott, A., Grobbee, D. E., van der Meché, F. G., & Hofman, A. (1995). Prevalence of Parkinson's disease in the elderly: The Rotterdam Study. *Neurology*, *45*(12), 2143–2146. doi:10.1212/WNL.45.12.2143 PMID:8848182

de Roodt, A. R., Lago, N. R., Salomón, O. D., Laskowicz, R. D., de Román, L. E. N., López, R. A., ... Vega, V. D. V. (2009). A new venomous scorpion responsible for severe envenomation in Argentina: Tityus confluens. *Toxicon*, *53*(1), 1–8. doi:10.1016/j.toxicon.2008.10.003 PMID:18983868

De Ruyck, K., De Boevre, M., Huybrechts, I., & De Saeger, S. (2015). Dietary mycotoxins, co-exposure, and carcinogenesis in humans: Short review. *Mutation Research/Reviews in Mutation Research*, *766*, 32–41. doi:10.1016/j.mrrev.2015.07.003

De Vega, G. R., Fernández-Sánchez, M. L., Pisonero, J., Eiró, N., Vizoso, F., & Sanz-Medel, A. (2017). Quantitative Bioimaging of Ca, Fe, Cu and Zn in Breast Cancer Tissues by LA-ICP-MS. *Journal of Analytical Atomic Spectrometry*, *32*(3), 671–677. doi:10.1039/C6JA00390G

de Vries, Y. A., Roest, A. M., Burgerhof, J. G. M., & de Jonge, P. (2018). Initial severity and antidepressant efficacy for anxiety disorders, obsessive-compulsive disorder, and posttraumatic stress disorder: An individual patient data meta-analysis. *Depression and Anxiety*, 1–8. doi:10.1002/da.22737 PMID:29659102

De Wit, C. A., Herzke, D., & Vorkamp, K. (2010). Brominated flame retardants in the Arctic environment trends and new candidates. *The Science of the Total Environment*, *408*(15), 2885–2918. doi:10.1016/j.scitotenv.2009.08.037

Dean, B., Lam, L. Q., Scarr, E., & Duce, J. A. (2019). Cortical biometals: Changed levels in suicide and with mood disorders. *Journal of Affective Disorders*, *243*, 539–544. doi:10.1016/j.jad.2018.09.026 PMID:30292148

Dear, K., Woodruff, R., Hanigan, I. and McMichael, A. (2007) The impacts of climate change on three health outcomes: temperature-related mortality and hospitalisations, salmonellosis and other bacterial gastroenteritis, and population at risk from dengue. *Garnaut Climate Change Review*, 1-47.

Dehesa-Dávila, M. (1989). Epidemiological characteristics of scorpion sting in Leon, Guanajuato, Mexico. *Toxicon, 27*(3), 281–286. doi:10.1016/0041-0101(89)90175-X PMID:2728020

Dehesa-Dávila, M., & Possani, L. D. (1994). Scorpionism and serotherapy in Mexico. *Toxicon, 32*(9), 1015–1018. doi:10.1016/0041-0101(94)90383-2 PMID:7801335

del Blanco, B., & Barco, A. (2018). Impact of environmental conditions and chemicals on the neuronal epigenome. *Current Opinion in Chemical Biology, 45*, 157–165. doi:10.1016/j.cbpa.2018.06.003 PMID:29913328

Del Vento, S., & Dachs, J. (2007). Atmospheric occurrence and deposition of polycyclic aromatic hydrocarbons in the northeast tropical and subtropical Atlantic Ocean. *Environmental Science & Technology, 41*(16), 5608–5613. doi:10.1021/es0707660 PMID:17874762

Delcour, I., Spanoghe, P., & Uyttendaele, M. (2015). Literature review : Impact of climate change on pesticide use. *Food Research International Journal, 68*, 7–15. doi:10.1016/j.foodres.2014.09.030

Delille, D., Bassères, A., Dessommes, A., & Rosiers, C. (1998). Influence of daylight on potential biodegradation of diesel and crude oil in Antarctic seawater. *Marine Environmental Research, 45*(3), 249–258. doi:10.1016/S0141-1136(97)00129-3

DeLorenzo, M. E., Wallace, S. C., Danese, L. E., & Baird, T. D. (2009). Temperature and salinity effects on the toxicity of common pesticides to the grass shrimp, *Palaemonetes pugio. Journal of Environmental Science and Health. Part. B, Pesticides, Food Contaminants, and Agricultural Wastes, 44*(5), 455–460. doi:10.1080/03601230902935121 PMID:20183050

Delpla, I., Jung, A. V., Baures, E., Clement, M., & Thomas, O. (2009). Impacts of climate change on surface water quality in relation to drinking water production. *Environment International, 35*(8), 1225–1233. doi:10.1016/j.envint.2009.07.001 PMID:19640587

DeMaagd, G., & Philip, A. (2015). Parkinson's Disease and Its Management: Part 1: Disease Entity, Risk Factors, Pathophysiology, Clinical Presentation, and Diagnosis. *P & T : A Peer-Reviewed Journal for Formulary Management, 40*(8), 504–532. Retrieved from http://www.ncbi.nlm.nih.gov/pubmed/26236139

Demers, A., Ayotte, P., Brisson, J., Dodin, S., Robert, J., & Dewailly, E. (2000). Risk and aggressiveness of breast cancer in relation to plasma organochlorine concentration. *Cancer Epidemiology, Biomarkers & Prevention, 9*, 161–166.

Demers, A., Ayotte, P., Brisson, J., Dodin, S., Robert, J., & Dewailly, E. (2002). Plasma concentrations of polychlorinated biphenyls and the risk of breast cancer: A congener-specific analysis. *American Journal of Epidemiology, 155*(7), 629–635. doi:10.1093/aje/155.7.629

Dent, R., Trudeau, M., Pritchard, K. I., Hanna, W. M., Kahn, H. K., Sawka, C. A., ... Narod, S. A. (2007). Triple-negative breast cancer: Clinical features and patterns of recurrence. *Clinical Cancer Research, 13*(15), 4429–4434. doi:10.1158/1078-0432.CCR-06-3045

Desantis, C., Siegel, R., & Jemal, A. (2014). Cancer Treatment & Survivorship Facts and Figures 2014 & 2015. American Cancer Society.

Desaulniers, D., Leingartner, K., Musicki, B., Cole, J., Li, M., Charbonneau, M., & Tsang, B. K. (2004). Lack of effects of postnatal exposure to a mixture of aryl hydrocarbon-receptor agonists on the development of methylnitrosourea-induced mammary tumors in Sprague- Dawley rats. *Journal of Toxicology and Environmental Health. Part A., 67*(18), 1457–1475. doi:10.1080/15287390490483818

Deshmukh, R. S., Chaudhary, R. K., & Roy, I. (2012). Effect of Pesticides on the Aggregation of Mutant Huntingtin Protein. *Molecular Neurobiology, 45*(3), 405–414. doi:10.100712035-012-8252-2 PMID:22415443

Desjeux, P. (2004). – Leishmaniasis: Current situation and new perspectives. *Comparative Immunology, Microbiology and Infectious Diseases*, *27*(5), 305–318. doi:10.1016/j.cimid.2004.03.004 PMID:15225981

Deutch, A.Y., Colbran, R.J., & Winder, D.J. (2007). Striatal plasticity and medium spiny neuron dendritic remodeling in Parkinsonism. *Parkinsonism Related Disorders, 13*, S251 S258.

Devinsky, O., Vezzani, A., Jette, N., Curtis, M. De, & Perucca, P. (2018). Epilepsy. *Nature Reviews, 3*. doi:10.1038/nrdp.2018.24

Dexter, D. T., Carayon, A., Javoy-Agid, F., Agid, Y., Wells, F. R., Daniel, S. E., ... Marsden, C. D. (1991). Alterations in the levels of iron, ferritin and other trace metals in Parkinson's disease and other neurodegenerative diseases affecting the basal ganglia. *Brain*, *114*(Pt 4), 1953–1975. doi:10.1093/brain/114.4.1953 PMID:1832073

Dexter, D. T., Wells, F. R., Agid, F., Agid, Y., Lees, A. J., Jenner, P., & Marsden, C. D. (1987). Increased nigral iron content in postmortem parkinsonian brain. *Lancet*, *330*(8569), 1219–1220. doi:10.1016/S0140-6736(87)91361-4 PMID:2890848

Dich, J., & Wiklund, K. (1998). Prostate cancer in pesticide applicators in swedish agriculture. *The Prostate*, *34*(2), 100–112. doi:10.1002/(SICI)1097-0045(19980201)34:2<100::AID-PROS4>3.0.CO;2-O PMID:9465941

Dich, J., Zahm, S., Hanberg, A., Adami, H., & Adami, H.-O. (1997). Pesticides and cancer. *Cancer Causes & Control*, *8*(3), 420–443. doi:10.1023/A:1018413522959

Dick, F. D. (2006). Parkinson's disease and pesticide exposures. *British Medical Bulletin*, *79–80*(1), 219–231. doi:10.1093/bmb/ldl018 PMID:17242039

Dickhut, R. M., & Gustafson, K. E. (1995). Atmospheric washout of polycyclic aromatic hydrocarbons in the southern chesapeake bay region. *Environmental Science & Technology*, *29*(6), 1518–1525. doi:10.1021/es00006a013 PMID:22276872

Dickin, S. K., Schuster-Wallace, C. J., Qadir, M., & Pizzacalla, K. (2016). A Review of Health Risks and Pathways for Exposure to Wastewater Use in Agriculture. *Environmental Health Perspectives*, *124*(7). doi:10.1289/ehp.1509995 PMID:26824464

Dietert, R. R., & Piepenbrink, M. S. (2006). Lead and immune function. *Critical Reviews in Toxicology*, *36*(4), 359–385. doi:10.1080/10408440500534297 PMID:16809103

Dietrich, W. D., Alonso, O., & Busto, R. (1993). Moderate Hyperglycemia Worsens Acute Blood-Brain Barrier Injury After Forebrain Ischemia in Rats. *Stroke*, 24. PMID:8418533

DiFiglia, M., Sapp, E., Chase, K., Schwarz, C., Meloni, A., Young, C., ... Aronin, N. (1995). Huntingtin is a cytoplasmic protein associated with vesicles in human and rat brain neurons. *Neuron*, *14*(5), 1075–1081. doi:10.1016/0896-6273(95)90346-1 PMID:7748555

Dinis-Oliveira, R. J., Remião, F., Carmo, H., Duarte, J. A., Navarro, A. S., Bastos, M. L., & Carvalho, F. (2006). Paraquat exposure as an etiological factor of Parkinson's disease. *Neurotoxicology*, *27*(6), 1110–1122. doi:10.1016/j.neuro.2006.05.012 PMID:16815551

Direction Regionale de l'Oriental, H. C. (2017). *La région de l'Oriental en.* Haut Commissariat Au Plan Direction Regionale de l'Oriental.

Divya, R. T., Sunil, B., & Latha, C. (2011). Physico- Chemical Analysis of Well Water at Eloor Industrial Area-Seasonal Study. *Current World Environment*, *6*(2), 259.

Dodds, W. K. (2006). Eutrophication and trophic state in rivers and streams. *Limnology and Oceanography*, *51*(1part2), 671–680. doi:10.4319/lo.2006.51.1_part_2.0671

Doha, S. A., & Samy, A. M. (2010). Bionomics of phlebotomine sand flies (Diptera: Psychodidae) in the province of Al-Baha, Saudi Arabia. *Memorias do Instituto Oswaldo Cruz*, *105*(7), 850–856. doi:10.1590/S0074-02762010000700002 PMID:21120352

Doi, H., Kikuchi, H., Murai, H., Kawano, Y., Shigeto, H., Ohyagi, Y., & Kira, J. (2006). Motor neuron disorder simulating ALS induced by chronic inhalation of pyrethroid insecticides. *Neurology*, *67*(10), 1894–1895. doi:10.1212/01.wnl.0000244489.65670.9f PMID:17130437

Dokhan, M. R., Kenawy, M. A., Doha, S. A., El-Hosary, S. S., Shaibi, T., & Annajar, B. B. (2016). Entomological studies of phlebotomine sand flies (Diptera: Psychodidae) in relation to cutaneous leishmaniasis transmission in Al Rabta, North West of Libya. *Acta Tropica*, *154*, 95–101. doi:10.1016/j.actatropica.2015.11.004 PMID:26589378

Dollear, F. G., & Goldblatt, L. A. (1969). *Aflatoxin, scientific background, control and implications*. New York: Academic Press.

Döll, P., & Siebert, S. (2002). Global modeling of irrigation water requirements. *Water Resources Research*, *38*(4), 8-1–8-10. doi:10.1029/2001WR000355

Domingo, J. L. (2012). Polybrominated diphenyl ethers in food and human dietary exposure: A review of the recent scientific literature. *Food and Chemical Toxicology*, *50*(2), 238–249. doi:10.1016/j.fct.2011.11.004

Donald, D. B., Cessna, A. J., Sverko, E., & Glozier, N. E. (2007). Pesticides in surface drinking-water supplies of the Northern Great Plains. *Environmental Health Perspectives*, *115*(8), 1183–1191. doi:10.1289/ehp.9435 PMID:17687445

Dongus, S. (2016). *Dynamics and sustainability of urban agriculture : Examples from sub-Saharan Africa*. Academic Press. doi:10.100711625-009-0097-x

Dong, W., Jiang, Y., & Yang, S. (2010). Response of the starting dates and the lengths of seasons in Mainland China to global warming. *Climatic Change*, *99*(1-2), 81–91. doi:10.100710584-009-9669-0

Donner, S. (2003). The impact of cropland cover on river nutrient levels in the Mississippi River Basin. *Global Ecology and Biogeography*, *12*(4), 341–355. doi:10.1046/j.1466-822X.2003.00032.x

Donner, S. D., & Kucharik, C. J. (2003). Evaluating the impacts of land management and climate variability on crop production and nitrate export across the Upper Mississippi Basin. *Global Biogeochemical Cycles*, *17*(3), n/a. doi:10.1029/2001GB001808

Dorgan, J. F., Brock, J. W., Rothman, N., Needham, L. L., Miller, R., Stephenson, H. E. Jr, ... Taylor, P. R. (1999). Serum organochlorine pesticides and PCBs and breast cancer risk: Results from a prospective analysis (USA). *Cancer Causes & Control*, *10*(1), 1–11. doi:10.1023/A:1008824131727

Dorsey, C. D., Lee, B.-K., Bolla, K. I., Weaver, V. M., Lee, S.-S., Lee, G.-S., ... Schwartz, B. S. (2006). Comparison of Patella Lead With Blood Lead and Tibia Lead and Their Associations With Neurobehavioral Test Scores. *Journal of Occupational and Environmental Medicine*, *48*(5), 489–496. doi:10.1097/01.jom.0000199678.86629.3b PMID:16688005

Dosemeci, M., Cocco, P., Gómez, M., Stewart, P. A., & Heineman, E. F. (1994). Effects of three features of a job-exposure matrix on risk estimates. *Epidemiology (Cambridge, Mass.)*, *5*(1), 124–127. doi:10.1097/00001648-199401000-00019 PMID:8117771

Dosemeci, M., Hoover, N., Blair, A., Figgs, W., Devesa, S., Grauman, D., ... Carolina, S. (1994). BRIEF Farming and Prostate Cancer for a grouping of three southeastern. *American Journal of Industrial Medicine*, *86*(22).

Douglas, E. M., Vogel, R. M., & Kroll, C. N. (2000). Trends in floods and low fows in the United States : Impact of spatial correlation. *Journal of Hydrology (Amsterdam)*, *240*(1-2), 90–105. doi:10.1016/S0022-1694(00)00336-X

Doull, J. (1972). The effect of physical environmental factors on drug response. In W. J. Hayes (Ed.), *Essays in Toxicology* (pp. 37–63). Elsevier.

Dowell, S. F. (2001). Seasonal variation in host susceptibility and cycles of certain infectious diseases. *Emerging Infectious Diseases*, *7*(3), 369–374. doi:10.3201/eid0703.017301 PMID:11384511

Driouech, F., Déqué, M., & Mokssit, A. (2009). Numerical simulation of the probability distribution function of precipitation over Morocco. *international journal. Climate Dynamics*, *32*(7-8), 1055–1063. doi:10.100700382-008-0430-6

Droogers, P., & Aerts, J. (2005). Adaptation strategies to climate change and climate variability: A comparative study between seven contrasting river basins. *Physics and Chemistry of the Earth Parts A/B/C*, *30*(6-7), 339–346. doi:10.1016/j.pce.2005.06.015

Drudge-Coates, L., & Turner, B. (2013). *Parasitic waterborne disease.* Academic Press.

Dssouli, K. (2006). *Treatment and reuse of wastewater in agriculture Marocoriental: Study of helminth parasites* (PhD in Social Sciences). Faculty of Sciences, Oujda.

Dssouli, K., El Halouani, H., & Berrichi, A. (2006). The health impact of the reuse of raw sewage from the city of Oujda in agriculture: A study of the parasite load in helminth eggs in some vegetable crops. *Biology & Health*, *6*, 1.

Duffus, J. H. (1997). *Carcinogenicity of Inorganic Substances: Risks from Occupational Exposure.* Woodhead Publishing.

Dunnwald, L. K., Rossing, M. A., & Li, C. I. (2007). Hormone receptor status, tumor characteristics, and prognosis: A prospective cohort of breast cancer patients. *Breast Cancer Research*, *9*(1), R6. doi:10.1186/bcr1639

Durack, P. J., Wijffels, S. E., & Matear, R. J. (2012). Ocean salinities reveal strong global water cycle intensification during 1950 to 2000. *Science*, *336*(6080), 455–458. doi:10.1126cience.1212222 PMID:22539717

Durr, A. (2010). Autosomal dominant cerebellar ataxias: Polyglutamine expansions and beyond. *Lancet Neurology*, *9*, 885–894.

Duvoisin, R. C. (1986). Etiology of Parkinson's disease: Current concepts. *Clinical Neuropharmacology*, *9*(Suppl 1), S3–S21. doi:10.1097/00002826-198609001-00003 PMID:2423234

Duyckaerts, C., Hauw, J.-J., Bastenaire, F., Piette, F., Poulain, C., Rainsard, V., ... Berthaux, P. (1986). Laminar distribution of neocortical senile plaques in senile dementia of the alzheimer type. *Acta Neuropathologica*, *70*(3–4), 249–256. doi:10.1007/BF00686079 PMID:3766125

Dyatlov, V. A., Platoshin, A. V., Lawrence, D. A., & Carpenter, D. O. (1998). Lead potentiates cytokine- and glutamate-mediated increases in permeability of the blood-brain barrier. *Neurotoxicology*, *19*(2), 283–291. Retrieved from http://www.ncbi.nlm.nih.gov/pubmed/9553965 PMID:9553965

Easterling, D. R., Meehl, G. A., Parmesan, C., Changnon, S. A., Karl, T. R., & Mearns, L. O. (2000). Climate extremes: observations, modeling, and impacts. *Science, 289*(5487), 2068-2074.

Easterling, W. E., Aggarwal, P. K., Batima, P., Brander, K. M., Erda, L., Howden, S. M., . . . Tubiello, F. N. (2007). Food, Fibre and Forest Products. In Climate Change 2007: Impacts, Adaptation and Vulnerability. Contribution of Working Group II to the Fourth Assessment Report of the Intergovernmental Panel on Climate Change (IPCC), Cambridge University Press.

Eaton, D. L., & Gallagher, E. P. (1994). Mechanisms of aflatoxins carcinogenesis. *Annual Review of Pharmacology and Toxicology*, *34*(1), 135–172. doi:10.1146/annurev.pa.34.040194.001031

Eaton, W. W., Kramer, M., Anthony, J. C., Dryman, A., Shapiro, S., & Locke, B. Z. (1989). The incidence of specific DIS/DSM-III mental disorders: Data from the NIMH Epidemiologic Catchment Area Program. *Acta Psychiatrica Scandinavica*, *79*(2), 163–178. doi:10.1111/j.1600-0447.1989.tb08584.x PMID:2784251

Ebi, K. L., Hartman, J., Chan, N., Mcconnell, J., Schlesinger, M., & Weyant, J. (2005). Climate suitability for stable malaria transmission in Zimbabwe under different climate change scenarios. *Climatic Change*, *73*(3), 375–393. doi:10.100710584-005-6875-2

Ebi, K. L., & Paulson, J. A. (2010). Climate change and child health in the United States. *Current Problems in Pediatric and Adolescent Health Care*, *40*(1), 2–18. doi:10.1016/j.cppeds.2009.12.001 PMID:20114000

Eckelman, B. P., Salvesen, G. S., & Scott, F. L. (2006). Human inhibitor of apoptosis proteins: Why XIAP isthe black sheep of the family. *EMBO Reports*, *7*(10), 988–994. doi:10.1038j.embor.7400795 PMID:17016456

Eddy, J. A. (2010). *The Sun, the Earth, and Near-Earth Space: A Guide to the Sun-Earth System.* NASA.

Ediriweera, D. S., Kasturiratne, A., Pathmeswaran, A., Gunawardena, N. K., Wijayawickrama, B. A., Jayamanne, S. F., ... Lalloo, D. G. (2016). Mapping the risk of snakebite in Sri Lanka-a national survey with geospatial analysis. *PLoS Neglected Tropical Diseases*, *10*(7), e0004813. doi:10.1371/journal.pntd.0004813 PMID:27391023

Eeles, R. A., Kote-Jarai, Z., Giles, G. G., Al Olama, A. A., Guy, M., Jugurnauth, S. K., ... Field, H. I. (2008). Multiple newly identified loci associated with prostate cancer susceptibility. *Nature Genetics*, *40*(3), 316–321. doi:10.1038/ng.90 PMID:18264097

Efstratiou, A., Ongerth, J. E., & Karanis, P. (2017). Waterborne transmission of protozoan parasites : Review of worldwide outbreaks - An update 2011 e 2016. *Water Research*, *114*, 14–22. doi:10.1016/j.watres.2017.01.036 PMID:28214721

Eganhouse, R. P., & Calder, J. A. (1976). The solubility of medium molecular weight aromatic hydrocarbons and the effects of hydrocarbon co-solutes and salinity. *Geochimica et Cosmochimica Acta*, *40*(5), 555–561. doi:10.1016/0016-7037(76)90223-4

Egorov, A. I., Naumova, E. N., Tereschenko, A. A., Kislitsin, V. A., & Ford, T. E. (2003). Daily variations in effluent water turbidity and diarrhoeal illness in a Russian city. *International Journal of Environmental Health Research*, *13*(1), 81–94. doi:10.1080/0960312021000071567 PMID:12745350

Eisenberg, J. N., Desai, M. A., Levy, K., Bates, S. J., Liang, S., Naumoff, K., & Scott, J. C. (2007). Environmental determinants of infectious disease: A framework for tracking causal links and guiding public health research. *Environmental Health Perspectives*, *115*(8), 1216–1223. doi:10.1289/ehp.9806 PMID:17687450

Eisenberg, J. N., Seto, E. Y., Colford, J. M. Jr, Olivieri, A., & Spear, R. C. (1998). An analysis of the Milwaukee cryptosporidiosis outbreak based on a dynamic model of the infection process. *Epidemiology (Cambridge, Mass.)*, *9*(3), 255–263. doi:10.1097/00001648-199805000-00008 PMID:9583416

Eisen, L., Monaghan, A. J., Lozano-Fuentes, S., Steinhoff, D. F., Hayden, M. H., & Bieringer, P. E. (2014). The impact of temperature on the bionomics of Aedes (Stegomyia) aegypti, with special reference to the cool geographic range margins. *Journal of Medical Entomology*, *51*(3), 496–516. doi:10.1603/ME13214

Ekmekyapar, F., Sabudak, T., & Seren, G. (2012). Assessment of heavy metal contamination in soil and wheat (Triticum aestivum L.) plant around the Corlu-Cerkezkoy highway in Thrace region. *Global NEST Journal*, *14*(4), 496.

El Aminn, E. O., & Berair, R. (1995). Piqûres de scorpion chez l'enfant. Expérience saoudienne. *Archives de Pediatrie*, *2*(8), 766–773. doi:10.1016/0929-693X(96)81248-6 PMID:7550843

El Amiri, B., Sibaoueih, M., Hadarbach, D., Cohen, N., & Chriyaa, A. (2009). Impact of wastewater use on cattle in the common Mzamza, Province of Settat. *Laboratory Technology, 17*.

El Hidan, M. A., Touloun, O., Bouazza, A., Laaradia, M. A., & Boumezzough, A. (2018). Androctonus genus species in arid regions: Ecological niche models, geographical distributions, and envenomation risk. *Veterinary World, 11*(3), 286–292. doi:10.14202/vetworld.2018.286-292 PMID:29657418

El Hidan, M. A., Touloun, O., & Boumezzough, A. (2015). An epidemiological study on scorpion envenomation in the Zagora oases (Morocco). *Journal of Coastal Life Medicine, 3*(9), 704–707. doi:10.12980/jclm.3.2015j5-86

El-Bayoumy, K., Chae, Y. H., Upadhyaya, P., Meschter, C., Cohen, L. A., & Reddy, B. S. (1992). Inhibition of 7,12-dimethylbenz(a)anthracene-induced tumors and DNA adduct formation in the mammary glands of female Sprague-Dawley rats by the synthetic organoselenium compound, 1,4- phenylenebis(methylene)selenocyanate. *Cancer Research, 52*, 2402–2407. PMID:1568209

El-Borie, M. A., Abd-Elzaher, M., & Al Shenawy, A. (2012). Solar and Geomagnetic Activity Effects on Global Surface Temperatures. *American Journal of Environmental Engineering, 2*(4), 80–85. doi:10.5923/j.ajee.20120204.02

El-Din, H. G., Ghafar, N. A., Saad, N. E., Aziz, M., Rasheed, D., & Hassan, E. M. (2010). Relationship between codon 249 mutation in exon 7 of p53 gene and diagnosis of hepatocellular carcinoma. *Archives of Medical Science : AMS, 6*(3), 348–355. doi:10.5114/aoms.2010.14254 PMID:22371770

Eldridge, S. R., Gould, M. N., & Butterworth, B. E. (1992). Genotoxicity of environmental agents in human mammary epithelial cells. *Cancer Research, 52*, 5617–5621.

El-Fawal, H. A., Waterman, S. J., De Feo, A., & Shamy, M. Y. (1999). Neuroimmunotoxicology: Humoral assessment of neurotoxicity and autoimmune mechanisms. *Environmental Health Perspectives, 107*(Suppl 5), 767–775. PMID:10502543

Elghany, N. A., Schumacher, M. C., Slattery, M. L., West, D. W., & Lee, J. S. (1990). Occupation, cadmiumexposure, and prostate cancer. *Epidemiology (Cambridge, Mass.), 1*(2), 107–115. doi:10.1097/00001648-199003000-00005 PMID:2073496

Elhafi, A. (2012). *Changement Climatique et impact sur la sécurité alimentaire de la région d'Afrique du Nord*. Academic Press.

El-Harouny, M. A., El-Morsi, D. A., Ahmed, B. R. A., & El-Atta, H. M. A. (2011). Chronic Toxicity of Some Heavy Metals and Breast Cancer in Egyptian Females. *Journal of Clinical Toxicology, 1*, 109.

Elissalde, M. H., & Clark, D. E. (1979). Testosterone metabolism by hexachlorobenzene-induced hepatic microsomal enzymes. *American Journal of Veterinary Research, 40*(12), 1762–1766.

Elroy-Stein, O., Bernstein, Y., & Groner, Y. (1986). Overproduction of human Cu/Zn-superoxide dismutase in transfected cells: Extenuation of paraquat-mediated cytotoxicity and enhancement of lipid peroxidation. *The EMBO Journal, 5*(3), 615–622. doi:10.1002/j.1460-2075.1986.tb04255.x PMID:3011416

Elsner, J. B., & Kavlakov, S. P. (2001). Hurricane intensity changes associated geomagnetic variation. *Atmospheric Science Letters, 2*(1-4), 86–93. doi:10.1006/asle.2001.0040

Emre, M., Aarsland, D., Albanese, A., Byrne, E. J., Deuschl, G., De Deyn, P. P., ... Lane, R. (2004). Rivastigmine for Dementia Associated with Parkinson's Disease. *The New England Journal of Medicine, 351*(24), 2509–2518. doi:10.1056/NEJMoa041470 PMID:15590953

Eniang, E. A., Akani, G. C., Rugiero, L., Vignoli, L., & Luiselli, L. (2013). Ecological data of Nigerian Crotaphopeltishotamboeia (Colubridae) populations. *The Herpetological Journal, 23*(1), 5–9.

Ensink, J. H. J., Blumenthal, U. J., & Brooker, S. (2009). *Europe PMC Funders Group Wastewater quality and the risk of intestinal nematode infection in sewage farming families in Hyderabad, India*. Academic Press.

Ensink, J. H. J., Mahmood, T., van der Hoek, W., Raschid-Sally, L., & Amerasinghe, F. P. (2004). A nation-wide assessment of wastewater use in Pakistan: An obscure activity or a vitally important one? *Water Policy*, *6*(3), 197–206. doi:10.2166/wp.2004.0013

Environmental Protection Agency. (2000). *Exposure and human health reassessment of 2,3,7,8-tetrachlorodibenzo-p-dioxin (TCDD) and related compounds. Part II: Health assessment for 2,3,7,8-tetrachlorodibenzo- p-dioxin (TCDD) and related compounds*. Retrieved from http://www.epa.gov/ncea/pdfs/dioxin/nasreview/pdfs/part2/dioxin_pt2_ch08_dec2003.pdf

Environmental Protection Agency. (2009). *Atrazine science reevaluation: Potential health impacts document*. Retrieved from http://www. epa.gov/pesticides/reregistration/atrazine/atrazine_update.htm

Environmental Protection Agency. (2010). *Atrazine updates*. Retrieved from http://www.epa.gov/opp00001/reregistration/atrazine/ atrazine_update.htm#cancer

Epstein, M. M., Kasperzyk, J. L., Andren, O., Giovannucci, E. L., Wolk, A., Hakansson, N., ... Mucci, L. A. (2011). Dietary zinc and prostate cancer survival in a Swedish cohort. *The American Journal of Clinical Nutrition*, *93*(3), 586–593.

Eriksen, S., & Kelly, P. (2007). Developing Credible Vulnerability Indicators for Climate Adaptation Policy Assessment. *Mitigation and Adaptation Strategies for Global Change*, *12*(4), 495–524. doi:10.100711027-006-3460-6

Escobedo, A. A., Almirall, P., Robertson, J., Franco, R. M. B., Hanevik, K., Mørch, K., & Cimerman, S. (2010). *Giardiasis : The Ever-Present Threat of a Neglected Disease, (Ci)*. Academic Press.

Escrivá, L., Font, G., Manyes, L., & Berrada, H. (2017). Studies on the Presence of Mycotoxins in Biological Samples: An Overview. *Toxins*, *9*(8). doi:10.3390/toxins9080251 PMID:28820481

Escudero Augusto, D., Marqués Álvarez, L., & Taboada Costa, F. (2008). Actualización en hemorragia cerebral espontánea. *Medicina Intensiva*, *32*(6), 282–295. doi:10.1016/S0210-5691(08)70956-2 PMID:18601836

Esper, J., Franck, D., B¨untgen, U., Verstege, A., Luterbacher, J., & Xoplaki, E. (2007). Long-term drought severity variations in Morocco. *Geophysical Research Letters*, *34*(17), L17702. doi:10.1029/2007GL030844

Etewa, S. E., & Naglaa, S. A. A. (2016). Geohelminths distribution as affected by soil properties, physicochemical factors and climate in Sharkyia governorate. *Journal of Parasitic Diseases: Official Organ of the Indian Society for Parasitology*, *40*(2), 496–504. doi:10.100712639-014-0532-5 PMID:27413327

EU. (2009). *Common Implementation Strategy for the Water Framework Directive (2000/60/EC) Guidance document No. 24: River Basin Management in a Changing Climate (Technical Report — 2009–040)*. EU.

European Association for Study of Liver. (2012). EASL Clinical Practice Guidelines: Wilson's disease. *Journal of Hepatology*, *56*(3), 671–685. doi:10.1016/j.jhep.2011.11.007 PMID:22340672

European Commission. (2006). *Commission Regulation No. 1881/2006 of December 19th setting maximum levels of certain contaminants in foodstuffs*. Official Journal of the European Union No. L364/5.

European Commission. (2010). Commission Regulation (EC) No 165/2010 of 26 February 2010 amending Regulation (EC) No 1881/2006 setting maximum levels for certain contaminants in foodstuffs as regards aflatoxins. *Official Journal of the European Union, L*, *50*(8).

European Commission. (2012). Commission Regulation (EC) No 1058/2012 of 12 November 2012 amending Regulation (EC) No 1881/2006 setting maximum levels for certain contaminants in foodstuffs as regards maximum levels for aflatoxins in dried figs. *Official Journal of the European Union, L, 313*(14).

European Food Safety Authority (EFSA). (2006). Opinion of the scientific panel on contaminants in the food chain on a request from the commission related to ochratoxin A in food (Question No. EFSA-Q-2005-154). The EFSA Journal, 365.

Evans, G. W. (2019). Projected Behavioral Impacts of Global Climate Change. *Annual Review of Psychology, 70*(1). doi:10.1146/annurev-psych-010418-103023

Exley, C., & Esiri, M. M. (2006). Severe cerebral congophilic angiopathy coincident with increased brain aluminium in a resident of Camelford, Cornwall, UK. *Journal of Neurology, Neurosurgery, and Psychiatry, 77*(7), 877–879. doi:10.1136/jnnp.2005.086553 PMID:16627535

Eze, J. I., Scott, E. M., Pollock, K. G., Stidson, R., Miller, C. A., & Lee, D. (2013). The association of weather and bathing water quality on the incidence of gastrointestinal illness in the west of Scotland. *Epidemiology and Infection, 142*(06), 1289–1299. doi:10.1017/S0950268813002148 PMID:24007797

Eze, U. A., Routledge, M. N., Okonofua, F. E., Huntriss, J., & Gong, Y. Y. (2018). Mycotoxin exposure and adverse reproductive health outcomes in Africa: A review. *World Mycotoxin Journal*, 1–20.

Ezzine, H. (2015). *Regional Analysis of Desaster Loss Databases in Arab States.* Cairo, Egypt: UNISDR.

Fabre, I., & Gardiès, C. (2008). L'accès à l'information scientifique numérique: Organisation des savoirs et enjeu de pouvoir dans une communauté scientifique. *Science and Society, 75*, 85–99.

Fahn, S. (1988). Parkinsonism. In J. B. Wyngaarden & L. H. Smith Jr., (Eds.), *Cecil's textbook of medicine* (pp. 2143–2147). Philadelphia: Saunders.

Faiz, M. A., Ghose, A., Ahsan, M. F., Rahman, M. R., Amin, M. R., Hassan, M. M. U., ... Theakston, R. D. G. (2010). The greater black krait (Bungarusniger), a newly recognized cause of neuro-myotoxic snake bite envenoming in Bangladesh. *Brain, 133*(11), 3181–3193. doi:10.1093/brain/awq265 PMID:20855420

Fan, C. W., & Kao, S. J. (2008). Effects of climate events driven hydrodynamics on dissolved oxygen in a subtropical deep reservoir in Taiwan. *The Science of the Total Environment, 393*(2-3), 326–332. doi:10.1016/j.scitotenv.2007.12.018 PMID:18243280

Faneyte, I. F., Kristel, P. M., Maliepaard, M., Scheffer, G. L., Scheper, R. J., Schellens, J. H., & van de Vijver, M. J. (2002). Expression of the breast cancer resistance protein in breast cancer. *Clinical Cancer Research, 8*, 1068–1074.

Fan, W., Yanase, T., Morinaga, H., Gondo, S., Okabe, T., Nomura, M., ... Nawata, H. (2007). Atrazine-induced aromatase expression is SF-1 dependent: Implications for endocrine disruption in wildlife and reproductive cancers in humans. *Environmental Health Perspectives, 115*(5), 720–727. doi:10.1289/ehp.9758

FAO (Food Agriculture Organization). (2000). *The Energy and Agriculture Nexus.* Environment and Natural Resources Working Paper No. 4. Rome: FAO.

FAO, IFAD, UNICEF, WFP, & WHO. (2017). *The State of Food Security and Nutrition in the World 2017. Building resilience for peace and food security.* Rome: FAO.

FAO. (2012). *Statistical year book 2012.* Retrieved from http://www.fao.org /docrep/015 /i2490e/i2490e00.htm

FAO. (2012). *The state of food insecurity in the world 2012. Economic growth is necessary but not sufficient to accelerate reduction of hunger and malnutrition.* Rome: Food and Agriculture Organization of the United Nations.

FAO/World Health Organization (WHO) expert committee in 2012. (n.d.). Retrieved from http://www.who.int/foodsafety/micro/jemra/meetings/sep12/en/

Faraj, C., Ouahabi, S., Adlaoui, E., Boccolini, D., Romi, R., & El Aouad, R. (2008). Risque de reemergence du paludisme au Maroc étude de la capacité vectorielle d'*Anopheleslabranchiae*dans une zone rizicole au nord du pays. *Parasite (Paris, France)*, *15*(4), 605–610. doi:10.1051/parasite/2008154605 PMID:19202769

Farshid, G. (2014). The Normal Breast and Risk Factors for Breast Cancer. Pathobiology of Human Disease: A Dynamic Encyclopedia of Disease Mechanisms, 899-919.

Fauci, A. S., Braunwald, E., Kasper, D. L., & Hauser, S. L. (Eds.). (2008). *Harrison's Principles of Internal Medicine*. New York: McGraw-Hill Medical.

Faulde, M. K., Heyl, G., & Amirih, M. L. (2006). Zoonotic cutaneous leishmaniasis, Afghanistan. *Emerging Infectious Diseases*, *12*(10), 1623–1624. doi:10.3201/eid1210.060076 PMID:17176597

Faulde, M., Schrader, J., Heyl, G., & Amirih, M. (2008). Differences in transmission seasons as an epidemiological tool for characterization of anthroponotic and zoonotic cutaneous leishmaniasis in northern Afghanistan. *Acta Tropica*, *105*(2), 131–138. doi:10.1016/j.actatropica.2007.10.011 PMID:18068684

Fava, M., Uebelacker, L. A., Alpert, J. E., Nierenberg, A. A., Pava, J. A., & Rosenbaum, J. F. (1997). Major depressive subtypes and treatment response. *Biological Psychiatry*, *42*(7), 568–576. doi:10.1016/S0006-3223(96)00440-4 PMID:9376453

Fayer, R., Speer, C. A., & Dubey, J. P. (1997). The general biology of Cryptosporidium. In R. Fayer (Ed.), *Cryptosporidium and Cryptosporidiosis* (pp. 1–41). Boca Raton, FL: Taylor & Francis.

Fayer, R., Trout, J. M., Graczyk, T. K., & Lewis, E. J. (2000). Prevalence of Cryptosporidium, Eimeria and Giardia infections in post-weaned and adult cattle on three Maryland farms. *Veterinary Parasitology*, *93*(2), 103–112. doi:10.1016/S0304-4017(00)00356-3 PMID:11035228

Fay, P. A., & Schultz, M. J. (2009). Germination, survival, and growth of grass and forb seedlings: Effects of soil moisture variability. *Acta Oecologica*, *35*(5), 679–684. doi:10.1016/j.actao.2009.06.007

Feachem, R. G., Bradley, D. J., Garelick, H., & Mara, D. D. (1983). *Sanitation and disease health aspects of excreta and wastewater management*. Chichester, UK: Wiley.

Feigin, V. L., Forouzanfar, M. H., Krishnamurthi, R., Mensah, G. A., Connor, M., Bennett, D. A., ... Murray, C. (2014). Global and regional burden of stroke during 1990 – 2010 : Fi ndings from the Global Burden of Disease Study 2010. *Lancet*, *383*(9913), 245–255. doi:10.1016/S0140-6736(13)61953-4 PMID:24449944

Feigin, V. L., Lawes, C. M. M., Bennett, D. A., Barker-collo, S. L., & Parag, V. (2009). Worldwide stroke incidence and early case fatality reported in 56 population-based studies : A systematic review. *Lancet Neurology*, *8*(4), 355–369. doi:10.1016/S1474-4422(09)70025-0 PMID:19233729

Fei, Q., McCormack, A. L., Di Monte, D. A., & Ethell, D. W. (2008). Paraquat Neurotoxicity Is Mediated by a Bak-dependent Mechanism. *The Journal of Biological Chemistry*, *283*(6), 3357–3364. doi:10.1074/jbc.M708451200 PMID:18056701

Feliciangeli, M. D., & Rabinovich, J. E. (1998). Abundance of *Lutzomyiaovallesi*but not *Lu. gomezi*(Diptera: Psychodidae) correlated with cutaneous leishmaniasis incidence in north-central Venezuela. *Medical and Veterinary Entomology*, *12*(2), 121–131. doi:10.1046/j.1365-2915.1998.00072.x PMID:9622364

Feneley, M. R., Young, M. P., Chinyama, C., Kirby, R. S., & Parkinson, M. C. (1996). Ki-67 expression in early prostate cancer and associated pathological lesions. *Journal of Clinical Pathology*, *49*(9), 741–748.

Feng, Y., & Xiao, L. (2011). Zoonotic Potential and Molecular Epidemiology of Giardia Species and Giardiasis. *Clinical Microbiology Reviews*, *24*(1), 110–140. doi:10.1128/CMR.00033-10 PMID:21233509

Fenker, J., Tedeschi, L. G., Pyron, R. A., & Nogueira, C. D. C. (2014). Phylogenetic diversity, habitat loss and conservation in South American pitvipers (Crotalinae: Bothrops and Bothrocophias). *Diversity & Distributions*, *20*(10), 1108–1119. doi:10.1111/ddi.12217

Fenwick, A. (2006). Waterborne infectious diseases—could they be consigned to history? *Science*, *313*(5790), 1077–1081. doi:10.1126cience.1127184 PMID:16931751

Ferenci, P., Członkowska, A., Merle, U., Ferenc, S., Gromadzka, G., Yurdaydin, C., ... Stremmel, W. (2007). Late-onset Wilson's disease. *Gastroenterology*, *132*(4), 1294–1298. doi:10.1053/j.gastro.2007.02.057 PMID:17433323

Ferraiuolo, L., Kirby, J., Grierson, A. J., Sendtner, M., & Shaw, P. J. (2011). Molecular pathways of motor neuron injury in amyotrophic lateral sclerosis. *Nature Reviews. Neurology*, *7*(11), 616–630. doi:10.1038/nrneurol.2011.152 PMID:22051914

Ferrari, T. E. (2017). Cetacean beachings correlate with geomagnetic disturbances in Earth's magnetosphere: An example of how astronomical changes impact the future of life. *International Journal of Astrobiology*, *16*(02), 163–175. doi:10.1017/S1473550416000252

Ferreira, D., Nobre, A., Silva, M. L., Tulha, J., Faria-oliveira, F., Ferreira, C., & Lucas, C. (2013). encodes a xylose/H + symporter from the highly related yeast species Debaryomyces fabryi and Debaryomyces hansenii. *FEMS Yeast Research*, *13*(7), 585–596. doi:10.1111/1567-1364.12061 PMID:23809840

Feustel, A., Wennrich, R., Steiniger, D., & Klauss, P. (1982). Zinc and cadmium concentration in prostaticcarcinoma of different histological grading in comparison to normal prostate tissue andadenofibromyomatosis (BPH). *Journal of Urology and Research*, *10*(6), 301–303.

Few, R., Ahern, M., Matthies, F., & Kovats, S. (2004). Floods, health and climate change: a strategic review. *East.* Repéré à http://www.tyndall.ac.uk/publications/working_papers/wp63_summary.shtml%5Cnhttp://tyndall.ac.uk/sites/default/files/wp63.pdf

Fewtrell, L., Kay, D., Watkins, J., Davies, C., & Francis, C. (2011). The microbiology of urban UK floodwaters and a quantitative microbial risk assessment of flooding and gastrointestinal illness. *Journal of Flood Risk Management*, *4*(2), 77–87. doi:10.1111/j.1753-318X.2011.01092.x

Fichet-Calvet, E., Jommaa, I., Ben Ismail, R., & Ashford, R. W. (2003). *Leishmania major* infection in the fat sand rat *Psammomys obesus* in Tunisia: Interaction of host and parasite populations. *Annals of Tropical Medicine and Parasitology*, *97*(6), 593–603. doi:10.1179/000349803225001517 PMID:14511558

Field, C. B., Barros, V. R., Mach, K. J., Mastrandrea, M. D., van Aalst, M., Adger, W. N., & ... . (2014). Technical summary. In C. B. Field, V. R. Barros, D. J. Dokken, K. J. Mach, M. D. Mastrandrea, T. E. Bilir, & ... (Eds.), *Climate Change 2014: Impacts, Adaptation and Vulnerability. Part A: Global and Sectoral Aspects. Contribution of Working Group II to the Fifth Assessment Report of the Intergovernmental Panel on Climate Change* (pp. 35–94). Cambridge, UK: Cambridge University Press.

Filali, A., Betbeder, A. M., Baudrimont, I., Benayada, A., Soulaymani, R., & Creppy, E. E. (2002). Ochratoxin A in human plasma in Morocco: A preliminary survey. *Human and Experimental Toxicology*, *21*(5), 241–245. doi:10.1191/0960327102ht249oa

Filipov, N. M., Stewart, M. A., Carr, R. L., & Sistrunk, S. C. (2007). Dopaminergic toxicity of the herbicide atrazine in rat striatal slices. *Toxicology*, *232*(1–2), 68–78. doi:10.1016/j.tox.2006.12.007 PMID:17218051

Fischer, G., Tubiello, F. N., Van Velthuizen, H., & Wiberg, D. A. (2007). Climate change impacts on irrigation water requirements: Effects of mitigation, 1990–2080. *Technological Forecasting and Social Change*, *74*(7), 1083–1107. doi:10.1016/j.techfore.2006.05.021

Fitzgerald, K., Mikalunas, V., Rubin, H., McCarthey, R., Vanagunas, A., & Craig, R. M. (1999). Hypermanganesemia in patients receiving total parenteral nutrition. *JPEN. Journal of Parenteral and Enteral Nutrition*, *23*(6), 333–336. doi:10.1177/014860719902300633 PMID:10574481

Fitzmaurice, C., Allen, C., Barber, R. M., Barregard, L., Bhutta, Z. A., Brenner, H., ... Naghavi, M. (2017). Global, Regional, and National Cancer Incidence, Mortality, Years of Life Lost, Years Lived With Disability, and Disability-Adjusted Life-Years for 29 Cancer Groups, 1990 to 2016: A Systematic Analysis for the Global Burden of Disease Study. *JAMA Oncology*, *3*, 524. doi:10.1001/jamaoncol.2016.5688

Fleming, L., Mann, J. B., Bean, J., Briggle, T., & Sanchez-Ramos, J. R. (1994). Parkinson's disease and brain levels of organochlorine pesticides. *Annals of Neurology*, *36*(1), 100–103. doi:10.1002/ana.410360119 PMID:7517654

Florea, A. M., & Büsselberg, D. (2011). Metals and Breast Cancer: Risk Factors or Healing Agents? *Journal of Toxicology*.

Florkin, M. (1956). *Aspects biochimiques communs aux êtres vivants: introduction à la biochimie générale des organismes*. Masson.

Floyd, L. E., Prinz, D. K., Crane, P. C., & Herring, L. C. (2002). Solar UV irradiance variation during cycles 22 and 23. *Advances in Space Research*, *29*(12), 1957–1962. doi:10.1016/S0273-1177(02)00242-9

Foley, D. H., Klein, T. A., Kim, H. C., Wilkerson, R. C., & Rueda, L. M. (2008). *Malaria risk assessment for the Republic of Korea based on models of mosquito distribution*. Walter Reed Army Inst Of Research.

Foley, D. H., Weitzman, A. L., Miller, S. E., Faran, M. E., Rueda, L. M., & Wilkerson, R. C. (2008). The value of georeferenced collection records for predicting patterns of mosquito species richness and endemism in the Neotropics. *Ecological Entomology*, *33*(1), 12–23.

Foley, J. A., DeFries, R., Asner, G. P., Barford, C., Bonan, G., Carpenter, S. R., ... Helkowski, J. H. (2005). Global consequences of land use. *Science*, *309*(5734), 570–574. doi:10.1126cience.1111772 PMID:16040698

Folkman, J. (1984). Angiogenesis. In *Biology of endothelial cells* (pp. 412–428). Boston, MA: Springer. doi:10.1007/978-1-4613-2825-4_42

Folkman, J. (1996). Fighting cancer by attacking its blood supply. *Scientific American*, *275*(3), 150–154. doi:10.1038c ientificamerican0996-150 PMID:8701285

Fonseca, R. G., Resende, L. A., Silva, M. D., & Camargo, A. (1993). Chronic motor neuron disease possibly related to intoxication with organochlorine insecticides. *Acta Neurologica Scandinavica*, *88*(1), 56–58. doi:10.1111/j.1600-0404.1993.tb04187.x PMID:7690513

Fontaine, J. J., Decker, K. L., Skagen, S. K., & Van Riper, C. (2009). Spatial and temporal variation in climate change: A bird's eye view. *Climatic Change*, *97*(1-2), 305–311. doi:10.100710584-009-9644-9

Forbes, J. R., & Cox, D. W. (2000). Copper-dependent trafficking of Wilson disease mutant ATP7B proteins. *Human Molecular Genetics*, *9*(13), 1927–1935. doi:10.1093/hmg/9.13.1927 PMID:10942420

Forsell, Y., & Winblad, B. (1999). Incidence of major depression in a very elderly population. *International Journal of Geriatric Psychiatry*, *14*(5), 368–372. doi:10.1002/(SICI)1099-1166(199905)14:5<368::AID-GPS919>3.0.CO;2-Y PMID:10389040

Forsman, A. (1999). Variation in thermal sensitivity of performance among colour morphs of a pygmy grasshopper. *Journal of Evolutionary Biology*, *12*(5), 869–878. doi:10.1046/j.1420-9101.1999.00084.x

Forstl, & Burns, Levy, Cairns, Luthert, & Lantos. (1992). Neurologic Signs in Alzheimer ' s. *Archives of Neurology*, 49.

Förstner, U. (1981). Metal transfer between solid and aqueous phases. In *Metal Pollution in the Aquatic Environment* (pp. 197–270). Berlin: Springer Berlin Heidelberg. doi:10.1007/978-3-642-69385-4_5

Foster, P. L., Eisenstadt, E., & Miller, J. H. (1983). Base substitution mutations induced by metabolically activated aflatoxin B1. *Proceedings of the National Academy of Sciences of the United States of America*, *80*(9), 2695–2698. doi:10.1073/pnas.80.9.2695

Foster, W. G., Jarrell, J. F., Younglai, E. V., Wade, M. G., Arnold, D. L., & Jordan, S. (1996). An Overview of Some Reproductive Toxicology Studies Conducted At Health Canada. *Toxicology and Industrial Health*, *12*(3–4), 447–459. doi:10.1177/074823379601200316 PMID:8843561

Foufopoulos, J., Kilpatrick, A. M., & Ives, A. R. (2010). Climate change and elevated extinction rates of reptiles from Mediterranean islands. *American Naturalist*, *177*(1), 119–129. doi:10.1086/657624 PMID:21091198

Foulkes, W. D. (2008). Inherited susceptibility to common cancers. *The New England Journal of Medicine*, *359*(20), 2143–2153. doi:10.1056/NEJMra0802968

Fox, J. H., Kama, J. A., Lieberman, G., Chopra, R., Dorsey, K., Chopra, V., ... Hersch, S. (2007). Mechanisms of copper ion mediated Huntington's disease progression. *PLoS One*, *2*(3), e334. doi:10.1371/journal.pone.0000334 PMID:17396163

Franchignoni, F., Mora, G., Giordano, A., Volanti, P., & Chio, A. (2013). Evidence of multidimensionality in the ALSFRS-R scale: A critical appraisal on its measurement properties using Rasch analysis. *Journal of Neurology, Neurosurgery, and Psychiatry*, *84*(12), 1340–1345. doi:10.1136/jnnp-2012-304701 PMID:23516308

Franco, R., Li, S., Rodriguez-Rocha, H., Burns, M., & Panayiotidis, M. I. (2010). Molecular mechanisms of pesticide-induced neurotoxicity: Relevance to Parkinson's disease. *Chemico-Biological Interactions*, *188*(2), 289–300. doi:10.1016/j.cbi.2010.06.003 PMID:20542017

Frederick, K. D., & Schwarz, G. E. (1999). Socioeconomic impacts of climate change on U.S. water supplies. *Journal of the American Water Resources Association*, *35*(6), 1563–1583. doi:10.1111/j.1752-1688.1999.tb04238.x

Frederickson, C. J. (1989). Neurobiology of Zinc and Zinc-Containing Neurons. *International Review of Neurobiology*, *31*, 145–238. doi:10.1016/S0074-7742(08)60279-2 PMID:2689380

Freire, C., Amaya, E., Gil, F., Fernández, M. F., Murcia, M., Llop, S., ... Olea, N. (2017, November). Prenatal co-exposure to neurotoxic metals and neurodevelopment in preschool children: The Environment and Childhood (INMA) Project. *The Science of the Total Environment*, *621*, 340–351. doi:10.1016/j.scitotenv.2017.11.273 PMID:29190557

Freire, L., & Sant'Ana, A. S. (2018). Modified mycotoxins: An updated review on their formation, detection, occurrence, and toxic effects. *Food and Chemical Toxicology*, *111*, 189–205. doi:10.1016/j.fct.2017.11.021

Freire-Maia, L., Campos, J. A., & Amaral, C. F. S. (1994). Approaches to the treatment of scorpion envenoming. *Toxicon*, *32*(9), 1009–1014. doi:10.1016/0041-0101(94)90382-4 PMID:7801334

French, S. S., Johnston, G. I. H., & Moore, M. C. (2007). Immune activity suppresses reproduction in food-limited female tree lizards Urosaurusornatus. *Functional Ecology*, *21*(6), 1115–1122. doi:10.1111/j.1365-2435.2007.01311.x

Freyaldenhoven, T. E., Ali, S. F., & Hart, R. W. (1995). MPTP- and MPP+-induced effects on body temperature exhibit age- and strain-dependence in mice. *Brain Research*, *688*(1–2), 161–170. doi:10.1016/0006-8993(95)00529-Y PMID:8542303

Frey, S. D., Drijber, R., Smith, H., & Melillo, J. (2008). Microbial biomass, functional capacity, and community structure after 12 years of soil warming. *Soil Biology & Biochemistry*, *40*(11), 2904–2907. doi:10.1016/j.soilbio.2008.07.020

Frisvad, J. C., & Samson, R. A. (1991). Mycotoxins produced by species of Penicillium and Aspergillus occurring in cereals. In J. Chelkowski (Ed.), *Cereal Grain Mycotoxins, Fungi and Quality in Drying and Storage*. Amsterdam: Elsevier.

Fritioff, Å., Kautsky, L., & Greger, M. (2005). Influence of temperature and salinity on heavy metal uptake by submersed plants. *Environmental Pollution*, *133*(2), 265–274. doi:10.1016/j.envpol.2004.05.036 PMID:15519457

Fritzsche, K. (2017). *Vulnerability Sourcebook – Concept and Guidelines for standardised Vulnerability Assessments.* Bonn: Deutsche Gesellschaft für Internationale Zusammenarbeit (GIZ) GmbH.

Frohman, E. M., Racke, M. K., & Raine, C. S. (2006). Multiple sclerosis--the plaque and its pathogenesis. *The New England Journal of Medicine*, *354*(9), 942–955. doi:10.1056/NEJMra052130 PMID:16510748

Frölich, L., & Riederer, P. (1995). Free radical mechanisms in dementia of Alzheimer type and the potential for antioxidative treatment. *Arzneimittel-Forschung*, *45*(3A), 443–446. Retrieved from http://www.ncbi.nlm.nih.gov/pubmed/7763342 PMID:7763342

Fry, B. G., Vidal, N., Norman, J. A., Vonk, F. J., Scheib, H., Ramjan, S. F., ... Kochva, E. (2006). Early evolution of the venom system in lizards and snakes. *Nature*, *439*(7076), 584–588. doi:10.1038/nature04328 PMID:16292255

Frydman, M., Bonné-Tamir, B., Farrer, L. A., Conneally, P. M., Magazanik, A., Ashbel, S., & Goldwitch, Z. (1985). Assignment of the gene for Wilson disease to chromosome 13: Linkage to the esterase D locus. *Proceedings of the National Academy of Sciences of the United States of America*, *82*(6), 1819–1821. doi:10.1073/pnas.82.6.1819 PMID:3856863

Frye, R. E., Casanova, M. F., Fatemi, S. H., Folsom, T. D., Reutiman, T. J., Brown, G. L., ... Zimmerman, A. W. (2016). Neuropathological Mechanisms of Seizures in Autism Spectrum Disorder. *Frontiers in Neuroscience*, *10*(May), 1–9. doi:10.3389/fnins.2016.00192 PMID:27242398

Fucic, A., Guszak, V., & Mantovani, A. (2017). Transplacental exposure to environmental carcinogens: Association with childhood cancer risks and the role of modulating factors. *Reproductive Toxicology (Elmsford, N.Y.)*, *72*, 182–190. doi:10.1016/j.reprotox.2017.06.044 PMID:28624605

Fuhrimann, S., & Medlicott, K. (2017). *Sanitation safety planning as a tool for achieving safely managed sanitation systems and safe use of wastewater.* Academic Press. doi:10.4103/2224-3151.213790

Fuhrimann, S., Pham-duc, P., Cissé, G., Thuy Tram, N., Ha Thu, H., Do Trung, D., & .... (2016). Microbial contamination along the main open wastewater and storm water channel of Hanoi, Vietnam, and potential health risks for urban farmers. *The Science of the Total Environment*, *566–7*, 1014–1022. doi:10.1016/j.scitotenv.2016.05.080 PMID:27325013

Fujiwara, T., & Shigematsu, H. (2004). Etiologic factors and clinical features of symptomatic epilepsy : Focus on pediatric cases. *Psychiatry and Clinical Neurosciences*, *58*(3), 9–12. doi:10.1111/j.1440-1819.2004.01244_3.x

Fukushima, T., Yamada, K., Isobe, A., Shiwaku, K., & Yamane, Y. (1993). Mechanism of cytotoxicity of paraquat. *Experimental and Toxicologic Pathology*, *45*(5–6), 345–349. doi:10.1016/S0940-2993(11)80424-0 PMID:8312721

Fukuto, T. R. (1990). Mechanism of action of organophosphorus and carbamate insecticides. *Environmental Health Perspectives*, *87*, 245–254. doi:10.1289/ehp.9087245 PMID:2176588

Fumonisin Toxicosis. (2012). InWilson, D. A. (Ed.), *Clinical Veterinary Advisor* (pp. 214–215). St. Louis, MO: W.B. Saunders; doi:10.1016/B978-1-4160-9979-6.00147-1

Funari, E., Manganelli, M., & Sinisi, L. (2012). Impact of climate change on waterborne diseases. *Annali dell'Istituto Superiore di Sanita*, *48*(4), 473–487. doi:10.4415/ANN_12_04_13 PMID:23247142

Funk, C., Dettinger, M. D., Michaelsen, J. C., Verdin, J. P., Brown, M. E., Barlow, M., & Hoell, A. (2008). Warming of the Indian Ocean threatens eastern and southern African food security but could be mitigated by agricultural development. *Proceedings of the National Academy of Sciences of the United States of America*, *105*(32), 11081–11086. doi:10.1073/pnas.0708196105 PMID:18685101

Furieri, L. B., Galán, M., Avendaño, M. S., García-redondo, A. B., Aguado, A., Martínez, S., ... Salaices, M. (2011). Endothelial dysfunction of rat coronary arteries after exposure to low concentrations of mercury is dependent on reactive oxygen species. *British Journal of Pharmacology*, *162*(8), 1819–1831. doi:10.1111/j.1476-5381.2011.01203.x PMID:21232032

Furtado, C., Adak, G. K., Stuart, J. M., Wall, P. G., Evans, H. S., & Casemore, D. P. (1998). Outbreaks of waterborne infectious intestinal disease in England and Wales, 1992-5. *Epidemiology and Infection*, *121*(1), 109–119. doi:10.1017/S0950268898001083 PMID:9747762

Gadzała-Kopciuch, R., Cendrowski, K., Cesarz, A., Kiełbasa, P., & Buszewski, B. (2011). Determination of zearalenone and its metabolites in endometrial cancer by coupled separation techniques. *Analytical and Bioanalytical Chemistry*, *401*(7), 2069–2078. doi:10.100700216-011-5206-x

Gaeta, A., & Hider, R. C. (2005). The crucial role of metal ions in neurodegeneration: The basis for a promising therapeutic strategy. *British Journal of Pharmacology*, *146*(8), 1041–1059. doi:10.1038j.bjp.0706416 PMID:16205720

Gajecki, M., Przybyłowicz, M., Zielonka, L., Zwierzchowski, W., Obremski, K., Skorska-Wyszyńska, E., ... Jakimiuk, E. (2004). Preliminary results of monitoring research on zearalenone presence in blood of women with neoplastic lesions in reproductive system. *Polish Journal of Veterinary Sciences*, *7*, 153–156.

Galasso, S. L., & Dyck, R. H. (2007). The Role of Zinc in Cerebral Ischemia. *Molecular Medicine (Cambridge, Mass.)*, *13*(9), 380–387. doi:10.2119/2007 PMID:17622314

Galbraith, N. S., Barrett, N. J., & Stanwell-smith, R. (1987). Water and disease after croydon: A review of water-borne and water-associated disease in the UK 1937–86. *Water and Environment Journal: the Journal / the Chartered Institution of Water and Environmental Management*, *1*(1), 7–21. doi:10.1111/j.1747-6593.1987.tb01184.x

Gallagher, R. P., & Fleshner, N. (1998). Prostate cancer: 3. Individual risk factors. *CMAJ: Canadian Medical Association Journal*, *159*(7), 807. PMID:9805030

Gallez, B., Demeure, R., Baudelet, C., Abdelouahab, N., Beghein, N., Jordan, B., ... Roels, H. A. (2001). Non invasive quantification of manganese deposits in the rat brain by local measurement of NMR proton T1 relaxation times. *Neurotoxicology*, *22*(3), 387–392. doi:10.1016/S0161-813X(01)00020-1 PMID:11456339

Galloway, J. N., & Cowling, E. B. (1978). The effects of precipitation on aquatic and terrestrial ecosystems: A proposed precipitation chemistry network. *Journal of the Air Pollution Control Association*, *28*(3), 229–235. doi:10.1080/00022470.1978.10470594

Gallus, S., Foschi, R., Negri, E., Talamini, R., Franceschi, S., Montella, M., ... La Vecchia, C. (2007). Dietary zinc and prostate cancer risk: A case-control study from Italy. *European Urology*, *52*(4), 1052–1056. doi:10.1016/j.eururo.2007.01.094 PMID:17292532

Gammon, M. D., Wolff, M. S., Neugut, A. I., Eng, S. M., Teitelbaum, S. L., Britton, J. A., ... Santella, R. M. (2002). Environmental toxins and breast cancer on Long Island. II. Organochlorine compound levels in blood. *Cancer Epidemiology, Biomarkers & Prevention*, *11*, 686–697.

Ganesh, R., & Janakiraman, L. (2008). Serum Zinc Levels in Children With Simple Febrile Seizure. *Clinical Pediatrics*, *47*(2), 164–166. doi:10.1177/0009922807306165 PMID:17873242

Ganesh, R., Janakiraman, L., & Meenakshi, B. (2011). Serum zinc levels are low in children with simple febrile seizures compared with those in children with epileptic seizures and controls. *Annals of Tropical Paediatrics*, *345–349*. doi:10.1179/1465328111Y.0000000032 PMID:22041469

Gao, C., Du, H., Hua, Y., Keep, R. F., Strahle, J., & Xi, G. (2014). Role of red blood cell lysis and iron in hydrocephalus after intraventricular hemorrhage. *Journal of Cerebral Blood Flow and Metabolism*, *34*(6), 1070–1075. doi:10.1038/jcbfm.2014.56 PMID:24667910

Gao, H.-M., & Hong, J.-S. (2011). Gene-environment interactions: Key to unraveling the mystery of Parkinson's disease. *Progress in Neurobiology*, *94*(1), 1–19. doi:10.1016/j.pneurobio.2011.03.005 PMID:21439347

Gao, T., Wang, X. C., Chen, R., Ngo, H. H., & Guo, W. (2015). Disability adjusted life year (DALY): A useful tool for quantitative assessment of environmental pollution. *The Science of the Total Environment*, *511*, 268–287. doi:10.1016/j.scitotenv.2014.11.048 PMID:25549348

Gardella, C. (2001). Lead exposure in pregnancy: A review of the literature and argument for routine prenatal screening. *Obstetrical & Gynecological Survey*, *56*(4), 231–238. doi:10.1097/00006254-200104000-00024 PMID:11285436

Gargano, J. W., Freeland, A. L., Morrison, M. A., Stevens, K., Zajac, L., Wolkon, A., ... Brunkard, J. M. (2015). Acute gastrointestinal illness following a prolonged community-wide water emergency. *Epidemiology and Infection*, *143*(13), 2766–2776. doi:10.1017/S0950268814003501 PMID:25608522

Garnett, T., Appleby, M. C., Balmford, A., Bateman, I. J., Benton, T. G., Bloomer, P., ... Herrero, M. (2013). Sustainable intensification in agriculture: Premises and policies. *Science*, *341*(6141), 33–34. doi:10.1126cience.1234485 PMID:23828927

Garrett-Jones, C. (1964). Prognosis for interruption of malaria transmission through assessment of the mosquito's vectorial capacity. *Nature*, *204*(4964), 1173–1175. doi:10.1038/2041173a0

Gatto, M. P., Cabella, R., & Gherardi, M. (2016). Climate change: The potential impact on occupational exposure to pesticides. *Annali dell'Istituto Superiore di Sanita*, *52*, 374–385. PMID:27698296

Gauvreau, D. (1987). Le paradigme de la maladie d'Alzheimer. *Interface*, *8*, 16–21.

Gefen, E. (2008). Sexual dimorphism in desiccation responses of the sand scorpion Smeringurus mesaensis (Vaejovidae). *Journal of Insect Physiology*, *54*(5), 798–805. doi:10.1016/j.jinsphys.2008.02.004 PMID:18374353

Gefen, E., & Ar, A. (2005). The effect of desiccation on water management and compartmentalisation in scorpions: The hepatopancreas as a water reservoir. *The Journal of Experimental Biology*, *208*(10), 1887–1894. doi:10.1242/jeb.01568 PMID:15879069

Gelderblom, W. C. A., Jaskiewicz, K., Marasas, W. F. O., Thiel, P. G., Horak, R. M., Vleggaar, R., & ... . (1988). Fumonisins – Novel mycotoxins with cancerpromoting activity produced by Fusarium moniliforme. *Applied and Environmental Microbiology*, *54*, 1806–1811.

Gelderblom, W. C. A., Jaskiewicz, K., Marasas, W. F. O., Thiel, P. G., Horak, R. M., Vleggaar, R., & ... . (1988). Fumonisins-Novel mycotoxins with cancerpromoting activity produced by Fusarium moniliforme. *Applied and Environmental Microbiology*, *54*, 1806–1811.

Genton, B., & D'Acremont, V. (2001). *Clinical features of malaria in returning travelers and migrants. Travelers' malaria*. Hamilton, Canada: Decker.

Gersberg, R. M., Daft, D., & Yorkey, D. (2004). Temporal pattern of toxicity in runoff from the Tijuana River Watershed. *Water Research*, *38*(3), 559–568. doi:10.1016/j.watres.2003.11.002 PMID:14723924

Gertler, M., Dürr, M., Renner, P., Poppert, S., Askar, M., Breidenbach, J., ... Chalmers, R. (2015). Outbreak of Cryptosporidium hominis following river flooding in the city of Halle (Saale), Germany, August 2013. *BMC Infectious Diseases*, *15*(1), 1. doi:10.118612879-015-0807-1 PMID:25879490

Ghasemi, E., Rahdar, M., & Rostami, M. (2015). *Prevalence of Entamoeba histolytica / dispar in drinking water in the city of Shush, Khuzestan Province in 2011*. Academic Press.

Ghazani, M., FitzGerald, G., Hu, W., Toloo, G. S., & Xu, Z. (2018). Temperature variability and gastrointestinal infections: A review of impacts and future perspectives. *International Journal of Environmental Research and Public Health*, *15*(4), 766. doi:10.3390/ijerph15040766 PMID:29659519

Gholamrezaei, M., Mohebali, M., Hanafi-Bojd, A. A., Sedaghat, M. M., & Shirzadi, M. R. (2016). Ecological niche modeling of main reservoir hosts of zoonotic cutaneous leishmaniasis in Iran. *Acta Tropica*, *160*, 44–52. doi:10.1016/j.actatropica.2016.04.014 PMID:27150212

Giangaspero, A., Berrilli, F., & Brandonisio, O. B.-P. (2007). *Giardia and Cryptosporidium and public health: the epidemiological scenario from the Italian perspective*. Springer. Retrieved from https://link.springer.com/article/10.1007/s00436-007-0598-4

Giannandrea, F., & Fargnoli, S. (2017). Environmental Factors Affecting Growth and Occurrence of Testicular Cancer in Childhood: An Overview of the Current Epidemiological Evidence. *Children*, *4*(1), 1. doi:10.3390/children4010001 PMID:28067779

Gianutsos, G., & Murray, M. T. (1982). Alterations in brain dopamine and GABA following inorganic or organic manganese administration. *Neurotoxicology*, *3*, 75–81. PMID:6891761

Gibb, H. J., Lees, P. S., Pinsky, P. F., & Rooney, B. C. (2000). Lung cancer among workers in chromium chemical production. *American Journal of Industrial Medicine*, *38*(2), 115–126. doi:10.1002/1097-0274(200008)38:2<115::AID-AJIM1>3.0.CO;2-Y PMID:10893504

Gill, C. A. (1928). *The Genesis of Epidemics and the Natural History of Disease. An Introduction to the Science of Epidemiology based upon the Study of Epidemics of Malaria, Influenza, and Plague. Academic Press.*

Ginat, D. T., & Meyers, S. P. (2012). Intracranial lesions with high signal intensity on T1-weighted MR images: differential diagnosis. *Radiographics: A Review Publication of the Radiological Society of North America. Inc*, *32*(2), 499–516. doi:10.1148/rg.322105761

Girmay, G., Singh, B. R., Nyssen, J., & Borrosen, T. (2009). Runoff and sediment-associated nutrient losses under different land uses in Tigray, Northern Ethiopia. *Journal of Hydrology (Amsterdam)*, *376*(1-2), 70–80. doi:10.1016/j.jhydrol.2009.07.066

Githeko, A. K., Lindsay, S. W., Confalonieri, U. E., & Patz, J. A. (2000). Climate change and vector-borne diseases: A regional analysis. *Bulletin of the World Health Organization*, *78*, 1136–1147. PMID:11019462

Gleadow, R. M., Edwards, E. J., & Evans, J. R. (2009). Changes in nutritional value of cyanogenic Trifolium repens grown at elevated atmospheric CO2. *Journal of Chemical Ecology*, *35*(4), 476–478. doi:10.100710886-009-9617-5 PMID:19352773

Glunt, K. D., Blanford, J. I., & Paaijmans, K. P. (2013). Chemicals, climate, and control: Increasing the effectiveness of malaria vector control tools by considering relevant temperatures. *PLoS Pathogens*, *9*(10), e1003602. doi:10.1371/journal.ppat.1003602 PMID:24098111

Gocmen, A., Peters, H. A., Cripps, D. J., Bryan, G. T., & Morris, C. R. (1989). Hexachlorobenzene episode in Turkey. *Biomedical and Environmental Sciences : BES*, *2*(1), 36–43. PMID:2590490

Godfree, A., & Farrell, J. (2005). Processes for managing pathogens. *Journal of Environmental Quality*, *34*(1), 105–113. doi:10.2134/jeq2005.0105 PMID:15647539

Goldman, S. M. (2014). Environmental Toxins and Parkinson's Disease. *Annual Review of Pharmacology and Toxicology*, *54*(1), 141–164. doi:10.1146/annurev-pharmtox-011613-135937 PMID:24050700

Golovine, K., Makhov, P., Uzzo, R. G., Kutikov, A., Kaplan, D. J., Fox, E., & Kolenko, V. M. (2010). Cadmiumdown-regulates expression of XIAP at the post-transcriptional level in prostate cancer cellsthrough an NF-kappaB-independent, proteasome-mediated mechanism. *Molecular Cancer*, *9*(1), 183. doi:10.1186/1476-4598-9-183 PMID:20618956

Gomez-Couso, H., Mendez-Hermida, F., Castro-Hermida, J., & Ares-Mazas, E. (2005). Giardia in shellfish-farming areas: Detection in mussels, river water and wastewaters. *Veterinary Parasitology*, *133*(1), 13–18. doi:10.1016/j.vetpar.2005.04.043 PMID:15982819

Goncharov, A., Haase, R. F., Santiago-rivera, A., Morse, G., McCaffrey, R. J., Rej, R., & Carpenter, D. O. (2008). High serum PCBs are associated with elevation of serum lipids and cardiovascular disease in a Native American population $. *Environmental Research*, *106*(2), 226–239. doi:10.1016/j.envres.2007.10.006 PMID:18054906

Gonzalez, A., Peters, U., Lampe, J. W., & White, E. (2009). Zinc intake from supplements and diet and prostate cancer. *Nutrition and Cancer*, *61*(2), 206–215. doi:10.1080/01635580802419749 PMID:19235036

Goode, M. J., & Duvall, D. (1989). Body temperature and defensive behaviour of free-ranging prairie rattlesnakes, Crotalusviridisviridis. *Animal Behaviour*, *38*(2), 360–362. doi:10.1016/S0003-3472(89)80100-9

Gordon, P. H., Artaud, F., Aouba, A., Laurent, F., Meininger, V., & Elbaz, A. (2011a). Changing mortality for motor neuron disease in France (1968-2007): An age-period-cohort analysis. *European Journal of Epidemiology*, *26*(9), 729–737. doi:10.100710654-011-9595-0 PMID:21674216

Gorell, J. M., Peterson, E. L., Rybicki, B. A., & Johnson, C. C. (2004). Multiple risk factors for Parkinson's disease. *Journal of the Neurological Sciences*, *217*(2), 169–174. doi:10.1016/j.jns.2003.09.014 PMID:14706220

Gorgogietas, V. A., Tsialtas, I., Sotiriou, N., Laschou, V. C., Karra, A. G., Leonidas, D. D., ... Psarra, A. G. (2018). Potential interference of aluminum chlorohydrate with estrogen receptor signaling in breast cancer cells. *Journal of Molecular Biochemistry*, *7*, 1–13.

Gosling, S. N., & Arnell, N. W. (2016). A global assessment of the impact of climate change on water scarcity. *Climatic Change*, *134*(3), 371–385. doi:10.100710584-013-0853-x

Gosling, S. N., Lowe, J. A., McGregor, G. R., Pelling, M., & Malamud, B. D. (2009). Associations between elevated atmospheric temperature and human mortality: A critical review of the literature. *Climatic Change*, *92*(3-4), 299–341. doi:10.100710584-008-9441-x

Gouin, T., Armitage, J. M., Cousins, I. T., Muir, D. C., Ng, C. A., Reid, L., & Tao, S. (2013). Influence of global climate change on chemical fate and bioaccumulation: The role of multimedia models. *Environmental Toxicology and Chemistry*, *32*(1), 20–31. doi:10.1002/etc.2044 PMID:23136071

Goyffon, M., & Guette, C. (2005). Scorpions dangereux du Niger. *Bulletin de la Société de Pathologie Exotique*, *98*(4), 293–295. PMID:16402579

Goytain, A., Hines, R. M., & Quamme, G. A. (2008). Huntingtin-interacting Proteins, HIP14 and HIP14L, Mediate Dual Functions, Palmitoyl Acyltransferase and Mg2+ Transport. *The Journal of Biological Chemistry*, *283*(48), 33365–33374. doi:10.1074/jbc.M801469200 PMID:18794299

Grabemann, H. J., Grabemann, I., Herbers, D., & Müller, A. (2001). Effects of a specific climate scenario on the hydrography and transport of conservative substances in the Weser estuary, Germany: A case study. *Climate Research*, *18*, 77–87. doi:10.3354/cr018077

Graczyk, T., Fayer, R., & Cranfield, M. R. (1997). Zoonotic transmission of Cryptosporidium parvum: Implications for water-borne cryptosporidiosis. *Parasitol. Today*, *13*, 348–351. Retrieved from https://www.cell.com/partod/abstract/S0169-4758(97)01076-4

Graczyk, T. K., Fayer, R., & Cranfield, M. R. (1997a). Zoonotic potential of Cryptosporidium parvum: Implications for waterborne cryptosporidiosis. *Parasitology Today (Personal Ed.)*, *13*(9), 348–351. doi:10.1016/S0169-4758(97)01076-4 PMID:15275048

Graeff, F. G., Guimarães, F. S., De Andrade, T. G. C. S., & Deakin, J. F. W. (1996). Role of 5-HT in stress, anxiety, and depression. *Pharmacology, Biochemistry, and Behavior*, *54*(I), 129–141. doi:10.1016/0091-3057(95)02135-3 PMID:8728550

Graham, S. H., & Chen, J. (2001). Programmed Cell Death in Cerebral Ischemia. *Journal of Cerebral Blood Flow and Metabolism*, *21*(2), 99–109. doi:10.1097/00004647-200102000-00001 PMID:11176275

Greenamyre, J. T., Betarbet, R., & Sherer, T. B. (2003). The rotenone model of Parkinson's disease: Genes, environment and mitochondria. *Parkinsonism & Related Disorders*, *9*, 59–64. doi:10.1016/S1353-8020(03)00023-3 PMID:12915069

Greenlee, R. T., Murray, T., Bolden, S., & Wingo, P. A. (2000). Cancer statistics, 2000. *CA: a Cancer Journal for Clinicians*, *50*(1), 7–33. doi:10.3322/canjclin.50.1.7 PMID:10735013

Green, R. E., Cornell, S. J., Scharlemann, J. P., & Balmford, A. (2005). Farming and the fate of wild nature. *Science*, *307*(5709), 550–555. doi:10.1126cience.1106049 PMID:15618485

Greenwald, O. E. (1974). Thermal dependence of striking and prey capture by gopher snakes. *Copeia*, *1974*(1), 141–148. doi:10.2307/1443016

Gresham, L. S., Molgaard, C. A., Golbeck, A. L., & Smith, R. (1986). Amyotrophic lateral sclerosis and occupational heavy metal exposure: A case control study. *Neuroepidemiology*, *5*(1), 29–38. doi:10.1159/000110810 PMID:3748267

Griffin, A. C. (1979). Role of selenium in the chemoprevention of cancer. *Advances in Cancer Research*, *29*, 419–442. doi:10.1016/S0065-230X(08)60851-9 PMID:382777

Grigg, J. (2004). Environmental toxins; their impact on children's health. *Archives of Disease in Childhood*, *89*(3), 244–250. doi:10.1136/adc.2002.022202 PMID:14977703

Griggs, D. J., & Noguer, M. (2002). Climate change 2001: The scientific basis. Contribution of working group I to the third assessment report of the intergovernmental panel on climate change. *Weather*, *57*(8), 267–269. doi:10.1256/004316502320517344

Grimsrud, T. K., & Andersen, A. (2012). Unrecognized risks of nickel-related respiratory cancer among Canadian electrolysis workers. *Scandinavian Journal of Work, Environment & Health*, *38*(6), 503–515. doi:10.5271jweh.3274 PMID:22258056

Groopman, J. D., Egner, P. A., Schulze, K. J., Wu, L. S.-F., Merrill, R., Mehra, S., … Christian, P. (2014). Aflatoxin exposure during the first 1000 days of life in rural South Asia assessed by aflatoxin B1-lysine albumin biomarkers. *Food and Chemical Toxicology: An International Journal, 74*, 184-189. doi:10.1016/j.fct.2014.09.016

Gros, R. (2002). *Fonctionnement et qualité des sols soumis à des perturbations physiques et chimiques d'origines anthropiques: réponses du sol, de la flore et de la microflore bactérienne tellurique* (Doctoral dissertation). Université de Savoie.

Gross, R. E. (2000). Breast cancer: Risk factors, screening, and prevention. *Journal of Seminars in Oncology Nursing, 16*(3), 176–184. doi:10.1053onc.2000.8110

Grover-Kopec, E., Kawano, M., Klaver, R. W., Blumenthal, B., Ceccato, P., & Connor, S. J. (2005). An online operational rainfall-monitoring resource for epidemic malaria early warning systems in Africa. *Malaria Journal, 4*(1), 6. doi:10.1186/1475-2875-4-6

Gubler. (1998). Resurgent vector borne diseases as a global health problem. *EmergInfec Dis, 4*, 442-450.

Guernaoui, S., Boumezzough, A., Pesson, B., & Pichon, G. (2005). Entomological investigations in Chichaoua: An emerging epidemic focus of cutaneous leishmaniasis in Morocco. *Journal of Medical Entomology, 42*(4), 697–70. doi:10.1093/jmedent/42.4.697 PMID:16119562

Guerra, C. A., Gikandi, P. W., Tatem, A. J., Noor, A. M., Smith, D. L., Hay, S. I., & Snow, R. W. (2008). The limits and intensity of Plasmodium falciparum transmission: Emplications for malaria control and elimination worldwide. *PLoS Medicine, 5*(2), 300–312. doi:10.1371/journal.pmed.0050038

Guerra, C. A., Howes, R. E., Patil, A. P., Gething, P. W., Van Boeckel, T. P., Temperley, W. H., ... Baird, J. K. (2010). The international limits and population at risk of Plasmodium vivax transmission in 2009. *PLoS Neglected Tropical Diseases, 4*(8), e774. doi:10.1371/journal.pntd.0000774

Guerreiro, C. A. M. (2016). Epilepsy: Is there hope? *The Indian Journal of Medical Research, 144*(5), 657–660. doi:10.4103/ijmr.IJMR_1051_16 PMID:28361817

Guessous-Idrissi, N., Chiheb, S., Hamdani, A., Riyad, M., Bichichi, M., Hamdani, S., & Krimech, A. (1997). Cutaneous leishmaniasis: An emerging epidemic focus of *Leishmaniatropica* in north Morocco. *Transactions of the Royal Society of Tropical Medicine and Hygiene, 91*(6), 660–663. doi:10.1016/S0035-9203(97)90511-3 PMID:9509172

Guilarte, T. R. (2010). Manganese and Parkinson's Disease: A Critical Review and New Findings. *Environmental Health Perspectives, 118*(8), 1071–1080. doi:10.1289/ehp.0901748 PMID:20403794

Guilarte, T. R. (2013). Manganese neurotoxicity: New perspectives from behavioral, neuroimaging, and neuropathological studies in humans and non-human primates. *Frontiers in Aging Neuroscience, 5*. doi:10.3389/fnagi.2013.00023 PMID:23805100

Guilarte, T. R., Chen, M. K., McGlothan, J. L., Verina, T., Wong, D. F., Zhou, Y., ... Schneider, J. S. (2006). Dopamine system dysfunction and subtle motor deficits in manganese-exposed non-human primates. *Experimental Neurology, 202*(2), 381–390. doi:10.1016/j.expneurol.2006.06.015 PMID:16925997

Guimaraes, R. M., Asmus, C. I., & Meyer, A. (2007). DDT reintroduction for malaria control: The cost-benefit debate for public health. *Cadernos de Saude Publica, 23*(12), 2835–2844. doi:10.1590/S0102-311X2007001200004 PMID:18157325

Guinotte, J. M., & Fabry, V. J. (2008). Ocean acidification and its potential effects on marine ecosystems. *Annals of the New York Academy of Sciences, 1134*(1), 320–342. doi:10.1196/annals.1439.013 PMID:18566099

Gund, G.-B. M., Jagtap, P. N., Ingale, V. B., & Patil, R. Y. (2013). Stroke : A Brain Attack. *IOSR Journal Of Pharmacy*, *3*(8), 1–23. doi:10.9790/3013-038201-23 PMID:27024833

Gunnarsson, L. G., Bodin, L., Söderfeldt, B., & Axelson, O. (1992). A case-control study of motor neurone disease: Its relation to heritability, and occupational exposures, particularly to solvents. *Occupational and Environmental Medicine*, *49*(11), 791–798. doi:10.1136/oem.49.11.791 PMID:1463680

Gupta, N., Khan, D. K., & Santra, S. C. (2009). An assessment of heavy metal contamination in vegetables grown in wastewater-irrigated areas of Titagarh, West Bengal, India. *Bulletin of Environmental Contamination and Toxicology*, *80*(2), 115–118. doi:10.100700128-007-9327-z PMID:18165915

Gupta, N., Khan, D. K., & Santra, S. C. (2009). Prevalence of intestinal helminth eggs on vegetables grown in wastewater-irrigated areas of Titagarh West Bengal, India. *Food Control*, *20*(10), 942–945. doi:10.1016/j.foodcont.2009.02.003

Gupta, S. K., Shukla, V. K., Vaidya, M. P., Roy, S. K., & Gupta, S. (1991). Serum trace elements and Cu/Zn ratio in breast cancer patients. *Journal of Surgical Oncology*, *46*(3), 178–181. doi:10.1002/jso.2930460311

Gursoy-Yuzugullu, O., Yuzugullu, H., Yilmaz, M., & Ozturk, M. (2011). Aflatoxin genotoxicity is associated with a defective DNA damage response bypassing p53 activation. *Liver International*, *31*(4), 561–571. doi:10.1111/j.1478-3231.2011.02474.x

Gusella, J. F., MacDonald, M. E., Ambrose, C. M., & Duyao, M. P. (1994). Molecular Genetics of Huntington's Disease. *Archives of Neurology*, *50*(11), 1157–1163. doi:10.1001/archneur.1993.00540110037003 PMID:8215974

Gutiérrez, J. M. (2012). Snakebite envenoming: a public health perspective. In *Public health-methodology, environmental and systems issues*. InTech.

Gutiérrez, J. M., Theakston, R. D. G., & Warrell, D. A. (2006). Confronting the neglected problem of snake bite envenoming: The need for a global partnership. *PLoS Medicine*, *3*(6), e150. doi:10.1371/journal.pmed.0030150 PMID:16729843

Guyton, K. Z., Loomis, D., Grosse, Y., El Ghissassi, F., Benbrahim-Tallaa, L., Guha, N., ... Straif, K. (2015). Carcinogenicity of tetrachlorvinphos, parathion, malathion, diazinon, and glyphosate. *The Lancet Oncology*, *16*(5), 490–491. doi:10.1016/S1470-2045(15)70134-8 PMID:25801782

Guzman Herrador, B. R., de Blasio, B. F., MacDonald, E., Nichols, G., Sudre, B., Vold, L., ... Nygård, K. (2015). Analytical studies assessing the association between extreme precipitation or temperature and drinking water-related waterborne infections: A review. *Environmental Health*, *14*(1), 29. doi:10.118612940-015-0014-y PMID:25885050

Gwiazda, R. H., Lee, D., Sheridan, J., & Smith, D. R. (2002). Low cumulative manganese exposure affects striatal GABA but not dopamine. *Neurotoxicology*, *23*(1), 69–76. doi:10.1016/S0161-813X(02)00002-5 PMID:12164549

Haacke, E. M., Cheng, N. Y. C., House, M. J., Liu, Q., Neelavalli, J., Ogg, R. J., ... Obenaus, A. (2005). Imaging iron stores in the brain using magnetic resonance imaging. *Magnetic Resonance Imaging*, *23*(1), 1–25. doi:10.1016/j.mri.2004.10.001 PMID:15733784

Haber, G., Ahmed, N. U., & Pekovic, V. (2012). Family History of Cancer and Its Association With Breast Cancer Risk Perception and Repeat Mammography. *American Journal of Public Health*, *102*(12), 2322–2329. doi:10.2105/AJPH.2012.300786

Habib, F. K., Lee, I. R., Stitch, S. R., & Smith, P. H. (1976). Androgen levels in the plasma and prostatic tissuesof patients with benign hypertrophy and carcinoma of the prostate. *The Journal of Endocrinology*, *71*(1), 99–107. doi:10.1677/joe.0.0710099 PMID:62008

Hadipour, A., Rajaee, T., Hadipour, V., & Seidirad, S. (2016). Multi-criteria decision-making model for wastewater reuse application: A case study from Iran. *Desalination and Water Treatment*, *57*(30), 13857–13864. doi:10.1080/19443994.2015.1060905

Hadley, N. F. (1970). Water relations of the desert scorpion, Hadrurus arizonensis. *The Journal of Experimental Biology*, *53*(3), 547–558. PMID:5487163

Haga, A., Nagase, H., Kito, H., & Sato, T. (1996). Enhanced invasiveness of tumour cells after hostexposure to heavy metals. *European Journal of Cancer*, *32A*(13), 2342–2347. doi:10.1016/S0959-8049(96)00349-8 PMID:9038619

Hagstrum, D. W., & Workman, E. B. (1971). Interaction of temperature and feeding rate in determining the rate of development of larval Culex tarsalis (Diptera, Culicidae). *Annals of the Entomological Society of America*, *64*(3), 668–671. doi:10.1093/aesa/64.3.668

Hahn, M., Massen, O., Nencki, M., & Pawlow, J. (1893). Die Eck'sche Fistel zwischen der unteren Hohlvene und der Pfortader und ihre Folgen für den Organismus. *Archiv für experimentelle Pathologie und Pharmakologie*, *32*(3–4), 161–210. doi:10.1007/BF01995065

Haider, L., Simeonidou, C., Steinberger, G., Hametner, S., Grigoriadis, N., Deretzi, G., ... Frischer, J. M. (2014). Multiple sclerosis deep grey matter: The relation between demyelination, neurodegeneration, inflammation and iron. *Journal of Neurology, Neurosurgery, and Psychiatry*, *85*(12), 1386–1395. doi:10.1136/jnnp-2014-307712 PMID:24899728

Haines, A. (1998). Global warming and vector-borne diseases. *Lancet*, *351*(9117), 1737–1738. doi:10.1016/S0140-6736(05)77778-3 PMID:9734919

Hajjami, K., Ennaji, M. M., Fouad, S., Oubrim, N., & Cohen, N. (2013). Wastewater Reuse for Irrigation in Morocco: Helminth Eggs Contamination's Level of Irrigated Crops and Sanitary Risk (A Case Study of Settat and Soualem Regions). *Journal of Bacteriology & Parasitology*, *4*, 163. doi:10.4172/2155- 9597.1000163

Haley, B. J., Cole, D. J., & Lipp, E. K. (2009). Distribution, diversity, and seasonality of waterborne *Salmonellae* in a rural watershed. *Applied and Environmental Microbiology*, *75*(5), 1248–1255. doi:10.1128/AEM.01648-08 PMID:19124594

Halliwell, B. (1992). Reactive Oxygen Species and the Central Nervous System. *Journal of Neurochemistry*, *59*(5), 1609–1623. doi:10.1111/j.1471-4159.1992.tb10990.x PMID:1402908

Hallman, H., Lange, L., Olson, L., Strömberg, I., & Jonsson, G. (1985). Neurochemical and Histochemical Characterization of Neurotoxic Effects of 1-Methyl-4-Phenyl-1,2,3,6-Tetrahydropyridine on Brain Catecholamine Neurones in the Mouse. *Journal of Neurochemistry*, *44*(1), 117–127. doi:10.1111/j.1471-4159.1985.tb07120.x PMID:3880581

Hamilton, J. L., & Luffman, I. (2009). Precipitation, pathogens, and turbidity trends in the little river, tennessee. *Physical Geography*, *30*(3), 236–248. doi:10.2747/0272-3646.30.3.236

Hanagasi, H. A., Ayribas, D., Baysal, K., & Emre, M. (2005). Mitochondrial complex i, ii/iii, and iv activities in familial and sporadic parkinson's disease. *The International Journal of Neuroscience*, *115*(4), 479–493. doi:10.1080/00207450590523017 PMID:15809215

Hands, S. L., Mason, R., Sajjad, M. U., Giorgini, F., & Wyttenbach, A. (2010). Metallothioneins and copper metabolism are candidate therapeutic targets in Huntington's disease. *Biochemical Society Transactions*, *38*(2), 552–558. doi:10.1042/BST0380552 PMID:20298220

Häni, H.Häni. (1990). The Analysis of Inorganic and Organic Pollutants in Soil with Special Regard to Their Bioavailability. *International Journal of Environmental Analytical Chemistry*, *39*(2), 197–208. doi:10.1080/03067319008027697

Hanjra, M. A., Blackwell, J., Carr, G., Zhang, F., & Jackson, T. M. (2012). Wastewater irrigation and environmental health: Implications for water governance and public policy. *International Journal of Hygiene and Environmental Health, 215*(3), 255–269. doi:10.1016/j.ijheh.2011.10.003 PMID:22093903

Han, Q., Lu, J., Duan, J., Su, D., Hou, X., Li, F., ... Huang, B. (2008). Gcn5- and elp3-induced histone H3 acetylation regulates hsp70 gene transcription in yeast. *The Biochemical Journal, 409*(3), 779–788. doi:10.1042/BJ20070578 PMID:17910533

Hansen, J., Sato, M., & Ruedy, R. (2012). Perception of climate change. *Proceedings of the National Academy of Sciences of the United States of America, 109*(37), E2415–E2423. doi:10.1073/pnas.1205276109 PMID:22869707

Hansen, M. E., & Matsumura, F. (2001). Effects of heptachlor epoxide on components of various signal transduction pathways important in tumor promotion in mouse hepatoma cells. Determination of the most sensitive tumor promoter related effect induced by heptachlor epoxide. *Toxicology, 160*(1–3), 139–153.

Hanslmeier, A. (2004). *The Sun and Space Weather.* Dordrecht: Springer Netherlands.

Hansson, E., Sasa, M., Mattisson, K., Robles, A., & Gutiérrez, J. M. (2013). Using geographical information systems to identify populations in need of improved accessibility to antivenom treatment for snakebite envenoming in Costa Rica. *PLoS Neglected Tropical Diseases, 7*(1), e2009. doi:10.1371/journal.pntd.0002009 PMID:23383352

Hansson, M., Djerbi, M., Rabbani, H., Mellstedt, H., Gharibdoost, F., Hassan, M., ... Abedi-Valugerdi, M. (2005). Exposure to mercuric chloride during the induction phase and after the onset of collagen-induced arthritis enhances immune/autoimmune responses and exacerbates the disease in DBA/1 mice. *Immunology, 114*(3), 428–437. doi:10.1111/j.1365-2567.2005.02105.x PMID:15720444

Harbach, R. E. (2004). The classification of genus Anopheles (Diptera: Culicidae): a working hypothesis of phylogenetic relationships. *Bulletin of Entomological Research, 94*(6), 537–553. doi:10.1079/BER2004321

Harper, S. L., Edge, V. L., Schuster-Wallace, C. J., Berke, O., & McEwen, S. A. (2011). Weather, water quality and infectious gastrointestinal illness in two Inuit communities in Nunatsiavut, Canada: Potential implications for climate change. *EcoHealth, 8*(1), 93–108. doi:10.100710393-011-0690-1 PMID:21785890

Harrison, J. C., Carvajal, M., & Garner, R. C. (1993). Does aflatoxin exposure in the United Kingdom constitute a cancer risk? *Environmental Health Perspectives, 99*, 99–105. doi:10.1289/ehp.939999

Harrison, R. M., Smith, D. J. T., & Luhana, L. (1996). Source apportionment of atmospheric polycyclic aromatic hydrocarbons collected from an urban location in Birmingham, U.K. *Environmental Science & Technology, 30*(3), 825–832. doi:10.1021/es950252d

Hartwig, A. (2013). Cadmium and cancer. *Metal Ions in Life Sciences, 11*, 491–507. doi:10.1007/978-94-007-5179-8_15 PMID:23430782

Harvell, C. D., Mitchell, C. E., Ward, J. R., Altizer, S., Dobson, A. P., Ostfeld, R. S., & Samuel, M. D. (2002). Climate warming and disease risks for terrestrial and marine biota. *Science, 296*(5576), 2158–2162. doi:10.1126cience.1063699 PMID:12077394

Hashimoto, M., Hsu, L. J., Xia, Y., Takeda, A., Sisk, A., Sundsmo, M., & Masliah, E. (1999). Oxidative stress induces amyloid–like aggregate formation of NACP/α–synuclein in vitro. *Neuroreport, 10*(4), 717–721. doi:10.1097/00001756-199903170-00011 PMID:10208537

Hashizume, M., Armstrong, B., Hajat, S., Wagatsuma, Y., Faruque, A. S., Hayashi, T., & Sack, D. A. (2007). Association between climate variability and hospital visits for non-cholera diarrhoea in Bangladesh: Effects and vulnerable groups. *International Journal of Epidemiology*, *36*(5), 1030–1037. doi:10.1093/ije/dym148 PMID:17664224

Hatcher, J., Pennell, K., & Miller, G. (2008). Parkinson's disease and pesticides: A toxicological perspective. *Trends in Pharmacological Sciences*, *29*(6), 322–329. doi:10.1016/j.tips.2008.03.007 PMID:18453001

Haurie, L., Sceia, A. & Thénié, J. (2009). *Inland transport and climate change.* A Literature Review UNECE.

Hayes, T. B., Case, P., Chui, S., Chung, D., Haeffele, C., Haston, K., ... Tsui, M. (2006). Pesticide mixtures, endocrine disruption, and amphibian declines: Are we underestimating the impact? *Environmental Health Perspectives*, *114*(Suppl 1), 40–50. doi:10.1289/ehp.8051 PMID:16818245

Hayes, T., Haston, K., Tsui, M., Honag, A., Haeffele, C., & Vonk, A. (2003). Atrazine-Induced Hermaphroditism at 0.1 ppb in American Leopard Frogs (Ranapipiens): Laboratory and Field Evidence. *Environmental Health Perspectives*, *111*(4), 568–575. doi:10.1289/ehp.5932

Hay, S. I., Guerra, C. A., Tatem, A. J., Noor, A. M., & Snow, R. W. (2004). The global distribution and population at risk of malaria: Past, present, and future. *The Lancet Infectious Diseases*, *4*(6), 327–336. doi:10.1016/S1473-3099(04)01043-6

Hay, S. I., Sinka, M. E., Okara, R. M., Kabaria, C. W., Mbithi, P. M., Tago, C. C., ... Temperley, W. H. (2010). Developing global maps of the dominant Anopheles vectors of human malaria. *PLoS Medicine*, *7*(2), e1000209. doi:10.1371/journal.pmed.1000209

HCEFLCD. (2007). *Etude d'aménagement du parc National d'Iriqui.* Maroc: HCEFLCD-DREF SO.

HCEFLCD. (2018, June). *Le Haut Commissariat aux Eaux et Forêts et à la Lutte Contre la Désertification (HCEFLCD).* Retrieved from eaux et forets: http://www.eauxetforets.gov.ma

HCP. (2014). *Caractéristiques Démographiques et Socio-Economiques: REGION DRAA-TAFILALET. Errachidia: HCP.* Direction Régionale Drâa-Tafilalet.

HCP. (2015). *Morocco between Millennium Development Goals and Sustainable Development Goals Achievements and Challenges.* Kingdom of Morocco High Commission For Planning.

HCP. (2018, May). *High Commission for Planning (Morocco).* Retrieved from Haut commissariat au plan: https://www.hcp.ma/

Heath, D. F., & Schlesinger, B. M. (1986). The Mg 280-nm doublet as a monitor of changes in solar ultraviolet irradiance. *Journal of Geophysical Research*, *91*(D8), 8672–8682. doi:10.1029/JD091iD08p08672

Heck, W. W., & Dunning, J. A. (1967). The effects of ozone on Tobacco and Pinto Bean as conditioned by several ecological factors. *Journal of the Air Pollution Control Association*, *17*(2), 112–114. doi:10.1080/00022470.1967.10468950

Hellerman, L., & Perkins, M. E. (1935). Activation of Enzymes Iii. the Role of Metal Ions in the Activation of Arginase. the Hydrolysis of Arginine Induced by Certain Metal Ions with Urease. *The Journal of Biological Chemistry*, *112*(1), 175–194.

Helzlsouer, K., Alberg, A., Huang, H., Hoffman, S., Strickland, P., Brock, J., ... Yager, J. (1999). Serum concentrations of organochlorine compounds and the subsequent development of breast cancer. *Cancer Epidemiology, Biomarkers & Prevention*, *8*, 525–532.

Hem, L. (2008). Modeller for vurdering av vannkilden som hygienisk barriere (Models for evaluation of the hygienic barrier effect of water sources). *Vann*, *1*, 19–34.

Henderson, R. W., & Hoevers, L. G. (1977). The seasonal incidence of snakes at a locality in northern Belize. *Copeia, 1977*(2), 349–355. doi:10.2307/1443914

Hendrickse, R. G. (1997). Of sick Turkeys, Kwashiorkor, malaria, perinatal mortality, Herion addicts and food poisoning: Research on influence of aflatoxins on child health in the tropics. *Journal of Tropical Medicine and Parasitology, 91*, 87–93.

Hendricks, K. A., Simpson, J. S., & Larsen, R. D. (1999). Neural tube defect along the Texas-Mexico border, 1993-1995. *American Journal of Epidemiology, 149*(12), 119–127. doi:10.1093/oxfordjournals.aje.a009766

Heneka, M. T., Carson, M. J., El Khoury, J., Landreth, G. E., Brosseron, F., Feinstein, D. L., ... Kummer, M. P. (2015). Neuroinflammation in Alzheimer's disease. *Lancet Neurology, 14*(4), 388–405. doi:10.1016/S1474-4422(15)70016-5 PMID:25792098

Hennig, A., Fink-Gremmels, J., & Leistner, L. (1991). Mutagenicity and effects of ochratoxin A on the frequency of sister chromatid exchange after metabolic activation. *IARC Scientific Publications*, 255–260.

Hennion, S., & Hennion, S. (2017). *Neuropsychologie et neuroimagerie des troubles de cognition sociale dans l ' épilepsie du lobe temporal.* Academic Press.

Herndon, E. S., Hladik, C. L., Shang, P., Burns, D. K., Raisanen, J., & White, C. L. III. (2009). Neuroanatomic profile of polyglutamine immunoreactivity in Huntington disease brains. *Journal of Neuropathology and Experimental Neurology, 68*(3), 250–261. doi:10.1097/NEN.0b013e318198d320 PMID:19225411

Herrington, R., Johnson, B., & Hunter, F. (1997). Canada Country Study: Climate Impacts and Adaptation, Volume 3: Responding to Global Climate Change in the Prairies. Environment Canada.

Hertzman, C., Wiens, M., Bowering, D., Snow, B., & Calne, D. (1990). Parkinson's disease: A case-control study of occupational and environmental risk factors. *American Journal of Industrial Medicine, 17*(3), 349–355. doi:10.1002/ajim.4700170307 PMID:2305814

Hertz, P. E., Huey, R. B., & Garland, T. Jr. (1988). Time budgets, thermoregulation, and maximal locomotor performance: Are reptiles olympians or boy scouts? *American Zoologist, 28*(3), 927–938. doi:10.1093/icb/28.3.927

Herwaarden, A. E. V., Wagenaar, E., Karnekamp, B., Merino, G., Jonker, J. W., & Schinkel, A. H. (2005). Breast cancer resistance protein (Bcrp1/Abcg2) reduces systemic exposure of the dietary carcinogens aflatoxin B1, IQ and Trp-P-1 but also mediates their secretion into breast milk. *Carcinogenesis, 27*(1), 123–130. doi:10.1093/carcin/bgi176

He, T.-T., Zuo, A.-J., Wang, J.-G., & Zhao, P. (2017). Organochlorine pesticides accumulation and breast cancer: A hospital-based case–control study. *Tumour Biology, 39*(5), 1010428317699114. doi:10.1177/1010428317699114 PMID:28459199

Hettiaratchi, J., Amatya, P., & Joshi, R. (2002). Biotreatment of flare pit waste. *Journal of Canadian Petroleum Technology, 41*, 30–36.

Heugens, E. H. W., Hendriks, A. J., Dekker, T., Straalen, N. M. V., & Admiraal, W. (2001). A review of the effects of multiple stressors on aquatic organisms and analysis of uncertainty factors for use in risk assessment. *Critical Reviews in Toxicology, 31*(3), 247–284. doi:10.1080/20014091111695 PMID:11405441

Heussner, A. H., & Bingle, L. E. (2015). Comparative ochratoxin toxicity: A review of the available data. *Toxins, 7*(10), 4253–4282. doi:10.3390/toxins7104253

He, Y., Peng, L., Zhang, W., Liu, C., Yang, Q., Zheng, S., ... Wu, K. (2018). Adipose tissue levels of polybrominated diphenyl ethers and breast cancer risk in Chinese women: A case-control study. *Environmental Research, 167*, 160–168. doi:10.1016/j.envres.2018.07.009

Hickling, R., Roy, D. B., Hill, J. K., Fox, R., & Thomas, C. D. (2006). The distributions of a wide range of taxonomic groups are expanding polewards. *Global Change Biology*, *12*(3), 450–455. doi:10.1111/j.1365-2486.2006.01116.x

High Planning Commission of Morocco (HCP). (2014). *Report of General Census of Population and Housing of 2004.* Retrieved from http://www.hcp.ma/Recensement-General-de-la-Population-et-de-l-Habitat-2004_a92.html

Higley, E. B., Newsted, J. L., Zhang, X., Giesy, J. P., & Hecker, M. (2010). Assessment of chemical effects on aromatase activity using the H295R cell line. *Environmental Science and Pollution Research International*, *17*(5), 1137–1148. doi:10.100711356-009-0285-3

Hiller, W., Zaudig, M., & von Bose, M. (1989). The overlap between depression and anxiety on different levels of psychopathology. *Journal of Affective Disorders*, *16*(2-3), 223-31. Retrieved from http://www.ncbi.nlm.nih.gov/pubmed/2522121

Hinkel, J. (2011). "Indicators of vulnerability and adaptive capacity": Towards a clarification of the science–policy interface. *Global Environmental Change*, *21*(1), 198–208. doi:10.1016/j.gloenvcha.2010.08.002

Hinwood, A. L., Jolley, D. J., & Sim, M. R. (1999). Cancer incidence and high environmental arsenic concentrations in rural populations: Results of an ecological study. *International Journal of Environmental Health Research*, *9*, 131–141.

Hiraga, K., Yamamoto, S., Fukuda, H., Hamanaka, N., & Oda, K. (2005). Enniatin has a new function as an inhibitor of Pdr5p, one of the ABC transporters in Saccharomyces cerevisiae. *Biochemical and Biophysical Research Communications*, *328*(4), 1119–1125. doi:10.1016/j.bbrc.2005.01.075

Hmamouch, A., Amarir, F., Fellah, H., Karzaz, M., Bekhti, K., & Rhajaoui, M. (2014). Coexistence of *Leishmaniatropica* and *Leishmaniainfantum*in Sefrou province, Morocco. *Acta Tropica*, *130*, 94–99. doi:10.1016/j.actatropica.2013.10.012 PMID:24161534

Hnichi, M. (1989). Un parc national en perspective. *Le Matin du Maghreb*, *12*, 142.

Hobson, M., Milhouse, M., Rajanna, B., Natural, D., & Sciences, A. (1986). *~ S Effects of Cadmium on the Uptake of Dopamine and Norepinephrine in Rat Brain Synaptosomes.* Academic Press.

Hodges, R. E., & Elsner, J. B. (2011). Evidence linking solar variability with US hurricanes. *International Journal of Climatology*, *31*(13), 1897–1907. doi:10.1002/joc.2196

Hofer-Warbinek, R., Schmid, J. A., Stehlik, C., Binder, B. R., Lipp, J., & de Martin, R. (2000). Activation ofNF-kappa B by XIAP, the X chromosome-linked inhibitor of apoptosis, in endothelial cellsinvolves TAK1. *The Journal of Biological Chemistry*, *275*(29), 22064–22068. doi:10.1074/jbc.M910346199 PMID:10807933

Hoffner, G.1., Kahlem, P., & Djian, P. (2002). Perinuclear localization of huntingtin as a consequence of its binding to microtubules through an interaction with beta-tubulin: Relevance to Huntington's disease. *Journal of Cell Science*, *115*(Pt 5), 941–948. PMID:11870213

Hofmann, G., & Somero, G. (1995). Evidence for protein damage at environmental temperatures: Seasonal changes in levels of ubiquitin conjugates and hsp70 in the intertidal mussel Mytilus trossulus. *The Journal of Experimental Biology*, *198*(7), 1509–1518. PMID:9319406

Holford, T., Zheng, T., Mayne, S., Zahm, S., Tessari, J., & Boyle, P. (2000). Joint effects of nine polychlorinated biphenyl (PCB) congeners on breast cancer risk. *International Journal of Epidemiology*, *29*(6), 975–982. doi:10.1093/ije/29.6.975

Holmes, A. K., Koller, K. R., Kieszak, S. M., Sjodin, A., Calafat, A. M., Sacco, F. D., ... Rubin, C. H. (2014). Case-control study of breast cancer and exposure to synthetic environmental chemicals among Alaska Native women. *International Journal of Circumpolar Health*, *73*(1), 25760. doi:10.3402/ijch.v73.25760

Holmstrup, M., Bindesbøl, A.-M., Oostingh, G. J., Duschl, A., Scheil, V., Köhler, H.-R., ... Spurgeon, D. J. (2010). Interactions between effects of environmental chemicals and natural stressors: A review. *The Science of the Total Environment*, *408*(18), 3746–3762. doi:10.1016/j.scitotenv.2009.10.067 PMID:19922980

Homme, K., Kern, J. K., Geier, D. A., Bjørklund, G., King, P. G., Homme, K. G., … Geier, M. R. (2014). *Amalgams and Chronic Illness, Fatigue, Depression, Anxiety, and Suicide*. Academic Press.

Honisch, B., Ridgwell, A., Schmidt, D. N., Thomas, E., Gibbs, S. J., Sluijs, A., ... Williams, B. (2012). The geological record of ocean acidification. *Science*, *335*(6072), 1058–1063. doi:10.1126cience.1208277 PMID:22383840

Honjo, K., Black, S. E., & Verhoeff, N. P. L. G. (2012). Alzheimer's Disease, Cerebrovascular Disease, and the β-amyloid Cascade. *The Canadian Journal of Neurological Sciences*, *39*(06), 712–728. doi:10.1017/S0317167100015547 PMID:23227576

Hoogenraad, T. (1988). Dithiocarbamates and Parkinson's disease. *Lancet*, *1*(8588), 767. doi:10.1016/S0140-6736(88)91573-5 PMID:2895292

Hooper, M. J., Ankley, G. T., Cristol, D. A., Maryoung, L. A., Noyes, P. D., & Pinkerton, K. E. (2013). Interactions between chemical and climate stressors: A role for mechanistic toxicology in assessing climate change risks. *Environmental Toxicology and Chemistry*, *32*(1), 32–48. doi:10.1002/etc.2043 PMID:23136056

Hopenhayn-Rich, C., Stump, M. L., & Browning, S. R. (2002). Regional assessment of atrazine exposure and incidence of breast and ovarian cancers in Kentucky. *Archives of Environmental Contamination and Toxicology*, *42*, 127–136. doi:10.1007002440010300

Hoque, E., Hope, V., Scragg, R., Baker, M., & Shrestha, R. (2004). A descriptive epidemiology of giardiasis in New Zealand and gaps in surveillance data. *The New Zealand Medical Journal*, *117*, U1149. PMID:15570332

Ho, S.-M., Lee, M.-T., Lam, H.-M., & Leung, Y.-K. (2011). Estrogens and Prostate Cancer: Etiology, Mediators, Prevention, and Management. *Endocrinology and Metabolism Clinics of North America*, *40*(3), 591–614. doi:10.1016/j.ecl.2011.05.002 PMID:21889723

Houghton, M. (1997). *Global warming: the complete briefing* (2nd ed.). Cambridge, UK: Cambridge University Press. doi:10.1017/CBO9781139164252

Houghton, R. A. (2003). Revised estimates of the annual net flux of carbon to the atmosphere from changes in land use and land management 1850–2000. *Tellus. Series B, Chemical and Physical Meteorology*, *55*(2), 378–390.

House, E., Collingwood, J., Khan, A., Korchazkina, O., Berthon, G., & Exley, C. (2004). Aluminium, iron, zinc and copper influence the in vitro formation of amyloid fibrils of Aβ42 in a manner which may have consequences for metal chelation therapy in Alzheimer's disease. *Journal of Alzheimer's Disease*, *6*(3), 291–301. doi:10.3233/JAD-2004-6310 PMID:15201484

Houston, M. C. (2011). Role of Mercury Toxicity in Hypertension, Cardiovascular Disease, and Stroke. *Journal of Clinical Hypertension*, *13*(8), 621–627. doi:10.1111/j.1751-7176.2011.00489.x PMID:21806773

Hovey, R. C., Coder, P. S., Wolf, J. C., Sielken, R. L., Tisdel, M. O., & Breckenridge, C. B. (2011). Quantitative assessment of mammary gland development in female Long Evans rats following in utero exposure to atrazine. *Toxicological Sciences*, *119*(2), 380–390. doi:10.1093/toxsci/kfq337

Howard, G., Calow, R., Macdonald, A., & Bartram, J. (2016). *Climate Change and Water and Sanitation : Likely Impacts and Emerging Trends for Action*. Academic Press. . doi:10.1146/annurev-environ-110615-085856

Howden, S. M., Soussana, J. F., Tubiello, F. N., Chhetri, N., Dunlop, M., & Meinke, H. (2007). Adapting agriculture to climate change. *Proceedings of the National Academy of Sciences of the United States of America*, *104*(50), 19691–19696. doi:10.1073/pnas.0701890104 PMID:18077402

Howes, R. E., Reiner, R. C. Jr, Battle, K. E., Longbottom, J., Mappin, B., Ordanovich, D., ... Smith, D. L. (2015). Plasmodium vivax transmission in Africa. *PLoS Neglected Tropical Diseases*, *9*(11), e0004222. doi:10.1371/journal.pntd.0004222

Hoxie, N. J., Davis, J. P., Vergeront, J. M., Nashold, R. D., & Blair, K. A. (1997). Cryptosporidiosis-associated mortality following a massive waterborne outbreak in Milwaukee, Wisconsin. *American Journal of Public Health*, *87*(12), 2032–2035. doi:10.2105/AJPH.87.12.2032 PMID:9431298

Hoyer, A. P., Grandjean, P., Jorgensen, T., Brock, J. W., & Hartvig, H. B. (1998). Organochlorine exposure and risk of breast cancer. *Lancet*, *352*(9143), 1816–1820. doi:10.1016/S0140-6736(98)04504-8

Hoyer, A. P., Jørgensen, T., Rank, F., & Grandjea, P. (2001). Organo-Chlorine Exposures Influence on Breast Cancer Risk and Survival According to Estrogen Receptor Status: A Danish Cohort-Nested Case-Control Study. *Biomedical Central Cancer*, *1*, 8.

Hoyer, A., Gerdes, A., Jorgensen, T., Rank, F., & Hartvig, H. (2002a). Organochlorines, p53 mutations in relation to breast cancer risk and survival. *Breast Cancer Research and Treatment*, *71*(1), 59–65. doi:10.1023/A:1013340327099

Hoyer, A., Jorgensen, T., Grandjean, P., & Hartvig, H. (2000). Repeated measurements of organochlorine exposure and breast cancer risk (Denmark). *Cancer Causes & Control*, *11*(2), 177–184. doi:10.1023/A:1008926219539

Hoyer, P., Grandjean, P., Jorgensen, T., Brock, J. W., & Hartvig, H. B. (2002b). Dieldrin as a Risk Factor for Breast Cancer and for Increased Mortality Once Breast Cancer Is Detected. *Journal of Clinical Epidemiology*, *53*, 323–330.

Hsu, I. C., Metcalf, R. A., Sun, T., Welsh, J. A., Wang, N. J., & Harris, C. C. (1991). Mutational hot spot in the p53 gene in human hepatocellular carcinomas. *Nature*, *350*(6317), 427–428. doi:10.1038/350427a0

Huang, E., Ong, W. Y., & Connor, J. R. (2004). Distribution of divalent metal transporter-1 in the monkey basal ganglia. *Neuroscience*, *128*(3), 487–496. doi:10.1016/j.neuroscience.2004.06.055 PMID:15381278

Huang, T., Li, X., Rijnaarts, H., Grotenhuis, T., Ma, W., Sun, X., & Xu, J. (2014). Effects of storm runoff on the thermal regime and water quality of a deep, stratified reservoir in a temperate monsoon zone, in Northwest China. *The Science of the Total Environment*, *485-486*, 820–827. doi:10.1016/j.scitotenv.2014.01.008 PMID:24485908

Huang, Z., Footitt, S., Tang, A., & Finch-Savage, W. E. (2018). Predicted global warming scenarios impact on the mother plant to alter seed dormancy and germination behaviour in Arabidopsis. *Plant, Cell & Environment*, *41*(1), 187–197. doi:10.1111/pce.13082 PMID:29044545

Huffaker, C. B. (1944). The temperature relations of the immature stages of the malarial mosquito, Anopheles quadrimaculatus Say, with a comparison of the developmental power of constant and variable temperatures in insect metabolism. *Annals of the Entomological Society of America*, *37*(1), 1–27. doi:10.1093/aesa/37.1.1

Huff, J., Lunn, R. M., Waalkes, M. P., Tomatis, L., & Infante, P. F. (2007). Cadmium-induced cancers in animals and in humans. *International Journal of Occupational and Environmental Health*, *13*(2), 202–212. doi:10.1179/oeh.2007.13.2.202 PMID:17718178

Hulme, M., Wigley, T. M., Barrow, E., Raper, S., Centella, A., Smith, S., & Chipanshi, A. (2000). *Using a Climate Scenario Generator for Vulnerability and Adaptation Assessments: MAGICC and SCENGEN: Version 2.4 Workbook*. Climatic Research Unit.

Hung, A. S. M., Liang, Y., Chow, T. C. H., Tang, H. C., Wu, S. L. Y., Wai, M. S. M., & Yew, D. T. (2016). Mutated tau, amyloid and neuroinflammation in Alzheimer disease—A brief review. *Progress in Histochemistry and Cytochemistry*, *51*(1), 1–8. doi:10.1016/j.proghi.2016.01.001 PMID:26851150

Hung, D. Z., Liau, M. Y., & Lin-Shiau, S. Y. (2003). The clinical significance of venom detection in patients of cobra snakebite. *Toxicon*, *41*(4), 409–415. doi:10.1016/S0041-0101(02)00336-7 PMID:12657310

Hung, I. H., Suzuki, M., Yamaguchi, Y., Yuan, D. S., Klausner, R. D., & Gitlin, J. D. (1997). Biochemical characterization of the Wilson disease protein and functional expression in the yeast Saccharomyces cerevisiae. *The Journal of Biological Chemistry*, *272*(34), 21461–21466. doi:10.1074/jbc.272.34.21461 PMID:9261163

Hunter, P., Colford, J., LeChevallier, M., Binder, S., & Berger, P. (2001). Waterborne diseases. *Emerg Infect Dis, 7*, 544 – 545.

Hunter, P.R. (2003). Climate change and waterborne and vector-borne disease. *J Appl Microbiol, 94*, 37S-46S.

Hunter, P. (1997). *Waterborne disease: epidemiology and ecology*. John Wiley & Sons.

Hunter, P. R. (2003). Climate change and waterborne and vector-borne disease. *Journal of Applied Microbiology*, *94*(s1), 37–46. doi:10.1046/j.1365-2672.94.s1.5.x PMID:12675935

Hunter, P. R., & Thompson, R. C. (2005). The zoonotic transmission of *Giardia* and *Cryptosporidium. International Journal for Parasitology*, *35*(11-12), 1181–1190. doi:10.1016/j.ijpara.2005.07.009 PMID:16159658

Huntingford, C., Jones, P. D., Livina, V. N., Lenton, T. M., & Cox, P. M. (2013). No increase in global temperature variability despite changing regional patterns. *Nature*, *500*(7462), 327–330. doi:10.1038/nature12310 PMID:23883935

Huovinen, P. (2001). Resistance to trimethoprim-sulfamethoxazole. *Nephrology, Dialysis, Transplantation*, *32*, 1608–1614. PMID:11340533

Hurley, S., Reynolds, P., Goldberg, D., Nelson, D. O., Jeffrey, S. S., & Petreas, M. (2011). Adipose levels of polybrominated diphenyl ethers and risk of breast cancer. *Breast Cancer Research and Treatment*, *129*(2), 505–511. doi:10.100710549-011-1481-7

Hursh, J. B., Clarkson, T. W., Cherian, M. G., Vostal, J. J., & Mallie, R. V. (1976). Clearance of mercury vapor inhaled by human subjects. *Archives of Environmental Health*, *31*(6), 302–309. doi:10.1080/00039896.1976.10667240 PMID:999343

Huster, D. (2010). Wilson disease. *Best Practice & Research. Clinical Gastroenterology*, *24*(5), 531–539. doi:10.1016/j.bpg.2010.07.014 PMID:20955957

Hutchinson, M. L., Walters, L. D., Moore, T., Thomas, D. J. I., & Avery, S. M. (2005a). Fate of pathogens present in livestock wastes spread onto fescue lots. *Applied and Environmental Microbiology*, *71*(2), 691–696. doi:10.1128/AEM.71.2.691-696.2005 PMID:15691918

Hutchinson, M. L., Walters, L. D., Moore, T., Thomas, D. J. I., & Avery, S. M. (2005b). Decline of zoonotic agents in liquid livestock wastes stored in batches on-farm. *Letters in Applied Microbiology*, *99*(1), 58–65. doi:10.1111/j.1365-2672.2005.02585.x

Hutton, G. (2013). Global costs and benefits of reaching universal coverage of sanitation and drinking-water supply. *Journal of Water and Health*, *11*(1), 1–12. doi:10.2166/wh.2012.105 PMID:23428544

Huwe, J. K., Hakk, H., Smith, D. J., Diliberto, J. J., Richardson, V., Stapleton, H. M., & Birnbaum, L. S. (2008). Comparative absorption and bioaccumulation of polybrominated diphenyl ethers following ingestion via dust and oil in male rats. *Environmental Science & Technology*, *42*(7), 2694–2700. doi:10.1021/es702644k

IARC. (1990). *IARC Monographs on the Evaluation of Carcinogenic Risks to Humans: Chromium, Nickel and Welding.* International Agency for Research on Cancer.

IARC. (1991). Occupational exposures in insecticide application, and some pesticides. *IARC Monographs on the Evaluation of Carcinogenic Risks to Humans*, 53. PMID:1688189

IARC. (1999). Some chemicals that cause tumours of the kidney or urinary bladder in rodents and some other substances. *IARC Monographs on the Evaluation of Carcinogenic Risks to Humans*, *73*, 131–182. PMID:10804954

IARC. (2002). *Some traditional herbal medicines, some mycotoxins, naphthalene and styrene.* IARC..

IARC. (2004). Some drinking-water disinfectants and contaminants, including arsenic. *IARC Monographs on the Evaluation of Carcinogenic Risks to Humans*, *84*, 269–477. PMID:15645578

IARC. (2006). *Inorganic and organic lead compounds.* IARC.

IARC. (2012). review of human carcinogens, Part C: Arsenic, Metals, Fibres, and Dusts. *IARC Monographs, Lyon, France*, *100*, 196–211.

IARC. (2018). *Aflatoxins.* Available at https://monographs.iarc.fr/wp-content/uploads/2018/06/mono100F-23.pdf

Ibarluzea, J., Fernandez, M., Santa-Marina, L., Olea-Serrano, M., Rivas, A., Aurrekoetxea, J., ... Olea, N. (2004). Breast cancer risk and the combined effect of environmental estrogens. *Cancer Causes & Control*, *15*(6), 591–600. doi:10.1023/B:CACO.0000036167.51236.86

Idrisa, S., Belghyti, D., El Kharrim, K., & Yoro, C. (2010). *Parasitic contamination of the mint and turnip irrigated by untreated wastewater in Sidi Yahia Gharb.* Balwois, Ohrid, Republic of Macedonia.

Iguermia, S., Harmouche, T., Mikou, O., Amarti, A., & Mernissi, F. Z. (2011). Mucocutaneousleishmaniasis in Morocco, evidence of the parasite's ecological evolution. *Medecine et Maladies Infectieuses*, *41*, 47–48. doi:10.1016/j.medmal.2010.07.004 PMID:20947275

Ikeda, Y., Toda, S., Wang, M., & Nakazawa, S. (1994). Early changes of blood-brain barrier and superoxide scavenging activity in rat cryogenic brain injury. In *Brain Edema IX* (pp. 136–138). Vienna: Springer. doi:10.1007/978-3-7091-9334-1_36

Imaida, K., & Shirai, T. (2000). Endocrine disrupting chemicals and carcinogenesis--breast, testis and prostate cancer. Nihon Rinsho. *Japanese Journal of Clinical Medicine*, *58*(12), 2527–2532.

Imbahale, S. S., Mweresa, C. K., Takken, W., & Mukabana, W. R. (2011). Development of environmental tools for anopheline larval control. *Parasites & Vectors*, *4*(1), 130. doi:10.1186/1756-3305-4-130

Immerzeel, W., Droogers, P., Terink, W., Hoogeveen, J., Hellegers, P., Bierkens, M., & Beek, R. v. (2011). *Middle-East and Northern Africa Water Outlook.* World Bank, Future Water.

In, R. F., Intermittently, M., To, E., & Mercury, E. (2000). *Renal function in miners intermittently exposed to elemental mercury vapour.* Academic Press.

Ingre, C., Roos, P. M., Piehl, F., Kamel, F., & Fang, F. (2015). Risk factors for amyotrophic lateral sclerosis. *Clinical Epidemiology*, *7*, 181–193. doi:10.2147/CLEP.S37505 PMID:25709501

INSERM Collective Expertise Centre. (2000). Cancer: A methodological approach for studying the link between cancer and the environment. In *INSERM Collective Expert Reports.* Paris: Institut national de la santé et de la recherche médicale. Retrieved from http://www.ncbi.nlm.nih.gov/books/NBK7114/

Intergovernmental Panel on Climate Change (IPCC). (2014). IPCC Fifth Assessment Report - *The Synthesis Report*. Author.

International Agency for Research on Cancer (IARC). (1993). Evaluation of carcinogenic risks of chemical to humans. In Some naturally-occurring substances: Food Items and Constituent. IARC.

International Agency for Research on Cancer (IARC). (2006). Inorganic and Organic Lead Compounds. Lyon, France: Author.

International Agency for Research on Cancer. (1993). IARC Scientific Publications: Vol. 58. *Be, Cd, Hg, and Exposures in the Glass Manufacturing Industry, IARC Monographs on the Evaluation of Carcinogenic Risks of Chemicals to Humans*. Lyon, France: Author.

International Agency for Research on Cancer. (1997). *Polychlorinated dibenzo-para-dioxins and polychlorinated dibenzofurans. IARC monographs on the evaluation of carcinogenic risks to humans* (Vol. 69). Lyon, France: IARC.

International Agency for Research on Cancer. (1999). *Some chemicals that cause tumours of the kidney or urinary bladder in rodents and some other substances: Summary of data reported and evaluation. IARC monographs on the evaluation of carcinogenic risks to humans* (Vol. 73). Lyon, France: IARC.

Ionescu, J. G., Novotny, J., Stejskal, V., Latsch, A., Blaurock-Busch, E., & Eisenmann-Klein, M. (2006). Increased levels of transition metals in breast cancer tissue. *Neuroendocrinology Letters*, *27*, 36–39.

IPCC AR4. (2007). Climate Change 2007: Synthesis Report. Contribution of Working Groups I, II and III to the Fourth Assessment. Geneva, Switzerland: IPCC.

IPCC. (2001). *IPCC Third Assessment Report. Synthesis Report*. Cambridge, UK: Cambridge University.

IPCC. (2007). *Climate Change 2007: Synthesis report*. Cambridge, UK: Cambridge University Press.

IPCC. (2007). *Climate change 2007-impacts, adaptation and vulnerability: Working group II contribution to the fourth assessment report of the IPCC* (Vol. 4). Cambridge University Press.

IPCC. (2014). *Climate Change 2014: Impacts, Adaptation, and Vulnerability. Part A: Global and Sectoral Aspects. Contributions of Working Group II to the Fifth Assessment Report of the Intergovernmental Panel on Climate Change*. Cambridge, UK: Cambridge University Press.

Irvine, D. G., Schiefer, H. B., & Hader, W. J. (1988). Geotoxicology of multiple sclerosis: The Henribourg, Saskatchewan, cluster focus. II. The soil. *The Science of the Total Environment*, *77*(2-3), 175–188. doi:10.1016/0048-9697(88)90054-X PMID:3241961

Islam, M. S., Han, S., Ahmed, M. K., & Masunaga, S. (2014). Assessment of Trace Metal Contamination in Water and Sediment of Some Rivers in Bangladesh. *Journal of Water and Environment Technology*, *12*(2), 109–121. doi:10.2965/jwet.2014.109

Itoh, H., Iwasaki, M., Hanaoka, T., Kasuga, Y., Yokoyama, S., Onuma, H., ... Tsugane, S. (2009). Serum organochlorines and breast cancer risk in Japanese women: A case–control study. *Cancer Causes & Control*, *20*(5), 567–580. doi:10.100710552-008-9265-z PMID:19031103

Ittner, L. M., & Götz, J. (2011). Amyloid-β and tau — a toxic pas de deux in Alzheimer's disease. *Nature Reviews. Neuroscience*, *12*(2), 67–72. doi:10.1038/nrn2967 PMID:21193853

Iwasaki, M., Inoue, M., Sasazuki, S., Kurahashi, N., Itoh, H., Usuda, M., & Tsugane, S. (2008). Plasma organochlorine levels and subsequent risk of breast cancer among Japanese women: A nested case–control study. *The Science of the Total Environment*, *402*(2-3), 176–183. doi:10.1016/j.scitotenv.2008.05.009 PMID:18555519

Jackson, L. E., Wheeler, S. M., Hollander, A. D., Orlove, B. S., Six, J., Sumner, D. A., ... Tomich, T. P. (2011). Case study on potential agricultural responses to climate change in a California landscape. *Climatic Change*, *109*(S1), 407–427. doi:10.100710584-011-0306-3

Jagai, J. S., Castronovo, D. A., Monchak, J., & Naumova, E. N. (2009). Seasonality of cryptosporidiosis: A meta-analysis approach. *Environmental Research*, *109*(4), 465–478. doi:10.1016/j.envres.2009.02.008 PMID:19328462

Jagai, J. S., Sarkar, R., Castronovo, D., Kattula, D., McEntee, J., Ward, H., ... Naumova, E. N. (2012). Seasonality of rotavirus in South Asia: A meta-analysis approach assessing associations with temperature, precipitation, and vegetation index. *PLoS One*, *7*(5), e38168. doi:10.1371/journal.pone.0038168 PMID:22693594

Jaga, K., & Dharmani, C. (2007). The interrelation between organophosphate toxicity and the epidemiology of depression and suicide. *Reviews on Environmental Health*, *22*(1), 57–73. doi:10.1515/REVEH.2007.22.1.57 PMID:17508698

Jāhān, S. (2015). *Human Development Report 2015 - Work for Human Development*. New York: United Nations Development Programme.

Jallow, M., Awadh, D. J., Albaho, M. S., Devi, V. Y., & Thomas, B. M. (2017). Pesticide Knowledge and Safety Practices among Farm Workers in Kuwait: Results of a Survey. *International Journal of Environmental Research and Public Health*, *14*(4), 340. doi:10.3390/ijerph14040340 PMID:28338612

Jan, A. T., Azam, M., Siddiqui, K., Ali, A., Choi, I., & Haq, Q. M. R. (2015). Heavy metals and human health: Mechanistic insight into toxicity and counter defense system of antioxidants. *International Journal of Molecular Sciences*, *16*(12), 29592–29630. doi:10.3390/ijms161226183 PMID:26690422

Jankovic, J. (2008). Parkinson's disease: Clinical features and diagnosis. *Journal of Neurology, Neurosurgery, and Psychiatry*, *79*(4), 368–376. doi:10.1136/jnnp.2007.131045 PMID:18344392

Järup, L. (2003). Hazards of heavy metal contamination. *British Medical Bulletin*, *68*(1), 167–182. doi:10.1093/bmb/ldg032 PMID:14757716

Jarup, L., & Akesson, A. (2009). Current status of cadmium as an environmental health problem. *Toxicology and Applied Pharmacology*, *238*(3), 201–208. doi:10.1016/j.taap.2009.04.020

Jayakody, N., Harris, E. C., & Coggon, D. (2015). Phenoxy herbicides, soft-tissue sarcoma and non-Hodgkin lymphoma: A systematic review of evidence from cohort and case–control studies. *British Medical Bulletin*, *114*(1), 75–94. doi:10.1093/bmb/ldv008 PMID:25790819

Jean, H., Emard, J.-F., Thouez, J.-P., Houde, L., Robitaille, Y., Mathieu, J., ... Gauvreau, D. (1996). Alzheimer's disease: Preliminary study of spatial distribution at birth place. *Social Science & Medicine*, *42*(6), 871–878. doi:10.1016/0277-9536(95)00185-9 PMID:8778999

JECFA-Joint FAO/WHO Expert Committee on Food Additives. (2000). Zearalenone. In: JECFA (Ed.), Safety evaluation of certain food additives and contaminants. WHO.

JECFA-Joint FAO/WHO Expert Committee on Food Additives. (2001). *Safety evaluation of certain mycotoxins in food.* Prepared by the 56th Meeting of the Food Additives Series No. 47. Geneva.

Jenkitkasemwong, S., Wang, C.-Y., Mackenzie, B., & Knutson, M. D. (2012). Physiologic implications of metal-ion transport by ZIP14 and ZIP8. *Biometals : An International Journal on the Role of Metal Ions in Biology. Biochemical Medicine*, *25*(4), 643–655. doi:10.100710534-012-9526-x

Jenner, P. (2003). Oxidative stress in Parkinson's disease. *Annals of Neurology*, *53*(S3), S26–S38. doi:10.1002/ana.10483 PMID:12666096

Jestoi, M. (2008). Emerging *Fusarium*-mycotoxins fusaproliferin, beauvericin, enniatins, and moniliformin- a review. *Critical Reviews in Food Science and Nutrition*, *48*(1), 21–49. doi:10.1080/10408390601062021

Jett, D. A. (2011). Neurotoxic Pesticides and Neurologic Effects. *Neurologic Clinics*, *29*(3), 667–677. doi:10.1016/j.ncl.2011.06.002 PMID:21803217

Jimenez, R., & Alvarez, A. M. (2005). Control de la degradaci_on de suelos. Universidad Aut_onoma de Madrid, Madrid.

Jimenez, B., & Asano, T. (2008). *Water Reuse: An International Survey. Contrasts, Issues and Needs around the World.* London: IWA Publishing.

Jimenez-Sanchez, M., Licitra, F., Underwood, B. R., & Rubinsztein, D. C. (2017). Huntington's Disease: Mechanisms of Pathogenesis and Therapeutic Strategies. *Cold Spring Harbor Perspectives in Medicine*, *7*(7), a024240. doi:10.1101/cshperspect.a024240 PMID:27940602

Jin, M. W., Xu, S. M., An, Q., & Wang, P. (2016). A review of risk factors for childhood leukemia. *European Review for Medical and Pharmacological Sciences*, *20*(18), 3760–3764. PMID:27735044

Jiricka, A., Formayer, H., Schmidt, A., Völler, S., Leitner, M., Fischer, T. B., & Wachter, T. F. (2016). Consideration of climate change impacts and adaptation in EIA practice — Perspectives of actors in Austria and Germany. *Environmental Impact Assessment Review*, *57*, 78–88. doi:10.1016/j.eiar.2015.11.010

Jofre, J., Blanch, A. R., & Lucena, F. (2010) Waterborne infectious disease outbreaks associated with water scarcity and rainfall events. In Water Scarcity in the Mediterranean: Perspectives Under Global Change (pp. 147-159). Berlin: Springer Berlin Heidelberg.

Johanning, E., Gareis, M., Nielsen, K., Dietrich, R., & Märtlbauer, E. (2002). Airborne mycotoxin sampling and screening analysis. *Proceedings of the 9th International Conference on Indoor Air Quality and Climate (Indoor Air).*

Johansson, C., Castoldi, A. F., Onishchenko, N., Manzo, L., Vahter, M., & Ceccatelli, S. (2007). Neurobehavioural and molecular changes induced by methylmercury exposure during development. *Neurotoxicity Research*, *11*(3–4), 241–260. doi:10.1007/BF03033570 PMID:17449462

Jöhnk, K. D., Huisman, J., Sharples, J., Sommeijer, B., Visser, P. M., & Stroom, J. M. (2008). Summer heatwaves promote blooms of harmful cyanobacteria. *Global Change Biology*, *14*(3), 495–512. doi:10.1111/j.1365-2486.2007.01510.x

Johnson, F. O., & Atchison, W. D. (2008). Potential enhanced development of paralytic phenotype in SOD1-G93a mice by methylmercury. Environmental etiologies of neurological disorders: scientific, translational and policy implications. *25th International Neurotoxicology Conference.*

Johnson, M. A. (2001). High calcium intake blunts pregnancy-induced increases in maternal blood lead. *Nutrition Reviews*, *59*(5), 152-156. Retrieved from http://www.scopus.com/inward/record.url?eid=2-s2.0-0034989948&partnerID=tZOtx3y1

Johnson, J. M. F., Franzluebbers, A. J., Weyers, S. L., & Reicosky, D. C. (2007). Agricultural opportunities to mitigate greenhouse gas emissions. *Environmental Pollution*, *150*(1), 107–124. doi:10.1016/j.envpol.2007.06.030 PMID:17706849

Johnson, J., Weissman, M. M., & Klerman, G. L. (1992). Service utilization and social morbidity associated with depressive symptoms in the community. *Journal of the American Medical Association*, *267*(11), 1478–1483. doi:10.1001/jama.1992.03480110054033 PMID:1538538

Johnson, N. P., & Wang, K. (2003). Is ovarian surgery effective for androgenic symptoms of polycystic ovarian syndrome. *Journal of Obstetrics & Gynaecology*, *23*(6), 599–606.

Johnson, P. E., Milne, D. B., & Lykken, G. I. (1992). Effect of age and sex on copper absorption, biological half life and status in humans. *The American Journal of Clinical Nutrition*, *56*(5), 917–925. doi:10.1093/ajcn/56.5.917 PMID:1329483

Johnston, S. C., Mendis, S., & Mathers, C. D. (2009). Global variation in stroke burden and mortality : Estimates from monitoring, surveillance, and modelling. *Lancet Neurology*, *8*(4), 345–354. doi:10.1016/S1474-4422(09)70023-7 PMID:19233730

Jomova, K., & Valko, M. (2011). Advances in metal-induced oxidative stress and human disease. *Toxicology*, *283*(2-3), 65–87. doi:10.1016/j.tox.2011.03.001 PMID:21414382

Jomova, K., Vondrakova, D., Lawson, M., & Valko, M. (2010). Metals, oxidative stress and neurodegenerative disorders. *Molecular and Cellular Biochemistry*, *345*(1-2), 91–104. doi:10.100711010-010-0563-x PMID:20730621

Jones, K. E., Patel, N. G., Levy, M. A., Storeygard, A., Balk, D., Gittleman, J. L., & Daszak, P. (2008). Global trends in emerging infectious diseases. *Nature*, *451*(7181), 990–993. doi:10.1038/nature06536 PMID:18288193

Joo, N. S., Kim, S. M., Jung, Y. S., & Kim, K. M. (2009). Hair iron and other minerals' level in breast cancer patients. *Biological Trace Element Research*, *129*(1-3), 28–35. doi:10.100712011-008-8281-x

Joseph, P. (2009). Mechanisms of cadmium carcinogenesis. *Toxicology and Applied Pharmacology*, *238*(3), 272–279. doi:10.1016/j.taap.2009.01.011 PMID:19371617

Joshi, P. L., Bhattacharya, M., Yadava, R. L., Chand, B., Narasimham, M. V., Nigam, D. K., & Jain, C. L. (1996). A community-based study on the effect of hexachlorocyclohexane (HCH) exposure in spraymen and general population. *The Journal of Communicable Diseases*, *28*, 189–198.

Jouybari, L., Saei Ghare Naz, M., Sanagoo, A., Kiani, F., Sayehmiri, F., Sayehmiri, K., & Hasanpour Dehkordi, A. (2018). Toxic elements as biomarkers for breast cancer: A meta-analysis study. *Cancer Management and Research*, *10*, 69–79. doi:10.2147/CMAR.S151324

Jow, G. M., Chou, C. J., Chen, B. F., & Tsai, J. H. (2004). Beauvericin induces cytotoxic effects in human acute lymphoblastic leukemia cells through cytochrome c release, caspase 3 activation: The causative role of calcium. *Cancer Letters*, *216*(2), 165–173. doi:10.1016/j.canlet.2004.06.005

Juckett, G., & Hancox, J. G. (2002). Venomous snakebites in the United States: Management review and update. *American Family Physician*, *65*(7), 1367–1378. PMID:11996419

Ju, H., Arumugam, P., Lee, J., & Song, J. M. (2017). Impact of Environmental Pollutant Cadmium on the Establishment of a Cancer Stem Cell Population in Breast and Hepatic Cancer. *ACS Omega*, *2*(2), 563–572. doi:10.1021/acsomega.6b00181

Ju-Kun, S., Yuan, D. B., Rao, H. F., Chen, T. F., Luan, B. S., Xu, X. M., ... Zhu, J. G. (2016). Association between Cd exposure and risk of prostate Cancer: A PRISMA-compliant systematic review and meta-analysis. *Medicine; Analytical Reviews of General Medicine, Neurology, Psychiatry, Dermatology, and Pediatries*, *95*(6), e2708. doi:10.1097/MD.0000000000002708

Julin, B., Wolk, A., Johansson, J. E., Andersson, S. O., Andren, O., & Akesson, A. (2012). Dietary cadmium exposure and prostate cancer incidence: A population-based prospective cohort study. *British Journal of Cancer*, *107*(5), 895–900. doi:10.1038/bjc.2012.311

Jung, Y., Jeong, E., Kyung, E., Kim, Y., Sohn, S., Hwan, S., ... Moon, C. (2008). Cadmium induces apoptotic cell death through p38 MAPK in brain microvessel endothelial cells. *European Journal of Pharmacology*, *578*(1), 11–18. doi:10.1016/j.ejphar.2007.08.049 PMID:17961543

Jurczak, A., Brodowska, A., Szkup, M., Prokopowicz, A., Karakiewicz, B., Łój, B., ... Grochans, E. (2018). Influence of Pb and Cd levels in whole blood of postmenopausal women on the incidence of anxiety and depressive symptoms. *Annals of Agricultural and Environmental Medicine*, *25*(2), 219–223. doi:10.26444/aaem/85929 PMID:29936823

Justić, D., Turner, R. E., Rabalais, N. N., Justic, D., Turner, R. E., & Rabalais, N. N. (2018). *Climatic Influences on Riverine Nitrate Flux : Implications for Coastal Marine Eutrophication and Hypoxia*. Retrieved from https://www.jstor.org/stable/1353186

Kagara, N., Tanaka, N., Noguchi, S., & Hirano, T. (2007). Zinc and its transporter ZIP10 are involved in invasive behavior of breast cancer cells. *Cancer Science*, *98*(5), 692–697. doi:10.1111/j.1349-7006.2007.00446.x

Kahime, K., Bounoua, B., Messouli, M., & Boussaa, S. (in press). Evaluation of Eco-Adaptation Strategies of Health to Climate change: Case of Zoonotic Cutaneous Leishmaniasis (ZCL) as Vulnerability Indicator in Pre-Saharan region of Morocco. In *Environmental Change and Human Security in the Middle East and Africa. GECS2012 Conference Proceedings*. Springer.

Kahime, K., Bounoua, B., Messouli, M., Boussaa, S., & Boumezzough, A. (2016a). Evaluation of Eco-Adaptation Strategies of Health to Climate change: Case of Zoonotic Cutaneous Leishmaniasis (ZCL) as Vulnerability Indicator in Pre-Saharan region of Morocco. In *Environmental Change and Human Security in the Middle East and Africa. GECS2012 Conference Proceedings*. Springer.

Kahime, K., Boussaa, S., Ouanaimi, F., Boumezzough, A. (2015). Entomological Assessment of Human Visceral Leishmaniasis Risk in area with sporadic cases, Morocco. *ActaTropica*, *48*, 58-65. doi: tropica.2015.04.006 doi:10.1016/j.acta

Kahime, K., Boussaa, S., Bounoua, L., Fouad, O., Messouli, M., & Boumezzough, A. (2014). Leishmaniasis in Morocco: Diseases & vectors. *Asian Pacific Journal of Tropical Disease*, *4*, 530–534. doi:10.1016/S2222-1808(14)60671-X

Kahime, K., Boussaa, S., Laamrani, A., Nhammi, H., & Boumezzough, A. (2016b). Epidemiological study on acute cutaneous leishmaniasis in Morocco. *Journal of Acute Disease*, *5*(1), 41–45. doi:10.1016/j.joad.2015.08.004

Kahime, K., Boussaa, S., Ouanaimi, F., & Boumezzough, A. (2015). Species composition of phlebotomine sand fly fauna in an area with sporadic cases of *Leishmania infantum* human visceral leishmaniasis, Morocco. *Acta Tropica*, *148*, 58–65. doi:10.1016/j.actatropica.2015.04.006 PMID:25930188

Kahime, K., Denis, S., Lahouari, B., El Hidan, M., & Souad, B. (2018). Management of Leishmaniases in the Era of Climate Change in Morocco. *International Journal of Environmental Research and Public Health*, *15*(7), 1542. doi:10.3390/ijerph15071542 PMID:30037049

Kain, K. C., Harrington, M. A., Tennyson, S., & Keystone, J. S. (1998). Imported malaria: Prospective analysis of problems in diagnosis and management. *Clinical Infectious Diseases*, *27*(1), 142–149. doi:10.1086/514616

Kalkhoff, S. J., Hubbard, L. E., Tomer, M. D., & James, D. E. (2016). Effect of variable annual precipitation and nutrient input on nitrogen and phosphorus transport from two Midwestern agricultural watersheds. *The Science of the Total Environment*, *559*, 53–62. doi:10.1016/j.scitotenv.2016.03.127 PMID:27054493

Kallenborn, R., Halsall, C., Dellong, M., & Carlsson, P. (2012). The influence of climate change on the global distribution and fate processes of anthropogenic persistent organic pollutants. *Journal of environmental monitoring. JEM*, *14*, 2854–2869. PMID:23014859

Kalra, J., Kumar, P., Majeed, A. B. A., & Prakash, A. (2016). Modulation of LOX and COX pathways via inhibition of amyloidogenesis contributes to mitoprotection against β-amyloid oligomer-induced toxicity in an animal model of Alzheimer's disease in rats. *Pharmacology, Biochemistry, and Behavior*, *146–147*, 1–12. doi:10.1016/j.pbb.2016.04.002 PMID:27106205

Kamel, F., Tanner, C., Umbach, D., Hoppin, J., Alavanja, M., Blair, A., ... Sandler, D. (2006). Pesticide Exposure and Self-reported Parkinson's Disease in the Agricultural Health Study. *American Journal of Epidemiology*, *165*(4), 364–374. doi:10.1093/aje/kwk024 PMID:17116648

Kamel, F., Umbach, D. M., Bedlack, R. S., Richards, M., Watson, M., Alavanja, M. C., ... Sandler, D. P. (2012). Pesticide exposure and amyotrophic lateral sclerosis. *Neurotoxicology*, *33*(3), 457–462. doi:10.1016/j.neuro.2012.04.001 PMID:22521219

Kamel, F., Umbach, D. M., Hu, H., Munsat, T. L., Shefner, J. M., Taylor, J. A., & Sandler, D. P. (2005). Lead exposure as a risk factor for amyotrophic lateral sclerosis. *Neurodegenerative Diseases*, *2*(3-4), 195–201. doi:10.1159/000089625 PMID:16909025

Kamel, F., Umbach, D. M., Stallone, L., Richards, M., Hu, H., & Sandler, D. P. (2008). Association of lead exposure with survival in amyotrophic lateral sclerosis. *Environmental Health Perspectives*, *116*(7), 943–947. doi:10.1289/ehp.11193 PMID:18629318

Kamińska, M., Ciszewski, T., Łopacka-Szatan, K., Miotła, P., & Starosławska, E. (2015). Breast cancer risk factors. *Przeglad Menopauzalny Journal*, *14*, 196–202.

Kanaly, R. A., & Harayama, S. (2000). Biodegradation of high-molecular-weight polycyclic aromatic hydrocarbons by bacteria. *Journal of Bacteriology*, *182*(8), 2059–2067. doi:10.1128/JB.182.8.2059-2067.2000 PMID:10735846

Kanavouras, K., Tzatzarakis, M. N., Mastorodemos, V., Plaitakis, A., & Tsatsakis, A. M. (2011). A case report of motor neuron disease in a patient showing significant level of DDTs, HCHs and organophosphate metabolites in hair as well as levels of hexane and toluene in blood. *Toxicology and Applied Pharmacology*, *256*(3), 399–404. doi:10.1016/j.taap.2011.07.022 PMID:21851830

Kandlikar, M., & Risbey, J. (2000). Agricultural impacts of climate change: If adaptation is the answer, what is the question? *Climatic Change*, *45*(3-4), 529–539. doi:10.1023/A:1005546716266

Kaneko, T., & Colwell, R. R. (1975). Incidence of *Vibrio parahaemolyticus* in Chesapeake Bay. *Applied Microbiology*, *30*, 251–257. PMID:1164012

Kannan, K., Tanabe, S., Ramesh, A., Subramanian, A., & Tasukawa, R. (1992). Persistent organochlorine residues in food stuffs from India and their implications on human dietary exposure. *Journal of Agricultural and Food Chemistry*, *40*(3), 518–524. doi:10.1021/jf00015a032

Kanthasamy, A. G., Kitazawa, M., Kanthasamy, A., & Anantharam, V. (2005). Dieldrin-Induced Neurotoxicity: Relevance to Parkinson's Disease Pathogenesis. *Neurotoxicology*, *26*(4), 701–719. doi:10.1016/j.neuro.2004.07.010 PMID:16112328

Kaplay, R. D., & Patode, H. S. (2004). Groundwater pollution due to industrial effluent at Tuppa, New Nanded, Maharashtra, India. *Environ Geol.*, *46*(6-7), 871–882. doi:10.100700254-004-1095-z

Kappus, K., Lundgren, R. G., Jr., Juranek, D. D., Roberts, J. M., & Spencer, H. C. (1994). *Intestinal parasitism in the United States: update on a continuing problem.* ASTMH. Retrieved from http://ajtmh.org/content/journals/10.4269/ajtmh.1994.50.705

Kard, S. P., Mandal, S. K., Sahu, T., & Kole, R. K. (2008). Assessment of heavy metal pollution in surface water. *International Journal of Environmental Science and Technology*, *5*(1), 119–124. doi:10.1007/BF03326004

Karimi, G., Shahar, S., Homayouni, N., Rajikan, R., Abu Bakar, N. F., & Othman, M. S. (2012). Association between trace element and heavy metal levels in hair and nail with prostate cancer. *Asian Pacific Journal of Cancer Prevention*, *13*(9), 4249–4253. doi:10.7314/APJCP.2012.13.9.4249 PMID:23167323

Karkashan, A., Khallaf, B., Morris, J., Thurbon, N., Rouch, D., Smith, S. R., & Deighton, M. (2015). Comparison of methodologies for enumerating and detecting the viability of Ascaris eggs in sewage sludge by standard incubation-microscopy, the BacLight Live/Dead viability assay and other vital dyes. *Water Research*, *68*, 533–544. doi:10.1016/j.watres.2014.10.003 PMID:25462759

Karl, T. R., Knight, R. W., Easterling, D. R., & Quayle, R. G. (1996). Indices of climate change for the United States. *Bulletin of the American Meteorological Society*, *77*(2), 279–303. doi:10.1175/1520-0477(1996)077<0279:IOCCFT>2.0.CO;2

Karmaoui A, Ifaadassan I, Babqiqi A, Messouli M & Yacoubi Khebiza M. (2016c). Analysis of the water supply-demand relationship in the Draa valley basin, Morocco, under climate change and socio-economic scenarios. *Journal of Scientific Research & Reports*. Doi:10.9734/JSRR/2016/21536

Karmaoui A, Messouli M, Yacoubi Khebiza M & Ifaadassan I. (2014). Environmental Vulnerability to Climate Change and Anthropogenic Impacts in Dryland, (Pilot Study: Middle Draa Valley, South Morocco). *J Earth Sci Clim Change, S11*, 2. doi:. doi:10.4172/2157-7617.S11-002

Karmaoui, A. (2018). The Cutaneous Leishmaniasis Vulnerability index (CLVI). *Acta Ecologica Sinica*, *38*(4), 288–295. doi:10.1016/j.chnaes.2018.01.001

Karmaoui, A., Ifaadassan, I., Messouli, M., & Khebiza, M. Y. (2015b). Sustainability of the Moroccan oasean system (Case study: Middle Draa Valley). *Global J Technol Optim*, *6*(01), 170. doi:10.4172/2229-8711.1000170

Karmaoui, A., Ifaadassan, I., Messouli, M., & Yacoubi Khebiza, M. (2015d). Monetarization of Ecosystem Services of Oasean Biome (Case study: Provisioning services of Middle Draa Valley Oases). *British Journal of Applied Science and Technology*, *11*(6), 1–18. doi:10.9734/BJAST/2015/19689

Karmaoui, A., Messouli, M., & Yacoubi, M. (2015a). Vulnerability of Ecosystem Services to Climate Change and Anthropogenic Impacts in South East of Morocco (Case study: Drying up of Iriki lake). *The International Journal of Climate Change: Impacts and Responses.*, *7*(3), 1835–7156.

Karmaoui, A., & Zerouali, S. (2018). Modeling the vulnerability to Zoonotic Cutaneous Leishmaniasis at the local scale. *Asian Journal of Applied Sciences*.

Karpeta, A., Maniecka, A., & Gregoraszczuk, E. L. (2016). Different mechanisms of action of 2, 2′, 4, 4′-tetrabromodiphenyl ether (BDE-47) and its metabolites (5-OH-BDE-47 and 6-OH-BDE-47) on cell proliferation in OVCAR-3 ovarian cancer cells and MCF-7 breast cancer cells. *Journal of Applied Toxicology*, *36*(12), 1558–1567. doi:10.1002/jat.3316

Karvonen, A., Rintamäki, P., Jokela, J., & Valtonen, E. T. (2010). Increasing water temperature and disease risks in aquatic systems: Climate change increases the risk of some, but not all, diseases. *International Journal for Parasitology*, *40*(13), 1483–1488. doi:10.1016/j.ijpara.2010.04.015 PMID:20580904

Kashyap, M. P., Singh, A. K., Kumar, V., Yadav, D. K., Khan, F., Jahan, S., ... Pant, A. B. (2013). Pkb/Akt1 Mediates Wnt/GSK3β/β-Catenin Signaling-Induced Apoptosis in Human Cord Blood Stem Cells Exposed to Organophosphate Pesticide Monocrotophos. *Stem Cells and Development*, *22*(2), 224–238. doi:10.1089cd.2012.0220 PMID:22897592

Kassas, M. (1995). Desertification: A general review. *Journal of Arid Environments*, *30*(2), 115–128. doi:10.1016/S0140-1963(05)80063-1

Kasten, M., Chade, A., & Tanner, C. M. (2007). Epidemiology of Parkinson's disease. *Handbook of Clinical Neurology*, *83*, 129–151. doi:10.1016/S0072-9752(07)83006-5 PMID:18808913

Kasturiratne, A., Pathmeswaran, A., Fonseka, M. M. D., Lalloo, D. G., Brooker, S., & De Silva, H. J. (2005). Estimates of disease burden due to land-snake bite in Sri Lankan hospitals. *The Southeast Asian Journal of Tropical Medicine and Public Health*, *36*(3), 733. PMID:16124448

Kasturiratne, A., Wickremasinghe, A. R., de Silva, N., Gunawardena, N. K., Pathmeswaran, A., Premaratna, R., ... de Silva, H. J. (2008). The global burden of snakebite: A literature analysis and modelling based on regional estimates of envenoming and deaths. *PLoS Medicine*, *5*(11), e218. doi:10.1371/journal.pmed.0050218 PMID:18986210

Kato, S., Jenkins, M., Fogarty, E., & Bowman, D. (2004). *Cryptosporidium parvum oocyst inactivation in field soil and its relation to soil characteristics : analyses using the geographic information systems.* Academic Press. doi:10.1016/j.scitotenv.2003.08.027

Kattwinkel, M., Kuhne, J.V., Foit, K. & Liess, M. (2011) Climate change, agricultural insecticide exposure, and risk for freshwater communities. *Ecological Applications*, *21*, 2068-2081.

Kaushal, S. S., Likens, G. E., Jaworski, N. A., Pace, M. L., Sides, A. M., Seekell, D., ... Wingate, R. L. (2010). Rising stream and river temperatures in the United States. *Frontiers in Ecology and the Environment*, *8*(9), 461–466. doi:10.1890/090037

Kavlakov, S., Elsner, J. B., & Pérez-Peraza, J. (2008). A statistical link between tropical cyclone intensification and major geomagnetic disturbances. *Geofísica Internacional*, *47*(3), 207–213.

Kearney, M., Shine, R., & Porter, W. P. (2009). The potential for behavioral thermoregulation to buffer "cold-blooded" animals against climate warming. *Proceedings of the National Academy of Sciences of the United States of America*, *106*(10), 3835–3840. doi:10.1073/pnas.0808913106 PMID:19234117

Keen, C. L., & Zidenberg-Cherr, S. (2003). MANGANESE. In B. Caballero (Ed.), *Encyclopedia of Food Sciences and Nutrition* (2nd ed.; pp. 3686–3691). Oxford, UK: Academic Press; doi:10.1016/B0-12-227055-X/00732-X

Keffala, C., Harerimana, C., & Vasel, J. L. (2012). Oeufs d'helminthes dans les eaux us ées et les boues de station d' épuration: Enjeux sanitaires et int eret du traitement par lagunage. *Environ Risque Sante*, *11*, 511–520. doi:10.1684/ers.2012.0573

Kegel, K. B., Meloni, A. R., Yi, Y., Kim, Y. J., Doyle, E., Cuiffo, B. G., ... DiFiglia, M. (2002). Huntingtin is present in the nucleus, interacts with the transcriptional corepressor C-terminal binding protein, and represses transcription. *The Journal of Biological Chemistry*, *277*(9), 7466–7476. doi:10.1074/jbc.M103946200 PMID:11739372

Keifer, M. C., & Firestone, J. (2007). Neurotoxicity of pesticides. *Journal of Agromedicine*, *12*(1), 17–25. doi:10.1300/J096v12n01_03 PMID:18032333

Keller, B., & Shapiro, W. (1982). *Double Episodes.* Academic Press.

Kellinghaus, C., Engbring, C., Kovac, S., Möddel, G., Boesebeck, F., Fischera, M., ... Husstedt, I. W. (2008). Frequency of seizures and epilepsy in neurological HIV-infected patients. *Seizure*, *17*(1), 27–33. doi:10.1016/j.seizure.2007.05.017 PMID:17618132

Kelly-Hope, L., & Thomson, M. C. (2008). Climate and infectious diseases. In M. C. Thomson, R. Garcia-Herrera, & M. Beniston (Eds.), *Seasonal forecasts, climate change and human health* (p. 31). New York: Springer. doi:10.1007/978-1-4020-6877-5_3

Kennish, M. J. (2002). Environmental threats and environmental future of estuaries. *Environmental Conservation*, *29*(01), 78–107. doi:10.1017/S0376892902000061

Kepp, K. P. (2012). Bioinorganic Chemistry of Alzheimer's Disease. *Chemical Reviews*, *112*(10), 5193–5239. doi:10.1021/cr300009x PMID:22793492

Keraita, B., & Amoah, P. (2011). Fecal Exposure Pathways in Accra: A Literature Reviewwith Specific Focus on IWMI's Work on Wastewater Irrigated Agriculture (IWMI). Literature Review Report. Centre for Global SafeWater, Emory University.

Keraita, B., Drechsel, P., Huibers, F., & Raschid-Sally, L., (2002). Wastewater use in informal irrigation in urban and periurban areas of Kumasi, Ghana. *UA Magazine - Wastewater Reuse in Urban Agriculture, 8.*

Kesel, A. J., Sonnenbichler, I., Polborn, K., Gürtler, L., Klinkert, W. E., Modolell, M., ... Oberthür, W. (1999). A new antioxidative vitamin B6 analogue modulates pathophysiological cell proliferation and damage. *Bioorganic & Medicinal Chemistry*, *7*(2), 359–367. doi:10.1016/S0968-0896(98)00229-6 PMID:10218829

Ketelaar, M., Hofstra, R., & Hayden, M. (2012). What monozygotic twins discordant for phenotype illustrate about mechanisms influencing genetic forms of neurodegeneration. *Clinical Genetics*, *81*(4), 325–333. doi:10.1111/j.1399-0004.2011.01795.x PMID:21981075

Kettles, M., Browning, S., Prince, T., & Horstman, S. (1997). Triazine herbicide exposure and breast cancer incidence: An ecologic study of Kentucky counties. *Environmental Health Perspectives*, *105*(11), 1222–1227. doi:10.1289/ehp.971051222

Kew, M. C. (2013). Aflatoxins as a cause of hepatocellular carcinoma. *Journal of Gastrointestinal and Liver Diseases; JGLD*, *22*, 305–310.

Khaldi, S., Ratajczak, M., Gargala, G., Fournier, M., Berthe, T., Favennec, L., & Dupont, J. P. (2011). Intensive exploitation of a karst aquifer leads to Cryptosporidium water supply contamination. *Water Research*, *45*(9), 2906–2914. doi:10.1016/j.watres.2011.03.010 PMID:21477840

Khallayoun, K., Ziad, H., Lhadi, E. K., & Cabaret, J. (2009). Reuse of untreated wastewater in agriculture in Morocco: Potential contamination of crops and soil helminth eggs. In Collection Environment/Health aspects Waste Water/Treatment and reuse of wastewater: impact on health and the environment. IAV Edition.

Khallayoune, K. (2009). *Aspects of the sanitary wastewater reuse in agriculture, Treatment and reuse of wastewater: impact on health and the environment*. IAV Edition.

Khan, A., Dobson, J. P., & Exley, C. (2006). Redox cycling of iron by Aβ42. *Free Radical Biology & Medicine*, *40*(4), 557–569. doi:10.1016/j.freeradbiomed.2005.09.013 PMID:16458186

Khanjani, N., Jafarnejad, A. B., & Tavakkoli, L. (2017). Arsenic and breast cancer: A systematic review of epidemiologic studies. *Reviews on Environmental Health*, *32*(3), 267–277. doi:10.1515/reveh-2016-0068

Khan, M. A., Ahmed, S. A., Catalin, B., Khodadoust, A., Ajayi, O., & Vaughn, M. (2006). Effect of temperature on heavy metal toxicity to juvenile crayfish, *Orconectes immunis* (Hagen). *Environmental Toxicology*, *21*(5), 513–520. doi:10.1002/tox.20213 PMID:16944513

Kheradmand, A., Yarali, B., Zare, A., Pourpak, Z., Shams, S., & Comparison, A. (2014). Comparison of Serum Zinc and Copper levels in Children and adolescents with Intractable and Controlled Epilepsy. *Iranian Journal of Child Neurology*, *8*(3), 49–54. PMID:25143774

Kiernan, M. C., Vucic, S., Cheah, B. C., Turner, M. R., Eisen, A., Hardiman, O., ... Zoing, M. C. (2011). Amyotrophic lateral sclerosis. *Lancet*, *377*(9769), 942–955. doi:10.1016/S0140-6736(10)61156-7 PMID:21296405

Kikuchihara, Y., Abe, H., Tanaka, T., Kato, M., Wang, L., Ikarashi, Y., ... Shibutani, M. (2015). Relationship between brain accumulation of manganese and aberration of hippocampal adult neurogenesis after oral exposure to manganese chloride in mice. *Toxicology*, *331*, 24–34. doi:10.1016/j.tox.2015.02.005 PMID:25698507

Kilness, A. W., & Hochberg, F. H. (1977). Amyotrophic lateral sclerosis in a high selenium environment. *Journal of the American Medical Association*, *237*(26), 2843–2844. doi:10.1001/jama.1977.03270530051023 PMID:577250

Kilvington, S., & Beeching, J. (1995). Identification and epidemiological typing of *Naegleria fowleri* with DNA probes. *Applied and Environmental Microbiology*, *61*, 2071–2078. PMID:7793928

Kim, B. E., Nevitt, T., & Thiele, D. J. (2008). Mechanisms for copper acquisition, distribution and regulation. *Nature Chemical Biology*, *4*(3), 176–185. doi:10.1038/nchembio.72

Kimberly, D. A., & Salice, C. J. (2012). Understanding interactive effects of climate change and toxicants: Importance of evolutionary processes. *Integrated Environmental Assessment and Management*, *8*(2), 385–386. doi:10.1002/ieam.1286 PMID:22431356

Kimbrough, R. D., Gaines, T. B., & Linder, R. E. (1971). The Ultrastructure of Livers of Rats Fed DDT and Dieldrin. *Archives of Environmental Health: An International Journal*, *22*(4), 460–467. doi:10.1080/00039896.1971.10665879 PMID:5545497

Kim, J., Byun, H., Chung, E., Chung, H., & Bae, O. (2013). Loss of Integrity : Impairment of the Blood-brain Barrier in Heavy Metal-associated Ischemic Stroke. *Toxicological Research*, *29*(3), 157–164. doi:10.5487/TR.2013.29.3.157 PMID:24386515

Kim, K. H., Kabir, E., & Jahan, S. A. (2017). Exposure to pesticides and the associated human health effects. *The Science of the Total Environment*, *575*, 525–535. doi:10.1016/j.scitotenv.2016.09.009 PMID:27614863

Kim, M.-K., Pyo, K.-H., Hwang, Y.-S., Park, K. H., Hwang, I. G., Chai, J.-Y., & Shin, E.-H. (2012, August 13). temperature on embryonation of Ascaris suum eggs in an environmental chamber. *Korean Journal of Parasitology*, *50*(3), 239–242. doi:10.3347/kjp.2012.50.3.239 PMID:22949753

Kim, S., & Mountz, J. M. (2011). SPECT Imaging of Epilepsy : An Overview and Comparison with F-18 FDG PET. *International Journal OfMolecular Imaging*, *2011*. doi:10.1155/2011/813028 PMID:21785722

Kim, Y. M., Park, J. W., & Cheong, H. K. (2012). Estimated effect of climatic variables on the transmission of Plasmodium vivax malaria in the Republic of Korea. *Environmental Health Perspectives*, *120*(9), 1314–1319. doi:10.1289/ehp.1104577

Kishiwa, P., Nobert, J., Kongo, V., & Ndomba, P. (2018). Assessment of impacts of climate change on surface water availability using coupled SWAT and WEAP models : case of upper Pangani River Basin, Tanzania. Academic Press.

Kistemann, T., Claßen, T., Koch, C., Dangendorf, F., Fischeder, R., Gebel, J., ... Exner, M. (2002). Microbial load of drinking water reservoir tributaries during extreme rainfall and runoff. *Applied and Environmental Microbiology*, *68*(5), 2188–2197. doi:10.1128/AEM.68.5.2188-2197.2002 PMID:11976088

Kistemann, T., Dangendorf, F., & Exner, M. (2001). A Geographical Information System (CIS) as a tool for microbial risk assessment in catchment areas of drinking water reservoirs. *International Journal of Hygiene and Environmental Health*, *203*(3), 225–233. doi:10.1078/S1438-4639(04)70033-4 PMID:11279819

Kitzberger, R., Madl, C., & Ferenci, P. (2005). Wilson disease. *Metabolic Brain Disease*, *20*(4), 295–302. doi:10.100711011-005-7910-8 PMID:16382340

Klein, R. (2011). Adaptation to climate change: more than technology. In Climate: Global Change and Local Adaptation (pp. 157-168). NATO.

Klein, K. R., Woodward, C. S., Waller, E. K., Lechowicz, M. J., & Rosenthal, H. (2005). Effects of Mycotoxins on Mononuclear Cells (MNCs) in Normal Blood, T-Cell Leukemia and Lymphoma Cell Lines. *Blood*, *106*(11), 4828–4828.

Klerks, P. L., Xie, L., & Levinton, J. S. (2011). Quantitative genetics approaches to study evolutionary processes in ecotoxicology; a perspective from research on the evolution of resistance. *Ecotoxicology (London, England), 20*(3), 513–523. doi:10.100710646-011-0640-2 PMID:21516382

Knowles, N. & Cayan, D.R. (2002). Potential effects of global warming on the Sacramento/San Joaquin watershed and the San Francisco estuary. *Geophysical Research Letters, 29*, 38-31-38-34.

Kociba, R. J., Keyes, D. G., Beyer, J. E., Carreon, R. M., Wade, C. E., Dittenber, D. A., ... Humiston, C. G. (1978). Results of a two-year chronic toxicity and oncogenicity study of 2,3,7,8-tetrachlorodibenzo-p-dioxin in rats. *Toxicology and Applied Pharmacology, 46*(2), 279–303. doi:10.1016/0041-008X(78)90075-3

Kohzadi, S., Shahmoradi, B., Ghaderi, E., Loqmani, H., & Maleki, A. (2018). Concentration, Source, and Potential Human Health Risk of Heavy Metals in the Commonly Consumed Medicinal Plants. *Biological Trace Element Research.* doi:10.100712011-018-1357-3 PMID:29700699

Koleva, N. G., Schneider, U., & Tol, T. (2009). *The impact of weather variability and climate change on pesticide applications in the US - An empirical investigation: Research unit Sustainability and Global Change.* Hamburg University.

Kolonel, L. N., Yoshizawa, C. N., & Hankin, J. H. (1988). Diet and prostatic cancer: A case-control study in Hawaii. *American Journal of Epidemiology, 127*(5), 999–1012. doi:10.1093/oxfordjournals.aje.a114903 PMID:3358418

Kolstad, E. W., & Johansson, K. A. (2011). Uncertainties associated with quantifying climate change impacts on human health: A case study for diarrhea. *Environmental Health Perspectives, 119*(3), 299–305. doi:10.1289/ehp.1002060 PMID:20929684

Komulainen, H., & Tuomisto, J. (1980). Effect of Heavy Metals on Dopamine, Noradrenaline and Serotonin Uptake and Release in Rat Brain Synaptosomes. *Acta Pharmacologica et Toxicologica, 48*(3), 199–204. doi:10.1111/j.1600-0773.1981.tb01609.x

Kondoh, M., Araragi, S., Sato, K., Higashimoto, M., Takiguchi, M., & Sato, M. (2002). Cadmium induces apoptosis partly via caspase-9 activation in HL-60 cells. *Toxicology, 170*(1-2), 111–117. doi:10.1016/S0300-483X(01)00536-4 PMID:11750088

Konstantinova, S. G., Russanova, I. E., & Russanov, E. M. (1991). Do the copper complexes of histamine, histidine and two H2antagonists react with O2? *Free Radical Research Communications, 12-13*(Pt 1), 215–220. doi:10.3109/10715769109145789 PMID:1676975

Kontakos, N., & Stokes, J. (1999). Monograph Series on Aging-related Diseases: XII. Parkinson s Disease Recent Developments and New Directions. *Chronic Diseases and Injuries in Canada, 20*(2), 58–76. PMID:10455038

Kontur, P. J., & Fechter, L. D. (1988). Brain regional manganese levels and monoamine metabolism in manganese-treated neonatal rats. *Neurotoxicology and Teratology, 10*(4), 295–303. doi:10.1016/0892-0362(88)90031-1 PMID:3226372

Kopp, G., & Lean, J. L. (2011). A new, lower value of total solar irradiance: Evidence and climate significance. *Geophysical Research Letters, 38*(1), L01706. doi:10.1029/2010GL045777

Kosek, M., Bern, C., & Guerrant, R. L. (2003). The global burden of diarrhoeal disease, as estimated from studies published between 1992 and 2000. *Bulletin of the World Health Organization, 81*(3), 197–204. PMID:12764516

Kotloff, K.L., Nataro, J.P., Blackwelder, W.C., Nasrin, D., Farag, T.H., Panchalingam, S., … Faruque, A.S. (2013). Burden and aetiology of diarrhoeal disease in infants and young children in developing countries (the Global Enteric Multicenter Study, GEMS): a prospective, case-control study. *Lancet, 382*(9888), 209-222.

Kouri, K., Lemmens, M., & Lemmens-Gruber, R. (2003). Beauvericin-induced channels in ventricular myocytes and liposomes. *Biochimica et Biophysica Acta*, *1609*(2), 203–210. doi:10.1016/S0005-2736(02)00689-2

Koutros, S., Langseth, H., Grimsrud, T. K., Barr, D. B., Vermeulen, R., Portengen, L., ... Engel, L. S. (2015). Prediagnostic Serum Organochlorine Concentrations and Metastatic Prostate Cancer: A Nested Case–Control Study in the Norwegian Janus Serum Bank Cohort. *Environmental Health Perspectives*, *123*(9), 867–872. doi:10.1289/ehp.1408245 PMID:25734605

Kovac, S., & Walker, M. C. (2013). Neuropeptides Neuropeptides in epilepsy. *Neuropeptides*, *47*(6), 467–475. doi:10.1016/j.npep.2013.10.015 PMID:24210141

Kowall, N. W., Pendlebury, W. W., Kessler, J. B., Perl, D. P., & Beal, M. F. (1989). Aluminum-induced neurofibrillary degeneration affects a subset of neurons in rabbit cerebral cortex, basal forebrain and upper brainstem. *Neuroscience*, *29*(2), 329–337. doi:10.1016/0306-4522(89)90060-2 PMID:2725861

Koyama, S., Arawaka, S., Chang-Hong, R., Wada, M., Kawanami, T., Kurita, K., ... Kato, T. (2006). Alteration of familial ALS-linked mutant SOD1 solubility with disease progression: Its modulation by the proteasome and hsp70. *Biochemical and Biophysical Research Communications*, *343*(3), 719–730. doi:10.1016/j.bbrc.2006.02.170 PMID:16563356

Krewski, D., Barakat-Haddad, C., Donnan, J., Martino, R., Pringsheim, T., Tremlett, H., ... Cashman, N. R. (2017). (in press). Determinants of neurological disease: Synthesis of systematic reviews. *Neurotoxicology*, *2017*. doi:10.1016/j.neuro.2017.04.002 PMID:28410962

Krishnamurthy, P., Choularton, R., Betts, R., & Lewis, K. (2011). Assessing the impacts of climate risk on food security through a vulnerability index. *Vulnerability Workshop of GI Forum.* Salzburg.

Kristal, A. R., Stanford, J. L., Cohen, J. H., Wicklund, K., & Patterson, R. E. (1999). Vitamin and mineral supplement use is associated with reduced risk of prostate cancer. *Cancer Epidemiology, Biomarkers & Prevention*, *8*(10), 887–892. PMID:10548317

Kristie, L. (2014). Health in the New Scenarios for Climate Change Research. *International Journal of Environmental Research and Public Health.*

Krueger, W. H., Gonye, G. E., Madison, D. L., Murray, K. E., Kumar, M., Spoerel, N., & Pfeiffer, S. E. (1997). TPO1, a member of a novel protein family, is developmentally regulated in cultured oligodendrocytes. *Journal of Neurochemistry*, *69*(4), 1343–1355. doi:10.1046/j.1471-4159.1997.69041343.x PMID:9326262

Kuch, U., & Mebs, D. (2002). Envenomations by colubrid snakes in Africa, Europe, and the Middle East. *Journal of Toxicology. Toxin Reviews*, *21*(1-2), 159–179. doi:10.1081/TXR-120004745

Kuciel-Lisieska, G. R. A. Ż. Y. N. A., Obremski, K., Stelmachów, J. E. R. Z. Y., Gajecka, M., Zielonka, Ł. U. K. A. S. Z., Jakimiuk, E., & Gajecki, M. (2008). Presence of zearalenone in blood plasma in women with neoplastic lesions in the mammary gland. *Bulletin of the Veterinary Institute in Pulawy*, *52*, 671–674.

Kuhn, K. G., Campbell-Lendrum, D. H., & Davies, C. R. (2002). A continental risk map for malaria mosquito (Diptera: Culicidae) vectors in Europe. *Journal of Medical Entomology*, *39*(4), 621–630. doi:10.1603/0022-2585-39.4.621

Kuhn, W., Winkel, R., Woitalla, D., Meves, S., Przuntek, H., & Muller, T. (1998). High prevalence of parkinsonism after occupational exposure to lead-sulfate batteries. *Neurology*, *50*(6), 1885–1886. doi:10.1212/WNL.50.6.1885 PMID:9633752

Kujawa, M. (1994). Some Naturally Occurring Substances: Food Items and Constituents, Heterocyclic Aromatic Amines and Mycotoxins. Food/Nahrung, 38, 351-351.

Kumar, K. K., Hakeem, A. A., & Kumar, S. S. (2010). *Proceedings of the Twenty second Kerala Science Congress*. Kerala State Council for Science, Technology and Environment, Government of Kerala.

Kumar, N., Tischbeinb, B., Kuschec, J., Laux, P., Bege, M. K., & Bogardi, J. J. (2017). Impact of climate change on water resources of upper Kharun catchment in Chhattisgarh, India. *Journal of Hydrology: Regional Studies*, *13*, 189–207.

Kumar, R., Alam, S., Chaudhari, B. P., Dwivedi, P. D., Jain, S. K., Ansari, K. M., & Das, M. (2012). Ochratoxin A-induced cell proliferation and tumor promotion in mouse skin by activating the expression of cyclin-D1 and cyclooxygenase-2 through nuclear factor-kappa B and activator protein-1. *Carcinogenesis*, *34*(3), 647–657. doi:10.1093/carcin/bgs368 PMID:23172667

Kumar, R., Razab, S., Prabhu, K., Ray, S., & Prakash, B. (2017). Serum butyrylcholinesterase and zinc in breast cancer. *Journal of Cancer Research and Therapeutics*, *13*(2), 367–370. doi:10.4103/0973-1482.165869

Kuo, H. W., Chen, S. F., Wu, C. C., Chen, D. R., & Lee, J. H. (2002). Serum and tissue trace elements in patients with breast cancer in Taiwan. *Biological Trace Element Research*, *89*(1), 1–11. doi:10.1385/BTER:89:1:1

Kurlander, H. M., & Patten, B. M. (1979). Metals in spinal cord tissue of patients dying of motor neuron disease. *Annals of Neurology*, *6*(1), 21–24. doi:10.1002/ana.410060105 PMID:507754

Kuroda, K., Hibi, D., Ishii, Y., Yokoo, Y., Takasu, S., Kijima, A., ... Sakai, H. (2015). Role of p53 in the progression from ochratoxin A-Induced DNA damage to gene mutations in the kidneys of mice. *Toxicological Sciences*, *144*(1), 65–76. doi:10.1093/toxsci/kfu267

Kurukulasuriya, P., Mendelsohn, R., Hassan, R., Benhin, J., Deressa, T., Diop, M., ... Mahamadou, A. (2006). Will African agriculture survive climate change? *The World Bank Economic Review*, *20*(3), 367–388. doi:10.1093/wber/lhl004

Kuter, K., Nowak, P., Gołembiowska, K., & Ossowska, K. (2010). Increased Reactive Oxygen Species Production in the Brain After Repeated Low-Dose Pesticide Paraquat Exposure in Rats. A Comparison with Peripheral Tissues. *Neurochemical Research*, *35*(8), 1121–1130. doi:10.100711064-010-0163-x PMID:20369291

Kuz'mina, V. V., & Ushakova, N. V. (2007). Effects of temperature, pH, and heavy metals (copper, zinc) upon proteinase activities in digestive tract mucosa of typical and facultative piscivorous fish. *Journal of Ichthyology*, *47*(6), 473–480. doi:10.1134/S0032945207060070

La Fontaine, S., Theophilos, M. B., Firth, S. D., Gould, R., Parton, R. G., & Mercer, J. F. (2001). Effect of the toxic milk mutation (tx) on the function and intracellular localization of Wnd, the murine homologue of the Wilson copper ATPase. *Human Molecular Genetics*, *10*(4), 361–370. doi:10.1093/hmg/10.4.361 PMID:11157799

La Merrill, M., Harper, R., Birnbaum, L. S., Cardiff, R. D., & Threadgill, D. W. (2010). Maternal dioxin exposure combined with a diet high in fat increases mammary cancer incidence in mice. *Environmental Health Perspectives*, *118*(5), 596–601. doi:10.1289/ehp.0901047

Lacorte, L. M., Delella, F. K., Porto Amorim, E. M., Justulin, L. A. Jr, Godinho, A. F., Almeida, A. A., ... Felisbino, S. L. (2011). Early changes induced by short-term low-dose cadmium exposure in rat ventral and dorsolateral prostates. *Microscopy Research and Technique*, *74*(11), 988–997. doi:10.1002/jemt.20985 PMID:21287658

Laden, F., Collman, G., Iwamoto, K., Alberg, A., Berkowitz, G., Freudenheim, J., ... Hunter, D. J. (2001a). 1,1-Dichloro-2,2-bis(p-chlorophenyl)ethylene and polychlorinated biphenyls and breast cancer: Combined analysis of five U.S. studies. *Journal of the National Cancer Institute*, *93*(10), 768–776. doi:10.1093/jnci/93.10.768

Laden, F., Hankinson, S., Wolff, M., Colditz, G., Willett, W., Speizer, F., & Hunter, D. (2001b). Plasma organochlorine levels and the risk of breast cancer: An extended follow-up in the Nurses' Health Study. *International Journal of Cancer*, *91*(4), 568–574. doi:10.1002/1097-0215(200002)9999:9999<::AID-IJC1081>3.0.CO;2-W

Lafferty, K. D. (2009). The ecology of climate change and infectious diseases. *Ecology*, *90*(4), 888–900. doi:10.1890/08-0079.1 PMID:19449681

Lahouiti, K., Maniar, S., & Bekhti, K. (2013). Seasonal fluctuations of phlebotomines sand fly populations (Diptera: Psychodidae) in the Moulay Yacoub province, centre Morocco: Effect of ecological factors. *African Journal of Environmental Science and Technology*, *7*(11), 1028–1031.

Lai, T. (2002). *Perspectives of peri-urban vegetable production in Hanoi*. Paper prepared for the Action Planning Workshop of the CGIAR Strategic Initiative for Urban and Peri-urban Agriculture (SIUPA), Lima, Peru.

Lake, I. R., Bentham, G., Kovats, R. S., & Nichols, G. L. (2005). Effects of weather and river flow on cryptosporidiosis. *Journal of Water and Health*, *3*(4), 469–474. doi:10.2166/wh.2005.048 PMID:16459850

Lake, I. R., Pearce, J., & Savill, M. (2008). The seasonality of human cryptosporidiosis in New Zealand. *Epidemiology and Infection*, *136*(10), 1383–1387. doi:10.1017/S0950268807009922 PMID:18053274

Lal, A., Baker, M. G., Hales, S., & French, N. P. (2013). Potential effects of global environmental changes on cryptosporidiosis and giardiasis transmission. *Trends in Parasitology*, *29*(2), 83–90. doi:10.1016/j.pt.2012.10.005 PMID:23219188

Lal, A., Hales, S., Kirk, M., Baker, M. G., & French, N. P. (2016). Spatial and temporal variation in the association between temperature and salmonellosis in NZ. *Australian and New Zealand Journal of Public Health*, *40*(2), 165–169. doi:10.1111/1753-6405.12413 PMID:26260292

Lal, A., Ikeda, T., French, N., Baker, M. G., & Hales, S. (2013). Climate variability, weather and enteric disease incidence in new zealand: Time series analysis. *PLoS One*, *8*(12), e83484. doi:10.1371/journal.pone.0083484 PMID:24376707

Lamon, L., Dalla Valle, M., Critto, A., & Marcomini, A. (2009). Introducing an integrated climate change perspective in POPs modelling, monitoring and regulation. *Environmental Pollution*, *157*(7), 1971–1980. doi:10.1016/j.envpol.2009.02.016 PMID:19272683

Lanata, C.F. (2003). Studies of food hygiene and diarrhoeal disease. *Int. J. Environ. Health Res.*, *13*(Suppl. 1), S175-S183.

Lanata, C. F. (2003). Studies of food hygiene and diarrhoeal disease. *International Journal of Environmental Health Research*, *13*(sup1), 175–183. doi:10.1080/0960312031000102921 PMID:12775394

Landtblom, A. M., Flodin, U., Söderfeldt, B., Wolfson, C., & Axelson, O. (1996). Organic solvents and multiple sclerosis: A synthesis of the current evidence. *Epidemiology (Cambridge, Mass.)*, *7*(4), 429–433. doi:10.1097/00001648-199607000-00015 PMID:8793371

Lane, E. A., Canty, M. J., & More, S. J. (2015). Cadmium exposure and consequence for the health and productivity of farmed ruminants. *Research in Veterinary Science*, *101*, 132–139. doi:10.1016/j.rvsc.2015.06.004 PMID:26267103

Lane, S., & Lloyd, D. (2002). Current trends in research into the waterborne parasite Giardia. *Critical Reviews in Microbiology*, *28*(2), 123–147. doi:10.1080/1040-840291046713 PMID:12109771

Lang, A. E., & Obeso, J. A. (2004). Challenges in Parkinson's disease: Restoration of the nigrostriatal dopamine system is not enough. *Lancet Neurology*, *3*(5), 309–316. doi:10.1016/S1474-4422(04)00740-9 PMID:15099546

Langkulsen, U., & Rwodzi, D. (2018). Climate Change, Air Pollution, and Human Health in Bangkok. In *Climate Change and Air Pollution* (pp. 255–271). Cham: Springer. doi:10.1007/978-3-319-61346-8_16

Langston, J. W., & Ballard, P. A. Jr. (1983). Parkinson's disease in a chemist working with l-methyl-4-phenyl-l, 2, 5, 6-tetrahydropyridine. *The New England Journal of Medicine, 309*(5). PMID:6602944

Langston, J. W., Ballard, P., Tetrud, J. W., & Irwin, I. (1983). Chronic Parkinsonism in humans due to a product of meperidine-analog synthesis. *Science, 219*(4587), 979–980. doi:10.1126cience.6823561 PMID:6823561

Langston, J. W., Forno, L. S., Rebert, C. S., & Irwin, I. (1984). Selective nigral toxicity after systemic administration of 1-methyl-4-phenyl-1, 2, 5, 6-tetrahydropyrine (MPTP) in the squirrel monkey. *Brain Research, 292*(2), 390–394.

Lanska, D. J. (1997). The geographic distribution of Parkinson's disease mortality in the United States. *Journal of the Neurological Sciences, 150*(1), 63–70. doi:10.1016/S0022-510X(97)05371-9 PMID:9260859

Lappano, R., Malaguarnera, R., Belfiore, A., & Maggiolini, M. (2017). Recent advances on the stimulatory effects of metals in breast cancer. *Molecular and Cellular Endocrinology, 5*, 49–56. doi:10.1016/j.mce.2016.10.017

Larkin, N. K., & Harrison, D. E. (2005). Global seasonal temperature and precipitation anomalies during El Niño autumn and winter. *Geophysical Research Letters, 32*(16), L16705. doi:10.1029/2005GL022860

Lassmann, H., & van Horssen, J. (2011). The molecular basis of neurodegeneration in multiple sclerosis. *FEBS Letters, 585*(23), 3715–3723. doi:10.1016/j.febslet.2011.08.004 PMID:21854776

Lauer, K., & Firnhaber, W. (1985). Epidemiological investigations into multiple sclerosis in Southern Hesse. III. The possible influence of occupation on the risk of disease. *Acta Neurologica Scandinavica, 72*(4), 397–402. doi:10.1111/j.1600-0404.1985.tb00890.x PMID:4082904

Lawler, J. J., Ruesch, A. S., Olden, J. D., & McRae, B. H. (2013). Projected climate-driven faunal movement routes. *Ecology Letters, 16*(8), 1014–1022. doi:10.1111/ele.12132 PMID:23782906

Lawrence, E., Hill, D., Hoppin, J., Lubin, J., Lynch, C., Pierce, J., ... Alavanja, M. (2005). 'Pesticide Use and Breast Cancer Risk among Farmers' Wives in the Agricultural Health Study. *American Journal of Epidemiology, 161*(2), 121–135. doi:10.1093/aje/kwi022

Le Couteur, D. G., McLean, A. J., Taylor, M. C., Woodham, B. L., & Board, P. G. (1999). Pesticides and Parkinson's disease. *Biomedicine and Pharmacotherapy, 53*(3), 122–130. doi:10.1016/S0753-3322(99)80077-8 PMID:10349500

Le Galliard, J. F., Massot, M., Baron, J. P., &Clobert, J. (2012). Ecological effects of climate change on European reptiles. *Wildlife conservation in a changing climate*, 179-203.

Le Houérou, H. N. (1992). Vegetation and land use in the Mediterranean basin by the year 2050: A prospective study. *Climatic Change and the Mediterranean, 1*, 175–232.

Leach, R. M. Jr, & Muenster, A. M. (1962). Studies on the role of manganese in bone formation. I. Effect upon the mucopolysaccharide content of chick bone. *The Journal of Nutrition, 78*(1), 51–56. doi:10.1093/jn/78.1.51 PMID:14463214

Leahy, J. G., & Colwell, R. R. (1990). Microbial degradation of hydrocarbons in the environment. *Microbiological Reviews, 54*, 305–315. PMID:2215423

Lechpammer, M., Clegg, M. S., Muzar, Z., Huebner, P. A., Jin, L.-W., & Gospe, S. M. Jr. (2014). Pathology of inherited manganese transporter deficiency. *Annals of Neurology, 75*(4), 608–612. doi:10.1002/ana.24131 PMID:24599576

Lee, D., Lind, P. M. Jr, Salihovic, D. R. J., Bavel, S., Van, B., & Lind, L. (2012). Background exposure to persistent organic pollutants predicts stroke in the elderly. *Environment International, 47*, 115–120. doi:10.1016/j.envint.2012.06.009 PMID:22809777

Lee, J. D., Wu, S. M., Lu, L. Y., Yang, Y. T., & Jeng, S. Y. (2009). Cadmium concentration and metallothioneinexpression in prostate cancer and benign prostatic hyperplasia of humans. *Journal of the Formosan Medical Association*, *108*(7), 554–559. doi:10.1016/S0929-6646(09)60373-9 PMID:19586829

Lee, R., Woo, W., Kummer, A., Duminy, H., & Xu, Z. (2003). Zinc accumulation in *N*-methyl-*N*-nitrosourea-induced rat mammary tumors is accompanied by an altered expression of ZnT-1 and metallothionein. *Experimental Biology and Medicine*, *228*, 689–696.

Lee, T., Tseng, M., Chen, C., & Lin, J. (2009). NeuroToxicology Association of high body lead store with severe intracranial carotid atherosclerosis. *Neurotoxicology*, *30*(6), 876–880. doi:10.1016/j.neuro.2009.07.004 PMID:19616024

Leff, B., Ramankutty, N., & Foley, J. A. (2004). Geographic distribution of major crops across the world. *Global Biogeochemical Cycles*, *18*(1), n/a. doi:10.1029/2003GB002108

Leichsenring, F., & Leweke, F. (2006). Social Anxiety Disorder. *The New England Journal of Medicine*, *355*(10), 1029–1036. doi:10.1056/NEJMcp060145 PMID:16957148

Lemarchand, C., Tual, S., Boulanger, M., Levêque-Morlais, N., Perrier, S., Clin, B., ... Baldi, I. (2016). Prostate cancer risk among French farmers in the AGRICAN cohort. *Scandinavian Journal of Work, Environment & Health*, *42*(2), 144–152. doi:10.5271jweh.3552 PMID:26932763

Lemmens-Gruber, R., Rachoy, B., Steininger, E., Kouri, K., Saleh, P., Krska, R., ... Lemmens, M. (2000). The effect of the *Fusarium* metabolite beauvericin on electromechanical and physiological properties in isolated smooth and heart muscle preparations of guinea pigs. *Mycopathologia*, *149*(1), 5–12. doi:10.1023/A:1007293812007

Leplège, A., Ecosse, E., Verdier, A., & Perneger, T. V. (1998). The French SF-36 Health Survey: Translation, cultural adaptation and preliminary psychometric evaluation. *Journal of Clinical Epidemiology*, *51*(11), 1013–1023. doi:10.1016/S0895-4356(98)00093-6 PMID:9817119

Leret, M. L., San Millán, J. A., & Antonio, M. T. (2003). Perinatal exposure to lead and cadmium affects anxiety-like behaviour. *Toxicology*, *186*(1-2), 125–130. doi:10.1016/S0300-483X(02)00728-X PMID:12604176

Lettenmaier, D. P., Wood, A. W., Palmer, R. N., Wood, E. F., & Stakhiv, E. Z. (1999). Water resources implications of global warming: A U.S. regional perspective. *Climatic Change*, *43*(3), 537–579. doi:10.1023/A:1005448007910

Levine, R. S., Peterson, A. T., & Benedict, M. Q. (2004). Distribution of members of Anopheles quadrimaculatus Say sl (Diptera: Culicidae) and implications for their roles in malaria transmission in the United States. *Journal of Medical Entomology*, *41*(4), 607–613. doi:10.1603/0022-2585-41.4.607

Levine, R. S., Peterson, A. T., & Benedict, M. Q. (2004). Geographic and ecologic distributions of the Anopheles gambiae complex predicted using a genetic algorithm. *The American Journal of Tropical Medicine and Hygiene*, *70*(2), 105–109. doi:10.4269/ajtmh.2004.70.105

Levy, K., Hubbard, A. E., & Eisenberg, J. N. (2009). Seasonality of rotavirus disease in the tropics: A systematic review and meta-analysis. *International Journal of Epidemiology*, *38*(6), 1487–1496. doi:10.1093/ije/dyn260 PMID:19056806

Levy, K., Woster, A. P., Goldstein, R. S., & Carlton, E. J. (2016). Untangling the impacts of climate change on waterborne diseases: A systematic review of relationships between diarrheal diseases and temperature, rainfall, flooding, and drought. *Environmental Science & Technology*, *50*(10), 4905–4922. doi:10.1021/acs.est.5b06186 PMID:27058059

Lewinsohn, P. M., Holm-Denoma, J. M., Small, J. W., Seeley, J. R., & Joiner, T. E. Jr. (2008). Separation anxiety disorder in childhood as a risk factor for future mental illness. *Journal of the American Academy of Child and Adolescent Psychiatry*, *47*(5), 548–555. doi:10.1097/CHI.0b013e31816765e7 PMID:18356763

Lewis, D. R., Southwick, J. W., Ouellet-Hellstrom, R., Rench, J., & Calderon, R. L. (1999). Drinking water arsenic in Utah: A cohort mortality study. *Environmental Health Perspectives*, *107*(5), 359–365. doi:10.1289/ehp.99107359 PMID:10210691

Lewis, J. A., Gehman, E. A., Baer, C. E., & Jackson, D. A. (2013). Alterations in gene expression in Caenorhabditis elegans associated with organophosphate pesticide intoxication and recovery. *BMC Genomics*, *14*(1), 291. doi:10.1186/1471-2164-14-291 PMID:23631360

Lewis, M. B., MacQuillan, G., Bamford, J. M., & Howdle, P. D. (2000). Delayed myelopathic presentation of the acquired hepatocerebral degeneration syndrome. *Neurology*, *54*(4), 1011. doi:10.1212/WNL.54.4.1011 PMID:10691011

Leye, E. H. M., Makoumba, D., Ndiaye, F., Bassirou, D., Maiguizo, D., & Tahir, D. (2012). Effet de la mycorhization sur la croissance et le développement de Jatropha curcas L., cultivée sous serre. *Revue des Energies Renouvelables*, *6*, 1741–1760.

Li, Y. V, & Zhang, J. H. (2012). Metal Ions in Stroke Pathophysiology. *Metal Ion in Stroke*, 1–12. doi:10.1007/978-1-4419-9663-3

Liao, S. L., & Chen, C. J. (2001). Manganese stimulates stellation of cultured rat cortical astrocytes. *Neuroreport*, *12*(18), 3877–3881. doi:10.1097/00001756-200112210-00004 PMID:11742202

Li, C., Jian, L., Zhang, Q. J., Feng, J. J., & Gui, Z. H. (2011). Alterations of emotion, cognition and firing activity of the basolateral nucleus of the amygdala after partial bilateral lesions of the nigrostriatal pathway in rats. *Brain Research Bulletin*, *85*(6), 329–338. doi:10.1016/j.brainresbull.2011.05.009 PMID:21624440

Lidsky, T. I., & Schneider, J. S. (2003). Lead neurotoxicity in children: Basic mechanisms and clinical correlates. *Brain*, *126*(1), 5–19. doi:10.1093/brain/awg014 PMID:12477693

Lighton, J., Brownell, P., Joos, B., & Turner, R. (2001). Low metabolic rate in scorpions: Implications for population biomass and cannibalism. *The Journal of Experimental Biology*, *204*(3), 607–613. PMID:11171311

Liigant, A., Haldre, S., Oun, A., Linnamägi, Ü., Saar, A., & Asser, T. A.-E. K. (2001). Seizure Disorders in Patients with Brain Tumors. *European Neurology*, *45*, 46–51.

Li, J., Li, N., Ma, M., Giesy, J. P., & Wang, Z. (2008). In vitro profiling of the endocrine disrupting potency of organochlorine pesticides. *Toxicology Letters*, *183*, 65–71.

Lilensten, L., Dudok de Wit, T., Kretzschmar, L., Amblard, P.-O., Moussaoui, S., Aboudarham, J., & Auchère, F. (2008). Review on the solar spectral variability in the EUV for space weather purposes. *Annales Geophysicae*, *26*(2), 269–279. doi:10.5194/angeo-26-269-2008

Liljegren, G., Hardell, L., Lindstrom, G., Dahl, P., & Magnuson, A. (1998). Case-control study on breast cancer and adipose tissue concentrations of congener specific polychlorinated biphenyls, DDE and hexachlorobenzene. *European Journal of Cancer Prevention*, *7*, 135–140.

Lima, A., Lovin, D. D., Hickner, P. V., & Severson, D. W. (2016). Evidence for an overwintering population of Aedes aegypti in Capitol Hill neighborhood, Washington, DC. *The American Journal of Tropical Medicine and Hygiene*, *94*(1), 231–235. doi:10.4269/ajtmh.15-0351

Lim, J., Lee, S., Lee, S., & Ha, S. (2018). Serum persistent organic pollutants levels and stroke risk. *Environmental Pollution*, *233*, 855–861. doi:10.1016/j.envpol.2017.12.031 PMID:29248762

Lim, S.-Y., Fox, S. H., & Lang, A. E. (2009). Overview of the Extranigral Aspects of Parkinson Disease. *Archives of Neurology*, *66*(2). doi:10.1001/archneurol.2008.561 PMID:19204152

Lindell, L. E. (1997). Annual variation in growth rate and body condition of adders, Viperaberus: Effects of food availability and weather. *Canadian Journal of Zoology*, *75*(2), 261–270. doi:10.1139/z97-033

Lindow-López, L., Galíndez, G., Aparicio-González, M., Sühring, S., Rojas-Aréchiga, M., Pritchard, H. W., & Ortega-Baes, P. (2018). Effects of alternating temperature on cactus seeds with a positive photoblastic response. *Journal of Arid Environments*, *148*, 74–77. doi:10.1016/j.jaridenv.2017.10.006

Lin, M. T., & Beal, M. F. (2006). Mitochondrial dysfunction and oxidative stress in neurodegenerative diseases. *Nature*, *443*(7113), 787–795. doi:10.1038/nature05292 PMID:17051205

Linnman, L., Andersson, A., Nilsson, K. O., Lind, B., Kjellstrom, T., & Friberg, L. (1973). Cadmium uptake by wheat from sewage sludge used as a plant nutrient source. A comparative study using flameless atomic absorption and neutron activation analysis. *Archives of Environmental Health*, *27*(1), 45–47. doi:10.1080/00039896.1973.10666306 PMID:4709111

Lin, P. H., Lin, C. H., Huang, C. C., Chuang, M. C., & Lin, P. (2007). 2,3,7,8-Tetrachlorodibenzo- p-dioxin (TCDD) induces oxidative stress, DNA strand breaks, and poly(ADPribose) polymerase-1 activation in human breast carcinoma cell lines. *Toxicology Letters*, *172*(3), 146–158. doi:10.1016/j.toxlet.2007.06.003

Lin, T., Huang, Y., & Huang, S. (1996). Lipid Peroxidation in Liver of Rats Administrated with Methyl Mercuric Chloride. *Biological Trace Element Research*, *33*, 54. PMID:8862759

Lioi, M. B., Santoro, A., Barbieri, R., Salzano, S., & Ursini, M. V. (2004). Ochratoxin and zearalenone: A comparative study on genotoxic effects and cell death induced in bovine lymphocytes. *Mutation Research*, *557*(1), 19–24. doi:10.1016/j.mrgentox.2003.09.009

Lioubimtseva, E., & Henebry, G. M. (2009). Climate and environmental change in arid Central Asia: Impacts, vulnerability, and adaptations. *Journal of Arid Environments*, *73*(11), 963–977. doi:10.1016/j.jaridenv.2009.04.022

Liou, H. H., Tsai, M. C., Chen, C. J., Jeng, J. S., Chang, Y. C., Chen, S. Y., & Chen, R. C. (1997). Environmental risk factors and Parkinson's disease: A case-control study in Taiwan. *Neurology*, *48*(6), 1583–1588. doi:10.1212/WNL.48.6.1583 PMID:9191770

Lipinski, C. (2001, April). Avoiding investment in doomed drugs. *Current Drug Discovery*, 17-19.

Lipp, E. K., Farrah, S. A., & Rose, J. B. (2001). Assessment and impact of microbial fecal pollution and human enteric pathogens in a coastal community. *Marine Pollution Bulletin*, *42*(4), 286–293. doi:10.1016/S0025-326X(00)00152-1 PMID:11381749

Lisle, J. T., & Rose, J. B. (1995). Cryptosporidium contamination of water in the USA and UK: A mini-review. *Aqua*, *44*(3), 103–117.

Liu, L., Oza, S., Hogan, D., Chu, Y., Perin, J., & Zhu, J. (2016). *Global, regional, and national causes of under-5 mortality in 2000–15: an updated systematic analysis with implications for the Sustainable Development.* Elsevier. Retrieved from https://www.sciencedirect.com/science/article/pii/S0140673616315938

Liu, C., Hofstra, N., & Franz, E. (2013). Impacts of climate change on the microbial safety of pre-harvest leafy green vegetables as indicated by *Escherichia coli* O157 and *Salmonella spp. International Journal of Food Microbiology*, *163*(2-3), 119–128. doi:10.1016/j.ijfoodmicro.2013.02.026 PMID:23558195

Liu, J., Wu, S., Shen, H., Cui, J., Wang, Y., Xing, L., ... Zhang, X. (2015). Ochratoxin A induces DNA damage and G2 phase arrest in human esophageal epithelium Het-1A cells in vitro. *The Journal of Toxicological Sciences*, *40*(5), 657–665. doi:10.2131/jts.40.657

Liu, R., Wang, J., Shi, J., Chen, Y., Sun, C., Zhang, P., & Shen, Z. (2014). Runoff characteristics and nutrient loss mechanism from plain farmland under simulated rainfall conditions. *The Science of the Total Environment*, *468-469*, 1069–1077. doi:10.1016/j.scitotenv.2013.09.035 PMID:24095969

Liu, S. H., Fu, W. M., & Lin-Shiau, S. Y. (1989). Effects of sodium selenite on neuromuscular junction of the mouse phrenic nerve-diaphragm preparation. *Neuropharmacology*, *28*(7), 733–739. doi:10.1016/0028-3908(89)90159-7 PMID:2474766

Liu, S., Li, S., & Du, Y. (2010). Polychlorinated Biphenyls (PCBs) Enhance Metastatic Properties of Breast Cancer Cells by Activating Rho-Associated Kinase (ROCK). *PLoS One*, *5*(6), 11272. doi:10.1371/journal.pone.0011272

Liu, Y., & Wu, F. (2010). Global burden of aflatoxin-induced hepatocellular carcinoma: A risk assessment. *Environmental Health Perspectives*, *118*(6), 818–824. doi:10.1289/ehp.0901388

Li, X. M., Zhang, L., Li, J., Li, Y., Wang, H. L., Ji, G. Y., ... Zhao, X. J. (2005). Measurement of serum zinc improves prostate cancer detection efficiency in patients with PSA levels between 4 ng/mL and 10 ng/mL. *Asian Journal of Andrology*, *7*(3), 323–328. doi:10.1111/j.1745-7262.2005.00044.x PMID:16110361

Li, X., Baker, D. N., Kanekal, S. G., Looper, M., & Temerin, M. (2001). Long Term Measurements of Radiation Belts by SAMPEX and Their Variations. *Geophysical Research Letters*, *28*(20), 3827–3830. doi:10.1029/2001GL013586

Li, X., Yin, J., Cheng, C., Sun, J., Li, Z., & Wu, Y. (2005). Paraquat induces selective dopaminergic nigrostriatal degeneration in aging C57BL/6 mice. *Chinese Medical Journal*, *118*(16), 1357–1361. Retrieved from http://www.ncbi.nlm.nih.gov/pubmed/16157030 PMID:16157030

Li, Y., Millikan, R. C., Bell, D. A., Cui, L., Tse, C. K., Newman, B., & Conway, K. (2005). Polychlorinated biphenyls, cytochrome P450 1A1 (CYP1A1) polymorphisms, and breast cancer risk among African American women and white women in North Carolina: A population-based case-control study. *Breast Cancer Research*, *7*(1), 12–18. doi:10.1186/bcr941

Li, Z. H., Liu, X. Y., Wang, N., Chen, J. S., Chen, Y. H., Huang, J. T., ... Chen, D. J. (2012). Effects of decabrominated diphenyl ether (PBDE-209) in regulation of growth and apoptosis of breast, ovarian and cervical cancer cells. *Environmental Health Perspectives*, *120*(4), 541–546. doi:10.1289/ehp.1104051

Llabjani, V., Trevisan, J., Jones, K. C., Shore, R. F., & Martin, F. L. (2011). Derivation by infrared spectroscopy with multivariate analysis of bimodal contaminant-induced dose-response effects in MCF-7 cells. *Environmental Science & Technology*, *45*(14), 6129–6135. doi:10.1021/es200383a

Lloret, F., Penuelas, J., & Estiarte, M. (2004). Experimental evidence of reduced diversity of seedlings due to climate modification in a Mediterranean-type community. *Global Change Biology*, *10*(2), 248–258. doi:10.1111/j.1365-2486.2004.00725.x

Lodovici, M., & Bigagli, E. (2011). *Oxidative Stress and Air Pollution Exposure* [Research article]. doi:10.1155/2011/487074

Lo, E. H., Dalkara, T., & Moskowitz, M. A. (2003). Mechanisms, challenges and opportunities in stroke. *Nature Reviews. Neuroscience*, *4*. doi:10.1038/nrn1106 PMID:12728267

Logrieco, A., Rizzo, A., Ferracane, R., & Ritieni, A. (2002). Occurrence of beauvericin and enniatins in wheat affected by Fusarium avenaceum head blight. *Applied and Environmental Microbiology*, *68*(1), 82–85. doi:10.1128/AEM.68.1.82-85.2002

Logroscino, G., Gao, X., Chen, H., Wing, A., & Ascherio, A. (2008). Dietary Iron Intake and Risk of Parkinson's Disease. *American Journal of Epidemiology*, *168*(12), 1381–1388. doi:10.1093/aje/kwn273 PMID:18945687

Logroscino, G., Traynor, B. J., Hardiman, O., Chio, A., Couratier, P., Mitchell, J. D., ... Beghi, E. (2008). Descriptive epidemiology of amyotrophic lateral sclerosis: New evidence and unsolved issues. *Journal of Neurology, Neurosurgery, and Psychiatry*, *79*(1), 6–11. doi:10.1136/jnnp.2006.104828 PMID:18079297

Loomis, D., Guyton, K., Grosse, Y., El Ghissasi, F., Bouvard, V., Benbrahim-Tallaa, L., ... Straif, K. (2015). Carcinogenicity of lindane, DDT, and 2, 4-dichlorophenoxyacetic acid. *The Lancet Oncology*, *16*(8), 891–892. doi:10.1016/S1470-2045(15)00081-9 PMID:26111929

López-Alcaide, S., & Macip-Ríos, R. (2011). Effects of climate change in amphibians and reptiles. In *Biodiversity loss in a changing planet*. InTech. doi:10.5772/24663

López-Carrillo, L., Hernández-Ramírez, R. U., Gandolfi, A. J., Ornelas Aguirre, J. M., Torres-Sánchez, L., & Cebrian, M. E. (2014). Arsenic methylation capacity is associated with breast cancer in northern Mexico. *Toxicology and Applied Pharmacology*, *280*(1), 53–59. doi:10.1016/j.taap.2014.07.013

Lopez-Carrillo, L., Lopez-Cervantes, M., Torres-Sanchez, L., Blair, A., Cebrain, M., & Garcia, R. (2002). Serum levels of betahexachlorocyclohexane, hexachlorobenzene and polychlorinated biphenyls and breast cancer in Mexican women. *European Journal of Cancer Prevention, 11*(2), 129–135. doi:10.1097/00008469-200204000-00004

Lopez-Cervantes, M., Torres-Sanchez, L., Tobias, A., & Lopez-Carrillo, L. (2004). Dichlorodiphenyldichloroethane burden and breast cancer risk: A meta-analysis of the epidemiologic evidence. *Environmental Health Perspectives*, *112*(2), 207–214. doi:10.1289/ehp.112-1241830

Lopez, O., Hernandez, A., Rodrigo, L., Gil, F., Pena, G., Serrano, J., ... Pla, A. (2007). Changes in antioxidant enzymes in humans with long-term exposure to pesticides. *Toxicology Letters*, *171*(3), 146–153. doi:10.1016/j.toxlet.2007.05.004 PMID:17590542

Loreto, F., Ciccioli, P., Cecinato, A., Brancaleoni, E., Frattoni, M., & Tricoli, D. (1996). Influence of environmental factors and air composition on the emission of [alpha]-pinene from quercus ilex leaves. *Plant Physiology*, *110*(1), 267–275. doi:10.1104/pp.110.1.267 PMID:12226182

Lorincz, M. T. (2010). Neurologic Wilson's disease. *Annals of the New York Academy of Sciences*, *1184*(1), 173–187. doi:10.1111/j.1749-6632.2009.05109.x PMID:20146697

Lourenço, W. (1991). Biogéographie évolutive, écologies et les stratégies biodémographiques chez les scorpions néotropicaux. *Compte rendu des séances de la société de biogéographie, 67*(4), 171-190.

Lourenço, W. R. (1994). Diversity and endemism in tropical versus temperate scorpion communities. *Compte rendu des séances de la société de biogéographie, 70*(3), 155-160.

Lourenço, W. R. (1988). Peut-on parler d'une biogéographie du scorpionisme. *CR Seances Soc Biogéogr*, *64*(4), 137–143.

Lourenço, W. R., & Cuellar, O. (1995). Scorpions, scorpionism, life history strategies and parthenogenesis. *The Journal of Venomous Animals and Toxins*, *1*(2), 51–62. doi:10.1590/S0104-79301995000200002

Lourengo, W. (1991). Biogéographie évolutive, écologies et les stratégies biodémographiques chez les scorpions néotropicaux. *Compte rendu des séances de la société de biogéographie, 67*(4), 171-190.

Lowenstein, D. H. (2009). Posttraumatic Epilepsy: Epilepsy after head injury : An overview. *Epilepsia*, *50*, 4–9. doi:10.1111/j.1528-1167.2008.02004.x PMID:19187288

Lozano, R., Naghavi, M., Foreman, K., & Lim, S. (n.d.). *Global and regional mortality from 235 causes of death for 20 age groups in 1990 and 2010: a systematic analysis for the Global Burden of Disease Study 2010.* Elsevier. Retrieved from https://www.sciencedirect.com/science/article/pii/S0140673612617280

Lublin, F. (2005). History of modern multiple sclerosis therapy. *Journal of Neurology*, *252*(S3), iii3–iii9. doi:10.100700415-005-2010-6 PMID:16170498

Lu, C. S., Huang, C. C., Chu, N. S., & Calne, D. B. (1994). Levodopa failure in chronic manganism. *Neurology*, *44*(9), 1600–1602. doi:10.1212/WNL.44.9.1600 PMID:7936281

Lu, C., Fenske, R. A., Simcox, N. J., & Kalman, D. (2000). Pesticide exposure of children in an agricultural community: Evidence of household proximity to farmland and take home exposure pathways. *Environmental Research*, *84*(3), 290–302. doi:10.1006/enrs.2000.4076 PMID:11097803

Lucena, R., Allam, M., Costabeber, I., Villarejo, M., & Navajas, R. (2001). Breast cancer risk factors: PCB congeners. *European Journal of Cancer Prevention*, *10*(1), 117–119. doi:10.1097/00008469-200102000-00018

Ludolph, A. C., He, F., Spencer, P. S., Hammerstad, J., & Sabri, M. (1991). 3- Nitropropionic acid exogenous animal neurotoxin and possible human striatal toxin. *The Canadian Journal of Neurological Sciences*, *18*(04), 492–498. doi:10.1017/S0317167100032212 PMID:1782616

Lueth, F. X. (1941). Effects of temperature on snakes. *Copeia*, *1941*(3), 125–132. doi:10.2307/1437733

Luiselli, L. (2001). The ghost of a recent invasion in the reduced feeding rates of spitting cobras during the dry season in a rainforest region of tropical Africa? *Acta Oecologica*, *22*(5-6), 311–314. doi:10.1016/S1146-609X(01)01113-4

Luiselli, L. (2006). Site occupancy and density of sympatric Gaboon viper (Bitisgabonica) and nose-horned viper (Bitisnasicornis). *Journal of Tropical Ecology*, *22*(5), 555–564. doi:10.1017/S0266467406003397

Lukaszewicz-hussain, A. (2010). Role of oxidative stress in organophosphate insecticide toxicity – Short review. *Pesticide Biochemistry and Physiology*, *98*(2), 145–150. doi:10.1016/j.pestbp.2010.07.006

Lukiw, W. J., Percy, M. E., & Kruck, T. P. (2005). Nanomolar aluminum induces pro-inflammatory and pro-apoptotic gene expression in human brain cells in primary culture. *Journal of Inorganic Biochemistry*, *99*(9), 1895–1898. doi:10.1016/j.jinorgbio.2005.04.021 PMID:15961160

Lundh, T., Wennberg, M., Bergdahl, I. A., Stegmayr, B., Vessby, B., Skerfving, S., & Stro, U. (2007). And risk of stroke in northern Sweden. *British Journal of Nutrition*, 1038–1045. doi:10.1017/S0007114507756519

Luo, T., Young, R., & Reig, P. (2015). *Aqueduct Projected Stress Country Rankings*. Retrieved from http://www.wri.org/blog/2015/08/ranking-world's-most-water-stressed-countries-2040

Lutz, C., Erken, M., Noorian, P., Sun, S., & McDougald, D. (2013). Environmental reservoirs and mechanisms of persistence of *Vibrio cholerae*. *Frontiers in Microbiology*, *4*, 375. doi:10.3389/fmicb.2013.00375 PMID:24379807

Ly, D., Forman, D., Ferlay, J., Brinton, L. A., & Cook, M. B. (2013). An international comparison of male and female breast cancer incidence rates. *International Journal of Cancer*, *132*(8), 1918–1926. doi:10.1002/ijc.27841

Lyimo, E. O., & Takken, W. (1993). Effects of adult body size on fecundity and the pre-gravid rate of Anopheles gambiae females in Tanzania. *Medical and Veterinary Entomology*, *7*(4), 328–332. doi:10.1111/j.1365-2915.1993.tb00700.x

Lyons, G., & Watterson, A. (2010). *Review of the Role Pesticides Play in Some Cancers: Children, Farmers and Pesticide Users at Risk*. CHEMTrust Report.

Lysenko, A. J., & Beljaev, A. E. (1969). An analysis of the geographical distribution of Plasmodium ovale. *Bulletin of the World Health Organization*, *40*(3), 383.

Mabey, N., Gulledge, J., Finel, B., & Silverthorne, K. (2011). *Degrees of risk: Defining a risk management framework for climate security*. London: Third Generation Environmentalism.

Mac Kenzie, W. R., Hoxie, N. J., Proctor, M. E., Gradus, M. S., Blair, K. A., Peterson, D. E., ... Davis, J. P. (1994). A massive outbreak in Milwaukee of cryptosporidium infection transmitted through the public water supply. *The New England Journal of Medicine*, *331*(3), 161–167. doi:10.1056/NEJM199407213310304 PMID:7818640

Macdonald, J. F., Xiong, Z., & Jackson, M. F. (2006). Paradox of Ca 2 C signaling, cell death and stroke. *Neurosciences*, *29*(2). doi:10.1016/j.tins.2005.12.001

MacDonald, M.Huntington's Disease Collaborative Research Group. (1993). A novel gene containing a trinucleotide repeat that is expanded and unstable on Huntington's disease chromosomes. *Cell*, *72*(6), 971–983. doi:10.1016/0092-8674(93)90585-E PMID:8458085

Macdonald, R. W., Harner, T., & Fyfe, J. (2005). Recent climate change in the Arctic and its impact on contaminant pathways and interpretation of temporal trend data. *The Science of the Total Environment*, *342*(1-3), 5–86. doi:10.1016/j.scitotenv.2004.12.059 PMID:15866268

Macdonald, R. W., Mackay, D., Li, Y. F., & Hickie, B. (2003). How will global climate change affect risks from long-range transport of persistent organic pollutants? *Human and Ecological Risk Assessment*, *9*(3), 643–660. doi:10.1080/713609959

Maciorowski, K. G., Herrera, P., Jones, F. T., Pillai, S. D., & Ricke, S. C. (2007). Effects on poultry and livestock of feed contamination with bacteria and fungi. *Animal Feed Science and Technology*, *133*(1-2), 109–136. doi:10.1016/j.anifeedsci.2006.08.006

MacKenzie, W. R., Schell, W. L., Blair, K. A., Addiss, D. G., Peterson, D. E., Hoxie, N. J., ... Davis, J. P. (1995). Massive outbreak of waterborne cryptosporidium infection in Milwaukee, Wisconsin: Recurrence of illness and risk of secondary transmission. *Nephrology, Dialysis, Transplantation*, *21*, 57–62. PMID:7578760

Macneale, K. H., Kiffney, P. M., & Scholz, N. L. (2010). Pesticides, aquatic food webs, and the conservation of Pacific salmon. *Frontiers in Ecology and the Environment*, *8*(9), 475–482. doi:10.1890/090142

Mac, T. L., Tran, D., Quet, F., Odermatt, P., Preux, P., & Tan, C. T. (2007). Epidemiology, aetiology, and clinical management of epilepsy in Asia : A systematic review. *Lancet Neurology*, *6*(6), 533–543. doi:10.1016/S1474-4422(07)70127-8 PMID:17509488

Madhun, Y. A., & Freed, V. H. (1990). Impact of pesticides on the environment. In Pesticides in the soil environment. Soil Science Society of America.

Mahad, D. H., Trapp, B. D., & Lassmann, H. (2015). Pathological mechanisms in progressive multiple sclerosis. *Lancet Neurology*, *14*(2), 183–193. doi:10.1016/S1474-4422(14)70256-X PMID:25772897

Mahé, G., & Olivry, J. C. (1995). Variations des précipitations et des écoulements en Afrique de l'Ouest et centrale de 1951 à 1989. *Science et changements planétaires/Sécheresse*, *6*(1), 109-117.

Mahé, G., Olivry, J. C., Dessouassi, R., Orange, D., Bamba, F., & Servat, E. (2000). Relations eaux de surface–eaux souterraines d'une rivière tropicale au Mali. *Comptes Rendus de l'Académie des Sciences-Series IIA-Earth and Planetary Science*, *330*(10), 689–692.

Mahjour, J., Akalay, O., & Saddiki, A. (1992). Les leishmanioses au Maroc de l'analyse écoépidémiologie à la prévention. *DEPS Bul. Epidém., 7.*

Maine, M. A., Sune, N. L., & Lagger, S. C. (2004). Chromium bioaccumulation: Comparison of the capacity of two floating aquatic macrophytes. *Water Research*, *38*(6), 1494–1501. doi:10.1016/j.watres.2003.12.025 PMID:15016526

Ma, J., Hung, H., & Blanchard, P. (2004). How do climate fluctuations affect persistent organic pollutant distribution in North America? Evidence from a decade of air monitoring. *Environmental Science & Technology*, *38*(9), 2538–2543. doi:10.1021/es0349610 PMID:15180048

Ma, J., Qiu, X., Zhang, J., Duan, X., & Zhu, T. (2012). State of polybrominated diphenyl ethers in China: An overview. *Chemosphere*, *88*(7), 769–778. doi:10.1016/j.chemosphere.2012.03.093

Malek, A. M., Barchowsky, A., Bowser, R., Youk, A., & Talbott, E. O. (2012). Pesticide exposure as a risk factor for amyotrophic lateral sclerosis: a meta-analysis of epidemiological studies: pesticide exposure as a risk factor for ALS. *Environmental Research*, *117*, 112–119. doi:10.1016/j.envres.2012.06.007 PMID:22819005

Malina, T., Krecsak, L., & Warrell, D. A. (2008). Neurotoxicity and hypertension following European adder (*Viperaberus-berus*) bites in Hungary: Case report and review. *QJM*, *101*(10), 801–806. doi:10.1093/qjmed/hcn079 PMID:18647797

Mallin, M. A., Johnson, V. L., & Ensign, S. H. (2009). Comparative impacts of stormwater runoff on water quality of an urban, a suburban, and a rural stream. *Environmental Monitoring and Assessment*, *159*(1-4), 475–491. doi:10.100710661-008-0644-4 PMID:19067207

Mandarapu, R., & Prakhya, B. M. (2015). In vitro myelotoxic effects of cypermethrin and mancozeb on human hematopoietic progenitor cells. *Journal of Immunotoxicology*, *12*(1), 48–55. doi:10.3109/1547691X.2014.880535 PMID:24499300

Mandrioli, J., Faglioni, P., Merelli, E., & Sola, P. (2003). The epidemiology of ALS in Modena, Italy. *Neurology*, *60*(4), 683–689. doi:10.1212/01.WNL.0000048208.54755.78 PMID:12601112

Manguin, S., Carnevale, P., Mouchet, J., Coosemans, M., Julvez, J., Richard-Lenoble, D., & Sircoulon, J. (2008). *Biodiversity of malaria in the world*. Academic Press.

Manguin, S., Garros, C., Dusfour, I., Harbach, R. E., & Coosemans, M. (2008). Bionomics, taxonomy, and distribution of the major malaria vector taxa of Anopheles subgenus Cellia in Southeast Asia: An updated review. *Infection, Genetics and Evolution*, *8*(4), 489–503. doi:10.1016/j.meegid.2007.11.004

Mannello, F., Tonti, G. A., Medda, V., Simone, P., & Darbre, P. D. (2011). Analysis of aluminium content and iron homeostasis in nipple aspirate fluids from healthy women and breast cancer-affected patients. *Journal of Applied Toxicology*, *31*(3), 262–269. doi:10.1002/jat.1641

Manning, H. C., Buck, J. R., & Cook, R. S. (2016). Mouse models of breast cancer: platforms for discovering precision imaging diagnostics and future cancer medicine. *Journal of Nuclear Medicine*, *57*, 60S–68S.

Manning-BoĞ, A. B., & Langston, J. W. (2007). Model fusion: The next phase in developing animal models for parkinson's disease. *Neurotoxicity Research*, *11*(3–4), 219–240. doi:10.1007/BF03033569 PMID:17449461

Manolova, Y., Manolov, G., Parvanova, L., Petkova-Bocharova, T., Castegnaro, M., & Chernozemsky, I. N. (1990). Induction of characteristic chromosomal aberrations, particularly X-trisomy, in cultured human lymphocytes treated by ochratoxin A, a mycotoxin implicated in Balkan endemic nephropathy. *Mutation Research. Fundamental and Molecular Mechanisms of Mutagenesis*, *231*(2), 143–149. doi:10.1016/0027-5107(90)90021-U

Mansouri, M. T., Muñoz-Fambuena, I., & Cauli, O. (2018). Cognitive impairment associated with chronic lead exposure in adults. *Neurology, Psychiatry & Brain Research*, *30*(March), 5–8. doi:10.1016/j.npbr.2018.04.001

Manthripragada, A. D., Costello, S., Cockburn, M. G., Bronstein, J. M., & Ritz, B. (2010). Paraoxonase 1, Agricultural Organophosphate Exposure, and Parkinson Disease. *Epidemiology (Cambridge, Mass.)*, *21*(1), 87–94. doi:10.1097/EDE.0b013e3181c15ec6 PMID:19907334

Mantle, P., Kulinskaya, E., & Nestler, S. (2005). Renal tumourigenesis in male rats in response to chronic dietary ochratoxin A. *Food Additives and Contaminants*, *22*(sup1), 58–64. doi:10.1080/02652030500358431

Manuwald, U., Velasco-Garrido, M., Berger, J., Manz, A., & Baur, X. (2012). Mortality study of chemical workers exposed to dioxins: Follow-up 23 years after chemical plant closure. *Occupational and Environmental Medicine*, *69*(9), 636–642. doi:10.1136/oemed-2012-100682

Marasas, W. F. O., Jaskiewicz, K., Venter, F. S., & Van Schalkwyk, D. J. (1988). Fusarium moniliforme contamination of maize in esophageal cancer aereas in Transkei. *South African Medical Journal*, *74*, 110–114.

Marasas, W. F., Kellerman, T. S., Gelderblom, W. C., Coetzer, J. A., Thiel, P. G., & van der Lugt, J. J. (1988). Leukoencephalomalacia in a horse induced by fumonisin B1 isolated from Fusarium moniliforme. *The Onderstepoort Journal of Veterinary Research*, *55*, 197–203.

Marcovecchio, J. E., Botté, S. E., & Freije, R. H. (2007). Heavy metals, major metals, trace elements. Handbook of Water Analysis, 2, 275-311.

Marcussen, H., Dalsgaard, A., & Holm, P. E. (2008). Content, distribution and fate of 33 elements in sediments of rivers receiving wastewater in Hanoi, Vietnam. *Environmental Pollution*, *155*(1), 41–51. doi:10.1016/j.envpol.2007.11.001 PMID:18164523

Maret, W. (2017). The bioinorganic chemistry of lead in the context of its toxicity. *Metal Ions in Biological Systems.* doi:10.1515/9783110434330-001

Marin, B., Couratier, P., Preux, P. M., & Logroscino, G. (2011). Can mortality data be used to estimate amyotrophic lateral sclerosis incidence? *Neuroepidemiology*, *36*(1), 29–38. doi:10.1159/000321930 PMID:21088432

Marin, S., Ramos, A. J., Cano-Sancho, G., & Sanchis, V. (2013). Mycotoxins: Occurrence, toxicology, and exposure assessment. *Food and Chemical Toxicology*, *60*, 218–237. doi:10.1016/j.fct.2013.07.047

Markesbery, W. R., Ehmann, W. D., Candy, J. M., Ince, P. G., Shaw, P. J., Tandon, L., & Deibel, M. A. (1995). Neutron activation analysis of trace elements in motor neuron disease spinal cord. *Neurodegeneration*, *4*(4), 383–390. doi:10.1006/neur.1995.0046 PMID:8846231

Marques dos Santos, D., Aschner, M., & Marreilha dos Santos, A. P. (2017). Manganese and Neurodegeneration. In A. R. White, M. Aschner, L. G. Costa, & A. I. Bush (Eds.), *Biometals in Neurodegenerative Diseases* (pp. 117–151). Academic Press. doi:10.1016/B978-0-12-804562-6.00008-7

Marrie, R. A. (2004). Environmental risk factors in multiple sclerosis aetiology. *Lancet Neurology*, *3*(12), 709–718. doi:10.1016/S1474-4422(04)00933-0 PMID:15556803

Martens, W. J., Jetten, T. H., & Focks, D. A. (1997). Sensitivity of malaria, schistosomiasis and dengue to global warming. *Climatic Change*, *35*(2), 145–156. doi:10.1023/A:1005365413932

Marthy, P., LeFichoux, Y., Pratlong, F., Rioux, J. A., Rostain, G., & Lacour, J. P. (1989). Cutaneous leishmaniasis due to *Leishmaniatropica*in a young Moroccan child observed in Nice, France. *Transactions of the Royal Society of Tropical Medicine and Hygiene*, *83*(4), 510. doi:10.1016/0035-9203(89)90268-X PMID:2617601

Martin, M. B., Reiter, R., Pham, T., Avellanet, Y. R., Camara, J., Lahm, M., ... Stoica, A. (2003). Estrogen like activity of metals in MCF-7 breast cancer cells. *The Journal of Endocrinology*, *144*(6), 2425–2436. doi:10.1210/en.2002-221054

Martin, M. B., Voeller, H. J., Gelmann, E. P., Lu, J., Stoica, E. G., Hebert, E. J., ... Stoica, A. (2002). Role of cadmium in the regulation of AR gene expressionand activity. *Endocrinology*, *143*(1), 263–275. doi:10.1210/endo.143.1.8581 PMID:11751618

Martins, L. F., & Peixoto, R. S. (2012). Biodegradation of petroleum hydrocarbons in hypersaline environments. *Brazilian Journal of Microbiology*, *43*(3), 865–872. doi:10.1590/S1517-83822012000300003 PMID:24031900

Martynyuk, A. E., Glushakov, A. V., Sumners, C., Laipis, P. J., Dennis, D. M., & Seubert, C. N. (2005). Impaired glutamatergic synaptic transmission in the PKU brain. *Molecular Genetics and Metabolism*, *86*, 34–42. doi:10.1016/j.ymgme.2005.06.014 PMID:16153867

Marx, A., Kumar, R., Thober, S., Rakovec, O., Wanders, N., Zink, M., ... Samaniego, L. (2018). Climate change alters low flows in Europe under global warming of 1.5, 2, and 3 °C. *Hydrology and Earth System Sciences*, *22*(2), 1017–1032. doi:10.5194/hess-22-1017-2018

Mas–Coma, S., Valero, M. A., & Bargues, M. D. (2009). Climate change effects in trematodiases, with emphasis on zoonotic fasciolasis and schistosomiasis. *Veterinary Parasitology*, *163*(4), 264–280. doi:10.1016/j.vetpar.2009.03.024 PMID:19375233

Mason, L. H., Harp, J. P., & Han, D. Y. (2014). Pb Neurotoxicity: Neuropsychological Effects of Lead Toxicity. *BioMed Research International*, *2014*, 1–8. doi:10.1155/2014/840547 PMID:24516855

Massanyi, P., Lukac, N., Slivkova, J., Kovacik, J., Makarevich, A. V., Chrenek, P., ... Stawarz, R. (2007). Mercury-induced alterations in rat kidneys and testes in vivo. *Journal of Environmental Science and Health Part A*, *42*(7), 865–870. doi:10.1080/10934520701370410 PMID:17558766

Mateo-Sagasta, J., Medlicott, J., Qadir, M., Raschid-Sally, L., Drechsel, P., & Liebe, J. (2013). *Proceedings of the UN-Water project on the safe use of wastewater in agriculture*. UN-Water Decade Programme on Capacity Development Proceedings Series No. 11.

Mateo-Sagasta, J., Medlicott, K., Qadir, M., Rashid-Sally, L., Dreschel, P., & Liebe, J. (2013). *Proceedings of the UN-Water project on the Safe Use of Wastewater in Agriculture*. Retrieved from http://collections.unu.edu/eserv/UNU:2661/proceedings-no-11_WEB.pdf

Mathee, A., Singh, E., Mogotsi, M., Timothy, G., Maduka, B., Olivier, J., & Ing, D. (2009). Lead-based paint on playground equipment in public children's parks in Johannesburg, Tshwane and Ekurhuleni. *South African Medical Journal*, *99*(11), 819–821. PMID:20218484

Mathews, A. O. C., & Kakulu, S. (2012). Investigation of heavy metal levels in roadside agricultural soil and plant samples in Adogo, Nigeria. *Acad J Enviro Sci.*, *1*(2), 31.

Matin, M. A., Jaffery, F. N., & Siddiqui, R. A. (1981). A Possible Neurochemical Basis of the Central Stimulatory Effects of pp ' DDT. *Neurochemistry*, *36*(3), 1000–1005. doi:10.1111/j.1471-4159.1981.tb01692.x PMID:7205252

Matsubara, K., Kobayashi, S., Kobayashi, Y., Yamashita, K., Koide, H., Hatta, M., ... Kimura, K. (1995). beta-Carbolinium cations, endogenous MPPplus analogs, in the lumbar cerebrospinal fluid of patients with Parkinson's disease. *Neurology*, *45*(12), 2240–2245. doi:10.1212/WNL.45.12.2240 PMID:8848200

Matsumura, H., Takahata, R., & Hayaishi, O. (1991). Inhibition of sleep in rats by inorganic selenium compounds, inhibitors of prostaglandin D synthase. *Proceedings of the national avademy of sciences of the united states of America*, *88*, 9046-50.10.1073/pnas.88.20.9046

Matthys, B., Vounatsou, P., Raso, G., Tschannen, A. B., Becket, E. G., Gosoniu, L., ... Utzinger, J. (2006). Urban farming and malaria risk factors in a medium-sized town in Côte d'Ivoire. *The American Journal of Tropical Medicine and Hygiene*, *75*(6), 1223–1231. doi:10.4269/ajtmh.2006.75.1223 PMID:17172397

Matyja, E. (2000). Aluminum enhances glutamate-mediated neurotoxicity in organotypic cultures of rat hippocampus. *Folia Neuropathologica*, *38*(2), 47–53. Retrieved from http://www.ncbi.nlm.nih.gov/pubmed/11057035 PMID:11057035

Mazzotti, L., & Bravo-Becherelle, M. A. (1961). Scorpionism in the Mexican republic. *Revista del Instituto de Salubridad y Enfermedades Tropicales*, *21*, 3. PMID:14471889

Mbundi, L., Gallar-Ayala, H., Khan, M. R., Barber, J. L., Losada, S., & Busquets, R. (2014). Advances in the analysis of challenging food contaminants: Nanoparticles, bisphenols, mycotoxins, and brominated flame retardants. In J. C. Fishbein & J. M. Heilman (Eds.), *Advances in Molecular Toxicology* (Vol. 8, pp. 35–105). Elsevier.

Mccarren, H. S., & Jr, J. H. M. (2016). Anticonvulsant discovery through animal models of status epilepticus induced by organophosphorus nerve agents and pesticides. *Annals of the New York Academy of Sciences*, 1–7. doi:10.1111/nyas.13092 PMID:27258770

McCarthy, D. T., Hathaway, J. M., Hunt, W. F., & Deletic, A. (2012). Intra-event variability of *Escherichia coli* and total suspended solids in urban stormwater runoff. *Water Research*, *46*(20), 6661–6670. doi:10.1016/j.watres.2012.01.006 PMID:22321764

McCarthy, S., Somayajulu, M., Sikorska, M., Borowy-Borowski, H., & Pandey, S. (2004). Paraquat induces oxidative stress and neuronal cell death; neuroprotection by water-soluble Coenzyme Q10. *Toxicology and Applied Pharmacology*, *201*(1), 21–31. doi:10.1016/j.taap.2004.04.019 PMID:15519605

McCombe, P. A., & Henderson, R. D. (2010). Effects of gender in amyotrophic lateral sclerosis. *Gender Medicine*, *7*(6), 557–570. doi:10.1016/j.genm.2010.11.010 PMID:21195356

McConoughey, S. J., Basso, M., Niatsetskaya, Z. V., Sleiman, S. F., Smirnova, N. A., Langley, B. C., ... Ratan, R. R. (2010). Inhibition of transglutaminase 2 mitigates transcriptional dysregulation in models of Huntington disease. *EMBO Molecular Medicine*, *2*(9), 349–370. doi:10.1002/emmm.201000084 PMID:20665636

McCormack, A. L., Thiruchelvam, M., Manning-Bog, A. B., Thiffault, C., Langston, J. W., Cory-Slechta, D. A., & Di Monte, D. A. (2002). Environmental Risk Factors and Parkinson's Disease: Selective Degeneration of Nigral Dopaminergic Neurons Caused by the Herbicide Paraquat. *Neurobiology of Disease*, *10*(2), 119–127. doi:10.1006/nbdi.2002.0507 PMID:12127150

McCready, D., Aronson, K., Chu, W., Fan, W., Vesprini, D., & Narod, S. (2004). Breast tissue organochlorine levels and metabolic genotypes in relation to breast cancer risk Canada. *Cancer Causes & Control*, *15*(4), 399–418. doi:10.1023/B:CACO.0000027505.32564.c2

McCreesh, N., & Booth, M. (2013). Challenges in predicting the effects of climate change on Schistosoma mansoni and Schistosoma haematobium transmission potential. *Trends in Parasitology*, *29*(11), 548–555. doi:10.1016/j.pt.2013.08.007 PMID:24064438

McCreesh, N., Nikulin, G., & Booth, M. (2015). Predicting the effects of climate change on Schistosomamansoni transmission in eastern Africa. *Parasites & Vectors*, *8*(1), 4. doi:10.118613071-014-0617-0 PMID:25558917

McDaniel, M. D., Kaye, J. P., & Kaye, M. W. (2013). Increased temperature and precipitation had limited effects on soil extracellular enzyme activities in a post-harvest forest. *Soil Biology & Biochemistry*, *56*, 90–98. doi:10.1016/j.soilbio.2012.02.026

McDermott, J. R., Smith, A. I., Iqbal, K., & Wisniewski, M.D., H. M. (. (1979). Brain aluminum in aging and Alzheimer disease. *Neurology*, *29*(6), 809–809. doi:10.1212/WNL.29.6.809 PMID:572003

McDonald, T. A. A. (2002). Perspective on the potential health risks of PBDEs. *Chemosphere*, *46*(5), 745–755. doi:10.1016/S0045-6535(01)00239-9

McDowell, I., Hill, G., Lindsay, J., Helliwell, B., Costa, L., Beattie, L., ... Parboosingh, J. (1994). The Canadian Study of Health and Aging: Risk factors for Alzheimer's disease in Canada. *Neurology*, *44*(11), 2073–2080. doi:10.1212/WNL.44.11.2073 PMID:7969962

McElroy, J. A., Gangnon, R. E., Newcomb, P. A., Kanarek, M. S., Anderson, H. A., Brook, J. V., ... Remington, P. L. (2007). Risk of breast cancer for women living in rural reas from adult exposure to atrazine from well water in Wisconsin. *Journal of Exposure Science & Environmental Epidemiology*, *17*(2), 207–214. doi:10.1038j.jes.7500511

McElroy, J. A., Shafer, M. M., Trentham-Dietz, A., Hampton, J. M., & Newcomb, P. A. (2006). Cadmium exposure and breast cancer risk. *Journal of the National Cancer Institute*, *98*(12), 869–873. doi:10.1093/jnci/djj233

McGechan, M. B., & Lewis, D. R. (2002). SW—Soil and Water: Transport of particulate and colloid-sorbed contaminants through Soil, Part 1: General Principles. *Biosystems Engineering*, *83*(3), 255–273. doi:10.1006/bioe.2002.0125

McGrath, K. G. (2003). An earlier age of breast cancer diagnosis related to more frequent use of antiperspirants/deodorants and underarm shaving. *European Journal of Cancer Prevention*, *12*(6), 479–485. doi:10.1097/00008469-200312000-00006

McGuire, V., Longstreth, W. T., Nelson, L. M., Koepsell, T. D., Checkoway, H., Morgan, M. S., & van Belle, G. (1997). Occupational exposures and amyotrophic lateral sclerosis. A population based case-control study. *American Journal of Epidemiology*, *145*(12), 1076–1088. doi:10.1093/oxfordjournals.aje.a009070 PMID:9199537

McKinlay, R., Plant, J. A., Bell, J. N. B., & Voulvoulis, N. (2008). Endocrine disrupting pesticides: Implications for risk assessment. *Environment International*, *34*(2), 168–183. doi:10.1016/j.envint.2007.07.013 PMID:17881056

McLean, D., Mannetje, A., Dryson, E., Walls, C., McKenzie, F., & Maule, M. (2009). Leukaemia and occupation: A New Zealand Cancer Registry-based case–control Study. *International Journal of Epidemiology*, *38*(2), 594–606. doi:10.1093/ije/dyn220 PMID:18953052

Mclusky, D. S., Bryant, V., & Campbell, R. (1986). The effect of temperature and salinity on the toxicity of heavy metals to marine and estuarine invertebrates. *Oceanography and Marine Biology - an Annual Review*, *24*, 481–520.

McMichael, A. J., Campbell Lendrum, D. H., Corvalán, C. F., Ebi, K. L., Githeko, A., Scheraga, J. D., & Woodward, A. (2003). *Climate change and human health—Risks and responses* (p. 322). Geneva: World Health Organization.

McMurry, J., & Fay, R. C. (2004). Hydrogen, Oxygen and Water. In K. P. Hamann (Ed.), *McMurry Fay Chemistry* (4th ed.; pp. 575–599). Pearson Education.

Meca, G., Sospedra, I., Soriano, J. M., Ritieni, A., Valero, M. A., & Mañes, J. (2009). Isolation, purification and antibacterial effects of fusaproliferin produced by Fusarium subglutinans in submerged culture. *Food and Chemical Toxicology*, *47*(10), 2539–2543. doi:10.1016/j.fct.2009.07.014

Meca, G., Zinedine, A., Blesa, J., Font, G., & Mañes, J. (2010). Further data on the presence of Fusarium emerging mycotoxins enniatins, fusaproliferin and beauvericin in cereals available on the Spanish markets. *Food and Chemical Toxicology*, *48*(5), 1412–1416. doi:10.1016/j.fct.2010.03.010

Meddich, A., El Mokhtar, M. A., Bourzik, W., Mitsui, T., Baslam, M., & Hafidi, M. (2018). Optimizing Growth and Tolerance of Date. *Root Biology*, 239.

Medema, G. J., & Schijven, J. F. (2001). Modeling the sewage discharge and dispersion of Cryptosporidium and Giardia in surface water. *Water Research*, *18*(18), 4307–4316. doi:10.1016/S0043-1354(01)00161-0 PMID:11763032

Medjakovic, S., Zoechling, A., Gerster, P., Ivanova, M. M., Teng, Y., Klinge, C. M., ... Jungbauer, A. (2014). Effect of nonpersistent pesticides on estrogen receptor, androgen receptor, and aryl hydrocarbon receptor. *Environmental Toxicology, 29*(10), 1201–1216. doi:10.1002/tox.21852 PMID:23436777

Meerts, I. A., Letcher, R. J., Hoving, S., Marsh, G., Bergman, A., Lemmen, J. G., ... Brouwer, A. (2001). *In vitro* estrogenicity of polybrominated diphenyl ethers, hydroxylated PDBEs, and polybrominated bisphenol A compounds. *Environmental Health Perspectives, 109*(4), 399–407. doi:10.1289/ehp.01109399

Megateli, S., Semsari, S., & Couderchet, M. (2009). Toxicity and removal of heavy metals (Cadmium, Copper, and Zinc) by Lemna gibba. *Ecotoxicology and Environmental Safety, 72*(6), 1774–1780. doi:10.1016/j.ecoenv.2009.05.004 PMID:19505721

Mehta, K. (2017). Impact of temperature on contaminants toxicity in fish fauna: A review. *Indian Journal of Science and Technology, 10*, 1–6.

Meier, M. M., & Matthia, D. (2014). A space weather index for the radiation field at aviation altitudes. *Journal of Space Weather Space Climate, 4*, A13. doi:10.1051wsc/2014010

Meinardi, H., Scott, R. A., Reis, R., & Sander, J. W. A. S. (2001). The Treatment Gap in Epilepsy : The Current Situation and Ways Forward. *Epilepsia, 42*(1), 136–149. doi:10.1046/j.1528-1157.2001.32800.x PMID:11207798

Mekonnen, M. M., & Hoekstra, Y. A. (2016). Four Billion People Experience Water Scarcity. *Science Advances, 2*(2), 1–7. doi:10.1126ciadv.1500323 PMID:26933676

Meli, S., Maurizio, M., Belligno, A., Bufo, S. A., Mazzatura, A., & Scopa, A. (2002). Influence of irrigation with lagooned urban wastewater on chemical and microbial soil parameters in a citrus orchard under Mediterranean condition. *The Science of the Total Environment, 285*(1-3), 69–77. doi:10.1016/S0048-9697(01)00896-8 PMID:11874049

Mendelsohn, R., Dinar, A., & Dalfelt, A. (2000). Climate change impacts on African agriculture. Preliminary analysis prepared for the World Bank.

Mendie, U. E. (2005). *The theory and practice of clean water production for domestic and industrial use: Purified and package water*. Lagos: Lacto-Medals Publishers.

Meredith, G. E., Halliday, G. M., & Totterdell, S. (2004). A critical review of the development and importance of proteinaceous aggregates in animal models of Parkinson's disease: New insights into Lewy body formation. *Parkinsonism & Related Disorders, 10*(4), 191–202. doi:10.1016/j.parkreldis.2004.01.001 PMID:15120093

Merhi, M., Raynal, H., Cahuzac, E., Vinson, F., Cravedi, J. P., & Gamet-Payrastre, L. (2007). Occupational exposure to pesticides and risk of hematopoietic cancers: Meta-analysis of case–control studies. *Cancer Causes & Control, 18*(10), 1209–1226. doi:10.100710552-007-9061-1 PMID:17874193

Merilä, J., & Hendry, A. P. (2014). Climate change, adaptation, and phenotypic plasticity: The problem and the evidence. *Evolutionary Applications, 7*(1), 1–14. doi:10.1111/eva.12137 PMID:24454544

Merritt, W. S., Alila, Y., Barton, M., Taylor, B., & Cohen, S. J. (2003). Exploring Impacts of Climate Change on the Hydrology of the Okanagan Basin. Proc. of the Canadian Water Resources Association.

Messing, P. G., Farenhorst, A., Waite, D. T., McQueen, D. A. R., Sproull, J. F., Humphries, D. A., & Thompson, L. L. (2011). Predicting wetland contamination from atmospheric deposition measurements of pesticides in the Canadian Prairie Pothole region. *Atmospheric Environment, 45*(39), 7227–7234. doi:10.1016/j.atmosenv.2011.08.074

Messouli, M. (2014). *Anticipation et gestion des risques d'evenements climatiques extremes et de catastrophes naturelles au maroc.* Rabat: Institut royal des études stratégiques (IRES).

Messouli, M., Ben Salem, A., & Rochdane, S. (2013). *Développement d'un concept d'Indice de Vulnérabilité au Changement Climatique et Environnementale au Niveau Régional.* Rabat: Département du développement durable (MNSSD), Région Moyen-Orient et Afrique du Nord, Banque Mondiale.

Mhirit, O., Khattabi, A., Rouchdi, M., Yessef, M., & Chouraichi, M. (2011). *Adaptation du Programme d'Action National de Lutte Contre la Désertification aux Spécificités Zonales.* Rabat: Le Haut Commissariat aux Eaux et Forêts et à la Lutte Contre la Désertification (HCEFLCD).

Miki, Y., Swensen, J., Shattuck-Eidens, D., Futreal, P. A., Harshman, K., Tavtigian, S., ... Skolnick, M. H. (1994). A strong candidate for the breast and ovarian cancer susceptibility gene BRCA1. *Science, 266*(5182), 66–71. doi:10.1126cience.7545954

Milatovic, D., Zaja-Milatovic, S., Gupta, R. C., Yu, Y., & Aschner, M. (2009). Oxidative damage and neurodegeneration in manganese-induced neurotoxicity. *Toxicology and Applied Pharmacology, 240*(2), 219–225. doi:10.1016/j.taap.2009.07.004 PMID:19607852

Mili, A. F., Saghrouni, F., BenSaid, Z., Saadi-BenAoun, Y., Guizani, I., & BenSaid, M. (2012). Retrospective Analysis of Leishmaniasis in Central Tunisia: An Update on Emerging Epidemiological Trends. In Current Topics in Tropical Medicine. InTech.

Miller, D. C., Hafez, K. S., Stewart, A., Montie, J. E., & Wei, J. T. (2003). Prostate carcinoma presentation, diagnosis, and staging. *Cancer, 98*(6), 1169–1178. doi:10.1002/cncr.11635 PMID:12973840

Miller, D., & Marasas, W. (2002). Ecology of mycotoxins in maize and groundnuts. *LEISA-LEUSDEN, 17*, 23–24.

Miller, K. (1982). Effect of temperature on sprint performance in the frog Xenopus laevis and the salamander Necturus maculosus. *Copeia, 1982*(3), 695–698. doi:10.2307/1444671

Miller, W. A., Lewis, D. J., Lennox, M., Pereira, M. G., Tate, K. W., Conrad, P. A., & Atwill, E. R. (2007). Climate and on-farm risk factors associated with *Giardia duodenalis* cysts in storm runoff from California coastal dairies. *Applied and Environmental Microbiology, 73*(21), 6972–6979. doi:10.1128/AEM.00100-07 PMID:17873066

Millikan, R., DeVoto, E., Duell, E., Tse, C., Savitz, D., Beach, J., ... Newman, B. (2000). Dichlorodiphenyldichloroethene, polychlorinated biphenyls, and breast cancer among African-American and white women in North Carolina. *Cancer Epidemiology, Biomarkers & Prevention, 9*, 1233–1240.

Mills, P. K., & Yang, R. (2003). Prostate cancer risk in California farm workers. *Journal of Occupational and Environmental Medicine, 45*(3), 249–258. doi:10.1097/01.jom.0000058339.05741.0c PMID:12661182

Mills, P. K., & Yang, R. (2005). Breast Cancer Risk in Hispanic Agricultural Workers in California. *International Journal of Occupational and Environmental Health, 11*(2), 123–131. doi:10.1179/oeh.2005.11.2.123

Mills, P. K., & Yang, R. (2006). Regression analysis of pesticide use and breast cancer incidence in California Latinas. *Journal of Environmental Health, 68*, 15–22.

Minai-Tehrani, D., Minoui, S., & Herfatmanesh, A. (2009). Effect of salinity on biodegradation of polycyclic aromatic hydrocarbons (PAHs) of heavy crude oil in soil. *Bulletin of Environmental Contamination and Toxicology, 82*(2), 179–184. doi:10.100700128-008-9548-9 PMID:18777147

Ministère de l'Aménagement du Territoire, l. (2005). *Rapport national sur la prevention des desastres conference de kobe- hyogo. Royaume du Maroc, Ministère de l'Aménagement du Territoire, de l'Eau et de l'Environnement.* Département de l'Environnement.

Ministère de l'Equipement, d. T. (2014). *Impact de la Houle Exceptionnelle du 06-07 Janvier 2014 sur les infrastructures Portuaires.* Royaume du Maroc, Ministère de l'Equipement, du Transport, de la Logistique.

Min, J., Cho, J., Lee, K., Park, J., Park, S., Yong, J., & Min, K. (2011). Potential role for organochlorine pesticides in the prevalence of peripheral arterial diseases in obese persons : Results from the National Health and Nutrition Examination Survey 1999 – 2004. *Atherosclerosis, 218*(1), 200–206. doi:10.1016/j.atherosclerosis.2011.04.044 PMID:21620405

Minton, S. A. (1990). Venomous bites by nonvenomous snakes: An annotated bibliography of colubrid envenomation. *Wilderness & Environmental Medicine, 1*(2), 119–127. doi:10.1580/0953-9859-1.2.119

Mir, M. A., Al-Baradie, R. S., & Alhussainawi, M. D. (2014). *Pathophysiology of Strokes. Recent Advances in Stroke Therapeutics.* Retrieved from https://www.researchgate.net/publication/273061843_Pathophysiology_of_Strokes?enrichId=rgreq-b11151049e7ecb01203e75e2f76d2998-XXX&enrichSource=Y292ZXJQYWdlOzI3MzA2MTg0MztBUzoyMTA4MjY5NTE2OTYzODdAMTQyNzI3NjQ3MDY1NQ%3D%3D&el=1_x_2&_esc=publicationCoverPdf

Misihairabgwi, J. M., Ezekiel, C. N., Sulyok, M., Shephard, G. S., & Krska, R. (2017). Mycotoxin contamination of foods in Southern Africa: A 10-year review (2007–2016). *Critical Reviews in Food Science and Nutrition*, 1–16. doi:10.1080/10408398.2017.1357003

Misra, A. K. (2014). Climate change and challenges of water and food security. *International Journal of Sustainable Built Environment, 3*(1), 153–165. doi:10.1016/j.ijsbe.2014.04.006

Mitchell, J. D., East, B. W., Harris, I. A., & Pentland, B. (1991). Manganese, selenium and other trace elements in spinal cord, liver and bone in motor neurone disease. *European Neurology, 31*(1), 7–11. doi:10.1159/000116626 PMID:2015841

Mitra, J., Vasquez, V., Hegde, P. M., Boldogh, I., Mitra, S., Kent, T. A., & Kosagi, S. (2014). Neurological Research and Therapy Stroke : Therapeutic Potential. *Neurol Res Ther, 1*(2), 1–8. doi:10.14437/NRTOA-1-107.Revisiting PMID:25717476

Miyake, Y., Tanaka, K., Fukushima, W., Sasaki, S., Kiyohara, C., Tsuboi, Y., ... Nagai, M. (2011). Dietary intake of metals and risk of Parkinson's disease: A case-control study in Japan. *Journal of the Neurological Sciences, 306*(1–2), 98–102. doi:10.1016/j.jns.2011.03.035 PMID:21497832

Miyoshi, Y., Morioka, A., & Misawa, H. (2000). Long Term Modulation of Low Altitude Proton Radiation Belt by the Earth's Atmosphere. *Geophysical Research Letters, 27*(14), 2169–2172. doi:10.1029/1999GL003721

Mnif, W., Hassine, A. I. H., Bouaziz, A., Bartegi, A., Thomas, O., & Roig, B. (2011). Effect of Endocrine Disruptor Pesticides: A Review. *International Journal of Environmental Research and Public Health, 8*(6), 2265–2303. doi:10.3390/ijerph8062265 PMID:21776230

Modgil, S., Lahiri, D. K., Sharma, V. L., & Anand, A. (2014). Role of early life exposure and environment on neurodegeneration: Implications on brain disorders. *Translational Neurodegeneration, 29*, 3–9. doi:10.1186/2047-9158-3-9 PMID:24847438

Moe, S.J., De Schamphelaere, K., Clements, W.H., Sorensen, M.T., Van den Brink, P.J. & Liess, M. (2013). Combined and interactive effects of global climate change and toxicants on populations and communities. *Environmental Toxicology and Chemistry / Setac, 32*, 49-61.

Moffett, A., Shackelford, N., & Sarkar, S. (2007). Malaria in Africa: Vector species' niche models and relative risk maps. *PLoS One, 2*(9), e824. doi:10.1371/journal.pone.0000824

Moffett, A., Strutz, S., Guda, N., González, C., Ferro, M. C., Sánchez-Cordero, V., & Sarkar, S. (2009). A global public database of disease vector and reservoir distributions. *PLoS Neglected Tropical Diseases*, *3*(3), e378. doi:10.1371/journal.pntd.0000378

Mo, H., Yoo, D., Bae, Y. J., & Cho, K. (2013). Effects of water temperature on development and heavy metal toxicity change in two midge species of *Chironomus riparius* and *C. yoshimatsui* in an era of rapid climate change. *Entomological Research*, *43*(2), 123–129. doi:10.1111/1748-5967.12014

Mohammadi, M., Riyahi Bakhtiari, A., & Khodabandeh, S. (2014). Concentration of Cd, Pb, Hg, and Se in different parts of human breast cancer tissues. *Journal of Toxicology*, *2014*, 1–5. doi:10.1155/2014/413870

Mokarizadeh, A., Faryabi, M. R., Rezvanfar, M. A., & Abdollahi, M. (2015). A comprehensive review of pesticides and the immune dysregulation: Mechanisms, evidence and consequences. *Toxicology Mechanisms and Methods*, *25*(4), 258–278. doi:10.3109/15376516.2015.1020182 PMID:25757504

Mokssit, A. (2012). *Impact des Changements Climatiques au Maroc.* IRES-Rabat: Ministère des Energies, des Mines, de l'Eau et de l'Environnement.

Molesworth, A. M., Harrison, R., David, R., Theakston, G., & Lalloo, D. G. (2003). Geographic information system mapping of snakebite incidence in northern Ghana and Nigeria using environmental indicators: A preliminary study. *Transactions of the Royal Society of Tropical Medicine and Hygiene*, *97*(2), 188–192. doi:10.1016/S0035-9203(03)90115-5 PMID:14584375

Molina-Holgado, F., Hider, R. C., Gaeta, A., Williams, R., & Francis, P. (2007). Metals ions and neurodegeneration. *Biometals*, *20*(3-4), 639–654. doi:10.100710534-006-9033-z PMID:17294125

Moline, M. A., Karnovsky, N. J., Brown, Z., Divoky, G. J., Frazer, T. K., Jacoby, C. A., ... Fraser, W. R. (2008). High latitude changes in ice dynamics and their impact on polar marine ecosystems. *Annals of the New York Academy of Sciences*, *1134*(1), 267–319. doi:10.1196/annals.1439.010 PMID:18566098

Mollah, M. A. H., Rakshit, S. C., Anwar, K. S., Arslan, M. I., Saha, N., Ahmed, S., ... Hassan, T. (2008). Zinc concentration in serum and cerebrospinal fluid simultaneously decrease in children with febrile seizure : Findings from a prospective study in Bangladesh. *Acta Paediatrica (Oslo, Norway)*, *97*(1), 1707–1711. doi:10.1111/j.1651-2227.2008.01001.x PMID:18795906

Møller, L. B., Ott, P., Lund, C., & Horn, N. (2005). Homozygosity for a gross partial gene deletion of the C-terminal end of ATP7B in a Wilson patient with hepatic and no neurological manifestations. *American Journal of Medical Genetics. Part A*, *138*(4), 340–343. doi:10.1002/ajmg.a.30977 PMID:16222684

Montgomery, E. B. Jr, & Gale, J. T. (2008). Mechanisms of action of deep brain stimulation (DBS). *Neuroscience and Biobehavioral Reviews*, *32*(3), 388–407. doi:10.1016/j.neubiorev.2007.06.003 PMID:17706780

Montgomery, M. P., Kamel, F., Saldana, T. M., Alavanja, M. C. R., & Sandler, D. P. (2008). Original Contribution Incident Diabetes and Pesticide Exposure among Licensed Pesticide Applicators : Agricultural Health Study, 1993 – 2003. *American Journal of Epidemiology*, *167*(10), 1235–1246. doi:10.1093/aje/kwn028 PMID:18343878

Montz, B. E., & Gruntfest, E. (2002). Flash flood mitigation: Recommendations for research and applications. *Environmental Hazards*, *4*(1), 15–22. doi:10.3763/ehaz.2002.0402

Moors, E., Singh, T., Siderius, C., Balakrishnan, S., & Mishra, A. (2013). Climate change and waterborne diarrhoea in northern India: Impacts and adaptation strategies. *The Science of the Total Environment*, *468-469*, S139–S151. doi:10.1016/j.scitotenv.2013.07.021 PMID:23972324

Mor Siobhan, M., & Tzipori, S. (2008). Cryptosporidiosis in children in sub-Saharan Africa: A lingering challenge. *Clinical Infectious Diseases*, *47*(7), 915–921. doi:10.1086/591539 PMID:18715159

Morahan, J. M., & Pamphlett, R. (2006). Amyotrophic lateral sclerosis and exposure to environmental toxins: An Australian case-control study. *Neuroepidemiolog*, *27*(3), 130–135. doi:10.1159/000095552 PMID:16946624

Mordecai, E. A., Paaijmans, K. P., Johnson, L. R., Balzer, C., Ben-Horin, T., de Moor, E., ... Lafferty, K. D. (2013). Optimal temperature for malaria transmission is dramatically lower than previously predicted. *Ecology Letters*, *16*(1), 22–30. doi:10.1111/ele.12015

Moreno-Rueda, G., Pleguezuelos, J. M., Pizarro, M., & Montori, A. (2012). Northward shifts of the distributions of Spanish reptiles in association with climate change. *Conservation Biology*, *26*(2), 278–283. doi:10.1111/j.1523-1739.2011.01793.x PMID:22098311

Moretti, A., Mulè, G., Ritieni, A., & Logrieco, A. (2007). Further data on the production of beauvericin, enniatins and fusaproliferin and toxicity to Artemia salina by Fusarium species of Gibberella fujikuroi species complex. *International Journal of Food Microbiology*, *18*(2), 158–163. doi:10.1016/j.ijfoodmicro.2007.07.004

Moretto, A., & Colosio, C. (2011). Biochemical and toxicological evidence of neurological effects of pesticides: The example of Parkinson's disease. *Neurotoxicology*, *32*(4), 383–391. doi:10.1016/j.neuro.2011.03.004 PMID:21402100

Morgagni, G. B. (1820). *De sedibus et causis morborum per anatomen indagatis.* Typographia Remondiniana. Retrieved from http://archive.org/details/desedibusetcaus06adelgoog

Morgan, J. A. T., Dejong, R. J., Snyder, S. D., Mkoji, G. M., & Loker, E. S. (2001). Schistosomamansoni and Biomphalaria: Past history and future trends. *Parasitology*, *123*(7), 211–228. doi:10.1017/S0031182001007703 PMID:11769285

Morgan-Ryan, U. M., Fall, A., Ward, L. A., Hijjawi, N., Sulaiman, N., Fayer, R., ... Xiao, L. (2002). Cryptosporidium hominis n. sp. (Apicomplexa: Cryptosporidiidae) from Homo sapiens. *The Journal of Eukaryotic Microbiology*, *49*(6), 433–440. doi:10.1111/j.1550-7408.2002.tb00224.x PMID:12503676

Mori, H., Kawai, K., Ohbayashi, F., Kuniyasu, T., Yamazaki, M., Hamasaki, T., & Williams, G. M. (1984). Genotoxicity of a variety of mycotoxins in the hepatocyte primary culture/DNA repair test using rat and mouse hepatocytes. *Cancer Research*, *44*, 2918–2923.

Moroccan Ministry of Health (Direction de la Planification et des Ressources Financière, Division de la Planification et des Etudes, Service des Etudes et de l'Information Sanitaire). (2016). Santé En Chiffre. Rabat: Moroccan Ministry of Health.

Moroccan Ministry of Health. (2004). Direction de l'épidémiologie et de lutte contre les maladies (1966–2004). Données épidémiologiques des maladies sous surveillance. Bulletin épidémiologique, ministère de la Santé, Maroc.

Moroccan Ministry of Health. (2011). *A report on progress of control programs against parasitic diseases.* Directorate of Epidemiology and Disease Control, Ministry of Health, Rabat, Morocco. Available from: http://www.sante.gov.ma/departements/delm/index-delm.htm

Moroccan Ministry of Health. (2012). *A report on State of Health of the Moroccan population.* Author.

Moroccan Ministry of Health. (2014). *A report on progress of control programs against parasitic diseases.* Directorate of Epidemiology and Disease Control, Ministry of Health, Rabat, Morocco. Available from: http://www.sante.gov.ma/departements/delm/index-delm.htm

Morrison, H., Savitz, D., Semenciw, R., Hulka, B., Mao, Y., Morison, D., & Wigle, D. (1993). Farming and prostate cancer mortality. *American Journal of Epidemiology*, *137*(3), 270-280

Morrison, H., Savitz, D., Semenciw, R., Hulka, B., Mao, Y., Morison, D., & Wigle, D. (1993). Farming and prostate cancer mortality. *American Journal of Epidemiology*, *137*(3), 270–280.

Mortimer, J. A., Borenstein, A. R., & Nelson, L. M. (2012). Associations of welding and manganese exposure with Parkinson disease: Review and meta-analysis. *Neurology*, *79*(11), 1174–1180. doi:10.1212/WNL.0b013e3182698ced PMID:22965675

Mortimer, L., & Chadee, K. (2010). The immunopathogenesis of Entamoeba histolytica. *Experimental Parasitology*, *126*(3), 366–380. doi:10.1016/j.exppara.2010.03.005 PMID:20303955

Mostafalou, S., & Abdollahi, M. (2013). Pesticides and human chronic diseases: Evidences, mechanisms, and perspectives. *Toxicology and Applied Pharmacology*, *268*(2), 157–177.

Most, P., Remppis, A., Pleger, S. T., Löffler, E., Ehlermann, P., Bernotat, J., ... Koch, W. J. (2003). Transgenic Overexpression of the Ca2+-binding Protein S100A1 in the Heart Leads to Increased in Vivo Myocardial Contractile Performance. *The Journal of Biological Chemistry*, *278*(36), 33809–33817. doi:10.1074/jbc.M301788200 PMID:12777394

Moszczynski, P. (1981). Organic solvents and T lymphocytes. Lancet, 1(8217), 438. Murray, T.J. (2009). The history of multiple sclerosis: The changing frame of the disease over the centuries. *Journal of the Neurological Sciences*, *277*(Suppl 1), S3–S8.

Mourino-Perez, R. R. (1998). Oceanography and the seventh cholera pandemic. *Epidemiology (Cambridge, Mass.)*, *9*, 355–357. doi:10.1097/00001648-199805000-00024 PMID:9583431

Moysich, K., Ambrosone, C., Vena, J., Shields, P., Mendola, P., Kostyniak, P., ... Freudenheim, J. (1998). Environmental organochlorine exposure and postmenopausal breast cancer risk. *Cancer Epidemiology, Biomarkers & Prevention*, *7*, 181–188.

Muchiri, J. M., Ascolillo, L., Mugambi, M., Mutwiri, T., Ward, H. D., Naumova, E. N., ... Griffiths, J. K. (2009). Seasonality of *Cryptosporidium* oocyst detection in surface waters of Meru, Kenya as determined by two isolation methods followed by PCR. *Journal of Water and Health*, *7*(1), 67–75. doi:10.2166/wh.2009.109 PMID:18957776

Mufti, A. R., Burstein, E., Csomos, R. A., Graf, P. C. F., Wilkinson, J. C., Dick, R. D., ... Duckett, C. S. (2006). XIAP Is a copper binding protein deregulated in Wilson's disease and other copper toxicosis disorders. *Molecular Cell*, *21*(6), 775–785. doi:10.1016/j.molcel.2006.01.033 PMID:16543147

Mufti, S., & Shah, G. N. (2011). Solar-geomagnetic activity influence on Earth's climate. *Journal of Atmospheric and Solar-Terrestrial Physics*, *73*(13), 1607–1615. doi:10.1016/j.jastp.2010.12.012

Muhammad, S., Müller, T., & Joergensen, R. G. (2008). Relationships between soil biological and other soil properties in saline and alkaline arable soils from the Pakistani Punjab. *Journal of Arid Environments*, *72*(4), 448–457. doi:10.1016/j.jaridenv.2007.06.016

Muirhead-Thomson, R. C. (1951). Mosquito Behaviour in Relation to Malaria Transmission and Control in the Tropics. London: Edward Arnold.

Muir, K., Rattanamongkolgul, S., Smallman-Raynor, M., Thomas, M., Downer, S., & Jenkinson, C. (2004). Breast cancer incidence and its possible spatial association with pesticide application in two counties of England. *Public Health*, *118*(7), 513–520. doi:10.1016/j.puhe.2003.12.019

Müller, R., Seeland, A., Jagodzinski, L. S., Diogo, J. B., Nowak, C., & Oehlmann, J. (2012). Simulated climate change conditions unveil the toxic potential of the fungicide pyrimethanil on the midge *Chironomus riparius*: A multigeneration experiment. *Ecology and Evolution*, *2*(1), 196–210. doi:10.1002/ece3.71 PMID:22408736

Müller, W. F., Hobson, W., Fuller, G. B., Knauf, W., Coulston, F., & Korte, F. (1978). Endocrine effects of chlorinated hydrocarbons in rhesus monkeys. *Ecotoxicology and Environmental Safety*, *2*(2), 161–172.

Mullins, J. K., & Loeb, S. (2012, March). Environmental exposures and prostate cancer. In Urologic Oncology: Seminars and Original Investigations (Vol. 30, No. 2, pp. 216-219). Elsevier doi:10.1016/j.urolonc.2011.11.014

Mullins, J. K., & Loeb, S. (2012). Environmental exposures and prostate cancer. *Urologic Oncology*, *30*(2), 216–219.

Mulware, S. J. (2013). Article Comparative Trace Elemental Analysis in Cancerous and Noncancerous Human Tissues Using PIXE. *Journal of Biophysics*.

Munga, S., Minakawa, N., Zhou, G., Barrack, O. O. J., Githeko, A. K., & Yan, G. (2006). Effects of larval competitors and predators on oviposition site selection of Anopheles gambiae sensu stricto. *Journal of Medical Entomology*, *43*(2), 221–224. doi:10.1093/jmedent/43.2.221

Murdoch, P. S., Baron, J. S., & Miller, T. L. (2000). Potential effects of climate change on surface-water quality in north america. *JAWRA Journal of the American Water Resources Association*, *36*(2), 347–366. doi:10.1111/j.1752-1688.2000.tb04273.x

Murphy, G. S., & Oldfield, E. C. III. (1996). Falciparum malaria. *Infectious Disease Clinics of North America*, *10*(4), 747–775. doi:10.1016/S0891-5520(05)70325-1

Murray, T. J. (2009). The history of multiple sclerosis: The changing frame of the disease over the centuries. *Journal of the Neurological Sciences*, *277*(Suppl 1), S3–S8. doi:10.1016/S0022-510X(09)70003-6 PMID:19200863

Muscat, J., Britton, J., Djordjevic, M., Citron, M., Kemeny, M., Busch-Devereaux, E., ... Stellman, S. (2003). Adipose concentrations of organochlorine compounds and breast cancer recurrence in Long Island, New York. *Cancer Epidemiology, Biomarkers & Prevention*, *12*, 1474–1478.

Mussalo-Rauhamaa, H., Hasanen, E., Pyysalo, H., Antervo, K., Kauppila, R., & Pantzar, P. (1990). Occurrence of beta-hexachlorocyclohexane in breast cancer patients. *Cancer*, *66*(10), 2124–2128. doi:10.1002/1097-0142(19901115)66:10<2124::AID-CNCR2820661014>3.0.CO;2-A

Mustafa, M., Salih, A., Illzam, E., Sharifa, A., Suleiman, M., & Hussain, S. (2016). Prostate Cancer: Pathophysiology, Diagnosis, and Prognosis. *IOSR Journal of Dental and Medical Sciences Ver. II*, *15*(6), 2279–2861.

Mutter, J., & Yeter, D. (2008). Kawasaki's disease, acrodynia, and mercury. *Current Medicinal Chemistry*, *15*(28), 3000–3010. doi:10.2174/092986708786848712 PMID:19075648

Muturi, E. J., Lampman, R., Costanzo, K., & Alto, B. W. (2011). Effect of temperature and insecticide stress on life-history traits of *Culex restuans* and *Aedes albopictus* (Diptera: Culicidae). *Journal of Medical Entomology*, *48*(2), 243–250. doi:10.1603/ME10017 PMID:21485359

Muzandu, K., Shaban, Z., Ishizuka, M., Kazusaka, A., & Fujita, S. (2005). Nitric oxide enhances catechol estrogen-induced oxidative stress in LNCaP cells. *Free Radical Research*, *39*(4), 389–398. doi:10.1080/10715760400029710 PMID:16028364

Nagels, J.W., Davies-Colley, R.J., Donnison, A.M., & Muirhead, R.W. (2002) Faecal contamination over flood events in a pastoral agricultural stream in New Zealand. *Water Science and Technology: A Journal of the International Association on Water Pollution Research, 45*, 45-52.

Nagvenkar, G. S., & Ramaiah, N. (2009). Abundance of sewage-pollution indicator and human pathogenic bacteria in a tropical estuarine complex. *Environmental Monitoring and Assessment*, *155*(1-4), 245–256. doi:10.100710661-008-0432-1 PMID:18633722

Nakamura, K., Yasunaga, Y., Ko, D., Xu, L. L., Moul, J. W., Peehl, D. M., ... Rhim, J. S. (2002). Cadmium-induced neoplastic transformation of human prostate epithelial cells. *International Journal of Oncology*, *20*(3), 543–547. PMID:11836566

Nallagouni, C. S. R., & Pratap, K., R. (2017). Aluminum and fluoride impacts cortex, hippocampus and dentate gyrus structure in rats: Protective role of resveratrol. *International Journal of Applied Biology and Pharmaceutical Technology*, *8*, 89–97.

Nam, J. K. (2014). Medieval European medicine and Asian spices. *Ui Sahak*, *23*(2), 319–342. doi:10.13081/kjmh.2014.23.319 PMID:25223223

Napier, M. D., Poole, C., Satten, G. A., Ashley-Koch, A., Marrie, R. A., & Williamson, D. M. (2016). Heavy metals, organic solvents, and multiple sclerosis: An exploratory look at gene-environment interactions. *Archives of Environmental & Occupational Health*, *71*(1), 26–34. doi:10.1080/19338244.2014.937381 PMID:25137520

Nasir, J., Floresco, S. B., O'Kusky, J. R., Borowski, A., Richman, J. M., Zeisler, J., ... Hayden, M. R. (1995). Targeted disruption of the HD gene results in embryonic lethality and behavioral and morphological changes in heterozygotes. *Cell*, *81*, 811–823. doi:10.1016/0092-8674(95)90542-1 PMID:7774020

National Research Council. (2000). *Health Effects of Excess Copper*. Author.

National Research Council. (2006). *Health risks from dioxin and related compounds: Evaluation of the EPA reassessment*. Washington, DC: The National Academies Press.

National Toxicology Program. (2004). *Toxicology and carcinogenesis studies of 2,3,7,8-tetrachlorodibenzo-p-dioxin (TCDD) (CAS No. 1746-01-6) in female Harlan Sprague-Dawley rats (gavage studies). TR-521*. Research Triangle Park, NC: NIEHS.

National Toxicology Program. (2006). *Toxicology and carcinogenesis studies of 2,3,7,8-tetrachlorodibenzo-p-dioxin (TCDD) (CAS No. 1746–01–6) in female Harlan Sprague-Dawley rats (gavage studies)*. TR-521. Research Triangle Park, NC: NIEHS. Retrieved from http://ntp.niehs.nih.gov/files/521_web.pdf

Naumova, E. N., Christodouleas, J., Hunter, P. R., & Syed, Q. (2005). Effect of precipitation on seasonal variability in cryptosporidiosis recorded by the North West England surveillance system in 1990–1999. *Journal of Water and Health*, *3*(2), 185–196. doi:10.2166/wh.2005.0017 PMID:16075943

Naumova, E. N., Egorov, A. I., Morris, R. D., & Griffiths, J. K. (2003). The elderly and waterborne cryptosporidium infection: Gastroenteritis hospitalizations before and during the 1993 Milwaukee outbreak. *Emerging Infectious Diseases*, *9*(4), 418–425. doi:10.3201/eid0904.020260 PMID:12702220

Navikas, V., & Link, H. (1996). Review: Cytokines and the pathogenesis of multiple sclerosis. *Journal of Neuroscience Research*, *45*(4), 322–333. doi:10.1002/(SICI)1097-4547(19960815)45:4<322::AID-JNR1>3.0.CO;2-B PMID:8872892

Naya, D. E., Veloso, C., & Bozinovic, F. (2008). Physiological flexibility in the Andean lizard Liolaemusbellii: Seasonal changes in energy acquisition, storage and expenditure. *Journal of Comparative Physiology. B, Biochemical, Systemic, and Environmental Physiology*, *178*(8), 1007–1015. doi:10.100700360-008-0292-6 PMID:18626649

Nebdi, H., Jodogne, J.-C., Crabbé, M., & Malcorps, H. (2002), *On the Identification of New Phenomena Observed on Safir System Measurements*. Paper presented at 2002' SAFIR Workshop, Budapest, Hungary.

Nebdi, H., Warnant, R., & Lejeune S. (2004). *Study of the correlation between small-scale TEC variability and geomagnetic activity*. Technical Report of Work Package 222. Solar Influences Data Centre Space Weather Pilot Project, ESA contract 16913/03/NL/LvH.

Nebdi, H. (2010). Study of geomagnetic activity during three solar cycles using a K-index data base. *Umm Al-Qura University Journal of Applied Science*, *2*(1), 33–41.

Nechifor, R., & Vink, M. (2011). Magnesium in the central nervous system. University of Adelaide Press.

Nedwell, D. B. (1999). Effect of low temperature on microbial growth: Lowered affinity for substrates limits growth at low temperature. *FEMS Microbiology Ecology*, *30*(2), 101–111. doi:10.1111/j.1574-6941.1999.tb00639.x PMID:10508935

Needleman, R. K., Neylan, I. P., & Erickson, T. (2018). Potential Environmental and Ecological Effects of Global Climate Change on Venomous Terrestrial Species in the Wilderness. *Wilderness & Environmental Medicine*, *29*(2), 226–238. doi:10.1016/j.wem.2017.11.004 PMID:29395962

Nelson, G. C., Rosegrant, M. W., Koo, J., Robertson, R., Sulser, T., Zhu, T., . . . Magalhaes, M. (2009). Climate change: Impact on agriculture and costs of adaptation (Vol. 21). Intl Food Policy Res Inst. (IFPRI).

Nelson, K. B., & Bauman, M. L. (2003). Thimerosal and Autism? *Pediatrics*, *111*(3), 674–679. doi:10.1542/peds.111.3.674 PMID:12612255

Nelson, K., Golnick, J., Korn, T., & Angle, C. (1993). Manganese encephalopathy: Utility of early magnetic resonance imaging. *British Journal of Industrial Medicine*, *50*(6), 510–513. PMID:8329316

Nelson, W. G., De Marzo, A. M., & Isaacs, W. B. (2003). Prostate cancer. *The New England Journal of Medicine*, *349*(4), 366–381. doi:10.1056/NEJMra021562 PMID:12878745

Nephron. (2018). In *Wikipedia*. Retrieved from https://en.wikipedia.org/w/index.php?title=Alzheimer_type_II_astrocyte&oldid=838733504

Nevers, M. B., & Whitman, R. L. (2011). Efficacy of monitoring and empirical predictive modeling at improving public health protection at Chicago beaches. *Water Research*, *45*(4), 1659–1668. doi:10.1016/j.watres.2010.12.010 PMID:21195447

Ngugi, A. K., Bottomley, C., Kleinschmidt, I., Sander, J. W., & Newton, C. R. (2010). Estimation of the burden of active and life-time epilepsy : A meta-analytic approach. *Epilepsia*, *51*(5), 883–890. doi:10.1111/j.1528-1167.2009.02481.x PMID:20067507

Nichols, G., Lane, C., Asgari, N., Verlander, N. Q., & Charlett, A. (2009). Rainfall and outbreaks of drinking water related disease and in England and Wales. *Journal of Water and Health*, *7*(1), 1–8. doi:10.2166/wh.2009.143 PMID:18957770

Nicoletti, A. (2010). Toxocariasis. *Épilepsies*, *22*(2).

Nicolopoulou-Stamati, P., Maipas, S., Kotampasi, C., Stamatis, P., & Hens, L. (2016). Chemical Pesticides and Human Health: The Urgent Need for a New Concept in Agriculture. *Frontiers in Public Health*, *4*. doi:10.3389/fpubh.2016.00148 PMID:27486573

Nieboer, E., Tsuji, L. J. S., Martin, I. D., & Liberda, E. N. (2013). Human biomonitoring issues related to lead exposure. *Environmental Science. Processes & Impacts*, *15*(10), 1824–1829. doi:10.1039/c3em00270e PMID:23970117

Niehoff, N. M., Nichols, H. B., White, A. J., Parks, C. G., D'Aloisio, A. A., & Sandler, D. P. (2016). Childhood and adolescent pesticide exposure and breast cancer risk. *Epidemiology (Cambridge, Mass.)*, *27*(3), 326–333. doi:10.1097/EDE.0000000000000451

Nilanonta, C., Isaka, M., Kittakoop, P., Palittapongarnpim, P., Kamchonwongpaisan, S., Pittayakhajonwut, D., ... Thebtaranonth, Y. (2000). Antimycobacterial and antiplasmodial cyclodepsipeptides from the insect pathogenic fungus Paecilomyces tenuipes BCC 1614. *Planta Medica*, *66*(8), 756–758. doi:10.1055-2000-9776

Nime, M. F., Casanoves, F., Vrech, D. E., & Mattoni, C. I. (2013). Relationship between environmental variables and surface activity of scorpions in the Arid Chaco ecoregion of Argentina. *Invertebrate Biology, 132*(2), 145–155. doi:10.1111/ivb.12019

Nissen, K. M., Leckebusch, G. C., Pinto, J. G., & Ulbrich, U. (2014). Mediterranean cyclones and windstorms in a changing climate. *Regional Environmental Change, 14*(5), 1873–1890. doi:10.100710113-012-0400-8

Nittoli, A. C., Costantini, S., Sorice, A., Capone, F., Ciarcia, R., Marzocco, S., ... Severino, L. (2018). Effects of α-zearalenol on the metabolome of two breast cancer cell lines by 1H-NMR approach. *Metabolomics, 14*(3), 33. doi:10.100711306-018-1330-3

Nizzetto, L., Lohmann, R., Gioia, R., Dachs, J., & Jones, K. C. (2010). Atlantic ocean surface waters buffer declining atmospheric concentrations of persistent organic pollutants. *Environmental Science & Technology, 44*(18), 6978–6984. doi:10.1021/es101293v PMID:20726587

Njung'e, K., & Handley, S. L. (1991). Effects of 5???HT uptake inhibitors, agonists and antagonists on the burying of harmless objects by mice; a putative test for anxiolytic agents. *British Journal of Pharmacology, 104*(1), 105–112. doi:10.1111/j.1476-5381.1991.tb12392.x PMID:1686200

Nobre, C. A., Marengo, J. A., Seluchi, M. E., Cuartas, L. A., & Alves, L. M. (2016). Some Characteristics and Impacts of the Drought and Water Crisis in Southeastern Brazil during 2014 and 2015. Academic Press. doi:10.4236/jwarp.2016.82022

Nolte, W., Wiltfang, J., Schindler, C. G., Unterberg, K., Finkenstaedt, M., Niedmann, P. D., ... Ramadori, G. (1998). Bright basal ganglia in T1-weighted magnetic resonance images are frequent in patients with portal vein thrombosis without liver cirrhosis and not suggestive of hepatic encephalopathy. *Journal of Hepatology, 29*(3), 443–449. doi:10.1016/S0168-8278(98)80063-9 PMID:9764992

Nomura, T., Yamasaki, M., Nomura, Y., & Mimata, H. (2005). Expression of the inhibitors of apoptosisproteins in cisplatin-resistant prostate cancer cells. *Oncology Reports, 14*(4), 993–997. PMID:16142363

Nordberg, G. F. (2004). Cadmium and health in the 21st century--historical remarks and trends for the future. *Biometals: An International Journal on the Role of Metal Ions in Biology, Biochemistry, and Medicine, 17*(5), 485-489. doi:10.1023/B:BIOM.0000045726.75367.85

Norenberg, M. D. (1987). The role of astrocytes in hepatic encephalopathy. *Neurochemical Pathology, 6*(1–2), 13–33. doi:10.1007/BF02833599 PMID:3306480

Nori, J., Carrasco, P. A., & Leynaud, G. C. (2014). Venomous snakes and climate change: Ophidism as a dynamic problem. *Climatic Change, 122*(1-2), 67–80. doi:10.100710584-013-1019-6

NourEddine, D., Miloud, S., & Abdelkader, A. (2005). Effect of lead exposure on dopaminergic transmission in the rat brain. *Toxicology, 207*(3), 363–368. doi:10.1016/j.tox.2004.10.016 PMID:15664264

Noyes, P. D., McElwee, M. K., Miller, H. D., Clark, B. W., Van Tiem, L. A., Walcott, K. C., ... Levin, E. D. (2009). The toxicology of climate change: Environmental contaminants in a warming world. *Environment International, 35*(6), 971–986. doi:10.1016/j.envint.2009.02.006 PMID:19375165

Nunes, C. S., Bevilacqua, P. D., & Jardim, C. C. G. (2000). Aspectos demográficos e espaciais dos acidentes escorpiônicos no Distrito Sanitário Noroeste, Município de Belo Horizonte, Minas Gerais, 1993 a 1996. *Cadernos de Saude Publica, 16*(1), 213–223. doi:10.1590/S0102-311X2000000100022 PMID:10738166

Nunomura, A., Moreira, P. I., Castellani, R. J., Lee, H., Zhu, X., Smith, M. A., & Perry, G. (2012). Oxidative Damage to RNA in Aging and Neurodegenerative Disorders. *Neurotoxicity Research*, *22*(3), 231–248. doi:10.100712640-012-9331-x PMID:22669748

Nutt, B. D. J. (2005). *Overview of Diagnosis and Drug Treatments of Anxiety Disorders*. Academic Press.

O'Driscoll, K., Mayer, B., Su, J., & Mathis, M. (2014). The effects of global climate change on the cycling and processes of persistent organic pollutants (POPs) in the North Sea. *Ocean Science*, *10*, 397–409. doi:10.5194/os-10-397-2014

O'Leary, M., Krailo, M., Anderson, J. R., & Reaman, G. H. (2008). Progress in Childhood Cancer: 50 Years of Research Collaboration, A Report from the Children's Oncology Group. *Seminars in Oncology*, *35*(5), 484–493. doi:10.1053/j.seminoncol.2008.07.008 PMID:18929147

O'Neal, S. L., & Zheng, W. (2015). Manganese Toxicity Upon Overexposure: A Decade in Review. *Current Environmental Health Reports*, *2*(3), 315–328. doi:10.100740572-015-0056-x PMID:26231508

O'Neil, J. M., Davis, T. W., Burford, M. A., & Gobler, C. J. (2012). The rise of harmful cyanobacteria blooms: The potential roles of eutrophication and climate change. *Harmful Algae*, *14*, 313–334. doi:10.1016/j.hal.2011.10.027

Oerke, E. C. (2005). Crop losses to pests. *The Journal of Agricultural Science*, *144*(01), 31–43. doi:10.1017/S0021859605005708

Ogawa, H., & Male, J. W. (1986). Simulating the Flood Mitigation Role of Wetlands. *Journal of Water Resources Planning and Management*, *112*(1), 114–128. doi:10.1061/(ASCE)0733-9496(1986)112:1(114)

Ogden, N. H., & Lindsay, L. R. (2016). Effects of climate and climate change on vectors and vector-borne diseases: Ticks are different. *Trends in Parasitology*, *32*(8), 646–656. doi:10.1016/j.pt.2016.04.015

Ogunlewe, J. O., & Osegbe, D. N. (1989). Zinc and cadmium concentrations in indigenous blacks withnormal, hypertrophic, and malignant prostate. *Cancer*, *63*(7), 1388–1392. doi:10.1002/1097-0142(19890401)63:7<1388::AID-CNCR2820630725>3.0.CO;2-M PMID:2465818

Ojcius, D. M., Zychlinsky, A., Zheng, L. M., & Young, J. D. E. (1991). Ionophore-induced apoptosis: Role of DNA fragmentation and calcium fluxes. *Experimental Cell Research*, *197*(1), 43–49. doi:10.1016/0014-4827(91)90477-C

Ojo, O., Oni, F., & Ogunkunle, O. (2003). Implications of climatic variability and climate change on water resources availability and water resources management in West Africa. *IAHS-AISH Publication*, *280*, 37–47.

Okamoto, K., Kihira, T., Kondo, T., Kobashi, G., Washio, M., Sasaki, S., ... Nagai, M. (2009b). Lifestyle Factors and Risk of Amyotrophic Lateral Sclerosis: A Case-Control Study in Japan. *Annals of Epidemiology*, *19*(6), 359–364. doi:10.1016/j.annepidem.2009.01.015 PMID:19342254

Okhuysen, P. C., Chappell, C. L., Crabb, J. H., Sterling, C. R., & Dupont, H. L. (1999). Virulence of three distinct Cryptosporidium parvum isolates for healthy adults. *The Journal of Infectious Diseases*, *180*(4), 1275–1281. doi:10.1086/315033 PMID:10479158

Okinaka, S., Yoshikawa, M., Mozai, T., Ogihara, K., Watanabe, H., Terao, T., ... Hirai, S. (1962). Studies on hepatocerebral disease. IV. Portal-systemic encephalopathy in chronic schistosomiasis japonica. *Archives of Neurology*, *7*(1), 1–9. doi:10.1001/archneur.1962.04210010007001 PMID:14481571

Olago, D., Marshall, M., Wandiga, S. O., Opondo, M., Yanda, P. Z., Kanalawe, R., ... Achola, P. (2007). Climatic, socioeconomic, and health factors affecting human vulnerability to cholera in the Lake Victoria basin, East Africa. *Ambio*, *36*(4), 350–358. doi:10.1579/0044-7447(2007)36[350:CSAHFA]2.0.CO;2 PMID:17626474

Olanow, C. W. (2004). Manganese Induced Parkinsonism and Parkinson's Disease. *Annals of the New York Academy of Sciences*, *1012*(1), 209–223. doi:10.1196/annals.1306.018 PMID:15105268

Olanow, C. W., Good, P. F., Shinotoh, H., Hewitt, K. A., Vingerhoets, F., Snow, B. J., ... Perl, D. P. (1996). Manganese intoxication in the rhesus monkey: A clinical, imaging, pathologic, and biochemical study. *Neurology*, *46*(2), 492–498. doi:10.1212/WNL.46.2.492 PMID:8614520

Olesen, J. E., Trnka, M., Kersebaum, K. C., Skjelvåg, A. O., Seguin, B., Peltonen-Sainio, P., ... Micale, F. (2011). Impacts and adaptation of European crop production systems to climate change. *European Journal of Agronomy*, *34*(2), 96–112. doi:10.1016/j.eja.2010.11.003

Oliver, A. M., Wright, K. D., Kakadekar, A., Pharis, S., Pockett, C., Bradley, T. J., ... Erlandson, M. C. (2018). Health anxiety and associated constructs in children and adolescents with congenital heart disease: A CHAMPS cohort study. *Journal of Health Psychology*. doi:10.1177/1359105318755263 PMID:29402140

Olson, M., O'Handley, R., Ralston, B., McAllister, T., & Thompson, R. (2004). Update on Cryptosporidium and Giardia infections in cattle. *Trends in Parasitology, 20*, 151-198. DOI: . pt.2004.01.015 doi:10.1016/j

Olson, M. E., Goh, J., Phillips, M., Guselle, N., & McAllister, T. A. (1999). Giardia cyst and Cryptosporidium oocyst survivaln in water, soil, and cattle feces. *Journal of Environmental Quality*, *28*(6), 1991–1996. doi:10.2134/jeq1999.00472425002800060040x

Olveda, R. M., Tallo, V., Olveda, D. U., Inobaya, M. T., Chau, T. N., & Ross, A. G. (2016). National survey data for zoonotic schistosomiasis in the Philippines grossly underestimates the true burden of disease within endemic zones: Implications for future control. *International Journal of Infectious Diseases*, *45*, 13–17. doi:10.1016/j.ijid.2016.01.011 PMID:26820760

Olyphant, G. A. (2005). Statistical basis for predicting the need for bacterially induced beach closures: Emergence of a paradigm? *Water Research*, *39*(20), 4953–4960. doi:10.1016/j.watres.2005.09.031 PMID:16290180

Olyphant, G. A., Thomas, J., Whitman, R. L., & Harper, D. (2003). Characterization and statistical modeling of bacterial (*Escherichia coli*) outflows from watersheds that discharge into southern Lake Michigan. *Environmental Monitoring and Assessment*, *81*(1/3), 289–300. doi:10.1023/A:1021345512203 PMID:12620022

Ongerth, J. E. (2013). The concentration of Cryptosporidium and Giardia in water--the role and importance of recovery efficiency. *Water Research*, *47*(7), 2479–2488. doi:10.1016/j.watres.2013.02.015 PMID:23490099

Ongerth, J. E. (2017). *Cryptosporidium* and *Giardia* in Water: Reassessment of Occurrence and Significance. *Journal of Environmental Engineering*, *143*(3), 04016084. doi:10.1061/(ASCE)EE.1943-7870.0001161

Onozuka, D. (2014). Effect of non-stationary climate on infectious gastroenteritis transmission in Japan. *Scientific Reports*, *4*(1), 5157. doi:10.1038rep05157 PMID:24889802

Onozuka, D., & Hashizume, M. (2011). Weather variability and paediatric infectious gastroenteritis. *Epidemiology and Infection*, *139*(09), 1369–1378. doi:10.1017/S0950268810002451 PMID:21044404

Onwuzuligbo, O., Hendricks, A. R., Hassler, J., Domanski, K., Goto, C., & Wolf, M. T. F. (2018). Mercury Intoxication as a Rare Cause of Membranous Nephropathy in a Child. *American Journal of Kidney Diseases*, *1-5*. doi:10.1053/j.ajkd.2018.05.013 PMID:30041877

Oreskes, N. (2004). The scientific consensus on climate change. *Science*, *306*(5702), 1686–1686. doi:10.1126cience.1103618 PMID:15576594

Orisakwe, O. E. (2014). The role of lead and cadmium in psychiatry. *North American Journal of Medical Sciences*, *6*(8), 370–376. doi:10.4103/1947-2714.139283 PMID:25210669

Orshan, L., Szekely, D., Khalfa, Z., & Bitton, S. (2010). Distribution and seasonality of Phlebotomus sand flies in cutaneous leishmaniasis foci, Judean Desert, Israel. *Journal of Medical Entomology*, *47*(3), 319–328. doi:10.1093/jmedent/47.3.319 PMID:20496578

Otero, R., Navio, E., Céspedes, F. A., Núñez, M. J., Lozano, L., Moscoso, E. R., ... Rodas, J. H. (2004). Scorpion envenoming in two regions of Colombia: Clinical, epidemiological and therapeutic aspects. *Transactions of the Royal Society of Tropical Medicine and Hygiene*, *98*(12), 742–750. doi:10.1016/j.trstmh.2003.12.018 PMID:15485705

Ottoson, J., Hansen, A., Bjorlenius, B., Norder, H., & Stenstrom, T. (2006). Removal of virus, parasitic protozoa and microbial indicators in conventional and membrane processes in a wastewater pilot plant. *Water Research*, *40*(7), 1449–1457. doi:10.1016/j.watres.2006.01.039 PMID:16533517

Ouanes, Z., Abid, S., Ayed, I., Anane, R., Mobio, T., Creppy, E., & Bacha, H. (2003). Induction of micronuclei by zearalenone in Vero monkey kidney cells and in bone marrow cells of mice: Protective effect of vitamin E. *Mutation Research*, *538*(1-2), 63–70. doi:10.1016/S1383-5718(03)00093-7

Overall, J. E. (1976). The Brief Psychiatric Rating Scale in psychopharmacologic research. *ECDEU Assessment Manual for Psychopharmocol*, *7*, 166–169.

Owolabi, L., Sale, S., Owolabi, S., Nalado, A., Umar, M., & Taura, A. (2018). Electroencephalography Abnormalities in Generalized Epilepsy and Their Predictors: A Multicenter Experience. *Annals of African Medicine*, *17*(2), 64. doi:10.4103/aam.aam_2_17 PMID:29536959

Ozkan, O., Uzun, R., Adiguzel, S., Cesaretli, Y., & Ertek, M. (2008). Evaluation of scorpion sting incidence in Turkey. *The Journal of Venomous Animals and Toxins Including Tropical Diseases*, *14*(1), 128–140. doi:10.1590/S1678-91992008000100010

Paaijmans, K. P., Heinig, R. L., Seliga, R. A., Blanford, J. I., Blanford, S., Murdock, C. C., & Thomas, M. B. (2013). Temperature variation makes ectotherms more sensitive to climate change. *Global Change Biology*, *19*(8), 2373–2380. doi:10.1111/gcb.12240 PMID:23630036

Paaijmans, K. P., Imbahale, S. S., Thomas, M. B., & Takken, W. (2010). Relevant microclimate for determining the development rate of malaria mosquitoes and possible implications of climate change. *Malaria Journal*, *9*(1), 196. doi:10.1186/1475-2875-9-196

Paaijmans, K. P., Jacobs, A. F. G., Takken, W., Heusinkveld, B. G., Githeko, A. K., Dicke, M., & Holtslag, A. A. M. (2008). Observations and model estimates of diurnal water temperature dynamics in mosquito breeding sites in western Kenya. *Hydrological Processes: An International Journal*, *22*(24), 4789–4801. doi:10.1002/hyp.7099

Paalman, M. A. A., Van Der Weijden, C. H., & Loch, J. P. G. (1994). Sorption of cadmium on suspended matter under estuarine conditions; competition and complexation with major sea-water ions. *Water, Air, and Soil Pollution*, *73*(1), 49–60. doi:10.1007/BF00477975

Pachauri, R. K., Allen, M. R., Barros, V. R., Broome, J., Cramer, W., Christ, R., & Dubash, N. K. (2014). *Climate change 2014: synthesis report. Contribution of Working Groups I, II and III to the fifth assessment report of the Intergovernmental Panel on Climate Change*. IPCC.

Pacini, F. (1854). Observazioni microscopiche e deduzioni patologiche sul cholera asiatico. *Gazzetta Medica Italiana*, *6*, 405–412.

Paerl, H. W., & Paul, V. J. (2012). Climate change: Links to global expansion of harmful cyanobacteria. *Water Research*, *46*(5), 1349–1363. doi:10.1016/j.watres.2011.08.002 PMID:21893330

Paeth, H., Born, K., Girmes, R., Podzun, R., & Jacob, D. (2009). Regional climate change in tropical and northern Africa due to greenhouse forcing and land use changes. *Journal of Climate*, *22*(1), 114–132. doi:10.1175/2008JCLI2390.1

Paillard, T., Rolland, Y., & de Souto Barreto, P. (2015). Protective Effects of Physical Exercise in Alzheimer's Disease and Parkinson's Disease: A Narrative Review. *Journal of Clinical Neurology (Seoul, Korea)*, *11*(3), 212. doi:10.3988/jcn.2015.11.3.212 PMID:26174783

Palikhe, B. R. (2007). *Relationship between pesticide use and climate change for crops*. Academic Press.

Palikhe, B.R. (2007). Relationship between pesticide use and climate change for crops. Academic Press.

Pall, H. A., Williams, A. C., Waring, R., & Elias, E. (1987). Motorneurone disease as manifestation of pesticide toxicity. *Lancet*, *ii*(8560), 685. doi:10.1016/S0140-6736(87)92468-8 PMID:2887964

Pal, P. B., Pal, S., Das, J., & Sil, P. C. (2012). Modulation of mercury-induced mitochondria-dependent apoptosis by glycine in hepatocytes. *Amino Acids*, *42*(5), 1669–1683. doi:10.100700726-011-0869-3 PMID:21373768

Pal, P. K., Samii, A., & Calne, D. B. (1999). Manganese neurotoxicity: A review of clinical features, imaging and pathology. *Neurotoxicology*, *20*, 227–238. PMID:10385886

Pamphlett, R., & Waley, P. (1998). Mercury in human spinal motor neurons. *Acta Neuropathologica*, *96*(5), 515–519. doi:10.1007004010050927 PMID:9829816

Parandin, R., Rassouli, M. B., Sisakhtnezhad, S., & Shahri, N. M. (2015). In vitro evaluation of the effects of zearalenone and α-zearalenol on MCF-7 and MDA-MB-468 cell lines of human breast cancer. *Razavi International Journal of Medicine, 3*.

Paraskevas, P. A., Giokas, D. L., & Lekkas, T. D. (2002). Wastewater management in coastal urban areas: The case of Greece. *Water Science and Technology*, *46*(8), 177–186. doi:10.2166/wst.2002.0174 PMID:12420981

Parent, M. É., & Siemiatycki, J. (2001). Occupation and prostate cancer. *Epidemiologic Reviews*, *23*(1), 138–143. doi:10.1093/oxfordjournals.epirev.a000779 PMID:11588838

Paris, M. P. K., Schweiger, W., Hametner, C., Stückler, R., Muehlbauer, G. J., Varga, E., & … . (2014). Zearalenone-16-O-glucoside: A new masked mycotoxin. *Journal of Agricultural and Food Chemistry*, *62*(5), 1181–1189. doi:10.1021/jf405627d

Parker, B. C., Ford, M. A., Gruft, H., & Falkinham, J. O. III. (1983). Epidemiology of infection by nontuberculous mycobacteria. IV. Preferential aerosolization of *Mycobacterium* intracellulare from natural waters. *The American Review of Respiratory Disease*, *128*, 652–656. PMID:6354024

Parkhouse, R. M. E., & Harrison, L. J. S. (2014). Helminth-Cestode: Taenia saginata and Taenia solium. *Encycl Food Saf*, *2*, 70–77. doi:10.1016/B978-0-12-378612-8.00148-7

Parkin, D., Bray, F., & Devesa, S. (2001). Cancer burden in the year 2000. The global picture. *European Journal of Cancer*, *37*, 4–66. doi:10.1016/S0959-8049(01)00267-2 PMID:11602373

Park, J. H., Lee, D. W., Park, K. S., & Joung, H. (2014). Serum trace metal levels in Alzheimer's disease and normal control groups. *American Journal of Alzheimer's Disease and Other Dementias*, *29*(1), 76–83. doi:10.1177/1533317513506778 PMID:24164932

Park, J. S., Linderholm, L., Charles, M. J., Athanasiadou, M., Petrik, J., Kocan, A., ... Hertz-Picciotto, I. (2007). Polychlorinated biphenyls their hydroxylated metabolites (OH-PCBS) in pregnant women from eastern Slovakia. *Environmental Health Perspectives*, *115*(1), 20–27. doi:10.1289/ehp.8913

Park, J.-H., Duan, L., Kim, B., Mitchell, M. J., & Shibata, H. (2010). Potential effects of climate change and variability on watershed biogeochemical processes and water quality in Northeast Asia. *Environment International*, *36*(2), 212–225. doi:10.1016/j.envint.2009.10.008 PMID:19926135

Parmesan, C., & Yohe, G. (2003). A globally coherent fingerprint of climate change impacts across natural systems. *Nature*, *421*(6918), 37–42. doi:10.1038/nature01286

Parrón, T., Requena, M., Hernández, A. F., & Alarcón, R. (2011). Association between environmental exposure to pesticides and neurodegenerative diseases. *Toxicology and Applied Pharmacology*, *256*(3), 379–385. doi:10.1016/j.taap.2011.05.006 PMID:21601587

Parry, M., Arnell, N., McMichael, T., Nicholls, R., Martens, P., Kovats, S., ... Fischer, G. (2001). Millions at risk: Defining critical climate change threats and targets. *Global Environmental Change*, *11*(3), 181–183. doi:10.1016/S0959-3780(01)00011-5

Partanen, H. A., El-Nezami, H. S., Leppänen, J. M., Myllynen, P. K., Woodhouse, H. J., & Vähäkangas, K. H. (2010). Aflatoxin B1 transfer and metabolism in human placenta. *Toxicological Sciences: An Official Journal of the Society of Toxicology*, *113*(1), 216–225. doi:10.1093/toxsci/kfp257 PMID:19875679

Pascual, M., Rodo, X., Ellner, S. P., Colwell, R., & Bouma, M. J. (2000). Cholera dynamics and El Nino-Southern Oscillation. *Science*, *289*(5485), 1766–1769. doi:10.1126cience.289.5485.1766 PMID:10976073

Pasture, O., & Onkia, O. (1994). Canadian Study of Health and Aging: Study methods and prevalence of dementia. *Canadian Medical Association Journal*, *150*(6), 899–913. PMID:8131123

Patel, Y. J., Payne, S. M. D., de Belleroche, J., & Latchman, D. S. (2005). Hsp27 and Hsp70 administered in combination have a potent protective effect against FALS-associated SOD1- mutant-induced cell death in mammalian neuronal cells. *Brain Research. Molecular Brain Research*, *134*(2), 256–274. doi:10.1016/j.molbrainres.2004.10.028 PMID:15836922

Paterson, D. L., Wright, H., & Harris, P. N. A. (2018, July). Health Risks of Flood Disasters. *Clinical Infectious Diseases*, 1–5. doi:10.1093/cid/ciy227

Paterson, R. R., & Lima, N. (2010). Toxicology of mycotoxins. *EXS*, *100*, 31–63. doi:10.1007/978-3-7643-8338-1_2

Patra, R. W., Chapman, J. C., Lim, R. P., Gehrke, P. C., & Sunderam, R. M. (2015). Interactions between water temperature and contaminant toxicity to freshwater fish. *Environmental Toxicology and Chemistry*, *34*(8), 1809–1817. doi:10.1002/etc.2990 PMID:26033197

Patyar, S., Pharm, M. S., & Patyar, R. R. (2015). Correlation between Sleep Duration and Risk of Stroke. *Journal of Stroke and Cerebrovascular Diseases*, 1–7. doi:10.1016/j.jstrokecerebrovasdis.2014.12.038 PMID:25817615

Patz, J. A., Campbell-Lendrum, D., Holloway, T., & Foley, J. A. (2005). Impact of regional climate change on human health. *Nature*, *438*(7066), 310–317. doi:10.1038/nature04188 PMID:16292302

Patz, J. A., Epstein, P. R., Burke, T. A., & Balbus, J. M. (1996). Global climate change and emerging infectious diseases. *Journal of the American Medical Association*, *275*(3), 217–223. doi:10.1001/jama.1996.03530270057032 PMID:8604175

Patz, J. A., Frumkin, H., Holloway, T., Vimont, D. J., & Haines, A. (2014). Climate change: Challenges and opportunities for global health. *Journal of the American Medical Association*, *312*(15), 1565–1580. doi:10.1001/jama.2014.13186 PMID:25244362

Patz, J. A., Githeko, A. K., McCarty, J. P., Hussein, S., Confalonieri, U., & De Wet, N. (1997). Climate Change and Human Health. *Population and Development Review*, *23*(1), 205. doi:10.2307/2137486

Patz, J. A., Graczyk, T. K., Geller, N., & Vittor, A. Y. (2000). Effects of environmental change on emerging parasitic diseases. *International Journal for Parasitology*, *30*(12), 1395–1405. doi:10.1016/S0020-7519(00)00141-7 PMID:11113264

Patz, J. A., McGeehin, M. A., Bernard, S. M., Ebi, K. L., Epstein, P. R., Grambsch, A., ... Trtanj, J. (2000). The potential health impacts of climate variability and change for the United States: Executive summary of the report of the health sector of the U.S. National Assessment. *Environmental Health Perspectives*, *108*(4), 367–376. doi:10.1289/ehp.00108367 PMID:10753097

Patz, J. A., Vavrus, S. J., Uejio, C. K., & McLellan, S. L. (2008). Climate change and waterborne disease risk in the Great Lakes region of the U.S. *American Journal of Preventive Medicine*, *35*(5), 451–458. doi:10.1016/j.amepre.2008.08.026 PMID:18929971

Paulo & Sergio. (2016). *Mycotoxins: Biochemical Approach*. Editora Albatroz.

Payment, P., Berte, A., Prevost, M., Menard, B., & Barbeau, B. (2000). Occurence of pathogenic microorganisims in the Saint Lawrence River (Canada) and comparison of health risks for populations using it as their source of drinking water. *Canadian Journal of Microbiology*, *46*(6), 565–576. doi:10.1139/w00-022 PMID:10913979

Pazaiti, A., Kontos, M., & Fentiman, I. S. (2012). ZEN and the art of breast health maintenance. *International Journal of Clinical Practice*, *66*(1), 28–36. doi:10.1111/j.1742-1241.2011.02805.x

Peasey, A., Blumenthal, U., Mara, D., & Ruiz-Palacios, G. (2000). *A review of policy and standards for wastewater reuse in agriculture: a Latin American perspective*. WELL Study. Retrieved from http://www.iboro.ac.uk/well/

Pedersen, U. B., Stendel, M., Midzi, N., Mduluza, T., Soko, W., Stensgaard, A. S., ... Kristensen, T. K. (2014). Modelling climate change impact on the spatial distribution of fresh water snails hosting trematodes in Zimbabwe. *Parasites & Vectors*, *7*(1), 536.

Pedersen, W. A., Fu, W., Keller, J. N., Markesbery, W. R., Appel, S., Smith, R. G., ... Mattson, M. P. (1998). Protein modification by the lipid peroxidation product 4-hydroxynonenal in the spinal cords of amyotrophic lateral sclerosis patients. *Annals of Neurology*, *44*(5), 819–824. doi:10.1002/ana.410440518 PMID:9818940

Pelling, M., & High, C. (2005). Understanding adaptation: What can social capital offer assessments of adaptive capacity? *Global Environmental Change*, *15*(4), 308–319. doi:10.1016/j.gloenvcha.2005.02.001

Peng, J., Peng, L., Stevenson, F. F., Doctrow, S. R., & Andersen, J. K. (2007). Iron and Paraquat as Synergistic Environmental Risk Factors in Sporadic Parkinson's Disease Accelerate Age-Related Neurodegeneration. *The Journal of Neuroscience*, *27*(26), 6914–6922. doi:10.1523/JNEUROSCI.1569-07.2007 PMID:17596439

Pentyala, S., Ruggeri, J., Veerraju, A., Yu, Z., & Bhatia, A. (2010). Microsomal Ca 2 + flux modulation as an indicator of heavy metal toxicity. *Indian Journal of Experimental Biology*, *48*(July), 737–743. PMID:20929057

Peraica, M., Richter, D., & Rasi, C. D. (2014). Mycotoxicoses in children. *Arhiv za Higijenu Rada i Toksikologiju*, *65*(4), 347–363. doi:10.2478/10004-1254-65-2014-2557

Perl, D. P., & Olanow, C. W. (2007). The neuropathology of manganese-induced Parkinsonism. *Journal of Neuropathology and Experimental Neurology*, *66*(8), 675–682. doi:10.1097/nen.0b013e31812503cf PMID:17882011

Perlroth, N. H., & Branco, C. W. C. (2017). Current knowledge of environmental exposure in children during the sensitive developmental periods. *Jornal de Pediatria*, *93*(1), 17–27. doi:10.1016/j.jped.2016.07.002 PMID:27821252

Perry, E., Perry, R., Blessed, G., & Tomlinson, B. (1977). Necropsy evidence of central cholinergic deficits in senile dementia. *Lancet*, *309*(8004), 189. doi:10.1016/S0140-6736(77)91780-9 PMID:64712

Pesatori, A. C., Consonni, D., Rubagotti, M., Grillo, P., & Bertazzi, P. A. (2009). Cancer incidence in the population exposed to dioxin after the "Seveso accident": Twenty years of follow-up. *Environmental Health*, *8*(1), 39. doi:10.1186/1476-069X-8-39

Pesticide Action Network UK. (1993). High residues in India. In *A history of HCH contamination*. Academic Press. Retrieved from http://www.pan-uk.org/ pestnews/pn21/pn21p15c.htm

Pestka, J. (2010). Deoxynivalenol: Mechanisms of action, human exposure, and toxicological relevance. *Archives of Toxicology*, *84*(9), 663–679. doi:10.100700204-010-0579-8

Peterson, T. C., McGuirk, M., Houston, T. G., Horvitz, A. H., & Wehner, M. F. (2008). Climate variability and change with implications for transportation. The National Research Council.

Peterson, E. W., & Cardoso, E. R. (1983). The blood-brain barrier following experimental subarachnoid hemorrhage. Part 2: Response to mercuric chloride infusion. *Journal of Neurosurgery*, *58*(3), 345–351. doi:10.3171/jns.1983.58.3.0345 PMID:6219189

Petrukhin, K., Fischer, S. G., Pirastu, M., Tanzi, R. E., Chernov, I., Devoto, M., ... Russo, J. J. (1993). Mapping, cloning and genetic characterization of the region containing the Wilson disease gene. *Nature Genetics*, *5*(4), 338–343. doi:10.1038/ng1293-338 PMID:8298640

Pettigrew, N. R., Churchill, J. H., Janzen, C. D., Mangum, L. J., Signell, R. P., Thomas, A. C., ... Xue, H. (2005). The kinematic and hydrographic structure of the Gulf of Maine Coastal Current. *Deep-sea Research. Part II, Topical Studies in Oceanography*, *52*(19-21), 2369–2391. doi:10.1016/j.dsr2.2005.06.033

Pfeiffer, R. F. (2007). Wilson's Disease. *Seminars in Neurology*, *27*(2), 123–132. doi:10.1055-2007-971173 PMID:17390257

Pfohl-Leszkowicz, A. (2009). Mycotoxines: Facteur de risque de cancers. *Journal africain du cancer/African. Journal of Cancer*, *1*(1), 42–55.

Pfohl-Leszkowicz, A., & Manderville, R. A. (2011). An update on direct genotoxicity as a molecular mechanism of ochratoxin A carcinogenicity. *Chemical Research in Toxicology*, *25*(2), 252–262. doi:10.1021/tx200430f

Pfohl-Leszkowicz, A., Petkova-Bocharova, T., Chernozemsky, I. N., & Castegnaro, M. (2002). Balkan endemic nephropathy and associated urinary tract tumors: A review on etiological causes and the potential role of mycotoxins. *Food Additives and Contaminants*, *19*(3), 282–302. doi:10.1080/02652030110079815

Pham Duc, P., Nguyen-Viet, H., Hattendorf, J., Zinsstag, J., Dac Cam, P., & Odermatt, P. (2011). Risk factors for Entamoeba histolytica infection in an agricultural community in Hanam province, Vietnam. *Parasites & Vectors*, *4*(1), 102. doi:10.1186/1756-3305-4-102 PMID:21663665

Phelps, T. (2010). *Old World Vipers, a natural history of the azemiopinae and viperinae*. Edition Chimaira.

Philips, T., & Robberecht, W. (2011). Neuroinflammation in amyotrophic lateral sclerosis: Role of glial activation in motor neuron disease. *Lancet Neurology*, *10*(3), 253–263. doi:10.1016/S1474-4422(11)70015-1 PMID:21349440

Piao, S., Ciais, P., Huang, Y., Shen, Z., Peng, S., Li, J., ... Fang, J. (2010). The impacts of climate change on water resources and agriculture in China. *Nature*, *467*(7311), 43–51. doi:10.1038/nature09364 PMID:20811450

Piazza, V., Gambardella, C., Canepa, S., Costa, E., Faimali, M., & Garaventa, F. (2016). Temperature and salinity effects on cadmium toxicity on lethal and sublethal responses of *Amphibalanus amphitrite nauplii. Ecotoxicology and Environmental Safety*, *123*, 8–17. doi:10.1016/j.ecoenv.2015.09.023 PMID:26410667

Piccinini, L., Borella, P., Bargellini, A., Medici, C. I., & Zoboli, A. (1996). A case-control study on selenium, zinc, and copper in plasma and hair of subjects affected by breast and lung cancer. *Biological Trace Element Research*, *51*(1), 23–30. doi:10.1007/BF02790144

Pichery, C., Bellanger, M., Zmirou-Navier, D., Glorennec, P., Hartemann, P., & Grandjean, P. (2011). Childhood lead exposure in France: Benefit estimation and partial cost-benefit analysis of lead hazard control. *Environmental Health*, *10*(1), 44. doi:10.1186/1476-069X-10-44 PMID:21599937

Pielke, R. A., Marland, G., Betts, R. A., Chase, T. N., Eastman, J. L., Niles, J. O., & Running, S. W. (2002). The influence of land-use change and landscape dynamics on the climate system: relevance to climate-change policy beyond the radiative effect of greenhouse gases. *Philosophical Transactions of the Royal Society of London A: Mathematical, Physical and Engineering Sciences, 360*(1797), 1705-1719.

Pielke, R. Jr, Prins, G., Rayner, S., & Sarewitz, D. (2007). Lifting the taboo on adaptation. *Nature*, *445*(7128), 597–598. doi:10.1038/445597a PMID:17287795

Pierce, D. W., Gleckler, P. J., Barnett, T. P., Santer, B. D., & Durack, P. J. (2012). The fingerprint of human-induced changes in the ocean's salinity and temperature fields. *Geophysical Research Letters*, 39.

Pierce-Ruhland, R., & Patten, B. M. (1981). Repeat study of antecedent events in motor neuron disease. *Annals of Clinical Research*, *13*, 102–107. PMID:7235610

Pillay, D., Chuturgoon, A. A., Nevines, E., Manickum, T., Deppe, W., & Dutton, M. F. (2002). The quantitative analysis of zearalenone and its derivatives in plasma of patients with breast and cervical cancer. *Clinical Chemistry and Laboratory Medicine*, *40*(9), 946–951. doi:10.1515/CCLM.2002.166

Pintilie, L., Torres, C. M., Teodosiu, C., & Castells, F. (2016). Urban wastewater reclamation for industrial reuse: An LCA case study. *Journal of Cleaner Production*, *139*, 1–14. doi:10.1016/j.jclepro.2016.07.209

Pipelzadeh, M. H., Jalali, A., Taraz, M., Pourabbas, R., & Zaremirakabadi, A. (2007). An epidemiological and a clinical study on scorpionism by the Iranian scorpion Hemiscorpius lepturus. *Toxicon*, *50*(7), 984–992. doi:10.1016/j.toxicon.2007.07.018 PMID:17854855

Pirkle, J., Wolfe, W., Patterson, D., Needham, L., Michalek, J., Miner, J., ... Philips, D. (1989). Estimates of the half-life of 2,3,7,8-tetrachlorodibenzo-*p*-dioxin in Vietnam veterans of Operation Ranch Hand. *Journal of Toxicology and Environmental Health*, *27*(2), 165–171. doi:10.1080/15287398909531288

Pittet, A. (1998). Natural occurrence of mycotoxins in foods and feeds—an updated review. *Revue de Medecine Veterinaire*, *6*, 479–492.

Pitt, J. I., Basilico, J. C., Abarca, M. L., & Lopez, C. (2000). Mycotoxins and toxigenic fungi. *Medical Mycology*, *38*(s1), 41–46. doi:10.1080/mmy.38.s1.41.46

Plaskon, L. A., Penson, D. F., Vaughan, T. L., & Stanford, J. L. (2003). Cigarette smoking and risk of prostatecancer in middle-aged men. *Cancer Epidemiology, Biomarkers & Prevention*, *12*(7), 604–609. PMID:12869398

Plato, C. C., Garruto, R. M., Galasko, D., Craig, U., Plato, M., Gamst, A., ... Wiederholt, W. (2003). Amyotrophic Lateral Sclerosis and Parkinsonism-Dementia Complex of Guam: Changing Incidence Rates during the Past 60 Years. *American Journal of Epidemiology*, *157*(2), 149–157. doi:10.1093/aje/kwf175 PMID:12522022

Plutzer, J., & Karanis, P. (2016). Neglected waterborne parasitic protozoa and their detection in water. *Water Res., 101*, 318-332.

Polis, G. A. (1990). *The biology of scorpions* (No. 595.46 B5). Academic Press.

Polis, G. A. (1980). Seasonal patterns and age-specific variation in the surface activity of a population of desert scorpions in relation to environmental factors. *Journal of Animal Ecology*, *49*(1), 1–18. doi:10.2307/4275

Polis, G. A., & Farley, R. D. (1980). Population biology of a desert scorpion: Survivorship, microhabitat, and the evolution of life history strategy. *Ecology*, *61*(3), 620–629. doi:10.2307/1937428

Polkowska, Z., Kozłowska, K., Namieśnik, J., & Przyjazny, A. (2004). Biological fluids as a source of information on the exposure of man to environmental chemical agents. *Critical Reviews in Analytical Chemistry*, *34*(2), 105–119. doi:10.1080/10408340490475911

Polsky, P., Neff, R., & Yarnal, B. (2007). Building comparable global change vulnerability assessments: The vulnerability scoping diagram. *Global Environmental Change*, *17*(3-4), 472–485. doi:10.1016/j.gloenvcha.2007.01.005

Pomier-Layrargues, G., Spahr, L., & Butterworth, R. F. (1995). Increased manganese concentrations in pallidum of cirrhotic patients. *Lancet (London, England)*, *345*(8951), 735. doi:10.1016/S0140-6736(95)90909-5 PMID:7885158

Portner, H. O., & Knust, R. (2007). Climate change affects marine fishes through the oxygen limitation of thermal tolerance. *Science*, *315*(5808), 95–97. doi:10.1126cience.1135471 PMID:17204649

Pough, F. H. (1980). The advantages of ectothermy for tetrapods. *American Naturalist*, *115*(1), 92–112. doi:10.1086/283547

Praline, J., Guennoc, A., Limousin, N., Hallak, H., de Toffol, B., & Corcia, P. (2007). Case report; ALS and mercury intoxication: A relationship? *Clinical Neurology and Neurosurgery*, *109*(10), 880–883. doi:10.1016/j.clineuro.2007.07.008 PMID:17719172

Prathumratana, L., Sthiannopkao, S., & Kim, K. W. (2008). The relationship of climatic and hydrological parameters to surface water quality in the lower Mekong River. *Environment International*, *34*(6), 860–866. doi:10.1016/j.envint.2007.10.011 PMID:18068783

Praticò, D. (2008). Oxidative stress hypothesis in Alzheimer's disease: A reappraisal. *Trends in Pharmacological Sciences*, *29*(12), 609–615. doi:10.1016/j.tips.2008.09.001 PMID:18838179

Pratlong, F., Rioux, J. A., Dereure, J., Mahjour, J., Gallego, M., Guilvard, E., ... Saddiki, A. (1991). *Leishmania tropica*au Maroc. IV. Diversité isozymiqueintrafocale. *Annales de Parasitologie Humaine et Comparee*, *66*(3), 100–104. doi:10.1051/parasite/1991663100 PMID:1776781

Premanandh, J. (2011). Factors affecting food security and contribution of modern technologies in food sustainability. *Journal of the Science of Food and Agriculture*, *91*(15), 2707–2714. doi:10.1002/jsfa.4666 PMID:22002569

Prezioso, D., Denis, L. J., Klocker, H., Sciarra, A., Reis, M., Naber, K., ... Griffiths, K. (2007). Estrogens and aspects of prostate disease. *International Journal of Urology*, *14*(1), 1–16. doi:10.1111/j.1442-2042.2006.01476.x PMID:17199853

Prikryl, P., Bruntz, R., Tsukijihara, T., Iwao, K., Muldrew, D. B., Rusin, V., ... Stastný, P. (2017). Tropospheric weather influenced by solar wind through atmospheric vertical coupling downward control. *Journal of Atmospheric and Solar-Terrestrial Physics*, *171*, 94–110. doi:10.1016/j.jastp.2017.07.023

Prikryl, P., Iwao, K., Muldrew, D., Rusin, V., Rybansky, M., & Bruntz, R. (2016). A link between high-speed solar wind streams and explosive extratropical cyclones. *Journal of Atmospheric and Solar-Terrestrial Physics*, *149*, 219–231. doi:10.1016/j.jastp.2016.04.002

Prikryl, P., Muldrew, D. B., & Sofko, G. J. (2009b). The influence of solar wind on extratropical cyclones – Part 2: A link mediated by auroral atmospheric gravity waves? *Annales Geophysicae*, *27*(1), 31–57. doi:10.5194/angeo-27-31-2009

Prikryl, P., Rusin, V., & Rybanský, M. (2009a). The influence of solar wind on extratropical cyclones – Part 1: Wilcox effect revisited. *Annales Geophysicae*, *27*(1), 1–30. doi:10.5194/angeo-27-1-2009

Prince, M., Wimo, A., Guerchet, M., Ali, G. C., Wu, Y. T., & Prina, M. (2015). *World Alzheimer Report 2015. The Global Impact of Dementia. Alzheimer's Disease International. Alzheimer's Disease International*. London: ADI.

Prins, G. S. (2008). Endocrine disruptors and prostate cancer risk. *Endocrine-Related Cancer*, *15*(3), 649–656. doi:10.1677/ERC-08-0043 PMID:18524946

Priyadarshi, A., Khuder, S. A., Schaub, E. A., & Priyadarshi, S. S. (2001). Environmental Risk Factors and Parkinson's Disease: A Metaanalysis. *Environmental Research*, *86*(2), 122–127. doi:10.1006/enrs.2001.4264 PMID:11437458

Priyadarshi, A., Khuder, S. A., Schaub, E. A., & Shrivastava, S. (2000). A meta-analysis of Parkinson's disease and exposure to pesticides. *Neurotoxicology*, *21*(4), 435–440. Retrieved from http://www.ncbi.nlm.nih.gov/pubmed/11022853 PMID:11022853

Provinciali, L., & Giovagnoli, A. R. (1990). Antecedent events in amyotrophic lateral sclerosis: Do they influence clinical onset and progression? *Neuroepidemiology*, *9*(5), 255–262. doi:10.1159/000110782 PMID:2087250

Pruss, A. (1998). Review of epidemiological studies on health effects from exposure to recreational water. *International Journal of Epidemiology*, *27*(1), 1–9. doi:10.1093/ije/27.1.1 PMID:9563686

Pruss, A., Kay, D., Fewtrell, L., & Bartram, J. (2002). Estimating the Burden of Disease from Water, Sanitation, Hygene at a Global Level. *Environmental Health Perspectives*, *110*(5), 537–542. doi:10.1289/ehp.02110537 PMID:12003760

Przybylowicz, A., Chesy, P., Herman, M., Parczewski, A., Walas, S., & Piekoszewski, W. (2012). Examination of distribution of trace elements in hair, fingernails and toenails as alternative biological materials. Application of chemometric methods. *Central European Journal of Chemistry*, *10*, 1590–1599.

Pullan, R. L., Smith, J. L., Jasrasaria, R., & Brooker, S. J. (2014). Global numbers of infection and disease burden of soil-transmitted helminth infections in 2010. *Parasites & Vectors*, *7*(1), 37. doi:10.1186/1756-3305-7-37 PMID:24447578

Putignani, L., & Menichella, D. (2010). Global distribution, public health and clinical impact of the protozoan pathogen Cryptosporidium. *Interdisciplinary Perspectives on Infectious Diseases*. PMID:20706669

Putnam, J. J. (1887a). On the Frequency with Which Lead Is Found in the Urine and on Certain Points in the Symptomatology of Chronic Lead Poisoning. *Boston Medical and Surgical Journal*, *117*(4), 73–76. doi:10.1056/NEJM188707281170401

Putnam, J. J. (1887b). On the Frequency with Which Lead Is Found in the Urine, and on Certain Points in the Symptomatology of Chronic Lead Poisoning. *Boston Medical and Surgical Journal*, *117*(5), 97–99. doi:10.1056/NEJM188708041170501

Qadir, M., Sharma, B. R., Bruggeman, A., Choukr-Allah, R., & Karajeh, F. (2007). Non-conventional water resources and opportunities for water augmentation to achieve food security in water scarce countries. *Agricultural Water Management*, *87*(1), 2–22. doi:10.1016/j.agwat.2006.03.018

Qayyum, M. A., & Shah, M. H. (2014). Comparative study of trace elements in blood, scalp hair and nails of prostate cancer patients in relation to healthy donors. *Biological Trace Element Research*, *162*(1-3), 46–57. doi:10.100712011-014-0123-4 PMID:25231348

Qin, H. P., Khu, S. T., & Yu, X. Y. (2010). Spatial variations of storm runoff pollution and their correlation with land-use in a rapidly urbanizing catchment in China. *The Science of the Total Environment*, *408*(20), 4613–4623. doi:10.1016/j.scitotenv.2010.07.021 PMID:20667581

Qishlaqi, A., Moore, F., & Forghani, G. (2008). *Impact of untreated wastewater irrigation on soils and crops in Shiraz suburban area*. SW Iran.

Qiu, C., Kivipelto, M., & von Strauss, E. (2009). Epidemiology of Alzheimer's disease: Occurrence, determinants, and strategies toward intervention. *Dialogues in Clinical Neuroscience*, *11*(2), 111–128. Retrieved from http://www.ncbi.nlm.nih.gov/pubmed/19585947 PMID:19585947

Queiroga, R. P. F. D., Sá, L. D. D., Nogueira, J. D. A., Lima, E. R. V. D., Silva, A. C. O., Pinheiro, P. G. O. D., & Braga, J. U. (2012). Distribuição espacial da tuberculose ea relação com condições de vida na área urbana do município de Campina Grande-2004 a 2007. *Revista Brasileira de Epidemiologia*, *15*(1), 222–232. doi:10.1590/S1415-790X2012000100020 PMID:22450507

Quintero-Betancourt, W., Peele, E. R., & Rose, J. B. (2002). Cryptosporidium parvum and Cyclospora cayetanensis: A review of laboratory methods for detection of these waterborne parasites. *Journal of Microbiological Methods*, *49*(3), 209–224. doi:10.1016/S0167-7012(02)00007-6 PMID:11869786

Quintero-Hernández, V., Jiménez-Vargas, J. M., Gurrola, G. B., Valdivia, H. H., & Possani, L. D. (2013). Scorpion venom components that affect ion-channels function. *Toxicon*, *76*, 328–342. doi:10.1016/j.toxicon.2013.07.012 PMID:23891887

Raaschou-Nielsen, O., Pavuk, M., Leblanc, A., Dumas, P., Weber, J., Olsen, A., ... Olsen, J. (2005). Adipose organochlorine concentrations and risk of breast cancer among postmenopausal Danish women. *Cancer Epidemiology, Biomarkers & Prevention*, *14*, 67–74.

Radhouane, L. (2013). Climate change impacts on North African countries and on some Tunisian economic sectors. *Journal of Agriculture and Environment for International Development*, *107*(1), 101–113.

Rahman, R., Faiz, M. A., Selim, S., Rahman, B., Basher, A., Jones, A., ... Milton, A. H. (2010). Annual incidence of snake bite in rural Bangladesh. *PLoS Neglected Tropical Diseases*, *4*(10), e860. doi:10.1371/journal.pntd.0000860 PMID:21049056

Rahmstorf, S., Cazenave, A., Church, J. A., Hansen, J. E., Keeling, R. F., Parker, D. E., & Somerville, R. C. (2007). Recent climate observations compared to projections. *Science*, *316*(5825), 709–709. doi:10.1126cience.1136843 PMID:17272686

Rajaganapa, V., Xavier, F., Sreekumar, D., & Mandal, P. K. (2011). Heavy Metal Contamination in Soil, Water and Fodder and their Presence in Livestock and Products : A Review. *Journal of Environmental Science and Technology*, *4*(3), 234–249. doi:10.3923/jest.2011.234.249

Rajan, S., Nadiah, N., Firdaus, M., & Appukutty, M., & Rama-, K. (2012). Effects of climate changes on dissolved heavy metal concentrations among recreational park tributaries in Pahang, Malaysia. *Biomedical Research*, *23*(1), 23–30.

Ralph, J. L., Orgebin-Crist, M.-C., Lareyre, J.-J., & Nelson, C. C. (2003). Disruption of Androgen Regulation in the Prostate by the Environmental Contaminant Hexachlorobenzene. *Environmental Health Perspectives*, *111*(4), 461–466. doi:10.1289/ehp.5919 PMID:12676599

Rama K. Maganti, MD; Paul Rutecki, M. (2013). EEG and Epilepsy Monitoring. *American Academy of Neurology*, (June), 598–622.

Ramankutty, N. (2004). Croplands in West Africa: A geographically explicit dataset for use in models. *Earth Interactions*, *8*(23), 1–22. doi:10.1175/1087-3562(2004)8<1:CIWAAG>2.0.CO;2

Ramankutty, N., Foley, J. A., Norman, J., & McSweeney, K. (2002). The global distribution of cultivable lands: Current patterns and sensitivity to possible climate change. *Global Ecology and Biogeography*, *11*(5), 377–392. doi:10.1046/j.1466-822x.2002.00294.x

Ramaoui, K., Guernaoui, S., & Boumezzough, A. (2008). Entomological and epidemiological study of a new focus of cutaneous leishmaniasis in Morocco. *Parasitology Research*, *103*(4), 859–863. doi:10.100700436-008-1068-3 PMID:18592272

Ramirez-Fuentes, E., Lucho-Constantino, C., Escamilla-Silva, E., & Dendooven, L. (2002). Characteristics, and carbon and nitrogen dynamics in soil irrigated with wastewater for different lengths of time. *Bioresource Technology*, *85*(2), 179–187. doi:10.1016/S0960-8524(02)00035-4 PMID:12227543

Randazzo, G., Fogliano, V., Ritieni, A., Mannina, L., Rossi, E., Scarallo, A., & Segre, A. L. (1993). Proliferin, a new sesterterpene from Fusarium proliferatum. *Tetrahedron*, *49*(47), 10883–10896. doi:10.1016/S0040-4020(01)80241-6

Ranjan, A. (2005). Risk factors for Indian kala-azar. *The American Journal of Tropical Medicine and Hygiene*, *73*(1), 74–78. doi:10.4269/ajtmh.2005.73.74 PMID:16014837

Ransohoff, R. M. (2016). How neuroinflammation contributes to neurodegeneration. *Science*, *353*(6301), 777–783. doi:10.1126cience.aag2590 PMID:27540165

Rao, D. G. V. P., & Khan, M. A. Q. (2000). *Zebra mussels*: Enhancement of copper toxicity by high temperature and its relationship with respiration and metabolism. *Water Environment Research*, *72*(2), 175–178. doi:10.2175/106143000X137257

Rappold, P. M., Cui, M., Chesser, A. S., Tibbett, J., Grima, J. C., & Duan, L., … Tieu, K. (2011). Paraquat neurotoxicity is mediated by the dopamine transporter and organic cation transporter-3. *Proceedings of the National Academy of Sciences*, *108*(51), 20766–20771. 10.1073/pnas.1115141108

Rashid, A., Haley, B. J., Rajabov, M., Ahmadova, S., Gurbanov, S., Colwell, R. R., & Huq, A. (2013). Detection of *Vibrio cholerae* in environmental waters including drinking water reservoirs of Azerbaijan. *Environmental Microbiology Reports*, *5*(1), 30–38. doi:10.1111/j.1758-2229.2012.00369.x PMID:23757128

Raynor, J. L., Enoch, R. R., & Fenton, S. E. (2005). Adverse Effects of Prenatal Exposure to Atrazine during a Critical Period of Mammary Gland Growth. *Toxicological Sciences*, *87*(1), 255–266. doi:10.1093/toxsci/kfi213

Razakandrainibe, R., Le, L., Gargala, G., & Favennec, L. (2014). Giardiose et cryptosporidiose : deux parasitoses à transmission hydrique. *RFL Rev. Francoph. des Lab.*, *459*, 51–56.

Ready, P.D. (2008). Leishmaniasis emergence and climate change. *Rev Sci Tech Off intEpiz, 27*, 399-412.

Ready, P. D. (2008). Leishmaniasis emergence and climate change. *Revue Scientifique et Technique (International Office of Epizootics)*, *27*, 399–412. doi:10.20506/rst.27.2.1803 PMID:18819668

Reaney, S. H., Bench, G., & Smith, D. R. (2006). Brain accumulation and toxicity of Mn(II) and Mn(III) exposures. *Toxicological Sciences*, *93*(1), 114–124. doi:10.1093/toxsci/kfl028 PMID:16740617

Reckziegel, G. C., & Pinto, V. L. (2014). Scorpionism in Brazil in the years 2000 to 2012. *The Journal of Venomous Animals and Toxins Including Tropical Diseases*, *20*(1), 46. doi:10.1186/1678-9199-20-46 PMID:25873937

Reddy, P. H. (2006). Amyloid precursor protein-mediated free radicals and oxidative damage: Implications for the development and progression of Alzheimer's disease. *Journal of Neurochemistry*, *96*(1), 1–13. doi:10.1111/j.1471-4159.2005.03530.x PMID:16305625

Regier, D. (1988). One-Month Prevalence of Mental Disorders in the United States. *Archives of General Psychiatry*, *45*(11), 977. doi:10.1001/archpsyc.1988.01800350011002 PMID:3263101

Reiner, A., Albin, R. L., Anderson, K. D., D'Amato, C. J., Penney, J. B., & Young, A. B. (1988). Differential loss of striatal projection neurons in Huntington disease. *Proceedings of the National Academy of Sciences of the United States of America*, *85*(15), 5733–5737. doi:10.1073/pnas.85.15.5733 PMID:2456581

Rein, J. O. (1993). Sting use in two species of Parabuthus scorpions (Buthidae). *The Journal of Arachnology*, 60–63.

Reinoso, R., Becares, E., & Smith, H. V. (2008). Effect of various environmental factors on the viability of Cryptosporidium parvum oocysts. *Journal of Applied Microbiology*, *104*(4), 980–986. doi:10.1111/j.1365-2672.2007.03620.x PMID:17973913

Reisen, W. K., Lothrop, H. D., & Hardy, J. L. (1995). Bionomics of Culex tarsalis (Diptera: Culicidae) in relation to arbovirus transmission in southeastern California. *Journal of Medical Entomology*, *32*(3), 316–327. doi:10.1093/jmedent/32.3.316

Reiter, P. (2001). Climate change and mosquito-borne disease. *Environmental Health Perspectives*, *109*, 141–161. PMID:11250812

Revich, E., Aksel, E., Ushakova, T., Ivanova, J., Zhuchenko, N., Klyuev, N., ... Sotskov, Y. (2001). Dioxin exposure and public health in Chapaevsk, Russia. *Chemosphere*, *43*(4-7), 951–966. doi:10.1016/S0045-6535(00)00456-2

Rhajaoui, M., Abedelmajeed, N., Fellah, H., Azmi, K., Amrir, F., Al-jawabreh, A., & .... (2007). Molecular typing reveals emergence of a new clinic-epidemiologic profile of cutaneous leishmanisis in Morocco. *Emerging Infectious Diseases*, *3*(9), 1358–1360. doi:10.3201/eid1309.060946 PMID:18252108

Rhajaoui, M., Fellah, H., Pratlong, F., Dedet, J. P., & Lyagoubi, M. (2004). Leishmaniasis due to *Leishmaniatropica*MON-102 in a new Moroccan focus. *Transactions of the Royal Society of Tropical Medicine and Hygiene*, *98*(5), 299–301. doi:10.1016/S0035-9203(03)00071-3 PMID:15109554

Rhallabi, N.., Moundib, R., Maaroufi, M., Marghich, M., & Khallayoune, K.H. (1990). Effects of irrigation with raw sewage and purified by the soil, crop yield of tomato and hygienic quality of the harvest. *Vet Inst Agron acts, 10*, 57-66.

Ribot, J., & Norton, A. (2009). Social dimensions of climate change: equity and vulnerability in a warming world. In Social Dimensions of Climate Change (pp. 47-75). Washington, DC: WB.

Rice, K. M., Walker, E. M. Jr, Wu, M., Gillette, C., & Blough, E. R. (2014). Environmental mercury and its toxic effects. *Journal of Preventive Medicine and Public Health*, *47*(2), 74–83. doi:10.3961/jpmph.2014.47.2.74 PMID:24744824

Richardson, K., Steffen, W., Schellnhuber, H. J., Alcamo, J., Barker, T., Kammen, D. M., . . . Waever, O. (2009) Synthesis report. Climate change: Global risks, challenges, and decisions. *Proceedings, United Nations Framework Convention on Climate Change.*

Richardson, D. B., Keil, A. P., Cole, S. R., & Dement, J. (2018). Asbestos standards: Impact of currently uncounted chrysotile asbestos fibers on lifetime lung cancer risk. *American Journal of Industrial Medicine*, *61*(5), 383–390. doi:10.1002/ajim.22836 PMID:29573442

Ricke, W. A., Wang, Y., & Cunha, G. R. (2007). Steroid hormones and carcinogenesis of the prostate: The role of estrogens. Differentiation. *Research in Biological Diversity*, *75*(9), 871–882. PMID:17924963

Riise, T., Kirkeleit, J., Aarseth, J. H., Farbu, E., & Midgard, R. (2011). Risk of MS is not associated with exposure to crude oil, but increases with low level of education. *Multiple Sclerosis*, *17*(7), 780–787. doi:10.1177/1352458510397686 PMID:21343231

Rijsberman, F. R. (2004). Water scarcity: Fact or fiction? *Proceedings of the 4th International Crop Science Congress, 80*, 5-22. 10.1016/j.agwat.2005.07.001

Rike, A. G., Schiewer, S., & Filler, D. M. (2008). Temperature effects on biodegradation of petroleum contaminants in cold soils. In D. L. Barnes, D. M. Filler, & I. Snape (Eds.), *Bioremediation of Petroleum Hydrocarbons in Cold Regions* (pp. 84–108). Cambridge, UK: Cambridge University Press. doi:10.1017/CBO9780511535956.007

Rimhanen-Finne, R., Vuorinen, A., Marmo, S., Malmberg, S., & Hänninen, M. L. (2004). Comparative analysis of Cryptosporidium, Giardia and indicator bacteria during sewage sludge hygienization in various composting processes. *Letters in Applied Microbiology, 38*(4), 301–305. doi:10.1111/j.1472-765X.2004.01487.x PMID:15214729

Rinkevich, F. D., Du, Y., Tolinski, J., Ueda, A., Wu, C. F., Zhorov, B. S., & Dong, K. (2015). Distinct roles of the DmNav and DSC1 channels in the action of DDT and pyrethroids. *Neurotoxicology, 47*, 99–106. doi:10.1016/j.neuro.2015.02.001 PMID:25687544

Rioux, J. A., Guilvard, E., Dereure, J., Lanotte, G., Denial, M., & Pratlong, F. (1986) Infestation naturelle de Phlebotomuspapatasi(Scopoli, 1786) par Leishmania major MON-25. A propos de 28 souches isolées dans un foyer du Sud Marocain. In:Leishmania. Taxinomie et Phylogenèse. Applications écoépidémiologiques. (Coll. Int. CNRS/ INSERM, 1984). IMEEE, 471–480.

Rioux, J. A. (2001). Trente ans de coopération franco–marocaine sur les leishmanioses: Dépistage et analyse des foyers. Facteurs de risque. Changements climatiques et dynamique nosogéographique. *Association des Anciens Elèves de l'Institut Pasteur, 168*, 90–101.

Rioux, J. A., Akalay, O., Périères, J., Dereure, J., & Mahjour, J. (1997). Evolution of Eco-epidemiological of 'leishmaniasis risk' at Atlantic Moroccan Sahara. Heuristic interest of the relationship 'sandflies bioclimates'. *Ecologia Mediterranea, 23*, 73–92.

Rioux, J. A., & de La Rocque, S. (2003). Climats, leishmanioses et trypanosomiases. Changements climatiques, maladies infectieuses et allergiques. *Annales de l'Institut Pasteur, 16*, 41–62.

Rioux, J. A., Mahjour, J., Gallego, M., Dereure, J., Périères, J., Laamrani, A., & ... . (1996). Leishmaniose cutanée humaine à *Leishmania infantum*MON-24 au Maroc. *Bulletin de la Société Française de Parasitologie, 14*, 179–183.

Rioux, J. A., Rispail, P., Lanotte, G., & Lepart, J. (1984). Relations Phlébotomes-bioclimats en écologie des leishmanioses Corollaires épidémiologiques. L'exemple du Maroc. *Bull Soc Bot Fr, 131*, 549–557.

Rishi, I., Baidouri, H., Abbasi, J. A., Bullard-Dillard, R., Kajdacsy-Balla, A., Pestaner, J. P., ... Bagasra, O. (2003). Prostate cancer in African American men is associated with downregulation of zinc transporters. *Applied Immunohistochemistry & Molecular Morphology, 11*(3), 253–260. doi:10.1097/00129039-200309000-00009 PMID:12966353

Rispail, P., Dereure, J., & Jarry, D. (2002). Risk zones of human leishmaniases in the western Mediterranean basin. Correlations between vectors sand flies, bioclimatology and phytosociology. *MemInstOswaldo Cruz Rio de Janeiro, 97*, 477-483.

Ritieni, A., Fogliano, V., Randazzo, G., Scarallo, A., Logrieco, A., Moretti, A., ... Bottalico, A. (1995). Isolation and characterization of fusaproliferin, a new toxic metabolite from Fusarium proliferatum. *Natural Toxins, 3*(1), 17–20. doi:10.1002/nt.2620030105

Ritieni, A., Monti, S. M., Randazzo, G., Logrieco, A., Moretti, A., Peluso, G., ... Fogliano, V. (1997). Teratogenic effects of fusaproliferin on chicken embryos. *Journal ofAgricultural and Food Chemistry, 45*(8), 3039–3043. doi:10.1021/jf960890v

Ritz, B., & Yu, F. (2000). Parkinson's disease mortality and pesticide exposure in California 1984–1994. *International Journal of Epidemiology, 29*(2), 323–329. doi:10.1093/ije/29.2.323 PMID:10817132

Robens, J., & Cardwell, K. (2003). The costs of mycotoxin management to the USA:management of aflatoxins in the United States. *Journal of Toxicology, 22*, 139–152.

Roberts, E. A., & Schilsky, M. L. (2008). Diagnosis and treatment of Wilson disease: An update. *Hepatology (Baltimore, Md.), 47*(6), 2089–2111. doi:10.1002/hep.22261 PMID:18506894

Roberts, J. R., & Karr, C. J. (2012). Pesticide Exposure in Children. *Pediatrics, 130*(6), e1765–e1788. doi:10.1542/peds.2012-2758 PMID:23184105

Robertson, L. J., Campbell, A. T., & Smith, H. V. (1992). Survival of *Cryptosporidium parvum* oocysts under various environmental pressures. *Applied and Environmental Microbiology, 58*, 3494–3500. PMID:1482175

Robertson, L. J., Hermansen, L., & Gjerde, B. K. (2006). Occurrence of *Cryptosporidium* Oocysts and Giardia Cysts in Sewage in Norway. *Applied and Environmental Microbiology, 72*(8), 5297–5303. doi:10.1128/AEM.00464-06 PMID:16885279

Robinson, S. A. (2009). Introduction: Climate change biology at the ends of the Earth-International Polar year special issue. *Global Change Biology, 15*(7), 1615–1617. doi:10.1111/j.1365-2486.2009.01977.x

Rodier, J. (1955). Manganese poisoning in Moroccan miners. *British Journal of Industrial Medicine, 12*(1), 21–35. PMID:14351643

Rodriguez, L., Ruiz, E., Alonso-Azcarate, J., & Rincon, J. (2009). Heavy metal distribution and chemical speciation in tailings and soils around a Pb–Zn mine in Spain. *Journal of Environmental Management, 90*(2), 1106–1116. doi:10.1016/j.jenvman.2008.04.007 PMID:18572301

Røe Utvik, T. I. (1999). Chemical characterisation of produced water from four offshore oil production platforms in the North Sea. *Chemosphere, 39*(15), 2593–2606. doi:10.1016/S0045-6535(99)00171-X

Roger, A., Nacher, M., Hanf, M., Drogoul, A. S., Adenis, A., Basurko, C., & Dufour, J. (2013). Climate and Leishmaniasis in French Guiana. *The American Journal of Tropical Medicine and Hygiene, 89*(3), 564–569. doi:10.4269/ajtmh.12-0771 PMID:23939706

Rogers, D. J., & Randolph, S. E. (1985). Population ecology of tsetse. *Annual Review of Entomology, 30*(1), 197–216. doi:10.1146/annurev.en.30.010185.001213

Roger, V. L., Go, A. S., Lloyd-Jones, D. M., Benjamin, E. J., Berry, J. D., & Borden, W. B., & American Heart Association Statistics Committee and Stroke Statistics Subcommittee. (2012). Heart disease and stroke statistics--2012 update: A report from the American Heart Association. *Circulation, 125*(1), e2–e220. doi:10.1161/CIR.0b013e31823ac046 PMID:22179539

Rolls, R. J., Leigh, C., & Sheldon, F. (2012). Mechanistic effects of low-flow hydrology on riverine ecosystems: Ecological principles and consequences of alteration. *Freshwater Science, 31*(4), 1163–1186. doi:10.1899/12-002.1

Romaniuk, A., Lyndin, M., Sikora, V., Lyndina, Y., Romaniuk, S., & Sikora, K. (2017). Heavy metals effect on breast cancer progression. *Journal of Occupational Medicine and Toxicology, 28*, 12-32.

Romero-Lankao, P., Smith, J. B., Davidson, D. J., Diffenbaugh, N. S., Kinney, P. L., Kirshen, R., … Villers Ruiz, L. (2014). North America. In Climate Change 2014: Impacts, Adaptation, and Vulnerability. Part B: Regional Aspects Contribution of Working Group II to the Fifth Assessment Report of the Intergovernmental Panel on Climate Change. Cambridge, UK: Cambridge University Press.

Romieu, I., Hernandez, M., Lazcano-Ponce, E., Weber, J. P., & Dewailly, E. (2000). Breast Cancer, Lactation History, and Serum Organochlorines. *American Journal of Epidemiology*, *152*(4), 363–370. doi:10.1093/aje/152.4.363

Ronckers, C. M., Erdmann, C. A., & Land, C. E. (2005). Radiation and breast cancer: A review of current evidence. *Breast Cancer Research*, *7*(1), 21–32. doi:10.1186/bcr970

Roodt, A. R. D., De Titto, E., Dolab, J. A., & Chippaux, J. P. (2013). Envenoming by coral snakes (Micrurus) in Argentina, during the period between 1979-2003. *Revista do Instituto de Medicina Tropical de São Paulo*, *55*(1), 13–18. doi:10.1590/S0036-46652013000100003 PMID:23328720

Rosas, H. D., Liu, A. K., Hersch, S., Glessner, M., Ferrante, R. J., Salat, D. H., ... Fischl, B. (2002). Regional and progressive thinning of the cortical ribbon in Huntington's disease. *Neurology*, *58*(5), 695–701. doi:10.1212/WNL.58.5.695 PMID:11889230

Roscoe, R. J., Ball, W., Curran, J. J., DeLaurier, C., Falken, M. C., Fitchett, R., ... Whittaker, S. G. (2002). Adult blood lead epidemiology and surveillance—United States, 1998–2001. *MMWR. Surveillance Summaries*, *13*, 1–10. PMID:12528812

Rosegrant, M. (2009). Changement climatique et sécurité alimentaire au Maroc. Rencontre internationale « Le changement climatique: enjeux et perspectives d'adaptation pour le Maroc ». Rabat: Institut Royal des Etudes Stratégiques (IRES).

Rose, J. B. (1997). Environmental ecology of *Cryptosporidium* and public health implications. *Annual Review of Public Health*, *18*(1), 135–161. doi:10.1146/annurev.publhealth.18.1.135 PMID:9143715

Rose, J. B., Daeschner, S., Easterling, D. R., Curriero, F. C., Lele, S., & Patz, J. A. (2000). Climate and waterborne disease outbreaks. *Journal - American Water Works Association*, *9*(9), 77–87. doi:10.1002/j.1551-8833.2000.tb09006.x

Rosenbaum, J. F., Pollock, R. A., Otto, M. W., & Pollack, M. H. (1995). Integrated treatment of panic disorder. *Bulletin of the Menninger Clinic, 59*(2 Suppl A), A4-26. Retrieved from http://www.ncbi.nlm.nih.gov/pubmed/7795571

Rosenblatt, A. (2007). Neuropsychiatry of Huntington's disease. *Dialogues in Clinical Neuroscience*, *9*, 191–197. PMID:17726917

Rosenblatt, A., Brinkman, R. R., Liang, K. Y., Almqvist, E. W., Margolis, R. L., Huang, C. Y., ... Ross, C. A. (2001). Familial influence on age of onset among siblings with Huntington disease. *American Journal of Medical Genetics*, *105*(5), 399–403. doi:10.1002/ajmg.1400 PMID:11449389

Rosenzweig, C., Iglesias, A., Yang, X. B. R., Epstein, P., & Chivian, E. (2001). Climate change and extreme weather events; implications for food production, plant diseases, and pests. *Global Change and Human Health*, *2*(2), 90–104. doi:10.1023/A:1015086831467

Rosenzweig, C., & Parry, M. L. (1994). Potential impact of climate change on world food supply. *Nature*, *367*(6459), 133–138. doi:10.1038/367133a0

Rosenzweig, C., & Tubiello, F. N. (1997). Impacts of Global Climate Change on Mediterranean Agrigulture: Current Methodologies and Future Directions. An Introductory Essay. *Mitigation and Adaptation Strategies for Global Change*, *1*(3), 219–232. doi:10.1007/BF00517804

Rose, S., Wynne, R., Frye, R. E., Melnyk, S., & James, S. J. (2015). Increased Susceptibility to Ethylmercury-Induced Mitochondrial Dysfunction in a Subset of Autism Lymphoblastoid Cell Lines. *Journal of Toxicology*, *2015*, 13. doi:10.1155/2015/573701 PMID:25688267

Rossignol, D. A., & Frye, R. E. (2011). Mitochondrial dysfunction in autism spectrum disorders : A systematic review and meta-analysis. *Molecular Psychiatry*, 1–25. doi:10.1038/mp.2010.136 PMID:21263444

Rowland, A. P., Lindley, D. K., Hall, G. H., Rossall, M. J., Wilson, D. R., Benham, D. G., ... Daniels, R. E. (2000). Effects of beach sand properties, temperature and rainfall on the degradation rates of oil in buried oil/beach sand mixtures. *Environmental Pollution*, *109*(1), 109–118. doi:10.1016/S0269-7491(99)00224-9 PMID:15092918

Rowley, W. A., & Graham, C. L. (1968). The effect of temperature and relative humidity on the flight performance of female Aedes aegypti. *Journal of Insect Physiology*, *14*(9), 1251–1257. doi:10.1016/0022-1910(68)90018-8

Royaume du Maroc. (2009). Ministère de l'Aménagement du territoire de l'habitat et de l'environnement. Second National Communication à la Convention Cadre des Nations Unis sur les Changements Climatiques.

Rozemeijer, J. C., & Broers, H. P. (2007). The groundwater contribution to surface water contamination in a region with intensive agricultural land use (Noord-Brabant, The Netherlands). *Environmental Pollution*, *148*(3), 695–706. doi:10.1016/j.envpol.2007.01.028 PMID:17418466

Rubinsztein, D. C., Leggo, J., Coles, R., Almqvist, E., Biancalana, V., Cassiman, J. J., ... Hayden, M. R. (1996). Phenotypic characterization of individuals with 30-40 CAG repeats in the Huntington disease (HD) gene reveals HD cases with 36 repeats and apparently normal elderly individuals with 36-39 repeats. *American Journal of Human Genetics*, *59*, 16–22. PMID:8659522

Rudant, J., Menegaux, F., Leverger, G., Baruchel, A., Nelken, B., Bertrand, Y., ... Clavel, J. (2007). Household Exposure to Pesticides and Risk of Childhood Hematopoietic Malignancies: The ESCALE Study (SFCE). *Environmental Health Perspectives*, *115*(12), 1787–1793. doi:10.1289/ehp.10596 PMID:18087601

Rudolph, A., Chang-Claude, J., & Schmidt, M. K. (2016). Gene–environment interaction and risk of breast cancer. *British Journal of Cancer*, *114*(2), 125–133. doi:10.1038/bjc.2015.439

Rush, A. J., & Weissenburger, J. E. (1994). Melancholic symptom features and DSM-IV. *The American Journal of Psychiatry*, *151*(4), 489–498. doi:10.1176/ajp.151.4.489 PMID:8147445

Rusiecki, J. A., De Roos, A., Lee, W. J., Dosemeci, M., Lubin, J. H., Hoppin, J. A., ... Alavanja, M. C. (2004). Cancer incidence among pesticide applicators exposed to atrazine in the Agricultural Health Study. *Journal of the National Cancer Institute*, *96*(18), 1375–1382. doi:10.1093/jnci/djh264 PMID:15367570

Rusiecki, J. A., Holford, T. R., Zahm, S. H., & Zheng, T. (2004). Polychlorinated biphenyls and breast cancer risk by combined estrogen and progesterone receptor status. *European Journal of Epidemiology*, *19*(8), 793–801. doi:10.1023/B:EJEP.0000036580.05471.31

Ruska, K. M., Sauvageot, J., & Epstein, J. I. (1998). Histology and cellular kinetics of prostatic atrophy. The American Journal of Surgical Pathology, 22(9), 1073–1077. Retrieved from Schimmer, A.D., Dalili, S., Batey, R.A., &Riedl, S.J., (2006).*Targeting XIAP for the treatment ofmalignancy. Cell Death and Differentiation*, *13*(2), 179–188.

Rybicki, B. A., Johnson, C. C., Uman, J., & Gorell, J. M. (1993). Parkinson's disease mortality and the industrial use of heavy metals in Michigan. *Movement Disorders*, *8*(1), 87–92. doi:10.1002/mds.870080116 PMID:8419812

Rychlik, M., Humpf, H. U., Marko, D., Danicke, S., Mally, A., Berthiller, F., ... Lorenz, N. (2014). Proposal of a comprehensive definition of modified and other forms of mycotoxins including "masked" mycotoxins. *Mycotoxin Research*, *30*(4), 197–205. doi:10.100712550-014-0203-5

Saad-Hussein, A., Shahy, E. M., Shaheen, W., Taha, M. M., Mahdy-Abdallah, H., Ibrahim, K. S., ... El-Shamy, K. A. (2016). Comparative Hepatotoxicity of Aflatoxin B1 among Workers Exposed to Different Organic Dust with Emphasis on Polymorphism Role of Glutathione S-Transferase Gene. *Open Access Macedonian Journal of Medical Sciences*, *4*(2), 312. doi:10.3889/oamjms.2016.051 PMID:27335608

Sabarwal, A., Kumar, K., & Singh, R. P. (2018). Hazardous Effects of Chemical Pesticides on Human Health-Cancer and Other Associated Disorders. *Environmental Toxicology and Pharmacology*, *63*, 103–114. doi:10.1016/j.etap.2018.08.018 PMID:30199797

Sáenz de Rodríguez, C. A. (1984). Environmental hormone contamination in Puerto Rico. *The New England Journal of Medicine*, *310*(26), 1741–1742. doi:10.1056/NEJM198406283102612

Safe, S. (2005). Clinical correlates of environmental endocrine disruptors. *Trends in Endocrinology and Metabolism*, *16*(4), 139–144. doi:10.1016/j.tem.2005.03.004

Safe, S., & Wormke, M. (2003). Inhibitory aryl hydrocarbon receptor-estrogen receptor alpha cross-talk and mechanisms of action. *Chemical Research in Toxicology*, *16*(7), 807–816. doi:10.1021/tx034036r

Safini, F., Bouskraoui, M., Pédiatrie, S. De, Mohammed, A. C. H. U., & Marrakech, V. I. (2009). *Epidémiologie de l ' épilepsie à l ' hôpital de jour pédiatrique*. Academic Press.

Sahathevan, R., Brodtmann, A., & Donnan, G. A. (2012). Systematic Review Dementia, stroke, and vascular risk factors; a review. *International Journal of Stroke*, *7*, 61–73. doi:10.1111/j.1747-4949.2011.00731.x PMID:22188853

Saleh, F., Behbehani, A., Asfar, S., Khan, I., & Ibrahim, G. (2011). Abnormal blood levels of trace elements and metals, DNA damage, and breast cancer in the state of Kuwait. *Biological Trace Element Research*, *141*(1-3), 96–109. doi:10.100712011-010-8724-z

Salem, A. B., Messouli, M., & Yacoubi-Khebiza, M. (2011). *Developing an oasis-based water management tool: Eco-hydrologic approach and weap software for a large arid catchment in Morocco*. Academic Press.

Salinas, L. M., Delaune, R. D., & Patrick, W. H. (1986). Changes occurring along a rapidly submerging coastal area: Louisiana, USA. *Journal of Coastal Research*, *2*(3), 269–284.

Sallon, S., Deckelbaum, R. J., Schmid, I. I., Harlap, S., Baras, M., & Spira, D. T. (1988). *Cryptosporidium*, malnutrition, and chronic diarrhea in children. *American Journal of Diseases of Children*, *142*, 312–315. PMID:3344720

Salomão, M. D., Santos, S. M. A., & Puorto, G. (1995). Activity pattern of Crotalusdurissus (Viperidae, Crotalinae): Feeding, reproduction and snakebite. *Studies on Neotropical Fauna and Environment*, *30*(2), 101–106. doi:10.1080/01650529509360946

Samanic, C. M., De Roos, A. J., Stewart, P. A., Rajaraman, P., Waters, M. A., & Inskip, P. D. (2008). Occupational exposure to pesticides and risk of adult brain tumors. *American Journal of Epidemiology*, *167*(8), 976–985. doi:10.1093/aje/kwm401 PMID:18299277

Samokhvalov, A. V., Irving, H., Mohapatra, S., & Rehm, J. (2010). Alcohol consumption, unprovoked seizures, and epilepsy : A systematic review and meta-analysis. *Epilepsia*, *51*(7), 1177–1184. doi:10.1111/j.1528-1167.2009.02426.x PMID:20074233

Sanchez-Santed, F., Colomina, M. T., & Herrero, H. E. (2016). Organophosphate pesticide exposure and neurodegeneration. *Cortex*, *74*, 417–426. doi:10.1016/j.cortex.2015.10.003 PMID:26687930

Sang, D. K., Njeru, W. K., & Ashford, R. W. (1992). A possible animal reservoir for Leishmania 331 tropica s.l. in Kenya. *Annals of Tropical Medicine and Parasitology*, *86*(3), 311–312. doi:10.1080/00034983.1992.11812670 PMID:1449281

Santos Pereira, L., Cordery, I., & Iacovides, I. (2009). *Coping with water scarcity: Addressing the challenges. Coping with Water Scarcity: Addressing the Challenges*. doi:10.1007/978-1-4020-9579-5

Santos, D., Batoreu, C., Mateus, L., dos Santos, A. M., & Aschner, M. (2014). Manganese in Human Parenteral Nutrition: Considerations for Toxicity and Biomonitoring. *Neurotoxicology, 43*, 36–45. doi:10.1016/j.neuro.2013.10.003 PMID:24184781

Sappino, A. P., Buser, R., Lesne, L., Gimelli, S., Bena, F., Belin, D., & Mandriota, S. J. (2012). Aluminium chloride promotes anchorage-independent growth in human mammary epithelial cells. *Journal of Applied Toxicology, 32*(3), 233–243. doi:10.1002/jat.1793

Sargeant, K., Sheridan, A., O'kelly, J., & Carnaghan, R. B. A. (1961). Toxicity associated with certain samples of groundnuts. *Nature, 192*(4807), 1096–1097. doi:10.1038/1921096a0

Sarkar, S., Malovic, E., Harischandra, D. S., Ngwa, H. A., Ghosh, A., Hogan, C., ... Kanthasamy, A. (2018). Manganese exposure induces neuroinflammation by impairing mitochondrial dynamics in astrocytes. *Neurotoxicology, 64*, 204–218. doi:10.1016/j.neuro.2017.05.009 PMID:28539244

Sarris, A. (2009*). Evolving Structure of World Agricultural Trade and Requirements for New World Trade Rules.* Paper presented at the Expert Meeting on "How to Feed the World in 2050", Rome, Italy.

Sathiakumar, N., MacLennan, P. A., Mandel, J., & Delzell, E. (2011). A review of epidemiologic studies of triazine herbicides and cancer. *Critical Reviews in Toxicology, 41*(sup1), 1–34.

Sathiakumar, N., MacLennan, P. A., Mandel, J., & Delzell, E. (2011). A review of epidemiologic studies of triazine herbicides and cancer. *Critical Reviews in Toxicology, 41*(sup1), 1–34. doi:10.3109/10408444.2011.554793

Sati, V. P. (2015). Climate change and socio-ecological transformation in high mountains: An empirical study of Garhwal Himalaya. *Change and Adaptation in Socio-Ecological Systems, 2*(1). doi:10.1515/cass-2015-0005

Sauer, R. M., Zook, B. C., & Garner, F. M. (1970). Demyelinating encephalomyelopathy associated with lead poisoning in nonhuman primates. *Science, 169*(3950), 1091–1093. doi:10.1126cience.169.3950.1091 PMID:5449319

Saupe, E. E., Papes, M., Selden, P. A., & Vetter, R. S. (2011). Tracking a medically important spider: Climate change, ecological niche modeling, and the brown recluse (Loxosceles reclusa). *PLoS One, 6*(3), e17731. doi:10.1371/journal.pone.0017731 PMID:21464985

Savettieri, G., Salemi, G., Arcara, A., Cassata, M., Castiglione, M. G., & Fierro, B. (1991). A case-control study of amyotrophic lateral sclerosis. *Neuroepidemiology, 10*(5-6), 242–245. doi:10.1159/000110279 PMID:1798425

Savitz, S. I., & Fisher, M. (2007). Future of Neuroprotection for Acute Stroke : In the Aftermath of the SAINT Trials. *Annals of Neurology, 61*(5), 396–402. doi:10.1002/ana.21127 PMID:17420989

Sayre, L. M., Perry, G., & Smith, M. A. (2008). Oxidative Stress and Neurotoxicity. *Chemical Research in Toxicology, 21*(1), 172–188. doi:10.1021/tx700210j PMID:18052107

SCF. (2000). *Opinion on Fusarium Toxins – Part 2: zearalenone (ZEA).* Retrieved from http://ec.europa.eu/food/fs/sc/scf/out65_en.pdf

Schapira, A. H. V. (2009). Neurobiology and treatment of Parkinson's disease. *Trends in Pharmacological Sciences, 30*(1), 41–47. doi:10.1016/j.tips.2008.10.005 PMID:19042040

Schelter, B. (2008). Seizure Prediction in Epilepsy. Academic Press.

Schenck, J. F., & Zimmerman, E. A. (2004). High-field magnetic resonance imaging of brain iron: Birth of a biomarker? *NMR in Biomedicine, 17*(7), 433–445. doi:10.1002/nbm.922 PMID:15523705

Schiedek, D., Sundelin, B., Readman, J. W., & Macdonald, R. W. (2007). Interactions between climate change and contaminants. *Marine Pollution Bulletin*, *54*(12), 1845–1856. doi:10.1016/j.marpolbul.2007.09.020 PMID:17963794

Schiff, K., & Bay, S. (2003). Impacts of stormwater discharges on the nearshore benthic environment of Santa Monica Bay. *Marine Environmental Research*, *56*(1-2), 225–243. doi:10.1016/S0141-1136(02)00332-X PMID:12648957

Schiff, P. L. Jr. (2006). Ergot and its alkaloids. *American Journal of Pharmaceutical Education*, *70*(5), 98. doi:10.5688/aj700598

Schijven, J.F., & de Roda Husman, A.M. (2005) Effect of climate changes on waterborne disease in The Netherlands. *Water Science and Technology: A Journal of the International Association on Water Pollution Research, 51*, 79-87.

Schilling, J., Freier, K. P., Hertig, E., & Scheffran, J. (2012). Climate change, vulnerability and adaptation in North Africa with focus on Morocco. *Agriculture, Ecosystems & Environment*, *156*, 12–26. doi:10.1016/j.agee.2012.04.021

Schlenker, W., & Lobell, D. B. (2010). Robust negative impacts of climate change on African agriculture. *Environmental Research Letters*, *5*(1), 014010. doi:10.1088/1748-9326/5/1/014010

Schmidhuber, J., & Tubiello, F. N. (2007). Global food security under climate change. *Proceedings of the National Academy of Sciences of the United States of America*, *104*(50), 19703–19708. doi:10.1073/pnas.0701976104 PMID:18077404

Schneider, S. H., Semenov, S., Patwardhan, A., Burton, I., Magadza, C. H., Oppenheimer, M., ... Yamin, F. (2007). Assessing key vulnerabilities and the risk from climate change. In *Climate Change 2007: Adaptation and Vulnerability Contribution of Working Group II to the Fourth Assessment Report of the Intergovernmental Panel on Climate Change* (pp. 779–810). Cambridge, UK: Cambridge University.

Schoenholtz, S. H., Van Miegroet, H., & Burger, J. A. (2000). A review of chemical and physical properties as indicators of forest soil quality: Challenges and opportunities. *Forest Ecology and Management*, *138*(1-3), 335–356. doi:10.1016/S0378-1127(00)00423-0

Scholz, N. L., Fleishman, E., Brown, L., Werner, I., Johnson, M. L., Brooks, M. L., ... Schlenk, D. (2012). A perspective on modern pesticides, pelagic fish declines, and unknown ecological resilience in highly managed ecosystems. *Bioscience*, *62*(4), 428–434. doi:10.1525/bio.2012.62.4.13

Schon, E. A., & Manfredi, G. (2003). Neuronal degeneration and mitochondrial dysfunction. *The Journal of Clinical Investigation*, *111*(3), 303–312. doi:10.1172/JCI200317741 PMID:12569152

Schöpfer, J., Drasch, G., & Schrauzer, G. N. (2010). Selenium and cadmium levels and ratios inprostates,livers, and kidneys of nonsmokers and smokers. *Biological Trace Element Research*, *134*(2), 180–187. doi:10.100712011-010-8636-y PMID:20174974

Schrauzer, G. N. (2008). Interactive effects of selenium and cadmium on mammary tumor developmentand growth in MMTV-infected female mice. A model study on the roles of cadmium andselenium in human breast cancer. *Biological Trace Element Research*, *123*(1-3), 27–34. doi:10.100712011-008-8091-1 PMID:18264682

Schuster, C.J., Ellis, A.G., Robertson, W.J., Charron, D.F., Aramini, J.J., Marshall, B.J., & Medeiros, D.T. (2005). Infectious disease outbreaks related to drinking water in Canada, 1974-2001. *Revue canadienne de sante publique, 96*, 254-258.

Schwartz, E., Sadetzki, S., Murad, H., & Raveh, D. (2001). Age as a risk factor for severe Plasmodium falciparum malaria in nonimmune patients. *Clinical Infectious Diseases*, *33*(10), 1774–1777. doi:10.1086/322522

Schwartz, G. G. (2002). Hypothesis: Does ochratoxin A cause testicular cancer? *Cancer Causes & Control*, *13*(1), 91–100. doi:10.1023/A:1013973715289

Schwartz, J., & Levin, R. (1999). Drinking water turbidity and health. *Epidemiology (Cambridge, Mass.)*, *10*(1), 86–89. doi:10.1097/00001648-199901000-00017 PMID:9888287

Schwartz, J., Levin, R., & Goldstein, R. (2000). Drinking water turbidity and gastrointestinal illness in the elderly of Philadelphia. *Journal of Epidemiology and Community Health*, *54*(1), 45–51. doi:10.1136/jech.54.1.45 PMID:10692962

Schwarz, S., Husstedt, I., Bertram, H. P., & Kuchelmeister, K. (1996). Amyotrophic lateral sclerosis after accidental injection of mercury. *Journal of Neurology, Neurosurgery, and Psychiatry*, *60*(6), 698. doi:10.1136/jnnp.60.6.698 PMID:8648348

Sciarillo, W. G., Alexander, G., & Farrell, K. P. (1992). Lead exposure and child behavior. *American Journal of Public Health*, *82*(10), 1356–1360. doi:10.2105/AJPH.82.10.1356 PMID:1415859

Scinicariello, F., & Buser, M. C. (2015). Blood cadmium and depressive symptoms in young adults (aged 20-39 years). *Psychological Medicine*, *45*(4), 807–815. doi:10.1017/S0033291714001883 PMID:25115444

Scott, L. (2008). *Climate variability and climate change: implications for chronic poverty*. Research Centre Working Paper, N 108.

Scott, C. A., Drechsel, P., Raschid-Sally, L., Bahri, A., Mara, D., Redwood, M., & … . (2010). Wastewater irrigation and health: challenges and outlook for mitigating risks in low-income countries. In P. Drechsel, C. A. Scott, L. Raschid-Sally, M. Redwood, & A. Bahri (Eds.), *Wastewater Irrigation and Health: Assessing and Mitigating Risk in Low-Income Countries* (pp. 381–394). London, UK: Earthscan.

Scott, C. A., Faruqui, N. I., & Raschid-Sally, L. (2004). *Wastewater use in irrigated agriculture: confronting the livelihood and environmental realities*. Wallingford: Cabi Publishing. doi:10.1079/9780851998237.0000

Seddiki, A. (2014, August 20). In Morocco youth unemployment is driving up inequality. The Guardian.

Seeland, A., Oehlmann, J., & Müller, R. (2012). Aquatic ecotoxicity of the fungicide pyrimethanil : Effect pro fi le under optimal and thermal stress conditions. *Environmental Pollution*, *168*, 161–169. doi:10.1016/j.envpol.2012.04.020 PMID:22622013

Segal, R. J., Reid, R. D., Courneya, K. S., Malone, S. C., Parliament, M. B., Scott, C. G., ... Wells, G. A. (2003). Resistance exercise in men receiving androgen deprivation therapy for prostate cancer. *Journal of Clinical Oncology*, *21*(9), 1653–1659. doi:10.1200/JCO.2003.09.534 PMID:12721238

Seguin, F., Leboulanger, C., Rimet, F., Druart, J. C., & Berard, A. (2001). Effects of atrazine and nicosulfuron on phytoplankton in systems of increasing complexity. *Archives of Environmental Contamination and Toxicology*, *40*, 198–208. doi:10.1007002440010164 PMID:11243322

Selvakumar, K., Bavithra, S., Ganesh, L., Krishnamoorthy, G., Venkataraman, P., & Arunakaran, J. (2013). Polychlorinated biphenyls induced oxidative stress mediated neurodegeneration in hippocampus and behavioral changes of adult rats: Anxiolyticlike effects of quercetin. *Toxicology Letters*, *222*(1), 45–54. doi:10.1016/j.toxlet.2013.06.237 PMID:23831945

Semenza, J.C., & Nichols, G. (2007) Cryptosporidiosis surveillance and water-borne outbreaks in Europe. *Euro Surveillance: Bulletin Europeen sur les maladies transmissibles, 12*, E13-14.

Semenza, J. C., Höuser, C., Herbst, S., Rechenburg, A., Suk, J. E., Frechen, T., & Kistemann, T. (2012). Knowledge mapping for climate change and food- and waterborne diseases. *Critical Reviews in Environmental Science and Technology*, *42*(4), 378–411. doi:10.1080/10643389.2010.518520 PMID:24771989

Semenza, J. C., & Menne, B. (2009). Climate change and infectious diseases in Europe. *The Lancet Infectious Diseases*, *9*(6), 365–375. doi:10.1016/S1473-3099(09)70104-5 PMID:19467476

Sene, M. (1994). *Etude de la Schistosomose intestinale à Schistosoma mansoni chez les rongeurs sauvages à Richard-Toil (Sénégal): Suivi de L'infestation naturelle et transmission expérimentale* (Dissertation). Université Cheikh Anta Diop de Dakar.

Sengupta, A., Mense, S. M., Lan, C., Zhou, M., Mauro, R. E., Kellerman, L., ... Zhang, L. (2007). Gene Expression Profiling of Human Primary Astrocytes Exposed to Manganese Chloride Indicates Selective Effects on Several Functions of the Cells. *Neurotoxicology*, *28*(3), 478–489. doi:10.1016/j.neuro.2006.10.005 PMID:17175027

Senhorst, H.A. & Zwolsman, J.J. (2005) Climate change and effects on water quality: a first impression. *Water Science and Technology, 51*, 53-59.

Sensenig, A. T., & Shultz, J. W. (2004). Elastic energy storage in the pedipalpal joints of scorpions and sun-spiders (Arachnida, Scorpiones, Solifugae). *The Journal of Arachnology*, *32*(1), 1–10. doi:10.1636/S02-73

Severn, D. J., & Ballard, G. (1990). Risk/benefit and regulations. In Pesticides in the soil environment. Soil Science Society of America Book.

Sevim, S., Aktekin, M., Dogu, O., Ozturk, H., & Ertas, M. (2002). Late Onset Polyneuropathy due to Organophosphate (DDVP) Intoxication. *The Canadian Journal of Neurological Sciences*, 75–78. PMID:12619790

Shachak, M., & Brand, S. (1983). The relationship between sit and wait foraging strategy and dispersal in the desert scorpion, Scorpio maurus palmatus. *Oecologia*, *60*(3), 371–377. doi:10.1007/BF00376854 PMID:28310698

Shaffer, L. R., Daniel, R., & Formanowicz, J. R. (1996). A cost of viviparity and parental care in scorpions: Reduced sprint speed and behavioural compensation. *Animal Behaviour*, *51*(5), 1017–1024. doi:10.1006/anbe.1996.0104

Shaffer, L. R., & Formanowicz, D. R. Jr. (2000). Sprint speeds of juvenile scorpions: Among family differences and parent offspring correlations. *Journal of Insect Behavior*, *13*(1), 45–54. doi:10.1023/A:1007711508847

Shah, H. K., Bhat, M. A., Sharma, T., Banerjee, B. D., & Guleria, K. (2018). Delineating Potential Transcriptomic Association with Organochlorine Pesticides in the Etiology of Epithelial Ovarian Cancer. *The Open Biochemistry Journal*, *12*(1), 16–28. doi:10.2174/1874091X01812010016 PMID:29576811

Shah, S. I. A., Minhas, U., & Khan, H. A. (2015). Trace minerals and heavy metals: Implications in prostate cancer. *World Cancer Research Journal*, *2*, e615.

Shaman, J., & Lipsitch, M. (2013). The El Nino-Southern Oscillation (ENSO)-pandemic influenza connection: Coincident or causal? *Proceedings of the National Academy of Sciences of the United States of America*, *110*(Suppl 1), 3689–3691. doi:10.1073/pnas.1107485109 PMID:22308322

Shamberger, R. J., Tytko, S. A., & Willis, C. E. (1976). Antioxidants and cancer. Part VI. Selenium and ageadjusted human cancer mortality. *Archives of Environmental Health*, *31*(5), 231–235. doi:10.1080/00039896.1976.10667225 PMID:973735

Shanks, G. D. (2012). Control and elimination of Plasmodium vivax. *Advances in Parasitology*, *80*, 301–341. doi:10.1016/B978-0-12-397900-1.00006-2

Shaw, I. (2010). Motor neurone disease - a methyl bromide exposure cluster points to a causal mechanism. *Human and Experimental Toxicology*, *29*(3), 241–242. doi:10.1177/0960327109359462 PMID:20071473

Shehata, M. G., Samy, A. M., Doha, S. A., Fahmy, A. R., Kaldas, R. M., & ... . (2009). First report of 335 Leishmania tropica from a classical focus of L. major in North-Sinai, Egypt. Am J Trop 336. *Medecine et Hygiene*, *81*, 213–218. PMID:19635872

Shelton, R. M. (1973). The effect of temperatures on development of eight mosquito species. *Mosquito News*, *33*(1), 1–12.

Shen, X.-L., Yu, J.-H., Zhang, D.-F., Xie, J.-X., & Jiang, H. (2014). Positive Relationship between Mortality from Alzheimer's Disease and Soil Metal Concentration in Mainland China. *Journal of Alzheimer's Disease*, *42*(3), 893–900. doi:10.3233/JAD-140153 PMID:25024310

Sherif, S. O., Salama, E. E., & Abdel-Wahhab, M. A. (2009). Mycotoxins and child health: The need for health risk assessment. *International Journal of Hygiene and Environmental Health*, *212*(4), 347–368. doi:10.1016/j.ijheh.2008.08.002

Shibayama, H., Kasahara, Y., & Kobayashi, H. (1986). Prevalence of dementia in a Japanese elderly population. *Acta Psychiatrica Scandinavica*, *74*(2), 144–151. doi:10.1111/j.1600-0447.1986.tb10598.x PMID:3490741

Shibayama, H., Kobayashi, H., Iwase, S., Nakagawa, M., Marui, Y., Kayukawa, Y., ... Takeuchi, T. (1986). Unusual Cases of Presenile Dementia with Fahr's Syndrome. *Psychiatry and Clinical Neurosciences*, *40*(1), 85–100. doi:10.1111/j.1440-1819.1986.tb01615.x PMID:3773356

Shields, P. G. (2006). Understanding population and individual risk assessment: The case of polychlorinated biphenyls. *Cancer Epidemiology, Biomarkers & Prevention*, *15*(5), 830–839. doi:10.1158/1055-9965.EPI-06-0222

Shi, J., Wang, G., Fang, Y., Ding, L., Huang, S., Hou, M., ... Li, P. (2017). A new moth-preying alpine pit viper species from Qinghai-Tibetan Plateau (Viperidae, Crotalinae). *Amphibia-Reptilia*, *38*(4), 517–532. doi:10.1163/15685381-00003134

Shimizu, K., Matsubara, K., Ohtaki, K., Fujimaru, S., Saito, O., & Shiono, H. (2003). Paraquat induces long-lasting dopamine overflow through the excitotoxic pathway in the striatum of freely moving rats. *Brain Research*, *976*(2), 243–252. doi:10.1016/S0006-8993(03)02750-1 PMID:12763259

Shine, R., & Madsen, T. (1996). Is thermoregulation unimportant for most reptiles? An example using water pythons (Liasisfuscus) in tropical Australia. *Physiological Zoology*, *69*(2), 252–269. doi:10.1086/physzool.69.2.30164182

Shine, R., & Shetty, S. (2001). Moving in two worlds: Aquatic and terrestrial locomotion in sea snakes (Laticaudacolubrina, Laticaudidae). *Journal of Evolutionary Biology*, *14*(2), 338–346. doi:10.1046/j.1420-9101.2001.00265.x

Shin, H. W., Jewells, V., Hadar, E., Fisher, T., & Hinn, A. (2014). Review of Epilepsy - Etiology, Diagnostic Evaluation and Treatment. *International Journal of Neurorehabilitation*, *1*(3), 1–8. doi:10.4172/2376-0281.1000129

Shin, H.-W., & Park, H. K. (2017). Recent Updates on Acquired Hepatocerebral Degeneration. *Tremor and Other Hyperkinetic Movements (New York, N.Y.)*, *7*, 463. doi:10.7916/D8TB1K44 PMID:28975044

Shinotoh, H., Snow, B. J., Hewitt, K. A., Pate, B. D., Doudet, D., Nugent, R., ... Calne, D. B. (1995). MRI and PET studies of manganese-intoxicated monkeys. *Neurology*, *45*(6), 1199–1204. doi:10.1212/WNL.45.6.1199 PMID:7783889

Shin, Y. C., Kim, E., Cheong, H.-K., Cho, S., Sakong, J., Kim, K. S., ... Kim, Y. (2007). High signal intensity on magnetic resonance imaging as a predictor of neurobehavioral performance of workers exposed to manganese. *Neurotoxicology*, *28*(2), 257–262. doi:10.1016/j.neuro.2006.03.014 PMID:16647136

Shipley, J. M., & Waxman, D. J. (2006). Aryl hydrocarbon receptor-independent activation of estrogen receptor-dependent transcription by 3-methylcholanthrene. *Toxicology and Applied Pharmacology*, *213*(2), 87–97. doi:10.1016/j.taap.2005.09.011

Shirzadi, M. R., Mollalo, A., & Yaghoobi-Ershadi, M. R. (2015). Dynamic relations between incidence of zoonotic cutaneous leishmaniasis and climatic factors in Golestan Province. *Iranian Journal of Arthropod-Borne Diseases*, *9*(2), 148. PMID:26623427

Shiue, I. (2015). Urinary heavy metals, phthalates and polyaromatic hydrocarbons independent of health events are associated with adult depression: USA NHANES, 2011–2012. *Environmental Science and Pollution Research International, 22*(21), 17095–17103. doi:10.100711356-015-4944-2 PMID:26126689

Shope, R. (1991). Global climate change and infectious diseases. *Environmental Health Perspectives, 96*, 171–174. doi:10.1289/ehp.9196171 PMID:1820262

Shotwell, O. L., Hesseltine, C. W., & Goulden, M. L. (1969). Ochratoxin A: Occurrence as natural contaminant of a corn sample. *Applied Microbiology, 17*, 765.

Shui-Ming, Z., Duo-Sen, L., Zong-Sheng, W., & Xing-Fa, M. (1993). A kinetic model describing the effect of temperature on the loss rate of pesticides in soil. *Ecological Modelling, 70*(1-2), 115–125. doi:10.1016/0304-3800(93)90075-4

Shuval, H. I., Adin, A., Fattal, B., Rawitz, E., & Yekutiel, P. (1986). *Wastewater Irrigation in Developing Countries: Health Effects and Technical Solutions*. World Bank Technical Paper Number 51, The World Bank.

Sian-Hülsmann, J., Mandel, S., Youdim, M. B. H., & Riederer, P. (2011). The relevance of iron in the pathogenesis of Parkinson's disease. *Journal of Neurochemistry, 118*(6), 939–957. doi:10.1111/j.1471-4159.2010.07132.x PMID:21138437

Siblerud, R. L., & Kienholz, E. (1994). Evidence that mercury from silver dental fillings may be an etiological factor in multiple sclerosis. *The Science of the Total Environment, 42*(3), 191–205. doi:10.1016/0048-9697(94)90327-1 PMID:8191275

Siblerud, R. L., Motl, J., & Kienholz, E. (1994). Psychometric evidence that mercury from silver dental fillings may be an etiological factor in depression, excessive anger, and anxiety. *Psychological Reports, 74*(36), 67–80. doi:10.2466/pr0.1994.74.1.67 PMID:8153237

Siddiqui, M. K., Jyoti, Singh, S., Mehrotra, P., Singh, K., & Sarangi, R. (2006). Comparison of some trace elements concentration in blood, tumor free breast and tumor tissues of women with benign and malignant breast lesions: An Indian study. *Environment International, 32*(5), 630–637. doi:10.1016/j.envint.2006.02.002

Sidhu, G. S. (2002). Mycotoxin genetics and gene clusters. *European Journal of Plant Pathology, 108*(7), 705–711. doi:10.1023/A:1020613413483

Sidoryk-Wegrzynowicz, M., & Aschner, M. (2013). Role of astrocytes in manganese mediated neurotoxicity. *BMC Pharmacology & Toxicology, 14*(1), 23. doi:10.1186/2050-6511-14-23 PMID:23594835

Siegel, R. L., Miller, K. D., & Jemal, A. (2017). Cancer statistics, 2017. *CA: a Cancer Journal for Clinicians, 67*(1), 7–30. doi:10.3322/caac.21387 PMID:28055103

Siegel, R., Naishadham, D., & Jemal, A. (2013). Cancer statistics, 2013. *CA: a Cancer Journal for Clinicians, 63*(1), 11–30. doi:10.3322/caac.21166 PMID:23335087

Signor, R. S., Roser, D. J., Ashbolt, N. J., & Ball, J. E. (2005). Quantifying the impact of runoff events on microbiological contaminant concentrations entering surface drinking source waters. *Journal of Water and Health, 3*(4), 453–468. doi:10.2166/wh.2005.052 PMID:16459849

Silva, M. P., Soave, D. F., Ribeiro-Silva, A., & Poletti, M. E. (2012). Trace elements as tumor biomarkers and prognostic factors in breast cancer: a study through energy dispersive x-ray fluorescence. *BMC Research Notes, 6.*

Silva-Pereira, L. C., Cardoso, P. C. dos S., Leite, D. S., Bahia, M. de O., Bastos, W. R., Smith, M. A. C., & Burbano, R. R. (2005). Cytotoxicity and genotoxicity of low doses of mercury chloride and methylmercury chloride on human lymphocytes in vitro. *Brazilian Journal of Medical and Biological Research, 38*(6), 901–907. doi:10.1590/S0100-879X2005000600012 PMID:15933784

Simpson, E. P., Henry, Y. K., Henkel, J. S., Smith, R. G., & Appel, S. H. (2004). Increased lipid peroxidation in sera of ALS patients: A potential biomarker of disease burden. *Neurology*, *62*(10), 1758–1765. doi:10.1212/WNL.62.10.1758 PMID:15159474

Simpson, S. L., Angel, B. M., & Jolley, D. F. (2004). Metal equilibration in laboratory-contaminated (spiked) sediments used for the development of whole-sediment toxicity tests. *Chemosphere*, *54*(5), 597–609. doi:10.1016/j.chemosphere.2003.08.007 PMID:14599505

Simpson, S. L., & Batley, G. E. (2003). Disturbances to metal partitioning during toxicity testing of iron(II)-rich estuarine pore waters and whole sediments. *Environmental Toxicology and Chemistry*, *22*(2), 424–432. doi:10.1002/etc.5620220225 PMID:12558176

Simpson, S. L., Pryor, I. D., Mewburn, B. R., Batley, G. E., & Jolley, D. (2002). Considerations for capping metal-contaminated sediments in dynamic estuarine environments. *Environmental Science & Technology*, *36*(17), 3772–3778. doi:10.1021/es025632v PMID:12322750

Sinert, R., Zehtabchi, S., Desai, S., Peacock, P., Altura, B. T., & Altura, B. M. (2007). Serum ionized magnesium and calcium levels in adult patients with seizures. *Scand J Clin Lab Invest*, 317–326. doi:10.1080/00365510601051441

Sinervo, B., & Adolph, S. C. (1994). Growth plasticity and thermal opportunity in Sceloporus lizards. *Ecology*, *75*(3), 776–790. doi:10.2307/1941734

Singh, A., & Fleurat, M. (2010). Pediatric emergency medicine practice acute gastroenteritis-an update. *Pediatric Emergency Medicine Practice*, *7*, 1–12.

Singh, J. S., & Gupta, S. R. (1977). Plant decomposition and soil respiration in terrestrial ecosystems. *Botanical Review*, *43*(4), 449–528. doi:10.1007/BF02860844

Singh, K. V. (1999). Studies on the role of climatological factors in the distribution of *Phlebotomine* sandflies (Diptera: Psychodidae) in semi-arid areas of Rajasthan, India. *Journal of Arid Environments*, *42*(1), 43–48. doi:10.1006/jare.1999.0499

Singh, S., Kaur, S., & Budhiraja, R. D. (2013). Chlorpyrifos-induced oxidative stress in rat's brain and protective effect of grape seed extract. *The Journal of Phytopharmacology*, *2*(3), 26–33.

Singh, V. K., Mehrotra, S., & Agarwal, S. S. (1999). The paradigm of Th1 and Th2 cytokines: Its relevance to autoimmunity and allergy. *Immunologic Research*, *20*(3), 147–161. doi:10.1007/BF02786470 PMID:10580639

Sinha, R. C., & Kanungo, M. S. (1967). Effect of starvation on the scorpion Palamnaeus bengalensis. *Physiological Zoology*, *40*(4), 386–390. doi:10.1086/physzool.40.4.30158456

Sinka, M. E., Bangs, M. J., Manguin, S., Chareonviriyaphap, T., Patil, A. P., Temperley, W. H., ... Hay, S. I. (2011). The dominant Anopheles vectors of human malaria in the Asia-Pacific region: Occurrence data, distribution maps and bionomic précis. *Parasites & Vectors*, *4*(1), 89. doi:10.1186/1756-3305-4-89

Sinka, M. E., Bangs, M. J., Manguin, S., Coetzee, M., Mbogo, C. M., Hemingway, J., ... Okara, R. M. (2010). The dominant Anopheles vectors of human malaria in Africa, Europe and the Middle East: Occurrence data, distribution maps and bionomic précis. *Parasites & Vectors*, *3*(1), 117. doi:10.1186/1756-3305-3-117

Sinka, M. E., Rubio-Palis, Y., Manguin, S., Patil, A. P., Temperley, W. H., Gething, P. W., ... Hay, S. I. (2010). The dominant Anopheles vectors of human malaria in the Americas: Occurrence data, distribution maps and bionomic précis. *Parasites & Vectors*, *3*(1), 72. doi:10.1186/1756-3305-3-72

Sin, S. N., Chua, H., Lo, W., & Ng, L. M. (2001). Assessment of heavy metal cations in sediments of Shing Mun River, Hong Kong. *Environment International*, *26*(5-6), 297–301. doi:10.1016/S0160-4120(01)00003-4 PMID:11392742

Sioen, I., Den Hond, E., Nelen, V., Van de Mieroop, E., Croes, K., Van Larebeke, N., ... Schoeters, G. (2013). Prenatal exposure to environmental contaminants and behavioural problems at age 7-8years. *Environment International, 59,* 225–231. doi:10.1016/j.envint.2013.06.014 PMID:23845936

Sisodia, R., & Moundiotiya, C. (2006). Assessment of the water quality index of wetland Kalakho Lake, Rajasthan, India. *J Environ Hydrology., 14*(23), 1.

Six, J., Bossuyt, H., Degryze, S., & Denef, K. (2004). A history of research on the link between (micro) aggregates, soil biota, and soil organic matter dynamics. *Soil & Tillage Research, 79*(1), 7–31. doi:10.1016/j.still.2004.03.008

Skoczyńska, A., Jedrejko, M., Martynowicz, H., Poreba, R., Affelska-Jercha, A., Steinmetz-Beck, A., ... Jedrychowska, I. (2010). The cardiovascular risk in chemical factory workers exposed to mercury vapor. *Medycyna Pracy, 61*(4), 381–391. Retrieved from http://www.ncbi.nlm.nih.gov/pubmed/20865852

Skrajnowska, D., Bobrowska-Korczak, B., Tokarz, A., Bialek, S., Jezierska, E., & Makowska, J. (2013). Copper and resveratrol attenuates serum catalase, glutathione peroxidase, and element values in rats with DMBA-induced mammary carcinogenesis. *Biological Trace Element Research, 156*(1-3), 271–278. doi:10.100712011-013-9854-x

Skyhøj Olsen. (2001). Post-stroke epilepsy. *Medicina, 78*(2), 86–90. doi:10.1016/j.neuint.2017.02.002 PMID:29659357

Slaoui, M., Mouh, F. Z., Ghanname, I., Razine, R., El Mzibri, M., & Amrani, M. (2016). Outcome of Breast Cancer in Moroccan Young Women Correlated to Clinic-Pathological Features, Risk Factors and Treatment: A Comparative Study of 716 Cases in a Single Institution. *PLoS One, 11*(10), e0164841. doi:10.1371/journal.pone.0164841

Slifko, T. R., Smith, H. V., & Rose, J. B. (2000). Emerging parasite zoonoses associated with water and food. *International Journal for Parasitology, 30*(12-13), 1379–1393. doi:10.1016/S0020-7519(00)00128-4 PMID:11113263

Smit, B., & Pilifosova, O. (2001). Adaptation to climate change in the context of sustainable development and equity. In Climate Change 2001: Impacts, Adaptation and Vulnerability. Cambridge University Press.

Smith, A., Reacher, M., Smerdon, W., Adak, G. K., Nichols, G., & Chalmers, R. M. (2006). Outbreaks of waterborne infectious intestinal disease in England and Wales, 1992–2003. *Epidemiology and Infection, 134*(06), 1141–1149. doi:10.1017/S0950268806006406 PMID:16690002

Smith, D. L., & McKenzie, F. E. (2004). Statics and dynamics of malaria infection in Anopheles mosquitoes. *Malaria Journal, 3*(1), 13. doi:10.1186/1475-2875-3-13

Smith, F. E., Scott, N. L., Keys, N., & Smith, F. T. (2017). The socio-economic vulnerability of the Australian east coast grazing sector to the impacts of climate change. *Regional Environmental Change,* 1185–1199.

Smith, J. E. Jr, Surampalli, R. Y., Reimers, R. S., Tyagi, R. D., & Lohani, B. N. (2008). Disinfection Processes and Stability Refinements to Biosolids Treatment Technologies. *Practice Periodical of Hazardous, Toxic, and Radioactive Waste Management, 12*(1), 10–17. doi:10.1061/(ASCE)1090-025X(2008)12:1(10)

Smith, L. M. (2003). *Playas of the Great Plains.* Austin, TX: University of Texas Press.

Smith, M. A., Richey Harris, P. L., Sayre, L. M., Beckman, J. S., & Perry, G. (1997). Widespread Peroxynitrite-Mediated Damage in Alzheimer's Disease. *The Journal of Neuroscience, 17*(8), 2653–2657. doi:10.1523/JNEUROSCI.17-08-02653.1997 PMID:9092586

Snedeker, S. M. (2001). Pesticides and breast cancer risk: A review of DDT, DDE, and dieldrin. *Environmental Health Perspectives, 109,* 35–47.

Snow, R. W., & Gilles, H. M. (2002). The epidemiology of malaria. *Essential Malariology, 4.*

Sobolev, D., & Begonia, M. F. T. (2008). Effects of Heavy Metal Contamination upon Soil Microbes : Lead-induced Changes in General and Denitrifying Microbial Communities as Evidenced by Molecular Markers. *International Journal of Environmental Research and Public Health*, *5*(5), 450–456. doi:10.3390/ijerph5050450 PMID:19151442

Soh, Y. C., Roddick, F., & van Leeuwen, J. (2008). The future of water in Australia: The potential effects of climate change and ozone depletion on Australian water quality, quantity and treatability. *The Environmentalist*, *28*(2), 158–165. doi:10.100710669-007-9123-7

Soliman, A. S., Wang, X., Digiovanni, J., Eissa, S., Morad, M., Vulimiri, S., ... Bondy, M. L. (2003). Serum organochlorine levels and history of lactation in Egypt. *Environmental Research*, *92*(2), 110–117. doi:10.1016/S0013-9351(02)00056-7

Soltani, Z., Fakoorziba, M. R., Moemenbellah-Fard, M. D., Kalantari, M., Akbarpour, M., Faramarzi, H., & Azizi, K. (2015). Phlebotomus papatasi (Diptera: Psychodidae) as the Vector of Leishmania Major in Kharameh District, Southern Iran. *Journal of Health Sciences and Surveillance System, 3*(4), 160-164.

Son, W. C., Kamino, K., Lee, Y. S., & Kang, K. S. (2003). Strain-specific mammary proliferative lesion development following lifetime oral administration of ochratoxin A in DA and Lewis rats. *International Journal of Cancer*, *105*(3), 305–311. doi:10.1002/ijc.11069

SOPAC. (2004). *The Environmental Vulnerability Index (EVI), Technical Report 384.* South Pacific Applied Geoscience Commission.

Soriano, J. M., González, L., & Catalá, A. I. (2005). Mechanism of action of sphingolipids and their metabolites in the toxicity of fumonisin B1. *Progress in Lipid Research*, *44*(6), 345–356. doi:10.1016/j.plipres.2005.09.001

Soto, A. M., Sonnenschein, C., Chung, K. L., Fernandez, M. F., Olea, N., & Serrano, F. O. (1995). The E-SCREEN assay as a tool to identify estrogens: An update on estrogenic environmental pollutants. *Environmental Health Perspectives*, *103*(suppl 7), 113–122. PMID:8593856

Souchon, Y., & Nicolas, V. (2011). *Barrages et seuils: principaux impacts environnementaux: Dynamique Physique.* ONEMA. Retrieved from www.irstea.fr/sites/default/files/

Spalding, M., Spalding, M. D., Ravilious, C., & Green, E. P. (2001). *World atlas of coral reefs*. Univ of California Press.

Spash, C. L. (2007). a. Climate change: Need for new economic thought. *Economic and Political Weekly*, 483–490.

Spash, C. L. (2007). b. The economics of climate change impacts à la Stern: Novel and nuanced or rhetorically restricted? *Ecological Economics*, *63*(4), 706–713. doi:10.1016/j.ecolecon.2007.05.017

Specchio, N., Fusco, L., Claps, D., & Vigevano, F. (2010). Epileptic encephalopathy in children possibly related to immune-mediated pathogenesis. *Brain & Development*, *32*(1), 51–56. doi:10.1016/j.braindev.2009.09.017 PMID:19850427

Spencer, D. C., & Forno, L. S. (2000). February 2000: Dementia with motor dysfunction in a patient with liver disease. *Brain Pathology (Zurich, Switzerland)*, *10*(2), 315–316, 319. doi:10.1111/j.1750-3639.2000.tb00205.x PMID:10764052

Spencer, P. S., Palmer, V. S., & Ludolph, A. C. (2005). On the decline and etiology of high-incidence motor system disease in West Papua (southwest New Guinea). *Movement Disorders*, *12*(S12), S119–S126. doi:10.1002/mds.20552 PMID:16092101

Squitieri, F., Frati, L., Ciarmiello, A., Lastoria, S., & Quarrell, O. (2006). Juvenile Huntington's disease: Does a dosage-effect pathogenic mechanism differ from the classical adult disease? *Mechanisms of Ageing and Development*, *127*(2), 208–212. doi:10.1016/j.mad.2005.09.012 PMID:16274727

Srebotnjak, T., Carr, G., De Sherbinin, A., & Rickwood, C. (2012). A global Water Quality Index and hot-deck imputation of missing data. *Ecological Indicators*, *17*, 108–119. doi:10.1016/j.ecolind.2011.04.023

Sritharan, J., MacLeod, J., Harris, S., Cole, D. C., Harris, A., Tjepkema, M., ... Demers, P. A. (2018). Prostate cancer surveillance by occupation and industry: The Canadian Census Health and Environment Cohort (CanCHEC). *Cancer Medicine*, *7*(4), 1468–1478. doi:10.1002/cam4.1358 PMID:29493883

Stabler, L. B., Martin, C. A., & Brazel, A. J. (2005). Microclimates in a desert city were related to land use and vegetation index. *Urban Forestry & Urban Greening*, *3*(3-4), 137–147. doi:10.1016/j.ufug.2004.11.001

Stafstrom, C. E., & Carmant, L. (2015). *Seizures and Epilepsy : An Overview for Neuroscientists*. Cold Spring Harbor Laboratory Press.

Stahl, R. G. Jr, Hooper, M. J., Balbus, J. M., Clements, W., Fritz, A., Gouin, T., ... Moe, S. J. (2013). The influence of global climate change on the scientific foundations and applications of Environmental Toxicology and Chemistry: Introduction to a SETAC international workshop. *Environmental Toxicology and Chemistry*, *32*(1), 13–19. doi:10.1002/etc.2037 PMID:23097130

Stallings, W. C., Metzger, A. L., Pattridge, K. A., Fee, J. A., & Ludwig, M. L. (1991). Structure-function relationships in iron and manganese superoxide dismutases. *Free Radical Research Communications*, *12-13*(Pt 1), 259–268. doi:10.3109/10715769109145794 PMID:2071033

Stanford, B. D., Wright, B., Routt, J. C., & Khan, S. J. (2014). *Water quality impacts of extreme weather-related events*. Alexandria, VA: Academic Press.

Stanford, J. L., & Ostrander, E. (2001). Familial prostate cancer. *Epidemiologic Reviews*, *23*(1), 19–23. doi:10.1093/oxfordjournals.epirev.a000789 PMID:11588848

Stanzani, M., Orciuolo, E., Lewis, R., Kontoyiannis, D. P., Martins, S. L. R., St John, L. S., & Komanduri, K. V. (2005). Aspergillus fumigatus suppresses the human cellular immune response via gliotoxin-mediated apoptosis of monocytes. *Blood*, *105*(6), 2258–2265. doi:10.1182/blood-2004-09-3421 PMID:15546954

Starzl, T. E., Porter, K. A., Francavilla, A., & McDermott, W. V. Jr. (1983). THE ECK FISTULA IN ANIMALS AND HUMANS. *Current Problems in Surgery*, *20*(11), 687–752. doi:10.1016/S0011-3840(83)80010-0 PMID:6357642

Stauder, M., Vezzulli, L., Pezzati, E., Repetto, B., & Pruzzo, C. (2010). Temperature affects *Vibrio cholerae* O1 El Tor persistence in the aquatic environment via an enhanced expression of GbpA and MSHA adhesins. *Environmental Microbiology Reports*, *2*(1), 140–144. doi:10.1111/j.1758-2229.2009.00121.x PMID:23766009

Stearns, S. C. (1976). Life-history tactics: A review of the ideas. *The Quarterly Review of Biology*, *51*(1), 3–47. doi:10.1086/409052 PMID:778893

Steele, J. C. (2005). Parkinsonism-dementia complex of Guam. *Movement Disorders*, *12*(S12), S99–S107. doi:10.1002/mds.20547 PMID:16092098

Steenland, K., MacNeil, J., Vega, I., & Levey, A. (2009). Recent trends in Alzheimer disease mortality in the United States, 1999 to 2004. *Alzheimer Disease and Associated Disorders*, *23*(2), 165–170. doi:10.1097/WAD.0b013e3181902c3e PMID:19484918

Steffens, K., Larsbo, M., Moeys, J., Jarvis, N., & Lewan, E. (2013). Predicting pesticide leaching under climate change: Importance of model structure and parameter uncertainty. *Agriculture, Ecosystems & Environment*, *172*, 24–34. doi:10.1016/j.agee.2013.03.018

Steffens, K., Larsbo, M., Moeys, J., Kjellström, E., Jarvis, N., & Lewan, E. (2014). Modelling pesticide leaching under climate change: Parameter vs. climate input uncertainty. *Hydrology and Earth System Sciences*, *18*(2), 479–491. doi:10.5194/hess-18-479-2014

Steinbruner, J. D., Stern, P. C., & Husbands, J. L. (2012). Climate and Social Stress: Implications for Security Analysis. National Research Council.

Stejskal, J., & Stejskal, V. D. (1999). The role of metals in autoimmunity and the link to neuroendocrinology. *Neuroendocrinology Letters*, *20*(6), 351–364. PMID:11458198

Stenrød, M., Perceval, J., Benoit, P., Almvik, M., Bolli, R. I., Eklo, O. M., ... Kværner, J. (2008). Cold climatic conditions: Effects on bioavailability and leaching of the mobile pesticide metribuzin in a silt loam soil in Norway. *Cold Regions Science and Technology*, *53*(1), 4–15. doi:10.1016/j.coldregions.2007.06.007

Stenseth, N. C., Leirs, H., Skonhoft, A., Davis, S. A., Pech, R. P., Andreassen, H. P., ... Wan, X. R. (2003). Mice, rats, and people: The bio-economics of agricultural rodent pests. *Frontiers in Ecology and the Environment*, *1*(7), 367–375. doi:10.1890/1540-9295(2003)001[0367:MRAPTB]2.0.CO;2

Stenseth, N. C., Mysterud, A., Ottersen, G., Hurrell, J. W., Chan, K. S., & Lima, M. (2002). Ecological effects of climate fluctuations. *Science*, *297*(5585), 1292–1296. doi:10.1126cience.1071281 PMID:12193777

Stenström, T. A., Seidu, R., Nelson, E., & Christian, Z. (2011). *Microbial Exposure and Health Assessments in Sanitation Technologies and Systems*. (Repéré à www.ecosanres.org

Stephenson, L. S. (2009). Nutritional and economic implications of soiltransmitted helminths with special reference to Ascariasis. In F. Lifshitz (Ed.), *Clinical disorders in pediatric gastroenterology and nutrition* (pp. 390–409). New York: Marcel Dekker.

Stephens, R., Spurgeon, A., Calvert, I. A., Beach, J., Levy, L. S., Harrington, J. M., & Berry, H. (1995). Neuropsychological effects of long-term exposure to organophosphates in sheep dip. *Lancet*, *345*(8958), 1135–1139. doi:10.1016/S0140-6736(95)90976-1 PMID:7723544

Sterk, A., Schijven, J., de Nijs, T., & de Roda Husman, A. M. (2013). Direct and indirect effects of climate change on the risk of infection by water-transmitted pathogens. *Environmental Science & Technology*, *47*(22), 12648–12660. doi:10.1021/es403549s PMID:24125400

Stevens, J. J., Graham, B., Walker, A. M., Tchounwou, P. B., & Rogers, C. (2010). The effects of arsenic trioxide on DNA synthesis and genotoxicity in human colon cancer cells. *International Journal of Environmental Research and Public Health*, *7*(5), 2018–2032. doi:10.3390/ijerph7052018 PMID:20623008

Stob, M., Baldwin, R. S., Tuite, J., Andrews, F. N., & Gillette, K. G. (1962). Isolation of an anabolic, uterotrophic compound from corn infected with *Gibberella zeae*. *Nature*, *196*(4861), 1318. doi:10.1038/1961318a0

Stocker, T. F., Qin, D., Plattner, G. K., Tignor, M., Allen, S. K., Boschung, J., . . . Midgley, P. M. (2013). IPCC, 2013: Climate Change 2013: The Physical Science Basis. Contribution of Working Group I to the Fifth Assessment Report of the Intergovernmental Panel on Climate Change.

Stocker, T. F. (2014). *Climate change 2013: The physical science basis:Working group I contribution to the fifth assessment report of the intergovernmental panel on climate change*. Cambridge University Press.

Stockmann, R., & Ythier, E. (2010). *Scorpions of the world*. N.A.P. Editions, Verrières-le- Buisson.

Stohs, S. J., & Bagchi, D. (1995). Oxidative mechanisms in the toxicity of metal ions. *Free Radical Biology & Medicine*, *18*(2), 321–336. doi:10.1016/0891-5849(94)00159-H PMID:7744317

Stoica, A., Katzenellenbogen, B. S., & Martin, M. B. (2000). Activation of estrogen receptor-alpha by theheavy metal cadmium. *Molecular Endocrinology (Baltimore, Md.)*, *14*(4), 545–553. PMID:10770491

Stolk, W. A., Kulik, M. C., le Rutte, E. A., Jacobson, J., Richardus, J. H., & de Vlas, S. J. (2016). Between-country inequalities in the neglected tropical disease burden in 1990 and 2010, with projections for 2020. *PLoS Neglected Tropical Diseases*, *10*(5), e0004560. doi:10.1371/journal.pntd.0004560

Stoltenburg-Didinger, G. (1994). *Neuropathology of the hippocampus and its susceptibility to neurotoxic insult.* Retrieved from https://sci-hub.tw/https://europepmc.org/abstract/med/7854577

Stour, L., & Agoumi, A. (2009). *Climatic drought in Morocco during the last decades [Sécheresse climatique au Maroc Sécheresse climatique au Maroc].* Paris: Hydroécologie Appliquée.

Strande, L., Ronteltap, M., & Brdjanovic, D. (2014). *Faecal sludge management: systems approach for implementation and operation* (1st ed.). London: IWA Publishing. doi:10.26530/OAPEN_578132

Stresman, G. H. (2010). Beyond temperature and precipitation: Ecological risk factors that modify malaria transmission. *Acta Tropica*, *116*(3), 167–172. doi:10.1016/j.actatropica.2010.08.005

Striepen, B. (2013). Time to tackle cryptosporidiosis. *Nature*, *503*, 189–191.

Strong, K., Mathers, C., & Bonita, R. (2007). Preventing stroke : Saving lives around the world. *Lancet Neurology*, *6*(February), 182–187. doi:10.1016/S1474-4422(07)70031-5 PMID:17239805

Struewing, J. P., Hartge, P., Wacholder, S., Baker, S. M., Berlin, M., McAdams, M., ... Tucker, M. A. (1997). The Risk of Cancer Associated with Specific Mutations of BRCA1 and BRCA2 among Ashkenazi Jews. *The New England Journal of Medicine*, *336*(20), 1401–1408. doi:10.1056/NEJM199705153362001 PMID:9145676

Strumylaite, L., Bogusevicius, A., Abdrachmanovas, O., Baranauskiene, D., Kregzdyte, R., Pranys, D., & Poskiene, L. (2011). Cadmium concentration in biological media of breast cancer patients. *Breast Cancer Research and Treatment*, *125*(2), 511–517. doi:10.100710549-010-1007-8

Strunz, E. C., Addiss, D. G., Stocks, M. E., Ogden, S., Utzinger, J., & Freeman, M. C. (2014). Water, sanitation, hygiene, and soil-transmitted helminth infection: Asystematic review and meta-analysis. *PLoS Medicine*, *11*(3), e1001620. doi:10.1371/journal.pmed.1001620 PMID:24667810

Struve, M. F., McManus, B. E., Wong, B. A., & Dorman, D. C. (2007). Basal ganglia neurotransmitter concentrations in rhesus monkeys following subchronic manganese sulfate inhalation. *American Journal of Industrial Medicine*, *50*(10), 772–778. doi:10.1002/ajim.20489 PMID:17620281

Struzyiqska, L., & Gadamski, R. (1997). Lead-Induced Abnormalities in Blood-Brain Barrier Permeability in Experimental Chronic Toxicity. *Molecular and Chemical Neuropathology*, *31*(3), 207–224. doi:10.1007/BF02815125 PMID:9336764

Stutz, W. H., Bendeck, O., Macedo, E. M., Camargo, J. C. C., Oliveira, F. S., & Bonito, R. F. (2008). Bioensaio Visando Controle de Escorpionídeos (Tityus serrulatus), através do uso de Bendiocarb. *Deltamethrin e Lambda-cyhalothrin, 2003.*

Subrahmanyam, K., & Yadaiah, Y. (2001). - Assessment of the impact of industrial effluents on water quality in Patancheru and environs, Medak district, Andhra Pradesh, India. *Hydrology Journal*, *9*(3), 297.

Sugita, M. (1978). The biological half-time of heavy metals. *International Archives of Occupational and Environmental Health*, *41*(1), 25–40. doi:10.1007/BF00377797 PMID:627414

Suk, J. E., & Semenza, J. C. (2011). Future infectious disease threats to Europe. *American Journal of Public Health*, *101*(11), 2068–2079. doi:10.2105/AJPH.2011.300181 PMID:21940915

Su, L., Blamire, A. M., Watson, R., He, J., Hayes, L., & O'Brien, J. T. (2016). Whole-brain patterns of 1H-magnetic resonance spectroscopy imaging in Alzheimer's disease and dementia with Lewy bodies. *Translational Psychiatry*, *6*(8), e877–e877. doi:10.1038/tp.2016.140 PMID:27576166

Sullivan, A. C., & Meigh, J. (2007). Integration of the biophysical and social sciences using an indicator approach: Addressing water problems at different scales. *Water Resources Management*, 111–128. doi:10.100711269-006-9044-0

Su, M., Wakabayashi, K., Kakita, A., Ikuta, F., & Takahashi, H. (1998). Selective involvement of large motor neurons in the spinal cord of rats treated with methylmercury. *Journal of the Neurological Sciences*, *156*(1), 12–17. doi:10.1016/S0022-510X(98)00030-6 PMID:9559981

Summerskill, W. H., Davidson, E. A., Sherlock, S., & Steiner, R. E. (1956). The neuropsychiatric syndrome associated with hepatic cirrhosis and an extensive portal collateral circulation. *The Quarterly Journal of Medicine*, *25*(98), 245–266. PMID:13323252

Sunkireddy, P., Sriramoju, B., Roy, K., Kanwar, R. K., & Kanwar, J. R. (2018). Oxidative Stress in Ocular Disorders: Exploring the Link to Pesticide Exposure and Potential for Using Nanotechnology for Antioxidant Delivery. In Drug Delivery for the Retina and Posterior Segment Disease (pp. 151–167). Springer.

Sun, X. D., Su, P., & Shan, H. (2017). Mycotoxin contamination of rice in China. *Journal of Food Science*, *82*(3), 573–584. doi:10.1111/1750-3841.13631

Sun, X., Zhang, T., Deng, Q., Zhou, Q., Sun, X., Li, E., ... Zhong, C. (2018). Benzidine Induces Epithelial–Mesenchymal Transition of Human Bladder Cancer Cells through Activation of ERK5 Pathway. *Molecules and Cells*, *41*(3), 188. PMID:29463068

Supothina, S., Isaka, M., Kirtikara, K., Tanticharoen, M., & Thebtaranonth, Y. (2004). Enniatin production by the entomopathogenic fungus Verticillium hemipterigenum BCC 1449. *J. Antibiot.*, *57*(11), 732–738. doi:10.7164/antibiotics.57.732

Sutherst, R. B. (2004). Global Change and Human Vulnerability to Vector-Borne Diseases. *Clinical Microbiology Reviews*, 136–173.

Sutherst, R. W. (1998). Implications of global change and climate variability for vector-borne diseases: Generic approaches to impact assessments. *International Journal for Parasitology*, *28*(6), 935–945. doi:10.1016/S0020-7519(98)00056-3 PMID:9673872

Svenson, J. E., MacLean, J. D., Gyorkos, T. W., & Keystone, J. (1995). Imported malaria: Clinical presentation and examination of symptomatic travelers. *Archives of Internal Medicine*, *155*(8), 861–868. doi:10.1001/archinte.1995.00430080109013

Swart, R. & Biesbroek, R. (2008) *Adaptation of infrastructure to climate change: International inventory.* University and Research Centre Alterra/Wageningen.

Szostakowska, B., Kruminis-Lozowska, W., Racewicz, M., Knight, R., Tamang, L., Myjak, P., & Graczyk, T. K. (2004). Cryptosporidium parvum and Giardia lamblia recovered from flies on a cattle farm and in a landfill. *Applied and Environmental Microbiology*, *70*(6), 3742–3744. doi:10.1128/AEM.70.6.3742-3744.2004 PMID:15184182

Szuets, P., Mesterhazy, A., Falkay, G. Y., & Bartok, T. (1997). Early telarche symptoms in children and their relations to zearalenon contamination in foodstuffs. *Cereal Research Communications*, 429–436.

Tago, D., Andersson, H., & Treich, N. (2014). *Preference Measurement in Health Article information : Users who downloaded this article also downloaded.* Preference Measurement in Health.

Takken, W., & Scott, T. W. (Eds.). (2003). *Ecological aspects for application of genetically modified mosquitoes* (Vol. 2). Springer Science & Business Media.

Talbi, F. Z., El Ouali Lalami, A., Janati Idrissi, A., Sebti, F., & Faraj, C. (2015). Leishmaniasis in central Morocco: Seasonal fluctuations of phlebotomine sand fly in Aichoun locality, from Sefrou province. *Pathology Research International*. PMID:25741448

Tamburrini, A., & Pozio, E. (1999). Long-term survival of Cryptosporidium parvum oocysts in sea water and in experimentally infected mussels (Mytilus galloprovincialis). *International Journal for Parasitology*, *29*(5), 711–715. doi:10.1016/S0020-7519(99)00033-8 PMID:10404265

Tamimy, H. (2011). *The infantile visceral leishmaniasis (About 73 cases)*. PhD in Medicine. Sidi Mohammed Ben Abdellah University, Faculty of Medicine and Pharmacy. Available from: http://scolarite.fmp-usmba.ac.ma/cdim/mediatheque/e_theses/89-11.pdf

Tan, J., Zhang, T., Jiang, L., Chi, J., Hu, D., Pan, Q., … Zhang, Z. (2011). Regulation of intracellular manganese hemeostasis by kufor-rakeb syndrome associated ATP13A2. *Journal of Biological Chemistry*. doi:10.1074/jbc.M111.233874

Tan, C., & Chen, H. (2011). Screening of prostate cancer by analyzing trace elements in hair and chemometrics. *Biological Trace Element Research*, *144*(1-3), 97–108. doi:10.100712011-011-9038-5 PMID:21452047

Tan, N. H., Arunmozhiarasi, A., & Ponnudurai, G. (1991). A comparative study of the biological properties of Dendroaspis (mamba) snake venoms. *Comparative Biochemistry and Physiology. C. Comparative Pharmacology and Toxicology*, *99*(3), 463–466. doi:10.1016/0742-8413(91)90272-U PMID:1685421

Tanner, C. M., Kamel, F., Ross, G. W., Hoppin, J. A., Goldman, S. M., Korell, M., ... Langston, J. W. (2011). Rotenone, Paraquat, and Parkinson's Disease. *Environmental Health Perspectives*, *119*(6), 866–872. doi:10.1289/ehp.1002839 PMID:21269927

Tanner, C. M., Ottman, R., Goldman, S. M., Ellenberg, J., Chan, P., Mayeux, R., & Langston, J. W. (1999). Parkinson disease in twins: An etiologic study. *Journal of the American Medical Association*, *281*(4), 341–346. doi:10.1001/jama.281.4.341 PMID:9929087

Tansey, M. G., & Goldberg, M. S. (2010). Neuroinflammation in Parkinson's disease: Its role in neuronal death and implications for therapeutic intervention. *Neurobiology of Disease*, *37*(3), 510–518. doi:10.1016/j.nbd.2009.11.004 PMID:19913097

Tantaoui-Elaraki, A., Riba, A., Oueslati, S., & Zinedine, A. (2018). Toxigenic fungi and mycotoxin occurrence and prevention in food and feed in northern Africa – a review. Special issue: "Mycotoxins in Africa. *World Mycotoxin Journal*, *11*(3), 385–400. doi:10.3920/WMJ2017.2290

Tanya, S., Chutima, R., Andre, F. C., Michel, B., & Michael, M. (2018). Anxiety disorders: Sex differences in serotonin and tryptophan metabolism. *Current Topics in Medicinal Chemistry*, 1–46. doi:10.20944/preprints201808.0487.v1 PMID:30430940

Tanzi, R. E., Petrukhin, K., Chernov, I., Pellequer, J. L., Wasco, W., Ross, B., ... Brzustowicz, L. M. (1993). The Wilson disease gene is a copper transporting ATPase with homology to the Menkes disease gene. *Nature Genetics*, *5*(4), 344–350. doi:10.1038/ng1293-344 PMID:8298641

Tao, Z., Shi, A., Lu, C., Song, T., Zhang, Z., & Zhao, J. (2015). Breast Cancer: Epidemiology and Etiology. *Cell Biochemistry and Biophysics*, *72*(2), 333–338. doi:10.100712013-014-0459-6

Tattersall, J. (2009). Biomedical Sciences Department, Dstl Porton Down, Salisbury, Wiltshire, SP4 0JQ, UK. *Frontiers in Bioscience*, *14*, 3688–3711. doi:10.2741/3481

Tavtigian, S. V., Simard, J., Rommens, J., Couch, F., Shattuck-Eidens, D., Neuhausen, S., ... Goldgar, D. E. (1996). The complete BRCA2 gene and mutations in chromosome 13q-linked kindreds. *Nature Genetics*, *12*(3), 333–337. doi:10.1038/ng0396-333

Taylor, P., Rinsky, J. L., Hoppin, J. A., Blair, A., He, K., & Freeman, L. E. B. (2013). Article. *Journal of Toxicology and Environmental Health, Part A: Current Issues*, 37–41. doi:10.1080/15287394.2013.819308

Teklehaimanot, H. D., Schwartz, J., Teklehaimanot, A., & Lipsitch, M. (2004). Alert threshold algorithms and malaria epidemic detection. *Emerging Infectious Diseases*, *10*(7), 1220–1226. doi:10.3201/eid1007.030722 PMID:15324541

Teresi, J. A., Abrams, R., Holmes, D., Ramirez, M., Shapiro, C., & Eimicke, J. P. (2002). Influence of cognitive impairment, illness, gender, and African-American status on psychiatric ratings and staff recognition of depression. *The American Journal of Geriatric Psychiatry*, *10*(5), 506–514. doi:10.1097/00019442-200209000-00003 PMID:12213684

Ternes, T. A., Bonerz, M., Herrmann, N., Teiser, B., & Andersen, H. R. (2007). Irrigation of treated wastewater in Braunschweig, Germany: An option to remove pharmaceuticals and musk fragrances. *Chemosphere*, *66*(5), 894–904. doi:10.1016/j.chemosphere.2006.06.035 PMID:16872661

Terracini, B. (2011). Epidemiology of childhood cancer. *Environmental Health*, *10*(Suppl 1), S8. doi:10.1186/1476-069X-10-S1-S8 PMID:21489218

Tessier, A., Campbell, P. G. C., & Bisson, M. (1979). Sequential extraction procedure for the speciation of particulate trace metals. *Analytical Chemistry*, *51*(7), 844–851. doi:10.1021/ac50043a017

Thakur, J. S., Rao, B. T., Rajwanshi, A., Parwana, H. K., & Kumar, R. (2008). Epidemiological study of high cancer among rural agricultural community of Punjab in Northern India. *International Journal of Environmental Research and Public Health*, *5*(5), 399–407. doi:10.3390/ijerph5050399 PMID:19151435

The United Nations Framework Convention on Climate Change Portal. (n.d.). Retrieved from http://unfccc.int

Theorell, T., Hammarström, A., Aronsson, G., Träskman Bendz, L., Grape, T., Hogstedt, C., ... Hall, C. (2015). A systematic review including meta-analysis of work environment and depressive symptoms. *BMC Public Health*, *15*(1), 738. doi:10.118612889-015-1954-4 PMID:26232123

Theumer, M. G., Henneb, Y., Khoury, L., Snini, S. P., Tadrist, S., Canlet, C., ... Audebert, M. (2018). Genotoxicity of aflatoxins and their precursors in human cells. *Toxicology Letters*, *287*, 100–107. doi:10.1016/j.toxlet.2018.02.007

Thiruchelvam, M., Richfield, E. K., Baggs, R. B., Tank, A. W., & Cory-Slechta, D. A. (2000). The nigrostriatal dopaminergic system as a preferential target of repeated exposures to combined paraquat and maneb: Implications for Parkinson's disease. *The Journal of Neuroscience*, *20*(24), 9207–9214. doi:10.1523/JNEUROSCI.20-24-09207.2000 PMID:11124998

Thiruchelvam, M., Richfield, E. K., Goodman, B. M., Baggs, R. B., & Cory-Slechta, D. A. (2002). Developmental Exposure to the Pesticides Paraquat and Maneb and the Parkinson's Disease Phenotype. *Neurotoxicology*, *23*(4–5), 621–633. doi:10.1016/S0161-813X(02)00092-X PMID:12428734

Thomas, K. M., Charron, D. F., Waltner-Toews, D., Schuster, C., Maarouf, A. R., & Holt, J. D. (2006). A role of high impact weather events in waterborne disease outbreaks in Canada, 1975 - 2001. *International Journal of Environmental Health Research*, *16*(3), 167–180. doi:10.1080/09603120600641326 PMID:16611562

Thomas, R. H. (2012). seizures and Epilepsy: Pathophysiology and principles of Diagnosis. *Epilepsy*, *1*, 28–42. doi:10.1016/j.engfracmech.2017.04.040

Thompson, H. J., Wilson, A., Lu, J., Singh, M., Jiang, C., Upadhyaya, P., ... Ip, C. (1994). Comparison of the effects of an organic and an inorganic form of selenium on a mammary carcinoma cell line. *Carcinogenesis*, *15*(2), 183–186. doi:10.1093/carcin/15.2.183 PMID:8313506

Thompson, R. C. A. (2004). The zoonotic significance and molecular epidemiology of Giardia and giardiasis. *Veterinary Parasitology*, *126*(1-2), 15–35. doi:10.1016/j.vetpar.2004.09.008 PMID:15567577

Thompson, R., & Monis, P. (2012). Giardia – from genome to proteome. *Advances in Parasitology*, *78*, 57–95. doi:10.1016/B978-0-12-394303-3.00003-7 PMID:22520441

Thomson, M. C., Mason, S. J., Phindela, T., & Connor, S. J. (2005). Use of rainfall and sea surface temperature monitoring for malaria early warning in Botswana. *The American Journal of Tropical Medicine and Hygiene*, *73*(1), 214–221. doi:10.4269/ajtmh.2005.73.214

Thornton, P. K., van de Steeg, J., Notenbaert, A., & Herrero, M. (2009). The impacts of climate change on livestock and livestock systems in developing countries: A review of what we know and what we need to know. *Agricultural Systems*, *101*(3), 113–127. doi:10.1016/j.agsy.2009.05.002

Thurston-Enriquez, J. A., Gilley, J. E., & Eghball, B. (2005). Microbial quality of runoff following land application of cattle manure and swine slurry. *Journal of Water and Health*, *3*(2), 157–171. doi:10.2166/wh.2005.0015 PMID:16075941

Tilman, D., Fargione, J., Wolff, B., D'antonio, C., Dobson, A., Howarth, R., ... Swackhamer, D. (2001). Forecasting agriculturally driven global environmental change. *Science*, *292*(5515), 281–284. doi:10.1126cience.1057544 PMID:11303102

Tinfo, N. S., Hotchkiss, M. G., Buckalew, A. R., Zorrilla, L. M., Cooper, R. L., & Laws, S. C. (2011). Understanding the effects of atrazine on steroidogenesis in rat granulosa and H295R adrenal cortical carcinoma cells. *Reproductive Toxicology (Elmsford, N.Y.)*, *31*(2), 184–193. doi:10.1016/j.reprotox.2010.11.005

Tinoco-Veras, C. M., Bezerra Sousa, M. S., da Silva, B. B., Franciscato Cozzolino, S. M., Viana Pires, L., Coelho Pimentel, J. A., ... do Nascimento-Marreiro, D. (2011). Analysis of plasma and erythrocyte zinc levels in premenopausal women with breast cancer. *Nutrición Hospitalaria*, *26*, 293–297.

Tirado, M. C., Clarke, R., Jaykus, L. A., McQuatters-Gollop, A., & Frank, J. M. (2010). Climate change and food safety:A review. *Food Research International*, *43*(7), 1745–1765. doi:10.1016/j.foodres.2010.07.003

To, P.K., Do, M.H., Cho, Y.S., Kwon, S.Y., Kim, M.S., &Jung, C., (2018). Zinc Inhibits Expression of Androgen Receptor to Suppress Growth of Prostate Cancer Cells. *International Journal of Molecular Sciences*, *19*(10).

Toews, A. D., Kolber, A., Hayward, J., Krigman, M. R., & Morell, P. (1978). Experimental animals Isolation of capillaries and microvessels. *North-Holland Biomedical Press*, *147*, 131–138.

Tola, M., & Kebede, B. (2016). Occurrence, importance and control of mycotoxins: A review. *Cogent Food & Agriculture*, *2*(1), 1191103. doi:10.1080/23311932.2016.1191103

Tollis, D. (1996). Who was Charcot? *Nursing Times*, *92*, 56–57. PMID:8710677

Tomaszewski, J., Miturski, R., Semczuk, A., Kotarski, J., & Jakowicki, J. (1998). Tissue zearalenone concentration in normal, hyperplastic and neoplastic human endometrium. *Ginekologia Polska*, *69*, 363–366.

Tomoda, H., Huang, X.-H., Cao, J., Nishida, H., Nagao, R., Okuda, S., ... Inoue, K. (1992). Inhibition of acyl-CoA: Cholesterol acyltransferase activity by cyclodepsipeptide antibiotics. *J. Antibiot.*, *45*(10), 1626–1632. doi:10.7164/antibiotics.45.1626

Toppari, J., Larsen, J. C., Christiansen, P., Giwercman, A., Grandjean, P., Guillette, L. J., ... Skakkebaek, N. E. (1996). Male reproductive health and environmental xenoestrogens. *Environmental Health Perspectives*, *104*(suppl 4), 741–803. PMID:8880001

Tore, K., Bergdahl, I. A., Nilsson, T., & Ja, B. (2007). Occupational exposure to particulate air pollution and mortality due to ischaemic heart disease and cerebrovascular disease. *Occupational and Environmental Medicine*. doi:10.1136/oem.2006.029488 PMID:17303673

Toren, T. J. (1973). *Biology of the California Coast Range scorpion Vaejovis gertschi striatus Hjelle* (Unpublished Masters thesis). California State University at San Francisco.

Torres, A. D., Rai, A. N., & Hardiek, M. L. (2000). Mercury Intoxication and Arterial Hypertension : Report of Two Patients and Review of the Literature. *Pediatrics*, *105*(3), e34. doi:10.1542/peds.105.3.e34 PMID:10699136

Toumi, A., Chlif, S., Bettaieb, J., Alaya, N. B., Boukthir, A., Ahmadi, Z. E., & Salah, A. B. (2012). Temporal dynamics and impact of climate factors on the cases of zoonotic cutaneous Leishmaniasis in central Tunisia. *PLoS Neglected Tropical Diseases*, *6*(5), e1633. doi:10.1371/journal.pntd.0001633 PMID:22563513

Toz, S. (1999). PCR and the detection of microbial pathogens in water and wastewater. *Water Research*, *33*(17), 3545–3556. doi:10.1016/S0043-1354(99)00071-8

Trari, B., & Carnevale, P. (2011). De la préélimination à l'élimination du paludisme au Maroc. Quels risques pour l'avenir? *Bulletin de la Société de Pathologie Exotique*, *104*(4), 291–295. doi:10.100713149-011-0156-2 PMID:21638201

Trenberth, K. E. (2001). Climate variability and global warming. *Science*, *293*(5527), 48–49. doi:10.1126cience.293.5527.48

Trottier, Y., Biancalana, V., & Mandel, J. L. (1994). Instability of CAG repeats in Huntington's disease: Relation to parental transmission and age of onset. *Journal of Medical Genetics*, *31*(5), 377–382. doi:10.1136/jmg.31.5.377 PMID:8064815

Trottier, Y., Waithe, W. I., & Anderson, A. (1992). Kinds of mutations induced by aflatoxin B1 in a shuttle vector replicating in human cells transiently expressing cytochrome P450IA2 cDNA. *Molecular Carcinogenesis*, *6*(2), 140–147. doi:10.1002/mc.2940060209

Trpis, M. (1972). Development and predatory behavior of Toxorhynchites brevipalpis (Diptera: Culicidae) in relation to temperature. *Environmental Entomology*, *1*(5), 537–546. doi:10.1093/ee/1.5.537

Tsai, S. M., Wang, T. N., & Ko, Y. C. (1999). Mortality for certain diseases in areas with high levels of arsenic in drinking water. *Archives of Environmental Health*, *54*(3), 186–193. doi:10.1080/00039899909602258 PMID:10444040

Tudzynski, P., & Neubauer, L. (2014). Biosynthesis and Molecular Genetics of Fungal Secondary Metabolites. Academic Press. doi:10.1007/978-1-4939-1191-2_14

Tu, J. V. (2010). Reducing the global burden of stroke : INTERSTROKE Public health research funding : independence is important The Diet and Health Research Industry Club (DRINC) is. *Lancet*, *376*(9735), 74–75. doi:10.1016/S0140-6736(10)60975-0 PMID:20561674

Túnez, I., Drucker-Colín, R., Jimena, I., Medina, F. J., Muñoz Mdel, C., Peña, J., & Montilla, P. (2006). Transcranial magnetic stimulation attenuates cell loss and oxidative damage in the striatum induced in the 3-nitropropionic model of Huntington's disease. *Journal of Neurochemistry*, *97*(3), 619–630. doi:10.1111/j.1471-4159.2006.03724.x PMID:16524377

Tun-Lin, W., Burkot, T. R., & Kay, B. H. (2000). Effects of temperature and larval diet on development rates and survival of the dengue vector Aedes aegypti in north Queensland, Australia. *Medical and Veterinary Entomology*, *14*(1), 31–37. doi:10.1046/j.1365-2915.2000.00207.x

Turner, E. L., Lane, J. A., Donovan, J. L., Davis, M. J., Metcalfe, C., Neal, D. E., ... Martin, R. M. (2011). Association of diabetes mellitus with prostate cancer: Nested case-control study (Prostate testing for cancer and Treatment study). *International Journal of Cancer*, *128*(2), 440–446. doi:10.1002/ijc.25360 PMID:20473853

Turner, M. C., Wigle, D. T., & Krewski, D. (2010). Residential Pesticides and Childhood Leukemia: A Systematic Review and Meta-Analysis. *Environmental Health Perspectives*, *118*(1), 33–41. doi:10.1289/ehp.0900966 PMID:20056585

Turner, P. C., Collinson, A. C., Cheung, Y. B., Gong, Y., Hall, A. J., Prentice, A. M., & Wild, C. P. (2007). Aflatoxin exposure in utero causes growth faltering in Gambian infants. *International Journal of Epidemiology*, *36*(5), 1119–1125. doi:10.1093/ije/dym122 PMID:17576701

Tuteja, N., & Gill, S. S. (2016). *Abiotic Stress Response in Plants*. John Wiley & Sons. doi:10.1002/9783527694570

Ueda, M., Imai, T., Takizawa, T., Onodera, H., Mitsumori, K., Matsui, T., & Hirose, M. (2005). Possible Enhancing Effects of Atrazine on Growth of 7, 12-Dimethylbenz(a) Anthracene-Induced Mammary Tumors in Ovariectomized Sprague-Dawley Rats. *Cancer Science*, *96*(1), 19–25. doi:10.1111/j.1349-7006.2005.00008.x

Uetz, P., Freed, P., & Hošek, J. (2018). *The Reptile Database*. Retrieved from http://www.reptile-database.org

Uhlig, S., Ivanova, L., Petersen, D., & Kristensen, R. (2009). Structural studies on minor enniatins from Fusarium sp. VI 03441: Novel N-methyl-threonine containing enniatins. *Toxicon*, *53*(7-8), 734–742. doi:10.1016/j.toxicon.2009.02.014

Uhlig, S., Torp, M., & Heier, B. T. (2006). Beauvericin and enniatin A, A1, B and B1 in Norwegian grain: A survey. *Food Chemistry*, *94*(2), 193–201. doi:10.1016/j.foodchem.2004.11.004

Ukraintseva, Sloan, Arbeev, & Yashin. (2015). *Increasing Rates of Dementia at Time of Declining Mortality From Stroke*. American Heart Association. doi:10.1161/01.STR.0000217971.88034.e9

Ullah, R., Malik, R. N., & Qadir, A. (2009). Assessment of groundwater contamination in an industrial city, Sialkot, Pakistan. *African Journal of Environmental Science and Technology*, *3*(12), 429.

Ulrich, A. C., Guigard, S. E., Foght, J. M., Semple, K. M., Pooley, K., Armstrong, J. E., & Biggar, K. W. (2009). Effect of salt on aerobic biodegradation of petroleum hydrocarbons in contaminated groundwater. *Biodegradation*, *20*(1), 27–38. doi:10.100710532-008-9196-0 PMID:18437506

Ulukanligil, M., Seyrek, A., Aslan, G., Ozbilge, H., & Atay, S. (2001). Environmental pollution with soil-transmitted helminthes in Sanliurfa, Turkey. *Memorias do Instituto Oswaldo Cruz*, *96*(7), 903–909. doi:10.1590/S0074-02762001000700004 PMID:11685253

Unc, A., & Goss, M. J. (2003). Movement of faecal bacteria through the Vadose zone. *Water, Air, and Soil Pollution*, *149*(1/4), 327–337. doi:10.1023/A:1025693109248

Underwood, G. (1979). Classification and distribution of venomous snakes in the world. In *Snake venoms* (pp. 15–40). Berlin: Springer. doi:10.1007/978-3-642-66913-2_2

Underwood, G., & Kochva, E. (1993). On the affinities of the burrowing asps Atractaspis (Serpentes: Atractaspididae). *Zoological Journal of the Linnean Society*, *107*(1), 3–64. doi:10.1111/j.1096-3642.1993.tb01252.x

UNFCCC. (1992). *United Nations framework convention on climate change*. New York: United Nations.

Uppala, P. T., Roy, S. K., Tousson, A., Barnes, S., Uppala, G. R., & Eastmond, D. A. (2005). Induction of cell proliferation, micronuclei and hyperdiploidy/polyploidy in the mammary cells of DDT- and DMBA-treated pubertal rats. *Environmental and Molecular Mutagenesis*, *46*(1), 43–52. doi:10.1002/em.20131

Urdinguio, R. G., Sanchez-Mut, J. V., & Esteller, M. (2009). Epigenetic mechanisms in neurological diseases: Genes, syndromes, and therapies. *Lancet Neurology*, *8*(11), 1056–1072. doi:10.1016/S1474-4422(09)70262-5 PMID:19833297

Uren, A. G., Beilharz, T., O'Connell, M. J., Bugg, S. J., van Driel, R., Vaux, D. L., & Lithgow, T. (1999). Role foryeast inhibitor of apoptosis (IAP)-like proteins in cell division. *Proceedings of the National Academy of Sciences of the United States of America*, *96*(18), 10170–10175. doi:10.1073/pnas.96.18.10170 PMID:10468581

Urry, W. H., Wehrmeister, H. L., Hodge, E. B., & Hidy, P. H. (1966). The structure of zearalenone. *Tetrahedron Letters*, *7*(27), 3109–3114. doi:10.1016/S0040-4039(01)99923-X

US Environmental. (2007), *What is a pesticide?*. epa.gov.

Valenta, J., Stach, Z., & Svítek, M. (2010). Acute pancreatitis after viperid snake cerastescerastes envenoming: A case report. *Prague Medical Report*, *111*(1), 69–75. PMID:20359440

Van Berkum, F. H., Huey, R. B., & Adams, B. A. (1986). Physiological consequences of thermoregulation in a tropical lizard (Ameiva festiva). *Physiological Zoology*, *59*(4), 464–472. doi:10.1086/physzool.59.4.30158599

Van Broeckhoven, C. L. (1995). Molecular genetics of Alzheimer disease: Identification of genes and gene mutations. *European Neurology*, *35*(1), 8–19. doi:10.1159/000117083 PMID:7737252

Van der Cruijsen-Koeter, I. W., Vis, A. N., Roobol, M. J., Wildhagen, M. F., de Koning, H. J., van der Kwast, T. H., & Schroder, F. (2005). Comparison of screen detected and clinically diagnosed prostate cancer in the European randomized study of screening for prostate cancer, section Rotterdam. *The Journal of Urology*, *174*(1), 121–125. doi:10.1097/01.ju.0000162061.40533.0f PMID:15947595

Van der Merwe, K. J., Steyne, P. S., Fourie, L. F., Scott, D. B., & Theron, J. J. (1965). Ochratoxin A, a toxic metabolite produced by Aspergillus ochraceus Wilh. *Nature*, *205*(4976), 1112–1113. doi:10.1038/2051112a0

Van der Merwe, K. J., Steyn, P. S., & Fourie, L. (1965). Mycotoxins. II. The constitution of ochratoxins A, B, and C, metabolites of *Aspergillus ochraceus* Wilh. *Journal of the Chemical Society, Perkin Transactions 1: Organic and Bio-Organic Chemistry*, 7083–7088.

Van Der Valk, P., & De Groot, C. J. (2000). Staging of multiple sclerosis (MS) lesions: Pathology of the time frame of MS. *Neuropathology and Applied Neurobiology*, *26*(1), 2–10. doi:10.1046/j.1365-2990.2000.00217.x PMID:10736062

Van Egmond, H. P. (1989). *Mycotoxins in dairy products*. London: Elsevier Applied Science.

Van Lieshout, M., Kovats, R. S., Livermore, M. T. J., & Martens, P. (2004). Climate change and malaria: Analysis of the SRES climate and socio-economic scenarios. *Global Environmental Change*, *14*(1), 87–99. doi:10.1016/j.gloenvcha.2003.10.009

Van Maele-Fabry, G., Gamet-Payrastre, L., & Lison, D. (2018). Household exposure to pesticides and risk of leukemia in children and adolescents: Updated systematic review and meta-analysis. *International Journal of Hygiene and Environmental Health*. PMID:30268646

Van Maele-Fabry, G., Hoet, P., & Lison, D. (2013). Parental occupational exposure to pesticides as risk factor for brain tumors in children and young adults: A systematic review and meta-analysis. *Environment International*, *56*, 19–31. doi:10.1016/j.envint.2013.02.011 PMID:23567326

Van Maele-Fabry, G., Lantin, A.-C., Hoet, P., & Lison, D. (2010). Childhood leukaemia and parental occupational exposure to pesticides: A systematic review and meta-analysis. *Cancer Causes & Control: CCC*, *21*(6), 787–809. doi:10.100710552-010-9516-7 PMID:20467891 van Vliet, M. T. H., & Zwolsman, J. J. G. (2008). Impact of summer droughts on the water quality of the Meuse river. *Journal of Hydrology (Amsterdam)*, *353*(1-2), 1–17. doi:10.1016/j.jhydrol.2008.01.001

Vanloon, G. W., & Duffy, S. J. (2009). The Hydrosphere. In *Environmental Chemistry: A Global Perspective* (2nd ed.; pp. 197–211). New York: Oxford University Press.

Vanselow, K. H., Jacobsen, S., Hall, C., & Garthe, S. (2017). Solar storms may trigger sperm whale strandings: Explanation approaches for multiple strandings in the North Sea in 2016. *International Journal of Astrobiology*, 1–9. doi:10.1017/S147355041700026X

Vautier, G., Bomford, A. B., Portmann, B. C., Metivier, E., Williams, R., & Ryder, S. D. (1999). p53 mutations in British patients with hepatocellular carcinoma: Clustering in genetic hemochromatosis. *Gastroenterology*, *117*(1), 154–160. doi:10.1016/S0016-5085(99)70562-7

Vela, N., Navarro, S., & Navarro García, G. (2007). Review. An overview on the environmental behaviour of pesticide residues in soils. *Spanish Journal of Agricultural Research*, *5*(3), 357–375. doi:10.5424jar/2007053-5344

Veldkamp, T. I. E., Wada, Y., Aerts, J. C. J. H., Döll, P., Gosling, S. N., Liu, J., ... Ward, P. J. (2017). Water scarcity hotspots travel downstream due to human interventions in the 20th and 21st century. *Nature Communications*, *8*, 15697. doi:10.1038/ncomms15697 PMID:28643784

Vella, V., Malaguarnera, R., Lappano, R., Maggiolini, M., & Belfiore, A. (2017). Recent views of heavy metals as possible risk factors and potential preventive and therapeutic agents in prostate cancer. *Molecular and Cellular Endocrinology*, *5*, 57–72. doi:10.1016/j.mce.2016.10.020

Venier, M., & Hites, R. A. (2014). DDT and HCH, two discontinued organochlorine insecticides in the Great Lakes region: Isomer trends and sources. *Environment International*, *69*, 159–165. doi:10.1016/j.envint.2014.03.028 PMID:24858704

Venkatesan, R., Park, Y. U., Ji, E., Yeo, E.-J., & Kim, S. Y. (2017). Malathion increases apoptotic cell death by inducing lysosomal membrane permeabilization in N2a neuroblastoma cells: A model for neurodegeneration in Alzheimer's disease. *Cell Death & Disease*, *3*, 17007. doi:10.1038/cddiscovery.2017.7 PMID:28487766

Venter, O., Sanderson, E. W., Magrach, A., Allan, J. R., Beher, J., Jones, K. R., ... Levy, M. A. (2016). Sixteen years of change in the global terrestrial human footprint and implications for biodiversity conservation. *Nature Communications*, *7*(1), 12558. doi:10.1038/ncomms12558 PMID:27552116

Vera, L., Martel, G., Gutierrez, J., & Marquez, M. (2006). Gesti on sostenible del agua residual en entornos rurales. Capitulo 3: Evaluaci ́on de los sistemas de depuraci ́on natural. Proyecto depuranat. Instituto Tecnol ́ogico de Canarias. Departamento de agua, Direcci ́on de investigaci ́on y desarrollo tecnol ́ogico.

Verchot, L. V., Van Noordwijk, M., Kandji, S., Tomich, T., Ong, C., Albrecht, A., ... Palm, C. (2007). Climate change: Linking adaptation and mitigation through agroforestry. *Mitigation and Adaptation Strategies for Global Change*, *12*(5), 901–918. doi:10.100711027-007-9105-6

Vergine, P., Saliba, R., Salerno, C., Laera, G., Berardi, G., & Pollice, A. (2015). Fate of the fecal indicator Escherichia coli in irrigation with partially treated wastewater. *Water Research*, *85*, 66–73. doi:10.1016/j.watres.2015.08.001 PMID:26302216

Vezér, T., Papp, A., Hoyk, Z., Varga, C., Náray, M., & Nagymajtényi, L. (2005). Behavioral and neurotoxicological effects of subchronic manganese exposure in rats. *Environmental Toxicology and Pharmacology*, *19*(3), 797–810. doi:10.1016/j.etap.2004.12.046 PMID:21783557

Vezzulli, L., Colwell, R. R., & Pruzzo, C. (2013). Ocean warming and spread of pathogenic vibrios in the aquatic environment. *Microbial Ecology, 65*(4), 817–825. doi:10.100700248-012-0163-2 PMID:23280498

Via, C. S., Nguyen, P., Niculescu, F., Papadimitriou, J., Hoover, D., & Silbergeld, E. K. (2003). Low-dose exposure to inorganic mercury accelerates disease and mortality in acquired murine lupus. *Environmental Health Perspectives, 111*(10), 1273–1277. doi:10.1289/ehp.6064 PMID:12896845

Vickery, W. L., & Bider, J. R. (1981). The influence of weather on rodent activity. *Journal of Mammalogy, 62*(1), 140–145. doi:10.2307/1380484

Victor, M., Adams, R. D., & Cole, M. (1965). The acquired (non-Wilsonian) type of chronic hepatocerebral degeneration. *Medicine, 44*(5), 345–396. doi:10.1097/00005792-196509000-00001 PMID:5318075

Vidal, L., Alfonso, M., Campos, F., Faro, L. R., Cervantes, R. C., & Durán, R. (2005). Effects of manganese on extracellular levels of dopamine in rat striatum: An analysis in vivo by brain microdialysis. *Neurochemical Research, 30*(9), 1147–1154. doi:10.100711064-005-7775-6 PMID:16292508

Vieira, C. L. Z., Janot-Pacheco, E., Lage, C., Pacini, A., Koutrakis, P., Cury, P. R., ... Saldiva, P. H. N. (2018). Long-term association between the intensity of cosmic rays and mortality rates in the city of Sao Paulo. *Environmental Research Letters, 13*(2). doi:10.1088/1748-9326/aaa27a

Vieira, L. E. A., & da Silva, L. A. (2006). Geomagnetic modulation of clouds effects in the Southern Hemisphere Magnetic Anomaly through lower atmosphere cosmic ray effects. *Geophysical Research Letters, 33*(14), L14802. doi:10.1029/2006GL026389

Viel, T., Dransart, E., Nemati, F., Henry, E., Thézé, B., Decaudin, D., ... Tavitian, B. (2008). In vivo tumor targeting by the B-subunit of shiga toxin. *Molecular Imaging, 7*(6), 239–247. doi:10.2310/7290.2008.00022

Vig, P.J., & Nath, R. (1991). *In vivo effects of cadmium on calmodulin and calmodulin regulated enzymes in rat brain.* Academic Press.

Vinceti, M., Rovesti, S., Bergomi, M.,& Vivoli, G. (2000). The epidemiology of selenium and human cancer. *Tumori Journal, 86*, 105-18. doi: . doi:10.1177/030089160008600201

Vinceti, M., Bottecchi, I., Fan, A., Finkelstein, Y., & Mandrioli, J. (2012). Are environmental exposures to selenium, heavy metals, and pesticides risk factors for amyotrophic lateral sclerosis? *Reviews on Environmental Health, 27*(1), 19–41. doi:10.1515/reveh-2012-0002 PMID:22755265

Vinceti, M., Guidetti, D., Pinotti, M., Rovesti, S., Merlin, M., Vescovi, L., ... Vivoli, G. (1996). Amyotrophic lateral sclerosis after long-term exposure to drinking water with high selenium content. *Epidemiology (Cambridge, Mass.), 7*(5), 529–532. doi:10.1097/00001648-199609000-00014 PMID:8862986

Vinceti, M., Maraldi, T., Bergomi, M., & Malagoli, C. (2009). Risk of chronic low-dose selenium overexposure in humans: Insights from epidemiology and biochemistry. *Reviews on Environmental Health, 24*(3), 231–248. doi:10.1515/REVEH.2009.24.3.231 PMID:19891121

Vinceti, M., Violi, F., Tzatzarakis, M., Mandrioli, J., Malagoli, C., Hatch, E. E., ... Tsatsakis, A. (2017). Pesticides, polychlorinated biphenyls and polycyclic aromatic hydrocarbons in cerebrospinal fluid of amyotrophic lateral sclerosis patients: A case-control study. *Environmental Research, 155*, 261–267. doi:10.1016/j.envres.2017.02.025 PMID:28242563

Vinet, F. (2017). Flood Impacts on Loss of Life and Human Health. *Floods*, 33-51. doi:10.1016/B978-1-78548-268-7.50003-1

Visser, A., Kroes, J. T. H., Van Vliet, M., Blenkinsop, S., Fowler, H. J., & Broers, H. P. (2012). Climate change impacts on the leaching of a heavy metal contamination in a small lowland catchment. *Journal of Contaminant Hydrology*, *127*, 47–64. doi:10.1016/j.jconhyd.2011.04.007 PMID:21684031

Vitt, L. J., & Caldwell, J. P. (2013). *Herpetology: an introductory biology of amphibians and reptiles*. Academic Press.

Vodela, J. K., Renden, J. A., Lenz, S. D., Mc Elhenney, W. H., & Kemppainen, B. W. (1997). Drinking Water Contaminants (Arsenic, Cadmium, Lead, Benzene, and Trichloroethylene). Interaction of Contaminants with Nutritional Status on General Performance and Immune Function in Broiler Chickens. *Poultry Science*, *76*(11), 1474–1492. doi:10.1093/ps/76.11.1474 PMID:9355140

Volirath, M., & Angst, J. (1989). Outcome of panic and depression in a seven???year follow???up: Results of the Zurich study. *Acta Psychiatrica Scandinavica*, *80*(6), 591–596. doi:10.1111/j.1600-0447.1989.tb03031.x PMID:2618782

Von Braun, J. (2007). *The world food situation: new driving forces and required actions*. Washington, DC: International Food Policy Research Institute.

von Campenhausen, S., Bornschein, B., Wick, R., Bötzel, K., Sampaio, C., Poewe, W., ... Dodel, R. (2005). Prevalence and incidence of Parkinson's disease in Europe. *European Neuropsychopharmacology : The Journal of the European College of Neuropsychopharmacology*, *15*(4), 473–490. doi:10.1016/j.euroneuro.2005.04.007 PMID:15963700

von Grebmer, K., von Grebmer, K., Ringler, C., Rosegrant, M. W., Olofinbiyi, T., Wiesmann, D., & Thompson, J. (2012). *Global hunger index: the challenge of hunger: Ensuring sustainable food security under land, water, and energy stresses* (Vol. 70). Intl Food Policy Res Inst.

Vonsattel, J. P., & DiFiglia, M. (1998). Huntington disease. *Journal of Neuropathology and Experimental Neurology*, *57*(5), 369–384. doi:10.1097/00005072-199805000-00001 PMID:9596408

Vonsattel, J. P., Keller, C., & Del Pilar Amaya, M. (2008). Neuropathology of Huntington's disease. *Handbook of Clinical Neurology*, *89*, 599–618. doi:10.1016/S0072-9752(07)01256-0 PMID:18631782

Vorderstrasse, B. A., Fenton, S. E., Bohn, A. A., Cundiff, J. A., & Lawrence, B. P. (2004). A novel effect of dioxin: Exposure during pregnancy severely impairs mammary gland differentiation. *Toxicological Sciences*, *78*(2), 248–257. doi:10.1093/toxsci/kfh062

Vrijheid, M., Casas, M., Bergström, A., Carmichael, A., Cordier, S., Eggesbø, M., ... Fernández-Somoano, A. (2011). European birth cohorts for environmental health research. *Environmental Health Perspectives*, *120*(1), 29–37. doi:10.1289/ehp.1103823 PMID:21878421

Waalkes, M. P. (2000). Cadmium carcinogenesis in review. *Journal of Inorganic Biochemistry*, *79*(1-4), 241–244. doi:10.1016/S0162-0134(00)00009-X PMID:10830873

Waalkes, M. P., & Rehm, S. (1994). Cadmium and prostate cancer. *Journal of Toxicology and Environmental Health*, *43*(3), 251–269. doi:10.1080/15287399409531920 PMID:7966437

Wade, T.J., Sams, E., Brenner, K.P., Haugland, R., Chern, E., Beach, M., … Dufour, A.P. (2010). Rapidly measured indicators of recreational water quality and swimming-associated illness at marine beaches: A prospective cohort study. *Environmental Health: A Global Access Science Source, 9*, 66.

Wade, T. J., Calderon, R. L., Brenner, K. P., Sams, E., Beach, M., Haugland, R., ... Dufour, A. P. (2008). High sensitivity of children to swimming-associated gastrointestinal illness: Results using a rapid assay of recreational water quality. *Epidemiology (Cambridge, Mass.)*, *19*(3), 375–383. doi:10.1097/EDE.0b013e318169cc87 PMID:18379427

Wagner, S. E., Burch, J. B., Hussey, J., Temples, T., Bolick-Aldrich, S., Mosley-Broughton, C., ... Hebert, J. R. (2009). Soil zinc content, groundwater usage, and prostate cancer incidence in South Carolina. *Cancer causes &control. CCC*, *20*(3), 345–353. PMID:18949566

Waisberg, M., Joseph, P., Hale, B., & Beyersmann, D. (2003). Molecular and cellular mechanisms of cadmium carcinogenesis. *Toxicology*, *192*(2), 95–117. doi:10.1016/S0300-483X(03)00305-6 PMID:14580780

Walck, J. L., Baskin, J. M., & Baskin, C. C. (1997). A comparative study of the seed germination biology of a narrow endemic and two geographically-widespread species of Solidago (Asteraceae). 1. Germination phenology and effect of cold stratification on germination. *Seed Science Research*, *7*(1), 47–58. doi:10.1017/S0960258500003366

Walck, J. L., Hidayati, S. N., Dixon, K. W., Thompson, K. E. N., & Poschlod, P. (2011). Climate change and plant regeneration from seed. *Global Change Biology*, *17*(6), 2145–2161. doi:10.1111/j.1365-2486.2010.02368.x

Waldbaum, S., & Patel, M. (2010). Mitochondria, oxidative stress, and temporal lobe epilepsy. *Epilepsy Research*, *88*(1), 23–45. doi:10.1016/j.eplepsyres.2009.09.020 PMID:19850449

Waldrop, M., & Firestone, M. (2006). Response of microbial community composition and function to soil climate change. *Microbial Ecology*, *52*(4), 716–724. doi:10.100700248-006-9103-3 PMID:17061172

Waldschmidt, S., & Tracy, C. R. (1983). Interactions between a lizard and its thermal environment: Implications for sprint performance and space utilization in the lizard Uta stansburiana. *Ecology*, *64*(3), 476–484. doi:10.2307/1939967

Walker, C. H. (2006). Ecotoxicity testing of chemicals with particular reference to pesticides. *Pest Management Science*, *62*(7), 571–583. doi:10.1002/ps.1218 PMID:16671067

Walt, M. (1996). Source and loss processes for radiation belt particles. In Radiation Belts: Model & Standards. Academic Press.

Walter, F. G., Stolz, U., Shirazi, F., & McNally, J. (2009). Epidemiology of severe and fatal rattlesnake bites published in the American Association of Poison Control Center's Annual Reports. *Clinical Toxicology (Philadelphia, PA)*, *47*(7), 663–669. doi:10.1080/15563650903113701 PMID:19640239

Walter, F. G., Stolz, U., Shirazi, F., & Mcnally, J. (2010). Temporal analyses of coral snakebite severity published in the American Association of Poison Control Centers' Annual Reports from 1983 through 2007. *Clinical Toxicology*, *48*(1), 72–78. doi:10.3109/15563650903430944 PMID:20070180

Walther, G. R., Post, E., Convey, P., Menzel, A., Parmesan, C., Beebee, T. J., ... Bairlein, F. (2002). Ecological responses to recent climate change. *Nature*, *416*(6879), 389–395. doi:10.1038/416389a PMID:11919621

Walton, J. R. (2006). Aluminum in hippocampal neurons from humans with Alzheimer's disease. *Neurotoxicology*, *27*(3), 385–394. doi:10.1016/j.neuro.2005.11.007 PMID:16458972

Walton, J. R. (2009a). Brain lesions comprised of aluminum-rich cells that lack microtubules may be associated with the cognitive deficit of Alzheimer's disease. *Neurotoxicology*, *30*(6), 1059–1069. doi:10.1016/j.neuro.2009.06.010 PMID:19591863

Walton, J. R. (2009b). Functional impairment in aged rats chronically exposed to human range dietary aluminum equivalents. *Neurotoxicology*, *30*(2), 182–193. doi:10.1016/j.neuro.2008.11.012 PMID:19109991

Wandiga, S. O., Opondo, M., Olago, D., Githeko, A., Githui, F., Marshall, M., ... Achola, P. (2010). Vulnerability to epidemic malaria in the highlands of Lake Victoria basin: The role of climate change/variability, hydrology and socio-economic factors. *Climatic Change*, *99*(3-4), 473–497. doi:10.100710584-009-9670-7

Wang, T., Guo, X., Hong, Y., Han, H., Cao, X., & Gene, Y. H.. (n.d.). *Comparison of apoptosis between adult worms of Schistosoma japonicum from susceptible (BALB/c mice) and less-susceptible (Wistar rats) hosts.* Elsevier. Retrieved from https://www.sciencedirect.com/science/article/pii/S0378111916305911

Wang, A., Costello, S., Cockburn, M., Zhang, X., Bronstein, J., & Ritz, B. (2011). Parkinson's disease risk from ambient exposure to pesticides. *European Journal of Epidemiology*, *26*(7), 547–555. doi:10.100710654-011-9574-5 PMID:21505849

Wang, B., & Du, Y. (2013). Cadmium and Its Neurotoxic Effects. *Oxidative Medicine and Cellular Longevity*, *2013*, 1–12. doi:10.1155/2013/898034 PMID:23997854

Wang, F., David, E., Lonial, S., Arellano, M., Khoury, H. J., Langston, A., ... Waller, N. (2007). Patulin, a Mycotoxin, Has Potent Cytotoxic Activity on Actue Lymphoblastic Leukemia and Induce Apoptosis through Intrinsic and Extrinsic Pathways. *Blood*, *110*(11), 4334–4334.

Wang, J. S., & Groopman, J. D. (1999). DNA damage by mycotoxins. *Mutation Research. Fundamental and Molecular Mechanisms of Mutagenesis*, *424*(1-2), 167–181. doi:10.1016/S0027-5107(99)00017-2

Wang, S., Qian, X., Han, B.-P., Luo, L.-C., & Hamilton, D. P. (2012). Effects of local climate and hydrological conditions on the thermal regime of a reservoir at Tropic of Cancer, in southern China. *Water Research*, *46*(8), 2591–2604. doi:10.1016/j.watres.2012.02.014 PMID:22391017

Wang, T., Gavin, H. M., Arlt, V. M., Lawrence, B. P., Fenton, S. E., Medina, D., & Vorderstrasse, B. A. (2011). Aryl hydrocarbon receptor activation during pregnancy, and in adult nulliparous mice, delays the subsequent development of DMBA-induced mammary tumors. *International Journal of Cancer*, *128*(7), 1509–1523. doi:10.1002/ijc.25493

Wang, T., Zaar, M., Arvedsen, S., Vedel-Smith, C., & Overgaard, J. (2002). Effects of temperature on the metabolic response to feeding in Python molurus. *Comparative Biochemistry and Physiology. Part A, Molecular & Integrative Physiology*, *133*(3), 519–527. doi:10.1016/S1095-6433(02)00250-7 PMID:12443910

Wang, Y., Jiang, G., Lam, P. K., & Li, A. (2007). Polybrominated diphenyl ether in the East Asian environment: A critical review. *Environment International*, *33*(7), 963–973. doi:10.1016/j.envint.2007.03.016

Wang, Y., Wang, L., Liu, F., Wang, Q., Selvaraj, J. N., Xing, F., ... Liu, Y. (2016). Ochratoxin A producing fungi, biosynthetic pathway and regulatory mechanisms. *Toxins*, *8*(3), 83. doi:10.3390/toxins8030083

Wani, A. L., Ara, A., & Usmani, J. A. (2015). Lead toxicity: A review. *Interdisciplinary Toxicology*, *8*(2), 55–64. doi:10.1515/intox-2015-0009

Warburg, M. R., & Polis, G. A. (1990). Behavioral responses, rhythms, and activity patterns. *The Biology of Scorpions*, *224*, 246.

Ward, D. M., & Brock, T. D. (1978). Hydrocarbon degradation in hypersaline environments. *Applied and Environmental Microbiology*, *35*, 353–359. PMID:16345276

War, G. (2008). Challenges to 'Acetylcholinesterase inhibitors and Gulf War illnesses'. *Proceedings of the National Academy of Sciences of the United States of America*, 33.

Waring, S. C., Esteban-Santillan, C., Reed, D. M., Craig, U., Labarthe, D. R., Petersen, R. C., & Kurland, L. T. (2004). Incidence of Amyotrophic Lateral Sclerosis and of the Parkinsonism-Dementia Complex of Guam, 1950-1989. *Neuroepidemiology*, *23*(4), 192–200. doi:10.1159/000078505 PMID:15272222

Warner, M., Eskenazi, B., Mocarelli, P., Gerthoux, P. M., Samuels, S., Needham, L., ... Brambilla, P. (2002). Serum dioxin concentrations and breast cancer risk in the Seveso Women's Health Study. *Environmental Health Perspectives*, *110*(7), 625–628. doi:10.1289/ehp.02110625

Warner, M., Mocarelli, P., Samuels, S., Needham, L. L., Brambilla, P., & Eskenazi, B. (2011). Dioxin exposure and cancer risk in the Seveso Women's Health Study. *Environmental Health Perspectives, 119*(12), 1700–1705. doi:10.1289/ehp.1103720

Warner, T. T., & Schapira, A. H. V. (2003). Genetic and environmental factors in the cause of Parkinson's disease. *Annals of Neurology, 53*(S3), S16–S25. doi:10.1002/ana.10487 PMID:12666095

Warrell, D. A. (1993). Venomous bites and stings in the tropical world. *The Medical Journal of Australia, 159*, 773–779. PMID:8264466

Warrell, D. A. (2010). Snake bite. *Lancet, 375*(9708), 77–88. doi:10.1016/S0140-6736(09)61754-2 PMID:20109866

Warrell, D. A., Ormerod, L. D., & Davidson, N. M. (1976). Bites by the night adder (*Caususmaculatus*) and burrowing vipers (genus Atractaspis) in Nigeria. *The American Journal of Tropical Medicine and Hygiene, 25*(3), 517–524. doi:10.4269/ajtmh.1976.25.517 PMID:945703

Warrick, J. A., DiGiacomo, P. M., Weisberg, S. B., Nezlin, N. P., Mengel, M., Jones, B. H., ... Farnsworth, K. L. (2007). River plume patterns and dynamics within the Southern California Bight. *Continental Shelf Research, 27*(19), 2427–2448. doi:10.1016/j.csr.2007.06.015

Washburn, L., McClure, K. A., Jones, B. H., & Bay, S. M. (2003). Spatial scales and evolution of stormwater plumes in Santa Monica Bay. *Marine Environmental Research, 56*(1-2), 103–125. doi:10.1016/S0141-1136(02)00327-6 PMID:12648952

Wasserberg, G., Yarom, I., & Warburg, A. (2003). Seasonal abundance patterns of the sandfly Phlebotomus papatasi in climatically distinct foci of cutaneous leishmaniasis in Israeli deserts. *Medical and Veterinary Entomology, 17*(4), 452–456. doi:10.1111/j.1365-2915.2003.00461.x PMID:14651661

Water, U. N. (2006). *Coping with Water Scarcity–A Strategic Issue and Priority for System-Wide Action*. Geneva, Switzerland: UN-Water.

Watkins, K. (2006). *Human Development Report 2006 - Beyond scarcity: Power, poverty and the global water crisis*. New York: United Nations Development Programme. doi:10.1016/S1352-0237(02)00387-8

Watson, R. T., & Albritton, D. L. (Eds.). (2001). *Climate change 2001: Synthesis report: Third assessment report of the Intergovernmental Panel on Climate Change*. Cambridge University Press.

Watt, G., Padre, L., Tuazon, M. L., Theakston, R. D. G., & Laughlin, L. (1988). Bites by the Philippine cobra (Najanajaphilippinensis): Prominent neurotoxicity with minimal local signs. *The American Journal of Tropical Medicine and Hygiene, 39*(3), 306–311. doi:10.4269/ajtmh.1988.39.306 PMID:3177741

Weber, F. A., Hofacker, A. F., Voegelin, A., & Kretzschmar, R. (2010). Temperature dependence and coupling of iron and arsenic reduction and release during flooding of a contaminated soil. *Environmental Science & Technology, 44*(1), 116–122. doi:10.1021/es902100h PMID:20039741

Wedler, F. C., Denman, R. B., & Roby, W. G. (1982). Glutamine synthetase from ovine brain is a manganese(II) enzyme. *Biochemistry, 21*(25), 6389–6396. doi:10.1021/bi00268a011 PMID:6129892

Wegrzynowicz, M., Holt, H. K., Friedman, D. B., & Bowman, A. B. (2012). Changes in the striatal proteome of YAC128Q mice exhibit geneenvironment interactions between mutant huntingtin and manganese. *Journal of Proteome Research, 1*(2), 1118–1132. doi:10.1021/pr200839d PMID:22191580

Weichenthal, S., Moase, C., Chan, P., Environmental, S., Perspectives, H., August, N., ... Chan, P. (2014). A Review of Pesticide Exposure and Cancer Incidence in the Agricultural Health Study Cohort All use subject to JSTOR Terms and Conditions Review A Review of Pesticide Exposure and Cancer Agricultural Health Study Cohort in the. *Environmental Health Perspectives*, *118*(8), 1117–1125. doi:10.1289/ehp.0901731 PMID:20444670

Weinshenker, B. C. (1996). Epidemiology of multiple sclerosis. *Neurologic Clinics*, *142*, 1–308.

Weinstein, S. A. (2017). Non-Front-Fanged Colubroid Snakes. In *Critical Care Toxicology* (pp. 1–41). Cham: Springer.

Weinstein, S. A., & Kardong, K. V. (1994). Properties of Duvernoy's secretions from opisthoglyphous and aglyphous colubrid snakes. *Toxicon*, *32*(10), 1161–1185. doi:10.1016/0041-0101(94)90347-6 PMID:7846688

Weiss, K. H. (1993). Wilson Disease. In M. P. Adam, H. H. Ardinger, R. A. Pagon, S. E. Wallace, L. J. Bean, K. Stephens, & A. Amemiya (Eds.), *GeneReviews®*. Seattle, WA: University of Washington. Retrieved from http://www.ncbi.nlm.nih.gov/books/NBK1512/

Weisskopf, M. G., Morozova, N., O'Reilly, E. J., McCullough, M. L., Calle, E. E., Thun, M. J., & Ascherio, A. (2009). Prospective study of chemical exposures and amyotrophic lateral sclerosis. *Journal of Neurology, Neurosurgery, and Psychiatry*, *80*(5), 558–561. doi:10.1136/jnnp.2008.156976 PMID:19372290

Wei, T., Yang, S., Moore, J. C., Shi, P., Cui, X., Duan, Q., ... Dong, W. (2012). Developed and developing world responsibilities for historical climate change and CO(2) mitigation. *Proceedings of the National Academy of Sciences of the United States of America*, *109*(32), 12911–12915. doi:10.1073/pnas.1203282109 PMID:22826257

Weniger, B. G., Blaser, M. J., Gedrose, J., Lippy, E. C., & Juranek, D. D. (1983). An outbreak of waterborne giardiasis associated with heavy water runoff due to warm weather and volcanic ashfall. *American Journal of Public Health*, *73*(8), 868–872. doi:10.2105/AJPH.73.8.868 PMID:6869640

Wen, J., Mu, P., & Deng, Y. (2016). Mycotoxins: Cytotoxicity and biotransformation in animal cells. *Toxicology Research*, *5*(2), 377–387. doi:10.1039/C5TX00293A PMID:30090353

Wenstrup, D., Ehman, W. D., & Markesbery, W. R. (1990). Trace element imbalances in isolated subcellular fractions of Alzheimer's disease brains. *Brain Research*, *533*(1), 125–131. doi:10.1016/0006-8993(90)91804-P PMID:2085723

Werner-Seidler, A., Perry, Y., Calear, A. L., Newby, J. M., & Christensen, H. (2017). School-based depression and anxiety prevention programs for young people: A systematic review and meta-analysis. *Clinical Psychology Review*, *51*, 30–47. doi:10.1016/j.cpr.2016.10.005 PMID:27821267

Wheeler, T., & Von Braun, J. (2013). Climate change impacts on global food security. *Science*, *341*(6145), 508–513. doi:10.1126cience.1239402 PMID:23908229

Whitaker, H. C., Kote-Jarai, Z., Ross-Adams, H., Warren, A. Y., Burge, J., George, A., ... Neal, D. E. (2010). The rs10993994 Risk Allele for Prostate Cancer Results in Clinically Relevant Changes in Microseminoprotein-Beta Expression in Tissue and Urine. *PLoS One*, *5*(10), e13363. doi:10.1371/journal.pone.0013363 PMID:20967219

White, N.J., Pukrittayakamee, S., Hien, T.T., Faiz, M.A., Mokuolu, O.A., & Dondorp, A. M. (2014). Malaria. *Lancet*, *383*(9918), 723-35. doi: . doi:10.1016/S0140-6736(13)60024-0

White, A. C., Robinson, P., Okhuysen, P. C., Lewis, D. E., Shahab, I., Lahoti, S., ... Chappell, C. L. (2000). Interferon-gamma expression in jejunal biopsies in experimental human cryptosporidiosis correlates with prior sensitization and control of oocyst excretion. *The Journal of Infectious Diseases*, *181*(2), 701–709. doi:10.1086/315261 PMID:10669358

White, A. J., Teitelbaum, S. L., Wolff, M. S., Stellman, S. D., Neugut, A. I., & Gammon, M. D. (2013). Exposure to fogger trucks and breast cancer incidence in the Long Island Breast Cancer Study Project: A case-control study. *Environmental Health*, *12*(1), 24. doi:10.1186/1476-069X-12-24

Whitehead, P. G., Wilby, R. L., Battarbee, R. W., Kernan, M., & Wade, A. J. (2009). A review of the potential impacts of climate change on surface water quality. *Hydrological Sciences Journal*, *54*(1), 101–123. doi:10.1623/hysj.54.1.101

White, J. (2018). Snakebite as a neglected disease. *ToxicologieAnalytiqueet Clinique*, *30*(3), 161.

WHO. (1996). *The World Health Report 1996. Fighting Disease Fostering Development*. Geneva: World Health Organization.

WHO. (1998). World Health Organization, Aluminum. In *Guidelines for Drinking-Water Quality* (2nd ed.; pp. 3–13). Geneva: World Health Organization.

WHO. (2004). *Guidelines for Drinking Water Quality* (3rd ed.). Geneva: World Health Organization.

WHO. (2006). *Guidelines for drinking water quality*. Geneva: World Health Organization.

WHO. (2009). *Global Malaria Action Plan World Health Organization*. Geneva, Switzerland: World Health Organization.

WHO. (2012). *Atlas of health and climate*. Joint publication World Health Organization and meteorogical world organization. Available at http://www.who.int/globalchange/publications/atlas/en/

WHO. (2017). *World Malaria report 2017*. Geneva: World Health Organization. Available http//www.who.int/malaria/publication/world-malaria-report-2017/en/

WHO. (2018). *Neglected tropical diseases: snakebite*. Retrieved from http://www.who.int/neglected _diseases/diseases/snakebites/en/index.html

Wiedner, C., Rücker, J., Brüggemann, R., & Nixdorf, B. (2007). Climate change affects timing and size of populations of an invasive *Cyanobacterium* in temperate regions. *Oecologia*, *152*(3), 473–484. doi:10.100700442-007-0683-5 PMID:17375336

Wierzbicki, R., & Praz, M. (2002). Disorders in Blood Coagulation in Humans Occupationally Exposed to Mercuric Vapors. *The Journal of Trace Elements in Experimental Medicine*, *29*, 21–29. doi:10.1002/jtra.1055

Wigle, D. T., Turner, M. C., & Krewski, D. (2009). A Systematic Review and Meta-analysis of Childhood Leukemia and Parental Occupational Pesticide Exposure. *Environmental Health Perspectives*, *117*(10), 1505–1513. doi:10.1289/ehp.0900582 PMID:20019898

Wijesundera, M. D. S. (1988). Malaria outbreaks in new foci in Sri Lanka. *Parasitology Today (Personal Ed.)*, *4*(5), 147–150. doi:10.1016/0169-4758(88)90193-7

Wijngaard, R. R., van der Perk, M., van der Grift, B., de Nijs, T. C. M., & Bierkens, M. F. P. (2017). The impact of climate change on metal transport in a lowland catchment. *Water, Air, and Soil Pollution*, *228*(3), 107. doi:10.100711270-017-3261-4 PMID:28260820

Wilbur, H. M. (1977). Propagule size, number, and dispersion pattern in Ambystoma and Asclepias. *American Naturalist*, *111*(977), 43–68. doi:10.1086/283137

Wild, C. P., Miller, J. D., & Groopman, J. D. (2015). *Effects of aflatoxins and fumonisins on child growth*. International Agency for Research on Cancer. Retrieved from https://www.ncbi.nlm.nih.gov/books/NBK350566/

Wild, C. P., & Kleinjans, J. (2003). Children and increased susceptibility to environmental carcinogens: Evidence or empathy? *Cancer Epidemiology, Biomarkers & Prevention, 12*(12), 1389–1394. PMID:14693727

William Langston, J., Forno, L. S., Rebert, C. S., & Irwin, I. (1984). Selective nigral toxicity after systemic administration of 1-methyl-4-phenyl-1,2,5,6-tetrahydropyrine (MPTP) in the squirrel monkey. *Brain Research, 292*(2), 390–394. doi:10.1016/0006-8993(84)90777-7 PMID:6607092

Williams, M., Todd, G. D., Roney, N., Crawford, J., Coles, C., McClure, P. R., … Citra, M. (2012). *Relevance to public health.* Agency for Toxic Substances and Disease Registry (US). Retrieved from https://www.ncbi.nlm.nih.gov/books/NBK158871/

Williams, D., Gutiérrez, J. M., Harrison, R., Warrell, D. A., White, J., Winkel, K. D., & Gopalakrishnakone, P. (2010). The Global Snake Bite Initiative: An antidote for snake bite. *Lancet, 375*(9708), 89–91. doi:10.1016/S0140-6736(09)61159-4 PMID:20109867

Williamson, C. E., Dodds, W., Kratz, T. K., & Palmer, M. A. (2008). Lakes and streams as sentinels of environmental change in terrestrial and atmospheric processes. *Frontiers in Ecology and the Environment, 6*(5), 247–254. doi:10.1890/070140

Wills, A. M., Cronin, S., Slowik, A., Kasperaviciute, A., Van Es, A. M., Morahan, J. M., ... Brown, R. H. (2009). A large-scale international meta-analysis of paraoxonase gene polymorphisms in sporadic ALS. *Neurology, 73*(1), 16–24. doi:10.1212/WNL.0b013e3181a18674 PMID:19321847

Wilson, S.R., Solomon, K.R. & Tang, X. (2007) Changes in tropospheric composition and air quality due to stratospheric ozone depletion and climate change. *Photochemical & Photobiological Sciences, 6*, 301-310.

Wilson's disease (a volume in the major problems in internal medicine series) By I. H. Scheinberg and I. Sternlieb Philadelphia, W. B. Saunders, 1984 192 pp, illustrated, $35.00. (1984). *Annals of Neurology, 16*(5), 626–626. doi:10.1002/ana.410160531

Wilson, D. M., & King, J. K. (1995). Production of aflatoxins B_1, B_2, G_1, and G_2 in pure and mixed cultures of *Aspergillus parasiticus* and *Aspergillus flavus*. *Food Additives and Contaminants, 12*(3), 521–525. doi:10.1080/02652039509374339

Wilson, D. M., Mubatanhema, W., & Jurjevic, Z. (2002). Biology and ecology of mycotoxigenic aspergillus species as related to economic and health concerns. In J. W. De Vries, M. W. Trucksess, & L. S. Jackson (Eds.), *Mycotoxins and Food Safety*. New York: Kluwer Academic / Plenum Publishers. doi:10.1007/978-1-4615-0629-4_2

Winpenny, J., Heinz, I., Koo-Oshima, S., Salgot, M., Collado, J., & Hernandex, F. (2010). *The Wealth of Waste: The Economics of Wastewater Use in Agriculture.* Rome, Italy: Food and Agriculture Organization of the United Nations. Available http://www.fao.org/docrep/012/i1629e/i1629e.pdf

Wirdefeldt, K., Adami, H.-O., Cole, P., Trichopoulos, D., & Mandel, J. (2011). Epidemiology and etiology of Parkinson's disease: A review of the evidence. *European Journal of Epidemiology, 26*(S1), 1–58. doi:10.100710654-011-9581-6 PMID:21626386

Wojcik, D. P., Godfrey, M. E., Christie, D., & Haley, B. E. (2006). Mercury toxicity presenting as chronic fatigue, memory impairment and depression: Diagnosis, treatment, susceptibility, and outcomes in a New Zealand general practice setting (1994-2006). *Neuroendocrinology Letters, 27*(4), 415–423. Retrieved from http://www.ncbi.nlm.nih.gov/pubmed/16891999 PMID:16891999

Wolfe, M. F., Schwartz, G. J. B., Singaram, S., Mielbrecht, E. E., Tjeerdema, R. S., & Sowby, M. L. (1998). Effects of salinity and temperature on the bioavailability of dispersed petroleum hydrocarbons to the golden-brown algae, *Isochrysis galbana*. *Archives of Environmental Contamination and Toxicology, 35*, 268–273. doi:10.1007002449900375 PMID:9680519

Wolff, M., Zeleniuch-Jacquotte, A., Dubin, N., & Toniolo, P. (2000). Risk of breast cancer and organochlorine exposure. *Cancer Epidemiology, Biomarkers & Prevention*, *9*, 271–277.

Wolff, N. A., Garrick, M. D., Zhao, L., Garrick, L. M., Ghio, A. J., & Thévenod, F. (2018). A role for divalent metal transporter (DMT1) in mitochondrial uptake of iron and manganese. *Scientific Reports*, *8*(1), 211. doi:10.103841598-017-18584-4 PMID:29317744

Wong, M. L., & Licinio, J. (2004). From monoamines to genomic targets: A paradigm shift for drug discovery in depression. *Nature Reviews. Drug Discovery*, *3*(2), 136–151. doi:10.1038/nrd1303 PMID:15040578

Woodall, C. J. (2009). Waterborne diseases - What are the primary killers? *Desalination*, *248*(1-3), 616–621. doi:10.1016/j.desal.2008.05.110

Woodruff, T., & Morello-Frosch, R. (2011). Communicating about chemical body burden, with Tracey Woodruff and Rachel Morello-Frosch. *Environmental Health Perspectives*, 119.

Woolcott, C. G., Aronson, K. J., Hanna, W. M., SenGupta, S. K., McCready, D. R., Sterns, E., & Miller, A. (2001). Organochlorines and breast cancer risk by receptor status, tumor size, and grade (Canada). *Cancer Causes & Control*, *12*(5), 395–404. doi:10.1023/A:1011289905751

World Health Organization (WHO). (2006). *Guidelines for the Safe Use of Wastewater Excreta and Greywater* (Vol. 4). Geneva: World Health Organization.

World Health Organization (WHO). (2007). *Report of the Sixtieth Worldwide Assembly on Health*. Geneva, Switzerland: WHO.

World Health Organization (WHO). (2010). *Control of the Leishmaniasis: Report of a Meeting of the WHO Expert Committee on the Control of Leishmaniases*. World Health Organization Technical Report Series No. 949. WHO.

World Health Organization (WHO). (2012 a). *Accelerating work to overcome the global impact of neglected tropical diseases: a roadmap for implementation*. Geneva, Switzerland: WHO.

World Health Organization (WHO). (2015). Investing To Overcome The Global ImpactOf Neglected Tropical Diseases Third Who Report On Neglected TropicalDiseases. WHO Document Production Services (WHO/HTM/NTD/2015.1).

World Health Organization (WHO). (2015). *Media centre, Malaria Fact sheet N°94*. Retrieved from http://www.who.int/mediacentre/factsheets/fs094/en/

World Health Organization. (1990). Severe and complicated malaria. *Transactions of the Royal Society of Tropical Medicine and Hygiene*, *84*(suppl 2), S1–S65. doi:10.1016/0035-9203(90)90363-J

World Health Organization. (1991). *Inorganic mercury: environ- mental health criteria 118*. Geneva, Switzerland: World Health Organization.

World Health Organization. (1995). Social and Economic Research Reports TAR/SER/PRS, Identification of high-risk communities for control of schistosomiasis in Africa: A multicommunity study. WHO.

World Health Organization. (2000). Severe falciparum malaria. *Transactions of the Royal Society of Tropical Medicine and Hygiene*, *94*(suppl 1), S1–S90. doi:10.1016/S0035-9203(00)90300-6

World Health Organization. (2005). *Ecosystems and human well-being : health synthesis : a report of the Millennium Ecosystem Assessment*. Author.

World Health Organization. (2012). UN-Water Global Analysis and Assessment of Sanitation and Drinking-Water: The Challenge of Extending and Sustaining Services. *UN Water Report*, 1-112.

World Health Organization. (2012b). *World malaria report*. Geneva: World Health Organization. Retrieved from http://www.who.int/malaria/publications/world_malaria_report_2012/report/en

World Health Organization. (2012c). *Élimination de la schistosomiaseau Maroc Une réalité et un succèsaprès trois décennies de lutte*. WHO.

World Health Organization. (2015). *Leishmaniasis Fact Sheet No 375*. Available online: http://www.who.int/mediacentre/factsheets/fs375/en/

World Health Organization. (2015). *World Malaria Report 2015*. Geneva, Switzerland: World Health Organization.

World Malaria Report. (2011). Retrieved from http://www.who.int/malaria/world_malaria_report_2011/9789241564403_eng.pdf

Worrall, F., Burt, T., & Shedden, R. (2003). Long term records of riverine dissolved organic matter. *Biogeochemistry*, *64*(2), 165–178. doi:10.1023/A:1024924216148

Wray, B. B., & O'Steen, K. G. (1975). Mycotoxin-producing fungi from house associated with leukemia. *Archives of Environmental Health*, *30*(12), 571–573. doi:10.1080/00039896.1975.10666781 PMID:1239242

Wright, L. D., Elia, C. F. D., & Nichols, C. R. (2019). *Tomorrow's Coasts: Complex and Impermanent* (Vol. 27). Springer International Publishing. doi:10.1007/978-3-319-75453-6

Wright, J., & Gundry, S. W. (2009). Household characteristics associated with home water treatment: An analysis of the Egyptian demographic and health survey. *Journal of Water and Health*, *7*(1), 21–29. doi:10.2166/wh.2009.056 PMID:18957772

Wright, S. J., & Calderón, O. (2006). Seasonal, El Nino and longer term changes in flower and seed production in a moist tropical forest. *Ecology Letters*, *9*(1), 35–44. PMID:16958866

Wu, H. C., & Santella, R. (2012). The role of aflatoxins in hepatocellular carcinoma. *Hepatitis Monthly, 12*(10).

Wu, F., Wang, J., Pu, C., Qiao, L., & Jiang, C. (2015). Wilson's disease: A comprehensive review of the molecular mechanisms. *International Journal of Molecular Sciences*, *16*(3), 6419–6431. doi:10.3390/ijms16036419 PMID:25803104

Wu, L., Long, T.-y., Liu, X., & Guo, J. (2012). Impacts of climate and land-use changes on the migration of non-point source nitrogen and phosphorus during rainfall-runoff in the Jialing River Watershed, China. *Journal of Hydrology (Amsterdam)*, *475*, 26–41. doi:10.1016/j.jhydrol.2012.08.022

Wu, M. M., Kuo, T. L., Hwang, Y. H., & Chen, C. J. (1989). Dose-response relation between arsenic concentration in well water and mortality from cancers and vascular diseases. *American Journal of Epidemiology*, *130*(6), 1123–1132. doi:10.1093/oxfordjournals.aje.a115439 PMID:2589305

Wu, S. N., Chen, H., Liu, Y. C., & Chiang, H. T. (2002). Block of L-type Ca2+ current by beauvericin, a toxic cyclopeptide in the NG108-15 neuronal cell line. *Chemical Research in Toxicology*, *15*(6), 854–860. doi:10.1021/tx020003k

Wu, W., Wu, P., Yang, F., Sun, D., Zhang, D.-X., & Zhou, Y.-K. (2018). Assessment of heavy metal pollution and human health risks in urban soils around an electronics manufacturing facility. *The Science of the Total Environment*, *630*, 53–61. doi:10.1016/j.scitotenv.2018.02.183 PMID:29475113

Wu, X., Tang, J., & Xie, M. (2015). Serum and hair zinc levels in breast cancer: A meta-analysis. *Scientific Reports*, *5*(1), 12249. doi:10.1038rep12249

Xiao, X. (1997). *Transient climate change and potential croplands of the world in the 21st century Report*. Cambridge, MA: Massachusetts Institute of Technology.

Xue, D., Zhou, C.X., Shi, Y.B., Lu, H., &He, X.Z., (2015). Decreased expression of ferroportin in prostate cancer. *Oncology Letters*, *10*, 913-916.

Xu, X., Dailey, A. B., Talbott, E. O., Ilacqua, V. A., Kearney, G., & Asal, N. R. (2009). Associations of serum concentrations of organochlorine pesticides with breast cancer and prostate cancer in US adults. *Environmental Health Perspectives*, *118*(1), 60–66. doi:10.1289/ehp.0900919 PMID:20056587

Xu, Y., Tokar, E. J., & Waalkes, M. P. (2014). Arsenic-induced cancer cell phenotype in human breast epithelia is estrogen receptor-independent but involves aromatase activation. *Archives of Toxicology*, *88*(2), 263–274. doi:10.100700204-013-1131-4

Yaghoobi Ershadi, M. R., Akhavan, A. A., Zahraei Ramazani, A. V., Abai, M. R., Ebrahimi, B., Vafaei Nezhad, R., ... Jafari, R. (2003). *Epidemiological study in a new focus of cutaneous leishmaniasis in the Islamic Republic of Iran.* Academic Press.

Yaghoobi-Ershadi, M. R., Akhavan, A. A., Zahraei-Ramazani, A. R., Jalali-Zand, A. R., & Piazak, N. (2005). Bionomics of Phlebotomus papatasi (Diptera: Psychodidae) in an endemic focus of zoonotic cutaneous leishmaniasis in central Iran. *Journal of Vector Ecology*, *30*(1), 115. PMID:16007964

Yaghoobi-Ershadi, M. R., Hakimiparizi, M., Zahraei-Ramazani, A. R., Abdoli, H., Akhavan, A. A., Aghasi, M., ... Ranjbar, A. A. (2010). Sand fly surveillance within an emerging epidemic focus of cutaneous leishmaniasis in southeastern Iran. *Journal of Arthropod-Borne Diseases*, *4*(1), 17. PMID:22808384

Yaghoobi-Ershadi, M. R., Marvi-Moghadam, N., Jafari, R., Akhavan, A. A., Solimani, H., Zahrai-Ramazani, A. R., ... Dehghan-Dehnavi, A. R. (2015). Some epidemiological aspects of cutaneous leishmaniasis in a new focus, central Iran. *Dermatology Research and Practice*. PMID:26483838

Yamaguchi, Y., Heiny, M. E., & Gitlin, J. D. (1993). Isolation and characterization of a human liver cDNA as a candidate gene for Wilson disease. *Biochemical and Biophysical Research Communications*, *197*(1), 271–277. doi:10.1006/bbrc.1993.2471 PMID:8250934

Yañez-Arenas, C., Peterson, A. T., Rodríguez-Medina, K., & Barve, N. (2016). Mapping current and future potential snakebite risk in the new world. *Climatic Change*, *134*(4), 697–711. doi:10.100710584-015-1544-6

Yang, G. Q., Wang, S. Z., Zhou, R. H., & Sun, S. Z. (1983). Endemic selenium intoxication of humans in China. *The American Journal of Clinical Nutrition*, *37*(5), 872–881. doi:10.1093/ajcn/37.5.872 PMID:6846228

Yang, G., Sau, C., Lai, W., Cichon, J., & Li, W. (2015). Article. *HHS Public Access*, *344*(6188), 1173–1178. doi:10.1126cience.1249098.Sleep

Yangin, C., Yilmaz, S., Altinbas, M., & Ozturk, I. (2002). A new process for the combined treatment of municipal wastewaters and land leachates in coastal areas. *Water Science and Technology*, *46*(8), 111–118. doi:10.2166/wst.2002.0156 PMID:12420973

Yang, K., LeJeune, J., Alsdorf, D., Lu, B., Shum, C. K., & Liang, S. (2012). Global distribution of outbreaks of water-associated infectious diseases. *PLoS Neglected Tropical Diseases*, *6*(2), e1483. doi:10.1371/journal.pntd.0001483 PMID:22348158

Yang, M., Zhou, H., Kong, R. Y., Fong, W. F., Ren, L. Q., Liao, X. H., ... Yang, S. (1997). Mutations at codon 249 of p53 gene in human hepatocellular carcinomas from Tongan, China. *Mutation Research*, *381*(1), 25–29. doi:10.1016/S0027-5107(97)00142-5 PMID:9403027

Yang, Y., Fang, S., Jensen, J. P., Weissman, A. M., & Ashwell, J. D. (2000). Ubiquitin protein ligase activity of IAPs and their degradation in proteasomes in response to apoptotic stimuli. *Science*, *288*(5467), 874–877. doi:10.1126cience.288.5467.874 PMID:10797013

Yang, Y., Jin, X. M., Yan, C. H., Tian, Y., Tang, J. Y., & Shen, X. M. (2008). Urinary level of nickel and acute leukaemia in Chinese children. *Toxicology and Industrial Health*, *24*(9), 603–610. doi:10.1177/0748233708100091 PMID:19106127

Yang, Y.-W., Liou, S.-H., Hsueh, Y.-M., Lyu, W.-S., Liu, C.-S., Liu, H.-J., ... Chung, C.-J. (2018). Risk of Alzheimer's disease with metal concentrations in whole blood and urine: A case–control study using propensity score matching. *Toxicology and Applied Pharmacology*, *356*, 8–14. doi:10.1016/j.taap.2018.07.015 PMID:30025849

Yap, M. B. H., Martin, P. D., & Jorm, A. F. (2018). Online parenting guidelines to prevent adolescent depression and anxiety: Evaluating user characteristics and usefulness. *Early Intervention in Psychiatry*, *12*(5), 951–958. doi:10.1111/eip.12503 PMID:29076246

Yates, T. L., Mills, J. N., Parmenter, C. A., Ksiazek, T. G., Parmenter, R. R., Vande Castle, J. R., ... Peters, C. J. (2002). The ecology and evolutionary history of an emergent disease: Hantavirus pulmonary syndrome. *Bioscience*, *52*(11), 989–998. doi:10.1641/0006-3568(2002)052[0989:TEAEHO]2.0.CO;2

Yazar, S., & Omurtag, G. Z. (2008). Fumonisins, trichothecenes and zearalenone in cereals. *International Journal of Molecular Sciences*, *9*(11), 2062–2090. doi:10.3390/ijms9112062

Yela, J. L., & Holyoak, M. (1997). Effects of moonlight and meteorological factors on light and bait trap catches of noctuid moths (Lepidoptera: Noctuidae). *Environmental Entomology*, *26*(6), 1283–1290. doi:10.1093/ee/26.6.1283

Yeo, H.-G., Choi, M., Chun, M.-Y., & Sunwoo, Y. (2003). Concentration distribution of polychlorinated biphenyls and organochlorine pesticides and their relationship with temperature in rural air of Korea. *Atmospheric Environment*, *37*(27), 3831–3839. doi:10.1016/S1352-2310(03)00456-4

Yip, K. Y., Wan, M. L. Y., Wong, A. S. T., Korach, K. S., & El-Nezami, H. (2017). Combined low-dose zearalenone and aflatoxin B_1 on cell growth and cell-cycle progression in breast cancer MCF-7 cells. *Toxicology Letters*, *281*, 139–151. doi:10.1016/j.toxlet.2017.09.022

Yoder, J.S., Gargano, J.W., Wallace, R.M., & Beach, M.J. (2012a). Giardiasis surveillancee United States, 2009e2010. *MMWR Surveill. Summ.*, *61*(5), 13-23.

Yoder, J.S., Wallace, R.M., Collier, S.A., Beach, M.J., & Hlavsa, M.C. (2012b). Centers for Disease Control and Prevention (CDC), Cryptosporidiosis surveillanced United States, 2009-2010. *MMWR Surveill. Summ.*, *61*(5), 1-12.

Yoshida, S., Uebayashi, Y., Kihira, T., Kohmoto, J., Wakayama, I., Taguchi, S., & Yase, Y. (1998). Epidemiology of motor neuron disease in the Kii Peninsula of Japan, 1989–1993: Active or disappearing focus? *Journal of the Neurological Sciences*, *155*(2), 146–155. doi:10.1016/S0022-510X(97)00300-6 PMID:9562259

Young, I., Smith, B. A., & Fazil, A. (2015). A systematic review and meta-analysis of the effects of extreme weather events and other weather-related variables on *Cryptosporidium* and *Giardia* in fresh surface waters. *Journal of Water and Health*, *13*(1), 1–17. doi:10.2166/wh.2014.079 PMID:25719461

Young, J. F., Mufson, L., & Davies, M. (2006). Impact of comorbid anxiety in an effectiveness study of interpersonal psychotherapy for depressed adolescents. *Journal of the American Academy of Child and Adolescent Psychiatry*, *45*(8), 904–912. doi:10.1097/01.chi.0000222791.23927.5f PMID:16865032

Younker, J. M., & Walsh, M. E. (2015). Impact of salinity and dispersed oil on adsorption of dissolved aromatic hydrocarbons by activated carbon and organoclay. *Journal of Hazardous Materials*, *299*, 562–569. doi:10.1016/j.jhazmat.2015.07.063 PMID:26259095

Yousef, A. E., & Marth, E. H. (1985). Degradation of aflatoxin M1 in milk by ultraviolet energy. *Journal of Food Protection*, *48*(8), 697–698. doi:10.4315/0362-028X-48.8.697

Yousefinejad, V., Mansouri, B., Ramezani, Z., Mohammadzadeh, N., & Akhlaghi, M. (2018). Evaluation of heavy metals in tobacco and hookah water used in coffee houses in Sanandaj city in 2017. *Majallah-i Ilmi-i Danishgah-i Ulum-i Pizishki-i Kurdistan*, *22*(6), 96–106.

Yuan, G. L., Liu, C., Chen, L., & Yang, Z. (2011). Inputting history of heavy metals into the inland lake recorded in sediment profiles: Poyang Lake in China. *Journal of Hazardous Materials*, *185*(1), 336–345. doi:10.1016/j.jhazmat.2010.09.039 PMID:20932644

Yücel, I., Arpaci, F., Ozet, A., Döner, B., Karayilanoğlu, T., Sayar, A., & Berk, O. (1994). Serum copper and zinc levels and copper/zinc ratio in patients with breast cancer. *Biological Trace Element Research*, *40*(1), 31–38. doi:10.1007/BF02916818

Yu, Z.-z., & Wang, L.-l. (2011). Factors influencing thermal structure in a tributary bay of Three Gorges Reservoir. *Journal of Hydrodynamics. Ser B*, *23*, 407–415.

Yu, Z., Zhang, L., Wu, D., & Liu, F. (2005). Anti-apoptotic action of zearalenone in MCF-7 cells. *Ecotoxicology and Environmental Safety*, *62*(3), 441–446. doi:10.1016/j.ecoenv.2004.10.003

Zachek, C. M., Miller, M. D., Hsu, C., Schiffman, J. D., Sallan, S., Metayer, C., & Dahl, G. V. (2015). Children's cancer and environmental exposures: Professional attitudes and practices. *Journal of Pediatric Hematology/Oncology*, *37*(7), 491–497. doi:10.1097/MPH.0000000000000416 PMID:26334434

Zahm, S. H., & Ward, M. H. (1998). Pesticides and childhood cancer. *Environmental Health Perspectives*, *106*(Suppl 3), 893. PMID:9646054

Zain, M. E. (2011). Impact of mycotoxins on humans and animals. *Journal of Saudi Chemical Society*, *15*(2), 129–144. doi:10.1016/j.jscs.2010.06.006

Zartman, R. E., Ramsey, R. H., Evans, P. W., Koenig, G., Truky, C., & Kamara, L. (1996). Outerbasin, annulus, and playa basin infiltration studies. *Texas Journal of Agriculture and Natural Resources*, *9*, 23–32.

Zecca, L., Fariello, R., Riederer, P., Sulzer, D., Gatti, A., & Tampellini, D. (2002). The absolute concentration of nigral neuromelanin, assayed by a new sensitive method, increases throughout the life and is dramatically decreased in Parkinson's disease. *FEBS Letters*, *510*(3), 216–220. doi:10.1016/S0014-5793(01)03269-0 PMID:11801257

Zecca, L., Youdim, M. B. H., Riederer, P., Connor, J. R., & Crichton, R. R. (2004). Iron, brain ageing and neurodegenerative disorders. *Nature Reviews. Neuroscience*, *5*(11), 863–873. doi:10.1038/nrn1537 PMID:15496864

Zeitlin, S., Liu, J. P., Chapman, D. L., Papaioannou, V. E., & Efstratiadis, A. (1995). *Increased apoptosis and early embryonic lethality in mice nullizygous for the HD gene*. Academic Press.

Zeng, F., Lerro, C., Lavoué, J., Huang, H., Siemiatycki, J., Zhao, N., ... Zhang, Y. (2017). Occupational Exposure to Pesticides and Other Biocides and Risk of Thyroid Cancer. *Occupational and Environmental Medicine*, *74*(7), 502–510. doi:10.1136/oemed-2016-103931 PMID:28202579

Zepnik, H., Völkel, W., & Dekant, W. (2003). Toxicokinetics of the mycotoxin ochratoxin A in F 344 rats after oral administration. *Toxicology and Applied Pharmacology*, *192*(1), 36–44. doi:10.1016/S0041-008X(03)00261-8

Zhang, L. C., Yu, H. N., Sun, S. L., Yang, J. G., He, G. Q., Ruan, H., & Shen, S. R. (2008). Investigations of thecytotoxicity of epigallocatechin-3-gallate against PC-3 cells in the presence of Cd2+ invitro. *Toxicology In Vitro*, *22*(4), 953–960. doi:10.1016/j.tiv.2008.02.005 PMID:18359184

Zhang, Q., Xu, C.-y., Becker, S., & Jiang, T. (2006). Sediment and runoff changes in the Yangtze River basin during past 50 years. *Journal of Hydrology (Amsterdam)*, *331*(3-4), 511–523. doi:10.1016/j.jhydrol.2006.05.036

Zhang, S., Lei, P., Liu, X., Li, X., Walker, K., Kotha, L., ... Safe, S. (2009). The aryl hydrocarbon receptor as a target for estrogen receptor-negative breast cancer chemotherapy. *Endocrine-Related Cancer*, *16*(3), 835–844. doi:10.1677/ERC-09-0054

Zhang, W., Jiang, F., & Ou, J. (2011). Global pesticide consumption and pollution: With China as a focus. *Proceedings of the International Academy of Ecology and Environmental Sciences*, *1*(2), 125–144.

Zhang, W., Parker, K. M., Luo, Y., Wan, S., Wallace, L. L., & Hu, S. (2005). Soil microbial responses to experimental warming and clipping in a tallgrass prairie. *Global Change Biology*, *11*(2), 266–277. doi:10.1111/j.1365-2486.2005.00902.x

Zhang, Y. P., & Ji, X. (2004). The thermal dependence of food assimilation and locomotor performance in southern grass lizards, Takydromus sexlineatus (Lacertidae). *Journal of Thermal Biology*, *29*(1), 45–53. doi:10.1016/j.jtherbio.2003.10.007

Zhang, Y., Coogan, P., Palmer, J. R., Strom, B. L., & Rosenberg, L. (2009). Vitamin and mineral use and risk of prostate cancer: The case-control surveillance study. *Cancer causes & control. CCC*, *20*(5), 691–698. PMID:19093214

Zhang, Y., & Pardridge, W. M. (2001). Rapid transferrin ef ¯ ux from brain to blood across the blood ± brain barrier. *Journal of Neurochemistry*, *76*(5), 1597–1600. doi:10.1046/j.1471-4159.2001.00222.x PMID:11238745

Zhao, F., Xi, G., Liu, W., Keep, R. F., & Hua, Y. (2016). Minocycline Attenuates Iron-Induced Brain Injury Materials and Methods. *Acta Neurochirurgica*, (S121). doi:10.1007/978-3-319-18497-5

Zheng, T., Holford, T., Mayne, S., Owens, P., Ward, B., Carter, D., ... Tessari, J. (1999). Beta-benzene hexachloride in breast adipose tissue and risk of breast carcinoma. *Cancer*, *85*(10), 2212–2218. doi:10.1002/(SICI)1097-0142(19990515)85:10<2212::AID-CNCR16>3.0.CO;2-F

Zheng, W., Aschner, M., & Ghersi-egea, J. (2003). Brain barrier systems : A new frontier in metal neurotoxicological research. *Toxicology and Applied Pharmacology*, *192*(1), 1–11. doi:10.1016/S0041-008X(03)00251-5 PMID:14554098

Zheng, W., & Monnot, A. D. (2012). Regulation of brain iron and copper homeostasis by brain barrier systems: Implication in neurodegenerative diseases. *Pharmacology & Therapeutics*, *133*(2), 177–188. doi:10.1016/j.pharmthera.2011.10.006 PMID:22115751

Zhou, E., & Crawford, R. L. (1995). Effects of oxygen, nitrogen, and temperature on gasoline biodegradation in soil. *Biodegradation*, *6*(2), 127–140. doi:10.1007/BF00695343 PMID:7772939

Zhou, X. N., Yang, G. J., Yang, K., Wang, X. H., Hong, Q. B., Sun, L. P., ... Utzinger, J. (2008). Potential impact of climate change on schistosomiasis transmission in China. *The American Journal of Tropical Medicine and Hygiene*, *78*(2), 188–194. doi:10.4269/ajtmh.2008.78.188 PMID:18256410

Zhou, X., Zhou, Y., Chen, R., Ma, W., Deng, H., & Kan, H. (2013). High temperature as a risk factor for infectious diarrhea in Shanghai, China. *Journal of Epidemiology*, *23*(6), 418–423. doi:10.2188/jea.JE20130012 PMID:23994865

Zhu, Z., Ge, T., Liu, S., Hu, Y., Ye, R., Xiao, M., ... Wu, J. (2018). Rice rhizodeposits affect organic matter priming in paddy soil: The role of N fertilization and plant growth for enzyme activities, CO2 and CH 4 emissions. *Soil Biology & Biochemistry*, *116*, 369–377. doi:10.1016/j.soilbio.2017.11.001

Zic, L., & Targa, A. (2014). Risk Nexus: Inondations au Maroc de 2014: quels enseignements tirer de Guelmim et Sidi Ifni. Zurich: Zurich Insurance Company Ltd.

Ziervogel, G., & Ericksen, P. J. (2010). Adapting to climate change to sustain food security. *Wiley Interdisciplinary Reviews: Climate Change*, *1*(4), 525–540. doi:10.1002/wcc.56

Zinedine, A., & Ruiz, M. J. (2014). Zearalenone. In *Mycotoxins and their implications in food safety*. Future Science Ltd. Retrieved from www.future-science.com

Zinedine, A., Gonzales-Osnaya, L., Soriano, J. M., Moltó, J. C., Idrissi, L., & Mañes, J. (2007a). Presence of aflatoxin M1 in pasteurized milk from Morocco. *International Journal of Food Microbiology*, *114*(1), 25–29. doi:10.1016/j.ijfoodmicro.2006.11.001

Zinedine, A., & Mañes, J. (2009). Occurrence and legislation of mycotoxins in food and feed from Morocco. *Food Control*, *20*(4), 334–344. doi:10.1016/j.foodcont.2008.07.002

Zinedine, A., Meca, G., Mañes, J., & Font, G. (2011). Further data on the occurrence of Fusarium emerging mycotoxins enniatins (A, A1, B, B1), fusaproliferin and beauvericin in raw cereals commercialized in Morocco. *Food Control*, *22*(1), 1–5. doi:10.1016/j.foodcont.2010.05.002

Zinedine, A., Soriano, J. M., Moltó, J. C., & Mañes, J. (2007b). Review on the toxicity, occurrence, metabolism, detoxification, regulations and intake of zearalenone: An oestrogenic mycotoxin. *Food and Chemical Toxicology*, *45*(1), 1–18. doi:10.1016/j.fct.2006.07.030

Zougari-Elwedi, B., Sanaa, M., Labidi, S., & Sahraoui, A. L. H. (2012). Évaluation de l'impact de la mycorhization arbusculaire sur la nutrition minérale des plantules de palmier dattier (Phœnix dactylifera L. var. Deglet Nour). *Étude Gest. des Sols*, *19*(3), 193-202.

Zou, X., Jiang, S., Wu, Z., Shi, Y., Cai, S., Zhu, R., & Chen, L. (2017). Effectiveness of deferoxamine on ferric chloride-induced epilepsy in rats. *Brain Research*, *1658*, 25–30. doi:10.1016/j.brainres.2017.01.001 PMID:28063856

Zufiria, M., Gil-Bea, F. J., Fernandez-Torron, R., Poza, J. J., Munoz-Blanco, J. L., Rojas-García, R., ... López de Munain, A. (2016). ALS: A bucket of genes, environment, metabolism and unknown ingredients. *Progress in Neurobiology*, *142*, 104–129. doi:10.1016/j.pneurobio.2016.05.004 PMID:27236050

Zumstein, V., & Regli, F. (1982). Amyotrophic lateral sclerosis: Epidemiologic study. *Revue swisse de médecine praxis*, *14*, 1424–7.

About the Contributors

Kholoud Kahime born in Marrakech, she obtained her Baccalaureate in "Experimental Sciences". Afterword, she started his high university studies and was graduated in 2008 from Cadi Ayyad University (Morocco) with a bachelor degree in Biology, Environment option. She was graduated from Ibn Zohr University with Master's degree in "Environment and Health Sciences" and she won university merit award. She got her PhD in December 2016 on Ecology, Epidemiology and Climate Change from Cadi Ayyad University. Pr. Kholoud Kahime is currently a lecturer-researcher at the School of Technology, Essaouira, Cadi Ayyad University of Marrakech. Her research focuses on environment, climate change, health, food security and sustainable development. Her research has resulted in 22 articles published in indexed and prominent, influential journals, one of which is cited in the IPCC report, (2014), alongside more than 40 oral and Poster presentations was presented in France, Tanzania, UK, USA, Spain, Germany, Poland, Monaco, Morocco, Turkey, Brazil. She is a consultant for German development agency GIZ, the Friedrich Naumann Foundation and IRD. In 2015, she was awarded the Prince Albert II of Monaco Scholarship and undertook specialist training at NASA. In 2016, was awarded the best researcher on climate change study in the Cadi Ayyad University. In 2018, she was nominated "Femmes leaders scientifiques du Maroc". Currently, she is President of the International Center for Research and Capacity Building (CI2RC).

Moulay Abdelmonaim El Hidan was born in 1985 in Marrakech-Safi state (Morocco). In 2004 he obtained his baccalaureate, option; "Experimental Sciences". Later he continued his university studies and was graduated in 2007 from Cadi Ayyad University (Morocco) with a bachelor degree in Biology. 2 years later, Pr. El Hidan was graduated from the same university with Master's degree. From 2011 to 2018, Pr. EL Hidan held the position of high school professor of Biology. At the same time he completed his PhD project and was graduated PhD in 2016, from Cadi Ayyad University, specialty Ecophysiology and epidemiology. Since 2018, Mr. EL Hidan was nominated as professor in the Ibn Zohr University (Morocco) and associate investigator in Cadi Ayyad University. Pr. EL Hidan is currently supervising several PhD and Master's projects with different research interests aiming the scorpion venom toxicology and climate change effects on Vector-Borne diseases spread with more than 20 papers and book chapters in numerous international journals as well as being reviewer for several scientific papers in addition to around 40 communications in prestigious international congresses and symposiums.

Omar El Hiba was born in 12 September 1985 in Youssoufia, Marrakech-Safi state (Morocco) where he finished his high school studies with Baccalaureate in "Experimental Sciences" in 2004. One year later, he started his high university studies and was graduated in 2007 from Cadi Ayyad University (Morocco) with a bachelor degree in Biology. In 2009, Pr. El Hiba was graduated from the same university with Master's degree in "Life and Health Sciences". He began his professional carrier as professor of Biology in the high school during 5 years. Concomitantly, he started his PhD in Neuroscience, receiving his PhD in 2015 on the "Physiopathology of Hepatic Encephalopathy" from Cadi Ayyad University. Since 2016, Mr. EL Hiba holds the position of professor in the department of Biology where he is currently permanent member in the laboratory of Biotechnology, Biochemistry and Nutrition (Nutrition and Food Sciences team), while he is associate to the Nutritional Physiopathologies Team, as well, he is associate member to laboratory of Innovation in Science, Technologies and Modeling, physics department, Faculty of Sciences, Chouaib Doukkali University, El Jadida (Morocco). In the faculty of Sciences Semlalia, Cadi Ayyad University, Pr. El Hiba is associate investigator to Neurosciences, Pharmacology and Environment Team, where he is the assistant of the research team leader and currently supervising several PhD projects with different research interests that span both heavy metals-induced brain disorders and hepatic encephalopathy. Much of his works shed light on the role of environmental pollutants, especially heavy metals, as key elements in the pathogenesis of different brain disorders including neurodegenerative diseases, in particular Parkinson's disease, while he is, since 2014, member of the International Parkinson and Movement Disorders Society (MDS), Milwaukee, USA. As well, his research area focuses on new drug discovery from medicinal plants with neuroprotective potential which have been subjected to several publications in different international journals to which he was solicited as reviewer for several papers. Besides, Pr. EL Hiba has more than 24 international communications in several international congresses and symposiums (Egypt, France, Canada, India, Sweden, Denmark…) with more then 18 international awards, grants and scholarships (IBRO, MDS, CNRST, ISHEN, NeuroMed, GDRI).

Lahouari Bounoua was born in Oran, Algeria. He completed his education at Florida State University with a Master degree in 1990 and a PhD in 1992, specialty "atmospheric sciences". Dr. Bounoua joined the National Aeronautics and Space Administration (NASA) at Goddard Space Flight Center (GSFC) in 1993 with an affiliation as a research associate with the University Space Research Association. In 1995 he was affiliated as an assistant professor with the University of Maryland College Park, and then, as an associate professor in 2002. to 2005 Dr. Bounoua joined NASA as a civil servant where he is now a Senior Research Scientist and the Associate Chief for the Biospheric Sciences Laboratory. Dr. Bounoua's studies revolve around terrestrial ecosystems and their interactions with the atmosphere using multi-scale remote sensing, modeling, and advanced analytical techniques. Dr. Bounoua leads a group focusing on land surface processes modeling, including land cover and land use change, climate impacts and feedbacks including health, water and food security and sustainable development. Another part of the research studies the impact of anthropogenic activities such as emission of carbon dioxide in the atmosphere and the quantification of the role of vegetation in climate regulation. Recently Dr. Bounoua conducted high impact work linking climate to incidence of the zoonotic cutaneous leishmaniasis (L. major) in the North African and developed a line of research in modeling vector borne diseases. Dr. Bounoua has more than 50 peer reviewed articles in high impact journals, several books and book chapters and has delivered several scientific seminars at national and international conferences. Dr. Bounoua was awarded the "Hydrospheric and Biospheric Sciences Award for Outstanding Education and Outreach" in 2009 and "University Space Research Association Excellence in Research" in 1996.

* * *

Abdelmohcine Aimrane is a researcher on human health and environment. He graduated from the university Abdelmalik Essâadi, with a master degree in Neuroscience and biotechnology in 2013. After graduation he worked as a high school teacher in Essaouira. He is currently working on a PhD in biological sciences and human health in Cadi Ayyad university. A reader by morning, an experimentalist by day and a writer by night. He has conducted and supervised bachelor students in the same university since 2015.

Mehdi Ait Laaradia obtained his bachelor's degree from Semlalia Science faculty, Cadi Ayyad University in 2016. He is currently a Ph. D. student under the supervision of Prof. Abderrahman Chait. His research is focused on studying physiological effects of scorpion venom

Hassan Alahyane was born in Morocco in 1987. He received the Matser degree from Cadi Ayyad University, Marrakech, in 2014. In 2015 until now, he joined the laboratory of Molecular Modeling & Ecophysiology, Department of Biology, to finish his doctorales studies in biological control of cirtus pests.

Mohamed Anli is a PhD Student, Faculty of Sciences Semalalia, Cadi Ayyad University of Marrakesh.

Fatima Zahra Azzaoui prior to her current post as assistant professor at the Faculty of Science, Kenitra (Morocco), Fatima-Zahra AZZAOUI served for 4 years in the same position at the Faculty of Science Ben M'Sik, Casablanca, (Morocco). She is a young researcher on Neurotoxicology and Epidemiology fields. She earned her PhD in 2009 from the IBN TOFAIL University. She worked on neurotoxicity of xenobiotics (heavy metals, pesticides..) on behavioral and cognitive health among animal and human models during her PhD project and her postdoctoral activity. She also investigated the environmental toxicity of these xenobiotics in water-soil-plant system in the North-West of Morocco, on her master's thesis. Also, she completed in 2007, a Universitary Diploma on Epidemiology and Clinical research, from Faculty of Medicine and Pharmacy, SIDI MOHAMED BEN ABDELLAH University, Fes, Morocco. She was an ex-alumnus of IBRO, she participated in several international trainings, conferences and congress, and she had several publications in scientific journals.

Abdelkrim Ben Salem is a PhD researcher on environment sciences at the University of Marrakesh. His current research interest is on "Vulnerability and Adaptation to Climatical Changes in the Biosphere Reserve of the Oasis in Southern Morocco." This work that was proposing has an objective, to methodically, clearly and objectively assess scientific, technical and socio-economic information which is necessary and useful for understanding the scientific basis of risks related to climate change, identify more precisely the possible consequences of this change and consider possible strategies for adaptation and mitigation. He is also interested in ecosystem services valuation and water management and usage.

Abdellah Bouazza is a PhD a Marrakech University, herpetologist currently focusing "on some ecophysiological (thermoregulation, reproduction) and biogeographical aspects (distribution and conservation) of amphibians and reptiles in Morocco and other Mediterranean areas.

Hanane Boutaj is a PhD student at Cadi Ayyad University faculty of Sciences Semlalia.

Bouimeja Btissam is PhD student at the Faculty of Sciences Sultan Moulay Slimane University, Beni Mellal in collaboration between the LPRD-FP of Beni Mellal, and the laboratory "Ecology and Environment" of the Faculty of Sciences Semlalia of Marrakech.

Hicham Chatoui was born in Beni Mellal (Tagzirt), Morocco, in 1984. He received He obtained a Master's degree in Research in Life Sciences and Health, option: Anthropobiology, From de Faculty of Science Semlalia, University of Cady Ayyad, Marrakesh, Morocco, in 2009. He is a doctoral candidate in neuroscience at the Semlalia Faculty of Sciences, Cadi Ayyad University, Marrakesh, Morocco. He is a pedagogical director and teacher researcher at the school of engineering of the Health and Agroindustry of the Private University of Marrakech since 2015. He is current research interests include hydric stress, neurotransmitter: serotonin, dopamine; glial and ependymocytic markers; behavior; anxiety; locomotor activity; motivation; stress, in the laboratory of Neurosciences, Department of Biology, Faculty of Science Semlalia, University of Cadi Ayyad, Marrakesh, Morocco.

Mohamed Echchakery is a Dr at Faculty of Sciences Semlalia University Cadi Ayyad, Morocco. He has coauthored over 10 publications including Climate change, Vector-Borne diseases, and scorpion venoms. The current research interests of Dr. Echchakery focuses on the ecology, phylogeny, systematics, and biogeography of scorpion species in Morocco.

Samira El Akhdari is a Professor at the Ibn Khatib Lycee, Ministry of Education of Morocco, and was a Researcher at the Agronomic and Veterinary Medicine Institute in Rabat. She is a member of The Moroccan Society for Mycotoxicolgy (MSM). She co-authored scientific papers and presented communications at National and International events. His research is mainly focused on Environmental microbiology, safety and hygiene.

Soraia El Baz is a Doctor in Microbiology and Toxicology Environmental in 2015 from the Faculty of Sciences Semlalia, participate in international congress with oral presentation or poster, and also articles published in international journal.

Mohamed Ait El Mokhtar works in Plant Biotechnology and Physiology Laboratory at the Faculty of Science Semlalia Cadi Ayyad University, Marrakesh, Morocco. El Mokhtar is interested in the effect of biofertilizers on improving plant tolerance to biotic and abiotic stresses in relation to climate change.

Ismail El Moudden's interests span Machine Learning, Genes Selection, Mining Cancer Behavior, and its applications to domains where one has to draw inferences from observing the complex, real-world systems that evolve over time. He enjoys challenging and complex data analysis, data mining, machine learning and data visualization tasks. He is now using deep Feature Selection and Extraction based on Neural Network for Cancer Markers Identification.

Abdeljalil Elgot is an Associated Professor on Neurosciences at the Higher Institute of Health sciences, University Hassan First, of Settat, Morocco.

Khalid Elkalat is currently Professor-researcher on Environment Studies at School of Technology, Cadi Ayyad University, Marrakech, Morocco.

Rokia Ghchime is a PhD in surgical and medical sciences, option neurophysiology, Faculty of Medicine and Pharmacy, Mohamed V University, Rabat.

Ahmed Karmaoui started his academic career as Research Fellow at Cadi Ayyad University and awarded Ph.D. from Faculty of Semlalia; and as an associate editor at the International Journal of Climate Change: Impacts and Responses, editor at several International Journals such as Journal of Research in Chemistry & Environment (SRI: JRCE), the International Journal of Earth & Environmental Sciences, Nanomedicine and Nanoscience Research Journal; as International Advisory Board member of the International Journal of ICT Research and Development in Africa (IJICTRDA), and the International Journal of Social Ecology and Sustainable Development (IJSESD). KARMAOUI has reviewed for 58 international journals. He has published various peer-reviewed international journal articles and book chapters on a range of environmental topics. He is currently the president of the Southern Center for Culture and Science, Morocco. Broad areas of his research interests include ecosystem services assessment, climate change impacts, response, adaptation, disaster risk reduction, risk and vulnerability, water management.

Ahmed Khadra is PhD in Cadi Ayyad University, Faculty of Sciences Semlalia Marrakesh, Department of Biology Marrakesh. Ahmed does research in Solid Waste Management, composting, Environmental Chemistry and Agro-Ecology.

Youssef Khamsi was born in Texas and grew up in Casablanca, Morocco. He attended the University of Texas at El Paso where he earned his Bachelor of Science in Physics. He then completed medical physics graduate school at The University of Arizona, earning a professional master's degree.

Mohammed Messouli has nearly 30 years of professional experience in the fields of environment, climate change and sustainable development. He is currently the Director of the Research Group on Impact, Vulnerability and Adaptation to Climate Change (GRIVAC) at the Faculty of Sciences of Marrakech. He has also acted as contributing author to the IPCC 5th Assessment Report, Working group III (2010-2014).

Hamid Nebdi received his Master degree (1997) and his PhD (2000) in atomic physics at University Catholic of Louvain (Belgium). He joined the Royal Meteorological Institute (RMI) of the Belgian Space Pole in 2001 as First Assistant Researcher and became the Acting Head of Section Ionospheic Profiles in 2003. He participated to many national and international research projects such as ESA Space Weather Pilot Project and Galileo Joint Undertaking. He was Belgian delegate in many Management Committees of COST actions and member in the Ionospheric Effects Group of the Space Weather Working Team of the European Space Agency. He was one of the initiators of the project: Solar-Terrestrial Centre of Excellence and a research valorisation project: Magnetic Valley. He worked four years as Assistant Professor of Physics at Umm Al-Qura University conjointly in College of Applied Sciences, Medicine Colleges and College of Engineering and Islamic Architecture in Saudi Arabia. And he participated actively in many projects and committees. His research has been extensively published in peer-reviewed scientific journals, reported in invited talks, oral and poster communications, and published in national/international conference proceedings. He has authored and co-authored several papers, chapters, book of abstracts, and a special issue. He has served as Conference Chair, Co-Chair, Program Committee, Scientific Committee and Organizing Committee member in several national and international conferences, workshops and congresses in different fields of the physics of radiation-matter interactions (atomic physics, laser physics, space weather, ionosphere …). He teaches more than 24 courses of physics and mathematics in Arabic, French and English in 3 universities. He is now Associate Professor of Physics at Chouaïb Doukkali University where he began as Assistant Professor in 2012.

Makbal Rachida is a PhD student from Faculty of Sciences Semelalia, University Cadi Ayyad Marrakech, Morocco. I have got the opportunity to train inside Tsukuba University in Japan (2016-2017) and worked on the antiproliferative effect of some Moroccan endemic plants extracts against skin cancer cell lines.

Zakaria Tazart is a PhD student at the Department of Biology the Faculty of Sciences Semlalia, Cadi Ayyad University of Marrakesh, Morocco and member of the Laboratory of Biology and Biotechnology of Microorganisms (LBBM).

Mounir Tilaoui received his Ph.D. in Pharmacology-Biochemistry from the Faculty of Science and Technology of Beni Mellal in July 2012. His thesis projects were focused on the antitumor activity of natural product: chemical, cellular and molecular analysis. After completion of his degree, he was appointed as a Postdoctoral Fellow in the department of chemical and life sciences, at the Waterford Institute of Technology. Ireland. His research focused on studying the effects of chemopreventive agents on activation of the intrinsic apoptotic pathway in cancer. He then continued at the Sultan Moulay Slimane University another postdoctoral researcher focused on natural product-based anticancer drug discovery / gene expression profiles for triple-negative breast cancer in Moroccan population. His current work is mainly focused on anticancer drug discovery.

Oulaid Touloun is currently a Professor at the University of Sultan Moulay Slimane, Beni Mellal. He has a PhD degree in Ecology and Epidemiology from the University of Cadi Ayyad in Marrakesh. Dr. Touloun has contributed extensively to expanding knowledge on Moroccan scorpions. His research activities include ecology, biogeography, and epidemiology. His passion, and his extensive experience in the field and in scorpions systematics have led him to discover and describe three new scorpion species. Oulaid Touloun is a reviewer in numerous specialized scientific journals and is also an active member of several associations involved in scientific research.

Mohammed Yacoubi Khebiza is a Professor at the University of Marrakesh with a PhD in the Ecology and groundwater quality, senior researcher on groundwater/surface water and climate change specifically on assessing impact adaptation to climate change in various land uses in Morocco. He will evaluate the impacts of climate change and climate variability on Moroccan waters ecosystems in selected sites and help in developing and test a set of methods for the assessment of vulnerability of relevant development topics resulting from impacts of climate change over waters ecosystems and implement of the methodology to assess vulnerability on the development topics. He will also initiate a science-policy dialogue in and across the pilot regions to test the practicality of the developed methods, to disseminate practical and useful information on reducing vulnerability through adaptive forest management, and to build capacity in developing countries through training sessions.

Siham Zerouali has ten years experience in teaching biological sciences. She is a Research Fellow at Southern Center for Culture and Science, Morocco. She has published various peer-reviewed articles and book chapters on a range of environmental topics.

Abdellah Zinedine is a Professor at the Faculty of Sciences, Chouaib Doukkali University, and was a Researcher Scientist at the National Institute of Health of Morocco (1998-2013). He is co-founder and the General Secretary of The Moroccan Society for Mycotoxicolgy (MSM). He was a National Expert and member of the scientific committees of the CNRST in Morocco (2011-2014). He is a member of the Executive Committee of SAMEFOOD Network (IAMZ-Zaragoza-Spain). He published book chapters and more than 50 scientific publications. He coordinated National projects and involved in International collaborative projects with European and African Teams. His research is mainly focused on microbiology, safety and food hygiene.

Index

F

G

H

I

K

L

M

N

P

R

S

www.ingramcontent.com/pod-product-compliance
Lightning Source LLC
LaVergne TN
LVHW060251011026
849161LV00011B/2688